SHRIVER & ATKINS
Inorganic Chemistry

SHRIVER & ATKINS
Inorganic Chemistry

FOURTH EDITION

Peter Atkins, Tina Overton,
Jonathan Rourke, Mark Weller,
Fraser Armstrong

OXFORD
UNIVERSITY PRESS

OXFORD

UNIVERSITY PRESS

Great Clarendon Street, Oxford OX2 6DP

Oxford University Press is a department of the University of Oxford.
It furthers the University's objective of excellence in research, scholarship,
and education by publishing worldwide in

Oxford New York

Auckland Cape Town Dar es Salaam Hong Kong Karachi
Kuala Lumpur Madrid Melbourne Mexico City Nairobi
New Delhi Shanghai Taipei Toronto

With offices in

Argentina Austria Brazil Chile Czech Republic France Greece
Guatemala Hungary Italy Japan Poland Portugal Singapore
South Korea Switzerland Thailand Turkey Ukraine Vietnam

Oxford is a registered trade mark of Oxford University Press
in the UK and in certain other countries

Published in the United States and Canada
by W. H. Freeman and Company, New York

British Library Cataloguing in Publication Data

Data available

Library of Congress Cataloging in Publication Data

Data available

Typeset by Newgen Imaging Systems (P) Ltd., Chennai, India
Printed in Italy
on acid-free paper by
Lito Terrazzi s.r.l.

ISBN 0–19–926463–5   978–0–19–926463–6

1 3 5 7 9 10 8 6 4 2

Preface

Our aim is to provide a comprehensive and contemporary introduction to the diverse and fascinating discipline of inorganic chemistry. Inorganic chemistry deals with the properties of all of the elements in the periodic table. These elements range from highly reactive metals, such as sodium, to noble metals, such as gold. The nonmetals include solids, liquids, and gases and range from the aggressive oxidizing agent fluorine to unreactive gases such as helium. Although this variety and diversity are features of any study of inorganic chemistry, there are underlying patterns and trends which enrich and enhance our understanding of the discipline. These trends in reactivity, structure, and properties of the elements and their compounds provide an insight into the landscape of the periodic table and provide a foundation on which to build understanding.

Inorganic compounds vary from ionic solids, which can be described by simple applications of classical electrostatics, to covalent compounds and metals, which are best described by models that have their origin in quantum mechanics. We can rationalize and interpret the properties of most inorganic compounds by using qualitative models that are based on quantum mechanics, such as atomic orbitals and their use to form molecular orbitals. The text builds upon similar qualitative bonding models that should already be familiar from introductory chemistry courses. Although qualitative models of bonding and reactivity clarify and systematize the subject, inorganic chemistry is essentially an experimental subject. New areas of inorganic chemistry are constantly being explored and new and often unusual inorganic compounds are constantly being synthesized and identified. These new inorganic syntheses continue to enrich the field with compounds that give us new perspectives on structure, bonding, and reactivity.

Inorganic chemistry has considerable impact on our everyday lives and on other scientific disciplines. The chemical industry is strongly dependent on it. Inorganic chemistry is essential to the formulation and improvement of modern materials such as catalysts, semiconductors, optical devices, superconductors, and advanced ceramic materials. The environmental and biological impact of inorganic chemistry is also huge. Current topics in industrial, biological, and environmental chemistry are mentioned throughout the book and are developed more thoroughly in later chapters.

In this new edition we have refined the presentation, organization, and visual representation. Most of the book has been rewritten and there is much completely new material. We have written with the student in mind, and we have added new pedagogical features and have enhanced others.

The topics in Part 1, *Foundations*, have been largely rewritten to make them more accessible to the reader with more qualitative explanation accompanying the more mathematical treatments. There is a completely new chapter on techniques of inorganic chemistry which provides a useful toolkit to spectroscopic and associated methods used by the inorganic chemist to probe the structures of inorganic molecules.

Part 2, *The elements and their compounds*, has been expanded and each group in the periodic table now has its own chapter. The section starts with hydrogen and proceeds across the periodic table from the *s*-block metals, across the *p*-block, and finishing with the *d*- and *f*-block elements. Organometallic compounds have been moved into this section of the text in recognition of their importance in the study of inorganic chemistry. The chemical properties of each group of elements and their compounds are described and enriched with descriptions of current applications. The patterns and trends that emerge are rationalized by drawing on the principles introduced in Part 1.

Part 3, *Frontiers*, takes the reader to the edge of knowledge in several areas of current research. These chapters explore specialized subjects that are of importance to industry, materials, and biology and include catalysis, nanomaterials, and bioinorganic chemistry.

All the illustrations and the marginal structures—nearly 1500 in all—have been redrawn and are now presented in full colour. We have used colour systematically rather than just for decoration, and have ensured that it serves a pedagogical purpose.

We are confident that this text will serve the undergraduate chemist well. It provides the theoretical building blocks upon which to build knowledge and understanding of inorganic chemistry. It should help to rationalize the sometimes bewildering diversity of descriptive chemistry. It also takes the student to the forefront of the discipline and should therefore complement many courses taken in the later stages of a programme.

The authors
September 2005

About the book

Inorganic chemistry is an extensive subject that at first sight can be daunting. We have made every effort to help by organizing the information systematically and by providing a lot of support material. Whether you work through the book systematically or dip in at an appropriate point of your studies there are several features that should make the book easier to use, more engaging, and help to develop deeper understanding. We have also provided further resources on the web in the accompanying Online Resource Centre.

Organizing the information

The book is divided into three sections. Part 1, *Foundations*, sets out the underlying principles of structure and thermodynamics that are used to systematize the discussion throughout the rest of the book. Part 2, *The elements and their compounds*, is a systematic journey through the periodic table that illustrates how the underlying principles govern the properties of the elements. Part 3, *Frontiers*, describes areas of chemistry that are currently at the cutting edge of research.

Key points

3.10 The rationalization of structures

The thermodynamic properties of such ionic solids can be treated very simply in terms of the ionic model. However, a model of a solid in terms of charged spheres interacting Coulombically is likely to be fairly crude and we should expect significant departures from its predictions because many solids are more covalent than ionic. Even conventional 'good' ionic solids, such as the alkali metal halides, have some covalent character. Nevertheless, the ionic model provides an attractively simple and effective scheme for correlating many properties.

(a) Ionic radii

Key point: The sizes of ions, ionic radii, generally increase down a group, decrease across a period, increase with coordination number, and decrease with increasing charge number.

Most sections of the text begin with a statement of one or two key points. This feature summarizes the most important aspects of the discussion in the section and can be used as a preparation for what is to come, to check understanding, to aid review, and to navigate around a chapter.

Boxes

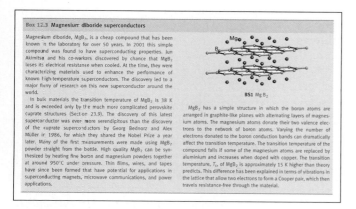

The numerous *Boxes* illustrate the diversity of inorganic chemistry and its applications to advanced materials, industrial processes, environmental chemistry, and everyday life, without obscuring the principles set out in the text itself.

Further information

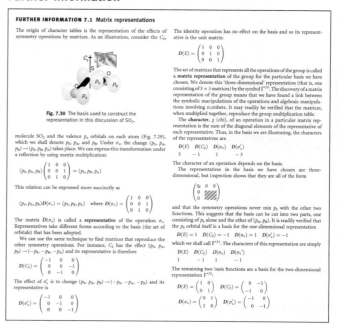

Where we judge it helpful to have more information without overburdening the exposition in the chapter itself, we provide it in the *Further information* sections at the end of the chapters.

Further reading

FURTHER READING

P.A. Cox, *Introduction to quantum theory and atomic structure.* Oxford University Press (1996). An introduction to the subject.

P. Atkins and J. de Paula, *Physical chemistry.* Oxford University Press and W.H. Freeman & Co. (2006). Chapters 8 and 9 give an account of quantum theory and atomic structure.

J. Emsley, *Nature's building blocks.* Oxford University Press (2003). A interesting guide to the elements.

D.M.P Mingos, *Essential trends in inorganic chemistry.* Oxford University Press (1998). Includes a detailed discussion of the important horizontal, vertical, and diagonal trends in the properties of the atoms.

P.A. Cox, *The elements: their origin, abundance, and distribution.* Oxford University Press (1989). Examines the origin of the elements, the factors controlling their widely differing abundances, and their distributions in the Earth, the solar system, and the universe.

J.A. Cleverty and N.G. Connelly, *Nomenclature of inorganic chemistry II. Recommendations.* Royal Society of Chemistry (2001). This book outlines the conventions for the periodic table and inorganic substances. It is known colloquially as the 'Red Book' on account of its distinctive red cover.

Each chapter lists some sources where more information can be found. We have tried to ensure that these sources are easily available, such as text books and monographs, rather than the primary research literature and have indicated the type of information each one provides.

Resource section

Resource section 6:
Tanabe–Sugano diagrams

This section collects together the Tanabe–Sugano diagrams for octahedral complexes with electron configurations d^2 to d^8. The diagrams, which were introduced in Section 19.4, show the dependence of the term energies on ligand-field strength. The term energies E are expressed as the ratio E/B, where B is a Racah parameter, and the ligand-field splitting Δ_O is expressed as Δ_O/B. Terms of different multiplicity are included in the same diagram by making specific, plausible choices about the value of the Racah parameter C, and these choices are given for each diagram. The term energy is always measured from the lowest energy term, and so there are discontinuities of slope where a low-spin term displaces a high-spin term at sufficiently high ligand-field strengths for d^4 to d^7 configurations. Moreover, the noncrossing rule requires terms of the same symmetry to mix rather than to cross, and this mixing accounts for the curved rather than the straight lines in a number of cases. The term labels are those of the point group O_h.

The diagrams were first introduced by Y. Tanabe and S. Sugano, *J. Phys. Soc. Japan*, 1954, **9**, 753. They may be used to find the parameters Δ_O and B by fitting the ratios of the energies of observed transitions to the lines. Alternatively, if the ligand-field parameters are known, then the ligand-field spectra may be predicted.

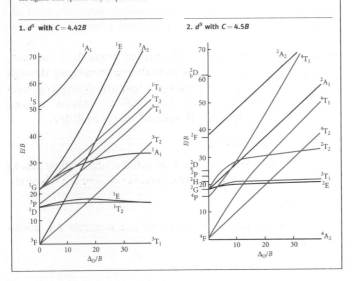

1. d^1 with $C = 4.42B$

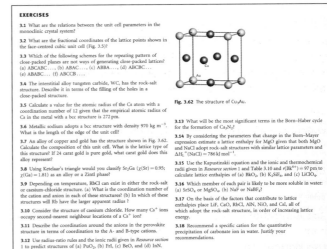

2. d^3 with $C = 4.5B$

At the back of the book is a collection of resources, including more extensive listings of data and information relating to group theory and spectroscopy.

Problem solving

Worked examples

Example 14.6 Determining the chain length of a polyphosphoric acid

A sample of a polyphosphoric acid was dissolved in water and titrated with dilute NaOH(aq). Two stoichiometric points were observed at 16.8 and 28.0 cm³. Determine the chain length of the polyphosphate.

Answer The strongly acidic OH groups are titrated by the first 16.8 cm³. The two terminal OH groups are titrated by the remaining $28.0 - 16.8\ \text{cm}^3 = 11.2\ \text{cm}^3$. Because the concentrations of analyte and titrant are such that each OH group requires 5.6 cm³ of the titrant, we conclude that there are (16.8 cm³)/(5.6 cm³) = 3 strongly acidic OH groups per molecule. A molecule with two terminal OH groups and three further OH groups is a tripolyphosphate.

Self-test 14.6 When titrated against base a sample of polyphosphate gave end points at 30.4 and 45.6 cm³. Calculate the chain length.

We have provided numerous *Worked examples* throughout the text. Each one illustrates an important aspect of the topics under discussion or provides practice with calculations and problems.

Self-test

Self-test 14.6 When titrated against base a sample of polyphosphate gave end points at 30.4 and 45.6 cm³. Calculate the chain length.

A *Self-test* follows each *Worked example*. Their role is to help you judge your understanding of the Example. Each *Self-test* is similar to the preceding *Worked example*, but extends the problem or takes a slightly different approach. Answers to *Self-tests* will be found in the *Answers* section near the end of the book.

Exercises

EXERCISES

3.1 What are the relations between the unit cell parameters in the monoclinic crystal system?

3.2 What are the fractional coordinates of the lattice points shown in the face-centred cubic unit cell (Fig. 3.5)?

3.3 Which of the following schemes for the repeating pattern of close-packed planes are not ways of generating close-packed lattices? (a) ABCABC..., (b) ABAC..., (c) ABBA..., (d) ABCBC... (e) ABABC... (f) ABCCB....

3.4 The interstitial alloy tungsten carbide, WC, has the rock-salt structure. Describe it in terms of the filling of the holes in a close-packed structure.

3.5 Calculate a value for the atomic radius of the Cs atom with a coordination number of 12 given that the empirical atomic radius of Cs in the metal with a bcc structure is 272 pm.

3.6 Metallic sodium adopts a bcc structure with density 970 kg m⁻³. What is the length of the edge of the unit cell?

3.7 An alloy of copper and gold has the structure shown in Fig. 3.62. Calculate the composition of this unit cell. What is the lattice type of this structure? If 24 carat gold is pure gold, what carat gold does this alloy represent?

3.8 Using Ketelaar's triangle would you classify Sr₂Ga (χ(Sr) = 0.95; χ(Ga) = 1.81) as an alloy or a Zintl phase?

3.9 Depending on temperature, RbCl can exist in either the rock-salt or caesium-chloride structure. (a) What is the coordination number of the cation and anion in each of these structures? (b) In which of these structures will Rb have the larger apparent radius?

3.10 Consider the structure of caesium chloride. How many Cs⁺ ions occupy second-nearest neighbour locations of a Cs⁺ ion?

3.11 Describe the coordination around the anions in the perovskite structure in terms of coordination to the A- and B-type cations.

3.12 Use radius-ratio rules and the ionic radii given in *Resource section* 1 to predict structures of (a) PuO₂, (b) FrI, (c) BeO, and (d) InN.

Fig. 3.62 The structure of Cu₃Au.

3.13 What will be the most significant terms in the Born–Haber cycle for the formation of Ca₃N₂?

3.14 By considering the parameters that change in the Born–Mayer expression estimate a lattice enthalpy for MgO given that both MgO and NaCl adopt rock-salt structures with similar lattice parameters and $\Delta H_L^\circ(\text{NaCl}) = 786\ \text{kJ mol}^{-1}$.

3.15 Use the Kapustinskii equation and the ionic and thermochemical radii given in *Resource section* 1 and Table 3.10 and $r(\text{Bk}^{4+}) = 97$ pm to calculate lattice enthalpies of (a) BkO₂, (b) K₂SiF₆, and (c) LiClO₄.

3.16 Which member of each pair is likely to be more soluble in water: (a) SrSO₄ or MgSO₄, (b) NaF or NaBF₄?

3.17 On the basis of the factors that contribute to lattice enthalpies place LiF, CaO, RbCl, AlN, NiO, and CsI, all of which adopt the rock-salt structure, in order of increasing lattice energy.

3.18 Recommend a specific cation for the quantitative precipitation of carbonate ion in water. Justify your recommendations.

There are many brief *Exercises* at the end of each chapter. Answers will be found in the *Answers* section and fully worked answers are available in the separate *Solutions manual* (see below). The *Exercises* are all reasonably straightforward and can be used to check your understanding, gain experience and practice in tasks such as balancing equations, predicting and drawing structures, and manipulating data.

Problems

PROBLEMS

3.1 Draw a cubic unit cell (a) in projection (showing the fractional heights of the atoms) and (b) as a three-dimensional representation that has atoms at the following positions: Ti at (½, ½, ½), O at (½, ½, 0), (½, ½, 0), and (½, 0, ½), and Ba at (0,0,0). Remember that a cubic unit cell with an atom on the cell face, edge, or corner will have equivalent atoms displaced by the unit cell repeat in any direction. Which structure type is the cell?

3.2 Draw a tetragonal unit cell and mark on it a set of points that would define (a) a face-centred lattice and (b) a body-centred lattice. Demonstrate, by considering two adjacent unit cells, that a tetragonal face-centred lattice of dimensions a and c can always be redrawn as a body-centred tetragonal lattice with dimensions $a/2^{1/2}c$.

3.3 Draw one layer of close-packed spheres. On this layer mark the positions of the centres of the B layer atoms using the symbol ⊗ and, with the symbol ○, mark the positions of the centres of the C layer atoms of an fcc lattice.

3.4 In the structure of MoS₂, the S atoms are arranged in close-packed layers that repeat themselves in the sequence AAA.... The Mo atoms occupy holes with coordination number 6. Show that each Mo atom is surrounded by a trigonal prism of S atoms.

3.5 The ReO₃ structure is cubic with an Re atom at each corner of the unit cell and one O atom on each unit cell edge midway between the Re atoms. Sketch this unit cell and determine the coordination numbers of the ions and (b) the identity of the structure type that would be generated if a cation were inserted in the centre of each ReO₃ unit.

3.5 Consider the structure of rock salt. (a) How many Na⁺ ions occupy second-nearest neighbour locations of an Na⁺ ion? (b) Pick out the closest-packed plane of Cl⁻ ions. (*Hint*: this hexagonal plane will be perpendicular to a threefold axis.)

3.6 Imagine the construction of an MX₂ structure from the CsCl structure by removal of half the Cs⁺ ions to leave tetrahedral coordination around each Cl⁻ ion. Identify this MX₂ structure.

The *Problems* are more demanding in content and in style and are often based on a research paper or other additional source of information. Problems generally require a discursive response and there may not be a single correct answer. They may be used as essay type questions or for classroom discussion.

About the Online Resource Centre

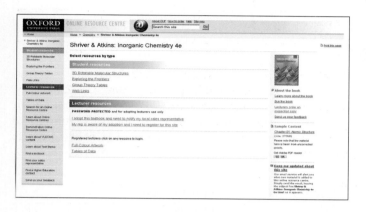

The Online Resource Centre provides teaching and learning resources. It is free of charge, complements the textbook, and offers additional materials which can be downloaded. The resources it provides are fully customisable and can be incorporated into a virtual learning environment.

Go to:

www.oxfordtextbooks.co.uk/orc/ichem4e/

Lecturer resources

Artwork

All the illustrations are available and can be used for lectures without charge (but not for commercial purposes without specific permission).

Tables of data

All the tables of data that appear in the chapter text are available and may be used under the same conditions as the figures.

Student resources

3D rotatable molecular structures

All the numbered molecular structures will be found in a three-dimensional, viewable, rotatable form along with many of the crystal structures and bioinorganic molecules. These have been produced in collaboration with Karl Harrison, University of Oxford.

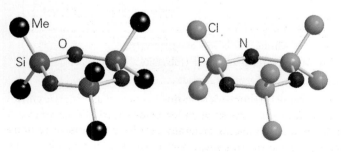

Web links

There is a huge network of information available about inorganic chemistry, and it can be bewildering to navigate through it. These links to web sites, which contain educational resources and further information are annotated to provide a guide through the large amount of material that is available.

Exploring the frontiers

Links to recent news and research articles which allow you to explore each of the topics introduced in the 'frontier' chapters.

Group theory tables

Comprehensive group theory tables are available for downloading.

Videos of chemical reactions

Video clips showing demonstrations of inorganic chemistry reactions.

Solutions manual

A *Solutions manual* to accompany this book, by Michael Hagerman, Union College, Christopher Schnabel, Eckerd College, and Kandalam Ramanujachary, Rowan University, contains complete solutions to the *Self-tests* and end-of-chapter *Exercises* (978–019–928859–5).

Acknowledgements

We have taken care to ensure that the text is free of errors. This is difficult in a rapidly changing field, where today's knowledge is soon replaced by tomorrow's. We would particularly like to thank Jennifer Armstrong and Steve Ogden, University of Southampton; Sandra Dann, University of Loughborough; Peter Taylor and Peter Moore, University of Warwick; Leslie Dutton, University of Philadelphia; Stuart Ferguson, University of Oxford; Judy Hirst and Edmund Kunji, Medical Research Council Dunn Human Nutrition Unit, Cambridge, for their guidance and advice. We acknowledge and thank all those colleagues who so unstintingly gave their time and expertise to a careful reading of a variety of draft chapters.

Jörgen Albertsson, Chalmers University of Technology

Angelo Amoroso, Cardiff University

James Anderson, University of Aberdeen

Stephen Best, University of Melbourne

Rolf Berger, Uppsala University

Henrik Birkedal, University of Aarhus

Adam Bridgeman, University of Hull

Alice Bruce, University of Maine

Andrew Burrows, University of Bath

Houston Byrd, University of Montevallo

David Cardin, University of Reading

Neil Champness, University of Nottingham

Patricia Chernovitz, Park University

Ann Chippendale, University of Reading

Peter Cragg, University of Brighton

Leroy Cronin, University of Glasgow

Michael Crowder, Miami University

Sandra Dann, Loughborough University

Bas de Bruin, Radboud University Nijmegen

Roger de la Rosa, St. Louis University

Nancy Dervisi, Cardiff University

Simon Duckett, University of York

Margaret Farago, Imperial College London

Walter Flomer, Northwestern State University

Kevin Flower, University of Manchester

David Geiger, SUNY College at Geneseo

John Goodwin, Coastal Carolina University

Gary Gray, University of Alabama at Birmingham

Tim Greene, University of Exeter

Gerald Gries, University of St. Thomas

Malcolm Halcrow, University of Leeds

Joan Halfpenny, Nottingham Trent University

Anders Hammershøi, University of Copenhagen

Justin Hargreaves, University of Glasgow

Andrew Harrison, University of Edinburgh

William Harrison, University of Aberdeen

Catherine Hinga Haustein, Central College

Anthony Haynes, University of Sheffield

Richard Henderson, University of Newcastle-upon-Tyne

Simon Higgins, University of Liverpool

Gary Hix, Nottingham Trent University

Richard Hotz, College of Mount St. Joseph

Alan Howard, University of Southampton

Michael Hudson, University of Reading,

Carl Hultman, Gannon University

John Irvine, University of St Andrews

David Jackson, University of Glasgow

Stuart James, Queen's University, Belfast

Robert Janes, Open University

Roger Jewsbury, University of Huddersfield

Alan Jircitano, Pennsylvania State Erie

Richard Jones, Keele University

Frederick Keen, Clarion University of Pennsylvania

Paul Kelly, Loughborough University

Lars Kloo, Royal Institute of Technology, Göteborg

Brian Knettle, University of Virginia College at Wise

Randolf Köhn, University of Bath

Donald Krogstad, Concordia College-Moorhead

Simon Lancaster, University of East Anglia

Art Landis, Emporia State University

Scott Luaders, Quincy University

Laura Lee, Lock Haven University

Christopher Levy, Kansas State University

Oliver Lindqvist, Göteborg University

Paul Low, University of Durham

Charles Mahler, Lycoming College

Mark McClure, The University of North Carolina at Pembroke

Claude Mertzenich, Luther College

Gregory Moehring, Governors State University

Karen Moss, Nottingham Trent University

John Nelson, University of Nevada

Nicolas Norman, University of Bristol

Lars-Johan Norrby, University of Stockholm

Ei-Ichiro Ochiai, Juniata College

Mark Ogden, Curtin University of Technology

Thomas Palstra, University of Groningen

Itai Panas, Chalmers University of Technology

Jason Powell, Ferrum College

Stephen Ralph, University of Wollongong

Peter Sadler, University of Edinburgh

Graham Saunders, Queen's University, Belfast

Milo Shaffer, Imperial College London

Stephen Skinner, Imperial College London

Robert Slade, University of Surrey

Peter Slater, University of Surrey

Peter Smith, Westminster College

Lothar Stahl, University of North Dakota

Alexander Steiner, University of Liverpool

Daniel Strömberg, Göteborg University

Gunnar Svensson, Stockholm University

Duane Swank, Pacific Lutheran University

Jan Teuben, University of Groningen

Hans Toftlund, University of Southern Denmark

Edman Tsang, University of Reading

Jens Ulstrup, Technical University of Denmark

John Vincent, University of Alabama

Kevin Wainwright, Flinders University

Richard Walton, University of Exeter

John Webb, Murdoch University

Colin White, University of Sheffield

Derk Andrew Wierda, Saint Anselm College

Derek Woollins, University of St Andrews

Many of the figures in Chapter 26 were produced using PyMOL software; for more information see DeLano, W.L. *The PyMOL Molecular Graphics System* (2002), De Lano Scientific, San Carlos, CA, USA.

Finally we wish to thank our publishers, Oxford University Press and W.H. Freeman, especially Ruth Hughes, for the understanding and assistance they have provided at all stages of the intricate and time-consuming task of producing a book of such complexity as this. We owe a considerable debt to their patience, wisdom, and understanding.

The authors

Summary of contents

Contents

PART 3 Frontiers   587

Glossary of chemical abbreviations

Ac	acetyl, CH_3CO
acac	acetylacetonato
aq	aqueous solution species
bipy	2,2′-bipyridine
bn	2,3-butanediamine
cod	1,5-cyclooctadiene
cot	cyclooctatetraene
Cy	cyclohexyl
Cp	cyclopentadienyl
Cp*	pentamethylcyclopentadienyl
cyclam	tetraazacyclotetradecane
dien	diethylenetriamine
DMSO	dimethylsulfoxide
DMF	dimethylformamide
η	hapticity
edta	ethylenediaminetetraacetato
en	ethylenediamine (1,2-diaminoethane)
Et	ethyl
gly	glycinato
Hal	halide
^{i}Pr	isopropyl
KCP	$K_2Pt(CN)_4Br_{0.3} \cdot 3H_2O$
L	a ligand
μ	signifies a bridging ligand
M	a metal
Me	methyl
mes	mesityl, 2,4,6-trimethylphenyl
Ox	an oxidized species
ox	oxalato
Ph	phenyl
phen	phenanthroline
py	pyridine
pyz	pyrazine
Sol	solvent, or a solvent molecule
soln	nonaqueous solution species
^{t}Bu	tertiary butyl
THF	tetrahydrofuran
TMEDA	*N, N, N′, N′*-tetramethylethylenediamine
trien	2,2′,2″-triaminotriethylene
X	generally halogen, also a leaving group or an anion
Xyl	xylyl, 2,6-dimethylphenyl
Y	an entering group

PART 1
Foundations

The eight chapters in this part of the book lay the foundations of inorganic chemistry. The first three chapters develop an understanding of the structures of atoms, molecules, and solids. Chapter 1 introduces the structure of atoms in terms of quantum theory and describes important periodic trends in their properties. Chapter 2 develops molecular structure in terms of increasingly sophisticated models of covalent bonding. Chapter 3 describes ionic bonding and the structures and properties of a range of typical ionic solids.

The next two chapters focus on two major types of reactions. Chapter 4 introduces the definitions of acids and bases and uses their properties to systematize many inorganic reactions. Chapter 5 describes oxidation and reduction and demonstrates how electrochemical data can be used to predict and explain the outcomes of redox reactions.

Chapter 6 provides a toolbox for inorganic chemistry: it describes a wide range of the instrumental techniques that are used to identify and determine the structures of compounds. Chapter 7 shows how a systematic consideration of the symmetry of molecules can be used to discuss the bonding and structure of molecules and help interpret the techniques described in Chapter 6.

Chapter 8 describes the coordination compounds of the elements. We discuss bonding, structure, and reactions of complexes and see how symmetry considerations can provide useful insight into this important class of compounds.

Atomic structure

1

This chapter lays the foundations for the explanation of the trends in the physical and chemical properties of all inorganic compounds. To understand the behaviour of molecules and solids we need to understand atoms. Our study of inorganic chemistry must therefore begin with a review of the structures and properties of atoms. We begin with discussion of the origin of matter in the solar system and then consider the development of our understanding of atomic structure and the behaviour of electrons in atoms. We introduce quantum theory qualitatively and use the results to rationalize properties such as atomic radii, ionization energy, electron affinity, and electronegativity. An understanding of these properties allows us to begin to rationalize the diverse chemical properties of the 111 elements known today.

The observation that the universe is expanding has led to the current view that about 15 billion years ago the currently visible universe was concentrated into a point-like region that exploded in an event called the **Big Bang**. With initial temperatures immediately after the Big Bang thought to be about 10^9 K, the fundamental particles produced in the explosion had too much kinetic energy to bind together in the forms we know today. However, the universe cooled as it expanded, the particles moved more slowly, and they soon began to adhere together under the influence of a variety of forces. In particular, the **strong force**, a short-range but powerful attractive force between nucleons (protons and neutrons), bound these particles together into nuclei. As the temperature fell still further, the **electromagnetic force**, a relatively weak but long-range force between electric charges, bound electrons to nuclei to form atoms.

Table 1.1 summarizes the properties of the only subatomic particles that we need to consider in chemistry. The 111 known elements that are formed from these subatomic particles are distinguished by their **atomic number**, Z, the number of protons in the nucleus of an atom of the element. Many elements have a number of **isotopes**, which are atoms with the same atomic number but different atomic masses. These isotopes are distinguished by the **mass number**, A, which is the total number of protons and neutrons in the nucleus. The mass number is also sometimes termed more appropriately the 'nucleon number'. Hydrogen, for instance, has three isotopes. In each case $Z = 1$, indicating that the nucleus contains one proton. The most abundant isotope has $A = 1$, denoted ^{1}H, its nucleus consisting of a single proton. Far less abundant (only 1 atom in 6000) is deuterium, with $A = 2$. This mass number indicates that, in addition to a proton, the nucleus contains one neutron. The formal designation of deuterium is ^{2}H, but it is commonly denoted D. The third, short-lived, radioactive isotope of hydrogen is tritium, ^{3}H or T. Its nucleus consists of one proton and two neutrons. In certain cases it is helpful to display the atomic number of the element as a left suffix; so the three isotopes of hydrogen would then be denoted $^1_1$H, $^2_1$H, and $^3_1$H.

The origin of the elements

About two hours after the start of the universe, the temperature had fallen so much that most of the matter was in the form of H atoms (89 per cent) and He atoms (11 per cent). In one sense, not much has happened since then for, as Fig. 1.1 shows, hydrogen and

Table 1.1 Subatomic particles of relevance to chemistry

Particle	Symbol	Mass/u*	Mass number	Charge/$e^{\dagger}$	Spin
electron	e^-	5.486×10^{-4}	0	−1	$\frac{1}{2}$
proton	p	1.0073	1	+1	$\frac{1}{2}$
neutron	n	1.0087	1	0	$\frac{1}{2}$
photon	γ	0	0	0	1
neutrino	ν	≈ 0	0	0	$\frac{1}{2}$
positron	e^+	5.486×10^{-4}	0	+1	$\frac{1}{2}$
α particle	α	[$^4_2$He^{2+} nucleus]	4	+2	0
β particle	β	[e^- ejected from nucleus]	0	−1	$\frac{1}{2}$
γ photon	γ	[electromagnetic radiation from nucleus]	0	0	1

* Masses are expressed in atomic mass units, u, with 1 u = 1.6605×10^{-27} kg.
† The elementary charge e is 1.602×10^{-19} C.

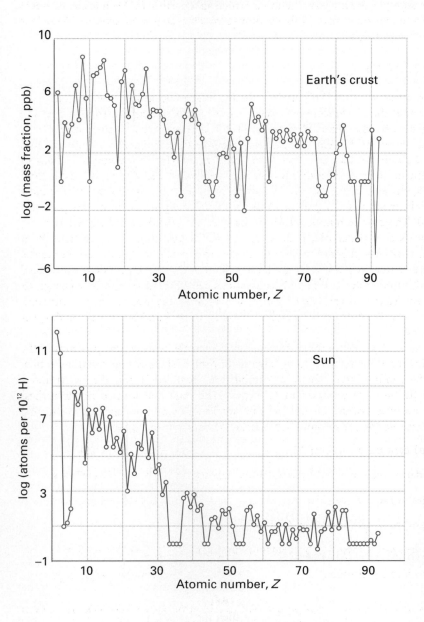

Fig. 1.1 The abundances of the elements in the universe. Elements with odd Z are less stable than their neighbours with even Z. The abundances refer to the number of atoms of each element relative to Si taken as 10^6.

helium remain overwhelmingly the most abundant elements in the universe. However, nuclear reactions have formed a wide assortment of other elements and have immeasurably enriched the variety of matter in the universe, and thus given rise to the whole area of chemistry.

1.1 The nucleosynthesis of light elements

Key points: The light elements were formed by nuclear reactions in stars formed from primeval hydrogen and helium; total mass number and overall charge are conserved in nuclear reactions; a large binding energy signifies a stable nucleus.

The earliest stars resulted from the gravitational condensation of clouds of H and He atoms. The compression of these clouds under the influence of gravity gave rise to high temperatures and densities within them, and fusion reactions began as nuclei merged together. The earliest nuclear reactions are closely related to those now being studied in connection with the development of controlled nuclear fusion.

Energy is released when light nuclei fuse together to give elements of higher atomic number. For example, the nuclear reaction in which an α particle (a ^{4}He nucleus with two protons and two neutrons) fuses with a carbon-12 nucleus to give an oxygen-16 nucleus and a γ-ray photon (γ) is

$$^{12}_{6}C + ^{4}_{2}\alpha \rightarrow ^{16}_{8}O + \gamma$$

This reaction releases 7.2 MeV of energy.[1] Nuclear reactions are very much more energetic than normal chemical reactions because the strong force is much stronger than the electromagnetic force that binds electrons to nuclei. Whereas a typical chemical reaction might release about 10^3 kJ mol^{-1}, a nuclear reaction typically releases a million times more energy, about 10^9 kJ mol^{-1}. In this nuclear equation, the **nuclide**, a nucleus of specific atomic number Z and mass number A, is designated ^A_ZE, where E is the chemical symbol of the element. Note that, in a balanced nuclear equation, the sum of the mass numbers of the reactants is equal to the sum of the mass numbers of the products ($12 + 4 = 16$). The atomic numbers sum similarly ($6 + 2 = 8$) provided an electron, e$^-$, when it appears as a β particle, is denoted $^0_{-1}$e and a positron, e$^+$, is denoted $^0_1$e. A positron is a positively charged version of an electron: it has zero mass number and a single positive charge. When it is emitted, the mass number of the nuclide is unchanged but the atomic number decreases by 1 because the nucleus has lost one positive charge. Its emission is equivalent to the conversion of a proton in the nucleus into a neutron: $^1_1p \rightarrow ^1_0n + e^+ + \nu$. A neutrino, ν (nu), is electrically neutral and has a very small (possibly zero) mass.

Elements up to $Z = 26$ were formed inside stars. Such elements are the products of the nuclear fusion reactions referred to as 'nuclear burning'. The burning reactions, which should not be confused with chemical combustion, involved H and He nuclei and a complicated fusion cycle catalysed by C nuclei. (The stars that formed in the earliest stages of the evolution of the cosmos lacked C nuclei and used noncatalysed H-burning reactions.) Some of the most important nuclear reactions in the cycle are

Proton (p) capture by carbon-12 : $^{12}_{6}C + ^1_1p \rightarrow ^{13}_{7}N + \gamma$

Positron decay accompanied by neutrino(ν) emission : $^{13}_{7}N \rightarrow ^{13}_{6}C + e^+ + \nu$

Proton capture by carbon-13 : $^{13}_{6}C + ^1_1p \rightarrow ^{14}_{7}N + \gamma$

Proton capture by nitrogen-14 : $^{14}_{7}N + ^1_1p \rightarrow ^{15}_{8}O + \gamma$

Positron decay, accompanied by neutrino emission : $^{15}_{8}O \rightarrow ^{15}_{7}N + e^+ + \nu$

Proton capture by nitrogen-15 : $^{15}_{7}N + ^1_1p \rightarrow ^{12}_{6}C + ^4_2\alpha$

---

[1] An electronvolt (1 eV) is the energy required to move an electron through a potential difference of 1 V. It follows that 1 eV $= 1.602 \times 10^{-19}$ J, which is equivalent to 96.48 kJ mol^{-1}; 1 MeV $= 10^6$ eV.

The net result of this sequence of nuclear reactions is the conversion of four protons (four ^{1}H nuclei) into an α particle (a ^{4}He nucleus):

$$4\,^1_1\text{p} \rightarrow\,^4_2\alpha + 2\,\text{e}^+ + 2\,\nu + 3\,\gamma$$

The reactions in the sequence are rapid at temperatures between 5 and 10 MK (where 1 MK $= 10^6$ K). Here we have another contrast between chemical and nuclear reactions, because chemical reactions take place at temperatures a hundred-thousand times lower. Moderately energetic collisions between species can result in chemical change, but only highly vigorous collisions can provide the energy required to bring about most nuclear processes.

Heavier elements are produced in significant quantities when hydrogen burning is complete and the collapse of the star's core raises the density there to 10^8 kg m^{-3} (about 10^5 times the density of water) and the temperature to 100 MK. Under these extreme conditions, helium burning becomes viable. The low abundance of beryllium in the present-day universe is consistent with the observation that $^8_4$Be formed by collisions between α particles goes on to react with more α particles to produce the more stable carbon nuclide, $^{12}_6$C:

$$^8_4\text{Be} + \,^4_2\alpha \rightarrow\,^{12}_6\text{C} + \gamma$$

Thus, the helium-burning stage of stellar evolution does not result in the formation of beryllium as a stable end product; for similar reasons, low concentrations of lithium and boron are also formed. The nuclear reactions leading to these three elements are still uncertain, but they may result from the fragmentation of C, N, and O nuclei by collisions with high-energy particles.

Elements can also be produced by nuclear reactions such as neutron (n) capture accompanied by proton emission:

$$^{14}_7\text{N} + \,^1_0\text{n} \rightarrow\,^{14}_6\text{C} + \,^1_1\text{p}$$

This reaction still continues in our atmosphere as a result of the impact of cosmic rays and contributes to the steady-state concentration of radioactive carbon-14 on Earth.

The high abundance of iron and nickel in the universe is consistent with these elements having the most stable of all nuclei. This stability is expressed in terms of the **binding energy**, which represents the difference in energy between the nucleus itself and the same numbers of individual protons and neutrons. This binding energy is often presented in terms of a difference in mass between the nucleus and its individual protons and neutrons because, according to Einstein's theory of relativity, mass and energy are related by $E = mc^2$, where c is the speed of light. Therefore, if the mass of a nucleus differs from the total mass of its components by $\Delta m = m_{\text{nucleons}} - m_{\text{nucleus}}$, then its binding energy is $E_{\text{bind}} = (\Delta m)c^2$. The binding energy of ^{56}Fe, for example, is the difference in energy between the ^{56}Fe nucleus and 26 protons and 30 neutrons. A positive binding energy corresponds to a nucleus that has a lower, more favourable, energy (and lower mass) than its constituent nucleons (Box 1.1).

Figure 1.2 shows the binding energy per nucleon, E_{bind}/A (obtained by dividing the total binding energy by the number of nucleons), for all the elements. Iron and nickel occur at the maximum of the curve, showing that their nucleons are bound more strongly than in any other nuclide. Harder to see from the graph is an alternation of binding energies as the atomic number varies from even to odd, with even-Z nuclides slightly more stable than their odd-Z neighbours. There is a corresponding alternation in cosmic abundances, with nuclides of even atomic number being marginally more abundant than those of odd atomic number.

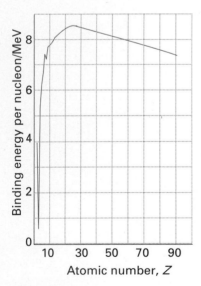

Fig. 1.2 Nuclear binding energies. The greater the binding energy, the more stable the nucleus.

1.2 The nucleosynthesis of heavy elements

Key point: Heavier nuclides are formed by processes that include neutron capture and subsequent β decay.

Because nuclei close to iron are the most stable, heavier elements are produced by a variety of energy-consuming processes. These processes include the capture of free

Box 1.1 Nuclear fusion and fission

If two nuclei with mass numbers lower than 56 merge to produce a new nucleus with a larger nuclear binding energy, E_{bind}/A (Fig. 1.2), the 'excess' binding energy is released. This process is called **fusion**. For example, two neon-20 nuclei may fuse to give a calcium-40 nucleus:

$$2\,^{20}_{10}\mathrm{Ne} \rightarrow\,^{40}_{20}\mathrm{Ca}$$

The value of E_{bind}/A for Ne is approximately 8.0 MeV. Therefore, the total binding energy on the left-hand side of the equation is $2 \times 20 \times 8.0 = 320$ MeV. The value of E_{bind}/A for Ca is around 8.6 MeV and the total energy on the right-hand side is $40 \times 8.6 = 344$ MeV. Thus, the difference in the binding energies of the products and reactants is 34 MeV.

For nuclei with mass numbers greater than 56, binding energy can be released when they split into lighter products with higher values of E_{bind}/A. This process is called **fission**. For example, uranium-236 can undergo fission into xenon-140 and strontium-93 nuclei:

$$^{236}_{92}\mathrm{U} \rightarrow\,^{140}_{54}\mathrm{Xe} +\,^{93}_{38}\mathrm{Sr} + 3\,\mathrm{n}$$

The values of E_{bind}/A for ^{236}U, ^{140}Xe, and ^{93}Sr nuclei are 7.6, 8.4, and 8.7 MeV, respectively. Therefore, the energy released in this reaction is $(140 \times 8.4) + (93 \times 8.7) - (236 \times 7.6) = 191.5$ MeV for the fission of each ^{236}U nucleus.

Fission can also be induced by bombarding heavy elements with neutrons:

$$^{235}_{92}\mathrm{U} + \mathrm{n} \rightarrow \text{fission products} + \text{neutrons}$$

For each neutron consumed, approximately 2.5 neutrons are produced. These go on to cause fission of other ^{235}U nuclei. Most of the released energy occurs as kinetic energy of the fission products and the neutrons. The kinetic energy is rapidly converted into thermal motion ('heat') through collisions with other atoms. In a nuclear reactor, the chain reaction is controlled by absorbing some of the neutrons in material such as graphite. An equilibrium state is reached in which one neutron survives for each fission event. In nuclear weapons, no attempt is made to control the chain reaction and the resulting energy release is explosively fast.

The kinetic energy of fission products from ^{235}U is about 165 MeV, that of the neutrons is about 5 MeV, and the γ-rays produced have an energy of about 7 MeV. The fission products are themselves radioactive and decay by β-, γ-, and X-radiation, releasing about 23 MeV. The neutrons that are not consumed by fission are captured in the reactor with the release of about 10 MeV. The energy produced is offset by about 10 MeV, which escapes from the reactor as radiation, and about 1 MeV as undecayed fission products remaining in the spent fuel. Therefore, the total energy produced for one fission event is about 200 MeV, or 32 pJ. It follows that about 1 W of reactor heat (where $1\,\mathrm{W} = 1\,\mathrm{J\,s}^{-1}$) corresponds to about 3.1×10^{10} fission events per second. A nuclear reactor producing 3 GW of heat will have an electrical output of approximately 1 GW and corresponds to the fission of 3 kg of ^{235}U per day.

neutrons, which are not present in the earliest stages of stellar evolution but are produced later in reactions such as

$$^{23}_{10}\mathrm{Na} +\,^{4}_{2}\alpha \rightarrow\,^{26}_{12}\mathrm{Mg} +\,^{1}_{0}\mathrm{n}$$

Under conditions of intense neutron flux, as in a supernova (the explosion of a star), a given nucleus may capture a succession of neutrons and become a progressively heavier isotope. However, there comes a point at which the nucleus will eject an electron from the nucleus as a β particle (a high-velocity electron, e^-). Because β decay leaves the mass number of the nuclide unchanged but increases its atomic number by 1 (the nuclear charge increases by 1 unit when an electron is ejected), a new element is formed. An example is

Neutron capture : $\quad^{98}_{42}\mathrm{Mo} +\,^{1}_{0}\mathrm{n} \rightarrow\,^{99}_{42}\mathrm{Mo} + \gamma$

Followed by β decay accompanied by neutrino emission : $\quad^{99}_{42}\mathrm{Mo} \rightarrow\,^{99}_{43}\mathrm{Tc} + e^- + \nu$

The **daughter nuclide**, the product of a nuclear reaction ($^{99}_{43}\mathrm{Tc}$, an isotope of technetium, in this example), can absorb another neutron, and the process can continue, gradually building up the heavier elements.

Example 1.1 Balancing equations for nuclear reactions

Synthesis of heavy elements occurs in the neutron-capture reactions believed to take place in the interior of cool 'red giant' stars. One such reaction is the conversion of $^{68}_{30}\mathrm{Zn}$ to $^{69}_{31}\mathrm{Ga}$ by neutron capture to form $^{69}_{30}\mathrm{Zn}$, which then undergoes β decay. Write balanced nuclear equations for this process.

Answer Neutron capture increases the mass number of a nuclide by 1 but leaves the atomic number (and hence the identity of the element) unchanged:

$$^{68}_{30}\mathrm{Zn} +\,^{1}_{0}\mathrm{n} \rightarrow\,^{69}_{30}\mathrm{Zn} + \gamma$$

The excess energy is carried away as a photon. The loss of an electron from the nucleus by β decay leaves the mass number unchanged but increases the atomic number by 1. Because zinc has atomic number 30, the daughter nuclide has $Z = 31$, corresponding to gallium. Therefore, the nuclear reaction is

$$^{69}_{30}\text{Zn} \rightarrow {}^{69}_{31}\text{Ga} + \text{e}^-$$

In fact, a neutrino is also emitted, but this cannot be inferred from the data as a neutrino is effectively massless and electrically neutral.

Self-test 1.1 Write the balanced nuclear equation for neutron capture by $^{80}_{35}\text{Br}$.

1.3 The classification of the elements

Some substances that we now recognize as chemical elements have been known since antiquity: they include carbon, sulfur, iron, copper, silver, gold, and mercury. The alchemists and their immediate successors, the early chemists, had added about another 18 elements by 1800. By that time, the precursor of the modern concept of an element had been formulated as a substance that consists of only one type of atom. (Now, of course, by 'type' of atom we mean an atom with a particular atomic number.) By 1800 many experimental techniques were available for converting oxides and other compounds into elements. These techniques were considerably enhanced by the introduction of electrolysis. The list of elements grew rapidly in the later nineteenth century. This growth was in part a result of the development of atomic spectroscopy, in which thermally excited atoms of a particular element are observed to emit electromagnetic radiation with a unique pattern of frequencies. These spectroscopic observations made it much easier to detect previously unknown elements.

(a) Patterns and periodicity

Key points: The elements are broadly divided into metals, nonmetals, and metalloids according to their physical and chemical properties; the organization of elements into the form resembling the modern periodic table is accredited to Mendeleev.

A useful broad division of elements is into **metals** and **nonmetals**. Metallic elements (such as iron and copper) are typically lustrous, malleable, ductile, electrically conducting solids at about room temperature. Nonmetals are often gases (oxygen), liquids (bromine), or solids that do not conduct electricity appreciably (sulfur). The chemical implications of this classification should already be clear from introductory chemistry:

1 Metallic elements combine with nonmetallic elements to give compounds that are typically hard, nonvolatile solids (for example, sodium chloride).

2 When combined with each other, the nonmetals often form volatile molecular compounds (for example, phosphorus trichloride).

3 When metals combine (or simply mix together) they produce alloys that have most of the physical characteristics of metals (for example, brass from copper and zinc).

Some elements have properties that make it difficult to classify them as metals or nonmetals. These elements are called **metalloids**. Examples of metalloids are silicon, germanium, arsenic, and tellurium.

A more detailed classification of the elements is the one devised by Dmitri Mendeleev in 1869; this scheme is familiar to every chemist as the **periodic table**. Mendeleev arranged the known elements in order of increasing atomic weight (molar mass). This arrangement resulted in families of elements with similar chemical properties, which he arranged into the groups of the periodic table. For example, the fact that C, Si, Ge, and Sn all form hydrides of the general formula EH_4 suggests that they belong to the same group. That N, P, As, and Sb all form hydrides with the general formula EH_3 suggests that

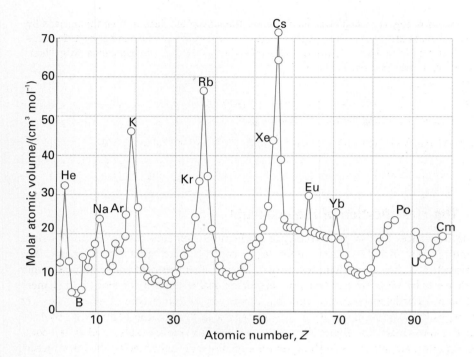

Fig. 1.3 The periodic variation of molar volume with atomic number.

they belong to a different group. Other compounds of these elements show family similarities, as in the formulas CF_4 and SiF_4 in the first group, and NF_3 and PF_3 in the second.

Mendeleev concentrated on the chemical properties of the elements. At about the same time Lothar Meyer in Germany was investigating their physical properties, and found that similar values repeated periodically with increasing molar mass. Figure 1.3 shows a classic example, where the molar volume of the element (its volume per mole of atoms) in its normal form is plotted against atomic number.

Mendeleev provided a spectacular demonstration of the usefulness of the periodic table by predicting the general chemical properties, such as the numbers of bonds they form, of unknown elements corresponding to gaps in his original periodic table. The same process of inference from periodic trends is still used by inorganic chemists to rationalize trends in the physical and chemical properties of compounds and to suggest the synthesis of previously unknown compounds. For instance, by recognizing that carbon and silicon are in the same family, the existence of alkenes $R_2C{=}CR_2$ suggests that $R_2Si{=}SiR_2$ ought to exist too. Compounds with silicon-silicon double bonds (disilaethenes) do indeed exist, but it was not until 1981 that chemists succeeded in isolating one.

(b) The modern periodic table

Key points: The periodic table is divided into periods and groups; the groups belong to four major blocks; the main-group elements are those in the *s* and *p* blocks.

The general structure of the modern periodic table will be familiar from previous chemistry courses (Fig. 1.4), and the following is a review. The elements are listed in order of atomic number, not atomic weight, because the atomic number tells us the number of electrons in the atom and is therefore a more useful quantity. The horizontal rows of the table are called **periods** and the vertical columns are called **groups**. The numbering system used for groups in the illustration follows the IUPAC recommendation. We often use the group number to designate the general position of an element, as in 'gallium is in Group 13'; alternatively, the lightest element in the group is used to designate the group, as in 'gallium is a member of the boron group'. The members of the same group as a given element are called the **congeners** of that element. Thus, sodium and potassium are congeners of lithium.

The periodic table is divided into four **blocks**. The members of the *s*- and *p*-blocks are collectively called the **main-group elements**. The *d*-block elements (often with the

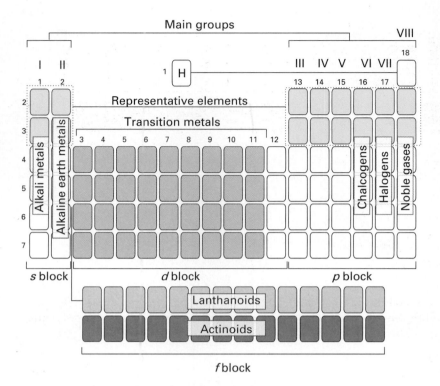

Fig. 1.4 The general structure of the periodic table. Compare this template with the complete table inside the front cover for the identities of the elements that belong to each block.

exception of Group 12, zinc, cadmium, and mercury) are also referred to collectively as the **transition elements**. The *f*-block elements are divided into the lighter series (atomic numbers 57–71) called the **lanthanoids** (more commonly, still the 'lanthanides') and the heavier series (atomic numbers 89–103) called the **actinoids** (commonly, the 'actinides'). To save space, the *f*-block is normally removed from its 'true' position and placed below the rest of the elements. Consequently, this arrangement is sometimes referred to as the **short form** of the periodic table. The **representative elements** are the members of the first three periods of the main-group elements (from hydrogen to argon).

In the illustration we show both the traditional numbering of the main groups (with the roman numerals from I to VIII) and the current IUPAC recommendation, in which the groups of the *s*-, *d*-, and *p*-blocks are numbered from 1 to 18. The groups of the *f*-block are not numbered because each row forms a highly homogeneous family.

The structures of hydrogenic atoms

The organization of the periodic table is a direct consequence of periodic variations in the electronic structure of atoms. Initially, we consider hydrogen-like or **hydrogenic atoms**, which have only one electron and so are free of the complicating effects of electron–electron repulsions. Hydrogenic atoms include ions such as He^+ and C^{5+} (found in stellar interiors) as well as the hydrogen atom itself. Then we use the concepts these atoms introduce to build up an approximate description of the structures of **many-electron atoms**, which are atoms with more than one electron.

1.4 Spectroscopic information

Key points: Spectroscopic observations on hydrogen atoms suggest that an electron can occupy only certain energy levels and that the emission of discrete frequencies of electromagnetic radiation occurs when an electron makes a transition between these levels.

Electromagnetic radiation is emitted when an electric discharge is passed through hydrogen gas. When passed through a prism or diffraction grating, this radiation is found to consist of a series of components, one in the ultraviolet region, one in the visible

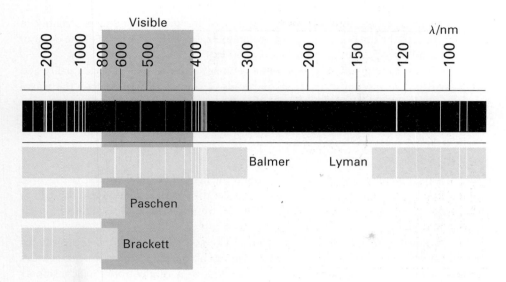

Fig. 1.5 The spectrum of atomic hydrogen and its analysis into series.

region, and several in the infrared region of the electromagnetic spectrum (Fig. 1.5). The nineteenth-century spectroscopist Johann Rydberg found that all the wavelengths (λ, lambda) can be described by the expression

$$\frac{1}{\lambda} = R\left(\frac{1}{n_1^2} - \frac{1}{n_2^2}\right) \tag{1.1}$$

where R is the **Rydberg constant**, an empirical constant with the value $1.097 \times 10^7 \, \text{m}^{-1}$. The n are integers, with $n_1 = 1, 2, \ldots$ and $n_2 = n_1 + 1, n_1 + 2, \ldots$. The series with $n_1 = 1$ is called the *Lyman series* and lies in the ultraviolet. The series with $n_1 = 2$ lies in the visible region and is called the *Balmer series*. The infrared series include the *Paschen series* ($n_1 = 3$) and the *Brackett series* ($n_1 = 4$).

The structure of the spectrum is explained if it is supposed that the emission of radiation takes place when an electron makes a transition from a state of energy $-hcR/n_2^2$ to a state of energy $-hcR/n_1^2$ and that the difference, which is equal to $hcR(1/n_1^2 - 1/n_2^2)$, is carried away as a photon of energy hc/λ. By equating these two energies, and cancelling hc, we obtain eqn 1.1.

The question these observations raise is why the energy of the electron in the atom is limited to the values $-hcR/n^2$ and why R has the value observed. An initial attempt to explain these features was made by Niels Bohr in 1913 using an early form of quantum theory in which he supposed that the electron could exist in only certain circular orbits. Although he obtained the correct value of R, his model was later shown to be untenable as it conflicted with the version of quantum theory developed by Erwin Schrödinger and Werner Heisenberg in 1926.

1.5 Some principles of quantum mechanics

Key points: Electrons can behave as particles or as waves; solution of the Schrödinger equation gives wavefunctions, which describe the location and properties of electrons in atoms. The probability of finding an electron at a given location is proportional to the square of the wavefunction. Wavefunctions generally have regions of positive and negative amplitude, and may undergo constructive or destructive interference with one another.

In 1924, Louis de Broglie suggested that because electromagnetic radiation could be considered to consist of particles called photons yet at the same time exhibit wave-like properties, such as interference and diffraction, then the same might be true of electrons. This dual nature is called **wave–particle duality**. An immediate consequence of duality is that it is impossible to know the linear momentum (the product of mass and velocity) and the location of an electron (and any particle) simultaneously. This restriction is the content of Heisenberg's **uncertainty principle**, that the product of the uncertainty in momentum and the uncertainty in position cannot be less than a quantity of the order of Planck's constant (specifically, $\frac{1}{2}\hbar$, where $\hbar = h/2\pi$).

Schrödinger formulated an equation that took account of wave–particle duality and accounted for the motion of electrons in atoms. To do so, he introduced the **wavefunction**, ψ (psi), a mathematical function of the position coordinates x, y, and z. The **Schrödinger equation**, of which the wavefunction is a solution, is

$$\frac{\partial^2 \psi}{\partial x^2} + \frac{\partial^2 \psi}{\partial y^2} + \frac{\partial^2 \psi}{\partial z^2} + \frac{2m_e}{\hbar^2}(E - V)\psi = 0 \tag{1.2}$$

Here V is the potential energy of the electron in the field of the nucleus and E is its total energy. The Schrödinger equation is a second-order differential equation and difficult to solve for all except the simplest systems. However, we shall need only qualitative aspects of its solutions.

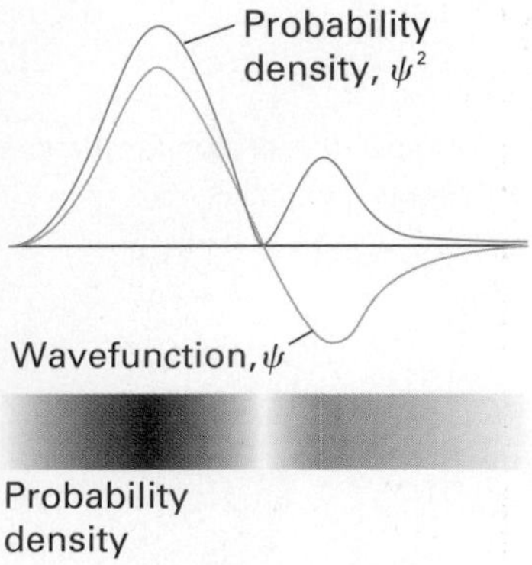

One crucial feature of eqn 1.2 is that physically acceptable solutions exist only for certain values of E. Therefore, the **quantization** of energy, the fact that an electron can possess only certain discrete energies in an atom, follows naturally from the Schrödinger equation, in addition to the imposition of certain requirements ('boundary conditions') that restrict the number of acceptable solutions.

A wavefunction contains all the dynamical information possible about the electron, including where it is and what it is doing. Specifically, the probability of finding an electron at a given location is proportional to the square of the wavefunction at that point, ψ^2. According to this interpretation, there is a high probability of finding the electron where ψ^2 is large, and the electron will not be found where ψ^2 is zero (Fig. 1.6). The quantity ψ^2 is called the **probability density** of the electron. It is a 'density' in the sense that the product of ψ^2 and the infinitesimal volume element $d\tau = dx\,dy\,dz$ (where τ is tau) is proportional to the probability of finding the electron in that volume. The probability is *equal* to $\psi^2 d\tau$ if the wavefunction is 'normalized' in the sense that

$$\int \psi^2 \, d\tau = 1 \tag{1.3}$$

where the integration is over all the space accessible to the electron. This expression simply states that the total probability of finding the electron somewhere must be 1. Any wavefunction can be made to fulfil this condition by multiplication by a *normalization constant, N*, a numerical constant that ensures that the integral in eqn 1.3 is indeed equal to 1.

Like other waves, wavefunctions in general have regions of positive and negative amplitude, or sign. The sign of the wavefunction is of crucial importance when two wavefunctions spread into the same region of space and interact. Then a positive region of one wavefunction may add to a positive region of the other wavefunction to give a region of enhanced amplitude. This enhancement is called **constructive interference** (Fig. 1.7a). It means that, where the two wavefunctions spread into the same region of space, such as occurs when two atoms are close together, there may be a significantly enhanced probability of finding the particles in that region. Conversely, a positive region of one wavefunction may be cancelled by a negative region of the second wavefunction (Fig. 1.7b). This **destructive interference** between wavefunctions will greatly reduce the probability that an electron will be found in that region. As we shall see, the interference of wavefunctions is of great importance in the explanation of chemical bonding. To keep track of the relative signs of different regions of a wavefunction in illustrations, we label regions of opposite sign with dark and light shading (sometimes white in the place of light shading).

1.6 Atomic orbitals

The wavefunction of an electron in a hydrogenic atom is called an **atomic orbital**. Hydrogenic atomic orbitals are central to the interpretation of inorganic chemistry, and we shall spend some time describing their shapes and significance.

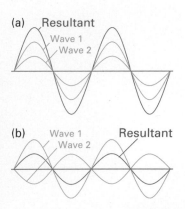

Fig. 1.6 The Born interpretation of the wavefunction is that its square is a probability density. There is zero probability density at a node. In the lower part of the illustration, the probability density is indicated by the density of shading.

Fig. 1.7 Wavefunctions interfere where they spread into the same region of space. (a) If they have the same sign in a region, they interfere constructively and the total wavefunction has an enhanced amplitude in the region. (b) If the wavefunctions have opposite signs, then they interfere destructively, and the resulting superposition has a reduced amplitude.

(a) Hydrogenic energy levels

Key points: The energy of the bound electron is determined by n, the principal quantum number, l specifies the orbital angular momentum, and m_l specifies the orientation of the angular momentum.

Each of the wavefunctions obtained by solving the Schrödinger equation for a hydrogenic atom is uniquely labelled by a set of three integers called **quantum numbers**. These quantum numbers are designated n, l, and m_l: n is called the **principal quantum number**, l is the **orbital angular momentum quantum number** (formerly the 'azimuthal quantum number'), and m_l is called the **magnetic quantum number**. Each quantum number specifies a physical property of the electron: n specifies the energy, l labels the orbital angular momentum, and m_l labels the orientation of that angular momentum. The value of n also indicates the size of the orbital, with high-energy orbitals more diffuse than compact, low-energy orbitals. The value of l also indicates the angular shape of the orbital, with the number of lobes increasing as l increases. The value of m_l also indicates the orientation of the orbital.

The allowed energies are specified by the principal quantum number, n. For a hydrogenic atom of atomic number Z, they are given by

$$E_n = -\frac{hcRZ^2}{n^2} \tag{1.4}$$

with $n = 1, 2, 3, \ldots$ and

$$R = \frac{m_e e^4}{8h^3 c\varepsilon_0^2} \tag{1.5}$$

(The fundamental constants in this expression are given inside the back cover.) The numerical value of R is $1.097 \times 10^7 \text{ m}^{-1}$, in excellent agreement with the empirical value determined spectroscopically. For future reference, the value hcR corresponds to 13.6 eV. The zero of energy (corresponding to $n = \infty$) corresponds to the electron and nucleus being widely separated and stationary. Positive values of the energy correspond to unbound states of the electron in which it may travel with any velocity and hence possess any energy. The energies given by eqn 1.4 are all negative, signifying that the energy of the electron in a bound state is lower than a widely separated stationary electron and nucleus. Finally, because the energy is proportional to $1/n^2$, the energy levels converge as the energy increases (becomes less negative, Fig. 1.8).

The value of l specifies the magnitude of the orbital angular momentum through $\{l(l+1)\}^{1/2}\hbar$, with $l = 0, 1, 2, \ldots$. We can think of l as indicating the rate at which the electron circulates around the nucleus. As we shall see shortly, the third quantum number m_l specifies the orientation of this momentum; for instance, whether the circulation is clockwise or anticlockwise.

(b) Shells, subshells, and orbitals

Key points: All orbitals with a given value of n belong to the same shell, all orbitals of a given shell with the same value of l belong to the same subshell, and individual orbitals are distinguished by the value of m_l.

In a hydrogenic atom, all orbitals with the same value of n have the same energy and are said to be **degenerate**. The principal quantum number therefore defines a series of **shells** of the atom, or sets of orbitals with the same value of n and hence with the same energy and approximately the same radial extent.

The orbitals belonging to each shell are classified into **subshells** distinguished by a quantum number l. For a given value of n, the quantum number l can have the values $l = 0, 1, \ldots, n-1$, giving n different values in all. For example, the shell with $n = 1$ consists of just one subshell with $l = 0$, the shell with $n = 2$ consists of two subshells, one with $l = 0$ and the other with $l = 1$, the shell with $n = 3$ consists of three subshells, with

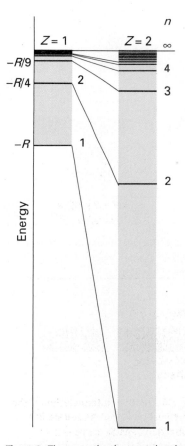

Fig. 1.8 The quantized energy levels of an H atom ($Z = 1$) and an He$^+$ ion ($Z = 2$). The energy levels of a hydrogenic atom are proportional to Z^2.

values of l of 0, 1, and 2. It is common practice to refer to each subshell by a letter:

Value of l	0	1	2	3	4	...
Orbital designation	s	p	d	f	g	...

For most purposes in chemistry we need consider only s, p, d, and f subshells.

A subshell with quantum number l consists of $2l+1$ individual orbitals. These orbitals are distinguished by the **magnetic quantum number**, m_l, which can have the $2l+1$ integer values from $+l$ down to $-l$. This quantum number specifies the component of orbital angular momentum around an arbitrary axis (commonly designated z) passing through the nucleus. So, for example, a d subshell of an atom ($l=2$) consists of five individual atomic orbitals that are distinguished by the values $m_l=+2, +1, 0, -1, -2$.

The practical conclusion for chemistry from these remarks is that there is only one orbital in an s subshell ($l=0$), the one with $m_l=0$: this orbital is called an **s orbital**. There are three orbitals in a p subshell ($l=1$), with quantum numbers $m_l=+1, 0, -1$; they are called **p orbitals**. The five orbitals of a d subshell ($l=2$) are called **d orbitals**, and so on (Fig. 1.9).

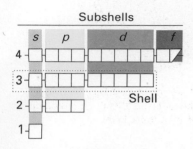

Fig. 1.9 The classification of orbitals into subshells (same value of l) and shells (same value of n).

Example 1.2 Identifying orbitals from quantum numbers

Which set of orbitals is defined by $n=4$ and $l=1$? How many orbitals are there in this set?

Answer The principal quantum number n identifies the shell; the subsidiary quantum number l identifies the subshell. The subshell with $l=1$ consists of p orbitals. The allowed values of $m_l=l, l-1, \ldots, -l$ give the number of orbitals of that type. In this case, $m_l=+1, 0$, and -1. There are therefore three $4p$ orbitals.

Self-test 1.2 Which set of orbitals is defined by the quantum numbers $n=3$ and $l=2$? How many orbitals are there in this set?

(c) Electron spin

Key points: The intrinsic spin angular momentum of an electron is defined by the two quantum numbers s and m_s. Four quantum numbers are needed to define the state of an electron in a hydrogenic atom.

In addition to the three quantum numbers required to specify the spatial distribution of an electron in a hydrogenic atom, two more quantum numbers are needed to define the state of an electron. These additional quantum numbers relate to the intrinsic angular momentum of an electron, its **spin**. This evocative name suggests that an electron can be regarded as having an angular momentum arising from a spinning motion, rather like the daily rotation of a planet as it travels in its annual orbit around the sun. However, spin is a purely quantum mechanical property and differs considerably from its classical namesake.

Spin is described by two quantum numbers, s and m_s. The former is the analogue of l for orbital motion but it is restricted to the single, unchangeable value $s=\frac{1}{2}$. The magnitude of the spin angular momentum is given by the expression $\{s(s+1)\}^{1/2}\hbar$, so for an electron this magnitude is fixed at $\frac{1}{2}\sqrt{3}\hbar$, for any electron. The second quantum number, the **spin magnetic quantum number**, m_s, may take only two values, $+\frac{1}{2}$ (anticlockwise spin, imagined from above) and $-\frac{1}{2}$ (clockwise spin). The two states are often represented by the two arrows ↑ ('spin-up', $m_s=+\frac{1}{2}$) and ↓ ('spin-down', $m_s=-\frac{1}{2}$) or by the Greek letters α and β, respectively.

Because the spin state of an electron must be specified if the state of the atom is to be specified fully, it is common to say that the state of an electron in a hydrogenic atom is characterized by four quantum numbers, namely n, l, m_l, and m_s (the fifth quantum number, s, is fixed at $\frac{1}{2}$).

(d) The radial variation of atomic orbitals

Key points: Regions where wavefunctions pass through zero are called nodes; an s orbital has a nonzero amplitude at the nucleus; all other orbitals (those with $l>0$) vanish at the nucleus.

Chemists generally find it adequate to use visual representations of atomic orbitals rather than mathematical expressions. However, we need to be aware of the mathematical expressions that underlie these representations.

Table 1.2 gives the mathematical expressions for some of the hydrogenic atomic orbitals. Because the potential energy of an electron in the field of a nucleus is spherically symmetric (it is proportional to Z/r and independent of orientation relative to the nucleus), the orbitals are best expressed in terms of the spherical polar coordinates defined in Fig. 1.10. In these coordinates, the orbitals all have the form

$$\psi_{nlm_l} = R_{nl}(r)\,Y_{lm_l}(\theta, \phi) \tag{1.6}$$

This expression and the entries in the table may look somewhat complicated, but they express the simple idea that a hydrogenic orbital can be written as the product of a function $R(r)$ of the radius and a function $Y(\theta, \phi)$ of the angular coordinates. The radial wavefunction expresses the variation of the orbital with distance from the nucleus. The angular wavefunction expresses the orbital's angular shape. The locations where the radial wavefunction passes *through* zero (not simply becoming zero) are called **radial nodes**. The planes on which the angular wavefunction passes through zero are called **angular nodes** or **nodal planes**.

Figures 1.11 and 1.12 show the radial variation of atomic orbitals. A $1s$ orbital, the wavefunction with $n = 1$, $l = 0$, and $m_l = 0$, decays exponentially with distance from the nucleus and never passes through zero. All orbitals decay exponentially at sufficiently great distances from the nucleus, but some orbitals oscillate through zero close to the nucleus and thus have one or more radial nodes before beginning their final exponential decay. An orbital with quantum numbers n and l in general has $n - l - 1$ radial nodes. This oscillation is evident in the $2s$ orbital, the orbital with $n = 2$, $l = 0$, and $m_l = 0$, which passes through zero once and hence has one radial node. A $3s$ orbital passes through zero twice and so has two radial nodes. A $2p$ orbital (one of the three orbitals with $n = 2$ and $l = 1$) has no radial nodes because its radial wavefunction does not pass through zero anywhere. However, a $2p$ orbital, like *all* orbitals other than s orbitals, is zero at the nucleus. Although an electron in an s orbital may be found at the nucleus, an electron in

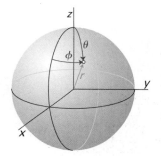

Fig. 1.10 Spherical polar coordinates: r is the radius, θ (theta) the colatitude, and ϕ (phi) the azimuth.

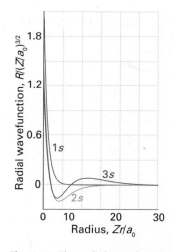

Fig. 1.11 The radial wavefunctions of the $1s$, $2s$, and $3s$ hydrogenic orbitals. Note that the number of radial nodes is 0, 1, and 2, respectively. Each orbital has a nonzero amplitude at the nucleus (at $r = 0$).

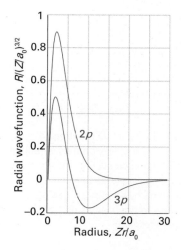

Fig. 1.12 The radial wavefunctions of the $2p$ and $3p$ hydrogenic orbitals. Note that the number of radial nodes is 0 and 1, respectively. Each orbital has zero amplitude at the nucleus (at $r = 0$).

Table 1.2 Hydrogenic orbitals

(a) Radial wavefunctions

$R_{nl}(r) = f(r)(Z/a_0)^{3/2}e^{-\rho/2}$

where a_0 is the Bohr radius (53 pm) and $\rho = 2Zr/na_0$

n	L	$f(r)$
1	0	2
2	0	$(1/2\sqrt{2})(2 - \rho)$
2	1	$(1/2\sqrt{6})\rho$
3	0	$(1/9\sqrt{3})(6 - 6\rho + \rho^2)$
3	1	$(1/9\sqrt{6})(4 - \rho)\rho$
3	2	$(1/9\sqrt{30})\rho^2$

(b) Angular wavefunctions

$Y_{l,m_l}(\theta, \phi) = (1/4\pi)^{1/2}y(\theta, \phi)$

l	m_l	$y(\theta, \phi)$
0	0	1
1	0	$3^{1/2}\cos\theta$
1	± 1	$3(3/2)^{1/2}\sin\theta e^{\pm i\phi}$
2	0	$(5/4)^{1/2}(3\cos^2\theta - 1)$
2	± 1	$3(15/4)^{1/2}\cos\theta\sin\theta e^{\pm i\phi}$
2	± 2	$(15/8)^{1/2}\sin^2\theta e^{\pm 2i\phi}$

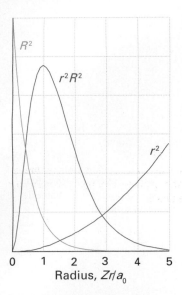

Fig. 1.13 The radial distribution function of a hydrogenic 1s orbital. The product of $4\pi r^2$ (which increases as r increases) and ψ^2 (which decreases exponentially) passes through a maximum at $r = a_0/Z$.

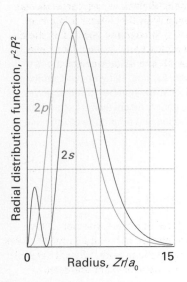

Fig. 1.14 The radial distribution functions of hydrogenic orbitals. Although the 2p orbital is *on average* closer to the nucleus (note where its maximum lies), the 2s orbital has a high probability of being close to the nucleus on account of the inner maximum.

any other type of orbital will not be found there. We shall soon see that this apparently minor detail, which is a consequence of the absence of orbital angular momentum when $l = 0$, is one of the key concepts for understanding the periodic table.

(e) The radial distribution function

Key point: A radial distribution function gives the probability that an electron will be found at a given distance from the nucleus, regardless of the direction.

The Coulombic force that binds the electron is centred on the nucleus, so it is often of interest to know the probability of finding an electron at a given distance from the nucleus regardless of its direction. This information enables us to judge how tightly the electron is bound. The total probability of finding the electron in a spherical shell of radius r and thickness dr is the integral of $\psi^2 d\tau$ over all angles. This result is often written $P(r)dr$, where $P(r)$ is called the **radial distribution function**. In general,

$$P(r) = r^2 R(r)^2 \tag{1.7}$$

(For s orbitals, this expression is the same as $P = 4\pi r^2 \psi^2$.) If we know the value of P at some radius r, then we can state the probability of finding the electron somewhere in a shell of thickness dr at that radius simply by multiplying P by dr. In general, a radial distribution function for an orbital in a shell of principal quantum number n has $n-1$ peaks, the outermost peak being the highest.

Because the wavefunction of a 1s orbital decreases exponentially with distance from the nucleus, and r^2 increases, the radial distribution function of a 1s orbital goes through a maximum (Fig. 1.13). Therefore, there is a distance at which the electron is most likely to be found. In general, this most probable distance decreases as the nuclear charge increases (because the electron is attracted more strongly to the nucleus), and specifically

$$r_{max} = \frac{a_0}{Z} \tag{1.8}$$

where a_0 is the **Bohr radius**, $a_0 = \varepsilon_0 \hbar^2 \pi m_e e^2$, a quantity that appeared in Bohr's formulation of his model of the atom; its numerical value is 52.9 pm. The most probable distance increases as n increases because the higher the energy, the more likely it is that the electron will be found far from the nucleus.

Example 1.3 Interpreting radial distribution functions

Figure 1.14 shows the radial distribution functions for 2s and 2p hydrogenic orbitals. Which orbital gives the electron a greater probability of close approach to the nucleus?

Answer The radial distribution function of a 2p orbital approaches zero near the nucleus faster than that of a 2s electron. This difference is a consequence of the fact that a 2p orbital has zero amplitude at the nucleus on account of its orbital angular momentum. Thus, the 2s electron has a greater probability of close approach to the nucleus.

Self-test 1.3 Which orbital, 3p or 3d, gives an electron a greater probability of being found close to the nucleus?

(f) The angular variation of atomic orbitals

Key points: The boundary surface of an orbital indicates the region of space within which the electron is most likely to be found; orbitals with the quantum number l have l nodal planes.

An s orbital has the same amplitude at a given distance from the nucleus whatever the angular coordinates of the point of interest: that is, an s orbital is spherically symmetrical. The orbital is normally represented by a spherical surface with the nucleus at its centre. The surface is called the **boundary surface** of the orbital, and defines the region of space within which there is a high (typically 75 per cent) probability of finding the electron. The boundary surface of any s orbital is spherical (Fig. 1.15).

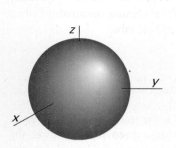

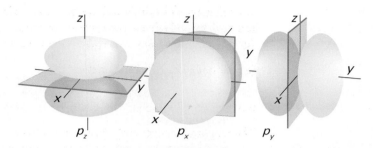

Fig. 1.15 The spherical boundary surface of an *s* orbital.

Fig. 1.16 The boundary surfaces of *p* orbitals. Each orbital has one nodal plane running through the nucleus. For example, the nodal plane of the p_z orbital is the *xy*-plane. The lightly shaded lobe has a positive amplitude, the more darkly shaded one is negative.

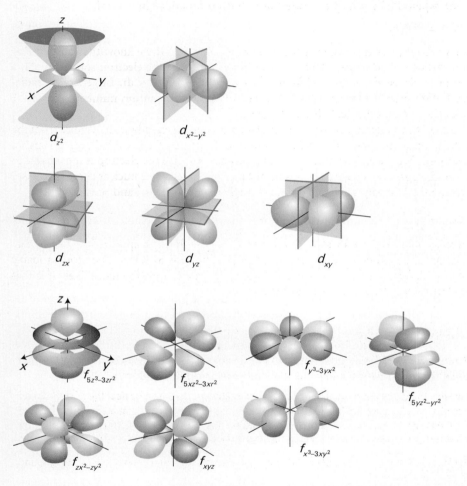

Fig. 1.17 One representation of the boundary surfaces of the *d* orbitals. Four of the orbitals have two perpendicular nodal planes that intersect in a line passing through the nucleus. In the d_{z^2} orbital, the nodal surface forms two cones that meet at the nucleus.

Fig. 1.18 One representation of the boundary surfaces of the *f* orbitals. Other representations (with different shapes) are also sometimes encountered.

All orbitals with $l > 0$ have amplitudes that vary with the angle. In the most common graphical representation, the boundary surfaces of the three *p* orbitals of a given shell are identical apart from the fact that their axes lie parallel to each of the three different Cartesian axes centred on the nucleus, and each one possesses a nodal plane passing through the nucleus (Fig. 1.16). This representation is the origin of the labels p_x, p_y, and p_z, which are alternatives to the use of m_l to label the individual orbitals. Each *p* orbital has a single nodal plane. A p_z orbital, for instance, is proportional to $\cos\theta$ (see Table 1.2), so its wavefunction vanishes everywhere on the plane corresponding to $\theta = 90°$ (the *xy*-plane). An electron will not be found anywhere on a nodal plane. A nodal plane cuts through the nucleus and separates the regions of positive and negative sign of the wavefunction.

The boundary surfaces and labels we use for the *d* and *f* orbitals are shown in Figs 1.17 and 1.18, respectively. Note that a typical *d* orbital has two nodal planes that intersect at the nucleus, and a typical f orbital has three nodal planes. In general, an orbital with the quantum number l has l nodal planes.

Many-electron atoms

As remarked at the start of the chapter, a 'many-electron atom' is an atom with more than one electron, so even He, with two electrons, is technically a many-electron atom. The exact solution of the Schrödinger equation for an atom with N electrons would be a function of the $3N$ coordinates of all the electrons. There is no hope of finding exact formulas for such complicated functions; however, it is straightforward to perform numerical computations by using widely available software to obtain precise energies and probability densities. The price of numerical precision, though, is the loss of the ability to visualize the solutions. For most of inorganic chemistry we rely on the **orbital approximation**, in which each electron occupies an atomic orbital that resembles those found in hydrogenic atoms. When we say that an electron 'occupies' an atomic orbital, we mean that it is described by the corresponding wavefunction.

1.7 Penetration and shielding

Key points: The ground-state electron configuration is a specification of the orbital occupation of an atom in its lowest energy state. The exclusion principle forbids more than two electrons to occupy a single orbital. The nuclear charge experienced by an electron is reduced by shielding by other electrons. As a result of the combined effects of penetration and shielding, the order of energy levels in a shell of a many-electron atom is $s < p < d < f$.

It is quite easy to account for the electronic structure of the helium atom in its **ground state**, its state of lowest energy. According to the orbital approximation, we suppose that both electrons occupy an atomic orbital that has the same spherical shape as a hydrogenic $1s$ orbital. However, the orbital will be more compact because, as the nuclear charge of helium is greater than that of hydrogen, the electrons are drawn in towards the nucleus more closely than is the one electron of an H atom. The ground-state **configuration** of an atom is a statement of the orbitals its electrons occupy in its ground state. For helium, with two electrons in the $1s$ orbital, the ground-state configuration is denoted $1s^2$.

As soon as we come to the next atom in the periodic table, lithium ($Z = 3$), we encounter several major new features. The configuration $1s^3$ is forbidden by a fundamental feature of nature known as the **Pauli exclusion principle**:

No more than two electrons may occupy a single orbital and, if two do occupy a single orbital, then their spins must be paired.

By 'paired' we mean that one electron spin must be ↑ and the other ↓; the pair is denoted ↑↓. Another way of expressing the principle is to note that, because an electron in an atom is described by four variable quantum numbers, n, l, m_l, and m_s, no two electrons can have the same four quantum numbers. The Pauli principle was introduced originally to account for the absence of certain transitions in the spectrum of atomic helium.

Because the configuration $1s^3$ is forbidden by the Pauli exclusion principle, the third electron must occupy an orbital of the next higher shell, the shell with $n = 2$. The question that now arises is whether the third electron occupies a $2s$ orbital or one of the three $2p$ orbitals. To answer this question, we need to examine the energies of the two subshells and the effect of the other electrons in the atom. Although $2s$ and $2p$ orbitals have the same energy in a hydrogenic atom, spectroscopic data and calculations show that that is not the case in a many-electron atom.

In the orbital approximation we treat the repulsion between electrons in an approximate manner by supposing that the electronic charge is distributed spherically around the nucleus. Then each electron moves in the attractive field of the nucleus and experiences an average repulsive charge from the other electrons. According to classical electrostatics, the field that arises from a spherical distribution of charge is equivalent to the field generated by a single point charge at the centre of the distribution (Fig. 1.19). This negative charge reduces the actual charge of the nucleus, Ze, to $Z_{eff}e$, where Z_{eff} (more precisely, $Z_{eff}e$) is called the **effective nuclear charge**. This effective nuclear charge depends on the values of n and l of the electron of interest, because electrons in different shells and subshells approach the nucleus to different extents. The reduction of the true nuclear charge to the effective nuclear charge by the other electrons is called **shielding**. The effective nuclear charge is sometimes expressed in terms of the true nuclear charge and an empirical **shielding constant**, σ, by writing

$$Z_{eff} = Z - \sigma$$

The shielding constant can be determined by fitting hydrogenic orbitals to those computed numerically.

The closer to the nucleus that an electron can approach, the closer is the value of Z_{eff} to Z itself, because the electron is repelled less by the other electrons present in the atom. With this point in mind, consider a $2s$ electron in the Li atom. There is a nonzero probability that the $2s$ electron can be found inside the $1s$ shell and experience the full nuclear charge (Fig. 1.20). The presence of an electron inside shells of other electrons is called **penetration**. A $2p$ electron does not penetrate through the **core**, the filled inner shells of electrons, so effectively because its wavefunction goes to zero at the nucleus. As a consequence, it is more fully shielded from the nucleus by the core electrons. We can conclude that a $2s$ electron has a lower energy (is bound more tightly) than a $2p$ electron, and therefore that the $2s$ orbital will be occupied before the $2p$ orbitals, giving a ground-state electron configuration for Li of $1s^2 2s^1$. This configuration is commonly denoted $[He]2s^1$, where $[He]$ denotes the atom's helium-like $1s^2$ core.

The pattern of energies in lithium, with $2s$ lower than $2p$, and in general ns lower than np, is a general feature of many-electron atoms. This pattern can be seen from Table 1.3, which gives the values of Z_{eff} for a number of valence-shell atomic orbitals in the ground-state electron configuration of atoms. The typical trend in effective nuclear charge is an increase across a period, for in most cases the increase in nuclear charge in successive groups is not cancelled by the additional electron. The values in the table also confirm that an s electron in the outermost shell of the atom is generally less shielded than a p electron of that shell. So, for example, $Z_{eff} = 5.13$ for a $2s$ electron in an F atom, whereas for a $2p$ electron $Z_{eff} = 5.10$, a lower value. Similarly, the effective nuclear charge is larger for an electron in an np orbital than for one in an nd orbital.

As a result of penetration and shielding, the order of energies in many-electron atoms is typically

$$ns < np < nd < nf$$

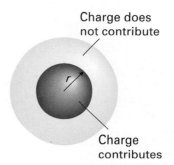

Charge does not contribute

Charge contributes

Fig. 1.19 The electron at the radius r experiences a repulsion from the total charge within the sphere of radius r; charge outside that radius has no net effect.

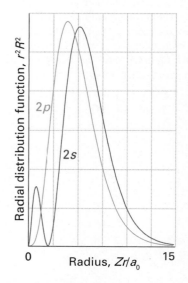

Fig. 1.20 The penetration of a $2s$ electron through the inner core is greater than that of a $2p$ electron because the latter vanishes at the nucleus. Therefore, the $2s$ electrons are less shielded than the $2p$ electrons.

Table 1.3 Effective nuclear charges, Z_{eff}

	H							He
Z	1							2
$1s$	1.00							1.69
	Li	Be	B	C	N	O	F	Ne
Z	3	4	5	6	7	8	9	10
$1s$	2.69	3.68	4.68	5.67	6.66	7.66	8.65	9.64
$2s$	1.28	1.91	2.58	3.22	3.85	4.49	5.13	5.76
$2p$			2.42	3.14	3.83	4.45	5.10	5.76
	Na	Mg	Al	Si	P	S	Cl	Ar
Z	11	12	13	14	15	16	17	18
$1s$	10.63	11.61	12.59	13.57	14.56	15.54	16.52	17.51
$2s$	6.57	7.39	8.21	9.02	9.82	10.63	11.43	12.23
$2p$	6.80	7.83	8.96	9.94	10.96	11.98	12.99	14.01
$3s$	2.51	3.31	4.12	4.90	5.64	6.37	7.07	7.76
$3p$			4.07	4.29	4.89	5.48	6.12	6.76

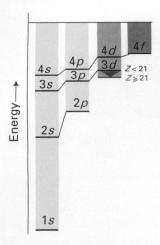

Fig. 1.21 A schematic diagram of the energy levels of a many-electron atom with $Z < 21$ (as far as calcium). There is a change in order for $Z \geq 21$ (from scandium onwards). This is the diagram that justifies the building-up principle, with up to two electrons being allowed to occupy each orbital.

because, in a given shell, s orbitals are the most penetrating and f orbitals are the least penetrating. The overall effect of penetration and shielding is depicted in the energy-level diagram for a neutral atom shown in Fig. 1.21.

Figure 1.22 summarizes the energies of the orbitals through the periodic table. The effects are quite subtle, and the order of the orbitals depends strongly on the numbers of electrons present in the atom and may change on ionization. For example, the effects of penetration are very pronounced for $4s$ electrons in K and Ca, and in these atoms the $4s$ orbitals lie lower in energy than the $3d$ orbitals. However, from Sc through Zn, the $3d$ orbitals in the neutral atoms lie close to but lower than the $4s$ orbitals. In atoms from Ga ($Z = 31$) onwards, the $3d$ orbitals lie well below the $4s$ orbital in energy, and the outermost electrons are unambiguously those of the $4s$ and $4p$ subshells.

1.8 The building-up principle

The ground-state electron configurations of many-electron atoms are determined experimentally by spectroscopy and are summarized in *Resource section* 2. To account for them, we need to consider both the effects of penetration and shielding on the energies of the orbitals and the role of the Pauli exclusion principle. The **building-up principle** (which is also known as the *Aufbau principle*, and is described below) is a procedure that leads to plausible ground-state configurations. It is not infallible, but it is an excellent starting point for the discussion. Moreover, as we shall see, it provides a theoretical framework for understanding the structure and implications of the periodic table.

(a) Ground-state electron configurations

Key points: The order of occupation of atomic orbitals follows the order 1s, 2s, 2p, 3s, 3p, 4s, 3d, 4p, ... Degenerate orbitals are occupied singly before being doubly occupied; certain modifications of the order of occupation occur for d and f orbitals.

According to the building-up principle, orbitals of neutral atoms are treated as being occupied in the order determined in part by the principal quantum number and in part by penetration and shielding:

Order of occupation: $1s\,2s\,2p\,3s\,3p\,4s\,3d\,4p\ldots$

Each orbital can accommodate up to two electrons. Thus, the three orbitals in a p subshell can accommodate a total of six electrons and the five orbitals in a d subshell can accommodate up to ten electrons. The ground-state configurations of the first five elements are therefore expected to be

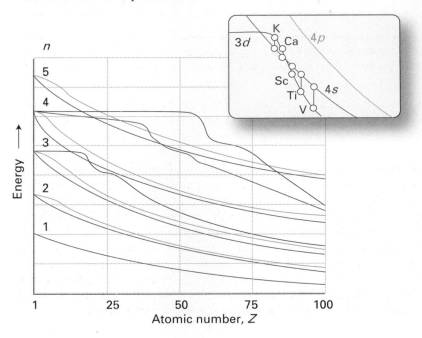

Fig. 1.22 A more detailed portrayal of the energy levels of many-electron atoms in the periodic table. The inset shows a magnified view of the order near $Z = 20$, where the $3d$ series of elements begin.

H	He	Li	Be	B
$1s^1$	$1s^2$	$1s^2 2s^1$	$1s^2 2s^2$	$1s^2 2s^2 2p^1$

This order agrees with experiment. When more than one orbital of the same energy is available for occupation, such as when the $2p$ orbitals begin to be filled in B and C, we adopt **Hund's rule**:

> *When more than one orbital has the same energy, electrons occupy separate orbitals and do so with parallel spins (↑↑).*

The occupation of separate orbitals (such as a p_x orbital and a p_y orbital) can be understood in terms of the weaker repulsive interactions that exist between electrons occupying different regions of space (electrons in different orbitals) than between those occupying the same region of space (electrons in the same orbital). The requirement of parallel spins for electrons that do occupy different orbitals is a consequence of a quantum mechanical effect called **spin correlation**, the tendency for two electrons with parallel spins to stay apart from one another and hence to repel each other less. One consequence of this effect is that half-filled shells of electrons with parallel spins are particularly stable. For example, the ground state of the chromium atom is $4s^1 3d^5$ rather than $4s^2 3d^4$. Further examples of the effect of spin correlation will be seen later in the chapter.

It is arbitrary which of the p orbitals of a subshell is occupied first because they are degenerate, but it is common to adopt the alphabetical order p_x, p_y, p_z. It then follows from the building-up principle that the ground-state configuration of C is $1s^2 2s^2 2p_x^1 2p_y^1$, or, more simply, $1s^2 2s^2 2p^2$. If we recognize the helium-like core ($1s^2$), an even briefer notation is $[He]2s^2 2p^2$, and we can think of the electronic valence structure of the atom as consisting of two paired $2s$ electrons and two parallel $2p$ electrons surrounding a closed helium-like core. The electron configurations of the remaining elements in the period are similarly

C	N	O	F	Ne
$[He]2s^2 2p^2$	$[He]2s^2 2p^3$	$[He]2s^2 2p^4$	$[He]2s^2 2p^5$	$[He]2s^2 2p^6$

The $2s^2 2p^6$ configuration of neon is another example of a **closed shell**, a shell with its full complement of electrons. The configuration $1s^2 2s^2 2p^6$ is denoted [Ne] when it occurs as a core.

Example 1.4 Accounting for trends in effective nuclear charge

Refer to Table 1.3. Suggest a reason why the increase in Z_{eff} for a $2p$ electron is smaller between N and O than between C and N given the configurations of the atoms listed above.

Answer On going from C to N, the additional electron occupies an empty $2p$ orbital. On going from N to O, the additional electron must occupy a $2p$ orbital that is already occupied by one electron. It therefore experiences stronger electron–electron repulsion and this has an effect on the Z_{eff}.

Self-test 1.4 Account for the larger increase in effective nuclear charge for a $2p$ electron on going from B to C compared with a $2s$ electron on going from Li to Be.

The ground-state configuration of Na is obtained by adding one more electron to a neon-like core, and is $[Ne]3s^1$, showing that it consists of a single electron outside a completely filled $1s^2 2s^2 2p^6$ core. Now a similar sequence of filling subshells begins again, with the $3s$ and $3p$ orbitals complete at argon, with configuration $[Ne]3s^2 3p^6$, which can be denoted [Ar]. Because the $3d$ orbitals are so much higher in energy, this configuration is effectively closed. Moreover, the $4s$ orbital is next in line for occupation, so the configuration of K is analogous to that of Na, with a single electron outside a noble-gas core: specifically, it is $[Ar]4s^1$. The next electron, for Ca, also enters the $4s$ orbital, giving $[Ar]4s^2$, which is the analogue of Mg. However, in the next element, scandium, the added electron occupies a $3d$ orbital, and the d block of the periodic table begins.

In the d block, the d orbitals of the atoms are in the process of being occupied (according to the rules of the building-up principle). However, the energy levels in Figs 1.21 and 1.22 are for individual atomic orbitals and do not fully take into account repulsion between electrons. For most of the d block, the determination of actual ground states by spectroscopy and calculation shows that it is advantageous to occupy orbitals predicted to be *higher* in energy (the $4s$ orbitals). The explanation for this order is that the occupation of orbitals of higher energy can result in a reduction in the repulsions between electrons that would occur if the lower-energy $3d$ orbitals were occupied. It is essential when assessing the total energy of the electrons to consider all contributions to the energy of a configuration, not merely the one-electron orbital energies. Spectroscopic data show that the ground-state configurations of d-block atoms are mostly of the form $3d^n4s^2$, with the $4s$ orbitals fully occupied despite individual $3d$ orbitals being lower in energy.

An additional feature, another consequence of spin correlation, is that in some cases a lower total energy may be obtained by forming a half-filled or filled d subshell, even though that may mean moving an s electron into the d subshell. Therefore, close to the centre of the d block the ground-state configuration is likely to be d^5s^1 and not d^4s^2 (as for Cr). Close to the right of the d block the configuration is likely to be $d^{10}s^1$ rather than d^9s^2 (as for Cu). A similar effect occurs in the f block, where f orbitals are being occupied, and a d electron may be moved into the f subshell so as to achieve an f^7 or an f^{14} configuration, with a net lowering of energy. For instance, the ground-state electron configuration of Gd is $[Xe]4f^75d^16s^2$ and not $[Xe]4f^65d^26s^2$.

The complication of the orbital energy not being a reliable guide to the total energy disappears when the $3d$ orbital energies fall well below that of the $4s$ orbitals, for then the competition is less subtle. The same is true of the cations of the d-block elements, where the removal of electrons reduces the complicating effects of electron–electron repulsions. Consequently, all d-block cations have d^n configurations and no electrons in the s orbitals. For example, the configuration of Fe is $[Ar]3d^64s^2$ whereas that of Fe^{2+} is $[Ar]3d^6$. For the purposes of chemistry, the electron configurations of the d-block ions are more important than those of the neutral atoms. In later chapters (starting in Chapter 18), we shall see the great significance of the configurations of the d-metal ions, for the subtle modulations of their energies provide the basis for the explanations of important properties of their compounds.

Example 1.5 Deriving an electron configuration

Predict the ground-state electron configurations of Ti and Ti^{3+}.

Answer For the atom, for which $Z = 22$, we must add 22 electrons in the order specified above, with no more than two electrons in any one orbital. This procedure results in the configuration $[Ar]4s^23d^2$, with the two $3d$ electrons in different orbitals with parallel spins. However, because the $3d$ orbitals lie below the $4s$ orbitals for elements beyond Ca, it is appropriate to reverse the order in which they are written. The configuration is therefore reported as $[Ar]3d^24s^2$. The configuration of the cation is obtained formally by removing first the s electrons and then as many d electrons as required. We must remove three electrons in all, two s electrons and one d electron. The configuration of Ti^{3+} is therefore $[Ar]3d^1$.

Self-test 1.5 Predict the ground-state electron configurations of Ni and Ni^{2+}.

(b) The format of the periodic table

Key points: The blocks of the periodic table reflect the identity of the orbitals that are occupied last in the building-up process. The period number is the principal quantum number of the valence shell. The group number is related to the number of valence electrons.

The layout of the periodic table reflects the electronic structure of the atoms of the elements. We can now see, for instance, that the **blocks** of the table indicate the type of subshell currently being occupied according to the building-up principle. Each **period**, or row, of the table corresponds to the completion of the s and p subshells of a given shell.

The period number is the value of the principal quantum number n of the shell currently being occupied in the main groups of the table. For example, Period 2 corresponds to the $n = 2$ shell and the filling of the $2s$ and $2p$ subshells.

The group numbers are closely related to the number of electrons in the **valence shell**, the outermost shell of the atom. The precise relation depends on the group number G and the numbering system adopted. In the '1−18' numbering system recommended by IUPAC:

Block:		s, d	p
Number of electrons in valence shell:		G	$G - 10$

For the purpose of this expression, the 'valence shell' of a d-block element consists of the ns and $(n - 1)d$ orbitals, so a scandium atom has three valence electrons (two $4s$ and one $3d$ electron). The number of valence electrons for the p-block element selenium (Group 16) is $16 - 10 = 6$, which gives the configuration s^2p^4. Alternatively, in the Roman numeral system, the group number is equal to the number of s and p valence electrons for the s and p blocks. Thus, selenium belongs to Group VI; hence it has six valence (s and p) electrons. Thallium belongs to Group III, so it has three valence s and p electrons.

Example 1.6 Placing elements within the periodic table.

State to which period, group, and block of the periodic table the element with the electron configuration $1s^2 2s^2 2p^6 3s^2 3p^4$ belongs.

Answer The valence electrons have $n = 3$. Therefore, the element is in Period 3 of the periodic table. The six valence electrons identify the element as a member of Group 16. The outer most electrons are p electrons, so the element is in the p block. (The element is sulfur.)

Self-test 1.6 State to which period, group, and block of the periodic table the element with the electron configuration $1s^2 2s^2 2p^6 3s^2 3p^6 4s^2$ belongs.

1.9 Atomic parameters

Certain characteristic properties of atoms, particularly their radii and the energies associated with the removal and addition of electrons, show regular periodic variations with atomic number. These atomic properties are of considerable importance when trying to understand many of the chemical properties of the elements. A knowledge of these trends enables chemists to rationalize observations and predict likely chemical and structural behaviour without resort to tabulated data for each element.

(a) Atomic and ionic radii

Key points: Atomic radii increase down a group and, within the s and p blocks, decrease from left to right across a period. The lanthanide contraction results in a decrease in atomic radius for elements following the f block. All anions are larger than their parent atoms and all cations are smaller.

One of the most useful atomic characteristics of an element is the size of its atoms and ions. As we shall see in later chapters, geometrical considerations are central to explaining the structures of many solids and individual molecules. In addition, the average distance of electrons from the nucleus of an atom correlates with the energy needed to remove it in the process of forming a cation.

An atom does not have a precise radius because at large distances the wavefunction of the electrons falls off exponentially with increasing distance from the nucleus. However, we can expect atoms with numerous electrons to be larger, in some sense, than atoms that have only a few electrons. Such considerations have led chemists to propose a variety of definitions of atomic radius on the basis of empirical considerations.

The **metallic radius** of a metallic element is defined as half the experimentally determined distance between the centres of nearest-neighbour atoms in the solid

(Fig. 1.23a, but see Section 3.7 for a refinement of this definition). The **covalent radius** of a nonmetallic element is similarly defined as half the internuclear distance between neighbouring atoms of the same element in a molecule (Fig. 1.23b). We shall refer to metallic and covalent radii jointly as **atomic radii** (Table 1.4). The periodic trends in metallic and covalent radii can be seen from the data in the table and are illustrated in Fig. 1.24. As will be familiar from introductory chemistry, atoms may be linked by single, double, and triple bonds, with multiple bonds shorter than single bonds between the same two elements. The **ionic radius** (Fig. 1.23c) of an element is related to the distance between the centres of neighbouring cations and anions. An arbitrary decision has to be taken on how to apportion the cation–anion distance between the two ions. There have been many suggestions: in one common scheme, the radius of the O^{2-} ion is taken to be 140 pm (Table 1.5; see Section 3.10a for a refinement of this definition). For example, the ionic radius of Mg^{2+} is obtained by subtracting 140 pm from the internuclear distance between adjacent Mg^{2+} and O^{2-} ions in solid MgO.

The data in Table 1.4 show that *atomic radii increase down a group*, and that they *decrease from left to right across a period*. These trends are readily interpreted in terms of the electronic structure of the atoms. On descending a group, the valence electrons are found in orbitals of successively higher principal quantum number. The atoms within the group have a greater number of completed shells of electrons in successive periods and hence their radii increase down the group. Across a period, the valence electrons enter

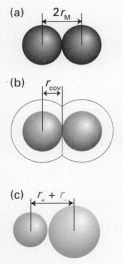

Fig. 1.23 A representation of (a) metallic radius, (b) covalent radius, and (c) ionic radius.

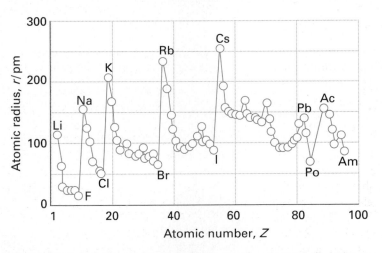

Fig. 1.24 The variation of atomic radii through the periodic table. Note the contraction of radii following the lanthanoids in Period 6. Metallic radii have been used for the metallic elements and covalent radii have been used for the nonmetallic elements.

Table 1.4 Atomic radii, r/pm*

Li	Be											B	C	N	O	F
157	112											88	77	74	66	64
Na	**Mg**											**Al**	**Si**	**P**	**S**	**Cl**
191	160											143	118	110	104	99
K	**Ca**	**Sc**	**Ti**	**V**	**Cr**	**Mn**	**Fe**	**Co**	**Ni**	**Cu**	**Zn**	**Ga**	**Ge**	**As**	**Se**	**Br**
235	197	164	147	135	129	137	126	125	125	128	137	153	122	121	117	114
Rb	**Sr**	**Y**	**Zr**	**Nb**	**Mo**	**Tc**	**Ru**	**Rh**	**Pd**	**Ag**	**Cd**	**In**	**Sn**	**Sb**	**Te**	**I**
250	215	182	160	147	140	135	134	134	137	144	152	167	158	141	137	133
Cs	**Ba**	**Lu**	**Hf**	**Ta**	**W**	**Re**	**Os**	**Ir**	**Pt**	**Au**	**Hg**	**Tl**	**Pb**	**Bi**		
272	224	172	159	147	141	137	135	136	139	144	155	171	175	182		

*The values refer to coordination number 12 (see Section 3.2).

Table 1.5 Ionic radii, r/pm*

Li^+	Be^{2+}	B^{3+}			N^{3-}	O^{2-}	F^-
59(4)	27(4)	12(4)			132	135(2)	128(2)
76(6)						138(4)	131(4)
						140(6)	133(6)
						142(8)	
Na^+	Mg^{2+}	Al^{3+}			P^{3-}	S^{2-}	Cl^-
99(4)	49(4)	39(4)			212	184(6)	167(6)
102(6)	72(6)	53(6)					
116(8)	89(8)						
K^+	Ca^{2+}	Ga^{3+}			As^{3-}	Se^{2-}	Br^-
138(6)	100(6)	62(6)			222	198(6)	196(6)
151(8)	112(8)						
159(10)	128(10)						
160(12)	135(12)						
Rb^+	Sr^{2+}	In^{3+}	Sn^{2+}	Sn^{4+}		Te^{2-}	I^-
149(6)	116(6)	79(6)	83(6)	74(6)		221(6)	206(6)
160(8)	125(8)	92(8)	93(8)				
173(12)	144(12)						
Cs^+	Ba^{2+}	Tl^{3+}					
167(6)	149(6)	88(6)					
174(8)	156(8)	Tl^+					
188(12)	175(12)	164(6)					

*Numbers in parentheses are the coordination number of the ion. For more values, see *Resource section* 1.

orbitals of the same shell; however, the increase in effective nuclear charge across the period draws in the electrons and results in progressively more compact atoms. The general increase in radius down a group and decrease across a period should be remembered as they correlate well with trends in many chemical properties.

Period 6 shows an interesting and important modification to these otherwise general trends. We see from Fig. 1.24 that the metallic radii in the third row of the d block are very similar to those in the second row, and not significantly larger as might be expected given their considerably larger numbers of electrons. For example, the radii of molybdenum ($Z = 42$) and tungsten ($Z = 74$) are 140 and 141 pm, respectively, despite the latter having many more electrons. The reduction of radius below that expected on the basis of a simple extrapolation down the group is called the **lanthanide contraction**. The name points to the origin of the effect. The elements in the third row of the d block (Period 6) are preceded by the elements of the first row of the f block, the lanthanoids, in which the $4f$ orbitals are being occupied. These orbitals have poor shielding properties and so the valence electrons experience more attraction from the nuclear charge than might be expected. The repulsions between electrons being added on crossing the f block fail to compensate for the increasing nuclear charge, so Z_{eff} increases from left to right across a period. The dominating effect of the latter is to draw in all the electrons and hence to result in a more compact atom. A similar contraction is found in the elements that follow the d block for the same reasons. For example, although there is a substantial increase in atomic radius between boron and aluminium (88 and 143 pm, respectively), the atomic radius of gallium (153 pm) is only slightly greater than that of aluminium.

There is an additional contraction of the heavier elements from Period 6 onwards that is ascribed to relativistic effects. According to the theory of relativity, the mass of an electron increases as its speed approaches the speed of light. For light elements, relativistic effects can be neglected but for the heavier elements with high atomic numbers they become significant and can result in an approximately 20 per cent contraction in the size of the atom. Relativistic effects are particularly apparent for $6s$ electrons, which penetrate close to the nucleus and are subject to very strong accelerating forces.

Another general feature apparent from Table 1.5 is that all anions are larger than their parent atoms and all cations are smaller (in some cases, markedly so). The increase in radius of an atom on anion formation is a result of the greater electron–electron repulsions that occur when an additional electron is added to form an anion. There is also an associated decrease in the value of Z_{eff}. The smaller radius of a cation compared with its parent atom is a consequence not only of the reduction in electron–electron repulsions that follows electron loss but also of the fact that cation formation typically results in the loss of the valence electrons and an increase in Z_{eff}. That loss often leaves behind only the much more compact closed shells of electrons. Once these gross differences are taken into account, the variation in ionic radii through the periodic table mirrors that of the atoms.

Although small variations in atomic radii may seem of little importance, in fact atomic radius plays a central role in the chemical properties of the elements, and small changes can have profound consequences.

(b) Ionization energy

Key points: First ionization energies are lowest at the lower left of the periodic table (near caesium) and greatest near the upper right (near helium). Successive ionizations of a species require higher energies.

The ease with which an electron can be removed from an atom is measured by its **ionization energy**, I, the minimum energy needed to remove an electron from a gas-phase atom:

$$A(g) \rightarrow A^+(g) + e^-(g) \qquad I = E(A^+, g) - E(A, g) \tag{1.9}$$

The **first ionization energy**, I_1, is the energy required to remove the least tightly bound electron from the neutral atom, the **second ionization energy**, I_2, is the energy required to remove the least tightly bound electron from the resulting cation, and so on. Ionization energies are conveniently expressed in electronvolts (eV), but are easily converted into kilojoules per mole by using $1 \, eV = 96.485 \, kJ \, mol^{-1}$. The ionization energy of the H atom is 13.6 eV, so to remove an electron from an H atom is equivalent to dragging the electron through a potential difference of 13.6 V.

In thermodynamic calculations it is often more appropriate to use the **ionization enthalpy**, the standard enthalpy of the process in eqn 1.9, typically at 298 K. The molar ionization enthalpy is larger by $\frac{5}{2}RT$ than the ionization energy. This difference stems from the change from $T = 0$ (assumed implicitly for I) to the temperature T (typically 298 K) to which the enthalpy value refers, and the replacement of 1 mol of gas particles by 2 mol. However, because RT is only $2.5 \, kJ \, mol^{-1}$ (corresponding to 0.026 eV) at room temperature and ionization energies are of the order of $10^2–10^3 \, kJ \, mol^{-1}$ (1–10 eV), the difference between ionization energy and enthalpy can often be ignored.

To a large extent, the first ionization energy of an element is determined by the energy of the highest occupied orbital of its ground-state atom. First ionization energies vary systematically through the periodic table (Table 1.6), being smallest at the lower left (near caesium) and greatest near the upper right (near helium). The variation follows the pattern of effective nuclear charge, and (as Z_{eff} itself shows) there are some subtle modulations arising from the effect of electron–electron repulsions within the same subshell. A useful approximation is that for an electron from a shell with principal quantum number n

$$I \propto \frac{Z_{eff}^2}{n^2}$$

Table 1.6 First and second (and some higher) ionization energies of the elements, $I/(\text{kJ mol}^{-1})$

H							He
1312							2373
							5259
Li	**Be**	**B**	**C**	**N**	**O**	**F**	**Ne**
513	899	801	1086	1402	1314	1681	2080
7297	1757	2426	2352	2855	3386	3375	3952
11809	14844	3660	4619	4577	5300	6050	6122
		25018					
Na	**Mg**	**Al**	**Si**	**P**	**S**	**Cl**	**Ar**
495	737	577	786	1011	1000	1251	1520
4562	1476	1816	1577	1903	2251	2296	2665
6911	7732	2744	3231	2911	3361	3826	3928
		11574					
K	**Ca**	**Ga**	**Ge**	**As**	**Se**	**Br**	**Kr**
419	589	579	762	947	941	1139	1351
3051	1145	1979	1537	1798	2044	2103	3314
4410	4910	2963	3302	2734	2974	3500	3565
Rb	**Sr**	**In**	**Sn**	**Sb**	**Te**	**I**	**Xe**
403	549	558	708	834	869	1008	1170
2632	1064	1821	1412	1794	1795	1846	2045
3900	4210	2704	2943	2443	2698	3197	3097
Cs	**Ba**	**Tl**	**Pb**	**Bi**	**Po**	**At**	**Rn**
375	502	590	716	704	812	926	1036
2420	965	1971	1450	1610	1800	1600	
3400	3619	2878	3080	2466	2700	2900	

Ionization energies also correlate strongly with atomic radii, and elements that have small atomic radii generally have high ionization energies. The explanation of the correlation is that in a small atom an electron is close to the nucleus and experiences a strong Coulombic attraction, making it difficult to remove. Therefore, as the atomic radius increases down a group, the ionization energy decreases and the decrease in radius across a period is accompanied by a gradual increase in ionization energy.

Some deviation from this general trend in ionization energy can be explained quite readily. An example is the observation that the first ionization energy of boron is smaller than that of beryllium despite the former's higher nuclear charge. This anomaly is readily explained by noting that, on going to boron, the outermost electron occupies a $2p$ orbital and hence is less strongly bound than if it had occupied a $2s$ orbital. As a result, the value of I_1 falls back to a lower value. The decrease between nitrogen and oxygen has a slightly different explanation. The configurations of the two atoms are

N $[He]2s^2 2p_x^1 2p_y^1 2p_z^1$ O $[He]2s^2 2p_x^2 2p_y^1 2p_z^1$

We see that, in an O atom, two electrons are present in a single $2p$ orbital. They repel each other strongly, and this strong repulsion offsets the greater nuclear charge. Another contribution to the difference is the lower energy of the O^+ ion on account of its $2s^2 2p^3$ configuration: as we have seen, a half-filled subshell has a relatively low energy (Fig. 1.25).

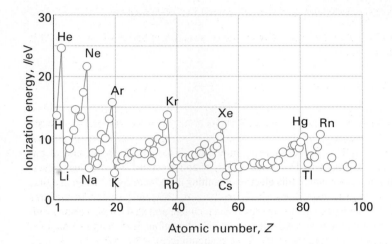

Fig. 1.25 The periodic variation of first ionization energies.

Example 1.7 Accounting for a variation in ionization energy

Account for the decrease in first ionization energy between phosphorus and sulfur.

Answer The ground-state configurations of the two atoms are

P $[\text{Ne}]3s^2 3p_x^1 3p_y^1 3p_z^1$ S $[\text{Ne}]3s^2 3p_x^2 3p_y^1 3p_z^1$

As in the analogous case of N and O, in the ground state of S, two electrons are present in a single $3p$ orbital. They are so close together that they repel each other strongly, and this increased repulsion offsets the effect of the greater nuclear charge of S compared with P. As in the difference between N and O, the half-filled subshell of S^+ also contributes to the lowering of energy of the ion and hence to the smaller ionization energy.

Self-test 1.7 Account for the decrease in first ionization energy between fluorine and chlorine.

When considering fluorine and neon, the last electrons enter orbitals that are already half full, and continue the trend in ionization energy from oxygen. The higher values of the ionization energies of these two elements reflect the high value of Z_{eff}. The value of I_1 falls back sharply from neon to sodium as the outermost electron occupies the next shell with an increased principal quantum number and is, hence, further from the nucleus.

Another important pattern is that successive ionizations require increasingly higher energies (Fig. 1.26). Thus, the second ionization energy of an element (the energy needed to remove an electron from the cation, E^+) is higher than its first ionization energy, and its third ionization energy is higher still. The explanation is that the higher the positive charge of a species, the greater the energy needed to remove an electron from the species. Moreover, when an electron is removed, Z_{eff} increases and the atom contracts. It is then more difficult to remove an electron from this smaller, more compact cation. The difference in ionization energy is greatly magnified when the electron is removed from a closed shell of the atom (as is the case for the second ionization energy of lithium and any of its congeners) because the electron must then be extracted from a compact orbital in which it interacts strongly with the nucleus. The first ionization energy of lithium, for instance, is 513 kJ mol^{-1}, but its second ionization energy is 7297 kJ mol^{-1}, more than ten times greater.

The pattern of successive ionization energies down a group is far from simple. Figure 1.26 shows the first, second, and third ionization energies of Group 13. Although they lie in the expected order $I_1 < I_2 < I_3$, there is no simple trend. The lesson to be drawn is that whenever an argument hangs on trends in small differences in ionization energies, it is always best to refer to actual numerical values rather than to guess a likely outcome.

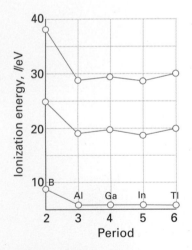

Fig. 1.26 The first, second, and third ionization energies of the elements of Group 13. Successive ionization energies increase, but there is no clear pattern of ionization energies down the group.

Example 1.8 Accounting for values of successive energies of ionization

Rationalize the following values for successive ionization energies of boron, where $\Delta_{ion}H(N)$ is the Nth enthalpy of ionization:

N	1	2	3	4	5
$\Delta_{ion}H(N)/(MJ\ mol^{-1})$	0.807	2.433	3.666	25.033	32.834

Answer The electron configuration of boron is $1s^22s^22p^1$. The first ionization energy corresponds to removal of the electron in the p orbital. This electron is shielded from nuclear charge by the core $1s$ shell and the $2s$ subshell. The second value corresponds to removal of the first $2s$ electron from the B^+ cation. This electron is more difficult to remove on account of the increased effective nuclear charge and the lower energy of an s electron. The effective nuclear charge increases further on removal of this electron, resulting in an increase between $\Delta_{ion}H(2)$ and $\Delta_{ion}H(3)$. There is a large increase between $\Delta_{ion}H(3)$ and $\Delta_{ion}H(4)$ because the $1s$ shell lies at very low energy as it experiences almost the full nuclear charge and also it lies in a shell with $n=1$. The final electron to be removed experiences no shielding of nuclear charge so $\Delta_{ion}H(5)$ is very high, and is given by $hcRZ^2$ with $Z=5$, corresponding to $(13.6\ eV) \times 25 = 340$ eV ($32.834\ MJ\ mol^{-1}$).

Self-test 1.8 Study the values listed below of the first five ionization energies of an element and deduce to which group of the periodic table the element belongs. Give your reasoning.

N	1	2	3	4	5
$\Delta_{ion}H(N)/(MJ\ mol^{-1})$	1.093	2.359	4.627	6.229	37.838

(c) Electron affinity

Key point: Electron affinities are highest for elements near fluorine in the periodic table.

The **electron-gain enthalpy**, $\Delta_{eg}H^\circ$, is the change in standard molar enthalpy when a gaseous atom gains an electron:

$$A(g) + e^-(g) \rightarrow A^-(g)$$

Electron gain may be either exothermic or endothermic. Although the electron-gain enthalpy is the thermodynamically appropriate term, much of inorganic chemistry is discussed in terms of a closely related property, the **electron affinity**, E_a, of an element (Table 1.7), which is the difference in energy between the gaseous atoms and the gaseous ions at $T=0$.

$$E_a = E(A, g) - E(A^-, g) \tag{1.10}$$

Table 1.7 Electron affinities of the main-group elements, $E_a/(kJ\ mol^{-1})$*

H							He
72							−48
Li	**Be**	**B**	**C**	**N**	**O**	**F**	**Ne**
60	≤0	27	122	−8	141	328	−116
					−780		
Na	**Mg**	**Al**	**Si**	**P**	**S**	**Cl**	**Ar**
53	≤0	43	134	72	200	349	−96
					−492		
K	**Ca**	**Ga**	**Ge**	**As**	**Se**	**Br**	**Kr**
48	2	29	116	78	195	325	−96
Rb	**Sr**	**In**	**Sn**	**Sb**	**Te**	**I**	**Xe**
47	5	29	116	103	190	295	−77

* The first values refer to the formation of the ion X^- from the neutral atom; the second value to the formation of X^{2-} from X^-.

Although the precise relation is $\Delta_{eg}H^{\ominus} = -E_a - \frac{5}{2}RT$, the contribution $\frac{5}{2}RT$ is commonly ignored. A positive electron affinity indicates that the ion A^- has a lower, more negative energy than the neutral atom, A.[2] The second electron-gain enthalpy, the enthalpy change for the attachment of a second electron to an initially neutral atom, is invariably positive because the electron repulsion outweighs the nuclear attraction.

Example 1.9 Accounting for the variation in electron affinity

Account for the large decrease in electron affinity between Li and Be despite the increase in nuclear charge.

Answer The electron configurations of Li and Be are [He]$2s^1$ and [He]$2s^2$, respectively. The additional electron enters the $2s$ orbital of Li but the $2p$ orbital of Be, and hence is much less tightly bound. In fact, the nuclear charge is so well shielded in Be that electron gain is endothermic.

Self-test 1.9 Account for the decrease in electron affinity between C and N.

The electron affinity of an element is largely determined by the energy of the *lowest unfilled* (or half-filled) orbital of the ground-state atom. This orbital is one of the two **frontier orbitals** of an atom, the other one being the *highest filled* atomic orbital. The frontier orbitals are the sites of many of the changes in electron distributions when bonds form, and we shall see more of their importance as the text progresses. An element has a high electron affinity if the additional electron can enter a shell where it experiences a strong effective nuclear charge. This is the case for elements towards the top right of the periodic table, as we have already explained. Therefore, elements close to fluorine (specifically O and Cl, but not the noble gases) can be expected to have the highest electron affinities as their Z_{eff} is large and it is possible to add electrons to the valence shell. Nitrogen has very low electron affinity because there is a high electron repulsion when the incoming electron enters an orbital that is already half full.

(d) Electronegativity

Key points: The electronegativity of an element is the power of an atom of the element to attract electrons when it is part of a compound; there is a general increase in electronegativity across a period and a general decrease down a group.

The **electronegativity**, χ (chi), of an element is the power of an atom of the element to attract electrons to itself when it is part of a compound. If an atom has a strong tendency to acquire electrons, it is said to be 'highly electronegative' (like the elements close to fluorine). If it has a tendency to lose electrons (like the alkali metals), it is said to be 'electropositive'. Electronegativity is a very useful concept in chemistry and has numerous applications, which include a rationalization of bond energies and the types of reactions that substances undergo and the prediction of the polarities of bonds and molecules (Chapter 2).

Periodic trends in electronegativity can be related to the size of the atoms and electron configuration. If an atom is small and has an almost closed shell of electrons, then it is more likely to attract an electron to itself than a large atom with few valence electrons. Consequently, the electronegativities of the elements typically increase left to right across a period and decrease down a group.

Quantitative measures of electronegativity have been defined in many different ways. Linus Pauling's original formulation (which results in the values denoted χ_P in Table 1.8 and illustrated in Fig. 1.27) draws on concepts relating to the energetics of bond formation, which will be dealt with in Chapter 2.[3] A definition more in the spirit of this

---

[2] Be alert to the fact that some people use the terms electron affinity and electron-gain enthalpy interchangeably, so, somewhat unhelpfully and misleadingly, a positive electron affinity then indicates that A^- has a higher, less favourable, energy than A.

[3] Pauling values of electronegativity are used throughout the subsequent chapters.

Table 1.8 Pauling χ_P, Mulliken, χ_M, and **Allred–Rochow**, χ_{AR}, electronegativities

H							He
2.20							5.5
3.06							
2.20							
Li	**Be**	**B**	**C**	**N**	**O**	**F**	**Ne**
0.98	1.57	2.04	2.55	3.04	3.44	3.98	
1.28	1.99	1.83	2.67	3.08	3.22	4.43	4.60
0.97	1.47	2.01	2.50	3.07	3.50	4.10	5.10
Na	**Mg**	**Al**	**Si**	**P**	**S**	**Cl**	**Ar**
0.93	1.31	1.61	1.90	2.19	2.58	3.16	
1.21	1.63	1.37	2.03	2.39	2.65	3.54	3.36
1.01	1.23	1.47	1.74	2.06	2.44	2.83	3.30
K	**Ca**	**Ga**	**Ge**	**As**	**Se**	**Br**	**Kr**
0.82	1.00	1.81	2.01	2.18	2.55	2.96	3.0
1.03	1.30	1.34	1.95	2.26	2.51	3.24	2.98
0.91	1.04	1.82	2.02	2.20	2.48	2.74	3.10
Rb	**Sr**	**In**	**Sn**	**Sb**	**Te**	**I**	**Xe**
0.82	0.95	1.78	1.96	2.05	2.10	2.66	2.6
0.99	1.21	1.30	1.83	2.06	2.34	2.88	2.59
0.89	0.99	1.49	1.72	1.82	2.01	2.21	2.40
Cs	**Ba**	**Tl**	**Pb**	**Bi**			
0.79	0.89	2.04	2.33	2.02			
0.70	0.90	1.80	1.90	1.90			
0.86	0.97	1.44	1.55	1.67			

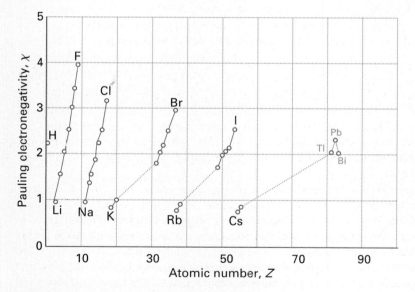

Fig. 1.27 The periodic variation of Pauling electronegativities.

chapter, in the sense that it is based on the properties of individual atoms, was proposed by Robert Mulliken. He observed that, if an atom has a high ionization energy, I, and a high electron affinity, E_a, then it will be likely to acquire rather than lose electrons when it is part of a compound, and hence be classified as highly electronegative. Conversely, if its ionization energy and electron affinity are both low, then the atom will tend to lose electrons rather than gain them, and hence be classified as electropositive. These observations motivate the definition of the **Mulliken electronegativity**, χ_M, as the average value of the ionization energy and the electron affinity of the element (both expressed in electronvolts):

$$\chi_M = \tfrac{1}{2}(I + E_a) \tag{1.11}$$

The complication in the definition of the Mulliken electronegativity is that the ionization energy and electron affinity in the definition relate to the **valence state**, the electron configuration the atom is supposed to have when it is part of a molecule. Hence, some calculation is required because the ionization energy and electron affinity to be used in calculating χ_M are mixtures of values for various actual spectroscopically observable states of the atom. We need not go into the calculation, but the resulting values given in Table 1.8 may be compared with the Pauling values. The two scales give similar values and show the same trends. One reasonably reliable conversion between the two is

$$\chi_P = 1.35(\chi_M)^{1/2} - 1.37 \tag{1.12}$$

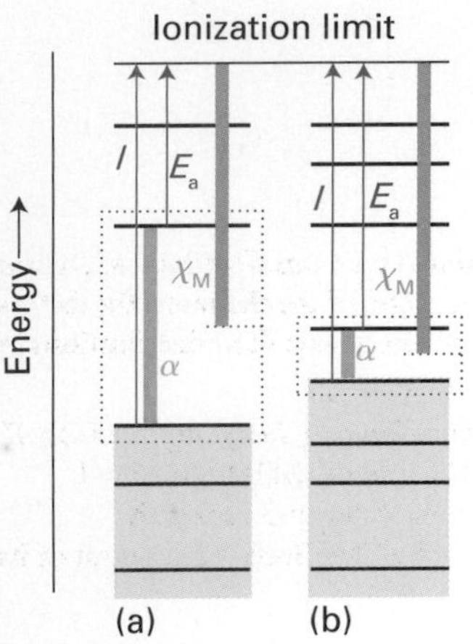

Because the elements near fluorine have high ionization energies and appreciable electron affinities, these elements have the highest Mulliken electronegativities. Because χ_M depends on atomic energy levels—and in particular on the location of the highest filled and lowest empty orbitals (Fig. 1.28)—the electronegativity of an element is high if the two frontier orbitals of its atoms are low in energy.

Various alternative 'atomic' definitions of electronegativity have been proposed. A widely used scale, suggested by A.L. Allred and E. Rochow, is based on the view that electronegativity is determined by the electric field at the surface of an atom. As we have seen, an electron in an atom experiences an effective nuclear charge Z_{eff}. The Coulombic potential at the surface of such an atom is proportional to Z_{eff}/r, and the electric field there is proportional to Z_{eff}/r^2. In the **Allred–Rochow definition** of electronegativity, χ_{AR} is assumed to be proportional to this field, with r taken to be the covalent radius of the atom:

$$\chi_{AR} = 0.744 + \frac{35.90 Z_{eff}}{(r/\mathrm{pm})^2} \tag{1.13}$$

The numerical constants have been chosen to give values comparable to Pauling electronegativities. According to the Allred–Rochow definition, elements with high electronegativity are those with high effective nuclear charge and the small covalent radius: such elements lie close to fluorine. The Allred–Rochow values parallel closely those of the Pauling electronegativities and are useful for discussing the electron distributions in compounds.

(e) Polarizability

Key points: A polarizable atom or ion is one with orbitals that lie close in energy; large, heavy atoms and ions tend to be highly polarizable.

The **polarizability**, α, of an atom is its ability to be distorted by an electric field (such as that of a neighbouring ion). An atom or ion (most commonly, an anion) is highly **polarizable** if its electron distribution can be distorted readily, which is the case if unfilled atomic orbitals lie close to the highest-energy filled orbitals. That is, the polarizability is likely to be high if the separation of the frontier orbitals is small and the polarizability will be low if the separation of the frontier orbitals is large (see Fig. 1.28). Closely separated frontier orbitals are typically found for large, heavy atoms and ions, such as the atoms and ions of the heavier alkali metals and the heavier halogens, so these atoms and ions are the most polarizable. Small, light atoms, such as the atoms and ions near fluorine, typically have widely spaced energy levels, so these atoms and ions are least polarizable. Species that effectively distort the electron distribution of a neighbouring atom or anion are described as having **polarizing ability**.

Fig. 1.28 The interpretation of the electronegativity and polarizability of an element in terms of the energies of the frontier orbitals (the highest filled and lowest unfilled atomic orbitals). (a) Low electronegativity and polarizability; (b) high electronegativity and polarizability.

We shall see the consequences of polarizability when considering the nature of bonding in Section 2.2, but it is appropriate to anticipate here that extensive polarization leads to covalency. **Fajan's rules** summarize the factors that affect polarization:

- Small, highly charged cations have polarizing ability.
- Large, highly charged anions are easily polarized.
- Cations that do not have a noble-gas electron configuration are easily polarized.

The last rule is particularly important for the *d*-block elements.

Example 1.10 Identifying polarizable species

Which would be the more polarizable, an F^- ion or an I^- ion?

Answer An F^- ion is small and singly charged. An I^- ion has the same charge but is large. Therefore, an I^- ion is likely to be the more polarizable.

Self-test 1.10 Which would be more polarizing, Na^+ or Cs^+?

FURTHER READING

P.A. Cox, *Introduction to quantum theory and atomic structure*. Oxford University Press (1996). An introduction to the subject.

P. Atkins and J. de Paula, *Physical chemistry*. Oxford University Press and W.H. Freeman & Co. (2006). Chapters 8 and 9 give an account of quantum theory and atomic structure.

J. Emsley, *Nature's building blocks*. Oxford University Press (2003). An interesting guide to the elements.

D.M.P Mingos, *Essential trends in inorganic chemistry*. Oxford University Press (1998). Includes a detailed discussion of the important horizontal, vertical, and diagonal trends in the properties of the atoms.

P.A. Cox, *The elements: their origin, abundance, and distribution*. Oxford University Press (1989). Examines the origin of the elements, the factors controlling their widely differing abundances, and their distributions in the Earth, the solar system, and the universe.

J.A. Cleverty and N.G. Connelly, *Nomenclature of inorganic chemistry II. Recommendations*. Royal Society of Chemistry (2001). This book outlines the conventions for the periodic table and inorganic substances. It is known colloquially as the 'Red Book' on account of its distinctive red cover.

EXERCISES

1.1 Write balanced equations for the following nuclear reactions (show emission of excess energy as a photon of electromagnetic radiation, γ): (a) $^{14}N + {}^4He$ to produce ^{17}O; (b) $^{12}C + p$ to produce ^{13}N; (c) $^{14}N + n$ to produce ^3H and ^{12}C. (The last reaction produces a steady state concentration of radioactive ^3H in the upper atmosphere.)

1.2 One possible source of neutrons for the neutron-capture processes mentioned in the text is the reaction of ^{22}Ne with α particles to produce ^{25}Mg and neutrons. Write the balanced equation for the nuclear reaction.

1.3 Without consulting reference material, draw the form of the periodic table with the numbers of the groups and the periods and identify the *s*, *p*, and *d* blocks. Identify as many elements as you can. (As you progress through your study of inorganic chemistry, you should learn the positions of all the *s*-, *p*-, and *d*-block elements and associate their positions in the periodic table with their chemical properties.)

1.4 What is the ratio of the energy of a ground-state He^+ ion to that of a Be^{3+} ion?

1.5 The ionization energy of H is 13.6 eV. What is the difference in energy between the $n = 1$ and $n = 6$ levels?

1.6 Calculate the wavenumber ($\tilde{v} = 1/\lambda$) and wavelength of the first transition in the visible region of the atomic spectrum of hydrogen.

1.7 What is the relation of the possible angular momentum quantum numbers to the principal quantum number?

1.8 How many orbitals are there in a shell of principal quantum number *n*? (Hint: begin with $n = 1$, 2, and 3 and see if you can recognize the pattern.)

1.9 Complete the following table:

n	*l*	m_l	Orbital designation	Number of orbitals
2			2*p*	
3	2			
			4*s*	
4		$+3, +2, \ldots, -3$		

1.10 Use sketches of 2*s* and 2*p* orbitals to distinguish between (a) the radial wavefunction, (b) the radial distribution function, and (c) the angular wavefunction.

1.11 Compare the first ionization energy of calcium with that of zinc. Explain the difference in terms of the balance between shielding with increasing numbers of *d* electrons and the effect of increasing nuclear charge.

1.12 Compare the first ionization energies of strontium, barium, and radium. Relate the irregularity to the lanthanide contraction.

1.13 The second ionization energies of some Period 4 elements are

Ca	Sc	Ti	V	Cr	Mn
1145	1235	1310	1365	1592	1509 kJ mol^{-1}

Identify the orbital from which ionization occurs and account for the trend in values.

1.14 Give the ground-state electron configurations of (a) C, (b) F, (c) Ca, (d) Ga^{3+}, (e) Bi, (f) Pb^{2+}.

1.15 Give the ground-state electron configurations of (a) Sc, (b) V^{3+}, (c) Mn^{2+}, (d) Cr^{2+}, (e) Co^{3+}, (f) Cr^{6+}, (g) Cu, (h) Gd^{3+}.

1.16 Give the ground-state electron configurations of (a) W, (b) Rh^{3+}, (c) Eu^{3+}, (d) Eu^{2+}, (e) V^{5+}, (f) Mo^{4+}.

1.17 Identify the elements that have the ground-state electron configurations: (a) [Ne]$3s^23p^4$, (b) [Kr]$5s^2$, (c) [Ar]$4s^23d^3$, (d) [Kr]$5s^24d^5$, (e) [Kr]$5s^24d^{10}5p^1$, (f) [Xe]$6s^24f^6$.

1.18 Account for the trends, element by element, across Period 3 in (a) ionization energy, (b) electron affinity, and (c) electronegativity.

1.19 Account for the fact that the two Group 5 elements niobium (Period 5) and tantalum (Period 6) have the same atomic radii.

1.20 Discuss the trend in electronegativites across Period 2 from lithium to fluorine. Can you account for the difference in details from the trend in ionization energy?

1.21 Identify the frontier orbitals of a Be atom.

1.22 Compare the broad trends in ionization energy, atomic radius, and electronegativity. Account for the parallels.

PROBLEMS

1.1 Show that an atom with the configuration ns^2np^6 is spherically symmetrical. Is the same true of an atom with the configuration ns^2np^3?

1.2 According to the Born interpretation, the probability of finding an electron in a volume element dτ is proportional to ψ^2dτ. (a) What is the most probable location of an electron in an H atom in its ground state? (b) What is its most probable distance from the nucleus, and why is this different? (c) What is the most probable distance of a $2s$ electron from the nucleus?

1.3 The ionization energies of rubidium and silver are 4.18 and 7.57 eV, respectively. Calculate the ionization energies of an H atom with its electron in the same orbitals as in these two atoms and account for the differences in values.

1.4 When 58.4 nm radiation from a helium discharge lamp is directed on a sample of krypton, electrons are ejected with a velocity of 1.59×10^6 m s^{-1}. The same radiation ejects electrons from rubidium atoms with a velocity of 2.45×10^6 m s^{-1}. What are the ionization energies (in eV) of the two elements?

1.5 Survey the early and modern proposals for the construction of the periodic table. You should consider attempts to arrange the elements on helices and cones as well as the more practical two-dimensional surfaces. What, in your judgement, are the advantages and disadvantages of the various arrangements?

1.6 The decision about the which elements should be identified as belonging to the f block has been a matter of some controversy. A view has been expressed by W.B. Jensen (*J. Chem. Educ.*, 1982, **59**, 635). Summarize the controversy and Jensen's arguments.

1.7 The natural abundances of adjacent elements in the periodic table usually differ by a factor of ten or more. Explain this phenomenon.

1.8 Balance the following nuclear reaction:

$$^{246}_{96}\text{Cm} + ^{12}_{6}\text{C} \rightarrow ? + ^{1}_{0}\text{n}$$

1.9 Explain how you would determine, using data you would look up in tables, whether or not the nuclear reaction in Problem 1.8 corresponds to a release of energy.

1.10 Consider the process of shielding in atoms, using Be as an example. What is being shielded? What is it shielded from? What is doing the shielding?

1.11 In general, ionization energies increase across a period from left to right. Explain why the second ionization energy of Cr is higher, not lower, than that of Mn.

1.12 Draw pictures of the two d orbitals in the *xy*-plane as flat projections in the plane of the paper. Label each drawing with the appropriate mathematical function, and include a labelled pair or Cartesian coordinate axes. Label the orbital lobes correctly with + and − signs.

1.13 During 1999 several papers appeared in the scientific literature claiming that d orbitals of Cu_2O had been observed experimentally. In his paper 'Have orbitals really been observed?' (*J. Chem. Educ.*, 2000, **77**, 1494), Eric Scerri reviews these claims and discusses whether orbitals can be observed physically. Summarize his arguments briefly.

1.14 At various times the following two sequences have been proposed for the elements to be included in Group 3: (a) Sc, Y, La, Ac; (b) Sc, Y, Lu, Lr. Because ionic radii strongly influence the chemical properties of the metallic elements, it might be thought that ionic radii could be used as one criterion for the periodic arrangement of the elements. Use this criterion to describe which of these sequences is preferred.

1.15 In the paper 'Ionization energies of atoms and atomic ions' (P.F. Lang and B.C. Smith, *J. Chem. Educ.*, 2003, **80**, 938) the authors discuss the apparent irregularities in the first and second ionization energies of d- and f-block elements. Describe how these inconsistencies are rationalized.

Molecular structure and bonding

2

The interpretation of structures and reactions in inorganic chemistry is often based on semiquantitative models. In this chapter we review the use of pictorial representations of electrons in molecules and methods for predicting their shapes. In addition, we examine the development of models of molecular structure in terms of the concepts of valence bond and molecular orbital theory. This chapter introduces concepts that will be used throughout the text to explain the structures and reactions of a wide variety of species. The chapter also illustrates the importance of the interplay between qualitative models, experiment, and calculation.

A great deal of inorganic chemistry depends on the relationship between the chemical properties of compounds and their electronic structures. The most elementary discussion of covalent bonding, the topic of this chapter, is in terms of shared pairs of electrons. That approach was introduced by G.N. Lewis in 1916. Since that time, our understanding of bonding has been greatly enriched both experimentally and theoretically, particularly through the development of molecular orbital theory. In this chapter we review Lewis's elementary but useful theory and show how modern theories capture some of its spirit but go far beyond it.

Lewis structures

Lewis proposed that a **covalent bond** is formed when two neighbouring atoms share an electron pair. A **single bond**, a shared electron pair (A:B), is denoted A—B; likewise, a **double bond**, two shared electron pairs (A::B), is denoted A=B, and a **triple bond**, three shared pairs of electrons (A:::B), is denoted A≡B. An unshared pair of valence electrons on an atom (A:) is called a **lone pair**. Although lone pairs do not contribute directly to the bonding, they do influence the shape of the molecule and its chemical properties.

2.1 The octet rule

Key point: Atoms share electron pairs until they have acquired an octet of valence electrons.

Lewis found that he could account for the existence of a wide range of molecules by proposing the **octet rule**:

> *Each atom shares electrons with neighbouring atoms to achieve a total of eight valence electrons (an 'octet').*

As we saw in Section 1.8, a closed-shell, noble-gas configuration is achieved when eight electrons occupy the s and p subshells of the valence shell. One exception is the hydrogen atom, which fills its valence shell, the $1s$ orbital, with two electrons (a 'duplet').

The octet rule provides a simple way of constructing a **Lewis structure**, a diagram that shows the pattern of bonds and lone pairs in a molecule. In most cases we can construct a Lewis structure in three steps.

1 Decide on the number of electrons that are to be included in the structure by adding together the numbers of all the valence electrons provided by the atoms.

Each atom provides all its valence electrons (thus, H provides one electron and O, with the configuration $[He]2s^2 2p^4$, provides six). Each negative charge on an ion corresponds to an additional electron; each positive charge corresponds to one electron less.

2 Write the chemical symbols of the atoms in the arrangement that shows which atoms are bonded together.

In most cases we know the arrangement or can make an informed guess. The less electronegative element is usually the central atom of a molecule, as in CO_2 and SO_4^{2-}, but there are plenty of well-known exceptions (H_2O and NH_3 among them).

3 Distribute the electrons in pairs so that there is one pair of electrons forming a single bond between each pair of atoms bonded together, and then supply electron pairs (to form lone pairs or multiple bonds) until each atom has an octet.

Each bonding pair (:) is then represented by a single line (—). The net charge of a polyatomic ion is assumed to be possessed by the ion as a whole, not by a particular individual atom.

Example 2.1 Writing a Lewis structure

Write a Lewis structure for the BF_4^- ion.

Answer The atoms supply $3 + (4 \times 7) = 31$ valence electrons; the single negative charge of the ion reflects the presence of an additional electron. We must therefore accommodate 32 electrons in 16 pairs around the five atoms. One solution is (**1**). The negative charge is ascribed to the ion as a whole, not to a particular individual atom.

Self-test 2.1 Write a Lewis structure for the PCl_3 molecule.

Table 2.1 gives examples of Lewis structures of some common molecules and ions. Except in simple cases, a Lewis structure does not portray the shape of the species, but only the pattern of bonds and lone pairs: it shows the number of the links, not the geometry of the molecule. For example, the BF_4^- ion is in fact tetrahedral (**2**), not planar, and PF_3 is trigonal pyramidal (**3**).

(a) Resonance

Key points: Resonance between Lewis structures lowers the calculated energy of the molecule and distributes the bonding character of electrons over the molecule; Lewis structures with similar energies provide the greatest resonance stabilization.

A single Lewis structure is often an inadequate description of the molecule. The Lewis structure of O_3 (**4**), for instance, suggests incorrectly that one O—O bond is different from the other, whereas in fact they have identical lengths (128 pm) intermediate between those of typical single O—O and double O=O bonds (148 and 121 pm, respectively). This deficiency of the Lewis description is overcome by introducing the concept of **resonance**, in which the actual structure of the molecule is taken to be a superposition, or average, of all the feasible Lewis structures corresponding to a given atomic arrangement.

Resonance is indicated by a double-headed arrow:

$$:\ddot{O}—\ddot{O}=\ddot{O} \quad \longleftrightarrow \quad \ddot{O}=\ddot{O}—\ddot{O}:$$

At this stage we are not indicating the shape of the molecule. Resonance should be pictured as a *blending* of structures, not a flickering alternation between them. In

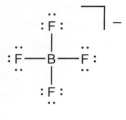

1 BF_4^-

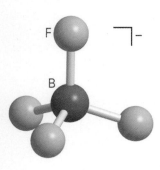

2 BF_4^-

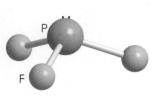

3 PF_3

$$:\ddot{O}—\ddot{O}=\ddot{O}$$

4 O_3

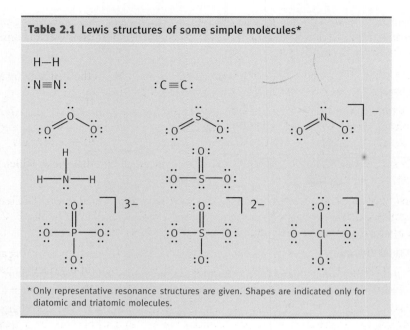

Table 2.1 Lewis structures of some simple molecules*

*Only representative resonance structures are given. Shapes are indicated only for diatomic and triatomic molecules.

quantum mechanical terms, the electron distribution of each structure is represented by a wavefunction, and the actual wavefunction, ψ, of the molecule is the superposition of the individual wavefunctions for each contributing structure:[1]

$$\psi = \psi(\text{O—O}{=}\text{O}) + \psi(\text{O}{=}\text{O—O})$$

The overall wavefunction is written as a superposition with equal contributions from both structures because the two structures have identical energies. The *blended* structure of two or more Lewis structures is called a **resonance hybrid**. Note that resonance occurs between structures that differ only in the allocation of electrons; resonance does not occur between structures in which the atoms themselves lie in different positions. For instance, there is no resonance between the structures SOO and OSO.

Resonance has two main effects:

1 Resonance averages the bond characteristics over the molecule.

2 The energy of a resonance hybrid structure is lower than that of any single contributing structure.

The energy of the O_3 resonance hybrid, for instance, is lower than that of either individual structure alone. Resonance is most important when there are several structures of identical energy that can be written to describe the molecule, as for O_3. In such cases, all the structures of the same energy contribute equally to the overall structure.

Structures with different energies may also contribute to an overall resonance hybrid but, in general, the greater the energy difference between two Lewis structures, the less the higher energy structure contributes. The BF_3 molecule, for instance, could be regarded as a resonance hybrid of the structures shown in (**5**), but the first structure dominates even though the octet is incomplete. Consequently, BF_3 is regarded *primarily* as having that structure with a small admixture of double-bond character. By contrast, for the NO_3^- ion (**6**), the last three structures dominate, and we treat the ion as having partial double-bond character.

---

[1] This wavefunction is not normalized (Section 1.5). We shall often omit normalization constants from linear combinations so as to clarify their structure. The wavefunctions themselves are formulated in the valence-bond theory, which is described later.

5 BF$_3$

6 NO$_3^-$

(b) Formal charge

Key points: The formal charge is the charge an atom would have if electron pairs were shared equally; Lewis structures with low formal charges typically have the lowest energy.

The decision about how best to arrange electrons to give the Lewis structure with the lowest energy can be put on a simple quantitative footing by assessing the **formal charge**, f, on each atom, the charge it would have if the bonding were perfectly covalent, with each shared pair of electrons shared equally between the two bonded atoms. That is, each atom is assumed to 'own' one electron of a bonding pair. Each lone pair of electrons belongs wholly to the atom on which the lone pair resides. The formal charge is then the net charge of the atom based on this 'purely covalent, perfect sharing' model of the bonding and is a measure of the extent to which the atom has gained or lost electrons in reaching the Lewis structure. More precisely,

$$f = V - L - \tfrac{1}{2}P \tag{2.1}$$

where V is the number of valence electrons on the parent atom, L is the number of lone pair electrons on that atom in the Lewis structure of the molecule, and P is the number of shared electrons. Thus, the formal charge is the difference between the number of valence electrons in the free atom and the number that the atom has in the molecule, assuming it owns one electron of each shared pair and both electrons of any lone pair (Fig. 2.1). For example, consider nitrogen in the ammonium ion, NH$_4^+$. Nitrogen has five valence electrons, there are no lone pairs and eight electrons shared in the four N—H bonds. Therefore, the formal charge on nitrogen is $5 - 0 - 4 = +1$. The formal charge represents (in some idealized sense) the number of electrons that an atom gains or loses when it enters into perfect covalent bonding with other atoms. The sum of the formal charges on a Lewis structure is equal to the total charge of the species (and is therefore zero for an electrically neutral molecule).

It is commonly the case that the lowest energy Lewis structure is the one with the lowest formal charges on the atoms (because such a structure corresponds to the least extensive rearrangement of electrons) and the structure in which the more electronegative element is assigned a formal negative charge and the less electronegative element is assigned a formal positive charge.

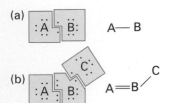

Fig. 2.1 A pictorial representation of the calculation of formal charge. The lines show how the bonding electrons and lone pair electrons are apportioned to each atom (a) in a diatomic molecule A—B and (b) in a triatomic molecule A═B—C. The formal charge on each atom is the difference between the number of electrons obtained in this way and the number in the free, neutral atom.

Example 2.2 Writing resonance structures and assessing their formal charges

Write resonance structures for NO$_2$F and identify the dominant structure.

Answer Four Lewis structures and their formal charges are shown in (**7**). It is very unlikely that a low energy will be achieved with a positive charge on an electronegative F atom. Each structure has a positive charge on the N atom but the structure with the highest charge is likely to have the highest energy. So the two structures with N═O bonds are most likely to dominate in the resonance.

Self-test 2.2 Write resonance structures for the nitrite ion, NO_2^-, and identify the formal charge on the atoms.

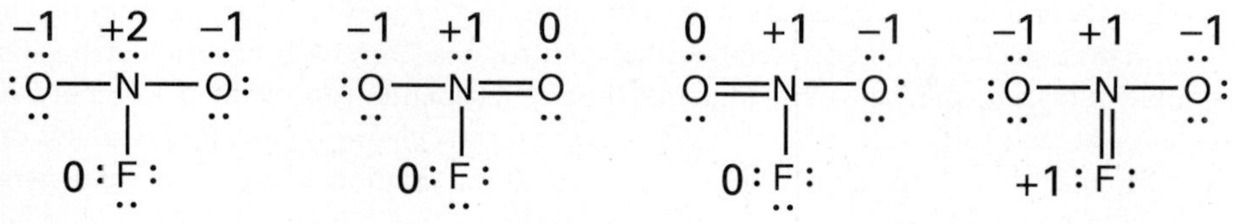

7 NO_2F

(c) Oxidation states

Key point: Oxidation numbers are assigned by applying the rules set out in Table 2.2.

Formal charge is a parameter derived by exaggerating the *covalent* character of a bond. The **oxidation number**, ω (omega),[2] is a parameter obtained by exaggerating the *ionic* character of a bond. It can be regarded as the charge that an atom would have if the more electronegative atom in a bond acquired the two electrons of the bond completely. The **oxidation state** is the physical state of the element corresponding to its oxidation number. Thus, an atom may *be assigned* an oxidation number and be *in* the corresponding oxidation state.[3] Each O atom in a compound is regarded as an O^{2-} ion (provided an O—F bond is not present), and hence is ascribed an oxidation number of -2. Likewise, the exaggerated ionic structure of NO_3^- is $N^{5+}(O^{2-})_3$, so the oxidation number of nitrogen in this compound is $+5$, which is denoted either $N(V)$ or $N(+5)$. These conventions may be used even if the oxidation number is negative, so

Table 2.2 The determination of oxidation number*

	Oxidation number
1. The sum of the oxidation numbers of all the atoms in the species is equal to its total charge	
2. For atoms in their elemental form	0
3. For atoms of Group 1	$+1$
For atoms of Group 2	$+2$
For atoms of Group 13 (except B)	$+3(EX_3)$, $+1(EX)$
For atoms of Group 14 (except C, Si)	$+4(EX_4)$, $+2(EX_2)$
4. For hydrogen	$+1$ in combination with nonmetals
	-1 in combination with metals
5. For fluorine	-1 in all its compounds
6. For oxygen	-2 unless combined with F
	-1 in peroxides (O_2^{2-})
	$-\frac{1}{2}$ in superoxides (O_2^-)
	$-\frac{1}{3}$ in ozonides (O_3^-)
7. Halogens	-1 in most compounds, unless the other elements include oxygen or more electronegative halogens

* To determine an oxidation number, work through the following rules in the order given. Stop as soon as the oxidation number has been assigned. These rules are not exhaustive, but they are applicable to a wide range of common compounds.

[2] There is no formally agreed symbol for oxidation number.

[3] In practice, inorganic chemists use the terms 'oxidation number' and 'oxidation state' interchangeably, but in this text we shall preserve the distinction.

oxygen has oxidation number -2, denoted $O(-2)$ or more rarely $O(-II)$, in most of its compounds.

In practice, oxidation numbers are assigned by applying a set of simple rules (Table 2.2). These rules reflect the consequences of electronegativity for the 'exaggerated ionic' structures of compounds and match the increase in the degree of oxidation that we would expect as the number of oxygen atoms in a compound increases (as in going from NO to NO_3^-). This aspect of oxidation number is taken further in Chapter 5. Many elements, for example nitrogen, the halogens, and the d-block elements, can exist in a variety of oxidation states (Table 2.2). Note that whereas the formal charge on an atom depends on the particular Lewis structure being considered, the oxidation state of an element is independent of the Lewis structure.

Example 2.3 Assigning an oxidation number to an element

What is the oxidation number of (a) S in hydrogen sulfide, H_2S, (b) Mn in the permanganate ion, MnO_4^-?

Answer Work through the steps set out in Table 2.2, in the order given. (a) The overall charge of the species is 0; so $2\omega(H) + \omega(S) = 0$. Because $\omega(H) = +1$ in combination with a nonmetal, it follows that $\omega(S) = -2$. (b) The sum of the oxidation numbers of all the atoms is -1, so $\omega(Mn) + 4\omega(O) = -1$. Because $\omega(O) = -2$, it follows that $\omega(Mn) = -1 - 4(-2) = +7$. That is, MnO_4^- is a compound of Mn(VII). Its formal name is tetraoxomanganese(VII) ion.

Self-test 2.3 What is the oxidation number of (a) O in O_2^+, (b) P in PO_4^{3-}?

(d) Hypervalence

Key point: Hypervalence and octet expansion occur for elements following Period 2.

The elements of Period 2, Li through Ne, obey the octet rule quite well, but elements of later periods show deviations from it. For example, the bonding in PCl_5 requires the P atom to have 10 electrons in its valence shell, one pair for each P—Cl bond (**8**). Similarly, in SF_6 the S atom must have 12 electrons if each F atom is to be bound to the central S atom by an electron pair (**9**). Species of this kind, which demand the presence of more than an octet of electrons around at least one atom, are called **hypervalent**. Species with Lewis structures that include expanded octets but which do *not necessarily* have more than eight valence electrons are not regarded as hypervalent. Thus, SO_4^{2-} is not hypervalent because a Lewis structure can be drawn with an octet of electrons around the S atom even though some contributing Lewis structures have 12 electrons in the valence shell of S.

The traditional explanation of hypervalence (for SF_6, for instance) and of octet expansion in general (for certain Lewis structures of SO_4^{2-}, for instance) invokes the availability of low-lying unfilled d orbitals, which can accommodate the additional electrons. According to this explanation, a P atom can accommodate more than eight electrons if it uses its vacant $3d$ orbitals. In PCl_5, with its five pairs of bonding electrons, at least one $3d$ orbital must be used in addition to the four $3s$ and $3p$ orbitals of the valence shell. The rarity of hypervalence in Period 2 is then ascribed to the absence of $2d$ orbitals. However, the real reason for the rarity of hypervalence in Period 2 may be the geometrical difficulty of packing more than four atoms around a small central atom and may in fact have little to do with the availability of d orbitals. The molecular orbital theory of bonding, which is described later in this chapter, suggests that the traditional explanation overemphasizes the role of d orbitals in hypervalent compounds (Section 2.11b).

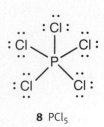

8 PCl_5

9 SF_6

2.2 Structure and bond properties

Certain properties of bonds are approximately the same in different compounds of the elements. Thus, if we know the strength of an O—H bond in H_2O, then with some confidence we can use the same value for the O—H bond in CH_3OH. At this stage we confine our attention to two of the most important characteristics of a bond: its length and its strength.

(a) Bond length

Key points: The equilibrium bond length in a molecule is the separation of the centres of the two bonded atoms; covalent radii vary through the periodic table in much the same way as metallic and ionic radii.

The **equilibrium bond length** in a molecule is the distance between the centres of the two bonded atoms. A wealth of useful and accurate bond length information is available in the literature, most of it obtained by X-ray diffraction on solids (Section 6.1). Equilibrium bond lengths of molecules in the gas phase are usually determined by infrared or microwave spectroscopy, or more directly by electron diffraction. Some typical values are given in Table 2.3.

To a reasonable first approximation, equilibrium bond lengths can be partitioned into contributions from each atom of the bonded pair. The contribution of an atom to a covalent bond is called the **covalent radius** of the element (**10**). We can use covalent radii in Table 2.4 to predict, for example, that the length of a P—N bond is 110 pm + 74 pm = 184 pm; experimentally, this bond length is close to 180 pm in a number of compounds. Experimental bond lengths should be used whenever possible, but covalent radii are useful for making cautious estimates when experimental data are not available.

Covalent radii vary through the periodic table in much the same way as metallic and ionic radii (Section 1.9a), for the same reasons, and are smallest close to F. Covalent radii are approximately equal to the separation of nuclei when the cores of the two atoms are in contact: the valence electrons draw the two atoms together until the repulsion between the cores starts to dominate. A covalent radius expresses the closeness of approach of *bonded* atoms; the closeness of approach of *nonbonded* atoms in neighbouring molecules that are in contact is expressed in terms of the **van der Waals radius** of the element, which is the internuclear separation when the *valence* shells of the two atoms are in nonbonding contact (**11**). van der Waals radii are of paramount importance for understanding the packing of molecular compounds in crystals, the conformations adopted by small but flexible molecules, and the shapes of biological macromolecules (Chapter 26).

(b) Bond strength

Key points: The strength of a bond is measured by its dissociation enthalpy; mean bond enthalpies are used to make estimates of reaction enthalpies.

A convenient thermodynamic measure of the strength of an AB bond is the **bond dissociation enthalpy**, $\Delta H^{\ominus}(A-B)$, the standard reaction enthalpy for the process

$$AB(g) \rightarrow A(g) + B(g)$$

The **mean bond enthalpy**, B, is the average bond dissociation enthalpy taken over a series of A—B bonds in different molecules (Table 2.5).

Mean bond enthalpies can be used to estimate reaction enthalpies. However, thermodynamic data on actual species should be used whenever possible in preference to mean values because the latter can be misleading. For instance, the Si—Si bond enthalpy ranges from 226 kJ mol^{-1} in Si_2H_6 to 322 kJ mol^{-1} in $Si_2(CH_3)_6$. The values in Table 2.5 are best considered as data of last resort: they may be used to make rough estimates of reaction enthalpies when enthalpies of formation or actual bond enthalpies are unavailable.

Example 2.4 Making estimates using mean bond enthalpies

Estimate the reaction enthalpy for the production of $SF_6(g)$ from $SF_4(g)$ given that the mean bond enthalpies of F_2, SF_4, and SF_6 are 158, 343, and 327 kJ mol^{-1}, respectively, at 25°C.

Answer The reaction is

$$SF_4(g) + F_2(g) \rightarrow SF_6(g)$$

Table 2.3 Bond lengths, R_e/pm

H_2^+	106
H_2	74
HF	92
HCl	127
HBr	141
HI	160
N_2	109
O_2	121
F_2	144
Cl_2	199
I_2	267

Table 2.4 Covalent radii, r_{cov}/pm*

H			
37			
C	**N**	**O**	**F**
77 (1)	74 (1)	66 (1)	64
67 (2)	65 (2)	57 (2)	
60 (3)	54 (3)		
70 (a)			
Si	**P**	**S**	**Cl**
118	110	104 (1)	99
		95 (2)	
Ge	**As**	**Se**	**Br**
122	121	117	114
	Sb	**Te**	**I**
	141	137	133

*Values are for single bonds except where otherwise stated (in parentheses); (a) denotes aromatic.

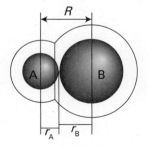

10 Covalent radius

Table 2.5 Mean bond enthalpies, B/(kJ mol^{-1})*

	H	C	N	O	F	Cl	Br	I	S	P	Si
H	436										
C	412	348 (1)									
		612 (2)									
		837 (3)									
		518 (a)									
N	388	305 (1)	163 (1)								
		613 (2)	409 (2)								
		890 (3)	946 (3)								
O	463	360 (1)	157	146 (1)							
		743 (2)		497 (2)							
F	565	484	270	185	155						
Cl	431	338	200	203	254	242					
Br	366	276				219	193				
I	299	238				210	178	151			
S	338	259	464	523	343	250	212		264		
P	322 (1)									201	
										480 (3)	
Si	318			466							226

*Values are for single bonds except where otherwise stated (in parentheses); (a) denotes aromatic.

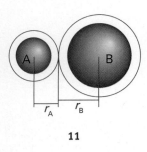

11

In this reaction, 1 mol F—F bonds and 4 mol S—F bonds (in SF$_4$) must be broken, corresponding to an enthalpy change of 158 kJ + (4 × 343 kJ) = +1530 kJ. This enthalpy change is positive because energy will be used in breaking bonds. Then 6 mol S—F bonds (in SF$_6$) must be formed, corresponding to an enthalpy change of 6 × (−327 kJ) = −1962 kJ. This enthalpy change is negative because energy is released when the bonds are formed. The net enthalpy change is therefore

$$\Delta H^{\ominus} = +1530 \text{ kJ} - 1962 \text{ kJ} = -432 \text{ kJ}$$

Hence, the reaction is strongly exothermic. The experimental value for the reaction is − 434 kJ, which is in excellent agreement with the estimated value.

Self-test 2.4 Estimate the enthalpy of formation of H$_2$S from S$_8$ (a cyclic molecule) and H$_2$.

(c) Bond enthalpy trends in the *p* block

Key points: For an element E that has no lone pairs, the E—X bond enthalpy decreases down the group; for an element that has lone pairs, the E—X bond enthalpy typically increases between Periods 2 and 3 and then decreases down the group.

The trend in mean bond enthalpies in the *p* block can be summarized as follows:

For an element E that has no lone pairs, the E—X bond enthalpy decreases down a group.

Some examples of this trend are

	B/(kJ mol^{-1})		B/(kJ mol^{-1})		B/(kJ mol^{-1})
C—C	347	C—H	412	C—Cl	327
Si—Si	222	Si—H	328	Si—Cl	391
Ge—Ge	188	Ge—H	289	Ge—Cl	342

Smaller atoms form stronger bonds because the shared electrons are closer to each of the atomic nuclei. The high value for the Si—Cl bond is attributed the large difference in electronegativity between Si and Cl. Another general trend is:

> For an element that has lone pairs, the bond enthalpy decreases down a group, but the value for an element at the head of a group is anomalous and smaller than the bond enthalpy of an element in Period 3, with the exception of the hydride.

Some examples of this trend are:

	$B/(\text{kJ mol}^{-1})$		$B/(\text{kJ mol}^{-1})$		$B/(\text{kJ mol}^{-1})$
N—N	165	N—Cl	194	N—H	388
P—P	200	P—Cl	319	P—H	322
As—As	180	As—Cl	317	As—H	247

The relative weakness of single bonds between Period 2 elements that have lone pairs is often ascribed to the closeness of the lone pairs on neighbouring atoms and the repulsion between them. In Period 2, the lone pairs of electrons on neighbouring atoms are close together and electron–electron repulsion leads to a decrease in the bond enthalpy. This reduction is observed for E—E bonds and E—X bonds if X also has lone pairs. Note that this trend is not observed for E—H because an H atom does not have a lone pair of electrons.

A number of features of the *p*-block elements can be interpreted (and remembered) with the aid of arguments based on bond enthalpies. For example, the bond dissociation enthalpy of gaseous BO is 788 kJ mol^{-1} whereas the B—O single bond enthalpy is 523 kJ mol^{-1}. It follows that the boron–oxygen bond in BO must be at least double and perhaps triple. Although acyclic alkanes with less than four C atoms are thermodynamically stable with respect to decomposition into their elements, their silicon analogues the silanes, $\text{Si}_n\text{H}_{2n+2}$, are unstable with respect to Si(s) and H_2(g). This difference is a consequence of the strength of the C—C bond compared to the Si—Si bond and the weakness of the Si—H bond compared to that of the H—H bond. The strength of the C—C bond also accounts for extensive **catenation** in carbon compounds, the formation of chains of atoms of the same element (Section 13.5b).

The trend in bond strengths for atoms with lone pairs is illustrated by the P—P single bond, for which $B = 200$ kJ mol^{-1}, compared with the N—N single bond, for which $B = 165$ kJ mol^{-1}. However, nitrogen forms much stronger multiple bonds than phosphorus. This difference can account for the fact that phosphorus is found as P_4 molecules under normal conditions (**12**) whereas nitrogen is found as N_2 molecules (:N≡N:, Section 14.1). The considerable difference in single and triple bond enthalpies of nitrogen also explains the rarity of catenation in its compounds. Thus, hydrazine, $\text{H}_2\text{N—NH}_2$ (the analogue of ethane) is quite strongly endoergic[4] at room temperature ($\Delta_f G^\ominus = +149$ kJ mol^{-1}) and analogues of higher alkanes are unknown.

Similar differences are found in Group 16 (Chapter 15). The S—S single bond enthalpy is larger than the O—O single bond enthalpy (263 and 142 kJ mol^{-1}, respectively). However, the bond enthalpies of the doubly bonded species O_2 (O=O) is significantly larger than that of S_2 (498 and 431 kJ mol^{-1}, respectively). This difference explains why elemental sulfur forms rings or chains with S—S single bonds, whereas oxygen exists as diatomic molecules. Similarly, sulfur catenation leads to polysulfides of formula $[\text{S—S—S}]^{2-}$ and $[\text{S—S—S—S}]^{2-}$ but polyoxygen anions beyond O_3^- are unknown.

One enlightening application of bond enthalpy arguments concerns **subvalent compounds**, or compounds in which fewer bonds are formed than valence rules suggest, such as PH_2. Although this compound is thermodynamically stable with respect to dissociation into the constituent atoms, it is unstable with respect to the disproportionation

$$3\,\text{PH}_2(\text{g}) \rightarrow 2\,\text{PH}_3(\text{g}) + \tfrac{1}{4}\text{P}_4(\text{s})$$

In a disproportionation reaction, a reactant produces two products, one in a lower and one in a higher oxidation state than the reactant. The origin of the spontaneity of this

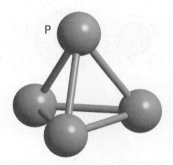

P

12 P_4

---

[4] An endoergic compound is one with a positive standard Gibbs energy of formation, and is therefore thermodynamically unstable with respect to its elements.

reaction is the strength of the P—P bonds in solid phosphorus, P_4. There are the same number (six) of P—H bonds in the reactants as there are in the products, but the reactants have no P—P bonds.

Although enthalpy changes are a reasonably reliable guide to stability, entropy also plays a role, particularly when gaseous species are involved. Judgements are therefore more reliable when based on the Gibbs energy rather than the enthalpy alone.

(d) Electronegativity and bond enthalpy

Key points: The Pauling scale of electronegativity is useful for estimating bond enthalpies and for assessing the polarities of bonds.

The concept of electronegativity was introduced in Section 1.9d, where it was defined as the power of an atom of the element to attract electrons to itself when it is part of a compound. The greater the difference in electronegativity between two elements A and B, the greater the ionic character of the A—B bond.

Linus Pauling's original formulation of electronegativity drew on concepts relating to the energetics of bond formation. For example, in the formation of AB from the diatomic A_2 and B_2 molecules,

$$A_2(g) + B_2(g) \rightarrow 2\ AB(g)$$

He argued that the excess energy, ΔE, of the A—B bond over the average energy of A—A and B—B bonds can be attributed to the presence of ionic contributions to the covalent bonding. He defined the difference in electronegativity as

$$|\chi_P(A) - \chi_P(B)| = 0.102(\Delta E/\text{kJ mol}^{-1})^{1/2} \qquad (2.2a)$$

where

$$\Delta E = B(A\text{—}B) - \tfrac{1}{2}\{B(A\text{—}A) + B(B\text{—}B)\} \qquad (2.2b)$$

with $B(A\text{—}B)$ the mean A—B bond enthalpy. Thus, if the A—B bond enthalpy is significantly greater than the average of the nonpolar A—A and B—B bonds, then it is presumed that there is a substantial ionic contribution to the wavefunction, and hence a large difference in electronegativity between the two atoms. Pauling electronegativities increase with increasing oxidation number of the element and the values in Table 1.8 are for the most common oxidation state.

Pauling electronegativities are useful for estimating the enthalpies of bonds between elements of different electronegativity and to make qualitative assessments of the polarities of bonds. Binary compounds in which the difference in electronegativity between the two elements is greater than about 1.7 can generally be regarded as being predominantly ionic. However, this crude distinction was refined by Anton van Arkel and Jan Ketelaar in the 1940s, when they drew a triangle with vertices representing ionic, covalent, and metallic bonding. The **Ketelaar triangle** (more appropriately, the *van Arkel–Ketelaar triangle*) has been elaborated by Gordon Sproul, who constructed a triangle based on the difference in electronegativities ($\Delta\chi$) of the elements in a binary compound and their average electronegativity (χ_{mean}, Fig. 2.2). As we work through the following chapters, we shall see how this basic concept can be enriched and used to classify a wide range of compounds of different kinds.

Ionic bonding is characterized by a large difference in electronegativity. Because a large difference indicates that the electronegativity of one element is high and that of the other is low, the average electronegativity must be intermediate in value. The compound CsF, for instance, with $\Delta\chi = 3.19$ and $\chi_{\text{mean}} = 2.38$, lies at the 'ionic' apex of the triangle. Covalent bonding is characterized by a small difference in electronegativities. Such compounds lie at the base of the triangle. Binary compounds that are predominantly covalently bonded are typically formed between nonmetals, which have high electronegativities. It follows that the covalent region of the triangle is the lower, right-hand corner. This corner of the triangle is occupied by F_2, which has $\Delta\chi = 0$ and $\chi_{\text{mean}} = 3.98$ (the maximum value of any Pauling electronegativity). Metallic bonding is also characterized by a small electronegativity difference, and also lies towards the base of the

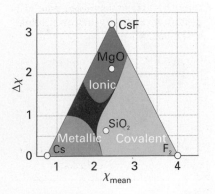

Fig. 2.2 A Ketelaar triangle, showing how a plot of average electronegativity against electronegativity difference can be used to classify the bond type for binary compounds.

triangle. In metallic bonding, however, electronegativities are low, the average values are therefore also low, and consequently metallic bonding occupies the lower, left-hand corner of the triangle. The outer corner is occupied by Cs, which has $\Delta\chi = 0$ and $\chi_{mean} = 0.79$ (the lowest value of Pauling electronegativity). The advantage of using a Ketelaar triangle over simple electronegativity difference is that it allows us to distinguish between covalent and metallic bonding, which are both indicated by a small electronegativity difference.

A Ketelaar triangle can be used to predict what type of bonding is likely to dominate in a binary compound. For example, in MgO, $\Delta\chi = 3.44 - 1.31 = 2.13$ and $\chi_{mean} = 2.38$. These values place MgO in the ionic region of the triangle. By contrast, for SiO_2, $\Delta\chi = 2.58 - 1.90 = 0.68$ and $\chi_{mean} = 2.24$. These values place SiO_2 lower on the triangle compared to MgO and in the covalent bonding region.

Example 2.5 Using a Ketelaar triangle

Using the electronegativity values in Table 1.8 and the Ketelaar triangle in Fig. 2.2 predict what type of bonding is likely to dominate in CaO.

Answer In CaO, $\Delta\chi = 3.44 - 1.00 = 2.44$ and $\chi_{mean} = 2.22$. These values place CaO in the ionic region of the triangle.

Self-test 2.5 Predict what type of bonding is likely to dominate in P_2O_5.

2.3 The VSEPR model

The **valence-shell electron pair repulsion model** (VSEPR model) of molecular shape is a simple extension of Lewis's ideas and is surprisingly successful for predicting the shapes of polyatomic molecules. The theory stems from suggestions made by Nevil Sidgwick and Herbert Powell in the years up to 1940 and later extended and put into a more modern context by Ronald Gillespie and Ronald Nyholm.

(a) The basic shapes

Key points: In the VSEPR model, regions of enhanced electron density take up positions as far apart as possible, and the shape of the molecule is identified by referring to the locations of the atoms in the resulting structure.

The primary assumption of the VSEPR model is that regions of enhanced electron density, by which we mean bonding pairs, lone pairs, or the concentrations of electrons associated with multiple bonds, take up positions as far apart as possible so that the repulsions between them are minimized. For instance, four such regions of electron density will lie at the corners of a regular tetrahedron, five will lie at the corners of a trigonal bipyramid, and so on (Table 2.6).

Although the arrangement of regions of electron density, both bonding regions and regions associated with lone pairs, governs the shape of the molecule, the *name* of the shape is determined by the arrangement of *atoms*, not the arrangement of the regions of electron density (Table 2.7). For instance, the NH_3 molecule has four electron pairs that are disposed tetrahedrally, but as one of them is a lone pair the molecule itself is classified as trigonal pyramidal. One apex of the pyramid is occupied by the lone pair. Similarly, H_2O has a tetrahedral arrangement of its electron pairs but, as two of the pairs are lone pairs, the molecule is classified as angular.

To apply the VSEPR model systematically, we first write down the Lewis structure for the molecule or ion and identify the central atom. Next, we count the number of atoms and lone pairs carried by that atom, because each atom (whether it is singly or multiply bonded to the central atom), and each lone pair, counts as one region of high electron density. To achieve lowest energy, these regions take up positions as far apart as possible, so we identify the basic shape they adopt by referring to Table 2.6. Finally, we note which locations correspond to atoms and identify the shape of the molecule from Table 2.7.

Table 2.6 The basic arrangement of regions of electron density according to the VSEPR model

Number of electron regions	Arrangement
2	Linear
3	Trigonal planar
4	Tetrahedral
5	Trigonal bipyramidal
6	Octahedral

Table 2.7 The description of molecular shapes

Linear	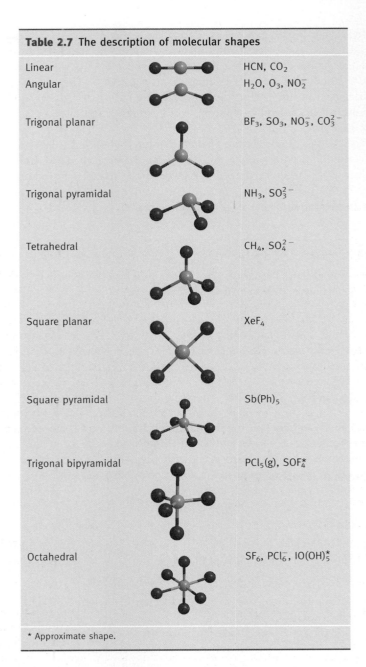	HCN, CO_2
Angular		H_2O, O_3, NO_2^-
Trigonal planar		BF_3, SO_3, NO_3^-, CO_3^{2-}
Trigonal pyramidal		NH_3, SO_3^{2-}
Tetrahedral		CH_4, SO_4^{2-}
Square planar		XeF_4
Square pyramidal		$Sb(Ph)_5$
Trigonal bipyramidal		$PCl_5(g)$, $SOF_4^{\star}$
Octahedral		SF_6, PCl_6^-, $IO(OH)_5^{\star}$

* Approximate shape.

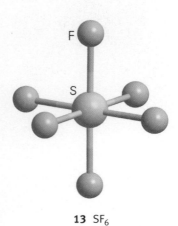

13 SF_6

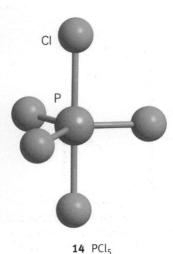

14 PCl_5

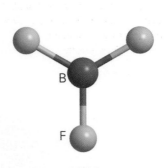

15 BF_3

Thus, an SF_6 molecule, with six single bonds around the central S atom, is predicted (and found) to be octahedral (**13**), and a PCl_5 molecule, with five single bonds and therefore five regions of electron density around the central atom, is predicted (and found) to be trigonal bipyramidal (**14**).

Example 2.6 Using the VSEPR model to predict shapes

Predict the shape of (a) a BF_3 molecule, (b) an SO_3^{2-} ion.

Answer (a) The Lewis structure of BF_3 is shown in (**5**). To the central B atom there are attached three F atoms but no lone pairs. The basic arrangement of three regions of electron density is trigonal planar. Because each location carries an F atom, the shape of the molecule is also trigonal planar (**15**). (b) Two Lewis structures for SO_3^{2-} are shown in (**16**): these two are representative of a variety of structures that contribute to the overall resonance structure. In each case there are three atoms attached to the central S atom and one lone pair, corresponding to four regions of electron density. The basic arrangement of these regions is tetrahedral.

Three of the locations correspond to atoms, so the shape of the ion is trigonal pyramidal (**17**). Note that the shape deduced in this way is independent of which resonance structure is being considered.

16 SO_3^{2-}

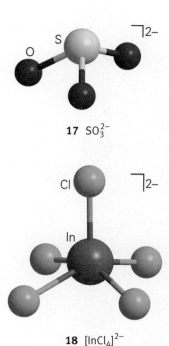

17 SO_3^{2-}

18 $[InCl_4]^{2-}$

Self-test 2.6 Predict the shape of (a) an H_2S molecule, (b) an XeO_4 molecule.

The VSEPR model is highly successful, but sometimes runs into difficulty when there is more than one basic shape of similar energy. For example, with five electron pairs attached to the central atom, a square-pyramidal arrangement of bonds is only slightly higher in energy than a trigonal-bipyramidal arrangement, and there are several examples of the former (**18**). Similarly, the basic shapes for seven regions of electron density are less readily predicted than others, partly because so many different conformations correspond to similar energies. However, in the *p* block, 7-coordination is dominated by pentagonal-bipyramidal structures. For example, IF_7 is pentagonal bipyramidal and XeF_5, with five bonds and two lone pairs, is pentagonal planar. Lone pairs are stereochemically less influential when they belong to heavy *p*-block elements. The SeF_6^{2-} and $TeCl_6^{2-}$ ions, for instance, are octahedral despite the presence of a lone pair on the Se and Te atoms. Lone pairs that do not influence the molecular geometry are said to be **stereochemically inert**.

(b) Modifications of the basic shapes

Key point: Lone pairs repel other pairs more strongly than bonding pairs do.

Once the basic shape of a molecule has been identified, adjustments are made by taking into account the differences in electrostatic repulsion between bonding regions and lone pairs. These repulsions are assumed to lie in the order

lone pair/lone pair > lone pair/bonding region > bonding region/bonding region

In elementary accounts, the greater repelling effect of a lone pair is explained by supposing that the lone pair is on average closer to the nucleus than a bonding pair and therefore repels other electron pairs more strongly. However, the true origin of the difference is obscure. An additional detail about this order of repulsions is that, given the choice between an axial and an equatorial site for a lone pair in a trigonal-bipyramidal array, the lone pair occupies the equatorial site. Whereas in the equatorial site the lone pair is repelled by the two bonding pairs at 90° (Fig. 2.3), in the axial position the lone pair is repelled by three bonding pairs at 90°. In an octahedral basic shape, a single lone pair can occupy any position but a second lone pair will occupy the position directly *trans* to the first, which results in a square-planar structure.

In a molecule with two adjacent bonding pairs and one or more lone pairs, the bond angle is decreased relative to that expected when all pairs are bonding Thus, the HNH angle in NH_3 is reduced from the tetrahedral angle (109.5°) of the underlying basic shape to a smaller value. This decrease is consistent with the observed HNH angle of 107°. Similarly, the HOH angle in H_2O is decreased from the tetrahedral value as the two lone pairs move apart. This decrease is in agreement with the observed HOH bond angle of 104.5°. A deficiency of the VSEPR model, however, is that it cannot be used to predict the actual bond angle adopted by the molecule.[5]

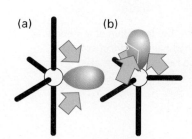

(a)   (b)

Fig. 2.3 In the VSEPR model a lone pair in (a) the equatorial position of a trigonal-bipyramidal arrangement interacts strongly with two bonding pairs, but in (b) an axial position it interacts strongly with three bonding pairs. The former arrangement is generally lower in energy.

---

[5] There are also problems with hydrides and fluorides. These are discussed by Gillespie. See *Further reading*.

19 SF_4

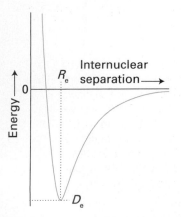

20 SF_4

Fig. 2.4 The formation of a σ bond. A σ bond has cylindrical symmetry around the internuclear axis.

Fig. 2.5 A molecular potential energy curve showing how the total energy of a molecule varies as the internuclear separation is changed.

> **Example 2.7 Accounting for the effect of lone pairs on molecular shape.**
>
> Predict the shape of an SF_4 molecule.
>
> **Answer** The Lewis structure of SF_4 is shown in (**19**). The central S atom has four F atoms attached to it and one lone pair. The basic shape adopted by these five regions is trigonal bipyramidal. The potential energy is least if the lone pair occupies an equatorial site to give a molecular shape that resembles a see-saw, with the axial bonds forming the 'plank' of the see-saw and the equatorial bonds the 'pivot'. The S—F bonds then bend away from the lone pair (**20**).
>
> **Self-test 2.7** Predict the shape of an XeF_2 molecule.

Valence-bond theory

The **valence-bond theory** (VB theory) of bonding was the first quantum mechanical theory of bonding to be developed and can be regarded as a way of expressing Lewis's concepts in terms of wavefunctions. Valence-bond theory considers the interaction of separate atoms as they are brought together to form a molecule. The computational techniques involved have been largely superseded by molecular orbital theory, but much of the language and some of the concepts of VB theory still remain.

2.4 The hydrogen molecule

Key points: In the valence-bond theory, the wavefunction of an electron pair is formed by superimposing the wavefunctions for the separated fragments of the molecule; a molecular potential energy curve shows the variation of the molecular energy with internuclear separation.

The two-electron wavefunction for two widely separated H atoms is $\psi = \phi_A(1)\phi_B(2)$, where ϕ_J is an H1s orbital on atom J. When the atoms are close, it is not possible to know whether it is electron 1 that is on A or electron 2. An equally valid description is therefore $\psi = \phi_A(2)\phi_B(1)$, in which electron 2 is on A and electron 1 is on B. When two outcomes are equally probable, quantum mechanics instructs us to describe the true state of the system as a superposition of the wavefunctions for each possibility, so a better description of the molecule than either wavefunction alone is the linear combination of the two possibilities.

$$\psi = \phi_A(1)\phi_B(2) + \phi_A(2)\phi_B(1) \tag{2.3}$$

This function is the (unnormalized) VB wavefunction for an H—H bond. The formation of the bond can be pictured as being due to the high probability that the two electrons will be found between the two nuclei and hence will bind them together (Fig. 2.4). More formally, the wave pattern represented by the term $\phi_A(1)\phi_B(2)$ interferes constructively with the wave pattern represented by the contribution $\phi_A(2)\phi_B(1)$ and there is an enhancement in the amplitude of the wavefunction in the internuclear region. For technical reasons stemming from the Pauli principle, only electrons with paired spins can be described by a wavefunction of the type written in eqn 2.3, so only paired electrons can contribute to a bond in VB theory. We say, therefore, that a VB wavefunction is formed by **spin pairing** of the electrons in the two contributing atomic orbitals. The electron distribution described by the wavefunction in eqn 2.3 is called a σ **bond**. As shown in Fig. 2.4, a σ bond has cylindrical symmetry around the internuclear axis, and the electrons in it have zero orbital angular momentum about that axis.

The **molecular potential energy curve** for H_2, a graph showing the variation of the energy of the molecule with internuclear separation, is calculated by changing the internuclear separation R and evaluating the energy at each selected separation (Fig. 2.5). The energy falls below that of two separated H atoms as the two atoms are brought within bonding distance and each electron is free to migrate to the other atom. However, the resulting lowering of energy is counteracted by an increase in energy from the Coulombic repulsion between the two positively charged nuclei. This positive contribution to the energy becomes large as R becomes small. Consequently, the total potential energy curve passes through a minimum and then climbs to a strongly positive value at small internuclear separations. The depth of the minimum of the curve is denoted D_e. The deeper the minimum, the more strongly the atoms are bonded together. The steepness of the well shows how rapidly the energy of the molecule rises as the bond is stretched or compressed. The steepness of the curve, an indication of the *stiffness* of the bond, therefore governs the vibrational frequency of the molecule (Section 6.4).

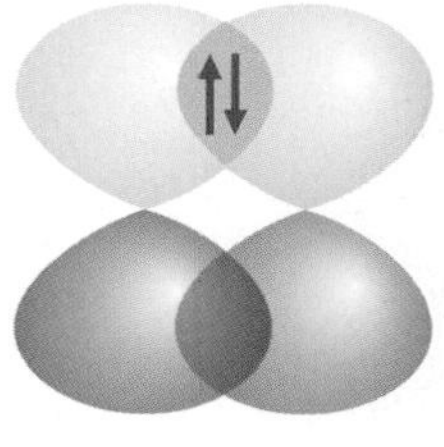

Fig. 2.6 The formation of a π bond.

2.5 Homonuclear diatomic molecules

Key point: Electrons in atomic orbitals of the same symmetry but on neighbouring atoms are paired to form σ and π bonds.

A similar description can be applied to more complex molecules, and we begin by considering **homonuclear diatomic molecules**, diatomic molecules in which both atoms belong to the same element (dinitrogen, N_2, is an example). To construct the VB description of N_2, we consider the valence-electron configuration of each atom, which from Section 1.8 we know to be $2s^2 2p_x^1 2p_y^1 2p_z^1$. It is conventional to take the z-axis to be the internuclear axis, so we can imagine each atom as having a $2p_z$ orbital pointing towards a $2p_z$ orbital on the other atom, with the $2p_x$ and $2p_y$ orbitals perpendicular to the axis. A σ bond is then formed by spin pairing between the two electrons in the opposing $2p_z$ orbitals. Its spatial wavefunction is still given by eqn 2.3, but now ϕ_A and ϕ_B stand for the two $2p_z$ orbitals. A simple way of identifying a σ bond is to envisage rotation of the bond around the internuclear axis: if the wavefunction remains unchanged, the bond is classified as σ.

The remaining $2p$ orbitals cannot merge to give σ bonds as they do not have cylindrical symmetry around the internuclear axis. Instead, the electrons in them merge to form two **π bonds**. A π bond arises from the spin pairing of electrons in two p orbitals that approach side-by-side (Fig. 2.6). The bond is so called because, viewed along the internuclear axis, it resembles a pair of electrons in a p orbital. More precisely, an electron in a π bond has one unit of orbital angular momentum about the internuclear axis. A simple way of identifying a π bond is to envisage rotation of the bond through 180° around the internuclear axis. If the signs (as indicated by the shading) of the lobes of the orbital are interchanged, then the bond is classified as π.

There are two π bonds in N_2, one formed by spin pairing in two neighbouring $2p_x$ orbitals and the other by spin pairing in two neighbouring $2p_y$ orbitals. The overall bonding pattern in N_2 is therefore a σ bond plus two π bonds (Fig. 2.7), which is consistent with the Lewis structure :N≡N:. Analysis of the total electron density in a triple bond shows that it has cylindrical symmetry around the internuclear axis, with the four electrons in the two π bonds forming a ring of electron density around the central σ bond.

Fig. 2.7 The VB description of N_2. (a) Two electrons form a σ bond and another two pairs form two π bonds. (b) In linear molecules, where the x and y axes are not specified, the electron density of π bonds is cylindrically symmetrical around the internuclear axis.

2.6 Polyatomic molecules

Key points: Each σ bond in a polyatomic molecule is formed by the spin pairing of electrons in any neighbouring atomic orbitals with cylindrical symmetry about the relevant internuclear axis; π bonds are formed by pairing electrons that occupy neighbouring atomic orbitals of the appropriate symmetry.

To introduce polyatomic molecules we consider the VB description of H_2O. The valence-electron configuration of a hydrogen atom is $1s^1$ and that of an O atom is $2s^2 2p_x^2 2p_y^1 2p_z^1$. The two unpaired electrons in the O2p orbitals can each pair with an electron in an H1s orbital, and each combination results in the formation of a σ bond (each bond has cylindrical symmetry about the respective O—H internuclear axis). Because the $2p_y$ and $2p_z$ orbitals lie at 90° to each other, the two σ bonds also lie at 90° to each other (Fig. 2.8).

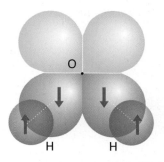

Fig. 2.8 The VB description of H_2O. There are two σ bonds formed by pairing electrons in O2p and H1s orbitals. This model predicts a bond angle of 90°.

We can predict, therefore, that H_2O should be an angular molecule, which it is. However, the theory predicts a bond angle of $90°$ whereas the actual bond angle is $104.5°$. Similarly, to predict the structure of an ammonia molecule, NH_3, we start by noting that the valence-electron configuration of an N atom given previously suggests that three H atoms can form bonds by spin pairing with the electrons in the three half-filled $2p$ orbitals. The latter are perpendicular to each other, so we predict a trigonal-pyramidal molecule with a bond angle of $90°$. An NH_3 molecule is indeed trigonal pyramidal, but the experimental bond angle is $107°$.

Another deficiency of the VB theory presented so far is its inability to account for carbon's tetravalence, its ability to form four bonds. The ground-state configuration of C is $2s^2 2p_x^1 2p_y^1$, which suggests that a C atom should be capable of forming only two bonds, not four. Clearly, something is missing from the VB approach.

These two deficiencies—the failure to account for bond angles and the valence of carbon—are overcome by introducing two new features, *promotion* and *hybridization*.

(a) Promotion

Key point: Promotion of electrons may occur if the outcome is to achieve more or stronger bonds and a lower overall energy.

Promotion is the excitation of an electron to an orbital of higher energy in the course of bond formation. Although electron promotion requires an investment of energy, that investment is worthwhile if the energy can be more than recovered from the greater strength or number of bonds that it allows to be formed. Promotion is not a 'real' process in which an atom somehow becomes excited and then forms bonds: it is a contribution to the overall energy change that occurs when bonds form.

In carbon, for example, the promotion of a $2s$ electron to a $2p$ orbital can be thought of as leading to the configuration $2s^1 2p_x^1 2p_y^1 2p_z^1$, with four unpaired electrons in separate orbitals. These electrons may pair with four electrons in orbitals provided by four other atoms, such as four $H1s$ orbitals if the molecule is CH_4, and hence form four σ bonds. Although energy was required to promote the electron, it is more than recovered by the atom's ability to form four bonds in place of the two bonds of the unpromoted atom. Promotion, and the formation of four bonds, is a characteristic feature of carbon and of its congeners in Group 14 (Chapter 13) because the promotion energy is quite small: the promoted electron leaves a doubly occupied ns orbital and enters a vacant np orbital, hence significantly relieving the electron–electron repulsion it experiences in the ground state.

(b) Hybridization

Key points: Hybrid orbitals are formed when atomic orbitals on the same atom interfere; specific hybridization schemes correspond to each local molecular geometry.

The description of the bonding in AB_4 molecules of Group 14 is still incomplete because it appears to imply the presence of three σ bonds of one type (formed from ϕ_B and ϕ_{A2p} orbitals) and a fourth σ bond of a distinctly different character (formed from ϕ_B and ϕ_{A2s}), whereas all the evidence (bond lengths, strengths, shape) point to the equivalence of all four A—B bonds, as in CH_4, for example.

This problem is overcome by realizing that the electron density distribution in the promoted atom is equivalent to the electron density in which each electron occupies a **hybrid orbital** formed by interference, or 'mixing', between the $A2s$ and the $A2p$ orbitals. The origin of the hybridization can be appreciated by thinking of the four atomic orbitals, which are waves centred on a nucleus, as being like ripples spreading from a single point on the surface of a lake: the waves interfere destructively and constructively in different regions, and give rise to four new shapes.

The specific linear combinations that give rise to four equivalent hybrid orbitals are

$$h_1 = s + p_x + p_y + p_z \quad h_2 = s - p_x - p_y + p_z$$
$$h_3 = s - p_x + p_y - p_z \quad h_4 = s + p_x - p_y - p_z \tag{2.4}$$

As a result of the interference between the component orbitals, each hybrid orbital consists of a large lobe pointing in the direction of one corner of a regular tetrahedron and a smaller lobe pointing in the opposite direction (Fig. 2.9). The angle between the axes of the hybrid orbitals is the tetrahedral angle, 109.47°. Because each hybrid is built from one s orbital and three p orbitals, it is called an sp^3 **hybrid orbital**.

It is now easy to see how the VB description of a CH_4 molecule is consistent with a tetrahedral shape with four equivalent C—H bonds. Each hybrid orbital of the promoted carbon atom contains a single unpaired electron; an electron in ϕ_{H1s} can pair with each one, giving rise to a σ bond pointing in a tetrahedral direction. Because each sp^3 hybrid orbital has the same composition, all four σ bonds are identical apart from their orientation in space.

A further feature of hybridization is that a hybrid orbital has pronounced directional character, in the sense that it has enhanced amplitude in the internuclear region. This directional character arises from the constructive interference between the s orbital and the positive lobes of the p orbitals. As a result of the enhanced amplitude in the internuclear region, the bond strength is greater than for an s or p orbital alone. This increased bond strength is another factor that helps to repay the promotion energy.

Hybrid orbitals of different compositions are used to match different molecular geometries and to provide a basis for their VB description. For example, sp^2 hybridization is used to reproduce the electron distribution needed for trigonal-planar species, such as on B in BF_3 and N in NO_3^-, and sp hybridization reproduces a linear distribution. Table 2.8 gives the hybrids needed to match the geometries of a variety of electron distributions.

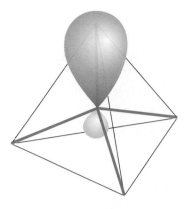

Fig. 2.9 One of the four equivalent sp^3 hybrid orbitals. Each one points towards the vertex of a regular tetrahedron.

(c) Isolobality

Key points: Structurally analogous fragments of molecules are described as isolobal; groups of isolobal molecular fragments are used to suggest patterns of bonding.

The concept of hybridization can be used to identify analogies in the structures of apparently unrelated molecules. Thus, we may view $N(CH_3)_3$ as derived from NH_3 by substitution of a CH_3 fragment for each H atom. In current terminology, the structurally analogous fragments are said to be **isolobal**, and the relationship is expressed by the symbol ⁻◊⁻. The origin of the name is the lobe-like shape of a hybrid orbital in a molecular fragment. Two fragments are isolobal if their highest-energy orbitals have the same symmetry (such as the σ symmetry of the H1s and a Csp^3 hybrid orbital), similar energies, and the same electron occupation (one in each case in H1s and Csp^3). As will become clear, the concept of isolobality is a helpful device for rationalizing observations.

Table 2.8 Some hybridization schemes		
Coordination number	Arrangement	Composition
2	Linear	sp, pd, sd
	Angular	sd
3	Trigonal planar	sp^2, p^2d
	Unsymmetrical planar	spd
	Trigonal pyramidal	pd^2
4	Tetrahedral	sp^3, sd^3
	Irregular tetrahedral	spd^2, p^3d, pd^3
	Square planar	p^2d^2, sp^2d
5	Trigonal bipyramidal	sp^3d, spd^3
	Tetragonal pyramidal	sp^2d^2, sd^4, pd^4, p^3d^2
	Pentagonal planar	p^2d^3
6	Octahedral	sp^3d^2
	Trigonal prismatic	spd^4, pd^5
	Trigonal antiprismatic	p^3d^3

A simple VB viewpoint can be adopted for many of the applications of isolobality. This viewpoint allows us to identify families of isolobal fragments, such as

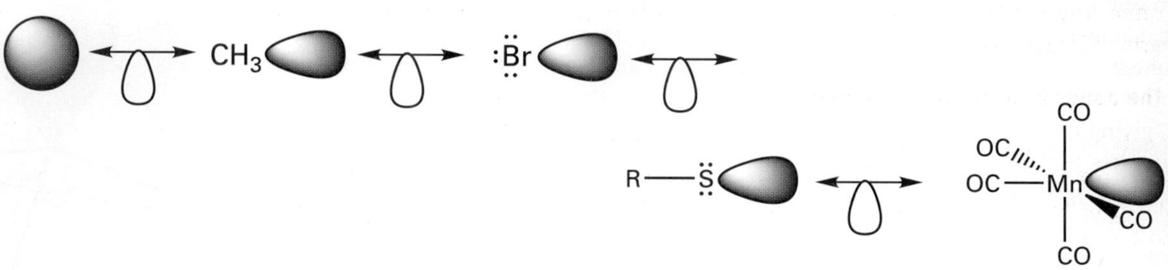

where, in this case, there is a single electron in each hybrid lobe. The recognition of this family permits us to anticipate by analogy with H—H that molecules such as H_3C—Br and $(OC)_5Mn$—CH_3 can be formed. It is also possible to identify isolobal fragments with two available singly occupied orbitals:

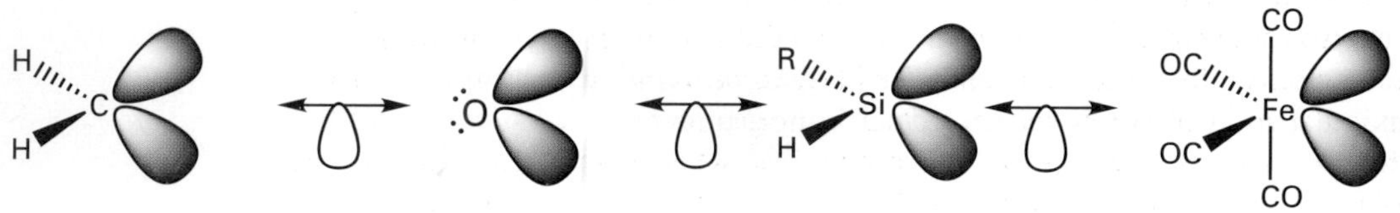

Some three-orbital isolobal fragments can also be identified:

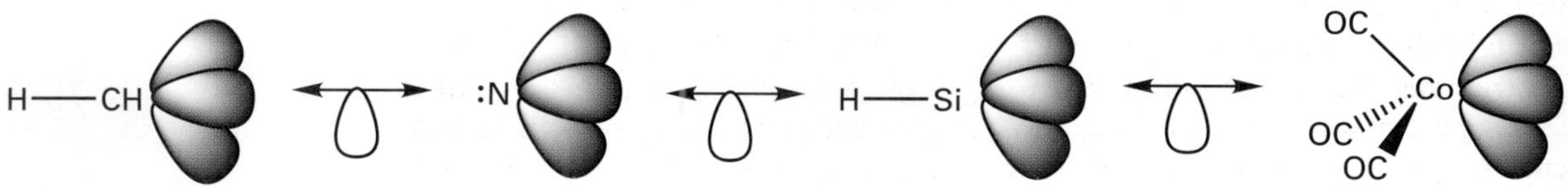

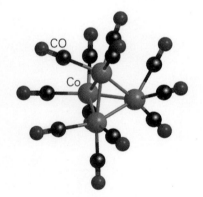

21 $Co_4(CO)_{12}$

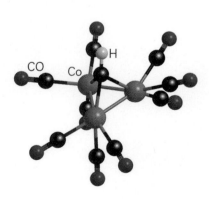

22 $Co_3(CO)_9CH$

The existence of these families suggests that we should expect to encounter molecules such as *cyclo*-C_4H_8, $O(CH_3)_2$, and $N(CH_3)_3$, which may all be built from these and similar fragments: these species are all known. The complexes $Co_4(CO)_{12}$ (**21**) and $Co_3(CO)_9CH$ (**22**) are two more examples. However, isolobal analogies must be used with care, as they may also tempt us to postulate the existence of $(OC)_5Mn$—O—$Mn(CO)_5$ and $(OC)_4Fe$=$Fe(CO)_4$, but both are currently unknown. Isolobal analogies provide useful correlations and hints for the synthesis of new molecules, but they are no substitute for experimental facts. The application of isolobal analogies to metal clusters is discussed in Section 21.3c.

Molecular orbital theory

We now generalize the *atomic* orbital description of atoms in a very natural way to a **molecular orbital** (MO) description of molecules in which electrons spread over *all* the atoms in a molecule and bind them all together. In the spirit of this chapter, we continue to treat the concepts qualitatively and to give a sense of how inorganic chemists discuss the electronic structures of molecules by using MO theory. Almost all qualitative discussions and calculations on inorganic molecules and ions are now carried out within the framework of MO theory.

2.7 An introduction to the theory

We begin by considering homonuclear diatomic molecules and diatomic ions formed by two atoms of the same element. The concepts these species introduce are readily extended to heteronuclear diatomic molecules formed between two atoms or ions of different elements. They are also easily extended to polyatomic molecules and solids composed of huge numbers of atoms and ions. In parts of this section we shall include molecular fragments in the discussion, such as the SF diatomic group in the SF_6 molecule or the OO diatomic group in H_2O_2, as similar concepts also apply to pairs of atoms bound together as parts of larger molecules.

(a) The approximations of the theory

Key points: Molecular orbitals are constructed as linear combinations of atomic orbitals; there is a high probability of finding electrons in atomic orbitals that have large coefficients in the linear combination; each molecular orbital can be occupied by up to two electrons.

As in the description of the electronic structures of atoms, we set out by making the **orbital approximation**, in which we assume that the wavefunction, Ψ, of the N electrons in the molecule can be written as a product of one-electron wavefunctions: $\Psi = \psi(1)\psi(2) \cdots \psi(N)$. The interpretation of this expression is that electron 1 is described by the wavefunction $\psi(1)$, electron 2 by the wavefunction $\psi(2)$, and so on. These one-electron wavefunctions are the **molecular orbitals** of the theory. As for atoms, the square of a one-electron wavefunction gives the probability distribution for that electron in the molecule: an electron in a molecular orbital is likely to be found where the orbital has a large amplitude, and will not be found at all at any of its nodes.

The next approximation is motivated by noting that, when an electron is close to the nucleus of one atom, its wavefunction closely resembles an atomic orbital of that atom. For instance, when an electron is close to the nucleus of an H atom in a molecule, its wavefunction is like a $1s$ orbital of that atom. Therefore, we may suspect that we can construct a reasonable first approximation to the molecular orbital by superimposing atomic orbitals contributed by each atom. This modelling of a molecular orbital in terms of contributing atomic orbitals is called the **linear combination of atomic orbitals** (LCAO) approximation. A 'linear combination' is a sum with various weighting coefficients. In simple terms, we combine the atomic orbitals of contributing atoms to give molecular orbitals that extend over the entire molecule.

In the most elementary form of MO theory, only the valence-shell atomic orbitals are used to form molecular orbitals. Thus, the molecular orbitals of H_2 are approximated by using two hydrogen $1s$ orbitals, one from each atom:

$$\psi = c_A\phi_A + c_B\phi_B \tag{2.5}$$

In this case the **basis set**, the atomic orbitals ϕ from which the molecular orbital is built, consists of two H$1s$ orbitals, one on atom A and the other on atom B. The coefficients c in the linear combination show the extent to which each atomic orbital contributes to the molecular orbital: the greater the value of c^2, the greater the contribution of that orbital to the molecular orbital.

Because H_2 is a homonuclear diatomic molecule, its electrons are equally likely to be found near each nucleus, so the linear combination that gives the lowest energy will have equal contributions from each $1s$ orbital ($c_A^2 = c_B^2$). Specifically, the coefficients in the unnormalized molecular orbital are $c_A = c_B = 1$, and

$$\psi_+ = \phi_A + \phi_B \tag{2.6}$$

The combination that corresponds to the next higher energy orbital also has equal contributions from each $1s$ orbital ($c_A^2 = c_B^2$), but the coefficients have opposite signs ($c_A = +1, c_B = -1$)

$$\psi_- = \phi_A - \phi_B \tag{2.7}$$

The relative signs of coefficients in LCAOs play a very important role in determining the energies of the orbitals. As we shall see, the relative signs determine whether atomic orbitals interfere constructively or destructively in different regions of the molecule and hence lead to an accumulation or a reduction of electron density in those regions.

Two more preliminary points should be noted. We see from this discussion that *two* molecular orbitals may be constructed from *two* atomic orbitals. In due course, we shall see the importance of the general point that N molecular orbitals can be constructed

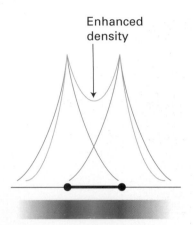

Enhanced
density

Fig. 2.10 The enhancement of electron density in the internuclear region arising from the constructive interference between the atomic orbitals on neighbouring atoms.

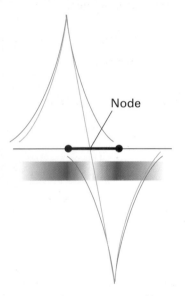

Node

Fig. 2.11 The destructive interference that arises if the overlapping orbitals have opposite signs. This interference leads to a nodal surface in an antibonding molecular orbital.

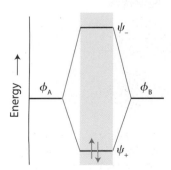

Fig. 2.12 The molecular orbital energy level diagram for H_2 and analogous molecules.

from a basis set of N atomic orbitals. For example, if we use all four valence orbitals on each O atom in O_2, then from the total of eight atomic orbitals we can construct eight molecular orbitals. In addition, as in atoms, the Pauli exclusion principle implies that each molecular orbital may be occupied by up to two electrons; if two electrons are present, then their spins must be paired. Thus, in a diatomic molecule constructed from two Period 2 atoms and in which there are eight molecular orbitals available for occupation, up to 16 electrons may be accommodated before all the molecular orbitals are full. The same rules that are used for filling atomic orbitals with electrons (the building-up principle, Section 1.8) apply to filling molecular orbitals with electrons.

The general pattern of the energies of molecular orbitals formed from N atomic orbitals is that one molecular orbital lies below that of the parent atomic energy levels, one lies higher in energy than they do, and the remainder are distributed between these two extremes.

(b) Bonding and antibonding orbitals

Key points: A bonding orbital arises from the constructive interference of neighbouring atomic orbitals; an antibonding orbital arises from their destructive interference, as indicated by a node between the atoms.

The orbital ψ_+ is an example of a **bonding orbital**. It is so called because the energy of the molecule is lowered relative to that of the separated atoms if this orbital is occupied by electrons. In elementary discussions of the chemical bond, the bonding character of ψ_+ is ascribed to the constructive interference between the two atomic orbitals and the enhanced amplitude this causes between the two nuclei (Fig. 2.10). An electron that occupies ψ_+ has an enhanced probability of being found in the internuclear region, and can interact strongly with both nuclei. Hence orbital overlap, the spreading of one orbital into the region occupied by another, leading to enhanced probability of electrons being found in the internuclear region, is taken to be the origin of the strength of bonds.

The orbital ψ_- is an example of an **antibonding orbital**. It is so called because, if it is occupied, the energy of the molecule is higher than for the two separated atoms. The greater energy of an electron in this orbital arises from the destructive interference between the two atomic orbitals, which cancels their amplitudes and gives rise to a nodal plane between the two nuclei (Fig 2.11). Electrons that occupy ψ_- are largely excluded from the internuclear region and are forced to occupy energetically less favourable locations. It is generally true that the energy of a molecular orbital in a polyatomic molecule is higher the more internuclear nodes it has. The increase in energy reflects an increasingly complete exclusion of electrons from the regions between nuclei. Note that an antibonding orbital is slightly more antibonding than its partner bonding orbital is bonding: the asymmetry arises partly from the details of the electron distribution and partly from the fact that internuclear repulsion pushes the entire diagram upwards.

The energies of the two molecular orbitals in H_2 are depicted in Fig. 2.12, which is an example of a **molecular orbital energy level diagram**, a diagram depicting the relative energies of molecular orbitals. The two electrons occupy the lower energy molecular orbital. An indication of the size of the energy gap between the two molecular orbitals is the observation of a spectroscopic absorption in H_2 at 11.4 eV (in the ultraviolet at 109 nm), which can be ascribed to the transition of an electron from the bonding orbital to the antibonding orbital. The dissociation energy of H_2 is 4.5 eV (434 kJ mol^{-1}), which gives an indication of the location of the bonding orbital relative to the separated atoms.

The Pauli exclusion principle limits to two the number of electrons that can occupy any molecular orbital and requires that those two electrons be paired ($\uparrow\downarrow$). The exclusion principle is the origin of the importance of the electron pair in bond formation in MO theory just as it is in VB theory: in the context of MO theory, two is the maximum number of electrons that can occupy an orbital that contributes to the stability of the molecule. The H_2 molecule, for example, has a lower energy than that of the separated atoms because two electrons can occupy the orbital ψ_+ and both can contribute to the lowering of its energy (as shown in Fig. 2.12). A weaker bond can be expected if only one electron is present in a bonding orbital, but, nevertheless, H_2^+ is known as a transient gas-phase ion; its dissociation energy is 2.6 eV (250.8 kJ mol^{-1}). Three electrons (as in H_2^-) are less effective than two electrons because the third electron must occupy the antibonding orbital ψ_- and hence destabilize the molecule. With four electrons, the antibonding effect of two electrons in ψ_- overcomes the bonding effect of two electrons in ψ_+. There is then no net bonding. It follows that a four-electron molecule with only 1s orbitals available for bond formation, such as He_2, is not expected to be stable relative to dissociation into its atoms.

So far, we have discussed interactions of atomic orbitals that give rise to molecular orbitals that are lower in energy (bonding) and higher in energy (antibonding) than the separated atoms. In addition, it is possible to generate a molecular orbital that has the same energy as the initial atomic orbitals. In this case, occupation of this orbital neither stabilizes nor destabilizes the molecule and so it is described as a **nonbonding orbital.** Typically, a nonbonding orbital is a molecular orbital that consists of a single orbital on one atom, perhaps because there is no atomic orbital of the correct symmetry for it to overlap on a neighbouring atom.

2.8 Homonuclear diatomic molecules

Although the structures of diatomic molecules can be calculated effortlessly by using commercial software packages, the validity of any such calculations must, at some point, be confirmed by experimental data. Moreover, elucidation of molecular structure can often be achieved by drawing on experimental information. One of the most direct portrayals of electronic structure is obtained from ultraviolet photoelectron spectroscopy (Section 6.8) in which electrons are ejected from the orbitals they occupy in molecules and their energies determined. Because the peaks in a photoelectron spectrum correspond to the various kinetic energies of photoelectrons ejected from different orbitals of the molecule, the spectrum gives a vivid portrayal of the molecular orbital energy levels of a molecule (Fig. 2.13).

(a) The orbitals

Key points: Molecular orbitals are classified as σ, π, or δ according to their rotational symmetry about the internuclear axis, and (in centrosymmetric species) as g or u according to their symmetry with respect to inversion.

Our task is to see how molecular orbital theory can account for the features revealed by photoelectron spectroscopy and the other techniques, principally absorption spectroscopy, that are used to study diatomic molecules. We are concerned predominantly with valence orbitals, rather than core orbitals. As with H_2, the starting point in the theoretical discussion is the **minimal basis set**, the smallest set of atomic orbitals from which useful molecular orbitals can be built. In Period 2 diatomic molecules, the minimal basis set consists of the one valence s orbital and three valence p orbitals on each atom, giving eight atomic orbitals in all. We shall now see how the minimal basis set of eight valence-shell atomic orbitals (four from each atom, one s and three p) is used to construct eight molecular orbitals. Then we shall use the Pauli principle to predict the ground-state electron configurations of the molecules.

The energies of the atomic orbitals that form the basis set are shown on either side of the molecular orbital diagram in Fig. 2.14. We form **σ orbitals** by allowing overlap between atomic orbitals that have cylindrical symmetry around the internuclear axis, which (as remarked earlier) is conventionally labelled z. The notation σ signifies that the orbital has cylindrical symmetry; atomic orbitals that can form σ orbitals include the 2s and $2p_z$ orbitals on the two atoms (Fig. 2.15). From these four orbitals (the 2s and the $2p_z$ orbitals on atom A and the corresponding orbitals on atom B) with cylindrical symmetry we can construct four σ molecular orbitals, two of which arise predominantly from interaction of the 2s orbitals, and two from interaction of the $2p_z$ orbitals. These molecular orbitals are labelled $1\sigma_g$, $1\sigma_u$, $2\sigma_g$, and $2\sigma_u$, respectively. Their energies

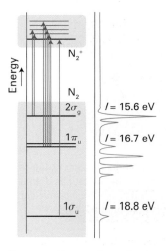

Fig. 2.13 The UV photoelectron spectrum of N_2. The fine structure in the spectrum arises from excitation of vibrations in the cation formed by photoejection of an electron.

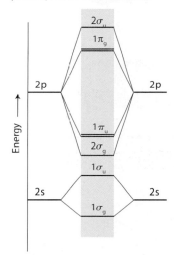

Fig. 2.14 The molecular orbital energy level diagram for the later Period 2 homonuclear diatomic molecules. This diagram should be used for O_2 and F_2.

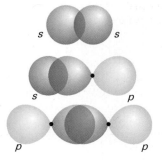

Fig. 2.15 A σ orbital can be formed in several ways, including s,s overlap, s,p overlap, and p,p overlap, with the p orbitals directed along the internuclear axis.

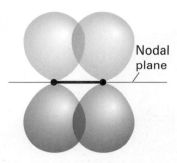

Fig. 2.16 Two *p* orbitals can overlap to form a π orbital. The orbital has a nodal plane passing through the internuclear axis, shown here from the side.

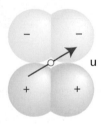

23 σ_g

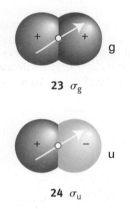

24 σ_u

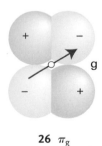

25 π_u

26 π_g

resemble those shown in Fig. 2.14 but it is difficult to predict the precise locations of the central two orbitals.

The remaining two $2p$ orbitals on each atom, which have a nodal plane containing the *z*-axis, overlap to give **π orbitals** (Fig. 2.16). Bonding and antibonding π orbitals can be formed from the mutual overlap of the two $2p_x$ orbitals, and also from the mutual overlap of the two $2p_y$ orbitals. This pattern of overlap gives rise to the two pairs of doubly degenerate energy levels (two energy levels of the same energy) shown in Fig. 2.14 and labelled $1\pi_u$ and $1\pi_g$.

For homonuclear diatomics, it is sometimes convenient (particularly for spectroscopic discussions) to signify the symmetry of the molecular orbitals with respect to their behaviour under inversion through the centre of the molecule. The operation of **inversion** consists of starting at an arbitrary point in the molecule, travelling in a straight line to the centre of the molecule, and then continuing an equal distance out on the other side of the centre. The orbital is designated g (for *gerade*, even) if it is identical under inversion, and u (for *ungerade*, odd) if it changes sign. Thus, a bonding σ orbital is g (**23**) and an antibonding σ orbital is u (**24**). On the other hand, a bonding π orbital is u (**25**) and an antibonding π orbital is g (**26**). Note that the σ_u orbitals are numbered separately from the σ_u orbitals, and similarly for the π orbitals.

The procedure can be summarized as follows:

1 From a basis set of *N* atomic orbitals, *N* molecular orbitals are constructed. In Period 2, $N = 8$.

2 The eight orbitals can be classified by symmetry into two sets: there are four σ orbitals and four π orbitals.

3 The four π orbitals form one doubly degenerate pair of bonding orbitals and one doubly degenerate pair of antibonding orbitals.

4 The four σ orbitals span a range of energies, one being strongly bonding and another strongly antibonding, with the remaining two σ orbitals lying between these extremes.

5 To establish the actual location of the energy levels, it is necessary to use electronic absorption spectroscopy, photoelectron spectroscopy, or detailed computation.

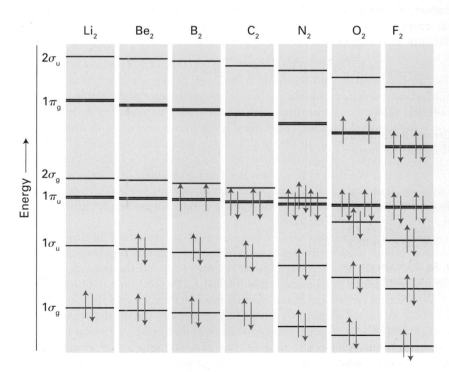

Fig. 2.17 The variation of orbital energies for Period 2 homonuclear diatomic molecules from Li_2 to F_2.

Photoelectron spectroscopy and detailed computation (the numerical solution of the Schrödinger equation for the molecules) enable us to build the orbital energy schemes shown in Fig. 2.17. As we see there, from Li_2 to N_2 the arrangement of orbitals is that shown in Fig. 2.18, whereas for O_2 and F_2 the order of the σ and π orbitals is reversed and the array is that shown in Fig. 2.14. The reversal of order can be traced to the increasing separation of the $2s$ and $2p$ orbitals that occurs on going to the right across Period 2. A general principle of quantum mechanics is that the mixing of wavefunctions is strongest if their energies are similar; therefore, as the s and p energy separation increases, the molecular orbitals become more purely s-like and p-like. When the s,p energy separation is small, each molecular orbital is a more extensive mixture of s and p character on each atom.

When considering species containing two neighbouring d-block atoms, as in Hg_2^{2+} and $[Cl_4ReReCl_4]^{2-}$, we should also allow for the possibility of forming bonds from d orbitals. A d_{z^2} orbital has cylindrical symmetry with respect to the internuclear (z) axis, and hence can contribute to the σ orbitals that are formed from s and p_z orbitals. The d_{yz} and d_{zx} orbitals both look like p orbitals when viewed along the internuclear axis, and hence can contribute to the π orbitals formed from p_x and p_y. The new feature is the role of $d_{x^2-y^2}$ and d_{xy}, which have no counterpart in the orbitals discussed up to now. These two orbitals can overlap with matching orbitals on the other atom to give rise to doubly degenerate pairs of bonding and antibonding δ orbitals (Fig. 2.19). As we shall see in Chapter 19, δ orbitals are important for the discussion of bonds between d-metal atoms, in d-metal complexes, and in organometallic compounds.

(b) The building-up principle for molecules

Key points: The building-up principle is used to predict the ground-state electron configurations by accommodating electrons in the array of molecular orbitals summarized in Fig. 2.14 or 2.18 and recognizing the constraints of the Pauli principle.

We use the building-up principle in conjunction with the molecular orbital energy level diagram in the same way as for atoms. The order of occupation of the orbitals is the order of increasing energy as depicted in Fig. 2.14 or 2.18. Each orbital can accommodate up to two electrons. If more than one orbital is available for occupation (because they happen to have identical energies, as in the case of pairs of π orbitals), then the orbitals are occupied separately. In that case, the electrons in the half-filled orbitals adopt parallel spins ($\uparrow\uparrow$), just as is required by Hund's rule for atoms (Section 1.8a). With very few exceptions, these rules lead to the actual ground-state configuration of the Period 2 diatomic molecules. For example, the electron configuration of N_2, with 10 valence electrons, is

$$N_2: \quad 1\sigma_g^2 1\sigma_u^2 1\pi_u^4 2\sigma_g^2$$

Molecular orbital configurations are written like those for atoms: the orbitals are listed in order of increasing energy, and the number of electrons in each one is indicated by a superscript. Note that π^4 is shorthand for the occupation of two different π orbitals.

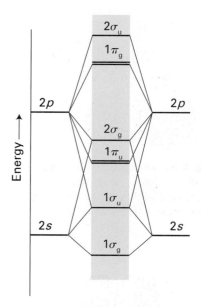

Fig. 2.18 The molecular orbital energy level diagram for Period 2 homonuclear diatomic molecules from Li_2 to N_2.

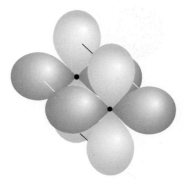

Fig. 2.19 The formation of δ orbitals by d-orbital overlap. The orbital has two mutually perpendicular nodal planes that intersect along the internuclear axis.

Example 2.8 Predicting the electron configurations of diatomic molecules

Predict the ground-state electron configurations of the oxygen molecule, O_2, the superoxide ion, O_2^-, and the peroxide ion, O_2^{2-}.

Answer An O_2 molecule has 12 valence electrons. The first ten electrons recreate the N_2 configuration except for the reversal of the order of the $1\pi_u$ and $2\sigma_g$ orbitals (see Fig. 2.17). Next in line for occupation are the doubly degenerate $1\pi_g$ orbitals. The last two electrons enter these orbitals separately, and have parallel spins. The configuration is therefore

$$O_2: \quad 1\sigma_g^2 1\sigma_u^2 2\sigma_g^2 1\pi_u^4 1\pi_g^2$$

The O_2 molecule is interesting because the lowest energy configuration has two unpaired electrons in different π orbitals. Hence, O_2 is paramagnetic (tends to be attracted into a magnetic field). The next two electrons can be accommodated in the $1\pi_g$ orbitals, giving

$$O_2^-: \quad 1\sigma_g^2 1\sigma_u^2 2\sigma_g^2 \pi_u^4 1\pi_g^3$$
$$O_2^{2-}: \quad 1\sigma_g^2 1\sigma_u^2 2\sigma_g^2 1\pi_u^4 1\pi_g^4$$

We are assuming that the orbital order does not change; this might not be the case.

Self-test 2.8: Write the valence-electron configuration for S_2^{2-} and Cl_2^-.

The **highest occupied molecular orbital** (HOMO) is the molecular orbital that, according to the building-up principle, is occupied last. The **lowest unoccupied molecular orbital** (LUMO) is the next higher molecular orbital. In Fig. 2.17, the HOMO of F_2 is $1\pi_g$ and its LUMO is $2\sigma_u$; for N_2 the HOMO is $2\sigma_g$ and the LUMO is $1\pi_g$. We shall increasingly see that these **frontier orbitals**, the LUMO and the HOMO, play special roles in the interpretation of structural and kinetic studies. The term SOMO, denoting a **singly occupied molecular orbital**, is sometimes encountered, and is of crucial importance for the properties of radical species.

2.9 Heteronuclear diatomic molecules

The molecular orbitals of heteronuclear diatomic molecules differ from those of homonuclear diatomic molecules in having unequal contributions from each atomic orbital. Each molecular orbital has the form

$$\psi = c_A\phi_A + c_B\phi_B + \cdots \tag{2.8}$$

as in homonuclear molecules. The unwritten orbitals include all the other orbitals of the correct symmetry for forming σ or π bonds but which typically make a smaller contribution than the two valence-shell orbitals we are considering. In contrast to orbitals for homonuclear species, the coefficients c_A and c_B are not necessarily equal in magnitude. If $c_A^2 > c_B^2$, the orbital is composed principally of ϕ_A and an electron that occupies the molecular orbital is more likely to be found near atom A than atom B. The opposite is true for a molecular orbital in which $c_B^2 > c_A^2$. In heteronuclear diatomic molecules, the more electronegative element makes the larger contribution to bonding orbitals, and the less electronegative element makes the greater contribution to the antibonding orbitals.

(a) Heteronuclear molecular orbitals

Key points: Heteronuclear diatomic molecules are polar; bonding electrons tend to be found on the more electronegative atom and antibonding electrons on the less electronegative atom.

The greater contribution to a bonding molecular orbital normally comes from the more electronegative atom: the bonding electrons are then likely to be found close to that atom and hence be in an energetically favourable location. The extreme case of a polar covalent bond, a covalent bond formed by an electron pair that is unequally shared by the two atoms, is an ionic bond. In an ionic bond, one atom gains complete control over the electron pair. The less electronegative atom normally contributes more to an antibonding orbital (Fig. 2.20). That is, antibonding electrons are more likely to be found in an energetically unfavourable location, close to the less electronegative atom.

A second difference between homonuclear and heteronuclear diatomic molecules stems from the energy mismatch in the latter between the two sets of atomic orbitals. We have already remarked that two wavefunctions interact less strongly as their energies diverge. This dependence on energy separation implies that the lowering of energy as a result of the overlap of atomic orbitals on different atoms in a heteronuclear molecule is less pronounced than in a homonuclear molecule, in which the orbitals have the same energies. However, we cannot necessarily conclude that A—B bonds are weaker than A—A bonds, because other factors (including orbital size and closeness of approach) are also important. The heteronuclear CO molecule, for example, which is isoelectronic with its homonuclear counterpart N_2, has an even higher bond enthalpy (1070 kJ mol^{-1}) than N_2 (946 kJ mol^{-1}).

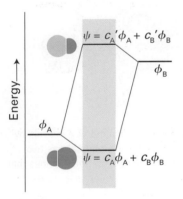

Fig. 2.20 A molecular orbital energy level diagram arising from interaction of two atomic orbitals with different energies. The lower molecular orbital is primarily composed of the lower energy atomic orbital, and vice versa. The shift in energies of the two levels is less than if the atomic orbitals had the same energy.

(b) Hydrogen fluoride

Key points: In hydrogen fluoride the bonding orbital is more concentrated on the F atom and the antibonding orbital is more concentrated on the H atom.

As an illustration of these general points, consider a simple heteronuclear diatomic molecule, HF. The five valence orbitals available for molecular orbital formation are the $1s$ orbital of H and the $2s$ and $2p$ orbitals of F; there are $1 + 7 = 8$ valence electrons to accommodate in the five molecular orbitals that can be constructed from the five basis orbitals.

The σ orbitals for HF can be constructed by allowing an H$1s$ orbital to overlap the F$2s$ and F$2p_z$ orbitals (z being the internuclear axis). These three atomic orbitals combine to give three σ molecular orbitals of the form $\psi = c_1\phi_{H1s} + c_2\phi_{F2s} + c_3\phi_{F2p_z}$. This procedure leaves the F$2p_x$ and F$2p_y$ orbitals unaffected as they have π symmetry and there is no valence H orbital of that symmetry. These π orbitals are therefore examples of the nonbonding orbitals mentioned earlier, and are molecular orbitals confined to a single atom. Note that, because there is no centre of inversion in a heteronuclear diatomic molecule, we do not use the g,u classification for its molecular orbitals.

Figure 2.21 shows the resulting energy level diagram. The 1σ bonding orbital is predominantly F$2s$ in character (in accord with the high electronegativity of fluorine). It is confined mainly to the F atom and is nonbonding. The 2σ orbital is more bonding than the 1σ orbital and has both H$1s$ and F$2s$ character. The 3σ orbital is antibonding, and principally H$1s$ in character: the $1s$ orbital has a relatively high energy (compared with the fluorine orbitals) and hence contributes predominantly to the high-energy antibonding molecular orbital.

Two of the eight valence electrons enter the 2σ orbital, forming a bond between the two atoms. Six more enter the 1σ and 1π orbitals; these two orbitals are largely nonbonding and confined mainly to the F atom. All the electrons are now accommodated, so the configuration of the molecule is $1\sigma^2 2\sigma^2 1\pi^4$. One important feature to note is that all the electrons occupy orbitals that are predominantly on the F atom. It follows that we can expect the HF molecule to be polar, with a partial negative charge on the F atom, which is found experimentally.

(c) Carbon monoxide

Key points: The HOMO of a carbon monoxide molecule is an almost nonbonding σ orbital largely localized on C; the LUMO is an antibonding π orbital.

The molecular orbital energy level diagram for carbon monoxide is a somewhat more complicated example than HF because both atoms have $2s$ and $2p$ orbitals that can participate in the formation of σ and π orbitals. The energy level diagram is shown in Fig. 2.22. The ground-state configuration is

CO: $1\sigma^2 2\sigma^2 1\pi^4 3\sigma^2$

The HOMO in CO is 3σ, which is a largely nonbonding lone pair on the C atom. The LUMO is the doubly degenerate pair of antibonding π orbitals, with mainly C$2p$ orbital character (Fig. 2.23). This combination of frontier orbitals—a full σ orbital largely localized on C and a pair of empty π orbitals—is one reason why metal carbonyls are such a characteristic feature of the d metals: in d-metal carbonyls, the HOMO lone pair orbital of CO participates in the formation of a σ bond, and the LUMO antibonding π orbital participates in the formation of π bonds to the metal atom (Chapter 19).

Although the difference in electronegativity between C and O is large, the experimental value of the electric dipole moment of the CO molecule (0.1 D) is small. Moreover, the negative end of the dipole is on the C atom despite that being the less electronegative atom. This odd situation stems from the fact that the lone pairs and bonding pairs have a complex distribution. It is wrong to conclude that, because the bonding electrons are mainly on the O atom, O is the negative end of the dipole, as this ignores the balancing effect of the lone pair on the C atom. The inference of polarity from electronegativity is particularly unreliable when antibonding orbitals are occupied.

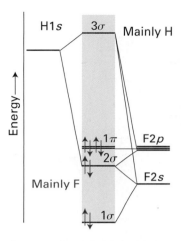

Fig. 2.21 The molecular orbital energy level diagram for HF. The relative positions of the atomic orbitals reflect the ionization energies of the atoms.

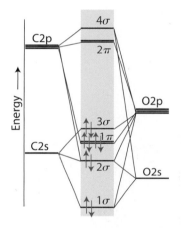

Fig. 2.22 The molecular orbital energy level diagram for CO.

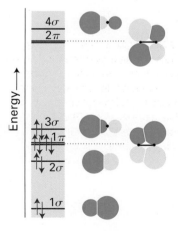

Fig. 2.23 A schematic illustration of the molecular orbitals of CO, with the size of the atomic orbital indicating the magnitude of its contribution to the molecular orbital.

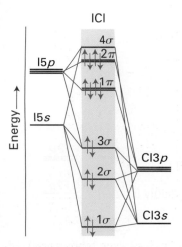

ICl

4σ

I5p

2π

1π

I5s

3σ

Cl3p

2σ

1σ

Cl3s

Energy →

Fig. 2.24 A schematic illustration of the energies of the molecular orbitals of ICl.

Example 2.9 Accounting for the structure of a heteronuclear diatomic molecule

The halogens form compounds among themselves. One of these 'interhalogen' compounds is iodine monochloride, ICl, in which the order of orbitals is 1σ, 2σ, 3σ, 1π, 2π, 4σ (from calculation). What is the ground-state electron configuration of ICl?

Answer First, we identify the atomic orbitals that are to be used to construct molecular orbitals: these are the $3s$ and $3p$ valence-shell orbitals of Cl and the $5s$ and $5p$ valence-shell orbitals of I. As for Period 2 elements, an array of σ and π orbitals can be constructed, and is shown in Fig. 2.24. The bonding orbitals are predominantly Cl in character (because that is the more electronegative element) and the antibonding orbitals are predominantly I in character. There are $7 + 7 = 14$ valence electrons to accommodate, which results in the ground-state electron configuration $1\sigma^2 2\sigma^2 3\sigma^2 1\pi^4 2\pi^4$

Self-test 2.9 Predict the ground-state electron configuration of the hypochlorite ion, ClO$^-$.

2.10 Bond properties

We now bring the discussion full circle, and show that, although seemingly very different, MO theory does elucidate many features of the Lewis description of molecules. We have already seen the origin of the importance of the electron pair: two electrons is the maximum number that can occupy a bonding orbital and hence contribute to a chemical bond. We now extend this concept by introducing the concept of 'bond order'.

(a) Bond order

Key points: The bond order assesses the net number of bonds between two atoms in the molecular orbital formalism; the greater the bond order between a given pair of atoms, the greater is the bond strength.

The **bond order**, b, identifies a shared electron pair as counting as a 'bond' and an electron pair in an antibonding orbital as an 'antibond' between two atoms. More precisely, the bond order is defined as

$$b = \tfrac{1}{2}(n - n^*) \tag{2.9}$$

where n is the number of electrons in bonding orbitals and n^* is the number in antibonding orbitals.

For example, F_2 has the configuration $1\sigma_g^2 1\sigma_u^2 2\sigma_g^2 1\pi_u^4 1\pi_g^4$ and, because $1\sigma_g$, $1\pi_u$, and $2\sigma_g$ orbitals are bonding but $1\sigma_u$ and $1\pi_g$ are antibonding (on the basis of whether they have an internuclear node), $b = \tfrac{1}{2}(2 + 2 + 4 - 2 - 4) = 1$. The bond order of F_2 is 1, which is consistent with the Lewis structure F—F and the conventional description of the molecule as having a single bond. N_2 has the configuration $1\sigma_g^2 1\sigma_u^2 1\pi_u^4 2\sigma_g^2$ and $b = \tfrac{1}{2}(2 + 4 + 2 - 2) = 3$. A bond order of 3 corresponds to a triply bonded molecule, which is in line with the Lewis structure :N≡N:. The high bond order is reflected in the high bond enthalpy of the molecule (946 kJ mol^{-1}), one of the highest for any molecule.

The bond order of the isoelectronic CO molecule is also 3, in accord with the analogous Lewis structure :C≡O:. However, this method of assessing bonding is primitive, especially for heteronuclear species. For instance, inspection of the computed molecular orbitals in Fig. 2.23 suggests that 2σ and 3σ are best regarded as nonbonding orbitals largely localized on O and C, and hence should really be disregarded in the calculation of b. The resulting bond order is unchanged by this modification. The lesson is that the definition of bond order provides a useful indication of the multiplicity of the bond, but any interpretation of contributions to b needs to be done in the light of guidance from the properties of computed orbitals.

Electron loss from N_2 leads to the formation of the transient species N_2^+ in which the bond order is reduced from 3 to 2.5. This reduction in bond order is accompanied by a corresponding decrease in bond strength (from 946 to 855 kJ mol^{-1}) and increase in the bond length from 109 pm for N_2 to 112 pm for N_2^+.

The definition of bond order in eqn 2.9 allows for the possibility that an orbital is only singly occupied. The bond order in O_2^-, for example, is 1.5, because three electrons occupy the $1\pi_g$ antibonding orbitals. Isoelectronic molecules and ions have the same bond order, so F_2 and O_2^{2-} both have bond order 1, and CO, N_2, and NO^+ all have bond order 3.

(b) Bond correlations

Key point: For a given pair of elements, bond strength increases and bond length decreases as bond order increases.

The strengths and lengths of bonds correlate quite well with each other and with the bond order. For a given pair of atoms:

Bond enthalpy increases as bond order increases.
Bond length decreases as bond order increases.

These trends are illustrated in Figs 2.25 and 2.26. The strength of the dependence varies with the elements. In Period 2 the correlation is relatively weak for CC bonds, with the result that a C=C double bond is less than twice as strong as a C—C single bond. This difference has profound consequences in organic chemistry, particularly for the reactions of unsaturated compounds. It implies, for example, that it is energetically favourable (but slow in the absence of a catalyst) for ethene and ethyne to polymerize: in this process, C—C single bonds form at the expense of the appropriate numbers of multiple bonds.

Familiarity with carbon's properties, however, must not be extrapolated without caution to the bonds between other elements. An N=N double bond (409 kJ mol^{-1}) is more than twice as strong as an N—N single bond (163 kJ mol^{-1}), and an N≡N triple bond (946 kJ mol^{-1}) is more than five times as strong. It is on account of this trend that NN multiply bonded compounds are stable relative to polymers or three-dimensional compounds having only single bonds. The same is not true of phosphorus, where the P—P, P=P, and P≡P bond enthalpies are 200, 310, and 490 kJ mol^{-1}, respectively. For phosphorus, single bonds are stable relative to the matching number of multiple bonds. Thus, phosphorus exists in a variety of solid forms in which P—P single bonds are present, including the tetrahedral P_4 molecules of white phosphorus. Diphosphorus molecules, P_2, are found only at high temperatures and low pressures.

The two correlations with bond order taken together imply that, for a given pair of elements,

Bond enthalpy increases as bond length decreases.

This correlation is illustrated in Fig. 2.27: it is a useful feature to bear in mind when considering the stabilities of molecules because bond lengths may be readily available from independent sources.

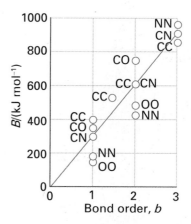

Fig. 2.25 The correlation between bond strength and bond order.

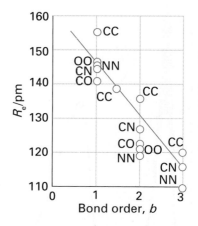

Fig. 2.26 The correlation between bond length and bond order.

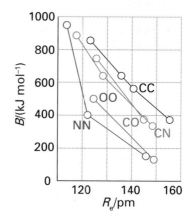

Fig. 2.27 The correlation between bond length and bond strength.

Example 2.10 Predicting correlations between bond order, bond length, and bond strength

Determine the bond order of the oxygen molecule, O_2, the superoxide ion, O_2^-, and the peroxide ion, O_2^{2-}, and predict the relative bond lengths and strengths of the species.

Answer The configurations of the three species are:

O_2: $1\sigma_g^2 1\sigma_u^2 2\sigma_g^2 1\pi_u^4 1\pi_g^2$

O_2^-: $1\sigma_g^2 1\sigma_u^2 2\sigma_g^2 1\pi_u^4 1\pi_g^3$

O_2^{2-}: $1\sigma_g^2 1\sigma_u^2 2\sigma_g^2 1\pi_u^4 1\pi_g^4$

These configurations correspond to bond orders of 2 for O_2, 1.5 for O_2^-, and 1 for O_2^{2-}. As bond enthalpy increases as bond order increases, we expect the bond enthalpies to increase in the order $O_2^{2-} < O_2^- < O_2$. Bond length decreases as the bond enthalpy increases, so bond length should follow the opposite trend; $O_2^{2-} > O_2^- > O_2$. These predictions are supported by the gas phase bond enthalpies of O—O bonds (146 kJ mol^{-1}) and O=O bonds (496 kJ mol^{-1}) and the associated bond lengths of 132 and 121 pm, respectively.

Self-test 2.10 Predict the order of bond enthalpies and bond lengths for C—N, C=N, and C≡N bonds.

Molecular orbitals of polyatomic molecules

Molecular orbital theory can be used to discuss in a uniform manner the electronic structures of triatomic molecules, finite groups of atoms, and the almost infinite arrays of atoms in solids. In each case the molecular orbitals resemble those of diatomic molecules, the only important difference being that the orbitals are built from a more extensive basis set of atomic orbitals. As remarked earlier, a key point to bear in mind is that from N atomic orbitals it is possible to construct N molecular orbitals.

We saw in Section 2.8 that the general structure of molecular orbital energy level diagrams can be derived by grouping the orbitals into different sets, the σ and π orbitals, according to their shapes. The same procedure is used in the discussion of the molecular orbitals of polyatomic molecules. However, because their shapes are more complex than diatomic molecules, we need a more powerful approach. The discussion of polyatomic molecules will therefore be carried out in two stages. In this chapter we use intuitive ideas about molecular shape to construct molecular orbitals. In Chapter 7 we discuss the shapes of molecules and the use of their symmetry characteristics to construct molecular orbitals and account for other properties. That chapter rationalizes the procedures presented here.

2.11 The construction of molecular orbitals

The photoelectron spectrum of NH_3 (Fig. 2.28) indicates some of the features that a theory of the structure of polyatomic molecules must elucidate. The spectrum shows two bands. The one with the lower ionization energy (in the region of 11 eV) has considerable vibrational structure. This structure indicates that the orbital from which the electron is ejected plays a considerable role in the determination of the molecule's shape. The broad band in the region of 16 eV arises from electrons that are bound more tightly.

(a) Polyatomic molecular orbitals

Key points: Molecular orbitals are formed from linear combinations of atomic orbitals of the same symmetry; their energies can be determined experimentally from gas-phase photoelectron spectra and interpreted in terms of the pattern of orbital overlap.

The features that have been introduced in connection with diatomic molecules are present in all polyatomic molecules. In each case, we write the molecular orbital of a given symmetry (such as the σ orbitals of a linear molecule) as a sum of *all* the atomic orbitals that can overlap to form orbitals of that symmetry:

$$\psi = \sum_i c_i \phi_i \tag{2.10}$$

In this linear combination, the ϕ_i are atomic orbitals (usually, the valence orbitals of each atom in the molecule) and the index i runs over all the atomic orbitals that have the appropriate symmetry. From N atomic orbitals we can construct N molecular orbitals. Then:

1 The greater the number of nodes in a molecular orbital, the greater the antibonding character and the higher the orbital energy.

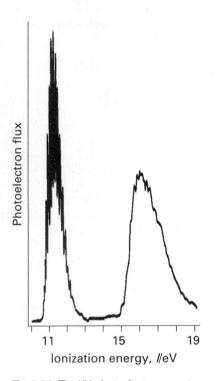

Photoelectron flux

Ionization energy, I/eV
11   15   19

Fig. 2.28 The UV photoelectron spectrum of NH_3, using He 21 eV radiation.

2 Interactions between non-nearest neighbour atoms are weakly bonding (lower the energy slightly) if the orbital lobes on these atoms have the same sign (and interfere constructively). They are weakly antibonding if the signs are opposite (and interfere destructively).

3 Orbitals constructed from lower energy atomic orbitals lie lower in energy. (So atomic *s* orbitals typically produce lower energy molecular orbitals than atomic *p* orbitals of the same shell.)

For instance, to account for the features in the photoelectron spectrum of NH_3, we need to build molecular orbitals that will accommodate the eight valence electrons in the molecule. Each molecular orbital is the combination of seven atomic orbitals: the three H1*s* orbitals, the N2*s* orbital, and the three N2*p* orbitals. It is possible to construct seven molecular orbitals from these seven atomic orbitals (Fig. 2.29).

It is not always strictly appropriate to use σ and π in polyatomic molecules because the labels apply to a linear molecule. However, it is often convenient to continue to use the notation when concentrating on the *local* form of an orbital, its shape relative to the internuclear axis between two neighbouring atoms. The correct procedure for labelling orbitals in polyatomic molecules according to their symmetry is described in Chapter 7. For our present purposes all we need know of this more appropriate procedure is the following:

> *a*, *b* denote a nondegenerate orbital.
> *e* denotes a doubly degenerate orbital.
> *t* denotes a triply degenerate orbital.

Subscripts and superscripts are sometimes added to these letters, as in a_1, b'', e_g, and t_2 because it is sometimes necessary to distinguish different *a*, *b*, *e*, and *t* orbitals according to a more detailed analysis of their symmetries.

The formal rules for the construction of the orbitals are described in Chapter 7, but it is possible to obtain a sense of their origin by imagining viewing the NH_3 molecule along its threefold axis (designated *z*). The N2p_z and N2*s* orbitals both have cylindrical symmetry about that axis. If the three H1*s* orbitals are superimposed with the same sign relative to each other (that is, so that all have the same size and tint in the diagram, Fig. 2.29), then they match this cylindrical symmetry. It follows that we can form molecular orbitals of the form

$$\psi = c_1\phi_{N2s} + c_2\phi_{Npz} + c_3\{\phi_{H1sA} + \phi_{H1sB} + \phi_{H1sC}\} \tag{2.11}$$

From these *three* basis orbitals (each specific combination of H1*s* orbitals counts as a single 'symmetry adapted' basis orbital), it is possible to construct three molecular orbitals (with different values of the coefficients *c*). The orbital with no nodes between the N and H atoms is the lowest in energy, that with a node between all the NH neighbours is the highest in energy, and the third orbital lies between the two. The three orbitals are nondegenerate and are labelled $1a^1$, $2a^1$, and $3a^1$ in order of increasing energy.

The N2p_x and N2p_y orbitals have π symmetry with respect to the *z*-axis, and can be used to form orbitals with combinations of the H1*s* orbitals that have a matching symmetry. For example, one such superposition will have the form

$$\psi = c_1\phi_{N2px} + c_2\{\phi_{H1sA} - \phi_{H1sB}\} \tag{2.12}$$

As can be seen from Fig. 2.29, the signs of the H1*s* orbital combination match those of the N2p_x orbital. The N2*s* orbital cannot contribute to this superposition, so only *two* combinations can be formed, one without a node between the N and H orbitals and the other with a node. The two orbitals differ in energy, the former being lower. A similar combination of orbitals can be formed with the N2p_y orbital, and it turns out (by the symmetry arguments that we use in Chapter 7) that the two orbitals are degenerate with the two we have just described. The combinations are examples of *e* orbitals (because they form doubly degenerate pairs), and are labelled 1*e* and 2*e* in order of increasing energy.

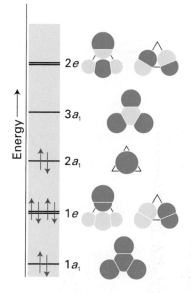

Fig. 2.29 A schematic illustration of the molecular orbitals of NH_3 with the size of the atomic orbital indicating the magnitude of its contribution to the molecular orbital. The view is along the *z*-axis.

Energy →

2*e*

3a_1

2a_1

1*e*

1a_1

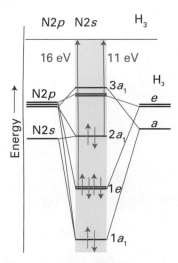

Fig. 2.30 The molecular orbital energy level diagram for NH_3 when the molecule has the observed bond angle (107°) and bond length.

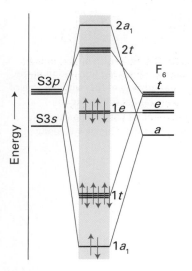

Fig. 2.31 A schematic molecular orbital energy level diagram for SF_6.

The general form of the molecular orbital energy level diagram is shown in Fig. 2.30. The actual location of the orbitals (particularly the relative positions of the a set and the e set), can be found only by detailed computation or by identifying the orbitals responsible for the photoelectron spectrum. We have indicated the probable assignment of the 11 eV and 16 eV peaks, which fixes the locations of two of the occupied orbitals. The third occupied orbital is out of range of the 21 eV radiation used to obtain the spectrum.

The photoelectron spectrum is consistent with the need to accommodate eight electrons in the orbitals. The electrons enter the molecular orbitals in increasing order of energy, starting with the orbital of lowest energy, and taking note of the requirement of the exclusion principle that no more than two electrons can occupy any one orbital. The first two electrons enter $1a$ and fill it. The next four enter the doubly degenerate $1e$ orbitals and fill them. The last two enter the $2a$ orbital, which calculations show is almost nonbonding and localized on the N atom. The resulting overall ground-state electron configuration is therefore $1a^2 1e^4 2a^2$. No antibonding orbitals are occupied, so the molecule has a lower energy than the separated atoms. The conventional description of NH_3 as a molecule with a lone pair is also mirrored in the configuration: the HOMO is $2a$, which is largely confined to the N atom and makes only a small contribution to the bonding. We have already seen that lone pair electrons play a considerable role in determining the shapes of molecules. The extensive vibrational structure in the 11 eV band of the photoelectron spectrum is consistent with this observation, as photoejection of a $2a$ electron removes the effectiveness of the lone pair and the shape of the ionized molecule is considerably different from that of NH_3 itself. Photoionization therefore results in extensive vibrational structure in the spectrum.

(b) Hypervalence in the context of molecular orbitals

Key point: The delocalization of molecular orbitals means that an electron pair can contribute to the bonding of more than two atoms.

A slightly more complicated example, but one that makes an important point, is the octahedral molecule SF_6. A simple basis set that is adequate to illustrate the point we want to make consists of the valence-shell s and p orbitals of the S atom and one p orbital of each of the six F atoms and pointing towards the S atom. We use the F2p orbitals rather than the F2s orbitals because they match the S orbitals more closely in energy. From these ten atomic orbitals it is possible to construct ten molecular orbitals. Calculations indicate that four of the orbitals are bonding and four are antibonding; the two remaining orbitals are nonbonding (Fig. 2.31).

There are 12 electrons to accommodate. The first two can enter $1a$ and the next six can enter $1t$. The remaining four fill the nonbonding pair of orbitals, resulting in $1a^2 1t^6 e^4$. As we see, none of the antibonding orbitals ($2a$ and $2t$) is occupied. The theory, therefore, accounts for the formation of SF_6, with four bonding orbitals and two nonbonding orbitals occupied.

The important point is that the bonding in SF_6 can be explained without using S3d orbitals to expand the octet. This does not mean that d orbitals cannot participate in the bonding, but it does show that they are not *necessary* for bonding six F atoms to the central S atom. The limitation of valence-bond theory is the assumption that each atomic orbital on the central atom can participate in the formation of only one bond. Molecular orbital theory takes hypervalence into its stride by having available plenty of orbitals, not all of which are antibonding. Therefore, the question of when hypervalence can occur appears to depend on factors other than d-orbital availability.

(c) Electron deficiency

Key point: The existence of electron-deficient species is explained by the delocalization of the bonding influence of electrons over several atoms.

The formation of molecular orbitals by combining several atomic orbitals accounts effortlessly for the existence of **electron-deficient compounds**, which are compounds for which there are not enough electrons for a Lewis structure to be written. This point can be illustrated most easily with diborane, B_2H_6 (**27**). There are only 12 valence electrons, but at least seven electron pairs are needed to bind eight atoms together if Lewis's model is correct. The terminal B—H bonds are $2c,2e$ bonds whereas the bridging B—H—B bonds are $3c,2e$ bonds in which three electrons are 'shared' between the three atoms. The bridging bonds in B_2H_6 are longer and weaker than the terminal ones.

The problem is solved by molecular orbital theory. The eight atoms of this molecule contribute a total of 14 valence orbitals (four orbitals from each B atom, making eight, and one orbital each from the six H atoms). These 14 atomic orbitals can be used to construct 14 molecular orbitals. About seven of these molecular orbitals will be bonding or nonbonding, which is more than enough to accommodate the 12 valence electrons provided by the atoms. In MO terms, there is nothing mysterious about the existence of this molecule.

A slightly different viewpoint is achieved by concentrating on the two BHB fragments and the molecular orbitals that can be formed by combining the atomic orbitals contributed by the three atoms (Fig. 2.32). This example is the first of several in which we construct molecular orbitals from the orbitals of molecular fragments: the approach is a half-way stage between the atomic orbitals and the final molecular orbitals, and is very useful for drawing analogies between molecules. The bonding orbital spanning these three atoms can accommodate two electrons and pull the molecule together. That two electrons can bind two pairs of atoms (BH and HB) in this way is not particularly remarkable, but it is an insurmountable difficulty for the Lewis approach. Electron deficiency is in fact well developed not only in boron (where it was first clearly recognized) but also in carbocations and a variety of other classes of compounds that we encounter later in the text.

(d) Localization

Key points: Localized and delocalized descriptions of bonds are mathematically equivalent, but one description may be more suitable for a particular property, as summarized in Table 2.9.

A striking feature of the Lewis approach to chemical bonding is its accord with chemical instinct, as it identifies something that can be called 'an A—B bond'. Both OH bonds in H_2O, for instance, are treated as localized, equivalent, structures, because each one consists of an electron pair shared between O and H. This feature appears to be absent from molecular orbital theory because molecular orbitals are delocalized, and the electrons that occupy them bind all the atoms together, not just a specific pair of neighbouring atoms. The concept of an A—B bond as existing independently of other bonds in the molecule, and of being transferable from one molecule to another, seems to have been lost. However, we shall now show that the molecular orbital description is mathematically almost equivalent to a localized description of the overall electron distribution. The demonstration hinges on the fact that linear combinations of molecular orbitals can be formed that result in the same overall electron distribution, but the individual orbitals are distinctly different.

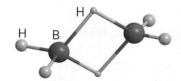

27 Diborane, B_2H_6

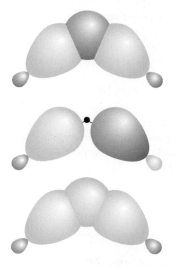

Fig. 2.32 The molecular orbital formed between two B atoms and one H atom lying between them, as in B_2H_6. Two electrons occupy the bonding combination and hold all three atoms together.

Table 2.9 A general indication of the properties for which localized and delocalized descriptions are appropriate

Localized appropriate	Delocalized appropriate
Bond strengths	Electronic spectra
Force constants	Photoionization
Bond lengths	Electron attachment
Brønsted acidity*	Magnetism
VSEPR description of molecular geometry	Walsh description of molecular geometry
	Standard potentials[†]

*Chapter 4. [†]Chapter 5.

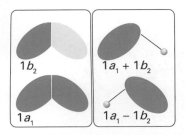

Fig. 2.33 The two occupied $1a_1$ and $1b_2$ orbitals of the H_2O molecule and their sum $1a_1 + 1b_2$ and difference $1a_1 - 1b_2$. In each case we form an almost fully delocalized orbital between a pair of atoms.

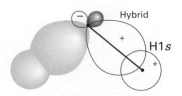

Fig. 2.34 The formation of localized O—H orbitals in H_2O by the overlap of hybrid orbitals on the O atom and H1s orbitals. The hybrid orbitals are a close approximation to the sp^3 hybrids shown in Fig. 2.9.

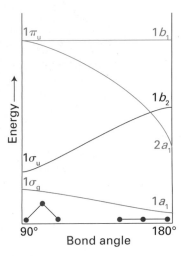

Fig. 2.35 The Walsh diagram for XH_2 molecules. Only the bonding and nonbonding orbitals are shown.

Consider the H_2O molecule. The two occupied bonding orbitals of the delocalized description, $1a_1$ and $1b_2$, are shown in Fig. 2.33. If we form the sum $1a_1 + 1b_2$, the negative half of $1b_2$ cancels half the $1a_1$ orbital almost completely, leaving a localized orbital between O and the other H. Likewise, when we form the difference $1a_1 - 1b_2$, the other half of the $1a_1$ orbital is cancelled almost completely, so leaving a localized orbital between the other pair of atoms. That is, by taking sums and differences of delocalized orbitals, localized orbitals are created (and vice versa). Because these are two equivalent ways of describing the same overall electron population, one description cannot be said to be better than the other.

Table 2.9 suggests when it is appropriate to select a delocalized description or a localized description. In general, a delocalized description is needed for dealing with global properties of the entire molecule. Such properties include electronic spectra (UV and visible transitions) (Section 6.3), photoionization spectra, ionization and electron attachment energies (Section 1.9), and reduction potentials (Section 5.1). By contrast, a localized description is most appropriate for dealing with properties of a fragment of a total molecule. Such properties include bond strength, bond length, bond force constant, and some aspects of reactions (such as acid–base character): in these aspects the localized description is more appropriate because it focuses attention on the distribution of electrons in and around a particular bond.

(e) Localized bonds and hybridization

Key point: Hybrid atomic orbitals are sometimes used in the discussion of localized molecular orbitals.

The localized molecular orbital description of bonding can be taken a stage further by invoking the concept of hybridization. Strictly speaking, hybridization belongs to VB theory, but it is commonly invoked in simple qualitative descriptions of molecular orbitals.

We have seen that in general a molecular orbital is constructed from all atomic orbitals of the appropriate symmetry. However, it is sometimes convenient to form a mixture of orbitals on one atom (the O atom in H_2O, for instance), and then to use these hybrid orbitals to construct localized molecular orbitals. In H_2O, for instance, each OH bond can be regarded as formed by the overlap of an H1s orbital and a hybrid orbital composed of O2s and O2p orbitals (Fig. 2.34).

We have already seen that the mixing of s and p orbitals on a given atom results in hybrid orbitals that have a definite direction in space, as in the formation of tetrahedral hybrids. Once the hybrid orbitals have been selected, a localized molecular orbital description can be constructed. For example, four bonds in CF_4 can be formed by building bonding and antibonding localized orbitals by overlap of each hybrid and one F2p orbital directed towards it. Similarly, to describe the electron distribution of BF_3, we could consider each localized BF σ orbital as formed by the overlap of an sp^2 hybrid with an F2p orbital. A localized orbital description of a PCl_5 molecule would be in terms of five PCl σ bonds formed by overlap of each of the five trigonal-bipyramidal sp^3d hybrid orbitals with a 2p orbital of a Cl atom. Similarly, where we wanted to form six localized orbitals in a regular octahedral arrangement (for example, in SF_6), we would need two d orbitals: the resulting six sp^3d^2 hybrids point in the required directions.

2.12 Molecular shape in terms of molecular orbitals

Key point: In the Walsh model, the shape of a molecule is predicted on the basis of the occupation of molecular orbitals that, in a correlation diagram, show a strong dependence on bond angle.

The VSEPR model is based on a localized picture of electron distributions. However, in MO theory, the electrons responsible for bonding are delocalized over the entire molecule. Current *ab initio* and semiempirical molecular orbital calculations, which are easily carried out with commercial software, are able to predict the shapes of even quite complicated molecules with high reliability. Nevertheless, there is still a need to understand the qualitative factors that contribute to the shape of a molecule within the framework of MO theory.

Figure 2.35 shows the **Walsh diagram** or **correlation diagram** for an H_2X molecule. A Walsh diagram adopts a simple pictorial approach to the task of analysing molecular shape in terms of delocalized molecular orbitals and was devised by A.D. Walsh in a classic series of papers published in 1953. Such diagrams play an important role in understanding the shapes, spectra, and reactions of polyatomic molecules. The diagram has been constructed by considering how the composition and energy of each molecular orbital changes as the bond angle is varied from 90° to 180°.

The molecular orbitals to consider in the angular molecule are[6]

$$\psi_{a_1} = c_1\psi_{2s} + c_2\phi_{2p_z} + c_3\phi_+$$
$$\psi_{b_1} = \phi_{2p_x} \tag{2.13}$$
$$\psi_{b_2} = c_4\phi_{2p_y} + c_5\phi_-$$

The linear combinations of $H1s$ orbitals ϕ_+ and ϕ_- are illustrated in Fig. 2.36. There are three a_1 orbitals and two b_2 orbitals; the lowest energy orbitals of each type are shown on the left of Fig. 2.37. In the linear molecule, the molecular orbitals are

$$\psi_{\sigma_g} = c_1\phi_{2s} + c_2\phi_+$$
$$\psi_{\pi_u} = \phi_{2p_y} \text{ and } \phi_{2p_z} \tag{2.14}$$
$$\psi_{\sigma_u} = c_3\phi_{2p_y} + c_4\phi_-$$

These orbitals are shown on the right in Fig. 2.37.

The lowest energy molecular orbital in 90° H_2X is the one labelled $1a_1$, which is built from the overlap of the $X2p_z$ orbital (the $2p_z$ orbital on X) with the ϕ_+ combination of $H1s$ orbitals. As the bond angle changes to 180° the energy of this orbital increases in part because the H—H overlap decreases and in part because the reduced involvement of the $X2p_z$ orbital decreases the overlap with ϕ_- (Fig 2.37). The energy of the $1b_2$ orbital is lowered because the $H1s$ orbitals move into a better position for overlap with the $X2p_y$ orbital. The weakly antibonding H—H contribution is also decreased. The biggest change occurs for the $2a_1$ orbital. It has considerable $X2s$ character in the 90° molecule, but the corresponding molecular orbital in the 180° molecule has pure $X2p_z$ orbital character. Hence, it shows a steep rise in energy as the bond angle increases. The $1b_1$ orbital is a nonbonding $X2p$ orbital perpendicular to the molecular plane in the 90° molecule and remains nonbonding in the linear molecule. Hence, its energy barely changes with angle.

The principal feature that determines whether or not the molecule is angular is whether $2a_1$ is occupied. This is the orbital that has considerable $X2s$ character in the angular molecule but not in the linear molecule. Hence, a lower energy is achieved if, when it is occupied, the molecule is angular. The shape adopted by an XH_2 molecule therefore depends on the number of electrons that occupy the orbitals.

The simplest XH_2 molecule in Period 2 is the transient gas-phase BeH_2 molecule (BeH_2 normally exists as a polymeric solid, with four-coordinate Be atoms). There are four valence electrons in a BeH_2 molecule, which occupy the lowest two molecular orbitals. If the lowest energy is achieved with the molecule angular, then that will be its shape. We can decide whether the molecule is likely to be angular by accommodating the electrons in the two lowest-energy orbitals corresponding to an arbitrary bond angle in Fig. 2.35. We then note that the HOMO decreases in energy on going to the left of the diagram (in the direction of increasing bond angle) and that the lowest total energy is obtained when the molecule is linear. Hence, BeH_2 is predicted to be linear and to have the configuration $1\sigma^2 2\sigma^2$. In CH_2, which has two more electrons than BeH_2, three of the molecular orbitals must be occupied. In this case, the lowest energy is achieved if the molecule is angular and has configuration $1a_1^2 2a_1^2 1b_2^2$.

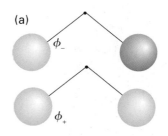

(a)

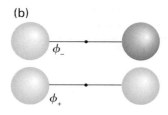

(b)

Fig. 2.36 The combination of $H1s$ orbitals that are used to construct molecular orbitals in (a) angular and (b) linear XH_2.

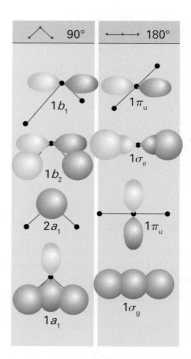

Fig. 2.37 The composition of the molecular orbitals of an XH_2 molecule at the two extremes of the correleation diagram shown in Fig. 2.35.

---

[6] We continue to use the letters a and b to label nondegenerate orbitals, and will explain their full significance in Chapter 7.

In general, any XH_2 molecule with from five to eight valence electrons is predicted to be angular. The observed bond angles are

BeH_2	BH_2	CH_2	NH_2	OH_2
$180°$	$131°$	$136°$	$103°$	$105°$

These experimental observations are qualitatively in line with Walsh's approach, but for quantitative predictions we have to turn to detailed molecular orbital calculations. However, for most qualitative considerations of molecular shape in inorganic chemistry, VSEPR serves very well.

Example 2.11 Using a Walsh diagram to predict a shape

Predict the shape of an H_2O molecule on the basis of a Walsh diagram for an XH_2 molecule.

Answer We choose an intermediate bond angle along the horizontal axis of the XH_2 diagram in Fig. 2.37 and accommodate eight electrons. The resulting configuration is $1a_1^2 2a_1^2 1b_2^2 1b_1^2$. The $2a_1$ orbital is occupied, so we expect the nonlinear molecule to have a lower energy than the linear molecule.

Self-test 2.11 Is any XH_2 molecule, in which X denotes an atom of a Period 3 element, expected to be linear? If so, which?

Walsh applied his approach to molecules other than compounds of hydrogen, but the correlation diagrams soon become very complicated. His approach represents a valuable complement to the VSEPR model because it traces the influences on molecular shapes of the occupation of orbitals spreading over the entire molecule and concentrates less on localized repulsions between pairs of electrons. Correlation diagrams like those introduced by Walsh are frequently encountered in contemporary discussions of the shapes of complex molecules, and we shall see a number of examples in later chapters. They illustrate how inorganic chemists can sometimes identify and weigh competing influences by considering two extreme cases (such as linear and 90° XH_2 molecules), and then rationalize the fact that the state of a molecule is a compromise intermediate between the two extremes. However, there is still no simple method for predicting the *numerical* value of bond angles even in simple molecules, except where the shape is governed by symmetry.

FURTHER READING

R.J. Gillespie and I. Hargittai, *The VSEPR model of molecular geometry.* Prentice Hall (1992). An excellent introduction to modern attitudes to VSEPR theory.

R.J. Gillespie and P.L.A. Popelier, *Chemical bonding and molecular geometry: from Lewis to electron densities.* Oxford University Press (2001). A comprehensive survey of modern theories of chemical bonding and geometry.

T. Albright, *Orbital interactions in chemistry.* Wiley, New York (2005). This text covers the application of molecular orbital theory to organic, organometallic, inorganic, and solid-state chemistry.

D.M.P. Mingos, *Essential trends in inorganic chemistry.* Oxford University Press (1998). An overview of inorganic chemistry from the perspective of structure and bonding.

K. Bansal, *Molecular structure and orbital theory.* Campus Books International (2000).

J.N. Murrell, S.F.A. Kettle, and J.M. Tedder, *The chemical bond.* Wiley, New York (1985).

T. Albright and J.K. Burdett, *Problems in molecular orbital theory.* Oxford University Press (1993).

J.K. Burdett, *An introduction to molecular orbitals.* Oxford University Press (1993).

EXERCISES

2.1 Write Lewis electron dot structures for (a) $GeCl_3^-$, (b) FCO_2^-, (c) CO_3^{2-}, (d) $AlCl_4^-$, (e) FNO. Where more than one resonance structure is important, give examples of all major contributors.

2.2 Construct Lewis structures of typical resonance contributions of (a) ONC^- and (b) NCO^- and assign formal charges to each atom. Which resonance structure is likely to be the dominant contribution in each case?

2.3 (a) Write Lewis structures for the major resonance forms of NO_2^-. (b) Assign formal charges. (c) Assign oxidation numbers to the atoms. (d) Describe whether oxidation numbers or formal charges are appropriate for the following applications: (i) predominant resonance Lewis dot structure among several resonance forms, (ii) determining whether there is the possibility of nitrogen being oxidized or reduced, (iii) determining the charge on the N atom.

2.4 Write Lewis structures for (a) XeF_4, (b) PF_5, (c) BrF_3, (d) $TeCl_4$, (e) ICl_2^-.

2.5 What shapes would you expect for the species (a) H_2S, (b) BF_4^-, (c) NH_4^+?

2.6 What shapes would you expect for the species (a) SO_3, (b) SO_3^{2-}, (c) IF_5?

2.7 What shapes would you expect for the species (a) ClF_3, (b) ICl_4^-, (c) I_3^-?

2.8 Solid phosphorus pentachloride is an ionic solid composed of PCl_4^+ cations and PCl_6^- anions, but the vapour is molecular. What are the shapes of the ions in the solid?

2.9 Use the covalent radii in Table 2.4 to calculate the bond lengths in (a) CCl_4 (177 pm), (b) $SiCl_4$ (201 pm), (c) $GeCl_4$ (210 pm). (The values in parentheses are experimental bond lengths and are included for comparison.)

2.10 Given that $B(Si=O) = 640$ kJ mol^{-1}, show that bond enthalpy considerations predict that silicon–oxygen compounds are likely to contain networks of tetrahedra with Si—O single bonds and not discrete molecules with Si=O double bonds.

2.11 The common forms of nitrogen and phosphorus are $N_2(g)$ and $P_4(s)$, respectively. Account for the difference in terms of the single and multiple bond enthalpies.

2.12 Use the data in Table 2.5 to calculate the standard enthalpy of the reaction $2 H_2(g) + O_2(g) \rightarrow 2 H_2O(g)$. The experimental value is -484 kJ mol^{-1}. Account for the difference between the estimated and experimental values.

2.13 Predict the standard enthalpies of the reactions

(a) $S_2^{2-}(g) + \frac{1}{4}S_8(g) \rightarrow S_4^{2-}(g)$

(b) $O_2^{2-}(g) + O_2(g) \rightarrow O_4^{2-}(g)$

by using mean bond enthalpy data. Assume that the unknown species O_4^{2-} is a singly bonded chain analogue of S_4^{2-}.

2.14 Four elements arbitrarily labelled A, B, C, and D have electronegativities 3.8, 3.3, 2.8, and 1.3, respectively. Place the compounds AB, AD, BD, and AC in order of increasing covalent character.

2.15 Use the Ketelaar triangle in Fig. 2.2 and the electronegativity values in Table 1.8 to predict what type of bonding is likely to dominate in (a) BCl_3, (b) KCl, (c) BeO.

2.16 Use molecular orbital diagrams to determine the number of unpaired electrons in (a) O_2^-, (b) O_2^+, (c) BN, and (d) NO^-.

2.17 Use Fig. 2.17 to write the electron configurations of (a) Be_2, (b) B_2, (c) C_2^-, and (d) F_2^+ and sketch the form of the HOMO in each case.

2.18 Determine the bond orders of (a) S_2, (b) Cl_2, and (c) NO^- from their molecular orbital configurations and compare the values with the bond orders determined from Lewis structures. (NO has orbitals like those of O_2.)

2.19 What are the expected changes in bond order and bond distance that accompany the following ionization processes? (a) $O_2 \rightarrow O_2^+ + e^-$, (b) $N_2 + e^- \rightarrow N_2^-$, (c) $NO \rightarrow NO^+ + e^-$.

2.20 (a) How many independent linear combinations are possible for four $1s$ orbitals? (b) Draw pictures of the linear combinations of H$1s$ orbitals for a hypothetical linear H_4 molecule. (c) From a consideration of the number of nonbonding and antibonding interactions, arrange these molecular orbitals in order of increasing energy.

2.21 (a) Construct the form of each molecular orbital in linear $[HHeH]^{2+}$ using $1s$ basis atomic orbitals on each atom and considering successive nodal surfaces. (b) Arrange the MOs in increasing energy. (c) Indicate the electron population of the MOs. (d) Should $[HHeH]^{2+}$ be stable in isolation or in solution? Explain your reasoning.

2.22 (a) Based on the MO discussion of NH_3 in the text, find the average NH bond order in NH_3 by calculating the net number of bonds and dividing by the number of NH groups.

2.23 From the relative atomic orbital and molecular orbital energies depicted in Fig. 2.31, describe the character as mainly F or mainly S for the frontier orbitals e (the HOMO) and $2t$ (the LUMO) in SF_6. Explain your reasoning.

2.24 Classify the hypothetical species (a) square H_4^{2+}, (b) angular O_3^{2-} as electron precise or electron deficient. Explain your answer and decide whether either of them is likely to exist.

2.25 Identify (a) the hydrogen–nitrogen molecule or molecular fragment that is isolobal with CH_3^-, (b) the hydrogen–boron molecule or molecular fragment that is isolobal with an O atom, (c) a nitrogen-containing species that is isolobal with $[Mn(CO)_5]^-$.

PROBLEMS

2.1 Use the concepts from Chapter 1, particularly the effects of penetration and shielding on the radial wavefunction, to account for the variation of single bond covalent radii with position in the periodic table.

2.2 Develop an argument based on bond enthalpies for the importance of SiO bonds in substances common in the Earth's crust in preference to SiSi or SiH bonds. How and why does the behaviour of silicon differ from that of carbon?

2.3 The van Arkel–Ketelaar triangle has been in use since the 1940s. A quantitative treatment of the triangle was carried out by Gordon Sproul in 1994 (*J. Phys. Chem.*, 1994, **98**, 6699). How many scales of electronegativity and how many compounds did Sproul investigate? What criteria were used to select compounds for the study? Which two electronegativity scales were found to give the best separation between areas of the triangle? What were the theoretical bases of these two scales?

2.4 When an He atom absorbs a photon to form the excited configuration $1s^1 2s^1$ (here called He*) a weak bond forms with another He atom to give the diatomic molecule HeHe*. Construct a molecular orbital description of the bonding in this species.

2.5 Construct an approximate molecular orbital energy diagram for a hypothetical planar form of NH_3. You may refer to *Resource section 4* to determine the form of the appropriate orbitals on the central N atom and on the triangle of H_3 atoms. From a consideration of the atomic energy levels, place the N and H_3 orbitals on either side of a molecular orbital energy-level diagram. Then use your judgement about the effect of bonding and antibonding interactions and energies of the parent orbitals to construct the molecular orbital energy levels in the centre of your diagram and draw lines indicating the contributions of the atomic orbitals to each molecular orbital. Atomic orbital energy level data are $H1s = -13.6$ eV, $N2s = -26.0$ eV, $N2p = -13.4$ eV.

2.6 (a) Use an extended Hückel molecular orbital program or input and output from software supplied by your instructor to construct a molecular orbital energy-level diagram to correlate the MO (from the output) and AO (from the input) energies and indicate the occupancy of the MOs (in the manner of Fig. 2.17) for one of the following molecules: HF (bond length 92 pm), HCl (127 pm), or CS (153 pm); (b) Use the output to sketch the form of the occupied orbitals, showing signs of the AO lobes by shading and their amplitudes by means of size of the orbital.

2.7 Perform an extended Hückel MO calculation on H_3 by using the H energy given in Problem 2.4 and H—H distances from NH_3 (N—H length 102 pm, HNH bond angle 107°) and then carry out the same type of calculation for NH_3. Use energy data for $N2s$ and and $N2p$ orbitals from Problem 2.4. From the output plot the molecular orbital energy levels with proper symmetry labels and correlate them with the N orbitals and H_3 orbitals of the appropriate symmetries. Compare the results of this calculation with the qualitative description in Problem 2.4.

2.8 Assign the lines in the UV photoelectron spectrum of CO shown in Fig. 2.38 and predict the appearance of the UV photoelectron spectrum of the SO molecule (see Section 6.8).

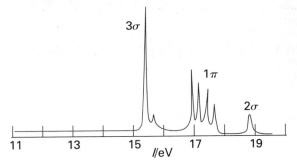

Fig. 2.38 The ultraviolet photoelectron spectrum of CO using 21 eV radiation.

The structures of simple solids

3

An understanding of the chemistry of compounds in the solid state is central to the study of many important inorganic materials, such as alloys, simple metal salts, inorganic pigments, nanostructured materials, zeolites, and high-temperature superconductors. This chapter surveys the structures adopted by atoms and ions in simple solids and explores why one arrangement may be preferred to another. We begin with the simplest model, in which atoms are represented by hard spheres and the structure of the solid is the outcome of stacking these spheres densely together. This 'close-packed' arrangement provides a good description of many metals and alloys and is a useful starting point for the discussion of numerous ionic solids. These simple solid structures can then be considered as building blocks for the construction of more complex inorganic materials. Introduction of partial covalent character into the bonding influences the choice of structure and thus trends in the adopted structural type correlate with the electronegativities of the constituent atoms. The chapter also describes some of the energy considerations that can be used to rationalize the trends in structure and reactivity. These arguments also systematize the discussion of the thermal stabilities and solubilities of ionic solids formed by the elements of Groups 1 and 2. Finally the electronic structures of materials are discussed in terms of an extension of molecular orbital theory to the almost infinite arrays of atoms found in solids. The classification of inorganic solids as conductors, semiconductors, and insulators is described in terms of this theory.

The majority of inorganic compounds exist as solids and comprise ordered arrays of atoms, ions, or molecules. Some of the simplest solids are the metals, whose structures can be described in terms of regular, space-filling arrangements of the metal atoms. These metal centres interact together through **metallic bonding** and a type of bonding that can be described in two similar ways. One view is that bonding occurs in metals when each atom loses one or more electrons to a common 'sea'. The strength of the bonding results from the combined attractions between all these freely moving electrons and the resulting cations. An alternative view is that metals are effectively enormous molecules with a multitude of atomic orbitals that interact to produce molecular orbitals extending throughout the sample. The familiar properties of metals stem from this type of bonding and the delocalization of electrons throughout the solid. Thus, metals are malleable and ductile because the electrons can adjust rapidly to relocation of the metal atom nuclei and there is no directionality in the bonding; they are lustrous because the electrons can respond almost freely to an incident wave of electromagnetic radiation and reflect it. Metallic bonding is characteristic of elements with low ionization energies, such as those on the left of the periodic table, through the d-block, and into part of the p-block close to the d-block. Most of the elements are metals, and metallic bonding also occurs in many other solid compounds, especially those of the d-metals, such as their oxides and sulfides.

In **ionic bonding** ions of different elements are held together in rigid, symmetrical arrays as a result of the attraction between their opposite charges. Ionic bonding also depends on electron loss and gain, so it is found typically in compounds of metals with electronegative elements. However, there are plenty of exceptions: not all compounds of metals are ionic and some compounds of nonmetals (such as ammonium nitrate) contain features of ionic bonding as well as covalent interactions. There are also materials that exhibit features of both ionic and metallic bonding.

Both ionic and metallic bonding are nondirectional, so structures where these types of bonding occur are most easily understood in terms of space-filling models that maximize, for example, the number and strength of the Coulombic interactions between the ions. The regular arrays of atoms, ions, or molecules in solids that produce these structures are best represented in terms of the repeating units that are produced as a result of the efficient methods of space-filling.

The description of the structures of solids

The arrangement of atoms or ions in simple solid structures can often be represented by different arrangements of hard spheres. The spheres used to describe metallic solids represent neutral atoms because each cation is still surrounded by its full complement of electrons. The spheres used to describe ionic solids represent the cations and anions because there has been a substantial transfer of electrons from one type of atom to the other.

3.1 Unit cells and the description of crystal structures

A crystal of an element or compound can be regarded as constructed from regularly repeating structural elements, which may be atoms, molecules, or ions. The 'crystal lattice' is the pattern formed by the points and used to represent the positions of these repeating structural elements.

(a) Lattice and unit cells

Key point: A unit cell is a subdivision of a crystal that, when stacked together without rotation or reflection, reproduces the crystal.

More formally, the **lattice** is a three-dimensional, infinite array of points, the **lattice points**, each of which is surrounded in an identical way by neighbouring points, and which defines the basic repeating structure of the crystal. In some cases the structural unit may be centred on the lattice point, but that is not necessary. The **crystal structure** itself is obtained by associating one or more identical structural units (such as molecules or ions) with each lattice point.

A **unit cell** of the crystal is an imaginary parallel-sided region (a 'parallelepiped') from which the entire crystal can be built up by purely translational displacements; unit cells so generated fit perfectly together with no space excluded. The angles (α, β, γ) and lengths (a, b, c) used to define the size and shape of a unit cell are the **unit cell parameters** (the 'lattice parameters'); by convention the angle between a and b is γ, that between b and c is α, and that between a and c is β. Unit cells may be chosen in a variety of ways but it is generally preferable to choose the smallest cell that exhibits the greatest symmetry. Thus, in the two-dimensional pattern in Fig. 3.1, a variety of unit cells may be chosen, each of which translates to repeat exactly the contents of the box. Two possible choices of repeating unit are shown but (b) would be preferred to (a) because it is half the area. The relationship between the lattice parameters in three dimensions as a result of the symmetry of the structure gives rise to the seven **crystal systems** (Table 3.1 and Fig. 3.2).

A **primitive** unit cell has just one lattice point in the unit cell (Fig. 3.3). More complex lattice types are **body-centred** and **face-centred** with two and four lattice points in each unit cell, respectively, and additional translational symmetry beyond that of the unit cell (Figs 3.4 and 3.5). The additional translational symmetry in the **body-centred cubic** (bcc) lattice, equivalent to the displacement ($+\frac{1}{2}$, $+\frac{1}{2}$, $+\frac{1}{2}$) from the unit cell origin at (0,0,0), produces a lattice point at the unit cell centre; note that the surroundings of each lattice point are identical, consisting of eight other lattice points at the corners of a cube. Centred lattices are sometimes preferred to primitive (though it is always possible to use

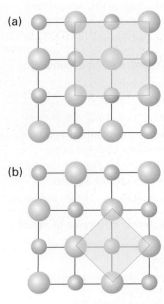

Fig. 3.1 A two-dimensional solid and two choices of a unit cell. The entire crystal is reproduced by translational displacements of either unit cell, but (b) is generally preferred to (a) because it is smaller.

(a)

(b)

Table 3.1 The seven crystal systems

System	Relations between lattice parameters		Unit cell defined by	Essential symmetries
Triclinic	$a \neq b \neq c$	$\alpha \neq \beta \neq \gamma \neq 90°$	$a\ b\ c\ \alpha\beta\gamma$	None
Monoclinic	$a \neq b \neq c$	$\alpha \neq \gamma \neq 90°\ \beta = 90°$	$a\ b\ c\ \beta$	One twofold rotation axis and/or a mirror plane
Orthorhombic	$a \neq b \neq c$	$\alpha = \beta = \gamma = 90°$	$a\ b\ c$	Three perpendicular twofold axes and/or mirror planes
Rhombohedral	$a = b = c$	$\alpha = \beta = \gamma \neq 90°$		One threefold rotation axis
Tetragonal	$a = b \neq c$	$\alpha = \beta = \gamma = 90°$	$a\ c$	One fourfold rotation axis
Hexagonal	$a = b \neq c$	$\gamma = 120°$	$a\ c$	One sixfold rotation axis
Cubic	$a = b = c$	$\alpha = \beta = \gamma = 90°$	a	Four threefold rotation axes tetrahedrally arranged

Fig. 3.3 Lattice points describing the translational symmetry of a primitive cubic unit cell.

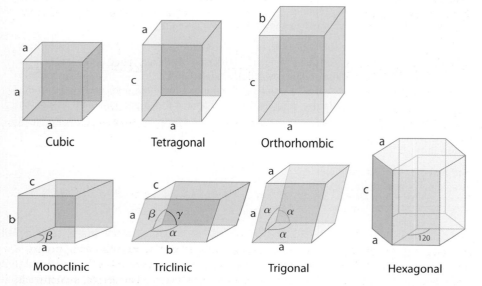

Fig. 3.2 The seven crystal systems.

Cubic   Tetragonal   Orthorhombic

Monoclinic   Triclinic   Trigonal   Hexagonal

Fig. 3.4 Lattice points describing the translational symmetry of a body-centred cubic unit cell.

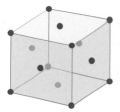

Fig. 3.5 Lattice points describing the translational symmetry of a face-centred cubic unit cell.

a primitive lattice for any structure) for with them all the essential structural symmetry of the cell is more apparent.

(b) Fractional atomic coordinates and projections

Key point: Structures may be drawn in projection denoting atom positions using fractional coordinates.

The position of an atom in a unit cell is normally described in terms of **fractional coordinates**, coordinates expressed as a fraction of the length of a side of the unit cell. Thus, the position of an atom located at xa parallel to a, yb parallel to b, and zc parallel to c is denoted (x,y,z), with $0 \leq x, y, z \leq 1$.

Three-dimensional representations of complex structures are often difficult to draw and to interpret. A clearer and faster method of representing three-dimensional structures is to draw the structure in projection and to view the unit cell down one direction, typically one of the axes of the unit cell. The positions of the atoms relative to the projection plane are denoted by the fractional coordinate above the base plane and written next to the symbol defining the atom in the projection. If two atoms lie above each other, then both fractional coordinates are noted in parentheses. For example, the structure of body-centred tungsten metal, shown in three dimensions in Fig. 3.6a, is represented in projection in Fig. 3.6b.

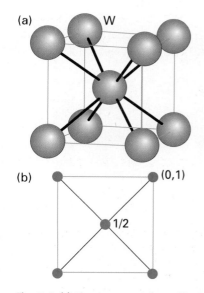

Fig. 3.6 (a) The structure of metallic tungsten and (b) its projection representation.

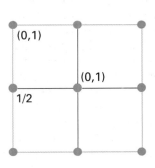

Fig. 3.7 The projection representation of an fcc unit cell.

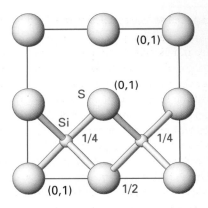

Fig. 3.8 The structure of silicon sulfide (SiS_2).

Example 3.1 Drawing a structure in projection

Convert the face-centred cubic lattice shown in Fig. 3.5 into a projection diagram.

Answer The faces of the cubic unit cell are square, so the projection diagram viewed from directly above the unit cell is a square. There is a lattice point at each corner of the unit cell, so the points at the corners of the square projection are labelled (0,1). There is a lattice point on each vertical face, which projects to points at fractional coordinate $\frac{1}{2}$ on each edge of the projection square. There is a lattice point on the lower and on the upper horizontal face of the unit cell, which projects to two points at the centre of the square at 0 and 1, respectively, so we place a final point in the centre of a square and label it (0,1). The resulting projection is shown in Fig. 3.7.

Self-test 3.1 Convert the projection diagram of the unit cell of SiS_2, shown in Fig. 3.8, into a three-dimensional representation.

3.2 The close packing of spheres

Key points: The close packing of identical spheres can result in a variety of polytypes, of which hexagonal and cubic close-packed structures are the most common.

Many metallic and ionic solids can be regarded as constructed from entities, such as atoms and ions, represented as hard spheres. If there is no directional covalent bonding, these spheres are free to pack together as closely as geometry allows and hence adopt a **close-packed structure**, a structure in which there is least unfilled space. The **coordination number** (C.N.) of a sphere in a close-packed arrangement (the 'number of nearest neighbours') is 12, the greatest number that geometry allows. When directional bonding is important, the resulting structures are no longer close-packed and the coordination number is less than 12.

Consider first a single layer of identical spheres (Fig. 3.9). The greatest number of immediate neighbours is 6 and there is only one way of constructing this close-packed layer.[1] A second close-packed layer of spheres is formed by placing spheres in the dips between the spheres of the first layer. (Note that only half the dips in the original layer are occupied, as there is insufficient space to place spheres into all the dips.) The third close-packed layer can be laid in either of two ways and hence can give rise to either of two **polytypes**, or structures that are the same in two dimensions (in this case, in the planes) but different in the third. Later we shall see that many different polytypes can be formed but those described here are two very important special cases.

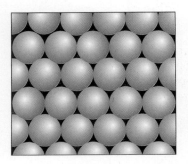

Fig. 3.9 A close-packed layer of hard spheres.

---

[1] A good way of showing this yourself is to get a number of identical coins and push them together on a flat surface; the most efficient arrangement for covering the area is with six coins around each coin. This simple modelling approach can be extended to three dimensions by using any collection of identical spherical objects such as balls or oranges.

In one polytype, the spheres of the third layer lie directly above the spheres of the first. This ABAB... pattern of layers, where A denotes layers that have spheres directly above each other and likewise for B, gives a structure with a hexagonal unit cell and hence is said to be **hexagonally close-packed** (hcp, Figs 3.10a and 3.11). In the second polytype, the spheres of the third layer are placed above the gaps in the first layer. The second layer covers half the holes in the first layer and the third layer lies above the remaining holes. This arrangement results in an ABCABC... pattern, where C denotes a layer that has spheres not directly above spheres of the A or the B layer positions (but they will be directly above another C type layer). This pattern corresponds to a structure with a cubic unit cell and hence it is termed **cubic close-packed** (ccp, Figs 3.10b and 3.12). Because each ccp unit cell has a sphere at one corner and one at the centre of each face, a ccp unit cell is sometimes referred to as **face-centred cubic** (fcc).[2]

The unoccupied space in a close-packed structure amounts to 26 per cent of the total volume (see Example 3.2). However, this unoccupied space is not empty in a real solid because electron density does not end as abruptly as the hard-sphere model suggests. The type and distribution of holes are important because many structures, including those of some alloys and many ionic compounds, can be regarded as formed from an expanded close-packed arrangement in which additional atoms or ions occupy all or some of the holes.

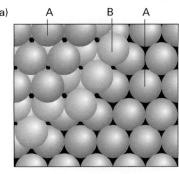
(a)

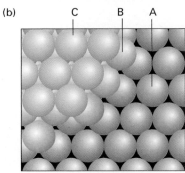
(b)

Fig. 3.10 The formation of two close-packed polytypes. (a) The third layer reproduces the first to give an ABA structure. (b) The third layer lies above the gaps in the first layer, giving an ABC structure. The different colours identify the different layers of identical spheres.

Example 3.2 Calculating the unoccupied space in a close-packed array

Calculate the percentage of unoccupied space in a close-packed arrangement of identical spheres.

Answer Because the space occupied by hard spheres is the same in the ccp and hcp arrays, we can choose the geometrically simpler structure, ccp, for the calculation. Consider Fig. 3.13. The side of such a cell is $\sqrt{8}r$, where r is the radius of each sphere, so its volume is $(\sqrt{8}r)^3 = (\sqrt{8})^3 r^3$. The unit cell contains $\frac{1}{8}$ of a sphere at each corner (for $8 \times \frac{1}{8} = 1$ in all) and half a cell on each face (for $6 \times \frac{1}{2} = 3$ in all), for a total of 4. Because the volume of each sphere is $\frac{4}{3}\pi r^3$, the total volume occupied by the spheres themselves is $4(\frac{4}{3}\pi r^3)$. The occupied fraction is therefore $4(\frac{4}{3}\pi r^3)/((\sqrt{8})^3 r^3) = (\frac{16}{3})\pi/(\sqrt{8})^3 = 0.740$. The unoccupied fraction is therefore 0.260, corresponding to 26.0 per cent.

Self-test 3.2 Calculate the fraction of space occupied by identical spheres in a primitive cubic unit cell.

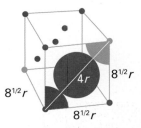

Fig. 3.13 The dimensions involved in the calculation of the packing fraction in a close-packed arrangement of identical spheres of radius r.

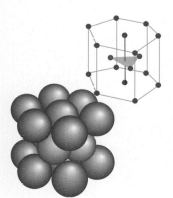

Fig. 3.11 The hexagonal close-packed (hcp) unit cell of the ABAB... polytype. The colours of the spheres correspond to layers in Fig. 3.10a.

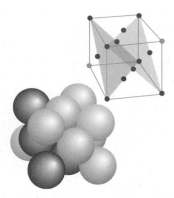

Fig. 3.12 The cubic close-packed (fcc) unit cell of the ABC... polytype. The colours of the spheres correspond to the layers in Fig. 3.10b.

---

[2] The descriptions ccp and fcc are often used interchangeably, although strictly ccp refers only to a close-packed arrangement whereas fcc refers to the lattice type of the common representation of ccp. Throughout this text the term ccp will be used to describe this close-packing arrangement. It will be drawn as the cubic unit cell, with the fcc lattice type, as this representation is easiest to visualize.

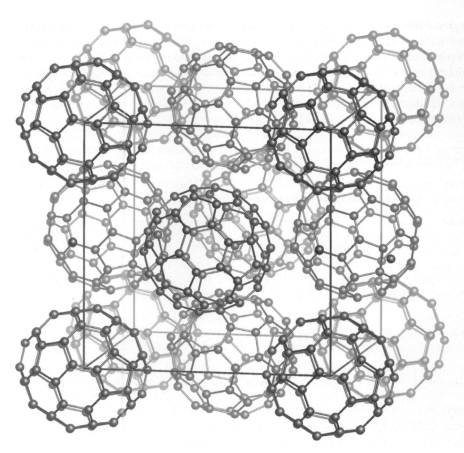

Fig. 3.14 The structure of solid C_{60}.

The ccp and hcp arrangements are the most efficient simple ways of filling space with identical spheres. They differ only in the stacking sequence of the close-packed layers and other, more complex, close-packed layer sequences may be formed by locating successive planes in different positions relative to their neighbours (Section 3.4). Any collection of atoms, such as those in the simple picture of a metal, or approximately spherical molecules is likely to adopt one of these close-packed structures unless there are additional energetic reasons—specifically covalent interactions—for choosing an alternative arrangement. Indeed, many metals adopt such close-packed structures (Section 3.4), as do the solid forms of the noble gases (which are ccp). Almost spherical molecules, such as C_{60}, in the solid state (Fig. 3.14) also adopt the ccp arrangement and so do many small molecules that rotate on their lattice sites and thus appear spherical, such as H_2, F_2, and one form of solid oxygen, O_2.

The proof that the ccp and hcp arrangements are the most efficient methods of packing identical spheres eluded mathematicians until the end of the twentieth century. Nonspherical shapes, such as ellipsoids, and mixtures of large and small spheres, can pack to occupy space even more efficiently than the 74 per cent achieved by spherical close packing.

3.3 Holes in close-packed structures

Key points: The structures of many solids can be discussed in terms of close-packed arrangements of one atom type in which the tetrahedral or octahedral holes are occupied by other atoms or ions. The ratio of spheres to octahedral holes to tetrahedral holes in a close-packed structure is 1:1:2.

The feature of a close-packed structure that enables us to extend the concept to describe structures more complicated than elemental metals is the existence of two types of **hole**, or unoccupied space between the spheres. An **octahedral hole** lies between two planar

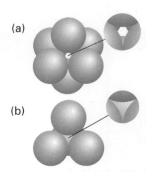

Fig 3.15 (a) An octahedral hole and (b) a tetrahedral hole formed in an arrangement of close-packed spheres.

(a)

(b)

triangles of spheres in adjoining layers (Fig. 3.15a). For a crystal consisting of N spheres in a close-packed structure, there are N octahedral holes. The distribution of these holes in an hcp unit cell is shown in Fig. 3.16. This illustration also shows that the hole has local octahedral symmetry in the sense that it is surrounded by six nearest-neighbour spheres with their centres at the corners of an octahedron. If each hard sphere has radius r, then each octahedral hole can accommodate another hard sphere with a radius no larger than $0.414r$ if the close-packed spheres are to remain in contact.

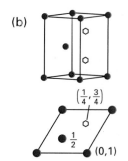

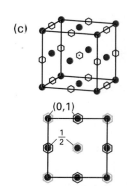

Example 3.3 Calculating the size of an octahedral hole

Calculate the maximum radius of a sphere that may be accommodated in an octahedral hole in a close-packed solid composed of spheres of radius r.

Answer The structure of a hole, with the top spheres removed, is shown in Fig. 3.17a. If the radius of a sphere is r and that of the hole is r_h, it follows from Pythagoras' theorem that $(r+r_h)^2 + (r+r_h)^2 = (2r)^2$ and therefore that $(r+r_h)^2 = 2r^2$, which implies that $r + r_h = \sqrt{2}r$. That is, $r_h = (\sqrt{2} - 1)r = 0.414r$. Note that this is the maximum size while keeping the close-packed spheres in contact; if the spheres are allowed to separate slightly while maintaining their relative positions, then the hole can accommodate a larger sphere.

Self-test 3.3 Show that the maximum radius of a sphere that can fit into a tetrahedral hole (see below) is $r_h = 0.225r$; base your calculation on Fig. 3.17b.

A **tetrahedral hole**, T, Fig. 3.15b, is formed by a planar triangle of touching spheres capped by a single sphere lying in the dip between them. The tetrahedral holes in any close-packed solid can be divided into two sets: in one the apex of the tetrahedron is directed up (T) and in the other the apex points down (T'). In an arrangement of N close-packed spheres there are N tetrahedral holes of each set and $2N$ tetrahedral holes in all. In a close-packed structure of spheres of radius r, a tetrahedral hole can accommodate another hard sphere of radius no greater than $0.225r$ (see Self-test 3.3). The location of tetrahedral holes, and the four nearest-neighbour spheres for one hole, in a hcp lattice is shown in Fig. 3.18. Individual tetrahedral holes in ccp and hcp structures are identical (because they are properties of two neighbouring close-packed layers) but in the hcp arrangement neighbouring T and T' holes share a common tetrahedral face and are so close together that they are never occupied simultaneously.

Fig. 3.16 (a) The locations of octahedral holes relative to the atoms arranged on a close-packed lattice. (b) The location (represented by a hexagon) of the two octahedral holes in the hcp unit cell and (c) the locations (represented by hexagons) of the octahedral holes in the ccp unit cell.

The structures of metals and alloys

X-ray diffraction studies (Section 6.1) reveal that many metallic elements have close-packed structures, indicating that the bonds between the atoms have little directional covalent character (Table 3.2, Fig 3.19). One consequence of this close packing is that metals often have high densities because the most mass is packed into the smallest volume. Indeed, the elements deep in the d block, near iridium and osmium, include the densest solids known under normal conditions of temperature and pressure.

3.4 Polytypism

Key point: Complex polytypes, different three-dimensional structural arrangements of the same stoichiometry, are sometimes observed for close-packed structures.

Which of the common close-packed polytypes, hcp or ccp, a metal adopts depends on the details of the electronic structure of its atoms, the extent of interaction between second-nearest-neighbours, and the residual effects of some directional character in their bonds. Indeed, a close-packed structure need not be either of the common ABAB... or ABCABC... polytypes. An infinite range of close-packed polytypes can in fact occur, as the layers may stack in a more complex repetition of A, B, and C layers or even in an almost random sequence. The sequences cannot be a completely random choice of A, B,

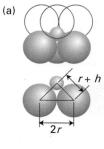

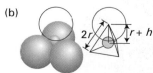

Fig. 3.17 The distances used to calculate the size of (a) an octahedral hole, (b) a tetrahedral hole.

Table 3.2 The crystal structures adopted by metals under normal conditions

Crystal structure	Element
Hexagonal close-packed (hcp)	Be, Cd, Co, Mg, Ti, Zn
Cubic close-packed (ccp)	Ag, Al, Au, Ca, Ca, Cu, Ni, Pb, Pt
Body-centred cubic (bcc)	Ba, Cr, Fe, W, alkali metals
Primitive cubic (cubic-P)	Po

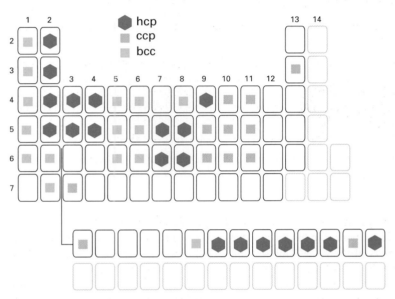

Fig. 3.19 The structures of the metallic elements at room temperature. Elements with more complex structures are left blank.

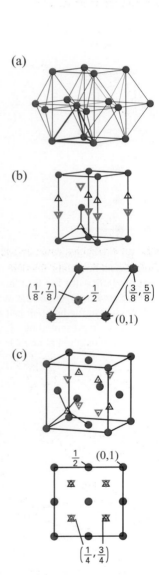

(a)

(b)

$\left(\frac{1}{8}, \frac{7}{8}\right)$ $\frac{1}{2}$ $\left(\frac{3}{8}, \frac{5}{8}\right)$

$(0,1)$

(c)

$\frac{1}{2}$ $(0,1)$

$\left(\frac{1}{4}, \frac{3}{4}\right)$

Fig. 3.18 (a) The location of a tetrahedral hole between close-packed layers (b) the locations (represented by triangles) of the tetrahedral holes in the hcp unit cell (the T and T′ sites are distinguished in green and black) (c) the locations (represented by triangles) of the tetrahedral holes in the ccp unit cell.

and C positions, however, because adjacent layers cannot have exactly the same sphere positions; for instance, AA and CC cannot occur because spheres in one layer must occupy dips in the adjacent layer.

Cobalt is an example of this more complex polytypism. Above 500°C, cobalt is ccp but it undergoes a transition when cooled. The structure that results is a nearly randomly stacked set (for instance, ABACBABABC . . .) of close-packed layers. In some samples of cobalt the polytypism is not random, as the sequence of planes repeats after several hundred layers. The long-range repeat may be a consequence of a spiral growth of the crystal that requires several hundred turns before a stacking pattern is repeated.

3.5 Non-close-packed structures

Key points: A common non-close-packed metal structure is body-centred cubic; a primitive cubic structure is occasionally observed. Metals that have structures more complex than those described so far can sometimes be regarded as slightly distorted versions of simple structures.

Not all metals are close-packed, and some other packing patterns use space nearly as efficiently. Even metals that are close-packed may undergo a phase transition to a less closely packed structure when they are heated and their atoms undergo large-amplitude vibrations.

One common structure is the **body-centred cubic** (cubic-I or bcc) structure in which there is a sphere at the centre of a cube with spheres at each corner (Fig. 3.20a). Metals with this structure have a coordination number of 8, because the central atom is in contact with the atoms at the corners of the unit cell. Although a bcc structure is less closely packed than the ccp and hcp structures (for which the coordination number is 12), the difference is not very great because the central atom has six second-nearest neighbours, at the centres of the adjacent unit cells, only 15 per cent further away. This arrangement leaves 32 per cent of the space unfilled compared with 26 per cent in the close-packed structures (see Example 3.4). A bcc structure is adopted by fifteen of the elements under standard conditions, including all the alkali metals and the metals in Groups 5 and 6. Accordingly, this simple arrangement of atoms is sometimes referred to as the 'tungsten type'.

Example 3.4 Calculating the volume filled in a body-centred cubic structure

Demonstrate that the fractional volume occupied by the spheres in a body-centred cubic arrangement is 0.68.

Answer Each unit cell contains the equivalent of two spheres of total volume $2 \times (4\pi/3)r^3$, where r is the radius of each sphere. Three spheres are in contact along the body diagonal that cuts across the unit cell from corner to corner for a total length of $4r$. The length of the body diagonal of a cube of side a is $\sqrt{3}a$. Therefore, the side of the unit cell and the radius of the close-packed spheres are related by $a = 4r/\sqrt{3}$. The volume of the cube is $a^3 = (4r/\sqrt{3})^3$. The fraction of space filled is therefore $(8/3)\pi r^3/(4r/\sqrt{3})^3 = \sqrt{3}\pi/8 = 0.68$.

Self-test 3.4 Calculate the maximum size of a sphere that can occupy the distorted octahedral site in the face of a bcc unit cell composed of spheres of radius r.

The least common metallic structure is the **primitive cubic** (cubic-P) structure (Fig. 3.21), in which spheres are located at the lattice points of a primitive cubic lattice, taken as the corners of the cube. The coordination number of a cubic-P structure is 6. One form of polonium (α-Po) is the only example of this structure among the elements under normal conditions. Solid mercury (α-Hg), however, has a closely related structure: it is obtained from the cubic-P arrangement by stretching the cube along one of its body diagonals (Fig. 3.22a); a second form of solid mercury (β-Hg) has a structure based on the body-centred cubic arrangement but compressed along one cell direction (Fig. 3.22b). Although antimony and bismuth normally have structures based on layers of atoms, both convert to a cubic-P structure under pressure and then to close-packed structures at even higher pressures.

Metals that have structures more complex than those described so far can sometimes be regarded, like solid mercury, as having slightly distorted versions of simple structures. Zinc and cadmium, for instance, have almost hcp structures, but the planes of close-packed atoms are separated by a slightly greater distance than in perfect hcp. This difference suggests stronger bonding between the close-packed atoms in the plane than between the planes: the bonding draws these atoms together and, in doing so, squeezes out the atoms of the neighbouring layers.

3.6 Polymorphism of metals

Key points: Polymorphism is a common consequence of the low directionality of metallic bonding. At high temperatures a bcc structure is common for metals that are close-packed at low temperatures on account of the increased amplitude of atomic vibrations.

The low directionality of the bonds that metal atoms may form accounts for the wide occurrence of **polymorphism**, the ability to adopt different crystal forms under different conditions of pressure and temperature. It is often, but not universally, found that the most closely packed phases are thermodynamically favoured at low temperatures and that the less closely packed structures are favoured at high temperatures.

The polymorphs of metals are generally (but not always systematically) labelled α, β, γ, . . . with increasing temperature. Some metals revert to a low-temperature form at higher temperatures. Iron, for example, shows several solid–solid phase transitions; α-Fe,

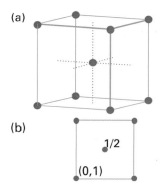

Fig. 3.20 (a) A bcc structure unit cell and (b) its projection representation.

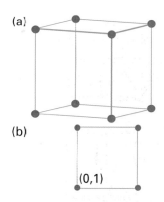

Fig. 3.21 (a) A primitive cubic unit cell and (b) its projection representation.

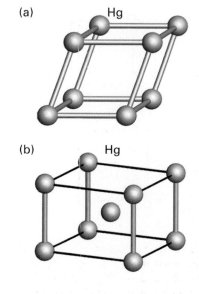

Fig 3.22 The structures of (a) α-mercury and (b) β-mercury that are closely related to the unit cells with primitive cubic- and body-centred cubic lattices, respectively.

which is bcc, occurs up to 906°C, γ-Fe, which is ccp, occurs up to 1401°C, and then α-Fe occurs again up to the melting point at 1530°C. The hcp polymorph, β-Fe, is formed at high pressures.

The bcc structure is common at high temperatures for metals that are close-packed at low temperatures because the increased amplitude of atomic vibrations in the hotter solid results in a less close-packed structure. For many metals (among them calcium, titanium, and manganese), the transition temperature is above room temperature; for others (among them lithium and sodium), the transition temperature is below room temperature. It is also found empirically that a bcc structure is favoured by metals with a small number of valence electrons per orbital.

3.7 Atomic radii of metals

Key point: The Goldschmidt correction converts atomic radii of metals to the value they would have in a close-packed structure with 12-fold coordination.

An informal definition of the atomic radius of a metallic element was given in Section 1.8 as half the distance between the centres of adjacent atoms in the solid. However, it is found that this distance generally increases with the coordination number of the lattice. The same atom in structures with different coordination numbers may therefore appear to have different radii, and an atom of an element with coordination number 12 appears bigger than one with coordination number 8. In an extensive study of internuclear separations in a wide variety of polymorphic elements and alloys, V. Goldschmidt found that the average relative radii are related as shown in Table 3.3.

It is desirable to put all elements on the same footing when comparing trends in their characteristics; that is, when comparing the intrinsic properties of their atoms rather than the properties that stem from their environment. Therefore, it is common to adjust the empirical internuclear separation to the value that would be expected if the element were in fact close-packed (with coordination number 12). Thus, the empirical atomic radius of Na is 185 pm, but that is for a structure in which the coordination number is 8. To adjust to 12-coordination we multiply this radius by $1/0.97 = 1.03$ and obtain 191 pm as the radius that a Na atom would have if it were in a close-packed structure.

Goldschmidt radii of the elements were in fact the ones listed in Table 1.4 as 'metallic radii' and used in the discussion of the periodicity of atomic radius (Section 1.9). The essential features of that discussion to bear in mind now, with 'atomic radius' interpreted as Goldschmidt-corrected metallic radius in the case of metallic elements, are that metallic radii generally increase down a group and decrease from left to right across a period. As remarked in Section 1.8, trends in atomic radii reveal the presence of the lanthanide contraction in Period 6, with atomic radii of the elements that follow the lanthanoids found to be smaller than simple extrapolation from earlier periods would suggest. As also remarked there, this contraction can be traced to the poor shielding effect of *f* electrons. A similar contraction occurs across each row of the *d* block.

3.8 Alloys

An **alloy** is a blend of metallic elements prepared by mixing the molten components and then cooling the mixture to produce a solid that exhibits metallic properties. Alloys may be homogeneous solid solutions, in which the atoms of one metal are distributed randomly among the atoms of the other, or they may be compounds with a definite composition and internal structure. The behaviour of mixtures of species in the solid state is an important one where the distribution of atoms or ions over an infinite lattice needs to be understood. Consideration of alloys provides a useful introduction to this chemistry and we shall use the concepts introduced here to describe the properties of ionic compounds and many important inorganic materials. Alloys typically form from two electropositive metals that have very similar electronegativities, so they are likely to be located towards the bottom left-hand corner of a Ketelaar triangle (Fig. 3.23).

Table 3.3 The variation of radius with coordination number

Coordination number	Relative radius
12	1
8	0.97
6	0.96
4	0.88

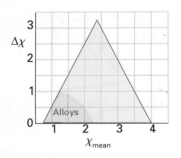

Fig. 3.23 The Ketelaar triangle and the approximate locations of alloys.

Solid solutions are classified as either 'substitutional' or 'interstitial'. A **substitutional solid solution** is a solid solution in which atoms of the solute metal occupy some of the locations of the solvent metal atoms (Fig. 3.24a). An **interstitial solid solution** is a solid solution in which the solute atoms occupy the interstices (the holes) between the solvent atoms (Fig. 3.24b). However, this distinction is not particularly fundamental, because interstitial atoms often lie in a definite array (Fig. 3.24c), and hence can be regarded as a substitutional version of another structure. Some of the classic examples of alloys are brass (up to 38 atom per cent Zn in Cu), bronze (a metal other than zinc or nickel in copper; casting bronze, for instance, is 10 atom per cent Sn and 5 atom per cent Pb), and stainless steel (over 12 atom per cent Cr in Fe).

(a) Substitutional solid solutions

Key point: A substitutional solid solution involves the replacement of one type of metal atom in a structure by another.

Substitutional solid solutions are generally formed if three criteria are fulfilled:

1 The atomic radii of the elements are within about 15 per cent of each other.

2 The crystal structures of the two pure metals are the same; this similarity indicates that the directional forces between the two types of atom are compatible with each other.

3 The electropositive characters of the two components are similar; otherwise compound formation would be more likely.

Thus, although sodium and potassium are chemically similar and have bcc structures, the atomic radius of sodium (191 pm) is 19 per cent smaller than that of potassium (235 pm), and the two metals do not form a solid solution. Copper and nickel, however, two neighbours late in the d block, have similar electropositive character, similar crystal structures (both ccp), and similar atomic radii (Ni 125 pm, Cu 128 pm, only 2.3 per cent different), and form a continuous series of solid solutions, ranging from pure nickel to pure copper. Zinc, copper's other neighbour in Period 4, has a similar atomic radius (137 pm, 7 per cent larger), but it is hcp, not ccp. In this instance, zinc and copper are partially miscible and form solid solutions known as 'α-brasses' of composition $Cu_{1-x}Zn_x$ with $0 < x < 0.38$ with the same structure type as pure copper.

(b) Interstitial solid solutions of nonmetals

Key point: In an interstitial solid solution, additional small atoms occupy holes within the lattice of the original metal structure.

Interstitial solid solutions are often formed between metals and small atoms (such as boron, carbon, and nitrogen) that can inhabit the interstices in the metal. The small atoms enter the host solid with preservation of the crystal structure of the original metal and without the transfer of electrons and formation of ionic species. There is either a simple whole-number ratio of metal and interstitial atoms (as in tungsten carbide, WC) or the small atoms are distributed randomly in the available holes in the metal. The former substances are true compounds and the latter can be considered as interstitial solid solutions or nonstoichiometric compounds.

Considerations of size can help to decide where the formation of an interstitial solid solution is likely to occur. Thus, the largest solute atom that can enter a close-packed solid without distorting the structure appreciably is one that just fits an octahedral hole, which as we have seen has radius $0.414r$. For small atoms such as B, C, or N the atomic radii of the possible host metal atom structures include those of the d metals such as iron, nickel, and cobalt, although there may be some additional bonding in these compounds. One important class of materials of this type consists of the carbon steels in which C atoms occupy some of the octahedral holes in the bcc lattice of iron. Carbon steels typically contain between 0.2 and 1.6 atom per cent C. With increasing carbon content they become harder and stronger but less malleable (Box 3.1).

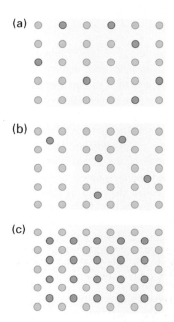

Fig. 3.24 (a) Substitutional and (b) interstitial alloys. (c) In some cases, an interstitial alloy may be regarded as a substitutional alloy derived from another lattice.

Box 3.1 Steels

Steels are alloys of iron, carbon, and other elements. They are classified as mild and medium-or high-carbon steels according to the percentage of carbon they contain. Mild steels contain up to 0.25 per cent (all percentages here are atom percentages), medium-carbon steels 0.25 to 0.45 per cent, and high-carbon steels 0.45 to 1.50 per cent. The addition of other metals to these carbon steels can have a major effect on the structure, properties, and therefore applications of the steel. Examples of metals added to carbon steels, so forming 'stainless steels', are listed in the table. Stainless steels are also classified by their crystalline structures, which are controlled by factors such as the rate of cooling following their formation in a furnace and the type of added metal. Thus pure iron adopts different polymorphs (Section 3.6) depending on temperature and some of these high-temperature structures can be stabilized at room temperature in steels or by quenching (cooling very rapidly).

Stainless steels with the *austenite* structure comprise over 70 per cent of total stainless steel production. Austenite is a solid solution of carbon and iron that exists in steel above 723°C and is ccp Fe with about 2 per cent of the octahedral holes filled with carbon. As it cools, it breaks down into other materials including *ferrite* and *martensite* as the solubility of carbon in the iron drops to below 1 per cent. The rate of cooling determines the relative proportions of these materials and therefore the mechanical properties (for example, hardness and tensile strength) of the steel. The addition of certain other metals, such as manganese, nickel, and chromium can allow the austenitic structure to survive cooling to room temperature. These steels contain a maximum of 0.15 per cent C and typically 10–20 per cent Cr plus Ni or Mn as a substitutional solid solution; they can retain an austenitic structure at all temperatures from the cryogenic region to the melting point of the alloy. A typical composition, 18 per cent Cr and 8 per cent Ni, is known as *18/8 stainless*.

Ferrite is α-Fe with only a very small level of carbon, less than 0.1 per cent, with a bcc iron crystal structure. Ferritic stainless steels are highly corrosion resistant, but far less durable than austenitic grades. They contain between 10.5 and 27 per cent chromium and some Mo, Al, or W. Martensitic stainless steels are not as corrosion resistant as the other two classes, but are extremely strong and tough as well as highly machineable, and can be hardened by heat treatment. They contain 11.5 to 18 per cent chromium and 1–2 per cent C which is trapped in the iron structure as a result of quenching compositions with the austenite structure type. The martensitic crystal structure is closely related to that of ferrite but the unit cell is tetragonal rather than cubic.

Metal	Percentage added	Effect on properties
Copper	0.2–1.5	Improves atmospheric corrosion resistance
Nickel	0.1–1	Benefits surface quality
Niobium	0.02	Increases tensile strength and yield point
Nitrogen	0.003–0.012	Improves strength
Manganese	0.2–1.6	Improves strength
Vanadium	Up to 0.12	Increases strength

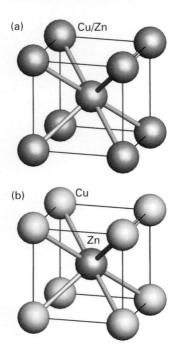

Fig. 3.25 The structure of β-brass with (a) disordered Cu and Zn and (b) ordered Cu and Zn.

(c) Intermetallic compounds

Key point: Intermetallic compounds are alloys in which the structure adopted is different from the structures of either component metal.

In contrast to the interstitial solid solutions of metals and nonmetals, where geometrical considerations are consistent with intuition, there is a class of material formed between two metals that are better regarded as actual compounds despite the similarity of the electronegativities of the metals. For instance, when some liquid mixtures of metals are cooled, they form phases with definite structures that are often unrelated to the parent structure. These phases are called **intermetallic compounds**. They include β-brass (CuZn), Fig. 3.25a, and compounds of composition $MgZn_2$, Cu_3Au, $NaTl$, and Na_5Zn_{21}. Note that some of these intermetallic compounds contain a very electropositive metal in combination with a less electropositive metal (for example Na, $\chi = 0.93$ and Zn, $\chi = 1.65$), and in a Ketelaar triangle lie above the true alloys (Fig. 3.26). Such combinations are called **Zintl phases**. These compounds are not fully ionic (although they are often brittle) and have some metallic properties, including lustre. A classic example of a Zintl phase is KGe with the structure shown in Fig. 3.27.

Example 3.5 Identifying the brass structure

What structure type is the copper/zinc alloy (β-brass, CuZn, Fig. 3.25a) formed by cooling a molten 1:1 mixture of zinc and copper?

Answer See Section 3.5. The structure has a body-centred cubic arrangement with the Cu and Zn atoms randomly distributed over the sites at the centre and corners of the unit cell.

Self-test 3.5 What is the structure type of the brass of composition CuZn if the metal atoms become ordered with Cu atoms on the cell corners and Zn atoms at the unit cell centre (Fig. 3.25b)?

Ionic solids

Key points: The ionic model treats a solid as an assembly of oppositely charged spheres that interact by nondirectional Coulombic forces; if the thermodynamic properties of the solid calculated on this model agree with experiment, then the compound is normally considered to be ionic.

Ionic solids, such as sodium chloride and potassium nitrate, are often recognized by their brittleness because the electrons made available by cation formation are localized to form a neighbouring anion instead of contributing to an adaptable, mobile electron sea. Ionic solids also commonly have high melting points and most are soluble in polar solvents, particularly water. However, there are exceptions: calcium fluoride, CaF_2, for example, is a high-melting ionic solid but it is insoluble in water. Ammonium nitrate, NH_4NO_3, is ionic in terms of its interactions between the ammonium and nitrate ions, but melts at 170°C. Binary ionic materials typically form for combinations of elements with large electronegativity differences, typically $\Delta\chi > 3$, and such compounds are therefore likely to be found at the top corner of a Ketelaar triangle.

The classification of a solid as ionic is based on comparison of its properties with those of the **ionic model**, which treats the solid as an assembly of oppositely charged spheres that interact primarily by nondirectional Coulombic forces together with repulsions between the complete shells of the ions in contact. If the thermodynamic properties of the solid calculated on this model agree with experiment, then the solid may be ionic. However, it should be noted that many examples of coincidental agreement with the ionic model are known, so numerical agreement alone does not imply ionic bonding. The nondirectional nature of interactions between ions in an ionic solid contrast with those present in a covalent solid, where the symmetries of atomic orbitals play a strong part in determining the geometries of molecules.

We start by describing some common ionic structures in terms of the packing of hard spheres, but in this case the spheres have different sizes and opposite charges. After that, we see how to rationalize the structures in terms of the energetics of crystal formation. The structures described were obtained by using X-ray diffraction and were among the first substances to be examined in this way.

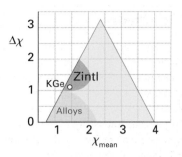

Fig. 3.26 The Ketelaar triangle and the approximate locations of Zintl phases. The point marks the location of one examplar, KGe.

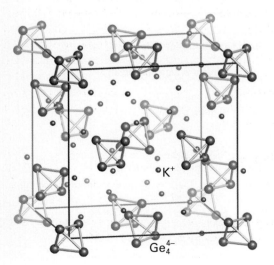

Fig. 3.27 The structure of the Zintl phase KGe showing the Ge_4^{4-} tetrahedral units and interspersed K^+ ions.

3.9 Characteristic structures of ionic solids

The ionic structures described in this section are prototypes of a wide range of solids. For instance, although the rock-salt structure takes its name from a mineral form of NaCl, it is characteristic of numerous other solids (Table 3.4). Many of the structures can be regarded as derived from arrays in which the larger of the ions, usually the anions, stack together in ccp or hcp patterns and the smaller counter-ions (usually the cations) occupy the octahedral or tetrahedral holes in the lattice (Table 3.5). Throughout the following discussion, it will be helpful to refer back to Figs 3.16 and 3.18 to see how the structure being described is related to the hole patterns shown there. The close-packed layer usually needs to expand to accommodate the counter-ions and avoid the strong repulsion between the identically charged ions, but this expansion is often a minor perturbation of the anion arrangement. Hence, the close-packed structure is often a good starting point.

(a) Binary phases, AX_n

Key points: Important structures that can be expressed in terms of the occupation of holes include the rock-salt, caesium-chloride, sphalerite, fluorite, wurtzite, nickel-arsenide, and rutile structures.

The simplest ionic compounds contain just one type of cation and one type of anion present in various ratios covering compositions such as AX and AX_2. Several different structures may exist for each of these compositions, depending on the relative sizes of the cations and anions.

Table 3.4 The crystals structures of compounds

Crystal structure	Example*
Antifluorite	K_2O, K_2S, Li_2O, Na_2O, Na_2Se, Na_2S
Caesium chloride	**CsCl**, CaS, TlSb, CsCN, CuZn
Fluorite	**CaF$_2$**, UO_2, $BaCl_2$, HgF_2, PbO_2
Nickel arsenide	**NiAs**, NiS, FeS, PtSn, CoS
Perovskite	**CaTiO$_3$**, $SrTiO_3$, $PbZrO_3$, $LaFeO_3$, $LiSrH_3$, $KMnF_3$
Rock salt	**NaCl**, KBr, RbI, AgCl, AgBr, MgO, CaO, TiO, FeO, NiO, SnAs, UC, ScN
Rutile	**TiO$_2$**, MnO_2, SnO_2, WO_2, MgF_2, NiF_2
Sphalerite (zinc blende)	**ZnS**, CuCl, CdS, HgS, GaP, InAs
Spinel	**MgAl$_2$O$_4$**, $ZnFe_2O_4$, $ZnCr_2S_4$
Wurtzite	**ZnS**, ZnO, BeO, MnS, AgI, AlN, SiC, NH_4F

* The substance in bold type is the one that gives its name to the structure.

Table 3.5 The relation of structure to the filling of holes

Close-packing type	Hole filling	Structure type (exemplar)
Cubic (ccp)	All octahedral	Rock salt (NaCl)
	All tetrahedral	Fluorite (CaF_2)
	Half tetrahedral	Sphalerite (ZnS)
Hexagonal (hcp)	All octahedral	Nickel arsenide (NiAs); with some distortion from perfect hcp
	Half octahedral	Rutile (TiO_2); with some distortion from perfect hcp
	All tetrahedral	No structure exists: tetrahedral holes share faces
	Half tetrahedral	Wurtzite (ZnS)

The **rock-salt structure** is based on a ccp array of bulky anions with cations in all the octahedral holes (Fig. 3.28). Alternatively, it can be viewed as a structure in which the anions occupy all the octahedral holes in a ccp array of cations. Because each ion is surrounded by an octahedron of six counterions, the coordination number of each type of ion is 6 and the structure is said to have **(6,6)-coordination**. In this notation, the first number in parentheses is the coordination number of the cation, and the second number is the coordination number of the anion. The rock-salt structure can still be described as having a face-centred cubic lattice after this hole filling because the translational symmetry demanded by this lattice type is preserved when all the octahedral sites are occupied.

To visualize the local environment of an ion in the rock-salt structure, we should note that the six nearest neighbours of the central ion of the cell shown in Fig. 3.28 lie at the centres of the faces of the cell and form an octahedron around the central ion. All six neighbours have a charge opposite to that of the central ion. The 12 second-nearest neighbours of the central ion, those next further away, are at the centres of the edges of the cell, and all have the same charge as the central ion. The eight third-nearest neighbours are at the corners of the unit cell, and have a charge opposite to that of the central ion.

We use the following rules to count the ions of each type in a unit cell:

1 An ion in the body of a cell belongs entirely to that cell and counts as 1.

2 An ion on a face is shared by two cells and contributes $\frac{1}{2}$ to the cell in question.

3 An ion on an edge is shared by four cells and hence contributes $\frac{1}{4}$.

4 An ion at a corner is shared by eight cells that share the corner, and so contributes $\frac{1}{8}$.

In the unit cell shown in Fig. 3.28, there are the equivalent of four Na^+ ions $(8 \times \frac{1}{8} + 6 \times \frac{1}{2})$ and four Cl^- ions $(12 \times \frac{1}{4} + 1)$. Hence, each unit cell contains four NaCl formula units. The number of formula units present in the unit cell is commonly denoted Z, so in this case $Z = 4$.

The rock-salt arrangement is not just formed for simple monatomic species such as M^+ and X^- but also for many 1:1 compounds in which the ions are complex units such as $[Co(NH_3)_6][TlCl_6]$. The structure of this compound can be considered as an array of close-packed octahedral $[TlCl_6]^{3-}$ ions with $[Co(NH_3)_6]^{3+}$ ions in all the octahedral holes. Similarly, compounds such as CaC_2, CsO_2, KCN, and FeS_2 all adopt structures closely related to the rock-salt structure with alternating cations and complex anions $(C_2^{2-}, O_2^-, CN^-,$ and S_2^{2-}, respectively), although the orientation of these linear diatomic species can lead to elongation of the unit cell and elimination of the cubic symmetry (Fig. 3.29). Further compositional flexibility, while retaining a rock-salt type of structure, can come from having more than one cation or anion type while maintaining the overall 1:1 ratio between opposite ions of opposite charge. For example, compounds such as $LiNiO_2$, equivalent to $Li_{1/2}Ni_{1/2}O$, have a mixture of Li^+ and Ni^{2+} ions distributed in an ordered manner over the octahedral sites in a close-packed O^{2-} ion array. This ordering consists of alternating planes of Li and Ni atoms between the close-packed O^{2-} ions; although $LiNiO_2$ does not have a strictly rock-salt structure, its relationship is easily recognized.

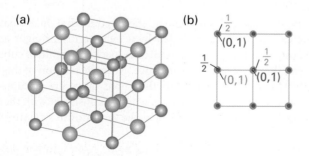

(a) (b)

Fig. 3.28 (a) The rock-salt structure and (b) its projection representation. Note the relation of this structure to the fcc structure in Fig. 3.16 with an atom in each octahedral hole.

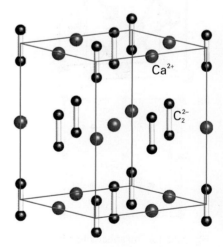

Fig. 3.29 The structure of CaC_2 is based on the rock-salt structure but is elongated in the direction parallel to the axes of the C_2^{2-} ions.

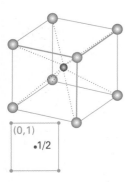

Fig. 3.30 The caesium-chloride structure. The corner lattice points, which are shared by eight neighbouring cells, are surrounded by eight nearest-neighbour lattice points. The cation occupies a cubic hole in a primitive cubic lattice.

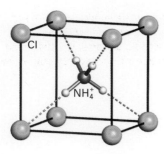

Fig. 3.31 The structure of ammonium chloride, NH_4Cl, reflects the ability of the tetrahedral NH_4^+ ion to form hydrogen bonds to the tetrahedral array of Cl^- ions around it.

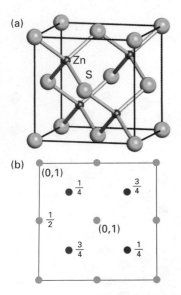

Fig. 3.32 (a) The sphalerite (zinc-blende) structure and (b) its projection representation. Note its relation to the ccp lattice in Fig. 3.18a with half the tetrahedral holes occupied by Zn^{2+} ions.

Much less common than the rock-salt structure is the **caesium-chloride structure** (Fig. 3.30), which is possessed by CsCl, CsBr, and CsI, as well as some other compounds formed of ions of similar radii to these, including TlI (see Table 3.4). The caesium-chloride structure has a cubic unit cell with each corner occupied by an anion and a cation occupying the 'cubic hole' at the cell centre (or vice versa); as a result, $Z = 1$. An alternative view of this structure is as two interlocking primitive cubic cells, one of Cs^+ and the other of Cl^-. The coordination number of both types of ion is 8, so the structure is described as having (8,8)-coordination. The radii are so similar that this energetically highly favourable coordination is feasible, with numerous counter-ions adjacent to a given ion. Note that NH_4Cl also forms this structure despite the relatively small size of the NH_4^+ ion because the cation can form hydrogen bonds with four of the Cl^- ions at the corners of the cube (Fig. 3.31). Many 1:1 alloys, such as AlFe and CuZn, have a caesium-chloride arrangement of the two metal atom types.

The **sphalerite structure** (Fig. 3.32), which is also known as the **zinc-blende structure**, takes its name from a mineral form of ZnS. Like the rock-salt structure it is based on an expanded ccp anion lattice but now the cations occupy one type of tetrahedral hole, one half the tetrahedral holes present in a close-packed structure. Each ion is surrounded by four neighbours and so the structure has (4,4)-coordination.

Example 3.6 Counting the number of ions in a unit cell

How many ions are there in the unit cell shown in the sphalerite structure shown in Fig. 3.32?

Answer According to the counting rules set out above, in the sphalerite structure, the count is:

Location (share)	Number of cations	Number of anions	Contribution
Body (1)	4×1	0	4
Face ($\frac{1}{2}$)	0	$6 \times \frac{1}{2}$	3
Edge ($\frac{1}{4}$)	0	0	0
Vertex ($\frac{1}{8}$)	0	$8 \times \frac{1}{8}$	1
Total:	4	4	8

Note that there are four cations and four anions in the unit cell. This ratio is consistent with the chemical formula ZnS, with $Z = 4$.

Self-test 3.6 Count the ions in the CsCl unit cell shown in Fig. 3.30.

The **fluorite structure** takes its name from its exemplar, the naturally occurring mineral fluorite, CaF_2. In fluorite, the Ca^{2+} ions lie in an expanded ccp array and the F^- ions occupy all the tetrahedral holes (Fig. 3.33). In this description the cations are close-packed because the F^- anions are very small. The **antifluorite** structure is the inverse of the fluorite structure in the sense that the locations of cations and anions are reversed. The latter structure is shown by some alkali metal oxides, including Li_2O. In it, the cations (which are twice as numerous as the anions) occupy both types of tetrahedral hole of the ccp array.

In the fluorite structure, the anions in their tetrahedral holes have four nearest neighbours. The cation site is surrounded by a cubic array of eight anions. From an alternative viewpoint, the anions form a non-close-packed, primitive cubic lattice and the cations occupy half the cubic holes in this lattice. Note the relation of this structure to the caesium-chloride structure, in which all the cubic holes are occupied. From either viewpoint, the lattice has (4,8)-coordination, which is consistent with there being twice as many anions as cations. The coordination in the antifluorite structure is the opposite, namely (8,4).

The **wurtzite structure** (Fig. 3.34) takes its name from another polymorph of zinc sulfide. It differs from the sphalerite structure in being derived from an expanded hcp anion array rather than a ccp array, but as in sphalerite the cations occupy one type of tetrahedral hole. This structure, which has (4,4)-coordination, is adopted by ZnO, AgI, and one polymorph of SiC, as well as several other compounds (Table 3.4). The local symmetries of the cations and anions are identical towards their nearest neighbours in wurtzite and sphalerite but differ at the second-nearest neighbours.

The **nickel-arsenide structure** (NiAs, Fig. 3.35) is also based on an expanded, distorted hcp anion array, but the Ni atoms now occupy the octahedral holes and each As atom lies at the centre of a trigonal prism of Ni atoms. This structure is adopted by NiS, FeS, and a number of other sulfides. The nickel-arsenide structure is typical of MX compounds that contain polarizable ions and with smaller electronegativity differences than ions adopting the rock-salt structure, and such compounds lie in the 'polarized ionic salt area' of a Ketelaar triangle (Fig. 3.36). There is also potential for some degree of metal-metal bonding between metal atoms in adjacent layers and this structure type (or distorted forms of it) is also common for a large number of alloys based on d- and p-block elements.

The **rutile structure** (Fig. 3.37) takes its name from rutile, a mineral form of titanium(IV) oxide, TiO_2. The structure can also be considered an example of hole filling in an hcp anion arrangement but now with the cations occupying only half the octahedral holes in an ordered manner and there is considerable buckling of the close-packed anion layers. This arrangement results in a structure that reflects the strong tendency of a Ti^{4+} ion to acquire octahedral coordination. Each Ti^{4+} atom is surrounded by six O atoms and each O atom is surrounded by three Ti^{4+} ions; hence the rutile structure has (6,3)-coordination. The principal ore of tin, cassiterite SnO_2, has the rutile structure, as do a number of metal difluorides (Table 3.4).

(b) Ternary phases $A_aB_bX_n$

Key point: The perovskite and spinel structures are adopted by many compounds with the stoichiometries ABO_3 and AB_2O_4, respectively.

The structural possibilities increase very rapidly once the compositional complexity is increased to three ionic species. Unlike binary compounds, it is difficult to predict the most likely structure type based on the ion sizes and preferred coordination numbers. This section describes two important structures formed by ternary oxides; the O^{2-} ion is the most common anion, so oxide chemistry is central to a significant part of solid-state chemistry.

The mineral perovskite, $CaTiO_3$, is the structural prototype of many ABX_3 solids (Table 3.4), particularly oxides. In its ideal form, the **perovskite structure** is cubic with each A cation surrounded by twelve X anions and each B cation surrounded by six X anions (Fig. 3.38). In fact, the perovskite structure may also be described as a close-packed array of A and oxide ions (arranged such that each A-type cation is surrounded by twelve O^{2-} ions from the original close-packed layers) with B cations in all the octahedral holes that are formed from six of the oxide ions, giving $B_{n/4}[AO_3]_{n/4}$ equivalent to ABO_3.

In oxides, X = O and the sum of the charges on the A and B ions must be +6, but that can be achieved in several ways ($A^{2+}B^{4+}$ and $A^{3+}B^{3+}$ among them), including the

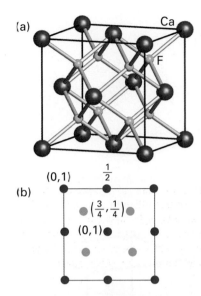

Fig. 3.33 (a) The fluorite structure and (b) its projection representation. This structure has a ccp array of cations and all the tetrahedral holes are occupied by anions.

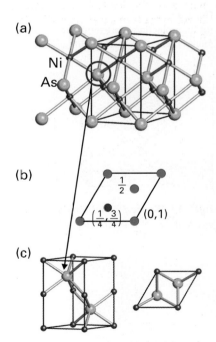

Fig. 3.35 (a) The nickel-arsenide structure, (b) the projection representation of the unit cell and (c) the trigonal prismatic coordination around As.

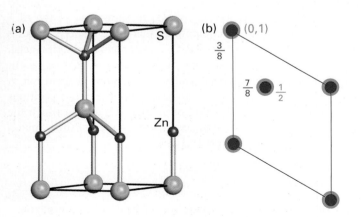

Fig. 3.34 (a) The wurtzite structure and (b) its projection representation.

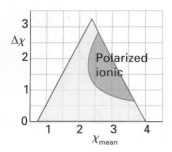

Fig. 3.36 The location of polarized ionic salts in the Ketelaar triangle.

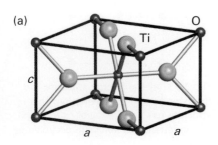

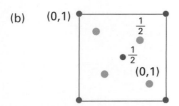

Fig. 3.37 (a) The rutile structure and (b) its projection representation. Rutile itself is one polymorph of TiO_2.

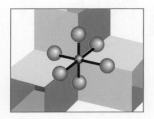

Fig. 3.39 The local coordination environment of a Ti atom in perovskite.

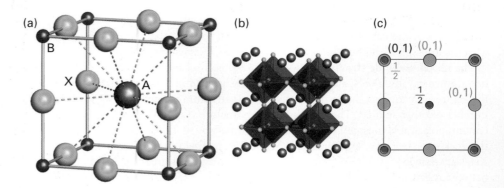

Fig. 3.38 (a) The perovskite structure, ABX_3, (b) a display that emphasizes the octahedral shape of the sites, and (c) its projection representation.

possibility of mixed oxides of formula $A(B_{0.5}B'_{0.5})O_3$, as in $La(Ni_{0.5}Ir_{0.5})O_3$. The A-type cation in perovskites is therefore usually a large ion of lower charge, such as Ba^{2+} or La^{3+} (ionic radius > 110 pm), and the B cation is a small ion of higher charge, such as Ti^{4+}, Nb^{5+}, or Fe^{3+} (ionic radius < 100 pm).

The perovskite structure is closely related to the materials that show interesting electrical properties, such as piezoelectricity, ferroelectricity, and high-temperature superconductivity (see Chapter 23).

Example 3.7 Determining coordination numbers

Demonstrate that the coordination number of the Ti^{4+} ion in the perovskite $CaTiO_3$ is 6.

Answer We need to imagine eight of the unit cells shown in Fig. 3.38 stacked together with a Ti atom shared by them all. A local fragment of the structure is shown in Fig. 3.39; it shows that there are six O^{2-} ions around the central Ti^{4+} ion, so the coordination number of Ti in perovskite is 6. An alternative way of viewing the perovskite structure is as BO_6 octahedra sharing all vertices in three orthogonal directions with the A cations at the centres of the cubes so formed (Fig. 3.39).

Self-test 3.7 What is the coordination number of a Ti^{4+} site and the O^{2-} site in rutile?

Spinel itself is $MgAl_2O_4$, and oxide spinels, in general, have the formula AB_2O_4. The **spinel structure** consists of a ccp array of O^{2-} ions in which the A cations occupy one-eighth of the tetrahedral holes and the B cations occupy half of the octahedral holes (Fig. 3.40). Spinels are sometimes denoted $A[B_2]O_4$, the square brackets denoting the cation type (normally the smaller, higher charged ion of A and B) that occupies the octahedral holes. So, for example, $ZnAl_2O_4$ can be written $Zn[Al_2]O_4$ to show that all the Al^{3+} cations occupy octahedral sites. Examples of compounds that have spinel structures include many ternary oxides with the stoichiometry AB_2O_4 that contain a 3d-series metal, such as $NiCr_2O_4$ and $ZnFe_2O_4$, and some simple binary d-block oxides, such as Fe_3O_4, Co_3O_4, and Mn_3O_4; note that in these structures A and B are the same element but in different oxidation states, as in $Fe^{2+}[Fe^{3+}]_2O_4$. There are also a number of compositions termed **inverse spinels**, in which the cation distribution is $B[AB]O_4$ and in which the more abundant cation is distributed over both tetrahedral and octahedral sites. Spinels and inverse spinels are discussed again in Sections 19.1 and 23.8.

Example 3.8 Predicting possible ternary phases

What ternary oxides with the perovskite or spinel structures might it be possible to synthesize that contain the cations Ti^{4+}, Zn^{2+}, In^{3+}, Pb^{2+}?

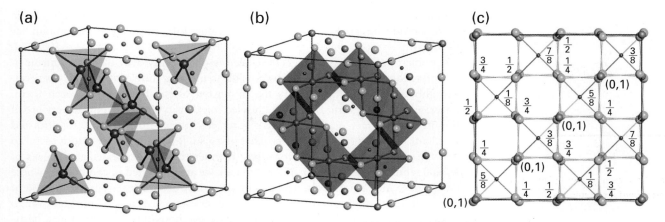

(a) **(b)** **(c)**

Fig. 3.40 (a) The spinel structure showing the tetrahedral environment around the O atoms, (b) showing their octahedral environment, and (c) its projection representation with only the cation locations specified.

Answer $PbTiO_3$ (perovskite), $TiZn_2O_4$ (spinel), $ZnIn_2O_4$ (spinel). $ZnTiO_3$ does not exist as a perovskite as the Zn^{2+} ion is too small for the A-type site; $PbIn_2O_4$ does not adopt the spinel structure as the Pb^{2+} cation is too large for the tetrahedral sites.

Self-test 3.8 What additional oxide perovskite composition(s) might be obtained if La^{3+} is added to this list of cations?

3.10 The rationalization of structures

The thermodynamic properties of such ionic solids can be treated very simply in terms of the ionic model. However, a model of a solid in terms of charged spheres interacting Coulombically is likely to be fairly crude and we should expect significant departures from its predictions because many solids are more covalent than ionic. Even conventional 'good' ionic solids, such as the alkali metal halides, have some covalent character. Nevertheless, the ionic model provides an attractively simple and effective scheme for correlating many properties.

(a) Ionic radii

Key point: The sizes of ions, ionic radii, generally increase down a group, decrease across a period, increase with coordination number, and decrease with increasing charge number.

A difficulty that confronts us at the outset is the meaning of the term 'ionic radius'. As remarked in Section 1.8, it is necessary to apportion the single internuclear separation of nearest-neighbour ions between the two different species (for example, an Na^+ ion and a Cl^- ion in contact). The most direct way to solve the problem is to make an assumption about the radius of one ion, and then to use that value to compile a set of self-consistent values for all other ions. The O^{2-} ion has the advantage of being found in combination with a wide range of elements. It is also reasonably unpolarizable, so its size does not vary much as the identity of the accompanying cation is changed. In a number of compilations, therefore, the values are based on $r(O^{2-}) = 140$ pm. However, this value is by no means sacrosanct: a set of values compiled by Goldschmidt was based on $r(O^{2-}) = 132$ pm and other values use the F^- ion as the basis.

For certain purposes (such as for predicting the sizes of unit cells) ionic radii can be helpful, but they are reliable only if they are all based on the same fundamental choice (such as the value 140 pm for O^{2-}). If values of ionic radii are used from different sources, it is essential to verify that they are based on the same convention. An additional complication that was first noted by Goldschmidt is that, as we have already seen for metals,

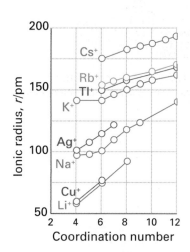

Fig. 3.41 The variation of ionic radius with coordination number.

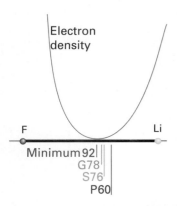

Fig. 3.42 The variation in electron density along the Li—F axis in LiF. The point P denotes the Pauling radii of the ions, G the original (1927) Goldschmidt radii, and S the Shannon radii.

ionic radii increase with coordination number (Fig. 3.41). Hence, when comparing ionic radii, we should compare like with like, and use values for a single coordination number (typically 6).

The problems of the early workers have been resolved only partly by developments in X-ray diffraction (Section 6.1). It is now possible to measure the electron density between two neighbouring ions and identify the minimum as the boundary between them. However, as can be seen from Fig. 3.42, the electron density passes through a very broad minimum, and its exact location may be very sensitive to experimental uncertainties and to the identities of the two neighbours. That being so, it is still probably more useful to express the sizes of ions in a self-consistent manner than to seek calculated values of individual radii in certain combinations. After all, we are interested in compounds where we always have interactions between pairs of ions so we are consistent in the method of determination and application of ionic radii. Very extensive lists of self-consistent values that have been compiled by analysing X-ray data on thousands of compounds, particularly oxides and fluorides, exist and some are given in Table 1.5 and *Resource section* 1.

The general trends for ionic radii are the same as for atomic radii. Thus:

1 Ionic radii increase down a group. (The lanthanide contraction, Section1.8, restricts the increase between the 4*d*- and 5*d*-series metal ions.)

2 The radii of ions of the same charge decrease across a period.

3 When an ion can occur in environments with different coordination numbers, its radius increases as the coordination number increases.

4 If an element can form cations with different charge numbers, then for a given coordination number its ionic radius decreases with increasing charge number.

5 Because a positive charge indicates a reduced number of electrons, and hence a more dominant nuclear attraction, cations are smaller than anions for elements with similar atomic numbers.

(b) The radius ratio

Key point: The radius ratio indicates the likely coordination numbers of the ions in a compound.

A parameter that figures widely in the literature of inorganic chemistry, particularly in introductory texts, is the 'radius ratio', ρ (rho), of the ions. The **radius ratio** is the ratio of the radius of the smaller ion (r_{small}) to that of the larger (r_{large}):

$$\rho = \frac{r_{small}}{r_{large}} \tag{3.1}$$

In most cases, r_{small} is the cation radius and r_{large} is the anion radius. The minimum radius ratio that can support a given coordination number is then calculated by considering the geometrical problem of packing together spheres of different sizes (Table 3.6). It is argued that, if the radius ratio falls below the minimum given, then ions of opposite charge will not be in contact and ions of like charge will touch. According to a simple electrostatic argument, the lower coordination number, in which the contact of oppositely charged ions is restored, then becomes favourable. As the ionic radius of the M$^+$ ion increases, more anions can pack around it; compare CsCl, having (8,8)-coordination, with NaCl, and its (6,6)-coordination.

Table 3.6 The correlation of structural type with radius ratio

Radius ratio (ρ)	Coordination numbers for 1:1 and 1:2 stoichiometries	Binary AB structure type	Binary AB$_2$ structure type
1	12	None known	None known
0.732–1	8:8 and 8:4	CsCl	CaF$_2$
0.414–0.732	6:6 and 6:3	NaCl (ccp), NiAs (hcp)	TiO$_2$
0.225–0.414	4:4	ZnS (ccp and hcp)	

As we have seen, a cation of radius between $0.225r$ and $0.414r$ can occupy a tetrahedral hole in a close-packed or slightly expanded close-packed array of anions of radius r. However, once a cation is larger than $0.414r$, the anions are forced so far apart that octahedral coordination becomes possible and most favourable. Note that $0.225r$ represents the size of the smallest ion that will fit in a tetrahedral hole and that cations between $0.225r$ and $0.414r$ will push the anions apart. However, the coordination number cannot increase to 6 with good contacts between cation and anions until the radius goes above $0.414r$. Similar arguments apply for the tetrahedral holes that can be filled by ions with sizes up to $0.225r$.

Concepts of ion packing based on radius ratios can often be used to predict which structure is most favourable for any particular choice of cation and anion (Table 3.6). In practice, the radius ratio is most reliable when the cation coordination number is 8, and less reliable with six- and four-coordinate cations because directional covalent bonding becomes more important for these lower coordination numbers.

Example 3.9 Predicting structure using radius ratio rules

What crystal structure is TlCl expected to adopt ?

Answer From *Resource section* 1, the ionic radii are $r(Tl^+) = 164$ pm and $r(Cl^-) = 167$ pm, giving $\rho = 0.98$. We can therefore predict that TlCl is likely to adopt a caesium chloride structure with eightfold coordination. That is the structure found in practice.

Self-test 3.9 Predict the likely structure of UO_2 using $r(U^{4+}) = 114$ pm

---

The ionic radii used in these calculations are those obtained by consideration of structures under normal conditions. At high pressures, different structures may be preferred, especially those with higher coordination numbers and greater density. Thus many simple compounds transform between the simple (4,4)-, (6,6)-, and (8,8)-coordination structures under pressure. Examples include most of the lighter alkali metal halides that undergo phase transitions from a rock-salt structure (6,6) to a caesium-chloride structure (8,8) under 5 kbar (the rubidium halides) or 10–20 kbar (the sodium and potassium halides). The ability to predict the structures of compounds under pressure is important for understanding the behaviour of ionic compounds under such conditions. Calcium oxide, for instance, is predicted to transform from the rock-salt to the caesium-chloride structure at around 600 kbar, the pressure in the Earth's lower mantle.

(c) Structure maps

Key point: A structure map is a representation of the variation in crystal structure with the character of the bonding.

Even though the use of radius ratios is not totally reliable, it is still possible to rationalize structures by collecting enough information empirically and looking for patterns. This approach has motivated the compilation of 'structure maps'. A **structure map** is an empirically compiled map that depicts the dependence of crystal structure on the electronegativity difference between the elements present and the average principal quantum number of the valence shells of the two atoms.[3] As such, a structure map can be regarded as an extension of the ideas introduced in Chapter 2 in relation to Ketelaar's triangle. As we have seen, binary ionic salts are formed for large differences in electronegativity $\Delta\chi$, but as this difference is reduced, polarized ionic salts and more covalently bonded networks become preferred. Now we can focus on this region of the triangle and explore how small changes in electronegativity and polarizability affect the choice of ion arrangement.

The ionic character of a bond increases with $\Delta\chi$, so moving from left to right along the horizontal axis of a structure map correlates with an increase in ionic character in the bonding. The principal quantum number is an indication of the radius of an ion, so moving up the vertical axis corresponds to an increase in the average radius of the ions.

---

[3] Structure maps were introduced by E. Mooser and W.B. Pearson, *Acta Crystallogr.*, 1959, **12**, 1015.

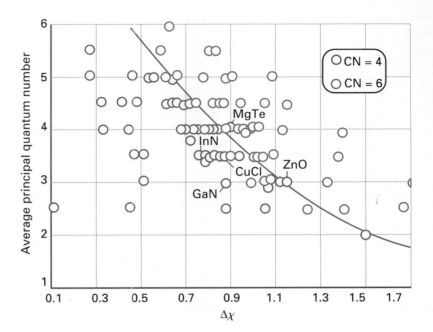

Fig. 3.43 A structure map for compounds of formula MX. A point is defined by the electronegativity difference ($\Delta\chi$) between M and X and their average principal quantum number n. The location in the map indicates the coordination number expected for that pair of properties. (Based on E. Mooser and W.B. Pearson, *Acta Crystallogr.*, 1959, **12**, 1015.)

Because atomic energy levels also become closer as the atom expands, the polarizability of the atom increases too (Section 1.9e). Consequently, the vertical axis of a structure map corresponds to increasing size and polarizability of the bonded atoms. Figure 3.43 is an example of a structure map for MX compounds. We see that the structures we have been discussing for MX compounds fall in distinct regions of the map. Elements with large $\Delta\chi$ have (6,6)-coordination, such as is found in the rock-salt structure; elements with small $\Delta\chi$ (and hence where there is the expectation of covalence) have a lower coordination number. In terms of a structure map representation, GaN is in a more covalent region of Fig. 3.43 than ZnO because $\Delta\chi$ is appreciably smaller.

Example 3.10 Using a structure map

What type of crystal structure should be expected for magnesium sulfide, MgS?

Answer The electronegativities of magnesium and sulfur are 1.3 and 2.6 respectively, so $\Delta\chi = 1.3$. The average principal quantum number is 3 (both elements are in Period 3). The point $\Delta\chi = 1.3$, $n = 3$ lies just in the 6-fold coordination region of the structure map in Fig. 3.43. This location is consistent with the observed rock-salt structure of MgS.

Self-test 3.10 Predict the coordination numbers of the cations and anions in rubidium chloride, RbCl.

The energetics of ionic bonding

A compound tends to adopt the crystal structure that corresponds to the lowest Gibbs energy. Therefore, if for the process

$$M^+(g) + X^-(g) \rightarrow MX(s)$$

the change in standard molar Gibbs energy, $\Delta G^\ominus$, is more negative for the formation of a structure A rather than B, then the transition from B to A is spontaneous under the prevailing conditions, and we can expect the solid to be found with structure A.

The process of solid formation from the gas of ions is so exothermic that at and near room temperature the contribution of the entropy to the change in Gibbs energy (as in $\Delta G^\ominus = \Delta H^\ominus - T\Delta S^\ominus$) may be neglected; this neglect is rigorously true at $T = 0$. Hence,

Table 3.7 Lattice enthalpies of some simple inorganic solids

Compound	Structure type	$\Delta H_{L}^{exp}/$ $(kJ\,mol^{-1})$	Compound	Structure type	$\Delta H_{L}^{exp}/$ $(kJ\,mol^{-1})$
LiF	Rock salt	1030	SrCl$_2$	Fluorite	2125
LiI	Rock salt	757	LiH	Rock salt	858
NaF	Rock salt	923	NaH	Rock salt	782
NaCl	Rock salt	786	KH	Rock salt	699
NaBr	Rock salt	747	RbH	Rock salt	674
NaI	Rock salt	704	CsH	Rock salt	648
KCl	Rock salt	719	BeO	Wurtzite	4293
KI	Rock salt	659	MgO	Rock salt	3795
CsF	Rock salt	744	CaO	Rock salt	3414
CsCl	Caesium chloride	657	SrO	Rock salt	3217
CsBr	Caesium chloride	632	BaO	Rocksalt	3029
CsI	Caesium chloride	600	Li$_2$O	Antifluorite	2799
MgF$_2$	Rutile	2922	TiO$_2$	Rutile	12150
CaF$_2$	Fluorite	2597	CeO$_2$	Fluorite	9627

discussions of the thermodynamic properties of solids normally focus, initially at least, on changes in enthalpy. That being so, we look for the structure that is formed most exothermically and identify it as the thermodynamically most stable form. Some typical values of lattice enthalpies are given in Table 3.7 for a number of simple ionic compounds.

3.11 Lattice enthalpy and the Born–Haber cycle

Key points: Lattice enthalpies are determined from enthalpy data by using a Born–Haber cycle; the most stable crystal structure of the compound is commonly the structure with the greatest lattice enthalpy under the prevailing conditions.

The **lattice enthalpy**, $\Delta H_{L}^{\ominus}$, is the standard molar enthalpy change accompanying the formation of a gas of ions from the solid:

$$MX(s) \rightarrow M^{+}(g) + X^{-}(g) \qquad \Delta H_{L}^{\ominus}$$

Because lattice disruption is always endothermic, lattice enthalpies are always positive and their positive signs are normally omitted from their numerical values. As remarked above, if entropy considerations are neglected, the most stable crystal structure of the compound is the structure with the greatest lattice enthalpy under the prevailing conditions.

Lattice enthalpies are determined from enthalpy data by using a **Born–Haber cycle**, a closed path of steps that includes lattice formation as one stage, such as that shown in Fig. 3.44. The standard enthalpy of decomposition of a compound into its elements in their reference states (their most stable states under the prevailing conditions) is the negative of its standard enthalpy of formation, $\Delta_{f}H^{\ominus}$:

$$M(s) + X(s) \rightarrow MX(s) \qquad \Delta_{f}H^{\ominus}$$
$$MX(s) \rightarrow M(s) + X(s) \qquad -\Delta_{f}H^{\ominus}$$

Likewise, the standard enthalpy of lattice formation from the gaseous ions is the negative of the lattice enthalpy:

$$MX(s) \rightarrow M^{+}(g) + X^{-}(g) \qquad \Delta H_{L}^{\ominus}$$
$$M^{+}(g) + X^{-}(g) \rightarrow MX(s) \qquad -\Delta H_{L}^{\ominus}$$

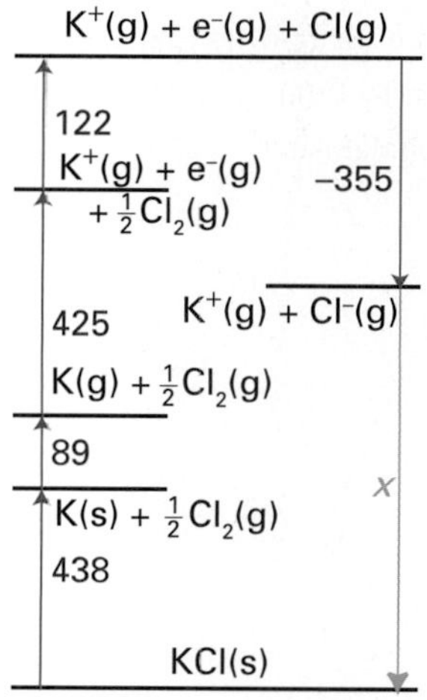

$K^{+}(g) + e^{-}(g) + Cl(g)$

122

$K^{+}(g) + e^{-}(g)$ $+ \frac{1}{2}Cl_2(g)$ −355

425    $K^{+}(g) + Cl^{-}(g)$

$K(g) + \frac{1}{2}Cl_2(g)$

89

$K(s) + \frac{1}{2}Cl_2(g)$ x

438

$KCl(s)$

Fig. 3.44 The Born–Haber cycle for KCl. The lattice enthalpy is equal to $-x$. All numerical values are in kilojoules per mole.

For a solid element, the standard enthalpy of atomization, $\Delta_{atom}H^{\ominus}$, is the standard enthalpy of sublimation, as in the process

$$K(s) \rightarrow K(g) \qquad \Delta_{sub}H^{\ominus} = +89\,kJ\,mol^{-1}$$

For a gaseous element, the standard enthalpy of atomization is the standard enthalpy of dissociation, as in

$$Cl_2(g) \rightarrow 2\,Cl(g) \qquad \Delta_{dis}H^{\ominus} = +244\,kJ\,mol^{-1}$$

The standard enthalpy for the formation of ions from their neutral atoms is the enthalpy of ionization (for the formation of cations) and the electron-gain enthalpy (for anions). Two examples are

$$K(g) \rightarrow K^+(g) + e^-(g) \qquad \Delta_{ion}H^{\ominus} = +425\,kJ\,mol^{-1}$$
$$Cl(g) + e^-(g) \rightarrow Cl^-(g) \qquad \Delta_{eg}H^{\ominus} = -355\,kJ\,mol^{-1}$$

The value of the lattice enthalpy—the only unknown in a well-chosen cycle—is found from the requirement that the sum of the enthalpy changes round a complete cycle is zero (because enthalpy is a state property).[4] The value of the lattice enthalpy obtained from a Born–Haber cycle depends on the accuracy of all the measurements being combined, and as a result there can be significant variations, typically $\pm 10\,kJ\,mol^{-1}$, in tabulated values.

	$\Delta H/(kJ\,mol^{-1})$
Sublimation of K(s)	+89
Ionization of K(g)	+425
Dissociation of $Cl_2(g)$	+244
Electron gain by Cl(g)	−355
Formation of KCl(s)	−438

	$\Delta H/(kJ\,mol^{-1})$
Sublimation of Mg(s)	+148
Ionization of Mg(g) to $Mg^{2+}(g)$	+2187
Vaporization of $Br_2(l)$	+31
Dissociation of $Br_2(g)$	+193
Electron gain by Br(g)	−331
Formation of $MgBr_2(s)$	−524

Example 3.11 Using a Born–Haber cycle to determine a lattice enthalpy

Calculate the lattice enthalpy of KCl(s) using a Born–Haber cycle and the information in the margin.

Answer The required cycle is shown in Fig. 3.44. The sum of the enthalpy changes around the cycle is zero, so

$$\Delta H_L^{\ominus} = (+438 + 89 + 425 + 122 - 355)\,kJ\,mol^{-1} = 719\,kJ\,mol^{-1}$$

Note that the calculation becomes more obvious by drawing an energy level diagram showing the signs of the various steps of the cycle; all lattice enthalpies are positive. Also as only one chlorine atom from $Cl_2(g)$ is required to produce KCl, half the dissociation energy of Cl_2 (in $kJ\,mol^{-1}$) is used in the calculation.

Self-test 3.11 Calculate the lattice enthalpy of magnesium bromide from the data shown in the margin.

3.12 Calculation of lattice enthalpies

Once the lattice enthalpy is known, it can be used to judge the character of the bonding in the solid. If the value calculated on the assumption that the lattice consists of ions interacting Coulombically is in good agreement with the measured value, then it may be appropriate to adopt a largely ionic model of the compound. A discrepancy indicates a degree of covalence. As mentioned earlier, it is important to remember that numerical coincidences can be misleading in this assessment.

(a) The Born–Mayer equation

Key points: The Born–Mayer equation is used to estimate lattice enthalpy for an ionic lattice. The Madelung constant reflects the effect of the geometry of the lattice on the strength of the net Coulombic interaction.

---

[4] Note that when the lattice enthalpy is known from calculation, a Born–Haber cycle may be used to determine the value of another elusive quantity, the electron-gain enthalpy (and hence the electron affinity).

To calculate the lattice enthalpy of a supposedly ionic solid we need to take into account several contributions to its enthalpy, including the attractions and repulsions between the ions. This calculation yields the **Born–Mayer equation** for the lattice enthalpy at $T=0$:

$$\Delta H_L = \frac{N_A |z_A z_B| e^2}{4\pi\epsilon_0 d_0} \left(1 - \frac{d}{d_0}\right)\mathcal{A} \tag{3.2}$$

where $d_0 = r_+ + r_-$ is the distance between centres of neighbouring cations and anions, and hence a measure of the 'scale' of the unit cell (for the derivation, see *Further information* 3.1). In this expression N_A is Avogadro's constant, z_A and z_B the charge numbers of the cation and anion, e the fundamental charge, ϵ_0 the vacuum permittivity, and d a constant (typically 34.5 pm) used to represent the repulsion between ions at short range. The quantity $\mathcal{A}$ is called the **Madelung constant**, and depends on the structure (specifically, on the relative distribution of ions, Table 3.8). The Born–Mayer equation in fact gives the lattice energy as distinct from the lattice enthalpy, but the two are identical at $T=0$ and the difference may be disregarded in practice at normal temperatures.

Table 3.8 Madelung constants

Structural type	$\mathcal{A}$
Caesium chloride	1.763
Fluorite	2.519
Rock salt	1.748
Rutile	2.408
Sphalerite	1.638
Wurtzite	1.641

Example 3.12 Using the Born–Mayer equation

Estimate the lattice enthalpy of sodium chloride.

Answer For sodium chloride $z_{Na}^+ = +1$, $z_{Cl}^- = -1$, from Table 3.8 $\mathcal{A} = 1.748$, and from Table 1.5 $d_C = r_{Na^+} + r_{Cl^-} = 282$ pm; hence (using fundamental constants from inside the back cover):

$$\Delta H_L = \frac{(6.023 \times 10^{23}\ \text{mol}^{-1}) \times |(+1)(-1)| \times (1.602 \times 10^{-19}\text{C})^2}{4\pi \times (8.854 \times 10^{-12}\ \text{J}^{-1}\text{C}^2\text{m}^{-1}) \times (2.82 \times 10^{-10}\text{m})} \times \left(1 - \frac{34.5\text{pm}}{282\text{pm}}\right) \times 1.748$$

$$= 7.56 \times 10^5\ \text{J mol}^{-1}$$

or 757 kJ mol^{-1}. This value compares reasonably well with the experimental value from the Born–Haber cycle, 788 kJ mol^{-1}.

Self-test 3.12 Estimate the lattice enthalpy of CsCl using the experimental $d_0 = 356$ pm.

The form of the Born–Mayer equation for lattice enthalpies allows us to account for their dependence on the charges and radii of the ions in the solid. Thus, the heart of the equation is

$$\Delta H_L \propto \frac{|z_A z_B|}{d_0}$$

Therefore, a large value of d_0 results in a low lattice enthalpy, whereas high ionic charges result in a high lattice enthalpy. This dependence is seen in some of the values given in Table 3.7. For the alkali metal halides, the lattice enthalpies decrease from LiF to LiI and from LiF to CsF as the halide and alkali metal ion radii increase, respectively. We also note that the lattice enthalpy of MgO ($|z_+z_-| = 4$) is almost four times that of NaCl ($|z_+z_-| = 1$) due to the increased charges on the ions for a similar value of d_0, noting that the Madelung constant is the same.

The Madelung constant typically increases with coordination number. For instance, $\mathcal{A} = 1.748$ for the (6,6) rock-salt structure but $\mathcal{A} = 1.763$ for the (8,8) caesium-chloride structure and 1.638 for the (4,4) sphalerite structure. This dependence reflects the fact that a large contribution comes from nearest neighbours, and such neighbours are more numerous when the coordination number is large. However, a high coordination number does not necessarily mean that the interactions are stronger in the caesium-chloride structure because the potential energy also depends on the scale of the lattice. Thus, d_0 may be so large in lattices with ions big enough to adopt eightfold-coordination that the separation of the ions reverses the effect of the small increase in the Madelung constant and results in a smaller lattice enthalpy.

(b) Other contributions to lattice enthalpies

Key point: Non-Coulombic contributions to the lattice enthalpy include van der Waals interactions, particularly the dispersion interaction.

Another contribution to the lattice enthalpy is the **van der Waals interaction** between the ions and molecules, the weak intermolecular interactions that are responsible for the formation of condensed phases of electrically neutral species. An important and sometimes dominant contribution of this kind is the **dispersion interaction** (the 'London interaction'). The dispersion interaction arises from the transient fluctuations in electron density (and, consequently, instantaneous electric dipole moment) on one molecule driving a fluctuation in electron density (and dipole moment) on a neighbouring molecule, and the attractive interaction between these two instantaneous electric dipoles. The molar potential energy of this interaction, V, is expected to vary as

$$V = -\frac{N_A C}{d_0^6} \tag{3.3}$$

The constant C depends on the substance. For ions of low polarizability, this contribution is only about 1 per cent of the Coulombic contribution and is ignored in elementary lattice enthalpy calculations of ionic solids. However, for highly polarizable ions, such as I^- and Tl^+, such terms can make significant contributions of several per cent.

3.13 Comparison of experimental and theoretical values of lattice enthalpy

Key points: For compounds formed from elements with $\Delta\chi > 2$, the ionic model is generally valid and lattice enthalpy values derived using the Born–Mayer equation and Born–Haber cycles are similar. For structures formed with small electronegativity differences and polarizable ions there may be additional, nonionic contributions to the bonding.

The agreement between the experimental lattice enthalpy and the value calculated using the ionic model of the solid (in practice, from the Born–Mayer equation) provides a measure of the extent to which the solid is ionic. Table 3.9 lists some calculated and measured lattice enthalpies. The ionic model is reasonably valid if $\Delta\chi > 2$, but the bonding becomes increasingly covalent if $\Delta\chi < 2$. However, it should be remembered that the electronegativity criterion ignores the role of polarizability of the ions. Thus, the alkali metal halides give fairly good agreement with the ionic model, the best with the least polarizable halide ions (F^-) formed from the highly electronegative F atom, and the worst with the highly polarizable halide ions I^- formed from the less electronegative I atom. This trend is also seen in the lattice enthalpy data for the silver halides in Table 3.9. The discrepancy between experimental and theoretical values is largest for the iodide, which indicates major deficiencies in the ionic model for this compound. Overall the agreement is much poorer with silver than with lithium as the electronegativity of silver ($\chi = 1.93$) is much higher than that of lithium ($\chi = 0.98$) and significant covalency in the bonding would be expected.

It is not always clear whether it is the electronegativity of the atoms or the polarizability of the resultant ions that should be used as a criterion. The worst agreement with the ionic model is for polarizable-cation–polarizable-anion combinations that are substantially covalent. Here again, though, the difference between the electronegativities of the parent elements is small and it is not clear whether electronegativity or polarizability provides the better criterion.

Table 3.9 Comparison of experimental and theoretical lattice enthalpies for rock-salt structures

	$\Delta H_L^{calc}/$ (kJ mol^{-1})	$\Delta H_L^{exp}/$ (kJ mol^{-1})	$(\Delta H_L^{exp} - \Delta H_L^{calc})/$ (kJ mol^{-1})
LiF	1029	1030	6
LiCl	834	853	19
LiBr	788	807	19
LiI	730	757	27
AgF	920	953	33
AgCl	832	903	71
AgBr	815	895	80
AgI	777	882	105

Example 3.13 Using the Born–Mayer equation to decide the theoretical stability of unknown compounds

Decide whether solid ArCl is likely to exist.

Answer Consideration of a Born–Haber cycle for the synthesis of ArCl would show two unknowns, the enthalpy of formation of ArCl and its lattice enthalpy. We can estimate the lattice enthalpy of a purely ionic ArCl by using the Born–Mayer equation assuming the radius of Ar^+ to be midway between that of Na^+ and K^+. That is, the lattice enthalpy is somewhere between the values for NaCl and KCl at about 745 kJ mol^{-1}. So, taking values of the enthalpy of atomization of Ar and Cl_2 as 0 and 122 kJ mol^{-1}, respectively, the ionization enthalpy of Ar as 1524 kJ mol^{-1}, and the electron affinity of Cl as 356 kJ mol^{-1} gives $\Delta_f H^{\ominus}(ArCl, s) = +545$ kJ mol^{-1}. That is, the compound is predicted to be very unstable with respect to its elements, mainly because the large ionization enthalpy of Ar is not compensated by the lattice enthalpy.

Self-test 3.13 Predict whether $CsCl_2$ with the fluorite structure is likely to exist.

Calculations like that in Example 3.13 were used to predict the stability of the first noble gas compounds. The ionic compound $(O_2)^+PtF_6^-$ had been obtained from the reaction of oxygen with PtF_6. Consideration of the ionization energies of O_2 (1176 kJ mol^{-1}) and Xe (1169 kJ mol^{-1}) showed them to be almost identical and the sizes of Xe^+ and O_2^{2+} would be expected to be similar, implying similar lattice enthalpies for their compounds. Hence once O_2 had been found to react with platinum hexafluoride, it could be predicted that Xe should too, as indeed it does, to give an ionic compound which is believed to contain $XeF^+Pt_6^-$. Similar calculations may be used to predict the stability, or otherwise, of a wide variety of compounds—for example the stability of alkaline earth monohalides, such as MgCl. Calculations based on Born–Mayer lattice enthalpies and Born–Haber cycles show that such a compound would not be expected to be stable but to disproportionate into Mg and $MgCl_2$. Note that calculations of this type only provide an estimate of the enthalpies of formation of ionic compounds and, hence, some idea of the thermodynamic stability of a compound. It may still be possible to isolate a thermodynamically unstable compound if its decomposition is very slow.

3.14 The Kapustinskii equation

Key point: The Kapustinskii equation is used to estimate lattice enthalpies of ionic compounds and to give a measure of the thermochemical radii of the constituent ions.

A.F. Kapustinskii observed that, if the Madelung constants for a number of structures are divided by the number of ions per formula unit, n, then approximately the same value is obtained for them all. He also noted that the value so obtained increases with the coordination number. Therefore, because ionic radius also increases with coordination number, the variation in A/nd_0 from one structure to another can be expected to be fairly small. This observation led Kapustinskii to propose that there exists a hypothetical rock-salt structure that is energetically equivalent to the true structure of any ionic solid and therefore that the lattice enthalpy can be calculated by using the rock-salt Madelung constant and the appropriate ionic radii for (6,6)-coordination. The resulting expression is called the **Kapustinskii equation**:

$$\Delta H_L = \frac{n|z_A z_B|}{d_0} \left(1 - \frac{d}{d_0}\right) \mathcal{K} \tag{3.4}$$

In this equation $\mathcal{K} = 1.21 \times 10^5$ kJ pm mol^{-1}

The Kapustinskii equation can be used to ascribe numerical values to the 'radii' of nonspherical molecular ions, as their values can be adjusted until the calculated value of the lattice enthalpy matches that obtained experimentally from the Born–Haber cycle. The self-consistent parameters obtained in this way are called **thermochemical radii** (Table 3.10). They may be used to estimate lattice enthalpies, and hence enthalpies of formation, of a wide range of compounds without needing to know the structure, assuming that the bonds are essentially ionic.

Example 3.14 Using the Kapustinskii equation

Estimate the lattice enthalpy of potassium nitrate, KNO_3.

Answer To use the Kapustinskii equation we need the number of ions per formula unit ($n = 2$), their charge numbers $z_{K^+} = +1$ and $z_{NO_3^-} = -1$, and the sum of their thermochemical radii 138 pm + 189 pm = 327 pm. Then, with $d = 34.5$ pm,

$$\Delta H_L = \frac{2 \times |(+1)(-1)|}{327 \text{pm}} \times \left(1 - \frac{34.5 \text{pm}}{327 \text{pm}}\right) \times (1.21 \times 10^5 \text{ kJ pm mol}^{-1})$$

$$= 622 \text{ kJ mol}^{-1}$$

Self-test 3.14 Estimate the lattice enthalpy of calcium sulfate, $CaSO_4$.

3.15 Consequences of lattice enthalpies

The Born–Mayer equation shows that, for a given lattice type (a given value of A), the lattice enthalpy increases with increasing ion charge numbers (as $|z_A z_B|$). The lattice enthalpy also increases as the ions come closer together and the scale of the lattice decreases. Energies that vary as the **electrostatic parameter**, ξ (xi),

$$\xi = \frac{|z_A z_B|}{d_0} \tag{3.5}$$

(which is often written more succinctly as $\xi = z^2/d$) are widely adopted in inorganic chemistry as indicative that an ionic model is appropriate. In this section we consider three consequences of lattice enthalpy and its relation to the electrostatic parameter.

(a) Thermal stabilities of ionic solids

Key point: Lattice enthalpies may be used to explain the chemical properties of many ionic solids including their thermal decomposition.

The particular aspect we consider here is the temperature needed to bring about thermal decomposition of carbonates (although the arguments can easily be extended to many inorganic solids):

$$MCO_3(s) \rightarrow MO(s) + CO_2(g)$$

Table 3.10 The thermochemical radii of ions, r/pm

Main-group elements

BeF_4^{2-}	BF_4^-	CO_3^{2-}	NO_3^-	OH^-	
245	228	185	189	140	
		CN^-	NO_2^-	O_2^{2-}	
		182	155	180	
			PO_4^{3-}	SO_4^{2-}	ClO_4^-
			238	230	236
			AsO_4^{3-}	SeO_4^{2-}	
			248	243	
			SbO_4^{3-}	TeO_4^{2-}	IO_4^-
			260	254	249
					IO_3^-
					182

Complex ions   *d-Metal oxoanions*

$[TiCl_6]^{2-}$	$[IrCl_6]^{2-}$	$[SiF_6]^{2-}$	$[GeCl_6]^{2-}$		CrO_4^{2-}	MnO_4^-
248	254	194	243		230	240
$[TiBr_6]^{2-}$	$[PtCl_6]^{2-}$	$[GeF_6]^{2-}$	$[SnCl_6]^{2-}$		MoO_4^{2-}	
261	259	201	247		254	
$[ZrCl_6]^{2-}$			$[PbCl_6]^{2-}$			
247			248			

Source: A.F. Kaputinskii, *Q. Rev. Chem. Soc.*, 1956, **10**, 283.

Table 3.11 Decomposition data for carbonates*

	$MgCO_3$	$CaCO_3$	$SrCO_3$	$BaCO_3$
$\Delta G^{\ominus}$/(kJ mol^{-1})	+48.3	+130.4	+183.8	+218.1
$\Delta H^{\ominus}$/(kJ mol^{-1})	+100.6	+178.3	+234.6	+269.3
$\Delta S^{\ominus}$/(J K^{-1} mol^{-1})	+175.0	+160.6	+171.0	+172.1
θ_{decomp}/°C	300	840	1100	1300

* Data are for the reaction $MCO_3(s) \rightarrow MO(s) + CO_2(g)$ at 298 K. θ is the temperature required to reach $p(CO_2) = 1$ bar and has been estimated from the thermodynamic data at 293 K.

Magnesium carbonate, for instance, decomposes when heated to about 300°C, whereas calcium carbonate decomposes only if the temperature is raised to over 800°C. The decomposition temperatures of thermally unstable compounds (such as carbonates) increase with cation radius (Table 3.11). In general, large cations stabilize large anions (and vice versa).

The stabilizing influence of a large cation on an unstable anion can be explained in terms of trends in lattice enthalpies. First, we note that the decomposition temperatures of solid inorganic compounds can be discussed in terms of their Gibbs energies of decomposition into specified products. The standard Gibbs energy for the decomposition of a solid, $\Delta G^{\ominus} = \Delta H^{\ominus} - T\Delta S^{\ominus}$, becomes negative when the second term on the right exceeds the first, which is when the temperature exceeds

$$T = \frac{\Delta H^{\ominus}}{\Delta S^{\ominus}} \tag{3.6}$$

In many cases it is sufficient to consider only trends in the reaction enthalpy, as the reaction entropy is essentially independent of M because it is dominated by the formation

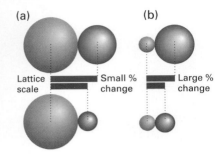

Fig. 3.45 A greatly exaggerated representation of the change in lattice parameter d for cations of different radii. (a) When the anion changes size (as when CO_3^{2-} decomposes into O^{2-} and CO_2, for instance) and the cation is large, the lattice parameter changes by a small amount. (b) If the cation is small, however, the relative change in lattice parameter is large and decomposition is thermodynamically more favourable.

of gaseous CO_2. The standard enthalpy of decomposition of the solid is

$$\Delta H^{\ominus} \approx \Delta_{decomp}H + \Delta H_L(MCO_3, s) - \Delta H_L(MO, s)$$

where $\Delta_{decomp}H$ is the enthalpy change for the gas-phase decomposition of CO_3^{2-}:

$$CO_3^{2-}(g) \rightarrow O^{2-}(g) + CO_2(g)$$

Because $\Delta_{decomp}H$ is large and positive, the overall reaction enthalpy is positive (decomposition is endothermic), but it is less strongly positive if the lattice enthalpy of the oxide is markedly greater than that of the carbonate because then $\Delta H_L(MCO_3, s) - \Delta H_L(MO, s)$ is negative. It follows that the decomposition temperature will be low for oxides that have relatively high lattice enthalpies compared with their parent carbonates. The compounds for which this is true are composed of small, highly charged cations, such as Mg^{2+}, which explains why a small cation increases the lattice enthalpy of an oxide more than that of a carbonate.

Figure 3.45 illustrates why a small cation has a more significant influence on the change in the lattice enthalpy as the cation size is varied. The *change* in separation is relatively small when the parent compound has a large cation initially. As the illustration shows in an exaggerated way, when the cation is very big, the change in size of the anion barely affects the scale of the lattice. Therefore, with a given unstable polyatomic anion, the lattice enthalpy difference is more significant and favourable to decomposition when the cation is small than when it is large.

The difference in lattice enthalpy between MO and MCO_3 is magnified by a larger charge on the cation as $\Delta H_L \propto z_A z_B/d$. As a result, thermal decomposition of a carbonate will occur at lower temperatures if it contains a higher charged cation. One consequence of this dependence on cation charge is that alkaline earth carbonates (M^{2+}) decompose at lower temperatures than the corresponding alkali metal carbonates (M^+).

Example 3.15 Assessing the dependence of stability on ionic radius

Present an argument to account for the fact that, when they burn in oxygen, lithium forms the oxide Li_2O but sodium forms the peroxide Na_2O_2.

Answer Because the small Li^+ ion results in Li_2O having a more favourable lattice enthalpy (in comparison with M_2O_2) than Na_2O, the decomposition reaction $M_2O_2(s) \rightarrow M_2O(s) + \frac{1}{2}O_2(g)$ is thermodynamically more favourable for Li_2O_2 than for Na_2O_2.

Self-test 3.15 Predict the order of decomposition temperatures of alkaline earth metal sulfates in the reaction $MSO_4(s) \rightarrow MO(s) + SO_3(g)$.

The use of a large cation to stabilize a large anion that is otherwise susceptible to decomposition, forming a smaller anionic species, is widely used by inorganic chemists to prepare compounds that are otherwise thermodynamically unstable. For example, the interhalogen anions, such as ICl_4^-, are obtained by the oxidation of I^- ions by Cl_2 but are susceptible to the reverse reaction:

$$MI(s) + 2Cl_2(g) \rightleftharpoons MICl_4(s) \rightarrow MCl(s) + ICl(g) + Cl_2(g)$$

To disfavour the decomposition, a large cation is used to reduce the lattice enthalpy difference between $MICl_4$ and MCl/MI. The larger alkali metals such as potassium, rubidium, and caesium can be used in some cases but it is even better to use a really bulky alkylammonium ion, such as $N^tBu_4^+$.

(b) The stabilities of oxidation states

Key point: The relative stabilities of different oxidation states in solids can be predicted from lattice enthalpy considerations.

A similar argument can be used to account for the general observation that high oxidation states are stabilized by small anions. In particular, fluorine has a greater ability than the other halogens to stabilize the high oxidation states of metals. Thus, the only known halides of Ag(II), Co(III), and Mn(IV) are the fluorides. Another sign of the decrease in stability of the heavier halides of metals in high oxidation states is that the iodides of Cu(II) and Fe(III) decompose on standing at room temperature (to CuI and FeI$_2$). Oxygen is also a very effective species for stabilizing the highest oxidation states of elements because of the high charge and small size of the O^{2-} ion.

To explain these observations consider the reaction

$$MX(s) + \tfrac{1}{2}X_2(g) \rightarrow MX_2(s)$$

where X is a halogen. The aim is to show why this reaction is most strongly spontaneous for X = F. If we ignore entropy contributions, we must show that the reaction is most exothermic for fluorine.

One contribution to the reaction enthalpy is the conversion of $\tfrac{1}{2}X_2$ to X$^-$. Despite fluorine having a lower electron affinity than chlorine, this step is more exothermic for X = F than for X = Cl because the bond enthalpy of F$_2$ is lower than that of Cl$_2$. The lattice enthalpies, however, play the major role. In the conversion of MX to MX$_2$, the charge number of the cation increases from +1 to +2, so the lattice enthalpy increases. As the radius of the anion increases, however, this difference in the two lattice enthalpies diminishes, and the exothermic contribution to the overall reaction decreases too. Hence, both the lattice enthalpy and the X$^-$ formation enthalpy lead to a less exothermic reaction as the halogen changes from F to I. Provided entropy factors are similar, which is plausible, we expect an increase in thermodynamic stability of MX relative to MX$_2$ on going from X = F to X = I down Group 17. Thus many iodides do not exist for metals in their higher oxidation states and compounds such as Cu^{2+}(I$^-$)$_2$, Tl^{3+}(I$^-$)$_3$ and VI$_5$ are unknown, whereas the corresponding fluorides CuF$_2$, TlF$_3$, and VF$_5$ are easily obtained. In effect, the high-oxidation-state metal oxidizes I$^-$ ions to I$_2$, leading to formation of a lower metal oxidation state such as Cu(I), Tl(I), and V(III) in the iodides of these metals.

(c) Solubility

Key point: The solubilities of salts can be rationalized by considering lattice and hydration enthalpies.

Lattice enthalpies play a role in solubilities, as the dissolution involves breaking up the lattice, but the trend is much more difficult to analyse than for decomposition reactions. One rule that is reasonably well obeyed is that *compounds that contain ions with widely different radii are soluble in water.* Conversely, the least water-soluble salts are those of ions with similar radii. That is, in general, *difference in size favours solubility in water.* It is found empirically that an ionic compound MX tends to be very soluble when the radius of M$^+$ is smaller than that of X$^-$ by about 80 pm.

Two familiar series of compounds illustrate these trends. In gravimetric analysis, Ba^{2+} is used to precipitate SO$_4^{2-}$, and the solubilities of the Group 2 sulfates decrease from MgSO$_4$ to BaSO$_4$. By contrast, the solubility of the Group 2 hydroxides increases down the group: Mg(OH)$_2$ is the sparingly soluble 'milk of magnesia' but Ba(OH)$_2$ can be used as a soluble hydroxide for preparation of solutions of OH$^-$. The first case shows that a large anion requires a large cation for precipitation. The second case shows that a small anion requires a small cation for precipitation.

Before attempting to rationalize the observations, we should note that the solubility of an ionic compound depends on the Gibbs energy for the process

$$MX(s) \rightarrow M^+(aq) + X^-(aq)$$

In the process, the interactions responsible for the lattice enthalpy of MX are replaced by hydration (and by solvation in general) of the ions. However, the exact balance of enthalpy and entropy effects is delicate and difficult to assess, particularly because the entropy change also depends on the degree of order of the solvent molecules that is brought about by the presence of the dissolved solute. The data in Fig. 3.46 suggest that enthalpy considerations are important in some cases at least, as the graph shows that there is a correlation between the enthalpy of solution of a salt and the difference in hydration enthalpies of the two ions. If the cation has a larger hydration enthalpy than its

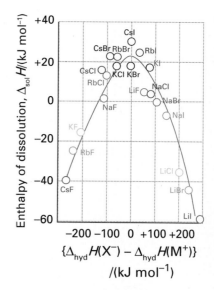

Fig. 3.46 The correlation between enthalpies of solution of halides and the differences between the hydration enthalpies of the ions. Dissolution is most exothermic when the difference is large.

anion partner (reflecting the difference in their sizes) or vice versa, then the dissolution of the salt is exothermic (reflecting the favourable solubility equilibrium).

The variation in enthalpy can be explained using the ionic model. The lattice enthalpy is inversely proportional to the distance between the centres of the ions:

$$\Delta H_L \propto \frac{1}{r_+ + r_-}$$

However, the hydration enthalpy, with each ion being hydrated individually, is the sum of individual ion contributions:

$$\Delta_{hyd} H \propto \frac{1}{r_+} + \frac{1}{r_-}$$

If the radius of one ion is small, the term in the hydration enthalpy for that ion will be large. However, in the expression for the lattice enthalpy one small ion cannot make the denominator of the expression small by itself. Thus, one small ion can result in a large hydration enthalpy but not necessarily lead to a high lattice enthalpy, so ion size asymmetry can result in exothermic dissolution. If both ions are small, then both the lattice enthalpy and the hydration enthalpy may be large, and dissolution might not be very exothermic.

Example 3.16 Accounting for trends in the solubility of s-block compounds

What is the trend in the solubilities of the Group 2 metal carbonates?

Answer The CO_3^{2-} anion has a large radius and is of the same charge level (2) as the cations M^{2+} of the Group 2 elements. The least soluble carbonate of the group is predicted to be that of the largest cation, Ra^{2+}. The most soluble is expected to be the carbonate of the smallest cation, Mg^{2+} (beryllium has too much covalent character in its bonding for it to be included in this analysis). Although magnesium carbonate is more soluble than radium carbonate, it is still only sparingly soluble: its solubility constant (its solubility product, K_{sp}) is only 3×10^{-8}.

Self-test 3.16 Which can be expected to be more soluble in water, $NaClO_4$ or $KClO_4$?

The electronic structures of solids

The previous sections have introduced concepts associated with the structures and energetics of ionic solids in which it was necessary to consider almost infinite arrays of ions and the interactions between them. Similarly, an understanding of the electronic structures of solids, and the derived properties such as electric conductivity, magnetism, and many optical effects, needs to consider the interactions of electrons with each other and extended arrays of ions. One simple approach is to regard a solid as a single huge molecule and to extend the ideas of molecular orbital theory introduced in Chapter 2 to very large numbers of orbitals. Similar concepts are used in later chapters to understand other key properties of large three-dimensional arrays of electronically interacting centres such as ferromagnetism, superconductivity, and the colours of solids.

3.16 The conductivities of inorganic solids

Key points: A metallic conductor is a substance with an electric conductivity that decreases with increasing temperature; a semiconductor is a substance with an electric conductivity that increases with increasing temperature.

The molecular orbital theory of small molecules can be extended to account for the properties of solids, which are aggregations of an almost infinite number of atoms. This approach is strikingly successful for the description of metals; it can be used to explain their characteristic lustre, their good electrical and thermal conductivity, and their malleability. All these properties arise from the ability of the atoms to contribute electrons to a common 'sea'. The lustre and electrical conductivities stem from the mobility of these electrons in response to either the oscillating electric field of an incident ray of light or to a potential difference. The high thermal conductivity is also a consequence of electron mobility, because an electron can collide with a vibrating atom, pick up its energy, and transfer it to another atom elsewhere in the solid. The ease with which metals can be mechanically deformed is another aspect of electron mobility, because the electron sea can quickly readjust to a deformation of the solid and continue to bind the atoms together.

Electronic conduction is also a characteristic of semiconductors. The criterion for distinguishing between a metal and a semiconductor is the temperature dependence of the electric conductivity (Fig. 3.47):

A **metallic conductor** is a substance with an electric conductivity that *decreases* with increasing temperature.

A **semiconductor** is a substance with an electric conductivity that *increases* with increasing temperature.

It is also generally the case (but not the criterion for distinguishing them) that the conductivities of metals at room temperature are higher than those of semiconductors. Typical values are given in Fig. 3.47. A solid **insulator** is a substance with a very low electrical conductivity. However, when that conductivity can be measured, it is found to increase with temperature, like that of a semiconductor. For some purposes, therefore, it is possible to disregard the classification 'insulator' and to treat all solids as either metals or semiconductors. **Superconductors** are a special class of materials that have zero electrical resistance below a critical temperature.

3.17 Bands formed from overlapping atomic orbitals

The central idea underlying the description of the electronic structure of solids is that the valence electrons supplied by the atoms spread through the entire structure. This concept is expressed more formally by making a simple extension of MO theory in which the solid is treated like an indefinitely large molecule. In solid-state physics, this approach is called the **tight-binding approximation**. The description in terms of delocalized electrons can also be used to describe nonmetallic solids. We therefore begin by showing how metals are described in terms of molecular orbitals. Then we go on to show that the same principles can be applied, but with a different outcome, to ionic and molecular solids.

(a) Band formation by orbital overlap

Key point: The overlap of atomic orbitals in solids gives rise to bands of energy levels separated by energy gaps.

The overlap of a large number of atomic orbitals in a solid leads to a large number of molecular orbitals that are closely spaced in energy and so form an almost continuous **band** of energy levels (Fig. 3.48). Bands are separated by **band gaps**, which are values of the energy for which there is no molecular orbital.

The formation of bands can be understood by considering a line of atoms, and supposing that each atom has an *s* orbital that overlaps the *s* orbitals on its immediate neighbours (Fig. 3.49). When the line consists of only two atoms, there is a bonding and an antibonding molecular orbital. When a third atom joins them, there are three orbitals. The central orbital of the set is nonbonding and the outer two are at low energy and high energy, respectively. As more atoms are added, each one contributes an atomic orbital, and hence one more molecular orbital is formed. When there are *N* atoms in the line, there are *N* molecular orbitals. The orbital of lowest energy has no nodes between neighbouring atoms. The orbital of highest energy has a node between every pair of neighbours. The remaining orbitals have successively 1, 2, . . . internuclear nodes and a corresponding range of energies between the two extremes.

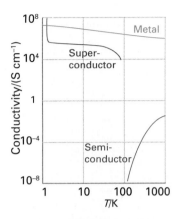

Fig. 3.47 The variation of the electrical conductivity of a substance with temperature is the basis of the classification of the substance as a metallic conductor, a semiconductor, or a superconductor.

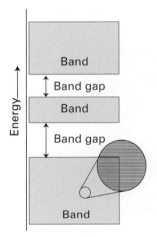

Fig. 3.48 The electronic structure of a solid is characterized by a series of bands of orbitals separated by gaps at energies where orbitals do not occur.

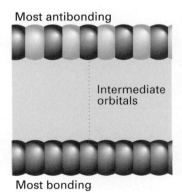

Fig. 3.49 A band can be thought of as formed by bringing up atoms successively to form a line of atoms. *N* atomic orbitals give rise to *N* molecular orbitals.

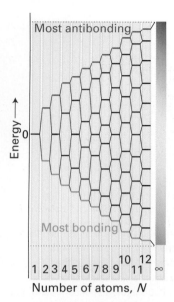

Fig. 3.50 The energies of the orbitals that are formed when *N* atoms are brought up to form a one-dimensional array.

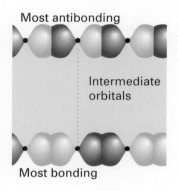

Fig. 3.51 An example of a *p* band in a one-dimensional solid.

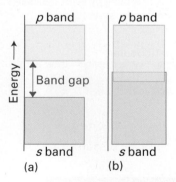

Fig. 3.52 (a) The *s* and *p* bands of a solid and the gap between them. Whether there is in fact a gap depends on the separation of the *s* and *p* orbitals of the atoms and the strength of the interaction between them in the solid. (b) If the interaction is strong, the bands are wide and may overlap.

The total width of the band, which remains finite even as *N* approaches infinity (as shown in Fig. 3.50), depends on the strength of the interaction between neighbouring atoms. The greater the strength of interaction (in broad terms, the greater the degree of overlap between neighbours), the greater the energy separation of the non-node orbital and the all-node orbital. However, whatever the number of atomic orbitals used to form the molecular orbitals, there is only a finite spread of orbital energies (as depicted in Fig. 3.50). It follows that the separation in energy between neighbouring orbitals must approach zero as *N* approaches infinity, otherwise the range of orbital energies could not be finite. That is, a band consists of a countable number but near-continuum of energy levels.

The band just described is built from *s* orbitals and is called an **s band**. If there are *p* orbitals available, a **p band** can be constructed from their overlap as shown in Fig. 3.51. Because *p* orbitals lie higher in energy than *s* orbitals of the same valence shell, there is often an energy gap between the *s* band and the *p* band (Fig. 3.52). However, if the bands span a wide range of energy and the atomic *s* and *p* energies are similar (as is often the case), then the two bands overlap. The **d band** is similarly constructed from the overlap of *d* orbitals. Note that the formation of bands is not restricted to simple atoms. Overlap of *s*, *p*, and *d* orbitals can occur in many solid compounds, particularly for the more spatially extensive valence orbitals, forming bands. Thus the description of energy levels in any solid material can be expressed in terms of bands.

Example 3.17 Identifying orbital overlap

Decide whether any *d* orbitals on titanium in TiO (with the rock-salt structure) can overlap to form a band.

Answer See Fig. 3.53, which shows one face of the rock-salt structure with the d_{xy} orbital drawn in on each the Ti atoms. The lobes of these orbitals point directly towards each other and will overlap to give a band. In a similar fashion the d_{zx} and d_{yz} orbitals overlap in the directions perpendicular to the *xz* and *yz* faces.

Self-test 3.17 Which *d* orbitals have the appropriate shape to overlap in a metal having a primitive structure?

(b) The Fermi level

Key point: The Fermi level is the highest occupied state in a solid at *T* = 0.

At *T* = 0, electrons occupy the individual molecular orbitals of the bands in accord with the building-up principle. If each atom supplies one *s* electron, then at *T* = 0 the lowest $\frac{1}{2}N$ orbitals are occupied. The highest occupied orbital at *T* = 0 is called the **Fermi level**; it lies near the centre of the band (Fig. 3.54). When the band is not completely full, the electrons close to the Fermi level can easily be promoted to nearby empty levels. As a result, they are mobile, and can move relatively freely through the solid and the substance is an electrical conductor.

The solid is in fact a *metallic* conductor. We have seen that the criterion of metallic conduction is the decrease of electrical conductivity with increasing temperature. This behaviour is the opposite of what we might expect if the conductivity were governed by thermal promotion of electrons above the Fermi level. The competing effect can be identified once we recognize that the ability of an electron to travel smoothly through the solid in a conduction band depends on the uniformity of the arrangement of the atoms. An atom vibrating vigorously at a site is equivalent to an impurity that disrupts the orderliness of the orbitals. This decrease in uniformity reduces the ability of the electron to travel from one edge of the solid to the other, so the conductivity of the solid is less than at *T* = 0. If we think of the electron as moving through the solid, then we would say that it was 'scattered' by the atomic vibration. This carrier scattering increases with increasing temperature as the lattice vibrations increase, and the increase accounts for the observed inverse temperature dependence of the conductivity of metals.

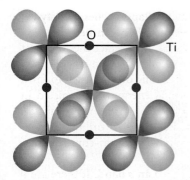

Fig. 3.53 One face of the TiO rock-salt structure showing how orbital overlap can occur for the d_{xy}, d_{yz}, and d_{zx} orbitals.

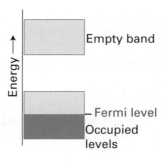

Fig. 3.54 If each of the N atoms supplies one s electron, then at $T=0$ the lower $\frac{1}{2}N$ orbitals are occupied and the Fermi level lies near the centre of the band.

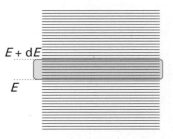

Fig. 3.55 The density of states is the number of energy levels in an infinitesimal range of energies between E and $E+\mathrm{d}E$.

(c) Densities of states

Key point: The density of states is not uniform across a band: in most cases, the states are densest close to the centre of the band.

The number of energy levels in an energy range divided by the width of the range is called the **density of states**, ρ (Fig. 3.55). The density of states is not uniform across a band because the energy levels are packed together more closely at some energies than at others. This variation is apparent even in one dimension, for—compared with its edges— the centre of the band is relatively sparse in orbitals (as can be seen in Fig. 3.50). In three dimensions, the variation of density of states is more like that shown in Fig. 3.56, with the greatest density of states near the centre of the band and the lowest density at the edges. The reason for this behaviour can be traced to the number of ways of producing a particular linear combination of atomic orbitals. There is only one way of forming a fully bonding molecular orbital (the lower edge of the band) and only one way of forming a fully antibonding orbital (the upper edge). However, there are many ways (in a three-dimensional array of atoms) of forming a molecular orbital with an energy corresponding to the interior of a band.

The density of states is zero in the band gap itself—there is no energy level in the gap. In certain special cases, however, a full band and an empty band might coincide in energy but with a zero density of states at their conjunction (Fig. 3.57). Solids with this band structure are called **semimetals**.[5] One important example is graphite, which is a semi- metal in directions parallel to the sheets of carbon atoms.

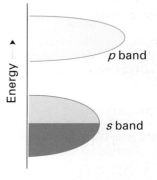

Fig. 3.56 Typical densities of states for two bands in a three-dimensional metal.

(d) Insulators

Key point: A solid insulator is a semiconductor with a large band gap.

A solid is an insulator if enough electrons are present to fill a band completely and there is a considerable energy gap before an empty orbital becomes available (Fig. 3.58). In a sodium chloride crystal, for instance, the $N\,\mathrm{Cl}^-$ ions are nearly in contact and their $3s$ and three $3p$ valence orbitals overlap to form a narrow band consisting of $4N$ levels. The Na^+ ions are also nearly in contact and also form a band. The electronegativity of chlorine is so much greater than that of sodium that the chlorine band lies well below the sodium band, and the band gap is about 7 eV. A total of $8N$ electrons are to be accommodated (seven from each Cl atom, one from each Na atom). These $8N$ electrons enter the lower chlorine band, fill it, and leave the sodium band empty. Because $kT\approx0.03$ eV at room temper- ature, very few electrons occupy the orbitals of the sodium band.

We normally think of an ionic or molecular solid as consisting of discrete ions or molecules. According to the picture we have just described, however, it seems that they should be regarded as having a band structure. The two pictures can be reconciled, because it is possible to show that a *full* band is equivalent to a sum of localized electron

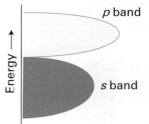

Fig. 3.57 The densities of states in a semimetal.

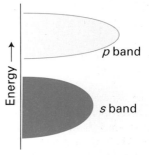

Fig. 3.58 The structure of a typical insulator: there is a significant gap between the filled and empty bands.

---

[5] This use of the term semimetal should be distinguished from its other use as a synonym for metalloid. In this text we avoid the latter usage.

Table 3.12 Some typical band gaps at 298 K

Material	E_g/eV
Carbon (diamond)	5.47
Silicon carbide	3.00
Silicon	1.11
Germaniun	0.66
Gallium arsenide	1.35
Indium arsenide	0.36

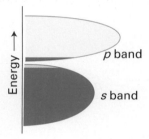

Fig. 3.59 In an intrinsic semiconductor, the band gap is so small that the Fermi distribution results in the population of some orbitals in the upper band.

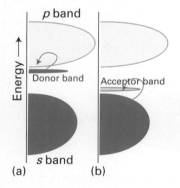

Fig. 3.60 The band structure in (a) an n-type semiconductor, (b) a p-type semiconductor.

densities. In sodium chloride, for example, a full band built from Cl orbitals is equivalent to a collection of discrete Cl^- ions.

3.18 Semiconduction

The characteristic physical property of a semiconductor is that its electrical conductivity increases with increasing temperature. At room temperature, the conductivities of semiconductors are typically intermediate between those of metals and insulators. The dividing line between insulators and semiconductors is a matter of the size of the band gap (Table 3.12); the conductivity itself is an unreliable criterion because, as the temperature is increased, a given substance may have in succession a low, intermediate, and high conductivity. The values of the band gap and conductivity that are taken as indicating semiconduction rather than insulation depend on the application being considered.

(a) Intrinsic semiconductors

Key point: The band gap in a semiconductor controls the temperature dependence of the conductivity through an Arrhenius-like expression.

In an **intrinsic semiconductor**, the band gap is so small that thermal energy results in some electrons populating the empty upper band (Fig. 3.59). This occupation of the conduction band introduces **positive holes**, equivalent to an absence of electrons, into the lower band, and as a result the solid is conducting. A semiconductor at room temperature generally has a much lower conductivity than a metallic conductor because only very few electrons and holes can act as charge carriers. The strong, increasing temperature dependence of the conductivity follows from the exponential Boltzmann-like temperature dependence of the electron population in the upper band.

It follows from the exponential form of the population of the conduction band that the conductivity of a semiconductor should show an Arrhenius-like temperature dependence of the form

$$\sigma = \sigma_0 e^{-E_g/2kT} \tag{3.7}$$

That is, the conductivity of a semiconductor can be expected to be Arrhenius-like with an activation energy equal to half the band gap, $E_a \approx \frac{1}{2} E_g$. This is found to be the case in practice.

(b) Extrinsic semiconductors

Key points: p-Type semiconductors are solids doped with atoms that remove electrons from the valence band; n-type semiconductors are solids doped with atoms that supply electrons to the conduction band.

An **extrinsic semiconductor** is a substance that is a semiconductor on account of the presence of intentionally added impurities. The number of electron carriers can be increased if atoms with more electrons than the parent element can be introduced by the process called **doping**. Remarkably low levels of dopant concentration are needed—only about one atom per 10^9 of the host material—so it is essential to achieve very high purity of the parent element initially.

If arsenic atoms ($[Ar]4s^2 4p^3$) are introduced into a silicon crystal ($[Ne]3s^2 3p^2$), one additional electron will be available for each dopant atom that is substituted. Note that the doping is *substitutional* in the sense that the dopant atom takes the place of an Si atom in the silicon structure. If the donor atoms, the As atoms, are far apart from each other, their electrons will be localized and the donor band will be very narrow (Fig. 3.60a). Moreover, the foreign atom levels will lie at higher energy than the valence electrons of the host structure and the filled dopant band is commonly near the empty conduction band. For $T > 0$, some of its electrons will be thermally promoted into the empty conduction band. In other words, thermal excitation will lead to the transfer of an electron from an As atom into the empty orbitals on a neighbouring Si atom. From there it will be able to migrate through the structure in the band formed by Si—Si overlap. This process gives rise to **n-type semiconductivity**, the 'n' indicating that the charge carriers are negatively charged (that is, electrons).

An alternative substitutional procedure is to dope the silicon with atoms of an element with fewer valence electrons on each atom, such as gallium ($[Ar]4s^2 4p^1$). A dopant atom of this kind effectively introduces holes into the solid. More formally, the dopant atoms form a very narrow, empty **acceptor band** that lies above the full Si band (Fig. 3.60b). At $T = 0$ the acceptor band is empty but at higher temperatures it can accept thermally excited electrons from the Si valence band. By doing so, it introduces holes into the latter and hence allows the remaining electrons in the band to be mobile. Because the charge carriers are now effectively positive holes in the lower band, this type of semiconductivity is called **p-type semiconductivity**.

Several d-metal oxides, including ZnO and Fe_2O_3, are n-type semiconductors. In their case, the property is due to small variations in stoichiometry and a small deficit of O atoms. The electrons that should be in localized O atomic orbitals (giving a very narrow oxide band, essentially localized individual O^{2-} ions) occupy a previously empty conduction band formed by the metal orbitals (Fig. 3.61b). The electrical conductivity decreases after the solids have been heated in oxygen and cooled slowly back to room temperature because the deficit of O atoms is partly replaced and, as the atoms are added, electrons are withdrawn from the conduction band to form oxide ions. However, when measured at high temperatures the conductivity of ZnO increases as further oxygen is lost from the structure, so increasing the number of electrons in the conduction band.

p-Type semiconduction is observed for some low oxidation number d-metal chalcogenides and halides, including Cu_2O, FeO, FeS, and CuI. In these compounds, the loss of electrons can occur through a process equivalent to the oxidation of some of the metal atoms, with the result that holes appear in the predominantly metal band (Fig. 3.61b). The conductivity increases when these compounds are heated in oxygen (or sulfur and halogen sources for FeS and CuI, respectively) because more holes are formed in the metal band as oxidation progresses. n-Type semiconductivity, however, tends to occur for oxides of metals in higher oxidation states, as the metal can be reduced to a lower oxidation state by occupation of a conduction band formed from the metal orbitals. Thus typical n-type semiconductors include Fe_2O_3, MnO_2, and CuO. By contrast, p-type semiconductivity occurs when the metal is in a low oxidation state, such as MnO and Cr_2O_3.

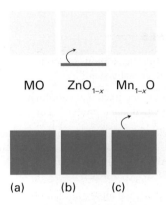

Fig. 3.61 The band structure in (a) a stoichiometric oxide MO, (b) an anion-deficient oxide, (c) an anion-excess oxide.

Example 3.18 Predicting extrinsic semiconducting properties

Decide which of the oxides WO_3, MgO, and CdO is likely to show p- or n-type extrinsic semiconductivity.

Answer WO_3, as W(VI), is readily reduced, losing oxygen, so n-type semiconductivity is expected. CdO is like ZnO and again would be predicted to be an n-type semiconductor. MgO does not lose or gain even small quantities of oxygen and so is an insulator.

Self-test 3.18 Predict p- or n-type extrinsic semiconductivity for V_2O_5 and CoO.

FURTHER READING

R.D. Shannon in *Encyclopaedia of inorganic chemistry* (ed. R.B. King). Wiley, New York (1994). A survey of ionic radii and their determination.

A.F. Wells, *Structural inorganic chemistry*. Oxford University Press (1985). The standard reference book, which surveys the structures of a huge number of inorganic solids.

J.K. Burdett, *Chemical bonding in solids*. Oxford University Press (1995). Further details of the electronic structures of solids.

Some introductory texts on solid-state inorganic chemistry are
U. Müller, *Inorganic structural chemistry*. Wiley, New York (1993).

A.R. West, *Basic solid state chemistry*. Wiley, New York (1999).

S.E. Dann, *Reactions and characterization of solids*. Royal Society of Chemistry, Cambridge (2000).

L.E. Smart and E.A. Moore, *Solid state chemistry: an introduction*. Taylor and Francis, CRC Press (2005).

P.A. Cox, *The electronic structure and chemistry of solids*. Oxford University Press (1987).

Two very useful texts on the application of thermodynamic arguments to inorganic chemistry are:
W.E. Dasent, *Inorganic energetics*. Cambridge University Press (1982).

D.A. Johnson, *Some thermodynamic aspects of inorganic chemistry*. Cambridge University Press (1982).

EXERCISES

3.1 What are the relations between the unit cell parameters in the monoclinic crystal system?

3.2 What are the fractional coordinates of the lattice points shown in the face-centred cubic unit cell (Fig. 3.5)?

3.3 Which of the following schemes for the repeating pattern of close-packed planes are not ways of generating close-packed lattices? (a) ABCABC..., (b) ABAC..., (c) ABBA..., (d) ABCBC... (e) ABABC... (f) ABCCB....

3.4 The interstitial alloy tungsten carbide, WC, has the rock-salt structure. Describe it in terms of the filling of the holes in a close-packed structure.

3.5 Calculate a value for the atomic radius of the Cs atom with a coordination number of 12 given that the empirical atomic radius of Cs in the metal with a bcc structure is 272 pm.

3.6 Metallic sodium adopts a bcc structure with density 970 kg m^{-3}. What is the length of the edge of the unit cell?

3.7 An alloy of copper and gold has the structure shown in Fig. 3.62. Calculate the composition of this unit cell. What is the lattice type of this structure? If 24 carat gold is pure gold, what carat gold does this alloy represent?

3.8 Using Ketelaar's triangle would you classify Sr$_2$Ga (χ(Sr) = 0.95; χ(Ga) = 1.81) as an alloy or a Zintl phase?

3.9 Depending on temperature, RbCl can exist in either the rock-salt or caesium-chloride structure. (a) What is the coordination number of the cation and anion in each of these structures? (b) In which of these structures will Rb have the larger apparent radius ?

3.10 Consider the structure of caesium chloride. How many Cs$^+$ ions occupy second-nearest neighbour locations of a Cs$^+$ ion?

3.11 Describe the coordination around the anions in the perovskite structure in terms of coordination to the A- and B-type cations.

3.12 Use radius-ratio rules and the ionic radii given in *Resource section* 1 to predict structures of (a) PuO$_2$, (b) FrI, (c) BeO, and (d) InN.

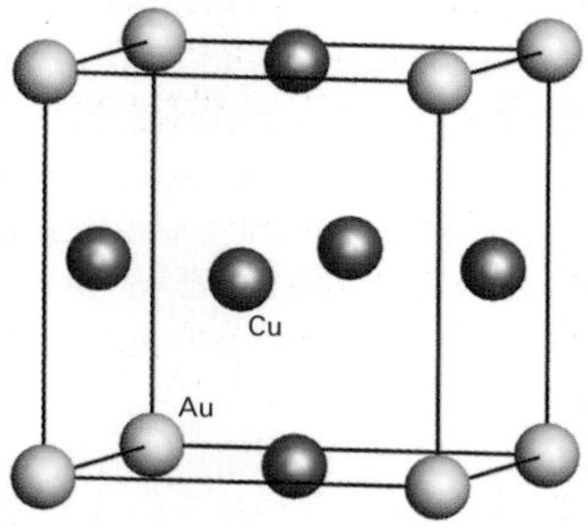

Fig. 3.62 The structure of Cu$_3$Au.

3.13 What will be the most significant terms in the Born–Haber cycle for the formation of Ca$_3$N$_2$?

3.14 By considering the parameters that change in the Born–Mayer expression estimate a lattice enthalpy for MgO given that both MgO and NaCl adopt rock-salt structures with similar lattice parameters and $\Delta H_L^{\ominus}$(NaCl) = 786 kJ mol^{-1}.

3.15 Use the Kapustinskii equation and the ionic and thermochemical radii given in *Resource section* 1 and Table 3.10 and r(Bk^{4+}) = 97 pm to calculate lattice enthalpies of (a) BkO$_2$, (b) K$_2$SiF$_6$, and (c) LiClO$_4$.

3.16 Which member of each pair is likely to be more soluble in water: (a) SrSO$_4$ or MgSO$_4$, (b) NaF or NaBF$_4$?

3.17 On the basis of the factors that contribute to lattice enthalpies place LiF, CaO, RbCl, AlN, NiO, and CsI, all of which adopt the rock-salt structure, in order of increasing lattice energy.

3.18 Recommend a specific cation for the quantitative precipitation of carbonate ion in water. Justify your recommendations.

PROBLEMS

3.1 Draw a cubic unit cell (a) in projection (showing the fractional heights of the atoms) and (b) as a three-dimensional representation that has atoms at the following positions: Ti at ($\frac{1}{2}, \frac{1}{2}, \frac{1}{2}$), O at ($\frac{1}{2}, \frac{1}{2}, 0$), ($\frac{1}{2}, \frac{1}{2}, 0$), and ($\frac{1}{2}, 0, \frac{1}{2}$), and Ba at (0,0,0). Remember that a cubic unit cell with an atom on the cell face, edge, or corner will have equivalent atoms displaced by the unit cell repeat in any direction. Which structure type is the cell?

3.2 Draw a tetragonal unit cell and mark on it a set of points that would define (a) a face-centred lattice and (b) a body-centred lattice. Demonstrate, by considering two adjacent unit cells, that a tetragonal face-centred lattice of dimensions a and c can always be redrawn as a body-centred tetragonal lattice with dimensions $a/2^{1/2}c$.

3.3 Draw one layer of close-packed spheres. On this layer mark the positions of the centres of the B layer atoms using the symbol $\otimes$ and, with the symbol $\bigcirc$, mark the positions of the centres of the C layer atoms of an fcc lattice.

3.4 In the structure of MoS$_2$, the S atoms are arranged in close-packed layers that repeat themselves in the sequence AAA.... The Mo atoms occupy holes with coordination number 6. Show that each Mo atom is surrounded by a trigonal prism of S atoms.

3.5 The ReO$_3$ structure is cubic with an Re atom at each corner of the unit cell and one O atom on each unit cell edge midway between the Re atoms. Sketch this unit cell and determine (a) the coordination numbers of the ions and (b) the identity of the structure type that would be generated if a cation were inserted in the centre of each ReO$_3$ unit.

3.5 Consider the structure of rock salt. (a) How many Na$^+$ ions occupy second-nearest neighbour locations of an Na$^+$ ion? (b) Pick out the closest-packed plane of Cl$^-$ ions. (*Hint*: this hexagonal plane will be perpendicular to a threefold axis.)

3.6 Imagine the construction of an MX$_2$ structure from the CsCl structure by removal of half the Cs$^+$ ions to leave tetrahedral coordination around each Cl$^-$ ion. Identify this MX$_2$ structure.

3.7 Obtain formulae (MX_n or M_nX) for the following structures derived from hole filling in close-packed arrays with (a) half the octahedral holes filled, (b) one quarter of the tetrahedral holes filled, and (c) all the octahedral and tetrahedral holes filled. What are the average coordination numbers of M and X in (a) and (b) ?

3.8 Given the following data for the length of a side of the unit cell for compounds that crystallize in the rock-salt structure, determine the cation radii: MgSe (545 pm), CaSe (591 pm), SrSe (623 pm), BaSe (662 pm). (*Hint*: To determine the radius of Se^{2-}, assume that the Se^{2-} ions are in contact in MgSe.)

3.9 Use the structure map in Fig. 3.43 to predict the coordination numbers of the cations and anions in (a) LiF, (b) RbBr, (c) SrS, and (d) BeO. The observed coordination numbers are (6,6) for LiF, RbBr, and SrS and (4,4) for BeO. Propose a possible reason for the discrepancies.

3.10 Using the accepted ionic radius of the ammonium ion, NH_4^+, NH_4Br is predicted to have a rock-salt structure with (6,6)-coordination. However, at room temperature NH_4Br has a caesium-chloride structure. Explain this observation.

3.11 (a) Calculate the enthalpy of formation of the hypothetical compound KF_2 assuming a CaF_2 structure. Use the Born–Mayer equation to obtain the lattice enthalpy and estimate the radius of K^{2+} by extrapolation of trends in Table 1.5 and *Resource section* 1. Ionization enthalpies and electron gain enthalpies are given in Tables 1.6 and 1.7. (b) What factor prevents the formation of this compound despite the favourable lattice enthalpy?

3.12 The common oxidation number for an alkaline earth metal is +2. Using the Born–Mayer equation and a Born–Haber cycle, show that CaCl is an exothermic compound. Use a suitable analogy to estimate an ionic radius for Ca^+. The sublimation enthalpy of Ca(s) is 176 kJ mol^{-1}. Show that an explanation for the nonexistence of CaCl can be found in the enthalpy change for the reaction $2\,CaCl(s) \rightarrow Ca(s) + CaCl_2(s)$.

3.13 The Coulombic attraction of nearest-neighbour cations and anions accounts for the bulk of the lattice enthalpy of an ionic compound. With this fact in mind, estimate the order of increasing lattice enthalpy of (a) MgO, (b) NaCl, and (c) AlN, all of which crystallize in the rock-salt structure. Give your reasoning.

3.14 There are two common polymorphs of zinc sulfide: cubic and hexagonal. Based on the analysis of Madelung constants alone, predict which polymorph should be more stable? Assume that the Zn—S distances in the two polymorphs are identical.

3.15 (a) Explain why lattice energy calculations based on the Born–Mayer equation reproduce the experimentally determined values to within 1 per cent for LiCl but only 10 per cent for AgCl given that both compounds have the rock-salt structure. (b) Identify a pair of compounds containing M^{2+} ions that might be expected to show similar behaviour.

3.16 Which of the following pairs of isostructural compounds are likely to undergo thermal decomposition at lower temperature? Give your reasoning. (a) $MgCO_3$ and $CaCO_3$ (decomposition products $MO + CO_2$). (b) CsI_3 and $N(CH_3)_4I_3$ (both compounds contain I_3^-; decomposition products $MI + I_2$; the radius of $N(CH_3)_4^+$ is much greater than that of Cs^+).

3.17 The Kapustinskii equation shows that lattice enthalpies are inversely proportional to the ion separations. Recent work has shown that further simplification of the Kapustinskii equation allows lattice enthalpies to be estimated from the molecular (formula) unit volume (the unit cell volume divided by the number of formula units, Z, it contains) or the mass density (see, for example, H.D.B. Jenkins and D. Tudela, *J. Chem. Educ.*, 2003, **80** 1482). How would you expect the lattice enthalpy to vary as a function of (a) the molecular unit volume and (b) the density? Given the following unit cell volumes (all in Å^3, $1\,Å = 10^{-10}$ m) for the alkaline earth carbonates MCO_3 and oxides, predict the observed decomposition behaviour of the carbonates.

MgCO$_3$ 47	CaCO$_3$ 61	SrCO$_3$ 64	BaCO$_3$ 76
MgO 19	CaO 28	SrO 34	BaO 42

3.18 By considering the rock-salt structure and the distances and charges around one central ion show that the first six terms of the Na^+ Madelung series are

$$+\frac{6}{\sqrt{1}} - \frac{12}{\sqrt{2}} + \frac{8}{\sqrt{3}} - \frac{6}{\sqrt{4}} + \frac{24}{\sqrt{5}} - \frac{24}{\sqrt{6}}$$

Discuss methods for showing this series converges to 1.748 by reference to R.P. Grosso, J.T. Fermann, and W.J. Vining, *J. Chem. Educ.*, 2001, **78**, 1198.

FURTHER INFORMATION 3.1 The Born–Mayer equation

Consider a one-dimensional line of alternating cations A and anions B of charges $+e$ and $-e$ separated by a distance d_0. The Coulomb potential energy of a single cation is the sum of its interactions with all the other ions:

$$V = \frac{e^2}{4\pi\varepsilon_0}\left(-\frac{2}{d_0} + \frac{2}{2d_0} - \frac{2}{3d_0} + \cdots\right) = -\frac{2e^2}{4\pi\varepsilon_0 d_0}\left(1 - \frac{1}{2} + \frac{1}{3} - \cdots\right)$$

The sum of the series in parentheses is ln 2, so for this arrangement of ions

$$V = -\frac{2e^2 \ln 2}{4\pi\varepsilon_0 d_0}$$

The total molar contribution of all the ions is this potential energy multiplied by Avogadro's constant N_A (to convert to a molar value) and divided by 2 (to avoid counting each interaction twice):

$$V = -\frac{N_A e^2}{4\pi\varepsilon_0 d_0}\mathcal{A}$$

The factor $\mathcal{A} = \ln 2$ is an example of a Madelung constant, a constant that represents the geometrical distribution of the ions (here, a straight line of constant separation). Two- and three-dimensional arrays of ions may be treated similarly, and give the values of $\mathcal{A}$ listed in Table 3.8.

The total molar potential energy includes the repulsive interaction between the ions. We can model that by a short-range exponential function of the form $Be^{-d_0/d}$, with d a constant that defines the range of the repulsive interaction and B a constant that defines its magnitude. The total molar potential energy of interaction is therefore

$$V = -\frac{N_A e^2}{4\pi\varepsilon_0 d_0}\mathcal{A} + Be^{-d_0/d}$$

This potential energy passes through a minimum when $dV/dd_0 = 0$, which occurs at

$$\frac{dV}{dd_0} = \frac{N_A e^2}{4\pi\varepsilon_0 d_0^2}\mathcal{A} - \frac{B}{d}e^{-d_0/d} = 0$$

It follows that, at the minimum,

$$Be^{d/d_0} = \frac{N_A e^2 d}{4\pi\varepsilon_0 d_0^2}\mathcal{A}$$

This relation can be substituted into the expression for V, to give

$$V = -\frac{N_A e^2}{4\pi\varepsilon_0 d_0}\left(1 - \frac{d}{d_0}\right)\mathcal{A}$$

On identifying $-V$ with the lattice enthalpy (more precisely, with the lattice energy at $T=0$), we obtain the Born–Mayer equation, eqn 3.2, for the special case of singly charged ions. The generalization to other charge types is straightforward. Note that if a different expression for the repulsive interaction between the ions is used then this expression will be modified. One alternative is to use an expression such as $1/r^n$ with a large n, typically $6 \le n \le 12$, which then gives rise to a slightly different expression for the lattice enthalpy known as the **Born–Landé equation**:

$$V = -\frac{N_A e^2}{4\pi\varepsilon_0 d_0}\left(1 - \frac{1}{n}\right)\mathcal{A}$$

The semiempirical Born–Mayer expression, with $d = 34.5\,\text{pm}$ determined from the best agreement with experimental data, is generally preferred to the Born–Landé equation.

Acids and bases

4

This chapter focuses on the wide variety of species that are classified as acids and bases. The acids and bases described in the first part of the chapter take part in proton transfer reactions. Proton transfer equilibria can be discussed quantitatively in terms of acidity constants, which are a measure of the tendency for species to donate protons. In the second part of the chapter, we broaden the definition of acids and bases to include reactions that involve electron-pair sharing between a donor and an acceptor. This broadening enables us to extend our discussion of acids and bases to species that do not contain protons and to nonaqueous media. Because of the greater diversity of these species, a single scale of strength is not appropriate. Therefore, we describe two approaches: in one, acids and bases are classified as 'hard' or 'soft'; in the other, thermochemical data are used to obtain a set of parameters characteristic of each species.

The original distinction between acids and bases was based, hazardously, on criteria of taste and feel: acids were sour and bases felt soapy. A deeper chemical understanding of their properties emerged from Arrhenius's (1884) conception of an acid as a compound that produced hydrogen ions in water. The modern definitions that we consider in this chapter are based on a broader range of chemical reactions. The definition due to Brønsted and Lowry focuses on proton transfer, and that due to Lewis is based on the interaction of electron pair acceptor and electron pair donor molecules and ions.

Brønsted acidity

Key points: A Brønsted acid is a proton donor and a Brønsted base is a proton acceptor; a simple representation of a hydrogen ion in water is as the hydronium ion, H_3O^+.

Johannes Brønsted in Denmark and Thomas Lowry in England proposed (in 1923) that the essential feature of an acid–base reaction is the transfer of a hydrogen ion, H^+, from one species to another. In the context of this definition, a hydrogen ion is often referred to as a proton. They suggested that any substance that acts as a proton donor should be classified as an acid, and any substance that acts as a proton acceptor should be classified as a base. Substances that act in this way are now called 'Brønsted acids' and 'Brønsted bases', respectively:

A **Brønsted acid** is a proton donor.
A **Brønsted base** is a proton acceptor.

The definitions make no reference to the environment in which proton transfer occurs, so they apply to proton transfer behaviour in any solvent and even in no solvent at all.

An example of a Brønsted acid is hydrogen fluoride, HF, which can donate a proton to another molecule, such as H_2O, when it dissolves in water:

$$HF(g) + H_2O(l) \rightarrow H_3O^+(aq) + F^-(aq)$$

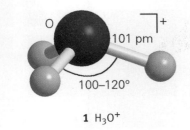

1 H_3O^+

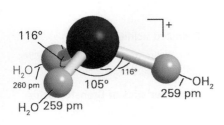

2 $H_9O_4^+$

An example of a Brønsted base is ammonia, NH_3, which can accept a proton from a proton donor:

$$H_2O(l) + NH_3(aq) \rightarrow NH_4^+(aq) + OH^-(aq)$$

As these two examples show, water is an example of an **amphiprotic** substance, a substance that can act as both a Brønsted acid and a Brønsted base.

When an acid donates a proton to a water molecule, the latter is converted into a *hydronium ion*, H_3O^+ (**1**; the dimensions are taken from the crystal structure of $H_3O^+ClO_4^-$). However, the entity H_3O^+ is almost certainly an oversimplified description of the proton in water, for it participates in extensive hydrogen bonding, and a better representation is $H_9O_4^+$ (**2**). Gas-phase studies of water clusters using mass spectrometry suggest that a cage of H_2O molecules can condense around one H_3O^+ ion in a regular pentagonal dodecahedral arrangement, resulting in the formation of the species $H^+(H_2O)_{21}$. As these structures indicate, the most appropriate description of a proton in water varies according to the environment and the experiment under consideration; for simplicity, we shall use the representation H_3O^+ throughout.

4.1 Proton transfer equilibria in water

Proton transfer between acids and bases is fast in both directions, so the dynamic equilibria

$$HF(aq) + H_2O(l) \rightleftharpoons H_3O^+(aq) + F^-(aq)$$
$$H_2O(1) + NH_3(aq) \rightleftharpoons NH_4^+(aq) + OH^-(aq)$$

give a more complete description of the behaviour of the acid HF and the base NH_3 in water than the forward reaction alone. The central feature of Brønsted acid–base chemistry in aqueous solution is that of rapid attainment of equilibrium in the proton transfer reaction, and we concentrate on this aspect.

(a) Conjugate acids and bases

Key points: When a species donates a proton, it becomes the conjugate base; when a species gains a proton, it becomes the conjugate acid. Conjugate acids and bases are in equilibrium in solution.

The structure of the two forward and reverse reactions given above, both of which depend on the transfer of a proton from an acid to a base, is expressed by writing the general Brønsted equilibrium as

$$Acid_1 + Base_2 \rightleftharpoons Acid_2 + Base_1$$

The species $Base_1$ is called the **conjugate base** of $Acid_1$, and $Acid_2$ is the **conjugate acid** of $Base_2$. The conjugate base of an acid is the species that is left after a proton is lost. The conjugate acid of a base is the species formed when a proton is gained. Thus, F^- is the conjugate base of HF and H_3O^+ is the conjugate acid of H_2O. There is no *fundamental* distinction between an acid and a conjugate acid or a base and a conjugate base: a conjugate acid is just another acid and a conjugate base is just another base.

Example 4.1 Identifying acids and bases

Identify the Brønsted acid and its conjugate base in the following reactions:

  (a) $HSO_4^-(aq) + OH^-(aq) \rightarrow H_2O(l) + SO_4^{2-}(aq)$

  (b) $PO_4^{3-}(aq) + H_2O(l) \rightarrow HPO_4^{2-}(aq) + OH^-(aq)$

Answer (a) The hydrogensulfate ion, HSO_4^-, transfers a proton to hydroxide; it is therefore the acid and the SO_4^{2-} ion produced is its conjugate base. (b) The H_2O molecule transfers a proton to the phosphate ion acting as a base; thus H_2O is the acid and the OH^- ion is its conjugate base.

Self-test 4.1 Identify the acid, base, conjugate acid, and conjugate base in the reactions (a) $HNO_3(aq) + H_2O(l) \rightarrow H_3O^+(aq) + NO_3^-(aq)$, (b) $CO_3^{2-}(aq) + H_2O(l) \rightarrow HCO_3^-(aq) + OH^-(aq)$, (c) $NH_3(aq) + H_2S(aq) \rightarrow NH_4^+(aq) + HS^-(aq)$.

(b) The strengths of Brønsted acids

Key points: The strength of a Brønsted acid is measured by its acidity constant, and the strength of a Brønsted base is measured by its basicity constant; the stronger the base, the weaker is its conjugate acid.

The strength of a Brønsted acid, such as HF, in aqueous solution is expressed by its **acidity constant** (or 'acid ionization constant'), K_a:

$$HF(aq) + H_2O(l) \rightleftharpoons H_3O^+(aq) + F^-(aq) \qquad K_a = \frac{[H_3O^+][F^-]}{[HF]}$$

or more generally:

$$HX(aq) + H_2O(l) \rightleftharpoons H_3O^+ + X^-(aq) \qquad K_a = \frac{[H_3O^+][X^-]}{[HX]}$$

In this definition, $[X^-]$ denotes the numerical value of the molar concentration of the species X^- (so, if the molar concentration of HF molecules is 0.001 mol dm^{-3}, then $[HF] = 0.001$).[1] A value $K_a \ll 1$ implies that $[HX]$ is large with respect to $[X^-]$, and so proton retention by the acid is favoured. The experimental value of K_a for hydrogen fluoride in water is 3.5×10^{-4}, indicating that under normal conditions, only a very small fraction of HF molecules are deprotonated in water. The actual fraction deprotonated can be calculated as a function of acid concentration from the numerical value of K_a.

The proton transfer equilibrium characteristic of a base, such as NH_3, in water can also be expressed in terms of an equilibrium constant, the **basicity constant**, K_b:

$$NH_3(aq) + H_2O(l) \rightleftharpoons NH_4^+(aq) + OH^-(aq) \qquad K_b = \frac{[NH_4^+][OH^-]}{[NH_3]}$$

or generally:

$$B(aq) + H_2O(l) \rightleftharpoons BH^+(aq) + OH^-(aq) \qquad K_b = \frac{[BH^+][OH^-]}{[B]}$$

If $K_b \ll 1$, then $[BH^+] \ll [B]$ and only a small fraction of B molecules are protonated. Therefore, the base is a weak proton acceptor and its conjugate acid is present in low concentration in solution. The experimental value of K_b for ammonia in water is 1.8×10^{-5}, indicating that under normal conditions, only a very small fraction of NH_3 molecules are protonated in water. As for the acid calculation, the actual fraction of base protonated can be calculated from the numerical value of K_b.

Because water is amphiprotic, a proton transfer equilibrium exists even in the absence of added acids or bases. The proton transfer from one water molecule to another is called **autoprotolysis** (or 'autoionization'). The extent of autoprotolysis and the composition of the solution at equilibrium is described by the **autoprotolysis constant** (or 'ionic product') of water:

$$2\,H_2O(l) \rightleftharpoons H_3O^+(aq) + OH^-(aq) \qquad K_w = [H_3O^+][OH^-]$$

The experimental value of K_w is 1.00×10^{-14} at 25°C, indicating that only a very tiny fraction of water molecules are present as ions in pure water.

An important role for the autoprotolysis constant of a solvent is that it enables us to express the strength of a base in terms of the strength of its conjugate acid. Thus, the value of K_b for the ammonia equilibrium in which NH_3 acts as a base is related to the value of K_a for the equilibrium

$$NH_4^+(aq) + H_2O(l) \rightleftharpoons H_3O^+(aq) + NH_3(aq)$$

---

[1] In precise work, K_a is expressed in terms of the activity of X, $a(X)$, its effective thermodynamic concentration.

in which its conjugate acid acts as an acid by

$$K_a K_b = K_w \tag{4.1}$$

This relation may be verified by multiplying together the expressions for the acidity constant of NH_4^+ and the basicity constant of NH_3. The implication of eqn 4.1 is that the larger the value of K_b, the smaller the value of K_a. That is, the stronger the base, the weaker is its conjugate acid. A further implication of eqn 4.1 is that the strengths of bases may be reported in terms of the acidity constants of their conjugate acids.

Because molar concentrations and acidity constants span many orders of magnitude, it is convenient to report them as their common logarithms (logarithms to the base 10) by using

$$pH = -\log[H_3O^+] \qquad pK = -\log K \tag{4.2}$$

where K may be any of the constants we have introduced. At 25°C, for instance, $pK_w = 14.00$. It follows from this definition and the relation in eqn 4.1 that

$$pK_a + pK_b = pK_w \tag{4.3}$$

A similar expression applies to the strengths of conjugate acids and bases in any solvent, with pK_w replaced by the appropriate autoprotolysis constant of the solvent, pK_{sol}.

Example 4.2 Relating acidity and basicity constants

The K_b of ammonia in water is 1.8×10^{-5}. Calculate K_a of the conjugate acid NH_4^+.

Answer The acidity and basicity constants are related to the autoprotolysis constant for water by the expression in eqn 4.1. Therefore, because $K_a \times 1.8 \times 10^{-5} = 1.00 \times 10^{-14}$, $K_a = 5.56 \times 10^{-10}$.

Self-test 4.2 The value of K_b for pyridine, C_5H_5N, is 1.8×10^{-9}. Calculate K_a for the conjugate acid, $C_5H_5NH^+$.

(c) Strong and weak acids and bases

Key points: An acid or base is classified as either weak or strong depending on the size of its acidity constant.

Table 4.1 lists the acidity constants of some common acids and conjugate acids of some bases. A substance is classified as a **strong acid** if the proton transfer equilibrium lies strongly in favour of donation of a proton to water. Thus, a substance with $pK_a < 0$ (corresponding to $K_a > 1$ and usually to $K_a \gg 1$) is a strong acid. Such acids are commonly regarded as being fully deprotonated in solution (but it must never be forgotten

---

Table 4.1 Acidity constants for species in aqueous solution at 25°C

Acid	HA	A⁻	K_a	pK_a	Acid	HA	A⁻	K_a	pK_a
Hydriodic	HI	I⁻	10^{11}	−11	Pyridinium ion	$HC_5H_5N^+$	C_5H_5N	5.6×10^{-6}	5.25
Perchloric	$HClO_4$	ClO_4^-	10^{10}	−10	Carbonic	H_2CO_3	HCO_3^-	4.3×10^{-7}	6.37
Hydrobromic	HBr	Br⁻	10^9	−9	Hydrogen sulfide	H_2S	HS^-	9.1×10^{-8}	7.04
Hydrochloric	HCl	Cl⁻	10^7	−7	Boric acid*	$B(OH)_3$	$B(OH)_4^-$	7.2×10^{-10}	9.14
Sulfuric	H_2SO_4	HSO_4^-	10^2	−2	Ammonium ion	NH_4^+	NH_3	5.6×10^{-10}	9.25
Hydronium ion	H_3O^+	H_2O	1	0.0	Hydrocyanic	HCN	CN⁻	4.9×10^{-10}	9.31
Chloric	$HClO_3$	ClO_3^-	10^{-1}	1	Hydrogencarbonate ion	HCO_3^-	CO_3^{2-}	4.8×10^{-11}	10.32
Sulfurous	H_2SO_3	HSO_3^-	1.5×10^{-2}	1.81	Hydrogenarsenate ion	$HAsO_4^{2-}$	AsO_4^{3-}	3.0×10^{-12}	11.53
Hydrogensulfate ion	HSO_4^-	SO_4^{2-}	1.2×10^{-2}	1.92	Hydrogensulfide ion	HS^-	S^{2-}	1.1×10^{-19}	19
Phosphoric	H_3PO_4	$H_2PO_4^-$	7.5×10^{-3}	2.12	Hydrogenphosphate ion	HPO_4^{2-}	PO_4^{3-}	2.2×10^{-13}	12.67
Hydrofluoric	HF	F⁻	3.5×10^{-4}	3.45	Dihydrogenphosphate ion	$H_2PO_4^-$	HPO_4^{2-}	6.2×10^{-8}	7.21

*The proton transfer equilibrium is $B(OH)_3(aq) + 2\,H_2O(l) \rightarrow H_3O^+(aq) + B(OH)_4^-(aq)$.

that that is only an approximation). For example, hydrochloric acid is regarded as a solution of H_3O^+ and Cl^- ions, and a negligible concentration of HCl molecules. A substance with $pK_a > 0$ (corresponding to $K_a < 1$) is classified as a **weak acid**; for such species, the proton transfer equilibrium lies in favour of nonionized acid. Hydrogen fluoride is a weak acid in water, and hydrofluoric acid consists of hydronium ions, fluoride ions, and a high proportion of HF molecules.

A **strong base** is a species that is almost virtually fully protonated in water. An example is the oxide ion, O^{2-}, which is immediately converted into OH^- ions in water. A **weak base** is only partially protonated in water. An example is NH_3, which is present almost entirely as NH_3 molecules in water, with a small proportion of NH_4^+ ions. The conjugate base of any strong acid is a weak base, because it is thermodynamically unfavourable for such a base to accept a proton.

(d) Polyprotic acids

Key points: A polyprotic acid loses protons in succession, and successive deprotonations are progressively less favourable; a distribution diagram summarizes how the fraction of each species present depends on the pH of the solution.

A **polyprotic acid** is a substance that can donate more than one proton. An example is hydrogen sulfide, H_2S, a diprotic acid. For a diprotic acid, there are two successive proton donations and two acidity constants:

$$H_2S(aq) + H_2O(1) \rightleftharpoons HS^- + H_3O^+(aq) \qquad K_{a1} = \frac{[H_3O^+][HS^-]}{[H_2S]}$$

$$HS^-(aq) + H_2O(1) \rightleftharpoons S^{2-}(aq) + H_3O^+(aq) \qquad K_{a2} = \frac{[H_3O^+][S^{2-}]}{[HS^-]}$$

From Table 4.1, $K_{a1} = 9.1 \times 10^{-8}$ ($pK_{a1} = 7.04$) and $K_{a2} \approx 10^{-19}$ ($pK_{a2} = 19$). The second acidity constant, K_{a2}, is almost always smaller than K_{a1} (and hence pK_{a2} is generally larger than pK_{a1}). The decrease in K_a is consistent with an electrostatic model of the acid in which, in the second deprotonation, a proton must separate from a centre with one more negative charge than in the first deprotonation. Because additional electrostatic work must be done to remove the positively charged proton, the deprotonation is less favourable.

The clearest representation of the concentrations of the species that are formed in the successive proton transfer equilibria of polyprotic acids is a **distribution diagram**, a diagram showing the fraction of solute present as a specified species X, $\alpha(X)$, plotted against the pH. Consider, for instance, the triprotic acid H_3PO_4, which releases three protons in succession to give $H_2PO_4^-$, HPO_4^{2-}, and PO_4^{3-}, then the fraction of solute present as intact H_3PO_4 molecules, for instance, is

$$\alpha(H_3PO_4) = \frac{[H_3PO_4]}{[H_3PO_4] + [H_2PO_4^-] + [HPO_4^{2-}] + [PO_4^{3-}]} \tag{4.4}$$

Figure 4.1 shows the fraction of all four solute species as a function of pH and hence summarizes the relative importance of each acid and its conjugate base at each pH. Conversely, the diagram indicates the pH of the solution that contains a particular fraction of the species. We see, for instance, that if $pH < pK_{a1}$, corresponding to high hydronium ion concentrations, then the dominant species is the fully protonated H_3PO_4 molecule. However, if $pH > pK_{a3}$, corresponding to low hydronium ion concentrations, then the dominant species is the fully deprotonated PO_4^{3-} ion. The intermediate species are dominant when pH values lie between the relevant pK_as.

4.2 Solvent levelling

Key point: A solvent with a large autoprotolysis constant can be used to discriminate between a wide range of acid and base strengths.

An acid that is weak in water may appear strong in a solvent that is a more effective proton acceptor, and vice versa. Indeed, in sufficiently basic solvents (such as liquid

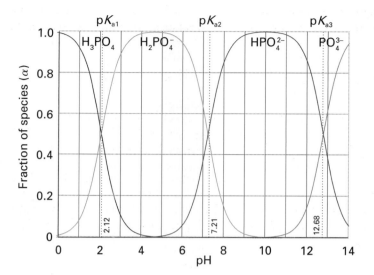

Fig. 4.1 The distribution diagram for the various forms of the triprotic acid phosphoric acid as a function of pH.

ammonia), it may not be possible to discriminate between their strengths, because all of them will be fully deprotonated. Similarly, bases that are weak in water may appear strong in a more strongly proton-donating solvent (such as anhydrous acetic acid). It may not be possible to arrange a series of bases according to strength, for all of them will be effectively fully protonated in acidic solvents. We shall now see that the autoprotolysis constant of a solvent plays a crucial role in determining the range of acid or base strengths that can be distinguished for species dissolved in it.

Any acid stronger than H_3O^+ in water donates a proton to H_2O and forms H_3O^+. Consequently, no acid significantly stronger than H_3O^+ can remain protonated in water. No experiment conducted in water can tell us which of HBr and HI is the stronger acid because both transfer their protons essentially completely to give H_3O^+. In effect, solutions of the strong acids HX and HY behave as though they are solutions of H_3O^+ ions regardless of whether HX is intrinsically stronger than HY. Water is therefore said to have a **levelling effect** that brings all stronger acids down to the acidity of H_3O^+. The strengths of such acids can be distinguished by using a less basic solvent. For instance, although HBr and HI have indistinguishable acid strengths in water, in acetic acid HBr and HI behave as weak acids and their strengths can be distinguished: in this way it is found that HI is a stronger proton donor than HBr.

The levelling effect can be expressed in terms of the pK_a of the acid. An acid such as HCN dissolved in a solvent, HSol, is classified as strong if $pK_a < 0$, where K_a is the acidity constant of the acid in the solvent Sol:

$$HCN(sol) + HSol(l) \rightleftharpoons H_2Sol^+(sol) + CN^-(sol) \qquad K_a = \frac{[H_2Sol^+][CN^-]}{[HCN]}$$

That is, all acids with $pK_a < 0$ (corresponding to $K_a > 1$) display the acidity of H_2Sol^+ when they are dissolved in the solvent HSol.

An analogous effect can be found for bases in water. Any base that is strong enough to undergo complete protonation by water produces an OH^- ion for each molecule of base added. The solution behaves as though it contains OH^- ions. Therefore, we cannot distinguish the proton-accepting power of such bases, and we say that they are levelled to a common strength. Indeed, the OH^- ion is the strongest base that can exist in water because any species that is a stronger proton acceptor immediately forms OH^- ions by proton transfer from water. For this reason, we cannot study NH_2^- or CH_3^- in water by dissolving alkali metal amides or methides because both anions generate OH^- ions and are fully protonated to NH_3 and CH_4:

$$KNH_2(s) + H_2O(l) \rightarrow K^+(aq) + OH^-(aq) + NH_3(aq)$$

$$Li_4(CH_3)_4(s) + 4\,H_2O(l) \rightarrow 4\,Li^+(aq) + 4\,OH^-(aq) + 4\,CH_4(g)$$

The base levelling effect can be expressed in terms of the pK_b of the base. A base dissolved in HSol is classified as strong if $pK_b < 0$, where K_b is the basicity constant of the base in HSol:

$$NH_3(sol) + HSol(l) \rightleftharpoons NH_4^+(sol) + Sol^-(sol) \qquad K_b = \frac{[NH_4^+][Sol^-]}{[NH_3]}$$

That is, all bases with $pK_b < 0$ (corresponding to $K_b > 1$) display the basicity of Sol^- in the solvent HSol. Now, because $pK_a + pK_b = pK_{sol}$, this criterion for levelling may be expressed as follows: all bases with $pK_a > pK_{sol}$ will give a negative value for pK_b and behave like Sol^- in the solvent HSol.

It follows from this discussion of acids and bases in a common solvent HSol that, because any acid is levelled if $pK_a < 0$ in HSol and any base is levelled if $pK_a > pK_{sol}$ in the same solvent, then the window of strengths that are not levelled in the solvent is from $pK_a = 0$ to pK_{sol}. For water, $pK_w = 14$. For liquid ammonia, the autoprotolysis equilibrium is

$$2\,NH_3(l) \rightleftharpoons NH_4^+(am) + NH_2^-(am) \qquad pK_{am} = 33$$

(The 'am' signifies solution in liquid ammonia.) It follows from these figures that acids and bases are discriminated much less in water than they are in ammonia. The discrimination windows of a number of solvents are shown in Fig. 4.2. The window for dimethylsulfoxide (DMSO, $(CH_3)_2SO$) is wide because $pK_{dmso} = 37$. Consequently, DMSO can be used to study a wide range of acids (from H_2SO_4 to PH_3). Water has a narrow window compared to some of the other solvents shown in the illustration. One reason is the high relative permittivity of water, which favours the formation of H_3O^+ and OH^- ions.

Example 4.3 Differentiating acidities in different solvents

Which of the solvents given in Fig. 4.2 could be used to differentiate the acidities of HCl ($pK_a \approx -6$) and HBr ($pK_a \approx -9$)?

Answer The table shows the windows for effective acid–base discrimination for various solvents. The only solvent for which the window covers the range -6 to -9 is methanoic (formic) acid, HCOOH

Self-test 4.3 Which of the solvents given in Fig. 4.2 could be used to discriminate the acidities of PH_3 ($pK_a \approx 27$) and GeH_4 ($pK_a \approx 25$)?

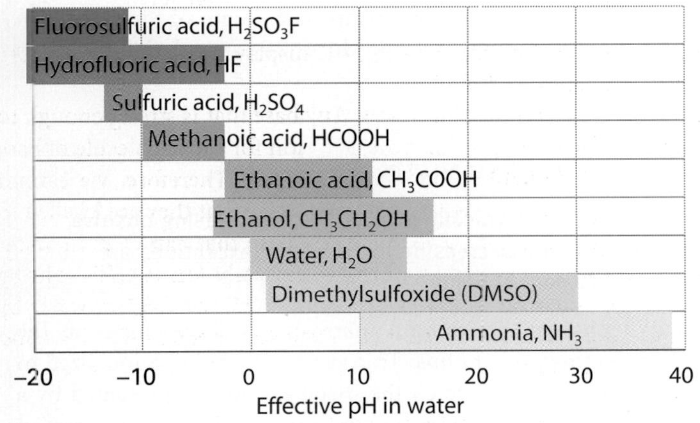

Fig. 4.2 The acid–base discrimination window for a variety of solvents. The width of each window is proportional to the autoprotolysis constant of the solvent.

Characteristics of Brønsted acids

Key point: Aqua acids, hydroxoacids, and oxoacids are typical of specific regions of the periodic table.

From now on we shall concentrate on Brønsted acids and bases in water. The largest class of acids in water consists of species that donate protons from an —OH group attached to a central atom. A donatable proton of this kind is called an **acidic proton** to distinguish it from other protons that may be present in the molecule, such as the nonacidic methyl protons in CH_3COOH.

There are three classes of acids to consider:

1 An **aqua acid**, in which the acidic proton is on a water molecule coordinated to a central metal ion.

$$E(OH_2)(aq) + H_2O(l) \rightleftharpoons E(OH)^-(aq) + H_3O^+(aq)$$

An example is

$$[Fe(OH_2)_6]^{3+}(aq) + H_2O(l) \rightleftharpoons [Fe(OH_2)_5(OH)]^{2+}(aq) + H_3O^+(aq)$$

The aqua acid, the hexaaquairon(III) ion, is shown as (**3**).

2 A **hydroxoacid**, in which the acidic proton is on a hydroxyl group without a neighbouring oxo group (=O).
An example is $Te(OH)_6$ (**4**).

3 An **oxoacid**, in which the acidic proton is on a hydroxyl group with an oxo group attached to the same atom.

Sulfuric acid, H_2SO_4 ($O_2S(OH)_2$; **5**), is an example of an oxoacid. The three classes of acids can be regarded as successive stages in the deprotonation of an aqua acid:

aqua acid   hydroxoacid   oxoacid

An example of these successive stages is provided by a *d*-block metal in an intermediate oxidation state, such as Ru(IV):

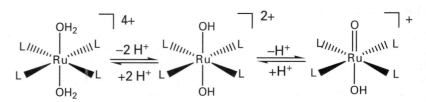

Aqua acids are characteristic of central atoms in low oxidation states, of *s*- and *d*-block metals, and of metals on the left of the *p*-block. Oxoacids are commonly found where the central element is in a high oxidation state. An element from the right of the *p*-block in one of its intermediate oxidation states may also produce an oxoacid ($HClO_2$ is an example).

4.3 Periodic trends in aqua acid strength

Key points: The strengths of aqua acids typically increase with increasing positive charge of the central metal ion and with decreasing ionic radius; exceptions are commonly due to the effects of covalent bonding.

The strengths of aqua acids typically increase with increasing positive charge of the central metal ion and with decreasing ionic radius. This variation can be rationalized to some extent in terms of an ionic model, in which the metal cation is represented by a sphere of radius r_+ carrying z positive charges. Because protons are more easily removed from the vicinity of cations of high charge and small radius, the model predicts that the acidity should increase with increasing z and with decreasing r_+.

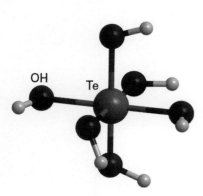

3 $[Fe(OH_2)_6]^{3+}$

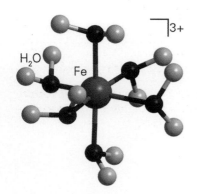

4 $B(OH)_3$

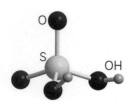

5 $O_2S(OH)_2$, H_2SO_4

The validity of the ionic model of acid strengths can be judged from Fig. 4.3. Aqua ions of elements that form ionic solids (principally those from the *s* block) have pK_a values that are quite well described by the ionic model. Several *d*-block ions (such as Fe^{2+} and Cr^{3+}) lie reasonably near the same straight line, but many ions (particularly those with low pK_a, corresponding to high acid strength) deviate markedly from it. This deviation indicates that the metal ions repel the departing proton more strongly than is predicted by the ionic model. This enhanced repulsion can be rationalized by supposing that the cation's positive charge is not confined to the central ion but is delocalized over the ligands and hence is closer to the departing proton. The delocalization is equivalent to attributing covalence to the element–oxygen bond. Indeed, the correlation is worst for ions that are disposed to form covalent bonds.

For the later *d*-block and the *p*-block metals (such as Cu^{2+} and Sn^{2+}, respectively), the strengths of the aqua acids are much greater than the ionic model predicts. For these species, covalent bonding is more important than ionic bonding and the ionic model is unrealistic. The overlap between metal orbitals and the orbitals of an oxygen ligand increases from left to right across a period. It also increases down a group, so aqua ions of heavier *d*-block metals tend to be stronger acids.

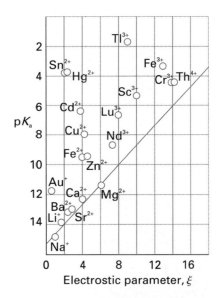

Fig. 4.3 The correlation between acidity constant and the electrostatic parameter ξ of aqua ions.

Example 4.4 Accounting for trends in aqua acid strength

Account for the trend in acidity $[Fe(OH_2)_6]^{2+} < [Fe(OH_2)_6]^{3+} < [Al(OH_2)_6]^{3+} \approx [Hg(OH_2)]^{2+}$

Answer The weakest acid is the Fe^{2+} complex on account of its relatively large ionic radius and low charge. The increase of charge to +3 increases the acid strength. The greater acidity of Al^{3+} can be explained by its smaller radius. The anomalous ion in the series is the Hg^{2+} complex. This complex reflects the failure of an ionic model, because in the complex there is a large transfer of positive charge to oxygen as a result of covalent bonding.

Self-test 4.4 Arrange $[Na(OH_2)_6]^+$, $[Sc(OH_2)_6]^{3+}$, $[Mn(OH_2)_6]^{2+}$, and $[Ni(OH_2)_6]^{2+}$ in order of increasing acidity.

4.4 Simple oxoacids

The simplest oxoacids are the **mononuclear acids**, which contain one atom of the parent element. They include H_2CO_3, HNO_3, H_3PO_4, and H_2SO_4. These oxoacids are formed by the electronegative elements at the upper right of the periodic table and by other elements in high oxidation states (Table 4.2). One interesting feature in the table is the occurrence of planar $B(OH)_3$, H_2CO_3, and HNO_3 molecules but not their analogues in later periods. As we saw in Chapter 2, π bonding is more important among the Period 2 elements, so their atoms are more likely to be constrained to lie in a plane. Deprotonation of an oxoacid leaves an oxoanion that is stabilized by resonance (Section 2.1a).

(a) Substituted oxoacids

Key points: Substituted oxoacids have strengths that may be rationalized in terms of the electron-withdrawing power of the substituent; in a few cases, a nonacidic H atom is attached directly to the central atom of an oxoacid.

One or more −OH groups of an oxoacid may be replaced by other groups to give a series of substituted oxoacids, which include fluorosulfuric acid, $O_2SF(OH)$, and aminosulfuric acid, $O_2S(NH_2)OH$ (**6**). Because fluorine is highly electronegative, it withdraws electrons from the central S atom and confers on S a higher effective positive charge. As a result, the substituted acid is stronger than $O_2S(OH)_2$. Another electron acceptor substituent is −CF_3, as in the strong acid trifluoromethylsulfonic acid, CF_3SO_3H (that is, $O_2S(CF_3)(OH)$). By contrast, the −NH_2 group, which has lone pair electrons, can donate electron density to S by π bonding. This transfer of charge reduces the positive charge of the central atom and weakens the acid.

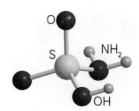

6 $O_2S(NH_2)OH$

Table 4.2 The structure and pK_a values of oxoacids*

$p = 0$	$p = 1$		$p = 2$	$p = 3$
HO—Cl	O‖Cl(HO)(OH)		O‖N(O)(OH)	
7.2	3.6		−1.4	
Si(OH)(HO)(OH)(OH)	P(O)(HO)(OH)(OH)	O(Cl)(OH)	S(O)(O)(OH)(OH)	Cl(O)(OH)(O)
10	2.1, 7.4, 12.7	2.0	−2.0, 1.9	−10
Te(OH)(HO)(HO)(OH)(OH)(OH)	I(O)(HO)(HO)(OH)(OH)(OH)	P(O)(OH)(HO)(H)	Cl(O)(OH)(O)	
7.8, 11.2	1.6, 7.0	1.8, 6.6	−1.0	
B(OH)(HO)(OH)	As(O)(HO)(OH)(OH)	Se(O)(OH)(OH)		
9.1†	2.3, 6.9, 11.5	2.6, 8.0		

* p is the number of nonprotonated O atoms.
† Boric acid is a special case; see Section 12.5.

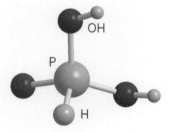

7 OPH(OH)$_2$, H$_3$PO$_3$

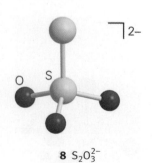

8 S$_2$O$_3^{2-}$

A trap for the unwary is that not all oxoacids follow the familiar structural pattern of a central atom surrounded by OH and O groups. Occasionally an H atom is attached directly to the central atom, as in phosphonic acid, H$_3$PO$_3$. Phosphonic acid is in fact only a *di*protic acid, as the substitution of two OH groups leaves a P—H bond (**7**) and consequently a nonacidic proton. This structure is consistent with NMR and vibrational spectra, and the structural formula is OPH(OH)$_2$. Substitution for an oxo group (as distinct from a hydroxyl group) is another example of a structural change that can occur. An important example is the thiosulfate ion, S$_2$O$_3^{2-}$ (**8**), in which an S atom replaces an O atom of a sulfate ion.

(b) Pauling's rules

Key point: The strengths of a series of oxoacids containing a specific central atom with a variable number of oxo and hydroxyl groups are summarized by Pauling's rules.

For a series of mononuclear oxoacids of an element E, the strength of the acids increases with increasing number of oxygen atoms. This trend can be explained qualitatively by considering the electron-withdrawing properties of oxygen. The oxygen atoms withdraw electrons making the O—H bond weaker. Consequently, protons are more readily released. In general, for any series of oxoacids, the one with the most oxygen is the most completely deprotonated. For example, the acids strength of the oxoacids of chlorine decrease in the order HOCl$_4$ > HClO$_3$ > HClO$_2$ > HClO. Similarly, H$_2$SO$_4$ is stronger than H$_2$SO$_3$ and HNO$_3$ is stronger than HNO$_2$.

The trends can be systematized semiquantitatively by using two empirical rules devised by Linus Pauling:

1 For the oxoacid $O_pE(OH)_q$, $pK_a \approx 8-5p$
2 The successive pK_a values of polyprotic acids (those with $q > 1$), increase by 5 units for each successive proton transfer.

Neutral hydroxoacids with $p = 0$ have $pK_a \approx 8$, acids with one oxo group have $pK_a \approx 3$, and acids with two oxo groups have $pK_a \approx -2$. For example, sulfuric acid, $O_2S(OH)_2$, has $p = 2$ and $q = 2$, and $pK_{a1} \approx -2$ (signifying a strong acid). Similarly, pK_{a2} is predicted to be $+3$, which is reasonably close to the experimental value of 1.9.

The success of these simple rules may be gauged by inspection of Table 4.2, in which acids are grouped according to p, the number of oxo groups. That the estimates are good to about ± 1 is surprising. The variation in strengths down a group is not large, and the complicated and perhaps cancelling effects of changing structures allow the rules to work moderately well. The more important variation across the periodic table from left to right and the effect of change of oxidation number are taken into account by the number of oxo groups characteristic of the molecular acids. In Group 15, the oxidation number $+5$ requires one oxo group (as in $OP(OH)_3$) whereas in Group 16 the oxidation number $+6$ requires two (as in $O_2S(OH)_2$).

(c) Structural anomalies

Key point: In certain cases, notably H_2CO_3 and H_2SO_3, a simple molecular formula misrepresents the composition of aqueous solutions of nonmetal oxides.

An interesting use of Pauling's rules is to detect structural anomalies. For example, carbonic acid, $OC(OH)_2$, is commonly reported as having $pK_{a1} = 6.4$, but the rules predict $pK_{a1} = 3$. The anomalously low acidity indicated by the experimental value is the result of treating the concentration of dissolved CO_2 as if it were all H_2CO_3. However, in the equilibrium

$$CO_2(aq) + H_2O(1) \rightleftharpoons OC(OH)_2(aq)$$

only about 1 per cent of the dissolved CO_2 is present as $OC(OH)_2$, so the actual concentration of acid is much less than the concentration of dissolved CO_2. When this difference is taken into account, the true pK_{a1} of H_2CO_3 is about 3.6, as Pauling's rules predict.

The experimental value $pK_{a1} = 1.8$ reported for sulfurous acid, H_2SO_3, suggests another anomaly, this time acting in the opposite way. In fact, spectroscopic studies have failed to detect the molecule $OS(OH)_2$ in solution, and the equilibrium constant for

$$SO_2(aq) + H_2O(l) \rightleftharpoons H_2SO_3(aq)$$

is less than 10^{-9}. The equilibria of dissolved SO_2 are complex, and a simple analysis is inappropriate. The ions that have been detected include HSO_3^- and $S_2O_5^{2-}$, and there is evidence for an SH bond in the solid salts of the hydrogensulfite ion.

This discussion of the composition of aqueous solutions of CO_2 and SO_2 calls attention to the important point that not all nonmetal oxides react fully with water to form acids. Carbon monoxide is another example: although it is formally the anhydride of methanoic acid, HCOOH, carbon monoxide does not in fact react with water at room temperature to give the acid. The same is true of some metal oxides: OsO_4, for example, can exist as dissolved neutral molecules.

Example 4.5 Using Pauling's rules

Identify the structural formulas that are consistent with the following pK_a values: H_3PO_4, 2.12; H_3PO_3, 1.80; H_3PO_2, 2.0.

Answer All three values are in the range that Pauling's first rule associates with one oxo group. This observation suggests the formulas $(HO)_3P{=}O$, $(HO)_2HP{=}O$, and $(HO)H_2P{=}O$. The second and the third formulas are derived from the first by replacement of —OH by H bound to P (as in structure **7**).

Self-test 4.5 Predict the pK_a values of (a) H_3PO_4, (b) $H_2PO_4^-$, (c) HPO_4^{2-}.

4.5 Anhydrous oxides

We have treated oxoacids as derived by deprotonation of their parent aqua acids. It is also useful to take the opposite viewpoint and to consider aqua acids and oxoacids as derived by hydration of the oxides of the central atom. This approach emphasizes the acid and base properties of oxides and their correlation with the location of the element in the periodic table.

(a) Acidic and basic oxides

Key points: Metallic elements typically form basic oxides; nonmetallic elements typically form acidic oxides.

An **acidic oxide** is an oxide that, on dissolution in water, binds an H_2O molecule and releases a proton to the surrounding solvent:

$$CO_2(g) + H_2O(l) \rightleftharpoons OC(OH)_2(aq)$$
$$OC(OH)_2(aq) + H_2O(l) \rightleftharpoons O_2C(OH)^-(aq) + H_3O^+(aq)$$

An equivalent interpretation is that an acidic oxide is an oxide that reacts with an aqueous base (an alkali):

$$CO_2(g) + OH^-(aq) \rightarrow O_2C(OH)^-(aq)$$

A **basic oxide** is an oxide to which a proton is transferred when it dissolves in water:

$$BaO(s) + H_2O(l) \rightarrow Ba^{2+}(aq) + 2\ OH^-(aq)$$

The equivalent interpretation in this case is that a basic oxide is an oxide that reacts with an acid:

$$BaO(s) + 2\ H_3O^+(aq) \rightarrow Ba^{2+}(aq) + 3\ H_2O(l)$$

Because acidic and basic oxide character often correlates with other chemical properties, a wide range of properties can be predicted from a knowledge of the character of oxides. In a number of cases the correlations follow from the basic oxides being largely ionic and of acidic oxides being largely covalent. For instance, an element that forms an acidic oxide is likely to form volatile, covalent halides. By contrast, an element that forms a basic oxide is likely to form solid, ionic halides. In short, the acidic or basic character of an oxide is a chemical indication of whether an element should be regarded as a metal or a nonmetal. Generally, metals form basic oxides and nonmetals form acidic oxides.

(b) Amphoterism

Key points: The frontier between metals and nonmetals in the periodic table is characterized by the formation of amphoteric oxides; amphoterism also varies with the oxidation state of the element.

An **amphoteric oxide** is an oxide that reacts with both acids and bases.[2] Thus, aluminium oxide reacts with acids and alkalis:

$$Al_2O_3(s) + 6\ H_3O^+(aq) + 3\ H_2O(l) \rightarrow 2\ [Al(OH_2)_6]^{3+}(aq)$$
$$Al_2O_3(s) + 2\ OH^-(aq) + 3\ H_2O(l) \rightarrow 2\ [Al(OH)_4]^-(aq)$$

Amphoterism is observed for the lighter elements of Groups 2 and 13, as in BeO, Al_2O_3, and Ga_2O_3. It is also observed for some of the d-block elements in high oxidation states,

---

[2] The word 'amphoteric' is derived from the Greek word for 'both'.

such as MoO_3 and V_2O_5, and some of the heavier elements of Groups 14 and 15, such as SnO_2 and Sb_2O_5.

Figure 4.4 shows the location of elements that in their characteristic group oxidation states have amphoteric oxides. They lie on the frontier between acidic and basic oxides, and hence serve as an important guide to the metallic or nonmetallic character of an element. The onset of amphoterism correlates with a significant degree of covalent character in the bonds formed by the elements, either because the metal ion is strongly polarizing (as for Be) or because the metal ion is polarized by the O atom attached to it (as for Sb).

An important issue in the *d*-block is the oxidation number necessary for amphoterism. Figure 4.5 shows the oxidation number for which an element in the first row of the block has an amphoteric oxide. We see that on the left of the block, from titanium to manganese and perhaps iron, oxidation state +4 is amphoteric (with higher values on the border of acidic and lower values of the border of basic). On the right of the block, amphoterism occurs at lower oxidation numbers: the oxidation states +3 for cobalt and nickel and +2 for copper and zinc are fully amphoteric. There is no simple way of predicting the onset of amphoterism. However, it presumably reflects the ability of the metal cation to polarize the oxide ions that surround it—that is, to introduce covalence into the metal—oxygen bond. The degree of covalence typically increases with the oxidation number of the metal as the increasingly positively charged cation becomes more strongly polarizing (Section 1.9e).

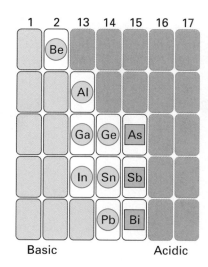

Fig. 4.4 The location of elements having amphoteric oxides. The circled elements have amphoteric oxides in all oxidation states. The elements in boxes have acidic oxides in the highest oxidation state and amphoteric oxides in lower oxidation states.

Example 4.6 Using oxide acidity in qualitative analysis

In the traditional scheme of qualitative analysis, a solution of metal ions is oxidized and then aqueous ammonia is added to raise the pH. The ions Fe^{3+}, Ce^{3+}, Al^{3+}, Cr^{3+}, and V^{3+} precipitate as hydrous oxides. The addition of H_2O_2 and NaOH redissolves the aluminium, chromium, and vanadium oxides. Discuss these steps in terms of the acidities of oxides.

Answer When the oxidation number of the metal is +3, all the metal oxides are sufficiently basic to be insoluble in a solution with $pH \approx 10$. Aluminium(III) oxide is amphoteric and redissolves in alkaline solution to give aluminate ions, $[Al(OH)_4]^-$. Vanadium(III) and chromium(III) oxides are oxidized by H_2O_2 to give vanadate ions, $[VO_4]^{3-}$, and chromate ions, $[CrO_4]^{2-}$, which are the anions derived from the acidic oxides V_2O_5 and CrO_3, respectively.

Self-test 4.6 If Ti(IV) ions were present in the sample, how would they behave?

4.6 Polyoxo compound formation

Key points: Acids containing the OH group condense to form polyoxoanions; polycation formation from simple aqua cations occurs with the loss of H_2O.

As the pH of a solution is increased, the aqua ions of metals that have basic or amphoteric oxides generally undergo polymerization and precipitation. One application of this is to the separation of metal ions, because the precipitation occurs quantitatively at a pH characteristic of each metal.

With the exception of Be^{2+} (which is amphoteric), the elements of Groups 1 and 2 have no important solution species beyond the aqua ions $M^+(aq)$ and $M^{2+}(aq)$. By contrast, the solution chemistry of the elements becomes very rich as the amphoteric region of the periodic table is approached. The two most common examples are polymers formed by Fe(III) and Al(III), both of which are abundant in the Earth's crust. In acidic solutions, both form octahedral hexaaquaions, $[Al(OH_2)_6]^{3+}$ and $[Fe(OH_2)_6]^{3+}$. In solutions of pH > 4, both precipitate as gelatinous hydrous oxides:

$$[Fe(OH_2)_6]^{3+}(aq) + (3+n)\,H_2O(l) \rightarrow Fe(OH)_3 \cdot nH_2O(s) + 3\,H_3O^+(aq)$$

$$[Al(OH_2)_6]^{3+}(aq) + (3+n)\,H_2O(l) \rightarrow Al(OH)_3 \cdot nH_2O(s) + 3\,H_3O^+(aq)$$

The precipitated polymers, which are often of colloidal dimensions, slowly crystallize to stable mineral forms. The extensive network structure of aluminium polymers, which are

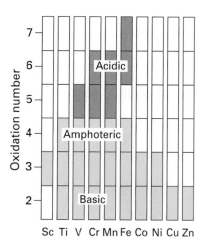

Fig. 4.5 The oxidation states for which elements in 3*d*-series elements have amphoteric oxides. Predominantly acidic oxides are shaded pink and predominantly basic oxides are shaded blue.

9 $P_2O_7^{4-}$

10 $P_2O_7^{4-}$

11 $P_4O_{12}^{4-}$

12 ATP^{4-}

13 ADP^{3-}

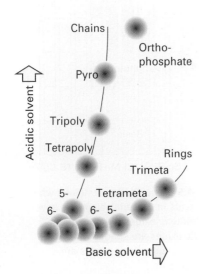

Fig. 4.6 A representation of a two-dimensional paper chromatogram of a complex mixture of phosphates formed by condensation reactions. The sample spot was placed at the lower left corner. Basic solvent separation was used first, followed by acidic solvent perpendicular to the basic one. This separates open chains from rings. The upper spot sequence corresponds to linear polymers and the lower sequence corresponds to rings.

neatly packed in three dimensions, contrasts with the linear polymers of their iron analogues. Aluminium polycations and similar ions are used constructively in water treatment to precipitate anions (such as F^-) that are present as pollutants in effluents from aluminium refining plants.

Polyoxoanion formation from oxoanions occurs by protonation of an O atom and its departure as H_2O:

$$2\,[CrO_4]^{2-}(aq) + 2\,H_3O^+(aq) \rightarrow [O_3CrOCrO_3]^{2-}(aq) + 3\,H_2O(l)$$

The importance of polyoxo anions can be judged by the fact that they account for most of the mass of oxygen in the Earth's crust, as they include almost all silicate minerals. They also include the phosphate polymers (such as ATP) used for energy transfer in living cells.

The formation of polyoxoanions is important for early d-block ions, particularly V(V), Mo(VI), W(VI), and (to a lesser extent) Nb(V), Ta(V), and Cr(VI); see Section 18.8. They are formed when base is added to aqueous solutions of the ions or oxides in high oxidation states. Polyoxoanions are also formed by some nonmetals, but their structures are different from those of their d-metal analogues. The common species in solution are rings and chains. The silicates are very important examples of polymeric oxoanions, and we discuss them in detail in Chapter 13. One example of a polysilicate mineral is $MgSiO_3$, which contains an infinite chain of SiO_3^{2-} units. In this section we illustrate some features of polyoxoanions using phosphates as examples.

The simplest condensation reaction, starting with the orthophosphate ion, PO_4^{3-}, is

$$2\,PO_4^{2-} + 2\,H^+ \longrightarrow \left[\, O{-}P{-}O{-}P{-}O \,\right]^{4-} + H_2O$$

The elimination of water consumes protons and decreases the average charge number of each P atom to -2. If each phosphate group is represented as a tetrahedron with the O atoms located at the corners, the diphosphate ion, $P_2O_7^{4-}$ (**9**) can be drawn as (**10**). Phosphoric acid can be prepared by hydrolysis of the solid phosphorus(V) oxide, P_4O_{10}. An initial step using a limited amount of water produces a metaphosphate ion with the formula $P_4O_{12}^{4-}$ (**11**). This reaction is only the simplest among many, and the separation of products from the hydrolysis of phosphorus(V) oxide by chromatography reveals the presence of chain species with from one to nine P atoms. Higher polymers are also present and can be removed from the column only by hydrolysis. Figure 4.6 is a schematic representation of a two-dimensional paper chromatogram: the upper spot sequence corresponds to linear polymers and the lower sequence corresponds to rings. Chain polymers of formula P_n with $n = 10$ to 50 can be isolated as mixed amorphous glasses analogous to those formed by silicates (Section 13.13).

The biological importance of polyphosphates was mentioned at the outset. At physiological pH (close to 7.4), the P—O—P bond is unstable with respect to hydrolysis. Consequently, its hydrolysis can serve as a mechanism for providing the energy to drive a reaction (the Gibbs energy). Similarly, the formation of the P—O—P bond is a means of storing Gibbs energy. The key to energy exchange in metabolism is the hydrolysis of adenosine triphosphate, ATP (**12**), to adenosine diphosphate, ADP (**13**):

$$ATP^{4-} + 2 H_2O \rightarrow ADP^{3-} + HPO_4^{2-} + H_3O^+ \Delta_r G^\ominus = -41 \text{ kJ mol}^{-1} \text{at pH} = 7.4$$

Energy flow in metabolism depends on the subtle construction of pathways to make ATP from ADP. The energy is used metabolically by pathways that have evolved to exploit the delivery of a thermodynamic driving force resulting from the hydrolysis of ATP.

Lewis acidity

Key points: A Lewis acid is an electron pair acceptor; a Lewis base is an electron pair donor.

The Brønsted—Lowry theory of acids and bases focuses on the transfer of a proton between species. Whereas this concept is more general than any that preceded it, it still fails to take into account reactions between substances that show similar features but in which no proton is transferred. This deficiency was remedied by a more general theory of acidity introduced by G.N. Lewis in the same year as Brønsted and Lowry introduced theirs (1923). Lewis's approach became influential only in the 1930s.

A **Lewis acid** is a substance that acts as an electron pair acceptor. A **Lewis base** is a substance that acts as an electron pair donor.[3] We denote a Lewis acid by A and a Lewis base by :B, often omitting any other lone pairs that may be present. The fundamental reaction of Lewis acids and bases is the formation of a **complex** (or adduct), A—B, in which A and :B bond together by sharing the electron pair supplied by the base.

4.7 Examples of Lewis acids and bases

Key points: Brønsted acids and bases exhibit Lewis acidity and basicity; the Lewis definition can be applied to aprotic systems.

A proton is a Lewis acid because it can attach to an electron pair, as in the formation of NH_4^+ from NH_3. It follows that any Brønsted acid, as it provides protons, exhibits Lewis acidity too. Note that the Brønsted acid HA is the complex formed by the Lewis acid H^+ with the Lewis base A^-. We say that a Brønsted acid *exhibits* Lewis acidity rather than that a Brønsted acid *is* a Lewis acid. All Brønsted bases are Lewis bases, because a proton acceptor is also an electron pair donor: an NH_3 molecule, for instance, is a Lewis base as well as a Brønsted base. Therefore, the whole of the material presented in the preceding sections of this chapter can be regarded as a special case of Lewis's approach. However, because the proton is not essential to the definition of a Lewis acid or base, a wider range of substances can be classified as acids and bases in the Lewis scheme than can be classified in the Brønsted scheme.

We meet many examples of Lewis acids later, but we should be alert to the following possibilities:

1 A molecule with an incomplete octet can complete its octet by accepting an electron pair.

   A prime example is $B(CH_3)_3$, which can accept the lone pair of NH_3 and other donors:

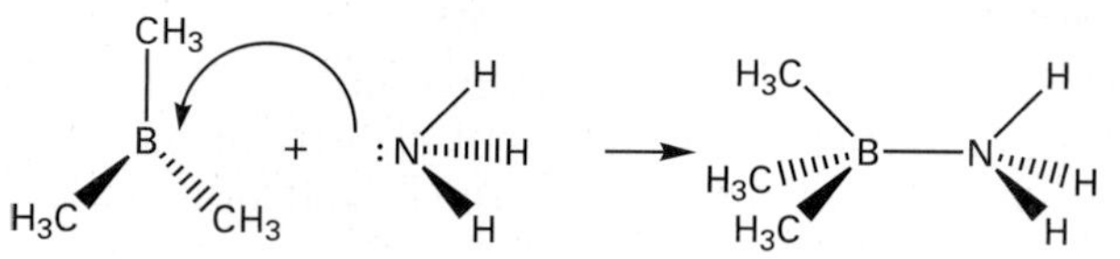

Hence, $B(CH_3)_3$ is a Lewis acid.

---

[3] The terms Lewis acid and base are used in discussions of the equilibrium properties of reactions. In the context of reaction rates, an electron pair donor is called a *nucleophile* and an electron acceptor is called an *electrophile*.

2 A metal cation can accept an electron pair supplied by the base in a coordination compound.

This aspect of Lewis acids and bases is treated at length in Chapters 8 and 19. An example is the hydration of Co^{2+}, in which the lone pairs of H_2O (acting as a Lewis base) donate to the central cation to give $[Co(OH_2)_6]^{2+}$. The Co^{2+} cation is therefore the Lewis acid.

3 A molecule or ion with a complete octet may be able to rearrange its valence electrons and accept an additional electron pair.

For example, CO_2 acts as a Lewis acid when it forms HCO_3^- by accepting an electron pair from an O atom in an OH^- ion:

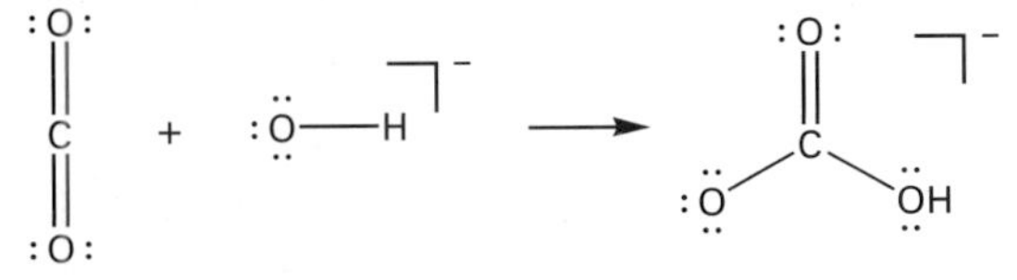

4 A molecule or ion may be able to expand its valence shell (or simply be large enough) to accept another electron pair. An example is the formation of the complex $[SiF_6]^{2-}$ when two F^- ions (the Lewis bases) bond to SiF_4 (the acid).

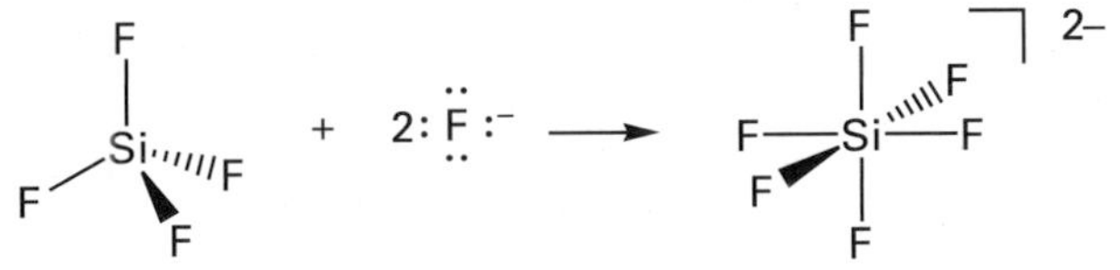

This type of Lewis acidity is common for the halides of the heavier p-block elements, such as SiX_4, AsX_3, and PX_5 (with X a halogen).

Example 4.7 Identifying Lewis acids and bases

Identify the Lewis acids and bases in the reactions (a) $BrF_3 + F^- \rightarrow BrF_4^-$, (b) $KH + H_2O \rightarrow KOH + H_2$.

Answer (a) The acid BrF_3 adds the base F^-. (b) The ionic hydride complex KH provides the base H^- to displace the acid H^+ from water to give H_2 together with KOH, in which the base OH^- is combined with the very weak acid K^+.

Self-test 4.7 Identify the acids and bases in the reactions (a) $FeCl_3 + Cl^- \rightarrow FeCl_4^-$, (b) $I^- + I_2 \rightarrow I_3^-$.

4.8 Group characteristics of Lewis acids

The chemical properties of many compounds can be understood in terms of Lewis acidity and basicity. Here we consider the Lewis acid and base properties of compounds of the main group elements.

(a) Lewis acids and bases of the s-block elements

The existence of hydrated alkali metal ions in water can be regarded as an aspect of their Lewis acid character, with H_2O the Lewis base. Alkali metal ions do not act as Lewis bases but their fluorides act as a source of the Lewis base F^- and form fluoride complexes with Lewis acids, such as SF_4.

$$CsF + SF_4 \rightarrow Cs^+[SF_5]^-$$

The Be atom in beryllium dihalides acts as a Lewis acid by forming a polymeric chain structure in the solid state (**14**). In this structure, a σ bond is formed when a lone pair of electrons of a halide ion, acting as a Lewis base, is donated into an empty sp^3 hybrid orbital on the Be atom. The Lewis acidity of beryllium chloride is also demonstrated by the formation of adducts such as $BeCl_4^{2-}$ (**15**).

(b) Group 13 Lewis acids

Key points: The ability of boron trihalides to act as Lewis acids generally increases in the order $BF_3 < BCl_3 < BBr_3$; aluminium halides are dimeric in the gas phase and are used as catalysts in solution.

The planar molecules BX_3 and AlX_3 have incomplete octets, and the vacant p orbital perpendicular to the plane (**16**) can accept a lone pair from a Lewis base:

The acid molecule becomes pyramidal as the complex is formed and the B—X bonds bend away from their new neighbours.

The order of thermodynamic stability of complexes of $:N(CH_3)_3$ with BX_3 is $BF_3 < BCl_3 < BBr_3$. This order is opposite to that expected on the basis of the relative electronegativities of the halogens: an electronegativity argument would suggest that fluorine, the most electronegative halogen, ought to leave the B atom in BF_3 most electron deficient and hence able to form the strongest bond to the incoming base. The currently accepted explanation is that the halogen atoms in the BX_3 molecule can form π bonds with the empty B$2p$ orbital (**17**), and that these π bonds must be disrupted to make the acceptor orbital available for complex formation. The π bond also favours the planar structure of the molecule, a structure that must be converted into tetrahedral in the adduct. The small F atom forms the strongest π bonds with the B$2p$ orbital: recall that p–p π bonding is strongest for Period 2 elements, largely on account of the small atomic radii of these elements and the significant overlap of their compact $2p$ orbitals. Thus, the BF_3 molecule has the strongest π bond to be broken when the amine forms an N—B bond.

Boron trifluoride is widely used as an industrial catalyst. Its role there is to extract bases bound to carbon and hence to generate carbocations:

Boron trifluoride is a gas, but it dissolves in diethyl ether to give a solution that is convenient to use. This dissolution is also an aspect of Lewis acid character because, as BF_3 dissolves, it forms a complex with the :O atom of a solvent molecule.

Aluminium halides are dimers in the gas phase; aluminium chloride, for example, has molecular formula Al_2Cl_6 in the vapour (**18**). Each Al atom acts as an acid towards a Cl atom initially belonging to the other Al atom. Aluminium chloride is widely used as a Lewis acid catalyst for organic reactions. The classic examples are Friedel—Crafts alkylation (the attachment of R^+ to an aromatic ring) and acylation (the attachment of RCO) during which $AlCl_4^-$ is formed. The catalytic cycle is shown in Fig. 4.7.

(c) Group 14 Lewis acids

Key points: The elements other than carbon exhibit hypervalence and act as Lewis acids by becoming five- or six-coordinate; tin(II) chloride is both a Lewis acid and a Lewis base.

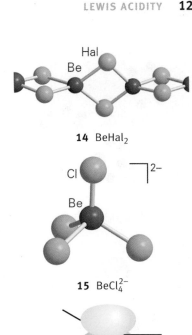

14 $BeHal_2$

15 $BeCl_4^{2-}$

16 AlX_3 and BX_3

17

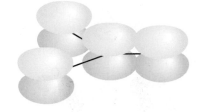

18 Al_2Cl_6

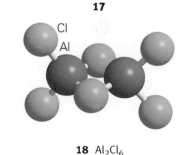

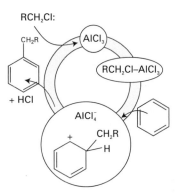

Fig. 4.7 The catalytic cycle for the Friedel–Crafts alklation reaction.

19 $[Si(C_6H_5)(OC_6H_4O)_2]^-$

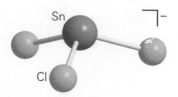

20 $SnCl_3^-$

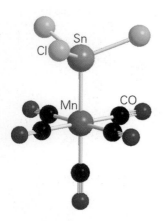

21 $[Mn(SnCl_3)(CO)_5]$

Unlike carbon, a Si atom can expand its valence shell (or is simply large enough) to become hypervalent:

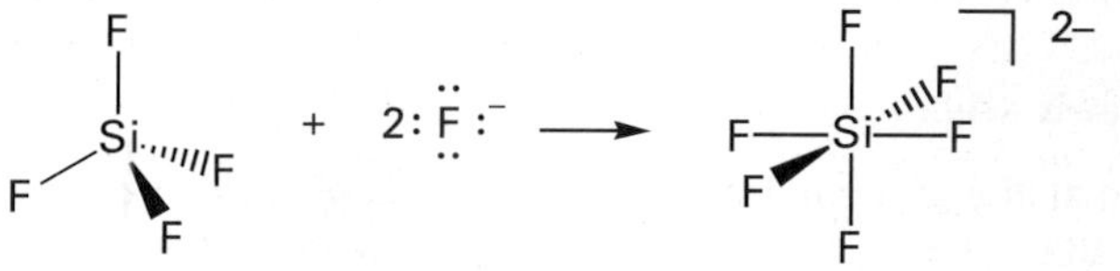

$$\text{SiF}_4 + 2\,:\!\ddot{\text{F}}\!:^- \longrightarrow [\text{SiF}_6]^{2-}$$

Germanium and tin can react similarly. Because the Lewis base F^-, aided by a proton, can displace O^{2-} from silicates, hydrofluoric acid is corrosive towards glass (SiO_2). The trend in acidity for SiX_4, which follows the order $SiI_4 < SiBr_4 < SiCl_4 < SiF_4$, correlates with the increase in the electron-withdrawing power of the halogen from I to F and is the reverse of that for BX_3.

Coordination number 6, as in $[SiF_6]^-$, is not the only state of coordination for silicon above 4. For example, a five-coordinate trigonal bipyramidal structure is possible (**19**).

Tin(II) chloride is both a Lewis acid and a Lewis base. As an acid, $SnCl_2$ combines with Cl^- to form $SnCl_3^-$ (**20**). This complex retains a lone pair, and it is sometimes more revealing to write its formula as $:SnCl_3^-$. It acts as a base to give metal–metal bonds, as in the complex $(CO)_5Mn$—$SnCl_3$ (**21**). Compounds containing metal–metal bonds are currently the focus of much attention in inorganic chemistry, as we see later in the text (Section 18.11). Tin (IV) halides are Lewis acids. They react with halide ions to form SnX_6^{2-}:

$$SnCl_4 + 2\ Cl^- \rightarrow SnCl_6^{2-}$$

The strength of the Lewis acidity follows the order $SnF_4 > SnCl_4 > SnBr_4 > SnI_4$.

Example 4.8 Predicting the relative Lewis basicity of compounds

Rationalize the relative Lewis basicities (a) $(H_3Si)_2O < (H_3C)_2O$; (b) $(H_3Si)_3N < (H_3C)_3N$.

Answer Nonmetallic elements in Period 3 and below can use available d orbitals to expand their valence shells by delocalization of the O or N lone pairs, so the silyl ether and silyl amine are the weaker Lewis bases in each pair.

Self-test 4.8 Given that π bonding between Si and the lone pairs of N is important, what difference in structure between $(H_3Si)_3N$ and $(H_3C)_3N$ do you expect?

(d) Group 15 Lewis acids

Key points: Oxides and halides of the heavier Group 15 elements act as Lewis acids.

Phosphorus pentafluoride is a strong Lewis acid and forms complexes with ethers and amines. The heavier elements of the nitrogen group (Group 15) form some of the most important Lewis acids, SbF_5 being one of the most widely studied compounds. This Lewis acid can be used to produce some of the strongest Brønsted acids, as in the reaction

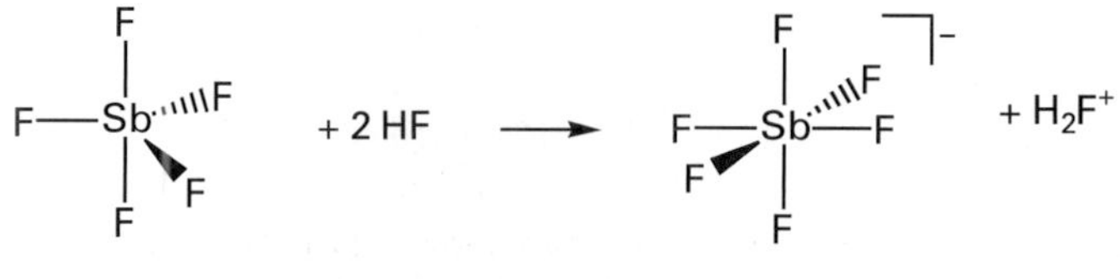

$$\text{SbF}_5 + 2\,HF \longrightarrow [\text{SbF}_6]^- + H_2F^+$$

A **superacid** is a mixture that can protonate almost any organic compound. The most common superacids are formed by dissolving SbF_5 in HSO_3F or anhydrous HF (Box 4.1).

(e) Group 16 Lewis acids

Key points: Sulfur dioxide can act as a Lewis acid by the formation of a complex; to act as a Lewis base, the SO_2 molecule can donate either its S or its O lone pair to a Lewis acid.

Box 4.1 Superacids

A *superacid* is a substance that is a more efficient proton donor than 100 mass per cent H_2SO_4. Superacids are typically viscous, corrosive liquids and can be up to 10^{18} times more acidic than 100 per cent H_2SO_4. They are formed when a powerful Lewis acid is dissolved in a powerful Brønsted acid. The most common superacids are formed when SbF_5 is dissolved in fluorosulfonic acid, HSO_3F, or anhydrous HF. An equimolar mixture of SbF_5 and HSO_3F is known as *Magic Acid*, so named because of its ability to dissolve candle wax. The enhanced acidity is due to the formation of a solvated proton, which is a better proton donor than the acid:

$$SbF_5(l) - 2\,HSO_3F(l) \rightarrow H_2SO_3F^+(sol) + SbF_5SO_3F^-(sol)$$

An even stronger superacid is formed when SbF_5 is added to anhydrous HF:

$$SbF_5(l) + 2\,HF(l) \rightarrow H_2F^+(sol) + SbF_6^-(sol)$$

Other pentafluorides also form superacids in HSO_3F and HF and the acidity of these compounds decreases in the order $SbF_5 > AsF_5 > TaF_5 > NbF_5 > PF_5$.

In the 1960s, George Olah and his colleagues at Case Western Reserve University found that carbonium ions were stabilized when hydrocarbons were dissolved in superacids. Carbocations could not be studied before Olah's experiments, and he won the 1994 Nobel Prize for Chemistry for this work. In inorganic chemistry, superacids have been used to observe a wide variety of reactive cations such as S_8^{2+}, $H_3O_2^+$, Xe_2^+, and HCO^+, some of which have been isolated for structural characterization.

Further reading

G.A. Olah, G.K. Prakash, and J. Sommer, *Superacids*. Wiley, New York (1985).

R.J. Gillespie and J. Laing, Superacid solutions in hydrogen fluoride. *J. Am. Chem. Soc.*, 1988, **110**, 6053.

Sulfur dioxide is both a Lewis acid and a Lewis base. Its Lewis acidity is illustrated by the formation of a complex with a trialkylamine acting as a Lewis base:

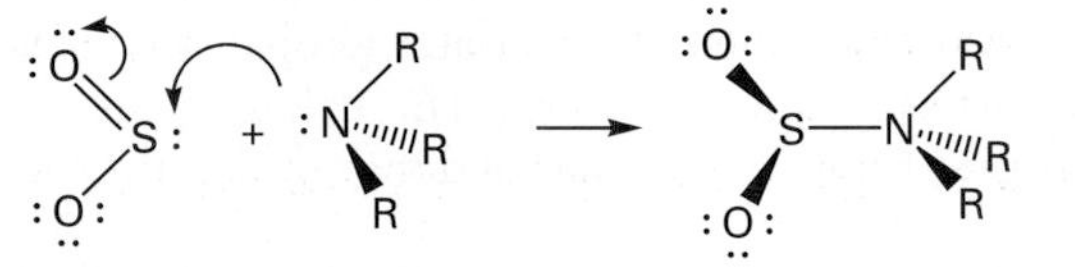

To act as a Lewis base, the SO_2 molecule can donate either its S or its O lone pair to a Lewis acid. When SbF_5 is the acid, the O atom of SO_2 acts as the electron pair donor, but when Ru(II) is the acid, the S atom acts as donor (**22**).

Sulfur trioxide is a strong Lewis acid and a very weak (O donor) Lewis base. Its acidity is illustrated by the reaction

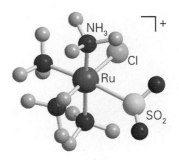

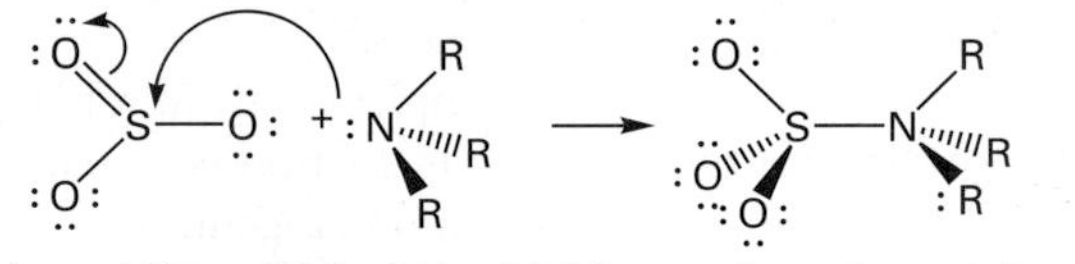

22 $[RuCl(NH_3)_4(SO_2)]^+$

A classic aspect of the acidity of SO_3 is its highly exothermic reaction with water in the formation of sulfuric acid. The resulting problem of having to remove large quantities of heat from the reactor used for the commercial production of the acid is alleviated by using a two-stage process that makes use of another aspect of the trioxide's Lewis acidity. Before dilution, sulfur trioxide is dissolved in sulfuric acid to form the mixture known as *oleum*. This reaction is an example of Lewis acid—base complex formation:

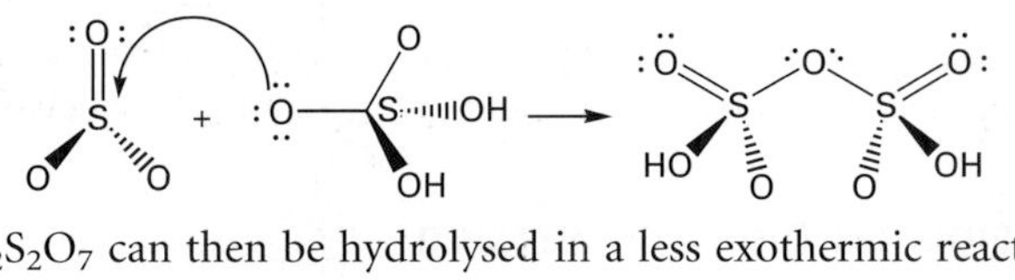

The resulting $H_2S_2O_7$ can then be hydrolysed in a less exothermic reaction:

$$H_2S_2O_7 + H_2O \rightarrow 2\,H_2SO_4$$

(f) Lewis acids of the halogens

Key point: Bromine and iodine molecules act as mild Lewis acids.

Lewis acidity is expressed in an interesting and subtle way by Br_2 and I_2, which are both strongly coloured. The strong visible absorption spectra of Br_2 and I_2 arise from

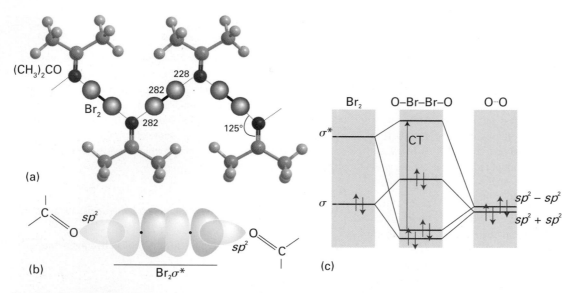

Fig. 4.8 The interaction of Br_2 with the carbonyl groups of two propanone molecules. (a) The structure of $(CH_3)_2COBr_2$ shown by X-ray diffraction. (b) The orbital overlap responsible for the complex formation. (c) A partial molecular orbital energy level diagram for the σ and σ^* orbitals on the Br_2 with the appropriate combinations of the sp^2 orbitals on the two O atoms. The charge transfer transition is labelled CT.

transitions to low-lying unfilled orbitals. The colours of the species therefore suggest that the empty orbitals may be low enough in energy to serve as acceptor orbitals in Lewis acid–base complex formation.[4] Iodine is violet in the solid and gas phases and in non-donor solvents such as trichloromethane. In water, propanone (acetone), or ethanol, all of which are Lewis bases, iodine is brown. The colour changes because a solvent–solute complex is formed from the lone pair of donor molecule O atoms and a low-lying σ^* orbital of the dihalogen.

The interaction of Br_2 with the carbonyl group of propanone is shown in Fig. 4.8. The illustration also shows the transition responsible for the new absorption band observed when a complex is formed. The orbital from which the electron originates in the transition is predominantly the lone pair orbital of the base (the ketone). The orbital to which the transition occurs is predominantly the LUMO of the acid (the dihalogen). Thus, to a first approximation, the transition transfers an electron from the base to the acid and is therefore called a **charge-transfer transition.**

The triiodide ion, I_3^-, is an example of a complex between a halogen acid (I_2) and a halide base (I^-). One of the applications of its formation is to render molecular iodine soluble in water so that it can be used as a titration reagent:

$$I_2(s) + I^-(aq) \rightarrow I_3^-(aq) \qquad K = 725$$

The triiodide ion is one example of a large class of polyhalide ions (Section 16.7).

Reactions and properties of Lewis acids and bases

Lewis acids and bases undergo a variety of characteristic reactions. We review these reactions here, and then show how to characterize the strengths of Lewis acids and bases.

4.9 The fundamental types of reaction

The simplest Lewis acid–base reaction in the gas phase or noncoordinating solvents is **complex formation:**

$$A + :B \rightarrow A{-}B$$

---

[4] The terms *donor–acceptor complex* and *charge-transfer complex* were at one time used to denote these complexes. However, the distinction between these complexes and the more familiar Lewis acid–base complexes is arbitrary and in the current literature the terms are used more or less interchangeably.

Two examples are

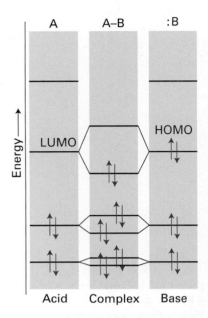

Fig. 4.9 The molecular orbital representation of the orbital interactions responsible for formation of a complex between Lewis acid A and Lewis base :B.

Both reactions involve Lewis acids and bases that are independently stable in the gas phase or in solvents that do not form complexes with them. Consequently, the individual species (as well as the complexes) may be studied experimentally.

Figure 4.9 shows the interaction of orbitals responsible for bonding in Lewis complexes. The exothermic character of the formation of the complex stems from the fact that the newly formed bonding orbital is populated by the two electrons supplied by the base whereas the newly formed antibonding orbital is left unoccupied. As a result, there is a net lowering of energy when the bond forms.

(a) Displacement reactions

Key point: In a displacement reaction, an acid or base drives out another acid or base from a Lewis complex.

A **displacement** of one Lewis base by another is a reaction of the form

$$B\text{—}A + :B' \rightarrow B: + A\text{—}B'$$

An example is

All Brønsted proton transfer reactions are of this type, as in

$$HS^-(aq) + H_2O(l) \rightarrow S^{2-}(aq) + H_3O^+(aq)$$

In this reaction, the Lewis base H_2O displaces the Lewis base S^- from its complex with the acid H^+. Displacement of one acid by another,

$$A' + B\text{—}A \rightarrow A'\text{—}B + A$$

is also possible, as in the reaction

In the context of d-metal complexes, a displacement reaction in which one ligand is driven out of the complex and is replaced by another is generally called a **substitution reaction** (Section 20.1).

(b) Metathesis reactions

Key point: A metathesis reaction is a displacement reaction assisted by the formation of another complex.

A **metathesis reaction** (or 'double displacement reaction') is an interchange of partners:[5]

$$A\text{—}B + A'\text{—}B' \rightarrow A\text{—}B' + A'\text{—}B$$

---

[5] The name metathesis comes from the Greek word for exchange.

The displacement of the base :B by :B′ is assisted by the extraction of :B by the acid A′. An example is the reaction

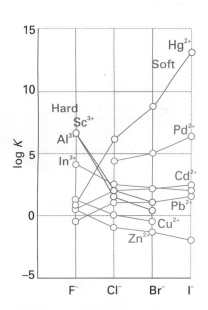

Here the base Br⁻ displaces I⁻, and the extraction is assisted by the formation of the less soluble AgI.

4.10 Hard and soft acids and bases

The proton (H^+) was the key electron pair acceptor in the discussion of Brønsted acid and base strengths. When considering Lewis acids and bases we must allow for a greater variety of acceptors and hence more factors that influence the interactions between electron pair donors and acceptors in general.

(a) The classification of acids and bases

Key points: Hard and soft acids and bases are identified empirically by the trends in stabilities of the complexes that they form: hard acids tend to bind to hard bases and soft acids tend to bind to soft bases.

It proves helpful when considering the interactions of Lewis acids and bases containing elements drawn from throughout the periodic table to consider at least two main classes of substance. The classification of substances as 'hard' and 'soft' acids and bases was introduced by R.G. Pearson; it is a generalization—and a more evocative renaming—of the distinction between two types of behaviour that were originally named simply 'class *a*' and 'class *b*' respectively, by S. Ahrland, J. Chatt, and N.R. Davies.

The two classes are identified empirically by the opposite order of strengths (as measured by the equilibrium constant, K_f, for the formation of the complex) with which they form complexes with halide ion bases:

- Hard acids bond in the order: $I^- < Br^- < Cl^- < F^-$
- Soft acids bond in the order: $F^- < Cl^- < Br^- < I^-$

Figure 4.10 shows the trends in K_f for complex formation with a variety of halide ion bases. The equilibrium constants increase steeply from F^- to I^- when the acid is Hg^{2+}, indicating that Hg^{2+} is a soft acid. The trend is less steep but in the same direction for Pb^{2+}, which indicates that this ion is a borderline soft acid. The trend is in the opposite direction for Zn^{2+}, so this ion is a borderline hard acid. The steep downward slope for Al^{3+} indicates that it is a hard acid. A useful rule of thumb is that small cations, which are not easily polarized, are hard and form complexes with small anions. Large cations are more polarizable and are soft.

For Al^{3+}, the binding strength increases as the electrostatic parameter ($\xi = z^2/r$) of the anion increases, which is consistent with an ionic model of the bonding. For Hg^{2+}, the binding strength increases with increasing polarizability of the anion. These two correlations suggest that hard acid cations form complexes in which simple Coulombic, or ionic, interactions are dominant, and that soft acid cations form more complexes in which covalent bonding is important.

A similar classification can be applied to neutral molecular acids and bases. For example, the Lewis acid phenol forms a more stable complex by hydrogen bonding to $(C_2H_5)_2O$: than to $(C_2H_5)_2S$:. This behaviour is analogous to the preference of Al^{3+} for F^- over Cl^-. By contrast, the Lewis acid I_2 forms a more stable complex with $(C_2H_5)_2S$:. We can conclude that phenol is hard whereas I_2 is soft.

In general, acids are identified as hard or soft by the thermodynamic stability of the complexes they form, as set out for the halide ions above and for other species as follows:

- Hard acids bond in the order: $R_3P \ll R_3N, R_2S \ll R_2O$
- Soft acids bond in the order: $R_2O \ll R_2S, R_3N \ll R_3P$

Fig. 4.10 The trends in stability constants for complex formation with a variety of halide ion bases. Hard ions are indicated by the blue lines, soft ions by the red line. Borderline hard or borderline soft ions are indicated by green lines.

Table 4.3 The classification of Lewis acids and bases*

	Hard	Borderline	Soft
Acids	H^+, Li^+, Na^+, K^+	Fe^{2+}, Co^{2+}, Ni^{2+}	Cu^+, Au^+, Ag^+, Tl^+, Hg^+
	Be^{2+}, Mg^{2+}, Ca^{2+}	Cu^{2+}, Zn^{2+}, Pb^{2+}	Pd^{2+}, Cd^{2+}, Pt^{2+}, Hg^{2+}
	Cr^{2+}, Cr^{3+}, Al^{3+}	SO_2, BBr_3	BH_3
	SO_3, BF_3		
Bases	F^-, OH^-, H_2O, NH_3	NO_2^-, SO_3^{2-}, Br^-	H^-, R^-, $\underline{C}N^-$, CO, I^-
	CO_3^{2-}, NO_3^-, O^{2-}	N_3^-, N_2	$\underline{S}CN^-$, R_3P, C_6H_5
	SO_4^{2-}, PO_4^{3-}, ClO_4^-	C_6H_5N, SCN^-	R_2S

* The underlined element is the site of attachment to which the classification refers.

Bases can also be defined as soft or hard. Bases such as halides and oxoanions are classified as hard because ionic bonding will be predominant in most of the complexes they form. Many soft bases bond through a carbon atom, such as CO and CN⁻. In addition to donating electron density to the metal through a σ interaction, these species are able to accept electron density through the π orbitals present on the base. The bonding is, consequently, predominantly covalent in character. As these soft bases are able to accept electron density into π orbitals they are known as **π acids**. The nature of this bonding will be explored in Chapter 19.

It follows from the definition of hardness that

- Hard acids tend to bind to hard bases.
- Soft acids tend to bind to soft bases.

When species are analysed with these rules in mind, it is possible to identify the classification summarized in Table 4.3.

(b) Interpretation of hardness

Key points: Hard acid–base interactions are predominantly electrostatic; soft acid–base interactions are predominantly covalent.

The bonding between hard acids and bases can be described approximately in terms of ionic or dipole–dipole interactions. Soft acids and bases are more polarizable than hard acids and bases, so the acid–base interaction has a more pronounced covalent character.

Although the type of bond formation is a major reason for the distinction between the two classes, we must not forget that there are other contributions to the Gibbs energy of complex formation and hence to the equilibrium constant. Among these contributions are:

1 The rearrangement of the substituents of the acid and base that may be necessary to permit formation of the complex.

2 Steric repulsion between substituents on the acid and the base.

3 Competition with the solvent in reactions in solution.

Later in the chapter we shall see that these additional contributions can have a marked effect on the outcome of a reaction.

(c) Chemical consequences of hardness

Key points: Hard–hard and soft–soft interactions help to systematize complex formation but must be considered in the light of other possible influences on bonding.

The concepts of hardness and softness help to rationalize a great deal of inorganic chemistry. For instance, they are useful for choosing preparative conditions and predicting the directions of reactions, and they help to rationalize the outcome of metathesis reactions. However, the concepts must always be used with due regard for other factors that may affect the outcome of reactions. This deeper understanding of chemical reactions will grow in the course of the rest of the book. For the time being we shall limit the discussion to a few straightforward examples.

The classification of molecules and ions as hard or soft acids and bases helps to clarify the terrestrial distribution of the elements described in Chapter 1. The tendency of soft acids to bond to soft bases and of hard acids to bond to hard bases explains certain aspects of the **Goldschmidt classification** of the elements into four types, a scheme widely used in geochemistry. Two of the classes are the **lithophile elements** and the **chalcophile elements**. The lithophile elements, which are found primarily in the Earth's crust (the lithosphere) in silicate minerals, include lithium, magnesium, titanium, aluminium, and chromium (as their cations). These cations are hard, and are found in association with the hard base O^{2-}. The chalcophile elements, however, are often found in combination with sulfide (and selenide and telluride) minerals, and include cadmium, lead, antimony, and bismuth. These elements (as their cations) are soft, and are found in association with the soft base S^{2-} (or Se^{2-} and Te^{2-}). Zinc cations are borderline hard, but softer than Al^{3+} and Cr^{3+}, and zinc is also often found as its sulfide.

Example 4.9 Explaining the Goldschmidt classification

The common ores of nickel and copper are sulfides. By contrast, aluminium is obtained from the oxide and calcium from the carbonate. Can these observations be explained in terms of hardness?

Answer From Table 4.3 we know that O^{2-} and CO_3^{2-} are hard bases; S^{2-} is a soft base. The table also shows that the cations Ni^{2+} and Cu^{2+} are considerably softer acids than Al^{3+} or Ca^{2+}. Hence the hard–hard and soft–soft rule accounts for the sorting observed.

Self-test 4.9 Of the metals cadmium, rubidium, chromium, lead, strontium, and palladium, which might be expected to be found in aluminosilicate minerals and which in sulfides?

Table 4.4 Drago–Wayland parameters for some acids and bases*

	E	C
Acids		
Antimony pentachloride	15.1	10.5
Boron trifluoride	20.2	3.31
Iodine	2.05	2.05
Iodine monochloride	10.4	1.70
Phenol	8.86	0.90
Sulfur dioxide	1.88	1.65
Trichloromethane	6.18	0.32
Trimethylboron	12.6	3.48
Bases		
Acetone	2.02	4.67
Ammonia	2.78	7.08
Benzene	0.57	1.21
Dimethylsulfide	0.70	15.26
Dimethylsulfoxide	2.76	5.83
Methylamine	2.66	12.00
p-Dioxane	2.23	4.87
Pyridine	2.39	13.10
Trimethylphosphine	17.2	13.40

*E and C parameters are often reported to give ΔH in kcal mol^{-1}; we have multiplied both by $\sqrt{(4.184)}$ to obtain ΔH in kJ mol^{-1}.

Polyatomic anions may contain donor atoms of different hardnesses. For example, the SCN^- ion is a base as it comprises both the harder N atom and the softer S atom. The ion binds to the hard Si atom through N. However, with a soft acid, such as a metal ion in a low oxidation state, the ion bonds through S. Platinum(II), for example, forms Pt—SCN in the complex $[Pt(SCN)_4]^{2-}$.

4.11 Thermodynamic acidity parameters

Key points: The standard enthalpies of complex formation are reproduced by the *E* and *C* parameters of the Drago–Wayland equation that reflect, in part, the ionic and covalent contributions to the bond in the complex.

An important alternative to the hard–soft classification of acids and bases makes use of an approach in which electronic, structural rearrangement, and steric effects are incorporated into a small set of parameters. The standard reaction enthalpies of complex formation

$$A(g) + B(g) \rightarrow A\text{—}B(g) \quad \Delta_r H^{\ominus}(A\text{—}B)$$

can be reproduced by the **Drago–Wayland equation**:

$$-\Delta_r H^{\ominus}(A\text{—}B)/(kJ\ mol^{-1}) = E_A E_B + C_A C_B \tag{4.5}$$

The parameters *E* and *C* were introduced with the idea that they represent 'electrostatic' and 'covalent' factors, respectively, but in fact they must accommodate all factors except solvation. The compounds for which the parameters are listed in Table 4.4 satisfy the equation with an error of less than ± 3 kJ mol^{-1}, as do a much larger number of examples in the original papers.

The Drago–Wayland equation is semiempirical but very successful and useful. In addition to providing estimates of the enthalpies of complex formation for over 1500 complexes, these enthalpies can be combined to calculate the enthalpies of displacement and metathesis reactions. Moreover, the equation is useful for reactions of acids and bases in nonpolar, noncoordinating solvents as well as for reactions in the gas phase. The major limitation is that the equation is restricted to substances that can conveniently be studied in the gas phase or in noncoordinating solvents; hence, in the main it is limited to neutral molecules.

4.12 Solvents as acids and bases

Most solvents are either electron-pair acceptors or donors and hence are either Lewis acids or bases. The chemical consequences of solvent acidity and basicity are considerable, as they help to account for the differences between reactions in aqueous and nonaqueous media (see Box 4.2). It follows that a displacement reaction often occurs

---

Box 4.2 Nonaqueous solvents

A good example of a synthetically useful nonaqueous solvent is tetrahydrofuran, THF (**B1**), a cyclic nonpolar ether that boils at 66°C. This weak hard-base solvent can readily be dried and deoxygenated by distillation from sodium strips in a nitrogen atmosphere. After use in syntheses (including preparation of air-sensitive compounds) it can easily be removed from a reaction mixture by heating under reduced pressure. For these reasons, it is one of the most widely used solvents for the synthesis of organometallic compounds. The Lewis basicity of the O atom is sometimes important, as it can coordinate to cations and metal atoms. For example, the preparation of substituted metal carbonyls, $[M(CO)_5L]$ (L = phosphine, amine, etc.) can be accomplished by photolysis of $[M(CO)_6]$ in THF to give $[M(CO)_5(thf)]$ and carbon monoxide. This intermediate is then allowed to react with the entering ligand L. Other useful polar aprotic solvents for inorganic substances include acetonitrile (methyl cyanide), CH_3CN, and dimethylsulfoxide, $(CH_3)_2SO$ (DMSO).

B4.1 Tetrahydrofuran

Ammonia is a strong hard base; this property is very important for its ability to coordinate to d-block acids and protons. It should not be forgotten that although ammonia is commonly regarded as a base, its own protons can act as centres of Lewis acidity because of their ability to participate in hydrogen bonding. Liquid ammonia (bp −33°C) is readily exploited as a solvent by using a Dewar flask. Despite a somewhat lower relative permittivity ($\varepsilon_r = 22$) than that of water, many salts of alkali metal cations with large anions are reasonably soluble in liquid ammonia. Organic compounds are often more soluble in liquid ammonia than they are in water.

One of the most remarkable reactions in liquid ammonia is the result of dissolution of alkali metals. Solutions of alkali metals are strongly reducing. Electron paramagnetic resonance spectra show that the solutions contain unpaired electrons. The blue colour typical of the solutions is the outcome of a very broad optical absorption band in the near IR with a maximum near 1500 nm.

The metal is ionized in ammonia solution to give 'solvated electrons':

$$Na(s) + NH_3(l) \rightarrow Na^+(am) + e^-(am)$$

where 'am' denotes ammonia solution. The blue solutions survive for long times at low temperature but decompose slowly to give hydrogen and sodium amide, $NaNH_2$. The exploitation of the blue solutions to produce compounds called 'electrides' is discussed in Section 10.9.

Liquid hydrogen fluoride (bp 19.5°C) is an acidic solvent with considerable Brønsted acid strength and a relative permittivity comparable to that of water. It is a good solvent for ionic substances. However, as it is both highly reactive and toxic, it presents handling problems, including its ability to etch glass. In practice, hydrogen fluoride is usually contained in polytetrafluoroethylene and polychlorotrifluoroethylene vessels.

Although the conjugate base of HF is formally F^-, the ability of HF to form a strong hydrogen bond to F^- means that the conjugate base is better regarded as the bifluoride ion, FHF^-. Many fluorides are soluble in hydrogen fluoride as a result of the formation of this ion; for example

$$LiF(s) + HF(l) \rightarrow Li^+(hf) + [FHF]^-(hf)$$

where 'hf' denotes solution in hydrogen fluoride. Because HF is quite acidic, protonation of the solute commonly accompanies bifluoride ion formation:

$$CH_3OH(l) + 2\,HF(l) \rightarrow [CH_2OH_2]^+(hf) + [FHF]^-(hf)$$

$$CH_3COOH(l) + 2\,HF(l) \rightarrow [CH_3C(OH)_2]^+(hf) + [FHF]^-(hf)$$

The second reaction is striking because acetic acid, an acid in water, is here acting as a base.

One of the most widely used and inexpensive nonaqueous solvent is supercritical CO_2. A *supercritical fluid* is a substance above its critical point, which is the highest temperature and pressure at which its vapour and liquid can exist in equilibrium. The critical point for CO_2 is 31.1°C and 7.382 MPa. The densities of supercritical fluids change dramatically with small changes in temperature or pressure and their viscosities are less than those of typical solvents. These properties and the fact that CO_2 is environmentally benign (as there is no net gain in the amount of CO_2) make supercritical CO_2 a very attractive solvent to industry. It is used in food processing for the extraction of organic compounds from natural products. For example, it is used in the decaffeination of tea and coffee and to extract flavours and fragrances from seeds, roots, leaves, and flowers. In the pharmaceutical industry supercritical CO_2 can be used to extract taxol, a promising anticancer drug, from the bark of *Taxus brevifolia*, the Pacific yew tree. It is used as a solvent for cleaning precision machinery as the solubilities of greases and oils are high in supercritical CO_2 and it does not leave a residue. Supercritical CO_2^- soluble dyes for the textile industry have been developed that result in almost 100 per cent uptake by the fabric and eliminate the production of dye-laden waste water.

Further reading

J. Chipperfield, *Non-aqueous solvents*. Oxford University Press (1999). A readable introduction.

N. Corcoran, *Chemistry in non-aqueous solvents*. Kluwer Academic Publishers (2003). A comprehensive account.

when a solute dissolves in a solvent, and that the subsequent reactions of the solution are also usually either displacements or metatheses. For example, when antimony pentafluoride dissolves in bromine trifluoride, the following displacement reaction occurs:

$$SbF_5 + BrF_3(l) \rightarrow BrF_2^+ + SbF_6^-$$

In the reaction, the strong Lewis acid SbF_5 abstracts F^- from BrF_3. A more familiar example of the solvent as participant in a reaction is in Brønsted theory. In this theory, the acid (H^+) is always regarded as complexed with the solvent, as in H_3O^+ if the solvent is water, and reactions are treated as the transfer of the acid, the proton, from a basic solvent molecule to another base. Only the saturated hydrocarbons among common solvents lack significant Lewis acid or base character.

(a) Basic solvents

Key points: Basic solvents are common; they may form complexes with the solute and participate in displacement reactions.

Solvents with Lewis base character are common. Most of the well-known polar solvents, including water, alcohols, ethers, amines, dimethylsulfoxide (DMSO, $(CH_3)_2SO$), dimethylformamide (DMF, $(CH_3)_2NCHO$), and acetonitrile (CH_3CN), are hard Lewis bases. Dimethylsulfoxide is an interesting example of a solvent that is hard on account of its O donor atom and soft on account of its S donor atom. Reactions of acids and bases in these solvents are generally displacements:

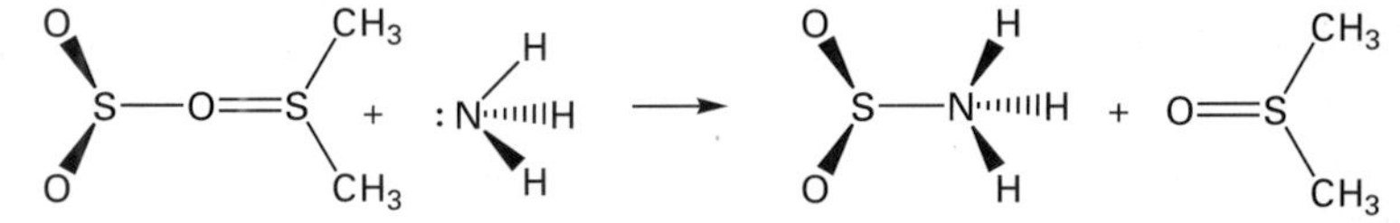

Example 4.10 Accounting for properties in terms of the Lewis basicity of solvents

Silver perchlorate, $AgClO_4$, is significantly more soluble in benzene than in alkane solvents. Account for this observation in terms of Lewis acid–base properties.

Answer The π electrons of benzene, a soft base, are available for complex formation with the empty orbitals of the cation Ag^+, a soft acid. The species $[Ag-C_6H_6]^+$ is the complex of the acid Ag^+ with π electrons of the weak base benzene.

Self-test 4.10 Boron trifluoride, BF_3, a hard acid, is often used in the laboratory as a solution in diethyl ether, $(C_2H_5)_2O:$, a hard base. Draw the structure of the complex that results from the dissolution of $BF_3(g)$ in $(C_2H_5)_2O(l)$.

(b) Acidic and neutral solvents

Key points: Hydrogen bond formation is an example of Lewis complex formation; other solvents may also show Lewis acid character.

Hydrogen bonding can be regarded as an example of complex formation. The 'reaction' is between A—H (the Lewis acid) and :B (the Lewis base) and gives the complex conventionally denoted A—H $\cdots$ B. Hence, many solutes that form hydrogen bonds with a solvent can be regarded as dissolving because of complex formation. A consequence of this view is that an acidic solvent molecule is displaced when proton transfer occurs:

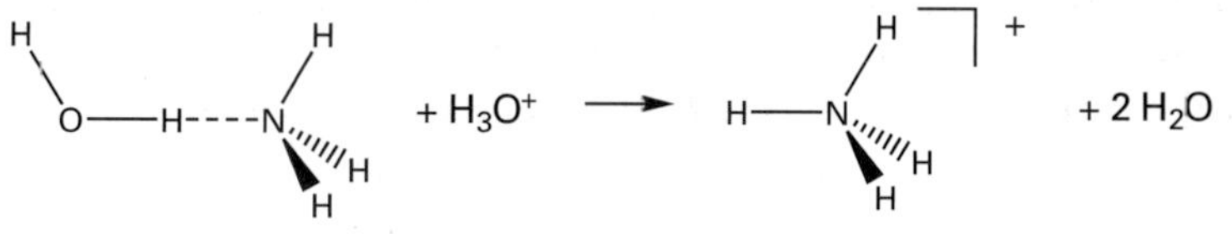

$$\text{A—B} \qquad \text{A'B'} \qquad \text{A'—B} \qquad \text{AB'}$$

Liquid sulfur dioxide is a good soft acidic solvent for dissolving the soft base benzene. Unsaturated hydrocarbons may act as acids or bases by using their π or π^* orbitals as frontier orbitals. Alkanes with electronegative substituents, such as haloalkanes (e.g. $CHCl_3$), are significantly acidic at the hydrogen atom.

4.13 Heterogeneous acid–base reactions

Key point: The surfaces of many catalytic materials and minerals have Brønsted and Lewis acid sites.

Some of the most important reactions involving the Lewis and Brønsted acidity of inorganic compounds occur at solid surfaces. For example, **surface acids**, which are solids with a high surface area and Lewis acid sites, are used as catalysts in the petrochemical industry for the interconversion of hydrocarbons. The surfaces of many materials that are important in the chemistry of soil and natural waters also have Brønsted and Lewis acid sites.

Silica surfaces do not readily produce Lewis acid sites because —OH groups remain tenaciously attached at the surface of SiO_2 derivatives; as a result, Brønsted acidity is dominant. The Brønsted acidity of silica surfaces themselves is only moderate (and comparable to that of acetic acid). However, as we have already remarked, aluminosilicates display strong Brønsted acidity. When surface OH groups are removed by heat treatment, the aluminosilicate surface possesses strong Lewis acid sites.

Surface reactions carried out using the Brønsted acid sites of silica gels are used to prepare thin coatings of a wide variety of organic groups using surface modification reactions such as

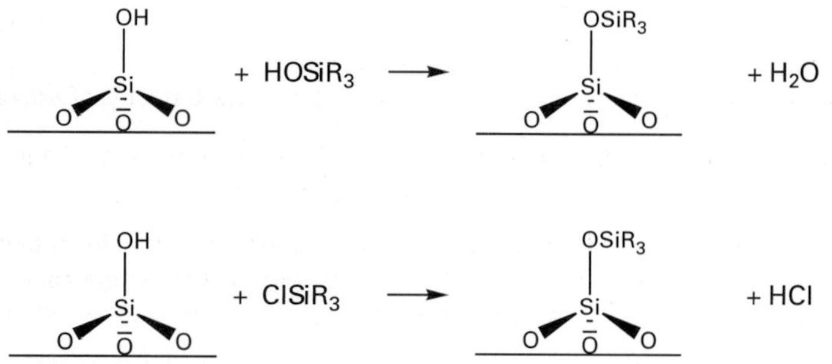

Thus, silica gel surfaces can be modified to have affinities for specific classes of molecules. This procedure greatly expands the range of stationary phases that can be used for chromatography. The surface —OH groups on glass can be modified similarly, and glassware treated in this manner is sometimes used in the laboratory when proton-sensitive compounds are being studied.

FURTHER READING

W. Stumm and J.J. Morgan, *Aquatic chemistry: chemical equilibria and rates in natural waters*. Wiley, New York (1995). The classic text on the chemistry of natural waters.

N. Corcoran, *Chemistry in non-aqueous solvents*. Kluwer Academic Publishers (2003). A comprehensive account.

J. Chipperfield, *Non-aqueous solvents*. Oxford University Press (1999). A readable introduction to the topic.

J. Burgess, *Ions in solution*. Ellis Horwood, Chichester (1988). A readable account of solvation with an introduction to acidity and polymerization.

J. Burgess, *Ions in solution: basic principles of chemical interactions*. Ellis Horwood, Chichester (1999).

EXERCISES

4.1 Sketch an outline of the s and p blocks of the periodic table and indicate on it the elements that form (a) strongly acidic oxides and (b) strongly basic oxides, and (c) show the regions for which amphoterism is common.

4.2 Identify the conjugate bases corresponding to the following acids: $[Co(NH_3)_5(OH_2)]^{3+}$, HSO_4^-, CH_3OH, $H_2PO_4^-$, $Si(OH)_4$, HS^-.

4.3 Identify the conjugate acids of the bases C_5H_5N (pyridine), HPO_4^{2-}, O^{2-}, CH_3COOH, $[Co(CO)_4]^-$, CN^-.

4.4 The K_a of ethanoic acid, CH_3COOH, in water is 1.8×10^{-5}. Calculate K_b of the conjugate base, $CH_3CO_2^-$.

4.5 Draw the structures of chloric and chlorous acid and predict their pK_a values using Pauling's rules.

4.6 Aided by Fig. 4.2 (taking solvent levelling into account), identify which bases from the following lists are (a) too strong to be studied experimentally; (b) too weak to be studied experimentally; or (c) of directly measurable base strength. (i) CO_3^{2-}, O^{2-}, ClO_4^-, and NO_3^- in water; (ii) HSO_4^-, NO_3^-, ClO_4^- in H_2SO_4.

4.7 The aqueous solution pK_a values for HOCN, H_2NCN, and CH_3CN are approximately 4, 10.5, and 20 (estimated), respectively. Explain the trend in these —CN derivatives of binary acids and compare them with H_2O, NH_3, and CH_4. Is —CN electron donating or withdrawing?

4.8 The pK_a value of $HAsO_4^{2-}$ is 11.6. Is this value consistent with Pauling's rules?

4.9 Use Pauling's rules to place the following acids in order of increasing acid strength: HNO_2, H_2SO_4, $HBrO_3$, and $HClO_4$ in a nonlevelling solvent.

4.10 Draw the structures and indicate the charges of the tetraoxoanions of $X = Si$, P, S, and Cl. Summarize and account for the trends in the pK_a values of their conjugate acids.

4.11 Which member of the following pairs is the stronger acid? Give reasons for your choice. (a) $[Fe(OH_2)_6]^{3+}$ or $[Fe(OH_2)_6]^{2+}$, (b) $[Al(OH_2)_6]^{3+}$ or $[Ga(OH_2)_6]^{3+}$, (c) $Si(OH)_4$ or $Ge(OH)_4$, (d) $HClO_3$ or $HClO_4$, (e) H_2CrO_4 or $HMnO_4$, (f) H_3PO_4 or H_2SO_4.

4.12 Arrange the oxides Al_2O_3, B_2O_3, BaO, CO_2, Cl_2O_7, SO_3 in order from the most acidic through amphoteric to the most basic.

4.13 Arrange the acids HSO_4^-, H_3O^+, H_4SiO_4, CH_3GeH_3, NH_3, HSO_3F in order of increasing acid strength.

4.14 The ions Na^+ and Ag^+ have similar radii. Which aqua ion is the stronger acid? Why?

4.15 Which of the elements Al, As, Cu, Mo, Si, B, Ti form oxide polyanions and which form oxide polycations?

4.16 When a pair of aqua cations forms an M—O—M bridge with the elimination of water, what is the general rule for the change in charge per M atom on the ion?

4.17 Write a balanced equation for the formation of $P_2O_7^{4-}$ from PO_4^{3-}. Write a balanced equation for the dimerization of $[Fe(OH_2)_6]^{3+}$ to give $[(H_2O)_4Fe(OH)_2Fe(OH_2)_4]^{4+}$.

4.18 Write balanced equations for the main reaction occurring when (a) H_3PO_4 and Na_2HPO_4 and (b) CO_2 and $CaCO_3$ are mixed in aqueous media.

4.19 Sketch the p block of the periodic table. Identify as many elements as you can that act as Lewis acids in one of their lower oxidation states and give the formula of a representative Lewis acid for each element.

4.20 For each of the following processes, identify the acids and bases involved and characterize the process as complex formation or acid–base displacement. Identify the species that exhibit Brønsted acidity as well as Lewis acidity.

(a) $SO_3 + H_2O \rightarrow HSO_4^- + H^+$

(b) $CH_3[B_{12}] + Hg^{2+} \rightarrow [B_{12}]^+ + CH_3Hg^+$; $[B_{12}]$ designates the Co-porphyrin, vitamin B_{12}.

(c) $KCl + SnCl_2 \rightarrow K^+ + [SnCl_3]^-$

(d) $AsF_3(g) + SbF_5(l) \rightarrow [AsF_2]^+[SbF_6]^-(s)$

(e) Ethanol dissolves in pyridine to produce a nonconducting solution.

4.21 Select the compound on each line with the named characteristic and state the reason for your choice.

(a) Strongest Lewis acid:
   BF_3 BCl_3 BBr_3
   $BeCl_2$ BCl_3
   $B(n\text{-}Bu)_3$ $B(t\text{-}Bu)_3$

(b) More basic towards $B(CH_3)_3$
   Me_3N Et_3N
   $2\text{-}CH_3C_5H_4N$ $4\text{-}CH_3C_5H_4N$

4.22 Using hard–soft concepts, which of the following reactions are predicated to have an equilibrium constant greater than 1? Unless otherwise stated, assume gas-phase or hydrocarbon solution and 25°C.

(a) $R_3PBBr_3 + R_3NBF_3 \rightleftharpoons R_3PBF_3 + R_3NBBr_3$

(b) $SO_2 + (C_6H_5)_3POHC(CH_3)_3 \rightleftharpoons (C_6H_5)_3PSO_2 + HOC(CH_3)_3$

(c) $CH_3HgI + HCl \rightleftharpoons CH_3HgCl + HI$

(d) $[AgCl_2]^-(aq) + 2CN^-(aq) \rightleftharpoons [Ag(CN)_2]^-(aq) + 2Cl^-(aq)$

4.23 The molecule $(CH_3)_2N$—PF_2 has two basic atoms, P and N. One is bound to B in a complex with BH_3, the other to B in a complex with BF_3. Decide which is which and state your reason.

4.24 The enthalpies of reaction of trimethylboron with NH_3CH_3, $NH_2(CH_3)_2NH$, and $(CH_3)_3N$ are -58, -74, -81, and -74 kJ mol^{-1}, respectively. Why is trimethylamine out of line?

4.25 With the aid of the table of E and C values (Table 4.4), discuss the relative basicity in (a) acetone and dimethylsulfoxide, (b) dimethylsulfide and dimethylsulfoxide. Comment on a possible ambiguity for dimethylsulfoxide.

4.26 Give the equation for the dissolution of SiO_2 glass by HF and interpret the reaction in terms of Lewis and Brønsted acid–base concepts.

4.27 Aluminium sulfide, Al_2S_3, gives off a foul odour characteristic of hydrogen sulfide when it becomes damp. Write a balanced chemical equation for the reaction and discuss it in terms of acid–base concepts.

4.28 Describe the solvent properties that would (a) favour displacement of Cl^- by I^- from an acid centre, (b) favour basicity of R_3As over R_3N, (c) favour acidity of Ag^+ over Al^{3+}, (d) promote the reaction $2\,FeCl_3 + ZnCl_2 \rightarrow Zn^{2+} + 2[FeCl_4]^-$. In each case, suggest a specific solvent that might be suitable.

4.29 Why are strongly acidic solvents (e.g. SbF_5/HSO_3F) used in the preparation of cations such as I_2^+ and Se_8^+, whereas strongly basic solvents are needed to stabilize anionic species such as S_4^{2-} and Pb_9^{4-}?

4.30 The Lewis acid $AlCl_3$ catalysis of the acylation of benzene was described in Section 4.8b. Propose a mechanism for a similar reaction catalysed by an alumina surface.

4.31 Use acid–base concepts to comment on the fact that the only important ore of mercury is cinnabar, HgS, whereas zinc occurs in nature as sulfides, silicates, carbonates, and oxides.

4.32 Write balanced Brønsted acid–base equations for the dissolution of the following compounds in liquid hydrogen fluoride: (a) CH_3CH_2OH, (b) NH_3, (c) C_6H_5COOH.

4.33 Is the dissolution of silicates in HF a Lewis acid–base reaction, a Brønsted acid–base reaction, or both?

4.34 The f-block elements are found as M(III) lithophiles in silicate minerals. What does this indicate about their hardness?

PROBLEMS

4.1 In analytical chemistry a standard trick for improving the detection of the stoichiometric point in titrations of weak bases with strong acids is to use acetic acid as a solvent. Explain the basis of this approach.

4.2 In the gas phase, the base strength of amines increases regularly along the series $NH_3 < CH_3NH_2 < (CH_3)_2NH < (CH_3)_3N$. Consider the role of steric effects and the electron-donating ability of CH_3 in determining this order. In aqueous solution, the order is reversed. What solvation effect is likely to be responsible?

4.3 The hydroxoacid $Si(OH)_4$ is weaker than H_2CO_3. Write balanced equations to show how dissolving a solid M_2SiO_4 can lead to a reduction in the pressure of CO_2 over an aqueous solution. Explain why silicates in ocean sediments might limit the increase of CO_2 in the atmosphere.

4.4 The precipitation of $Fe(OH)_3$ discussed in the chapter is used to clarify waste waters, because the gelatinous hydrous oxide is very efficient at the coprecipitation of some contaminants and the entrapment of others. The solubility constant of $Fe(OH)_3$ is $K_s = [Fe^{3+}][OH^-]^3 \approx 1 \times 10^{-38}$. As the autoprotolysis constant of water links $[H_3O^+]$ to $[OH^-]$ by $K_w = [H_3O^+][OH^-] = 1.0 \times 10^{-14}$, we can rewrite the solubility constant by substitution as $[Fe^{3+}]/[H^+]^3 = 10^4$. (a) Balance the chemical equation for the precipitation of $Fe(OH)_3$ when iron(III) nitrate is added to water. (b) If 6.6 kg of $Fe(NO_3)_3.9H_2O$ is added to 100 dm^3 of water, what is the final pH of the solution and the molar concentration of Fe^{3+}, neglecting other forms of dissolved Fe(III)? Give formulas for two Fe(III) species that have been neglected in this calculation.

4.5 The frequency of the symmetrical M—O stretching vibration of the octahedral aqua ions $[M(OH_2)_6]^{2+}$ increases along the series, $Ca^{2+} < Mn^{2+} < Ni^{2+}$. How does this trend relate to acidity?

4.6. An electrically conducting solution is produced when $AlCl_3$ is dissolved in the basic polar solvent CH_3CN. Give formulas for the most probable conducting species and describe their formation using Lewis acid–base concepts.

4.7 The complex anion $[FeCl_4]^-$ is yellow whereas $[Fe_2Cl_6]$ is reddish. Dissolution of 0.1 mol $FeCl_3(s)$ in 1 dm^3 of either $POCl_3$ or $PO(OR)_3$ produces a reddish solution that turns yellow on dilution. Titration of red solutions in $POCl_3$ with Et_4NCl solutions leads to a sharp colour change (from red to yellow) at a 1:1 mole ratio of $FeCl_3/Et_4NCl$. Vibrational spectra suggest that oxochloride solvents form adducts with typical Lewis acids by coordination of oxygen. Compare the following two sets of reactions as possible explanations of the observations.

(a) $Fe_2Cl_6 + 2\,POCl_3 \rightleftharpoons 2[FeCl_4]^- + 2[POCl_2]^+$
$POCl_2^+ + Et_4NCl \rightleftharpoons Et_4N^+ + POCl_3$

(b) $Fe_2Cl_6 + 4\,POCl_3 \rightleftharpoons [FeCl_2(OPCl_3)_4]^+ + [FeCl_4]^-$

Both equilibria are shifted to products by dilution.

4.8 In the traditional scheme for the separation of metal ions from solution that is the basis of qualitative analysis, ions of Au, As, Sb, and Sn precipitate as sulfides but redissolve on addition of excess ammonium polysulfide. By contrast, ions of Cu, Pb, Hg, Bi, and Cd precipitate as sulfides but do not redissolve. In the language of this chapter, the first group is amphoteric for reactions involving SH^- in place of OH^-. The second group is less acidic. Locate the amphoteric boundary in the periodic table for sulfides implied by this information. Compare this boundary with the amphoteric boundary for hydrous oxides in Fig. 4.4. Does this analysis agree with describing S^{2-} as a softer base than O^{2-}?

4.9 The compounds SO_2 and $SOCl_2$ can undergo an exchange of radioactively labelled sulfur. The exchange is catalysed by Cl^- and $SbCl_5$. Suggest mechanisms for these two exchange reactions with the first step being the formation of an appropriate complex.

4.10 In the reaction of t-butyl bromide with $Ba(NCS)_2$, the product is 91 per cent S-bound t-Bu-SCN. However, if $Ba(NCS)_2$ is impregnated into solid CaF_2, the yield is higher and the product is 99 per cent t-Bu-NCS. Discuss the effect of alkaline earth metal salt support on the hardness of the ambident nucleophile SCN^-. (See T. Kimura, M. Fujita, and T. Ando, *J. Chem Soc., Chem. Commun.*, 1990, 1213.)

4.11 Pyridine forms a stronger Lewis acid–base complex with SO_3 than with SO_2. However, pyridine forms a weaker complex with SF_6 than with SF_4. Explain the difference.

4.12 Predict whether the equilibrium constants for the following reactions should be greater than 1 or less than 1:

(a) $CdI_2(s) + CaF_2(s) \rightleftharpoons CdF_2(s) + CaI_2(s)$

(b) $[CuI_4]^{2-}(aq) + [CuCl_4]^{3-}(aq) \rightleftharpoons [CuCl_4]^{2-}(aq) + [CuI_4]^{3-}(aq)$

(c) $NH_2^-(aq) + H_2O(l) \rightleftharpoons NH_3(aq) + OH^-(aq)$

4.13 For parts (a), (b), and (c), state which of the two solutions has the lower pH:

(a) 0.1 M $Fe(ClO_4)_2(aq)$ or 0.1 M $Fe(ClO_4)_3(aq)$

(b) 0.1 M $Ca(NO_3)_2(aq)$ or 0.1 M $Mg(NO_3)_2$ (aq)

(c) 0.1 M $Hg(NO_3)_2(aq)$ or 0.1 M $Zn(NO_3)_2(aq)$

4.14 A paper by Gillespie and Liang entitled 'Superacid solutions in hydrogen fluoride' (*J. Am. Chem. Soc.*, 1988, **110**, 6053) discusses the acidity of various solutions of inorganic compounds in HF.
(a) Give the order of acid strength of the pentafluorides determined during the investigation. (b) Give the equations for the reactions of SbF_5 and AsF_5 with HF. (c) SbF_5 forms a dimer, $Sb_2F_{11}^-$, in HF. Give the equation for the equilibrium between the monomeric and the dimeric species.

4.15 In their paper 'The strengths of the hydrohalic acids' (*J. Chem. Educ.*, 2001, **78**, 116), R. Schmid and A. Miah discuss the validity of literature values of the pK_as for HF, HCl, HBr, and HI. (a) On what basis have the literature values been estimated? (b) To what is the low acid strength of HF relative to HCl usually attributed? (c) What reason do the authors suggest for the high acid strength of HCl?

Oxidation and reduction

5

Many elements form compounds in which they exist in one or more different oxidation states; as a result, they can participate in redox reactions in which electrons are transferred from one species to another. In this chapter we analyse these reactions by considering mainly their thermodynamic aspects and, through this, develop an understanding of the inorganic chemistry of the elements with respect to their oxidation and reduction. We discuss the procedures for analysing redox reactions in solution and see that the electrode potentials of electrochemically active species provide thermodynamic data in a useful form. In particular, we use diagrammatic techniques to summarize trends in the stabilities of various oxidation states, including the influence of pH. Applications of these data in environmental chemistry, chemical analysis, and inorganic synthesis are presented. The discussion concludes with a thermodynamic examination of the conditions needed for some major industrial oxidation and reduction processes. This analysis is exemplified by considering how the elements are extracted, particularly the metals, including those processes carried out at high temperatures and electrochemically.

A very large class of reactions of inorganic compounds can be regarded as occurring by the transfer of electrons from one species to another. Indeed many of the fascinating aspects of inorganic chemistry, including many applications of inorganic compounds, derive from changes in oxidation state. Electron gain is called **reduction** and electron loss is called **oxidation**; the joint process is called a **redox reaction**. The species that supplies electrons is the **reducing agent** (or 'reductant') and the species that removes electrons is the **oxidizing agent** (or 'oxidant'). The transfer of electrons is often accompanied by the transfer of atoms and it is sometimes difficult to keep track of where electrons have come from and gone to. It is therefore safest, and simplest, to analyse redox reactions according to a set of formal rules expressed in terms of oxidation numbers (Section 2.1) and not to think in terms of actual electron transfers. Oxidation then corresponds to the increase in oxidation number of an element and reduction corresponds to the decrease in the oxidation number. A redox reaction is a chemical reaction in which there are changes in oxidation number of at least one of the elements involved.

The simplest oxidation and reductions involve the formation of cations and anions from the elements. Examples include the oxidation of lithium to Li^+ ions when it burns in air to form Li_2O and the reduction of chlorine to Cl^- when it reacts with calcium to form $CaCl_2$. For the Group 1 and 2 elements the only oxidation numbers found are those of the element (0) and of the ions, +1 and +2, respectively. However, many of the other elements form compounds in more than one oxidation state. Thus lead is commonly found in its compounds as Pb(II), as in PbO, and as Pb(IV), as in PbO_2. The ability to exhibit multiple oxidation numbers is seen at its fullest in d-metal compounds, particularly in Groups 6, 7, and 8; osmium, for instance, forms compounds that span oxidation numbers between 0 (osmium metal) and +8, as in OsO_4. Because the oxidation state of an element is often reflected in the properties of its compounds, the ability to express the tendency of an element to form a compound in a particular oxidation state quantitatively is very useful in inorganic chemistry. In this chapter we demonstrate that the electrochemical potentials associated with electron transfer provide a simple and quantitative method of measuring the tendency for redox reactions to occur. Similar arguments help us to rationalize the processes required to extract the elements from their naturally occurring forms.

Reduction potentials

The driving force for a reduction reaction is measured by the potential difference required to reduce ions in solution to the metal. This quantitative measure of the tendency for a metal to be reduced can be extended to similar reduction reactions involving many other species in solution. In the following sections we introduce these concepts and show how they can be applied to understanding many redox processes including the reactions that occur between inorganic compounds.

5.1 Redox half-reactions

Key point: A redox reaction can be expressed as the difference of two reduction half-reactions.

It is convenient to think of a redox reaction as the combination of two conceptual **half-reactions** in which the electron loss (oxidation) and gain (reduction) are displayed explicitly. In a reduction half-reaction, a substance gains electrons, as in

$$2 H^+(aq) + 2 e^- \rightarrow H_2(g)$$

In an oxidation half-reaction, a substance loses electrons, as in

$$Zn(s) \rightarrow Zn^{2+}(aq) + 2 e^-$$

A state is not ascribed to the electrons in the equation of a half-reaction; moreover, the division into half-reactions is only conceptual and may not correspond to an actual physical separation of the two processes. Nor does the half-reaction represent any part of the real physical behaviour in an electron transfer between reagents in a reaction. The oxidized and reduced species in a half-reaction constitute a **redox couple**. A couple is written with the oxidized species before the reduced, as in H^+/H_2 and Zn^{2+}/Zn, and typically the phases are not shown.

Because an oxidation is the reverse of a reduction, we can write the second of the two half-reactions above as the reduction

$$Zn^{2+}(aq) + 2 e^- \rightarrow Zn(s)$$

The overall reaction

$$Zn(s) + 2 H^+ \rightarrow Zn^{2+}(aq) + H_2(g)$$

is the *difference* of the two reduction half-reactions. In other cases it may be necessary to multiply each half-reaction by a factor to ensure that the numbers of electrons match.

Example 5.1 Combining half-reactions

Form a balanced equation for the oxidation of Sn^{2+} by permanganate (MnO_4^-).

Answer The two reduction half-reactions are

$$MnO_4^-(aq) + 8 H^+(aq) + 5 e^- \rightarrow Mn^{2+}(aq) + 4 H_2O(l)$$
$$Sn^{4+}(aq) + 2 e^- \rightarrow Sn^{2+}(aq)$$

To balance the number of electrons in the two half-reactions the first is multiplied by 2 and the second by 5 to give 10 e^- in each case. Then subtracting one half-reaction from the other gives

$$2 MnO_4^-(aq) + 16 H^+(aq) + 5 Sn^{2+} \rightarrow 2 Mn^{2+}(aq) + 8 H_2O(l) + 5 Sn^{4+}(aq)$$

Self-test 5.1 Use reduction half-reactions to write a balanced equation for the oxidation of zinc metal by permanganate ions.

5.2 Standard potentials

Key point: A reaction is favourable, in the sense $K > 1$, if $E^\ominus > 0$, where $E^\ominus$ is the difference of the standard potentials corresponding to the half-reactions into which the overall reaction may be divided.

Thermodynamic arguments can be used to identify which reactions are spontaneous (that is, have a natural tendency to occur). The thermodynamic criterion of spontaneity is that, at constant temperature and pressure, the reaction Gibbs energy, $\Delta_r G$, is negative. It is usually sufficient to consider the standard reaction Gibbs energy, $\Delta_r G^\ominus$, which is related to the equilibrium constant through

$$\Delta_r G^\ominus = -RT \ln K \tag{5.1}$$

A negative value of $\Delta_r G^\ominus$ corresponds to $K > 1$ and therefore to a 'favourable' or spontaneous reaction. Because the overall chemical equation is the difference of two reduction half-reactions, the standard Gibbs energy of the overall reaction is the difference of the standard Gibbs energies of the two half-reactions. The overall reaction is favourable (in the sense that $K > 1$) and will proceed in the direction that corresponds, through eqn 5.1, to a negative value of the resulting overall $\Delta_r G^\ominus$.

As reduction half-reactions must always occur in pairs in any actual reaction, only the difference in their standard Gibbs energies has any significance. Therefore, we can choose one half-reaction to have $\Delta_r G^\ominus = 0$, and report all other values relative to it. By convention, the specially chosen half-reaction is the reduction of hydrogen ions:

$$\text{H}^+(\text{aq}) + \text{e}^- \rightarrow \tfrac{1}{2}\text{H}_2(\text{g}) \qquad \Delta_r G^\ominus = 0$$

at all temperatures. With this choice, the standard Gibbs energy for the reduction of Zn^{2+} ions, for example, is found by determining experimentally that

$$\text{Zn}^{2+}(\text{aq}) + \text{H}_2(\text{g}) \rightarrow \text{Zn}(\text{s}) + 2\,\text{H}^+(\text{aq}) \qquad \Delta_r G^\ominus = +147\text{ kJ mol}^{-1}$$

Then, because the H^+ reduction half-reaction makes zero contribution to the reaction Gibbs energy (according to our convention), it follows that

$$\text{Zn}^{2+}(\text{aq}) + 2\,\text{e}^- \rightarrow \text{Zn}(\text{s}) \qquad \Delta_r G^\ominus = +147\text{ kJ mol}^{-1}$$

Standard reaction Gibbs energies may be measured by setting up a galvanic cell (an electrochemical cell in which a chemical reaction is used to generate an electric current) in which the reaction driving the electric current through the external circuit is the reaction of interest (Fig. 5.1). The potential difference between its electrodes is then measured. In practice, we must ensure that the cell is acting reversibly in a thermodynamic sense, which means that the potential difference must be measured with negligible current flowing and, if desired, converted to a Gibbs energy by using $\Delta_r G = -\nu F E$, where ν is the stoichiometric coefficient of the electrons transferred when the half-reactions are combined[1] and F is Faraday's constant ($F = 96.48\text{ kC mol}^{-1}$). Tabulated values, normally for standard conditions,[2] are usually kept in the units in which they were measured, namely volts (V).

The potential that corresponds to the $\Delta_r G^\ominus$ of a half-reaction is written $E^\ominus$, with

$$\Delta_r G^\ominus = -\nu F E^\ominus \tag{5.2}$$

The potential $E^\ominus$ is called the **standard potential** (or 'standard reduction potential', to emphasize that the half-reaction is a reduction). Because $\Delta_r G^\ominus$ for the reduction of

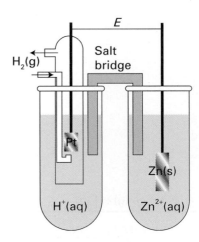

Fig. 5.1 A schematic diagram of a galvanic cell. The standard potential, $E^\ominus$, is the potential difference when the cell is not generating current and all the substances are in their standard states.

---

[1] Although we use ν (nu) for the stoichiometric coefficient of the electron, electrochemical equations in inorganic chemistry are also commonly written with n in its place; we use ν to emphasize that it is a dimensionless number, not an amount in moles.

[2] Standard conditions are all substances at 1 bar and unit activity. For reactions involving H^+ ions, standard conditions correspond to pH = 0, approximately 1 M acid. Pure solids and liquids have unit activity.

H^+ is arbitrarily set at zero, the standard potential of the H^+/H_2 couple is also zero at all temperatures:

$$H^+(aq) + e^- \rightarrow \tfrac{1}{2}H_2(g) \qquad E^\ominus(H^+, H_2) = 0$$

Similarly, for the Zn^{2+}/Zn couple, for which $v = 2$, it follows from the measured value of $\Delta_r G^\ominus$ that at 25°C:

$$Zn^{2+}(aq) + 2\,e^- \rightarrow Zn(s) \qquad E^\ominus(Zn^{2+}, Zn) = -0.76V$$

Because the standard reaction Gibbs energy is the difference of the $\Delta_r G^\ominus$ values for the half-reactions, $E^\ominus$ for an overall reaction is also the difference of the two standard potentials of the reduction half-reactions into which the overall reaction can be divided. Thus, from the half-reactions given above it follows that the difference is

$$2\,H^+(aq) + Zn(s) \rightarrow Zn^{2+}(aq) + H_2(g) \qquad E^\ominus = +0.76V$$

Note that the $E^\ominus$ values for couples (and their half-reactions) are called standard potentials and that their difference is called the **standard electromotive force** (emf) of the corresponding galvanic cell.[3] The consequence of the negative sign in eqn 5.2 is that a reaction is favourable (in the sense $K > 1$) if the corresponding standard emf is positive. Because $E^\ominus > 0$ for the reaction above ($E^\ominus = +0.76\,V$), we know that zinc has a thermodynamic tendency to reduce H^+ ions under standard conditions (acidic solution, pH = 0, and Zn^{2+} at unit activity); that is, zinc metal dissolves in acids. The same is true for any metal that has a couple with a negative standard potential.

Example 5.2 Calculating an emf

Use the following standard potentials to calculate the standard emf of a copper–zinc cell.

$$Zn^{2+}(aq) + 2\,e^- \rightarrow Zn(s) \qquad E^\ominus(Zn^{2+}, Zn) = -0.76\,V$$
$$Cu^{2+}(aq) + 2\,e^- \rightarrow Cu(s) \qquad E^\ominus(Cu^{2+}, Cu) = +0.34\,V$$

Answer For the reaction $Cu^{2+}(aq) + Zn(s) \rightarrow Zn^{2+}(aq) + Cu(s)$

$$E^\ominus = +0.34\,V - (-0.76\,V) = +1.10\,V$$

The cell will produce a potential of 1.1 V (under standard conditions).

Self-test 5.2 Is copper metal expected to dissolve in dilute hydrochloric acid?

5.3 Trends in standard potentials

Key point: The atomization and ionization of a metal and the hydration enthalpy of its ions all contribute to the value of the standard potential.

The factors that contribute to the standard potential $E^\ominus$ for a specific couple M^+/M can be identified by consideration of a thermodynamic cycle and the corresponding changes in Gibbs energy that contribute to the overall reaction (Fig. 5.2):

$$M^+(aq) + \tfrac{1}{2}H_2(g) \rightarrow H^+(aq) + M(s)$$

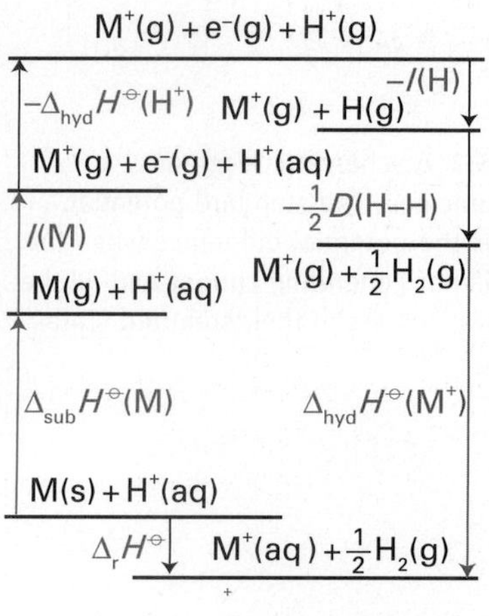

The reaction entropy is largely independent of the identity of M. It corresponds to a value of $T\Delta S^\ominus$ in the region of -20 to $-40\,kJ\,mol^{-1}$, which is small in comparison with the reaction enthalpy, the difference between the standard enthalpies of formation of $H^+(aq)$ and $M^+(aq)$. These calculations use the absolute values of the enthalpies of formation of M^+ and H^+ derived from a combination of thermodynamic measurements and judicious estimates. Thus, although by convention $\Delta_f H^\ominus(H^+, aq) = 0$, for the analysis in the illustration we use its actual value, approximately $+445\,kJ\,mol^{-1}$, which is obtained by considering the formation of hydrogen atoms from $H_2(g)$ ($+218\,kJ\,mol^{-1}$), ionization to $H^+(g)$ ($+1312\,kJ\,mol^{-1}$), and hydration of $H^+(g)$ (approximately $-1085\,kJ\,mol^{-1}$).

Fig. 5.2 A thermodynamic cycle showing the properties that contribute to the standard potential of a metal couple.

---

[3] IUPAC deprecates 'emf' because a force is not involved; the recommended term is *zero-current cell potential*.

Table 5.1 Thermodynamic contributions to $E^{\ominus}$ for a selection of metals at 298 K

	Li	Na	Cs	Ag
$\Delta_{sub}H^{\ominus}/(kJ\ mol^{-1})$	+161	+109	+79	+284
$I/(kJ\ mol^{-1})$	526	502	382	735
$\Delta_{hyd}H^{\ominus}/(kJ\ mol^{-1})$	−520	−406	−264	−468
$\Delta_f H^{\ominus}(M^+,aq)/(kJ\ mol^{-1})$	+167	+205	+197	+551
$\Delta_r H^{\ominus}/(kJ\ mol^{-1})$	+278	+240	+248	−106
$T\Delta_r S^{\ominus}/(kJ\ mol^{-1})$	−16	−22	−34	−29
$\Delta_r G^{\ominus}/(kJ\ mol^{-1})$	+294	+262	+282	−77
$E^{\ominus}/V$	−3.04	−2.71	−2.92	+0.80

$\Delta_{hyd}H^{\ominus}(H^+,aq) = +445\ kJ\ mol^{-1}$.

The analysis of the cell emf into its thermodynamic contributions allows us to account for trends in the standard potentials. For instance, the variation of standard potential down Group 1 seems contrary to expectation based on electronegativities in that highly electropositive caesium ($\chi = 0.79$, $E^{\ominus} = -2.94\ V$) has a less negative standard potential than lithium ($\chi = 2.20$, $E^{\ominus} = -3.04\ V$). Lithium has a higher enthalpy of sublimation and ionization energy than caesium, and in isolation this difference would imply a less negative standard potential as these more positive terms make formation of the ion less favourable. However, Li^+ has a large negative enthalpy of hydration, which results from its small size (its ionic radius is 90 pm) compared with Cs^+ (181 pm) and its consequent strong electrostatic interaction with water molecules. Overall, this term outweighs those derived from the formation of $Li^+(g)$ and gives rise to a more negative standard potential. It should be noted that this variation is a result of defining standard potentials in aqueous solution; in a less polar solvent, with lower solvation enthalpies, Cs would have the more negative standard potential. The relatively low standard potential for sodium ($-2.71\ V$) in comparison with the rest of Group 1 (close to $-2.9\ V$) can be explained in terms of a combination of a fairly high sublimation enthalpy and moderate hydration enthalpy (Table 5.1).

The value of $E^{\ominus}(Na^+, Na) = -2.71\ V$ may also be compared with that for $E^{\ominus}(Ag^+, Ag) = +0.80\ V$. The ionic radii of these ions ($r_{Na^+} = 116\ pm$ and $r_{Ag^+} = 129\ pm$) are similar, and consequently their ionic hydration enthalpies are similar too. However, the much higher enthalpy of sublimation of silver, and particularly its high ionization energy, which is due to the poor screening by the $4d$ electrons, results in a positive standard potential. This difference is reflected in the very different behaviours of the metals when treated with a dilute acid, when sodium reacts and dissolves explosively producing hydrogen whereas silver is unreactive. Similar arguments can be used to explain many of the trends observed in the standard potentials given in Table 5.2. For example, the positive potentials characteristic of the noble metals result from their very high sublimation enthalpies.

5.4 The electrochemical series

Key points: The oxidized member of a couple is a strong oxidizing agent if $E^{\ominus}$ is positive and large the reduced member is a strong reducing agent if $E^{\ominus}$ is negative and large.

A negative standard potential signifies a couple in which the reduced species (the Zn in Zn^{2-}/Zn) is a reducing agent for H^+ ions under standard conditions in aqueous solution. That is, if $E^{\ominus}(Ox, Red) < 0$, then the substance 'Red' is a strong enough reducing agent to reduce H^+ ions (in the sense that $K > 1$ for the reaction). A short compilation of $E^{\ominus}$ values at 25°C is given in Table 5.2. The list is arranged in the order of the **electrochemical series**:

Ox/Red couple with strongly positive $E^{\ominus}$ [Ox is strongly oxidizing]

⋮

Ox/Red couple with strongly negative $E^{\ominus}$ [Red is strongly reducing]

Table 5.2 Selected standard potentials at 298 K; further values are included in *Resource section 2*

Couple	$E^{\ominus}$/V
$F_2(g) + 2\,e^- \rightarrow 2\,F^-(aq)$	+2.87
$Ce^{4+}(aq) + e^- \rightarrow Ce^{3+}(aq)$	+1.76
$MnO_4^-(aq) + 8\,H^+(aq) + 5\,e^- \rightarrow Mn^{2+}(aq) + 4\,H_2O(l)$	+1.51
$Cl_2(g) + 2\,e^- \rightarrow 2\,Cl^-(aq)$	+1.36
$O_2(g) + 4\,H^+(aq) + 4\,e^- \rightarrow 2\,H_2O(l)$	+1.23
$[IrCl_6]^{2-}(aq) + e^- \rightarrow [IrCl_6]^{3-}(aq)$	+0.87
$Fe^{3+}(aq) + e^- \rightarrow Fe^{2+}(aq)$	+0.77
$[PtCl_4]^{2-}(aq) + 2\,e^- \rightarrow Pt(s) + 4\,Cl^-(aq)$	+0.60
$I_3^-(aq) + 2\,e^- \rightarrow 3\,I^{-1}(aq)$	+0.54
$[Fe(CN)_6]^{3-}(aq) + e^- \rightarrow [Fe(CN)_6]^{4-}(aq)$	+0.36
$AgCl(s) + e^- \rightarrow Ag(s) + Cl^-(aq)$	+0.22
$2\,H^+(aq) + 2\,e^- \rightarrow H_2(g)$	0
$AgI(s) + e^- \rightarrow Ag(s) + I^-(aq)$	−0.15
$Zn^{2+}(aq) + 2\,e^- \rightarrow Zn(s)$	−0.76
$Al^{3+}(aq) + 3\,e^- \rightarrow Al(s)$	−1.68
$Ca^{2+}(aq) + 2\,e^- \rightarrow Ca(s)$	−2.84
$Li^+(aq) + e^- \rightarrow Li(s)$	−3.04

An important feature of the electrochemical series is that the reduced member of a couple has a thermodynamic tendency to reduce the oxidized member of any couple that lies above it in the series. Note that the classification refers only to the thermodynamic aspect of the reaction—its spontaneity under standard conditions, not its rate. Thus even reactions that are found to be thermodynamically favourable from the electrochemical series may not progress, or only progress extremely slowly, if the kinetics of the process are unfavourable. Such reactions include those that have high activation energies.

Example 5.3 Using the electrochemical series

Among the couples in Table 5.2 is the permanganate ion, MnO_4^-, the common analytical reagent used in redox titrations of iron. Which of the ions Fe^{2+}, Cl^-, and Ce^{3+} can permanganate oxidize in acidic solution?

Answer The standard potential of the couple MnO_4^-/Mn^{2+} in acidic solution is +1.51 V. The standard potentials for Fe^{2+}, Cl^-, and Ce^{3+} are +0.77, +1.36, and +1.76 V, respectively. It follows that permanganate ions are sufficiently strong oxidizing agents in acidic solution (pH = 0) to oxidize the first two ions, which have less positive standard potentials. Permanganate ions cannot oxidize Ce^{3+}, which has a more positive standard potential. It should be noted that the presence of other ions in the solution can modify the potentials and the conclusions (Section 5.9); this variation with conditions is particularly important in the case of hydrogen ions, and the influence of pH is discussed in Section 5.12. The ability of permanganate to oxidize Cl^- means that HCl cannot be used to acidify redox reactions involving permanganate but instead H_2SO_4 is used.

Self-test 5.3 Another common analytical oxidizing agent is an acidic solution of dichromate ions, $Cr_2O_7^{2-}$, for which $E^{\ominus}(Cr_2O_7^{2-}, Cr^{3+}) = +1.38$ V. Is the solution useful for a redox titration of Fe^{2+} to Fe^{3+}? Could there be a side reaction when Cl^- is present?

5.5 The Nernst equation

Key point: The emf of a cell at an arbitrary composition of the reaction mixture is given by the Nernst equation.

To judge the tendency of a reaction to run in a particular direction at an arbitrary composition, we need to know the sign and value of $\Delta_r G$ at that composition. For this information, we use the thermodynamic result that

$$\Delta_r G = \Delta_r G^{\ominus} + RT \ln Q \tag{5.3a}$$

where Q is the reaction quotient[4]

$$a \; Ox_A + b \; Red_B \rightarrow a' \; Red_A + b' \; Ox_B \qquad Q = \frac{[Red_A]^{a'} [Ox_B]^{b'}}{[Ox_A]^{a} [Red_B]^{b}} \tag{5.3b}$$

Note that the reaction quotient, Q, is used rather than the equilibrium constant, K, as we are considering a general system that might not have reached equilibrium. The reaction is spontaneous under the stated conditions if $\Delta_r G < 0$. This criterion can be expressed in terms of the emf of the corresponding cell by substituting $E = - \Delta_r G / \nu F$ and $E^{\ominus} = -\Delta_r G^{\ominus} / \nu F$ into eqn 5.3a, which gives the **Nernst equation**:

$$E = E^{\ominus} - \frac{RT}{\nu F} \ln Q \tag{5.4}$$

A reaction is spontaneous if, under the prevailing conditions, $E > 0$, for then $\Delta_r G < 0$. At equilibrium $E = 0$ and $Q = K$, so eqn 5.4 implies the following very important relation between the standard potential of a reaction and its equilibrium constant at a temperature T:

$$\ln K = \frac{\nu F E^{\ominus}}{RT} \tag{5.5}$$

Table 5.3 lists the value of K that corresponds to emfs in the range -2 to $+2$ V, with $\nu = 1$ and at 25°C. The table shows that, although electrochemical data are often compressed into the range -2 to $+2$ V, that narrow range corresponds to 68 orders of magnitude in the value of the equilibrium constant for $\nu = 1$.

If we regard the emf E as the difference of two potentials, just as $E^{\ominus}$ is the difference of two standard potentials, then the potential of each couple can be written as in eqn 5.4, but with

$$a \; Ox + \nu \, e^- \rightarrow a' \; Red \qquad Q = \frac{[Red]^{a'}}{[Ox]^{a}}$$

By convention, the electrons do not appear in the expression for Q.

Table 5.3 The relation between K and $E^{\ominus}$

$E^{\ominus}$/V	K
+2	10^{34}
+1	10^{17}
0	1
−1	10^{-17}
−2	10^{-34}

Example 5.4 Using the Nernst equation

What is the dependence of the potential of the H^+/H_2 couple on pH when the hydrogen pressure is 1 bar and the temperature is 25°C?

Answer The reduction half-reaction is $2 \, H^+(aq) + 2 \, e^- \rightarrow H_2(g)$ with $\nu = 2$. We can expect a decrease in hydrogen ion concentration (an increase in pH) to lower the tendency of the reaction to form products and hence to lead to an increase in reaction Gibbs energy and therefore a decrease in potential. The Nernst equation for the couple is

$$E(H^+, H_2) = E^{\ominus}(H^+, H_2) - \frac{RT}{2F} \ln \frac{p_{H_2}}{[H^+]^2}$$

$$= \frac{RT}{F} \ln[H^+] = \frac{RT \ln 10}{F} \log[H^+] \tag{5.6}$$

$$= -\frac{RT \ln 10}{F} pH$$

We have used $\ln (1/x) = -\ln x$ and $\ln x = \ln 10 \times \log x$. On substituting numerical values we obtain

$$E(H^+, H_2) = -\frac{(8.3145 \; J \; K^{-1} mol^{-1}) \times (298.15 \; K) \times \ln 10}{9.6485 \times 10^4 \; C \; mol^{-1}} \times pH = -59.16 \; mV \times pH$$

---

[4] For reactions involving gas-phase species, the molar concentrations of the latter are replaced by partial pressures relative to $p^{\ominus} = 1$ bar.

where we have used $1\,J = 1\,C\,V$ and $10^{-3}\,V = 1\,mV$. That is, the potential becomes more negative by $59\,mV$ for each unit increase in pH. In neutral solution (pH = 7), $E(H^+, H_2) = -0.41\,V$.

Self-test 5.4 What is the potential for reduction of MnO_4^- to Mn^{2+} in neutral (pH = 7) solution?

Example 5.5 Analysing a fuel cell

A *fuel cell* can convert a chemical fuel, such as hydrogen, methane, or methanol, directly into electrical power (see Box 9.1). At the anode a fuel is oxidized and at the cathode the oxidant is reduced; the resulting flow of electrons from anode to cathode is used as a power source. Calculate the cell emf produced by a fuel cell that burns hydrogen with oxygen under standard conditions.

Answer The reduction half-reactions are

(a) $O_2(g) + 4\,H^+(aq) + 4\,e^- \rightarrow 2\,H_2O(l)$ $E^{\ominus} = +1.23\,V$

(b) $2\,H^+(aq) + 2\,e^- \rightarrow H_2(g)$ $E^{\ominus} = 0$

Overall $(a - 2b)$: $2\,H_2(g) + O_2(g) \rightarrow 2\,H_2O(l)$

The standard emf of the cell is therefore

$E^{\ominus} = (+1.23\,V) - 0 = +1.23\,V$

Self-test 5.5 What cell emf would be produced in a fuel cell operating with oxygen and hydrogen, with both gases at 5.0 bar?

Redox stability

When assessing the thermodynamic stability of a species in solution, we must bear in mind all possible reactants: the solvent, other solutes, the species itself, and dissolved oxygen. In the following discussion, we focus on the types of reaction that result from the thermodynamic instability of a solute. We also comment briefly on kinetic factors, but the trends they show are generally less systematic than those shown by stabilities.

5.6 Reactions with water

Water may act as an oxidizing agent, when it is reduced to H_2:

$H_2O(l) + e^- \rightarrow \tfrac{1}{2}H_2(g) + OH^-(aq)$

For the equivalent reduction of hydronium ions in water at any pH (and partial pressure of H_2 of 1 bar) we have seen that the Nernst equation gives

$H^+(aq) + e^- \rightarrow \tfrac{1}{2}H_2(g)$ $E = -0.059\,V \times pH$

This is the reaction that chemists typically have in mind when they refer to 'the reduction of water'. Water may also act as a reducing agent, when it is oxidized to O_2:

$O_2(g) + 4\,H^+(aq) + 4\,e^- \rightarrow 2\,H_2O(l)$

When the partial pressure of O_2 is 1 bar, the Nernst equation for this half-reaction becomes

$$E = E^{\ominus} - \frac{RT}{4F}\ln\frac{1}{[H^+]^4} = 1.23\,V - (0.059\,V \times pH) \tag{5.7}$$

The variation of these two potentials with pH is shown in Fig. 5.3 (the additional features shown in this illustration are discussed in the following sections).

(a) Oxidation by water

Key point: For metals with large, negative standard potentials, reaction with aqueous acids leads to the production of hydrogen unless a passivating oxide layer is formed.

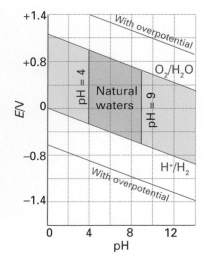

Fig. 5.3 The variation of the reduction potentials of water with pH. The additional features are discussed in Section 5.6: the blue lines are the boundaries when the overpotential is taken into account and the vertical lines represent the normal range for natural waters. The blue area is the stability field of water.

The reaction of a metal with water or aqueous acid is in fact the oxidation of the metal by water or hydrogen ions, because the overall reaction is one of the following processes (and their analogues for more highly charged metal ions):

$$M(s) + H_2O(l) \rightarrow M^+(aq) + \tfrac{1}{2}H_2(g) + OH^-(aq)$$

$$M(s) + H^+(aq) \rightarrow M^+(aq) + \tfrac{1}{2}H_2(g)$$

These reactions are thermodynamically favourable when M is an s-block metal or a $3d$-series metal from Group 3 to at least Group 8 or 9 and beyond (Ti, V, Cr, Mn, Ni). An example from Group 3 is

$$2\ Sc(s) + 6\ H^+(aq) \rightarrow 2\ Sc^{3+}(aq) + 3\ H_2(g)$$

When the standard potential for the reduction of a metal ion to the metal is negative, the metal should undergo oxidation in 1 M acid with the evolution of hydrogen.

Although the reactions of magnesium and aluminium with moist air are spontaneous, both metals can be used for years in the presence of water and oxygen. They survive because they are **passivated**, or protected against reaction, by an impervious film of oxide. Magnesium oxide and aluminium oxide both form a protective skin on the parent metal beneath. A similar passivation occurs with iron, copper, and zinc. The process of 'anodizing' a metal, in which the metal is made an anode (the site of oxidation) in an electrolytic cell, is one in which partial oxidation produces a smooth, hard passivating film on its surface. Anodizing is especially effective for the protection of aluminium by the formation of an inert, cohesive, and impenetrable Al_2O_3 layer.

(b) Reduction by water

Key point: Water can act as a reducing agent; that is, be oxidized by other species.

The strongly positive potential of the O_2, H^+/H_2 couple (eqn 5.7) shows that acidified water is a poor reducing agent except towards strong oxidizing agents. An example of the latter is $Co^{3+}(aq)$, for which $E^{\ominus}(Co^{3+}, Co^{2+}) = +1.92$ V. It is reduced by water with the evolution of O_2 and Co^{3+} does not survive in aqueous solution:

$$4\ Co^{3+}(aq) + 2\ H_2O(l) \rightarrow 4\ Co^{2+}(aq) + O_2(g) + 4\ H^+(aq) \quad E^{\ominus} = +0.69\ V$$

Because H^+ ions are produced in the reaction, lower acidity favours the oxidation; lowering the concentration of H^+ ions encourages the formation of the products.

Only a few oxidizing agents (Ag^{2+} is another example) can oxidize water rapidly enough to give appreciable rates of O_2 evolution. Standard potentials greater than $+1.23$ V occur for several redox couples that are regularly used in aqueous solution, including Ce^{4+}/Ce^{3+} ($+1.76$ V), the acidified dichromate ion couple $Cr_2O_7^{2-}/Cr^{3+}$ ($+1.38$ V), and the acidified permanganate couple MnO_4^-/Mn^{2+} ($+1.51$ V). The origin of the barrier to reaction is a kinetic one, stemming from the need to transfer four electrons and to form an oxygen–oxygen double bond.

Given that the rates of redox reactions are often controlled by the slow rate at with which an oxygen–oxygen bond can be formed, it remains a challenge for inorganic chemists to find good catalysts for O_2 evolution. Some progress has been made by using Ru complexes. Existing catalysts include the relatively poorly understood coatings that are used in the anodes of cells for the commercial electrolysis of water. They also include the enzyme system found in the O_2 evolution apparatus of the plant photosynthetic centre. This system is based on a special cofactor containing four Mn atoms and one Ca atom (Section 26.[10]). Although Nature is elegant and efficient, it is also complex, and the photosynthetic process is only slowly being elucidated by biochemists and bioinorganic chemists.

(c) The stability field of water

Key point: The stability field of water shows the region of pH and reduction potential where couples are neither oxidized by nor reduce hydrogen ions.

A reducing agent that can reduce water to H_2 rapidly, or an oxidizing agent that can oxidize water to O_2 rapidly, cannot survive in aqueous solution. The **stability field** of water, which is shown tinted blue in Fig. 5.3, is the range of values of potential and pH for which water is thermodynamically stable towards both oxidation and reduction.

The upper and lower boundaries of the stability field are identified by finding the dependence of E on pH for the relevant half-reactions. Thus, any species with a potential higher than that given in eqn 5.7 can oxidize water with the production of O_2. Hence, this expression defines the upper boundary of the stability field. Similarly, the reduction of $H^+(aq)$ to H_2 can be caused by any couple with a potential that is more negative than that given in eqn 5.6. This expression therefore gives the lower boundary of the stability field.

Couples that are thermodynamically unstable in water lie outside the limits defined by the sloping lines in Fig. 5.3. The couple consists either of too strong a reducing agent (below the H_2 production line) or too strong an oxidizing agent (above the O_2 production line). The stability field in 'natural' water is represented by the addition of two vertical lines at pH = 4 and pH = 9, which mark the limits on pH that are commonly found in lakes and streams. A diagram like that shown in the illustration is known as a **Pourbaix diagram** and is widely used in environmental chemistry, as we shall see in Section 5.13.

5.7 Oxidation by atmospheric oxygen

Key point: The oxygen present in air and dissolved in water can oxidize metals and metal ions in solution.

The possibility of reaction between the solutes and dissolved O_2 must be considered when a solution is contained in an open beaker or is otherwise exposed to air. As an example, consider a solution containing Fe^{2+}. Because $E^{\ominus}(Fe^{3+}, Fe^{2+}) = +0.77\,V$, which lies within the stability field of water, we expect Fe^{2+} to survive in water. Moreover, we can also infer that the oxidation of metallic iron by $H^+(aq)$ should not proceed beyond Fe(II), because further oxidation to Fe(III) is unfavourable (by 0.77 V) under standard conditions. However, the picture changes considerably in the presence of O_2. In fact, Fe(III) is the most common form of iron in the Earth's crust, and most iron in sediments that have been deposited from aqueous environments is present as Fe(III). The reaction

$$4\,Fe^{2+}(aq) + O_2(g) + 4\,H^+(aq) \rightarrow 4\,Fe^{3+}(aq) + 2\,H_2O(l)$$

is the difference of the following two half-reactions:

$$O_2(g) + 4\,H^+(aq) + 4\,e^- \rightarrow 2\,H_2O(l) \qquad E = +1.23\,V - (0.059\,V \times pH)$$
$$Fe^{3+}(aq) + e^- \rightarrow Fe^{2+}(aq) \qquad E^{\ominus} = +0.77\,V$$

which implies that

$$\begin{aligned} E^{\ominus} &= \{1.23\,V - (0.059\,V \times pH)\} - 0.77\,V \\ &= +0.46\,V - (0.059\,V \times pH) \end{aligned} \tag{5.8}$$

for the overall reaction (where the standard state symbol refers to all species except H^+), and therefore that $E^{\ominus} = +0.46\,V$ at pH = 0. The oxidation of Fe^{2+} by O_2 is therefore spontaneous ($K > 1$) at pH = 0. However, +0.46 V is not large and atmospheric oxidation of Fe(II) in aqueous solution is slow in the absence of catalysts. At higher pH, the ion species present in aqueous solution differ (Section 5.13), but solutions of Fe^{3+} may be used in the laboratory at pH close to 7.

Example 5.6 Judging the importance of atmospheric oxidation

The oxidation of copper roofs to a green substance (typically 'basic copper carbonate') is an example of atmospheric oxidation in a damp environment. Estimate the potential for oxidation of copper by oxygen in aqueous solution.

Answer The reduction half-reactions are

$$O_2(g) + 4\,H^+(aq) + 4\,e^- \rightarrow 2\,H_2O(l) \qquad E = +1.23\,V - (0.059\,V \times pH)$$
$$Cu^{2+}(aq) + 2\,e^- \rightarrow Cu(s) \qquad E^{\ominus} = +0.34\,V$$

The difference is

$$E^\ominus = \{1.23\ V - (0.059\ V \times pH)\} - 0.34\ V$$
$$= 0.89\ V - (0.059\ V \times pH) \tag{5.9}$$

Therefore, $E^\ominus = +0.89\ V$ at $pH = 0$ and $+0.48\ V$ at $pH = 7$, so atmospheric oxidation by the reaction

$$2\ Cu(s) + O_2(g) + 4\ H^+(aq) \rightarrow 2\ Cu^{2+}(aq) + 2\ H_2O(l)$$

is spontaneous (in the sense $E^\ominus > 0$ and $K > 1$) in both neutral and acid environments. Nevertheless, copper roofs do last for more than a few minutes: their familiar green surface is a passive layer of an almost impenetrable hydrated copper(II) carbonate, sulfate or, near the sea, chloride. These compounds are formed from oxidation in the presence of atmospheric CO_2, SO_2, or salt water and the anion is also involved in the redox chemistry.

Self-test 5.6 The standard potential for the conversion of sulfate ions, $SO_4^{2-}(aq)$, to $SO_2(aq)$ by the reaction $SO_4^{2-}(aq) + 4\ H^+(aq) + 2\ e^- \rightarrow SO_2(aq) + 2\ H_2O(l)$ is $+0.16\ V$. What is the thermodynamically expected fate of SO_2 emitted into fog or clouds?

5.8 Disproportionation and comproportionation

Key point: Standard potentials can be used to define the inherent stability and instability of different oxidation states in terms of disproportionation and comproportionation.

Because $E^\ominus(Cu^+, Cu) = +0.52\ V$ and $E^\ominus(Cu^{2+}, Cu^+) = +0.16\ V$, and both potentials lie within the stability field of water, Cu^+ ions neither oxidize nor reduce water. Despite this, Cu(I) is not stable in aqueous solution because it can undergo **disproportionation**, a redox reaction in which the oxidation number of an element is simultaneously raised and lowered. In other words, the element undergoing disproportionation serves as its own oxidizing and reducing agent:

$$2\ Cu^+(aq) \rightarrow Cu^{2+}(aq) + Cu(s)$$

This reaction is the difference of the following two half-reactions:

$$Cu^+(aq) + e^- \rightarrow Cu(s) \qquad E^\ominus = +0.52\ V$$
$$Cu^{2+}(aq) + e^- \rightarrow Cu^+(aq) \qquad E^\ominus = +0.16\ V$$

Because $E^\ominus = 0.52\ V - 0.16\ V = +0.36\ V$ for the disproportionation reaction, $K = 1.3 \times 10^6$ at 298 K, so the reaction is highly favourable. Hypochlorous acid also undergoes disproportionation:

$$5\ HClO(aq) \rightarrow 2\ Cl_2(g) + ClO_3^-(aq) + 2\ H_2O(l) + H^+(aq)$$

This redox reaction is the difference of the following two half-reactions:

$$4\ HClO(aq) + 4\ H^+(aq) + 4\ e^- \rightarrow 2\ Cl_2(g) + 4\ H_2O(l) \qquad E^\ominus = +1.63\ V$$
$$ClO_3^-(aq) + 5\ H^+(aq) + 4\ e^- \rightarrow HClO(aq) + 2\ H_2O(l) \qquad E^\ominus = +1.43\ V$$

So overall $E^\ominus = 1.63\ V - 1.43\ V = +0.20\ V$, and $K = 3 \times 10^{13}$ at 298 K.

Example 5.7 Assessing the likelihood of disproportionation

Show that Mn(VI) is unstable with respect to disproportionation into Mn(VII) and Mn(II) in acidic aqueous solution.

Answer The overall reaction

$$5\ HMnO_4^-(aq) + 3\ H^+(aq) \rightarrow 4\ MnO_4^-(aq) + Mn^{2+}(aq) + 4\ H_2O(l)$$

is the difference of the following two half-reactions

$$HMnO_4^-(aq) + 7\ H^+(aq) + 4\ e^- \rightarrow Mn^{2+}(aq) - 4\ H_2O(l) \qquad E^\ominus = +1.63\ V$$
$$4\ MnO_4^-(aq) + 4\ H^+(aq) + 4\ e^- \rightarrow 4\ HMnO_4^-(aq) \qquad E^\ominus = +0.90\ V$$

The difference of the standard potentials is $+0.73$ V, so the disproportionation is essentially complete ($K = 10^{50}$ at 298 K). A practical consequence of the disproportionation is that high concentrations of $HMnO_4^-$ ions cannot be obtained in acidic solution; they can, however, be obtained in basic solution, as we see in Section 5.12.

Self-test 5.7 The standard potentials for the couples Fe^{2+}/Fe and Fe^{3+}/Fe^{2+} are -0.41 V and $+0.77$ V, respectively. Should we expect Fe^{2+} to disproportionate in aqueous solution?

In **comproportionation**, the reverse of disproportionation, two species with the same element in different oxidation states form a product in which the element is in an intermediate oxidation state. An example is

$$Ag^{2+}(aq) + Ag(s) \rightarrow 2\,Ag^+(aq) \qquad E^{\ominus} = +1.18\text{ V}$$

The large positive potential indicates that Ag(II) and Ag(0) are completely converted to Ag(I) in aqueous solution ($K = 1 \times 10^{20}$ at 298 K).

5.9 The influence of complexation

Key points: The formation of a more thermodynamically stable complex when the metal is in the higher oxidation state of a couple favours oxidation and makes the standard potential more negative; the formation of a more stable complex when the metal is in the lower oxidation state of the couple favours reduction and the standard potential becomes more positive.

The formation of metal complexes (see Chapter 8) affects standard potentials because the ability of the complex ML_6 to accept or release an electron is different from that of the corresponding aqua ion with $L = H_2O$. An example is

$$[Fe(OH_2)_6]^{3+}(aq) + e^- \rightarrow [Fe(OH_2)_6]^{2+}(aq) \qquad E^{\ominus} = +0.77\text{ V}$$

compared with

$$[Fe(CN)_6]^{3-}(aq) + e^- \rightarrow [Fe(CN)_6]^{4-}(aq) \qquad E^{\ominus} = +0.36\text{ V}$$

We see that the hexacyanoferrate(III) complex is more resistant to reduction than the aqua complex. To analyse this difference, we treat the reduction of the cyano complex as the result of the three reactions shown in Fig. 5.4. The replacement of aqua ligands by cyano ligands stabilizes the hexacyanoferrate(III) complex more than it stabilizes the hexacyanoferrate(II) complex, so the net effect is to favour the reactants in the overall reaction

$$[Fe(CN)_6]^{3-}(aq) + e^- \rightarrow [Fe(CN)_6]^{4-}(aq)$$

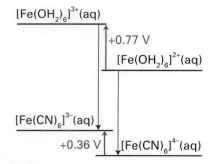

Fig. 5.4 The effect of complexation with CN^- on the standard potential of the Fe^{3+}/Fe^{2+} couple by changing the Gibbs energies (the vertical axis) of the various species involved.

Example 5.8 Assessing the effect of complexation on potential

In general, CN^- forms complexes that are thermodynamically more stable than those formed by Br^-. Which complex, $[Ni(CN)_4]^{2-}$ or $[NiBr_4]^{2-}$, is expected to have the more negative standard potential for reduction to Ni(s)?

Answer The reduction half-reaction is

$$[NiX_4]^{2-}(aq) + 2\,e^- \rightarrow Ni(s) + 4\,X^-(aq)$$

The more stable the complex, the higher its resistance to reductive decomposition. Because $[Ni(CN)_4]^{2-}$ is more stable, it is more difficult to reduce than $[NiBr_4]^{2-}$, and therefore it will have the more negative standard potential.

Self-test 5.8 Which couple has the more negative standard potential, $[Ni(en)_3]^{2+}/Ni$ or $[Ni(NH_3)_6]^{2+}/Ni$?

The diagrammatic presentation of potential data

There are several useful diagrammatic summaries of the relative stabilities of different oxidation states in aqueous solution. 'Latimer diagrams' are useful for summarizing quantitative data for individual elements. 'Frost diagrams' are useful for the qualitative portrayal of the relative and inherent stabilities of oxidation states of a range of elements. We use them frequently in this context in the following chapters to convey the sense of trends in the redox properties of the members of a group.

5.10 Latimer diagrams

In a **Latimer diagram** for an element, the numerical value of the standard potential (in volts) is written over a horizontal line connecting species with the element in different oxidation states. The most highly oxidized form of the element is on the left, and in species to the right the element is in successively lower oxidation states. A Latimer diagram summarizes a great deal of information in a compact form and (as we explain) shows the relationships between the various species in a particularly clear manner.

(a) Construction

Key points: In a Latimer diagram, oxidation numbers decrease from left to right and the numerical values of $E^{\ominus}$ in volts are written above the line joining the species involved in the couple.

The Latimer diagram for chlorine in acidic solution, for instance, is

$$\underset{+7}{ClO_4^-} \xrightarrow{+1.20} \underset{+5}{ClO_3^-} \xrightarrow{+1.18} \underset{+3}{HClO_2} \xrightarrow{+1.65} \underset{+1}{HClO} \xrightarrow{+1.67} \underset{0}{Cl_2} \xrightarrow{+1.36} \underset{-1}{Cl^-}$$

As in this example, oxidation numbers are sometimes written under (or over) the species. The notation

$$ClO_4^- \xrightarrow{+1.20} ClO_3^-$$

denotes

$$ClO_4^-(aq) + 2\,H^+(aq) + 2\,e^- \to ClO_3^-(aq) + H_2O(l) \qquad E^{\ominus} = +1.20\ V$$

Similarly,

$$HClO \xrightarrow{+1.67} Cl_2$$

denotes

$$2\,HClO(aq) + 2\,H^+(aq) + 2\,e^- \to Cl_2(g) + 2\,H_2O(l) \qquad E^{\ominus} = +1.67\ V$$

As in this example, the conversion of a Latimer diagram to a half-reaction often involves balancing elements by including the predominant species present in acidic aqueous solution (H^+ and H_2O). Electronic charge is then balanced by the inclusion of the appropriate number of electrons. The standard state for this couple includes the condition that pH $= 0$. Note that the reactant (HClO) and product (Cl_2) differ in the number of oxygen atoms and therefore, because the balancing oxygen species occurs in combination with hydrogen (as OH^-, H_2O, or H_3O^+), the standard potentials of oxoanion reduction reactions depend on the pH.

In basic aqueous solution (corresponding to pOH $= 0$ and therefore pH $= 14$), the Latimer diagram for chlorine is

$$ClO_4^- \xrightarrow{+0.37} ClO_3^- \xrightarrow{+0.30} HClO_2 \xrightarrow{+0.68} \underset{\underset{+0.89}{\underline{\qquad\qquad\qquad}}}{HClO \xrightarrow{+0.42} Cl_2} \xrightarrow{+1.36} Cl^-$$

Note that the value for the Cl_2/Cl^- couple is the same as in acidic solution because its half-reaction does not involve the transfer of protons. In basic solution, the predominant species present in solution are OH^- and H_2O, so these species are used to balance the equations for the half-reactions. As an example, the half-reaction for the ClO^-/Cl_2 couple in basic solution is

$$2 \ ClO^-(aq) + 2 \ H_2O(l) + 2 \ e^- \rightarrow Cl_2(g) + 4 \ OH^-(aq)$$
$$E^{\ominus} = +0.42 \ V \ at \ pH = 14$$

(b) Nonadjacent species

Key points: The standard potential of a couple that is the combination of two other couples is obtained by combining the standard Gibbs energies, not the standard potentials, of the half-reactions.

The Latimer diagram given above includes the standard potential for two nonadjacent species (the couple ClO^-/Cl^-). This information is redundant in the sense that it can be inferred from the data on adjacent species, but it is often included for commonly used couples as a convenience. To derive the standard potential of a nonadjacent couple when it is not listed explicitly we cannot in general just add the $E^{\ominus}$ values but must make use of eqn 5.2 ($\Delta_r G^{\ominus} = -\nu F E^{\ominus}$) and the fact that the overall $\Delta_r G^{\ominus}$ for two successive steps a and b is the sum of the individual values:

$$\Delta_r G^{\ominus}(a+b) = \Delta_r G^{\ominus}(a) + \Delta_r G^{\ominus}(b)$$

To find the standard potential of the composite process, we convert the individual $E^{\ominus}$ values to $\Delta_r G^{\ominus}$ by multiplication by the relevant factor $-\nu F$, add them together, and then convert the sum back to $E^{\ominus}$ for the nonadjacent couple by division by $-\nu F$ for the overall electron transfer:

$$-\nu F E^{\ominus}(a+b) = -\nu(a)F E^{\ominus}(a) - \nu(b)F E^{\ominus}(b)$$

Because the factors $-F$ cancel and $\nu = \nu(a) + \nu(b)$, the net result is

$$E^{\ominus}(a+b) = \frac{\nu(a)E^{\ominus}(a) + \nu(b)E^{\ominus}(b)}{\nu(a) + \nu(b)} \tag{5.10}$$

Example 5.9 Extracting $E^{\ominus}$ for nonadjacent oxidation states

Use the Latimer diagram to calculate the value of $E^{\ominus}$ for the $HClO/Cl^-$ couple in aqueous acidic solution.

Answer The oxidation number of chlorine changes from $+1$ to 0 in one step and from 0 to -1 in the subsequent step. From the Latimer diagram given in the text above, we can write

$$HClO(aq) + H^+(aq) + e^- \rightarrow \tfrac{1}{2} Cl_2(g) + H_2O(l) \qquad E^{\ominus}(a) = +1.63 \ V$$
$$\tfrac{1}{2} Cl_2(g) + e^- \rightarrow Cl^-(aq) \qquad\qquad\qquad E^{\ominus}(b) = +1.36 \ V$$

We see that $\nu(a) = 1$ and $\nu(b) = 1$. It follows that the standard potential of the $HClO/Cl^-$ couple is

$$E^{\ominus} = \frac{(1)(1.63 \ V) + (1)(1.36 \ V)}{2} = +1.50 \ V$$

Self-test 5.9 Calculate $E^{\ominus}$ for the reduction of ClO_3^- to $HClO$ at $pH = 0$.

(c) Disproportionation

Key point: A species has a tendency to disproportionate into its two neighbours in a Latimer diagram if the potential on the right of the species is higher than that on the left.

Consider the disproportionation

$$2 \ M^+(aq) \rightarrow M(s) + M^{2+}(aq)$$

This reaction has $K > 1$ if $E^{\ominus} > 0$. To analyse this criterion in terms of a Latimer diagram, we express the overall reaction as the difference of two half-reactions:

$$M^{+}(aq) + e^{-} \rightarrow M(s) \qquad E^{\ominus}(R)$$
$$M^{2+}(aq) + e^{-} \rightarrow M^{+}(aq) \qquad E^{\ominus}(L)$$

The designations L and R refer to the relative positions, left and right respectively, of the couples in a Latimer diagram (recall that the more highly oxidized species lies to the left). The standard potential for the overall reaction is $E^{\ominus} = E^{\ominus}(R) - E^{\ominus}(L)$, which is positive if $E^{\ominus}(R) > E^{\ominus}(L)$. We can conclude that a species has a tendency to disproportionate into its two neighbours if the potential on the right of the species is higher than the potential on the left.

Example 5.10 Identifying a tendency to disproportionate

A part of the Latimer diagram for oxygen is

$$O_2 \xrightarrow{+0.70} H_2O_2 \xrightarrow{+1.76} H_2O$$
$$\underset{+1.23}{\underline{\qquad\qquad\qquad}}$$

Does hydrogen peroxide have a tendency to disproportionate in acid solution?

Answer The potential to the right of H_2O_2 is higher than that to its left, so we can anticipate that H_2O_2 does have a tendency to disproportionate into its two neighbours under acid conditions. From the two half-reactions

$$2\,H^{+}(aq) + 2\,e^{-} + H_2O_2(aq) \rightarrow 2\,H_2O \qquad E^{\ominus} = +1.76\text{ V}$$
$$O_2 + 2\,H^{+}(aq) + 2\,e^{-} \rightarrow H_2O_2(aq) \qquad E^{\ominus} = +0.70\text{ V}$$

we can conclude that for the overall reaction

$$2\,H_2O_2(aq) \rightarrow 2\,H_2O(l) + O_2(g) \qquad E^{\ominus} = +1.06\text{ V}$$

and therefore that the reaction is spontaneous under standard conditions (and $K > 1$).

Self-test 5.10 Use the following Latimer diagram (acid solution) to discuss whether (a) Pu(IV) disproportionates to Pu(III) and Pu(V) in aqueous solution; (b) Pu(V) disproportionates into Pu(VI) and Pu(IV).

$$PuO_2^{2+} \xrightarrow{+1.02} PuO_2^{+} \xrightarrow{+1.04} Pu^{4+} \xrightarrow{+1.01} Pu^{3+}$$

5.11 Frost diagrams

A **Frost diagram** of an element X is a plot of $NE^{\ominus}$ for the couple $X(N)/X(0)$ against the oxidation number, N, of the element:

$$X(N) + Ne^{-} \rightarrow X(0) \qquad E^{\ominus}$$

(a) Construction

Key point: The most stable oxidation state of an element in aqueous solution corresponds to the species that lies lowest in its Frost diagram.

Figure 5.5 is an example of a Frost diagram. Because $NE^{\ominus}$ is proportional to the standard reaction Gibbs energy for the conversion of the species $X(N)$ to the element (explicitly, $NE^{\ominus} = -\Delta_r G^{\ominus}/F$, where $\Delta_r G^{\ominus}$ is the standard reaction Gibbs energy for the half-reaction given above), a Frost diagram can also be regarded as a plot of standard reaction Gibbs energy against oxidation number. It follows that the most stable oxidation state of an element in aqueous solution corresponds to the species that lies lowest in its Frost diagram (Fig. 5.6).

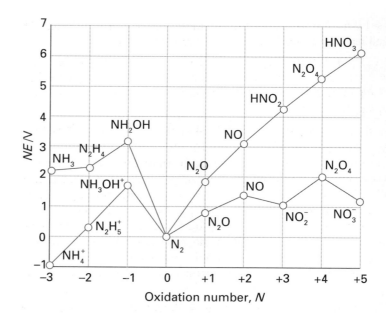

Fig. 5.5 The Frost diagram for nitrogen: the steeper the slope of a line, the higher the standard potential for the couple. The red line refers to standard (acid) conditions (pH = 0), the blue line refers to pH = 14.

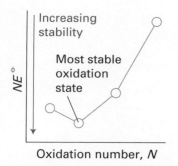

Fig. 5.6 Oxidation state stability in a Frost diagram.

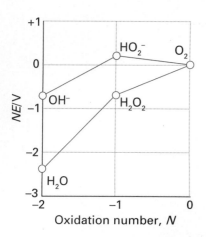

Fig. 5.7 The Frost diagram for oxygen in acidic solution (red line, pH = 0) and alkaline solution (blue line, pH = 14).

Example 5.11 Constructing a Frost diagram

Construct a Frost diagram for oxygen from the Latimer diagram in Example 5.10.

Answer For the reduction of O_2 to H_2O_2 (for which $N = -1$), $E = +0.70$ V, so $NE^{\ominus} = -0.70$ V. Because the oxidation number of O in H_2O is -2, $N = -2$, and $E^{\ominus}$ for the O_2/H_2O couple is $+1.23$ V, $NE^{\ominus} = -2.46$ V. These results are plotted in Fig. 5.7.

Self-test 5.11 Construct a Frost diagram from the Latimer diagram for Tl:

$$Tl^{3+} \xrightarrow{+1.25} Tl^+ \xrightarrow{-0.34} Tl$$
$$\underset{+0.72}{\underline{\qquad\qquad\qquad}}$$

(b) Interpretation

Key point: Frost diagrams can be used to gauge the inherent stabilities of different oxidation states of an element and to decide whether particular species are good oxidizing or reducing agents.

To interpret the qualitative information contained in a Frost diagram it will be useful to keep the following features in mind.

1 The slope of the line joining any two points in a Frost diagram is equal to the standard potential of the couple formed by the two species the points represent (Fig. 5.8).

As an illustration, refer to the oxygen diagram in Fig. 5.7. At the point corresponding to $N = -1$ (for H_2O_2), $(-1) \times E^{\ominus} = -0.70$ V, and at $N = -2$ (for H_2O), $(-2) \times E^{\ominus} = -2.46$ V. The difference of the two values is -1.76 V. The change in oxidation number of oxygen on going from H_2O_2 to H_2O is -1. Therefore, the slope of the line is $(-1.76 \text{ V})/(-1) = +1.76$ V, in accord with the value for the H_2O_2/H_2O couple in the Latimer diagram. It follows that *the steeper the line joining two points in a Frost diagram, the higher the standard potential of the corresponding couple* (Fig. 5.9a). Therefore, we can infer the magnitude of K (the 'spontaneity') for the reaction between any two couples by comparing the slopes of the corresponding lines. In particular (Fig. 5.9b):

2 The oxidizing agent in the couple with the more positive slope (the more positive $E^{\ominus}$) is liable to undergo reduction.

3 The reducing agent of the couple with the less positive slope (the most negative $E^{\ominus}$) is liable to undergo oxidation.

For example, the steep slope connecting HNO_3 to lower oxidation numbers in Fig. 5.5 shows that nitric acid is a good oxidizing agent under standard conditions.

We saw in the discussion of Latimer diagrams that a species is liable to undergo disproportionation if the potential for its reduction from $X(N)$ to $X(N-1)$ is greater than its potential for oxidation from $X(N)$ to $X(N+1)$. The same criterion can be expressed in terms of a Frost diagram (Fig. 5.9c):

4 A species in a Frost diagram is unstable with respect to disproportionation if its point lies above the line connecting the two adjacent species.

When this criterion is satisfied, the standard potential for the couple to the left of the species is greater than that for the species on the right. A specific example is NH_2OH; as can be seen in Fig. 5.5, this compound is unstable with respect to disproportionation into NH_3 and N_2. The origin of this rule is illustrated in Fig. 5.9d, where we show geometrically that the reaction Gibbs energy of a species with intermediate oxidation number lies above the average value for the two species on either side. As a result, there is a tendency for the intermediate species to disproportionate into the two other species.

The criterion for comproportionation to be spontaneous can be stated analogously (Fig. 5.9e):

5 Two species will tend to comproportionate into an intermediate species that lies below the straight line joining the terminal species.

A substance that lies below the line connecting its neighbours in a Frost diagram is more stable than they are because their average molar Gibbs energy is higher (Fig. 5.9f) and hence comproportionation is thermodynamically favourable. The nitrogen in NH_4NO_3, for instance, has two ions with oxidation numbers -3 (NH_4^+) and $+5$ (NO_3^-). Because N_2O, in which the oxidation number of nitrogen is the average of these two values, $+1$, lies below the line joining NH_4^+ to HNO_3, their comproportionation is spontaneous:

$$NH_4^+(aq) + NO_3^-(aq) \rightarrow N_2O(g) + 2\ H_2O(l)$$

However, although the reaction is expected to be spontaneous on thermodynamic grounds under standard conditions, the reaction is kinetically inhibited in solution and does not ordinarily occur. The corresponding reaction

$$NH_4NO_3(s) \rightarrow N_2O(g) + 2\ H_2O(g)$$

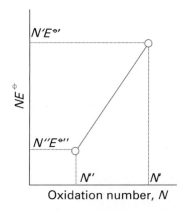

Fig. 5.8 The general structure of a region of a Frost diagram used to establish the relation between the slope of a line and the standard potential of the corresponding couple.

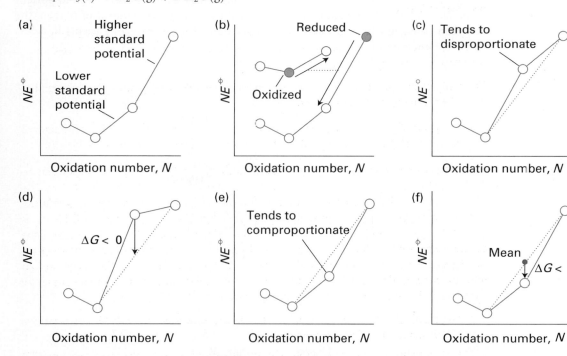

Fig. 5.9 The interpretation of a Frost diagram to gauge (a) reduction potential, (b) tendency towards oxidation and reduction, (c, d) disproportionation, and (e, f) comproportionation.

in the solid state is both thermodynamically spontaneous ($\Delta_r G^\ominus = -168\,\text{kJ mol}^{-1}$) and, once initiated by a detonation, explosively fast. Indeed, ammonium nitrate is often used in place of dynamite for blasting rocks.

Example 5.12 Using a Frost diagram to judge the thermodynamic stability of ions in solution

Figure 5.10 shows the Frost diagram for manganese. Comment on the stability of Mn^{3+} in acidic aqueous solution.

Answer Because Mn^{3+} lies above the line joining Mn^{2+} to MnO_2, it should disproportionate into these two species. The chemical reaction is

$$2\,Mn^{3+}(aq) + 2\,H_2O(l) \rightarrow Mn^{2+}(aq) + MnO_2(s) + 4\,H^+(aq)$$

Self-test 5.12 What is the oxidation number of Mn in the product when MnO_4^- is used as an oxidizing agent in aqueous acid?

5.12 The dependence of stability on pH

Key points: Modified Latimer and Frost diagrams summarize potential data under specified conditions of pH; their interpretation is the same as for pH = 0, but oxoanions often display markedly different thermodynamic stabilities; all oxoanions are stronger oxidizing agents in acidic than in basic solution.

Frost diagrams can equally well be constructed for other conditions. The potentials at pH = 14 are denoted $E_B^\ominus$ and the blue line in Fig. 5.5 is a 'basic Frost diagram' for nitrogen. The important difference from the behaviour in acidic solution is the stabilization of NO_2^- against disproportionation: its point in the basic Frost diagram no longer lies above the line connecting its neighbours. The practical outcome is that metal nitrites are stable in neutral and basic solutions and can be isolated, whereas HNO_2 cannot (although solutions of HNO_2 have some short-term stability as their decomposition is kinetically slow). In some cases, there are marked differences between strongly acidic and basic solutions, as for the phosphorus oxoanions. This example illustrates an important general point about oxoanions: when their reduction requires removal of oxygen, the reaction consumes H^+ ions, and all oxoanions are stronger oxidizing agents in acidic than in basic solution.

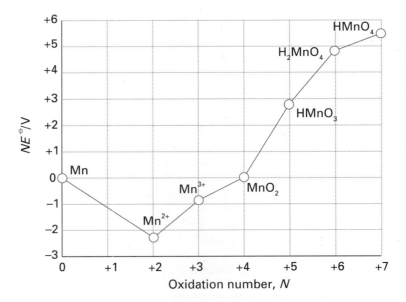

Fig. 5.10 The Frost diagram for manganese in acidic solution (pH = 0).

Standard potentials in neutral solution (pH = 7) are denoted $E_W^{\ominus}$. These potentials are particularly useful in biochemical discussions because cell fluids are buffered near pH = 7, and pH = 7 (with unit activity for the other electroactive species present) corresponds to the so-called **biological standard state**.

Example 5.13 Application of Frost diagrams at different pH

Potassium nitrite is stable in basic solution but, when the solution is acidified, a gas is evolved that turns brown on exposure to air. What is the reaction?

Answer Figure 5.5 shows that the point representing NO_2^- ion in basic solution lies below the line joining NO to NO_3^-; the ion therefore is not liable to disproportionation. On acidification, the HNO_2 point rises and the straightness of the line through NO, HNO_2, and N_2O_4 (dimeric NO_2) implies that all three species are present at equilibrium. The brown gas is NO_2 formed from the reaction of NO evolved from the solution with air. In solution, the species of oxidation number +2 (NO) tends to disproportionate. However, the escape of NO from the solution prevents its disproportionation to N_2O and HNO_2.

Self-test 5.13 By reference to Fig. 5.5, compare the strength of NO_3^- as an oxidizing agent in acidic and basic solution.

5.13 Pourbaix diagrams

Key point : A Pourbaix diagram is a map of the conditions of potential and pH under which species are stable in water.

Diagrams can also be used for discussing the general relations between redox activity and Brönsted acidity. In a **Pourbaix diagram**, the regions indicate the conditions of pH and potential under which a species is thermodynamically stable. The diagrams were introduced by Marcel Pourbaix in 1938 as a convenient way of discussing the chemical properties of species in natural waters.

Figure 5.11 is a simplified Pourbaix diagram for iron, omitting such low concentration species as oxygen-bridged dimers. This diagram is useful for the discussion of iron species in natural water because the total iron concentration is low; at high concentrations complex multinuclear iron species can form. We can see how the diagram has been constructed by considering some of the reactions involved. The reduction half-reaction

$$Fe^{3+}(aq) + e^- \rightarrow Fe^{2+}(aq) \qquad E^{\ominus} = +0.77 \text{ V}$$

does not involve H^+ ions, and so its potential is independent of pH, giving a horizontal line on the diagram.[5] If the environment contains a couple with a potential above this line (a more positive, oxidizing couple), then the oxidized species, Fe^{3+}, will be the major species. Hence, the horizontal line towards the top left of the diagram is a boundary that separates the regions where Fe^{3+} and Fe^{2+} dominate.

Another reaction to consider is

$$Fe^{3+}(aq) + 3 H_2O(l) \rightarrow Fe(OH)_3(s) + 3 H^+(aq)$$

This reaction is not a redox reaction (there is no change in oxidation number of any element), and the boundary between the region where Fe^{3+} is soluble and the region where $Fe(OH)_3$ precipitates and limits $Fe^{3+}(aq)$ to very low concentration is independent of any redox couples. However, this boundary does depend on pH, with $Fe^{3+}(aq)$ favoured by low pH and the $Fe(OH)_3$ favoured by high pH. We adopt the convention that Fe^{3+} is the dominant species in the solution if its concentration exceeds 10 μmol dm^{-3} (a typical freshwater value). The equilibrium concentration of Fe^{3+} varies with pH, and the vertical boundary represents the pH at which Fe^{3+} becomes dominant according to this definition. A similar consideration of possible further reactions such as

$$Fe(OH)_3(s) + 3 H^+(aq) + e^- \rightarrow Fe^{2+}(aq) + 3 H_2O(l)$$

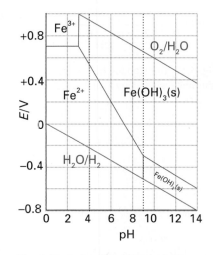

Fig. 5.11 A simplified Pourbaix diagram for some important naturally occurring compounds of iron. The broken black vertical lines represent the normal pH range in natural waters.

---

[5] The pH affects the activity of iron ions, so the potential does depend on pH even though H^+ does not appear in the reaction. We are ignoring this effect here and elsewhere.

allows the complete Pourbaix diagram for iron to be constructed. The presence of H^+ on the left of the half-reaction for a redox couple implies a downward slope of a boundary in a Pourbaix diagram.

5.14 Natural waters

Key point: Pourbaix diagrams show that iron can exist in solution as Fe^{2+} under acidic, reducing conditions such as in contact with soils rich in organic matter.

The chemistry of natural waters can be rationalized by using Pourbaix diagrams of the kind we have just constructed. Thus, where fresh water is in contact with the atmosphere, it is saturated with O_2, and many species may be oxidized by this powerful oxidizing agent ($E^\ominus = +1.23$ V). More fully reduced forms are found in the absence of oxygen, especially where there is organic matter to act as a reducing agent. The major acid system that controls the pH of the medium is $CO_2/H_2CO_3/HCO_3^-/CO_3^{2-}$, where atmospheric CO_2 provides the acid and dissolved carbonate minerals provide the base. Biological activity is also important, because respiration consumes oxygen and releases CO_2. This acidic oxide lowers the pH and hence makes the potential more positive. The reverse process, photosynthesis, consumes CO_2 and releases O_2. This consumption of acid raises the pH and makes the potential less negative. The condition of typical natural waters—their pH and the potentials of the redox couples they contain—is summarized in Fig. 5.12.

From Fig. 5.11 we see that Fe^{3+} can exist in water if the environment is oxidizing; hence, where O_2 is plentiful and the pH is low (below 4), iron will be present as Fe^{3+}. Because few natural waters are so acidic, Fe^{3+} is very unlikely to be found in the environment. The iron in insoluble Fe_2O_3 can enter solution as Fe^{2+} if it is reduced, which occurs when the condition of the water lies below the sloping boundary in the diagram. We should observe that, as the pH rises, Fe^{2+} can form only if there are strong reducing couples present, and its formation is very unlikely in oxygen-rich water. Figure 5.12 shows that iron will be reduced and dissolved in the form of Fe^{2+} in both bog waters and organic-rich waterlogged soils (at pH near 4.5 in both cases and with corresponding E values near $+0.03$ V and -0.1 V, respectively).

It is instructive to analyse a Pourbaix diagram in conjunction with an understanding of the physical processes that occur in water. As an example, consider a lake where the temperature gradient, cool at the bottom and warmer above, tends to prevent vertical mixing. At the surface, the water is fully oxygenated and the iron must be present in particles of the insoluble $Fe(OH)_3$; these particles tend to settle. At greater depth, the O_2 content is low. If the organic content or other sources of reducing agents are sufficient, the oxide will be reduced and iron will dissolve as Fe^{2+}. The $Fe(II)$ ions will then diffuse towards the surface where they encounter O_2 and be oxidized to insoluble $Fe(OH)_3$ again.

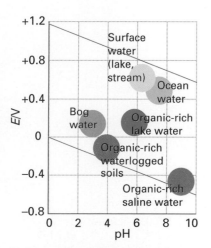

Fig. 5.12 The stability field of water showing regions typical of various natural waters.

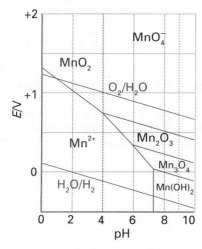

Fig. 5.13 A section of the Pourbaix diagram for manganese. The broken black vertical lines represent the normal pH range in natural waters.

Example 5.14 Using a Pourbaix diagram

Figure 5.13 is part of a Pourbaix diagram for manganese. Identify the environment in which the solid MnO_2 or its corresponding hydrous oxides are important.

Answer Manganese dioxide is thermodynamically favoured only if $E > +0.6$ V. Under mildly reducing conditions, the stable species is $Mn^{2+}(aq)$. Thus, MnO_2 is important in well-aerated waters (near the air–water boundary) where E approaches the value for the O_2/H_2O couple.

Self-test 5.14 Use Fig. 5.11 to evaluate the possibility of finding $Fe(OH)_3(s)$ in a waterlogged soil.

Chemical extraction of the elements

The original definition of 'oxidation' was a reaction in which an element reacts with oxygen and is converted to an oxide. 'Reduction' originally meant the reverse reaction, in which an oxide of a metal is converted to the metal. Both terms have been generalized and expressed in terms of electron transfer and changes in oxidation state, but these special cases are still the basis of a major part of chemical industry and laboratory chemistry. In the following sections we discuss the extraction of the elements in terms of changing their oxidation number from nonzero, that is the value in their naturally occurring compounds, to zero (corresponding to the element).

5.15 Chemical reduction

Oxygen has been a component of the atmosphere since its production from water by photosynthesis became a dominant process over 10^9 years ago, and many metals are found as their simple oxides, such as Fe_2O_3, or in more complex ternary phases, such as $FeTiO_3$. Metal sulfides also occur commonly, particularly in mineral veins where deposition occurs under water-free and oxygen-poor conditions. Copper could be extracted from its ores at temperatures attainable in the primitive hearths that became available about 6000 years ago, and the process of 'smelting' was discovered, in which ores are heated with a reducing agent such as carbon. Smelting is still used, but because many important ores of easily reduced metals are sulfides, it is often preceded by conversion of some of the sulfide to an oxide by 'roasting' in air, as in the reaction

$$2\ Cu_2S(s) + 3\ O_2(g) \rightarrow 2\ Cu_2O(s) + 2\ SO_2(g)$$

followed by smelting

$$Cu_2O(s) + C(s) \rightarrow 2\ Cu(s) + CO(g)$$

It was not until nearly 3000 years ago that the higher temperatures could be reached and less readily reduced elements, such as iron, could be extracted, leading to the Iron Age. Carbon remained the dominant reducing agent until the end of the nineteenth century, and metals that needed higher temperatures for their production remained unavailable even though their ores were reasonably abundant.

The technological breakthrough in the nineteenth century that resulted in the conversion of aluminium from a rarity into a major construction metal was the introduction of **electrolysis,** the driving of a nonspontaneous reaction (including the reduction of ores) by the passage of an electric current. The availability of electric power also expanded the scope of carbon reduction, because electric furnaces can reach much higher temperatures than carbon-combustion furnaces, such as the blast furnace. Thus, magnesium was a metal of the twentieth century because one of its modes of recovery, the *Pidgeon process*, involves the very high temperature, electrothermal reduction of the oxide by carbon:

$$MgO(s) + C(s) \xrightarrow{\Delta} Mg(l) + CO(g)$$

Note that the carbon is only oxidized to carbon monoxide as this is the product favoured thermodynamically at the very high reaction temperatures used.

(a) Thermodynamic aspects

Key points: An Ellingham diagram summarizes the temperature dependence of the standard Gibbs energies of formation of metal oxides and may be used to determine the temperature at which reduction by carbon or carbon monoxide becomes spontaneous.

As we have seen, the standard reaction Gibbs energy, $\Delta_r G^{\ominus}$, is related to the equilibrium constant, K, through $\Delta_r G^{\ominus} = -RT \ln K$, and a negative value of $\Delta_r G^{\ominus}$ corresponds to $K > 1$. It should be noted that equilibrium is rarely attained in commercial processes as many such systems involve dynamic stages where, for example, reactants and products are in contact only for short times. Furthermore, even a process at equilibrium for which $K < 1$ can be viable if the product is swept out of the reaction chamber and the reaction continues to chase the ever-vanishing equilibrium composition. In principle, we also need to consider rates when judging whether a reaction is feasible in practice, but reactions are often fast at high temperature and thermodynamically favourable reactions are likely to occur. A fluid phase (typically a gas or solvent) is usually required to facilitate what would otherwise be a sluggish reaction between coarse particles.

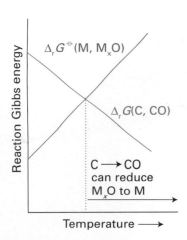

Fig. 5.14 The variation of the standard reaction Gibbs energies for the formation of a metal oxide and carbon monoxide with temperature. The formation of carbon monoxide from carbon can reduce the metal oxide to the metal at temperatures higher than the point of intersection of the two lines. More specifically, at the intersection the equilibrium constant changes from $K < 1$ to $K > 1$.

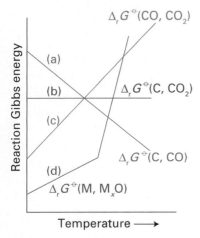

Fig. 5.15 Part of an Ellingham diagram showing the standard Gibbs energy for the formation of a metal oxide and the three carbon oxidation Gibbs energies. The slopes of the lines are determined largely by whether or not there is net gas formation or consumption in the reaction. A phase change generally results in a kink in the graph (because the entropy of the substance changes).

To achieve a negative $\Delta_r G^\ominus$ for the reduction of a metal oxide with carbon or carbon monoxide, one of the following reactions (balanced in terms of the amount of oxygen)

(a) $C(s) + \frac{1}{2}O_2(g) \rightarrow CO(g)$ \qquad $\Delta_r G^\ominus(C, CO)$

(b) $\frac{1}{2}C(s) + \frac{1}{2}O_2(g) \rightarrow \frac{1}{2}CO_2(g)$ \qquad $\Delta_r G^\ominus(C, CO_2)$

(c) $CO(g) + \frac{1}{2}O_2(g) \rightarrow CO_2(g)$ \qquad $\Delta_r G^\ominus(CO, CO_2)$

must have a more negative $\Delta_r G^\ominus$ than a reaction of the form

(d) $x\,M(s \text{ or } l) + \frac{1}{2}O_2(g) \rightarrow M_xO(s)$ \qquad $\Delta_r G^\ominus(M, M_xO)$

under the same reaction conditions. If that is so, then one of the reactions

(a – d) $M_xO(s) + C(s) \rightarrow x\,M(s \text{ or } l) + CO(g)$ \qquad $\Delta_r G^\ominus(C, CO) - \Delta_r G^\ominus(M, M_xO)$

(b – d) $M_xO(s) + \frac{1}{2}C(s) \rightarrow x\,M(s \text{ or } l) + \frac{1}{2}CO_2(g)$ \qquad $\Delta_r G^\ominus(C, CO_2) - \Delta_r G^\ominus(M, M_xO)$

(c – d) $M_xO(s) + CO(g) \rightarrow x\,M(s \text{ or } l) + CO_2(g)$ \qquad $\Delta_r G^\ominus(CO, CO_2) - \Delta_r G^\ominus(M, M_xO)$

will have a negative standard reaction Gibbs energy, and therefore have $K > 1$. The procedure followed here is similar to that adopted with half-reactions in aqueous solution (Section 5.1), but now all the reactions are written as oxidations, and the overall reaction is the difference of reactions with matching numbers of oxygen atoms. The relevant information is commonly summarized in an **Ellingham diagram**, which is a graph of $\Delta_r G^\ominus$ against temperature for the reactions (a) to (d) listed above (Fig. 5.14).

We can understand the appearance of an Ellingham diagram by noting that $\Delta_r G^\ominus = \Delta_r H^\ominus - T\Delta_r S^\ominus$ and using the fact that the enthalpy and entropy of reaction are, to a reasonable approximation, independent of temperature: the slope of a line in an Ellingham diagram should therefore be equal to $-\Delta_r S^\ominus$ for the relevant reaction. Because the standard molar entropies of gases are much larger than those of solids, the standard reaction entropy of (d), in which there is a net consumption of gas, is negative, and hence the plot in the Ellingham diagram should have a positive slope, as shown in Fig. 5.15. The kinks in the lines, where the slope of the metal oxidation line changes, are where the metal undergoes a phase change, particularly melting, and the reaction entropy changes accordingly.

The reaction entropy of (a), in which there is a net formation of gas (because 1 mol CO replaces $\frac{1}{2}$ mol O_2), is positive, and its line in the Ellingham diagram therefore has a negative slope. The standard reaction entropy of (b) is close to zero as there is no net change in the amount of gas, so its line is horizontal in the Ellingham diagram. Finally, (c) has a negative reaction entropy because $\frac{3}{2}$ mol of gas molecules is replaced by 1 mol CO_2; hence the line in the diagram has a positive slope.

At temperatures for which the C/CO line lies above the metal oxide lines in Fig. 5.15, $\Delta_r G^\ominus(M, M_xO)$ is more negative than $\Delta_r G^\ominus(C, CO)$. At these temperatures, $\Delta_r G^\ominus(C, CO) - \Delta_r G^\ominus(M, M_xO)$ is positive, so the reaction (a – d) has $K < 1$. However, for temperatures for which the C/CO line lies below the metal oxide line, the reduction of the metal oxide by carbon has $K > 1$. Similar remarks apply to the temperatures at which the other two carbon oxidation lines lie above or below the metal oxide lines in Fig. 5.15. In summary:

- For temperatures at which the C/CO line lies below the metal oxide line, carbon can be used to reduce the metal oxide and itself is oxidized to carbon monoxide.

- For temperatures at which the C/CO₂ line lies below the metal oxide line, carbon can be used to achieve the reduction, but is oxidized to carbon dioxide.

- For temperatures at which the CO/CO₂ line lies below the metal oxide line, carbon monoxide can reduce the metal oxide to the metal and is oxidized to carbon dioxide.

Figure 5.16 shows an Ellingham diagram for a selection of common metals. In principle, production of all the metals shown in the diagram, even magnesium and calcium, could be accomplished by **pyrometallurgy**, heating with a reducing agent. However, there are severe practical limitations. Efforts to produce aluminium by pyrometallurgy (most notably in Japan, where electricity is expensive) were frustrated by the volatility of

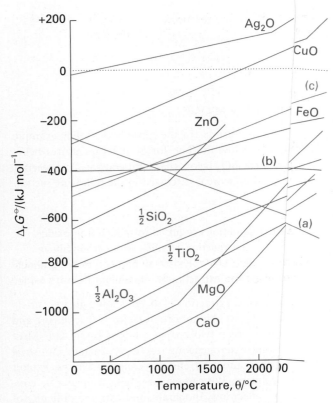

Fig. 5.16 An Ellingham diagram for the reduction of rdes.

Al$_2$O$_3$ at the very high temperatures required ficulty of a different kind is encountered in the pyrometallurgical extraction nium, where titanium carbide, TiC, is formed instead of the metal. In practice, p llurgical extraction of metals is confined principally to magnesium, iron, cobalt, inc, and a variety of ferroalloys (alloys with iron).

Example 5.15 Using an Ellingham diagram

What is the lowest temperature at which ZnO can b to zinc metal by carbon? What is the overall reaction at this temperature?

Answer The C,CO line in Fig. 5.16 lies below the Z pproximately 950°C; above this temperature, reduction of the metal oxide is spc The contributing reactions are reaction (a) and the reverse of

$$Zn(g) + \tfrac{1}{2} O_2(g) \rightarrow ZnO(s)$$

so the overall reaction is the difference, or

$$C(s) + ZnO(s) \rightarrow CO(g) + Zn(g)$$

The physical state of zinc is given as a gas beca ent boils at 907°C (the corres-ponding inflection in the ZnO line in the Ellingha n be seen in Fig. 5.16).

Self-test 5.15 What is the minimum temperatur n of MgO by carbon?

Similar principles apply to reductions usir cing agents. For instance, an Ellingham diagram can be used to explore wh M′ can be used to reduce the oxide of another metal M. In this case, we e diagram whether at a tem-perature of interest the M′/M′O line lies bel line, as M′ is now taking the place of C. When

$$\Delta_r G^\ominus = \Delta_r G^\ominus(M', M'O) - \Delta_r G^\ominus(M, M$$

is negative, where the Gibbs energies refer to the reactions

(a) $M'(s \text{ or } l) + \frac{1}{2}O_2(g) \to M'O(s)$ $\qquad \Delta_r G^\ominus(M', M'O)$

(b) $M(s \text{ or } l) + \frac{1}{2}O_2(g) \to MO(s)$ $\qquad \Delta_r G^\ominus(M, MO)$

the reaction

(a − b) $MO(s) + M'(s \text{ or } l) \cdot M(s \text{ or } l) + M'O(s)$

and its analogues for MO_2, also on) is feasible (in the sense $K > 1$). For example, because in Fig. 5.16 the line Mg lies below the line for Si at temperatures below 2200°C, magnesium may be u to reduce SiO_2 below that temperature. This reaction has in fact been used to produ ow-grade silicon as discussed in the following section.

(b) Survey of processes

Key points: A blast furnace p ces the conditions required to reduce iron oxides with carbon; impure silicon p ed from the reaction of silica with carbon above 1500°C may be purified by fu processing; electrolysis may be used to bring about a nonspontaneous reduction quired for the extraction of aluminium from its oxide.

Industrial processes for achie he reductive extraction of metals show a greater variety than the thermodynamic an might suggest. An important factor is that the ore and carbon are both solids, and ction between two coarse solids is rarely fast. Most processes exploit gas–solid id–solid heterogeneous reactions. Current industrial processes are varied in the gies they adopt to ensure economical rates, exploit materials, and avoid enviro al problems. We can explore these strategies by considering three important ex that reflect low, moderate, and extreme difficulty of reduction.

The least difficult reduct lude those of copper ores. Roasting and smelting are still widely used in the py urgical extraction of copper. However, some recent techniques seek to avoid th environmental problems caused by the production of the large quantity of SO_2, to the atmosphere, that accompanies roasting. One promising development is ometallurgical extraction of copper, the extraction of a metal by reduction of solutions of its ions, using H_2 or scrap iron as the reducing agent. In this p u^{2+} ions, leached from low-grade ores by acid or bacterial action, are reduc drogen in the reaction

$$Cu^{2+}(aq) + H_2(g) \to Cu H^+(aq)$$

or by a similar reduction n. This process is less harmful to the environment provided the acid by-pro ed or neutralized locally rather than contributing to acidic atmospheric pollu o allows economic exploitation of lower grade ores.

That extraction of iror mediate difficulty is shown by the fact that the Iron Age followed the Bron economic terms, iron ore reduction is the most important application of rometallurgy. In a blast furnace (Fig. 5.17), which is still the major source of t, the mixture of iron ores (Fe_2O_3, Fe_3O_4), coke (C), and limestone ($CaCO_3$) ith a blast of hot air. Combustion of coke in this air blast raises the temperat °C, and the carbon burns to carbon monoxide in the lower part of the furnac y of Fe_2O_3 from the top of the furnace meets the hot CO rising from below. oxide is reduced, first to Fe_3O_4 and then to FeO at 500–700°C, and the CC to CO_2. The final reduction to iron, from FeO, by carbon monoxide occu 00 and 1200°C in the central region of the furnace. Thus overall

$$Fe_2O_3(s) + 3\ CO(g) \quad 3\ CO_2(g)$$

The function of th formed by the thermal decomposition of calcium carbonate is to combin ates present in the ore to form a molten layer of slag in the hottest (lowest) nace. Slag is less dense than iron and can be drained away. The iron formed 400°C below the melting point of the pure metal on account of the dissolv ntains. The impure iron, the densest phase, settles to

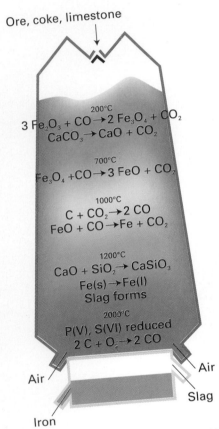

Ore, coke, limestone

200°C
$3\ Fe_2O_3 + CO \to 2\ Fe_3O_4 + CO_2$
$CaCO_3 \to CaO + CO_2$

700°C
$Fe_3O_4 + CO \to 3\ FeO + CO_2$

1000°C
$C + CO_2 \to 2\ CO$
$FeO + CO \to Fe + CO_2$

1200°C
$CaO + SiO_2 \to CaSiO_3$
$Fe(s) \to Fe(l)$
Slag forms

2000°C
P(V), S(VI) reduced
$2\ C + O_2 \to 2\ CO$

Air

Air

Slag

Iron

Fig. 5.17 A schematic diagram of a blast furnace showing the typical composition and temperature profile.

the bottom and is drawn off to solidify into 'pig iron', in which the carbon content is high (about 4 per cent by mass). The manufacture of steel is then a series of reactions in which the carbon content is reduced and other metals are used to form alloys with the iron (see Box 3.1).

More difficult than the extraction of either copper or iron is the extraction of silicon from its oxide: indeed, silicon is very much an element of the twentieth century. Silicon of 96 to 99 per cent purity is prepared by reduction of quartzite or sand (SiO_2) with high purity coke. The Ellingham diagram shows that the reduction is feasible only at temperatures in excess of about 1500°C. This high temperature is achieved in an electric arc furnace in the presence of excess silica (to prevent the accumulation of SiC):

$$SiO_2(l) + 2\ C(s) \xrightarrow{1500°C} Si(l) + 2\ CO(g)$$
$$2\ SiC(s) + SiO_2(l) \rightarrow 3\ Si(l) + 2\ CO(g)$$

Very pure silicon (for semiconductors) is made by converting crude silicon to volatile compounds, such as $SiCl_4$. These compounds are purified by exhaustive fractional distillation and then reduced to silicon with pure hydrogen. The resulting semiconductor-grade silicon is melted and large single crystals are pulled slowly from the cooled surface of a melt: this procedure is called the **Czochralski process**.

As we have remarked, the Ellingham diagram shows that the direct reduction of Al_2O_3 with carbon becomes feasible only above 2000°C, which makes it uneconomically expensive and wasteful in terms of any fossil fuels used to heat the system. However, the reduction can be brought about electrolytically (Section 5.17).

5.16 Chemical oxidation

Key points: Elements obtained by chemical oxidation include the heavier halogens, sulfur, and (in the course of their purification) certain noble metals.

As O_2 is available from fractional distillation of air, chemical methods of oxygen production are not necessary. Sulfur is an interesting mixed case. Elemental sulfur is either mined or produced by oxidation of the H_2S that is removed from 'sour' natural gas and crude oil. The oxidation is accomplished by the **Claus process**, which consists of two stages. In the first, some hydrogen sulfide is oxidized to sulfur dioxide:

$$2\ H_2S + 3\ O_2 \rightarrow 2\ SO_2 + 2\ H_2O$$

In the second stage, this sulfur dioxide is allowed to react in the presence of a catalyst with more hydrogen sulfide:

$$2\ H_2S + SO_2 \xrightarrow{Oxide\ catalyst, 300°C} 3\ S + 2\ H_2O$$

The catalyst is typically Fe_2O_3 or Al_2O_3. The Claus process is environmentally benign; otherwise it would be necessary to burn the toxic hydrogen sulfide to polluting sulfur dioxide.

The only important metals extracted in a process using an oxidation stage are the ones that occur in native form (that is, as the element). Gold is an example, because it is difficult to separate the granules of metal in low-grade ores by simple 'panning'. The dissolution of gold depends on oxidation, which is favoured by complexation with CN^- ions to form $[Au(CN)_2]^-$.

$$Au(s) + 2\ CN^-(aq) \rightarrow [Au(CN)_2]^-(aq)$$

This complex is then reduced to the metal by reaction with another reactive metal, such as zinc:

$$2\ [Au(CN)_2]^-(aq) + Zn(s) \rightarrow 2\ Au(s) + [Zn(CN)_4]^{2-}(aq)$$

However, because of the toxicity of cyanide, alternative methods of extracting gold have been used, involving oxidation of gold to form sulfate solutions followed by reduction with bacteria.

The lighter, strongly oxidizing halogens are extracted electrochemically, as described in Section 5.17. The more readily oxidizable halogens, Br_2 and I_2, are obtained by chemical oxidation of the aqueous halides with chlorine. For example,

$$2 NaBr(aq) + Cl_2(g) \rightarrow 2 NaCl(aq) + Br_2(l)$$

5.17 Electrochemical extraction

Key points: Elements obtained by electrochemical reduction include aluminium; those obtained by electrochemical oxidation include chlorine.

The extraction of metals from ores electrochemically is confined mainly to the more electropositive elements, as discussed in the case of aluminium later in this section. For other metals produced in bulk quantities, such as iron and copper, the more energy-efficient and cleaner routes used by industry in practice, and using chemical methods of reduction, were described in Section 5.15b. In some specialist cases, electrochemical reduction is used to isolate small quantities of platinum group metals. So, for example, treatment of spent catalytic converters with acids under oxidizing conditions produces a solution containing Pt^{2+} and other platinum group metals, which can then be reduced electrochemically. The metals are deposited at the cathode with an overall 80 per cent efficient extraction from the ceramic catalytic converter.

As we saw in Section 5.15, an Ellingham diagram shows that the reduction of Al_2O_3 with carbon becomes feasible only above 2000°C, which is uneconomically expensive. However, the reduction can be brought about electrolytically, and all modern production uses the electrochemical **Hall–Héroult process**, which was invented in 1886 independently by Charles Hall and Paul Héroult. The process requires pure aluminium hydroxide that is extracted from aluminium ores by using the **Bayer process**. In this process the bauxite ore used as a source of aluminium is a mixture of the acidic oxide SiO_2 and amphoteric oxides and hydroxides, such as Al_2O_3, AlOOH, and Fe_2O_3. The Al_2O_3 is dissolved in hot aqueous sodium hydroxide, which separates the aluminium from much of the less soluble Fe_2O_3, although silicates are also rendered soluble in these strongly basic conditions. Cooling the sodium aluminate solution results in the precipitation of $Al(OH)_3$, leaving the silicates in solution. The aluminium hydroxide is then dissolved in molten cryolite (Na_3AlF_6) and the melt is reduced electrolytically at a steel cathode with graphite anodes. The latter participate in the electrochemical reaction by reacting with the evolved oxygen atoms so that the overall process is

$$2 Al_2O_3 + 3 C \rightarrow 4 Al + 3 CO_2$$

As the power consumption of a typical plant is huge, aluminium is often produced where electricity is cheap (for example, from hydroelectric sources in Canada) and not where bauxite is mined (in Jamaica, for example).

The lighter halogens are the most important elements extracted by electrochemical oxidation. The standard reaction Gibbs energy for the oxidation of Cl^- ions in water

$$2 Cl^-(aq) + 2 H_2O(l) \rightarrow 2 OH^-(aq) + H_2(g) + Cl_2(g) \quad \Delta_r G^\ominus = +422 \text{ kJ mol}^{-1}$$

is strongly positive, which suggests that electrolysis is required. The minimum potential difference that can achieve the oxidation of Cl^- is about 2.2 V ($\Delta_r G^\ominus = -\nu F E^\ominus$ and $\nu = 2$)

It may appear that there is a problem with the competing reaction

$$2 H_2O(l) \rightarrow 2 H_2(g) + O_2(g) \quad \Delta_r G^\ominus = +414 \text{ kJ mol}^{-1}$$

which can be driven forwards by a potential difference of only 1.2 V (in this reaction, $\nu = 4$). However, the rate of oxidation of water is very slow at potentials at which it first becomes favourable thermodynamically. This slowness is expressed by saying that the reduction requires a high **overpotential**, η (eta), the potential that must be applied in addition to the equilibrium value before a significant rate of reaction is achieved. Consequently, the electrolysis of brine produces Cl_2, H_2, and aqueous NaOH, but not much O_2.

Oxygen, not fluorine, is produced if aqueous solutions of fluorides are electrolysed. Therefore, F_2 is prepared by the electrolysis of an anhydrous mixture of potassium fluoride and hydrogen fluoride, an ionic conductor that is molten above $72°C$.

FURTHER READING

A.J. Bard, R. Parsons, and R. Jordan, *Standard potentials in aqueous solution*. M. Dekker, New York (1985). A collection of cell potential data with discussion.

S.G. Bartsch, Standard electrode potentials in water and their temperature coefficients. *J. Phys. Chem. Ref. Data*, 1989, **18**, 1.

I. Barin, *Thermochemical data of pure substances*, Vols 1 and 2. VCH, Weinheim (1989). A comprehensive source of thermodynamic data for inorganic substances.

R.G. Compton and G.H.W. Sanders, *Electrode potentials*. Oxford University Press (1996). Equilibrium electrochemistry and the use of electrode potentials.

J. Emsley, *The elements*. Oxford University Press (1998). Excellent source of data on the elements including standard potentials.

A.G. Howard, *Aquatic environmental chemistry*. Oxford University Press (1998). Discussion of the compositions of freshwater and marine systems explaining the effects of oxidation and reduction processes.

M. Pourbaix, *Atlas of electrochemical equilibria in aqueous solution*. Pergamon Press, Oxford (1966). The original and still good source of Pourbaix diagrams.

W. Stumm and J.J. Morgan, *Aquatic chemistry*. Wiley, New York (1996). A standard reference on natural water chemistry.

T.W. Swaddle, *Inorganic chemistry*. Academic Press (1997). A good introductory account of the practical aspects of environmental chemistry, redox reactions in solution, and extractive metallurgy.

EXERCISES

5.1 Using data from *Resource section 1.2*, suggest chemical reagents that would be suitable for carrying out the following transformations and write balanced equations for the reactions: (a) oxidation of HCl to chlorine gas, (b) reduction of $Cr(III)(aq)$ to $Cr(II)(aq)$, (c) reduction of Ag^+ to $Ag(s)$, (d) reduction of I_2 to I^-.

5.2 Use standard potential data from *Resource section 1.2* as a guide to write balanced equations for the reactions that each of the following species might undergo in aerated aqueous acid. If the species is stable, write 'no reaction'. (a) Cr^{2+}, (b) Fe^{2+}, (c) Cl^-, (d) $HClO$, (e) $Zn(s)$.

5.3 Use the information in *Resources section 1.2* to write balanced equations for the reactions, including disproportionations, that can be expected for each of the following species in aerated acidic aqueous solution: (a) Fe^{2+}, (b) Ru^{2+}, (c) $HClO_2$, (d) Br_2.

5.4 Write the Nernst equation for

(a) the reduction of $O_2(g)$: $O_2(g) + 4H^+(aq) + 4e^- \rightarrow 2H_2O(l)$

(b) the reduction of $Fe_2O_3(s)$: $Fe_2O_3(s) + 6H^+(aq) + 6e^- \rightarrow 2Fe(s) + 3H_2O(l)$

and in each case express the formula in terms of pH. What is the potential for the reduction of oxygen at $pH = 7$ and $p(O_2) = 0.20$ bar (the partial pressure of oxygen in air)?

5.5 Given the following standard potentials in basic solution

$CrO_4^{2-}(aq) + 4H_2O(l) + 3e^- \rightarrow Cr(OH)_3(s) + 5OH^-(aq)$ $E^⊖ = -0.11V$

$[Cu(NH_3)_2]^+(aq) + e^- \rightarrow Cu(s) + 2NH_3(aq)$ $E^⊖ = -0.10V$

and assuming that a reversible reaction can be established on a suitable catalyst, calculate $E^⊖$, $\Delta_r G^⊖$, and K for the reductions of (a) CrO_4^{2-} and (b) $[Cu(NH_3)_2]^+$ in basic solution. Comment on why $\Delta_r G^⊖$ and K are so different between the two cases despite the values of $E^⊖$ being so similar.

5.6 Answer the following questions using the Frost diagram in Fig. 5.18. (a) What are the consequences of dissolving Cl_2 in aqueous basic solution? (b) What are the consequences of dissolving Cl_2 in

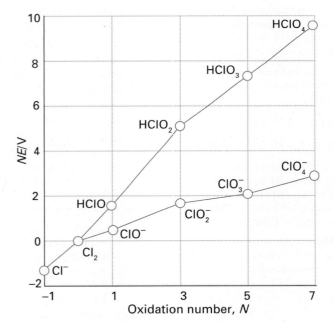

Fig. 5.18 A Frost diagram for chlorine. The red line refers to acid conditions ($pH = 0$) and the blue line to $pH = 14$.

aqueous acid? (c) Is the failure of $HClO_3$ to disproportionate in aqueous solution a thermodynamic or a kinetic phenomenon?

5.7 Use standard potentials as a guide to write equations for the main net reaction that you would predict in the following experiments: (a) N_2O is bubbled into aqueous NaOH solution, (b) zinc metal is added to aqueous sodium triiodide, (c) I_2 is added to excess aqueous $HClO_3$.

5.8 Characterize the condition of acidity or basicity that would most favour the following transformations in aqueous solution: (a) $Mn^{2+} \rightarrow MnO_4^-$, (b)$ClO_4^- \rightarrow ClO_3^-$, (c) $H_2O_2 \rightarrow O_2$, (d) $I_2 \rightarrow 2\,I^-$.

5.9 Use the Latimer diagram for chlorine to determine the potential for reduction of ClO_4^- to Cl_2. Write a balanced equation for this half-reaction.

5.10 Calculate the equilibrium constant of the reaction $Au^+(aq) + 2\,CN^-(aq) \rightarrow [Au(CN)_2]^-(aq)$ from the standard potentials

$$Au^+(aq) + e^- \rightarrow Au(s) \qquad\qquad E^\ominus = +1.68V$$

$$[Au(CN)_2]^-(aq) + e^- \rightarrow Au(s) + 2\,CN^-(aq) \quad E^\ominus = -0.6V$$

5.11 Use Fig. 5.12 to find the approximate potential of an aerated lake at pH = 6. With this information and Latimer diagrams from *Resource section 1.2*, predict the species at equilibrium for the elements (a) iron, (b) manganese, (c) sulfur.

5.12 The species Fe^{2+} and H_2S are important at the bottom of a lake where O_2 is scarce. If the pH = 6, what is the maximum value of E characterizing the environment?

5.13 The ligand edta forms stable complexes with hard acid centres. How will complexation with edta affect the reduction of M^{2+} to the metal in the $3d$ series?

5.14 In Fig. 5.11, which of the boundaries depend on the choice of Fe^{2+} concentration as 10^{-5} mol dm^{-3}?

5.15 Consult the Ellingham diagram in Fig. 5.16 and determine if there are any conditions under which aluminium might be expected to reduce MgO. Comment on these conditions.

PROBLEMS

5.1 Using standard potential data, suggest why permanganate is not a suitable oxidizing agent for the quantitative estimation of Fe^{2+} in the presence of HCl but becomes so if sufficient Mn^{2+} and phosphate ion are added to the solution. (*Hint:* Phosphate forms complexes with Fe^{3+}, thereby stabilizing it.)

5.2 R.A. Binstead and T.J. Meyer have described (*J. Am. Chem. Soc.*, 1987, **109**, 3287) the reduction of $[Ru^{IV}O(bipy)_2(py)]^{2+}$ to $[Ru^{III}(OH)(bipy)_2(py)]^{2+}$. The study revealed that the rate of the reaction is strongly influenced by the change of the solvent from H_2O to D_2O. What does this suggest about the mechanism of the reaction? Comment on the relation of this result to the cases of simple electron transfer and simple atom transfer described in the chapter.

5.3 It is often found that O_2 is a slow oxidizing agent. Suggest a mechanistic explanation that takes into account the two reduction potentials

$$O_2(g) + 4\,H^+(aq) + 4\,e^- \rightarrow 2\,H_2O(l) \quad E^\ominus = +1.23\text{ V}$$

$$O_2(g) + 2\,H^+(aq) + 2\,e^- \rightarrow H_2O_2(aq) \quad E^\ominus = +0.70\text{ V}$$

5.4 Some anaerobic bacteria use oxidizing agents other than O_2 as an energy source;. for example, SO_4^{2-}, NO_3^-, and Fe^{3+}. One half-reaction is $FeO(OH)(s) + HCO_3^-(aq) + 2\,H^+(aq) + e^- \rightarrow FeCO_3(s) + 2\,H_2O(l)$, for which $E^\ominus = +1.67$ V. What mass of iron gives the same change in Gibbs energy as 1.00 g of O_2?

5.5 Many of the tabulated data for standard potentials have been determined from thermochemical data rather than direct electrochemical measurements of cell potentials. Carry out a calculation to illustrate this approach for the half-reaction $Sc_2O_3(s) + 3\,H_2O(l) + 6\,e^- \rightarrow 2\,Sc(s) + 6\,OH^-(aq)$.

	$Sc^{3+}(aq)$	$OH^-(aq)$	$H_2O(l)$	$Sc_2O_3(s)$	$Sc(s)$
$\Delta_f H^\ominus/(\text{kJ mol}^{-1})$	−614.2	−230.0	−285.8	−1908.7	0
$S_m^\ominus/(\text{J K}^{-1}\text{ mol}^{-1})$	−255.2	−10.75	+69.91	+77.0	+34.76

5.6 The reduction potential of an ion such as OH^- can be strongly influenced by the solvent. (a) From the review article by D.T. Sawyer and J.L. Roberts, *Acc. Chem. Res.*, 1988, **21**, 469, describe the magnitude of the change in the potential of the OH/OH^- potential on changing the solvent from water to acetonitrile, CH_3CN. (b) Suggest a qualitative interpretation of the difference in solvation of the OH^- ion in these two solvents.

5.7 Reactions that involve atom transfer may be expressed either in terms of the redox language of changes in oxidation number or from the standpoint of nucleophilic substitution. Describe, from these two viewpoints, the mechanism of the reaction of NO_2 with ClO^- in water to produce NO_3^- and Cl. (See D.W. Johnson and D.W. Margerum, *Inorg. Chem.*, 1991, **30**, 4845.)

5.8 Balance the following redox reaction in acid solution: $MnO_4^- + H_2SO_3 \rightarrow Mn^{2+} + HSO_4^-$. Predict the qualitative pH dependence on the net potential for this reaction (i.e. increases, decreases, remains the same).

5.9 Draw a Frost diagram for mercury in acid solution given the following Latimer diagram:

$$Hg^{2+} \xrightarrow{\;0.911\;} Hg_2^{2+} \xrightarrow{\;0.796\;} Hg$$

Comment on the tendency of any of the species to act as an oxidizing agent, a reducing agent, or to undergo disproportionation.

5.10 From the following Latimer diagram, calculate the value of $E^\ominus$ for the reaction $2\,HO_2(aq) \rightarrow O_2(g) + H_2O_2(aq)$.

$$O_2 \xrightarrow{\;-0.125\;} HO_2 \xrightarrow{\;+1.510\;} H_2O_2$$

Comment on the thermodynamic tendency of HO_2 to undergo disproportionation.

5.11 Using the following aqueous acid solution reduction potentials $E^\ominus(Pd^{2+}, Pd) = +0.915$ V and $E^\ominus([PdCl_4]^{2-}, Pd) = +0.600$ V, calculate the equilibrium constant for $Pd^{2+}(aq) + 4\,Cl^-(aq) \rightleftharpoons [PdCl_4]^{2-}(aq)$ in 1 M HCl(aq).

5.12 Calculate the reduction potential at 25°C for the conversion of $MnO_4^-(aq)$ to $MnO_2(s)$ in aqueous solution at pH = 9.00 and 1 M $MnO_4^-(aq)$ given that $E^\ominus(MnO_4^-, MnO_2) = +1.69$ V.

5.13 How could a cyclic voltammogram be used to determine whether a metal ion in aqueous solution is present in a complex with ligands other than water?

5.14 Explain why water with high concentrations of dissolved carbon dioxide and open to atmospheric oxygen is very corrosive towards iron.

5.15 Discuss how the equilibrium $Cu^{2+}(aq) + Cu(s) \rightleftharpoons 2\,Cu^+(aq)$ can be shifted by complexation with chloride ions (see J. Malyyszko and M. Kaczor, *J. Chem. Educ.*, 2003, **80**, 1048).

5.16 In their article 'Variability of the cell potential of a given chemical reaction' (*J. Chem. Educ.*, 2004, **81**, 84), L.H. Berka and I. Fishtik conclude that $E^\ominus$ for a chemical reaction is not a state function because the half-reactions are arbitrarily chosen and may contain different numbers of transferred electrons. Discuss this objection.

Physical techniques in inorganic chemistry

6

Modern inorganic chemistry relies on a wide variety of physical techniques that use special instruments and apply sophisticated concepts and theories. In this chapter, we introduce the most important physical techniques that are used to investigate the atomic and electronic structures of inorganic compounds and study their reactions. In each case we discuss the basic principles of the measurements and their interpretation and recommend further reading material that will provide more comprehensive explanations and examples.

All the structures of the molecules and materials that we cover in this book have been determined by applying one or more kinds of investigation by a physical technique. The techniques and instruments available vary greatly in complexity and cost, as well as in their suitability for meeting particular challenges and solving problems. In this chapter we outline the most widely used techniques, presenting brief qualitative guides to the relevant concepts and providing examples of how the data are analysed to yield structures, rates, and other physical data. Many of the physical techniques used in contemporary inorganic research rely on the interaction of electromagnetic radiation with matter and there is hardly a section of the electromagnetic spectrum that is not used.

Diffraction methods

Diffraction techniques, particularly using X-rays, are the most important structure determination methods available to the inorganic chemist. X-ray diffraction has been responsible for determining the structures of a quarter of a million different substances including tens of thousands of purely inorganic compounds and many more organometallic compounds. The method allows for the unambiguous determination of the positions of the atoms and ions that make up a molecular or ionic compound and thus allows description of structures in terms of features such as bond lengths and angles and the relative positions of ions and molecules in a unit cell. The structural data obtained have been interpreted in terms of atomic and ionic radii, which then allow chemists to predict structure and explain trends in many properties.

6.1 X-ray diffraction

Key points: The scattering of radiation with wavelengths of about 100 pm from crystals gives rise to diffraction; the interpretation of these diffraction patterns allows the extraction of quantitative structural information and in many cases the complete molecular or ionic structure.

Diffraction is the interference between waves that occurs as a result of an object in their path. X-rays are scattered by the electrons in atoms, and diffraction can occur for a periodic array of scattering centres separated by distances similar to the wavelength of the radiation (about 100 pm), such as exists in a crystal. If we think of scattering as equivalent to reflection from two adjacent parallel planes separated by a distance d (Fig. 6.1) then the

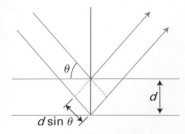

Fig. 6.1 The Bragg equation is derived by treating layers of atoms as reflecting planes; X-rays interfere constructively when the additional path length $2d \sin \theta$ is equal to the wavelength λ.

angle at which constructive interference occurs between waves of wavelength λ is given by the **Bragg equation**:

$$2d \sin \theta = \lambda \tag{6.1}$$

The intensity of the diffraction depends on the details of the crystal structure and the identities of the atoms. How well an atom scatters X-rays is related to how many electrons it possesses and its location in the unit cell. Thus measurement of diffraction angles and intensities allows us to work backwards to structural information.

There are two principal X-ray techniques: the **powder method**, in which the materials being studied are in polycrystalline form, consisting of thousands of crystallites of a few micrometres or less in dimension, and **single-crystal diffraction**, where the compound is available as a single crystal of dimensions of several tens of micrometres or larger.

(a) Powder X-ray diffraction

Key point: The powder X-ray diffraction method is used for phase identification and the determination of lattice parameters and lattice type.

A powder sample contains an enormous number of very small crystallites, typically 0.1 to 10 μm in dimension and orientated at random. An X-ray beam striking a polycrystalline sample is scattered in all possible directions. As a result, each plane of atoms separated by a different lattice spacing in the crystal gives rise to a cone of diffraction intensity. Each cone consists of a set of closely spaced dots, each one of which represents diffraction from a single crystallite within the powder sample. With a very large number of crystallites these dots join together to form the diffraction cone. To obtain powder X-ray diffraction data in a form useful for analysis, the positions of the various diffraction cones need to be determined. This can be achieved by using either photographic film or a detector sensitive to X-ray radiation. In each case the basic idea is to determine the diffraction angle, θ, of the various diffraction cones. A **powder diffractometer** uses an X-ray detector to measure the positions of the diffracted beams. Scanning the detector around the sample along the circumference of a circle cuts through the diffraction cones at the various diffraction maxima and the intensity of the X-rays detected is recorded as a function of the detector angle. A powder X-ray diffraction pattern obtained for a typical material is shown in Fig. 6.2.

The number and positions of the diffraction maxima, which are normally referred to as *reflections*, depend on the cell parameters, crystal system, lattice type, and wavelength used to collect the data; the peak intensities depend on the types of atoms present and their positions. Nearly all crystalline solids have a unique powder X-ray diffraction pattern in terms of the positions of the observed reflections and their intensities. In mixtures of compounds, each crystalline phase present contributes to the powder diffraction pattern its own unique set of lines and the relative intensities of line sets from mixtures depend on the amount present and the ability of a structure to scatter X-rays.

The effectiveness of powder X-ray diffraction has led to it becoming the major technique for the characterization of polycrystalline, solid inorganic materials (Table 6.1). Many of the powder diffraction data sets collected from inorganic, organometallic, and

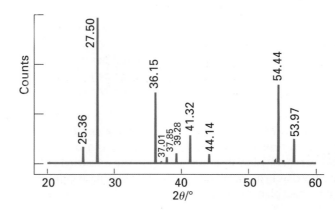

Fig. 6.2 A powder diffraction pattern obtained for a mixture of TiO_2 polymorphs (see Example 6.1).

Table 6.1 Application of powder X-ray diffraction

Application	Typical use and information extracted
Identification of unknown materials	Rapid identification of most crystalline phases
Determination of sample purity	Monitoring the progress of a chemical reaction occurring in the solid state
Determination and refinement of lattice parameters	Phase identification and monitoring structure as a function of composition
Investigation of phase diagrams/new materials	Mapping out composition and structure
Determination of crystallite size/stress	Particle size measurement and uses in metallurgy
Structure refinement	Extraction of crystallographic data from a known structure type
Structure determination	Structure determination (often at high precision) is possible from first principles
Phase changes/expansion coefficients	Studies as a function of temperature (cooling or heating typically in the range 100–1200 K). Observation of structural transitions

organic compounds have been compiled into a database by the Joint Committee on Powder Diffraction Standards (JCPDS). This database, which contains over 50 000 unique powder X-ray diffraction patterns, can be used like a fingerprint library to identify an unknown material from its powder pattern alone. Powder X-ray diffraction is used routinely in the investigation of phase formation and changes in structures of solids. The synthesis of a metal oxide can be verified by collecting a powder diffraction pattern and demonstrating that the data are consistent with a single pure phase of that material. Indeed, the progress of a chemical reaction is often monitored by observing the formation of the product phase at the expense of the reactants.

The presence or absence of certain reflections in the diffraction pattern permits the determination of the lattice type. In recent years the technique of fitting the intensities of the peaks in the diffraction pattern has become a popular method of extracting structural information such as atomic positions. The analysis, which is known as the **Rietveld method**, involves fitting a calculated diffraction pattern to the experimental trace. The technique is not as powerful as the single-crystal methods but has the advantage of not requiring the growth of crystals.

Example 6.1 Using powder X-ray diffraction to analyse polycrystalline compounds

Titanium dioxide exists in several polymorphic forms, the most common of which are anatase, rutile, and brookite. The experimental diffraction angles for six strongest lines observed in the powder diffraction pattern collected from each of these different polymorphs are summarized in the table in the margin.

The powder X-ray diffraction pattern collected (using 154 pm X-radiation) from a sample of white paint, known to contain TiO_2 in one or more of these polymorphic forms, showed the diffraction pattern in Fig. 6.2. Identify the TiO_2 polymorphs present.

Rutile	Anatase	Brookite
27.50	25.36	19.34
36.15	37.01	25.36
39.28	37.85	25.71
41.32	38.64	30.83
44.14	48.15	32.85
54.44	53.97	34.90

Answer The lines closely match those of rutile (strongest reflections) and anatase (a few weak reflections) so paint contains these phases, and rutile as the major TiO_2 phase.

Self-test 6.1 Chromium(IV) oxide also adopts the rutile structure. By consideration of the Bragg equation and the ionic radii of Ti^{4+} and Cr^{4+} (*Resource section* 1) predict the main features of the CrO_2 powder X-ray diffraction pattern.

(b) **Single-crystal X-ray diffraction**

Key point: The analysis of the diffraction patterns obtained from single crystals allows the full determination of the structure.

Analysis of the diffraction data obtained from single crystals is the most important method of obtaining the structures of inorganic solids. Provided a compound can be grown as a crystal of sufficient size and quality, the data provide definitive information on molecular or extended lattice structure.

The collection of diffraction data from a single crystal is normally carried out using a laboratory-based four-circle or area detector diffractometer (Fig. 6.3). For very small crystals, intense X-ray beams obtained at synchrotron sources can be used. A **four-circle diffractometer** uses a scintillation detector to measure the diffracted X-ray beam intensity as a function of the four angles shown in the illustration. An **area detector** uses an image plate that is sensitive to X-rays and so can measure a large number of diffraction maxima simultaneously; many new systems use this technology because the data can typically be collected in a few hours.

Analysis of the diffraction data from single crystals is formally a complex process involving many thousands of reflections with their intensities, but with increasing advances in computation power a skilled crystallographer can complete the structure determination of a small inorganic molecule in under an hour. Because the diffraction pattern is in effect the Fourier transform of the electron density in the crystal, the process of structure determination involves transforming the experimental diffraction data back to the three-dimensional array of atoms. Single-crystal X-ray diffraction can be used to determine the structures of most inorganic compounds provided they can be obtained as crystals of reasonable size. However, determining the positions of H atoms in many inorganic compounds that also contain heavy atoms such as the 4d- and 5d-metals can be difficult or impossible because X-rays are scattered by electrons and H atoms have very low electron densities.

(c) X-ray diffraction at synchrotron sources

Key point: High-intensity X-ray beams generated at synchrotron sources allow the structures of very complex molecules to be determined.

Whereas powder and single-crystal X-ray diffraction experiments are normally undertaken in chemistry laboratories using X-ray generators, much more intense X-ray beams can be obtained by using **synchrotron radiation**. Synchrotron radiation is produced by electrons undergoing acceleration around a storage ring and is typically several orders of magnitude more intense than laboratory X-ray sources. New, ever more intense, synchrotron sources are being constructed for research including diffraction studies of very small crystals. Diffraction equipment located on such an X-ray source permits the study of much smaller samples and crystals as small as $10 \times 10 \times 10\,\mu m$. Furthermore, data collection can be undertaken much more rapidly and more complex structures can be determined.

6.2 Neutron diffraction

Key point: The scattering of neutrons by crystals yields diffraction data that give additional information on structure, particularly on light atom positions.

Diffraction occurs from crystals for any particle with a velocity such that its associated wavelength (through the de Broglie relation, $\lambda = h/mv$) is comparable to the separations of the atom or ions in the crystal. Neutrons and electrons can have wavelengths of the order of 100–200 pm and thus undergo diffraction by crystalline inorganic compounds.

Neutron beams of the appropriate wavelength are generated by 'moderating' (slowing down) neutrons generated in nuclear reactors or through a process known as **spallation**, in which neutrons are chipped off the nuclei of heavy elements by accelerated beams of protons. The instrumentation used for collecting data and analysing single-crystal or powder neutron diffraction patterns is often similar to that used for X-ray diffraction. The scale is much larger, however, because neutron beam fluxes are much lower than laboratory X-ray sources.

The advantages of neutron diffraction stem from the fact that neutrons are scattered by nuclei rather than by the surrounding electrons. As a result, neutrons are sensitive to

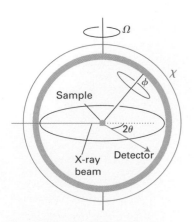

Fig. 6.3 The layout of a four-circle diffractometer. A computer controls the location of the detector as the four angles are changed systematically.

structural parameters that often complement those for X-rays. In particular, the scattering is not dominated by the heavy elements, which can be a problem with X-ray diffraction for most inorganic compounds. For example, locating the position of a light element such as H and Li in a material that also contains lead can be impossible with X-ray diffraction, as almost all the electron density is associated with the Pb atoms. With neutrons, in contrast, the scattering from light atoms is often similar to that of heavy elements; so neutron diffraction is frequently used in conjunction with X-ray diffraction techniques to define a structure more accurately. Typical applications include studies of the complex oxides of the heavier metals, such as the high-temperature superconductors (where accurate oxide ion positions are required in the presence of metals such as barium and thallium) and systems where hydrogen positions are of interest.

Absorption spectroscopy

Absorption spectroscopic methods make use of the absorption of electromagnetic radiation by a molecule or material at a characteristic frequency corresponding to the energy of a transition between vibrational or electronic energy levels. The intensity is related to the probability of the transition occurring, which in turn is determined by the symmetry rules described in Chapter 7.

6.3 Ultraviolet–visible spectroscopy

Key points: The energies and intensities of these transitions provide information on electronic structure and chemical environment; changes in spectral properties are used to monitor the progress of reactions.

Ultraviolet–visible spectroscopy (UV–visible spectroscopy) is the observation of the absorption of electromagnetic radiation in the visible and ultraviolet (UV) regions of the spectrum. It is sometimes known as **electronic spectroscopy** because the energy is used to excite species to higher electronic energy levels. UV–visible spectroscopy is among the most widely used techniques for studying inorganic compounds and their reactions and most laboratories possess a UV–visible spectrophotometer (Fig. 6.4). However, the theoretical framework is complicated; therefore, this section describes only basic principles and the ways in which UV–visible spectra are measured and used in investigations. These principles are extended in later chapters, particularly Chapter 19.

(a) Measuring a spectrum

The sample is usually a solution but may also be a gas. In either case, it is contained in a cell (a 'cuvette') constructed of an optically transparent material such as glass or, for examining UV spectra at wavelengths below 320 nm, silica. The cell is placed in the beam from a light source and the optical transmission is measured at a detector. Usually the incident light beam is split into two, one part of the beam passing through the sample and the other passing through a cell that is identical except that the sample is absent. The

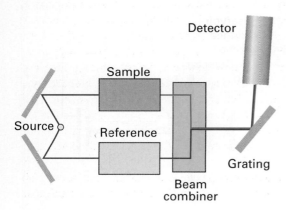

Fig. 6.4 The layout of a typical UV–visible absorption spectrometer.

sample and reference beams are compared at the detector (a photodiode). Conventional spectrometers sweep the wavelength of the incident beam by changing the angle of a diffraction grating, but it is now more common for the entire spectrum to be recorded at once by using a diode array detector.

The **absorbance**, A, of a sample is defined as

$$A = \log(I_0/I) \tag{6.2}$$

where I_0 is the incident intensity and I is the measured intensity after passing through the sample. Thus a sample that attenuates the light intensity by 10 per cent (so $I_0/I = 100/90$) has an absorbance of 0.05, one that attenuates it by 90 per cent (so $I_0/I = 100/10$) has an absorbance of 1, and one that attenuates it by 99 per cent an absorbance of 2, and so on. The detector is the limiting factor for strongly absorbing species because the measurement of low photon flux is unreliable.

The empirical **Beer–Lambert law** is used to relate the absorbance to the molar concentration [J] of the absorbing species J and optical pathlength (l).

$$A = \varepsilon[\text{J}]l \tag{6.3}$$

where ε is the **molar absorption coefficient** (still commonly referred to as the 'extinction coefficient'). Values of ε range from above 10^5 dm^3 mol^{-1} cm^{-1} for fully allowed transitions to less than 1 dm^3 mol^{-1} cm^{-1} for 'forbidden' transitions; in the latter case the absorbing species may be difficult to observe unless the concentration or pathlength are increased accordingly. The proportionality between absorbance and concentration provides a way to measure properties that depend on concentration, such as equilibria and rates of reaction.

(b) Spectroscopic monitoring of titrations and kinetics

When the emphasis is on measurement of absorbance rather than the energy of transitions, the spectroscopic investigation is usually called **spectrophotometry**. Provided at least one of the species involved has a suitable absorption band, it is usually straightforward to carry out a spectrophotometric titration in which the extent of reaction is monitored by measuring the concentrations of components. Measuring the UV–visible absorption spectra of species in solution also provides a method for monitoring the progress of reactions and determining rate constants.

The techniques that use UV–visible spectral monitoring range from those measuring reactions in the picosecond range (photochemically initiated by an ultrafast laser pulse) to the monitoring of slow reactions over hours and even days. The stopped-flow technique (Fig. 6.5) is commonly used to study reactions with half-lives of between 1 ms and 10 s that can be initiated by mixing. Two solutions, each containing one of the reactants, are mixed rapidly by a pneumatic impulse, then the flowing, reacting solution is brought to an abrupt stop by filling a 'stop-syringe', triggering the monitoring of absorbance. The reaction can be monitored at a single wavelength, or successive spectra can be measured very rapidly using a **diode array detector**.

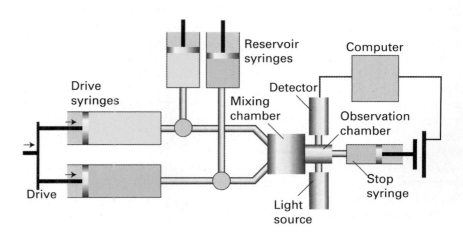

Fig 6.5 The structure of a stopped-flow instrument for studying fast reactions in solution.

The spectral changes incurred during a titration or the course of a reaction also provide information about the number of species that form during its progress. An important case is the appearance of one or more **isosbestic points**, wavelengths at which two species have equal values for their molar absorption coefficients, during a reaction or titration. At such an isosbestic point, it extremely unlikely that a third species will have the same molar absorption coefficient. Therefore, the retention of isosbestic points in a titration or during the course of a reaction is evidence for there being only two dominant species (reactant and product) in the solution (Fig. 6.6).

6.4 Infrared and Raman spectroscopy

Key points: Infrared and Raman spectroscopy are often complementary in that a particular type of vibration may be observed in one method but not the other; the information is used in many ways, ranging from structural determination to measuring reaction kinetics.

Vibrational spectroscopy is used to characterize compounds in terms of the strength and number of bonds that are present. It is used to detect the presence of known compounds (fingerprinting), to monitor changes in the concentration of a species during a reaction, to determine the components of an unknown compound (such as the presence of CO ligands), to determine a likely structure for a compound, and to measure properties of bonds (force constants).

(a) The energies of molecular vibrations

A bond in a molecule behaves like a spring: stretching it through a distance x produces a restoring force F. For small displacements, the restoring force is proportional to the displacement and $F = -kx$, where k is the **force constant** of the bond: the stiffer the bond, the greater the force constant. Such a system is known as a harmonic oscillator, and solution of the Schrödinger equation gives the energies

$$E_v = (v + \tfrac{1}{2})\hbar\omega \qquad \text{with } \omega = (k/\mu)^{1/2} \quad \text{and } v = 0, 1, 2, \dots \qquad (6.4a)$$

where v is the vibrational quantum number and μ is the effective mass of the oscillator. For a diatomic molecule composed of atoms of masses m_A and m_B,

$$\mu = \frac{m_A m_B}{m_A + m_B} \qquad (6.4b)$$

If $m_A \gg m_B$, then $\mu = m_B$ and only atom B moves appreciably during the vibration: in this case the vibrational energy levels are determined largely by m_B, the mass of the lighter atom. The frequency ω is therefore high when the force constant is large (a stiff bond) and the mass of the oscillator is low (only light atoms are moved during the vibration). Vibrational energies are usually expressed in terms of the wavenumber $\tilde{v}$ the reciprocal of the wavelength ($v = \omega/2\pi c$); typical values of v lie in the range 300–3800 cm^{-1} (Table 6.2).

A molecule consisting of N atoms can vibrate in $3N-6$ different ways if it is nonlinear and $3N-5$ different ways if it is linear. These different vibrations are called **normal modes**. For instance, a CO_2 molecule has four normal modes of vibration, two corresponding to stretching the bonds and two corresponding to bending the molecule in two perpendicular planes. Bending modes typically occur at lower frequencies than stretching modes and depend on the masses of the atoms in a complicated way that reflects the extents to which the various atoms move in each mode. The modes are labelled v_1, v_2, etc. and are sometimes given evocative names, such as *symmetric stretch* and *antisymmetric stretch*. Only normal modes that correspond to a changing electric dipole moment can interact with infrared radiation, so only these modes are active in producing an IR spectrum.

The lowest level ($v = 0$) of any normal mode corresponds to $E_0 = \tfrac{1}{2}\hbar\omega$, the so-called **zero-point energy**, which is the lowest energy that a vibrating bond can possess. In addition to the fundamental transitions with $\Delta v = +1$, vibrational spectra may also show bands arising from double quanta ($\Delta v = +2$ at $2\tilde{v}$) known as **overtones**, and

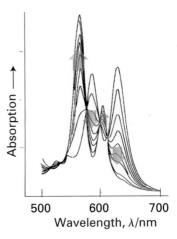

Fig. 6.6 Isosbestic points observed in the absorption spectral changes during reaction of HgTPP (TPP = tetraphenylporphyrin) with Zn^{2+} in which Zn replaces Hg in the macrocycle. The initial and final spectra are those of the reactant and product, which indicates that free TPP does not reach a detectable concentration during the reaction. (Adapted from C. Grant and P. Hambright, *J. Am. Chem. Soc.*, 1969, **91**, 4195.)

Table 6.2 Characteristic fundamental stretching wavenumbers of some common molecular species

Species	Range/cm^{-1}
CH	2900–3200
OH	3400–3600
CN$^-$	2000–2200
CO (terminal)	2000–2100
CO (bridging)	1800–1900
NO	1675–1870
O_2^-	920–1120
O_2^{2-}	800–900
Metal–metal bonds	120–400

combinations of two different vibrational modes (for example, $v_1 + v_2$). These special transitions may be helpful as they often arise even when the fundamental transition is not allowed by the selection rules (Section 7.8).

(b) The techniques

In **infrared spectroscopy** (IR spectroscopy) the vibrational spectrum of a compound is obtained by exposing the sample to infrared radiation and recording the variation of the transmission with frequency. In early spectrometers, the transmission was measured as the frequency was swept between two limits. Now the spectrum is extracted from an interferogram by **Fourier transformation**, which converts information in the time domain (based on the interference of waves travelling along paths of different lengths) to the frequency domain. The sample must be contained in a material that does not absorb infrared radiation, which means that glass cannot be used and aqueous solutions are unsuitable unless the spectral bands of interest occur at frequencies not absorbed by water. Optical windows are typically constructed from CsI. Traditional procedures of sample preparation in KBr pellets or paraffin mulls have largely been replaced by total internal reflectance devices in which the sample is simply placed in position.

In **Raman spectroscopy** the sample is exposed to intense laser radiation in the visible region of the spectrum. Most of the photons are scattered elastically (with no change of frequency) but some are scattered inelastically, having given up some of their energy to excite vibrations. These photons have frequencies different from that of the incident radiation (v_{in}) by amounts equivalent to vibrational frequencies (v_i) of the molecule. An advantage of Raman spectroscopy over IR spectroscopy is that aqueous solutions can be used, but a disadvantage is that linewidths are usually much greater. Conventional Raman spectroscopy involves the photon causing a transition to a 'virtual' excited state that then collapses back to a real lower state, emitting the detected photon in the process. The technique is not very sensitive but great enhancement is achieved if the species under investigation is coloured and the excitation laser is tuned to a real electronic transition. The latter technique is known as **resonance Raman spectroscopy** and is particularly valuable for studying the environment of d-metals in enzymes because only vibrations close to the chromophore are excited and the many thousands of bonds in the rest of the molecule are 'silent'.

Only vibrational modes that correspond to a changing polarizability of the molecule are active in Raman spectroscopy, so the technique is often complementary to IR spectroscopy. When the molecule has a centre of symmetry, no mode can be both IR and Raman active (but a mode may be inactive in both).

(c) Applications of infrared and Raman spectroscopy

A major use of IR and Raman spectroscopy is the study of numerous compounds of the d-block that contain carbonyl ligands. The CO group is a strong oscillator in the sense that it gives rise to intense vibrational absorption bands. Free CO absorbs at $2143 \, \text{cm}^{-1}$, but when coordinated in a compound the stretching frequency (and correspondingly the wavenumber) is lowered by an amount that depends on the extent to which electron density is transferred into the 2π orbital (the LUMO) by back donation from the metal (Section 21.5). The CO stretching absorption also allows distinction to be made between terminal and bridging ligands, with bridging ligands occurring at lower frequencies.

Raman and infrared spectroscopy are excellent methods for studying molecules that are formed and trapped in inert matrices, the technique known as **matrix isolation**. The principle is that highly unstable species that would not normally exist can be generated in an inert matrix such as solid xenon. The speed of data acquisition possible with Fourier-transform IR (FTIR) has meant that it can be incorporated into rapid kinetic techniques, including ultrafast laser photolysis and stopped-flow methods. This application has been exploited to investigate rates of reaction and to identify transient intermediates in reactions that involve changes in vibrational spectra, such as those involving the addition or loss of carbonyl ligands.

Resonance techniques

Several techniques of structural investigation depend on bringing energy level separations into resonance with electromagnetic radiation, with the separations in some cases controlled by the application of a magnetic field. Two of these techniques involve **magnetic resonance**: in one, the energy levels are those of magnetic nuclei (nuclei with nonzero spin); in the other, they are the energy levels of unpaired electrons.

6.5 Nuclear magnetic resonance

Key points: Nuclear magnetic resonance (NMR) is suitable for studying compounds containing elements with magnetic nuclei and it is particularly important for determining the structures of species containing hydrogen. Unlike X-ray diffraction, NMR gives information on dynamics, and it is an important tool for investigating rearrangement reactions occurring on the millisecond timescale.

Nuclear magnetic resonance (NMR) is the most powerful and widely used spectroscopic method for the determination of molecular structures in solution and pure liquids. In many cases, it provides information about shape and symmetry with greater certainty than is possible with other spectroscopic techniques, such as IR and Raman spectroscopy. It also provides information about the rate and nature of the interchange of ligands in fluxional molecules and can be used to follow reactions, in many cases providing exquisite mechanistic detail. The technique has been used to obtain the structures of protein molecules of up to 20 kDa, and has complemented the more static descriptions obtained with X-ray single-crystal diffraction. However, unlike X-ray diffraction, NMR studies of molecules in solution generally cannot provide detailed bond distance and angle information.

The sensitivity of NMR is dependent on several parameters, including the abundance of the isotope and the size of its nuclear magnetic moment. For example, ^{1}H, with 99.98 per cent natural abundance and a large magnetic moment, is easier to observe than ^{13}C, which has a smaller magnetic moment and only 1.1 per cent natural abundance. With modern multinuclear NMR techniques it is particularly easy to observe spectra for ^{1}H, ^{19}F, and ^{31}P and useful spectra can also be obtained for many other elements; Table 6.3 lists a selection of nuclei and their sensitivities. A common limitation for exotic nuclei is the presence of a nuclear quadrupole moment, a non-uniform distribution of electric charge (which is present for all nuclei with $I > \frac{1}{2}$), which broadens signals and degrades spectra. Nuclei with even atomic numbers and even mass numbers (such as ^{12}C and ^{16}O) have zero spin and are invisible in NMR.

(a) Observation of the spectrum

A nucleus of spin I can take up $2I + 1$ orientations relative to the direction of an applied magnetic field. Each orientation has a different energy (Fig. 6.7), with the lowest level the most highly populated. The energy separation of the two levels of a spin-$\frac{1}{2}$ nucleus (such as ^{1}H or ^{13}C) is

$$\Delta E = \hbar \gamma B_0 \tag{6.5}$$

where B_0 is the magnitude of the applied magnetic field and γ is the **magnetogyric ratio** of the nucleus, the ratio of its magnetic moment to its spin angular momentum. With modern superconducting magnets producing 5–20 T, resonance is achieved with electromagnetic radiation in the range 20–900 MHz. Because the difference in energy levels is relatively small, the population of the lowest level is only marginally more than higher levels, and thus the sensitivity of the NMR experiment is low but can be increased by using a stronger magnetic field, which increases the population difference and the absorption intensity.

Spectra were originally obtained in a continuous wave (CW) mode in which the sample is subjected to a scanning radiofrequency and the resonances recorded as the spectrum. Now the energy separations are identified by exciting nuclei in the sample with

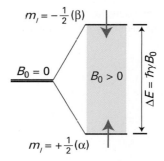

Fig. 6.7 When a nucleus with spin $I > 0$ is in a magnetic field, its $2I + 1$ orientations (designated m_I) have different energies. This diagram shows the energy levels of a nucleus with $I = \frac{1}{2}$ (as for ^{1}H, ^{13}C, ^{31}P).

Table 6.3 Nuclear spin characteristics of common nuclei

Nucleus	Natural abundance/%	Sensitivity*	Spin	NMR frequency/MHz[†]
^{1}H	99.98	5680	$\frac{1}{2}$	100.000
^{2}H	0.015	0.00821	1	15.351
^{7}Li	92.58	1540	$\frac{3}{2}$	38.863
^{11}B	80.42	754	$\frac{3}{2}$	32.072
^{13}C	1.11	1.00	$\frac{1}{2}$	25.145
^{15}N	0.37	0.0219	$\frac{1}{2}$	10.137
^{17}O	0.037	0.0611	$\frac{5}{2}$	13.556
^{19}F	100	4730	$\frac{1}{2}$	94.094
^{23}Na	100	525	$\frac{3}{2}$	26.452
^{29}Si	4.7	2.09	$\frac{1}{2}$	19.867
^{31}P	100	377	$\frac{1}{2}$	40.481
^{89}Y	100	0.668	$\frac{1}{2}$	4.900
^{103}Rh	100	0.177	$\frac{1}{2}$	3.185
^{109}Ag	48.18	0.276	$\frac{1}{2}$	4.654
^{183}W	14.4	0.0589	$\frac{1}{2}$	4.166
^{195}Pt	33.8	19.1	$\frac{1}{2}$	21.462
^{199}Hg	16.84	5.42	$\frac{1}{2}$	17.911

* Sensitivity is relative to ^{13}C = 1 and is the product of the relative sensitivity of the isotope and the natural abundance.
[†] At 2.349 T (a '100 MHz spectrometer').

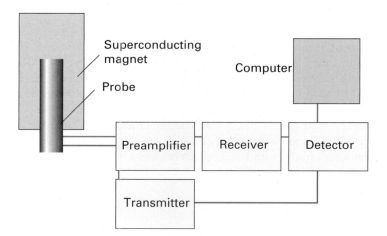

Fig. 6.8 The layout of a typical NMR spectrometer. The link between transmitter and detector is arranged so that only low-frequency signals are processed.

a sequence of radiofrequency pulses and then observing the return of the nuclear magnetization back to equilibrium. Fourier transformation then converts the time-domain data to the frequency domain with peaks at frequencies corresponding to transitions between the different nuclear energy levels. Figure 6.8 shows the experimental set-up of an NMR spectrometer.

(b) Chemical shifts

The frequency of an NMR transition depends on the local magnetic field experienced by the nucleus and is expressed in terms of the **chemical shift**, δ, the difference between the resonance frequency of nuclei in the sample and that of a reference compound:

$$\delta = \frac{\nu - \nu^{\circ}}{\nu^{\circ}} \times 10^6 \tag{6.6}$$

A common standard for ^{1}H, ^{13}C, or ^{29}Si spectra is tetramethylsilane Si(CH$_3$)$_4$, abbreviated as TMS. When $\delta < 0$ the nucleus is said to be **shielded** relative to the standard; $\delta > 0$ corresponds to a nucleus that is **deshielded** with respect to the reference. An

H atom bound to a closed-shell, low-oxidation-state, *d*-block element from Groups 6 to 10 (such as [HCo(CO)$_4$]) is generally found to be highly shielded whereas in an oxoacid (such as H$_2$SO$_4$) it is deshielded. From these examples it might be supposed that the higher the electron density around a nucleus, the greater its shielding. However, as several factors contribute to the shielding, a simple physical interpretation of chemical shifts in terms of electron density is generally not possible.

The chemical shifts of ^{1}H and other nuclei in various chemical environments are tabulated, so empirical correlations can often be used to identify compounds or the element to which the resonant nucleus is bound. For example the H chemical shift in CH$_4$ is only 0.1 because the H nuclei are in an environment similar to that in tetramethylsilane, but the H chemical shift is $\delta = 3.1$ for H in GeH$_4$ (Fig. 6.9). Chemical shifts are different for the same element in inequivalent positions within a molecule. Thus in ClF$_3$ the chemical shift of the equatorial ^{19}F nucleus is separated by $\Delta\delta = 120$ from that of the axial F nuclei (Fig. 6.10).

(c) Spin–spin coupling

Structural assignment is often helped by the observation of the **spin–spin coupling** of nuclei, which gives rise to multiplets in the spectrum. The strength of spin–spin coupling, which is reported as the **spin–spin coupling constant**, *J* (in hertz, Hz), decreases rapidly with distance through chemical bonds, and in many cases is greatest when the two atoms are directly bonded to each other. In **first-order spectra**, which are being considered here, the coupling constant is equal to the separation of adjacent lines in a multiplet. As can be seen in Fig. 6.9, $J(^1\text{H}-^{73}\text{Ge}) \approx 100$ Hz. The allowed transitions all occur at the same frequency when the nuclei are related by symmetry and any spin–spin coupling is invisible. Thus, a single ^{1}H signal is observed for the CH$_3$I molecule even though there is coupling between the H nuclei, because the three nuclei are related to each other by a threefold axis.

A multiplet of $2I + 1$ lines is obtained when a spin-$\frac{1}{2}$ nucleus (or a set of symmetry-related spin-$\frac{1}{2}$ nuclei) is coupled to a nucleus of spin *I*. In the spectrum of GeH$_4$ shown in Fig. 6.9, the single central line arises from the four equivalent H atoms in GeH$_4$ molecules that contain Ge isotopes with $I = 0$. This central line is flanked by 10 evenly spaced but less intense lines that arise from a small fraction of GeH$_4$ that contains the isotope ^{73}Ge, for which $I = \frac{9}{2}$, the four ^{1}H nuclei are coupled to the ^{73}Ge nucleus to yield a $2 \times \frac{9}{2} + 1 = 10$ line multiplet.

The coupling of the nuclear spins of different elements is called **heteronuclear coupling**, and the Ge–H coupling just discussed is an example. **Homonuclear coupling** between nuclei of the same element is detectable when the nuclei are unrelated by the symmetry operations of the molecule, as in the ^{19}F NMR spectrum of ClF$_3$ in Fig. 6.10. The signal ascribed to the two axial F nuclei (each with $I = \frac{1}{2}$) is split into a doublet by the single equatorial ^{19}F nucleus, and the latter is split into a triplet by the two axial ^{19}F nuclei (^{19}F is in 100 per cent abundance). Thus, the pattern of ^{19}F resonances readily distinguishes this unsymmetrical structure from trigonal-planar and trigonal-pyramidal structures, both of which would have equivalent F nuclei and hence a single ^{19}F resonance.

The sizes of ^{1}H—^{1}H homonuclear coupling constants in organic molecules are typically 18 Hz or less. By contrast, ^{1}H—X heteronuclear coupling constants can be several hundred hertz. Homonuclear and heteronuclear coupling between nuclei other than ^{1}H can lead to coupling constants of many kilohertz. The sizes of coupling constants are often related to the geometry of a molecule by noting empirical trends. In square-planar Pt(II) complexes, *J*(Pt—P) is sensitive to the group *trans* to a phosphine ligand and the value of *J*(Pt—P) increases in the following order of *trans* ligands:

$$\text{R}^- < \text{H}^- < \text{PR}_3 < \text{NH}_3 < \text{Br}^- < \text{Cl}^-$$

For example, *cis*-[PtCl$_2$(PEt$_3$)$_2$], where Cl$^-$ is *trans* to P, has *J*(Pt—P) = 3.5 kHz, whereas *trans*-[PtCl$_2$(PEt$_3$)$_2$], with P *trans* to P, has *J*(Pt—P) = 2.4 kHz. These systematic variations allow *cis* and *trans* isomers to be distinguished quite readily. The rationalization for the variation in the sizes of the coupling constants above stems from the fact that a ligand that exerts a large *trans* influence (Section 20.5) substantially weakens the bond *trans* to itself, causing a reduction in the NMR coupling between the nuclei.

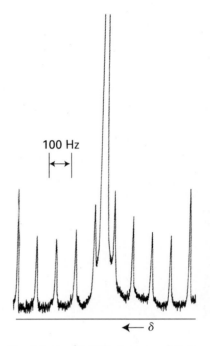

100 Hz

Fig. 6.9 The ^{1}H NMR spectrum of GeH$_4$.

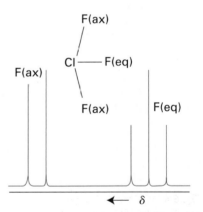

F(ax)

Cl——F(eq)

F(ax)

F(ax)

F(ax)

F(eq)

Fig. 6.10 The ^{19}F NMR spectrum of ClF$_3$.

(d) Intensities

The integrated intensity of a signal arising from a set of symmetry-related nuclei is proportional to the number of nuclei in the set. Provided sufficient time is allowed during spectrum acquisition for the full relaxation of the observed nucleus, integrated intensities ('integrals') can be used with confidence to aid spectral assignment. For instance, in the spectrum of ClF_3 the relative integrated intensities are 2:1 (for the doublet and triplet). This pattern is consistent with the structure and the splitting pattern because it indicates the presence of two symmetry-related F nuclei and one inequivalent F nucleus.

The relative intensities of the $2N+1$ lines in a multiplet that arises from coupling to N equivalent spin-$\frac{1}{2}$ nuclei are given by Pascal's triangle (**1**); thus three equivalent protons give a 1:3:3:1 quartet. Groups of nuclei with higher spin quantum numbers give different patterns. The ^1H NMR spectrum of HD, for instance, consists of three lines of equal intensity as a result of coupling to the ^2H nucleus ($I=1$, with $2I+1=3$ orientations).

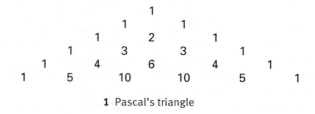

1 Pascal's triangle

(e) Fluxionality

The timescale of the NMR experiment is slow in the sense that structures can be resolved provided their lifetime is not less than a few milliseconds. For example, $Fe(CO)_5$ shows just one ^{13}C resonance, indicating that, on the NMR timescale, all five CO groups are equivalent. However, the IR spectrum (of timescale about 1 ps) shows distinct axial and equatorial CO groups, and by implication a trigonal-bipyramidal structure. The observed ^{13}C NMR spectrum of $Fe(CO)_5$ is the weighted average of these separate resonances.

Because the temperature at which NMR spectra are recorded can easily be changed, samples can often be cooled down to a temperature at which the rate of interconversion becomes slow enough for separate resonances to be observed. Figure 6.11, for instance, shows the idealized ^{31}P NMR spectra of $RhMe(PMe_3)_4$ at room temperature and at $-80°C$. Careful control allows determination of the temperature at which the spectrum changes from the high-temperature form to the low-temperature form ('the coalescence temperature') and thence to determine the barrier to interconversion.

(f) Solid-state NMR

Solid-state NMR rarely produces resolution that is as high as that from solution-state NMR. In part this difference is due to anisotropic interactions such as dipolar couplings and the fact that, in the solid state, chemically equivalent nuclei might be in different

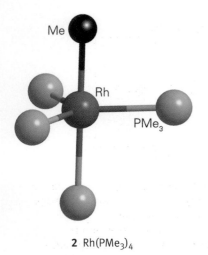

2 $Rh(PMe_3)_4$

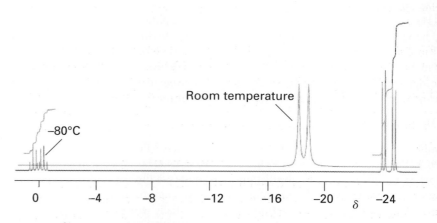

Fig. 6.11 ^{31}P NMR spectra of $Rh(PMe_3)_4Me$ (**2**) at room temperature and at $-80°C$.

magnetic environments and thus have different resonance frequencies. To average out anisotropic interactions, samples are spun at very high speeds (typically 10–25 kHz) at the 'magic angle' (54.7°) with respect to the field axis. This so-called **magic-angle spinning** (MAS) substantially reduces the effect of anisotropy but often leaves signals that are still significantly broadened compared to those from solution. The broadening of signals can sometimes be so great that signal widths are comparable to the chemical shift range for some nuclei. This is a particular problem for ^1H, which typically has a chemical shift range of $\Delta\delta = 10$. Broad signals are less of a problem for nuclei such as ^{195}Pt, for which the range in chemical shifts is $\Delta\delta = 16\,000$, although this large range can be reflected in large anisotropic linewidths. Quadrupolar nuclei (those with $I > \frac{1}{2}$) present additional problems as the peak position becomes field dependent and can no longer be identified with the chemical shift.

Despite these difficulties, developments in the technique have made possible the observation of high-resolution NMR spectra for solids and are of far-reaching importance in many areas of chemistry. An example is the use of ^{29}Si MAS-NMR to determine the environments of Si atoms in natural and synthetic aluminosilicates. Homonuclear and heteronuclear decoupling enhances the resolution of spectra and the use of multiple pulse sequences has allowed the observation of spectra with some difficult samples. The high-resolution technique CPMAS-NMR, a combination of MAS with **cross-polarization** (CP) usually with heteronuclear decoupling, has been used to study many compounds containing ^{13}C, ^{31}P, and ^{29}Si. The technique is also used to study molecular compounds in the solid state. For example, the ^{13}C CPMAS spectrum of $[Fe_2(C_8H_8)(CO)_5]$ at $-160°C$ indicates that all C atoms in the C_8 ring are equivalent on the timescale of the experiment. The interpretation of this observation is that the molecule is fluxional in the solid state.

Example 6.2 Interpreting NMR spectra

Explain why the ^{19}F NMR spectrum of SF_4 consists of two 1:2:1 triplets of equal intensity.

Answer An SF_4 molecule is trigonal bipyramidal with a lone pair occupying one of the equatorial sites. The two axial F atoms are chemically different from the two equatorial F atoms and give two signals of equal intensity. The signals are in fact 1:2:1 triplets as each distinct ^{19}F nucleus couples to the two chemically distinct ^{19}F nuclei.

Self-test 6.2 Use the isotope information in Table 6.3 to demonstrate why the ^1H resonance of the hydrido ligand in *cis*-[RhH(CO)(PMe$_3$)$_2$] consists of eight lines of equal intensity.

6.6 Electron paramagnetic resonance

Key points: EPR spectroscopy is used to study compounds possessing unpaired electrons, particularly those containing a *d* metal; it is often the technique of choice for identifying and studying metals such as Fe and Cu at the active sites of metalloenzymes.

Electron paramagnetic resonance (EPR; or electron spin resonance, ESR) spectroscopy is a technique for studying paramagnetic species such as organic and main-group radicals, although in inorganic chemistry, its major importance is for characterizing compounds and material containing the *d*- and *f*-block elements. The simplest case is for a species having one unpaired electron ($S = \frac{1}{2}$): by analogy with NMR, the application of an external magnetic field B_0 produces a difference in energy between the $m_s = -\frac{1}{2}$ and $+\frac{1}{2}$ states, with

$$\Delta E = g\mu_B B_0 \tag{6.7}$$

where μ_B is the Bohr magneton and g is the **g-value** (Fig. 6.12). The conventional method of recording EPR spectra is to use a continuous wave (CW) spectrometer, in which the sample is irradiated with a constant microwave frequency and the field is varied. The resonant frequency standard for most spectrometers is approximately 9 GHz, and

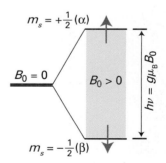

Fig. 6.12 When an unpaired electron is in a magnetic field, its two orientations (α, $m_s = +\frac{1}{2}$ and β, $m_s = -\frac{1}{2}$) have different energies. Resonance is achieved when the energy separation matches the energy of the incident microwave photons.

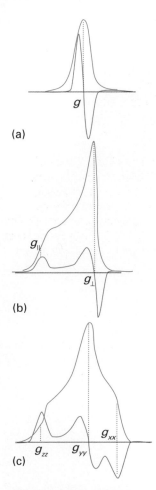

(a)

(b)

(c)

Fig. 6.13 The forms of EPR powder (frozen solution) spectra expected for different types of of g-value anisotropy. The green line is the absorption and the red line the first derivative of the absorption (its slope). For technical reasons related to the detection technique, the first derivative is normally observed in EPR spectrometers.

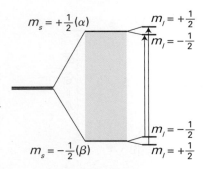

Fig. 6.14 When a magnetic nucleus is present, its $2I+1$ orientations give rise to a local magnetic field that splits each Zeeman level of an electron into $2I+1$ levels. The allowed transitions ($\Delta m_s = +1$, $\Delta m_I = 0$) give rise to the hyperfine structure of an EPR spectrum.

the instrument is then known as an 'X-band spectrometer'. Laboratories specializing in EPR spectroscopy often have a range of instruments, each operating at different fields. Thus an S-band (resonant frequency 3 GHz) spectrometer and particularly those operating at high fields, Q-band (35 GHz) and W-band (95 GHz) spectrometers, are used to complement the information gained with an X-band spectrometer.

Pulsed EPR spectrometers are becoming commercially available and offering new opportunities analogous to the way that pulsed Fourier-transform techniques have revolutionized NMR. Pulsed EPR techniques provide time resolution, making it possible to measure the dynamic properties of paramagnetic systems.

(a) The g-value

For a free electron $g = 2.0023$, but this value is altered by spin–orbit coupling. For many species, particularly d-metal complexes, g-values may be highly anisotropic so that the resonance condition depends on the angle that the paramagnetic species makes to the applied field. Figure 6.13 shows the EPR spectra of frozen solutions or 'glasses', expected for isotropic (all three g-values the same along perpendicular axes), axial (two the same), and rhombic (all three different) spin systems.

The sample (which is usually contained in a quartz tube) comprises the paramagnetic species in dilute form, either in the solid state (doped crystals or powders) or in solution. Relaxation is so efficient for d-metal ions that the spectra are often too broad to detect; consequently, liquid nitrogen and sometimes liquid helium are used to cool the sample. Frozen solutions behave as powders, so resonances are observed at all g-values, analogous to powder X-ray diffraction. Sometimes, more detailed studies are made with oriented single crystals. Provided relaxation is slow, EPR can also be observed at room temperature in liquids.

EPR spectra are also obtained for systems having more than one unpaired electron, but the theoretical background is much more complicated. Whereas species having an odd number of electrons are usually detectable, it can be difficult to observe spectra for systems having an even number of electrons. Table 6.4 shows the suitability of common paramagnetic species for EPR detection.

(b) Hyperfine coupling

The **hyperfine structure** of a spectrum is due to the magnetic coupling of the electron spin to any magnetic nuclei present. Coupling to a nucleus with spin I splits an EPR line into $2I+1$ lines of the same intensity (Fig. 6.14). A distinction is sometimes made between the hyperfine structure due to coupling to the nucleus of the atom on which the

Table 6.4 EPR detectability of common d-metal ions

Usually easy to study		Usually difficult to study or diamagnetic	
Species	Spin S	Species	Spin S
Ti(III)	$\frac{1}{2}$	Ti(II)	1
Cr(III)	$\frac{3}{2}$	It(IV)	0
V(IV)	$\frac{1}{2}$	Cr(II)	2
Fe(III)	$\frac{1}{2}, \frac{5}{2}$	V(III)	1
Co(II)	$\frac{3}{2}, \frac{1}{2}$	V(V)	0
Ni(III)	$\frac{3}{2}, \frac{1}{2}$	Fe(II)	1
Ni(I)	$\frac{1}{2}$	Co(III)	0
Cu(II)	$\frac{1}{2}$	Co(I)	0
Mo(V)	$\frac{1}{2}$	Ni(II)	1
W(V)	$\frac{1}{2}$	Cu(I)	0
		Mo(VI)	0
		Mo(IV)	1, 0
		W(VI)	0

unpaired electron is primarily located and the 'superhyperfine coupling', the coupling to ligand nuclei. Superhyperfine coupling to ligand nuclei is used to measure the extent of electron delocalization and covalence in metal complexes. The EPR spectrum of a frozen solution of a Cu(II) salt (d^9, one unpaired electron) in Fig. 6.15 illustrates the effects of both **g-value anisotropy** and **hyperfine structure**. The EPR spectrum is axial, showing that the electronic environment about the Cu(II) differs in one direction ($g_\parallel$) relative to the other two, and this resonance ($g_\parallel$) is subject to particularly strong hyperfine coupling with the $I = \frac{3}{2}$ copper nucleus.

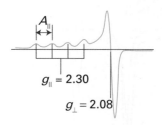

$g_\parallel = 2.30$

$g_\perp = 2.08$

Fig. 6.15 The EPR spectrum of Cu^{2+} (d^9, one unpaired electron) in frozen aqueous solution. The tetragonally distorted Cu^{2+} ion shows an axially symmetrical spectrum in which hyperfine coupling to Cu ($I = \frac{3}{2}$, four hyperfine lines) is clearly evident for the $g_\parallel$ component.

Example 6.3 Interpreting superhyperfine coupling

Using the data provided in Table 6.3 suggest how the EPR spectrum of a Co(II) complex might change its appearance when a single OH^- ligand is replaced by an F^- ligand.

Answer Fluorine-19 (100 per cent abundance) has nuclear spin $I = \frac{1}{2}$ whereas oxygen-16 (close to 100 per cent) has $I = 0$. Consequently, any part of the EPR spectrum might be split into two lines.

Self-test 6.3 Predict how you might show that an EPR signal that is characteristic of a new material arises from tungsten sites.

6.7 Mössbauer spectroscopy

Key point: Mössbauer spectroscopy is based on the resonant absorption of γ-radiation by nuclei and exploits the fact that nuclear energies are sensitive to the electronic environment.

The **Mössbauer effect** is the recoil-free emission of γ-radiation by a nucleus. To understand what is involved, we consider a ^{57}Co nucleus that decays by electron capture to produce an excited state of ^{57}Fe, denoted $^{57}Fe^{**}$ (Fig. 6.16). This nuclide decays to another excited state, denoted $^{57}Fe^*$, that lies 14.41 eV above the ground state and emits a γ-ray of energy 14.41 eV as it decays to the ground state. If the emitting nucleus recoils as it emits the γ-ray photon, its frequency is shifted by the Doppler effect. The recoil is avoided by ensuring that the $^{57}Fe^*$ nuclei are pinned down into a rigid lattice, so that the radiation is highly monochromatic.

If a sample containing ^{57}Fe (which occurs with 2 per cent natural abundance) is placed close to the ^{57}Co source, the monochromatic, recoil-free γ-ray emitted by $^{57}Fe^*$ can be expected to be absorbed resonantly by the ^{57}Fe nuclei. However, resonant absorption still might not occur because the electronic environment of the receiver ^{57}Fe is different from that of the $^{57}Fe^*$ transmitter. The two can be brought back into resonance by moving the target nucleus relative to the transmitter and relying on the Doppler shift to match the absorption frequency to the transmitted frequency as depicted in Fig. 6.16. A **Mössbauer spectrum** is a portrayal of the resonant absorption peaks that occur as the velocity of the sample is changed.

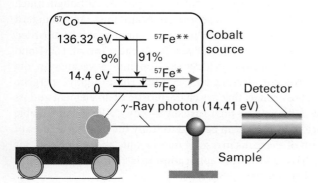

Fig. 6.16 The layout of a Mössbauer spectrometer. The speed of the carriage is adjusted until the Doppler shifted frequency of the emitted γ-ray matches the corresponding nuclear transition in the sample. The inset shows the nuclear transitions responsible for the emission of the γ-ray.

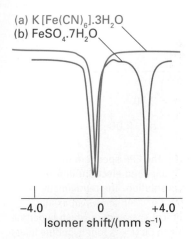

(a) K [Fe(CN)$_6$].3H$_2$O
(b) FeSO$_4$.7H$_2$O

−4.0	0	+4.0

Isomer shift/(mm s^{-1})

Fig. 6.17 Mössbauer spectra of Fe samples showing the isomer shift and quadrupole coupling: (a) a highly symmetric environment, (b) a nonsymmetric environment in which quadrupolar splitting is observed.

The Mössbauer spectrum of a sample containing iron can be expected to consist of a single signal due to absorption of radiation at the energy required (ΔE) to excite the nucleus from its ground state to the excited state. The difference between ΔE of the sample and that of metallic ^{57}Fe is called the **isomer shift** and is expressed in terms of the velocity (of the order of mm s^{-1}) that is required to achieve resonance by the Doppler shift. The value of ΔE depends on the electron density at the nucleus and although this effect is primarily due to s electrons (because their wavefunctions are nonzero at the nucleus), shielding effects cause ΔE to be sensitive to p and d electrons too. Oxidation states, such as Fe(II), Fe(III), and Fe(IV), can be distinguished as well as ionic and covalent bonding.

Although it is a rather specialized technique, the element most suited for study by Mössbauer spectroscopy is iron, which has great importance and is common in minerals and biological samples, for which other techniques are less effective. Mössbauer spectroscopy is also used to study some other suitable nuclei, including. ^{119}Sn, ^{129}I, and ^{197}Au. Because a ^{57}Fe* nucleus has $I = \frac{3}{2}$, it possesses an electric quadrupole moment. As a result, and providing the environment of the nucleus is not isotropic, the Mössbauer spectrum splits into two lines with separation ΔE_Q (Fig. 6.17). The splitting is a good guide as to the state of Fe in proteins and minerals as it depends on the oxidation state and the distribution of d-electron density.

Ionization-based techniques

Ionization techniques measure the energies of products, electrons, or molecular fragments generated when a sample is ionized by bombardment with high-energy radiation or particles.

6.8 Photoelectron spectroscopy

Key point: Photoelectron spectroscopy is used to determine the energies and order of orbitals in molecules and solids, by analyzing the kinetic energies of electrons (photoelectrons) that are emitted upon irradiation.

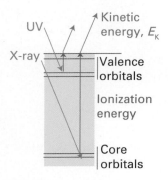

Fig. 6.18 In photoelectron spectroscopy, high-energy electromagnetic radiation (UV for the ejection of valence electrons, X-ray for core electrons) expels an electron from its orbital, and the kinetic energy of the photoelectron is equal to the difference between the photon energy and the ionization energy of the electron.

The basis of **photoelectron spectroscopy** (PES) is the measurement of the kinetic energies of electrons (photoelectrons) emitted by ionization of a sample that is irradiated with high-energy monochromatic radiation (Fig. 6.18). The kinetic energy of the photoelectrons, E_K, is related to the ionization energies (I) of electron in their original orbitals by the relation

$$E_K = h\nu - I \tag{6.8}$$

where ν is the frequency of the incident radiation. **Koopmans' theorem** states that the ionization energy is equal to the negative of the orbital energy, so the determination of the kinetic energies of photoelectrons can be used to determine orbital energies. The theorem assumes that the energy involved in electron reorganization after ionization is offset by the increase in electron–electron repulsion energy as the orbital contracts. This approximation is usually reasonably valid.

There are two major types of photoionization technique, **X-ray photoelectron spectroscopy** (XPS) and **ultraviolet photoelectron spectroscopy** (UPS). Although much more intense sources can be obtained using synchrotron beam lines, the standard laboratory source for XPS is usually a magnesium or aluminium anode that is bombarded by a high-energy electron beam. This bombardment results in radiation at 1.254 and 1.486 keV, respectively, due to the transition of a $2p$ electron into a vacancy in the $1s$ orbital caused by the bombardment ejecting an electron. These highly energetic photons cause ionizations from core orbitals that are characteristic of an element and its oxidation state. Because the linewidth is fairly high (usually 1–2 eV), XPS is not suitable for probing fine details of valence orbitals but can be used to study the band structures of solids. Because the mean free path of electrons in a solid may be only about 1 nm, XPS is suited for surface elemental analysis, and in this application it is commonly known as **electron spectroscopy for chemical analysis** (ESCA).

The source for UPS is typically a helium discharge lamp that emits radiation at 21.22 eV, known as He(I) radiation, or at 40.8 eV, known as He(II) radiation. Linewidths are much smaller than achieved with X-rays, so that the resolution is far greater. The technique is used to probe valence-shell energy levels and the vibrational fine structure that is revealed often provides important information on the bonding or antibonding character of the orbitals from which electrons are ejected (Fig. 6.19). When the electron is removed from a nonbonding orbital, the product is formed in its ground state, and a narrow line is observed. However, when the electron is removed from an antibonding orbital, the resulting ion is formed in several different vibrational levels and extensive fine structure is observed. Similarly, removal from a bonding orbital also gives a progression of vibrational levels. Bonding and antibonding orbitals can be distinguished by determining whether the vibrational frequencies in the resulting ion are higher or lower than for the original molecule.

Another useful aid is the comparison of photoelectron intensities for a sample irradiated with He(I) or He(II). The higher energy source preferentially excites electrons associated with d or f orbitals, allowing these contributions to be distinguished from s and p orbitals, for which He(I) causes higher intensities. The origin of this effect lies in differences in absorption cross-sections (see *Further reading*).

Photoelectrons emitted by a particular atom may be backscattered by adjacent atoms, and can result in an interference pattern that is detected as periodic variations in intensity at energies just above the absorption edge. In **extended X-ray absorption fine structure** (EXAFS) these variations are analysed to reveal the nature of nearby atoms (based on their electron density) and the distance between the target atom and the scattering atom. An advantage of this method is that it can provide bond distances in amorphous samples.

6.9 Mass spectrometry

Key point: Mass spectrometry is a technique for determining the mass of a molecule and of its fragments.

Mass spectrometry measures the mass-to-charge ratio of gaseous ions. The ions can be either positively or negatively charged, and it is normally trivial to infer the actual charge on an ion and hence the mass of a species.

The accuracy of measurement of the mass of the ions varies according to the use being made of the spectrometer. If all that is required is a crude measure of the mass, for instance, to within ± 1 u, then the resolution of the mass spectrometer need only be of the order of 1 part in 10^4. By contrast, to determine the mass of individual atoms so that the mass defect can be determined, the accuracy must approach 1 part in 10^{10}. With a mass spectrometer of this accuracy, molecules of nominally the same mass such as $^{12}C^{16}O$ (of mass 27.9949 u) can be distinguished from $^{14}N_2$ (of mass 28.0061 u) and the elemental and isotopic composition of ions of nominal mass less than 1000 u may be determined unambiguously. A simple experiment to determine the mass of an electron can be performed by comparing the mass of the Cl^- and Cl^+ ions, as the mass difference between the two ions is twice the mass of the electron.

(a) Ionization and detection methods

The major practical difficulty with mass spectrometry is to convert a sample (often liquid or solid and normally uncharged) into gaseous ions. Typically, less than a milligram of compound is used. Many different experimental arrangements have been devised to produce gas-phase ions but all suffer from a tendency to fragment the compound of interest. **Electron impact ionization** (EI) relies on bombarding a sample with high-energy electrons to cause both vaporization and ionization. The disadvantage is that EI tends to induce considerable decomposition in larger molecules. **Fast atom bombardment** (FAB) is similar to EI, but bombardment of the sample with fast neutral atoms is used to volatilize and ionize the sample; it induces less fragmentation than EI. **Matrix-assisted laser desorption/ionization** (MALDI) is similar to EI, but a short laser pulse is used to the same effect; this technique is particularly effective with polymeric samples. In **electrospray ionization** (ESI), charged droplets of solution are sprayed into a vacuum

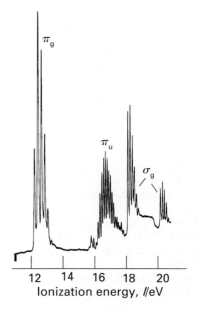

Fig. 6.19 UV photoelectron spectra of O_2. Loss of an electron from $2\sigma_g$ (see the MO energy-level diagram, Fig. 2.14) gives rise to two bands because the unpaired electron that remains can be parallel or antiparallel to the two unpaired electrons in the $1\pi_g$ orbitals.

chamber where solvent evaporation results in generation of individually charged ions; ESI mass spectrometry is becoming more widely used and is often the method of choice for ionic compounds in solution.

The traditional method of ion separation relies on the acceleration of ions with an electric field and then a magnetic field to deflect the moving ions: ions with a lower mass-to-charge ratio are deflected more than heavier ions. Thus scanning through a magnetic field will allow ions of different mass-to-charge ratios to be detected (Fig. 6.20). In a **time-of-flight** (TOF) mass spectrometer, the ions from a sample are accelerated by an electric field for a fixed time and then allowed to fly freely (Fig. 6.21). Because the force on all the ions of the same charge is the same, the lighter ions are accelerated to higher speeds than the heavier ions and strike a detector sooner. In an **ion cyclotron resonance** (ICR) mass spectrometer (often denoted FTICR, for Fourier transform) ions are collected in a small cyclotron cell inside a strong magnetic field. The ions circle round in the magnetic field, effectively behaving as an electrical current. Because an accelerated current generates electromagnetic radiation, the signal generated by the ions can be detected and used to establish the mass-to-charge ratio of the moving ions.

(b) Interpretation

A typical mass spectrum obtained in one of the ways described above is shown in Fig. 6.22. To interpret a mass spectrum is it helpful to detect a peak corresponding to the

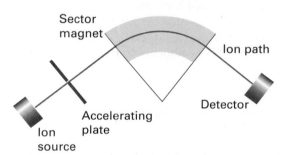

Fig. 6.20 A magnetic sector mass spectrometer. The molecular fragments are deflected according to their mass-to-charge ratio, allowing for separation at the detector.

Fig. 6.21 A time-of-flight (TOF) mass spectrometer. The molecular fragments are accelerated to different speeds by the potential difference and arrive at different times at the detector.

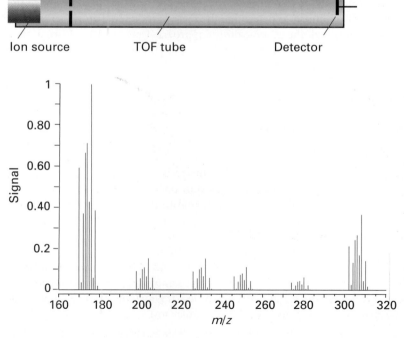

Fig. 6.22 The mass spectrum of $Mo(\eta^6\text{-}C_6H_6)(CO)_2PMe_3$.

singly charged, nonfragmented molecular ion. Sometimes a peak will occur at half the molecular mass and is ascribed to a doubly charged ion. Peaks from multiply charged ions are usually easy to identify because the separation between the peaks from the different isotopomers is no longer 1 u but fractions of an atomic mass unit; in a doubly charged ion isotopic peaks will be $\frac{1}{2}$ u apart, in a triply charged ion they will be $\frac{1}{3}$ u apart, and so on.

In addition to indicating the mass of the molecule (and hence its molar mass) or ion that is being studied, a mass spectrum also provides information about fragmentation pathways of molecules. This information can be used to confirm structural assignments. For example, complex ions will often lose ligands, and peaks representing the complete ion minus one or more ligands are found.

When an element present in the sample has a number of isotopes (for instance, chlorine is 75.5 per cent ^{35}Cl and 24.5 per cent ^{37}Cl), instead of a single peak representing an ion, multiple peaks will be present. Thus, for a molecule containing chlorine, the mass spectrum will show two peaks 2 u apart in the intensity ratio of about 3:1. For elements with a more complex isotopic composition, different patterns of peaks will occur and can be used to diagnose the presence of the elements in compounds of unknown composition. An Hg atom, for instance, has six isotopes in significant abundance (Fig. 6.23). The actual proportion of isotopes of an element varies according to the geographic source of that element, and this subtle aspect is easily seen with high-resolution mass spectrometers. Thus the precise determination of the proportions of isotopes within a sample can be used to determine the source of a sample.

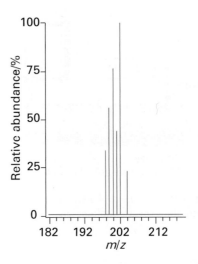

Fig. 6.23 The mass spectrum of a sample containing mercury showing the isotopic composition of the atoms.

Example 6.4 Interpreting a mass spectrum

Figure 6.22 shows part of the mass spectrum of $Mo(\eta^6\text{-}C_6H_6)(CO)_2PMe_3$. Assign all the main peaks.

Answer The complex $Mo(\eta^6\text{-}C_6H_6)(CO)_2PMe_3$ has an average molecular mass of 306 u, but because Mo has a large number of isotopes, a simple molecular ion is not seen. Instead, ten peaks centred on 306 u are seen; the most abundant isotope of Mo is ^{98}Mo (24 per cent) and the ion that contains this isotope has the highest intensity of the peaks of the molecular ion. In addition to the peaks representing the molecular ion, peaks at M^+-28, M^+-56, M^+-76, M^+-104, M^+-132, are seen. These peaks represent the loss of one CO, two CO, PMe_3+CO, and PMe_3+2 CO ligands, respectively, from the parent compound.

Self-test 6.4 Explain why the mass spectrum of $ClBr_3$ (Fig. 6.24) consists of five peaks separated by 2 u?

Chemical analysis

One of the classic applications of chemistry has been the determination of the elemental composition of compounds. The techniques now available are highly sophisticated and in many cases can be automated to achieve rapid, reliable results. In this section we include thermal techniques that can be used to follow the phase changes of substances without change of composition, as well as processes that result in changes of composition.

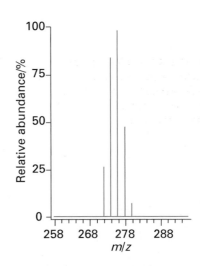

Fig. 6.24 The mass spectrum of $ClBr_3$.

6.10 Atomic absorption spectroscopy

Key point: Almost every metallic element can be determined quantitatively by using the absorption characteristics of atoms.

The principles of atomic absorption spectroscopy are similar to those of UV–visible spectroscopy except that the absorbing species are free atoms or ions. Unlike molecules, atoms and ions do not have rotational or vibrational energy and the only transitions that occur are between electronic energy levels. Consequently, atomic absorption spectra

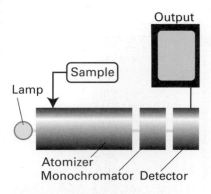

Fig. 6.25 The layout of a typical atomic absorption spectrophotometer.

consist of sharply defined lines rather than the broad bands typical in molecular spectroscopy.

Figure 6.25 shows the basic components of an atomic absorption spectrophotometer. The sample is exposed to radiation of a specific wavelength from a 'hollow cathode' lamp, which consists of a cathode constructed of a particular element and a tungsten anode in a sealed tube filled with neon. The energy of the photons emitted is exactly the same as can be absorbed by the unbound atoms or ions of the same element. A different lamp is required for each element that is to be analysed.

The major differences in instrumentation arise from the different methods used to convert the analyte (the substance being analysed) to free, unbound atoms or ions. In **flame atomization** the analyte solution is mixed with the fuel in a nebulizer, which creates an aerosol. The aerosol enters the burner where it passes into a fuel–oxidant flame. Typical fuel–oxidant mixtures are acetylene–air, which produces flame temperatures of up to 2500 K, and acetylene–nitrous oxide, which generates temperatures of up to 3000 K. A common type of **electrothermal atomizer** is the graphite furnace. The temperatures reached in the furnace are comparable to those attained in a flame atomizer but detection limits can be 1000 times better. The increased sensitivity is due to the ability to generate atoms quickly and keep them in the optical path for longer. Another advantage of the graphite furnace is that solid samples may be used. A monochromater is placed after the atomizer to isolate the desired wavelength for passage to the detector, as the ionization process may produce spectral lines from other components of the analyte.

Almost every metallic element can be analysed by using atomic absorption spectroscopy, although not all with high sensitivity or usefully low detection limit. For example, the detection limit for cadmium in a flame ionizer is 1 part per billion (1 ppb, 1 in 10^9) whereas that for mercury is only 500 ppb. Limits of detection using a graphite furnace can be as low as 1 part in 10^{15}. Direct determination is possible for any element for which hollow cathode lamp sources are available. Other species can be determined by indirect procedures. For example, PO_4^{3-} reacts with MoO_4^{2-} in acid conditions to form $H_3P(Mo_3O_{10})_4$, which can be extracted into an organic solvent and analysed for molybdenum. Generally, to analyse for a particular element, a set of calibration standards is prepared in a similar matrix to the sample, and the standards and the sample are analysed under the same conditions.

6.11 Elemental analysis

Key point: The carbon, hydrogen, nitrogen, oxygen, and sulfur content of a sample can be determined by high-temperature decomposition.

Elemental analysers convert organic compounds to gaseous molecules by high-temperature decomposition. Instruments are available that allow automated analysis of carbon, hydrogen, nitrogen, oxygen, and sulfur. Figure 6.26 shows the arrangement for an instrument that analyses for carbon, hydrogen, and nitrogen, sometimes referred to as **CHN analysis**. The sample is heated to 900°C, in oxygen and a mixture of carbon dioxide, carbon monoxide, water, nitrogen, and nitrogen oxides is produced. A stream

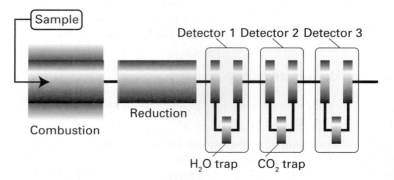

Fig. 6.26 The layout of the instrument used for CHN analysis.

of helium sweeps the products into a tube furnace at 750°C, where copper reduces nitrogen oxides to nitrogen and removes oxygen. Copper oxide converts carbon monoxide to carbon dioxide. The resulting mixture is analysed by passing it through a series of three thermal conductivity detectors. The first detector measures hydrogen and water is removed in a trap. At the second detector the carbon is measured and carbon dioxide is removed in a second trap. The remaining nitrogen is measured at the third detector. The data obtained from this technique are reported as mass per cent C, H, and N.

Oxygen may be analysed if the reaction tube is replaced with a quartz tube filled with platinized carbon. When the gaseous products are swept through this tube, the oxygen is converted to carbon monoxide, which is then converted to carbon dioxide by passage over hot copper oxide. The rest of the procedure is the same as described above. Sulfur can be measured if the sample is oxidized in a tube filled with copper oxide. Water is removed by trapping in a cool tube and the sulfur dioxide is determined at what is normally the hydrogen detector.

6.12 Thermal methods

Key points: In differential thermal analysis the temperature of the sample is monitored with respect to a reference sample; in differential scanning calorimetry the sample and the reference are maintained at the same temperature throughout the procedure.

Thermal analysis is the analysis of a change in a property of a sample, which is related to an imposed change in the temperature. The sample is usually in the solid state and the changes that occur on heating include melting, phase transition, sublimation, and decomposition.

The analysis of the change in the mass of a sample on heating is known as **thermogravimetric analysis** (TGA). The measurements are carried out using a thermobalance, which consists of an electronic microbalance, a temperature programmable furnace, and a controller, which enables the sample to be simultaneously heated and weighed (Fig. 6.27). The sample is weighed into a sample holder and then suspended from the balance within the furnace. The temperature of the furnace is usually increased linearly, but more complex heating schemes, isothermal heating, and cooling protocols can also be used. The balance and furnace are situated within an enclosed system so that the atmosphere can be controlled. The atmosphere used may be inert or reactive, depending on the nature of the investigation, and it can be static or flowing. A flowing atmosphere has the advantage of carrying away any evolved volatile or corrosive species and it also prevents the condensation of reaction products. In addition, species can be fed into a mass spectrometer to identify them.

Thermogravimetric analysis is most useful for desorption, decomposition, dehydration, and oxidation processes. For example, the thermogravimetric curve for $CuSO_4 \cdot 5H_2O$ from room temperature to 300°C shows three stepwise mass losses (Fig. 6.28), corresponding to the three stages in the dehydration to form first $CuSO_4 \cdot 4H_2O$, then $CuSO_4 \cdot H_2O$, and finally $CuSO_4$.

The most widely used thermal method of analysis is **differential thermal analysis** (DTA). In this technique the temperature of the sample is compared to that of a reference material while they are both being subjected to the same heating procedure. In a DTA instrument, the sample and reference are placed in low thermal conductivity sample holders that are then held within cavities in a block in the furnace. The temperature of the furnace is increased linearly and the difference in temperature between the sample and the reference is plotted against the furnace temperature. If an endothermic event takes place within the sample, the temperature of the sample will lag behind that of the reference and a minimum will be observed on the DTA curve. If an exothermal event takes place, then the temperature of the sample will exceed that of the reference and a maximum will be observed on the curve. The area under the endotherm or exotherm is related to the enthalpy of the thermal event, ΔH. Common reference samples for the analysis of inorganic compounds are alumina, Al_2O_3, and carborundum, SiC.

Carrier gas

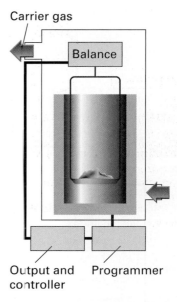

Fig. 6.27 A thermogravimetric analyser: the mass of the sample is monitored as the temperature is raised.

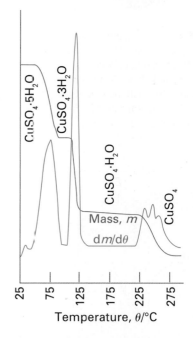

Fig. 6.28 The thermogravimetric curve obtained for $CuSO_4 \cdot 5H_2O$ as the temperature is raised from 25°C to 300°C. The red line is the mass of the sample and the green line is its first derivative (the slope of the red line).

A technique closely related to DTA is **differential scanning calorimetry** (DSC). In DSC, the sample and the reference are maintained at the same temperature throughout the heating procedure by using separate power supplies to the sample and reference holders. Any difference between the power supplied to the sample and reference is recorded against the furnace temperature. Thermal events appear as deviations from the DSC baseline as either endotherms or exotherms, depending on whether more or less power had to be supplied to the sample relative to the reference. In DSC, endothermic reactions are usually represented as positive deviations from the baseline corresponding to increased power supplied to the sample. Exothermic events are represented as negative deviations from the baseline.

The information obtained from DTA and DSC is very similar. However, DTA can be used up to higher temperatures and the quantitative data obtained from DSC are more reliable. Both DTA and DSC are often used for 'fingerprint' comparison of the results obtained from a sample with those of a reference material. Information about the temperatures and enthalpy changes of thermal events can also be extracted.

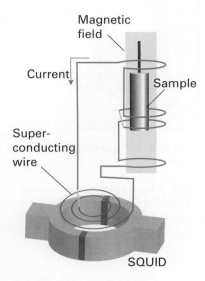

Fig. 6.29 The magnetic susceptibility of a sample is measured by using a SQUID: the sample, which is exposed to a magnetic field, is moved through the loops in small increments and the potential difference across the SQUID is monitored.

Magnetometry

Key point: Magnetometry is used to determine the characteristic response of a sample to the application of a magnetic field.

The classic way of monitoring the magnetic properties of a sample is to measure the attraction towards or repulsion away from a magnetic field by monitoring the change in the apparent weight of a sample when the magnetic field is turned on in a 'Gouy balance'. Measurements are now made using a **superconducting quantum interference device** (a SQUID, Fig. 6.29). A SQUID takes advantage of the quantization of magnetic flux and the property of current loops in superconductors that, as a part of the circuit, include a weakly conducting link through which electrons must tunnel. The current that flows in the loop in a magnetic field is determined by the value of the magnetic flux and SQUIDs can be exploited as very sensitive magnetometers.

Electrochemical techniques

Key point: Cyclic voltammetry is an electrochemical technique that measures the electrical currents due to reduction and oxidation of species in solution that react with an electrode.

Reduction potentials and the rates and coupled equilibria of redox reactions are measured by using an electrode placed in a solution of the sample. **Cyclic voltammetry** is a powerful method for surveying the redox reactions that a molecule is able to undergo. It provides direct information on reduction potentials and the stabilities of different products of oxidation or reduction. Electrochemical techniques are widely used in inorganic chemistry and cyclic voltammetry is a particularly useful method, monitoring the current flowing between the electrodes as the potential difference is changed cyclically. Cyclic voltammetry allows the investigator to obtain rapid qualitative insight into the redox properties of a compound and reliable quantitative information on its thermodynamic and kinetic properties.

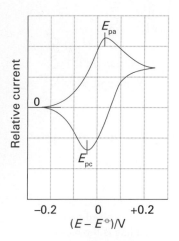

Fig. 6.30 The cyclic voltammogram for an electroactive species present in solution as the reduced form and displaying a reversible one-electron reaction at an electrode. The peak potentials E_{pa} and E_{pc} for oxidation and reduction, respectively, are separated by 0.06 V. The reduction potential is the mean of E_{pa} and E_{pc}.

The 'working electrode' at which the electrochemical reaction of interest occurs is usually constructed from platinum, silver, gold, or carbon. The reference electrode is normally a silver/silver-chloride electrode and the counter electrode is normally platinum. To understand what is involved, consider the redox couple $[Fe(CN)_6]^{3-} / [Fe(CN)_6]^{4-}$ where, initially, only the reduced form (the Fe(II) complex) is present (Fig. 6.30). The concentration of electroactive species is usually quite low (less than 0.001 M) and the solution contains a relatively high concentration of inert 'supporting' electrolyte (at concentrations greater than about 0.1 M) to provide conductivity. The potential difference is applied between the working electrode and the reference electrode and is scanned back and forth between two limits, tracing out a triangular waveform.

While the potential is low, no current flows. As it approaches the reduction potential of the Fe(III)/Fe(II) couple, the Fe(II) is oxidized at the working electrode and a current starts to flow. This current rises to a peak then decreases steadily because Fe(II) becomes depleted close to the electrode (the solution is unstirred) and must be supplemented by species diffusing from increasingly distant regions of the solution. Once the upper potential limit is reached, the potential sweep is reversed. Initially, Fe(II) diffusing to the electrode continues to be oxidized but eventually the potential difference becomes sufficiently negative to reduce the Fe(III) that had been formed; the current reaches a peak then decreases gradually to zero as the lower potential limit is reached.

The average of the two peak potentials is a good approximation to the reduction potential E under the prevailing conditions (E is not in general the standard potential because non-standard conditions are usually adopted). In the ideal case, the oxidation and reduction peaks are similar in magnitude and separated by a small potential increment, which is usually $(59\,mV)/v$ at $25°C$, where v is the number of electrons transferred during the reaction at the electrode. This case is known as 'reversible' electrochemistry, which means that electron transfer at the electrode is sufficiently fast that equilibrium is always maintained throughout the potential sweep. In such a case the current is usually limited by diffusion of electroactive species to the electrode.

Slow kinetic processes at the electrode result in a large separation of reduction and oxidation peaks that increases with increasing scan rate. This separation arises because an overpotential (effectively a driving force) is required to overcome barriers to electron transfer in each direction. Moreover, the peak due to a reduction or an oxidation in the initial part of the cycle process is often not matched by a corresponding peak in the reverse direction. This absence occurs because the species that is initially generated undergoes a further chemical reaction during the cycle and produces either a species with a different reduction potential or one that is not electroactive within the range of potential scanned. Inorganic chemists often refer to this behaviour as 'irreversible'. An electrochemical reaction followed by a chemical reaction is known as an **EC process**. By analogy, a **CE process** is a reaction in which the species able to undergo the electrochemical (E) reaction must first be generated by a chemical reaction. Thus, for a molecule that is suspected of decomposing upon oxidation, it may be possible to observe the initial unstable species formed by the E process provided the scan rate is sufficiently fast to re-reduce it before it undergoes further reaction. Consequently, by varying the scan rate, the kinetics of the chemical reaction can be determined.

Computational techniques

Computation has proved to be one of the most important techniques in chemistry. **Computer modelling** is the use of numerical models for exploring the structures and properties of individual molecules and materials. The methods used range from rigorous, and therefore computationally very time-consuming, treatments, known as '*ab initio* methods', based on numerical solution of the Schrödinger equation for the system, to the more rapid and necessarily less detailed 'semi-empirical techniques', which use approximate or 'effective functions' to describe the forces between particles.

6.13 Procedures

Key point: Computations use either *ab initio* methods or semi-empirical methods, in which experimental data are used to parameterize the calculations.

There are two principal approaches to solving the Schrödinger equation for many-electron polyatomic molecules. In the **semi-empirical methods**, integrals that occur in the formal solution of the Schrödinger equation are set equal to parameters that have been chosen to lead to the best fit to experimental quantities, such as enthalpies of formation. Semi-empirical methods are applicable to a wide range of molecules with an almost limitless number of atoms, and are widely popular. In the more fundamental ***ab initio*** methods, an attempt is made to calculate structures from first principles, using only the atomic numbers of the atoms present and their general arrangement in space. Such an approach is intrinsically more reliable than a semi-empirical procedure but much more demanding computationally.

Both types of procedure typically adopt a **self-consistent field** (SCF) procedure, in which an initial guess about the composition of the LCAO is successively refined until the solution remains unchanged in a cycle of calculation. The most common type of *ab initio* calculation is based on the **Hartree–Fock method** in which the primary approximation is applied to the electron–electron repulsion. Various methods of correcting for the explicit electron–electron repulsion, referred to as the **correlation problem**, are the Møller–Plesset perturbation theory (MP*n*, where *n* is the order of correction), the Generalized Valence Bond (GVB) method, Multi-Configurations Self-Consistent Field (MCSCF), Configuration Interaction (CI), and Coupled Cluster theory (CC).

A currently popular alternative to the *ab initio* method is **density functional theory** (DFT), in which the total energy is expressed in terms of the total electron density $\rho = |\psi|^2$ rather than the wavefunction ψ itself. When the Schrödinger equation is expressed in terms of ρ, it becomes a set of equations called the **Kohn–Sham equations**, which are solved iteratively starting from an initial estimate and continuing until they are self-consistent. The advantage of the DFT approach is that it is less demanding computationally, requires less computer time, and—in some cases, particularly *d*-metal complexes—gives better agreement with experimental values than is obtained from other procedures.

Semi-empirical methods are set up in the same general way as Hartree–Fock calculations but within this framework certain pieces of information, such as integrals representing the interaction between two electrons, are approximated by importing empirical data or simply ignored. To soften the effect of these approximations, parameters representing other integrals are adjusted so as to give the best agreement with experimental data. Semi-empirical calculations are much faster than the *ab initio* calculations but the quality of results is very dependent on using a reasonable set of experimental parameters that can be transferred from structure to structure. Thus semi-empirical calculations have been very successful in organic chemistry with just a few types of element and molecular geometries. Semi-empirical methods have also been devised specifically for the description of inorganic chemistry.

6.14 Representation of results and applications

Key point: Graphical techniques are used to display a variety of results of computation as an aid to interpretation and the development of insight into molecular properties.

The raw output of a molecular structure calculation is a list of the coefficients of the atomic orbitals in each molecular orbital and the energies of these orbitals. The graphical representation of a molecular orbital uses stylized shapes to represent the basis set and then scales their size to indicate the value of the coefficient in the LCAO. Different signs of the wavefunctions are typically represented by different colours. The total electron density at any point (the sum of the squares of the wavefunctions evaluated at that point) is commonly represented by an **isodensity surface**, a surface of constant total electron density (Fig. 6.31). An important aspect of a molecule other than its geometrical shape is the distribution of charge over its surface. A common procedure begins with calculation of the net electric charge at each point on an isodensity surface by subtracting the charge due to the electron density at that point from the charge due to the nuclei. Then the potential energy of interaction between each charge on the surface and a 'probe' charge located away from the molecule is computed. The result is an **electrostatic potential surface** (an 'elpot surface') in which net positive potential is shown in one colour and net negative potential is shown in another, with intermediate gradations of colour.

Computer modelling is applied to solids as well as to individual molecules and is useful for predicting the behaviour of a material; for example, for indicating which crystal structure of a compound is energetically most favourable, for predicting phase changes, for calculating thermal expansion coefficients, identifying preferred sites for a dopant

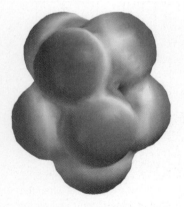

Fig. 6.31 The output of computations of the electronic structure of a molecule is conveyed in a variety of ways. Here we show the electric potential surface of SF_5CF_3, a molecule that has been found to act as a very powerful greenhouse gas but is of uncertain origin in the atmosphere. Red areas indicate regions of negative potential and green regions of positive potential.

ions, and calculating a diffusion pathway through a lattice. However, it is worth noting that very few aspects of inorganic chemistry can be computed exactly and although modelling computation can give a very useful insight into materials chemistry, it is not yet at a stage where it can be used reliably to predict the exact structure or properties of any complex compound.

FURTHER READING

A.K. Brisdon, *Inorganic spectroscopic methods.* Oxford Science Publications (1998).

R.P. Wayne, *Chemical instrumentation.* Oxford Science Publications (1994).

D.A. Skoog, F.J. Holler, and T.A. Nieman, *Principles of instrumental analysis.* Brooks Cole (1997).

R.S. Drago, *Physical methods in chemistry.* Saunders (1992).

E.I. Solomon and A.B.P. Lever, *Inorganic electronic structure and spectroscopy.* Wiley–Interscience (1999).

J.R. Ferraro and K. Nakamoto, *Introductory Raman spectroscopy.* Academic Press (1994).

K. Nakamoto, *Infrared and Raman spectra of inorganic and coordination compounds.* Wiley–Interscience (1997).

J.K.M. Saunders and B.K. Hunter, *Modern NMR spectroscopy, a guide for chemists.* Oxford University Press (1993).

J.A. Iggo, *NMR spectroscopy in inorganic chemistry.* Oxford University Press (1999).

J.W. Akitt and B.E. Mann, *NMR and chemistry.* Stanley Thornes, Cheltenham (2000).

M.E. Brown, *Introduction to thermal analysis.* Kluwer Academic Press (2001).

P.J. Haines, *Principles of thermal analysis and calorimetry.* Royal Society of Chemistry (2002).

A.J. Bard and L.R. Faulkner, *Electrochemical methods: fundamentals and approaches.* Wiley (2001).

O. Kahn, *Molecular magnetism.* VCH, New York (1993).

EXERCISES

6.1 How would you determine when the reaction of MgO and Al_2O_3 to produce $Mg_2Al_2O_4$ was complete?

6.2 The reaction of sodium carbonate, boron oxide, and silicon dioxide gives a borosilicate glass. Explain why the powder diffraction pattern of this product shows no diffraction maxima.

6.3 The minimum size of crystal that can typically be studied using a laboratory single crystal diffractometer is $50 \times 50 \times 50\,\mu m$. The X-ray flux from a synchrotron source is expected to be 10^6 times the intensity of a laboratory source. Calculate the minimum size of a cubic crystal that could be studied on a diffractometer by using this source.

6.4 Calculate the wavelength associated with a neutron moving at $2.20\,km\,s^{-1}$. Is this wavelength suitable for diffraction studies ($m_n = 1.675 \times 10^{-27}\,kg$)?

6.5 Suggest reasons for the order of stretching frequencies observed with diatomic species: $CN^- > CO > NO$.

6.6 Use the data in Table 6.2 to estimate the O—O stretching wavenumber expected for a compound believed to contain the oxygenyl species (O_2^+).

6.7 Figure 6.32 shows the UV photoelectron spectrum of NH_3. Explain why the band at approximately 11 eV shows such a long and sharply-resolved progression.

6.8 Explain why a Raman band assigned to the symmetric N—C stretching mode in $N(CH_3)_3$ shows a shift to lower frequency when ^{14}N is substituted by ^{15}N but no such shift is observed for the N—Si symmetric stretch in $N(SiH_3)_3$.

6.9 Explain why the ^{13}C NMR spectrum of $Co_2(CO)_9$ shows only a single peak at room temperature.

6.10 Explain the observation that the ^{19}F NMR spectrum of XeF_5^- consists of a central peak symmetrically flanked by two

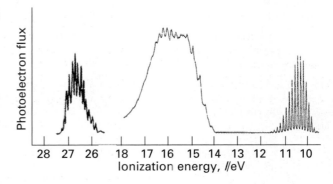

Fig. 6.32

peaks, each of which is roughly one-sixth of the intensity of the central peak.

6.11 Determine the g-values of the EPR spectrum shown in Fig. 6.33, measured for a frozen sample using a microwave frequency of 9.43 GHz.

6.12 Which technique is sensitive to the slowest processes, NMR or EPR?

6.13 For a paramagnetic compound of a d-metal compound having one unpaired electron, outline the main difference you would expect to see between an EPR spectrum measured in aqueous solution at room temperature and that recorded for a frozen solution.

6.14 How would you determine whether the compound $Fe_4[Fe(CN)_6]_3$ contains discrete Fe(II) and Fe(III) sites?

6.15 Suggest a reason why no resolved quadrupole splitting is observed in the ^{121}Sb ($I = \frac{5}{2}$) Mössbauer spectrum of solid SbF_5.

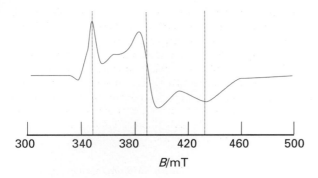

Fig. 6.33

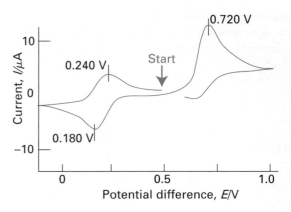

Fig. 6.34

6.16 Explain why, even though the average atomic mass of silver is 107.9 u, no peak at 108 u is observed in the mass spectrum of pure silver. What effect does this absence have on the mass spectra of silver compounds?

6.17 What peaks should you expect in the mass spectrum of $Cr(C_6H_6)(CO)_3$?

6.18 Interpret the cyclic voltammogram shown in Fig. 6.34, which has been recorded for an Fe (III) complex in aqueous solution.

6.19 A cobalt(II) salt was dissolved in water and reacted with excess acetylacetone (1,2-pentanedione, $CH_3COCHCOCH_3$) and hydrogen peroxide. A green solid was formed that gave the following results for elemental analysis: C, 50.4 per cent; H, 6.2 per cent; Co, 16.5 per cent (all by mass). Determine the ratio of cobalt to acetylacetonate ion in the product.

PROBLEMS

6.1 Discuss the importance of X-ray crystallography in inorganic chemistry. See for example: The history of molecular structure determination viewed through the Nobel Prizes. W.P. Jensen, G.J. Palenik, and I.-H. Suh, *J. Chem. Educ.*, 2003, **80**, 753.

6.2 Discuss why the length of an O—H bond obtained from X-ray diffraction experiments averages 85 pm while those obtained in neutron diffraction experiments average 96 pm.

6.3 Refer to Fig. 6.35. Discuss the likely reason for the shifts in λ_{max} for spectra of I_2 recorded in different solvent media: (a) heptane, (b) benzene, (c) diethylether/CCl_4, (d) pyridine/heptane, (e) triethylamine/heptane.

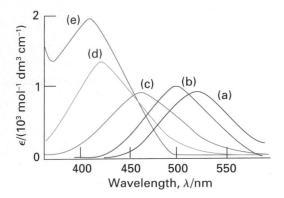

Fig. 6.35

6.4 Discuss how you would carry out the following analyses: (a) calcium levels in breakfast cereal, (b) mercury in shell fish, (c) number of organic ligands in a transition metal complex, (d) water of crystallization in an inorganic salt.

6.5 Write a review of the applications of thermal methods of analysis in inorganic chemistry.

6.6 Discuss the challenges involved in carrying out each of the following scenarios: (a) identification of inorganic pigments in a valuable painting, (b) analysis of evolved inorganic gases from a volcano, (c) determination of pollutants at the sea bed.

6.7 How may a consideration of ^{31}P chemical shifts and ^1H coupling constants be used to distinguish the isomers of the octahedral complex $[Rh(PMe_3)_3(CCR)_2(H)]$? (See J.P. Rourke, G. Stringer, D.S. Yufit, J.A.K. Howard, and T.B. Marder, *Organometallics*, 2002, **21**, 429.)

6.8 Ionization techniques in mass spectrometry frequently induce fragmentation and other undesirable reactions. However, under some circumstances these reactions can be helpful; using fullerenes as an example, discuss this phenomenon. (See M.M. Boorum, Y.V. Vasil'ev, T. Drewello, and L.T. Scott, *Science*, 2001, **294**, 828.)

6.9 The iron content of a dietary supplement tablet was determined using atomic absorption spectrometry. A tablet (0.4878 g) was ground to a fine powder and 0.1123 g was dissolved in dilute sulfuric acid and transferred to a 50 cm³ volumetric flask. A 10 cm³ aliquot of this solution was taken and made up to 100 cm³ in another volumetric flask. A series of standards was prepared that contained 1.00, 3.00, 5.00, 7.00, and 10.0 ppm iron. The absorptions of the standards and the sample solution were measured at the iron absorption wavelength:

Concentration/ppm	Absorbance
1.00	0.095
3.00	0.265
5.00	0.450
7.00	0.632
10.0	0.910
Sample	0.545

Calculate the mass of iron in the tablet.

6.10 A sample of water from a reservoir was analysed for copper content. The sample was filtered and diluted tenfold with deionized water. A set of standards was prepared with copper concentrations between 100 and 500 ppm. The standards and the sample were aspirated into an atomic absorption spectrometer and the absorbance was measured at the copper absorption wavelength. The results obtained are given below. Calculate the concentration of copper in the reservoir.

Concentration/ppm	Absorbance
100	0.152
200	0.388
300	0.590
400	0.718
500	0.865
Sample	0.751

6.11 A sample from an effluent stream was analysed for phosphate levels. Dilute hydrochloric acid and excess sodium molybdate were added to a 50 cm^3 sample of the effluent. The molybdophosphoric acid, $H_3PMo_{12}O_{40}$, that was formed was extracted into two 10 cm^3 portions of an organic solvent. A molybdenum standard with a concentration of 10 ppm was prepared in the same solvent. The combined extract and the standard were aspirated into an atomic absorption spectrometer that was set up to measure molybdenum. The extracts gave an absorbance of 0.573 and the standard gave an absorbance of 0.222. Calculate the concentration of phosphate in the effluent.

7

Molecular symmetry

Symmetry governs the physical and spectroscopic properties of molecules and provides hints about how reactions might occur. In this chapter we explore some of the consequences of molecular symmetry and introduce the systematic arguments of group theory. We shall see that symmetry considerations are essential for constructing molecular orbitals and analysing molecular vibrations. They also enable us to extract information about molecular and electronic structure from spectroscopic data.

The systematic, mathematical treatment of symmetry is called **group theory**. Group theory is a rich and powerful subject, but we shall confine our use of it at this stage to the classification of molecules in terms of their symmetry properties, the construction of molecular orbitals, and the analysis of molecular vibrations and the selection rules that govern their excitation. We shall also see that it is possible to draw some general conclusions about the properties of molecules without doing any calculations at all.

An introduction to symmetry analysis

That some molecules are 'more symmetrical' than others is intuitively obvious. However, our aim is to define the symmetries of individual molecules precisely, not just intuitively, and to provide a scheme for specifying and reporting these symmetries. It will become clear in later chapters that symmetry analysis is one of the most pervasive techniques in chemistry and provides the basis also for determining the structures and properties of materials containing an almost infinite number of repeating units.

7.1 Symmetry operations and symmetry elements

Key points: Symmetry operations are actions that leave the molecule apparently unchanged; each symmetry operation is associated with a symmetry element.

A fundamental concept of group theory is the **symmetry operation**, an action, such as rotation through a certain angle, that leaves the molecule apparently unchanged. An example is the rotation of an H_2O molecule by 180° around the bisector of the HOH angle (Fig. 7.1). Associated with each symmetry operation there is a **symmetry element**, a point, line, or plane with respect to which the symmetry operation is performed. Table 7.1 lists the most important symmetry operations and their corresponding elements. All these operations leave at least one point of the molecule unmoved, just as a rotation of a sphere leaves its centre unmoved, and hence they are the operations of **point-group symmetry**.

The **identity operation**, E, consists of doing nothing to the molecule. Every molecule has at least this operation and some have only this operation, so we need it if we are to classify all molecules according to their symmetry. The rotation of an H_2O molecule by 180° around a line bisecting the HOH angle (as in Fig. 7.1) is a symmetry operation, denoted C_2. In general, an ***n*-fold rotation** is a symmetry operation if the molecule appears unchanged after rotation by $360°/n$. The corresponding symmetry element is a line, an **n-fold rotation axis**, C_n, about which the rotation is performed. The trigonal-pyramidal NH_3 molecule has a threefold rotation axis, denoted C_3, but there are two operations associated with this axis, one a clockwise rotation by $120°$ and the other an anti-clockwise rotation by $120°$ (Fig. 7.2). The two operations are denoted C_3 and C_3^2 (because two successive clockwise rotations by $120°$ is equivalent to an anticlockwise rotation by $120°$), respectively.

There is only one rotation operation associated with a C_2 axis (as in H_2O) because clockwise and anticlockwise rotations by $180°$ are identical. The square-planar molecule XeF_4 also has twofold rotation axes that are perpendicular to the fourfold C_4 axis: one pair (C_2') passes through each *trans*-FXeF unit and the other pair (C_2'') passes through the bisectors of the FXeF angles (Fig. 7.3). By convention, the highest order rotational axis, which is called the **principal axis**, defines the z-axis.

The reflection of an H_2O molecule in either of the two planes shown in Fig. 7.4 is a symmetry operation; the corresponding symmetry element, the plane of the mirror, is a **mirror plane**, σ. The H_2O molecule has two mirror planes that intersect at the bisector of the HOH angle. Because the planes are 'vertical', in the sense of containing the rotational axis of the molecule, they are labelled with a subscript v, as in σ_v and σ_v'. An XeF_4 molecule has a mirror plane σ_h in the plane of the molecule. The subscript h signifies that the plane is 'horizontal' in the sense that the principal rotational axis of the molecule is

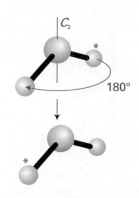

Fig. 7.1 An H_2O molecule may be rotated through any angle about the bisector of the HOH bond angle, but only a rotation of $180°$ (the C_2 operation) leaves it apparently unchanged.

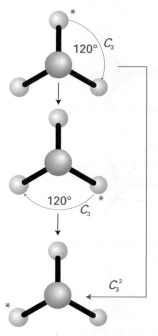

Fig. 7.2 A threefold rotation and the corresponding C_3 axis in NH_3. There are two rotations associated with this axis, one through $120°$ (C_3) and one through $240°$ (C_3^2).

Table 7.1 Symmetry operations and symmetry elements

Symmetry operation	Symmetry element	Symbol
Identity	'whole of space'	E
Rotation by $360°/n$	n-fold symmetry axis	C_n
Reflection	mirror plane	σ
Inversion	centre of inversion	i
Rotation by $360°/n$ followed by reflection in a plane perpendicular to the rotation axis	n-fold axis of improper rotation*	S_n

* Note the equivalences $S_1 = \sigma$ and $S_2 = i$.

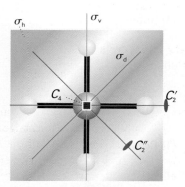

Fig. 7.3 Some of the symmetry elements of a square-planar molecule such as XeF_4. There are two pairs of twofold rotation axes that are perpendicular to the principal C_4 axis. One pair (C_2') passes through each *trans*-FXeF unit and the other pair (C_2'') passes through the bisectors of the FXeF angles. The horizontal reflection plane σ_h is in the plane of the paper and there are two sets of vertical reflection planes, σ_v and σ_d.

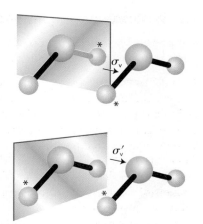

Fig. 7.4 The two vertical mirror planes σ_v and σ_v' in H_2O and the corresponding operations. Both planes cut through the C_2 axis.

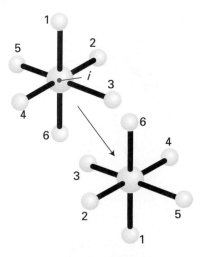

Fig. 7.5 The inversion operation and the centre of inversion *i* in SF$_6$.

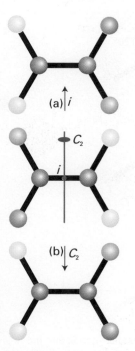

Fig. 7.6 Care must be taken not to confuse (a) an inversion operation with (b) a twofold rotation. Although the two operations may sometimes appear to have the same effect, that is not the case in general.

perpendicular to it. This molecule also has two more sets of two mirror planes that intersect the fourfold axis. The symmetry elements (and the associated operations) are denoted σ_v for the planes that pass through the F atoms and σ_d for the planes that bisect the angle between the F atoms. The d denotes 'dihedral' and signifies that the plane bisects the angle between two C_2' axes (the FXeF axes).

To understand the **inversion operation**, *i*, we need to imagine that each atom is projected in a straight line through a single point located at the centre of the molecule and then out to an equal distance on the other side (Fig. 7.5). In an octahedral molecule such as SF$_6$, with the point at the centre of the molecule, diametrically opposite pairs of atoms at the corners of the octahedron are interchanged. The symmetry element, the point through which the projections are made, is called the **centre of inversion**, *i*. For SF$_6$, the centre of inversion lies at the nucleus of the S atom: likewise, the molecule CO$_2$ has an inversion centre at the C nucleus. However, there need not be an atom at the centre of inversion: an N$_2$ molecule has a centre of inversion midway between the two nitrogen nuclei. An H$_2$O molecule does not possess a centre of inversion. No tetrahedral AB$_4$ molecule has a centre of inversion. Although an inversion and a twofold rotation may sometimes achieve the same effect (Fig. 7.6), that is not the case in general and the two operations must be distinguished.

An **improper rotation** consists of a rotation of the molecule through a certain angle around an axis followed by a reflection in the plane perpendicular to that axis (Fig. 7.7). The illustration shows a fourfold improper rotation of a CH$_4$ molecule. In this case, the operation consists of a 90° rotation about an axis bisecting two HCH bond angles, followed by a reflection through a plane perpendicular to the rotation axis. Neither the 90° operation nor the reflection alone is a symmetry operation for CH$_4$ but their overall effect is a symmetry operation. A fourfold improper rotation is denoted S_4. The symmetry element, the **improper-rotation axis**, S_n (S_4 in the example), is the corresponding combination of an *n*-fold rotational axis and a perpendicular mirror plane.

An S_1 axis, a rotation through 360° followed by a reflection in the perpendicular plane, is equivalent to a reflection alone, so S_1 and σ_h are the same; the symbol σ_h is generally used rather than S_1. Similarly, an S_2 axis, a rotation through 180° followed by a reflection in the perpendicular plane, is equivalent to an inversion, *i* (Fig. 7.8); the symbol *i* is employed rather than S_2.

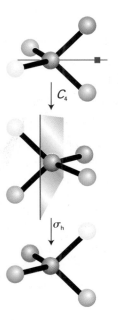

Fig. 7.7 A fourfold axis of improper rotation S_4 in the CH$_4$ molecule.

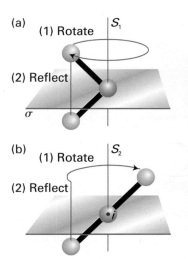

Fig. 7.8 (a) An S_1 axis is equivalent to a mirror plane and (b) an S_2 axis is equivalent to a centre of inversion.

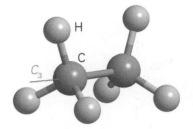

1 A C_3 axis

Example 7.1 Identifying symmetry elements

Which symmetry elements are present in (a) eclipsed and (b) staggered conformations of a CH_3CH_3 molecule?

Answer The eclipsed conformation (**1**) has the elements E, C_3, C_2, σ_h, σ_v, and S_3. The staggered conformation (**2**) has the elements E, C_3, C_2, σ_d, i, and S_6.

Self-test 7.1 Sketch the S_4 axis of an NH_4^+ ion. How many of these axes does the ion possess?

7.2 The point groups of molecules

Key point: The point group of a molecule is identified by noting its symmetry elements and comparing these elements with the elements that define each group.

The symmetry properties of a molecule that leave one point unchanged define its **point group** and are a basis for its symmetry classification. A point group is labelled by its **Schoenflies symbol**. If a molecule has only the identity element (CHBrClI is an example (**3**)), then we list its elements as E alone and look for the group that has only this element and identify it as belonging to the group C_1. The molecule CH_2BrCl belongs to a slightly richer group: it has the elements E (all groups have that element) and a mirror plane, and therefore belongs to the group C_s. This procedure can be continued, and molecules assigned to the group that matches the symmetry elements they possess. For example, molecules as diverse as H_2O, SO_2, and (by ignoring the H atoms) *cis*-$[CrCl_2(NH_3)_4]^+$ all belong to the point group C_{2v}.

The assignment of a molecule to its point group depends on identifying its symmetry elements and then referring to Table 7.2. In practice, the shapes in the table give a very good clue to the identity of the group to which the molecule belongs, at least in simple cases. The decision tree in Fig. 7.9 can also be used to assign most common point groups systematically by answering the questions at each decision point.

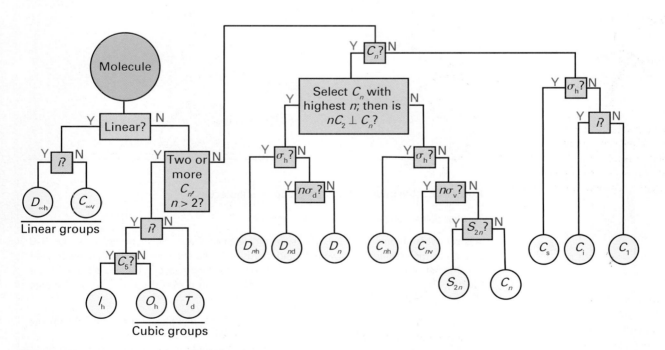

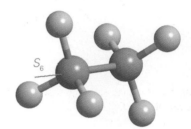

2 An S_6 axis

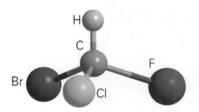

3 CHBrClF, C_1

Fig. 7.9 The decision tree for identifying a molecular point group. The symbols of each point refer to the symmetry elements (not the corresponding operations).

Table 7.2 The composition of some common groups

Point group	Symmetry elements	Shape	Examples
C_1	E		SiIClBrF
C_2	E, C_2		H_2O_2
C_s	E, σ		NHF_2
C_{2v}	E, C_2, σ_v, σ_v'		H_2O, SO_2Cl_2
C_{3v}	E, $2C_3$, $3\sigma_v$		NH_3, PCl_3, $POCl_3$
$C_{\infty v}$	E, C_2, $2C_\phi$, $\infty\sigma_v$		CO, HCl, OCS
D_{2h}	E, $3C_2$, i, 3σ		N_2O_4, B_2H_6
D_{3h}	E, $2C_3$, $3C_2$, σ_h, $2S_3$, $3\sigma_v$		BF_3, PCl_5
D_{4h}	E, $2C_4$, C_2, $2C_2'$, $2C_2''$, i, $2S_4$, σ_h, $2\sigma_v$, $2\sigma_d$		XeF_4, trans-$[MA_4B_2]$
$D_{\infty h}$	E, $2\infty C_2'$, $2C_\phi$, i, $\infty\sigma_v$, $2S_\phi$		H_2, CO_2, C_2H_2
T_d	E, $8C_3$, $3C_2$, $6S_4$, $6\sigma_d$		CH_4, $SiCl_4$
O_h	E, $8C_3$, $6C_2$, $6C_4$, $3C_2$, i, $6S_4$, $8S_6$, $3\sigma_h$, $6\sigma_d$		SF_6

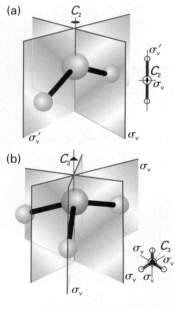

(a)

(b)

Fig. 7.10 The symmetry elements of (a) H_2O and (b) NH_3. The diagrams on the right are views from above and summarize the diagrams on the left.

Example 7.2 Identifying the point group of a molecule

To what point groups do H_2O and NH_3 belong?

Answer Work through Fig. 7.9. The symmetry elements are shown in Fig. 7.10. (a) H_2O possesses the identity (E), a twofold rotation axis (C_2), and two vertical mirror planes (σ_v and σ_v'). The set of elements (E, C_2, σ_v, σ_v') corresponds to the group C_{2v}. (b) NH_3 possesses the identity (E), a threefold axis (C_3), and three vertical mirror planes ($3\sigma_v$). The set of elements (E, C_3, $3\sigma_v$) identifies the point group as C_{3v}.

Self-test 7.2 Identify the point groups of (a) BF_3, a trigonal-planar molecule, and (b) the tetrahedral SO_4^{2-} ion.

It is very useful to be able to recognize immediately the point groups of some common molecules. Linear molecules with a centre of symmetry, such as H_2, CO_2 (**4**), and $HC{\equiv}CH$ belong to $D_{\infty h}$. A molecule that is linear but has no centre of symmetry, such as HCl, OCS (**5**), and NNO belongs to $C_{\infty v}$. Tetrahedral (T_d) and octahedral (O_h) molecules have more than one principal axis of symmetry (Fig. 7.11): a tetrahedral CH_4 molecule, for instance, has four C_3 axes, one along each CH bond. The O_h and T_d point groups are known as **cubic groups** because they are closely related to the symmetry of a cube. A closely related group, the **icosahedral group**, I_h, characteristic of the icosahedron, has 12 fivefold axes (Fig. 7.12). The icosahedral group is important for boron compounds and the C_{60} fullerene molecule.

The distribution of molecules among the various point groups is very uneven. Some of the most common groups for molecules are the low-symmetry groups C_1 and C_s, the groups for a number of other polar molecules C_{2v} (as in SO_2) and C_{3v} (as in NH_3), and the highly symmetrical tetrahedral and octahedral groups. There are many linear molecules, which belong to the groups $C_{\infty v}$ (HCl, OCS) and $D_{\infty h}$ (Cl_2 and CO_2), and a number of planar-trigonal molecules, D_{3h} (such as BF_3 (**6**)), trigonal-bipyramidal molecules (such as PCl_5 (**7**)), which are D_{3h}, and square-planar molecules, D_{4h} (**8**). So-called 'octahedral' molecules with two identical substituents opposite each other, as in (**9**), are also D_{4h}. The last example shows that the point-group classification of a molecule is more precise than the casual use of the terms 'octahedral' or 'tetrahedral' that indicate molecular *geometry*. For instance, a molecule may be called octahedral (that is, it has octahedral geometry) even if it has six different groups attached to the central atom. However, the 'octahedral molecule' belongs to the octahedral point group O_h only if all six groups and the lengths of their bonds to the central atom are identical and all angles are 90°.

Applications of symmetry

Important applications of symmetry in inorganic chemistry include the construction and labelling of molecular orbitals and the interpretation of spectroscopic data to determine structure. However, there are several simpler applications, one being to use group theory

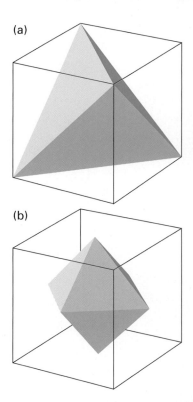

Fig. 7.11 Shapes having cubic symmetry. (a) The tetrahedron, point group T_d. (b) The octahedron, point group O_h.

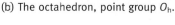
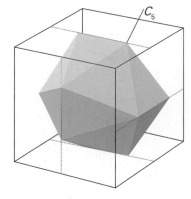

Fig. 7.12 The regular icosahedron, point group I_h, and its relation to a cube.

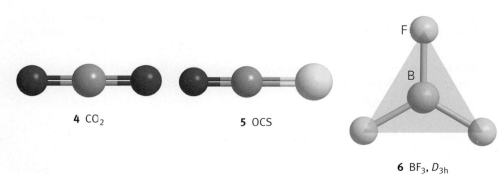

4 CO_2

5 OCS

6 BF_3, D_{3h}

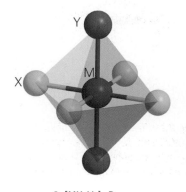

7 PCl_5, D_{3h}

8 $[PtCl_4]^{2-}$, D_{4h}

9 $[MX_4Y_2]$, D_{4h}

to decide whether a molecule is polar or chiral. In many cases the answer may be obvious and we do not need to use group theory. However, that is not always the case and the following examples illustrate the approach that can be adopted when the result is not obvious.

7.3 Polar molecules

Key point: A molecule cannot be polar if it belongs to any group that includes a centre of inversion, any of the groups *D* and their derivatives, the cubic groups (*T*, *O*), the icosahedral group (*I*), and their modifications.

A **polar molecule** is a molecule that has a permanent electric dipole moment. A molecule cannot be polar if it has a centre of inversion. Inversion implies that a molecule has matching charge distributions at all diametrically opposite points about a centre, which rules out a dipole moment. For the same reason, a dipole moment cannot lie perpendicular to any mirror plane or axis of rotation that the molecule may possess. For example, a mirror plane demands identical atoms on either side of the plane, so there can be no dipole moment across the plane. Similarly, a symmetry axis implies the presence of identical atoms at points related by the corresponding rotation, which rules out a dipole moment perpendicular to the axis.

In summary:

1 A molecule cannot be polar if it has a centre of inversion.

2 A molecule cannot have an electric dipole moment perpendicular to any mirror plane.

3 A molecule cannot have an electric dipole moment perpendicular to any axis of rotation.

Example 7.3 Judging whether a molecule can be polar

The ruthenocene molecule (**10**) is a pentagonal prism with the Ru atom sandwiched between two C_5H_5 rings. Can it be polar?

Answer We should decide whether the point group is *D* or cubic, because in neither case can it have a permanent electric dipole. Reference to Fig. 7.9 shows that a pentagonal prism belongs to the point group D_{5h}. Therefore, the molecule must be nonpolar.

Self-test 7.3 A conformation of the ferrocene molecule that lies 4 kJ mol^{-1} above the lowest energy configuration is a pentagonal antiprism (**11**). Is it polar?

7.4 Chiral molecules

Key point: A molecule cannot be chiral if it possesses an improper rotation axis (S_n).

A **chiral molecule** (from the Greek word for 'hand') is a molecule that cannot be superimposed on its own mirror image. An actual hand is chiral in the sense that the mirror image of a left hand is a right hand, and the two hands cannot be superimposed. A chiral molecule and its mirror image partner are called **enantiomers** (from the Greek word for 'both'). Chiral molecules that do not interconvert rapidly between enantiomeric forms are **optically active** in the sense that they can rotate the plane of polarized light. Enantiomeric pairs of molecules rotate the plane of polarization of light by equal amounts in opposite directions.

A chiral molecule cannot possess an improper rotation axis, S_n. A mirror plane is an S_1 axis of improper rotation and a centre of inversion is equivalent to an S_2 axis; therefore, molecules with either a mirror plane or a centre of inversion have axes of improper rotation and cannot be chiral. Groups in which S_n is present include D_{nh} (which include S_n), D_{nd}, and some of the cubic groups (specifically, T_d and O_h).

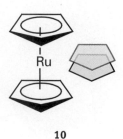

10

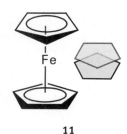

11

Therefore, molecules such as CH_4 and $Ni(CO)_4$ that belong to the group T_d are not chiral. That a 'tetrahedral' carbon atom leads to optical activity (as in CHClFBr) should serve as another reminder that group theory is stricter in its terminology than casual conversation. Thus CHBrClF (**3**) belongs to the group C_1, not to the group T_d; it has tetrahedral geometry but not tetrahedral symmetry.

When judging chirality, it is important to be alert for axes of improper rotation that might not be immediately apparent. Molecules with neither a centre of inversion nor a mirror plane (and hence with no S_1 or S_2 axes) are usually chiral, but it is important to verify that a higher-order improper-rotation axis is not also present. For instance, the quaternary ammonium ion (**12**) has neither a mirror plane (S_1) nor an inversion centre (S_2), but it does have an S_4 axis and so it is not chiral.

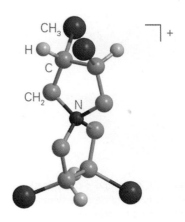

12

Example 7.4 Judging whether a molecule is chiral

The complex $[Mn(acac)_3]$, where acac denotes the acetylacetonato ligand ($CH_3COCHCOCH_3^-$), has the structure shown as (**13**). Is it chiral?

Answer We begin by identifying the point group. The chart in Fig. 7.9 shows that the ion belongs to the point group D_3, which consists of the elements (E, C_3, $3C_2$) and hence does not contain an improper-rotation axis either explicitly or in a disguised form. The complex ion is chiral and hence, because it is long-lived, optically active.

Self-test 7.4 Is the conformation of H_2O_2 shown in (**14**) chiral? The molecule can usually rotate freely about the O—O bond: comment on the possibility of observing optically active H_2O_2.

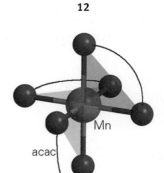

13 [Mn(acac)₃]

The symmetries of orbitals

We shall now see in more detail the significance of the labels used for molecular orbitals introduced in Sections 2.8 and 2.9 and gain more insight into their construction. At this stage the discussion will continue to be informal and pictorial, our aim being to give an elementary introduction to group theory but not the details of the calculations involved: those can be found in Sections 7.9 and 7.10 and in more detail in *Further information* 7.1 at the end of the chapter. The specific objective here is to show how to identify the symmetry label of an orbital from a drawing like those in *Resource section* 4 and, conversely, to appreciate the significance of a symmetry label. The arguments later in the book are all based on simply 'reading' molecular orbital diagrams qualitatively.

7.5 Symmetry labels and character tables

Key point: The systematic analysis of the symmetry properties of molecules is carried out by using character tables.

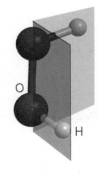

14 H_2O_2

Molecular orbitals of diatomic and linear polyatomic molecules are labelled σ, π, and so on. These labels refer to the symmetries of the orbitals with respect to rotations around the principal axis of the molecule, which in this case is the internuclear axis. The same designations can be used for specifying orbitals in nonlinear polyatomic molecules by noting the local symmetry with respect to the axis of a given bond, so we can speak, for instance, of σ and π orbitals in a more complex molecule such as benzene. A σ orbital does not change sign under a rotation through any angle about the internuclear axis, a π orbital changes sign when rotated by 180°, and so on (Fig. 7.13).

The labels σ and π are based on the rotational symmetry of orbitals with respect to an internuclear axis. The more elaborate labels a, a_1, e, e_g, and so on, which are used to label molecular orbitals in nonlinear molecules, are based on the behaviour of the orbitals under all the symmetry operations of the relevant molecular point group. The label is assigned by referring to the **character table** of the group, a table that lists the different symmetry types possible in a point group.

Fig. 7.13 The σ and π classification of orbitals is based on their symmetries with respect to rotations about an axis. A σ orbital is unchanged by any rotation; a π orbital changes sign when rotated by 180° about the internuclear axis.

The structure of a typical character table is shown in Table 7.3. The entries in the main part of the table are called **characters**, χ (chi). Each character shows how an object or mathematical function, such as an atomic orbital, is affected by the corresponding symmetry operation of the group (as explained more formally in *Further information* 7.1). For instance, a '1' shows that it is unchanged; a '−1' shows that it changes sign. A '0' signifies that the function undergoes a more complicated change. In some character tables, numbers such as '2' and '3' appear as characters: this feature is explained later. The **class** of an operation is a specific grouping of symmetry operations of the same general type: the two (clockwise and anticlockwise) threefold rotations about an axis form one class, reflections in a mirror plane form another, and so on. Each row of characters refers to a particular **irreducible representation** of the group, which has a technical meaning in group theory but, broadly speaking, is a fundamental type of symmetry in the group (like σ and π for linear molecules). The label in the first column is the **symmetry species** (essentially, a label, like σ and π) of that irreducible representation. The two columns on the right contain examples of functions that exhibit the characteristics of that symmetry species. The third column contains functions defined by a single axis, such as translations or p orbitals (x,y,z) or rotations (R_x, R_y, R_z), and the fourth column contains quadratic functions such as d orbitals (xy, etc.). Character tables for a selection of common point groups are given in *Resource section 3*.

As an example, the C_{2v} character table is given in Table 7.4, and the corresponding symmetry elements for the H_2O molecule are shown in Fig. 7.10. The symmetry operations are E, C_2 (a twofold rotation about the z-axis), and the two reflections σ_v and σ'_v (in each of the vertical planes, defined here as zx and yz, respectively). Now consider a $2p_z$ orbital on the O atom: this orbital is proportional to $zf(r)$ and its symmetry properties are those of the function z itself. All the characters are +1 because the function z, like p_z, is unchanged by all operations of the point group. The $O2p_z$ orbital is thus totally symmetric under all operations of the point group C_{2v} and therefore has symmetry species A_1. Now consider the $O2p_x$ orbital, which is proportional to the function $xf(r)$ and has the same symmetry as the function x. The character of this orbital under C_2 is −1, which means simply that the function x, like p_x, changes sign under a twofold rotation. A p_x orbital also changes sign (and therefore has character −1) when reflected in the yz plane (σ'_v), but is unchanged (character 1) when reflected in the xz plane (σ_v). It follows that the characters of an $O2p_x$ orbital are $(1, -1, 1, -1)$ and therefore that its symmetry species

Table 7.3 The components of a character table

Name of point group*	Symmetry operations R arranged by class (E, C_n, etc.)	Functions	Further functions	Order of group, h
Symmetry species (Γ)	Characters (χ)	Translations and components of dipole moments (x, y, z), of relevance to IR activity Rotations (about axes x, y, z)	Quadratic functions such as z^2, xy, etc., of relevance to Raman activity	

* Schoenflies symbol.

Table 7.4 The C_{2v} character table

C_{2v}	E	C_2	$\sigma(zx)$	$\sigma'(yz)$	$h=4$	
A_1	1	1	1	1	z	x^2, y^2, z^2
A_2	1	1	−1	−1	R_z	
B_1	1	−1	1	−1	x, R_y	xy
B_2	1	−1	−1	1	y, R_x	zx, yz

Table 7.5 The C_{3v} character table

C_{3v}	E	$2C_3$	$3\sigma_v$	$h=6$	
A_1	1	1	1	z	z^2
A_2	1	1	−1	R_z	
E	2	−1	0	(x,y) (R_x, R_y)	(zx, yz) $(x^2 - y^2, xy)$

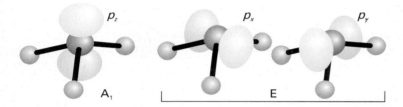

Fig. 7.14 The nitrogen $2p_z$ orbital in ammonia is symmetric under all operations of the C_{3v} point group and therefore has A_1 symmetry. The $2p_x$ and $2p_y$ orbitals behave identically under all operations (they cannot be distinguished) and are given the symmetry label E.

is B_1. An $O2s$ orbital is also totally symmetric (as it is for all point groups) so it also has symmetry species A_1.

The letter A used to label a symmetry species in the group C_{2v} means that the function to which it refers is symmetric with respect to rotation about the twofold axis. The label B indicates that the function changes sign under that rotation. The subscript 1 on A_1 means that the function to which it refers is also symmetric with respect to reflection in the principal vertical plane (for H_2O this is the plane that contains all three atoms). A subscript 2 is used to denote that the function changes sign under this reflection.

Now consider the slightly more complex example of NH_3, which belongs to the point group C_{3v} (Table 7.5). An NH_3 molecule has higher symmetry than H_2O. This higher symmetry is immediately apparent by noting the **order**, h, of the group, the total number of symmetry operations that can be carried out. For H_2O, $h=4$ and for NH_3, $h=6$. For highly symmetric molecules, h is large; for example $h=48$ for the point group O_h.

Inspection of the NH_3 molecule (Fig. 7.14) shows that whereas the $N2p_z$ orbital is unique (it has A_1 symmetry), the remaining two $N2p_x$ and $N2p_y$ orbitals belong to the same symmetry representation, E; in other words, they are degenerate (they have the same symmetry characteristics and hence have the same energy). In general, E always denotes double degeneracy and T denotes triple degeneracy.[1]

The characters in the column headed by the identity operation E give the degeneracy of the orbitals. Thus, in NH_3 or any other molecule having a C_3 axis, a symmetry label such as A_1 or A_2 (or B, etc.) refers to a nondegenerate irreducible representation that has a character of 1 in the column headed E. Similarly, any doubly degenerate pair of orbitals in a molecule with a C_3 axis must be labelled E and have a character 2 in the column labelled E. Degenerate irreducible representations also contain zero values for some operations. This character arises because it is the sum of the characters for the two (or more) orbitals of the set, and if one orbital changes sign but the other does not, then the total character is 0.

Example 7.5 Using a character table to judge degeneracy

Can there be triply degenerate orbitals in BF_3?

Answer The point group of the molecule is D_{3h}. Reference to the character table for this group (*Resource section* 3) shows that, because no character exceeds 2 in the column headed E, the maximum degeneracy is 2. Therefore, none of its orbitals can be triply degenerate.

Self-test 7.5 The SF_6 molecule is octahedral. What is the maximum possible degree of degeneracy of its orbitals?

---

[1] Care must be taken to distinguish the identity operation E (italic, a column heading) from the symmetry label E (roman, a row label).

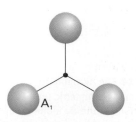

Fig. 7.15 The ϕ_1 symmetry-adapted linear combination of H1s orbitals in NH₃. This SALC has A₁ symmetry.

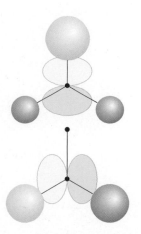

Fig. 7.16 The ϕ_2 and ϕ_3 symmetry-adapted linear combinations of H1s orbitals in NH₃. These SALCs have E symmetry (they are equal in energy).

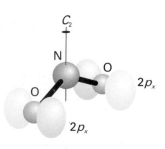

Fig. 7.17 The combination of O2p_x orbitals referred to in *Example 7.6*.

7.6 Symmetry-adapted linear combinations

A fundamental principle of the MO theory of diatomic molecules (Section 2.7) is that molecular orbitals are constructed from atomic obitals of the same symmetry. Thus, an s orbital may have nonzero overlap with another s orbital or with a p_z orbital on the second atom (where z is the internuclear direction), but not with a p_x or p_y orbital. Exactly the same principle applies in polyatomic molecules. Our task here is to group atomic orbitals, such as the three H1s orbitals of NH₃, together to form combinations of a particular symmetry and then to build molecular orbitals by allowing combinations of the same symmetry on different atoms, such as an N2s orbital and the appropriate combination of the three H1s orbitals, to overlap. Specific combinations of atomic orbitals that are used to build molecular orbitals of a given symmetry are called **symmetry-adapted linear combinations** (SALCs). A collection of commonly encountered SALCs of orbitals is shown in *Resource section* 4; it is usually simple to identify the symmetry of a combination of orbitals by comparing it with the diagrams provided there.

An example will make this clear. The three H1s orbitals in NH₃ give rise to three SALCs, one of which (Fig. 7.15) is

$$\phi_1 = \psi_{A1s} + \psi_{B1s} + \psi_{C1s}$$

This combination remains unchanged under a C_3 rotation and under any of the vertical reflections, so its characters are

$$E \quad 2C_3 \quad 3\sigma_v$$
$$1 \quad 1 \quad \quad 1$$

Comparison with the C_{3v} character table in Table 7.5 shows that ϕ_1 has symmetry species A₁. The other two SALCs that can be built from the three H1s orbitals in NH₃ are

$$\phi_2 = 2\psi_{A1s} - \psi_{B1s} - \psi_{C1s}$$
$$\phi_3 = \psi_{B1s} - \psi_{C1s}$$

Both SALCs have E symmetry (Fig. 7.16). How these SALCs are constructed is described in Section 7.10 after we have established some quantitative techniques of group theory.

Example 7.6 Identifying the symmetry labels of SALCs

Identify the symmetry label of the orbital $\phi = \psi_o - \psi_{o'}$ in the C_{2v} molecule NO₂, where ψ_o is an O2p_x orbital on one O atom and $\psi_{o'}$ is an O2p_x orbital on the other O atom.

Answer The combination is shown in Fig. 7.17. Under C_2, ϕ changes into itself, implying a character of 1. Under the reflection σ_v, both orbitals change sign, so ϕ is transformed into $-\phi$, implying a character of -1. Under σ_v', ϕ also changes sign, so the character for this operation is also -1. The characters are therefore

$$E \quad C_2 \quad \sigma_v \quad \sigma_v'$$
$$1 \quad 1 \quad -1 \quad -1$$

These values match the irreducible representation A₂.

Self-test 7.6 For a square-planar array of H atoms A, B, C, D, identify the symmetry label of the combination $\phi = \psi_{A1s} - \psi_{B1s} + \psi_{C1s} - \psi_{D1s}$.

7.7 The construction of molecular orbitals

Key point: Only atomic orbitals or combinations of orbitals of the same symmetry type can contribute to a molecular orbital of a given symmetry type.

Molecular orbitals are constructed from SALCs of the same symmetry type. Thus, as remarked above, in a linear molecule with the z-axis as the internuclear axis, an s orbital and a p_z orbital both have σ symmetry and so may combine to form σ molecular orbitals.

On the contrary, an s orbital and a p_x orbital have different symmetries (σ and π, respectively) and cannot contribute to the same molecular orbital (Fig. 7.18). As can be seen from the illustration, the physical interpretation for this difference is that the contribution from the region of constructive interference is cancelled by the contribution from the region of destructive interference.

The analogous argument for a nonlinear molecule can be illustrated by considering NH_3 again. We have seen (in Fig. 7.15) that the SALC ϕ_1 has A_1 symmetry. The N2s and N2p_z orbitals each have the same symmetry species and so all three contribute to molecular orbitals of the form

$$\psi = c_1\psi_{N2s} + c_2\psi_{N2p_z} + c_3\phi_1$$

in which the c_i are coefficients found by manipulating the Schrödinger equation (in practice, computationally). The symmetry species of this molecular orbital is A_1, like its components, and it is called an a_1 **orbital**. Note that the labels for molecular orbitals are italic, lower-case versions of the symmetry species of the orbital. Only three such linear combinations are possible (because the H combination ϕ_1 counts as a single orbital), and they are labelled $1a_1$, $2a_1$, and $3a_1$ in order of increasing energy (the order of increasing number of internuclear nodes).

We have also seen (and can confirm by referring to *Resource section* 4) that the symmetry-adapted combinations ϕ_2 and ϕ_3 have E symmetry in C_{3v}. The character table shows that the same is true of the N2p_x and N2p_y orbitals, and this identification is confirmed by noting that jointly the two $2p$ orbitals behave exactly like ϕ_2 and ϕ_3 (Fig. 7.19). It follows that ϕ_2 and ϕ_3 can combine with these two $2p$ orbitals to give doubly degenerate bonding and antibonding orbitals of the form

$$\psi = c_4\psi_{N2p_x} + c_5\phi_2, \quad c_6\psi_{N2p_y} + c_7\phi_3$$

These molecular orbitals have E symmetry and are therefore called e **orbitals**. The lower energy (bonding) pair is labelled $1e$ and the higher energy (antibonding) pair is labelled $2e$.

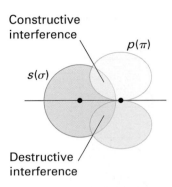

Fig. 7.18 An s orbital (with σ symmetry) has zero net overlap with a p orbital (with π symmetry) because the constructive interference between the parts of the atomic orbitals with the same sign exactly matches the destructive interference between the parts with opposite signs.

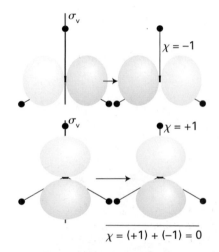

Fig. 7.19 An N2p_x orbital in NH_3 changes sign under a σ_v reflection but a N2p_y orbital is left unchanged. Hence the degenerate pair jointly has character 0 for this operation. The plane of the paper is the xy-plane.

Example 7.7 Identifying symmetry-adapted orbitals

The two H1s orbitals in H_2O form the SALCs $\phi_+ = \psi_{A1s} + \psi_{B1s}$ and $\phi_- = \psi_{A1s} - \psi_{B1s}$ (**15**). With what oxygen orbitals will they overlap to form molecular orbitals?

Answer Under C_2, ϕ_+ does not change sign but ϕ_- does; their characters are 1 and -1, respectively. Under the reflections, ϕ_+ does not change sign; ϕ_- changes sign under σ_v, so its character is -1 for this operation. The characters are therefore

	E	C_2	σ_v	σ_v'
ϕ_+	1	1	1	1
ϕ_-	1	-1	-1	1

This table identifies their symmetry labels as A_1 and B_2, respectively. The same conclusion could have been obtained more directly by referring to Resource section 3. According to the right of the character table, the O2s and O2p_z orbitals also have A_1 symmetry; O2p_y has B_2 symmetry. The linear combinations that can be formed are therefore

$$\psi = c_1\psi_{O2s} + c_2\psi_{O2p_z} + c_3\phi_+$$
$$\psi = c_4\psi_{O2p_y} + c_5\phi_-$$

The three a_1 orbitals are bonding, intermediate, and antibonding in character according to the relative signs of the coefficients c_1, c_2, and c_3. Similarly, depending on the relative signs of the coefficients c_4 and c_5, one of the two b_2 orbitals is bonding and the other is antibonding.

Self-test 7.7 What is the symmetry label of the SALC $\phi = \psi_{A1s} + \psi_{B1s} + \psi_{C1s} + \psi_{D1s}$ in CH_4, where ψ_{J1s} is an H1s orbital on atom J?

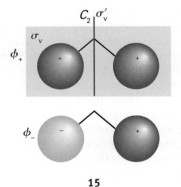

15

A symmetry analysis has nothing to say about the energies of orbitals other than to identify degeneracies. To calculate the energies, and even to arrange the orbitals in order, it is necessary to use quantum mechanics; to assess them experimentally it is necessary to

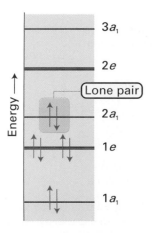

Fig. 7.20 A schematic molecular orbital energy level diagram for NH_3 and an indication of its ground-state electron configuration.

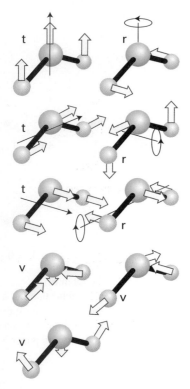

Fig. 7.21 An illustration of the counting procedure for displacements of the atoms in a nonlinear molecule. For a molecule with three atoms, there are nine atomic displacements in all. Three combinations of the displacements correspond to translation (t) of the molecule through space, and three correspond to rotations (r). Hence the three remaining combinations (v) correspond to vibrations of the molecule that leave its centre of mass and orientation unchanged.

use techniques such as photoelectron spectroscopy. In simple cases, though, we can use the general rules set out in Section 2.12 to judge the relative energies of the orbitals. For example, in NH_3, the $1a_1$ orbital, being composed of the low-lying N2s orbital, can be expected to lie lowest in energy and its antibonding partner, $3a_1$, will probably lie highest. The $1e$ bonding orbital is next higher in energy after $1a_1$, and is followed by the $2a_1$ orbital, which is largely nonbonding. This qualitative analysis leads to the energy level scheme shown in Fig. 7.20. These days, there is no difficulty in using one of the widely available software packages to calculate the energies of the orbitals directly by either an *ab initio* or a semi-empirical procedure; the energies given in Fig. 7.20 have in fact been calculated in this way. Nevertheless, the ease of achieving computed values should not be seen as a reason for disregarding the understanding of the energy level order that comes from investigating the structures of the orbitals.

The general procedure for constructing a molecular orbital scheme for a reasonably simple molecule can now be summarized as follows:

1 Assign a point group to the molecule.

2 Look up the shapes of the SALCs in *Resource section* 4.

3 Arrange the SALCs of each molecular fragments in increasing order of energy, first noting whether they stem from *s*, *p*, or *d* orbitals (and put them in the order $s < p < d$), and then their number of internuclear nodes.

4 Combine SALCs of the same symmetry type from the two fragments, and from *N* SALCs form *N* molecular orbitals.

5 Estimate the relative energies of the molecular orbitals from considerations of overlap and relative energies of the parent orbitals, and draw the levels on a molecular orbital energy level diagram (showing the origin of the orbitals).

6 Confirm, correct, and revise this qualitative order by carrying out a molecular orbital calculation by using commercial software.

The symmetries of molecular vibrations

The interpretation of the IR and Raman spectra introduced in Section 6.4 is greatly simplified by taking into account the symmetry of the molecule in question, and in this section we introduce the reasoning involved.

7.8 Symmetry considerations

Key points: If a molecule has a centre of inversion, none of its modes can be both IR and Raman active; a vibrational mode is IR active if it has the same symmetry as a component of the electric dipole vector; a vibrational mode is Raman active if it has the same symmetry as a component of the molecular polarizability.

A knowledge of the symmetry of a molecule can assist in the analysis of IR and Raman spectra. It is convenient to consider two aspects of symmetry. One is the information that can be obtained directly by knowing to which point group a molecule as a whole belongs. The other is the additional information that comes from knowing the symmetry species of each normal mode.

As we saw in Section 6.4, absorption of infrared radiation can occur when a vibration results in a change in the electric dipole moment of a molecule and a Raman transition can occur when the polarizability of a molecule changes during a vibration.

(a) The exclusion rule

It should be quite easy to see intuitively (and can be confirmed group theoretically) that all three vibrational displacements shown in Fig. 7.21 lead to a change in the dipole moment of H_2O. It follows that all three modes of this C_{2v} molecule are IR active. It is much more difficult to judge intuitively whether a mode is Raman active because it is hard to know whether a particular distortion of a molecule results in a change of polarizability (although modes that result in a swelling of the molecule such as the symmetric stretch of CO_2 are good prospects). This difficulty is partly overcome by the **exclusion rule**, which is sometimes very helpful:

- If a molecule has a centre of inversion, then none of its modes can be both IR and Raman active.

(A mode may be inactive in both.)

Example 7.8 Using the exclusion rule

Identify the normal modes of CO_2 that are IR and Raman active.

Answer The symmetric stretch v_1 leaves the electric dipole moment unchanged at zero and so it is IR inactive: it may therefore be Raman active (and is). By contrast, for the antisymmetric stretch, v_3, the C atom moves opposite to that of the two O atoms: as a result, the electric dipole moment changes from zero in the course of the vibration and the mode is IR active. Because the CO_2 molecule has a centre of inversion, it follows from the exclusion rule that this mode cannot be Raman active. Both bending modes cause a departure of the dipole moment from zero and are therefore IR active. It follows from the exclusion rule that the two bending modes are Raman inactive.

Self-test 7.8 The bending mode of N_2O is active in the IR. Can it also be Raman active?

(b) Information from the symmetries of normal modes

So far, we have remarked that it is often intuitively obvious whether a vibrational mode gives rise to a changing electric dipole and is therefore IR active. When intuition is unreliable (perhaps because the molecule is complex or the mode of vibration is difficult to visualize) a symmetry analysis can be used instead. We shall illustrate the procedure by considering the two square-planar species (**16**) and (**17**). The Pt analogues of these species (and the distinction between them) are of considerable social and practical significance because the *cis* isomer is used as a chemotherapeutic agent against certain cancers (whereas the *trans* isomer is therapeutically inactive; Chapter 26).

First, we note that the *cis* isomer (**16**) has C_{2v} symmetry, whereas the *trans* isomer (**17**) is D_{2h}. Both species have bands in the Pd—Cl stretching region between 200 and 400 cm^{-1}. We know immediately from the exclusion rule that the two modes of the *trans* isomer (which has a centre of symmetry) cannot be active in both IR and Raman. However, to decide which modes are IR active and which are Raman active we consider the characters of the modes themselves. It follows from the symmetry properties of dipole moments and polarizabilities (which we do not verify here) that:

> The symmetry label of the vibration must be the same as that of x, y, or z in the character table for the vibration to be IR active and the same as that of a quadratic function, such as xy or x^2, for it to be Raman active.

Our first task, therefore, is to classify the normal modes according to their symmetry species, and then to identify which of these modes have the same symmetry species as x, etc. and xy, etc. by referring to the final columns of the character table of the molecular point group.

Figure 7.22 shows the symmetric and antisymmetric stretches of the Pd—Cl bonds for each isomer, where the NH_3 group is treated as a single mass point. To classify them according to their symmetry species in their respective point groups we use an approach similar to the symmetry analysis of molecular orbitals in terms of SALCs.

Consider the *cis* isomer and its point group C_{2v} (Table 7.4). For the symmetric stretch, we see that the pair of displacement vectors representing the vibration is apparently unchanged by each operation of the group. For example, the twofold rotation simply interchanges two equivalent displacement vectors. It follows that the character of each operation is 1:

E	C_2	σ_v	σ_v'
1	1	1	1

16 *cis*-[PdCl$_2$(NH$_3$)$_2$]

17 *trans*-[PdCl$_2$(NH$_3$)$_2$]

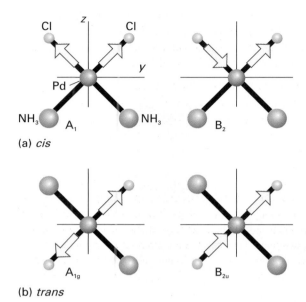

Fig. 7.22 Some of the Pd—Cl stretching modes of *cis* and *trans* forms of $[Pd(Cl)_2(NH_3)_2]$. The motion of the Pd atom (which preserves the centre of mass of the molecule) is not shown.

The symmetry of this vibration is therefore A_1. For the antisymmetric stretch, the identity E leaves the displacement vectors unchanged and the same is true of σ'_v, which lies in the plane containing the two Cl atoms. However, both C_2 and σ_v interchange the two oppositely directed displacement vectors, and so convert the overall displacement into -1 times itself:

E	C_2	σ_v	σ'_v
1	-1	-1	1

The C_{2v} character table identifies the symmetry type of this mode as B_2. A similar analysis of the *trans* isomer, but using the D_{2h} group, results in the labels A_g and B_{2u} for the symmetric and antisymmetric Pd—Cl stretches, respectively.

Example 7.9 Identifying the symmetry of vibrational displacements

The *trans* isomer in Fig. 7.22 has D_{2h} symmetry. Verify that the symmetry species of the antisymmetric stretch is B_{2u}.

Answer The elements of D_{2h} are E, $C_2(x)$, $C_2(y)$, $C_2(z)$, i, $\sigma(xy)$, $\sigma(yz)$, $\sigma(zx)$. Of these, E, $C_2(y)$, $\sigma(xy)$, and $\sigma(yz)$ leave the displacement vectors unchanged and so have characters 1. The remaining operations reverse the directions of the vectors, so giving characters of -1:

E	$C_2(x)$	$C_2(y)$	$C_2(z)$	i	$\sigma(xy)$	$\sigma(yz)$	$\sigma(zx)$
1	-1	1	-1	-1	1	1	-1

Comparison of this set of characters with the D_{2h} character table shows that the symmetry species is B_{2u}.

Self-test 7.9 Confirm that the symmetry species of the symmetric mode of the *trans* isomer is A_g.

As we have remarked, a vibrational mode is IR active if it has the same symmetry species as the displacements x, y, or z. In C_{2v}, z is A_1 and y is B_2. Both A_1 and B_2 vibrations of the *cis* isomer are therefore IR active. In D_{2h}, x, y, and z are B_{3u}, B_{2u}, and B_{1u}, respectively, and only these vibrations can be IR active. The antisymmetric Pd—Cl stretch of the *trans* isomer has symmetry B_{2u} and is IR active.

To determine the Raman activity, we note that in C_{2v} the quadratic forms xy, etc transform as A_1, A_2, B_1, and B_2 and therefore in the *cis* isomer the modes of symmetry A_1, A_2, B_1, and B_2 are Raman active. In D_{2h}, however, only A_g, B_{1g}, B_{2g}, and B_{3g} are Raman active.

The experimental distinction between the *cis* and *trans* isomers now emerges. In the Pd—Cl stretching region, the *cis* (C_{2v}) isomer has two bands in both the Raman and IR spectra. By contrast, the *trans* (D_{2h}) isomer has one band at a different frequency in each spectrum. The IR spectra of the two isomers are shown in Fig. 7.23.

(c) The assignment of molecular symmetry from vibrational spectra

An important application of vibrational spectra is the identification of molecular symmetry and hence shape and structure. An especially important example arises in metal carbonyls in which CO molecules are bound to a metal atom. Vibrational spectra are especially useful because the CO stretch is responsible for very strong characteristic absorptions between 1850 and 2200 cm^{-1}.

The first metal carbonyl to be characterized was the tetrahedral (T_d) molecule $Ni(CO)_4$. The vibrational modes of the molecule that arise from stretching motions of the CO groups are four combinations of the four CO displacement vectors. The problem of recognizing the appropriate linear combinations is similar to the problem of finding SALCs of atomic orbitals on the C atoms that are suitable for forming four σ bonds to the Ni atom. Reference to *Resource section* 4 shows that the four σ-donor orbitals transform as $A_1 + T_2$. The analogous linear combinations of the CO displacements are depicted in Fig. 7.24.

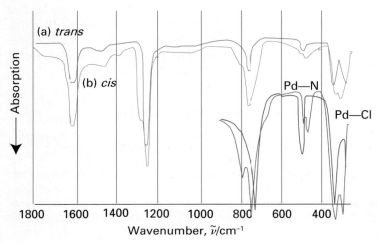

Fig. 7.23 The IR spectra of *cis* (red) and *trans* (blue) forms of $[Pd(Cl)_2(NH_3)_2]$. (R. Layton, D.W. Sink, and J.R. Durig, *J. Inorg. Nucl. Chem.*, 1966, **28**, 1965.)

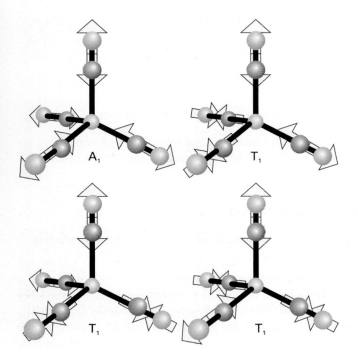

Fig. 7.24 The modes of $Ni(CO)_4$ that correspond to the stretching of CO bonds. The A_1 mode is nondegenerate whereas the T_2 modes are triply degenerate (they correspond to asymmetric displacements resolved along the three (*x, y,* or *z*) coordinates). (Compare the relative phases of the displacements with the relative phases of the SALCs of the same symmetry in *Resource section* 4.)

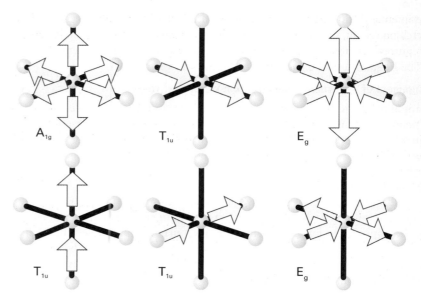

Fig. 7.25 The A_{1g} and T_{1u} M—L stretching modes of an octahedral ML_6 complex. The motion of the central metal atom M, which preserves the centre of mass of the molecule, is not shown (it is stationary in the A_{1g} mode).

At this stage we consult the character table for T_d. We see that the combination labelled A_1 transforms like $x^2 + y^2 + z^2$, indicating that it will be Raman active but not IR active. By contrast, x, y, and z and the products xy, yz, and zx transform as T_2, so the T_2 modes are both Raman and IR active. Consequently, a tetrahedral carbonyl molecule is recognized by one IR band and two Raman bands in the CO stretching region.

Example 7.10 Predicting the IR and Raman bands of an octahedral molecule

Consider an AB_6 molecule, such as SF_6, that belongs to the O_h point group. Sketch the normal modes of A—B stretches and comment on their activities in IR or Raman spectroscopy.

Answer We argue by analogy with the shapes of SALCs and identify the SALCs that can be constructed from s orbitals in an octahedral arrangement (*Resource section* 4). These orbitals are the analogues of the stretching displacements of the A—B bonds and the signs represent their relative phases. They have the symmetry species A_{1g}, E_g, and T_{1u}. The resulting linear combinations of stretches are illustrated in Fig 7.25. The A_{1g} (totally symmetric) and E_g modes are Raman active and the T_{1u} mode is IR active.

Self-test 7.10 Consider only bands due to stretching vibrations and predict how the IR and Raman spectra of SF_5Cl differ from those of SF_6.

Representations

We now move on to a more quantitative treatment and introduce three topics that are important for applying symmetry arguments in the treatment of molecular orbitals and spectroscopy in a systematic manner.

7.9 The reduction of a representation

Key point: A reducible representation can be resolved into its constituent irreducible representations by using the reduction formula.

We have seen that the three $H1s$ orbitals of NH_3 give rise to—the technical term is 'span'—two irreducible representations in C_{3v}, one of symmetry species A_1 and the other of symmetry species E. Similarly, we have seen that the stretching modes of SF_6 span A_{1g}, E_g, and T_{1u} in the group O_h. Here we present a systematic way for arriving at the identification of the symmetry species spanned by a set of orbitals or atom displacements.

The fact the three H1s orbitals of NH_3 span two particular irreducible representations is expressed formally by writing $\Gamma = A_1 + E$ where Γ stands for the reducible representation. In general, we write

$$\Gamma = c_1\Gamma_1 + c_2\Gamma_2 + \cdots \tag{7.1a}$$

where the Γ_i denote the various symmetry species of the group and the c_i tell us how many times each symmetry species appears in the reduction. A very deep theorem from group theory (see *Further reading*) provides an explicit formula for calculating the coefficients c_i in terms of the characters χ_i of the irreducible representation Γ_i and the corresponding characters χ of the original reducible representation Γ:

$$c_i = \frac{1}{h}\sum_R \chi_i(R)\chi(R) \tag{7.1b}$$

Here h is the order of the point group (the number of symmetry elements; that is given in the top row of the character table) and the sum is over each operation R of the group. How this expression is used is illustrated by the following example.

Example 7.11 Using the reduction formula

Consider the molecule *cis*-[$PdCl_2(NH_3)_2$], which, if we ignore the hydrogen atoms, belongs to the point group C_{2v}. What are the symmetry species spanned by the displacements of the atoms?

Answer To analyse this problem we consider the 15 displacements of the five non-hydrogen atoms (Fig. 7.26) and obtain the characters of what will turn out to be a reducible representation Γ by examining what happens when we apply the symmetry operations of the group. Then we use eqn 7.1b to identify the symmetry species of the irreducible representations into which that reducible representation can be reduced. To identify the characters of Γ we note that (based on the material set out in *Further information* 7.1) each displacement that moves to a new location under a particular symmetry operation contributes zero to the character of that operation; those that remain the same contribute 1; those that are reversed contribute -1. Thus, because all 15 displacements remain unmoved under the identity, $\chi(E) = 15$. A C_2 rotation leaves only the z displacement on Pd unchanged (contributing 1) but reverses the x and y displacements on Pd (contributing -2), so $\chi(C_2) = -1$. Under the reflection σ_v, the z and x displacements on Pd are unchanged (contributing 2) and the y displacement on Pd is reversed (contributing -1), so $\chi(\sigma_v) = 1$. Finally, for any reflection in a vertical plane passing through the plane of the atoms, the five z displacements on these atoms remain the same (contributing 5), so do the five y displacements (another 5), but the five x displacements are reversed (contributing -5); therefore $\chi(\sigma_v') = 5$. The characters of Γ are therefore

E	C_2	σ_v	σ_v'
15	-1	1	5

Now we use eqn 7.1b, noting that $h = 4$ for this group. To find how many times the symmetry species A_1 appears in the reducible representation, we write

$$c_1 = \tfrac{1}{4}\{1 \times 15 + 1 \times (-1) + 1 \times 1 + 1 \times 5\} = 5$$

By repeating this procedure for the other species, we find

$$\Gamma = 5A_1 + 2A_2 + 3B_1 + 5B_2$$

For C_{2v}, the translations of the entire molecule span $A_1 + B_1 + B_2$ (as given by the functions x, y, and z in the final column) and the rotations span $A_2 + B_1 + B_2$. By subtracting these symmetry species from the ones we have just found, we can conclude that the vibrations of the molecule span $4A_1 + A_2 + B_1 + 3B_2$.

Self-test 7.11 Determine the symmetries of all the vibration modes of $[PdCl_4]^{2-}$, a D_{4h} molecule.

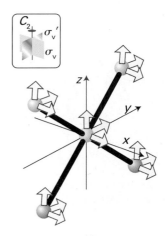

Fig. 7.26 The atomic displacements in *cis*-[$PdCl_2(NH_3)_3$] with the H atoms ignored.

Many of the modes of *cis*-[$PdCl_2(NH_3)_2$] found in Example 7.11 are complex motions that are not easy to visualize: they include Pd—N stretches and various buckling motions of the plane. However, even without being able to visualize them easily, we can infer at once that the A_1, B_1, and B_2 modes (because the functions x, y, and z, and hence the components of the electric dipole, span these symmetry species) are IR active and all the modes are Raman active (because the quadratic forms span all four species).

It is sometimes helpful to be able to derive the symmetry species of a function by regarding the function as the product of two other functions that span known irreducible representations. For instance, we can infer the symmetry species of xy from a knowledge of the symmetry species of x and y. A **direct product** is the (in general reducible) representation spanned by the product of two or more functions. To obtain the symmetry species spanned by the product of functions we simply multiply corresponding characters and, if necessary, use eqn 7.1b to reduce it to its component symmetry species. For instance, suppose one function spans A_2 in the group C_{2v} and another spans B_2. We draw up the following table

	E	C_2	σ_v	σ_v'
A_2	1	1	-1	-1
B_2	1	-1	-1	1
$A_2 \times B_2$	1	-1	1	-1

Here the direct product is irreducible $A_2 \times B_2 = B_1$, but that is not always the case. For example, in C_{3v}, $E \times E = A_1 + A_2 + E$. In Chapter 19 we shall use direct products to infer spectroscopic selection rules.

7.10 Projection operators

Key point: A projection operator is used to generate SALCs from a basis of orbitals.

To generate an unnormalized SALC of a particular symmetry species from an arbitrary set of basis atomic orbitals, we select any one of the set and form the following sum:

$$\phi = \sum_R \chi_i(R) R\psi \tag{7.2}$$

where $\chi_i(R)$ is the character of the operation R for the symmetry species of the SALC we want to generate. Once again, the best way to illustrate the use of this expression is with an example.

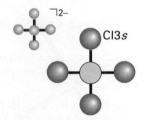

Fig. 7.27 The Cl orbital basis used to construct SALCs in $[PtCl_4]^{2-}$.

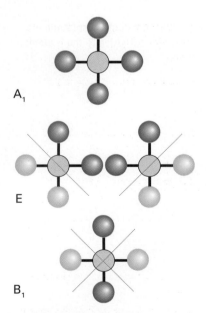

Fig. 7.28 The SALCs constructed from the basis in Fig. 7.27.

Example 7.12 Generating a SALC

Generate the SALC of Clσ orbitals for $[PtCl_4]^{2-}$. The basis orbitals are denoted ψ_1, ψ_2, ψ_3, and ψ_4 and are shown in Fig. 7.27.

Answer We choose one of the basis orbitals and subject it to all the symmetry operations of the D_{4h} point group, writing down the basis function $R\psi$ into which it is transformed. For example, the operation C_4 moves ψ_1 into the position occupied by ψ_2, C_2 moves it to ψ_3 and C_4^3 moves it to ψ_4. Continuing for all operations we obtain

Operation R:	E	C_4^1	C_4^3	C_2	C_2'	C_2'	C_2''	C_2''	i	S_4^1	S_4^3	σ_h	σ_v	σ_v	σ_d	σ_d
$R\psi_1$	ψ_1	ψ_2	ψ_4	ψ_3	ψ_1	ψ_3	ψ_2	ψ_4	ψ_3	ψ_2	ψ_4	ψ_1	ψ_1	ψ_3	ψ_2	ψ_4

We now add together all the new basis functions, and for each class of operation we multiply by the character $\chi_i(R)$ for the irreducible representation we are interested in. Thus, for A_{1g} (as all characters are 1) we obtain $4\psi_1 + 4\psi_2 + 4\psi_3 + 4\psi_4$. The (unnormalized) SALC is therefore

$$\phi(A_{1g}) = \psi_1 + \psi_2 + \psi_3 + \psi_4$$

As we continue down the character table using the various irreducible representations, the SALCs emerge as follows:

$$\phi(B_{1g}) = \psi_1 - \psi_2 + \psi_3 - \psi_4$$
$$\phi(E_u) = \psi_1 - \psi_3$$

Under all other irreducible representations the projection operators vanish (because no SALCs exist of those symmetries). Then choosing instead ψ_2 as our basis function, we obtain the same SALCs except for $\phi(E_u) = \psi_2 - \psi_4$. The forms of the SALCs are therefore $A_{1g} + B_{1g} + E_u$ (Fig. 7.28).

Self-test 7.12 Use projection operators in SF_6 to determine the SALCs for σ-bonding in an octahedral complex. (Use the O point group rather than O_h.)

FURTHER READING

P. Atkins and J. de Paula, *Physical chemistry*. Oxford University Press and W.H. Freeman & Co (2006). An account of the generation and use of character tables without too much mathematical background.

For more rigorous introductions, see.

J.S. Ogden, *Introduction to molecular symmetry*. Oxford University Press (2001).

P. Atkins and R. Friedman, *Molecular quantum mechanics*. Oxford University Press (2005).

F.A. Cotton, *Group theory for chemists*. Wiley, New York (1990).

EXERCISES

7.1 Draw sketches to identify the following symmetry elements: (a) a C_3 axis and a σ_v plane in the NH_3 molecule; (b) a C_4 axis and a σ_h plane in the square-planar $[PtCl_4]^{2-}$ ion.

7.2 Which of the following molecules and ions has (1) a centre of inversion, (2) an S_4 axis: (a) CO_2, (b) C_2H_2, (c) BF_3, (d) SO_4^{2-}?

7.3 Determine the symmetry elements and assign the point group of (a) NH_2Cl, (b) CO_3^{2-}, (c) SiF_4, (d) HCN, (e) $SiFClBrI$, (f) BF_4^-.

7.4 Determine the symmetry elements of objects with the same shape as the boundary surface of (a) an *s* orbital, (b) a *p* orbital, (c) a d_{xy} orbital, (d) a d_{z^2} orbital.

7.5 (a) Determine the symmetry group of an SO_3^{2-} ion. (b) What is the maximum degeneracy of a molecular orbital in this ion? (c) If the sulfur orbitals are 3*s* and 3*p*, which of them can contribute to molecular orbitals of this maximum degeneracy?

7.6 (a) Determine the point group of the PF_5 molecule. (Use VSEPR, if necessary, to assign geometry.) (b) What is the maximum degeneracy of its molecular orbitals? (c) Which phosphorus 3*p* orbitals contribute to a molecular orbital of this degeneracy?

7.7 Determine the point groups for SnF_4, SeF_4, and BrF_4^-.

7.8 Reaction of $AsCl_3$ with Cl_2 at low temperature yields a product, believed to be $AsCl_5$, which shows Raman bands at 437, 369, 295, 220, 213, and 83 cm^{-1}. Detailed analysis of the 369 and

295 cm^{-1} bands show these to arise from totally symmetric modes. Show that the Raman spectrum is consistent with a trigonal-bipyamidal geometry.

7.9 How many vibrational modes does an SO_3 molecule have (a) in the plane of the nuclei, (b) perpendicular to the molecular plane?

7.10 What are the symmetry species of the vibrations of (a) SF_6, (b) BF_3 that are both IR and Raman active?

7.11 What are the symmetry species of the vibrational modes of a C_{6v} molecule that are neither IR nor Raman active?

7.12 The $[AuCl_4]^-$ ion has D_{4h} symmetry. Determine the representation Γ of all $3N$ displacements and reduce it to obtain the irreducible representations.

7.13 Determine the following direct products: (a) $A_2 \times B_2$ in C_{2v}, (b) $E \times T_1$ in T_d, (c) $E_g \times E_u$ in D_{4h}.

7.14 Determine the normal modes of vibration for the PCl_4^+ ion and identify the modes that are IR or Raman active.

7.15 How could IR and Raman spectroscopy be used to distinguish between: (a) planar and pyramidal forms of PF_3, (b) planar and 90°-twisted forms of B_2F_4 (D_{2h} and D_{2d} respectively).

7.16 Use the projection operator method to determine the SALCs required for formation of σ-bonds in (a) BF_3, (b) PF_5, (c) BrF_5.

PROBLEMS

7.1 Consider a molecule IF_3O_2 (with I as the central atom). How many isomers are possible? Which is likely to be most stable? Assign point group designations to each isomer.

7.2 Group theory is often used by chemists as an aid in the interpretation of infrared spectra. For example, there are four N—H bonds in NH_4^+ and four stretching modes are possible: each one is a linear combination of the four stretching modes, and each one has a characteristic symmetry. There is the possibility that several vibrational modes occur at the same frequency, and hence are degenerate. A quick glance at the character table will tell if degeneracy is possible. (a) In the case of the tetrahedral NH_4^+ ion, is it necessary to consider the possibility of degeneracies? (b) Are degeneracies possible in any of the vibrational frequencies of $NH_2D_2^+$?

7.3 Figure 7.29 shows the energy levels for CH_3^+. What is the point group used for this illustration? What H1*s* orbital linear combination participates in a_1'? What C orbitals contribute to a_1'? What H linear combinations participate in the bonding e' pair? Which C orbitals are e'? Which C orbitals are a_2''? Is any H linear combination a_2''? Now add two H1*s* orbitals on the *z* axis (above and below the plane), modify the linear combinations of each symmetry type accordingly, and construct a new a_2'' linear combination. Are there bonding and nonbonding (or only

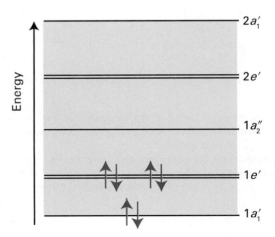

Fig. 7.29

weakly antibonding orbitals) that can accommodate ten electrons and allow carbon to become hypervalent?

7.4 Construct a Walsh diagram correlating the orbitals of a square-planar (D_{4h}) H_4 molecule with those of a linear ($D_{\infty h}$) H_4 molecule.

7.5 Construct a Walsh diagram correlating the orbitals of a trigonal-planar (D_{3h}) XH_3 molecule with those of a trigonal-pyramidal (C_{3v}) XH_3 molecule.

7.6 (a) Determine the point group of the most symmetric planar conformation of $B(OH)_3$ and the most symmetric nonplanar conformation of $B(OH)_3$. Assume that the B—O—H bond angles are all 109.5° in all conformations. (b) Sketch a conformation of $B(OH)_3$ that is chiral, once again keeping all three of the B—O—H bond angles equal to 109.5°.

7.7 How many isomers are there for 'octahedral' molecules with the formula MA_3B_3, where A and B are monoatomic ligands? What is the point group of each isomer? Are any of the isomers chiral? Repeat this exercise for molecules with the formula $MA_2B_2C_2$.

7.8 Determine whether the number of IR and Raman active stretching vibrations could be used to determine uniquely whether a sample of gas is BF_3, NF_3, ClF_3.

7.9 How many SALCs of 1s atomic orbitals will a square-planar array of four H atoms produce? Consider that the four H atoms lie directly on the x- and y-axes of a Cartesian coordinate system. Draw the SALC with no nodal planes and determine its symmetry type. Draw the SALC with two nodal planes and determine its symmetry type.

7.10 How many planes of symmetry does a benzene molecule possess? What chloro-substituted benzene of formula $C_6H_nCl_{6-n}$ has exactly four total planes of symmetry?

FURTHER INFORMATION 7.1 Matrix representations

The origin of character tables is the representation of the effects of symmetry operations by matrices. As an illustration, consider the C_{2v}

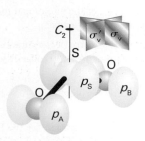

Fig. 7.30 The basis used to construct the representation in this discussion of SO_2.

molecule SO_2 and the valence p_x orbitals on each atom (Fig. 7.30), which we shall denote p_S, p_A, and p_B. Under σ_v, the change $(p_S, p_A, p_B) \rightarrow (p_S, p_B, p_A)$ takes place. We can express this transformation under a reflection by using matrix multiplication:

$$(p_S, p_A, p_B) \begin{pmatrix} 1 & 0 & 0 \\ 0 & 0 & 1 \\ 0 & 1 & 0 \end{pmatrix} = (p_S, p_B, p_A)$$

This relation can be expressed more succinctly as

$$(p_S, p_A, p_B)D(\sigma_v) = (p_S, p_B, p_A) \quad \text{where } D(\sigma_v) = \begin{pmatrix} 1 & 0 & 0 \\ 0 & 0 & 1 \\ 0 & 1 & 0 \end{pmatrix}$$

The matrix $D(\sigma_v)$ is called a **representative** of the operation σ_v. Representatives take different forms according to the basis (the set of orbitals) that has been adopted.

We can use the same technique to find matrices that reproduce the other symmetry operations. For instance, C_2 has the effect $(p_S, p_A, p_B) \rightarrow (-p_S, -p_B, -p_A)$ and its representative is therefore

$$D(C_2) = \begin{pmatrix} -1 & 0 & 0 \\ 0 & 0 & -1 \\ 0 & -1 & 0 \end{pmatrix}$$

The effect of σ_v' is to change $(p_S, p_A, p_B) \rightarrow (-p_S, -p_A, -p_B)$ and its representative is

$$D(\sigma_v') = \begin{pmatrix} -1 & 0 & 0 \\ 0 & -1 & 0 \\ 0 & 0 & -1 \end{pmatrix}$$

The identity operation has no effect on the basis and so its representative is the unit matrix:

$$D(E) = \begin{pmatrix} 1 & 0 & 0 \\ 0 & 1 & 0 \\ 0 & 0 & 1 \end{pmatrix}$$

The set of matrices that represents all the operations of the group is called a **matrix representation** of the group for the particular basis we have chosen. We denote this 'three-dimensional' representation (that is, one consisting of 3×3 matrices) by the symbol $\Gamma^{(3)}$. The discovery of a matrix representation of the group means that we have found a link between the symbolic manipulations of the operations and algebraic manipulations involving numbers. It may readily be verified that the matrices, when multiplied together, reproduce the group multiplication table.

The **character**, χ (chi), of an operation in a particular matrix representation is the sum of the diagonal elements of the representative of each representative. Thus, in the basis we are illustrating, the characters of the representatives are

$D(E)$	$D(C_2)$	$D(\sigma_v)$	$D(\sigma_v')$
3	−1	1	−3

The character of an operation depends on the basis.

The representatives in the basis we have chosen are three-dimensional, but inspection shows that they are all of the form

$$\begin{pmatrix} \blacksquare & 0 & 0 \\ 0 & & \\ 0 & & \blacksquare \end{pmatrix}$$

and that the symmetry operations never mix p_S with the other two functions. This suggests that the basis can be cut into two parts, one consisting of p_S alone and the other of (p_A, p_B). It is readily verified that the p_S orbital itself is a basis for the one-dimensional representation

$$D(E) = 1 \quad D(C_2) = -1 \quad D(\sigma_v) = 1 \quad D(\sigma_v') = -1$$

which we shall call $\Gamma^{(1)}$. The characters of this representation are simply

$D(E)$	$D(C_2)$	$D(\sigma_v)$	$D(\sigma_v')$
1	−1	1	−1

The remaining two basis functions are a basis for the two-dimensional representation $\Gamma^{(2)}$:

$$D(E) = \begin{pmatrix} 1 & 0 \\ 0 & 1 \end{pmatrix} \quad D(C_2) = \begin{pmatrix} 0 & -1 \\ -1 & 0 \end{pmatrix}$$

$$D(\sigma_v) = \begin{pmatrix} 0 & 1 \\ 1 & 0 \end{pmatrix} \quad D(\sigma_v') = \begin{pmatrix} -1 & 0 \\ 0 & -1 \end{pmatrix}$$

These matrices are the same as those of the original three-dimensional representation, except for the loss of the first row and column. We say that the original three-dimensional representation has been **reduced** to the **direct sum** of a one-dimensional representation spanned by p_S and a two-dimensional representation spanned by (p_A, p_B). This reduction is consistent with the common sense view that the central orbital plays a role different from the other two. The reduction is denoted symbolically by $\Gamma^{(3)} = \Gamma^{(1)} + \Gamma^{(2)}$.

The one-dimensional representation cannot be reduced any further, and is called an **irreducible representation** of the group. We can demonstrate that the two-dimensional representation is reducible (for this basis) by switching attention to the linear combinations $p_1 = p_A + p_B$ and $p_2 = p_A - p_B$.

The representatives in the new basis can be constructed from the old. For example, because under σ_v, $(p_A, p_B) \rightarrow (p_B, p_A)$, it follows (by applying these transformations to the linear combinations) that $(p_1, p_2) \rightarrow (p_1, -p_2)$. The same transformation is achieved by writing

$$(p_1, p_2)D(\sigma_v) \rightarrow (p_1, -p_2) \text{ with } D(\sigma_v) = \begin{pmatrix} 1 & 0 \\ 0 & -1 \end{pmatrix}$$

which gives us the representative $D(\sigma_v)$ in the new basis. The remaining three representatives may be found similarly, and the complete representation is

$$D(E) = \begin{pmatrix} 1 & 0 \\ 0 & 1 \end{pmatrix} \quad D(C_2) = \begin{pmatrix} -1 & 0 \\ 0 & -1 \end{pmatrix}$$

$$D(\sigma_v) = \begin{pmatrix} 1 & 0 \\ 0 & -1 \end{pmatrix} \quad D(\sigma_v') = \begin{pmatrix} -1 & 0 \\ 0 & -1 \end{pmatrix}$$

The new representatives are all in block diagonal form

$$D(R) = \begin{pmatrix} \blacksquare & 0 \\ 0 & \blacksquare \end{pmatrix}$$

and the two combinations are not mixed with one another by any operation of the group. We have therefore achieved the reduction of $\Gamma^{(2)}$ to the direct sum of two one-dimensional representations. Thus, p_1 spans

$$D(E) = 1 \quad D(C_2) = -1 \quad D(\sigma_v) = 1 \quad D(\sigma_v') = -1$$

which is the same one-dimensional representation as that spanned by p_S, and p_2 spans

$$D(E) = 1 \quad D(C_2) = 1 \quad D(\sigma_v) = -1 \quad D(\sigma_v') = -1$$

which is a different one-dimensional representation; we shall denote it $\Gamma^{(1)'}$. It is easy to check that either set of 1×1 matrices is a representation by multiplying pairs together and seeing that they reproduce the original group multiplication table.

Now we can make the final link to the material in the chapter. The character table of a group is the list of the characters of all its irreducible representations. At this point we have found two irreducible representations of the group C_{2v}. Their characters are

	E	C_2	σ_v	σ_v'
$\Gamma^{(1)}$	1	-1	1	-1
$\Gamma^{(1)'}$	1	1	-1	-1

The two irreducible representations are normally labelled B_1 and A_2, respectively. An A or a B is used to denote a one-dimensional representation; A is used if the character under the principal rotation is $+1$ and B is used if the character is -1. (The letter E denotes a two-dimensional irreducible representation and a T a three-dimensional irreducible representation: all the irreducible representations of C_{2v} are one-dimensional.) There are in fact only two more species of irreducible

representations of this group, for it is a surprising theorem of group theory that

Number of symmetry species = number of classes

In C_{2v} there are four classes (four columns in the character table), and so there are only four species of irreducible representation. The full character table for this group is shown in Table 7.4.

The most important applications in chemistry of group theory, the construction of symmetry-adapted orbitals and the analysis of selection rules, are based on the **little orthogonality theorem**:

$$\sum_C g(C)\chi^{(\Gamma)}(C)\chi^{(\Gamma')}(C) = 0$$

The sum is over the classes C of the operations of the group (the columns in the character table), $g(C)$ is the number of operations in each class (such as the 2 in $2C_3$ in the C_{3v} character table), and Γ and Γ' are two *different* irreducible representations. If Γ and Γ' are the same irreducible representations, then

$$\sum_C g(C)\chi^{(\Gamma)}(C)\chi^{(\Gamma)}(C) = h$$

where h is the order of the group (the number of elements).

To find out whether a reducible representation contains a given irreducible representation, we use an expression derived from the little orthogonality theorem. Thus, for a given operation, the character of a reducible representation is a linear combination of the characters of the irreducible representations of the group:

$$\chi(C) = \sum_\Gamma c_\Gamma \chi^{(\Gamma)}(C)$$

To find the coefficients for a given irreducible representation Γ', we multiply both sides by $g(C)\chi^{(\Gamma')}$ and sum over all the classes of operation:

$$\sum_C g(C)\chi^{(\Gamma')}(C)\chi(C) = \sum_C \sum_\Gamma c_\Gamma g(C)\chi^{(\Gamma')}(C)\chi^{(\Gamma)}(C)$$

When the right-hand side is summed over C, the little orthogonality theorem gives 0 for all terms for which Γ is not the same as Γ'. However, because we are summing over all Γ, a term with $\Gamma = \Gamma'$ is guaranteed to be present. Only that term contributes, and the right-hand side is then equal to $hc_{\Gamma'}$. It then follows that

$$c_\Gamma = \frac{1}{h}\sum_C g(C)\chi^{(\Gamma)}(C)\chi(C)$$

This formula is exceptionally important for finding the decomposition of a reducible representation, because Γ can be set equal to each irreducible representation in turn, and the coefficients c determined. It takes an even easier form if we only want to know if the totally symmetric irreducible representation (which for convenience we denote A_1) is present, because all the characters of that representation are 1, so

$$c_{A_1} = \frac{1}{h}\sum_C g(C)\chi(C)$$

To form an orbital of symmetry species Γ we form $P\psi$, where

$$P\psi = \sum_R \chi^{(\Gamma)}(R)R\psi$$

and R is an operation of the group. Note that the actual operations occur in the formula, not the classes as in the earlier expressions. The quantity P is called a **projection operator**. As an example of its form, to project out a B_1 symmetry-adapted linear combination in the group C_{2v} we would use

$$P = \chi^{(B_1)}(E)E + \chi^{(B_1)}(C_2)C_2 + \chi^{(B_1)}(\sigma_v)\sigma_v + \chi^{(B_1)}(\sigma_v')\sigma_v'$$
$$= E - C_2 + \sigma_v - \sigma_v'$$

An introduction to coordination compounds

Metal complexes, in which a single central metal atom or ion is surrounded by several ligands, play an important role in inorganic chemistry, especially for elements of the *d* block. In this chapter, we introduce the common structural arrangements for ligands around a central metal atom and the isomeric forms that are possible.

In the context of metal coordination chemistry, the term **complex** means a central metal atom or ion surrounded by a set of ligands. A **ligand** is an ion or molecule that can have an independent existence. An example of a complex is $[Co(NH_3)_6]^{3+}$, in which the Co^{3+} ion is surrounded by six NH_3 ligands. We shall use the term **coordination compound** to mean a neutral complex or an ionic compound in which at least one of the ions is a complex. Thus, $Ni(CO)_4$ (**1**) and $[Co(NH_3)_6]Cl_3$ (**2**) are both coordination compounds. A complex is a combination of a Lewis acid (the central metal atom) with a number of Lewis bases (the ligands). The atom in the Lewis base ligand that forms the bond to the central atom is called the **donor atom**, because it donates the electrons used in bond formation. Thus, O is the donor atom when H_2O acts as a ligand. The metal atom or ion, the Lewis acid in the complex, is the **acceptor atom**. All metals, from all blocks of the periodic table, form complexes.

The principal features of the geometrical structures of metal complexes were identified by Alfred Werner, whose training was in organic stereochemistry. Werner combined the interpretation of optical and geometrical isomerism, patterns of reactions, and conductance data in work that remains a model of how to use physical and chemical evidence effectively and imaginatively. The striking colours of many *d*- and *f*-metal coordination compounds, which reflect their electronic structures, were a mystery to Werner. This characteristic was clarified only when the description of electronic structure in terms of orbitals was applied to the problem in the period from 1930 to 1960. We look at the electronic structure of *d*-metal complexes in Chapter 19 and *f*-metal complexes in Chapter 22.

The geometrical structures of metal complexes can now be determined in many more ways than Werner had at his disposal. When single crystals of a compound can be grown, X-ray diffraction (Section 6.1) gives precise shapes, bond distances, and angles. Nuclear magnetic resonance (Section 6.5) can be used to study complexes with lifetimes longer than microseconds. Very short-lived complexes, those with lifetimes comparable to diffusional encounters in solution (a few nanoseconds), can be studied by vibrational and electronic spectroscopy. It is possible to infer the geometries of complexes with long lifetimes in solution (such as the classic complexes of Co(III), Cr(III), and Pt(II) and many organometallic compounds) by analysing patterns of reactions and isomerism. This method was originally exploited by Werner, and it still teaches us much about the synthetic chemistry of the compounds as well as helping to establish their structures.

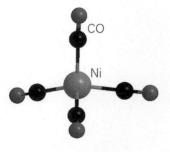

1 Ni(CO)$_4$

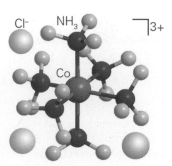

2 [Co(NH$_3$)$_6$]Cl$_3$

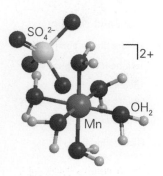

3 [Mn(OH$_2$)$_6$]SO$_4$

Constitution and geometry

Key points: In an inner-sphere complex, the ligands are attached directly to a central metal ion: the number of ligands depends on the size of the metal, the identity of the ligands, and the electronic interactions.

In what we normally understand as a complex, more precisely an **inner-sphere complex**, the ligands are attached directly to the central metal atom or ion. These ligands form the **primary coordination sphere** of the complex and their number is called the **coordination number** of the central metal atom. As in solids, a wide range of coordination numbers can occur, and the origin of the structural richness and chemical diversity of complexes is the ability of the coordination number to range up to 12.

Although we shall concentrate on inner-sphere complexes throughout this chapter, we should keep in mind that complex cations can associate electrostatically with anionic ligands (and, by other weak interactions, with solvent molecules) without displacement of the ligands already present. The product of this association is called an **outer-sphere complex**. With [Mn(OH$_2$)$_6$]$^{2+}$ and SO$_4^{2-}$ ions, for instance, the equilibrium concentration of the outer-sphere complex {[Mn(OH$_2$)$_6$]$^{2+}$SO$_4^{2-}$} (**3**) can, depending on the concentration, exceed that of the inner-sphere complex [Mn(OH$_2$)$_5$SO$_4$] in which the SO$_4^{2-}$ ligand is directly attached to the metal ion. It is worth remembering that most methods of measuring complex formation equilibria do not distinguish outer-sphere from inner-sphere complex formation but simply detect the sum of all bound ligands. Outer-sphere complexation should be suspected whenever the metal and ligands have opposite charges.

The coordination number of a metal atom or ion is not always evident from the composition of the solid, as solvent molecules and species that are potentially ligands may simply fill spaces within the structure and not have any direct bonds to the metal ion. For example, X-ray diffraction shows that CoCl$_2 \cdot 6$H$_2$O contains the neutral complex [Co(Cl)$_2$(OH$_2$)$_4$] and two uncoordinated H$_2$O molecules occupying well-defined positions in the crystal. Such additional solvent molecules are called **solvent of crystallization**.

Three factors govern the coordination number of a complex:

1 The size of the central atom or ion.

2 The steric interactions between the ligands.

3 Electronic interactions between the central atom or ion and the ligands.

In general, the large radii of atoms and ions lower down the periodic table favour higher coordination numbers. For similar steric reasons, bulky ligands often result in low coordination numbers, especially if the ligands are also charged (when unfavourable electrostatic interactions also come into play). High coordination numbers are also most common on the left of a period, where the ions have larger radii. They are especially common when the metal ion has only a few electrons because a small number of valence electrons means that the metal ion can accept more electrons from Lewis bases; one example is [Mo(CN)$_8$]$^{4-}$. Lower coordination numbers are found on the right of the *d* block, particularly if the ions are rich in electrons; an example is [PtCl$_4$]$^{2-}$. Such atoms are less able to accept electrons from any Lewis bases that are potential ligands. Low coordination numbers occur if the ligands can form multiple bonds with the central metal, as in MnO$_4^-$ and CrO$_4^{2-}$, as now the electrons provided by each ligand tend to exclude the attachment of more ligands. In Chapter 19 we consider these coordination number preferences in more detail.

8.1 Low coordination numbers

Key points: Two-coordinate complexes are found for Cu$^+$ and Ag$^+$; these complexes readily accommodate more ligands if they are available. Complexes may have coordination numbers higher than their empirical formulas suggest.

The best known complexes of metals with coordination number 2 that are formed in solution under ordinary laboratory conditions are linear species of the Group 11 ions. Linear two-coordinate complexes with two identical symmetric ligands have $D_{\infty h}$ symmetry. The complex $[AgCl_2]^-$, which is responsible for the dissolution of solid silver chloride in aqueous solutions containing excess Cl^- ions, is one example, dimethyl mercury, Me—Hg—Me, is another. A series of linear Au(I) complexes of formula LAuX, where X is a halogen and L is a neutral Lewis base such as a substituted phosphine, R_3P, or thioether, R_2S, are also known. Two-coordinate complexes often gain additional ligands to form three- or four-coordinate complexes.

A formula that suggests a certain coordination number in a solid compound might conceal a polymeric chain with a higher coordination number. For example, CuCN appears to have coordination number 1, but it in fact exists as linear —Cu—CN—Cu—CN— chains in which the coordination number of copper is 2.

Three-coordination is rare among metal complexes, but is found with bulky ligands such as tricyclohexylphosphine, as in $[Pt(PCy_3)_3]$ (**4**), where cy denotes cyclohexyl (—C_6H_{11}), with its trigonal arrangement of the ligands. MX_3 compounds, where X is a halogen, are usually chains or networks with a higher coordination number and shared ligands. Three-coordinate complexes with three identical symmetric ligands normally have D_{3h} symmetry.

8.2 Intermediate coordination numbers

Complexes of metal ions with the intermediate coordination numbers four, five, and six are the most important class of complex. They include the vast majority of complexes that exist in solution, and almost all the biologically important complexes.

(a) Four-coordination

Key points: Tetrahedral complexes are favoured over higher coordinate complexes if the central atom is small or the ligands large; square-planar complexes are typically observed for metals with d^8 configurations.

Four-coordination is found in an enormous number of compounds. Tetrahedral complexes of approximately T_d symmetry (**5**) are favoured over higher coordination numbers when the central atom is small and the ligands are large (such as Cl^-, Br^-, and I^-), because then ligand–ligand repulsions override the energy advantage of forming more metal–ligand bonds. Four-coordinate s- and p-block complexes with no lone pair on the central atom, such as $[BeCl_4]^{2-}$, $[BF_4]^-$, and $[SnCl_4]$, are almost always tetrahedral, and tetrahedral complexes are common for oxoanions of metal atoms on the left of the d block in high oxidation states, such as $[MoO_4]^{2-}$. Examples of tetrahedral complexes from Groups 5 through 11 are: $[VO_4]^{3-}$, $[CrO_4]^{2-}$, $[MnO_4]^-$, $[FeCl_4]^{2-}$, $[CoCl_4]^{2-}$, $[NiBr_4]^{2-}$, $[CuBr_4]^{2-}$.

Werner studied a series of four-coordinate Pt(II) complexes formed by the reactions of $PtCl_2$ with NH_3 and HCl. For a complex of formula MX_2L_2, only one isomer is expected if the species is tetrahedral, but two isomers are expected if the species is square planar, (**6**) and (**7**). Because Werner was able to isolate two nonelectrolytes of formula $[Pt(NH_3)_2Cl_2]$, he concluded that they could not be tetrahedral and were, in fact, square planar (**8**). The complex with like ligands on adjacent corners of the square is called a *cis* isomer (**6**, point group C_{2v}) and the complex with like ligands opposite is the *trans* isomer (**7**, D_{2h}). The existence of different spatial arrangements of the same ligands is called **geometric isomerism**. Geometric isomerism is far from being of only academic interest: platinum complexes are used in cancer chemotherapy, and it is found that only *cis*-Pt(II) complexes can bind to the bases of DNA for long enough to be effective. Square-planar complexes with four identical symmetric ligands have D_{4h} symmetry.

Square-planar complexes are rarely found for s- and p-block complexes but are abundant for d^8 complexes of the elements belonging to the 4d- and 5d-series metals such as Rh^+, Ir^+, Pd^{2+}, Pt^{2+}, and Au^{3+}, which are almost invariably square planar. For 3d-metals with d^8 configurations (for example, Ni^{2+}), square-planar geometry is favoured by ligands that can form π bonds by accepting electrons from the metal atom,

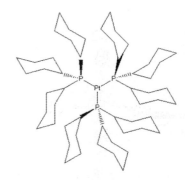

4 [Pt(Pcy$_3$)], cy=*cyclo*-C$_6$H$_{11}$

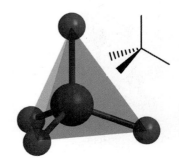

5 Tetrahedral complex, T_d

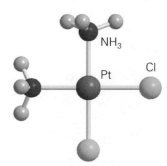

6 *cis*-[Pt(Cl)$_2$(NH$_3$)$_2$]

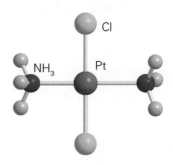

7 *trans*-[Pt(Cl)$_2$(NH$_3$)$_2$]

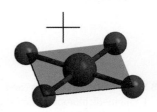

8 Square-planar complex, D_{4h}

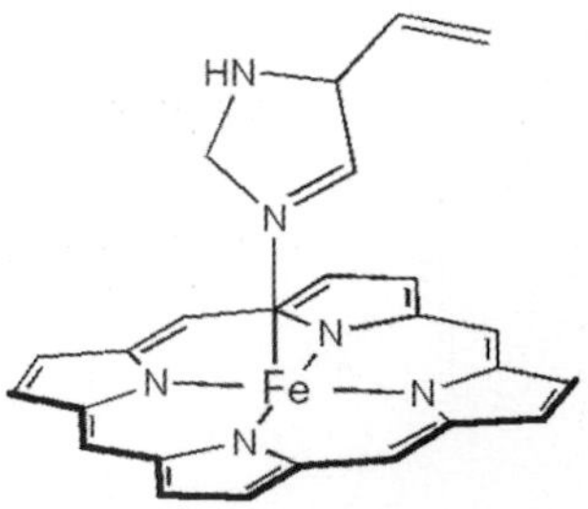

9

as in $[Ni(CN)_4]^{2-}$ (see Section 19.1). Examples of square-planar complexes from Groups 9, 10, and 11 are: $[Rh(PPh_3)_3Cl]$, trans-$[Ir(PMe_3)_2(CO)Cl]$, $[Ni(CN)_4]^{2-}$, $[PdCl_4]^{2-}$, $[Pt(NH_3)_4]^{2+}$, and $[AuCl_4]^-$. Square-planar geometry can also be forced on a central atom by complexation with a ligand that contains a rigid ring of four donor atoms, much as in the formation of a porphyrin complex (**9**).

Example 8.1 Identifying isomers from chemical evidence

Use the reactions indicated in Fig. 8.1 to show how the cis and trans geometries of a pair of platinum complexes may be assigned.

Answer The cis diamminedichloro isomer reacts with Ag_2O to lose Cl^-, and the product adds one oxalato dianion $(C_2O_4^{2-})$ at adjacent positions. The trans isomer loses Cl^-, but the product cannot displace the two OH^- ligands with only one $C_2O_4^{2-}$ anion. A reasonable explanation is that the $C_2O_4^{2-}$ anion cannot reach across the square plane to bridge two trans positions. This conclusion is supported by X-ray crystallography.

Self-test 8.1 The two square-planar isomers of $[Pt(PR_3)_2BrCl]$ (where PR_3 is a trialkylphosphine) have different ^{31}P NMR spectra (Fig. 8.2). For the sake of this exercise, we ignore coupling to ^{195}Pt ($I = \frac{1}{2}$ at 33 per cent abundance). One isomer (A) shows a single ^{31}P resonance; the other (B) shows two ^{31}P resonances, each of which is split into a doublet by the second ^{31}P nucleus. Which isomer is cis and which is trans?

(b) Five-coordination

Key points: In the absence of polydentate ligands that enforce the geometry, the energies of the various geometries of five-coordinate complexes differ little from one another and such complexes are often fluxional.

Five-coordinate complexes, which are less common than four- or six-coordinate complexes, are normally either square pyramidal or trigonal bipyramidal. A square-pyramidal

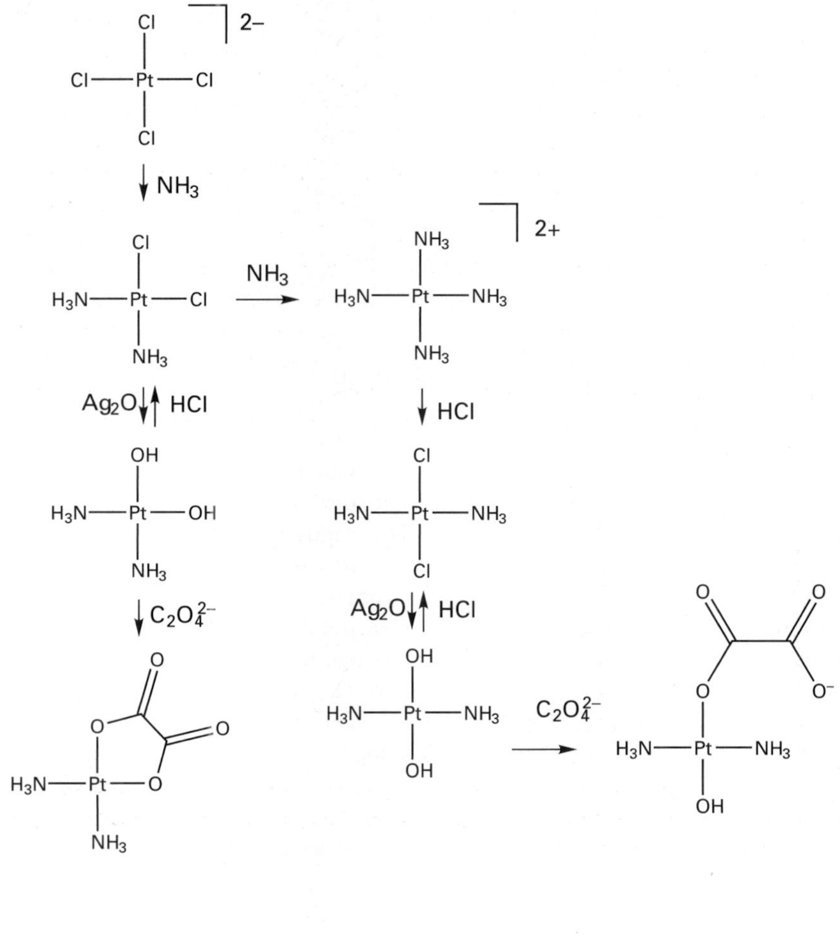

Fig. 8.1 The preparation of cis- and trans-diamminedichloroplatinum(II) and a chemical method for distinguishing the isomers.

complex would have C_{4v} symmetry if all the ligands were identical and the trigonal-bipyramidal complex would have D_{3h} symmetry with identical ligands. Distortions from these ideal geometries are common, and structures are known at all points between the two ideal geometries. A trigonal-bipyramidal shape minimizes ligand–ligand repulsions, but steric constraints on ligands that can bond through more than one site to a metal atom can favour a square-pyramidal structure. For instance, square-pyramidal five-coordination is found among the biologically important porphyrins, where the ligand ring enforces a square-planar structure and a fifth ligand attaches above the plane. Structure (**9**) shows the active centre of myoglobin, the oxygen transport protein; the location of the Fe atom above the plane of the ring is important to its function (Section 26.7). In some cases, five-coordination is induced by a polydentate ligand containing a donor atom that can bind to an axial location of a trigonal bipyramid, with its remaining donor atoms reaching down to the three equatorial positions (**10**). Ligands that force a trigonal-bipyramidal structure in this fashion are called **tripodal**.

The energies of the various geometries of five-coordinate complexes often differ little from one another. The delicacy of this balance is underlined by the fact that $[Ni(CN)_5]^{3-}$ can exist as both square-pyramidal (**11**) and trigonal-bipyramidal (**12**) conformations in the same crystal. In solution, trigonal-bipyramidal complexes with monodentate ligands are often highly fluxional (that is, able to twist into different shapes), so a ligand that is axial at one moment becomes equatorial at the next moment: the conversion from one stereochemistry to another may occur by a **Berry pseudorotation** (Fig. 8.3). The neutral complex $[Fe(CO)_5]$, for instance, is trigonal bipyramidal in the crystal; however, in solution the ligands exchange their axial and equatorial positions at a rate that is fast on an NMR timescale (so that all the ligands appear identical) but slow on an IR timescale (so that the different ligands are distinct).

(c) Six-coordination

Key point: The overwhelming majority of six-coordinate complexes are octahedral or have shapes that are small distortions of octahedral.

Six-coordination is the most common arrangement for metal complexes and is found in *s*-, *p*-, *d*-, and *f*-metal coordination compounds. Almost all six-coordinate complexes are octahedral (**13**), at least if we consider the ligands as represented by structureless points. A regular octahedral (O_h) arrangement of ligands is highly symmetric (Fig. 8.4). It is especially important, not only because it is found for many complexes of formula ML_6 but also because it is the starting point for discussions of complexes of lower symmetry, such as those shown in Fig. 8.5. The simplest deviation from O_h symmetry is tetragonal (D_{4h}), and occurs when two ligands along one axis differ from the other four; these two ligands, which are *trans* to each other, might be closer in than the other four or, more commonly, further away. For the d^9 configuration (particularly for Cu^{2+} complexes), a tetragonal distortion may occur even when all ligands are identical because of an inherent effect known as the Jahn–Teller distortion (Section 19.1). Rhombic (D_{2h}) distortions, in which a *trans* pair of ligands are close in and another *trans* pair are further out, can occur. Trigonal (D_{3d}) distortions occur when two opposite faces of the octahedron move away

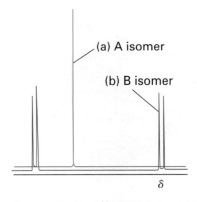

Fig. 8.2 Idealized ^{31}P NMR spectra of two isomers of $[PtBrCl(PR_3)_2]$. The fine structure due to Pt is not shown.

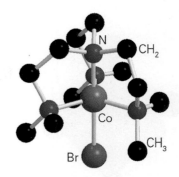

10 $[CoBrN(CH_2CH_2NMe_2)_3]$

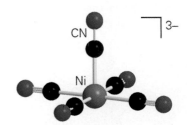

11 $[Ni(CN)_5]^{3-}$, square pyramidal

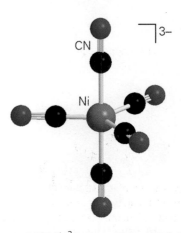

12 $[Ni(CN)_5]^{3-}$, trigonal bipyramidal

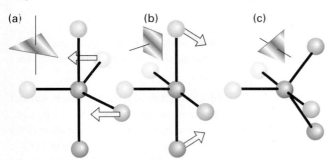

Fig. 8.3 A Berry pseudorotation in which (a) a trigonal-bipyramidal $Fe(CO)_5$ complex distorts into (b) a square-pyramidal isomer and then (c) becomes trigonal bipyramidal again, but with the two initially axial ligands now equatorial.

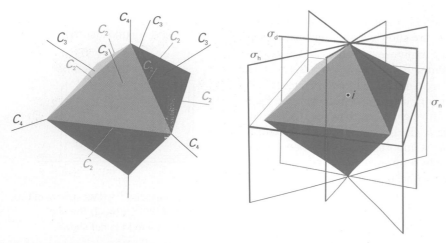

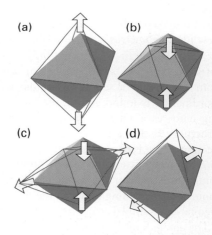

Fig. 8.4 The highly symmetric octahedral arrangement of six ligands around a central metal and the corresponding symmetry elements of an octahedron.

Fig. 8.5 Distortions of a regular octahedron: (a) and (b) tetragonal distortions, (c) rhombic distortion, (d) trigonal distortion.

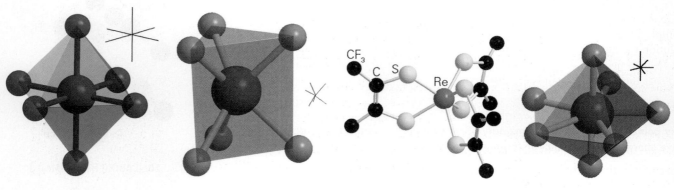

13 Octahedral complex, O_h **14** Trigonal prism, D_{3h} **15** [Re(SC(CF$_3$)=C(CF$_3$)S)$_3$]   **16** Pentagonal bipyramid, D_{5h}

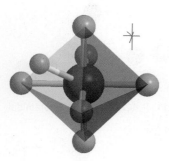

17 Capped octahedron

18 Capped trigonal prism

and give rise to a large family of structures that are intermediate between regular octahedral and trigonal prismatic (**14**).

Trigonal-prismatic (D_{3h}) complexes are rare, but have been found in solid MoS_2 and WS_2; the trigonal prism is also the shape of several complexes of formula $[M(S_2C_2R_2)_3]$ (**15**). Trigonal-prismatic d^0 complexes such as $[Zr(CH_3)_6]^{2-}$ have also been isolated. Such structures require either very small **σ-donor ligands**, ligands that bind by forming a σ bond to the central atom, or favourable ligand–ligand interactions that can constrain the complex into a trigonal prism; such ligand–ligand interactions are often provided by ligands that contain sulfur atoms, which can form covalent bonds to each other.[1]

8.3 Higher coordination numbers

Key points: Larger atoms and ions, particularly those of the *f* block, tend to form complexes with high coordination numbers; nine-coordination is particularly important in the *f* block.

Seven-coordination is encountered for a few 3*d* complexes and many more 4*d* and 5*d* complexes, where the larger central atom can accommodate more than six ligands. Seven-coordination resembles five-coordination in the similarity in energy of its various geometries. These limiting 'ideal' geometries include the pentagonal bipyramid (**16**), a capped octahedron (**17**), and a capped trigonal prism (**18**); in each of the latter two, the

---

[1] A general theoretical model for the structure of six-coordinate trigonal-prismatic complexes is advanced in K. Seppelt, *Acc. Chem. Res.*, 2003, **36**, 147.

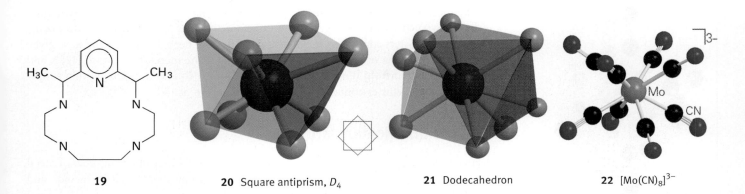

19 **20** Square antiprism, D_4 **21** Dodecahedron **22** $[Mo(CN)_8]^{3-}$

seventh capping ligand occupies one face. There are a number of intermediate structures, and interconversion between them is often rapid at room temperature. Examples include $[Mo(CNR)_7]^{2+}$, $[ZrF_7]^{3-}$, $[TaCl_4(PR_3)_3]$, and $[ReOCl_6]^{2-}$ from the d block and $[UO_2(OH_2)_5]^{2+}$ from the f block. A method to force seven-rather than six-coordination on the lighter elements is to synthesize a ring of five donor atoms (**19**) that then occupy the equatorial positions, leaving the axial positions free to accommodate two more ligands.

Stereochemical non-rigidity is also shown in eight-coordination; such complexes may be square antiprismatic (**20**) in one crystal but dodecahedral (**21**) in another. Two examples of complexes with these geometries are shown as (**22**) and (**23**), respectively. Cubic geometry (**24**) is rare.

Nine-coordination is important in the structures of f-block elements; their relatively large ions can act as host to a large number of ligands. A simple example of a nine-coordinate lanthanoid complex is $[Nd(OH_2)_9]^{3+}$. More complex examples arise with the MCl_3 solids, with M ranging from La to Gd, where a coordination number of nine is achieved through metal–halide–metal bridges (Section 22.4). An example of nine-coordination in the d block is $[ReH_9]^{2-}$ (**25**), which has small enough ligands for this coordination number to be feasible.

Ten- and twelve-coordination are encountered in complexes of the f-block M^{3+} ions. Examples include $[Ce(NO_3)_6]^{2-}$ (**26**), which is formed in the reaction of Ce(IV) salts with nitric acid. Each NO_3^- ligand is bonded to the metal atom by two O atoms. An example of a ten-coordinate complex is $[Th(ox)_4(OH_2)_2]^{4-}$, in which each oxalate ion ligand (ox, $C_2O_4^{2-}$) provides two O donor atoms. These high coordination numbers are rare with s-, p-, and d- block ions.

8.4 Polymetallic complexes

Key point: Polymetallic complexes are classified as metal clusters if they contain M–M bonds or as cage complexes if they contain ligand-bridged metal atoms.

Polymetallic complexes are complexes that contain more than one metal atom. In some cases, the metal atoms are held together by bridging ligands; in others there are direct metal–metal bonds; in yet others there are both types of link. The term **metal cluster** is usually reserved for polymetallic complexes in which there are direct metal–metal bonds that form triangular or larger closed structures. When no metal–metal bond is present, polymetallic complexes are referred to as **cage complexes** (or *cage compounds*).[2] Poly-metallic complexes are known for metals in every coordination number and geometry.

Cage complexes may be formed with a wide variety of anionic ligands. For example, two Cu^{2+} ions can be held together with acetate-ion bridges (**27**). Structure (**28**) is an example of a cubic structure formed from four Fe atoms bridged by RS^- ligands. This type of structure is of great biological importance as it is involved in a number of

---

[2] The term 'cage compound' has a variety of meanings in inorganic chemistry, and it is important to keep them distinct. For example, another use of the term is as a synonym of a clathrate compound (an inclusion compound, in which a species is trapped in a cage formed by molecules of another species).

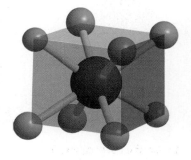

23 $[Zr(ox)_4]^{4-}$

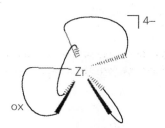

24 Cube

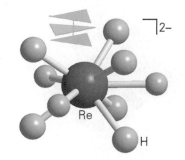

25 $[ReH_9]^{2-}$, D_{3h}

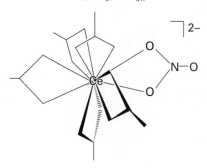

26 $[Ce(NO_3)_6]^{2-}$

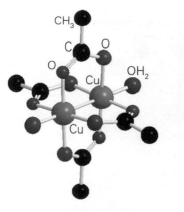

27 [(H$_2$O)Cu(μ-CH$_3$CO$_2$)$_4$Cu(OH$_2$)]

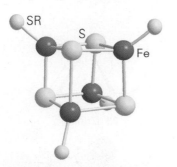

28 [Fe$_4$S$_4$(SR)$_4$]$^{2-}$

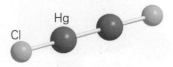

29 [Hg$_2$Cl$_2$] $D_{\infty h}$

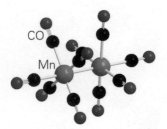

30 [(OC)$_5$MnMn(CO)$_5$]

biochemical redox reactions (Section 26.8). With the advent of modern structural techniques, such as automated X-ray diffractometers and multinuclear NMR, many polymetallic clusters containing metal–metal bonds have been discovered and have given rise to an active area of research. A simple example is the mercury(I) cation Hg$_2^{2+}$, and complexes derived from it, such as [Hg$_2$(Cl)$_2$] (**29**), which is commonly written simply Hg$_2$Cl$_2$. A metal cluster containing CO ligands is shown as (**30**).

Ligands and nomenclature

A large number of molecules and ions can behave as ligands, and a large number of metal ions form complexes. In this section we introduce some representative ligands and consider the basics of naming complexes.

8.5 Representative ligands

Key points: Polydentate ligands can form chelates; a bidentate ligand with a small bite angle can result in distortions from standard structures.

Table 8.1 gives the names and formulas of a number of common simple ligands and Table 8.2 gives the common prefixes used. Some of these ligands have only a single donor pair of electrons and will have only one point of attachment to the metal: such ligands are classified as **monodentate** (from the Latin meaning 'one-toothed'). Ligands that

Table 8.1 Typical ligands and their names

Name	Formula	Abbreviation	Donor atoms	Number of donors
Acetylacetonato		acac	O	2
Ammine	NH$_3$		N	1
Aqua	H$_2$O		O	1
2,2-Bipyridine		bipy	N	2
Bromo	Br$^-$		Br	1
Carbanato	CO$_3^{2-}$		O	1 or 2
Carbonyl	CO		C	1
Chloro	Cl$^-$		Cl	1
Cyano	CN$^-$		C	1
Diethylenetriamine	NH(CH$_2$CH$_2$NH$_2$)$_2$	dien	N	3
Bis(diphenylphosphino)ethane	Ph$_2$P〜PPh$_2$	dppe	P	2
Bis(diphenylphosphino)methane	Ph$_2$P⌒PPh$_2$	dppm	P	2
Cyclopentadienyl	C$_5$H$_5^-$	Cp	C	5
Ethylenediamine (1,2-diaminoethane)	NH$_2$CH$_2$CH$_2$NH$_2$	en	N	2
Ethylenediaminetetraacetato		edta	N, O	2N, 4O
Fluoro	F$^-$		F	1
Glycinato	NH$_2$CH$_2$CO$_2^-$	gly	N, O	1N, 1O
Hydrido	H$^-$		H	1
Hydroxo	OH$^-$		O	1
Iodo	I$^-$		I	1

Table 8.1 (*Continued*)

Name	Formula	Abbreviation	Donor atoms	Number of donors
Isothiocyanato	NCS^-		N	1
Nitrato	NO_3^-		O	1 or 2
Nitrito	NO_2^-		O	1
Nitro	NO_2^-		N	1
Oxo	O^{2-}		O	1
Oxalato		ox	O	2
Pyridine		py	N	1
Sulfido	S^{2-}		S	1
Tetraazacyclotetradecane		cyclam	N	4
Thiocyanato	SCN^-		S	1
Thiolato	RS^-		S	1
Triaminotriethylamine	$N(CH_2CH_2NH_2)_3$	tren	N	4
Tricyclohexylphosphine	$P(C_6H_{11})_3$	PCy_3	P	1
Triethylphosphine	$P(C_2H_5)_3$	PEt_3	P	1
Trimethylphosphine	$P(CH_3)_3$	PMe_3	P	1
Triphenylphosphine	$P(C_6H_5)_3$	PPh_3	P	1

Table 8.2 Prefixes used for naming complexes

Prefix	Meaning
mono-	1
di-, bis-	2
tri-, tris-	3
tetra-, tetrakis-	4
penta-	5
hexa-	6
hepta-	7
octa-	8
nona-	9
deca-	10
undeca-	11
dodeca-	12

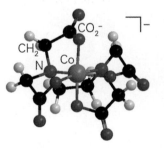

31 Ethylenediamine (en) ligand attached to M

have more than one point of attachment are classified as **polydentate**. Ligands that specifically have two points of attachment are known as **bidentate**, those with three, **tridentate**, and so on.

Ambidentate ligands have more than one different potential donor atoms. An example is the thiocyanate ion (NCS^-), which can attach to a metal atom either by the N atom, to give isothiocyanato complexes, or by the S atom, to give thiocyanato complexes. Another example of an ambidentate ligand is NO_2^-: as $M-NO_2^-$ the ligand is nitro and as $M-ONO^-$ it is nitrito.

Polydentate ligands can produce a **chelate** (from the Greek for 'claw'), a complex in which a ligand forms a ring that includes the metal atom. An example is the bidentate ligand ethylenediamine (en, $NH_2CH_2CH_2NH_2$), which forms a five-membered ring when both N atoms attach to the same metal atom (**31**). It is important to note that normal chelating ligands will only attach to the metal at two adjacent coordination sites, a *cis* fashion. The hexadentate ligand ethylenediaminetetraacetic acid, as its anion (edta, $(^-O_2CCH_2)_2NCH_2CH_2N(CH_2CO_2^-)_2$), can attach at six points (at two N atoms and four O atoms) and can form an elaborate complex with five five-membered rings (**32**). This ligand is used to trap metal ions, such as Ca^{2+} ions, in 'hard' water. Table 8.1 includes some of the most common chelating ligands.

In a chelate formed from a saturated organic ligand, such as ethylenediamine, a five-membered ring can fold into a conformation that preserves the tetrahedral angles within the ligand and yet still achieve an L—M—L angle of 90°, the angle typical of octahedral complexes. Six-membered rings may be favoured sterically or by electron delocalization through their π orbitals. The bidentate β-diketones, for example, coordinate as the anions of their enols in six-membered ring structures (**33**). An important example is the acetyl-acetonato anion (acac, $CH_3COCHCOCH_3^-$ (**34**)). Because biochemically important amino acids can form five- or six-membered rings, they also chelate readily.

32 $[Co(edta)]^-$

33

34

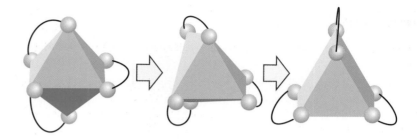

Fig. 8.6 A chelating ligand that permits only a small bite angle can distort an octahedral complex into trigonal-prismatic geometry.

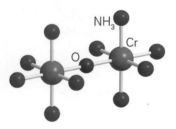

35

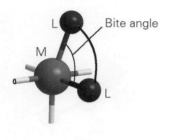

36 $[Re_2Cl_8]^{2-}$

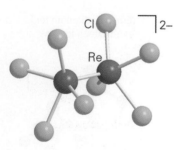

37 μ-Oxo-bis(pentaammine chromium(III)),
$[(H_3N)_5CrOCr(NH_3)_5]$

The degree of strain in a chelating ligand is often expressed in terms of the **bite angle**, the L—M—L angle in the chelate ring (**35**). A chelating ligand that permits only a small bite angle is one of the main causes of distortion from octahedral towards trigonal-prismatic geometry in six-coordinate complexes (Fig. 8.6).

8.6 Nomenclature

Key points: The cation and anion of a complexes are named according to a set of rules; cations are named first and ligands are named in alphabetical order.

Detailed guidance on nomenclature is beyond the scope of this book and only a general introduction is given here. In fact, the names of complexes often become so cumbersome that inorganic chemists often prefer to give the formula rather than spell out the entire name.

For compounds that consist of one or more ions, the cation is named first followed by the anion (as for simple ionic compounds), regardless of which is complex. Complex ions are named with their ligands in alphabetical order (ignoring any numerical prefixes). The ligand names are followed by the name of the metal with either its oxidation number in parentheses, as in hexaamminecobalt(III) for $[Co(NH_3)_6]^{3+}$, or with the overall charge on the complex specified in parentheses, as in hexaamminecobalt(3+). The suffix -ate is added to the name of the metal (sometimes in its Latin form) if the complex is an anion, as in the name hexacyanoferrate(II) for $[Fe(CN)_6]^{4-}$.

The number of a particular type of ligand in a complex is indicated by the prefixes mono-, di-, tri-, and tetra-. The same prefixes are used to state the number of metals atoms if more than one is present in a complex, as in octachlorodirhenate(III), $[Re_2Cl_8]^{2-}$ (**36**). Where confusion with the names of ligands is likely, perhaps because the name already includes a prefix, as with ethylenediamine, the alternative prefixes bis-, tris-, and tetrakis- are used, with the ligand name in parentheses. For example, dichloro- is unambiguous but tris(ethylenediamine) shows more clearly that there are three ethylenediamine ligands, as in tris(ethylenediamine)cobalt(II), $[Co(en)_3]^{2+}$. Ligands that bridge two metal centres are denoted by a prefix μ- added to the name of the relevant ligand, as in μ-oxo-bis(pentamminechromium(III)) (**37**). A subscript is used to indicate the number of metal atoms bridged (provided that number is greater than two); for instance, a hydride ligand bridging three metal atoms is denoted μ_3-H.

Square brackets are used to indicate which groups are bound to a metal atom, and should be used whether the complex is charged or not; however, in casual usage, neutral complexes and oxoanions are often written without brackets, as in $Ni(CO)_4$ for tetracarbonylnickel(0)[3] and MnO_4^- for tetraoxomanganate(VII) ('permanganate'). The metal symbol is given first, then the anionic ligands, and finally the neutral ligands, as in $[Co(Cl)_2(NH_3)_4]^+$ for tetraamminedichlorocobalt(III). (This order is sometimes varied to clarify which ligand is involved in a reaction.) Polyatomic ligand formulas are sometimes written in an unfamiliar sequence (as for OH_2 in $[Fe(OH_2)_6]^{2+}$ for hexaaquairon(II)) to place the donor atom adjacent to the metal atom and so help to make the structure of the complex clear. The donor atom of an ambidentate ligand is sometimes indicated by underlining it, for example $[Fe(\underline{N}CS)(HO_2)_5]^{2+}$.

---

[3] When assigning oxidation numbers in carbonyl complexes, CO is ascribed a net oxidation number of 0.

Example 8.2 Naming complexes

Name the complexes (a) $[Pt(Cl)_2(NH_3)_4]^{2+}$; (b) $[Ni(CO)_3(py)]$; (c) $[Cr(edta)]^-$; (d) $[Co(Cl)_2(en)_2]^+$; (e) $[Rh(CO)_2I_2]^-$.

Answer (a) The complex has two anionic ligands (Cl^-), four neutral ligands (NH_3) and an overall charge of +2; hence the oxidation number of platinum must be +4. According to the alphabetical order rules, the name of the complex is tetraamminedichloroplatinum(IV). (b) The ligands CO and py (pyridine) are neutral, so the oxidation number of nickel must be 0. It follows that the name of the complex is tricarbonylpyridinenickel(0). (c) This complex contains the hexadentate $edta^{4-}$ ion as the sole ligand. The four negative charges of the ligand result in a complex with a single negative charge if the central metal ion is Cr^{3+}. The complex is therefore ethylenediaminetetraacetatochromate(III). (d) This complex contains two anionic chloride ligands and two neutral en ligands. The overall charge of +1 must be the result of the cobalt having oxidation number +3. The complex is therefore dichlorobis(ethylenediamine)cobalt(III). (e) This complex contains two anionic iodide ligands and two neutral CO ligands. The overall charge of −1 must be the result of the rhodium having oxidation number +1. The complex is therefore dicarbonyldiiodorhodate(I).

Self-test 8.2 Write the formulas of the following complexes: (a) diaquadichloroplatinum(II); (b) diamminetetra(isothiocyanato)chromate(III); (c) tris(ethylenediamine)rhodium(III); (d) bromopentacarbonylmanganese(I); (e) chlorotris(triphenylphosphine)rhodium(I).

Isomerism and chirality

Key points: A molecular formula is not sufficient to identify a coordination compound: linkage, ionization, hydrate, and coordination isomerism are all possible for coordination compounds.

A molecular formula often does not give enough information to identify a compound unambiguously. We have already noted that the existence of ambidentate ligands gives rise to the possibility of **linkage isomerism**, in which the same ligand may link through different atoms. This type of isomerism accounts for the red and yellow isomers of the formula $[Co(NO_2)(NH_3)_5]^{2+}$. The red compound has a nitrito Co—O link (**38**); the yellow isomer, which forms from the unstable red form over time, has a nitro Co—N link (**39**). We can consider three further types of isomerism briefly, before looking at geometric and optical isomerism in more depth. **Ionization isomerism** occurs when a ligand and a counterion in one compound exchange places. An example is $[Pt(NH_3)_4Cl_2]Br_2$ and $[Pt(NH_3)_4Br_2]Cl_2$. If the compounds are soluble, the two isomers exist as different ionic species in solution (in this example, with free Br^- and Cl^- ions, respectively). Very similar to ionization isomerism is **hydrate isomerism**, which arises when one of the ligands is water; for example, there are three differently coloured hydration isomers of a compound with molecular formula $CrCl_3 \cdot 6H_2O$: $[Cr(H_2O)_6]Cl_3$, $[Cr(H_2O)_5Cl]Cl_2 \cdot H_2O$ and $[Cr(H_2O)_4Cl_2]Cl \cdot 2H_2O$. **Coordination isomerism** arises when there are different complex ions that can form from the same molecular formula, as in $[Co(NH_3)_6][Cr(CN)_6]$ and $[Cr(NH_3)_6][Co(CN)_6]$.

Once we have established which ligands bind to which metals, and through which donor atoms, we can then consider how to arrange these ligands, as the three-dimensional character of metal complexes can result in a multitude of possible arrangements of the ligands. These different arrangements can result in geometrical isomerism and optical isomerism. We now explore these varieties of isomerism by considering the permutations of ligand arrangement for each of the common complex geometries.

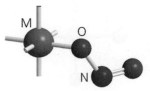

38 Nitrito ligand

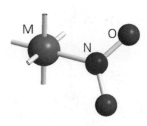

39 Nitro ligand

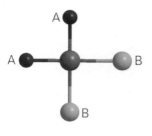

40 *cis*-[MA$_2$B$_2$]

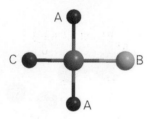

41 *trans*-[MA$_2$B$_2$]

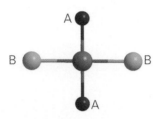

42 *cis*-[MA$_2$BC]

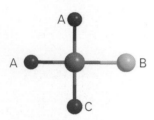

43 *trans*-[MA$_2$BC]

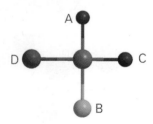

44 [MABCD], A *trans* to B

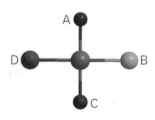

45 [MABCD], A *trans* to C

Example 8.3 Isomerism in metal complexes

What types of isomerism are possible for complexes with the following molecular formulas: (a) [PtSCN(PEt$_3$)$_3$]$^+$, (b) Co(NH$_3$)$_5$BrSO$_4$, (c) FeCl$_2$·6H$_2$O?

Answer (a) The complex contains the thiocyanate ligand, SCN$^-$, which can bind through either the S or the N atom to give rise to two linkage isomers: [Pt(<u>S</u>CN)(PEt$_3$)$_3$]$^+$ and [Pt(<u>N</u>CS)(PEt$_3$)$_3$]$^+$. (b) With an octahedral geometry and five coordinated ammonia ligands, it is possible to have two ionization isomers: [Co(NH$_3$)$_5$SO$_4$]Br and [Co(NH$_3$)$_5$Br]SO$_4$. (c) Hydrate isomerism occurs as complexes of formula [Fe(H$_2$O)$_6$]Cl$_2$ and [FeCl(H$_2$O)$_5$]Cl·H$_2$O are possible.

***Self-test* 8.3** Two types of isomerism are possible for the six-coordinate complex Cr(NO$_2$)$_2$·6H$_2$O. Identify all isomers.

8.7 Square-planar complexes

Key point: The only simple isomers of square-planar complexes are *cis*–*trans* isomers.

The simple case of two sets of two different monodentate ligands, as in [MA$_2$B$_2$], gives rise to two different geometric isomers (**40**) and (**41**). With three different ligands, as in [MA$_2$BC], the locations of the two A ligands also allow us to distinguish the geometric isomers as *cis* and *trans*, (**42**) and (**43**). When there are four different ligands, as in [MABCD], there are three different isomers and we have to specify the geometry more explicitly, as in (**44**), (**45**), and (**46**). Bidentate ligands with different endgroups, as in [M(AB)$_2$], can also give rise to geometrical isomers that can be classified as *cis* (**47**) and *trans* (**48**).

8.8 Tetrahedral complexes

Key point: The only simple isomers of tetrahedral complexes are optical isomers.

The only isomers of tetrahedral complexes normally encountered are those where either all four ligands are different or where there are two unsymmetrical bidentate chelating ligands. In both cases, (**49**) and (**50**), the molecules are **chiral**, not superimposable on their mirror image (Section 7.4). Two mirror-image isomers jointly make up an **enantiomeric pair**. The existence of a pair of chiral complexes that are each other's mirror image (like a right and left hand), and that have lifetimes that are long enough for them to be separable, is called **optical isomerism**. Optical isomers are so called because they are

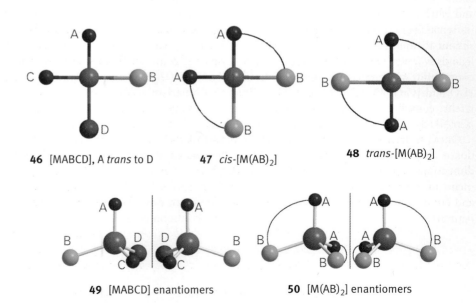

46 [MABCD], A *trans* to D **47** *cis*-[M(AB)$_2$] **48** *trans*-[M(AB)$_2$]

49 [MABCD] enantiomers **50** [M(AB)$_2$] enantiomers

51 [ML$_5$], trigonal bipyramid

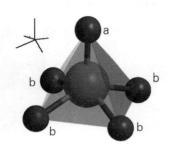

52 [ML$_5$], square pyramid

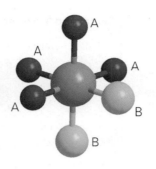

53 *cis*-[MA$_4$B$_2$]

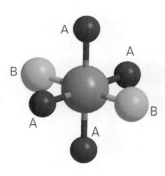

54 *trans*-[MA$_4$B$_2$]

optically active, in the sense that one enantiomer rotates the plane of polarized light in one direction and the other rotates it through an equal angle in the opposite direction.

8.9 Trigonal-bipyramidal and square-pyramidal complexes

Key points: Five-coordinate complexes are not stereochemically rigid; two chemically distinct coordination sites exist within both trigonal-bipyramidal and square-pyramidal complexes.

We noted previously that five-coordinate complexes are often distorted away from ideal shapes, and frequently interconvert very easily. Thus, although isomers of five-coordinate complexes do exist, they are commonly not separable. It is important to be aware that both trigonal-bipyramidal and square-pyramidal complexes have two chemically distinct sites: axial (a) and equatorial (e) for the trigonal bipyramid (**51**) and axial (a) and basal (b) for the square pyramid (**52**). Certain ligands will have preferences for the different sites because of their steric and electronic requirements, but we do not go into detail here.[4]

8.10 Octahedral complexes

Key points: *Cis* and *trans* isomers exist for octahedral complexes of formula [MA$_4$B$_2$], and *mer* and *fac* isomers are possible for complexes of formula [MA$_3$B$_3$]. More complicated ligand sets lead to further isomers.

There are huge numbers of complexes with nominally octahedral geometry, where in this context the nominal structure 'ML$_6$' is taken to mean a central metal atom surrounded by six ligands, not all of which are necessarily the same.

(a) Geometrical isomerism

Whereas there is only one way of arranging the ligands in octahedral complexes of general formula [MA$_6$] or [MA$_5$B], the two B ligands of an [MA$_4$B$_2$] complex may be placed on adjacent octahedral positions to give a *cis* isomer (**53**) or on diametrically opposite positions to give a *trans* isomer (**54**). Provided we treat the ligands as structureless points, the *trans* isomer has D_{4h} symmetry and the *cis* isomer has C_{2v} symmetry.

There are two ways of arranging the ligands in [MA$_3$B$_3$] complexes. In one isomer, three A ligands lie in one plane and three B ligands lie in a perpendicular plane (**55**). This complex is designated the *mer* isomer (for meridional) because each set of ligands can be regarded as lying on a meridian of a sphere. In the second isomer, all three A (and B) ligands are adjacent and occupy the corners of one triangular face of the octahedron (**56**); this complex is designated the *fac* isomer (for facial). Provided we treat the ligands as structureless points, the *mer* isomer has C_{2v} symmetry and the *fac* isomer has C_{3v} symmetry.

For a complex of composition [MA$_2$B$_2$C$_2$], there are five different geometrical isomers: an all-*trans* isomer (**57**); three different isomers where one pair of ligands are *trans* with the other two *cis*, as in (**58**), (**59**), and (**60**); and an enantiomeric pair of all-*cis*

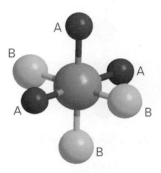

55 *mer*-[MA$_3$B$_3$]

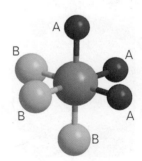

56 *fac*-[MA$_3$B$_3$]

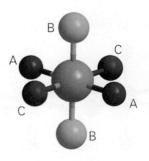

57 [MA$_2$B$_2$C$_2$]

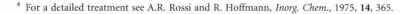

[4] For a detailed treatment see A.R. Rossi and R. Hoffmann, *Inorg. Chem.*, 1975, **14**, 365.

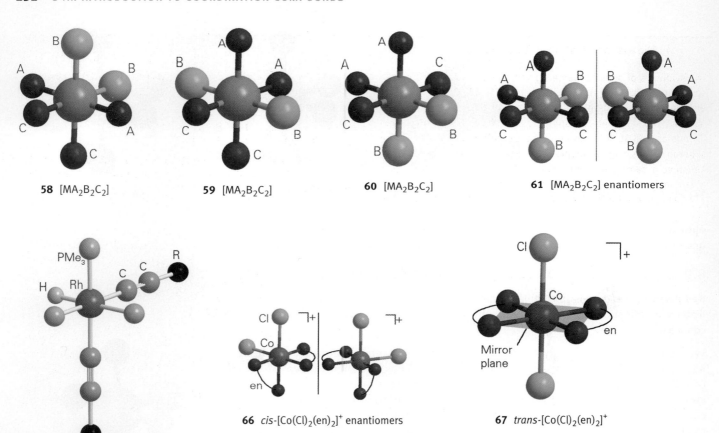

58 [MA$_2$B$_2$C$_2$]   **59** [MA$_2$B$_2$C$_2$]   **60** [MA$_2$B$_2$C$_2$]   **61** [MA$_2$B$_2$C$_2$] enantiomers

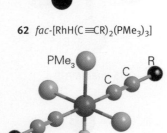

62 *fac*-[RhH(C≡CR)$_2$(PMe$_3$)$_3$]

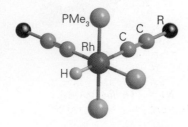

63 *mer-trans*-[RhH(C≡CR)$_2$(PMe$_3$)$_3$]

64 *mer-cis*-[RhH(C≡CR)$_2$(PMe$_3$)$_3$]

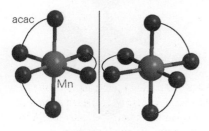

65 [Mn(acac)$_2$] enantiomers

66 *cis*-[Co(Cl)$_2$(en)$_2$]$^+$ enantiomers

67 *trans*-[Co(Cl)$_2$(en)$_2$]$^+$

isomers (**61**). More complicated compositions, such as [MA$_2$B$_2$CD] or [MA$_3$B$_2$C], result in more extensive geometrical isomerism. For instance the rhodium compound [RhH(C≡CR)$_2$(PMe$_3$)$_3$] exists as three different isomers: *fac* (**62**), *mer-trans* (**63**), and *mer-cis* (**64**).

Box 8.1 Summarizes an approach to the synthesis of specific isomers.

(b) Chirality and optical isomerism

Key points: A number of ligand arrangements at an octahedral centre give rise to chiral compounds; isomers are designated Δ or Λ depending on their configuration.

In addition to the many examples of geometrical isomerism shown by octahedral compounds, many are also chiral. A very simple example is [Mn(acac)$_3$] (**65**), where three bidentate acetylacetonato (acac) ligands result in the existence of enantiomers. One way of looking at the optical isomers that arise in complexes of this nature is to imagine looking down one of the threefold axes and see the ligand arrangement as a propellor, or screw thread.

Chirality can also exist for complexes of the formula [MA$_2$B$_2$C$_2$] when the ligands of each pair are *cis* to each other (**61**). In fact, many examples of optical isomerism are known for octahedral complexes with both monodentate and polydentate ligands, and we must always be alert to the possibility of optical isomerism.

As a further example of optical isomerism, consider the products of the reaction of Co(III) chloride and ethylenediamine in a 1:2 mole ratio. The product includes a pair of dichloro complexes, one of which is violet (**66**) and the other green (**67**); they are, respectively, the *cis* and *trans* isomers of dichlorobis(ethylenediamine)cobalt(III), [CoCl$_2$(en)$_2$]$^+$. As can be seen from their structures, the *cis* isomer cannot be superimposed on its mirror image. It is therefore chiral and hence (because the complexes are long-lived) optically active. The *trans* isomer has a mirror plane and can be superimposed on its mirror image; it is achiral and optically inactive.

The formal criterion of chirality is the absence of an axis of improper rotation (S_n, an *n*-fold axis in combination with a horizontal mirror plane, Section 7.4). The existence of

Box 8.1 The synthesis of specific isomers

The synthesis of specific isomers often requires subtle changes in synthetic conditions. For example, the most stable Co(II) complex in ammoniacal solutions of Co(II) salts, $[Co(NH_3)_6]^{2+}$, is only slowly oxidized. As a result, a variety of complexes containing other ligands as well as NH_3 can be prepared by bubbling air through a solution containing ammonia and a Co(II) salt. Starting with ammonium carbonate yields $[CoCO_3(NH_3)_4]^+$, in which CO_3^{2-} is a bidentate ligand that occupies two adjacent coordination positions. The complex cis-$[Co(NH_3)_4L_2]$ can be prepared by displacement of

the CO_3^{2-} ligand in acidic solution. When concentrated hydrochloric acid is used, the violet cis-$[CcCl_2(NH_3)_4]Cl$ compound (**B1**) can be isolated:

$$[Co(CO_3)(NH_3)_4]^+(aq) + 2\,H^+(aq) + 3\,Cl^-(aq)$$
$$\rightarrow cis\text{-}[CoCl_2(NH_3)_4]Cl(s) + H_2CO_3$$

By contrast, reaction of $[Co(NH_3)_6]^{3+}$ directly with a mixture of HCl and H_2SO_4 in air gives the bright green trans-$[CoCl_2(NH_3)_4]Cl$ isomer (**B2**).

B1   B2

such a symmetry element is implied by the presence of either a mirror plane through the central atom (which is equivalent to an S_1 axis) or a centre of inversion (which is equivalent to an S_2 axis), and if either of these elements is present then the complex is achiral. As remarked in Section 7.4, we must also be alert for higher-order axes of improper rotation (particularly S_4), because the presence of any S_n axis implies achirality. An example of a species with an S_4 axis but not an S_1 or S_2 axis was given as structure **12** in Chapter 7.

The absolute configuration of a chiral octahedral complex is described by imagining a view along a threefold rotation axis of the regular octahedron and noting the handedness of the helix formed by the ligands (Fig. 8.7). Clockwise rotation of the helix is then designated Δ (delta) whereas the anticlockwise rotation is designated Λ (lambda). The designation of the absolute configuration must be distinguished from the experimentally determined direction in which an isomer rotates polarized light: some Λ compounds rotate in one direction, others rotate in the opposite direction, and the direction may change with wavelength. The isomer that rotates the plane of polarization clockwise (when viewed into the oncoming beam) at a specified wavelength is designated the **d-isomer**, or the **(+)-isomer**; the one rotating the plane anticlockwise is designated the **l-isomer**, or the **(−)-isomer**. Box 8.2 describes how enantiomers of metal complexes may be separated.

Complexes with coordination numbers of greater than six have the potential for a great number of isomers, both geometrical and optical. As these complexes are often stereochemically nonrigid, the isomers are usually not separable, and we do not consider them further.

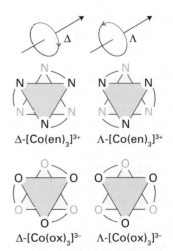

Fig. 8.7 Absolute configurations of $M(L\!-\!L)_3$ complexes. Δ is used to indicate clockwise rotation of the helix and Λ to indicate anticlockwise rotation.

Example 8.4 Identifying types of isomerism

When the four-coordinate square-planar complex $[IrCl(PMe_3)_3]$ (where PMe_3 is trimethylphosphine) reacts with Cl_2, two six-coordinate products of formula $[Ir(Cl)_3(PMe_3)_3]$ are formed. ^{31}P NMR spectra indicate one P environment in one of these isomers and two in the other. What isomers are possible?

Answer Because the complexes have the formula [MA$_3$B$_3$], we expect meridional and facial isomers. Structures (**68**) and (**69**) show the arrangement of the three Cl$^-$ ions in the *fac* and *mer* isomers, respectively. All P atoms are equivalent in the *fac* isomer and two environments exist in the *mer* isomer.

Self-test 8.4 When the anion of the amino acid glycine, H$_2$NCH$_2$CO$_2^-$ (gly$^-$), reacts with Co(III) oxide, both the N and an O atom of gly$^-$ coordinate and two Co(III) nonelectrolyte *mer* and *fac* isomers of [Co(gly)$_3$] are formed. Sketch the two isomers.

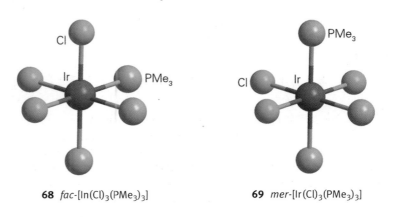

68 *fac*-[In(Cl)$_3$(PMe$_3$)$_3$]   **69** *mer*-[Ir(Cl)$_3$(PMe$_3$)$_3$]

Box 8.2 The resolution of enantiomers

Optical activity is the only physical manifestation of chirality for a compound with a single chiral centre. However, as soon as more than one chiral centre is present, other physical properties, such as solubility and melting points, are affected because they depend on the strengths of intermolecular forces, which are different between different isomers (just as there are different forces between a given nut and bolts with left- and right-handed threads). One method of separating a pair of enantiomers into the individual isomers is therefore to prepare diastereomers. As far as we need be concerned, *diastereomers* are isomeric compounds that contain two chiral centres, one being of the same absolute configuration in both components and the other being enantiomeric between the two components. An example of diastereomers is provided by the two salts of an enantiomeric pair of cations, A, with an optically pure anion, B, and hence of composition [Δ-A][Δ-B] and [Λ-A][Δ-B]. Because diastereomers differ in physical properties (such as solubility), they are separable by conventional techniques.

A classical chiral resolution procedure begins with the isolation of a naturally optically active species from a biochemical source (many naturally occurring compounds are chiral). A convenient compound is *d*-tartaric acid (**B3**), a carboxylic acid obtained from grapes. This molecule is a chelating ligand for complexation of antimony, so a convenient resolving agent is the potassium salt of the singly charged antimony *d*-tartrate anion. This anion is used for the resolution [Co(NO$_2$)$_2$(en)$_2$]$^+$ as follows.

The enantiomeric mixture of the cobalt(III) complex is dissolved in warm water and a solution of potassium antimony *d*-tartrate is added. The mixture is cooled immediately to induce crystal-

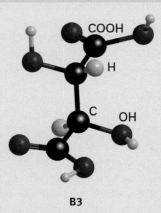

B3

lization. The less soluble diastereomer {*l*-[Co(NO$_2$)$_2$(en)$_2$]}{*d*-[SbOC$_4$H$_4$O$_6$]} separates as fine yellow crystals. The filtrate is reserved for isolation of the *d*-enantiomer. The solid diastereomer is ground with water and sodium iodide. The sparingly soluble compound *l*-[Co(NO$_2$)$_2$(en)$_2$]I separates, leaving sodium antimony tartrate in the solution. The *d*-isomer is obtained from the filtrate by precipitation of the bromide salt.

Further reading

A. von Zelewsky, *Stereochemistry of coordination compounds*. Wiley, Chichester (1996).

W.L. Jolly, *The synthesis and characterization of inorganic compounds*. Waveland Press, Prospect Heights (1991).

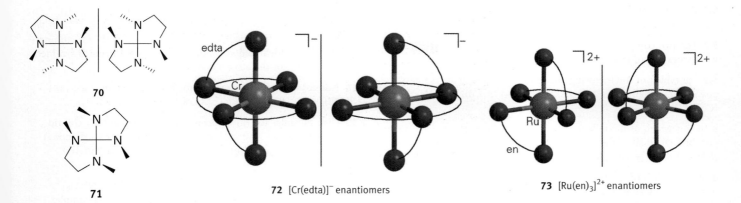

70

71

72 [Cr(edta)]⁻ enantiomers

73 [Ru(en)₃]²⁺ enantiomers

8.11 Ligand chirality

Key point: Coordination to a metal can stop a ligand inverting and hence lock it into a chiral configuration.

In certain cases, achiral ligands can become chiral on coordination to a metal, leading to a complex that is chiral. Usually the nonchiral ligand contains a donor that rapidly inverts as a free ligand, but that becomes locked in one configuration on coordination. An example is $MeNHCH_2CH_2NHMe$, where the two N atoms become chiral centres on coordination to a metal atom. For a square-planar complex, this imposed chirality results in three isomers, one pair of chiral enantiomers (**70**) and one complex that is not chiral (**71**).

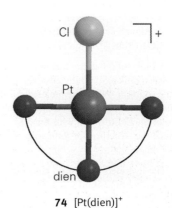

74 [Pt(dien)]⁺

Example 8.5 Recognizing chirality

Which of the complexes (a) [Cr(edta)]⁻, (b) [Ru(en)₃]²⁺, (c) [Pt(dien)Cl]⁺ are chiral?

Answer The complexes are shown schematically in (**72**), (**73**), and (**74**), respectively. Neither (**72**) nor (**73**) has a mirror plane or a centre of inversion; so both are chiral (they also have no higher S_n axis); (**74**) has a plane of symmetry and hence is achiral. (Although the CH_2 groups in a dien ligand are not in the mirror plane, they oscillate rapidly above and below it.)

Self-test 8.5 Which of the complexes (a) cis-[Cr(Cl)₂(ox)₂]³⁻, (b) trans-[Cr(Cl)₂(ox)₂]³⁻, (c) cis-[RhH(CO)(PR₃)₂] are chiral?

FURTHER READING

G.B. Kaufman, *Inorganic coordination compounds*. Wiley, New York (1981). A fascinating account of the history of structural coordination chemistry.

G.B. Kauffman, *Classics in coordination chemistry: I. Selected papers of Alfred Werner*. Dover, New York (1968). Provides translations of Werner's key papers.

A. von Zelewsky, *Stereochemistry of coordination compounds*. Wiley, Chichester (1996). A readable book that covers chirality in detail.

J.A. McCleverty and T.J. Meyer (eds), *Comprehensive coordination chemistry II*. Elsevier (2004).

N.G. Connelly and J.A. McCleverty, *Nomenclature of inorganic chemistry II. Recommendations 2000*. Royal Society of Chemistry, Cambridge (2001). Also known as 'The IUPAC red book', the definitive guide to naming inorganic compounds.

EXERCISES

8.1 (a) Sketch the two structures that describe most four-coordinate complexes. (b) In which structure are isomers possible for complexes of formula MA₂B₂?

8.2 Sketch the two structures that describe most five-coordinate complexes. Label the two different sites in each structure.

8.3 (a) Sketch the two structures that describe most six-coordinate complexes. (b) Which one of these is rare?

8.4 Explain the meaning of the terms *monodentate*, *bidentate*, and *quadridentate*.

8.5 What type of isomerism can arise with ambidentate ligands? Give an example.

8.6 Which of the following molecules could act as chelating ligands?

(a) P(OPh)₃ (b) Me₂P⌒PMe₂ (c) [structure] (d) [structure]

8.7 Draw the structures of representative complexes that contain the ligands (a) en, (b) ox, (c) phen, and (d) edta^{4-}.

8.8 The two compounds [RuBr(NH$_3$)$_5$]Cl and [RuCl(NH$_3$)$_5$]Br are what types of isomers?

8.9 For which of the following tetrahedral complexes are isomers possible? Draw all of the isomers. [CoBr$_2$Cl$_2$]$^-$, [CoBrCl$_2$(OH$_2$)], [CoBrClI(OH$_2$)].

8.10 For which of the following square-planar complexes are isomers possible? Draw all of the isomers. [Pt(ox)(NH$_3$)$_2$], [PdBrCl(PEt$_3$)$_2$], [IrH(CO)(PR$_3$)$_2$], [Pd(gly)$_2$].

8.11 For which of the following octahedral complexes are isomers possible? Draw all of the isomers. [FeCl(OH$_2$)$_5$]$^{2+}$, [Ir(Cl)$_3$(PEt$_3$)$_3$], [Ru(bipy)$_3$]$^{2+}$, [Co(Cl)$_2$(en)(NH$_3$)$_2$]$^+$, [W(CO)$_4$(py)$_2$].

8.12 Ignoring optical isomers, how many isomers are possible for an octahedral complex of general formula [MA$_2$BCDE]? How many isomers are possible *including* optical isomers?

8.13 Which of the following complexes are chiral? (a) [Cr(ox)$_3$]$^{3-}$, (b) *cis*-[Pt(Cl)$_2$(en)], (c) *cis*-[Rh(Cl)$_2$(NH$_3$)$_4$]$^+$, (d) [Ru(bipy)$_3$]$^{2+}$, (e) *fac*-[Co(NO$_2$)$_3$(dien)], (f) *mer*-[Co(NO$_2$)$_3$(dien)]. Draw the enantiomers of the complexes identified as chiral and identify the plane of symmetry in the structures of the achiral complexes.

8.14 Name and draw structures of the following complexes: (a) [Ni(CO)$_4$]; (b) [Ni(CN)$_4$]$^{2-}$; (c) [CoCl$_4$]$^{2-}$; (d) [Mn(NH$_3$)$_6$]$^{2+}$.

8.15 Give formulas for (a) chloropentaamminecobalt(III) chloride; (b) hexaaquairon(3+) nitrate, (c) *cis*-dichlorobis(ethylenediamine) ruthenium(II), (d) μ-hydroxobis(pentaamminechromium(III)) chloride.

8.16 Name the octahedral complex ions (a) *cis*-[CrCl$_2$(NH$_3$)$_4$]$^+$, (b) *trans*-[Cr(NCS)$_4$(NH$_3$)$_2$]$^-$, (c) [Co(C$_2$O$_4$)(en)$_2$]$^+$. Is complex (c) *cis* or *trans*?

PROBLEMS

8.1 The compound Na$_2$IrCl$_6$ reacts with triphenylphosphine in diethyleneglycol in an atmosphere of CO to give *trans*-[IrCl(CO)(PPh$_3$)$_2$], known as 'Vaska's compound'. Excess CO gives a five-coordinate species and treatment with NaBH$_4$ in ethanol gives [IrH(CO)$_2$(PPh$_3$)$_2$]. Draw and name the three complexes.

8.2 A pink solid has the formula CoCl$_3$·5NH$_3$·H$_2$O. A solution of this salt is also pink and rapidly gives three 3 mol AgCl on titration with silver nitrate solution. When the pink solid is heated, it loses 1 mol H$_2$O to give a purple solid with the same ratio of NH$_3$:Cl:Co. The purple solid, on dissolution and titration with AgNO$_3$, releases one of its chlorides slowly. Deduce the structures of the two octahedral complexes and draw and name them.

8.3 The hydrated chromium chloride that is available commercially has the overall composition CrCl$_3$·6H$_2$O. On boiling a solution, it becomes violet and has a molar electrical conductivity similar to that of [Co(NH$_3$)$_6$]Cl$_3$. By contrast, CrCl$_3$·5H$_2$O is green and has a lower molar conductivity in solution. If a dilute acidified solution of the green complex is allowed to stand for several hours, it turns violet. Interpret these observations with structural diagrams.

8.4 The complex first denoted β-[PtCl$_2$(NH$_3$)$_2$] was identified as the *trans* isomer. (The *cis* isomer was denoted α.) It reacts slowly with solid Ag$_2$O to produce [Pt(NH$_3$)$_2$(OH$_2$)$_2$]$^{2+}$. This complex does not react with ethylenediamine to give a chelated complex. Name and draw the structure of the diaqua complex. A third isomer of composition PtCl$_2$·2NH$_3$ is an insoluble solid that, when ground with AgNO$_3$, gives a solution containing [Pt(NH$_3$)$_4$](NO$_3$)$_2$ and a new solid phase of composition Ag$_2$[PtCl$_4$]. Give the structures and names of each of the three Pt(II) compounds.

8.5 Phosphane (phosphine) and arsane (arsine) analogues of [PtCl$_2$(NH$_3$)$_2$] were prepared in 1934 by Jensen. He reported zero dipole moments for the β isomers, where the designation β represents the product of a synthetic route analogous to that of the ammines. Give the structures of the complexes.

8.6 Air oxidation of Co(II) carbonate and aqueous ammonium chloride gives a pink chloride salt with a ratio of 4NH$_3$:Co. On addition of HCl to a solution of this salt, a gas is rapidly evolved and the solution slowly turns violet on heating. Complete evaporation of the violet solution yields CoCl$_3$·4NH$_3$. When this is heated in concentrated HCl, a green salt can be isolated with composition CoCl$_3$·4NH$_3$·HCl. Write balanced equations for all the transformations occurring after the air oxidation. Give as much information as possible concerning the isomerism occurring and the basis of your reasoning. Is it helpful to know that the form of [Co(Cl)$_2$(en)$_2$]$^+$ that is resolvable into enantiomers is violet?

8.7 When cobalt(II) salts are oxidized by air in a solution containing ammonia and sodium nitrite, a yellow solid, [Co(NO$_2$)$_3$(NH$_3$)$_3$], can be isolated. In solution it is nonconducting; treatment with HCl gives a complex that, after a series of further reactions, can be identified as *trans*-[Co(Cl)$_2$(NH$_3$)$_3$(OH$_2$)]$^+$. It requires an entirely different route to prepare *cis*-[Co(Cl)$_2$(NH$_3$)$_3$(OH$_2$)]$^+$. Is the yellow substance *fac* or *mer*? What assumption must you make to arrive at a conclusion?

8.8 The reaction of [ZrCl$_4$(dppe)] (dppe is a bidentate phosphine ligand) with Mg(CH$_3$)$_2$ gives [Zr(CH$_3$)$_4$(dppe)]. NMR spectra indicate that all methyl groups are equivalent. Draw octahedral and trigonal prism structures for the complex and show how the conclusion from NMR supports the trigonal prism assignment. (P.M. Morse and G.S. Girolami, *J. Am. Chem. Soc.*, 1989, **111**, 4114.)

8.9 The resolving agent *d*-*cis*[Co(NO$_2$)$_2$(en)$_2$]Br can be converted to the soluble nitrate by grinding in water with AgNO$_3$. Outline the use of this species for resolving a racemic mixture of the *d* and *l* enantiomers of K[Co(edta)]. (The *l*-[Co(edta)]$^-$ enantiomer forms the less soluble diastereomer. See F.P. Dwyer and F.L. Garvan, *Inorg. Synth.*, 1965, **6**, 192.)

8.10 Show how the coordination of two MeHNCH$_2$CH$_2$NH$_2$ ligands to a metal atom in a square-planar complex results in not only *cis* and *trans* but also optical isomers. Identify the isomer that does not have any mirror planes but is not chiral.

PART 2
The elements and their compounds

This part of the book describes the chemistry of the elements as they are set out in the periodic table. The descriptive chemistry of the elements is often viewed as a collection of unrelated facts with few underlying principles. In fact, a study of descriptive chemistry reveals a rich tapestry of patterns and trends, many of which can be rationalized and explained by application of the concepts developed in Part 1.

The first chapter in this part, Chapter 9, deals with the chemistry of the unique element hydrogen. The following eight chapters (Chapters 10–17) proceed systematically across the groups of the *s* and the *p* blocks. The chemistry of the elements of these groups demonstrates the diversity, intricacy, and fascinating nature of inorganic chemistry.

The chemistry of the *d*-block elements is so extensive that the remaining four chapters of the part are devoted to it. Chapter 18 reviews the descriptive chemistry of the three series of *d*-block elements. Chapters 19 and 20 describe how electronic structure affects the chemical and physical properties of the *d*-metal complexes and describes their reactions in solution. Chapter 21 describes the industrially vital *d*-metal organometallic compounds.

Our tour of the periodic table draws to a close in Chapter 22 with a description of the chemistry of the *f*-block elements.

Hydrogen

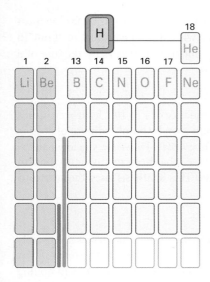

Hydrogen has very rich chemical properties despite its simple atomic structure; it was in fact at the forefront of much of the discussion in earlier chapters, especially in the description of acids and bases. Here we summarize some of the properties of the element and its binary compounds. The latter range from ionic compounds and solids with metallic properties to volatile, molecular compounds. We shall see that many of the properties of these compounds are understandable in terms of their ability to supply H^- or H^+ ions, and that it is often possible to anticipate which one is likely to dominate.

Hydrogen is the most abundant element in the universe and the tenth most abundant by mass on Earth, where it is found in minerals, the oceans, and in all living things. The partial depletion of hydrogen from the Earth reflects its volatility during the formation of the planet.

Because a hydrogen atom has only one electron, it might be thought that the element's chemical properties will be limited, but this is far from the case. Hydrogen has richly varied chemical properties, and despite its single electron, under certain circumstances it can bond to more than one atom simultaneously. Moreover, it ranges in character from a strong Lewis base (as the hydride ion, H^-) to a strong Lewis acid (as the hydrogen cation, H^+, the proton) (Section 4.7). In fact, hydrogen forms more compounds than any other element.

The element

Hydrogen does not fit neatly into the periodic table. It is sometimes placed at the head of the alkali metals in Group 1, with the rationale that hydrogen and the alkali metals have only one valence electron. That position, however, is not entirely satisfactory in terms of either the chemical or physical properties of the element. In particular, hydrogen is not a metal under normal conditions. Hydrogen may be metallic at the core of Jupiter, and the extreme pressures present there have been recreated in the laboratory to produce metallic hydrogen. Less frequently, hydrogen is placed above the halogens in Group 17, with the rationale that, like the halogens, it requires one electron to complete its valence shell. In this text, hydrogen is placed on its own at the head of the periodic table to emphasize its unique nature.

9.1 Nuclear properties

Key point: The three hydrogen isotopes H, D, and T have large differences in their atomic masses and different nuclear spins, which give rise to easily observed changes in IR, Raman, and NMR spectra of molecules containing these isotopes.

There are three isotopes of hydrogen: hydrogen itself (^{1}H), deuterium (D, ^{2}H), and tritium (T, ^{3}H); tritium is radioactive. The three isotopes are unique in having different names and symbols, reflecting the significant differences in their masses and the chemical properties that stem from mass, such as the rates of bond-cleavage reactions. The distinct properties of the isotopes make them useful as **tracers**, which are isotopes that can be followed through a series of reactions, such as by infrared (IR, Section 6.4) or nuclear magnetic resonance (NMR) spectroscopy (Section 6.5). Tritium is sometimes preferred as a tracer because it can be detected by its radioactivity, which can be a more sensitive probe than spectroscopy.

The lightest isotope, ^{1}H (which is very occasionally called protium), is by far the most abundant. Deuterium has variable natural abundance with an average value of about 16 atoms in 100 000. Neither ^{1}H nor ^{2}H is radioactive, but tritium decays by the loss of a β particle to yield a rare but stable isotope of helium:

$$^3_1\text{H} \rightarrow {}^3_2\text{He} + \beta^-$$

The half-life for this decay is 12.4 years. Tritium's abundance of 1 in 10^{21} hydrogen atoms in surface water reflects a steady state between its production by bombardment of cosmic rays on the upper atmosphere and its loss by radioactive decay. However, the low natural abundance of the isotope is increased artificially because it has been manufactured on an undisclosed scale for use in thermonuclear weapons. This synthesis uses the neutrons from a fission reactor with Li as a target:

$$^1_0\text{n} + {}^6_3\text{Li} \rightarrow {}^3_1\text{H} + {}^4_2\text{He}$$

The physical and chemical properties of isotopically substituted molecules are usually very similar. However, this is not true when D is substituted for H, as the mass of the substituted atom is doubled. For instance, Table 9.1 shows that the differences in boiling points and bond enthalpies are easily measurable for H_2 and D_2. The difference in boiling point between H_2O and D_2O reflects the greater strength of the O$\cdots$D—O hydrogen bond compared with that of the O$\cdots$H—O bond (Section 9.4b). The compound D_2O is known as 'heavy water' and is used as a moderator in the nuclear power industry; it slows down emitted neutrons and increases the rate of induced fission.

Reaction rates are often measurably different for processes in which E—H and E—D bonds, where E is another element, are broken, made, or rearranged. The detection of this **kinetic isotope effect** can often help to support a proposed reaction mechanism. Kinetic isotope effects are frequently observed when an H atom is transferred from one atom to another in an activated complex. For example, the electrochemical reduction of H^+(aq) to H_2(g) occurs with a substantial isotope effect, with H_2 being liberated more rapidly. A practical consequence of the difference in rates of formation of H_2 and D_2 is that D_2O may be concentrated electrolytically.

Because the frequencies of molecular vibrations depend on the masses of atoms, they are strongly influenced by substitution of D for H. The heavier isotope results in the lower frequency (Section 6.4). Inorganic chemists often take advantage of this isotope effect by observing the IR spectra of **isotopomers**, molecules differing in their isotopic

Table 9.1 The effect of deuteration on physical properties

	H_2	D_2	H_2O	D_2O
Normal boiling point/°C	−252.8	−249.7	100.0	101.4
Mean bond enthalpy/(kJ mol^{-1})	436.0	443.3	463.5	470.9

composition, to determine whether a particular infrared absorption involves significant motion of hydrogen.

Another important property of the hydrogen nucleus is its spin. The nucleus of hydrogen, a proton, has $I = \frac{1}{2}$; the nuclear spins of D and T are 1 and $\frac{1}{2}$, respectively. As explained in Chapter 6, proton NMR detects the presence of H nuclei in a compound and is a powerful method for determining structures of molecules, even proteins with molar masses as high as 20 kDa. Molecular hydrogen, H_2, exists in two forms that differ in the relative orientations of the two nuclear spins: in *ortho*-hydrogen the spins are parallel ($I = 1$), in *para*-hydrogen the spins are antiparallel ($I = 0$). At $T = 0$ hydrogen is 100 per cent *para*. As the temperature is raised, the proportion of the *ortho* form in a mixture at equilibrium increases until at room temperature there is approximately 75 per cent *ortho* and 25 per cent *para*. Most physical properties of the two forms are the same, but the melting and boiling points of *para*-hydrogen are about 0.1°C lower than those of normal hydrogen and the thermal conductivity of *para*-hydrogen is about 50 per cent greater than that of the *ortho* form.

9.2 Hydrogen atoms and ions

Key points: In combination with metals, hydrogen is often regarded as a hydride; hydrogen compounds with elements of similar electronegativity have low polarity; compounds in which H is bonded to a highly electronegative centre are usually acidic.

The H atom has a high ionization energy ($1310\ kJ\ mol^{-1}$) and a low but positive electron affinity ($77\ kJ\ mol^{-1}$). The Pauling electronegativity of hydrogen is 2.2. This value is similar to those of B, C, and Si, so E—H bonds involving these elements are not expected to be very polar.

Hydrogen forms compounds with metals that are referred to as 'metal hydrides'; however, only the compounds of the highly electropositive metals in Groups 1 and 2 can be regarded as containing the hydride ion, H^-. When combined with the highly electronegative elements from the right of the periodic table, the E—H bond is best regarded as covalent and polar, with the H atom carrying a very small partial positive charge. As can be seen by inspection of Fig. 9.1, strong acids have strongly deshielded proton NMR signals (Section 6.5). In accord with its electronegativity, hydrogen is normally assigned oxidation number -1 when in combination with metals (as in NaH and AlH_3) and $+1$ when in combination with nonmetals (as in NH_3 and HCl).

When an electric discharge is passed through hydrogen gas at low pressure, the molecules dissociate, ionize, and recombine to form a plasma that contains spectroscopically observable amounts of H, H^+, H_2^+, and H_3^+. The free hydrogen cation (H^+, the proton) has a very high charge/radius ratio and so it is not surprising to find that it is a very strong Lewis acid. In the gas phase it readily attaches to other molecules and atoms;

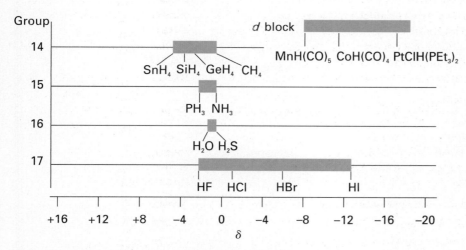

Fig. 9.1 Typical ^{1}H NMR chemical shifts. The tinted boxes group families of elements together.

it even attaches to He to form HeH^+. In the condensed phase, H^+ is always found in combination with a Lewis base, and its ability to transfer between Lewis bases gives it the special role in chemistry that we explored in detail in Chapter 4. The molecular cations H_2^+ and H_3^+ have only a transitory existence in the gas phase and are unknown in solution.

9.3 Properties and reactions of dihydrogen

The stable form of elemental hydrogen under normal conditions is dihydrogen, H_2, more commonly referred to simply as 'hydrogen'. The H_2 molecule has a high bond enthalpy

---

Box 9.1 The hydrogen cycle

Oxygen is an abundant oxidizing agent in nature and aerobic reactions are dominant on a global scale. However, in some environments anaerobic reactions become important. For example, in sediments, anaerobic reactions occur to a greater extent in lower layers. In these strata, carbon compounds, originally formed by photosynthesis or chemosynthesis, undergo decay reactions that are facilitated by bacteria. These reactions produce many reactive small molecules and ions, including H_2, CO, CO_2, HCO_2^-, and $CH_3CO_2^-$.

A schematic diagram of a fresh water environment is shown in Fig. B9.1. The inorganic compounds produced in the lower anaerobic strata of this environment provide raw materials that ultimately allow archaea (methanogens) to produce methane and bacteria (acetogens) to produce acetate ions. If there are sulfate ions in the environment, sulfate-reducing bacteria will convert sulfate to sulfide. Metalloenzymes with Fe-, Co-, Ni-, and Mo-based active sites play key roles in these reactions.

The methanogens produce methane, which diffuses towards the surface. As the methane diffuses up from the anaerobic sediments, at some point it encounters O_2 diffusing down from the surface. Where these two gases meet methanotrophs use the methane and oxygen to generate carbon compounds and release energy. The first step in methane oxidation is carried out by the methane mono-oxygenase (MMO) enzyme (Section 26.10), which depends on iron or copper and catalyses the formation of methanol. In anaerobic environments such as bogs, rice paddies, and the rumens of grazing animals, such as sheep and cows, methane may escape into the atmosphere without further biological conversion. In the atmosphere, CH_4 is a potent greenhouse gas, but, due to its reaction with OH radicals, has a limited lifetime ($t\frac{1}{2} \approx 12$ years). In very deep basins and at high, cold latitudes, CH_4 can form massive gas hydrates just below the surface layer. These gas hydrates are clathrate compounds, which may constitute the largest fossil fuel reservoir on Earth (Box 13.2).

In anaerobic environments many reactions occur in which molecular hydrogen plays a dramatic role (Fig. B9.2). In clostridia and other fermentative organisms, no strong oxidant, such as O_2, is available or usable and the electron acceptor involved in many reactions is the proton, so generating H_2. The enzyme responsible for the generation of H_2 is the Fe–Fe hydrogenase (Section 26.14), which has CO and CN^- ligands bound to two Fe atoms of a unique six Fe atom cluster. These strong-field ligands, which are generally toxic to most life, stabilize low-spin iron states that favour the evolution of hydrogen gas.

The H_2 generated by fermentation is used as a reductant by a wide variety of organisms in both the anaerobic and aerobic environment. These organisms use NiFe hydrogenases, which catalyse H_2 uptake and use it to execute diverse reaction chemistry. Thus, H_2 serves as a 'redox currency', and is used in the reduction of nitrate, sulfate, iron(III), dinitrogen, and sulfur in anaerobic habitats. Should H_2 reach the aerobic environment, H_2-oxidizing aerobes combine H_2 and O_2 to produce water. The processes that consume H_2 keep its level low enough that formation of H_2 by the fermentative organisms remains thermodynamically favourable.

B9.1 Species present in aerobic and anaerobic fresh water environments.

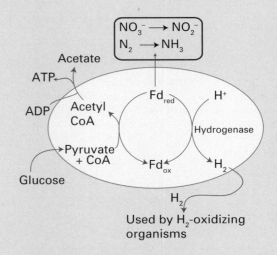

B9.2 The role of hydrogen in anaerobic environments.

($436\,\text{kJ mol}^{-1}$) and a short bond length (74 pm). Because it has so few electrons, the forces between H_2 molecules are weak, and at 1 atm the gas condenses to a liquid only when cooled to 20 K.

(a) Production

Key points: Much of the hydrogen for industry is produced by high-temperature reaction of H_2O with CH_4 or a similar reaction with coke. Hydrogen's major uses are in the processing of hydrocarbon fuels and synthesis of ammonia.

Molecular hydrogen is not present in significant quantities in the Earth's atmosphere or in underground gas deposits. However, there is a very high biological turnover, because various microorganisms use H^+ as an oxidant or H_2 as a fuel (Boxes 9.1 and 9.2).

Hydrogen is produced in huge quantities to satisfy the needs of industry. The main commercial process for the production of hydrogen is currently *steam reforming*, the catalysed reaction of water and hydrocarbons (typically methane from natural gas) at high temperatures:

$$CH_4(g) + H_2O(g) \xrightarrow{1000°C} CO(g) + 3\,H_2(g)$$

A similar reaction, but with coke as the reducing agent, is sometimes called the *water-gas reaction*:

$$C(s) + H_2O(g) \xrightarrow{1000°C} CO(g) + H_2(g)$$

This reaction was once a primary source of H_2 and it may become important again when natural hydrocarbons are depleted. Both reactions are generally followed by a second reaction, often called the *shift reaction*, in which water is reduced to hydrogen by reaction with carbon monoxide:

$$CO(g) + H_2O(g) \rightarrow CO_2(g) + H_2(g)$$

Hydrogen production is often integrated into chemical processes that require H_2 as a feedstock. As shown in Fig. 9.2, a major use of hydrogen is direct combination with N_2 to produce NH_3, the primary source of nitrogen-containing chemicals, plastics, and fertilizers. Another major chemical, methanol, is produced from the catalytic combination of H_2 and CO. Because of its high specific enthalpy (approximately three times that of a typical hydrocarbon), hydrogen is an excellent fuel for large rockets.

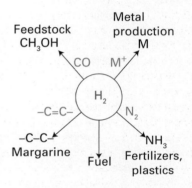

Fig. 9.2 The major uses of hydrogen.

(b) Reactions of hydrogen

Key points: Molecular hydrogen is activated by dissociation on a metal surface or by formation of a metal complex; reactions of hydrogen with oxygen and halogens commonly involve a radical chain mechanism.

Molecular hydrogen reacts slowly with most other elements, partly because of its high bond enthalpy and hence high activation energy for reaction. Under special conditions, however, the reactions are rapid. These conditions include:

1 The activation of the molecule by homolytic dissociation on a metal surface or a metal complex.[1]

2 Heterolytic dissociation by a surface or a metal ion.

3 Initiation of a radical chain reaction.

Two examples of homolytic dissociation are chemisorption on platinum and coordination to the Ir atom in a complex (Section 21.22). The former accounts for the use of finely divided platinum metal to catalyse the hydrogenation of alkenes and of platinum as the electrode material used in the electrolytic reduction of H^+ and the oxidation of

---

[1] In homolytic dissociation the bond breaks symmetrically to give one product. In heterolytic dissociation the bond breaks unsymmetrically to give two different products.

Box 9.2 Fuel cells

The use of hydrogen as a fuel has been analysed seriously since the early 1970s when petroleum prices rose sharply. Strategies have been devised for a 'hydrogen economy' in which hydrogen is the primary fuel, perhaps used in fuel cells to produce electricity. Aside from the eventual scarcity of fossil fuels, when hydrogen burns it produces water rather than greenhouse gases such as CO_2. Hydrogen fuel cells are already being used in a range of applications, from personal electronic equipment to motor vehicles. Challenges impeding their development are related to the efficient production of hydrogen and the development of hydrogen storage technologies. Hydrogen can be liberated from hydrocarbon fuels, but this method liberates greenhouse gases and does little to reduce society's reliance on fossil fuels. Hydrogen can be liberated from water, and the methods under development include electrolysis using energy from a renewable source, biomass gasification (in which agricultural waste is superheated to release hydrogen and other gases), and photobiological solutions in which enzymes in green algae produce hydrogen from water. Any application of hydrogen fuel cells must tackle the need to store hydrogen in as small a volume as possible. Liquefaction is a possibility but the cryogenic equipment required is expensive and requires energy. Metal hydrides have also been used as hydrogen storage media (Section 9.5). A more appealing solution is storage within microporous materials, and interest has focused on carbon-based materials such as nanotubes and microfibres.

Further reading

P. Hoffman. *Tomorrow's energy: hydrogen, fuel cells, and the prospects for a cleaner planet.* The MIT Press (2001).

For up-to-date information on the latest developments in fuel cell technology see the US Energy Department's website on energy efficiency and renewable energy, http://www.eere.energy.gov/hydrogenandfuelcells/.

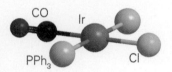

1 [IrClCO(PPh₃)₂]

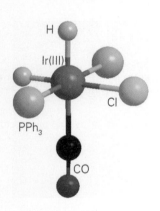

2 [IrCl(H)₂CO(PPh₃)₂]

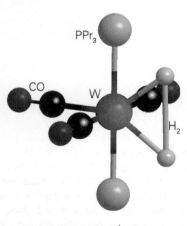

3 [W(CO)₃(H₂)(PⁱPr₃)₂]

H_2 in fuel cells (Box 9.2). The overpotential for H_2 formation is much smaller at a platinum electrode, where dissociative chemisorption occurs, than at a mercury electrode, where chemisorption is unfavourable. Similarly, H_2 readily coordinates to the Ir atom in complexes such as [IrClCO(PPh₃)₂] (**1**) to give a complex containing two hydrido (H^-) ligands (**2**). This type of reaction is common for Rh(I), Ir(I), and Pt(0) complexes. There are many d-block hydrido complexes; they are discussed in more detail in Chapter 21.

An H_2 molecule can coordinate to a metal atom without cleavage of the H—H bond. The first such compound to be identified was [W(CO)₃(H₂)(PⁱPr₃)₂] (**3**), where ⁱPr denotes isopropyl, CH(CH₃)₂, and well over a hundred compounds of this type are now known. The significance of these compounds is that they provide examples of species intermediate between molecular H_2 and a dihydrido complex. No dihydrogen complexes are known for the early d-block (Groups 3, 4, and 5), the f-block, or the p-block metals. The bonding of H_2 to d-block metals is discussed in detail in Section 21.7.

An example of heterolytic dissociation is the reaction of H_2 with a ZnO surface, which appears to produce a Zn(II)-bound hydride and an O-bound proton:

$$H_2 + Zn-O-Zn-O \longrightarrow Zn-O-Zn-O$$

This reaction is thought to be involved in the catalytic hydrogenation of carbon monoxide to methanol

$$CO(g) + 2\,H_2(g) \xrightarrow{Cu/ZnO/Al_2O_3} CH_3OH(g)$$

which is carried out on a very large scale worldwide as methanol is used as a feedstock in the industrial production of many organic chemicals. Another example is the sequence

$$H_2(g) + Cu^{2+}(aq) \rightarrow CuH^+(aq) + H^+(aq) \rightarrow Cu(s) + 2\,H^+(aq)$$

which is important in the hydrometallurgical reduction of Cu^{2+} (Chapter 5). The CuH^+ intermediate has only a transitory existence. Another example in which H_2 is dissociated into a hydride and a proton is during its oxidation at the active site of hydrogenases (see Section 26.14).

Radical chain mechanisms account for the thermally or photochemically initiated reactions between H_2 and the halogens. Initiation is by thermal or photochemical dissociation of the dihalogen molecules to give atoms that act as radical chain carriers in the propagation reaction:

$$\text{Initiation} \quad Br_2 \xrightarrow{\Delta\ or\ h\nu} Br\cdot + \cdot Br$$

$$\text{Propogation} \quad Br\cdot + H_2 \rightarrow HBr + H\cdot$$

$$H\cdot + Br_2 \rightarrow HBr + Br\cdot$$

The activation energy for radical attack is low because a new bond is formed as one bond is lost, so once initiated the formation and consumption of radicals is self-sustaining and the production of HBr is very rapid. Chain termination occurs when the radicals recombine:

$$\text{Termination} \quad H\cdot + \cdot H \rightarrow H_2$$

$$Br\cdot + \cdot Br \rightarrow Br_2$$

Termination becomes more important towards the end of the reaction when the concentrations of H_2 and Br_2 are low.

Despite the low reactivity of hydrogen due to its high bond enthalpy, one startling reaction is the combustion of hydrogen to form water. Mixtures of hydrogen gas and oxygen (or air) do not react until they are ignited with a spark or flame, when the resulting reaction is explosive:

$$2\,H_2 + O_2 \rightarrow 2\,H_2O \qquad \Delta_r H^\ominus = -244\,kJ\,mol^{-1}$$

The driving force for this reaction is the high O—H bond enthalpy ($464\,kJ\,mol^{-1}$), which ensures that the reaction is exoergic. That high bond enthalpy is in turn due, in part, to the coincidental near equality of the ionization energies of H and O, so their valence electrons have similar energies.

Compounds of hydrogen

Figure 9.3 summarizes the classification of the binary compounds of hydrogen and their distribution through the periodic table. The main point of devising this classification is to emphasize the principal trends in properties. In fact, there is range of structural types, and some elements form compounds with hydrogen that do not fall strictly into any one category. The three classes of binary hydrogen compounds that we consider are:

1 **Molecular hydrides**, binary compounds of an element and hydrogen in the form of individual, discrete molecules.
2 **Saline hydrides**, nonvolatile, electrically nonconducting, crystalline solids.
3 **Metallic hydrides**, nonstoichiometric, electrically conducting solids.

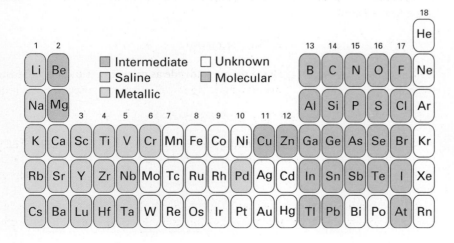

Fig. 9.3 Classification of the binary hydrogen compounds of the *s*-, *p*-, and *d*-block elements. Although some *d*-block elements, such as iron and ruthenium, do not form binary hydrides, they do form metal complexes containing the hydride ligand.

9.4 Molecular hydrides

Molecular compounds of hydrogen are common for the electronegative elements of Groups 13 to 17; some examples are B_2H_6, CH_4, NH_3, H_2O, and HF.

(a) Nomenclature and classification

Key point: Molecular compounds of hydrogen are classified as electron-rich, electron-precise, or electron-deficient.

The systematic names of the molecular hydrogen compounds are formed from the name of the element and the suffix -ane, as in phosphane for PH_3. The more traditional names, however, such as phosphine and hydrogen sulfide (H_2S, sulfane) are still widely used (Table 9.2). The common names ammonia and water are universally used rather than their systematic names azane and oxidane.

Molecular compounds of hydrogen are divided into three categories:

1 **Electron-precise**, in which all valence electrons of the central atom are engaged in bonds.

2 **Electron-deficient**, in which there are too few electrons to be able to write a Lewis structure for the molecule.

3 **Electron-rich**, in which there are more electron pairs on the central atom than are needed for bond formation (that is, there are lone pairs on the central atom).

The hydrocarbons, such as methane and ethane, are electron-precise; so too are silane, SiH_4, and germane, GeH_4 (Chapter 13). All these molecules are characterized by the presence of two-centre, two-electron bonds ($2c,2e$ bonds) and the absence of lone pairs on the central atom. Diborane, B_2H_6 (**4**), is an example of an electron-deficient compound. Its Lewis structure would require at least 14 valence electrons to bind the eight atoms together, but the molecule has only 12 valence electrons. The simple explanation of its structure is the presence of BHB three-centre, two-electron bonds ($3c,2e$; Section 2.11c) acting as bridges between the two B atoms, so two electrons can help to bind three atoms. These bridging B—H bonds are longer and weaker that the terminal B—H bonds. Electron-deficient hydrogen compounds are common for boron and aluminium and the

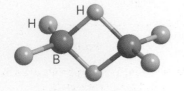

4 Diborane, B_2H_6

Table 9.2 Some molecular hydrogen compounds			
Group	Formula	Traditional name	IUPAC name
13	B_2H_6	Diborane	Diborane[6]
14	CH_4	Methane	Methane
	SiH_4	Silane	Silane
	GeH_4	Germane	Germane
	SnH_4	Stannane	Stannane
15	NH_3	Ammonia	Azane
	PH_3	Phosphine	Phosphane
	AsH_3	Arsine	Arsane
	SbH_3	Stibine	Stibane
16	H_2O	Water	Oxidane
	H_2S	Hydrogen sulfide	Sulfane
	H_2Se	Hydrogen selenide	Sellane
	H_2Te	Hydrogen telluride	Tellane
17	HF	Hydrogen fluoride	Hydrogen fluoride
	HCl	Hydrogen chloride	Hydrogen chloride
	HBr	Hydrogen bromide	Hydrogen bromide
	HI	Hydrogen iodide	Hydrogen iodide

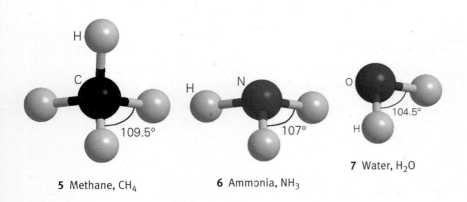

5 Methane, CH_4

6 Ammonia, NH_3

7 Water, H_2O

Table 9.3 Bond angles (in degrees) for Group 15 and 16 hydrogen compounds

NH_3	106.6	H_2O	104.5
PH_3	93.8	H_2S	92.1
AsH_3	91.8	H_2Se	91
SbH_3	91.3	H_2Te	89

compounds generally exhibit Lewis acidity. Electron-rich compounds are formed by the elements in Groups 15 to 17. Ammonia, NH_3, with one lone pair on nitrogen, and water, H_2O, with two lone pairs on oxygen, are examples. These types of molecules generally exhibit Lewis basicity. For example, the electron-rich compound trimethylamine, $N(CH_3)_3$, reacts with boron trifluoride to form the Lewis acid–base complex trimethylamine trifluoroborane, $(H_3C)_3NBF_3$. The hydrogen halides, HF, HCl, HBr, and HI, are another important group of electron-rich compounds (Section 16.2).

The shapes of the molecules of the electron-precise and electron-rich compounds can all be predicted by the VSEPR rules (Section 2.3). Thus, CH_4 is tetrahedral (**5**), NH_3 is trigonal pyramidal (**6**), H_2O is angular (**7**), and HF is (necessarily) linear. However, the simple VSEPR rules do not indicate the considerable change in bond angle between NH_3 and its heavier analogues or between H_2O and its analogues in Group 16. As noted in Table 9.3, the bond angles in NH_3 and H_2O are slightly less than the tetrahedral angle, but for their heavier analogues the bond angle is as small as 90°. The basic tetrahedral arrangement of electron pairs in these compounds was explained by the formation of hybrid sp^3 orbitals in Section 2.6. Hybridization relies on the effective overlap between s and p orbitals on the central atom and this overlap becomes less effective down a group.

(b) Hydrogen bonding

Key points: Hydrogen compounds of electronegative nonmetals with at least one lone pair often associate through hydrogen bonds; water, ice, and clathrate hydrates are examples of this association.

An important consequence of the simultaneous presence of highly electronegative atoms (N, O, and F) and lone pairs in the electron-rich compounds is the possibility of forming hydrogen bonds. The E—H bond between an electronegative element E and hydrogen is highly polar, $^{\delta-}E—H^{\delta+}$, and the partially positively charged H atom can interact with a lone pair on the electron-rich compound. Thus, a **hydrogen bond** consists of an H atom between atoms of more electronegative nonmetallic elements. This definition includes the widely recognized N—H···N and O—H···O hydrogen bonds but excludes the B—H—B bridges in boron hydrides because boron is not more electronegative than hydrogen. Striking evidence for hydrogen bonding is provided by the trends in boiling points, which are unusually high for the strongly hydrogen-bonded molecules water, ammonia, and hydrogen fluoride (Fig. 9.4). Some of the most persuasive evidence for hydrogen bonding comes from structural data, such as the open network structure of ice (Fig. 9.5), the existence of chains in solid HF that survive partially even in the vapour (**8**), and the location of hydrogen atoms in solids by diffraction and NMR techniques. As indicated by the trends in boiling points, H_2S, PH_3, HCl, and the heavier p-block molecular hydrides do not form strong hydrogen bonds.

Although hydrogen bonds are much weaker than conventional bonds (Table 9.4), they have important consequences for the properties of the electron-rich hydrogen compounds of Period 2, including their densities, viscosities, vapour pressures, and acid–base characters. Hydrogen bonding is readily detected by the shift to lower wavenumber and broadening of E—H stretching bands in infrared spectra (Fig. 9.6) and by unusual

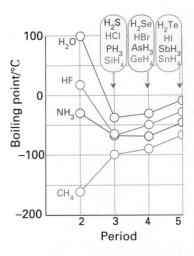

Fig. 9.4 Normal boiling points of p-block binary hydrogen compounds.

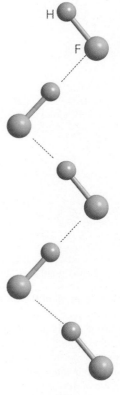

8 $(HF)_5$

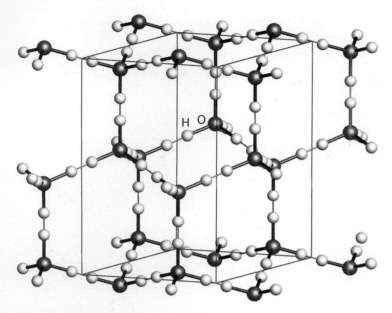

Fig. 9.5 The structure of ice. The structure shows all possible atom positions, but only half are actually occupied.

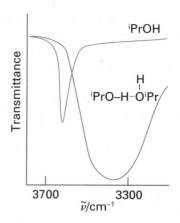

Fig. 9.6 Infrared spectra of 2-propanol. In the upper curve, 2-propanol is present as unassociated molecules in dilute solution. In the lower curve, the pure alcohol is associated through hydrogen bonds. The association lowers the frequency and broadens the O—H stretching absorption band. (From N.B. Colthup, L.H. Daly, and S.E. Wiberley, *Introduction to infrared and Raman spectroscopy.* Academic Press, New York (1975).)

Table 9.4 Comparison of hydrogen bond enthalpies with the corresponding E—H covalent bond enthalpies (kJ mol^{-1})

	Hydrogen bond		Covalent bond
HS—H$\cdots$SH$_2$	7	S—H	363
H$_2$N—H$\cdots$NH$_3$	17	N—H	386
HO—H$\cdots$OH$_2$	22	O—H	464
F—H$\cdots$FH	29	F—H	565
HO—H$\cdots$Cl$^-$	55	Cl—H	428
F$\cdots$H$\cdots$F$^-$	465	F—H	565

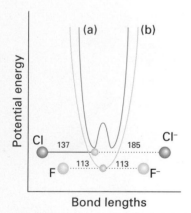

Fig. 9.7 The variation of the potential energy with the position of the proton between two atoms in a hydrogen bond. (a) The double minimum potential characteristic of a weak hydrogen bond. (b) The single minimum potential characteristic of the strong hydrogen bond in [FHF]$^-$.

proton chemical shifts in proton NMR. Hydrogen bonding may be symmetrical or unsymmetrical. In unsymmetrical hydrogen bonding, the H atom is not midway between the two nuclei, even when the heavier linked atoms are identical. For example, the [ClHCl]$^-$ ion is linear but the H atom is not midway between the Cl atoms (Fig. 9.7). By contrast, in the bifluoride ion, [FHF]$^-$, the H atom lies midway between the F atoms; the FF separation (226 pm) is significantly less than twice the van der Waals radius of the F atom (2 × 135 pm).

The structures of hydrogen-bonded complexes have been observed in the gas phase by microwave spectroscopy. The lone-pair orientation of electron-rich compounds implied by VSEPR theory shows good agreement with the HF orientation (Fig. 9.8). For example, HF is oriented along the threefold axis of NH$_3$ and PH$_3$, out of the H$_2$O plane in its complex with H$_2$O, and off the HF axis in the HF dimer. X-ray single crystal structure determinations often show the same patterns, as for example in the structure of ice and in solid HF, but packing forces in solids may have a strong influence on the orientation of the relatively weak hydrogen bond.

One of the most interesting manifestations of hydrogen bonding is the structure of ice. There are at least ten different phases of ice but only one is stable at ordinary pressures. The familiar low-pressure phase of ice, ice-I, crystallizes in a hexagonal unit cell with each O atom surrounded tetrahedrally by four others (as shown in Fig. 9.5). These O atoms are held together by hydrogen bonds with O—H···O and O···H—O bonds largely randomly distributed through the solid. The resulting structure is quite open, which accounts for the density of ice being lower than that of water. When ice melts, the network of hydrogen bonds partially collapses.

Water can also form **clathrate hydrates**, consisting of hydrogen-bonded cages of water molecules surrounding foreign molecules or ions. One example is the clathrate hydrate of composition $Xe_4(CCl_4)_8(H_2O)_{68}$ (Fig. 9.9). In this structure, the cages with O atoms defining their corners consist of 14-faced and 12-faced polyhedra in the ratio 3 : 2. These O atoms are held together by hydrogen bonds and guest molecules occupy the interiors of the polyhedra. Aside from their interesting structures, which illustrate the organization that can be enforced by hydrogen bonding, clathrate hydrates are often used as models for the way in which water appears to become organized around nonpolar groups, such as those in proteins. Methane clathrate hydrates occur in the Earth at high pressures, and it is estimated that huge quantities of natural gas are trapped in these formations (see Box 13.2).

Some ionic compounds form clathrate hydrates in which the anion is incorporated into the framework by hydrogen bonding. This type of clathrate is particularly common with the very strong hydrogen bond acceptors F^- and OH^-. One such example is $N(CH_3)_4F·4H_2O$ (9).

A very important aspect of hydrogen bonding is its role in determining the structures of proteins and nucleic acids. Specific hydrogen bonding between adenine and thymine and between guanine and cytosine is the basis of DNA replication; in proteins, hydrogen bonding between backbone peptide —NH and CO groups of different amino acid residues is the basis of the α helix and β sheet secondary structures (Section 26.2).

9.5 Saline hydrides

Key points: Hydrogen compounds of electropositive metals may be regarded as ionic hydrides, M^+H^-; they liberate H_2 in contact with acids and transfer H^- to electrophiles.

Group 1 hydrides have rock-salt structures (Section 3.9) and Group 2 hydrides have crystal structures like those of some heavy metal halides (Table 9.5). These structures (and their chemical properties) are the basis for classifying hydrogen compounds of the s-block elements other than beryllium as saline (salt-like). The ionic radius of H^- determined from X-ray diffraction varies from 126 pm in LiH to 154 pm in CsH. This wide variability reflects the loose control that the single charge of the proton has on its two surrounding electrons and the resulting high compressibility and polarizability of H^-.

The saline hydrides are insoluble in common nonaqueous solvents but they do dissolve in molten alkali halides. Electrolysis of these molten-salt solutions produces

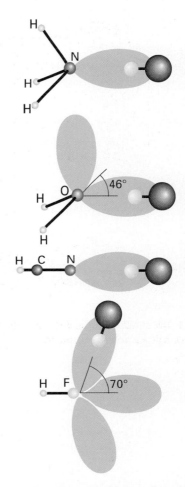

Fig. 9.8 The orientation of lone pairs as indicated by VSEPR theory compared with the orientation of HF in the gas-phase hydrogen-bonded complex.

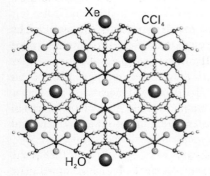

Fig. 9.9 The cages of water molecules in clathrate hydrates, such as $Xe_4(CCl_4)_8(H_2O)_{68}$.

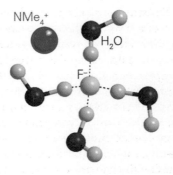

9 $N(CH_3)_4F·4H_2O$

Table 9.5 Structures of s-block hydrides	
Compound	Crystal structure
LiH, NaH, RbH, CsH	Rock salt
MgH_2	Rutile
CaH_2, SrH_2, BaH_2	Distorted $PbCl_2$

hydrogen gas at the anode (the site of oxidation):

$$2 \, H^-(\text{melt}) \rightarrow H_2(g) + 2 \, e^-$$

This reaction provides chemical evidence for the existence of H^-. The reaction of saline hydrides with water is dangerously vigorous:

$$NaH(s) + H_2O(l) \rightarrow H_2(g) + NaOH(aq)$$

Indeed, finely divided sodium hydride can ignite if it is left exposed to humid air. Such fires are difficult to extinguish because even carbon dioxide is reduced when it comes into contact with hot metal hydrides (water, of course, forms even more flammable hydrogen); they may, however, be blanketed with an inert solid, such as silica sand. The reaction between metal hydrides and water is used in the laboratory to remove traces of water from solvents and inert gases such as nitrogen and argon:

$$CaH_2(s) + 2 \, H_2O(g) \rightarrow Ca(OH)_2(s) + 2 \, H_2(g)$$

Calcium hydride is preferred in this application because it is the cheapest of the saline hydrides and is available in granular form, which is easy to handle. Large amounts of water should not be removed from a solvent in this manner because the strongly exothermic reaction evolves flammable hydrogen.

The absence of convenient solvents limits the use of saline hydrides as reagents, but this problem is partially overcome by the availability of commercial dispersions of finely divided NaH in oil. Even more finely divided and reactive alkali metal hydrides can be prepared from the metal alkyl and hydrogen. Alkali metal hydrides are convenient reagents for making other hydride compounds. For example, reaction with a trialkylboron compound yields a hydride complex that is soluble in polar organic solvents and therefore a useful reducing agent and source of hydride ions:

$$NaH(s) + B(C_2H_5)_3(\text{et}) \rightarrow Na[HB(C_2H_5)_3](\text{et})$$

where et denotes solution in diethyl ether.

Metal hydrides are useful deprotonating agents:

$$NaH(s) + H_2O(l) \rightarrow NaOH(aq) + H_2(g)$$
$$LiH(s) + CH_2Cl_2(l) \rightarrow LiCHCl_2(\text{sol}) + H_2(g)$$

Magnesium hydride is used as a hydrogen storage medium for fuel cells (Box 9.2). The magnesium hydride is dispersed in mineral oil and the resultant mixture can absorb up to 100 g of hydrogen per cubic decimetre.

9.6 Metallic hydrides

Key points: No stable binary metal hydrides are known for the metals in Groups 7 to 9; metallic hydrides have metallic conductivity and in many the hydrogen is very mobile.

Nonstoichiometric metallic hydrides are formed by all the *d*-block metals of Groups 3, 4, and 5 and by the *f*-block elements (Fig. 9.10). However, the only hydride in Group 6 is CrH, and no hydrides are known for the unalloyed metals of Groups 7, 8, and 9. The region of the periodic table covered by Groups 7 through to 9 is sometimes referred to as the **hydride gap** because few, if any, stable binary metal–hydrogen compounds are formed by these elements. Claims have been made, however, that hydrogen dissolves in iron at very high pressure and that iron hydride is abundant at the centre of the Earth.

Metallic hydrides (and the hydrides of alloys) have a metallic lustre and most are electrically conducting (hence their name). They are less dense than the parent metal and are brittle. Most metallic hydrides have variable composition. For example, at 550°C zirconium hydride exists over a composition range from $ZrH_{1.30}$ to $ZrH_{1.75}$; it has the fluorite structure (Fig. 3.33) with a variable number of anion sites unoccupied. The variable stoichiometry and metallic conductivity of these hydrides can be understood in terms of a model in which the band of delocalized orbitals responsible for the

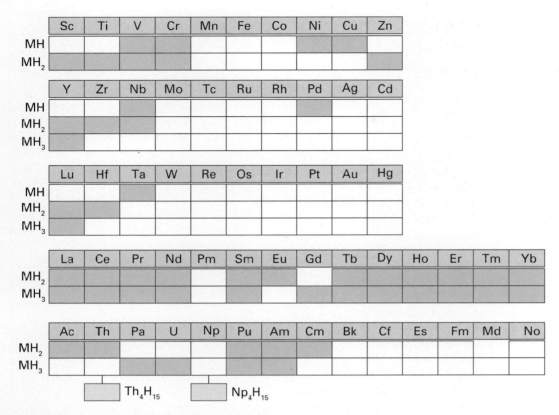

Fig. 9.10 Hydrides formed by *d*- and *f*-block elements. The formulas are limiting stoichiometries based on the structural type.

conductivity accommodates the electrons supplied by arriving H atoms. In this model, the H atoms as well as the metal atoms take up equilibrium positions in the electron sea. The conductivities of metallic hydrides typically vary with hydrogen content, and this variation can be correlated with the extent to which the conduction band is filled or emptied as hydrogen is added or removed. Thus, whereas CeH_{2-x} is a metallic conductor, CeH_3 (which has a full conduction band) is an insulator and is more like a saline hydride.

The Group 10 metals, especially nickel and platinum, are often used as hydrogenation catalysts in which surface hydride formation is thought to be involved (Chapter 25). However, somewhat surprisingly, at moderate pressures only palladium forms a stable bulk phase; its composition is PdH_x, with $x < 1$. Nickel forms hydride phases at very high pressures but platinum does not form any at all. Apparently the Pt—H bond enthalpy is sufficiently great to disrupt the H—H bond but not strong enough to offset the loss of Pt—Pt bonding, which would occur upon formation of a bulk platinum hydride. In agreement with this interpretation, the enthalpies of sublimation, which reflect M—M bond enthalpies, increase in the order Pd (378 kJ mol^{-1}) < Ni (430 kJ mol^{-1}) < Pt (565 kJ mol^{-1}) (Box 9.3).

A striking property of many metallic hydrides is the high mobility of hydrogen at slightly elevated temperatures. This mobility is used in the ultrapurification of H_2 by diffusion through a palladium–silver alloy tube (Fig. 9.11). The high mobility of the hydrogen they contain and their variable composition make the metallic hydrides potential hydrogen storage media. Palladium can absorb up to 900 times its own volume of hydrogen and is sometimes referred to a 'hydrogen sponge'. The intermetallic compound $LaNi_5$ forms a hydride phase with a limiting composition $LaNi_5H_6$, and at this composition it contains a greater density of hydrogen than liquid H_2. A less expensive system with the composition $FeTiH_x$ ($x < 1.95$) is now commercially available for low-pressure hydrogen storage and it has been tested as a source of energy in vehicle trials.

Box 9.3 Metal hydride batteries

A nickel metal hydride (or NiMH) battery is a type of rechargeable battery similar to the widely used nickel–cadmium battery (NiCad) battery. The main advantages of the metal hydride over the NiCad batteries are that they are more easily recycled and do not contain toxic cadmium. However, NiMH batteries have a high self-discharge rate of approximately 30 per cent per month. This is higher than that of NiCd batteries, which is around 20 per cent per month. Despite this, NiMH batteries are being investigated as possible power sources for electric vehicles. In contrast to vehicles powered by the internal combustion engine, electric vehicles are emission free (if the generation of electricity elsewhere is disregarded). In addition, the energy efficiency of generating electricity for vehicles is almost twice that of the internal combustion engine. Electric power also reduces society's reliance on oil and increases the opportunities for using coal, gas, and renewable energy sources.

The attractive properties of NiMH batteries include high power, long life, wide range of operating temperatures, short recharging times, and sealed, maintenance-free operation. The positive electrode is made from nickel hydroxide. The negative electrode is made from a mixed metal alloy at which the metal hydrides are formed reversibly. The reactions take place in a basic solution of 30 mass per cent KOH. The reactions taking place at the electrodes are

$$M + H_2O + e^- \underset{\text{recharge}}{\overset{\text{charge}}{\rightleftharpoons}} MH + OH^-$$

and

$$Ni(OH)_2 + OH^- \underset{\text{discharge}}{\overset{\text{charge}}{\rightleftharpoons}} NiOOH + H_2O + e^-$$

There is no net change in the electrolyte concentration over the charge–discharge cycle. In NiCad batteries, H_2O is generated at both electrodes during charge and is consumed during discharge.

The strength of the M—H bond in the metal hydride is crucial to the operation of the battery. The ideal bond enthalpy falls in the range $25–50 \text{ kJ mol}^{-1}$. If the bond enthalpy is too low, the hydrogen does not react with the metal alloy and hydrogen is evolved. If the bond enthalpy is too high, the reaction is not reversible. Other factors influence the choice of metal. For example, the alloy must not react with KOH solution, must be resistant to oxidation and corrosion, and must tolerate overcharge (during which O_2 is generated at the $Ni(OH)_2$ electrode) and overdischarge (during which H_2 is generated at the $Ni(OH)_2$ electrode). To satisfy these diverse requirements the alloys have disordered structures and use metals that would not be suitable if used alone, including Li, Mg, Al, Ca, V, Cr, Mn, Fe, Cu, and Zr. The number of hydrogen atoms per metal atom can be increased by using Mg, Ti, V, Zr, and Nb and the M—H bond enthalpy can be adjusted by using V, Mn, and Zr. The charge and discharge reactions are catalysed by Al, Mn, Co, Fe, and Ni and the corrosion resistance is improved by using Cr, Mo, and W. This wide range of properties allows NiMH battery performance to be optimized for different applications.

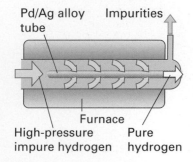

Fig. 9.11 Schematic diagram of a hydrogen purifier. Because of a pressure differential and the mobility of H atoms in palladium, hydrogen diffuses through the palladium–silver alloy tube as H atoms but impurities do not.

Example 9.1 Correlating the classification and properties of hydrogen compounds

Classify the compounds PH_3, CsH, and B_2H_6 and discuss their probable physical properties. For the molecular compounds specify their subclassification (electron-deficient, electron-precise, or electron-rich).

Answer The compound CsH is a compound of a Group 1 element, and so it can be expected to be a saline hydride typical of the s-block metals. It is an electrical insulator with the rock-salt structure. As with the hydrogen compounds of other p-block elements, the hydrides PH_3 and B_2H_6 are molecular with low molar masses and high volatilities. They are in fact gases under normal conditions. The Lewis structure indicates that PH_3 has a lone pair on the phosphorus atom and that it is therefore an electron-rich molecular compound. Diborane, B_2H_6, is an electron-deficient compound.

Self test 9.1 Give balanced equations (or NR, for no reaction) for (a) $Ca + H_2$, (b) $NH_3 + BF_3$, (c) $LiOH + H_2$.

Stability, synthesis, and reactions

Many of the chemical properties of hydrogen compounds can be rationalized by studying the periodic trends in their thermodynamic stabilities and the polarities of the E—H bond.

9.7 Stability

Key points: Compound formation between H_2 and Group 1 and 2 metals except Be is thermodynamically favourable; on going down a group in the *p* block, the

Table 9.6 Standard Gibbs energy of formation, $\Delta_f G^\ominus$/(kJ mol^{-1}), of binary s- and p-block hydrogen compounds at 25°C

Period	Group						
	1	2	13	14	15	16	17
2	LiH(s)	BeH$_2$(s)	B$_3$H$_6$(g)	CH$_4$(g)	NH$_3$(g)	H$_2$O(l)	HF(g)
	−68.4	(+20)	+86.7	−50.7	−16.5	−237.1	−273.2
3	NaH(s)	MgH$_2$(s)	AlH$_3$(s)	SiH$_4$(g)	PH$_3$(g)	H$_2$S(g)	HCl(g)
	−33.5	−35.9	(−1)	+56.9	+13.4	−33.6	−95.3
4	KH(s)	CaH$_2$(s)	Ga$_2$H$_6$(s)	GeH$_4$(g)	AsH$_3$(g)	H$_2$Se(g)	HBr(g)
	(−36)	−147.2	> 0	+113.4	+68.9	+15.9	−53.5
5	RbH(s)	SrH$_2$(s)		SnH$_4$(g)	SbH$_3$(g)	H$_2$Te(g)	HI(g)
	(−30)	(−141)		+188.3	+147.8	> 0	+1.7
6	CsH(s)	BaH$_2$(s)					
	(−32)	(−140)					

From *J. Phys. Chem. Ref. Data*, **11**, Supplement 2 (1982). Values in parentheses are based on $\Delta_f H^\ominus$ data from this source and entropy contributions.

element–hydrogen bond strengths decrease and the hydrides of the heavier elements become thermodynamically unstable.

A negative Gibbs energy of formation is a clue that the direct combination of hydrogen and an element may be the preferred synthetic route for a hydrogen compound. When a compound is thermodynamically unstable with respect to its elements, an indirect synthetic route from other compounds can often be found, but each step in the indirect route must be thermodynamically favourable.

The standard Gibbs energies of formation of the hydrogen compounds of s- and p-block elements reveal a regular variation in stability (Table 9.6). With the possible exception of BeH$_2$, for which good data are not available, all the s-block hydrides are exergonic ($\Delta_f G^\ominus < 0$) and therefore thermodynamically stable with respect to their elements at room temperature. The trend is erratic in Group 13, in that only AlH$_3$ is exergonic. In all the other groups of the p block, the simple hydrogen compounds of the first members of the groups (CH$_4$, NH$_3$, H$_2$O, and HF) are exergonic but become progressively less stable down the group. This general trend in stability is largely a reflection of the decreasing E—H bond strength down a group in the p block (Fig. 9.12). The weak bonds formed by the heavier elements are often attributed to poor overlap between the relatively compact H1s orbital and the more diffuse s and p orbitals of the heavier atoms in these groups. The heavier members become more stable on going from Group 14 across to the halogens. For example, SnH$_4$ is highly endergonic ($\Delta_f G^\ominus > 0$) whereas HI is barely so.

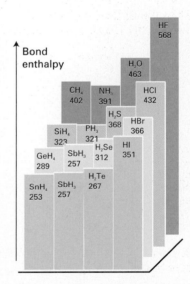

Fig. 9.12 Average bond enthalpies (in kilojoules per mole, at 298 K).

9.8 Synthesis

Key points: The general routes to hydrogen compounds are direct reaction of H$_2$ and the element, protonation of nonmetal anions, and metathesis between a hydride source and a halide or pseudohalide.

There are three common methods for synthesizing binary hydrogen compounds:

1 Direct combination of the elements:

$$2\,E + H_2(g) \rightarrow 2\,EH$$

2 Protonation of a Brønsted base:

$$E^- + H_2O(aq) \rightarrow EH + OH^-$$

3 Metathesis (double replacement) of a halide or pseudohalide with a hydride:[2]

$$E^+H^- + EX \rightarrow E^+X^- + EH$$

In such general equations, the symbol E could also denote an element with higher valence, with corresponding changes of detail in the formulas and stoichiometric coefficients.

Direct synthesis is used commercially for the synthesis of compounds that have negative Gibbs energies of formation, including NH_3 and the hydrides of lithium, sodium, and calcium. However, in some cases high pressure, high temperature, and a catalyst are necessary to overcome the unfavourable kinetics. The high temperature used for the lithium reaction is an example: it melts the metal and hence helps to break up the surface layer of hydride that would otherwise passivate it. This inconvenience is avoided in many laboratory preparations by adopting one of the alternative synthesis routes, which may also be used to prepare compounds with positive Gibbs energies of formation.

An example of protonation of a Brønsted base, such as a nitride ion, is

$$Li_3N(s) + 3\ H_2O(l) \rightarrow 3\ LiOH(aq) + NH_3(g)$$

Lithium nitride is too expensive for the reaction to be suitable for the commercial production of ammonia, but it is very useful in the laboratory for the preparation of ND_3 (by using D_2O in place of H_2O). The success of the reaction depends on the Brønsted acid being a better proton donor than the conjugate acid of the N^{3-} anion (NH_3 in this case). Water is a sufficiently strong acid to protonate the very strong base N^{3-}, but a stronger acid, such as H_2SO_4, is required to protonate the weak base Cl^-:

$$NaCl(s) + H_2SO_4(l) \rightarrow NaHSO_4(s) + HCl(g)$$

A synthesis by a metathesis reaction is

$$LiAlH_4 + SiCl_4 \rightarrow LiAlCl_4 + SiH_4$$

This reaction involves (at least formally) the exchange of Cl^- ions for H^- ions in the coordination sphere of the Si atom. Hydrides of the more electropositive elements (LiH, NaH, and AlH_4^-) are the most active H^- sources. The favourite sources are often the AlH_4^- and BH_4^-) ions in salts such as $LiAlH_4$ (lithium tetrahydridoaluminate, known commonly as lithium aluminium hydride) and $NaBH_4$ (sodium tetrahydridoborate, know commonly as sodium borohydride), which are soluble in ether solvents that solvate the alkali metal ion. Of these two anion complexes, AlH_4^- is much the stronger hydride donor.

9.9 Reactions

Three types of reaction can lead to the scission of an E—H bond:

1 Heterolytic cleavage by hydride transfer:

$$E-H \rightarrow E^+ + :H^-$$

2 Homolytic cleavage:

$$E-H \rightarrow E\cdot + H\cdot$$

3 Heterolytic cleavage by proton transfer:

$$E-H \rightarrow :E^- + :H^+$$

A free H^- or H^+ ion is unlikely to be formed in reactions that take place in solution. Instead, the ion is transferred by the formation of a complex that involves a hydrogen bridge.

Detailed experiments are required to establish which of these three primary processes occurs in practice. For example, any of them can lead to the addition of EH across

---

[2] A pseudohalide is a diatomic or polyatomic ion that resembles a halide ion chemically; an example is CN^- (Section 16.3).

a double bond:

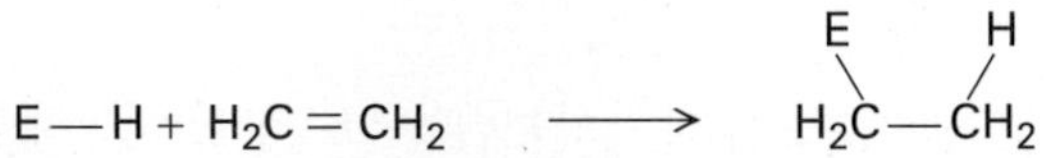

$$E-H + H_2C=CH_2 \longrightarrow \underset{\underset{H_2C-CH_2}{\diagdown}}{\overset{E\qquad H}{\diagup}}$$

However, there are some patterns of behaviour that at least suggest which one is likely to occur and these are discussed in the next section.

(a) Heterolytic cleavage by hydride transfer

Key points: Hydrides of active metals, such as LiH, are strongly hydridic: they react vigorously with proton sources to liberate H_2 and form a salt, and they may displace halides to form anionic hydride complexes, such as BH_4^-.

Compounds with **hydridic character** react vigorously with Brønsted acids by the transfer of an H^- ion with the evolution of H_2:

$$NaH(s) + H_2O(l) \rightarrow NaOH(aq) + H_2(g)$$

A compound that takes part in this reaction with a weak proton donor (water, in this example) is said to be 'strongly hydridic'. A compound that requires a strong proton donor is classified as 'weakly hydridic' (an example is germane, GeH_4). Hydridic character is most pronounced towards the left of a period where the element is most electropositive (in the s block) and decreases rapidly after Group 13. For example, the saline hydrides of Groups 1 and 2 are strongly hydridic, as demonstrated by their vigorous reactions with water and alcohols. In addition to protolysis, the saline hydrides display hydridic behaviour, by transferring an H^-, in the following synthetically useful reactions:

1 Metathesis with a halide, such as the reaction of finely divided lithium hydride with silicon tetrachloride in dry diethyl ether (et):

$$4\,LiH(s) + SiCl_4(et) \rightarrow 4\,LiCl(s) + SiH_4(g)$$

2 Addition to a Lewis acid:

$$LiH(s) + B(CH_3)_3(g) \rightarrow Li[BH(CH_3)_3](et)$$

3 Reaction with a proton source, to produce H_2:

$$NaH(s) + CH_3OH(l) \rightarrow NaOCH_3(s) + H_2(g)$$

The H atom of an OH group, but not the hydrogen bonded to the C atom, is sufficiently acidic to undergo this reaction. Combinations of these reactions are also common. For example, the following reaction may be thought of as metathesis (to form B_2H_6) to which H^- may then add (to form BH_4^-):

$$4\,LiH(s) + BF_3(et) \rightarrow LiBH_4(et) + 3\,LiF(s)$$

(b) Homolytic cleavage

Key point: Homolytic cleavage of an E—H bond to produce a radical E· and H_2 occurs most readily for the hydrides of the heavy p-block elements.

Homolytic cleavage occurs readily for the hydrogen compounds of some p-block elements, especially the heavier elements. For example, the use of a radical initiator greatly facilitates the reaction of trialkylstannanes, R_3SnH, with haloalkanes, RX, as a result of the formation of $R_3Sn\cdot$ radicals:

$$R_3SnH + R'X \rightarrow R'H + R_3SnX$$

The tendency towards radical reactions increases towards the heavier elements in each group, and SnH compounds are in general more prone to radical reactions than are SiH compounds. The ease of homolytic E—H bond cleavage correlates with the decrease in E—H bond strength down a group.

The order of reactivity for haloalkanes with trialkylstannanes is

$$RF < RCl < RBr < RI$$

Thus fluoroalkanes do not react with R_3SnH, chloroalkanes require heat, photolysis, or chemical radical initiators, and bromoalkanes and iodoalkanes react spontaneously at room temperature. This trend indicates that the initiation step is halogen abstraction.

(c) Heterolytic cleavage by proton transfer

Key point: Hydrogen attached to an electronegative element has protic character and the compound is typically a Brønsted acid.

Compounds reacting by deprotonation are said to show **protic behaviour**: in other words, they are Brønsted acids. We saw in Section 4.1 that Brønsted acid strength increases from left to right across a period in the p block and down a group. One striking example of this trend is the increase in acidity from CH_4 to HF. Binary hydrogen compounds of elements on the right of the periodic table typically undergo these reactions.

Example 9.2 Using hydrogen compounds in synthesis

Suggest a procedure for synthesizing lithium tetraethoxyaluminate, $Li[Al(OEt)_4]$, from $LiAlH_4$ and reagents and solvents of your choice.

Answer The reaction of the slightly acidic compound ethanol with the strongly hydridic lithium tetrahydroaluminate should yield the desired alkoxide and hydrogen. The reaction might be carried out by dissolving $LiAlH_4$ in tetrahydrofuran and dropping ethanol into this solution slowly:

$$LiAlH_4(thf) + 4\ C_2H_5OH(l) \rightarrow Li[Al(OEt)_4](thf) + 4\ H_2(g)$$

This type of reaction should be carried out slowly under a stream of inert gas (N_2 or Ar) to dilute the H_2, which is explosively flammable.

Self-test 9.2 Suggest a way of making triethylmethylstannane, $MeEt_3Sn$, from triethylstannane, Et_3SnH, and a reagent of your choice.

FURTHER READING

G.A. Jeffrey, *An introduction to hydrogen bonding*. Oxford University Press (1997).

G.A. Jeffrey, *Hydrogen bonds in biological systems*. Oxford University Press (1994).

R.B. King, *Inorganic chemistry of the main group elements*. Wiley (1994).

P. Enghag, *Encyclopedia of the elements*. Wiley (2004).

P. Ball, *H₂O: A biography of water*. Phoenix (2004). An entertaining look at the chemistry and physics of water.

EXERCISES

9.1 It has been suggested that hydrogen could be placed in Group 1, Group 14, or Group 17 of the periodic table. Give arguments for and against each of these positions.

9.2 Explain the relatively low reactivity of hydrogen.

9.3 Assign oxidation numbers to the elements in (a) H_2S, (b) KH, (c) $[ReH_9]^{2-}$, (d) H_2SO_4, (e) $H_2PO(OH)$.

9.4 Write balanced chemical equations for three major industrial preparations of hydrogen gas. Propose two different reactions that would be convenient for the preparation of hydrogen in the laboratory.

9.5 Preferably without consulting reference material, construct the periodic table, identify the elements, and (a) indicate positions of salt-like, metallic, and molecular hydrides, (b) add arrows to indicate trends in $\Delta_f G^\circ$ for the hydrogen compounds of the p-block elements, (c) delineate the areas where the molecular hydrides are electron-deficient, electron-precise, and electron-rich.

9.6 Describe the expected physical properties of water in the absence of hydrogen bonding.

9.7 Which hydrogen bond would you expect to be stronger, S—H$\cdots$O or O—H$\cdots$S? Why?

9.8 Name and classify the following hydrogen compounds: (a) BaH_2, (b) SiH_4, (c) NH_3, (d) AsH_3, (e) $PdH_{0.9}$, (f) HI.

9.9 Identify the compounds from Exercise 9.8 that provide the most pronounced example of the following chemical characteristics and give a balanced equation that illustrates each of the characteristics: (a) hydridic character, (b) Brønsted acidity, (c) variable composition, (d) Lewis basicity.

9.10 Divide the compounds in Exercise 9.8 into those that are solids, liquids, or gases at room temperature and pressure. Which of the solids are likely to be good electrical conductors?

9.11 Use Lewis structures and VSEPR theory to predict the shapes of H_2Se, P_2H_4, and H_3O^+ and to assign point groups. Assume a skew structure for P_2H_4.

9.12 Identify the reaction that is most likely to give the highest proportion of HD and give your reasoning: (a) $H_2 + D_2$ equilibrated over a platinum surface, (b) $D_2O + NaH$, (c) electrolysis of HDO.

9.13 Identify the compound in the following list that is most likely to undergo radical reactions with alkyl halides, and describe the reason for your choice: H_2O, NH_3, $(CH_3)_3SiH$, $(CH_3)_3SnH$.

9.14 Arrange H_2O, H_2S, and H_2Se in order of (a) increasing acidity, (b) increasing basicity towards a hard acid such as the proton.

9.15 Describe the three different common methods for the synthesis of binary hydrogen compounds and illustrate each one with a balanced chemical equation.

9.16 What is the trend in hydridic character of $[BH_4]^-$, $[AlH_4]^-$, and $[GaH_4]^-$? Which is the strongest reducing agent? Give the equations for the reaction of $[GaH_4]^-$ with excess 1 M $HCl(aq)$.

9.17 Describe the important physical differences and a chemical difference between each of the hydrogen compounds of the p-block elements in Period 2 with their counterparts in Period 3.

9.18 What type of substance is formed by the interaction of water and krypton at low temperatures and elevated krypton pressure? Describe the structure in general terms.

9.19 Sketch the approximate potential energy surfaces for the hydrogen bond between H_2O and the Cl^- ion and contrast this with the potential energy surface for the hydrogen bond in $[FHF]^-$.

PROBLEMS

9.1 What is the expected infrared stretching wavenumber of gaseous $^3H^{35}Cl$ given that the corresponding value for $^1H^{35}Cl$ is 2991 cm^{-1}?

9.2 Consult Chapter 6 and then sketch the qualitative splitting pattern and relative intensities within each set for the ^1H and ^{31}P NMR spectra of PH_3.

9.3 In his paper 'The proper place for hydrogen' (*J. Chem. Educ.*, 2003, **80**, 947), M.W. Cronyn argues that hydrogen should be placed at the head of Group 14 immediately above carbon. Summarize his reasoning.

9.4 (a) Sketch a qualitative molecular orbital energy level diagram for the HeH^+ molecule-ion and indicate the correlation of the molecular orbital levels with the atomic energy levels. The ionization energy of H is 13.6 eV and the first ionization energy of He is 24.6 eV. (b) Estimate the relative contribution of H1s and He1s orbitals to the bonding orbital and predict the location of the partial positive charge of the polar molecule. (c) Why do you suppose that HeH^+ is unstable on contact with common solvents and surfaces?

9.5 Hydrogen bonding can influence many reactions, including the rates of O_2 binding and dissociation from metalloproteins (G.D. Armstrong and A.G. Sykes, *Inorg. Chem.*, 1986, **25**, 3135). Describe the evidence for (or against) hydrogen bonding with O_2 in the metalloproteins haemerythrin, myoglobin, and haemocyanin (see Section 26.7).

9.6 Spectroscopic evidence has been obtained for the existence of $[Ir(C_5H_5)(H_3)(PR_3)]^+$, a complex in which one ligand is formally H_3^+. Devise a plausible molecular orbital scheme for the bonding in the complex, assuming that an angular H_3 unit occupies one coordination site and interacts with the e_g and t_{2g} orbitals of the metal. An alternative formulation of the structure of the complex, however, is as a trihydro species with very large coupling constants (see *J. Am. Chem. Soc.*, 1991, **113**, 6074 and the references therein, and especially *J. Am. Chem. Soc.*, 1990, **112**, 909 and 920). Review the evidence for this alternative formulation.

9.7 Correct the faulty statements in the following description of hydrogen compounds. 'Hydrogen, the lightest element, forms thermodynamically stable compounds with all of the nonmetals and most metals. The isotopes of hydrogen have mass numbers of 1, 2, and 3, and the isotope of mass number 2 is radioactive. The structures of the hydrides of the Group 1 and 2 elements are typical of ionic compounds because the H^- ion is compact and has a well-defined radius. The structures of the hydrogen compounds of the nonmetals are adequately described by VSEPR theory. The compound $NaBH_4$ is a versatile reagent because it has greater hydridic character than the simple Group 1 hydrides such as NaH. Heavy element hydrides such as the tin hydrides frequently undergo radical reactions, in part because of the low E—H bond energy. The boron hydrides are called electron-deficient compounds because they are easily reduced by hydrogen.'

10 The Group 1 elements

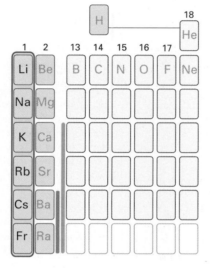

All the Group 1 elements are metallic. Unlike most metals, they have low densities and are very reactive. In this chapter we look at the occurrence of the elements in the natural environment and how they are extracted and used. The chapter also describes the trends in the properties of the simple binary compounds and the nature of complexes and organometallic compounds of the elements.

The Group 1 elements, the **alkali metals**, are lithium, sodium, potassium, rubidium, caesium (cesium), and francium. We shall not discuss francium, which exists naturally only in minute quantities and is highly radioactive. All the elements are metals and form simple, ionic compounds, most of which are soluble in water. The elements form a limited number of complexes and organometallic compounds.

The elements

Key point: The trends in the properties of the Group 1 metals and their compounds can be explained in terms of trends in their atomic radii.

All the Group 1 elements are metals with electron configuration ns^1. They conduct electricity and heat, are soft, and have low melting points that decrease down the group. Their softness and low melting points stem from the fact that their metallic bonding is weak because each atom contributes only one electron to the molecular orbital band (Section 3.17). They all adopt a body-centred cubic structure (Section 3.5) and because that structure is not close-packed, they have low densities. All the metals form alloys. Table 10.1 summarizes some important properties.

The chemical properties of the Group 1 elements correlate with the trend in their atomic radii (Fig. 10.1). The increase in atomic radius from lithium to francium leads to a decrease in first ionization energy down the group because the valence shell is increasingly distant from the nucleus (Fig. 10.2; see also Section 1.9). Because their first ionization energies are all low, the metals are reactive, and form M^+ ions readily and increasingly down the group. Thus, lithium reacts gently with water, sodium reacts

Table 10.1 Selected properties of the Group 1 elements

	Li	Na	K	Rb	Cs
Metallic radius/pm	152	186	231	244	262
Ionic radius/pm	60	95	133	148	169
Ionization energy/(kJ mol^{-1})	519	494	418	402	376
Standard potential/V	−3.04	−2.71	−2.94	−2.92	−3.03
Density/(g cm^{-3})	0.53	0.97	0.86	1.53	1.90
Melting point/°C	180	98	64	39	29
$\Delta_{hyd}H^{\ominus}$/(kJ mol^{-1})	−519	−406	−322	−301	−276
$\Delta_{sub}H^{\ominus}$/(kJ mol^{-1})	161	109	90	86	79

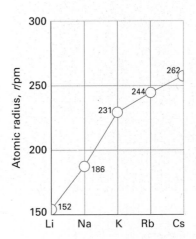

Fig. 10.1 The variation in atomic radius of the elements of Group 1.

vigorously, potassium reacts so exothermically that the hydrogen released is ignited, and rubidium and caesium react explosively:

$$2\ M(s) - 2\ H_2O(l) \rightarrow 2\ MOH(aq) + H_2(g)$$

This tendency to form M$^+$ in aqueous conditions is confirmed by the standard potentials of the couples M$^+$/M, which are all large and negative (Table 10.1), indicating that the metals are readily oxidized. The surprising uniformity of the standard potentials of the alkali metals can be explained by studying the thermodynamic cycle for the reduction half-reaction (Fig. 10.3). The enthalpies of sublimation and ionization both decrease down the group (making oxidation more favourable); however, this trend is counteracted by a smaller enthalpy of hydration as the radii of the ions increase (making oxidation less favourable).

All the elements must be stored under a hydrocarbon solvent to prevent reaction with atmospheric oxygen, although lithium, sodium, and potassium can be handled in air for short periods. Rubidium and caesium must be handled under an inert atmosphere at all times.

Most of the trends in the chemical properties of the elements within the periodic table are best discussed in terms of vertical trends within groups or horizontal trends across periods. However, the first element in each group also displays a **diagonal relationship** with the element to the lower right of it in the periodic table. Diagonal relationships arise because the atomic radii, and hence many of the chemical properties, of the two elements are similar (Table 1.4). Some examples of the similarities between lithium and magnesium are:

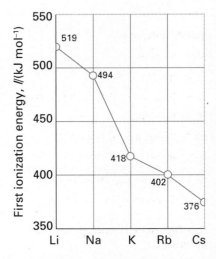

Fig. 10.2 The variation in first ionization energy of the elements of Group 1.

1 Many lithium and magnesium salts exhibit a high degree of covalent character in their bonding. This covalent character is due to the high polarizing power associated with high charge density (Section 1.9).

2 Lithium and magnesium form normal oxides, whereas other Group 1 elements form peroxides or superoxides.

3 Lithium is the only alkali metal to form a nitride, Li$_3$N. All Group 2 elements form nitrides.

4 Lithium carbonate, phosphate, and fluoride have very low solubilities in water. The corresponding compounds of Group 2 elements are insoluble.

5 Lithium forms organometallic compounds similar to those of magnesium.

6 The carbonates of lithium and magnesium decompose to give the metal oxide and carbon dioxide; the carbonates of the other alkali metals do not decompose when heated.

10.1 Occurrence and extraction

Key point: The Group 1 elements can be extracted by electrolysis.

The name lithium comes from the Greek *lithos* for stone. The natural abundance of lithium is low, the most abundant mineral being spodumene, LiAlSi$_2$O$_6$, from which

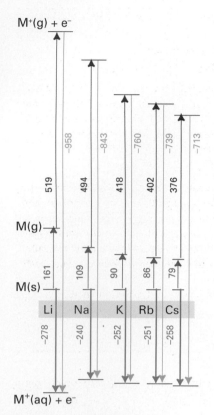

Fig. 10.3 Thermochemical cycle (standard enthalpy changes in kJ mol^{-1}) for the oxidation half-reaction $M(s) \rightarrow M^{+}(aq) + e^{-}$.

lithium is most commonly extracted, and lepidolite, which has the approximate formula $K_2Li_3Al_4Si_7O_{21}(F,OH)_3$. Spodumene is first converted to LiCl and then electrolysed to produce lithium metal.

Sodium occurs as the mineral rock salt (NaCl) and in salt lakes and seawater; it also makes up 2.6 per cent by mass of the biosphere. The metal is extracted by *Down's process*, the electrolysis of molten sodium chloride:

$$2 \text{ NaCl(l)} \rightarrow 2 \text{ Na(l)} + Cl_2(g)$$

This process is also used for the industrial production of chlorine.

Potassium occurs naturally as potash (KOH) and carnelite ($KCl \cdot MgCl_2 \cdot 6H_2O$). Natural potassium contains 0.012 per cent of the radioactive isotope ^{40}K, which undergoes β decay to ^{40}Ca and electron capture to ^{40}Ar. The ratio of ^{40}K and ^{40}Ar is used for dating rocks. In principle, potassium could be extracted electrolytically but the high reactivity of the element makes this far too hazardous. Instead, molten sodium and molten potassium chloride are heated together and potassium and sodium chloride are formed.

$$\text{Na(l)} + \text{KCl(l)} \rightleftharpoons \text{NaCl(l)} + \text{K(g)}$$

At the temperature of operation, potassium is a vapour and removing it from the system drives the equilibrium to the right.

Rubidium (from the Latin *rubidus* for deep red) and caesium (*caesius* for sky blue) were discovered by Robert Bunsen in 1861. Both elements occur as minor constituents of the mineral lepidolite, from which they are obtained as by-products of the extraction of lithium. Prolonged treatment of lepidolite with sulfuric acid forms the alums of the alkali metals, $M_2SO_4 \cdot Al_2(SO_4)_3 \cdot nH_2O$. The alums are separated by multiple fractional crystallizations and then converted to the hydroxide by reaction with $Ba(OH)_2$ and then to the chloride by ion exchange. The metals are obtained from the molten chloride by reduction with calcium or barium.

$$2 \text{ RbCl(l)} + \text{Ca(l)} \rightarrow CaCl_2(l) + 2 \text{ Rb(l)}$$

Caesium also occurs as the mineral pollucite, $Cs_4Al_4Si_9O_{26} \cdot H_2O$. The caesium is extracted from the mineral by leaching with sulfuric acid to form the alum $Cs_2SO_4 \cdot Al_2(SO_4)_3 \cdot 24H_2O$, which is then converted to the sulfate by roasting with carbon. The chloride is formed by ion exchange and is then reduced with calcium or barium as described above.

10.2 Uses of the elements and their compounds

Key points: Common uses of lithium are related to its low density; the most widely used compounds of Group 1 are sodium chloride and sodium hydroxide.

The applications of lithium are due to its low atomic mass and consequently low density. It is used in aircraft alloys and in several types of battery. The low density of lithium, only 5 per cent that of lead, coupled with the strongly negative standard potential of the Li^{+}/Li couple (Table 10.1), make lithium batteries an attractive alternative to lead–acid batteries (Box 10.1). Lithium carbonate is widely used to treat bipolar conditions (manic depression) and lithium stearate is a widely used lubricant in the automotive industry.

Sodium and potassium are essential for physiological function (Section 26.3). Sodium is used in the extraction of rarer metals, such as titanium from titanium(IV) chloride. The major uses of NaCl are road de-icing and production of NaOH for the chloralkali industry (Box 10.2). Sodium hydroxide is in the ten most important industrial chemicals in terms of annual tonnage produced. Other common applications of sodium and its compounds include the metal in street lamps, which gives a distinctive yellow glow when an electrical discharge is passed through its vapour, table salt, baking soda, and caustic soda (NaOH).

Potassium hydroxide is used in soap manufacture to make 'soft' liquid soaps. Potassium chloride and sulfate are used as fertilizers, the nitrate and chlorate are used in

Box 10.1 Lithium batteries

The very negative standard potential and low density of lithium make it an ideal anode material for batteries. Lithium batteries are commonplace, but there are many types based on different lithium compounds and reactions.

The lithium rechargeable battery used in portable computers and phones uses $LiCoO_2$ as the anode with a graphite cathode. Lithium ions are produced at the cathode and to maintain charge balance, Co(III) is oxidized to Co(IV) at the anode.

$$LiCoO_2(s) \rightarrow Li_{1-x}CoO_2 + x\, Li^+ \text{ (solvent) } + x\, e^-$$

The lithium ions are intercalated into the graphite and return to the $LiCoO_2$ electrode when the cell is discharged. The battery is rechargeable because both the cathode and the anode can act as host for the Li^+ ions, which can move back and forth between them when charging and discharging. There are many other lithium batteries using different electrode materials, mainly d-metal compounds, which take part in the redox reaction in a similar way to the cobalt.

Another popular lithium battery uses thionyl chloride, $SOCl_2$. This system produces a light, high-voltage cell with a constant energy output. The overall reaction in the battery is

$$2\, Li(s) + 3\, SOCl_2(l) \rightarrow LiCl(s) + S(s) + SO_2(l)$$

The battery requires no additional solvent as both $SOCl_2$ and SO_2 are liquids at the internal battery pressure. This battery is not rechargeable as both sulfur and LiCl are precipitated. It is used in military applications and in spacecraft. Another battery system is based on the reduction of SO_2:

$$2\, Li(s) + 2\, SO_2(l) \rightarrow Li_2S_2O_4(s)$$

The system is also not rechargeable as solid $Li_2S_2O_4$ deposits on the cathode. This battery uses acetonitrile (CH_3CN) as a co-solvent and the handling of this compound and the SO_2 present safety hazards. The batteries are hermetically sealed and not available to the general public. They are used in military communications and automated external defibrillators that are used to restore normal heart rhythm.

Rechargeable batteries are discussed further in Chapter 23.

Box 10.2 The chloralkali industry

The chloralkali industry has its roots in the industrial revolution when large quantities of alkali were required for the manufacture of soap, paper, and textiles. Today, sodium hydroxide is one of the top ten most important inorganic chemicals in terms of quantity produced and continues to be important in the manufacture of other inorganic chemicals and in the pulp and paper industries. Chlorine, the other product of the process, is also very important industrially and is used in the manufacture of PVC, in the extraction of titanium, and in the pulp and paper industries.

The industrial process is based on the electrolysis of an aqueous solution of sodium chloride. Water is reduced to hydrogen gas and hydroxide ions at the cathode, and chloride ions are oxidized to chlorine gas at the anode:

$$2\, H_2O(l) + 2\, e^- \rightarrow H_2(g) + 2\, OH^-(aq)$$
$$2\, Cl^-(aq) \rightarrow Cl_2(g) + 2\, e^-$$

There are three different types of cells that are used for the electrolysis. In a *diaphragm cell*, a diaphragm prevents the OH^- ions produced at the cathode from coming into contact with the chlorine gas produced at the anode. This diaphragm used to be made of asbestos, but it is now made of a polytetrafluoroethylene mesh. During the electrolysis, the solution at the cathode is removed continuously, and evaporated to crystallize out the sodium chloride impurities. The final sodium hydroxide typically contains approximately 1 per cent by mass NaCl.

A *membrane cell* functions like the diaphragm cell except that the anode and cathode solutions are separated by a microporous polymer membrane that is permeable only to the Na^+ ions. The sodium hydroxide solution produced using this cell typically contains approximately 50 ppm Cl^- ion. The disadvantage with this method is that the membrane is very expensive and can become clogged by trace impurities.

The *mercury cell* uses liquid mercury as the cathode. Chlorine gas is produced at the anode but sodium metal is produced at the cathode:

$$Na^+(aq) + e^- \rightarrow Na(Hg)$$

The sodium–mercury amalgam is reacted with water on a graphite surface:

$$2\, Na(Hg) + 2\, H_2O(l) \rightarrow 2\, NaOH(aq) + H_2(g)$$

The sodium hydroxide solution produced by this route is very pure and the mercury cell is the preferred source of high quality, solid sodium hydroxide. Unfortunately, the process is accompanied by discharge of mercury into the environment. Consequently, the chloralkali industry is under pressure to move away from the use of mercury electrodes.

fireworks. Potassium bromide has been used as an antaphrodisiac (a compound that reduces the libido).

Rubidium and caesium are often used in the same applications and one element may often be substituted for the other. The market for these elements is small but highly specialized. Applications include glasses for fibre-optic applications in the telecommunications industry, night vision equipment, and photoelectric cells. The 'caesium clock' (atomic clock) is used for the international standard measure of time and for the definition of the second and the metre.

Simple compounds

All the Group 1 elements form simple compounds with hydrogen and oxygen, and the halogens and also form salts of oxoacids. The compounds are predominantly ionic.

10.3 Hydrides

Key point: The hydrides of the Group 1 elements are ionic and contain the H^- ion.

Group 1 metals form ionic, saline hydrides with the rock-salt structure; the anion present is the *hydride ion*, H^-. These hydrides were discussed in some detail in Section 9.5.

The hydrides react violently with water:

$$NaH(s) + H_2O(l) \rightarrow NaOH(aq) + H_2(g)$$

Finely divided sodium hydride can ignite if it is left exposed to humid air. Such fires are difficult to extinguish because even carbon dioxide is reduced when it comes into contact with hot metal hydrides. Hydrides are useful as non-nucleophilic bases and reductants:

$$NaH(s) + NH_3(l) \rightarrow NaNH_2(am) + H_2(g)$$

where am denotes a solution in ammonia.

700
600
500
400
300
200
$-\Delta_f H^{\ominus}/(\text{kJ mol}^{-1})$
Li
Na
K
Rb
Cs
F
Cl
Br
I
-616
-574
-567
-558
-554
-409
-411
-437
-435
-443
-394
-395
-406
-351
-361
-328
-334
-347
-270
-288

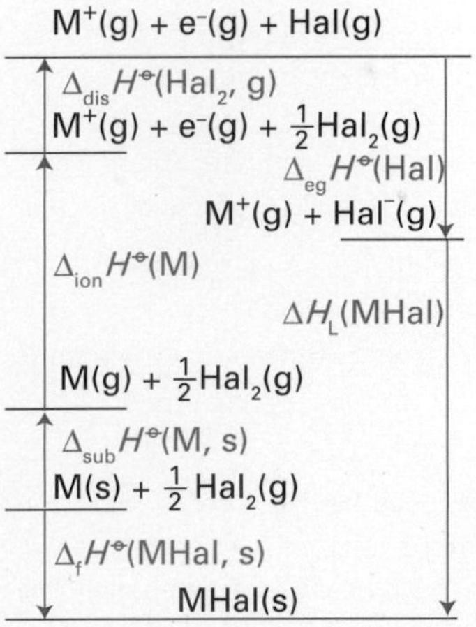

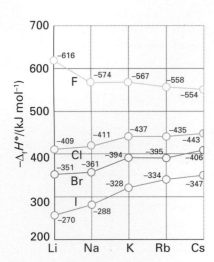

Fig. 10.4 The standard enthalpies of formation of the halides of Group 1 elements at 298 K.

10.4 Halides

Key point: On descending the group, the enthalpy of formation becomes less negative for the fluorides but more negative for the chlorides, bromides, and iodides.

All the Group 1 elements form halides, MX, by direct combination of the elements. Most of the halides have the (6,6)-coordinate rock-salt structure, but CsCl, CsBr, and CsI have the (8,8)-coordinate caesium-chloride structure (Section 3.9). The enthalpy of formation of all the halides is large and negative, becoming less negative from the fluoride to iodide for each element. The enthalpy of formation becomes less negative for the fluorides on descending the group but becomes more negative for the chlorides, bromides, and iodides (Fig. 10.4). These trends can be rationalized by considering a Born–Haber cycle for formation of the halides from the elements (Fig. 10.5).

As we saw in Section 3.11, the requirement that the sum of enthalpy changes round a Born–Haber cycle be zero implies that the enthalpy of formation of a compound is

$$\Delta_f H^{\ominus} = \Delta_{sub} H^{\ominus} + \Delta_{ion} H^{\ominus} + \Delta_{dis} H^{\ominus} + \Delta_{eg} H^{\ominus} - \Delta_L H^{\ominus} \tag{10.1}$$

$M^+(g) + e^-(g) + Hal(g)$
$\Delta_{dis}H^{\ominus}(Hal_2, g)$
$M^+(g) + e^-(g) + \frac{1}{2}Hal_2(g)$
$\Delta_{eg}H^{\ominus}(Hal)$
$M^+(g) + Hal^-(g)$
$\Delta_{ion}H^{\ominus}(M)$
$\Delta H_L(MHal)$
$M(g) + \frac{1}{2}Hal_2(g)$
$\Delta_{sub}H^{\ominus}(M, s)$
$M(s) + \frac{1}{2}Hal_2(g)$
$\Delta_f H^{\ominus}(MHal, s)$
MHal(s)

Fig. 10.5 The Born–Haber cycle for the formation of the Group 1 halides. The sum of the enthalpy changes round the cycle is zero.

The first two terms in the expression are constant for a series of halides of a given metal. The next two terms vary from fluoride to iodide and, as can be seen from the data in Table 10.2, their sum becomes less negative from F to I. The final term is the lattice enthalpy, which (from the Born–Mayer equation, as discussed in Section 3.12a) we know to be inversely proportional to the sum of the ionic radii:

$$\Delta_L H^{\ominus} \propto \frac{1}{r_+ + r_-}$$

As the radius of the anion increases from F^- to I^-, the lattice enthalpy becomes smaller. Consequently, $\Delta_f H^{\ominus}$ becomes less negative.

If we consider the formation of a series of Group 1 halides, the terms $\Delta_{dis} H^{\ominus}$ and $\Delta_{eg} H^{\ominus}$ are constant. The terms $\Delta_{sub} H^{\ominus}$ and $\Delta_{ion} H^{\ominus}$ vary with the metal and, as can be seen from Table 10.1, the sum of their values decreases down the group. The lattice enthalpy also decreases as the radius of the cation increases down a group. The trend in the enthalpy of formation depends upon the relative difference between these values, that is $(\Delta_{sub} H^{\ominus} + \Delta_{ion} H^{\ominus}) - \Delta_L H^{\ominus}$. For the chlorides, bromides, and iodides, the variation in values of $(\Delta_{sub} H^{\ominus} + \Delta_{ion} H^{\ominus})$ is greater than the variation in $\Delta_L H^{\ominus}$ and the enthalpies

Table 10.2 Selected data for the discussion of the stabilities of the Group 1 halides

	F	Cl	Br	I
Ionic radius/pm	136	181	195	216
$\frac{1}{2}\Delta_{dis}H^{\ominus}$/(kJ mol^{-1})	79	121	112	107
$\Delta_{eg}H^{\ominus}$/(kJ mol^{-1})	−328	−349	−325	−295
$(\frac{1}{2}\Delta_{dis}H^{\ominus} + \Delta_{eg}H^{\ominus})$/(kJ mol^{-1})	−249	−228	−213	−188

of formation become more negative down the group. However, for the fluorides, the small ionic radius of fluorine ensures that the differences in $\Delta_L H^{\ominus}$ are greater than those in $(\Delta_{sub}H^{\ominus} + \Delta_{ion}H^{\ominus})$ and the enthalpies of formation become less negative down the group.

Example 10.1 Calculating enthalpies of formation

Use data from Table 10.1 and 10.2 to calculate the enthalpies of formation of NaF and NaCl and comment on the values obtained.

Answer The lattice energies of the compounds can be calculated using the Kapustinskii equation (eqn 3.4), which gives 891 kJ mol^{-1} for NaF and 767 kJ mol^{-1} for NaCl. Then, from eqn 10.1,

$\Delta_f H^{\ominus}$(NaF) $= 109 + 494 + 79 + (-328) + (-891)$ kJ mol^{-1} $= -537$ kJ mol^{-1}

$\Delta_f H^{\ominus}$(NaCl) $= 109 + 494 + 121 + (-349) + (-767)$ kJ mol^{-1} $= -392$ kJ mol^{-1}

The enthalpy of formation for NaF is the more negative and therefore the fluoride would be expected to be more stable than the chloride. In this case, the most important term in the expression for $\Delta_f H^{\ominus}$ is the lattice enthalpy, $\Delta_L H^{\ominus}$, which is larger for NaF because of the smaller size of the anion.

Self-test 10.1 Calculate the enthalpies of formation for LiF and NaF and comment on the values obtained.

The halides are all soluble in water with the exception of LiF, which is only sparingly soluble. This low solubility of LiF can be traced to the fact that the high lattice enthalpy due to the small ionic radii is not offset by the enthalpy of hydration (see later in this chapter).

Sodium chloride occurs naturally as the mineral rock salt, the deposits of ancient dried saline lakes. Many of these deposits are underground and are mined in the conventional way. Alternatively, water may be pumped underground to dissolve the rock salt, which is then pumped out as saturated brine solution. Rock salt is used more than any other mineral in chemical manufacture and is also used extensively in road de-icing, where it reduces the freezing point of water. Potassium chloride is also recovered from ancient dried lakes. Its major use is as a fertilizer.

10.5 Oxides and related compounds

Key points: Only lithium forms a normal oxide on direct reaction with oxygen; sodium forms the peroxide and the heavier elements form the superoxides.

As mentioned previously, all the Group 1 elements react vigorously with oxygen. Only lithium reacts directly with oxygen to give the oxide, Li$_2$O.

$4\,Li(s) + O_2(g) \rightarrow 2\,Li_2O(s)$

Sodium reacts with oxygen to give the peroxide, Na$_2$O$_2$, which contains the peroxide ion, O_2^{2-}:

$2\,Na(s) + O_2(g) \rightarrow Na_2O_2(s)$

The other Group 1 elements form the superoxides, which contain the paramagnetic superoxide ion, O_2^-:

$$K(s) + O_2(g) \rightarrow KO_2(s)$$

All the oxides are basic and react with water to give the hydroxide ion:

$$Li_2O(s) + H_2O(l) \rightarrow 2\,Li^+(aq) + 2\,OH^-(aq)$$
$$Na_2O_2(s) + 2\,H_2O(l) \rightarrow 2\,Na^+(aq) + 2\,OH^-(aq) + H_2O_2(aq)$$
$$2\,KO_2(s) + 2\,H_2O(l) \rightarrow 2\,K^+(aq) + 2\,OH^-(aq) + H_2O_2(aq) + O_2(g)$$

The normal oxides of sodium, potassium, rubidium, and caesium can be prepared by heating the metal with a limited amount of oxygen or by thermal decomposition of the peroxide or superoxide.

$$Na_2O_2(s) \rightarrow Na_2O(s) + \tfrac{1}{2}\,O_2(g)$$

The stability of the peroxides and superoxides to this decomposition increases down the group, with Li_2O_2 being the least stable and Cs_2O_2 the most stable. Sodium peroxide is a widely used oxidizing agent as it provides a ready source of oxygen on warming. The tendency of the peroxide or superoxide to decompose to the oxide can be explained by examining the lattice enthalpies of the compounds. As discussed earlier, the lattice enthalpy is inversely proportional to the sum of the ionic radii. Consequently, as the O^{2-} ion is smaller than either the O_2^{2-} or the O_2^- ion, the lattice enthalpy of any oxide is larger than that of the corresponding peroxide or superoxide. On descending the group, the radius of the cations increase and the lattice enthalpies of both the oxide and peroxide (or superoxide) decrease. Therefore, the *difference* between the two lattice enthalpies decreases and the tendency to decompose is decreased.

Potassium superoxide, KO_2, absorbs carbon dioxide and liberates oxygen. This reaction is exploited to purify air in applications such as submarines and breathing apparatus.

$$4\,KO_2(s) + 2\,CO_2(g) \rightarrow 2\,K_2CO_3(s) + 3\,O_2(g)$$
$$K_2CO_3(s) + CO_2(g) + H_2O(g) \rightarrow 2\,KHCO_3(s)$$

Ozonides, compounds that contain the ozonide ion, O_3^-, exist for all the Group 1 metals. The ozonides of K, Rb, and Cs are obtained by heating the peroxide or super-oxides with ozone. Sodium and lithium ozonides may be prepared by ion exchange of CsO_3 in liquid ammonia. These compounds are very unstable and explode violently.

$$2\,KO_3(s) \rightarrow 2\,KO_2(s) + O_2(s)$$

Partial oxidation of rubidium and caesium yield suboxides of various compositions. Special conditions are needed to form these compounds in which the elements occur with oxidation numbers lower than $+1$. These compounds are formed only when air, water, and other oxidizing agents are rigorously excluded. For example, a series of metal-rich oxides is formed by the reaction of rubidium or caesium with a limited supply of oxygen. These compounds are dark, highly reactive metallic conductors with formulas such as Rb_6O, Rb_9O_2, Cs_4O_4, and Cs_7O. A clue to the nature of these compounds is that Rb_9O_2 consists of O atoms surrounded by octahedra of six Rb atoms, with two neighbouring octahedra sharing faces (Fig. 10.6). The metallic conduction of the compounds suggests that the valence electrons are delocalized beyond the individual Rb_9O_2 clusters.

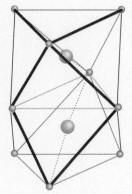

Fig. 10.6 The structure of Rb_9O_2. Each Rb atom is surrounded by an octahedron of O atoms, and neighbouring octahedral share triangular faces.

Example 10.2 Predicting stabilities of peroxides

The ionic radii of the O^{2-} and O_2^{2-} ions are 140 and 176 pm, respectively. Use the data in Table 10.1 and the Kapustinskii equation (eqn 3.4) to calculate the difference in lattice energies between Na_2O and Na_2O_2. Carry out the same calculation for Rb_2O and Rb_2O_2 and comment on the results.

Answer The Kapustinskii equation gives the following values of lattice energy:

Na_2O	Na_2O_2	Rb_2O	Rb_2O_2
1978	1782	1680	1536 $kJ\,mol^{-1}$

The difference between the values for sodium is 196 kJ mol^{-1}, and 144 kJ mol^{-1} for rubidium, confirming that the difference between the values decreases down the group resulting in decreased tendency for the peroxide to decompose.

Self-test 10.2 All the Group 1 ozonides are unstable. Predict whether that instability increases or decreases down the group.

10.6 Hydroxides

Key point: All Group 1 hydroxides absorb water from the atmosphere.

All the hydroxides are white, translucent, deliquescent solids. They absorb water from the atmosphere in an exothermic reaction. Lithium hydroxide, LiOH, forms the stable hydrate, LiOH·8H$_2$O. The solubility of the hydroxides makes them a ready source of OH$^-$ ions in the laboratory and in industry. Potassium hydroxide, KOH, is soluble in ethanol and this 'ethanolic KOH' is a useful reagent in organic synthesis.

Sodium hydroxide is produced by the chloralkali industry (see Box 10.2) and is used as a reagent in the organic chemical industry and in the preparation of other inorganic chemicals. It is also used in the papermaking industry and by the food industry to break down proteins. For example, olives are soaked in sodium hydroxide solution to make the skins soft enough to be edible. Domestic applications are based on the action of NaOH on grease and it is used extensively in oven and drain cleaners. In some 'foaming' drain cleaners it is mixed with aluminium powder. The aluminium reacts with the aqueous hydroxide ions to liberate hydrogen gas, which agitates the mixture and enhances the interaction with NaOH.

$$2\,Al(s) + 2\,OH^-(aq) + 6\,H_2O(l) \rightarrow 2\,[Al(OH)_4]^-(aq) + 3\,H_2(g)$$

10.7 Compounds of oxoacids

The Group 1 elements form salts with most oxoacids. The most industrially important Group 1 salts of oxoacids are sodium carbonate, commonly called *soda ash*, and sodium hydrogen carbonate, commonly known as *sodium bicarbonate*.

(a) Carbonates

Key point: The Group 1 carbonates are soluble and decompose to the oxide when heated.

The Group 1 elements form the only soluble carbonates (with the exception of the NH$_4^+$ ion), although lithium carbonate is only sparingly soluble.

Sodium carbonate has been produced by the Solvay process for many years. The overall reaction is

$$2\,NaCl(aq) + CaCO_3(s) \rightleftharpoons Na_2CO_3(s) + CaCl_2(aq)$$

The equilibrium lies to the left on account of the high lattice energy of CaCO$_3$, and the actual process uses a complex stepwise route. Ammonia and carbon dioxide are passed into a saturated sodium chloride solution to form soluble ammonium and hydrogencarbonate ions. When cooled to below 15°C, sodium hydrogencarbonate is precipitated and is filtered off and heated to produce sodium carbonate. The process is very energy intensive and produces large amounts of calcium chloride as a by-product. These problems mean that sodium carbonate is mined wherever sources of the mineral trona exist, which contains sodium sesquicarbonate, Na$_3$(CO$_3$)(HCO$_3$)·2H$_2$O.

The main uses of sodium carbonate are in glass manufacture, where it is heated with silica to form sodium silicate, Na$_2$O·xSiO$_2$, and as a water softener, where it removes Ca^{2+} ions as the calcium carbonate, the 'scale' formed in kettles in hard water areas. Potassium carbonate is produced by treating KOH with carbon dioxide and is used in glass and ceramics manufacture.

Lithium carbonate decomposes when heated:

$$Li_2CO_3(s) \xrightarrow{\Delta} Li_2O(s) + CO_2(g)$$

The carbonates of the heavier elements do not decompose significantly even when heated to very high temperature. This stabilizing influence of a large cation on a large anion can be explained in terms of trends in lattice energies and was discussed in Section 3.15.

Example 10.3 Predicting the thermal stabilities of carbonates

Use the Kapustinskii equation (eqn 3.4), the ionic radii given in Table 10.1, and the radii of the oxide and carbonate ions (140 pm and 164 pm, respectively) to calculate the difference between the lattice enthalpies of the carbonate and oxide of (a) sodium and (b) caesium. Comment on the values obtained.

Answer The Kapustinskii equation gives the following values for the lattice enthalpies of the compounds

	Na_2CO_3	Na_2O	Rb_2CO_3	Rb_2O
$\Delta_L H/(kJ\ mol^{-1})$	1620	1757	1380	1479

The difference between the carbonate and oxide for sodium and rubidium is 137 and 99 kJ mol^{-1}, respectively. This calculation confirms that the difference between the lattice enthalpies of the carbonate and the oxide decreases down the group.

Self-test 10.3 Sketch a thermochemical cycle for the decomposition of a Group 1 carbonate into the oxide and carbon dioxide.

(b) Hydrogencarbonates

Key points: Sodium hydrogencarbonate is less soluble than the carbonate and liberates CO_2 when heated.

Sodium hydrogencarbonate (sodium bicarbonate) is less soluble than sodium carbonate in water and can be prepared by bubbling carbon dioxide through a saturated solution of the carbonate.

$$Na_2CO_3(aq) + CO_2(g) + H_2O(l) \rightarrow 2\ NaHCO_3(s)$$

The reverse of this reaction occurs when the hydrogencarbonate is heated.

$$2\ NaHCO_3(s) \rightarrow Na_2CO_3(s) + CO_2(g) + H_2O(l)$$

This reaction provides the basis for the use of sodium hydrogencarbonate as a fire extinguisher. The powdered salt smothers the flames and it decomposes in the heat to liberate carbon dioxide and water, which themselves act as extinguishers. This reaction is also the basis for the use of sodium hydrogencarbonate in baking, when the carbon dioxide and water vapour released during the baking process cause the product to 'rise'. A more effective raising agent is baking powder in which sodium hydrogencarbonate is mixed with calcium dihydrogenphosphate:

$$2\ NaHCO_3(s) + Ca_2(H_2PO_4)_2(s) \rightarrow Na_2HPO_4(s) + CaHPO_4(s) + 2\ CO_2(g) + 2\ H_2O(l)$$

Potassium hydrogencarbonate is used as a buffer in wine production and in water treatment. It is used as a buffer in low pH liquid detergents, as an additive in soft drinks, and as an antacid to combat indigestion.

(c) Other oxosalts

Key point: The nitrates of Group 1 elements are used as fertilizers and explosives.

Sodium sulfate, Na_2SO_4, is very soluble and readily forms hydrates. The major commercial source of sodium sulfate is as a by-product of the production of hydrochloric acid from sodium chloride:

$$NaCl(aq) + H_2SO_4(aq) \rightarrow Na_2SO_4(aq) + 2\,HCl(aq)$$

It is also obtained as a by-product of several other industrial processes, including flue-gas desulfurization and the manufacture of rayon. The principal use of sodium sulfate is in processing wood pulp for making the tough, brown paper used in packaging and cardboard. During the process, the sodium sulfate is reduced to sodium sulfite, which dissolves the lignin in the wood. (The lignin is recovered from the pulp and used as an adhesive and binder.) It is also used in glass manufacture, in detergents, and as a mild laxative.

Sodium thiosulfate, $Na_2S_2O_3$, which contains the thiosulfate ion $S_2O_3^{2-}$ (1), is readily soluble in water and is a mild reducing agent.

$$S_4O_6^{2-}(aq) \rightarrow 2\,S_2O_3^{2-}(aq) + 2\,e^- \quad E^\ominus = +0.08\ V$$

Its major use is as a developer in photography, where it forms a soluble complex with silver ions. Sodium thionite, $Na_2S_2O_4$, which contains the thionite ion, $S_2O_4^{2-}$ (2), is used as a common reducing agent in biochemistry.

Sodium nitrate, $NaNO_3$, is deliquescent and is used in making other nitrate, fertilizers, and explosives. Potassium nitrate, KNO_3, occurs naturally as the mineral saltpetre. It is slightly soluble in cold water and very soluble in hot water. It has been used extensively in the manufacture of gunpowder since about the twelfth century and is used in explosives, fireworks, matches, and fertilizers.

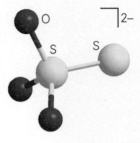

1 $S_2O_3^{2-}$

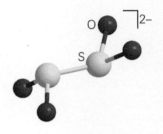

2 $S_2O_4^{2-}$

10.8 Nitrides

Key point: Only lithium forms a nitride by direct interaction with nitrogen.

Although lithium is the least reactive of the Group 1 metals, it is the only one that (like magnesium) forms a nitride by direct reaction with nitrogen:

$$6\,Li(s) + N_2(g) \rightarrow 2\,Li_3N(s)$$

Lithium nitride shows potential as a hydrogen storage material. It stores up to 11.5 per cent by mass when exposed to hydrogen at elevated temperatures and pressures. The Li_3N reacts with hydrogen to form $LiNH_2$ and LiH in a reversible reaction:

$$Li_3N(s) + 2\,H_2(g) \rightleftharpoons LiNH_2(s) + 2\,LiH(s)$$

When heated to 1700°C the $LiNH_2$ and LiH react together to form Li_3N and liberate hydrogen.

The other Group 1 elements form azides, which contain the N_3^- ion, on direct reaction of the elements:

$$2\,Na(s) + 3\,N_2(g) \rightarrow 2\,NaN_3(s)$$

Sodium nitride has recently been synthesized by deposition of Na and N atoms on a cooled sapphire surface at liquid nitrogen temperatures. The compound decomposes at 870°C.

10.9 Solubility and hydration

Key points: There is wide variation in the solubility of the common salts; only lithium and sodium form hydrated salts.

All the common salts of the Group 1 metals are soluble in water. The solubilities cover a wide range of values, some of the most soluble being those for which there is the greatest difference between the radii of the cation and anion. Thus, the solubilities of the lithium halides increase from the fluoride to the bromide whereas for caesium the trend is reversed. The explanation for these trends was discussed in Section 3.15.

Not all alkali metal salts occur as their hydrates. The lattice enthalpies of hydrated salts are lower than for the anhydrous salt because the radius of the cation is effectively increased by the hydration sphere and is further from its surrounding anions. The hydrated salt will be favoured if this decrease in lattice enthalpy is offset by the hydration enthalpy. The hydration enthalpy depends on the ion–dipole interaction between the cation and the polar water molecule. This interaction is greatest when the cation has high charge density. The Group 1 metal cations have low charge density on account of their large radii and low charge. Consequently, most of their salts are anhydrous. There are a few exceptions for the smaller lithium and sodium ions; for example, $LiOH \cdot 8H_2O$.

10.10 Solutions in liquid ammonia

Key points: Sodium dissolves in liquid ammonia to give a solution that is blue when dilute and bronze when concentrated.

Sodium dissolves in pure, anhydrous liquid ammonia (without hydrogen evolution) to give solutions that are deep blue when dilute. The colour of these **metal–ammonia solutions** originates from the tail of a strong absorption band that peaks in the near infrared.[1] The dissolution of sodium in liquid ammonia to give a very dilute solution is represented by the equation

$$Na(s) \rightarrow Na^+(am) + e^-(am)$$

These solutions survive for long periods at the temperature of boiling ammonia ($-33°C$) and in the absence of air. However, they are only metastable and their decomposition is catalysed by some d-block compounds:

$$Na^+(am) + e^-(am) + NH_3(l) \rightarrow NaNH_2(am) + \tfrac{1}{2}H_2(g)$$

Concentrated metal–ammonia solutions have a metallic bronze colour and have electrical conductance close to that of the metal. These solutions have been described as 'expanded metals' in which $e^-(am)$ associates with the ammoniated cation. This description is supported by the fact that, in saturated solutions, the ammonia to metal ratio is between 5 and 10, which corresponds to a reasonable coordination number for the metal.

The blue metal–ammonia solutions are excellent reducing agents. For example, the Ni(I) complex $[Ni_2(CN)_6]^{4-}$, in which nickel is in an unusually low oxidation state, may be prepared by the reduction of Ni(II) with potassium in liquid ammonia:

$$2\,K_2[Ni(CN)_4] + 2\,K^+(am) + 2\,e^-(am) \rightarrow K_4[Ni_2(CN)_6](am) + 2\,KCN$$

The reaction is performed in the absence of air in a vessel cooled to the boiling point of ammonia.

Alkali metals also dissolve in ethers and alkylamines to give solutions with absorption spectra that depend on the alkali metal. The dependence on the metal suggests that the spectrum is associated with charge transfer from an *alkalide ion*, M^- (such as a sodide ion, Na^-), to the solvent. Further evidence for the occurrence of alkalide ions is the diamagnetism associated with the species assigned as M^-, which would have the spin-paired s^2 valence electron configuration. Another observation in agreement with this interpretation is that, when sodium/potassium alloy is dissolved, the metal-dependent band is the same as for solutions of sodium itself. When ethylenediamine (1,2-diaminoethane, en) is used as a solvent, the metal-independent band is not observed, so the dissolution equation is written as

$$2\,Na(s) \rightarrow Na^+(en) + Na^-(en)$$
$$NaK(l) \rightarrow K^+(en) + Na^-(en)$$

Coordination and organometallic compounds

Group 1 elements form coordination compounds with ligands that contain O or N atoms. The most stable complexes are formed with chelating ligands and macrocycles. The organometallic compounds of Group 1 elements are pyrophoric and readily hydrolysed.

---

[1] Other electropositive metals with low enthalpies of sublimation, including calcium and europium, dissolve in liquid ammonia to give solutions with a blue colour that is independent of the metal.

10.11 Coordination compounds

Key point: The Group 1 elements form stable complexes with polydentate ligands.

The Group 1 metal ions are 'hard' Lewis acids (Section 4.15). Therefore, most of the complexes they form arise from Coulombic interactions with small, hard donors, such as those possessing O or N atoms. Monodentate ligands are only weakly bound on account of the weak Coulombic interactions and lack of significant covalent bonding by these ions. In the $M(OH_2)_n^+$ species the ligands readily exchange with the water solvent molecules. Chelating ligands such as the ethylenediaminetetraacetate ion, $(O_2C)_2NCH_2CH_2N(CO_2)_2^{4-}$, have much higher formation constants; but it is with macrocycles and related ligands that we see the strongest complexes formed. Crown ethers such as 18-crown-6 (**3**) form complexes with alkali metal ions that are reasonably stable in nonaqueous solution. Bicyclic cryptand ligands, such as 2.2.1 crypt (**4**) and 2.2.2 crypt (**5**), form complexes with alkali metals that are even more stable, and they can survive even in aqueous solution. These ligands are selective for a particular metal ion, the dominant factor being the fit between the cation and the cavity in the ligand that accommodates it (Fig. 10.7).

Another example of this fit between cation and ligand cavity is thought to be responsible for the transport of Na^+ and K^+ ions across cell membranes (Section 26.3). The ions cross the hydrophobic cell membrane by means of embedded protein molecules that contain cavities lined with donor atoms. The donor atoms are arranged to form a cavity, the size of which determines whether Na^+ or K^+ is bound. Such *ion channels* modulate the Na^+/K^+ concentration differential across the cell membrane that is essential for particular functions of the cell. The naturally occurring molecule valinomycin (**6**) is an antibiotic that selectively coordinates K^+: the resulting hydrophobic 1:1 complex transports K^+ through a bacterial cell membrane, depolarizing the ion differential and resulting in cell death.

The complexation of sodium with a cryptand can be used to prepare solid sodides, such as $[Na(2.2.2)]^+Na^-$, where (2.2.2) denotes the cryptand ligand. X-ray structure determination reveals the presence of $[Na(2.2.2)]^+$ and Na^- ions, with the latter located in a cavity of the crystal with an apparent radius larger than that of I^-. The precise nature of the products of this reaction vary with the ratio of sodium to cryptand. It is also

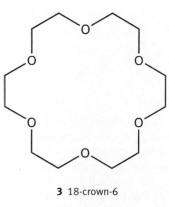

3 18-crown-6

4 2.2.1 crypt

5 2.2.2 crypt

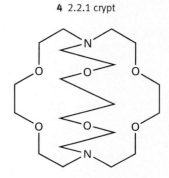

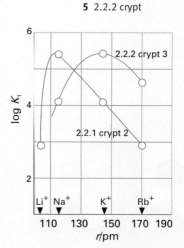

6 Valinomycin

Fig. 10.7 The formation constants of complexes of Group 1 metals with cryptand ligands plotted against cation radius. Note that the smaller 2.2.1 crypt favours complex formation with Na^+ and the larger 2.2.2 crypt favours K^+.

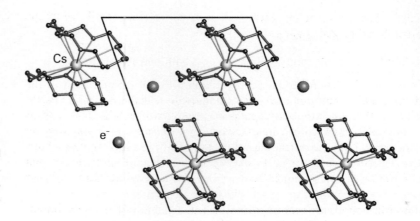

Fig. 10.8 The crystal structure of [Cs(18-crown-6)$_2$]$^+$e$^-$. The purple spheres mark the sites of highest electron density, and so indicate the locations of the 'anion' e$^-$. (From S.B. Dawes, D.L. Ward, R.H. Huang, and J.J. Dye, *J. Am. Chem Soc.*, 1986, **108**, 3534.)

possible to crystallize solids containing solvated electrons, the so-called **electrides**, and to obtain their X-ray crystal structures. Figure 10.8, for example, shows the inferred position of the maxima of the electron density in such a solid. The preparation of sodides and other alkalides demonstrates the powerful influence of solvents and complexing agents on the chemical properties of metals. A further example of these influences is the ability of crown ethers to produce reactive Cl$^-$ ions in organic solvents. When an aqueous solution of NaCl is shaken in a separating funnel with a solution of 18-crown-6 in organic solvent, the Na$^+$ ions cross into the organic phase drawing Cl$^-$ ions with them.

10.12 Organometallic compounds

Key point: The organometallic compounds of the Group 1 elements are unstable and pyrophoric.

Group 1 elements form a number of organometallic compounds that are unstable in the presence of water and pyrophoric in air. They are prepared in organic solvents, such as tetrahydrofuran (THF) or dichloromethane. Protic (proton-donating) organic compounds form ionic organometallic compounds with the Group 1 metals. For example, cyclopentadiene reacts with sodium metal in THF:

$$Na + C_5H_6 \rightarrow Na^+[C_5H_5]^- + \tfrac{1}{2}H_2$$

The resulting cyclopentadienide anion is an important intermediate in the synthesis of *d*-block organometallic compounds (Chapter 21). Sodium and potassium form intensely coloured compounds with aromatic species. The oxidation of the metal results in transfer of an electron to the aromatic system to produce a **radical anion**, an anion that possesses an unpaired electron:

Na + (naphthalene) ⟶ Na$^+$ [C$_{10}$H$_8$]$^-$

Sodium and potassium alkyls are colourless solids that are insoluble in organic solvents and, when stable, have fairly high melting temperatures. They are produced by a **transmetallation reaction**, which involves breaking a metal–carbon bond and forming a metal–carbon bond to a different metal. Alkylmercury compounds are often the starting materials in these reactions. For example, methylsodium is produced in the reaction between sodium metal and dimethylmercury in a hydrocarbon solvent:

$$Hg(CH_3)_2 + 2\,Na \rightarrow 2\,NaCH_3 + Hg$$

Lithium alkyls and aryls are by far the most important Group 1 organometallics. They are liquids or low melting solids, are the most thermally stable of the entire group, and are soluble in organic and nonpolar solvents such as THF. They can be synthesized from an

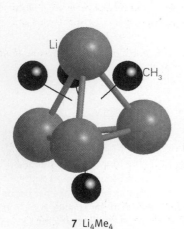

7 Li$_4$Me$_4$

alkyl halide and lithium metal or by reacting the organic species with butyllithium, Li(C₄H₉), commonly abbreviated to BuLi.

$$BuCl + 2\ Li \rightarrow BuLi + LiCl$$

$$BuLi + C_6H_6 \rightarrow Li(C_6H_5) + C_4H_{10}$$

A feature of many main-group organometallic compounds is the presence of bridging alkyl groups. When ethers are the solvent, methyl lithium exists as $Li_4(CH_3)_4$, with a tetrahedron of lithium atoms and bridging methyl groups (**7**). In hydrocarbon solvents, $Li_6(CH_3)_6$ (**8**) is formed; its structure is based upon an octahedral arrangement of lithium atoms. Other lithium alkyls adopt similar structures except when the alkyl groups become very bulky, as in the case of *t*-butyl, $-C(CH_3)_3$, when tetramers are the largest species formed.

Organolithium compounds are very important in organic synthesis, the most important reactions being those in which they act as nucleophiles and attack a carbonyl group:

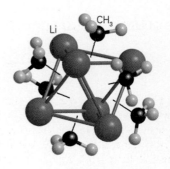

$Li_6(CH_3)_6$,square-pyramidal

8

Organolithium compounds are also used to convert *p*-block halides to organocompounds, as we see in later chapters. For example, boron trichloride reacts with butyllithium in THF to give the organoboron compound:

$$BCl_3 + 3\ BuLi \rightarrow Bu_3B + 3\ LiCl$$

The driving force for this and many other reactions of *s*- and *p*-block organometallic compounds is the formation of the halide compound of the more electropositive metal.

Alkyllithiums are important industrially in the stereospecific polymerization of alkenes to form synthetic rubber. Butyllithium is used as an initiator in solution polymerization to produce a wide range of elastomers and polymers. Organolithium compounds are also used in the synthesis of a range of pharmaceuticals including vitamins A and D, analgesics, antihistamines, antidepressants, and anticoagulants. Alkyllithiums can be used in the synthesis of other organometallic compounds. For example, they can be used to introduce alkyl groups into *d*-metal organometallic compounds (Section 21.8):

$$(C_5H_5)_2MoCl_2 + 2\ CH_3Li \rightarrow (C_5H_5)_2Mo(CH_3)_2 + 2\ LiCl$$

The reactivity and solubility of alkyllithium is enhanced by adding a chelating ligand such as tetramethylethylenediamine, TMEDA (**9**), which breaks up any tetramers to give complexes such as $[BuLi(TMEDA)]_2$.

9 TMEDA

FURTHER READING

R.B. King, *Inorganic chemistry of the main group elements*. Wiley (1994).

P. Enghag, *Encyclopedia of the elements*. Wiley (2004).

D.M.P Mingos, *Essential trends in inorganic chemistry*. Oxford University Press (1998). An overview of inorganic chemistry from the perspective of structure and bonding.

V.K. Grigorovich, *The metallic bond and the structure of metals*. Nova Science Publishers (1989).

N.C. Norman, *Periodicity and the s- and p-block elements*. Oxford University Press (1997). Includes coverage of essential trends and features of s-block chemistry.

A. Sapse and P.V. Schleyer (eds), *Lithium chemistry: a theoretical and experimental overview*. Wiley (1995).

EXERCISES

10.1 Why are Group 2 elements (a) strong reducing agents, (b) poor complexing agents?

10.2 Use the data given in Tables 10.1 and 10.2 to calculate the enthalpies of formation for the Group 2 fluorides and chlorides. Plot the data and comment on the trends observed.

10.3 Give equations for the synthesis of (a) C_2H_5Li and (b) C_2H_5Na.

10.4 Which of the following pairs are most likely to form the desired compound? Describe the periodic trend and the physical basis for your answer in each case. (a) Cs^+ or Mg^{2+}, form an acetate complex; (b) Li^+ or K^+, form a complex with crypt 2.2.2.

10.5 Identify compounds A, B, C, and D.

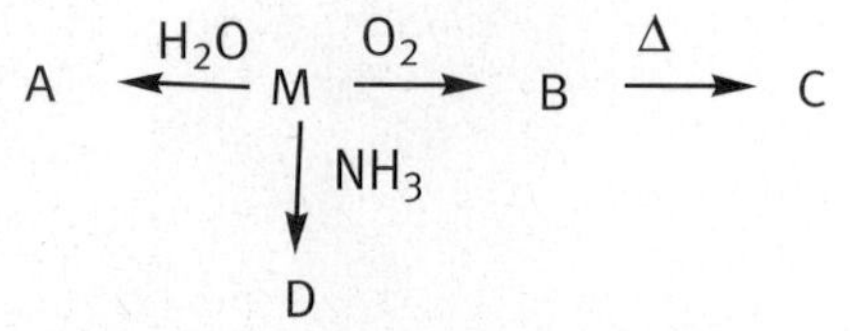

$$A \xleftarrow{H_2O} M \xrightarrow{O_2} B \xrightarrow{\Delta} C$$
$$M \xrightarrow{NH_3} D$$

10.6 Account for the fact that LiF and CsI have low solubility in water whereas LiI and CsF are very soluble.

10.7 Explain why LiH has greater thermal stability than the other Group 1 hydrides whereas Li_2CO_3 decomposes at a lower temperature than the other Group 1 carbonates.

10.8 Draw the structures of NaCl and CsCl and give the coordination number of the metal in each case. Explain why the compounds adopt different structures.

10.9 Explain how the nature of the alkyl group affects the structure of lithium alkyls.

10.10 Predict the products of the following reactions:

(a) $CH_3Br + Li \rightarrow$

(b) $MgCl_2 + LiC_2H_5 \rightarrow$

(c) $C_2H_5Li + C_6H_6 \rightarrow$

PROBLEMS

10.1 Describe the origin of the diagonal relationship between Li and Mg.

10.2 Identify the incorrect statement or statements in the following description and provide corrections and an explanation of the trend. (a) Sodium dissolves in ammonia and amines to produce the sodium cation and solvated electrons or the sodide ion. (b) Sodium dissolved in liquid ammonia will not react with NH_4^+ because of strong hydrogen bonding with the solvent.

10.3 Z. Jedlinski and M. Sokol describe the solubility of alkali metals in nonaqueous supramolecular systems (*Pure Appl. Chem.*, 1995, **67**, 587). They dissolved the metals in THF containing crown ethers or cryptands. Sketch the structure of the 18C6 ligand. Give the equations proposed for the dissolution process. Outline the two methods used to prepare the alkali metal solutions. What factors affect the stability of the solutions?

10.4 Alkali metal halides can be extracted from aqueous solution by solid-phase ditopic salt receptors. (See J.M. Mahoney, A.M. Beatty, and B.D. Smith, *Inorg. Chem.*, 2004, **43**, 7617.) (a) What is a ditopic receptor? (b) What is the order of selectivity to extraction of the alkali metal ions in aqueous solution? (c) What is the order of selectivity to extraction from the solid phase? (d) Explain the observed order of selectivity.

10.5 The molecular geometries of crown ether derivatives play an important role in capturing and transporting alkali metal ions. K. Okano and co-workers (see K. Okano, H. Tsukube, and K. Hori, *Tetrahedron*, 2004, **60**, 10877) studied stable conformations of 12-crown-O3N and its Li^+ complex in aqueous and acetonitrile solutions. (a) Which three programs did the authors use in their study and what did each program calculate? (b) Which Li^+ complex was found to be most stable in (i) aqueous and (ii) acetonitrile solutions?

The Group 2 elements

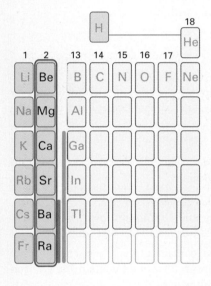

In this chapter we look at the occurrence and isolation of the Group 2 elements and study the chemical properties of their simple compounds, complexes, and organometallic compounds. Throughout the chapter we draw comparisons with the elements of Group 1. We see how the insolubility of some of the calcium compounds in particular leads to the existence of many inorganic minerals that provide the raw materials for the infrastructure of our built environment and provide the building blocks for the compounds from which many rigid biological structures are formed.

The elements calcium, strontium, barium, and radium are known as the **alkaline earth metals**, but the term is often applied to the whole group. All the elements are silvery white metals, but some aspects of the chemical properties of beryllium are more like those of a metalloid. The elements are harder, denser, and less reactive than the elements of Group 1 but are still more reactive than many typical metals. The elements form a limited number of complexes and organometallic compounds.

The elements

Key point: The most important factors influencing the chemical properties of the Group 2 elements are their atomic and ionic radii.

The increase in mechanical hardness of the Group 2 compared with the Group 1 elements indicates an increase in the strength of metallic bonding between Groups 1 and 2, which can be attributed to the increased number of electrons available (Section 3.16). The atomic radii of the Group 2 elements are smaller than those of Group 1. This reduction in atomic radius between the groups is responsible for the higher densities and ionization energies (Table 11.1). The ionization energies of the elements decrease down the group as the radius increases (Fig. 11.1, Section 1.9). The elements become more reactive and more electropositive down the group as it becomes easier to form the +2 ions. This decrease in ionization energy is reflected in the trend in standard potentials for the M^{2+}/M couples. Because the standard potentials become more negative down the group, the metals are more readily oxidized. Thus, whereas calcium, strontium, and barium

Table 11.1 Selected properties of the Group 2 elements

	Be	Mg	Ca	Sr	Ba	Ra
Metallic radius/pm	112	160	197	215	217	220
Ionic radius, $r(M^{2+})$/pm	31	65	99	113	135	140
Ionization energy, I/(kJ mol^{-1})	900	736	590	548	502	510
$E^{\ominus}(M^{2+}/M)$/V	−1.85	−2.38	−2.87	−2.89	−2.90	−2.92
Density, ρ/(g cm^{-3})	1.85	1.74	1.54	2.62	3.51	5.00
Melting point/°C	1280	650	850	768	714	700
$\Delta_{hyd}H^{\ominus}$/(kJ mol^{-1})	−2500	−1920	−1650	−1480	−1360	−
$\Delta_{sub}H^{\ominus}$/(kJ mol^{-1})	321	150	193	164	176	130

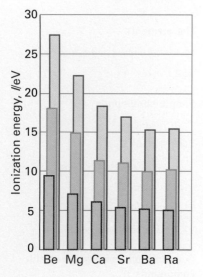

Fig. 11.1 The variation of first, second, and total ionization energies in Group 2.

react readily with cold water, magnesium reacts only with hot water:

$$M(s) + 2\,H_2O(l) \rightarrow M(OH)_2(aq) + H_2(g)$$

All the elements occur as M(II) in their compounds, which is consistent with their ns^2 valence-electron configuration. The compounds are predominantly ionic, with the exception of those of beryllium, which are predominantly covalent on account of the very small size of the Be^{2+} ions and the corresponding high charge density and polarizing power (Section 1.9).

There is a strong diagonal relationship between beryllium and aluminium:

1 Beryllium and aluminium form covalent hydrides, halides, and oxides; the analogous compounds of magnesium are predominantly ionic.

2 Beryllium oxide and aluminium oxide are amphoteric; magnesium oxide is basic.

3 In the presence of excess OH$^-$ ions, beryllium and aluminium form $[Be(OH)_4]^{2-}$ and $[Al(OH)_4]^-$, respectively; magnesium does not react with OH$^-$ ions.

4 Beryllium and aluminium form carbides that contain the C^{4-} ion and produce methane on reaction with water; the other Group 2 carbides contain the C_2^{2-} ion and liberate ethyne (acetylene) on reaction with water.

11.1 Occurrence and extraction

Key points: Magnesium is the only Group 2 element extracted on an industrial scale; magnesium, calcium, strontium, and barium can be extracted from the molten chloride.

Beryllium occurs naturally as the semiprecious mineral beryl, $Be_3Al_2(SiO_3)_6$, from which its name is taken. Beryl is the basis of the gemstone emerald, in which a small fraction of Al^{3+} is replaced by Cr^{3+}. Beryllium is extracted by heating beryl with sodium hexafluorosilicate, Na_2SiF_6, to produce BeF_2, which is then reduced to the element by magnesium.

Magnesium is the eighth most abundant element in the Earth's crust. It occurs naturally in a number of minerals such as dolomite, $CaCO_3\cdot MgCO_3$, and magnesite, $MgCO_3$, and is the third most abundant element in seawater, from which it is commercially extracted. A litre of seawater contains more than 1 g of magnesium ions. The extraction from seawater relies on the fact that magnesium hydroxide is less soluble than calcium hydroxide, because the solubility of the salts of mononegative anions increases down the group (Section 11.8). Either CaO (quicklime) or $Ca(OH)_2$ (slaked lime) is added to seawater and $Mg(OH)_2$ precipitates. The hydroxide is converted to the chloride by treatment with hydrochloric acid:

$$CaO(s) + H_2O(l) \rightarrow Ca^{2+}(aq) + 2\,OH^-(aq)$$

$$Mg^{2+}(aq) + 2\,OH^-(aq) \rightarrow Mg(OH)_2(s)$$

$$Mg(OH)_2(s) + 2\,HCl(aq) \rightarrow MgCl_2(aq) + 2\,H_2O(l)$$

The magnesium is then extracted by electrolysis of molten magnesium chloride:

Cathode: $Mg^{2+}(aq) + 2\ e^- \rightarrow Mg(s)$ Anode: $2\ Cl^-(aq) \rightarrow Cl_2(g) + 2\ e^-$

Magnesium is also extracted from dolomite. The dolomite is heated in air to give magnesium and calcium oxides. This mixture is heated with ferrosilicon (FeSi), which forms calcium silicate, Ca_2SiO_4, iron, and magnesium. Magnesium is a liquid at the high operating temperatures used in the process and can be removed by distillation.

A major problem in the production of magnesium is its high reactivity towards oxygen and water. Nitrogen is commonly used to provide an inert atmosphere for the production of many other reactive metals but cannot be used for magnesium because it reacts to form the nitride, Mg_3N_2. Sulfur hexafluoride or sulfur trioxide are used as alternatives to nitrogen, but each gives rise to environmental problems: SF_6 is a greenhouse gas and SO_3 contributes to acid rain. Although magnesium is very reactive towards oxygen and water, the metal can be handled safely on account of the presence of an inert passivating oxide film on its surface.

Calcium is the fifth most abundant element in the Earth's crust and occurs widely as limestone, $CaCO_3$. The name 'calcium' comes from the Latin *calx*, which means lime. The element is a major component of biominerals such as bone, shells, and teeth, and is central to cell signalling processes such as hormonal or electrical activation of enzymes in higher organisms (Section 26.4). The average adult human contains approximately 1 kg of calcium.

Calcium is extracted by electrolysis of the molten chloride, which is itself obtained as a by-product of the Solvay process for the production of sodium carbonate. Strontium is named after the Scottish village of Strontian where the strontium-containing ore was first found. It is extracted by electrolysis of molten $SrCl_2$ or by reduction of SrO with aluminium:

$$6\ SrO(s) + 2\ Al(s) \rightarrow 3\ Sr(s) + Sr_3Al_2O_6(s)$$

The metal reacts vigorously with water and as a finely divided powder ignites in air. Barium is extracted by electrolysis of the molten chloride or by reduction of BaO with aluminium. It reacts very vigorously with water and ignites very readily in air.

All the isotopes of radium are radioactive. They undergo α, β, and γ decay with half-lives that vary from 42 minutes to 1599 years. Radium was discovered by Pierre and Marie Curie in 1898 after the painstaking extraction from the uranium-bearing mineral pitchblende. Pitchblende is a complex mineral containing many elements: it contains approximately 1 g of Ra in 10 t of ore and the Curies took three years to isolate 0.1 g of $RaCl_2$.

11.2 Uses

Key points: Magnesium is very important for biological function; its major applications are pyrotechnics, alloys, and common medicines.

Beryllium is unreactive in air on account of a passivating layer of an inert oxide film on its surface, which makes beryllium very resistant to corrosion. This resistance to corrosion, and the fact that it is one of the lightest metals, led to its use in alloys to make precision instruments, aircraft, and missiles. Beryllium is also used as a moderator for nuclear reactions because the beryllium nucleus has a high neutron capture cross-section and the metal has a high melting point. However, the fact that soluble beryllium salts are highly toxic has restricted its industrial applications.

Magnesium is of great biological importance. Not only is it a component of chlorophyll but also it is coordinated by many other biologically important ligands, including ATP (adenosine triphosphate, Section 26.2). It is essential for human health, being responsible for the activity of many enzymes. The recommended adult human dose is approximately 0.3 g per day and the average adult contains about 25 g of magnesium.

Most of the applications of magnesium are based on the formation of light alloys, especially with aluminium, that are widely used in construction in applications where

Box 11.1 Cement and concrete

Cement is made by grinding together limestone and a source of aluminosilicates, such as clay, shale, or sand, and then heating the mixture to 1500°C in a rotary cement kiln. The first important reaction to occur in the lower temperature portion of the kiln (900°C) is the calcining of limestone (the process of heating to a high temperature to oxidize or decompose a substance and convert it to a powder), when calcium carbonate (limestone) decomposes to calcium oxide (lime) and carbon dioxide is driven off. At higher temperatures the calcium oxide reacts with the aluminosilicates and silicates to form molten Ca_2SiO_4, Ca_3SiO_5, and $Ca_3Al_2O_6$. The relative proportions of these compounds determines the properties of the final cement. As the compounds cool, they solidify into a form called *clinker*. The clinker is ground to a fine powder and a small amount of calcium sulfate (gypsum) is added to form Portland cement.

Concrete is produced by mixing cement with sand, gravel or crushed stone and water. Often small amounts of additives are added to achieve particular properties. For example, flow and dispersion are improved by adding polymeric materials such as phenolic resins and resistance to frost damage is improved by adding surfactants. When the water is added to the cement, complex hydration reactions occur that produce hydrates, such as $Ca_3Si_2O_7 \cdot H_2O$, $Ca_3Si_2O_7 \cdot 3H_2O$, and $Ca(OH)_2$:

$$2\ Ca_2SiO_4(s) + 2\ H_2O(l) \rightarrow Ca_3Si_2O_7 \cdot H_2O(s) + Ca(OH)_2(aq)$$
$$2\ Ca_2SiO_4(s) + 4\ H_2O(l) \rightarrow Ca_3Si_2O_7 \cdot 3H_2O(s) + Ca(OH)_2(aq)$$

These hydrates form a gel or slurry that coats the surfaces of the sand or aggregates and fills the voids to form the solid concrete. The properties of concrete are determined by the relative proportions of calcium silicates and calcium aluminosilicates in the cement used, the additives, and the amount of water, which determines the degree of hydration.

The raw materials for cement manufacture often contain traces of sodium and potassium sulfates, and sodium and potassium hydroxides are formed during the hydration process. These hydroxides are responsible for the cracking, swelling, and distortion of many ageing concrete structures. The hydroxides take part in a complex series of reactions with the aggregate material to form an alkali silicate gel. This gel is hygroscopic and expands as it absorbs water, producing stress in the concrete, which leads to cracking and deformation. The susceptibility of concrete to this 'alkali silicate reaction' is now monitored by calculating the total alkali levels in the concrete produced and strategies are in place to minimize its effects. For example, adding 'fly ash' to the mix, which is a waste product from coal-fired power stations, can reduce the problem.

weight is an issue, such as aircraft. A magnesium–aluminium alloy was previously used in warships but was discovered to be highly flammable when subjected to missile attack. Some of the uses of magnesium are based upon the fact that the metal burns in air with an intense white flame, and so it is used in fireworks and flares. Other applications are 'Milk of Magnesia', $Mg(OH)_2$, which is a common remedy for indigestion, and 'Epsom Salts', $MgSO_4 \cdot 7H_2O$, which is used for a variety of health treatments including a treatment for constipation, a purgative, and a soak for sprains and bruises. Magnesium oxide, MgO, is used as a refractory lining for furnaces. Organomagnesium compounds are widely used in organic synthesis as Grignard reagents (Section 11.10).

The compounds of calcium are much more useful than the element itself. Calcium oxide (as lime or quicklime) is a major component of mortar and cement (Box 11.1). It is also used in steelmaking and papermaking. Calcium carbonate is used in the Solvay process (Section 10.7) for the production of sodium carbonate (except in the USA, where sodium carbonate is mined as trona) and as the raw material for production of CaO. Calcium fluoride is insoluble and transparent over a wide range of wavelengths. It is used to make cells and windows for infrared and ultraviolet spectrometers.

Strontium is used in pyrotechnics (Box 11.2) and in glasses for colour television tubes. Barium compounds, with the large number of electrons possessed by the barium, are very effective at absorbing X-rays: they are used as 'barium meals' and 'barium enemas' to investigate the intestinal tract. Barium is highly toxic, so the sparingly soluble sulfate is used in this application. Barium carbonate is used in glassmaking and as a flux to aid the flow of glazes and enamels. It is also used as rat poison. The sulfide has been used as a depilatory, to remove unwanted body hair.

Soon after its discovery, radium was used to treat malignant tumours. Luminous radium paint was once widely used on clock and watch faces but has been replaced by less hazardous compounds.

Box 11.2 Fireworks and flares

Fireworks use exothermic reactions to produce heat, light, and sound. Common oxidants are nitrates and perchlorates, which decompose when heated to liberate oxygen. Common fuels are carbon, sulfur, powdered aluminium or magnesium, and organic materials such as poly(vinyl chloride)(PVC), starch, and gums. The most common constituent of fireworks is gunpowder or black powder, a mixture of potassium nitrate, sulfur, and charcoal and thus both an oxidant and a fuel. Special effects, such as colours, flashes, smoke, and noises, are provided by additives to the firework mixture. The Group 2 elements are used in fireworks to provide colour.

Barium compounds are added to fireworks to produce green flames. The species responsible for the colour is $BaCl^+$, which is produced when Ba^{2+} ions combine with Cl^- ions. The Cl^- ions are produced during decomposition of the perchlorate oxidant or during combustion of the PVC fuel:

$$KClO_4(s) \rightarrow KCl(s) + 2\ O_2(g)$$
$$KCl(s) \rightarrow K^+(g) + Cl^-(g)$$
$$Ba^{2+}(g) + Cl^-(g) \rightarrow BaCl^+(g)$$

Barium chlorate, $Ba(ClO_3)_2$, has been used instead of $KClO_4$ and a barium compound but is too unstable to shock and friction. Similarly, strontium nitrate and carbonate are used to produce a red colour on formation of $SrCl^+$. Strontium chlorate and perchlorate are effective at producing the red colour but are too unstable to shock and friction for routine use.

Distress flares also use strontium compounds. Strontium nitrate is mixed with sawdust, waxes, sulfur, and $KClO_4$ and packed into a waterproof tube. When ignited, the flares burn with an intense red flame for up to 30 minutes.

As well as being used as a fuel, powdered magnesium is added to fireworks and flares to maximize light output. As well as the magnesium producing an intense white light, illumination is increased by the incandescence of high-temperature MgO particles that are produced in the oxidation reaction.

Simple compounds

All the Group 2 elements form simple compounds with hydrogen, oxygen, and the halogens and also form a number of oxosalts. The properties of the compounds are predominantly ionic, with the exception of those of beryllium, which are covalent.

11.3 Hydrides

Key point: All the elements form saline hydrides with the exception of beryllium, which forms a polymeric, covalent compound.

Like the Group 1 elements, the Group 2 elements, with the exception of beryllium, form ionic, saline hydrides that contain the H^- ion (Section 9.5). They can be prepared by direct reaction between the metal and hydrogen. Beryllium hydride is covalent and must be prepared from alkylberyllium (Section 11.10). It has a chain structure with bridging hydrogen atoms. There is insufficient electron density for the formation of effective $2c,2e$ bonds, and the bonds formed are $3c,2e$ bonds, as indicated by relatively long bridging Be—H bonds (**1**).

The ionic hydrides of the heavier elements react violently with water to produce hydrogen gas:

$$MgH_2(s) + 2\ H_2O(l) \rightarrow Mg(OH)_2(aq) + 2\ H_2(g)$$

This reaction is not as violent as that for the Group 1 elements and the Group 2 hydrides are used as a convenient source of hydrogen in fuel cells (Box 9.1).

11.4 Halides

Key points: The halides of beryllium are covalent; all the fluorides, except BeF_2, are insoluble in water; all other halides are soluble.

All the halides of beryllium are covalent. The fluoride is a glassy solid that exists in several temperature-dependent phases similar to those of SiO_2 (Section 13.8). It is soluble in water, forming the hydrate $[Be(OH_2)_4]^{2+}$. It is prepared from the thermal decomposition of $(NH_4)_2BeF_4$. Beryllium chloride, $BeCl_2$, can be made from the oxide:

$$BeO(s) + C(s) + Cl_2(g) \rightarrow BeCl_2(s) + CO(g)$$

1 BeH_2

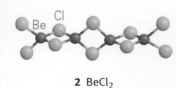

2 $BeCl_2$

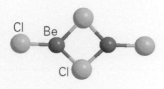

3 $BeCl_2(OEt_2)_2$

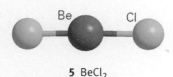

4 $(BeCl_2)_2$

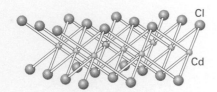

5 $BeCl_2$

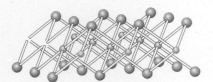

Fig. 11.2 The cadmium-chloride structure adopted by $MgCl_2$.

The chloride, as well as $BeBr_2$ and BeI_2, can also be prepared from direct reaction of the elements at elevated temperatures.

The structure of solid $BeCl_2$ is a polymeric chain (**2**). The local structure is an almost regular tetrahedron around the Be atom and the bonding can be considered to be based on sp^3 hybridization of Be. The structure resembles that of BeH_2 (Section 11.3) but the chloride has sufficient electron density for $2c,2e$ covalent bonding to take place. Beryllium chloride is a Lewis acid, readily forming adducts with electron-pair donors such as diethyl ether (**3**). In the vapour phase the compound tends to form a dimer based on sp^2 hybridization (**4**), and when the temperature is above 900°C linear monomers are formed, indicating sp hybridization (**5**).

The anhydrous halides of magnesium are prepared by the direct combination of the elements as preparation from aqueous solutions yields the hydrates, which are partially hydrolysed on heating. The anhydrous halides of the heavier Group 2 elements can be prepared by dehydration of the hydrates. All the fluorides except BeF_2 are sparingly soluble although the solubility increases slightly down the group. As the radius of the cation increases from Be to Ba, the cation coordination number increases from 4 to 8, with CaF_2, SrF_2, and BaF_2 adopting the fluorite structure (Section 3.9). The other halides of Group 2 form layer structures because of the increasing polarizability of the halide ions. Magnesium chloride adopts the cadmium-chloride layered structure in which the layers are arranged so that the chloride ions are cubic close packed (Fig. 11.2). MgI_2 and CaI_2 adopt the closely related cadmium-iodide structure in which the layers of iodide ions are hexagonal close packed.

The most important fluoride of the Group 2 elements is CaF_2. Its mineral form, fluorspar, is the only large-scale source of fluorine. Anhydrous hydrogen fluoride is prepared by the action of concentrated sulfuric acid on fluorspar:

$$CaF_2(s) + H_2SO_4(l) \rightarrow CaSO_4(s) + 2\,HF(l)$$

All the Group 2 chlorides are deliquescent and form hydrates. They have lower melting points than the fluorides. Magnesium chloride is the most important chloride for industry. It is extracted from seawater and then used in the production of magnesium metal. Calcium chloride is also of great importance and is produced on a massive scale industrially. Its deliquescence leads to its widespread use as a laboratory drying agent. It is also used to de-ice roads, where it is more effective than NaCl for two reasons. First, the dissolution is very exothermic

$$CaCl_2(s) \rightarrow Ca^{2+}(aq) + 2\,Cl^-(aq) \qquad \Delta_{sol}H^{\ominus} = -82\,\text{kJ mol}^{-1}$$

The heat that is generated helps to melt the ice. Second, the minimum freezing mixture of $CaCl_2$ in water has a freezing point of −55°C compared to −18°C for that of NaCl in water. The exothermic dissolution also leads to another application in instant heating packs and self-heating drink containers. Concentrated solutions of $CaCl_2$ have a very sticky consistency and they are sprayed onto unfinished roads to minimize dust production.

Example 11.1 Predicting the nature of the halides

Use the data in Table 1.8 and Fig. 2.2 to predict whether CaF_2 is predominantly ionic or covalent.

Answer The Pauling electronegativity values of Ca and F are 1.00 and 3.98, respectively. The average electronegativity is therefore 2.49 and the difference is 2.98. These values on the Ketelaar triangle in Fig. 2.2 indicate that CaF_2 is ionic.

Self-test 11.1 Predict whether (a) $BeCl_2$ and (b) $BaCl_2$ are predominantly ionic or covalent.

11.5 Oxides and hydroxides

The Group 2 elements react with oxygen to form the oxides. All the elements except beryllium also form unstable peroxides. The oxides react with water to form the basic hydroxides.

(a) Oxides and related compounds

Key points: All the elements form normal oxides with oxygen except barium, which forms the peroxide; all the peroxides decompose to the oxides, their stabilities increasing down the group.

Beryllium oxide is obtained by ignition of the metal in oxygen. It is a white, insoluble solid with the wurtzite structure (Section 3.9). Its high melting point (2570°C) leads to its use as a refractory material.

The oxides of the other Group 2 elements can be obtained by direct combination of the elements (except barium, which forms the peroxide) but they are more commonly obtained by decomposition of the carbonates:

$$MCO_3(s) \xrightarrow{\Delta} MO(s) + CO_2(g)$$

The oxides of the elements from Mg to Ba all adopt the rock-salt structure (Section 3.9). Their melting points decrease down the group as the lattice enthalpies decrease with increasing cation radius. Magnesium oxide is a high-melting-point solid (as is BeO) and is used as a refractory lining in industrial furnaces. Both BeO and MgO have very high thermal conductivity with low electrical conductivity. This combination of properties leads to their use for the heating elements of domestic appliances.

Calcium oxide (as lime or quicklime) is used in large quantities in the steel industry to remove phosphorus, silicon, and sulfur. When heated, CaO is thermoluminescent and emits a bright white light (hence 'limelight'). Calcium oxide is also used as a water softener to remove hardness by reacting with soluble carbonates and hydrogencarbonates to form the insoluble $CaCO_3$. It reacts with water to form $Ca(OH)_2$, which is sometimes known as *slaked lime* and is used in gardening to neutralize acidic soils.

Example 11.2 Explaining trends in s-block chemistry

Use simple bonding models to explain why barium forms a peroxide but beryllium does not.

Answer Large anions are generally stabilized by large cations (Section 3.15). Therefore, barium peroxide should be more stable than beryllium peroxide. In fact the peroxide compound forms spontaneously on exposure of barium to air.

Self-test 11.2 Use lattice enthalpy considerations to explain why MgO_2 is less stable than BaO_2.

The peroxides of magnesium, calcium, strontium, and barium exist and are prepared by a variety of routes; only SrO_2 and BaO_2 can be made by direct reaction of the elements, albeit at elevated temperatures and pressures. All the peroxides are strong oxidizing agents and decompose to the oxide:

$$MO_2(s) \rightarrow MO(s) + \tfrac{1}{2}O_2(g)$$

The stability of the peroxides towards thermal decomposition increases down the group as the radius of the cation increases. This trend is explained by considering the lattice enthalpies of the peroxide and the oxide and their dependence on the relative radii of the cations and anions. As the radius of O^{2-} is smaller than that of O_2^{2-}, the lattice enthalpy of the oxide is greater than that of the corresponding peroxide. The difference between the two lattice enthalpies decreases down the group as both values become smaller with increasing cation radius. Therefore, the tendency to decompose decreases. Magnesium peroxide, MgO_2, is consequently the least stable peroxide and is used as an *in situ* source of oxygen in a range of applications, including bioremediation to clean up polluted waterways.

The Group 2 elements also form a number of complex oxides such as perovskite, $CaTiO_3$, and spinel, $MgAl_2O_4$ (Section 3.9).

Example 11.3 Explaining the thermal stabilities of peroxides

Use the Kapustinskii equation (eqn 3.3), the ionic radii given in Table 11.1, and the radii of the oxide and peroxide ions (140 and 180 pm, respectively) to calculate the difference between the lattice enthalpies of the peroxide and oxide of (a) magnesium and (b) barium. Comment on the values obtained.

Answer The Kapustinskii equation gives the following values:

	MgO	MgO$_2$	BaO	BaO$_2$
$\Delta_L H/(\text{kJ mol}^{-1})$	3927	3395	3078	2736

The difference between the oxide and peroxide for magnesium and barium is 532 and 342 kJ mol^{-1}, respectively. This calculation confirms that the difference between the lattice enthalpies of the oxide and peroxide decreases down the group.

Self-test 11.3 Calculate the lattice enthalpies for CaO and CaO$_2$ and check that the above trend is confirmed.

(b) Hydroxides

Key point: The basicity and solubility of the hydroxides increases down the group.

All the hydroxides are formed by reaction of the oxides with water. The hydroxides become more basic down the group with increasing metal character of the element. Beryllium hydroxide, Be(OH)$_2$, is amphoteric. The solubility in water increases from Mg(OH)$_2$ to Ba(OH)$_2$. Magnesium hydroxide, Mg(OH)$_2$, is sparingly soluble and forms a mildly basic solution because a saturated solution contains a low concentration of the OH$^-$ ions. An aqueous suspension of magnesium hydroxide is used as an antacid ('milk of magnesia'). Calcium hydroxide, Ca(OH)$_2$, is more soluble than Mg(OH)$_2$ so a saturated solution contains a higher concentration of hydroxide ions and the solution is described as being moderately basic. A saturated solution of Ca(OH)$_2$ is called *limewater* and is used to test for the presence of CO$_2$. If CO$_2$ is bubbled through the limewater a white precipitate of CaCO$_3$ is formed, which then disappears on further reaction with CO$_2$ to form the hydrogencarbonate ion:

$$Ca(OH)_2 + CO_2(g) \rightarrow CaCO_3(s) + H_2O(l)$$
$$CaCO_3(s) + CO_2(g) + H_2O(l) \rightarrow Ca^{2+}(aq) + 2\,HCO_3^-(aq)$$

Barium hydroxide, Ba(OH)$_2$, is soluble and aqueous solutions are described as strongly basic.

Example 11.4 Calculating the pH of aqueous solutions of the hydroxides

Given that the solubility of Mg(OH)$_2$ is 1.54×10^{-4} mol dm^{-3}, calculate the pH of a saturated aqueous solution.

Answer As the solubility of Mg(OH)$_2$ is 1.54×10^{-4} mol dm^{-3}, the concentration of OH$^-$ ions in the saturated solution is 3.08×10^{-4} mol dm^{-3}. We know from Section 4.1b that [H$^+$][OH$^-$] = 1×10^{-14}, so [H$^+$] must be 3.2×10^{-11} mol dm^{-3}. This concentration corresponds to pH = 10.5.

Self-test 11.4 Calculate the pH of a saturated aqueous solution of Ca(OH)$_2$, given that its solubility is 2.11×10^{-3} mol dm^{-3}.

11.6 Carbides

Key point: All the Group 2 carbides react with water to produce a hydrocarbon gas.

All the Group 2 elements form carbides. Beryllium carbide, Be$_2$C, contains the methide ion, C^{4-}. It is an ionic, crystalline solid with the antifluorite structure (Fig. 11.3).

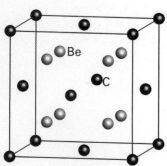

Fig. 11.3 The antifluorite structure adopted by Be$_2$C.

The carbides of Mg, Ca, Sr, and Ba have the formula MC_2 and contain the dicarbide (acetylide) anion, C_2^{2-}. They are prepared by heating the oxide or carbonate with carbon in a furnace at 2000°C:

$$MO(s) + 3\ C(s) \rightarrow MC_2(s) + CO_2(g)$$

$$2\ MCO_3(s) + 5\ C(s) \rightarrow 2\ MC_2(s) + 3\ CO_2(g)$$

All the carbides react with water to produce a hydrocarbon: beryllium carbide produces methane, whereas the other elements produce ethyne (acetylene):

$$Be_2C(s) + 4\ H_2O(l) \rightarrow 2\ Be(OH)_2(aq) + CH_4(g)$$

$$CaC_2(s) + 2\ H_2O(l) \rightarrow Ca(OH)_2(s) + C_2H_2(g)$$

Methane and ethyne are flammable and the latter burns with a bright light from the incandescent carbon particles that form in the flame. When this reaction was discovered in the late nineteenth century, calcium carbide found widespread use in vehicle lamps, enabling safe night-time driving for the first time. It was also used in miners' lamps and is still used today by some cavers and pot-holers as the lamps are reliable and produce an intense light. Ethyne is also a good feedstock for many organic syntheses and its production from CaC_2 was the foundation of the Union Carbide Company. The scale of industrial production of calcium carbide is falling as the chemical industries shift to petrochemicals for feedstocks in organic synthesis. The method is still used to produce ethyne (acetylene) for oxyacetylene welding.

11.7 Salts of oxoacids

The most important oxo compounds of the Group 2 elements are the carbonates, hydrogencarbonates, and the sulfates.

(a) Carbonates and hydrogencarbonates

Key points: All the carbonates are sparingly soluble in water, with the exception of $BeCO_3$; the carbonates decompose to the oxide on heating, most readily high in the group. The hydrogencarbonates are more soluble than the carbonates; magnesium and calcium hydrogencarbonates are responsible for temporary hardness of water.

Beryllium carbonate exists as a soluble hydrate that is susceptible to hydrolysis because of the high charge density on the Be^{2+} ion:

$$[Be(H_2O)_4]^{2+}(aq) + H_2O(l) \rightarrow [Be(H_2O)_3(OH)]^+(aq) + H_3O^+(aq)$$

This equation is greatly simplified; in fact several reactions occur and hydroxo-bridged species are present. The carbonates of the other Group 2 elements are all sparingly soluble and are decomposed to the oxide on heating:

$$MCO_3(s) \xrightarrow{\Delta} MO(s) + CO_2(g)$$

The temperature at which this decomposition occurs increases from 350°C for magnesium carbonate to 1360°C for barium carbonate (Fig. 11.4). The Group 2 carbonates are less thermally stable than the Group 1 carbonates. These trends can be explained in terms of trends in lattice enthalpies and this was discussed in detail in Section 3.15.

Calcium carbonate is the most important oxo compound of the Group 2 elements. It occurs widely in nature as limestone, chalk, marble, dolomite (with magnesium), and as coral, pearl, and seashell (Box 11.3). It has widespread use in construction and road building. It is also used as an antacid, as an abrasive in toothpaste, in chewing gum, and as a health supplement where it is taken to maintain bone density. Powdered limestone is known as *agricultural lime* and is used to neutralize acidic soil:

$$CaCO_3(s) + 2\ H^+(aq) \rightarrow Ca^{2+}(aq) + CO_2(g) + H_2O(l)$$

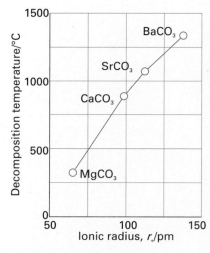

Fig. 11.4 The variation of the decomposition temperature of the Group 2 carbonates with ionic radius.

Box 11.3 The polymorphs of calcium carbonate

Calcium carbonate occurs as vast deposits of sedimentary rocks that are formed from the fossilized remains of marine creatures. The most common and stable form is the hexagonal *calcite*, which has been identified in more 300 different crystalline forms. Calcite comprises about 4 per cent by mass of the Earth's crust and is formed in many different geological environments. Calcite can form rocks of considerable mass and constitutes a significant part of all three major rock classification types, igneous, sedimentary, and metamorphic. Calcite is a major component in the igneous rock called carbonatite and forms the major portion of many hydrothermal veins. Limestone is the sedimentary form of calcite. Limestone metamorphizes to marble from the heat and pressure of metamorphic events, which increase the density of the rock and destroy any texture. Pure white marble is the result of metamorphism of very pure limestone. The characteristic swirls and veins of many coloured marble varieties are usually due to various mineral impurities such as clay, silt, sand, and iron oxides.

Iceland spar is a form of transparent, colourless calcite originally found in Iceland. It exhibits birefringence. Birefringence, or double refraction, is the division of a ray of light into two rays when it passes through certain types of material, depending on the polarization of the light. This behaviour is explained by assigning two different refractive indices to the material for different polarizations. As the two beams exit the crystal they are bent into two different angles of refraction.

Aragonite is a less abundant polymorph of calcite. Aragonite is orthorhombic with three crystals forms. It is less stable than calcite and converts into it at 400°C and, given enough time, aragonite in the environment will convert to calcite. Most bivalve animals, such as oysters, clams, mussels, and corals, secrete aragonite and their shells and pearls are composed of mostly aragonite. The pearlization and iridescent colours in seashells such as abalone are due to several layers of aragonite (Box 24.1). Other natural sources of aragonite include hot springs and cavities in volcanic rocks. Synthetic aragonite can be formed and is used as a filler in the paper industry, where its fine texture, whiteness, and absorbent properties add quality to the product. Powdered limestone is heated in a kiln to form CaO, lime. The lime is then slurried with water to form *milk of lime*. Carbon dioxide is then bubbled through the slurry until aragonite is formed.

$$CaCO_3(s) \xrightarrow{\Delta} CaO(s) + CO_2(g)$$
$$CaO(s) + H_2O(l) \rightarrow Ca(OH)_2(s)$$
$$Ca(OH)_2(s) + CO_2(g) \rightarrow CaCO_3(s) + H_2O(l)$$

The conditions, such as temperature and flow rate of carbon dioxide, determine the final particle size distribution and crystal type.

Travertine is a white, naturally occurring, very hard form of calcium carbonate. It is deposited from the water of hot mineral springs or streams containing CaO.

Calcium carbonate is sparingly soluble in water but its solubility is increased if CO_2 is dissolved in the water, as in rainwater. Thus, caverns have been eroded into limestone rock by the reaction to form the more soluble hydrogencarbonate:

$$CaCO_3(s) + H^+(aq) + HCO_3^-(aq) \rightarrow Ca^{2+}(aq) + 2\ HCO_3^-(aq)$$

This reaction is reversible and over time calcium carbonate stalactites and stalagmites are formed.

On account of the lower charge of the anion, the Group 2 hydrogencarbonates are more soluble in water than the corresponding carbonates. Temporary hardness of water is caused by the presence of magnesium and calcium hydrogencarbonates. These ions are precipitated as carbonate on boiling, when the equilibrium in the following reaction is moved to the right

$$Ca(HCO_3)_2(aq) \rightleftharpoons CaCO_3(s) + CO_2(g) + H_2O(l)$$

Temporary hardness may also be removed by adding $Ca(OH)_2$, which also precipitates the carbonate:

$$Ca(HCO_3)_2(aq) + Ca(OH)_2(aq) \rightleftharpoons 2\ CaCO_3(s) + 2\ H_2O(l)$$

(b) Sulfates

Key points: The most important sulfate is calcium sulfate, which occurs naturally as gypsum and alabaster; calcium and magnesium sulfates are responsible for permanent hardness of water.

Permanent hardness (so called because the hardness is not removed by boiling) is caused by magnesium and calcium sulfates. In this case the water is softened by passing through an ion-exchange resin, which replaces the Mg^{2+} and Ca^{2+} ions with Na^+ ions.

Calcium sulfate is the most important of the Group 2 sulfates. It occurs as gypsum, which is the dihydrate, $CaSO_4 \cdot 2H_2O$. Alabaster is another form of this compound, with a dense smooth form that resembles marble. When the dihydrate is heated above 150°C it loses water to form the hemihydrate, $CaSO_4 \cdot \frac{1}{2}H_2O$, which is also known as *plaster of Paris* as it was first mined in the Montmartre district of the city. When mixed with water, plaster of Paris expands as it forms the dihydrate and forms a strong structure that is used to make casts for broken limbs. Gypsum is mined and used as a building material. One application is in fireproof wallboard. In the event of fire the dihydrate will dehydrate to form the hemihydrate and release water vapour:

$$2\ CaSO_4 \cdot 2\ H_2O(s) \rightarrow 2\ CaSO_4 \cdot \tfrac{1}{2}H_2O(s) + 3\ H_2O(g)$$

The reaction is endothermic ($\Delta_r H^{\ominus} = +117\ \text{kJ mol}^{-1}$) so it absorbs heat from the fire. In addition, the water produced absorbs heat and evaporates and then the gaseous water provides an inert barrier, reducing the supply of oxygen to the fire.

11.8 Solubility and hydration

Key points: The large negative hydration enthalpies of the salts of mononegative ions, ensure that they are soluble. For salts of dinegative ions, the lattice enthalpies are more influential and the salts are insoluble.

The compounds of the Group 2 elements are generally much less soluble in water than those of the Group 1 elements even though the hydration enthalpies are more negative:

	Na^+	K^+	Mg^{2+}	Ca^{2+}
$\Delta_{hyd}H^{\ominus}/(\text{kJ mol}^{-1})$	−406	−322	−1920	−1650

With the exception of the fluorides, the salts of mononegative anions are usually soluble in water and those of dinegative anions (such as oxides) are usually sparingly soluble. For the compounds of dinegative anions, such as the carbonates and the sulfates, it is the high lattice enthalpy arising from the increased charge on the anion that is the deciding factor, outweighing the influence of the enthalpy of hydration. This insolubility is responsible for the enormous deposits of the magnesium- and calcium-containing minerals, such as limestone, gypsum, and dolomite, that are widely exploited in building. The fluorides are all insoluble in water because the small size of the fluoride anion leads to a high lattice enthalpy. The exception is BeF_2, which has a very high hydration enthalpy due to the high charge density on the small cation. Thus hydration enthalpy, rather than lattice enthalpy, is the most influential factor for BeF_2.

$$BeF_2(s) \rightarrow [Be(OH_2)_4]^{2+}(aq) + 2\ F^-(aq) \qquad \Delta_r H^{\ominus} = -2500\ \text{kJ mol}^{-1}$$

The hydrated Be^{2+} cation, $[Be(OH_2)_4]^{2+}$, acts as an acid in water.

$$[Be(OH_2)_4]^{2+}(aq) + H_2O(l) \rightarrow [Be(OH_2)_3(OH)]^+(aq) + H_3O^+(aq)$$

This reaction can be accredited to the high polarizing power of the small, doubly charged cations. Solutions of the hydrated salts of the heavier elements are neutral.

Example 11.5 Assessing factors effecting solubility

Use the data in Table 11.1 and the Kapustinskii equation (eqn 3.3) to calculate the lattice enthalpy of $MgCl_2$ and $MgCO_3$. The ionic radii of Cl^- and CO_3^{2-} are 181 and 164 pm, respectively. Comment on the likely effect on solubility.

Answer The lattice enthalpies for $MgCO_3$ and $MgCl_2$ are 3590 and 2537 kJ mol^{-1}, respectively. As the value for $MgCO_3$ is larger, it is more likely to offset the hydration enthalpy and the solubility will be reduced.

Self-test 11.5 Calculate the lattice enthalpy of MgF_2 and comment on how it will affect the solubility compared to $MgCl_2$.

Coordination and organometallic compounds

Only beryllium forms nonlabile complexes with simple ligands.[1] All the Group 2 elements form nonlabile complexes with polydentate, chelating ligands. The most important organometallic compounds of the group are the magnesium-based Grignard reagents.

11.9 Coordination compounds

Key points: Only beryllium forms coordination compounds with simple ligands; the most stable complexes are formed with polydentate, chelating ligands such as edta.

Compounds of beryllium show properties consistent with a greater covalent character than those of its congeners, and some of its complexes with ordinary ligands are stable. The complexes are usually tetrahedral although the coordination number of the Be atom can fall to 3 or 2 if the ligands are bulky. The most nonlabile complexes are formed with halide or chelating oxygen-donor ligands, such as oxalate, alkoxides, and diketonates. For example, basic beryllium acetate (beryllium oxoethanoate, $Be_4O(O_2CCH_3)_6$) consists of a central O atom surrounded by a tetrahedron of four Be atoms, which in turn are bridged by ethanoate ions. It can be prepared by the reaction of ethanoic acid with beryllium carbonate:

$$4\ BeCO_3(s) + 6\ CH_3COOH(l) \rightarrow 4\ CO_2(g) + 3\ H_2O(l) + Be_4O(O_2CCH_3)_6(s)$$

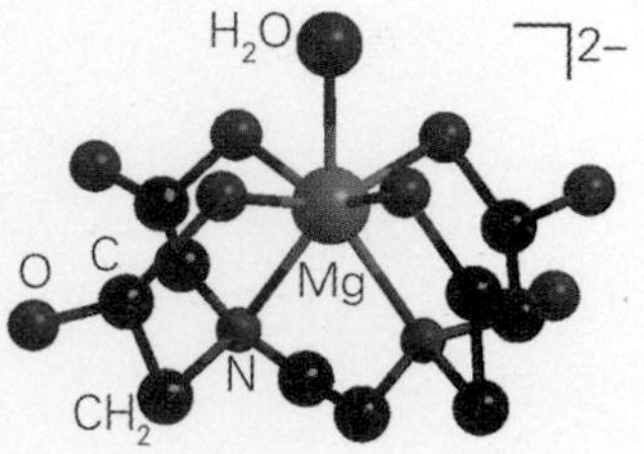

Basic beryllium acetate is a colourless, sublimable, molecular compound; it is soluble in chloroform, from which it can be recrystallized.

The Group 2 cations form complexes with crown and crypt ligands. The most non-labile of these complexes are formed with the larger Sr^{2+} and Ba^{2+} cations and all the complexes are more stable than these of the smaller Group 1 cations.

The most stable complexes are formed with charged polydentate ligands, such as the analytically important ethylenediaminetetraacetate ion (edta). The formation constants of edta complexes of the alkali metal complexes lie in the order $Ca^{2+} > Mg^{2+} > Sr^{2+} > Ba^{2+}$. In the solid state, the structure of the Mg^{2+} edta complex is seven-coordinate (**6**), with H_2O at one coordination site. The calcium complex is either seven- or eight- coordinate, depending on the counter ion, with one or two H_2O molecules serving as ligands.

Many complexes of Ca^{2+} and Mg^{2+} occur naturally. The most important macrocyclic complexes are the chlorophylls, which are porphyrin complexes of Mg involved in photosynthesis (Section 26.2). Magnesium is involved in phosphate transfer, and carbohydrate metabolism. Calcium is a component of biominerals and is also coordinated by proteins, notably those involved in cell signalling and muscle action (Section 26.3).

6 $[Mg(edta)(OH_2)]^{2-}$

11.10 Organometallic compounds

Key points: Alkylberyllium compounds polymerize in the solid phase; Grignard reagents are some of the most important main-group organometallic compounds.

Organometallic compounds of beryllium are pyrophoric in air and unstable in water. They can be prepared by *transmetallation* from methylmercury in a hydrocarbon solvent:

$$Hg(CH_3)_2(sol) + Be(s) \rightarrow Be(CH_3)_2(sol) + Hg(s)$$

Another synthetic route is by halogen exchange or metathesis reactions in which a beryllium halide reacts with an alkyllithium compound. The products are the lithium halide and an alkylberyllium compound. In this way, the halogen and organic groups are

---

[1] *Labile* is a kinetic term. A labile complex would be expected to undergo a rapid reaction. A nonlabile complex would not be expected to undergo a rapid reaction and may be described as *inert*. See Chapter 20.

transferred between the two metal atoms. The driving force for this and similar reactions is the formation of the halide of the more electropositive metal.

$$2\ BuLi(sol) + BeCl_2(sol) \rightarrow (Bu)_2Be(sol) + 2\ LiCl(s)$$

Grignard reagents in ether can also be used in the synthesis of organoberyllium compounds:

$$2\ RMgCl(sol) + BeCl_2(sol) \rightarrow R_2Be(sol) + 2\ MgCl_2(s)$$

Methylberyllium, $Be(CH_3)_2$, is predominantly a monomer in the vapour phase and in hydrocarbon solvents, where it adopts a linear structure, as expected from the VSEPR model. In the solid it forms polymeric chains in which the bridging methyl groups form $3c,2e$ bridging bonds (Section 9.11) (**7**). Bulkier alkyl groups lead to a lower degree of polymerization; ethylberyllium (**8**) is a dimer and t-butylberyllium (**9**) is a monomer.

An interesting organoberyllium compound is beryllocene, $(C_5H_5)_2Be$, which, although the formula suggests a similar structure to ferrocene (Sections 21.15 and 21.19), has a very different structure, with the Be atom positioned directly above the centre of one cyclopentadienyl ring and below a single C atom on the other ring (**10**). Low-temperature NMR at $-135°C$, however, suggests that the two rings are equivalent, indicating that even at this low temperature the Be atom rapidly oscillates between the two possible equivalent positions relative to the two cyclopentadienyl rings.

Alkyl- and arylmagnesium halides are very well known as **Grignard reagents** and are widely used in synthetic organic chemistry where they behave as a source of R^-. They are prepared from magnesium metal and an organohalide. As the surface of magnesium metal is covered by a passivating oxide film, the magnesium has to be activated before the reaction can proceed. A trace of iodine is usually added to the reactants, forming magnesium iodide. This compound is soluble in the solvent used and dissolves to expose an activated magnesium surface. The reaction is carried out in ether or tetrahydrofuran:

$$Mg(s) + RBr(sol) \rightarrow RMgBr(sol)$$

The structure of Grignard reagents is far from simple. The metal atom has a coordination number of 2 only in solution and when the alkyl group is bulky. Otherwise, it is solvated with a tetrahedral arrangement of solvent molecules around the Mg atom (**11**). In addition, complex equilibria in solution, known as **Schlenk equilibria**, lead to the presence of several species, the exact nature of which depend on temperature, concentration, and solvent. For example R_2Mg, $RMgX$, and MgX_2 have all been detected:

$$2\ RMgX(sol) \rightleftharpoons R_2Mg(sol) + MgX_2(sol)$$

Grignard reagents are widely used in the synthesis of organometallic compounds of other metals, as in the formation of alkylberyllium compounds mentioned above. They are also widely used in organic synthesis. One reaction is **organomagnesiation**, which involves addition of the Grignard reagent to an unsaturated bond:

$$R^1MgX(sol) + R^2R^3C{=}CR^4R^5(sol) \rightarrow R^1R^2R^3CCR^4R^5MgX(sol)$$

Grignard reagents undergo side reactions such as **Wurtz coupling** to form a carbon–carbon bond:

$$R^1MgX(sol) + R^2X(sol) \rightarrow R^1R^2(sol) + MgX_2(sol)$$

The organometallic compounds of calcium, strontium, and barium are generally ionic and very unstable. They all form analogues of Grignard reagents by direct interaction of the finely divided metal with organohalide.

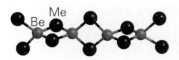

7 BeMe$_2$

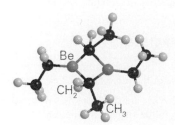

8 (BeEt$_2$)$_2$

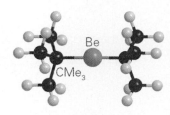

9 Be(^{t}Bu)$_2$

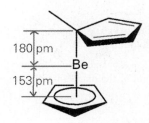

180 pm

153 pm

10 Be(Cp)$_2$

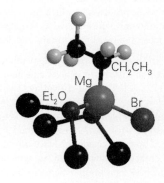

11 MgBrEt(OEt$_2$)$_2$

FURTHER READING

R.B. King, *Inorganic chemistry of the main group elements*. Wiley (1994).

P. Enghag, *Encyclopedia of the elements*. Wiley (2004).

D.M.P. Mingos, *Essential trends in inorganic chemistry*. Oxford University Press (1998). A survey of inorganic chemistry from the perspective of structure and bonding.

N.C. Norman, *Periodicity and the s- and p-block elements*. Oxford University Press (1997). Includes coverage of essential trends and features of *s*-block chemistry.

J.A.H. Oates, *Lime and limestone: chemistry and technology, production and uses*. Wiley (1998).

EXERCISES

11.1 Explain why compounds of beryllium are covalent whereas those of the other Group 2 elements are predominantly ionic.

11.2 Why are the properties of beryllium more similar to aluminium than to magnesium?

11.3 Identify the compounds A, B, C, and D of the Group 2 element M.

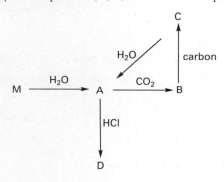

11.4 Give a balanced equation for the extraction of magnesium from dolomite.

11.5 Give a balanced equation for the extraction of barium by reduction of the oxide with aluminium.

11.6 Why is magnesium hydroxide a much more effective antacid than calcium or barium hydroxide?

11.7 Explain why Group 1 hydroxides are much more corrosive than Group 2 hydroxides.

11.8 How do Group 2 salts give rise to scaling from hard water?

11.9 Use the data in Table 1.1 and the Ketelaar triangle in Fig. 2.2 to predict the nature of the bonding in $BeCl_2$, MgI_2, and BaF_2.

11.10 Discuss the difference that you might expect between the structures of the two Grignard compounds C_2H_5MgBr and $2,4,6-(CH_3)_3C_6H_2MgBr$.

11.11 Predict the products of the following reactions:

(a) $MgCl_2 + LiC_2H_5 \rightarrow$

(b) $Mg + (C_2H_5)_2Hg \rightarrow$

(c) $Mg + C_2H_5HgCl \rightarrow$

PROBLEMS

11.1 Marble and limestone buildings are eroded by contact with acid rain. Define the term acid rain and discuss the origins of the acidity. Describe the processes by which the marble and limestone are attacked. List the compounds that are used as scrubbers in power plants to minimize emissions implicated in acid rain and describe how they are effective.

11.2 In their paper 'Noncovalent interaction of chemical bonding between alkaline earth cations and benzene?' (*Chem. Phys. Lett.*, 2001, **349**, 113), X.J. Tan and co-workers carried out theoretical calculation of complexes formed between beryllium, magnesium, and calcium ions and benzene. To which orbital interactions was the binding of the alkali metal to benzene attributed? How was the C—C bond length in benzene affected by this interaction? Place the M—C bonds in order of increasing bond enthalpy. How did the strength of the bonds compare to those formed between Group 1 elements and benzene? Sketch the geometry of the metal–benzene complexes.

11.3 P.C. Junk and J.W. Steed (*J. Chem. Soc., Dalton. Trans.*, 1999, 407) prepared crown ether complexes from the nitrates of Mg, Ca, Sr, and Ba.

Outline the general procedure that was used for the syntheses. Sketch the structures of the two crown ethers used. Comment on the structures of the complexes and how they change with different cations.

11.4 Some of the most successful superconducting materials contain barium. Write a review of the superconductivity of barium-containing compounds.

11.5 An experiment was carried out to determine the permanent and temporary hardness of domestic water. A few drops of pH = 10 buffer were added to a $100 \, cm^3$ sample of the water. This sample was titrated against 0.01 M edta(aq) using Eriochrome Black T indicator and gave a titre of $33.8 \, cm^3$. Under these circumstances both Mg^{2+} and Ca^{2+} ions react with edta. A second $100 \, cm^3$ sample was titrated against the edta solution after $5.0 \, cm^3$ of 0.1 M NaOH(aq) and a few drops of murexide indicator had been added. Under these conditions only the Ca^{2+} ions react with the edta and a titre of $27.5 \, cm^3$ was obtained. Determine the permanent and temporary hardness of the water sample in terms of Mg^{2+} and Ca^{2+} ions.

The Group 13 elements

12

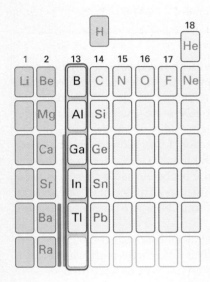

This chapter and the following five describe the general chemical properties of the *p*-block elements. The *p*-block is a very rich region of the periodic table; its members show a much greater variation in properties than those of the *s* and *d* blocks. There are some clear trends in chemical properties of the Group 13 elements that we shall see repeated in the other groups of the *p* block. In this chapter we look at the occurrence and isolation of each element in Group 13 and consider the chemical properties of the elements and their simple compounds, coordination compounds, and organometallic compounds. In addition, we introduce the extensive range of boron clusters.

The elements of Group 13, boron, aluminium, gallium, indium, and thallium, have interesting and diverse physical and chemical properties. The first member of the group, boron, is essentially nonmetallic whereas the properties of the heavier members of the group are distinctly metallic. Aluminium is the most important element commercially and is produced on a massive scale for a wide range of applications. Boron forms a large number of cluster compounds involving hydrogen, metals, and carbon.

The elements

Key points: Boron is the only nonmetal in the group; many of the compounds of the elements are electron deficient and act as Lewis acids.

The elements of Group 13 show a wide variation in abundance in crustal rocks, the oceans, and the atmosphere. Aluminium is abundant but the low cosmic and terrestrial abundance of boron, like that of lithium and beryllium, reflects how the light elements are sidestepped in nucleosynthesis (Section 1.1). The low abundance of heavier members of the group is in keeping with the progressive decrease in nuclear stability of the elements that follow iron.

Whereas the elements of the *s* and *d* blocks are all metallic, the elements of the *p* block range from nonmetals, through metalloids, to metals. This variety results in a diversity of chemical properties and some distinctive trends, many of which can be rationalized by consideration of atomic radii and electron configurations. There is an increase in metallic character from boron to thallium: boron is a nonmetal, aluminium is essentially metallic, although it is often classed as a metalloid on account of its amphoteric character, and

Table 12.1 Selected properties of the elements

	B	Al	Ga	In	Tl
Covalent radius/pm	80	125	125	150	155
Metallic radius/pm		143	141	166	171
Ionic radius, $r(M^{3+})$/pm	20	50	62	81	95
Melting point/°C	2300	660	30	157	304
Boiling point/°C	3930	2470	2400	2000	1460
First ionization energy, I_1/(kJ mol^{-1})	799	577	577	556	590
Second ionization energy, I_2/(kJ mol^{-1})	2427	1817	1979	1821	1971
Third ionization energy, I_3/(kJ mol^{-1})	3660	2745	2963	2704	2878
Electron affinity, E_a/(kJ mol^{-1})	26.7	42.5	28.9	28.9	
Pauling electronegativity	2.0	1.5	1.6	1.7	1.8
$E^{\ominus}(M^{3+},M)$/V	−0.89	−1.68	−0.53	−0.34	+1.26

gallium, indium, and thallium are metals. Associated with this trend is a variation from predominantly covalent to ionic bonding in the compounds of the elements that can be rationalized in terms of the increase in atomic radius and related decrease in ionization energy down the group (Table 12.1). Because the ionization energies of the heavier elements are low, the metals form cations increasingly readily down the group. In contrast to the expected trend in electronegativity (Section 1.9), gallium is more electronegative than aluminium. This anomaly in the general trend to lower electronegativity is known as the **alternation effect** and is a consequence of the increased effective nuclear charges of the $4p$ elements due to the presence of their poorly shielding $3d$ electrons. As we shall see, the alternation effect survives into Group 14, where germanium is more electronegative than silicon (Chapter 13).

As we have seen already in Groups 1 and 2, the first member of each group differs from its congeners on account of its small atomic radius. This difference is particularly evident in Group 13, where the chemical properties of boron are very distinct from those of the rest of the group. However, boron does have a pronounced diagonal relationship with silicon in Group 14:

1 Boron and silicon form acidic oxides, B_2O_3 and SiO_2; aluminium forms an amphoteric oxide.

2 Boron and silicon form many polymeric oxide structures.

3 Boron and silicon form flammable, gaseous hydrides; aluminium hydride is a solid.

The valence-electron configuration of the Group 13 elements is ns^2np^1 and, as this configuration suggests, all the elements adopt the +3 oxidation state in their compounds. However, the heavier elements of the group also form compounds with the metal in the +1 oxidation state and this state increases in stability down the group. In fact, the most common oxidation state of thallium is Tl(I). This trend is particularly evident within the halides. The relative stability of an oxidation state in which the oxidation number is 2 less than the group oxidation number is an example of the **inert pair effect** and it is a recurring theme within the p block. There is no simple explanation for this effect: it is often ascribed to the large energy that is needed to remove the ns^2 electrons after the np^1 electron has been removed; however, the sum of the first three ionization energies of thallium (5438 kJ mol^{-1}) is only slighter higher than the figure for gallium (5521 kJ mol^{-1}) and indium (5083 kJ mol^{-1}). Another contribution to the effect may be the low M—X bond enthalpies for the heavier p-block elements (Table 12.2).

The most striking feature of the Group 13 elements is the electron deficiency and the resulting Lewis acidity of their neutral compounds. The electron deficiency arises from the ns^2np^1 electron configuration, which contributes up to a maximum of six electrons in the valence shell when three covalent bonds are formed. This electron deficiency has a vital influence on the structure and reactions of the compounds of the elements.

Table 12.2 Mean bond enthalpies/(kJ mol^{-1})

	H	F	Cl	Br	I
B	334	757	536	423	220
Al	284	664	511	444	370
Ga	274	577	481	444	339
In	243	506	439	414	331
Tl	188	445	372	334	272

Boron exists in several hard and refractory allotropes. The three solid phases for which crystal structures are available contain the icosahedral (20-faced) B_{12} unit as a building block (Fig. 12.1). This icosahedral unit is a recurring motif in boron chemistry and we shall meet it again in the structures of metal borides and boron hydrides. The icosahedral unit is also found in some intermetallic compounds of other Group 13 elements, such as Al_5CuLi_3, $RbGa_7$, and K_3Ga_{13}.

12.1 Occurrence and recovery

Key points: Aluminium is highly abundant; thallium and indium are the least abundant of the Group 13 elements.

Boron occurs naturally as borax, $Na_2B_4O_5(OH)_4 \cdot 8H_2O$, and kernite, $Na_2B_4O_5(OH)_4 \cdot 2H_2O$, from which the impure element is obtained. The borax is converted to boric acid, $B(OH)_3$, and then to boron oxide, B_2O_3. The oxide is reduced with magnesium and washed with alkali and then hydrofluoric acid. Pure boron is produced by reduction of BBr_3 vapour with H_2:

$$2\ BBr_3(g) + 3\ H_2(g) \rightarrow 2\ B(g) + 6\ HBr(g)$$

Aluminium is the most abundant metallic element in the crustal regions of the Earth and makes up approximately 8 per cent by mass of crustal rocks. It occurs in numerous clays and aluminosilicate minerals but the commercially most important mineral is bauxite, a complex mixture of hydrated aluminium hydroxide and aluminium oxide, from which it is extracted by the Hall–Héroult process on an immense scale. The process is very expensive but this expense is offset by the scale of production and availability of the raw material. The oxide, alumina, occurs naturally as ruby, sapphire, corundum, and emery.

Gallium oxide occurs as an impurity in bauxite and is normally recovered as a by-product of the manufacture of aluminium. The process results in the concentration of gallium in the residues from which it is extracted by electrolysis. Indium is produced as a by-product of the extraction of lead and zinc and is isolated by electrolysis. Thallium compounds are found in flue dust, which is dissolved in dilute sulfuric acid; hydrochloric acid is then added to precipitate thallium(I) chloride and the metal is extracted by electrolysis.

12.2 Uses

Key points: The most useful compound of boron is borax; the most commercially important element is aluminium.

The main use of boron is in borosilicate glasses. Borax has many domestic uses; for example, as a water softener, cleaner, and mild pesticide. Boric acid, $B(OH)_3$, is used as a mild antiseptic. The technological uses of aluminium metal exploit its lightness, resistance to corrosion, and the fact that it is easily recycled. It is used in cans, foils, utensils, in construction, and in aircraft alloys (Box 12.1). Because the melting point of gallium is just above room temperature, it is used in high-temperature thermometers. Gallium and indium form a low-melting-point alloy that is used as the safety device in sprinkler systems. Gallium and indium are deposited on glass surfaces to form corrosion-resistant

(a)

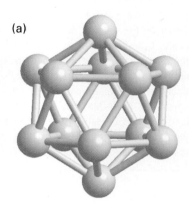

(b)

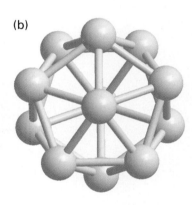

Fig. 12.1 A view of the B_{12} icosahedron in α-rhombohedral boron (a) along and (b) perpendicular to the threefold axis of the crystal. The individual icosahedra are linked by 3c,2e bonds.

mirrors, and a mixture of In_2O_3 and SnO_2 is used as a transparent, conducting coating for electronic displays and as a heat-reflective coating for light bulbs. Thallium compounds were once used to treat ringworm and as a rat and ant poison. However, this application has been banned because of their very high toxicity, which arises from the transport of Tl^+ ions across cell membranes together with K^+ ions (Section 26.3). Thallium is absorbed more efficiently by tumour cells and is used in nuclear medicine as an imaging agent.

Simple compounds of boron

As is typical of an element at the head of its group, the chemical properties of boron and its compounds are strikingly different from those of its congeners and will be treated separately.

12.3 Simple hydrides of boron

Boron forms an elaborate range of neutral and anionic hydrides with cage-like structures. These species are electron deficient and act as Lewis acids.

(a) Boranes

Key points: Diborane can be synthesized by metathesis between a boron halide and a hydride source; many of the higher boranes can be prepared by the partial pyrolysis of diborane; all the boron hydrides are flammable, sometimes explosively, and many of them are susceptible to hydrolysis.

The **boranes**, the binary hydrogen compounds of boron, were first isolated in pure form by the German chemist Alfred Stock in 1912. All the boron hydrides burn with a characteristic green flame and several of them ignite explosively on contact with air. We have already seen that the simplest member of the series, diborane, B_2H_6 (**1**), is electron-deficient and that its structure can be described in terms of $2c,2e$ and $3c,2e$ bonds (Section 2.11): bridging $3c,2e$ bonds are a recurring theme in borane chemistry. Diborane, B_2H_6, can be prepared in the laboratory by metathesis of a boron halide with either $LiAlH_4$ or $LiBH_4$ in ether:

$$3 \, LiBH_4(et) + 4 \, BF_3(et) \rightarrow 2 \, B_2H_6(g) + 3 \, LiBF_4(et)$$

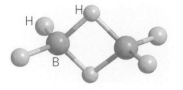

1 Diborane, B_2H_6

That this reaction is a metathesis (an exchange of partners) can be seen by writing it in the simplified form

$$\tfrac{3}{4}BH_4^-(et) + BF_3(et) \rightarrow BH_3(g) + \tfrac{3}{4}BF_4^-(et)$$

Both $LiBH_4$ and $LiAlH_4$, like LiH, are good reagents for the transfer of H^-, but they are generally preferred over LiH and NaH because they are soluble in ethers. The synthesis is carried out with the strict exclusion of air (typically in a vacuum line, because diborane ignites on contact with air). Diborane decomposes very slowly at room temperature, forming higher boron hydrides and a nonvolatile and insoluble yellow solid that consist of $B_{10}H_{14}$ and the polymeric species, BH_n.

Stock prepared six different boron hydrides and had good but inconclusive evidence for a seventh boron hydride, B_6H_{12}. The compounds fall into two classes. One class has the formula B_nH_{n+4} and the other, which is richer in hydrogen and less stable, has the formula B_nH_{n+6}. Examples include pentaborane(11) (**2**), B_5H_{11}, tetraborane(10) (**3**), B_4H_{10}, and pentaborane(9) (**4**), B_5H_9. The nomenclature, in which the number of B atoms is specified by a prefix and the number of H atoms is given in parentheses, should be noted. Thus, the systematic name for diborane is diborane(6); however, as there is no diborane(8), the simpler term 'diborane' is almost always used.

All the boranes are colourless and diamagnetic. They range from gases (B_2H_6 and B_4H_8) through volatile liquids (B_5H_9 and B_6H_{10} hydrides), to the sublimable solid $B_{10}H_{14}$. All the boron hydrides are flammable, and several of the lighter ones, including diborane, react spontaneously with air, often with explosive violence and a green flash (an emission from an excited state of the reaction intermediate BO). The final product of the reaction is the hydrated oxide:

$$B_2H_6(g) + 3\,O_2(g) \rightarrow 2\,B(OH)_3(s)$$

The lighter boranes are readily hydrolysed by water:

$$B_2H_6(g) + 6\,H_2O(l) \rightarrow 2\,B(OH)_3(aq) + 6\,H_2(g)$$

As described below, B_2H_6 is a Lewis acid, and the mechanism of this hydrolysis reaction involves coordination of H_2O acting as a Lewis base. Molecular hydrogen then forms as a result of the combination of the partially positively charged H atom on O with the partially negatively charged H atom on B.

(b) Lewis acidity

Key points: Soft and bulky Lewis bases cleave diborane symmetrically; more compact and hard Lewis bases cleave the hydrogen bridge unsymmetrically; although it reacts with many hard Lewis bases, diborane is best regarded as a soft Lewis acid.

As implied by the mechanism of hydrolysis, diborane and many other light boron hydrides act as Lewis acids and they are cleaved by reaction with Lewis bases. Two different cleavage patterns have been observed, namely, symmetric cleavage and unsymmetric cleavage. In **symmetric cleavage**, B_2H_6 is broken symmetrically into two BH_3 fragments, each of which forms a complex with a Lewis base:

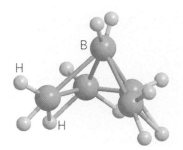

2 B_5H_{11}

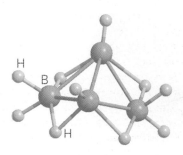

3 B_4H_{10}

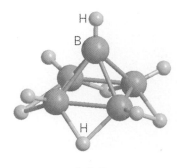

4 B_5H_9

$$H_3B—N(CH_3)_3 + 2\,N(CH_3)_3 \rightarrow 2\ H_3B—N(CH_3)_3$$

Many complexes of this kind exist. They are interesting partly because they are isoelectronic with hydrocarbons. For instance, the product of this reaction is isoelectronic with 2,2-dimethylpropane, $C(CH_3)_4$. Stability trends indicate that BH_3 is a soft Lewis acid, as illustrated by the reaction

$$H_3B—N(CH_3)_3 + F_3B—S(CH_3)_2 \rightarrow H_3B—S(CH_3)_2 + F_3B—N(CH_3)_3$$

in which BH_3 transfers to the soft S donor atom and the harder Lewis acid, BF_3, combines with the hard N donor atom.

The direct reaction of diborane and ammonia results in **unsymmetrical cleavage**, which is cleavage leading to an ionic product:

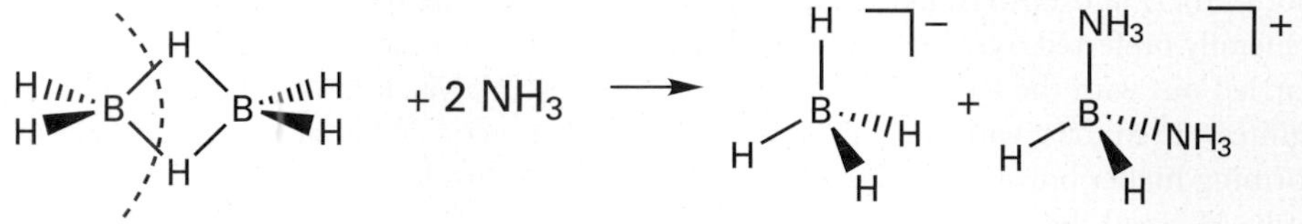

Unsymmetrical cleavage of this kind is generally observed when diborane and a few other boron hydrides react with strong, sterically uncrowded bases at low temperatures. The steric repulsion is such that only two small ligands can attack one B atom in the course of the reaction.

(c) Hydroboration

Key points: Hydroboration, the reaction of diborane with alkenes in ether solvent, produces organoboranes that are useful intermediates in synthetic organic chemistry.

An important component of a synthetic chemist's repertoire of reactions is **hydroboration**, the addition of HB across a multiple bond:

$$H_3B\!-\!OR_2 + H_2C\!=\!CH_2 \xrightarrow{\Delta,\,\text{ether}} CH_3CH_2BH_2 + R_2O$$

From the viewpoint of an organic chemist, the C—B bond in the primary product of hydroboration is an intermediate stage in the stereospecific formation of C—H or C—OH bonds, to which it can be converted. From the viewpoint of the inorganic chemist, the reaction is a convenient method for the preparation of a wide variety of organoboranes. The hydroboration reaction is one of a class of reactions in which EH adds across the multiple bond; hydrosilylation (Section 13.5) is another important example. Much of the driving force for these types of reaction stems from the strength of the C—H bond compared with B—H and Si—H bonds (Tables 12.2 and 13.1).

(d) The tetrahydroborate ion

Key point: The tetrahydroborate ion is a useful intermediate for the preparation of metal hydride complexes and borane adducts.

Diborane reacts with alkali metal hydrides to produce salts containing the tetrahydroborate ion, BH_4^-:

$$B_2H_6(\text{polyet}) + 2\,LiH(\text{polyet}) \rightarrow 2\,LiBH_4(\text{polyet})$$

Because of the sensitivity of diborane and LiH to water and oxygen, the synthesis must be carried out in the absence of air and in a nonaqueous solvent such as the short-chain polyether $CH_3OCH_2CH_2OCH_3$ (denoted here 'polyet'). We can view this reaction as another example of the Lewis acidity of BH_3 towards the strong Lewis basicity of H^-. The BH_4^- ion is isoelectronic with CH_4 and NH_4^+, and the three species show the following variation in chemical properties as the electronegativity of the central atom increases:

	BH_4^-	CH_4	NH_4^+
Character:	hydridic	–	protic

where 'protic' denotes Brønsted acid (proton donating) character; CH_4 is neither acidic nor basic under the conditions prevailing in aqueous solution.

Alkali metal tetrahydroborates are very useful laboratory and commercial reagents. They are often used as a mild source of H^- ions, as general reducing agents, and as precursors for most boron–hydrogen compounds. Most of these reactions are carried out in polar nonaqueous solvents. The preparation of diborane, mentioned previously,

$$3\,LiEH_4 + 4\,BF_3 \rightarrow 2\,B_2H_6 + 3\,LiEF_4 \qquad (E = B, Al)$$

is one example. Although BH_4^- is thermodynamically unstable with respect to hydrolysis, the reaction is very slow at high pH and some synthetic applications have been devised in water. For example, germane (GeH_4) can be prepared by dissolving GeO_2 and KBH_4 in aqueous potassium hydroxide and then acidifying the solution:

$$HGeO_3^- (aq) + BH_4^- (aq) + 2\ H^+(aq) \rightarrow GeH_4(g) + B(OH)_3(aq)$$

Aqueous BH_4^- also can serve as a simple reducing agent, as in the reduction of aqua ions such as Ni^{2+} or Cu^{2+} to the metal or metal boride. With halogen complexes of $4d$ and $5d$ elements that also have stabilizing ligands, such as phosphines, tetrahydroborate ions can be used to introduce a hydride ligand by a metathesis reaction in a nonaqueous solvent:

$$RuCl_2(PPh_3)_3 + NaBH_4 + PPh_3 \xrightarrow{\Delta\ benzene/ethanol} RuH_2(PPh)_4 + \text{other products}$$

It is probable that many of these metathesis reactions proceed through a transient BH_4^- complex. Indeed, many borohydride complexes are known, especially with highly electropositive metals: they include $Al(BH_4)_3$ (**5**), which contains a diborane-like double hydride bridge, and $Zr(BH_4)_4$ (**6**), in which triple hydride bridges are present. We see from these examples that many compounds can be described in terms of $3c,2e$ bonds.

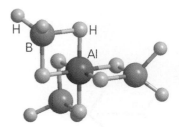

5 $Al(BH_4)_3$

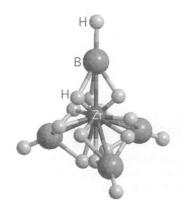

6 $Zr(BH_4)_4$

Example 12.1 Predicting the reactions of boron–hydrogen compounds

By means of a chemical equation, indicate the products resulting from the interaction of equal molar amounts of $[HN(CH_3)_3]Cl$ with $LiBH_4$ in tetrahydrofuran (THF).

Answer The interaction of the hydridic BH_4^- ion with the protic $[HN(CH_3)_3]^+$ ion will evolve hydrogen to produce trimethylamine and BH_3. In the absence of other Lewis bases, the BH_3 molecule would coordinate to THF; however, the stronger Lewis base trimethylamine is produced in the initial reactions, so the overall reaction will be

$$[HN(CH_3)_3]Cl + LiBH_4 \rightarrow H_2 + H_3BN(CH_3)_3 + LiCl$$

The boron atom in $H_3BN(CH_3)_3$ has four tetrahedrally attached groups.

Self-test 12.1 Write an equation for the reaction of $LiBH_4$ with propene in ether solvent and a 1:1 stoichiometry and another equation for its reaction with ammonium chloride in THF with the same stoichiometry.

12.4 Boron trihalides

Key points: Boron trihalides are useful Lewis acids, with BCl_3 stronger than BF_3, and important electrophiles for the formation of boron–element bonds; subhalides with B—B bonds, such as B_2Cl_4, are also known.

All the boron trihalides except BI_3 may be prepared by direct reaction between the elements. However, the preferred method for BF_3 is the reaction of B_2O_3 with CaF_2 in H_2SO_4. This reaction is driven in part by the reaction of the strong acid H_2SO_4 with oxides and the affinity of the hard B atom for fluorine:

$$B_2O_3(s) + 3\ CaF_2(s) + 6\ H_2SO_4(l) \rightarrow 2\ BF_3(g) + 3\ [H_3O][HSO_4](soln) + 3\ CaSO_4(s)$$

Boron trihalides consist of trigonal-planar BX_3 molecules. Unlike the halides of the other elements in the group, they are monomeric in the gas, liquid, and solid states. Boron trifluoride and boron trichloride are gases, the tribromide is a volatile liquid, and the triiodide is a solid (Table 12.3). This trend in volatility is consistent with the increase in strength of dispersion forces with the number of electrons in the molecules.

Boron trihalides have an incomplete octet and are Lewis acids. We have already drawn attention to the order of their strengths in this role, which is $BF_3 < BCl_3 \leq BBr_3$ and contrary to the order of electronegativity of the attached halogens (Section 4.8). The electron deficiency is partially removed by X—B π bonding between the halogen atoms and the boron giving rise to the partial occupation of the vacant p orbital on the B atom

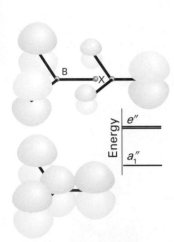

Table 12.3 Properties of the boron trihalides

	BF_3	BCl_3	BBr_3	BI_3
Melting point/°C	−127	−107	−46	50
Boiling point/°C	−100	13	91	210
Bond length/pm	130	175	187	210
$\Delta_f G^{\circ}/(\text{kJ mol}^{-1})$	−1112	−339	−232	+21

Fig. 12.2 The bonding π orbitals of boron trihalide are largely localized on the electronegative halogen atoms, but overlap with a p orbital of boron is significant in the a_1'' orbital.

by electrons donated by the halogen atoms (Fig. 12.2). The trend in Lewis acidity stems from more efficient X—B π bonding for the lighter, smaller halogens and the F—B bond is one of the strongest single bonds known. All the boron trihalides form simple Lewis complexes with suitable bases, as in the reaction

$$BF_3(g) + :NH_3(g) \rightarrow F_3B\text{—}NH_3(s)$$

However, boron chlorides, bromides, and iodides are susceptible to protolysis by mild proton sources such as water, alcohols, and even amines. As shown in Fig. 12.3, this reaction, together with metathesis reactions, is very useful in preparative chemistry. An example is the rapid hydrolysis of BCl_3 to give boric acid, $B(OH)_3$:

$$BCl_3(g) + 3\ H_2O(l) \rightarrow B(OH)_3(aq) + 3\ HCl(aq)$$

It is probable that a first step in this reaction is the formation of the complex $Cl_3B\text{—}OH_2$, which then eliminates HCl and reacts further with water.

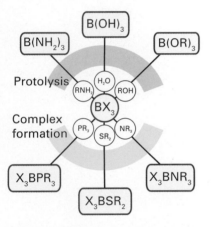

Fig. 12.3 The reactions of boron–halogen compounds (X = halogen).

Example 12.2 Predicting the products of reactions of the boron trihalides

Predict the likely products of the following reactions, and write the balanced chemical equations. (a) BF_3 and excess NaF in acidic aqueous solution; (b) BCl_3 and excess NaCl in acidic aqueous solution; (c) BBr_3 and excess $NH(CH_3)_2$ in a hydrocarbon solvent.

Answer (a) The F^- ion is a chemically hard and fairly strong base; BF_3 is a hard and strong Lewis acid with a high affinity for the F^- ion. Hence, the reaction should result in a complex:

$$BF_3(g) + F^-(aq) \rightarrow BF_4^-(aq)$$

Excess F^- and acid prevent the formation of hydrolysis products such as BF_3OH^-, which are formed at high pH. (b) Unlike B—F bonds, which are very strong and only mildly susceptible to hydrolysis, the other boron–halogen bonds are hydrolysed vigorously by water. We can anticipate that BCl_3 will undergo hydrolysis rather than coordinate to aqueous Cl^-:

$$BCl_3(g) + 3\ H_2O(l) \rightarrow B(OH)_3(aq) + 3\ HCl(aq)$$

(c) Boron tribromide will undergo protolysis with formation of a B—N bond:

$$BBr_3(g) + 6\ NH(CH_3)_2 \rightarrow B(N(CH_3)_2)_3 + 3\ [NH_2(CH_3)_2]Br$$

In this reaction the HBr produced by the protolysis protonates excess dimethylamine.

Self-test 12.2 Write and justify balanced equations for plausible reactions between (a) BCl_3 and ethanol, (b) BCl_3 and pyridine in hydrocarbon solution, (c) BBr_3 and $F_3BN(CH_3)_3$.

The tetrafluoroborate anion, BF_4^-, which is mentioned in Example 12.2, is used in preparative chemistry when a relatively large noncoordinating anion is needed. The tetrahaloborate anions BCl_4^- and BBr_4^- can be prepared in nonaqueous solvents. However, because of the ease with which B—Cl and B—Br bonds undergo solvolysis, they are stable in neither water nor alcohols.

Boron halides are the starting point for the synthesis of many boron–carbon and boron–pseudohalogen compounds (Section 16.3).[1] Examples include the formation of

---

[1] Pseudohalogens are species that resemble the halogens in their chemical properties. Cyanogen, $(CN)_2$, is a pseudohalogen and the cyanide ion, CN^-, is a pseudohalide.

alkylboron and arylboron compounds, such as trimethylboron by the reaction of boron trifluoride with a methyl Grignard reagent in ether solution:

$$BF_3 + 3\ CH_3MgI \rightarrow B(CH_3)_3 + magnesium\ halides$$

When an excess of the Grignard (or organolithium) reagent is present, tetraalkyl or tetraaryl borates are formed:

$$BF_3 + Li_4(CH_3)_4 \rightarrow Li[B(CH_3)_4] + 3\ LiF$$

Boron halides containing B—B bonds have been prepared. The best known of these compounds have the formula B_2X_4, with X = F, Cl, and Br, and the tetrahedral cluster compound B_4Cl_4. The B_2Cl_4 molecules are planar (**7**) in the solid state but staggered (**8**) in the gas. This conformational difference suggests that rotation about the B—B bond is quite easy, as is expected for a single bond.

One route to B_2Cl_4 is to pass an electric discharge through BCl_3 gas in the presence of a Cl atom scavenger, such as mercury vapour. Spectroscopic data indicate that BCl is produced by electron impact on BCl_3:

$$BCl_3(g) \xrightarrow{\text{electron impact}} BCl(g) + 2\ Cl(g)$$

The Cl atoms are scavenged by mercury vapour and removed as $Hg_2Cl_2(s)$, and the BCl fragment is thought to combine with BCl_3 to yield B_2Cl_4. Metathesis reactions can be used to make B_2X_4 derivatives from B_2Cl_4. The thermal stability of these derivatives increases with increasing tendency of the X group to form a π bond with B:

$$B_2Cl_4 < B_2F_4 < B_2(OR)_4 \ll B_2(NR_2)_4$$

It was thought for a long time that X groups with lone pairs were essential for the existence of B_2X_4 compounds, but diboron compounds with alkyl or aryl groups have been prepared. Compounds that survive at room temperature can be obtained when the groups are bulky, as in $B_2(^tBu)_4$.

A secondary product in the synthesis of B_2Cl_4 is B_4Cl_4, a pale yellow solid composed of molecules with the four B atoms forming a tetrahedron (**9**). Like B_2Cl_4, B_4Cl_4 does not have a formula analogous to those of the boranes (such as B_2H_6) discussed below. This difference may lie in the tendency of halogens to form π bonds with boron by the donation of lone electron pairs on the halide into the otherwise vacant p orbital on boron (as in Fig. 12.2).

12.5 Boron–oxygen compounds

Key point: Boron forms B_2O_3, polyborates, and borosilicate glasses.

The most important oxide of boron, B_2O_3, is prepared by dehydration of boric acid:

$$2\ B(OH)_3(s) \xrightarrow{\Delta} B_2O_3(s) + 3\ H_2O(g)$$

The vitreous form of the oxide consists of a network of partially ordered trigonal BO_3 units. Crystalline B_2O_3 consists of an ordered network of BO_3 units joined through O atoms. Metal oxides dissolve in molten B_2O_3 to give coloured glasses.

Boric acid, $B(OH)_3$, is a very weak Brønsted acid in aqueous solution. However, the equilibria are more complicated than the simple Brønsted proton transfer reactions characteristic of the later p-block oxoacids. Boric acid is in fact primarily a weak Lewis acid, and the complex it forms with H_2O, $H_2OB(OH)_3$, is the actual source of protons:

$$B(OH)_3(aq) + 2\ H_2O(l) \rightleftharpoons H_3O^+(aq) + [B(OH)_4]^-(aq) \qquad pK_a = 9.2$$

As is typical of many of the lighter elements of the p block, there is a tendency for the anion to polymerize by condensation, with the loss of H_2O. Thus, in concentrated neutral or basic solution, equilibria such as

$$3\ B(OH)_3(aq) \rightleftharpoons [B_3O_3(OH)_4]^-(aq) + H^+(aq) + 2\ H_2O(l) \qquad K = 1.4 \times 10^{-7}$$

occur to yield polynuclear anions (**10**).

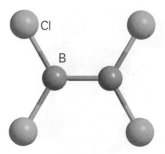

7 B_2Cl_4, D_{2h}

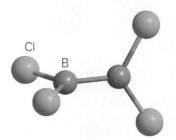

8 B_2Cl_4, D_{2d}

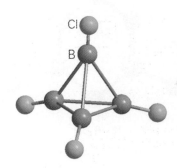

9 B_4Cl_4, T_d

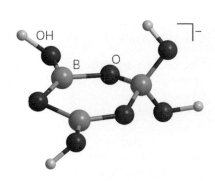

10 $[B_3O_3(OH)_4]^-$

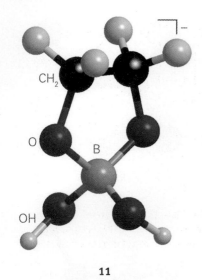

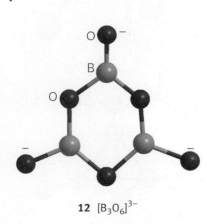

11

12 $[B_3O_6]^{3-}$

The reaction of boric acid with an alcohol in the presence of sulfuric acid leads to the formation of simple borate esters, which are compounds of the form $B(OR)_3$:

$$B(OH)_3 + 3\ CH_3OH \xrightarrow{H_2SO_4} B(OCH_3)_3 + 3\ H_2O$$

Borate esters are much weaker Lewis acids than the boron trihalides, presumably because the O atom acts as an intramolecular π donor, like the F atom in BF_3 (Section 4.8), and donates electron density to the p orbital of the B atom. Hence, judging from Lewis acidity, an O atom is more effective than an F atom as a π donor towards boron. 1,2-Diols have a particularly strong tendency to form borate esters on account of the chelate effect (Section 20.1), and they produce a cyclic borate ester (**11**).

As with silicates and aluminates, there are many polynuclear borates, and both cyclic and chain species are known. An example is the cyclic polyborate anion $[B_3O_6]^{3-}$ (**12**). A notable feature of borate formation is the possibility of both three-coordinate B atoms, as in (**11**), and four-coordinate B atoms, as in $[B(OH)_4]^-$ and for one of the B atoms in (**10**). Polyborates form by sharing one O atom with a neighbouring B atom, as in (**10**); structures in which two adjacent B atoms share two or three O atoms are unknown.

The rapid cooling of molten B_2O_3 or metal borates often leads to the formation of borate glasses. Although these glasses themselves have little technological significance, the fusion of sodium borate with silica leads to the formation of borosilicate glasses (such as Pyrex). Sodium perborate is used as a bleach in laundry powders, automatic dishwasher powders, and whitening toothpastes. Although the formula is often given as $NaBO_3 \cdot 4H_2O$, the compound contains the peroxide anion, O_2^{2-}, and is more accurately described as $Na_2[B_2(O_2)_2(OH)_4] \cdot 6H_2O$. The compound is preferred to hydrogen peroxide in many applications because it is more stable and liberates oxygen only at elevated temperatures.

12.6 Compounds of boron with nitrogen

Key points: Compounds containing BN, which is isoelectronic with CC, include the ethane analogue ammonia borane H_3NBH_3, the benzene analogue $H_3N_3B_3H_3$, and BN analogues of graphite and diamond.

There are many molecular compounds that contain BN bonds. When considering them, it is helpful to bear in mind that BN and CC are isoelectronic. The simplest compound of boron and nitrogen, BN, is easily synthesized by heating boron oxide with a nitrogen compound (Box 12.2):

$$B_2O_3(l) + 2\ NH_3(g) \xrightarrow{1200°C} 2\ BN(s) + 3\ H_2O(g)$$

The form of boron nitride produced by this reaction, and the thermodynamically stable phase under normal laboratory conditions, consists of planar sheets of atoms like those in graphite (Section 13.3). The planar sheets of alternating B and N atoms consist of edge-shared hexagons and, as in graphite, the B—N distance within the sheet (145 pm) is much shorter than the distance between the sheets (333 pm, Fig. 12.4). The difference between the structures of graphite and boron nitride, however, lies in the register of the atoms of neighbouring sheets: in BN, the hexagonal rings are stacked directly over

Box 12.2 Applications of boron nitride

Hexagonal boron nitride was first developed to meet the needs of the aerospace industry. It is stable in oxygen and is not attacked by steam until 900°C. It is a good thermal insulator, has low thermal expansion, and is resistant to thermal shock. These applications have led to its use in industry to make high-temperature crucibles. The powder is used as a mould release and thermal insulator. Boron nitride nanotubes have been formed by depositing boron and nitrogen on a tungsten surface under high vacuum. These nanotubes could be suitable for

high-temperature conditions under which carbon nanotubes would burn (Section 13.4).

The softness and sheen of powdered boron nitride has led to its widest application in the cosmetics and personal care industries. It is nontoxic and presents no known hazard and is added to many products up to around 10 per cent. It adds a pearlescent sheen to products such as nail polishes and lipsticks and is added to foundations to hide wrinkles. Its light reflective properties scatter the light, making wrinkles less noticeable.

each other, with B and a N atoms alternating in successive layers; in graphite, the hexagons are staggered. Molecular orbital calculations suggest that the stacking in BN stems from a partial positive charge on B and a partial negative charge on N. This charge distribution is consistent with the electronegativity difference of the two elements ($\chi_P(B) = 2.04$, $\chi_P(N) = 3.04$).

As with impure graphite, layered boron nitride is a slippery material that is used as a lubricant. Unlike graphite, however, it is a colourless electrical insulator, as there is a large energy gap between the filled and vacant π bands. The size of the band gap is consistent with its high electrical resistivity and lack of absorption in the visible spectrum. In keeping with this large band gap, BN forms a much smaller number of intercalation compounds than graphite (Section 13.3).

Layered boron nitride changes into a denser cubic phase at high pressures and temperatures (60 kbar and 2000°C, Fig. 12.5). This phase is a hard crystalline analogue of diamond but, as it has a lower lattice enthalpy, it has a slightly lower mechanical hardness (Fig. 12.6). Cubic boron nitride is manufactured and used as an abrasive for certain high-temperature applications in which diamond cannot be used because it forms carbides with the material being ground.

The fact that BN and CC are isoelectronic suggests that there might be analogies between these compounds and hydrocarbons. Many **amine-boranes**, the boron–nitrogen analogues of saturated hydrocarbons, can be synthesized by reaction between a nitrogen Lewis base and a boron Lewis acid:

$$\tfrac{1}{2} B_2H_6 + N(CH_3)_3 \rightarrow H_3BN(CH_3)_3$$

However, although amine-boranes are isoelectronic with hydrocarbons, their properties are significantly different, in large part due to the difference in electronegativities of B and N. For example, whereas ammoniaborane, H_3NBH_3, is a solid at room temperature with a vapour pressure of a few pascals, its analogue ethane, H_3CCH_3, is a gas that condenses at $-89°C$. This difference can be traced to the difference in polarity of the two molecules: ethane is nonpolar, whereas ammoniaborane has a large dipole moment of 5.2 D (**13**).

Several BN analogues of the amino acids have been prepared, including ammonia-carboxyborane, H_3NBH_2COOH, the analogue of propionic acid, CH_3CH_2COOH. These compounds display significant physiological activity, including tumour inhibition and reduction of serum cholesterol.

The simplest unsaturated boron–nitrogen compound is aminoborane, H_2NBH_2, which is isoelectronic with ethene. It has only a transient existence in the gas phase because it readily forms cyclic ring compounds such as a cyclohexane analogue (**14**). However, the aminoboranes do survive as monomers when the double bond is shielded from reaction by bulky alkyl groups on the N atom and by Cl atoms on the B atom (**15**). For instance, monomeric aminoboranes can be synthesized readily by the reaction of a dialkylamine and a boron halide:

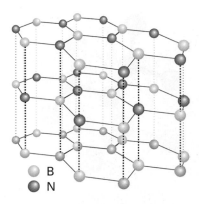

Fig. 12.4 The structure of layered hexagonal boron nitride. Note that the rings are in register between layers.

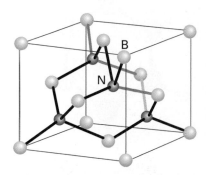

Fig. 12.5 The sphalerite structure of cubic boron nitride.

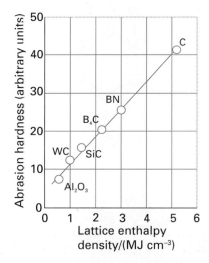

Fig. 12.6 The correlation of hardness with lattice enthalpy density (the lattice enthalpy divided by the molar volume of the substance). The point for carbon represents diamond; that for boron nitride represents the diamond-like sphalerite structure.

$((CH_3)_2CH)_2NH + BCl_3 \longrightarrow$ [Cl, Cl on B; CH(CH$_3$)$_2$, CH(CH$_3$)$_2$ on N; B=N] $+ HCl$

13 NH_3BH_3 **14** $N_3B_3H_{12}$

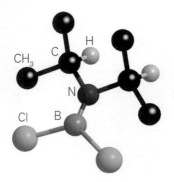

15 $Cl_2B=N(^iPr)_2$

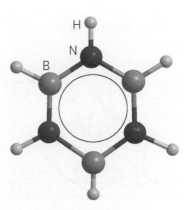

16 Borazine, $B_3N_3H_6$

17 $B_3N_3H_3Cl_3$

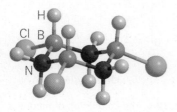

18 $B_3N_3H_9Cl_3$

The reaction also occurs with xylyl (2,4,6-trimethylphenyl) groups in place of isopropyl groups.

Apart from layered boron nitride, the best known unsaturated compound of boron and nitrogen is borazine, $H_3B_3N_3H_3$ (**16**), which is isoelectronic and isostructural with benzene. Borazine was first prepared in Alfred Stock's laboratory in 1926 by the reaction between diborane and ammonia. Since then, many symmetrically trisubstituted derivatives have been made by procedures that depend on the protolysis of BCl bonds of BCl_3 by an ammonium salt (**17**):

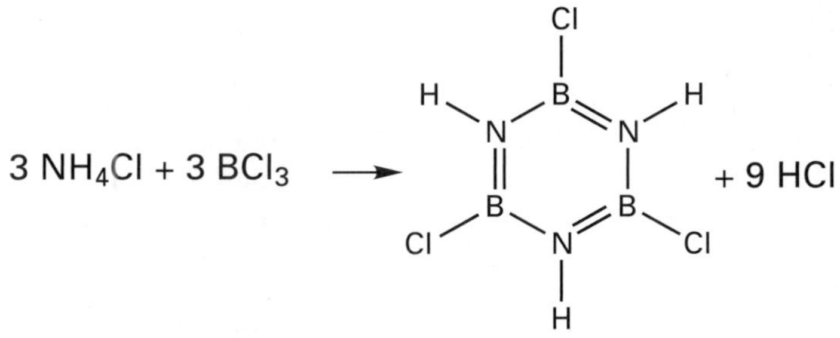

$$3\ NH_4Cl + 3\ BCl_3 \longrightarrow \text{(structure)} + 9\ HCl$$

The use of an alkylammonium chloride yields *N*-alkyl substituted *B*-trichloroborazines.

Despite their structural resemblance, there is little chemical resemblance between borazine and benzene. Once again, the difference in the electronegativities of boron and nitrogen is influential, and BCl bonds in trichloroborazine are much more labile than the CCl bonds in chlorobenzene. In the borazine compound, the π electrons are concentrated on the N atoms and there is a partial positive charge on the B atoms that leaves them open to electrophilic attack. A sign of the difference is that the reaction of a chloroborazine with a Grignard reagent or hydride source results in the substitution of Cl by alkyl, aryl, or hydride groups. Another example of the difference is the ready addition of HCl to borazine to produce a trichlorocyclohexane analogue (**18**):

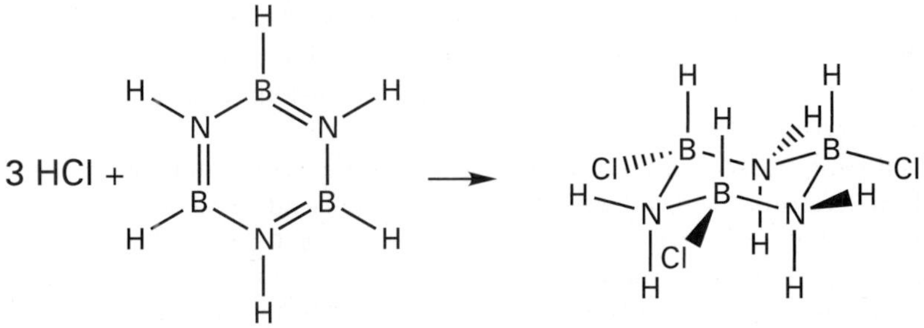

$$3\ HCl + \text{(structure)} \longrightarrow \text{(structure)}$$

The electrophile, H^+, in this reaction attaches to the partially negative N atom and the nucleophile Cl^- attaches to the partially positive B atom.

Example 12.3 Preparing borazine derivatives

Give balanced chemical equations for the synthesis of borazine starting with NH_4Cl and other reagents of your choice.

Answer The reaction of NH_4Cl with BCl_3 in refluxing chlorobenzene will yield the *B*-trichloroborazine:

$$3\ NH_4Cl + 3\ BCl_3 \rightarrow H_3N_3B_3Cl_3 + 9\ HCl$$

The Cl atoms in *B*-trichloroborazine can be displaced by hydride ions from reagents such as $LiBH_4$, to yield borazine:

$$3\ LiBH_4 + H_3N_3B_3Cl_3 \xrightarrow{THF} H_3B_3N_3H_3 + 3\ LiCl + 3\ THF \cdot BH_3$$

Self-test 12.3 Suggest a reaction or series of reactions for the preparation of *N, N', N''*-trimethyl-*B,B',B''*-trimethylborazine starting with methylamine and boron trichloride.

Boron clusters

The boron subhalide B_4Cl_4 discussed previously provides a hint about boron's ability to form cluster compounds. The first recognition of these clusters was the direct result of improvements in X-ray crystallography, which gave the first accurate indication of the structures of metal borides and of boranes more complex than B_2H_6. The neutral boranes and the anionic borohydrides were the first examples of molecular cluster compounds to be studied in depth.

12.7 Metal borides

Key points: Metal borides include boron anions as isolated B atoms, linked *closo*-boron polyhedra, and hexagonal boron networks.

The direct reaction of elemental boron and a metal at high temperatures provides a useful route to many metal borides. An example is the reaction of calcium and some other highly electropositive metals with boron to produce a phase of composition MB_6:

$$Ca(l) + 6\ B(s) \rightarrow CaB_6(s)$$

Metal borides are found with a wide range of compositions, as boron can occur in numerous types of structure, including isolated B atoms, chains, planar and puckered nets, and clusters. The simplest metal borides are metal-rich compounds that contain isolated B^{3-} ions. The most common examples of these compounds have the formula M_2B, where M may be one of the middle to late $3d$ metals (Mn to Ni) in low oxidation states. Another important class of metal borides contain planar or puckered hexagonal nets and have the composition MB_2 (Fig. 12.7). These compounds are formed primarily by electropositive metals, including magnesium, aluminium, the early d metals (from Sc to Mn, for instance, in Period 4), and uranium (Box 12.3).

The boron-rich borides, typically MB_6 and MB_{12}, where M is an electropositive metal, are of even greater structural interest. In them, the B atoms link to form an intricate network of interconnecting cages. In MB_6 compounds (which are formed by

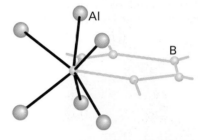

Fig. 12.7 The AlB_2 structure. To give a clear picture of the hexagonal layer, B atoms outside the unit cell are displayed.

Box 12.3 Magnesium diboride superconductors

Magnesium diboride, MgB_2, is a cheap compound that has been known in the laboratory for over 50 years. In 2001 this simple compound was found to have superconducting properties. Jun Akimitsu and his co-workers discovered by chance that MgB_2 loses its electrical resistance when cooled. At the time, they were characterizing materials used to enhance the performance of known high-temperature superconductors. The discovery led to a major flurry of research on this new superconductor around the world.

In bulk materials the transition temperature of MgB_2 is 38 K and is exceeded only by the much more complicated perovskite cuprate structures (Section 23.9). The discovery of this latest superconductor was even more serendipitous than the discovery of the cuprate superconductors by Georg Bednorz and Alex Müller in 1986, for which they shared the Nobel Prize a year later. Many of the first measurements were made using MgB_2 powder straight from the bottle. High quality MgB_2 can be synthesized by heating fine boron and magnesium powders together at around 950°C under pressure. Thin films, wires, and tapes have since been formed that have potential for applications in superconducting magnets, microwave communications, and power applications.

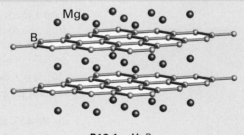

B12.1 MgB_2

MgB_2 has a simple structure in which the boron atoms are arranged in graphite-like planes with alternating layers of magnesium atoms. The magnesium atoms donate their two valence electrons to the network of boron atoms. Varying the number of electrons donated to the boron conduction bands can dramatically affect the transition temperature. The transition temperature of the compound falls if some of the magnesium atoms are replaced by aluminium and increases when doped with copper. The transition temperature, T_c, of MgB_2 is approximately 15 K higher than theory predicts. This difference has been explained in terms of vibrations in the lattice that allow two electrons to form a Cooper pair, which then travels resistance-free through the material.

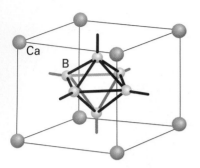

Fig. 12.8 The CaB_6 structure. Note that the B_6 octahedra are connected by a bond between vertices of adjacent B_6 octahedra. The crystal is a simple cube analogue of CsCl. Thus eight Ca atoms surround the central B_6 octahedron.

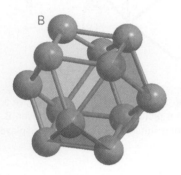

19 B_{12} cuboctahedron

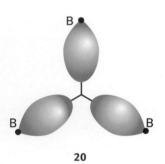

20

the electropositive s-block metals such as sodium, potassium, calcium, barium, and strontium), the B_6 octahedra are linked by their vertices to form a cubic framework (Fig. 12.8). The linked B_6 clusters bear a charge of -1, -2, or -3 depending on the cation with which they are associated. In the MB_{12} compounds the B-atom networks are based on linked cuboctahedra (**19**) rather than the more familiar icosahedron. This type of compound is formed by some of the heavier electropositive metals, particularly those of the f block.

12.8 Higher boranes and borohydrides

In addition to the simple hydrides discussed in Section 12.3, boron forms several series of polymeric cage-like boron–hydrogen compounds. Borohydrides are formed with up to twelve B atoms and fall into three classes.

(a) Bonding and structure

Key point: The bonding in boron hydrides and polyhedral borohydride ions can be approximated by conventional $2c,2e$ bonds together with $3c,2e$ bonds.

In this section we describe the structures and properties of the cage-like boranes and borohydrides, which include Stock's series B_nH_{n+4} and B_nH_{n+6} as well as the more recently discovered $B_nH_n^{2-}$ closed polyhedra. Boranes and borohydrides occur with a variety of shapes, some resembling nests and others (to the imaginative eye) butterflies and spiders' webs. They are all electron deficient in the sense that they cannot be described by simple Lewis structures.

The modern interpretation of the bonding in boranes and borohydrides derives from the work of Christopher Longuet-Higgins who, as an undergraduate at Oxford, published a seminal paper in which he introduced the concept of $3c,2e$ bonds (Section 2.11). He later developed a fully delocalized molecular orbital treatment for boron polyhedra and predicted the stability of the icosahedral ion $B_{12}H_{12}^{2-}$, which was subsequently verified. William Lipscomb and his students used single-crystal X-ray diffraction to determine the structures of a large number of boranes and borohydrides, and generalized the concept of multicentre bonding to these more complex species.

Boron cluster compounds are best considered from the standpoint of fully delocalized molecular orbitals containing electrons that contribute to the stability of the entire molecule. However, it is sometimes fruitful to identify groups of three atoms and to regard them as bonded together by versions of the $3c,2e$ bonds of the kind that occur in diborane itself (**1**). In the more complex boranes, the three centres of the $3c,2e$ bonds may be BHB bridge bonds, but they may also be bonds in which three B atoms lie at the corners of an equilateral triangle with their sp^3 hybrid orbitals overlapping at its centre (**20**). To reduce the complexity of the structural diagrams, the illustrations that follow will not in general indicate the $3c,2e$ bonds in the structures.

(b) Wade's rules

Key points: Boron hydride structures include simple polyhedral *closo* compounds and the progressively more open nido and arachno structures.

A correlation between the number of electrons (counted in a specific way), the formula, and the shape of the molecule was established by Kenneth Wade in the 1970s. These so-called **Wade's rules** apply to a class of polyhedra called **deltahedra** (because they are made up of triangular faces resembling Greek deltas, Δ) and can be used in two ways. For molecular and anionic boranes, they enable us to predict the general shape of the molecule or anion from its formula. However, because the rules are also expressed in terms of the number of electrons, we can extend them to analogous species in which there are atoms other than boron, such as carboranes and other p-block clusters. Here we concentrate on the boron clusters, where knowing the formula is sufficient for predicting the shape. However, so that we can cope with other clusters we shall show how to count the framework electrons too.

The building block from which the deltahedron is constructed is assumed to be one BH group (**21**) that contributes two electrons. The electrons in the B—H bond are ignored in the counting procedure, but all others are included whether or not it is obvious that they help to hold the skeleton together. By the 'skeleton' is meant the framework of the cluster with each BH group counted as a unit. If a B atom happens to carry two H atoms, only one of the B—H bonds is treated as a unit.[2] For instance, in B_5H_{11}, one of the B atoms has two 'terminal' H atoms but only one BH entity is treated as a unit, the other pair of electrons being treated as part of the skeleton and hence referred to as 'skeletal electrons'. A BH group makes two electrons available to the skeleton (the B atom provides three electrons and the H atom provides one but, of these four, two are used for the B—H bond).

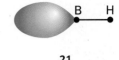

21

Example 12.4 Counting skeletal electrons

Count the number of skeletal electrons in B_4H_{10} (**3**).

Answer Four BH units contribute $4 \times 2 = 8$ electrons, and the six additional H atoms contribute a further six electrons, giving 14 in all. The resulting seven pairs are distributed as shown in (**22**): two are used for the additional terminal B—H bonds, four are used for the four B—H—B bridges, and one is used for the central B—B bond.

Self-test 12.4 How many skeletal electrons are present in B_5H_9?

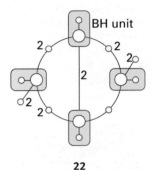

22

According to Wade's rules (Table 12.4), species of formula $[B_nH_n]^{2-}$ and $n + 1$ pairs of skeletal electrons have a **closo** structure, a name derived from the Greek for 'cage', with a B atom at each corner of a closed deltahedron and no B—H—B bonds. This series of anions is known for $n = 5$ to 12, and examples include the trigonal-bipyramidal $[B_5H_5]^{2-}$ ion, the octahedral $[B_6H_6]^{2-}$ ion, and the icosahedral $[B_{12}H_{12}]^{2-}$ ion. The *closo*-borohydrides and their carborane analogues (Section 12.10) are typically thermally stable and moderately unreactive.

Boron clusters of formula B_nH_{n+4} and $n + 2$ pairs of skeletal electrons have the **nido** structure, a name derived from the Latin for 'nest'. They can be regarded as derived from a *closo*-borane that has lost one vertex but have B—H—B bonds as well as B—B bonds. An example is B_5H_9. In general, the thermal stability of the *nido*-boranes is intermediate between that of *closo*- and the *arachno*-boranes described below.

Clusters of formula B_nH_{n+6} and $n+3$ skeletal electron pairs have an **arachno** structure, from the Greek for 'spider' (as they resemble untidy spiders' webs). They can be regarded as *closo*-borane polyhedra less two vertices (and must have B—H—B bonds). One example of an *arachno*-borane is pentaborane(11) (B_5H_{11}). As with most *arachno*-boranes, pentaborane(11) is thermally unstable at room temperature and is highly reactive.

Table 12.4 Classification and electron count of boron hydrides

Type	Formula*	Skeletal electron pairs	Examples
closo	$B_nH_n^{2-}$	$n + 1$	$B_5H_5^{2-}$ to $B_{12}H_{12}^{2-}$
nido	B_nH_{n+4}	$n + 2$	B_2H_6, B_5H_9, B_6H_{10}
arachno	B_nH_{n+6}	$n + 3$	B_4H_{10}, B_5H_{11}
hypho[†]	B_nH_{n+8}	$n + 4$	None[‡]

* In some cases, protons can be removed; thus $[B_5H_8]^-$ arises from the deprotonation of B_5H_9.
[†] The name comes from the Greek for 'net'.
[‡] Some derivatives are known.

[2] This rather odd feature is just a part of Wade's rules: it works. That is, counting electrons in this way provides a parameter that helps to correlate properties.

Example 12.5 Using Wade's rules

Infer the structure of $[B_6H_6]^{2-}$ from its formula and from its electron count.

Answer We note that the formula $[B_6H_6]^{2-}$ belongs to a class of borohydrides having the formula $[B_nH_n]^{2-}$, which is characteristic of a *closo* species. Alternatively, we can count the number of skeletal electron pairs and from that deduce the structural type. Assuming one B—H bond per B atom, there are six BH units to take into account and therefore twelve skeletal electrons plus two from the overall charge of -2: $6 \times 2 + 2 = 14$, or $2(n+1)$ with $n = 6$. This number is characteristic of *closo* clusters. The closed polyhedron must contain triangular faces and six vertices; therefore an octahedral structure is indicated.

Self-test 12.5 How many framework electron pairs are present in B_4H_{10} and to what structural category does it belong? Sketch its structure.

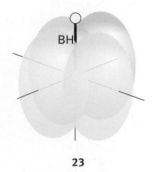

23

(c) The origin of Wade's rules

Key points: The molecular orbitals in a closo-borane can be constructed from BH units, each of which contributes one radial atomic orbital pointing towards the centre of the cluster and two perpendicular *p* orbitals that are tangential to the polyhedron.

Wade's rules have been justified by molecular orbital calculations. We shall indicate the kind of reasoning involved by considering the first of them (the $n + 1$ rule). In particular, we shall show that $[B_6H_6]^{2-}$ has a low energy if it has an octahedral *closo* structure, as predicted by the rules.

A B—H bond uses one electron and one orbital of the B atom, leaving three orbitals and two electrons for the skeletal bonding. One of these orbitals, which is called a **radial orbital**, can be considered to be a boron *sp* hybrid pointing towards the interior of the fragment (as in **21**). The remaining two boron *p* orbitals, the **tangential orbitals,** are perpendicular to the radial orbital (**23**). The shapes of the 18 symmetry-adapted linear combinations of these 18 orbitals in an octahedral B_6H_6 cluster can be inferred from the drawings in *Resource section* 4, and we show the ones with net bonding character in Fig. 12.9.

The lowest energy molecular orbital is totally symmetric (a_{1g}) and arises from in-phase contributions from all the radial orbitals. Calculations show that the next higher orbitals are the t_{1u} orbitals, each of which is a combination of four tangential and two radial orbitals. Above these three degenerate orbitals lie another three t_{2g} orbitals, which are tangential in character, giving seven bonding orbitals in all. Hence, there are seven

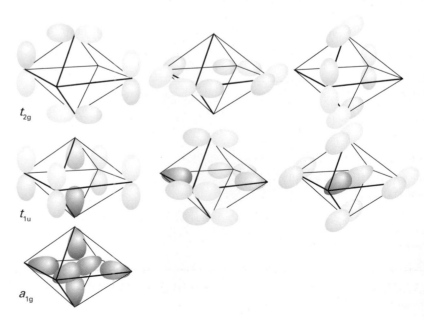

t_{2g}

t_{1u}

a_{1g}

Fig. 12.9 Radial and tangential bonding molecular orbitals for $B_6H_6^{2-}$. The relative energies are $a_{1g} < t_{1u} < t_{2g}$.

orbitals with net bonding character delocalized over the skeleton, and they are separated by a considerable gap from the remaining eleven largely antibonding orbitals (Fig. 12.10).

There are seven electron pairs to accommodate, one pair from each of the six B atoms and one pair from the overall charge (-2). These seven pairs can all enter and fill the seven bonding skeleton orbitals, and hence give rise to a stable structure, in accord with the $n+1$ rule. Note that the unknown neutral octahedral B_6H_6 molecule would have too few electrons to fill the t_{2g} bonding orbitals. Similar arguments can be used for all *closo* structures.

(d) Structural correlations

Key points: Conceptually, the *closo*, *nido*, and *arachno* structures are related by the successive removal of a BH fragment and addition of H or electrons.

A very useful structural correlation between *closo*, *nido*, and *arachno* species is based on the observation that clusters with the same numbers of skeletal electrons are related by removal of successive BH groups and the addition of the appropriate numbers of electrons and H atoms. This conceptual process provides a good way to think about the structures of the various boron clusters but does not represent how they are interconverted chemically.

The idea is amplified in Fig. 12.11, where the removal of a BH unit and two electrons and the addition of four H atoms converts the octahedral *closo*-$[B_6H_6]^{2-}$ anion to the square-pyramidal *nido*-B_5H_9 borane. A similar process (removal of a BH unit and addition of two H atoms) converts *nido*-B_5H_9 into a butterfly-like *arachno*-B_4H_{10} borane. Each of these three boranes has 14 skeletal electrons but, as the number of skeletal electrons per B atom increases, the structure becomes more open. A more systematic correlation of this type is indicated for many different boranes in Fig. 12.12.

(e) Synthesis of higher boranes and borohydrides

Key point: Pyrolysis followed by rapid quenching provides one method of converting small boranes to larger boranes.

As discovered by Stock and perfected by many subsequent workers, the controlled pyrolysis of B_2H_6 in the gas phase provides a route to most of the higher boranes and borohydrides, including B_4H_{10}, B_5H_9, and $B_{10}H_{14}$. A key first step in the proposed mechanism is the dissociation of B_2H_6 and the condensation of the resulting BH_3 fragment with borane fragments. For example, the mechanism of the formation of tetraborane(10) by the pyrolysis of diborane appears to be

$$B_2H_6 \rightarrow BH_3 + BH_3$$
$$B_2H_6 + BH_3 \rightarrow B_3H_7 + H_2$$
$$BH_3 + B_3H_7 \rightarrow B_4H_{10}$$

The synthesis of tetraborane(10), B_4H_{10}, is particularly difficult because it is highly unstable, in keeping with the instability of the B_nH_{n+6} (*arachno*) series. To improve the yield, the product that emerges from the hot reactor is immediately quenched on a cold surface. Pyrolytic syntheses to form species belonging to the more stable B_nH_{n+4} (*nido*) series proceed in higher yield, without the need for a rapid quench. Thus B_5H_9 and $B_{10}H_{14}$ are readily prepared by the pyrolysis reaction. More recently, these brute-force methods of pyrolysis have given way to more specific methods that are described below.

(f) Characteristic reactions of boranes and borohydrides

Key points: Characteristic reactions of boranes are cleavage of a BH_2 group from diborane and tetraborane by NH_3, deprotonation of large boron hydrides by bases, reaction of a boron hydride with a borohydride ion to produce a larger borohydride anion, and Friedel–Crafts-type substitution of an alkyl group for hydrogen in pentaborane and some larger boron hydrides.

The characteristic reactions of boron clusters with a Lewis base range from cleavage of BH_n from the cluster to deprotonation of the cluster, cluster enlargement, and abstraction of one or more protons.

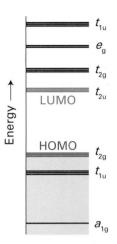

Fig. 12.10 Schematic molecular orbital energy levels of the B-atom skeleton of $[B_6H_6^{2-}]$. The form of the bonding orbitals is shown in Fig. 12.9.

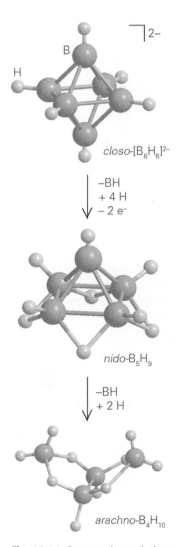

Fig. 12.11 Structural correlations between a B_6 *closo* octahedral structure, a B_5 *nido* square pyramid, and a B_4 *arachno* butterfly.

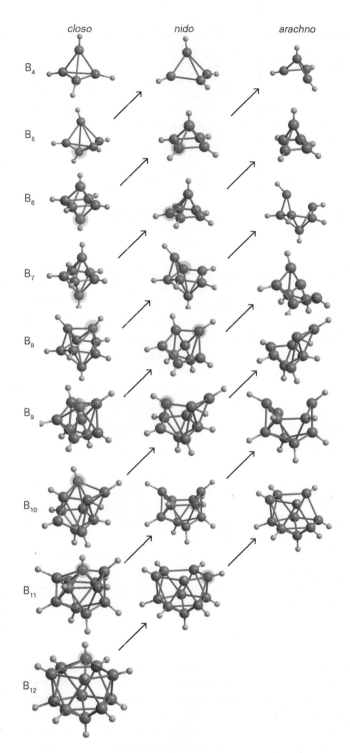

Fig. 12.12 Structural relations between *closo*-, *nido*-, and *arachno*-boranes and heteratomic boranes. Diagonal lines connect species that have the same number of skeletal electrons. Hydrogen atoms beyond those in the B—H framework and charges have been omitted. The blue tinted atom is removed first, and the red tinted atom is removed second. (Based on R.W. Rudolph, *Acc. Chem. Res.*, 1976, **9**, 446.)

Lewis base cleavage reactions have already been introduced in Section 12.3 in connection with diborane. With the robust higher borane B_4H_{10}, cleavage may break some BHB bonds leading to partial fragmentation of the cluster:

Deprotonation, rather than cleavage, occurs readily with the large borane $B_{10}H_{14}$:

$$B_{10}H_{14} + N(CH_3)_3 \rightarrow [HN(CH_3)_3]^+[B_{10}H_{13}]^-$$

The structure of the product anion indicates that deprotonation occurs from a $3c,2e$ B—H—B bridge, leaving the electron count on the boron cluster unchanged. This deprotonation of a B—H—B $3c,2e$ bond to yield a $2c,2e$ bond occurs without major disruption of the bonding:

$$B\diagdown_{H}\diagup B \longrightarrow \left[B-B \right]^- + H^+$$

The Brønsted acidity of boron hydrides increases approximately with size:

$$B_4H_{10} < B_5H_9 < B_{10}H_{14}$$

This trend correlates with the greater delocalization of charge in the larger clusters, in much the same way that delocalization accounts for the greater acidity of phenol than methanol. The variation in acidity is illustrated by the observation that, as shown above, the weak base trimethylamine deprotonates decaborane(10), but the much stronger base methyllithium is required to deprotonate B_5H_9:

$$+ \tfrac{1}{4}Li_4(CH_3)_4 \longrightarrow Li^+ \left[\right]^- + CH_4$$

Hydridic character is most characteristic of small anionic borohydrides. As an illustration, whereas BH_4^- readily surrenders a hydride ion in the reaction

$$BH_4^- + H^+ \rightarrow \tfrac{1}{2}B_2H_6 + H_2$$

the $[B_{10}H_{10}]^{2-}$ ion survives even in strongly acidic solution. Indeed, the hydronium salt $(H_3O)_2B_{10}H_{10}$ can even be crystallized.

The cluster-building reaction between a borane and a borohydride provides a convenient route to higher borohydride ions:

$$5\,K[B_9H_{14}] + 2\,B_5H_9 \xrightarrow{\text{polyether, }85°C} 5\,K[B_{11}H_{14}] + 9\,H_2$$

Similar reactions are used to prepare other borohydrides, such as $B_{10}H_{10}^{2-}$. This type of reaction has been used to synthesize a wide range of polynuclear borohydrides. Boron-11 NMR spectroscopy reveals that the boron skeleton in $B_{11}H_{14}^-$ consists of an icosahedron with a missing vertex (Fig. 12.13).

The **electrophilic displacement** of H^+ provides a route to alkylated and halogenated species. As with Friedel–Crafts reactions, the electrophilic displacement of H is catalysed by a Lewis acid, such as aluminium chloride, and the substitution generally occurs on the closed portion of the boron clusters:

$$+ CH_3Cl \xrightarrow{AlCl_3} + HCl$$

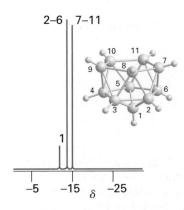

Fig. 12.13 The proton-decoupled ^{11}B NMR spectrum of $[B_{11}H_{14}]^-$. The *nido* structure (a truncated icosahedron) is indicated by the 1:5:5 pattern.

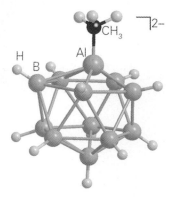

24 *closo*-[B$_{11}$H$_{11}$AlCH$_3$]$^{2-}$

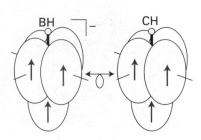

25 [Fe(CO)$_3$B$_4$H$_8$]

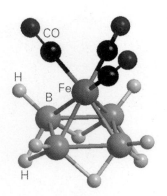

26

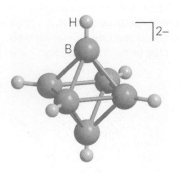

27 *closo*-[B$_5$H$_6$]$^{2-}$

Example 12.6 Proposing a structure for a boron-cluster reaction product

Propose a structure for the product of the reaction of B$_{10}$H$_{14}$ with LiBH$_4$ in a refluxing polyether, CH$_3$OC$_2$H$_4$OCH$_3$ (which boils at 162°C).

Answer The prediction of the probable outcome for the reactions of a boron cluster is difficult because several products are often plausible and the actual outcome is often sensitive to the conditions of the reaction. In the present case we note that an acidic borane, B$_{10}$H$_{14}$, is brought into contact with the hydridic anion BH$_4^-$ under rather vigorous conditions. Therefore, we might expect the evolution of hydrogen:

$$B_{10}H_{14} + Li[BH_4] \xrightarrow{\text{ether, R}_2\text{O}} Li[B_{10}H_{13}] + R_2OBH_3 + H_2$$

This set of products suggests the further possibility of condensation of the neutral BH$_3$ complex with B$_{10}$H$_{13}^-$ to yield a larger borohydride. That is in fact the observed outcome under these conditions:

$$Li[B_{10}H_{13}] + R_2OBH_3 \rightarrow Li[B_{11}H_{14}] + H_2 + R_2O$$

It turns out that, in the presence of excess LiBH$_4$, the cluster building continues to give the very stable icosahedral B$_{12}$H$_{12}^{2-}$ anion:

$$Li[\textit{nido}\text{-B}_{11}H_{14}] + Li[BH_4] \rightarrow Li_2[\textit{closo}\text{-B}_{12}H_{12}] + 3\ H_2$$

Self-test 12.6 Propose a plausible product for the reaction between Li[B$_{10}$H$_{13}$] and Al$_2$(CH$_3$)$_6$.

12.9 Metallaboranes

Key points: Main-group and *d*-block metals may be incorporated into boron hydrides through B—H—M bridges or more robust B—M bonds.

Many **metallaboranes**, or metal-containing boron clusters, have been characterized. In some cases the metal is appended to a borohydride ion through hydrogen bridges. A more common and generally more robust group of metallaboranes have direct metal–boron bonds. An example of a main-group metallaborane with an icosahedral framework is *closo*-[B$_{11}$H$_{11}$AlCH$_3$]$^{2-}$ (**24**). It is prepared by interaction of the acidic hydrogens in Na$_2$[B$_{11}$H$_{13}$] with trimethylaluminium:

$$2\ [B_{11}H_{13}]^{2-} + Al_2(CH_3)_6 \xrightarrow{\Delta} 2\ [B_{11}H_{11}AlCH_3]^{2-} + 4\ CH_4$$

When B$_5$H$_9$ is heated with Fe(CO)$_5$, a metallated analogue of pentaborane is formed (**25**).

12.10 Carboranes

Key points: When C—H is introduced in place of B—H in a polyhedral boron hydride, the charge of the resulting carboranes is one unit more positive; carborane anions are useful precursors of boron-containing organometallic compounds.

Closely related to the polyhedral boranes and borohydrides are the **carboranes** (more formally, the *carbaboranes*), a large family of clusters that contain both B and C atoms. Now we begin to see the full generality of Wade's electron counting rules, as BH$^-$ is isolectronic and isolobal with CH (**26**), and we can expect the polyhedral borohydrides and carboranes to be related. Thus, an analogue of B$_6$H$_6^{2-}$ (**27**) is the neutral carborane B$_4$C$_2$H$_6$ (**28**).

One entry into the interesting and diverse world of carboranes is the conversion of decaborane(14) to *closo*-1,2-B$_{10}$C$_2$H$_{12}$ (**29**). The first reaction in this preparation is the displacement of an H$_2$ molecule from decaborane by a thioether:

$$B_{10}H_{14} + 2\ SEt_2 \rightarrow B_{10}H_{12}(SEt_2)_2 + H_2$$

The loss of two H atoms in this reaction is compensated by the donation of electron pairs by the added thioethers, so the electron count is unchanged. The product of the reaction is then converted to the carborane by the addition of an alkyne:

$$B_{10}H_{12}(SEt_2)_2 + C_2H_2 \rightarrow B_{10}C_2H_{12} + 2\ SEt_2 + H_2$$

The four π electrons of ethyne displace two thioether molecules (two two-electron donors) and an H_2 molecule (which leaves with two additional electrons). The net loss of two electrons correlates with the change in structure from a *nido* starting material to the *closo* product. The C atoms are in adjacent (1,2) positions, reflecting their origin from ethyne. This *closo*-carborane survives in air and can be heated without decomposition. At 500°C in an inert atmosphere it undergoes isomerization into 1,7-$B_{10}C_2H_{12}$ (**30**), which in turn isomerizes at 700°C to the 1,12-isomer (**31**).

The H atoms attached to carbon in *closo*-$B_{10}C_2H_{12}$ are very mildly acidic, so it is possible to lithiate these compounds with butyllithium:

$$B_{10}C_2H_{12} + 2\,LiC_4H_9 \rightarrow B_{10}C_2H_2Li_2 + 2\,C_4H_{10}$$

These dilithiocarboranes are good nucleophiles and undergo many of the reactions characteristic of organolithium reagents (Section 10.12). Thus, a wide range of carborane derivatives can be synthesized. For example, reaction with CO_2 gives a dicarboxylic acid carborane:

$$B_{10}C_2H_{10}Li_2 \xrightarrow{(1)\,2\,CO_2;\ (2)\,2\,H_2O} B_{10}C_2H_{10}(COOH)_2$$

Similarly, I_2 leads to the diiodocarborane and NOCl yields $B_{10}C_2H_{10}(NO)_2$.

Although 1,2-$B_{10}C_2H_{12}$ is very stable, the cluster can be partially fragmented in strong base, and then deprotonated with NaH to yield *nido*-$B_9C_2H_{11}^{2-}$:

$$B_{10}C_2H_{12} + EtO^- + 2\,EtOH \rightarrow B_9C_2H_{50}^- + B(OEt)_3 + H_2$$
$$Na[B_9C_2H_{12}] + NaH \rightarrow Na_2[B_9C_2H_{11}] + H_2$$

The importance of these reactions is that *nido*-$B_9C_2H_{11}^{2-}$ (Fig. 12.14a) is an excellent ligand. In this role it mimics the cyclopentadienyl ligand ($C_5H_5^-$; Fig. 12.14b) which is widely used in organometallic chemistry:

$$2\,Na_2[B_9C_2H_{11}] + FeCl_2 \xrightarrow{THF} 2\,NaCl + Na_2[Fe(B_9C_2H_{11})_2]$$
$$2\,Na[C_5H_5] + FeCl_2 \xrightarrow{THF} 2\,NaCl + Fe(C_5H_5)_2$$

Although we shall not go into the details of their synthesis, a wide range of metal-coordinated carboranes can be synthesized. A notable feature is the ease of formation of multi-decker sandwich compounds containing carborane ligands, (**32**) and (**33**). The highly negative $B_3C_2H_5^{4-}$ ligand has a much greater tendency to form stacked sandwich compounds than the less negative and therefore poorer donor $C_5H_5^-$.

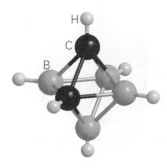

28 *closo*-1,2-$B_4C_2H_6$

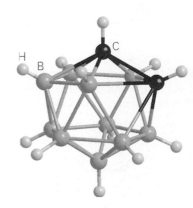

29 *closo*-1,2-$B_{10}C_2H_{12}$

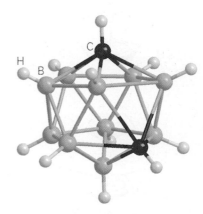

30 *closo*-1,7-$B_{10}C_2H_{12}$

Example 12.7 Planning the synthesis of a carborane derivative

Give balanced chemical equations for the synthesis of 1,2-$B_{10}C_2H_{10}(Si(CH_3)_3)_2$ starting with decaborane(10) and other reagents of your choice.

Answer The attachment of substituents to C atoms in 1,2-*closo*-$B_{10}C_2H_{12}$ is most readily carried out using the dilithium derivative $B_{10}C_2H_{10}Li_2$. We first prepare 1,2-$B_{10}C_2H_{12}$ from decaborane:

$$B_{10}H_{14} + 2\,SR_2 \rightarrow B_{10}H_{12}(SR_2)_2 + H_2$$
$$B_{10}H_{12}(SR_2)_2 + C_2H_2 \rightarrow B_{10}C_2H_{12} + 2\,SR_2$$

The product is then lithiated by lithium alkyl, where the alkyl carbanion abstracts the slightly acidic hydrogen atoms from $B_{10}C_2H_{50}$, replacing them with Li^+:

$$B_{10}C_2H_{12} + 2\,LiC_4H_9 \rightarrow B_{10}C_2H_{10}Li_2 + 2\,C_4H_{10}$$

The resulting carborane is then used in a nucleophilic displacement on $Si(CH_3)_3Cl$ to yield the desired product:

$$B_{10}C_2H_{10}Li_2 + 2\,Si(CH_3)_3Cl \rightarrow B_{10}C_2H_{10}(Si(CH_3)_3)_2 + 2\,LiCl$$

Self-test 12.7 Propose a synthesis for the polymer precursor 1,7-$B_{10}C_2H_{10}(Si(CH_3)_2Cl)_2$ from 1,2-$B_{10}C_2H_{12}$ and other reagents of your choice.

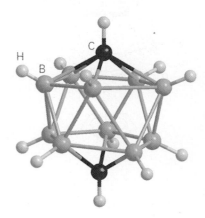

31 *closo*-1,12-$B_{10}C_2H_{12}$

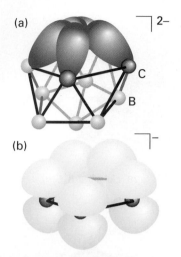

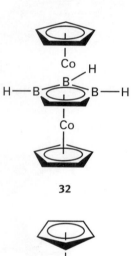

32

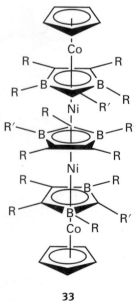

33

Fig. 12.14 The isolobal relation between (a) $[B_9C_2H_{11}]^{2-}$ and (b) $[C_5H_5]^-$. The H atoms have been omitted for clarity.

Compounds of aluminium through thallium

The elements aluminium, gallium, indium, and thallium are metals with many similarities in their chemical properties Like boron, they form electron-deficient compounds that act as Lewis acids.

12.11 The hydrides of aluminium and gallium

Key points: $LiAlH_4$ and $LiGaH_4$ are useful precursors of MH_3L_2 complexes; $LiAlH_4$ is also used as a source of H^- ions in the preparation of metalloid hydrides, such as SiH_4. The alkyl aluminium hydrides are used to couple alkenes.

Aluminium hydride, AlH_3, is a solid that is best regarded as saline, like the hydrides of the *s*-block metals. Unlike CaH_2 and NaH, which are more readily available commercially, AlH_3 has few applications in the laboratory. The alkylaluminium hydrides, such as $Al_2(C_2H_5)_4H_2$, are well-known molecular compounds and contain Al—H—Al $3c,2e$ bonds (Section 2.11). Hydrides of this kind are used to couple alkenes, the initial step being the addition of the AlH entity across the C=C double bond, as in hydroboration (Section 12.3). Pure Ga_2H_6 has been prepared only relatively recently but derivatives have been known for some time. The hydrides of indium and thallium are very unstable.

The metathesis of the halides with LiH leads to lithium tetrahydridoaluminate, $LiAlH_4$, or the analogous tetrahydrogallate, $LiGaH_4$:

$$4\,LiH + ECl_3 \xrightarrow{\Delta,\,ether} LiEH_4 + 3\,LiCl \quad (E = Al, Ga)$$

The direct reaction of lithium, aluminium, and hydrogen leads to the formation of either $LiAlH_4$ or Li_3AlH_6, depending on the conditions of the reaction. Their formal analogy with halide complexes such as $AlCl_4^-$ and AlF_6^{3-} should be noted.

The AlH_4^- and GaH_4^- ions are tetrahedral, and are much more hydridic than BH_4^-. The latter property is consistent with the higher electronegativity of boron compared with aluminium and gallium. For example, $NaAlH_4$ reacts violently with water but, as we saw earlier, basic aqueous solutions of $NaBH_4$ are useful in synthetic chemistry. They are also much stronger reducing agents; $LiAlH_4$ is commercially available and widely used as a strong hydride source and as a reducing agent. With the halides of many nonmetallic elements, AlH_4^- serves as a hydride source in metathesis reactions, such as the reaction of lithium tetrahydridoaluminate with silicon tetrachloride in tetrahydrofuran solution to produce silane (recall Section 9.8):

$$LiAlH_4 + SiCl_4 \xrightarrow{THF} LiAlCl_4 + SiH_4$$

The general rule in this important type of reaction is that H^- migrates from the element of lower electronegativity (Al in the example) to the element of greater electronegativity (Si).

Under conditions of controlled protolysis, both AlH_4^- and GaH_4^- lead to complexes of aluminium or gallium hydride:

$$LiEH_4 + [(CH_3)_3NH]Cl \rightarrow (CH_3)_3N—EH_3 + LiCl + H_2 \quad (E = Al, Ga)$$

In striking contrast to BH_3 complexes, these complexes will add a second molecule of base to form five-coordinate complexes of aluminium or gallium hydride :

$$(CH_3)_3N—EH_3 + N(CH_3)_3 \rightarrow ((CH_3)_3N)_2EH_3 \quad (E = Al, Ga)$$

This behaviour is consistent with the trend for Period 3 and heavier *p*-block elements to form five- and six-coordinate hypervalent compounds (Section 2.1).

12.12 Halides

All the elements form trihalides with the metal in its +3 oxidation state. However, as we expect from the inert pair effect, the +1 oxidation state becomes more common on descending the group and Tl forms stable monohalides.

(a) Trihalides

Key points: Aluminium, gallium, and indium all favour the +3 oxidation state, and their trihalides are Lewis acids

Although direct reaction of aluminium, gallium, or indium with a halogen yields a halide, these electropositive metals also react with HCl or HBr gas, and the latter is usually a more convenient route:

$$2\,Al(s) + 6\,HCl(g) \xrightarrow{100°C} 2\,AlCl_3(s) + 3\,H_2(g)$$

Because the F^- ion is so small, the trifluorides are mechanically hard ionic solids that have much higher melting points and sublimation enthalpies than the other halides. Their high lattice enthalpies also result in their having very limited solubility in most solvents, and they do not act as Lewis acids to simple donor molecules. However, despite their low reactivity towards most donors, AlF_3 and GaF_3 form salts of the type Na_3AlF_6 (cryolite) and Na_3GaF_6, which contain octahedral $[MF_6]^{3-}$ complex ions. Cryolite occurs naturally and molten synthetic cryolite is used as a solvent for bauxite in the industrial extraction of aluminium.

The heavier trihalides of aluminium, gallium, and indium are soluble in a wide variety of polar solvents and are excellent Lewis acids. The trigonal planar MX_3 monomer occurs only at elevated temperatures in the gas phase. Otherwise, they exist as M_2X_6 dimers in the vapour phase and in solution and the volatile solids are dimeric. An exception is $AlCl_3$, which has a six-coordinate layer structure in the solid phase and converts to four-coordinate molecular dimers at its melting point. The electron deficiency of the monomeric trihalides is removed to an extent by the formation of the dimers in which a coordinate M—X bond is formed involving a lone pair of electrons on the halogen (see Chapter 4, structure (**17**)), producing a tetrahedral arrangement of halogen atoms around each M atom.

The Lewis acidities of the halides reflect the relative chemical hardness of the Group 13 elements. Thus, towards a hard Lewis base (such as ethyl acetate, which is hard because of its O donor atoms), the Lewis acidities of the halides weaken as the softness of the acceptor element increases, so the Lewis acidities fall in the order:

$$BCl_3 > AlCl_3 > GaCl_3$$

By contrast, towards a soft Lewis base (such as dimethylsulfane, Me_2S, which is soft because of its S atom), the Lewis acidities strengthen as the softness of the acceptor element increases:

$$GaX_3 > AlX_3 > BX_3 \quad (X = Cl\ or\ Br)$$

Aluminium trichloride is a useful starting material for the synthesis of other aluminium compounds:

$$AlCl_3(sol) + 3\,LiR(sol) \rightarrow AlR_3(sol) + 3\,LiCl(s)$$

This reaction is an example of **transhalogenation**, which is important in the preparation of main-group organometallic compounds. In transhalogenation reactions, the halide formed is that of the more electronegative element and the high lattice enthalpy of that compound can be regarded as the 'driving force' of the reaction. The main industrial application of $AlCl_3$ is as a Friedel–Crafts catalyst in organic synthesis.

The thallium trihalides are much less stable than those of its lighter congeners. A trap for the unwary is that thallium triiodide is a compound of Tl(I) rather than Tl(III), as it contains the I_3^- ion, not I^-. This is confirmed by considering the standard potentials, which indicate that Tl(III) is rapidly reduced to Tl(I) by iodide:

$$Tl^{3+}(aq) + 2\,e^- \rightarrow Tl^+(aq) \quad E^{\ominus} = +1.26\ V$$

$$I_3^-(aq) + 2\,e^- \rightarrow 3\,I^-(aq) \quad E^{\ominus} = +0.55\ V$$

However, in excess iodide Tl(III) is stabilized by the formation of a complex:

$$TlI_3(s) + I^-(aq) \rightarrow TlI_4^-(aq)$$

In keeping with the general tendency towards higher coordination numbers for the heavier p-block elements, halides of aluminium and its heavier congeners may take on more than one Lewis base and become hypervalent:

$$AlCl_3 + N(CH_3)_3 \rightarrow Cl_3AlN(CH_3)_3$$

$$Cl_3AlN(CH_3)_3 + N(CH_3)_3 \rightarrow Cl_3Al(N(CH_3)_3)_2$$

(b) Low-oxidation-state halides

Key point: The $+1$ oxidation state becomes progressively more stable from aluminium to thallium.

The oxidation state M(I) increases in stability down the group because of the inert-pair effect. All the AlX compounds, GaF and InF are unstable, gaseous species that disproportionate in the solid phase:

$$3\ AlX(s) \rightarrow 2\ Al(s) + AlX_3(s)$$

The other monohalides of gallium, indium, and thallium are more stable. Gallium monohalides are formed by reacting GaX_3 with the metal in a 1:2 ratio:

$$GaX_3(s) + 2\ Ga(s) \rightarrow 3\ GaX(s) \quad (X = Cl, Br, or\ I)$$

The stability increases from the chloride to the iodide. The stability of the $+1$ oxidation state is increased by the formation of complexes such as $Ga[AlX_4]$. The apparently divalent $GaCl_2$ can be prepared by heating GaX_3 with gallium metal in a 2:1 ratio:

$$2\ GaX_3 + Ga \xrightarrow{\Delta} 3\ GaX_2 \quad (X = Cl, Br, or\ I)$$

The formula $GaCl_2$ is deceiving, as this solid and most other apparently divalent salts do not contain Ga(II); instead they are mixed-oxidation-state compounds containing Ga(I) and Ga(III). Mixed-oxidation-state halogen compounds are also known for the heavier metals, such as $InCl_2$ and $TlBr_2$. The presence of M^{3+} ions is indicated by the existence of MX_4^- complexes in these salts with short M—X distances, and the presence of M^+ ions is indicated by longer and less regular separation from the halide ions. There is in fact only a fine line between the formation of a mixed-oxidation-state ionic compound and the formation of a compound that contains M—M bonds. For example, mixing $GaCl_2$ with a solution of $[N(CH_3)_4]Cl$ in a nonaqueous solvent yields the compound $[N(CH_3)_4]_2[Cl_3Ga—GaCl_3]$, in which the anion has an ethane-like structure with a Ga—Ga bond.

Indium monohalides are prepared by direct interaction of the elements or by heating the metal with HgX_2. The stability increases from the chloride to the iodide and is enhanced by the formation of complexes such as $In[AlX_4]$.

Gallium(I) and indium(I) halides both disproportionate when dissolved in water:

$$3\ MX(s) \rightarrow 2\ M(s) + M^{3+}(aq) + 3\ X^-(aq) \quad (M = Ga, In; X = Cl, Br, I)$$

In contrast to the other elements in the group, Tl(I) is the most stable oxidation state of the halides. Thallium(I) is stable with respect to disproportionation in water because Tl^{3+} is difficult to achieve. Thallium(I) halides are prepared by the action of HX on an acidified solution of a soluble Tl(I) salt. Thallium(I) fluoride has a distorted rock-salt structure (Section 3.9) whereas TlCl and TlBr have the caesium-chloride structure (Section 3.9). Yellow TlI has an orthorhombic layer structure but, when pressure is applied, is converted to red TlI with a caesium-chloride structure. Thallium(I) iodide is used in photomultiplier tubes to detect ionizing radiation.

Other low-oxidation-state halides of thallium are known: TlX_2 is actually $Tl^I[Tl^{III}_3X_4]$ and Tl_2X_3 is $Tl^I_3[Tl^{III}X_6]$.

Example 12.8 Proposing reactions of Group 13 halides

Propose chemical equations (or indicate no reaction) for reactions between (a) $AlCl_3$ and $(C_2H_5)_3NGaCl_3$ in toluene, (b) $(C_2H_5)_3NGaCl_3$ and GaF_3 in toluene, and (c) TlCl and NaI in water.

Answer (a) Al(III) is a stronger and harder Lewis acid than Ga(III); therefore, the following reaction can be expected:

$$AlCl_3 + (C_2H_5)_3NGaCl_3 \rightarrow (C_2H_5)_3NAlCl_3 + GaCl_3$$

(b) No reaction, because GaF_3 has a very high lattice enthalpy and thus is not a good Lewis acid. (c) Tl(I) is chemically borderline soft, so it combines with the softer I^- ion rather than Cl^-:

$$TlCl(s) + NaI(aq) \rightarrow TlI(s) + NaCl(aq)$$

Like silver halides, Tl(I) halides have low solubility in water, so the reaction will probably proceed very slowly.

Self-test 12.8 Propose, with reasons, the chemical equation (or indicate no reaction) for reactions between (a) $(CH_3)_2SAlCl_3$ and $GaBr_3$ and (b) TlCl and formaldehyde (HCHO) in acidic aqueous solution. (*Hint*: formaldehyde is easily oxidized to CO_2 and H^+.)

12.13 Oxo compounds

Key points: Aluminium and gallium form α and β forms of the oxide in which the elements are in their +3 oxidation state; thallium forms an oxide, in which it is in its +1 oxidation state, and a peroxide.

The most stable form of Al_2O_3, α-alumina, is a very hard and refractory material. In its mineral form it is known as *corundum* and as a gemstone it is *sapphire* or *ruby, emerald*, or *amethyst*, depending on the metal ion impurities. The blue of sapphire arises from a charge-transfer transition from Fe^{2+} to Ti^{4+} ion impurities (Section 19.3). Ruby is α-alumina in which a small fraction of the Al^{3+} ions are replaced by Cr^{3+}. The structure of α-alumina and gallia, Ga_2O_3, consists of an hcp array of O^{2-} ions with the metal ions occupying two-thirds of the octahedral holes in an ordered array.

Dehydration of aluminium hydroxide at temperatures below 900°C leads to the formation of γ-alumina, which is a metastable polycrystalline form with a defect spinel structure (Section 3.9b) and a very high surface area. Partly because of its surface acid and base sites, this material is used as a solid phase in chromatography and as a heterogeneous catalyst and catalyst support (Section 25.11)

The α and γ forms of Ga_2O_3 have the same structures as their aluminium analogues. The metastable form is β-Ga_2O_3, which has a ccp structure with Ga(III) in distorted octahedral and tetrahedral sites. Half the Ga(III) ions are therefore four-coordinate despite its large radius (compared to Al(III)). This coordination may be due to the effect of the filled $3d^{10}$ shell of electrons, as remarked previously. Indium and thallium form In_2O_3. Thallium also forms the Tl(I) oxide and peroxide, Tl_2O and Tl_2O_2.

12.14 Salts of oxoacids

Key points: The elements all form a series of salts called alums.

The most important oxosalts of Group 13 are the *alums*, $MAl(SO_4)_2 \cdot 12H_2O$, where M is a univalent cation such as Na^+, K^+, Rb^+, Cs^+, Tl^+, or NH_4^+. Gallium and indium can also form analogous series of salts of this type but boron and thallium do not because of their respectively small and large size. The alums can be thought of as double salts containing the hydrated trivalent cation $[Al(H_2O)_6]^{3+}$. The remaining water molecules form hydrogen bonds between the cations and sulfate ions. The mineral *alum*, $KAl(SO_4)_2 \cdot 12H_2O$, from which aluminium takes its name, is the only common, water-soluble, aluminium-bearing mineral. It has been used since ancient times as a 'mordant' to fix dyes to textiles. The term alum is used widely to describe other compounds with the general formula $M^IM^{III}(SO_4)_2 \cdot 12H_2O$, where M^{III} is often a *d*-block metal ion, such as $KFe(SO_4)_2 \cdot 12H_2O$, ferric alum.

12.15 Sulfides

Key point: Gallium, indium, and thallium form many sulfides with a wide range of structures.

The only sulfide of aluminium is Al_2S_3, which is prepared by direct reaction of the elements at elevated temperatures:

$$2\ Al(s) + 3\ S(s) \xrightarrow{\Delta} Al_2S_3(s)$$

It is rapidly hydrolysed in aqueous solution:

$$Al_2S_3(s) + 6\ H_2O(l) \rightarrow 2\ Al(OH)_3(aq) + 3\ H_2S(g)$$

Aluminium sulfide exists in α, β, and γ forms. The structures of the α and β forms are based on the wurtzite structure (Section 3.9): in α-Al_2S_3 the S^{2-} ions are hexagonal close packed and the Al^{3+} ions occupy two-thirds of the tetrahedral sites in an ordered fashion; in β-Al_2S_3 the Al^{3+} ion occupy two-thirds of the tetrahedral sites randomly. The γ form adopts the same structure as γ-Al_2O_3.

The sulfides of gallium, indium, and thallium are more numerous and varied than those of aluminium and adopt many different structural types. Some examples are given in Table 12.5. Many of the sulfides are semiconductors, photoconductors, or light emitters and are used in electronic devices.

12.16 Compounds with Group 15 elements

Key point: Aluminium, gallium, and indium react with phosphorus, arsenic, and antimony to form materials that act as semiconductors.

The compounds formed between Group 13 and Group 15 elements (the nitrogen group) are important commercially and technologically as they are isoelectronic with silicon and germanium and act as semiconductors (Sections 14.4 and 23.18). The nitrides adopt the wurtzite structure and the phosphides, arsenides, and stibnides all adopt the zinc-blende (sphalerite) structure (Section 3.9). All the Group 13/15 binary compounds can be prepared by direct reaction of the elements at high temperature and pressure.

$$Ga(s) + As(s) \rightarrow GaAs(s)$$

The most widely used Group 13/15 (formerly, and still sometimes, Group III/V) semiconductor is gallium arsenide, GaAs, which is used to make devices such as integrated circuits, light-emitting diodes, and laser diodes. Its band gap is similar to that of silicon and larger than those of other Group 13/15 compounds (Table 12.6). Gallium arsenide is superior to silicon for such applications because it has higher electron mobility, allowing it to function at frequencies in excess of 250 GHz. GaAs devices also generate less electronic noise than silicon devices. One disadvantage of Group 13/15 semiconductors is that the compounds decompose in moist air and must be kept under an inert atmosphere, usually nitrogen, or be completely encapsulated.

Table 12.5 Selected sulfides of gallium, indium, and thallium

Sulfide	Structure
GaS	Layer structure with Ga—Ga bonds
α-Ga_2S_3	Defect wurtzite structure (hexagonal)
γ-Ga_2S_3	Defect sphalerite structure (cubic)
InS	Layer structure with In—In bonds
β-In_2S_3	Defect spinel (as γ-Al_2O_3)
TlS	Chains of edge-shared $Tl^{III}S_4$ tetrahedra
Tl_4S_3	Chains of $[Tl^{III}S_4]$ and $Tl^I[Tl^{III}S_3]$ tetrahedra

Organometallic compounds

The most important organometallic compounds of the Group 13 elements are those of boron and aluminium. Organoboron compounds are commonly treated as organometallic compounds even though boron is not a metal.

12.17 Organoborane compounds

Key points: Organoboron compounds are electron-deficient Lewis acids; tetraphenylborate is an important ion.

Organoboranes of the type BR_3 can be prepared by hydroboration of an alkane with diborane.

$$B_2H_6 + 6\ CH_2{=}CH_2 \rightarrow 2\ B(CH_2CH_3)_3$$

Alternatively, they can be produced from a Grignard reagent (Section 11.10):

$$(C_2H_5)_2O{:}BF_3 + 3\ RMgX \rightarrow BR_3 + 3\ MgXF + (C_2H_5)_2O$$

Alkylboranes are not hydrolysed but are pyrophoric. The aryl species are more stable. They are all monomeric and planar. Like other boron compounds, the organoboron species are electron deficient and consequently act as Lewis acids and form adducts easily.

An important anion is the tetraphenylborate ion, $[B(C_6H_5)_4]^-$, more commonly written BPh_4^-, analogous to the tetrahydridoborate ion, BH_4 (Section 12.3). The sodium salt can be obtained by a simple addition reaction:

$$BPh_3 + NaPh \rightarrow Na^+[BPh_4]^-$$

The sodium salt is soluble in water but the salts of most large, monopositive ions are insoluble. Consequently, the anion is useful as a precipitating agent and can be used in gravimetric analysis.

12.18 Organoaluminium compounds

Key points: Methyl- and ethylaluminium are dimers; bulky alkyl groups result in monomeric species.

Alkylaluminium compounds can be prepared on a laboratory scale by transmetallation of a mercury compound:

$$2\ Al + 3\ Hg(CH_3)_2 \rightarrow Al_2(CH_3)_6 + 3\ Hg$$

Trimethylaluminium is prepared commercially by the reaction of aluminium metal with chloromethane to give $Al_2Cl_2(CH_3)_4$. This intermediate is then reduced with sodium and the $Al_2(CH_3)_6$ (**34**) is removed by fractional distillation.

Alkylaluminium dimers are similar in structure to the analogous dimeric halides (Section 12.12) but the bonding is different. In the halides, the bridging Al—Cl—Al bonds are $2c,2e$ bonds; that is, each Al—Cl bond involves an electron pair. In the alkylaluminium dimers the Al—C—Al bonds are longer than the terminal Al—C bonds, which suggests that they are $3c,2e$ bonds, with one bonding pair shared across the Al—C—Al unit, somewhat analogous to the bonding in diborane, B_2H_6 (Section 12.3).

Triethylaluminium and higher alkyl compounds are prepared from the metal, an appropriate alkene and hydrogen gas at elevated temperatures and pressures.

$$2\ Al + 3\ H_2 + 6\ CH_2{=}CH_2 \xrightarrow{60-110^\circ C,\ 10-20\text{MPa}} Al_2(CH_2CH_3)_6$$

This route is relatively cost-effective and, as a result, alkylaluminium compounds have found many commercial applications. Triethylaluminium, often written as the monomer, $Al(C_2H_5)_3$, is an organometallic complex of aluminium of major industrial importance. It is used in the Ziegler–Natta polymerization catalyst (Section 25.17).

Steric factors have a powerful effect on the structures of alkylaluminiums. Where dimers are formed, the long weak bridging bonds are easily broken. This tendency

Table 12.6 Band gaps at 298 K	
	E_g/eV
GaAs	1.35
GaSb	0.67
InAs	0.36
InSb	0.163
Si	1.107

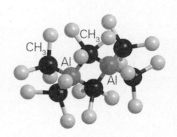

34 $Al_2(CH_3)_6$

increases with the bulkiness of the ligand. So, for example, triphenylaluminium is a dimer but trimesitylaluminium (where trimesityl is $2,4,6\text{-}(CH_3)_3C_6H_2\text{-}$) is a monomer.

FURTHER READING

R.B. King, *Inorganic chemistry of the main group elements.* Wiley (1994).

D.M.P. Mingos, *Essential trends in inorganic chemistry.* Oxford University Press (1998). A survey of inorganic chemistry from the perspective of structure and bonding.

N.C. Norman, *Periodicity and the s- and p-block elements.* Oxford University Press (1997). Includes coverage of essential trends and features of *s*-block chemistry.

R.B. King (ed.), *Encyclopedia of inorganic chemistry.* Wiley (2005)

C.E. Housecroft, *Boranes and metalloboranes.* Ellis Horwood, Chichester (2005). An introduction to borane chemistry.

EXERCISES

12.1 Give a balanced chemical equation and conditions for the recovery of boron.

12.2 Describe the bonding in (a) BF_3, (b) $AlCl_3$, (c) B_2H_6.

12.3 Arrange the following in order of increasing Lewis acidity: BF_3, BCl_3, $AlCl_3$. In the light of this order, write balanced chemical reactions (or no reaction) for (a) $BF_3N(CH_3)_3 + BCl_3 \rightarrow$, (b) $BH_3CO + BBr_3 \rightarrow$.

12.4 Thallium tribromide (1.11 g) reacts quantitatively with 0.257 g of NaBr to form a product A. Deduce the formula of A. Identify the cation and anion.

12.5 Identify compounds A, B, and C.

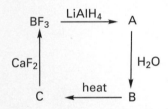

12.6 Does B_2H_6 survive in air? If not, write the equation for the reaction.

12.7 Using BCl_3 as a starting material and other reagents of your choice, devise a synthesis for the Lewis acid chelating agent, F_2B—C_2H_4—BF_2.

12.8 Given $NaBH_4$, a hydrocarbon of your choice, and appropriate ancillary reagents and solvents, give formulas and conditions for the synthesis of (a) $B(C_2H_5)_3$, (b) Et_3NBH_3.

12.9 Draw the B_{12} unit that is a common motif of boron structures; take a viewpoint along a C_2 axis.

12.10 Which boron hydride would you expect to be more thermally stable, B_6H_{10} or B_6H_{12}? Give a generalization by which the thermal stability of a borane can be judged.

12.11 (a) Give a balanced chemical equation (including the state of each reactant and product) for the air oxidation of pentaborane(9). (b) Describe the probable disadvantages, other than cost, for the use of pentaborane as a fuel for an internal combustion engine.

12.12 (a) From its formula, classify $B_{10}H_{14}$ as *closo*, *nido*, or *arachno*. (b) Use Wade's rules to determine the number of framework electron pairs for decaborane(14). (c) Verify by detailed accounting of valence electrons that the number of cluster valence electrons of $B_{10}H_{14}$ is the same as that determined in (b).

12.13 Starting with $B_{10}H_{14}$ and other reagents of your choice, give the equations for the synthesis of $[Fe(nido\text{-}B_9C_2H_{11})_2]^{2-}$, and sketch the structure of this species.

12.14 (a) What are the similarities and differences in structure of layered BN and graphite (Section 13.3)? (b) Contrast their reactivity with Na and Br_2. (c) Suggest a rationalization for the differences in structure and reactivity.

12.15 Devise a synthesis for the borazines (a) $Ph_3N_3B_3Cl_3$ and (b) $Me_3N_3B_3H_3$, starting with BCl_3 and other reagents of your choice. Draw the structures of the products.

12.16 Give the structures and names of B_4H_{10}, B_5H_9, and $1,2\text{-}B_{10}C_2H_{12}$.

12.17 Arrange the following boron hydrides in order of increasing Brønsted acidity, and draw a structure for the probable structure of the deprotonated form of one of them: B_2H_6, $B_{10}H_{14}$, B_5H_9.

PROBLEMS

12.1 Borane exists as the molecule B_2H_6 and trimethylborane exists as a monomer $B(CH_3)_3$. In addition, the molecular formulas of the compounds of intermediate compositions are observed to be $B_2H_5(CH_3)$, $B_2H_4(CH_3)_2$, $B_2H_3(CH_3)_3$, and $B_2H_2(CH_3)_4$. Based on these facts, describe the probable structures and bonding in the latter series.

12.2 ^{11}B NMR is an excellent spectroscopic tool for inferring the structures of boron compounds. With ^{11}B—^{11}B coupling ignored, it is possible to determine the number of attached H atoms by the multiplicity

of a resonance: BH gives a doublet, BH_2 a triplet, and BH_3 a quartet. B atoms on the closed side of *nido* and *arachno* clusters are generally more shielded than those on the open face. Assuming no B—B or B—H—B coupling, predict the general pattern of the ^{11}B NMR spectra of (a) BH_3CO, (b) $[B_{12}H_{12}]^{2-}$, and (c) B_4H_{10}.

12.3 Identify the incorrect statements in the following description of Group 13 chemistry and provide corrections along with explanations of principle or chemical generalization that applies. (a) All the elements in Group 13 are nonmetals. (b) The increase in chemical hardness on going down the group is illustrated by greater oxophilicity and fluorophilicity for the heavier elements. (c) The Lewis acidity increases for BX_3 from $X = F$ to Br and this may be explained by stronger Br—B π bonding. (d) *Arachno*-boron hydrides have a $2(n + 3)$ skeletal electron count and they are more stable than *nido*-boron hydrides. (e) In a series of *nido*-boron hydrides acidity increases with increasing size. (f) Layered boron nitride is similar in structure to graphite and because it has a small separation between HOMO and LUMO, it is a good electrical conductor.

12.4 Use suitable molecular orbital software to calculate the wavefunctions and energy levels for *closo*-$[B_6H_6]^{2-}$. From that output, draw a molecular orbital energy diagram for the orbitals primarily involved in B—B bonding and sketch the form of the orbitals. How do these orbitals compare qualitatively with the qualitative description for this anion in this chapter? Is B—H bonding neatly separated from the B—B bonding in the computed wavefunctions?

12.5 In his paper 'Covalent and ionic molecules: why are BeF_2 and AlF_3 high melting point solids whereas BF_3 and SiF_4 are gases?'(*J. Chem. Educ.*, 1998, **75** 923), R.J. Gillespie makes a case for the classification of the bonding in BF_3 and SiF_4 as predominantly ionic. Summarize his arguments and describe how these differ from the conventional view of bonding in gaseous molecules.

12.6 Nanotubes of C, N, and B have been synthesized by C. Colliex and co-workers (*Science*, 1997, **278**, 653). (a) What are the advantageous properties of these nanotubes over carbon analogues? (b) Outline the method used for the preparation of these compounds. (c) What was the main structural feature of the nanotubes and how could this be exploited in applications?

12.7 M. Montiverde discusses 'Pressure dependence of the superconducting temperature of MgB_2' (*Science*, 2001, **292**, 75). (a) Describe the theoretical bases of the two theories that have been postulated to explain the superconductivity of MgB_2. (b) How does the T_c of MgB_2 vary with pressure? What insight does this provide into the superconductivity?

12.8 Use the references in Z.W. Pan, Z.R. Dai, and Z.L. Wang (*Science*, 2001, **291**, 1947) as a starting point to write a review of wire-like nanomaterials of Group 13 elements. Indicate how In_2O_3 nanobelts were prepared and give the dimensions of a typical nanobelt.

The Group 14 elements

13

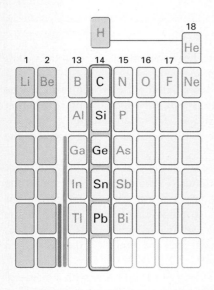

The elements of Group 14 are arguably the most important of all, carbon providing the basis for life on Earth and silicon being vital for the physical structure of the natural environment in the form of crustal rocks. The elements of this group exhibit great diversity in their properties, ranging as they do from the nonmetallic carbon to the well known metals tin and lead. All the elements form binary compounds with other elements. In addition, silicon forms a diverse range of network solids. Many of the organocompounds of the Group 14 elements are commercially important.

The elements of Group 14, carbon, silicon, germanium, tin, and lead, show considerable diversity in their chemical and physical properties. Carbon, of course, is the building block of life and central to organic chemistry. In this chapter, our focus with carbon is on its *inorganic* chemistry. Silicon is widely distributed in the natural environment and tin and lead find widespread applications in industry and manufacturing.

The elements

Key points: The lightest elements of the group are nonmetals; tin and lead are metals. All the elements except lead exist as several allotropes.

The elements of Group 14 (the carbon group) exhibit interesting and diverse physical and chemical properties as well as being of fundamental importance in industry and nature. Carbon forms many binary compounds with metals and nonmetals and a rich range of organometallic compounds. We discuss carbon in many contexts throughout this text, including organometallic compounds in Chapter 21 and catalysis in Chapter 25. In combination with oxygen and aluminium, silicon is a dominant component of minerals in the Earth's crust, just as carbon in combination with hydrogen and oxygen is dominant in the biosphere. Silicon and germanium are vital to modern high technology, particularly as semiconductors and optical fibres.

The lightest members of the group, carbon and silicon, are nonmetals, germanium is a metalloid, and tin and lead are metals. This increase in metallic properties on descending a group is a striking feature of the *p* block and can be understood in terms of the increasing ionic radius and associated decrease in ionization energy down the group

Table 13.1 Selected properties of the Group 14 elements

	C	Si	Ge	Sn	Pb
Melting point/°C	3730 (graphite sublimes)	1410	937	232	327
Atomic radius/pm	77	117	122	162	175
Ionic radius, $r(M^{n+})$/pm			93 (+2)	112 (+2)	120 (+2)
			53 (+4)	71 (+4)	84 (+4)
First ionization energy, I/(kJ mol^{-1})	1090	786	762	707	716
Pauling electronegativity	2.5	1.8	1.8	1.8	1.8
Electron affinity, E_a/(kJ mol^{-1})	122	134	116	116	
$E^{\circ}(M^{4+}, M^{2+})$/V				+0.15	+1.69
$E^{\circ}(M^{2+}, M)$/V				−0.14	−0.13

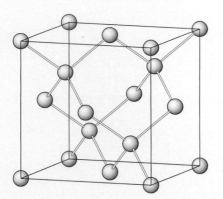

Fig. 13.1 The cubic diamond structure.

(Table 13.1). Because the ionization energies of the heavier elements are low, the metals form cations increasingly readily down the group.

As the configuration ns^2np^2 suggests, the +4 oxidation state is dominant in the compounds of the elements. The major exception is lead, for which the most common oxidation state is +2, two less than the group maximum. This relative stability of the low oxidation state is an example of the inert-pair effect, which is such a striking feature of the *p*-block elements.

The electronegativities of carbon and silicon are similar to that of hydrogen and they form many covalent hydrogen and alkyl compounds. Carbon and silicon are strong **oxophiles** and **fluorophiles**, in the sense that they have high affinities for the hard anions O^{2-} and F^-, respectively (Section 4.10). Their oxophilic character is evident in the existence of an extensive series of oxoanions, the carbonates and silicates. By contrast, lead forms more stable compounds with soft anions, such as I^- and S^{2-}, than with hard anions, and is therefore classified as chemically soft.

All the elements of the group except lead have at least one solid phase with a diamond structure (Fig. 13.1). The cubic phase of tin, which is called *grey tin* or α-*tin* (α-Sn), is not stable at room temperature. It converts to the more stable, common phase, *white tin* or β-*tin* (β-Sn). An atom of tin in the white allotrope has six nearest neighbours in a highly distorted octahedral array. When white tin is cooled to 13.2°C, it converts to grey tin. The effects of this transformation were first recognized on organ pipes in medieval European cathedrals, where it was believed to be due to the Devil's work. Legend has it that Napoleon's armies were defeated in Russia because, as the temperature fell, the white tin buttons on the soldiers' uniforms were converted to grey tin, which then crumbled away.

The gap between the valence and conduction bands (Section 3.17) decreases steadily from diamond, which is commonly regarded as an insulator, to grey tin, which behaves like a metal below its transition temperature.

13.1 Occurrence and recovery

Key points: Elemental carbon is mined as graphite and diamond; elemental silicon is recovered from SiO_2 by carbon-arc reduction; the much less abundant germanium is found in zinc ores.

Two reasonably pure forms of carbon, *diamond* and *graphite*, are mined. There are many less pure forms, such as *coke*, which is made by the pyrolysis of coal, and *lamp black*, which is the product of incomplete combustion of hydrocarbons. The 1996 Nobel Prize in Chemistry was awarded to Richard Smalley, Robert Curl, and Harold Kroto for the discovery of a new allotrope of carbon, C_{60}, named buckminsterfullerene after the geodesic domes designed by the architect Buckminster Fuller (see Section 13.4). Carbon occurs as carbon dioxide in the atmosphere and dissolved in natural waters and as the insoluble carbonates of calcium and magnesium.

Silicon occurs widely distributed in the natural environment and makes up 26 per cent by mass of the Earth's crust. It occurs as sand, quartz, amethyst, agate, and opal, and is also found in asbestos, feldspar, clays, and micas. Elemental silicon is produced from silica, SiO_2, by high-temperature reduction with carbon in an electric arc furnace:

$$SiO_2(s) + 2\ C(s) \rightarrow Si(s) + 2\ CO(g)$$

Germanium is low in abundance and generally not concentrated in nature. It is obtained by the reduction of GeO_2 with carbon monoxide or hydrogen. Tin is produced by the reduction of the mineral cassiterite, SnO_2, with coke in an electric furnace. Lead is obtained from its sulfide ores, which are converted to oxide and reduced by carbon in a blast furnace.

13.2 Uses

Key point: Elemental silicon and, to a lesser extent, germanium are produced for the manufacture of semiconductor devices.

Elemental carbon in the form of coal or coke is used as a fuel and reducing agent in the recovery of metals from their ores. Graphite is used as a lubricant and in pencils, and diamond is used in industrial cutting tools. The band gap and consequent semi-conductivity of silicon leads to its many applications in integrated circuits, computer chips, solar cells, and other electronic solid-state devices. Silica is the major raw material used to make glass. Germanium was the first widely used material for the construction of transistors because it was easier to purify than silicon and, having a smaller band gap than silicon (0.72 eV for Ge, 1.11 eV for Si), is a better intrinsic semiconductor.

Tin is resistant to corrosion and is used to plate steel for use in tin cans. Bronze is an alloy of tin and copper that typically contains less than 12 per cent by mass of tin; bronze with higher tin content is used to make bells. Solder is an alloy of tin and lead, and has been in use since Roman times. Window glass or float glass is made by floating molten glass on the surface of molten tin. The 'tin side' of window glass can be seen as a haze of tin(IV) oxide when viewed with ultraviolet radiation. Trialkyl and triaryltin compounds are in widespread use as fungicides and biocides.

The softness and malleability of lead has resulted in its use in plumbing, although this application is now illegal in many countries due to concerns over lead poisoning. Its low melting point leads to its use in solder and its high density ($11.34\ \mathrm{g\ cm}^{-3}$) leads to its use in ammunition and as shielding from ionizing radiation. Lead oxide is added to glass to raise its refractive index and form 'lead' or 'crystal' glass.

13.3 Diamond and graphite

Key points: Graphite consists of stacked two-dimensional carbon sheets; oxidizing agents or reducing agents may be intercalated between these sheets with concomitant electron transfer.

Diamond and graphite, the two common crystalline forms of elemental carbon, are strikingly different. Diamond is effectively an electrical insulator; graphite is a good conductor. Diamond is the hardest known natural substance and hence the ultimate abrasive; impure (partially oxidized) graphite is slippery and frequently used as a lubricant. Because of its durability, clarity, and high refractive index, diamond is one of the most highly prized gemstones; graphite is soft and black with a slightly metallic lustre and is neither durable nor particularly attractive. The origin of these widely different physical properties can be traced to the very different structures and bonding in the two polymorphs.

In diamond, each C atom forms single bonds of length 154 nm with four adjacent C atoms at the corners of a regular tetrahedron (Fig. 13.1); the result is a rigid, covalent, three-dimensional framework. Diamond has the highest known thermal conductivity because this structure distributes thermal motion in three dimensions very efficiently. Measurement of thermal conductivity is used to identify fake diamonds. Graphite

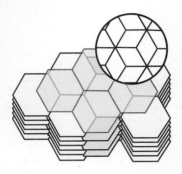

Fig. 13.2 The structure of graphite. The rings are in register in alternate planes, not adjacent planes.

consists of stacks of planar layers within which each C atom has three nearest neighbours at 142 nm (Fig. 13.2). The σ bonds between neighbours within the sheets are formed from the overlap of sp^2 hybrid orbitals, and the remaining perpendicular p orbitals overlap to form π bonds that are delocalized over the plane. The planes are widely separated from each other (at 335 nm), indicating that there are weaker forces between them. These forces are sometimes, but not very appropriately, called 'van der Waals forces' (because in the common impure form of graphite, graphitic oxide, they are weak, like intermolecular forces), and consequently the region between the planes is called the **van der Waals gap**. The ready cleavage of graphite parallel to the planes of atoms (which is largely due to the presence of impurities) accounts for its slipperiness. Diamond can be cleaved, but this ancient craft requires considerable expertise as the forces in the crystal are more symmetrical.

The conversion of diamond to graphite at room temperature and pressure is spontaneous ($\Delta_{trs}G^\ominus = -2.90\,\text{kJ mol}^{-1}$) but does not occur at an observable rate under ordinary conditions, and diamonds older than the solar system have been isolated from meteorites. Diamond is the denser phase, so it is favoured by high pressures, and large quantities of diamond abrasive are manufactured commercially by a d-metal-catalysed high-temperature, high-pressure process (Box 13.1).

The electrical conductivity and many of the chemical properties of graphite are closely related to the structure of its delocalized π bonds. Its electrical conductivity perpendicular to the planes is low ($5\,\text{S cm}^{-1}$ at 25°C) and increases with increasing temperature, signifying that graphite is a semiconductor in that direction. The electrical conductivity is much higher parallel to the planes ($30\,\text{kS cm}^{-1}$ at 25°C) but decreases as the temperature

---

Box 13.1 Synthetic diamonds

There has been interest in the synthesis of diamonds since Antoine Lavoisier discovered that they were composed entirely of carbon. The first reported synthesis was in 1880 by J.B. Henney. He claimed that he had made diamond from hydrocarbons, bone oil, and lithium but his results were never reproduced.

Diamonds were synthesized in 1955 after many failed attempts, using graphite and a d metal raised to 1500–2000 K and 7 GPa. The graphite and metal must both be molten for diamond to be produced, so the temperature of synthesis depends on the melting point of the metal. The d metal (typically nickel) dissolves the graphite and the less soluble diamond phase crystallizes from it. The size, shape, and colour of the diamonds depend upon the conditions. Low-temperature synthesis produces dark, impure crystals. High-temperature synthesis produces paler, purer crystals. Common impurities are species that can be accommodated into the diamond lattice with minimum distortion. The diamonds are often contaminated with graphite or the metal catalyst. For example, the lattice dimensions of nickel are similar to those of diamond and crystallites of nickel may be included in the diamond lattice.

The diamond crystals can be grown by seeding with small diamond crystals but the new growth is often uneven with gaps and inclusions. Better quality diamonds are formed when the source of carbon is diamond and the seed crystals are in a cooler part of the apparatus. The difference in the solubility with the change in temperature causes the diamond to crystallize in a slow, controlled way giving high quality diamonds. Diamonds up to 1 carat (200 mg) may take up to a week to cystallize this way.

Diamonds can be synthesized directly from graphite without a metal catalyst if the temperature and pressure are high enough. The shock synthesis method (the *Du Pont method*) exposes graphite to the intense pressure generated by a charge of high explosive. The graphite reaches a temperature of 1000 K and a pressure of 30 GPa for a few milliseconds and some of it is converted to diamond. The *static pressure method* heats graphite in high-pressure equipment by the discharge from a capacitor. Polycrystalline lumps of diamond are formed at 3300–4500 K and 13 GPa. Hydrocarbons may also be used as the carbon source in this method. Aromatic compounds such as naphthalene and anthracene produce graphite but aliphatic compounds such as paraffin wax and camphor produce diamond.

Because the high-pressure synthesis of diamond is costly and cumbersome, a low-pressure process would be highly attractive. It has in fact been known for a long time that microscopic diamond crystals mixed with graphite can be formed by depositing C atoms on a hot surface in the absence of air. The C atoms are produced by the pyrolysis of methane, and the atomic hydrogen also produced in the pyrolysis plays an important role in favouring diamond over graphite. One property of the atomic hydrogen is that it reacts more rapidly with the graphite than with diamond to produce volatile hydrocarbons, so the unwanted graphite is swept away. Although the process is not fully perfected, synthetic diamond films are already finding applications ranging from the hardening of surfaces subjected to wear, such as cutting tools and drills, to the construction of electronic devices. For example, boron-doped diamond films are very conducting and are used as electrodes in electrochemistry.

A promising new method of synthesis that is environmentally more friendly and cheaper than any of the high-temperature and high-pressure methods uses silicon carbide. Carbon is extracted as diamond under Cl_2 and H_2 gases at ambient pressure and the relatively low temperature of 1300 K.

is raised, indicating that graphite behaves as a metal, more precisely a semimetal,[1] in that direction. This effect is most striking in pyrolytic graphite, which is manufactured by the decomposition of a hydrocarbon gas at high temperature in a vacuum furnace. The resulting graphite is of very high purity with desirable mechanical, thermal, and electrical properties. Pyrolytic graphite is used in ion beam grids, thermal insulators, rocket nozzles, and heater elements.

Graphite can act as either an electron donor or an electron acceptor towards atoms and ions that penetrate between its sheets and give rise to an **intercalation compound**. Thus, K atoms reduce graphite by donating their valence electron to the empty orbitals of the π band and the resulting K^+ ions penetrate between the layers (Section 13.11a). The electrons added to the band are mobile, and therefore alkali metal graphite intercalates have high electrical conductivity. The stoichiometry of the compound depends on the quantity of potassium and the reaction conditions. The different stoichiometries are associated with an interesting series of structures, where the alkali metal ion may insert between neighbouring layers of carbon atoms, every other layer, and so on in a process known as *staging* (Fig. 13.3).

An example of an oxidation of graphite by removal of electrons from the π band is the formation of **graphite bisulfates** by heating graphite with a mixture of sulfuric and nitric acids. In this reaction, electrons are removed from the π band, and HSO_4^- ions penetrate between the sheets to give substances of approximate formula $(C_{24})^+SO_3(OH)^-$. In this oxidative intercalation reaction, the removal of electrons from the full π band leads to a higher conductivity than that of pure graphite. This process is analogous to the formation of p-type silicon by electron-accepting dopants (Section 3.18). When graphite bisulfates are treated with water, the layers are disrupted. When the water is subsequently removed at high temperatures, a highly flexible form of graphite is formed; this *graphite tape* is used to make sealing gaskets, valves, and brake linings.

The halogens show an alternation effect in their tendency to form intercalation compounds with graphite. Graphite reacts with fluorine to produce 'graphite fluoride', a nonstoichiometric species with formula $(CF)_n$ ($0.59 < n < 1$). This compound is black when n is low in its range and colourless when n approaches 1. It is used as a lubricant in high vacuum applications and as the cathode in lithium batteries. At elevated temperatures the products of the reaction also include C_2F and C_4F. Chlorine reacts slowly with graphite to form C_8Cl and iodine does not react at all. By contrast, bromine intercalates readily to give C_8Br, $C_{16}Br$, and $C_{20}Br$ in another example of staging.

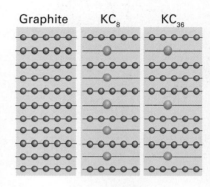

Fig. 13.3 Potassium graphite compounds showing two types of alternation of intercalated atoms.

13.4 Other forms of carbon

Diamond and graphite are not the only allotropes of carbon. The fullerenes (known informally as 'buckyballs') were discovered in the 1980s and have given rise to a new field within the inorganic chemistry of carbon.

(a) Carbon clusters

Key point: Fullerenes are formed when an electric arc is discharged between carbon electrodes in an inert atmosphere.

Metal and nonmetal cluster compounds have been known for decades, but the discovery of the soccer-ball shaped C_{60} cluster in the 1980s created great excitement in the scientific community and in the popular press. Much of this interest undoubtedly stemmed from the fact that carbon is a common element and there had seemed little likelihood that new molecular carbon structures would be found.

When an electric arc is struck between carbon electrodes in an inert atmosphere, a large quantity of soot is formed together with significant quantities of C_{60} and much smaller quantities of related **fullerenes** such as C_{70}, C_{76}, and C_{84}. The fullerenes can be dissolved in a hydrocarbon or halogenated hydrocarbon and separated by chromatography on an alumina column. The structure of C_{60} has been determined by X-ray

---

[1] A semimetal, remember (Section 3.17), is a substance in which two neighbouring bands have zero density of states at their edge but zero band gap between them.

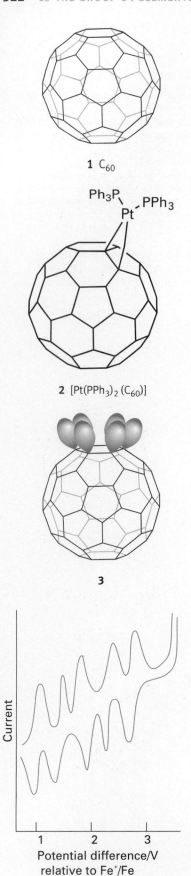

1 C_{60}

2 $[Pt(PPh_3)_2 (C_{60})]$

3

Fig. 13.5 The cyclic voltammogram of C_{60} in toluene at 25°C.

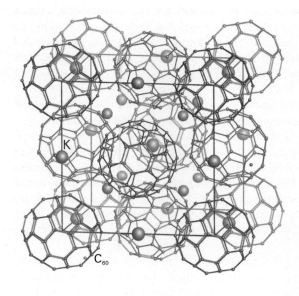

Fig. 13.4 The fcc structure of K_3C_{60}. Only a fragment of the complete unit cell is shown. The full cell is face-centred cubic. (The structure of solid C_{60} itself is shown in Fig. 3.13.)

crystallography on the solid at low temperature and electron diffraction in the gas phase. The molecule consists of five- and six-membered carbon rings, and the overall symmetry is icosahedral in the gas phase (**1**).

Fullerenes can be reduced to form [60]fulleride salts, C_{60}^{n-} ($n = 1$ to 12). Fullerides of alkali metals are solids having compositions such as K_3C_{60}. The structure of K_3C_{60} consists of a face-centred cubic array of C_{60} ions in which K^+ ions occupy the one octahedral and two tetrahedral sites available to each C_{60} ion (Fig. 13.4). The compound is metallic at room temperature and a superconductor below 18 K. Other superconducting salts include Rb_2CsC_{60}, which has a superconducting transition temperature (T_c) of 33 K, and Cs_3C_{60}, with $T_c = 40$ K. The superconductivity of E_3C_{60} compounds can be explained by considering that the conduction electrons are donated to the C_{60} molecules and are mobile because of overlapping C_{60} molecular orbitals (Section 23.19).

Some of the most interesting consequences of fullerene research have been the identification of *carbon nanotubes*. Nanotubes consist of one or more concentric cylindrical tubes formed of sheets of hexagonal arrangements of carbon atoms in which the ends may be closed by fullerene-like caps containing six five-membered rings of carbon atoms. The preparation of nanotubes has stimulated much research and the compounds could ultimately find a wide range of practical applications such as hydrogen storage and catalysis. They are treated more fully in Chapter 24.

(b) Fullerene–metal complexes

Key point: The polyhedral fullerenes undergo reversible multielectron reduction and form complexes with *d*-metal organometallic compounds and with OsO_4.

Reasonably efficient methods of synthesizing the fullerenes have been developed, and their redox and coordination chemistry is receiving serious investigation. In keeping with the formation of alkali metal fullerides described above, C_{60} undergoes five electrochemically reversible reduction and oxidation steps in nonaqueous solvents (Fig. 13.5). These observations suggest that the fullerenes might serve as either electrophiles or nucleophiles when paired with the appropriate metal.

One realization of this expectation is the attack of electron-rich platinum(0) phosphine complexes on C_{60}, yielding compounds such as (**2**), in which the Pt atom spans a pair of C atoms in the fullerene molecule. This reaction is analogous to the coordination of double bonds to platinum-phosphine complexes. Analogy with the η^6-benzenechromium complexes (Section 21.19) suggests that a metal atom might coordinate to a sixfold face of C_{60}. That such hexahapto complexes do not in fact form is attributed to the radial arrangement of the $C2p\pi$ orbitals (**3**) that results in poor overlap with *d* orbitals of a metal atom centred above a sixfold face of the molecule.

In contrast to the poor interaction of a fullerene sixfold face with a single metal atom, a larger array of metal atoms, the triruthenium cluster $Ru_3(CO)_{12}$, reacts to form a $Ru_3(CO)_9$ cap on a sixfold face of C_{60}. In the process, three CO ligands are displaced (**4**). The relatively large triangle of three metal atoms provides a favourable geometry for overlap with radially orientated $C2p\pi$ orbitals.

The chemical properties of C_{60} are not limited to its interaction with electron-rich metal complexes. Reaction with a strong electrophile and oxidant, OsO_4 in pyridine, yields an oxo bridge complex analogous to the adducts of OsO_4 with alkenes (**5**).

In addition to complexes formed with the metal atom outside the fullerene cage, **endohedral** fullerenes are formed in which one or more atoms are accommodated inside the C_{60} shell. Such complexes are denoted $M@C_{60}$, indicating that the M atom is inside the C_{60} cage. Small inert gas atoms and molecules may be driven inside the cage at high temperatures ($>600°C$) and pressures ($>2000\,atm$) to give, for example, $H_3@C_{60}$. Alternatively, the carbon cage can be formed around the endohedral atom by using a metal-doped carbon rod in an electric arc. Larger shells are often formed such as $La@C_{82}$ and $La_3@C_{106}$.

(c) Partially crystalline carbon

Key points: Amorphous and partially crystalline carbon in the form of small particles are used on a large scale as adsorbents and as strengthening agents for rubber; carbon fibres impart strength to polymeric materials.

There are many forms of carbon that have a low degree of crystallinity. These partially crystalline materials have considerable commercial importance; they include *carbon black*, *activated carbon*, and *carbon fibres*. Because single crystals suitable for complete X-ray analyses of these materials are not available, their structures are uncertain. However, what information there is suggests that their structures are similar to that of graphite, but the degree of crystallinity and shapes of the particles differ.

'Carbon black' is a very finely divided form of carbon. It is prepared (on a scale that exceeds 8 Mt annually) by the combustion of hydrocarbons under oxygen-deficient conditions. Planar stacks, like those of graphite, and multilayer balls, reminiscent of the fullerenes, have both been proposed for its structure (Fig. 13.6). Carbon black is used on a huge scale as a pigment, in printer's ink (as on this page), and as a filler for rubber goods, including tyres, where it greatly improves the strength and wear resistance of the rubber and helps to protect it from degradation by sunlight.

'Activated carbon' is prepared from the controlled pyrolysis of organic material, including coconut shells. It has a high surface area (in some cases exceeding $1000\,m^2\,g^{-1}$), that arises from the small particle size. It is therefore a very efficient adsorbent for molecules, including organic pollutants from drinking water, noxious gases from the air, and impurities from reaction mixtures. There is evidence that the parts of the surface defined by the edges of the hexagonal sheets are covered with oxidation products, including carboxyl and hydroxyl groups (**6**). This structure may account for some of its surface activity.

Carbon fibres are made by the controlled pyrolysis of asphalt fibres or synthetic fibres and are incorporated into a variety of high-strength plastic products, such as tennis rackets and aircraft components. Their structure bears a resemblance to that of graphite, but in place of the extended sheets the layers consist of ribbons parallel to the axis of the fibre. The strong in-plane bonds (which resemble those in graphite) give the fibre its very high tensile strength.

4 $[Ru_3(CO)_9(C_{60})]$

5 $[O_5(O)_2(py)_2(OC_{60}O)]$

6

Example 13.1 Comparing bonding in diamond and boron

Each B atom in elemental boron is bonded to five other B atoms but each C atom in diamond is bonded to four nearest neighbours. Suggest an explanation of this difference.

Answer The B and C atoms both have four orbitals available for bonding (one *s* and three *p*). However, a C atom has four valence electrons, one for each orbital, and it can therefore use all its electrons and orbitals in forming $2c,2e$ bonds with the four neighbouring C atoms.

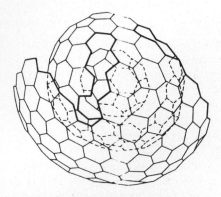

Fig. 13.6 A proposed structure for a soot particle resulting from imperfect closure of a curved C-atom network. Graphite-like structures have also been proposed.

By contrast, B has one less electron; therefore, to use all its orbitals, it forms $3c,2e$ bonds. The formation of these three-centre bonds brings another B atom into binding distance.

Self-test 13.1 Describe how the electronic structure of graphite is altered when it reacts with (a) potassium, (b) bromine.

Simple compounds

All the Group 14 elements form simple binary compounds with hydrogen, oxygen, the halogens, and nitrogen. Carbon and silicon also form carbides and silicides with metals.

13.5 Hydrides

All the Group 14 elements form a tetravalent hydride, EH_4. In addition, carbon and silicon form series of catenated molecular hydrides.

(a) Hydrocarbons

Key point: The stability of catenated hydrocarbons can be attributed to high C—C and C—H bond enthalpies.

Carbon forms an enormous range of hydrocarbon compounds that are best regarded from the viewpoint of organic chemistry. However, the simple hydrocarbons are discussed here briefly to allow comparisons with hydrides of the remainder of the group.

Carbon forms a series of simple hydrocarbons, alkanes, having the general formula C_nH_{2n+2}. The stability of the long-chain, catenated hydrocarbons can be attributed to the strong C—C and C—H bond enthalpies (Table 13.2). Carbon also forms strong multiple bonds (Table 13.2) in the unsaturated alkenes and alkynes. The stability of the C—C bond and the ability to form multiple bonds are largely responsible for the diversity and stability of carbon compounds.

Methane, CH_4, is the simplest hydrocarbon. It is an odourless, flammable gas. It is found in large natural underground deposits, from where it is extracted as natural gas and used as domestic and industrial fuel:

$$CH_4(g) + 2\,O_2(g) \rightarrow CO_2(g) + 2\,H_2O \qquad \Delta_{comb}H^{\ominus} = -882\,kJ\,mol^{-1}$$

Apart from this combustion reaction, methane is not very reactive. It is not hydrolysed by water (Box 13.2) and reacts with halogens only when exposed to ultraviolet radiation:

$$CH_4(g) + Cl_2(g) \xrightarrow{h\nu} CH_3Cl(g) + HCl(g)$$

The alkanes up to butane, C_4H_{10}, are gases, those containing from 5 to 17 carbon atoms are liquids, and the heavier hydrocarbons are solids.

(b) Silanes

Key points: Silane is a reducing agent; with a platinum complex as a catalyst it adds across carbon–carbon double bonds (hydrosilylation) and forms $Si(OR)_4$ with alcohols.

Table 13.2 Selected mean bond enthalpies, $B(X—Y)/(kJ\,mol^{-1})$

C—H	412	Si—H	318	Ge—H	288	Sn—H	250
C—O	360	Si—O	466	Ge—O	350		
C=O	743	Si=O	642				
C—C	348	Si—Si	326	Ge—Ge	186	Sn—Sn	150
C=C	612						
C≡C	837						
C—F	486	Si—F	584	Ge—F	466		
C—Cl	322	Si—Cl	390	Ge—Cl	344	Sn—Cl	320

Box 13.2 Methane clathrates

Methane clathrates are crystalline solids formed at low temperatures when ice crystallizes around methane molecules. Clathrates are also referred to as *methane hydrates* or *natural gas hydrates* and their formation has caused major problems in the past by clogging gas pipelines in cold climates. The hydrates may contain other small, gaseous molecules such as ethane and propene. Several different clathrate structures are known. The unit cell of the most common one, known as Structure I, contains 46 H_2O molecules and up to eight CH_4 molecules. Recently, clathrates have received attention as a possible energy source because $1\,m^3$ of clathrate liberates up to $164\,m^3$ of methane gas.

Clathrates have been found under sediments on the ocean floors. They are thought to form by migration of methane from beneath the ocean floor along geological faults, followed by crystallization on contact with cold seawater. The methane in clathrates is also generated by bacterial degradation of organic matter in low-oxygen environments at the ocean floor. Where sedimentation rates and organic carbon levels are high, the water in the pores of the sediment is low in oxygen and methane is produced by anaerobic bacteria. Below the zone of solid clathrates, large volumes of methane may occur as bubbles of free gas in the sediment. Methane hydrates are stable at low temperatures and high pressures. Because of these conditions and the need for relatively large amounts of organic matter for bacterial methanogenesis, clathrates are mainly restricted to high latitudes and along continental margins

in the oceans. On the continental margins the supply of organic material is high enough to generate enough methane and water temperatures are close to freezing. In polar regions the gas hydrates are commonly linked to the occurrence of permafrost. The permafrost reservoir of methane has been estimated at about 400 Gt of carbon in the Arctic but no estimates have been made of possible Antarctic reservoirs. The oceanic reservoir has been estimated to be about 10–11 Tt of carbon.

In recent years, many governments have become very interested in the possible use of methane hydrates as fossil fuels. The realization that huge reservoirs of methane hydrates occur on the ocean floor and in permafrost regions has led to exploration and investigation of how to use hydrates as an energy source. The USSR tried unsuccessfully to recover gas hydrates from permafrost reservoirs in the 1960s and 1970s. Not enough is known about how clathrate deposits occur in ocean sediments to be able to plan for their recovery and drilling has been carried out in very few places.

The potential recovery of methane from clathrates is not without serious implications. As methane is a greenhouse gas, the discharge of large amounts of it into the atmosphere would increase global warming. Methane levels in the atmosphere were lower during glacial periods than during interglacial periods. Disturbances could destabilize sea-floor methane hydrates, triggering submarine landslides and huge releases of methane.

The data in Table 13.2 illustrate how the E—E bond enthalpy decreases on descending the group. As a result, the tendency to catenation decreases from C to Pb. Silicon forms a series of compounds analogous to the alkanes, the *silanes*, but the longest chain contains just four Si atoms, as tetrasilane, Si_4H_{10}. The silanes, with their greater number of electrons and stronger intermolecular forces, are less volatile than their hydrocarbon analogues. Thus, whereas propane, C_3H_8, is a gas under normal conditions, its silicon analogue trisilane, Si_3H_8, is a liquid and boils at 53°C.

Silane is prepared commercially by the reduction of SiO_2 with aluminium under a high pressure of hydrogen in a molten salt mixture of NaCl and $AlCl_3$. An idealized equation for this reaction is

$$6\,H_2(g) + 3\,SiO_2(s) + 4\,Al(s) \rightarrow 3\,SiH_4(g) + 2\,Al_2O_3(s)$$

The silanes are much more reactive than the alkanes and their stability decreases with increasing chain length. Silane itself, SiH_4, is spontaneously flammable in air, reacts violently with halogens and is hydrolysed on contact with water. This increased reactivity compared to hydrocarbons is attributed to the large atomic radius of silicon, which leaves it open to attack by nucleophiles, the greater polarity of the Si—H bond, and the availability of low-lying d orbitals, which may facilitate the formation of adducts. Silane is a reducing agent in aqueous solution. For example, when silane is bubbled through an oxygen-free aqueous solution containing Fe^{3+}, it reduces the iron to Fe^{2+}.

Bonds between silicon and hydrogen are not readily hydrolysed in neutral water, but the reaction is rapid in strong acid or in the presence of traces of base. Similarly, alcoholysis is accelerated by catalytic amounts of alkoxide:

$$SiH_4 + 4\,ROH \xrightarrow{\Delta,\,OR^-} Si(OR)_4 + 4\,H_2$$

Kinetic studies indicate that the reaction proceeds through a structure in which OR^- attacks the Si atom while H_2 is being formed via a kind of H—H hydrogen bond between hydridic and protic hydrogen atoms.

Example 13.2 Investigating the formation of catenated species

Use the bond enthalpy data in Table 13.2 and the additional data given below to calculate the enthalpy of formation of C_2H_6 and Si_2H_6.

$$\Delta_{vap}H^{\ominus}(C, graphite) = 715 \text{ kJ mol}^{-1} \quad \Delta_{atom}H^{\ominus}(Si, s) = 439 \text{ kJ mol}^{-1}$$
$$B(H-H) = 436 \text{ kJ mol}^{-1}$$

Answer The enthalpy of formation can be calculated as the difference between the bonds broken and the bonds formed in a reaction:

$$2 \text{ C(graphite)} + 3 \text{ H}_2(g) \rightarrow C_2H_6(g)$$
$$\Delta_f H^{\ominus}(C_2H_6, g) = [2(715) + 3(436)] - [348 + 6(412)] \text{ kJ mol}^{-1}$$
$$= -82 \text{ kJ mol}^{-1}$$
$$2 \text{ Si(s)} + 3 \text{ H}_2(g) \rightarrow Si_2H_6(g)$$
$$\Delta_f H^{\ominus}(Si_2H_6, g) = [2(439) + 3(436)] - [226 + 6(318)] \text{ kJ mol}^{-1}$$
$$= +52 \text{ kJ mol}^{-1}$$

The negative value for ethane is due, to a large extent, to the greater C—C and C—H bond enthalpies compared to the Si—Si and Si—H values.

Self-test 13.2 Use the bond enthalpy data in Table 13.2 and above to calculate the enthalpy of formation of CH_4 and SiH_4.

The silicon analogue of hydroboration (Section 12.3c) is **hydrosilylation**, the addition of SiH across the multiple bonds of alkenes and alkynes. This reaction, which is used in both industrial and laboratory syntheses, can be carried out under conditions (300°C or ultraviolet irradiation) that produce a radical intermediate. In practice, it is usually performed under far milder conditions by using a platinum complex as catalyst:

$$CH_2{=}CH_2 + SiH_4 \xrightarrow{\Delta, H_2PtCl_6, \text{ isopropanol}} CH_3CH_2SiH_3$$

The current view is that this reaction proceeds through an intermediate in which both the alkene and silane are attached to the Pt atom.

Silane is used in the production of semiconductor devices such as solar cells and in the hydrosilylation of alkenes; it is prepared commercially by the high-pressure reaction of hydrogen, silicon dioxide, and aluminium.

(c) Germane, stannane, and plumbane

Key point: Thermal stability decreases from germane to stannane and plumbane.

The decreasing stability of the hydrides on going down the group severely limits the accessible chemical properties of stannanes and plumbane. Germane (GeH_4) and stannane (SnH_4) can be synthesized by the reaction of the appropriate tetrachloride with $LiAlH_4$ in tetrahydrofuran solution. Plumbane (PbH_4) has been synthesized in trace amounts by the protolysis of a magnesium/lead alloy. The presence of alkyl or aryl groups stabilizes the hydrides of all three elements. For example, trimethylplumbane, $(CH_3)_3PbH$, begins to decompose at −30°C, but it can survive for several hours at room temperature.

13.6 Compounds with halogens

The Group 14 elements react with all the halogens to form tetrahalides. Lead also forms stable dihalides.

(a) Halides of carbon

Key points: Nucleophiles displace halogens in carbon–halogen bonds; organometallic nucleophiles produce new M—C bonds; mixtures of polyhalocarbons and alkali metals are explosion hazards.

The tetrahalomethanes, the simplest halocarbons, vary from the highly stable and volatile CF_4 to the thermally unstable solid CI_4 (Table 13.3). These tetrahalomethanes and analogous partially halogenated alkanes provide a route to a wide variety of derivatives, mainly by nucleophilic displacement of one or more halogen atoms. Some useful and interesting reactions from an inorganic perspective are outlined in Fig. 13.7. Note in particular the metal–carbon bond-forming reactions, which take place either by complete displacement of halogen or by oxidative addition. The rates of nucleophilic displacement increase greatly from fluorine to iodine, and lie in the order $F \ll Cl < Br < I$. All tetrahalomethanes are thermodynamically unstable with respect to hydrolysis:

$$CX_4(l \text{ or } g) + 2\,H_2O(l) \rightarrow CO_2(g) + 4\,HX(aq)$$

However, the reaction for C—F bonds is extremely slow, and fluorocarbon polymers such as poly(tetrafluoroethene) are highly resistant to attack by water. Tetrahalomethanes can be reduced by strong reducing agents, such as alkali metals. For example, the reaction of carbon tetrachloride with sodium is highly exoergic:

$$CCl_4(l) + 4\,Na(s) \rightarrow 4\,NaCl(s) + C(s) \qquad \Delta_r G^{\ominus} = -249\,kJ\,mol^{-1}$$

This reaction can occur with explosive violence with CCl_4 and other polyhalocarbons, so alkali metals such as sodium should never be used to dry them. Analogous reactions occur on the surface of poly(tetrafluoroethene) when it is exposed to alkali metals or strongly reducing organometallic compounds. Fluorocarbons, together with other fluorine-containing molecules, exhibit many interesting properties, such as high volatility and strong electron-withdrawing character (Chapter 16).

The **carbonyl halides** (Table 13.4) are planar molecules and useful chemical intermediates. The simplest of these compounds, $OCCl_2$, phosgene (**7**), is a highly toxic gas. It is prepared on a large scale by the reaction of chlorine with carbon monoxide:

$$CO(g) + Cl_2(g) \xrightarrow{200°C,\,charcoal} Cl_2CO(g)$$

The utility of phosgene lies in the ease of nucleophilic displacement of chlorine to produce carbonyl compounds and isocyanates (Fig. 13.8). The fact that hydrolysis leads

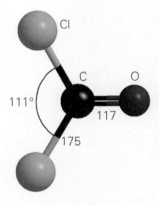

7 Phosgene, $OCCl_2$

Table 13.3 Properties of tetrahalomethanes

	CF_4	CCl_4	CBr_4	CI_4
Melting point/°C	−187	−23	90	171 *dec*
Boiling point/°C	−128	77	190	*sub*
$\Delta_f G^{\ominus}/(kJ\,mol^{-1})$	−879	−65	+48	>0

dec, decomposes; *sub*, sublimes.

Table 13.4 Properties of carbonyl halides

	COF_2	$COCl_2$	$COBr_2$
Melting point/°C	−114	−128	
Boiling point/°C	−83	8	65
$\Delta_f G^{\ominus}/(kJ\,mol^{-1})$	−619	−205	−111

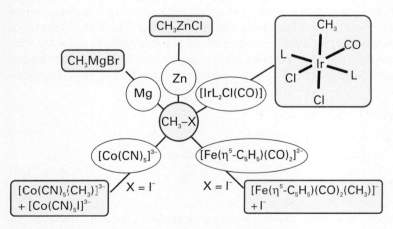

Fig. 13.7 Some characteristic reactions of carbon–halogen bonds (X = halogen).

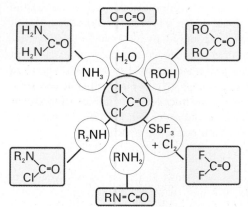

Fig. 13.8 Characteristic reactions of phosgene, Cl_2CO.

to carbon dioxide and hydrogen chloride rather than carbonic acid, $(HO)_2CO$, can be traced to the latter's instability.

(b) Compounds of silicon and germanium with halogens

Key point: Because silicon can form hypervalent transition states whereas carbon cannot, substitution reactions of silicon halides occur more readily than those of carbon halides.

The full range of tetrahalides is known for silicon and germanium; all of them are volatile molecular compounds. Germanium shows signs of an inert-pair effect in that it also forms nonvolatile dihalides. Among the silicon tetrahalides, the most important is the tetrachloride, which is prepared by direct reaction of the elements:

$$Si(s) + 2\ Cl_2(g) \rightarrow SiCl_4(l)$$

Silicon and germanium halides are mild Lewis acids and add one or two ligands to yield five- or six-coordinate complexes:

$$SiF_4(g) + 2\ F^-(aq) \rightarrow SiF_6^{2-}(aq)$$

$$GeCl_4(l) + N{\equiv}CCH_3(l) \rightarrow Cl_4GeN{\equiv}CCH_3(s)$$

The hydrolysis of the silicon and germanium tetrahalides is fast, and is represented schematically as follows:

$$EX_4 + 2\ H_2O \rightarrow EX_4(OH_2)_2 \rightarrow EO_2 + 4\ HX \qquad (E = Si\ or\ Ge, X = halogen)$$

The corresponding carbon tetrahalides are kinetically more resistant to hydrolysis because of the lack of access to the sterically shrouded C atom to form an intermediate aqua complex.

The substitution reactions of halosilanes have been studied extensively. The reactions are more facile than for their carbon analogues because a Si atom can readily expand its coordination sphere to accommodate the incoming nucleophile. The stereochemistry of these substitution reactions indicates that a five-coordinate intermediate is formed with the most electronegative substituents adopting the axial position. Moreover, substituents leave from the axial position. The H^- ion is a poor leaving group, and alkyl groups are even poorer:

Note that in these examples, the R^4 substituent replaces H with retention of configuration.

(c) Tin and lead halides

Key points: Tin forms dihalides and tetrahalides; for lead, only the dihalides are stable.

Evidence of the inert-pair effect becomes more prominent in the chemistry of tin and lead as the +2 oxidation state becomes increasingly stable. Aqueous and nonaqueous solutions of tin(II) salts are useful mild reducing agents, but they must be stored under an inert atmosphere because air oxidation is spontaneous and rapid:

$$Sn^{2+}(aq) + \tfrac{1}{2} O_2(g) + 2 H^+(aq) \rightarrow Sn^{4+}(aq) + H_2O(l) \quad E^{\ominus} = +1.08\,V$$

Tin dihalides and tetrahalides are both well known. The tetrachloride, tetrabromide, and tetraiodide are molecular compounds, but the tetrafluoride has a structure consistent with it being an ionic solid because the small F^- ion permits a six-coordinate structure. Lead tetrafluoride can be considered as an ionic solid but, as a manifestation of the inert-pair effect, $PbCl_4$ is an unstable, covalent, yellow oil that decomposes into $PbCl_2$ and Cl_2 at room temperature. Lead tetrabromide and tetraiodide are unknown, so the dihalides dominate the halogen compounds of lead. The arrangement of halogen atoms around the central metal atom in the dihalides of tin and lead often deviates from simple tetrahedral or octahedral coordination and is attributed to the presence of a stereochemically active lone pair. The tendency to achieve the distorted structure is more pronounced with the small F^- ion, and less distorted structures are observed with larger halides.

Both Sn(IV) and Sn(II) form a variety of complexes. Thus, $SnCl_4$ forms complex ions such as $SnCl_5^-$ and $SnCl_6^{2-}$ in acidic solution. In nonaqueous solution, a variety of donors interact with the moderately strong Lewis acid $SnCl_4$ to form complexes such as *cis*-$SnCl_4(OPMe_3)_2$. In aqueous and nonaqueous solutions Sn(II) forms trihalo complexes, such as $SnCl_3^-$, where the pyramidal structure indicates the presence of a stereochemically active lone pair (**8**). The $SnCl_3^-$ ion can act as a soft donor to *d*-metal ions. One unusual example of this ability is the red cluster compound $Pt_3Sn_8Cl_{20}$, which is trigonal bipyramidal (**9**).

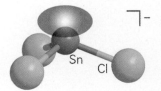

8 $SnCl_3^-$

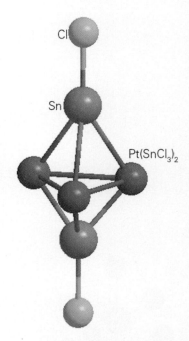

9 $[(SnCl)_2(Pt(SnCl_3)_2)_3]$

13.7 Compounds of carbon with oxygen and sulfur

Key points: Carbon monoxide is a key reducing agent in the production of iron and a common ligand in *d*-metal chemistry; carbon dioxide is much less important as a ligand and is the acid anhydride of carbonic acid; the sulfur compounds CS and CS_2 have similar structures to their oxygen analogues.

The two familiar oxides of carbon are CO and CO_2. Among the less familiar oxides is carbon suboxide, $O{=}C{=}C{=}C{=}O$. Physical data on all three compounds are summarized in Table 13.5. It should be noted that the bond length in carbon monoxide is short and strong (bond enthalpy 1076 kJ mol^{-1}) and the force constant high. These features are in accord with its possession of a triple bond, as in the Lewis structure $:C{\equiv}O:$.

The uses of CO include the reduction of metal oxides in a blast furnace (Section 5.15) and the shift reaction (Section 9.3a) for the production of H_2:

$$CO(g) + H_2O(g) \rightarrow CO_2(g) + H_2(g)$$

Table 13.5 Properties of some oxides of carbon

Oxide	m.p./°C	b.p./°C	$\bar{\nu}(CO)$/cm^{-1}		$k(CO)$/(N m^{-1})	Bond length/pm	
						CC	CO
CO	−199	−192	2145		1860		113
CO$_2$		−78	2449	1318	1550		116
OCCCO	−111	7	2200	2290		128	116

In Chapter 25, where we deal with catalysis, we describe the conversion of carbon monoxide to acetic acid and aldehydes. The CO molecule has very low Brønsted basicity and negligible Lewis acidity towards neutral electron-pair donors. Despite its weak Lewis acidity, however, CO is attacked by strong Lewis bases at high pressure and somewhat elevated temperatures. Thus, the reaction with hydroxide ions yields the formate ion, HCO_2^-:

$$CO(g) + OH^-(s) \rightarrow HCO_2^-(s)$$

Similarly, the reaction with methoxide ions (CH_3O^-) yields the acetate ion, $CH_3CO_2^-$.

Carbon monoxide is an excellent ligand towards metal atoms in low oxidation states (Section 21.18). Its well-known toxicity is an example of this behaviour; it binds to the Fe atom in haemoglobin, so excluding the attachment of O_2, and the victim suffocates. An interesting point is that H_3BCO can be prepared from B_2H_6 and CO at high pressures in a rare example of the coordination of CO to a simple Lewis acid. A complex of similar stability is not formed by BF_3; this observation is consistent with the classification of BH_3 as a soft acid and BF_3 as a hard acid.

Carbon dioxide, CO_2, shows a number of significant differences from carbon monoxide. The C—O bond is longer and the stretching force constants smaller in CO_2 than in CO, which is consistent with the bonds in CO_2 being double rather than triple. Carbon dioxide is only a very weak Lewis acid. For example, only a small fraction of molecules are complexed with water to form H_2CO_3 in acidic aqueous solution but, at higher pH, OH^- coordinates to the C atom, so forming the hydrogencarbonate (bicarbonate) ion, HCO_3^-. This reaction is very slow; yet the attainment of rapid equilibrium between CO_2 and HCO_3^- is so important to life that it is catalysed by a Zn-containing enzyme carbon dioxide hydratase (carbonic anhydrase, Section 26.8). The enzyme accelerates the reaction by a factor of about 10^9.

Carbon dioxide is one of several polyatomic molecules that are implicated in the **greenhouse effect**. In this effect, a polyatomic molecule in the atmosphere permits the passage of visible light but, because of its vibrational infrared absorptions, it blocks the immediate radiation of heat from the Earth. There is strong evidence for a significant increase in atmospheric CO_2 since the industrialization of society. In the past, nature has managed to stabilize the concentration of atmospheric CO_2, in part by precipitation of calcium carbonate in the deep oceans, but it seems that the rate of diffusion of CO_2 into the deep waters is too slow to compensate for the increased influx of CO_2 into the atmosphere (Box 13.3). There is convincing evidence for increasing concentrations of the greenhouse gases CO_2, CH_4, N_2O, and cholorofluorocarbons and it is clear that they are having an impact on global temperatures.

The principal chemical properties of CO_2 are summarized in Fig. 13.9. These properties are based on its mild Lewis acidity towards hard donors, as in the formation of CO_3^{2-} ions in strongly basic solution. The ability of CO_2 to form limestone and thereby to moderate the concentration of atmospheric CO_2 has been mentioned above. Similarly, carbonato complexes of metals can be formed in which CO_3^{2-} is a ligand (**10**). These complexes are often useful intermediates because CO_3^{2-} can be displaced in acidic solution to yield complexes that are otherwise difficult to prepare:

$$[Co(NH_3)_5(CO_3)]^+ + 2\ HF \rightarrow [Co(NH_3)_5F]^{2+} + CO_2 + H_2O + F^-$$

Carbonato complexes can be either monodentate, as shown above, or bidentate; in the latter case the CO_3^{2-} ion has a small bite angle.

From an economic perspective, an important reaction is CO_2 with ammonia to yield ammonium carbonate, $(NH_4)_2CO_3$, which at elevated temperatures is converted directly to urea, $CO(NH_2)_2$, a useful fertilizer, feed supplement for cattle, and chemical intermediate. In organic chemistry, a common synthetic reaction is that between CO_2 and carbanion reagents to produce carboxylic acids.

Metal complexes of CO_2 are known (**11**), but they are rare and far less important than the metal carbonyls. In its interaction with a low-oxidation state, electron-rich metal centre, the neutral CO_2 molecule acts as a Lewis acid and the bonding is dominated by electron donation from the metal atom into an antibonding π orbital of CO_2.

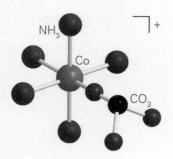

10 $[Co(CO_3)(NH_3)_5]^+$

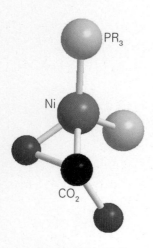

11 $[Ni(CO_2)(PR_3)_2]$

Box 13.3 The carbon cycle

The carbon cycle is of particular interest because of the increased levels of carbon dioxide in the Earth's atmosphere and its potential to cause global warming through an enhanced greenhouse effect. On a global scale, the biological carbon cycle cannot be discussed without also considering the oxygen cycle (Box 15.1). The intimate relationship of the two cycles is shown in Fig. B13.1.

Photosynthesis involves reduction of CO_2 to organic compounds and oxidation of H_2O to O_2 (Chapter 26). Oxygenic photosynthesis is present in the chloroplasts of higher plants, in a variety of algae, and in cyanobacteria. In effect, oxygenic photosynthesis produces O_2 as a side product of H_2O oxidation. When this process first occurred, the evolved O_2 would have been a toxin, producing reactive oxygen species capable of destroying most biomolecules.

Oxygen was not present when the Earth first cooled and liquid water first became available, and CO_2 was the principal atmospheric gas. Early organisms used photosynthesis or *chemolithotropy* (inorganic reactions) to produce the energy needed to reduce carbon dioxide or hydrogencarbonate ions to the organic molecules needed for cellular function. The electron donated in this reduction process did not come from water, the dominant electron donor of the past 2 Ga. However, the first photosynthetic organisms on the early Earth used far simpler forms of photosynthesis. Some of these processes persist in modern bacteria, which use molecules such as H_2S, S_8, thiosulfate, H_2, and organic acids to reduce CO_2. As these molecules are in limited supply (compared to water), non-O_2 evolving photosynthesis is capable of reducing only a small fraction of carbon dioxide. However, once oxygenic photosynthesis evolved, with H_2O serving as the electron donor, planetary biomass could be produced and sustained at levels two to three orders of magnitude larger than previously.

The mass balance of the biological cycle in Fig. B13.1 is not quantitatively complete. Whereas there is input of CO_2 from the eruption of volcanoes and consumption of CO_2 in the weathering of silicate solids, as far as oxygen and organic carbon are concerned,

there is no purely geochemical source. Therefore, for the cycle to be truly complete, no O_2 would ever accumulate: all the O_2 produced on the left side of the cycle by photosynthesis would be consumed on the right side by respiration and combustion. However, with each pass around the cycle, some of the reduced carbon biomass is buried in sediments, mostly land plants and algae in shallow marine basins and lakes. This small amount of buried biomass gradually becomes unavailable for oxidation, and some of it is transformed into hydrocarbon fossil fuels. Over geological time scales this buried reduced organic matter accumulates and is converted into the coal, shale, oil, and natural gas that constitute our fossil fuel reserves.

As this reduced carbon is buried, it is no longer available to be oxidized by O_2, which begins to accumulate in the atmosphere. Thus, over hundreds of millions of years, the process that created our fossil fuel reserves also formed the O_2 of the atmosphere and helped to decrease the initially high level of CO_2. The global accumulation of O_2 was slow on the early Earth on account of the vast amounts of iron(II) present in the oceans. This iron was oxidized by the O_2, yielding banded iron(III) formations. Once the iron(II) and reduced sulfur were consumed, O_2 began to accumulate in the atmosphere, achieving roughly modern levels of approximately 1 Ga ago.

Currently, we are extracting and burning fossil fuels on a geologically very short time scale, thereby disturbing the relationship between oxygen and carbon. The combustion reactions are obviously the major factor, but some oil or gas reaches the surface through natural and human activities. The oil or gas that is not burned can be biodegraded to produce CO_2 and complete the carbon cycle shown in Fig. B13.2. The biodegradation is carried out by aerobic organisms that, almost exclusively, use iron-dependent enzymes.

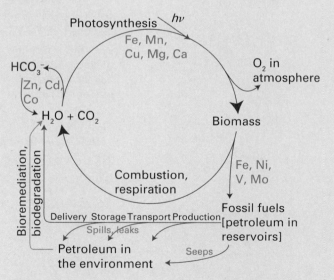

B13.1 The biological carbon cycle.

B13.2 The carbon cycle showing the impact of fossil fuel use.

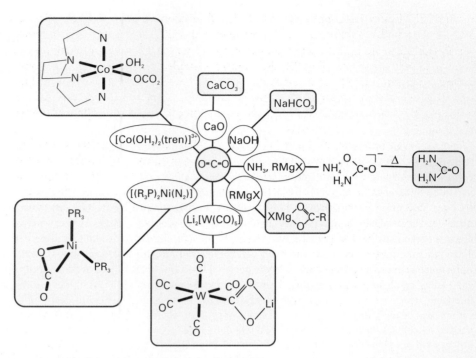

Fig. 13.9 Characteristic reactions of carbon dioxide.

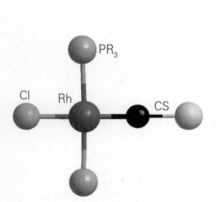

12 [RuCl(CS)(PR$_3$)$_2$]

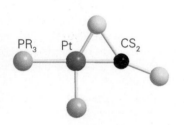

13 [Pt(CS$_2$)(PR$_3$)$_2$]

The sulfur analogues of carbon monoxide and carbon dioxide, CS and CS$_2$, are known. The former is an unstable transient molecule and the latter is endoergic ($\Delta_f G^{\ominus} = +65$ kJ mol^{-1}). Some complexes of CS (**12**) and CS$_2$ (**13**) exist, and their structures are similar to those formed by CO and CO$_2$. In basic aqueous solution, CS$_2$ undergoes hydrolysis and yields a mixture of carbonate ions, CO$_3^{2-}$, and trithiocarbonate ions, CS$_3^{2-}$.

Example 13.3 Proposing a synthesis that uses the reactions of carbon monoxide

Propose a synthesis of CH$_3$ ^{13}CO$_2^-$ that uses ^{13}CO, a primary starting material for many carbon-13 labelled compounds.

Answer We first note that CO$_2$ is readily attacked by strong nucleophiles such as LiCH$_3$ to produce acetate ions. Therefore, an appropriate procedure would be to oxidize ^{13}CO to ^{13}CO$_2$ and then to react the latter with LiCH$_3$. A strong oxidizing agent, such as solid MnO$_2$, can be used in the first step to avoid the problem of excess O$_2$ in the direct oxidation.

$$^{13}CO(g) + 2\,MnO_2(s) \xrightarrow{\Delta} {}^{13}CO_2(g) + Mn_2O_3(s)$$
$$4\,{}^{13}CO_2(g) + Li_4(CH_3)_4(et) \rightarrow 4\,Li[CH_3\,{}^{13}CO_2](et)$$

where et denotes solution in ether. (Another method involves the reaction of [Rh(I)$_2$(CO)$_2$]$^-$ with ^{13}CO. The basis of this reaction is discussed in Chapter 25.)

Self-test 13.3 Propose a synthesis of D^{13}CO$_2^-$ starting from ^{13}CO.

13.8 Simple compounds of silicon with oxygen

Key point: The Si—O—Si link is present in silica, a wide range of metal silicate minerals, and silicone polymers.

The high affinity of silicon for oxygen accounts for the existence of a vast array of silicate minerals and synthetic silicon–oxygen compounds, which are important in mineralogy, industrial processing, and the laboratory. Aside from rare high-temperature phases, the structures of silicates are confined to tetrahedral four-coordinate Si. The complicated silicate structures are often easier to comprehend if the SiO_4 unit is drawn as a tetrahedron with the Si atom at the centre and O atoms at the vertices. The representation is often cut to the bone by drawing the SiO_4 unit as a simple tetrahedron with the atoms omitted. Each terminal O atom contributes -1 to the charge of the SiO_4 unit, but each shared O atom contributes 0. Thus, orthosilicate is $[SiO_4]^{4-}$ (**14**), disilicate is $[O_3SiOSiO_3]^{6-}$ (**15**), and the SiO_2 unit of silica has no net charge because all the O atoms are shared.

With the above principles of charge balance in mind, it should be clear that an endless single-stranded chain or a ring of SiO_4 units, which has two shared O atoms for each Si atom, will have the formula and charge $[(SiO_3)^{2-}]_n$. An example of a compound containing such a cyclic metasilicate ion is the mineral beryl, $Be_3Al_2Si_6O_{18}$, which contains the $[Si_6O_{18}]^{12-}$ ion (**16**). Beryl is a major source of beryllium. The gemstone emerald is beryl in which Cr^{3+} ions are substituted for some Al^{3+} ions. A chain metasilicate is present in the mineral jadeite, $NaAl(SiO_3)_2$ (**17**), one of two different minerals sold as jade, the green colour arising from traces of iron impurities. In addition to other configurations for the single chain, there are double-chain silicates which include the family of minerals known commercially as asbestos (Box 13.4).

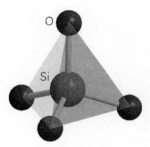

14 $[SiO_4]^{4-}$

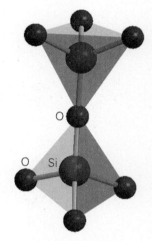

15 $[Si_2O_7]^{6-}$

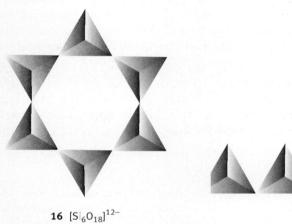

16 $[Si_6O_{18}]^{12-}$

17 $[SiO_3]_n$

Box 13.4 Asbestos

Asbestos is a generic term that applies to six types of naturally occurring mineral fibres. Only three of these have found commercial applications. White asbestos, or chrysotile, has the formula $Mg_3Si_2O_5(OH)_4$ and has a sheet silicate structure. When viewed under a microscope the fibres appear as a tangled mass. Brown asbestos, amosite, and blue asbestos, crocidolite, are amphiboles. Amphiboles have a double-chain silicate structure and appear as individual needles when viewed under a microscope.

All the fibres have properties that have made them very attractive in a range of industrial and domestic applications. These properties include thermal stability, heat resistance, nonbiodegradability, resistance to attack by a wide range of chemicals, and low electrical conductivity. Asbestos is most often used for its thermal properties or for matrix reinforcement. The first recorded use of asbestos was in 2000 BCE in Finland, where it was used to reinforce clay pottery. Marco Polo used asbestos fibres as a flame-resistant material. Demand for asbestos grew during the industrial revolution and by 1900 asbestos–cement sheets were being widely produced for use in construction. World production grew throughout the twentieth century until the 1960s, when health problems associated with exposure to asbestos led to a reduction or ban in its use. The health effects of asbestos are respiratory disease associated with inhalation of long, thin fibres. Asbestos is responsible for asbestosis, which is lung fibrosis resulting from long-term, high-level exposure, lung cancer, which often accompanies asbestosis, and mesothelioma, which is a rare cancer of the lining of the thoracic and abdominal cavities. The impact on health of each type of asbestos is not the same. The most harmful varieties are the needle-like amphiboles.

Where asbestos is still used today it is held within a matrix such as cement or organic resins. Asbestos substitutes in thermal insulation applications include glass fibre and vermiculite. In fibre–cement applications cellulose fibre or synthetic fibres such as polypropylene are being used.

Example 13.4 Determining the charge on a cyclic silicate

Draw the structure and determine the charge on the cyclic silicate anion $[Si_3O_9]^{n-}$.

Answer The ion is a six-membered ring with alternating Si and O atoms and six terminal O atoms, two on each Si atom. Because each terminal O atom contributes -1 to the charge, the overall charge is -6. From another perspective, the conventional oxidation numbers of silicon and oxygen, $+4$ and -2, respectively, also indicate a charge of -6 for the anion.

Self-test 13.4 Repeat the question for the cyclic anion $[Si_4O_{12}]^{n-}$.

Silica and many silicates crystallize slowly. Amorphous solids known as **glasses** can be obtained instead of crystals by cooling the melt at an appropriate rate. In some respects these glasses resemble liquids. As with liquids, their structures are ordered over distances of only a few interatomic spacings (such as within a single SiO_4 tetrahedron). Unlike liquids, however, their viscosities are very high, and for most practical purposes they behave like solids.

The composition of silicate glasses has a strong influence on their physical properties. For example, fused quartz (amorphous SiO_2) softens at about $1300°C$, borosilicate glass (which contains boron oxide, Section 12.5) softens at about $800°C$, and soda-lime glass softens at even lower temperatures. The variation in softening point can be understood by appreciating that the Si—O—Si links in silicate glasses form the framework that imparts rigidity. When basic oxides such as Na_2O and CaO are incorporated (as in soda-lime glass), they react with the SiO_2 melt and convert Si—O—Si links into terminal SiO groups and hence lower its softening temperature.

Very different properties are found for the —Si—O—Si— backbone of silicone polymers, which are described later in this chapter.

13.9 Oxides of germanium, tin, and lead

Key point: The $+2$ oxide becomes more stable on going down the group from Ge to Pb.

Germanium(II) oxide, GeO, is a reducing agent and disproportionates readily to Ge and GeO_2. Germanium(IV) oxide, GeO_2, resembles silica, SiO_2, and is based on tetrahedral four-coordinate GeO_4 units. It also exists in a six-coordinate crystalline form with a rutile-like structure and in a vitreous form that resembles fused silica. Germanium analogues of silicates and aluminosilicates are also known (Section 13.13).

Tin(II) oxide, SnO, exists as two polymorphs. In the blue–black form the Sn(II) ions are four-coordinate (Fig. 13.10), but the O^{2-} ions around the Sn(II) lie in a square to one side with the lone pair on Sn pointing away from the square. This structure can be rationalized by the presence of a stereochemically active lone pair on the metal atom and can be described as fluorite (Section 3.9) with alternate layers of anions missing. The red form of SnO has a similar structure and can be converted to the blue–black form by heat, pressure, and treatment with alkali. Both forms are readily oxidized to SnO_2 when heated in air.

When heated in the absence of air, SnO disproportionates into Sn and SnO_2. The latter occurs naturally as the mineral cassiterite and has a rutile structure (Section 3.9). It has low solubility in glasses and glazes and is used in large quantities as an opacifier and pigment carrier in ceramic glazes to make them less transparent.

The oxides of lead are very interesting from both fundamental and technological standpoints. The red form of PbO has the same structure as blue–black SnO with a stereochemically active lone pair (Fig. 13.10). Lead also forms mixed oxidation state oxides. The best known is 'red lead', Pb_3O_4, which contains Pb(IV) in an octahedral environment and Pb(II) in an irregular six-coordinate environment. The assignment of different oxidation numbers to the lead in these two sites is based on the shorter PbO distances for the atom identified as Pb(IV). The maroon form of lead(IV) oxide, PbO_2, crystallizes in the rutile structure. This oxide is a component of the cathode of a lead–acid battery (Box 13.5).

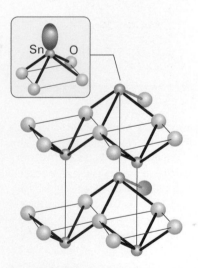

Fig. 13.10 The structure of blue–black SnO showing parallel layers of square-based pyramidal SnO_4 units.

The chemistry of the lead–acid battery is noteworthy because, as well as being the most successful rechargeable battery, it illustrates the role of both kinetics and thermodynamics in the operation of cells.

In its fully charged state, the active material on the cathode is PbO_2 and at the anode it is lead metal; the electrolyte is dilute sulfuric acid. One feature of this arrangement is that the lead-containing reactants and products at both electrodes are insoluble. When the cell is producing current, the reaction at the cathode is the reduction of Pb(IV) as PbO_2 to Pb(II), which in the presence of sulfuric acid is deposited on the electrode as insoluble $PbSO_4$:

$$PbO_2(s) + HSO_4^-(aq) + 3\,H^+(aq) + 2\,e^- \rightarrow PbSO_4(s) + 2\,H_2O(1)$$

At the anode, lead is oxidized to Pb(II), which is also deposited as the sulfate:

$$Pb(s) + SO_4^{2-}(aq) \rightarrow PbSO_4(s) + 2\,e^-$$

The overall reaction is

$$PbO_2(s) + 2\,HSO_4^-(aq) + 2\,H^+(aq) + Pb(s) \rightarrow 2\,PbSO_4(s) + 2\,H_2O(1)$$

The potential difference of about 2 V is remarkably high for a cell in which an aqueous electrolyte is used, and exceeds by far the potential for the oxidation of water to O_2, which is 1.23 V. The success of the battery hinges on the high overpotentials (and hence low rates) of oxidation of H_2O on PbO_2 and of reduction of H_2O on lead.

13.10 Compounds with nitrogen

Key points: The cyanide ion, CN^-, forms complexes with many *d*-metal ions; its coordination to the active sites of enzymes such as cytochrome *c* oxidase accounts for its high toxicity.

Hydrogen cyanide, HCN, is produced in large amounts by the high-temperature catalytic combination of methane and ammonia, and is used as an intermediate in the synthesis of many common polymers, such as poly(methyl methacrylate) and polyacrylonitrile. It is highly volatile (b.p. 26°C) and, like the CN^- ion, highly poisonous. In some respects the toxicity of the CN^- ion is similar to that of the isoelectronic CO molecule, because both form complexes with iron porphyrin molecules. However, whereas CO attaches to the Fe in haemoglobin and causes oxygen starvation, CN^- targets the active site of cytochrome *c* oxidase (the enzyme in mitochondria that reduces oxygen to water), which results in a rapid and catastrophic collapse of energy production.

Unlike the neutral ligand CO, the negatively charged CN^- ion is a strong Brønsted base ($pK_a = 9.4$) and a much poorer Lewis acid π acceptor. Its coordination chemistry is therefore mainly associated with metal ions in positive oxidation states, as with Fe^{2+} in the hexacyanoferrate(II) complex, $[Fe(CN)_6]^{4-}$.

The toxic, flammable gas cyanogen, $(CN)_2$ (**18**), is known as a **pseudohalogen** because of its similarity to a halogen. It dissociates to give $\cdot CN$ radicals and forms inter-pseudohalogen compounds. Similarly, CN^- is an example of a **pseudohalide ion** (Section 16.3).

The direct reaction of silicon and nitrogen gas at high temperatures produces silicon nitride, Si_3N_4. This substance is very hard and inert, and is used in high-temperature ceramic materials. Current industrial research projects focus on the use of suitable organosilicon–nitrogen compounds that might undergo pyrolysis to yield silicon nitride fibres and other shapes. When SiO_2 is heated with carbon, CO is evolved and silicon carbide, SiC, forms. This very hard material is widely used as the abrasive *carborundum*.

Trisilylamine, $(H_3Si)_3N$, the silicon analogue of trimethylamine, has very low Lewis basicity. It has a planar structure, or is fluxional with a very low barrier to inversion. The low basicity and planar structure have traditionally been attributed to *d*-orbital participation in bonding, allowing sp^2 hybridization around the nitrogen and delocalization of the lone pair through π-type bonding. However, recent quantum mechanical calculations indicate that whereas *d* orbitals play a role in delocalization, they are not responsible for the planar structure. Because the electronegativity of Si is lower than that of C, the Si—N bond is more polar than the C—N bond. This difference leads to long-range electrostatic repulsion between the silyl groups in trisilylamine and hence a planar structure.

N   C

18 Cyanogen, $(CN)_2$

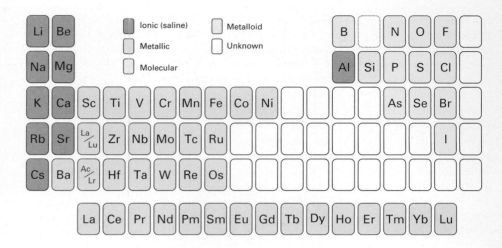

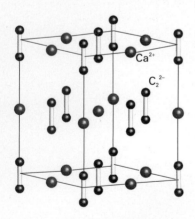

Fig. 13.12 In K_8C, a graphite intercalation compound, the potassium atoms lie in a symmetrical array between the sheets. (See Fig. 13.3 for a view parallel to the sheets.)

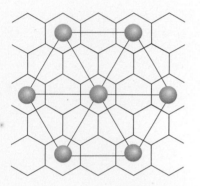

Fig. 13.13 The calcium carbide structure. Note that this structure bears a similarity to the rock-salt structure. Because C_2^{2-} is not spherical, the cell is elongated along one axis. This crystal is therefore tetragonal rather than cubic.

13.11 Carbides

The numerous binary compounds of carbon with metals and metalloids, the *carbides*, are classified as follows:

Saline carbides, which are largely ionic solids; they are formed by the elements of Groups 1 and 2 and by aluminium.

Metallic carbides, which have a metallic conductivity and lustre; they are formed by the *d*-block elements.

Metalloid carbides, which are hard covalent solids formed by boron and silicon.

Figure 13.11 summarizes the distribution of the different types in the periodic table; it also includes binary molecular compounds of carbon with electronegative elements, which are not normally regarded as carbides. This classification is very useful for correlating chemical and physical properties, but (as so often in inorganic chemistry) the borderlines are sometimes indistinct.

(a) Saline carbides

Key points: Metal–carbon compounds of highly electropositive metals are saline; nonmetal carbides are mechanically hard and are semiconductors.

Saline carbides of the Group 1 and 2 metals may be divided into three subcategories: **graphite intercalation compounds**, such as KC_8, **dicarbides** (or 'acetylides'), which contain the C_2^{2-} anion, and **methides**, which formally contain the C^{4-} anion.

Graphite intercalation compounds are formed by the Group 1 metals (Section 13.3). They are formed by a redox process, and specifically by the reaction of graphite with alkali metal vapour or with metal–ammonia solution. For example, contact between graphite and potassium vapour in a sealed tube at 300°C leads to the formation of KC_8 in which the alkali metal ions lie in an ordered array between the graphite sheets (Fig. 13.12). A series of alkali metal–graphite intercalation compounds can be prepared with different metal:carbon ratios, including KC_8 and KC_{16}.

The dicarbides are formed by a broad range of electropositive metals, including those from Groups 1 and 2 (Section 11.6) and the lanthanoids. The C_2^{2-} ion has a very short C—C distance in some dicarbides (for example, 119 pm in CaC_2), which is consistent with it being a triply bonded $[C{\equiv}C]^{2-}$ ion isoelectronic with $[C{\equiv}N]^-$ and $N{\equiv}N$. Some dicarbides have a structure related to rock salt, but replacement of the spherical Cl^- ion by the elongated $[C{\equiv}C]^{2-}$ ion leads to an elongation of the crystal along one axis, and a resulting tetragonal symmetry (Fig. 13.13). The C—C bond is significantly longer in the lanthanoid dicarbides, which suggests that for them the simple triply bonded structure is not a good approximation.

Carbides such as Be_2C and Al_4C_3 are borderline between saline and metalloid, and the isolated C ion is only formally C^{4-}. The existence of directional bonding to the C atom in carbides (as distinct from the nondirectional character expected of purely ionic bonding) is indicated by the crystal structures of methides, which are not those expected for the simple packing of spherical ions.

The principal synthetic routes to the saline carbides and acetylides of Groups 1 and 2 are very straightforward:

Direct reaction of the elements at high temperatures:

$$Ca(l) + 2\ C(s) \xrightarrow{>2000°C} CaC_2(s)$$

The formation of graphite intercalation compounds is another example of a direct reaction, but is carried out at much lower temperatures. The intercalation reaction is more facile because no CC covalent bonds are broken when an ion slips between the graphite layers.

Reaction of a metal oxide and carbon at a high temperature:

$$CaO(l) + 3\ C(s) \xrightarrow{2000°C} CaC_2(l) + CO(g)$$

Crude calcium carbide is prepared in electric arc furnaces by this method. The carbon serves both as a reducing agent to remove the oxygen and as a source of carbon to form the carbide.

Reaction of ethyne (acetylene) with a metal–ammonia solution:

$$2\ Na(am) + C_2H_2(g) \rightarrow Na_2C_2(s) + H_2(g)$$

This reaction occurs under mild conditions and leaves the carbon–carbon bonds of the starting material intact. As the ethyne molecule is a very weak Brønsted acid ($pK_a = 25$), the reaction can be regarded as a redox reaction between a highly active metal and a weak acid to yield H_2 (with H^+ the oxidizing agent) and the metal dicarbide.

The saline carbides have high electron density on carbon, so they are readily oxidized and protonated. For example, calcium carbide reacts with the weak acid water to produce ethyne:

$$CaC_2(s) + 2\ H_2O(l) \rightarrow Ca(OH)_2(s) + HC{\equiv}CH(g)$$

This reaction is readily understood as the transfer of a proton from a Brønsted acid (H_2O) to the conjugate base (C_2^{2-}) of a weaker acid ($HC{\equiv}CH$). Similarly, the controlled hydrolysis or oxidation of the graphite intercalation compound KC_8 restores the graphite and produces a hydroxide or oxide of the metal:

$$2\ KC_8(s) + 2\ H_2O(g) \rightarrow 16\ C(graphite) + 2\ KOH(s) + H_2(g)$$

(b) Metallic carbides

Key point: *d*-Metal carbides are often hard materials with the carbon atom octahedrally surrounded by metal atoms.

The *d* metals provide the largest class of carbides. They are sometimes referred to as **interstitial carbides** because their structures are often related to those of metals by the insertion of C atoms in octahedral holes. However, this nomenclature gives the erroneous impression that the metallic carbides are not legitimate compounds. In fact the hardness and other properties of metallic carbides demonstrate that strong metal–carbon bonding is present in them. Some of these carbides are economically useful materials. Tungsten carbide (WC), for example, is used for cutting tools and high-pressure apparatus such as that used to produce diamond. Cementite, Fe_3C, is a major constituent of steel and cast iron.

Metallic carbides of composition MC have an fcc or hcp arrangement of metal atoms with the C atoms in the octahedral holes. The fcc arrangement results in a rock-salt structure. The C atoms in carbides of composition M_2C occupy only half the octahedral holes between the close-packed metal atoms. A C atom in an octahedral hole is formally **hypercoordinate** (that is, has an untypically high coordination number) because it is surrounded by six metal atoms. However, the bonding can be expressed in terms of delocalized molecular orbitals formed from the C2*s* and C2*p* orbitals and the *d* orbitals (and perhaps other valence orbitals) of the surrounding metal atoms.

It has been found empirically that the formation of simple compounds in which the C atom resides in an octahedral hole of a close-packed structure occurs when $r_C/r_M < 0.59$, where r_C is the covalent radius of C and r_M the metallic radius of M. This relation also applies to metal compounds containing nitrogen or oxygen.

13.12 Silicides

Key points: Silicon–metal compounds (silicides) contain isolated Si, tetrahedral Si_4 units, or hexagonal nets of Si atoms.

Silicon, like its neighbours boron and carbon, forms a wide variety of binary compounds with metals. Some of these **silicides** contain isolated Si atoms. The structure of ferro-silicon, Fe_3Si, for instance, which plays an important role in steel manufacture, can be viewed as an fcc array of Fe atoms with some atoms replaced by Si. Compounds such as K_4Si_4 contain isolated tetrahedral cluster anions $[Si_4]^{4-}$ that are isoelectronic with P_4. Many of the f-block elements form compounds with the formula MSi_2 that have the hexagonal layers that adopt the AlB_2 structure shown in Fig. 12.8.

Extended silicon–oxygen compounds

As well as forming simple binary compounds with oxygen, silicon forms a wide range of extended network solids that find a range of applications in industry. Aluminosilicates occur naturally as clays, minerals, and rocks. Zeolite aluminosilicates are widely used as molecular sieves, catalysts, and catalyst support materials.

13.13 Aluminosilicates

Key point: Aluminium may replace silicon in a silicate framework to form an aluminosilicate.

Even greater structural diversity than that displayed by the silicates themselves is possible when Al atoms replace some of the Si atoms. The resulting **aluminosilicates** are largely responsible for the rich variety of the mineral world. We have already seen that in γ-alumina, Al^{3+} ions are present in both octahedral and tetrahedral holes (Section 3.3). This versatility carries over into the aluminosilicates, where Al may substitute for Si in tetrahedral sites, enter an octahedral environment external to the silicate framework, or, more rarely, occur with other coordination numbers. Because aluminium occurs as Al(III), its presence in place of Si(IV) in an aluminosilicate renders the overall charge negative by one unit. An additional cation, such as H^+, Na^+, or $\frac{1}{2}Ca^{2+}$, is therefore required for each Al atom that replaces an Si atom. As we shall see, these additional cations have a profound effect on the properties of the materials.

(a) Layered and three-dimensional aluminosilicates

Key point: The brittle layered aluminosilicates are the primary constituents of clay and some common minerals.

Many important minerals are varieties of layered aluminosilicates that also contain metals such as lithium, magnesium, and iron: they include clays, talc, and various micas. In one class of layered aluminosilicate the repeating unit consists of a silicate layer with the structure shown in Fig. 13.14. An example of a simple aluminosilicate of this type (simple, that is, in the sense of there being no additional elements) is the mineral kao-linite, $Al_2(OH)_4Si_2O_5$, which is used commercially as china clay. The electrically neutral layers are held together by rather weak hydrogen bonds, so the mineral readily cleaves and incorporates water between the layers.

A larger class of aluminosilicates has Al^{3+} ions sandwiched between silicate layers (Fig. 13.15). One such mineral is pyrophyllite, $Al_2(OH)_2Si_4O_{10}$. The mineral talc, $Mg_3(OH)_2Si_4O_{10}$, is obtained when three Mg^{2+} ions replace two Al^{3+} ions in the

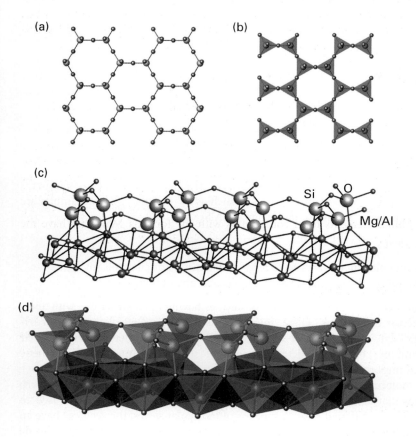

Fig. 13.14 (a) A net of SiO$_4$ tetrahedra with one O atom on top of each Si atom projecting towards the viewer and (b) its tetrahedral representations. (c) Edge view of the above net, with one O/Si incorporated into a net of MO$_6$ octahedra and (d) its polyhedral representation. This structure is for the mineral chrysotile, for which M is Mg. When M is Al^{3+} and each of the atoms on the bottom is replaced by an OH group this structure is close to that of the 1:1 clay mineral kaolinite.

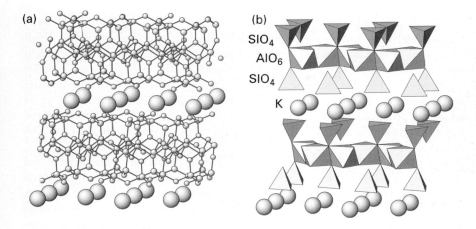

Fig. 13.15 (a) The structure of 2:1 clay minerals such as muscovite mica KAl$_2$(OH)$_2$Si$_3$AlO$_{10}$, in which K$^+$ resides between the charged layers (exchangeable cation sites), Si^{4+} resides in sites of coordination number 4, and Al^{3+} in sites of coordination number 6. (b) The polyhedral representation. In talc, Mg^{2+} ions occupy the octahedral sites and O atoms on the top and bottom are replaced by OH groups.

octahedral sites. In talc (and in pyrophyllite) the repeating layers are neutral, and as a result talc readily cleaves between them. This structure accounts for talc's familiar slippery feel.

Muscovite mica, KAl$_2$(OH)$_2$Si$_3$AlO$_{10}$, has charged layers because one Al(III) atom substitutes for one Si(IV) atom in the pyrophyllite structure. The resulting negative charge is compensated by a K$^+$ ion that lies between the repeating layers. Because of this electrostatic cohesion, muscovite is not soft like talc but it is readily cleaved into sheets. More highly charged layers with dipositive ions between the layers lead to greater hardness.

There are many minerals based on a three-dimensional aluminosilicate framework. The *feldspars*, for instance, which are the most important class of rock-forming minerals (and contribute to granite), belong to this class. The aluminosilicate frameworks of

Table 13.6 Some uses of zeolites

Function	Application
Ion exchange	Water softeners in detergents
Absorption of molecules	Selective gas separation
	Gas chromatography
Solid acid	Cracking high molar mass hydrocarbons for fuel and petrochemical intermediates
	Shape-selective alkylation and isomerization of aromatics for petroleum and polymer intermediates

feldspars are built up by sharing all vertices of SiO_4 or AlO_4 tetrahedra. The cavities in this three-dimensional network accommodate ions such as K^+ and Ba^{2+}. Two examples are the feldspars orthoclase, $KAlSi_3O_8$, and albite, $NaAlSi_3O_8$.

(b) Molecular sieves

Key point: Zeolite aluminosilicates have large open cavities or channels giving rise to useful properties such as ion exchange and molecular absorption.

The **molecular sieves** are crystalline aluminosilicates having open structures with apertures of molecular dimensions. These 'microporous' substances represent a major triumph of solid-state chemistry, for their synthesis and our understanding of their properties combine challenging determinations of structures, imaginative synthetic chemistry, and important practical applications.

The name 'molecular sieve' is prompted by the observation that these materials adsorb only molecules that are smaller than the aperture dimensions and so can be used to separate molecules of different sizes. A subclass of molecular sieves, the *zeolites*,[2] have an aluminosilicate framework with cations (typically from Groups 1 or 2) trapped inside tunnels or cages. In addition to their function as molecular sieves, zeolites can exchange their ions for those in a surrounding solution.

The cages are defined by the crystal structure, so they are highly regular and of precise size. Consequently, molecular sieves capture molecules with greater selectivity than high surface area solids such as silica gel or activated carbon, where molecules may be caught in irregular voids between the small particles. Zeolites are also used for shape-selective heterogeneous catalysis. For example, the molecular sieve ZSM-5 is used to synthesize 1,2-dimethylbenzene (*o*-xylene) for use as an octane booster in gasoline. The other xylenes are not produced because the catalytic process is controlled by the size and shape of the zeolite cages and tunnels. This and other applications are summarized in Table 13.6 and discussed in Chapter 23 and 25.

Synthetic procedures have added to the many naturally occurring zeolite varieties and have produced zeolites that have specific cage sizes and specific chemical properties within the cages. These synthetic zeolites are sometimes made at atmospheric pressure, but more often they are produced in a high-pressure autoclave. Their open structures seems to form around hydrated cations or other large cations such as NR_4^+ ions introduced into the reaction mixture. For example, a synthesis may be performed by heating colloidal silica to 100–200°C in an autoclave with an aqueous solution of tetrapropylammonium hydroxide. The microcrystalline product, which has the typical composition $[N(C_3H_7)_4]OH(SiO_2)_{48}$, is converted into the zeolite by burning away the C, H, and N of the quaternary ammonium cation at 500°C in air. Aluminosilicate zeolites are made by including high surface area alumina in the starting materials.

A wide range of zeolites has been prepared with varying cage and bottleneck sizes (Table 13.7). Their structures (Box 13.6) are based on approximately tetrahedral MO_4 units, which in the great majority of cases are SiO_4 and AlO_4. Because the structures involve many such tetrahedral units, it is common practice to abandon the polyhedral

---

[2] The name zeolite is derived from the Greek for 'boiling stone'. Geologists found that certain rocks seemed to boil when subjected to the flame of a blowpipe.

Table 13.7 Composition and properties of some molecular sieves

Molecular sieve	Composition	Bottleneck diameter/pm	Chemical properties
A	$Na_{12}[(AlO_2)_{12}(SiO_2)_{12}] \cdot xH_2O$	400	Absorbs small molecules; ion exchanger, hydrophilic
X	$Na_{86}[(AlO_2)_{86}(SiO_2)_{106}] \cdot xH_2O$	800	Absorbs medium-sized molecules; ion exchanger, hydrophilic
Chabazite	$Ca_2[(AlO_2)_4(SiO_2)_8] \cdot xH_2O$	400–500	Absorbs small molecules; ion exchanger, hydrophilic
ZSM-5	$Na_3[(AlO_2)_3(SiO_2)_{93}] \cdot xH_2O$	550	Moderately hydrophilic
ALPO-5	$AlPO_4 \cdot xH_2O$	800	Moderately hydrophilic
Silicalite	SiO_2	600	Hydrophobic

Box 13.6 Zeolite structures

Zeolites occur with a wide range of structures. The figure below shows polyhedral representations of six different zeolite frameworks; the three letter code given below each structure is used to label and distinguish these structures. In each case just the TO_4 tetrahedra that form the framework are shown in blue; non-framework atoms such as charge balancing cations and water molecules are omitted to show the porous natures of these fascinating and useful structures.

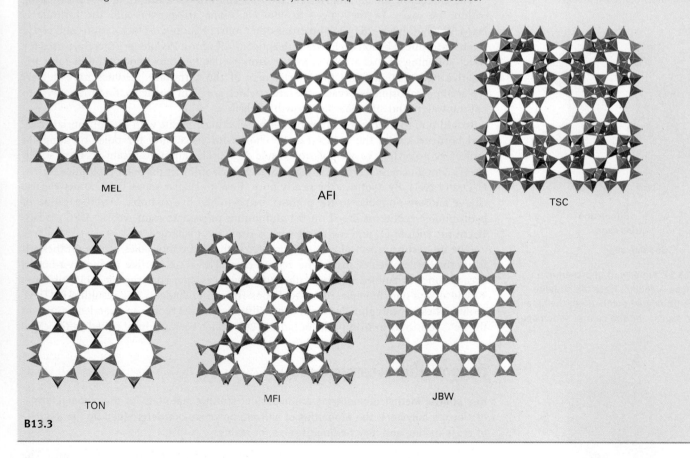

MEL   AFI   TSC

TON   MFI   JBW

B13.3

representation in favour of one that emphasizes the position of the Si and Al atoms. In this scheme, the Si or Al atom lies at the intersection of four line segments and the O atom bridge lies on the line segment (Fig. 13.16). This **framework representation** has the advantage of giving a clear impression of the shapes of the cages and channels in the zeolite.

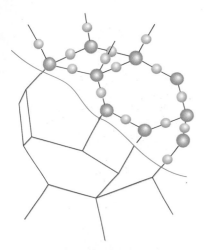

Fig. 13.16 Framework representation of a truncated octahedron (truncation perpendicular to the fourfold axes of the octahedron) and the relation of Si and O atoms to the framework. Note that a Si atom is at each vertex of the truncated octahedron and an O atom is approximately along each edge.

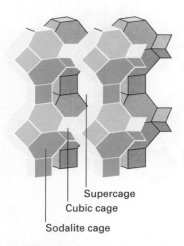

Supercage
Cubic cage
Sodalite cage

Fig. 13.17 Framework representation of a type-A zeolite. Note the sodalite cages (truncated octahedral), the small cubic cages, and the central supercage.

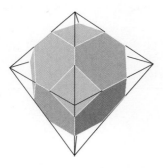

19 Truncated octahedron

A large and important class of zeolites is based on the **sodalite cage**. This cage is a truncated octahedron (Fig. 13.17), a shape that can be formed by slicing off each vertex of an octahedron (**19**). The truncation leaves a square face in the place of each vertex and the triangular faces of the octahedron are transformed into regular hexagons. The substance known as 'zeolite type A' is based on sodalite cages that are joined by O bridges between the square faces. Eight such sodalite cages are linked in a cubic pattern with a large central cavity called an **α cage**. The α cages share octagonal faces, with an open diameter of 420 pm. Thus H_2O or other small molecules can fill them and diffuse through octagonal faces. However, these faces are too small to permit the entrance of molecules with van der Waals diameters larger than 420 pm.

Example 13.5 Analysing the structure of the sodalite cage

Identify the fourfold and sixfold axes in the truncated octahedral polyhedron used to describe the sodalite cage.

Answer There is one fourfold axis running through each pair of opposite square faces for a total of six fourfold axes. Similarly, a set of four sixfold axes runs through opposite sixfold faces.

Self-test 13.5 How many Si and Al atoms are there in one sodalite cage?

The charge on the aluminosilicate zeolite framework is neutralized by cations lying within the cages. In the type-A zeolite, Na^+ ions are present and the formula is $Na_{12}(AlO_2)_{12}(SiO_2)_{12}\cdot xH_2O$. Numerous other ions, including d-block cations and NH_4^+, can be introduced by ion exchange with aqueous solutions. Zeolites are therefore used for water softening and as a component of laundry detergent to remove the di- and tripositive ions that decrease the effectiveness of the surfactant. Zeolites have in part replaced polyphosphates because the latter, which are plant nutrients, find their way into natural waters and stimulate the growth of algae.

In addition to the control of properties by selecting a zeolite with the appropriate cage and bottleneck size, the zeolite can be chosen for its affinity for polar or nonpolar molecules according to its polarity (Table 13.7). The aluminosilicate zeolites, which always contain charge-compensating ions, have high affinities for polar molecules such as H_2O and NH_3. By contrast, the nearly pure silica molecular sieves bear no net electric charge and are nonpolar to the point of being mildly hydrophobic. Another group of hydrophobic zeolites is based on the aluminium phosphate frameworks; $AlPO_4$ is isoelectronic with Si_2O_4 and the framework is similarly uncharged.

One interesting aspect of zeolite chemistry is that large molecules can be synthesized from smaller molecules inside the zeolite cage. The result is like a ship-in-a-bottle, because once assembled the molecule is too big to escape. For example, Na^+ ions in a Y-type zeolite may be replaced by Fe^{2+} ions (by ion exchange). The resulting Fe^{2+}—Y zeolite is heated with phthalonitrile, which diffuses into the zeolite and condenses around the Fe^{2+} ion to form iron phthalocyanine (**20**), which remains imprisoned in the cage.

Organosilicon compounds

Key points: Methylchlorosilanes are important starting materials for the manufacture of silicone polymers; the properties of silicone polymers are determined by the degree of cross-linking and may be liquids, gels, or resins.

All silicon tetraalkyls and tetraaryls are monomeric with a tetrahedral silicon centre. The carbon–silicon bond is strong and the compounds are fairly stable. They can be prepared in a variety of ways, examples of which are shown below.

$$SiCl_4 + 4\,RLi \rightarrow SiR_4 + 4\,LiCl$$
$$SiCl_4 + LiR \rightarrow RSiCl_3 + LiCl$$

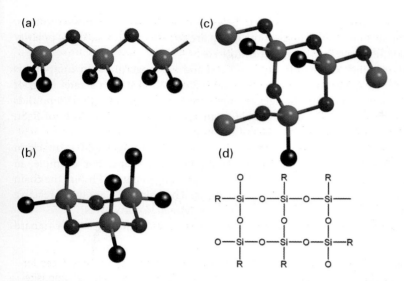

20

Fig. 13.18 The structure of (a) a chain, (b) a ring, and (c) cross-linked silicone; (d) the chemical formula of a fragment.

The *Rochow process* provides a cost-effective industrial route to methylchlorosilane, which is an important starting material in the manufacture of silicones:

$$n\, MeCl + Si/Cu \rightarrow Me_nSiCl_{4-n}$$

These methylchlorosilanes, Me_nSiCl_{4-n}, where $n = 1-3$, can be hydrolysed to form silicones or polysiloxanes:

$$Me_3SiCl + H_2O \rightarrow Me_3SiOH + HCl$$
$$2\, Me_3SiOH \rightarrow Me_3SiOSiMe_3 + H_2O$$

The reaction yields oligomers that contain the tetrahedral silicon group and oxygen atoms that form Si—O—Si bridges. Hydrolysis of $Me_2Si(OH)_2$ produces chains or rings and the hydrolysis of $MeSiCl_3$ produces a cross-linked polymer (Fig. 13.18). It is interesting to note that most silicon polymers are based upon a Si—O—Si backbone, whereas carbon polymers are generally based on a C—C backbone, reflecting the strengths of the Si—O and C—C bonds (Table 13.2).

Silicone polymers have a range of structures and uses. Their properties depend on the degree of polymerization and cross-linking, which are influenced by the choice and mix of reactants, and the use of dehydrating agents such as sulfuric acid and elevated temperatures. The liquid silicones are more stable than hydrocarbon oils. Moreover, unlike hydrocarbons, their viscosity changes only slightly with temperature. Thus silicones are used as lubricants and wherever inert fluids are needed; for example, in hydraulic braking systems. Silicones are very hydrophobic and are used in water-repellent sprays for shoes and other items. The lower molar mass silicones are essential in personal-care products such as shampoos, conditioners, shaving foams, hair gels, and toothpastes, and impart to them a 'silky' feel. At the other end of the spectrum of delicacy, silicone greases, oils, and resins are used as sealants, lubricants, varnishes, waterproofing, synthetic rubbers, and hydraulic fluids.

Organometallic compounds

Key points: Tin and lead form tetravalent organo compounds; organotin compounds are used as fungicides and pesticides.

Many organometallic compounds of Group 14 are of great commercial importance. Organotin compounds are used to stabilize poly (vinyl chloride)(PVC), as antifouling agents on ships, as wood preservatives, and as pesticides. Tetraethyllead has been

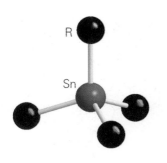

21 SnR$_4$

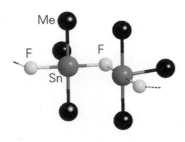

22 (SnFMe$_3$)$_n$

widely used as an antiknock agent in leaded fuels but is now banned in most countries. Generally, organometallic compounds of the group are tetravalent and have low polarity bonds. Their stability decreases from silicon to lead.

Organotin compounds differ from organosilicon and organogermanium compounds in several ways. There is a greater occurrence of the +2 oxidation state, a greater range of coordination numbers, and halide bridges are often present. Most organotin compounds are colourless liquids or solids that are stable to air and water. The structures of R$_4$Sn compounds are all similar, with a tetrahedral tin atom (**21**).

The halide derivatives, R$_3$SnX, often contain Sn—X—Sn bridges and form chain structures. The presence of bulky R groups may affect the shape. For example, in (SnFMe$_3$)$_n$ (**22**), the Sn—F—Sn backbone is in a zigzag arrangement, in Ph$_3$SnF the chain has straightened, and (Me$_3$SiC)Ph$_2$SnF is a monomer. The haloalkyls are more reactive than the tetraalkyls and are useful in the synthesis of tetraalkyl derivatives.

Alkyltin compounds may be prepared in a variety of ways, including via a Grignard reagent and by metathesis:

$$SnCl_4 + 4\ RMgBr \rightarrow SnR_4 + 4\ MgBrCl$$
$$3\ SnCl_4 + 2\ Al_2R_6 \rightarrow 3\ SnR_4 + 2\ Al_2Cl_6$$

Organotin compounds have the widest range of uses of all main-group organometallic compounds and the annual worldwide industrial production of organotin complexes probably exceeds 50 kt. Their major application is in the stabilization of PVC plastics. Without the additive, halogenated polymers are rapidly degraded by heat, light, and atmospheric oxygen to give discoloured, brittle products. The tin stabilizers scavenge

---

Box 13.7 Lead in the environment

Of all the toxic elements in the environment, lead is perhaps the most pervasive. It poisons thousands of people yearly, most of them children in urban areas. The key aspect of lead in the body is that Pb^{2+} can be absorbed through the intestine and is stored in bones, because Pb^{2+}, having a similar ionic radius to Ca^{2+}, can substitute for Ca^{2+} in hydroxyapatite, Ca$_5$(PO$_4$)$_3$(OH). Bone is continuously formed and resorbed, allowing Pb^{2+} to circulate in the blood, and reach critical targets in the blood-forming and nerve tissues. Several epidemiological studies have associated blood lead, down to quite low levels, with impairments in growth, hearing, and mental development.

Drinking water can be a significant source of lead exposure because of lead in pipes or in the solder used in fitting connections. On contact with oxygenated water, metallic lead can be oxidized and solubilized:

$$2\ Pb + O_2 + 4\ H^+ \rightarrow 2\ Pb^{2+} + 2\ H_2O$$

Because the reaction consumes protons, the dissolution rate depends on the pH and is greatest in areas of soft water (low in Ca^{2+} and Mg^{2+}), where there has been little neutralization of rainwater's natural acidity. Districts with soft water sometimes add phosphate to form a protective lead phosphate precipitate that limits further dissolution of lead.

$$3\ Pb^{2+} + 2\ PO_4^{3-} \rightarrow Pb_3(PO_4)_2$$

Modern pipes and fittings are lead free, but older lead-based stock is still in use.

The major exposure route is lead-laced dust, which is deposited on food crops or which is directly ingested, particularly by children, who play in the dust and sometimes eat it. The two major dust sources are leaded paint and leaded fuel. Lead has been added to

fuel in the form of tetraethyl or tetramethyl lead since the 1920s to suppress 'knocking', the tendency of the fuel–air mixture to pre-ignite when compressed in an engine's cylinder. The lead inhibits pre-ignition by catalysing recombination of radicals. To prevent lead from depositing in the cylinder, dichloroethene or dibromoethene is also added, so that during combustion the lead is volatilized as PbX$_2$ (X = Cl or Br). Once in the atmosphere, the PbX$_2$ condenses into fine particles. Although these particles can circulate widely, most of them settle out in the dust near roadways; lead levels in urban dust correlate strongly with traffic congestion.

Lead additives were phased out in the developed world in the early 1970s, when catalytic converters were introduced for control of emissions, because lead particles in the exhaust gas deactivate the catalytic surfaces. However, there are many areas of Africa, Asia, and South America where use of leaded fuels continues, and is even increasing as traffic increases. This is a public health issue, because it has become clear that blood lead levels in the population correlate strongly with the use of leaded fuel.

The remaining problem involves lead in paint. Lead salts are brightly coloured and have been widely used in pigments and paint bases: PbCrO$_4$ is yellow, Pb$_3$O$_4$ is red, and Pb$_3$(OH)$_2$CO$_3$ is white. As the paint wears away, the lead compound is dispersed in dust. Dust and paint chips are particularly hazardous indoors; lead was banned from indoor paint in 1927 throughout much of Europe but only in 1971 in the USA. Older housing, particularly in inner cities, still has leaded paint on interior walls. Leaded paints have also been widely used on building exteriors, and weathering can result in high lead levels in the adjacent soils, where children often play. In the USA, about one-fifth of children living in inner-city homes built before 1946 are estimated to have elevated blood lead levels.

labile Cl^- ions that initiate the loss of HCl, the first step in the degradation process. Organotin compounds also have a wide range of applications relating to their biocidal effects. They are used as fungicides, algaecides, wood preservative, and antifouling agents. However, their widespread use on boats to prevent fouling and attachment of barnacles has caused environmental concerns as high levels of organotin compounds kill some species of marine life and affect the growth and reproduction of others. Many nations now restrict the use of organotin compounds to vessels over 25 m long.

Tetraethyl lead used to be made on a huge scale as an anti-knock agent in petrol. However, concerns about the levels of lead in the environment has led to it being phased out (Box 13.7). Alkyllead compounds, R_4Pb, can be made in the laboratory by using a Grignard reagent or an organolithium compound:

$$2\ PbCl_2 + 4\ RLi \rightarrow R_4Pb + 4\ LiCl + Pb$$
$$2\ PbCl_2 + 4\ RMgBr \rightarrow 2\ R_4Pb + Pb + 4\ MgBrCl$$

They are all monomeric molecules with tetrahedral geometry around the Pb atom. The halide derivatives may contain bridging halide atoms to form chains. Monomers are favoured by more bulky organic substituents. For example, $Pb(CH_3)_3Cl$ exists as a chain structure with bridging Cl atoms (**23**) whereas the mesityl derivative $Pb(Me_3C_6H_2)_3Cl$ is a monomer.

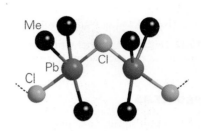

23 $(PbClMe_3)_n$

FURTHER READING

R.B. King, *Inorganic chemistry of the main group elements.* Wiley (1994).

D.M.P. Mingos, *Essential trends in inorganic chemistry.* Oxford University Press (1998). A survey of inorganic chemistry from the perspective of structure and bonding.

N.C. Norman, *Periodicity and the s- and p-block elements.* Oxford University Press (1997). Includes coverage of essential trends and features of *p*-block chemistry.

R.B. King (ed.), *Encyclopedia of inorganic chemistry.* Wiley (2005).

P.R. Birkett, A round-up of fullerene chemistry. *Educ. Chem.*, 1999, **36**, 24. A readable survey of fullerene chemistry.

J. Baggot, *Perfect symmetry: the accidental discovery of buckminsterfullerene.* Oxford University Press (1994). A general account of the story of the discovery of fullerene.

P.J.F. Harris, *Carbon nanotubes and related structures.* Cambridge University Press (2002).

N.N. Greenwood and A. Earnshaw, *Chemistry of the Elements.* Butterworth-Heinemann (1997).

EXERCISES

13.1 Silicon forms the chlorofluorides $SiCl_3F$, $SiCl_2F_2$, and $SiClF_3$. Sketch the structures of these molecules.

13.2 Explain why CH_4 burns in air whereas CF_4 does not. The enthalpy of combustion of CH_4 is $-888\ kJ\ mol^{-1}$ and the C—H and C—F bond enthalpies are -413 and $-489\ kJ\ mol^{-1}$, respectively.

13.3 SiF_4 reacts with $(CH_3)_4NF$ to form $[(CH_3)_4N][SiF_5]$. (a) Use the VSEPR rules to determine the shape of the cation and anion in the product. (b) Account for the fact that the ^{19}F NMR spectrum shows two fluorine environments.

13.4 Identify the compounds A to F:

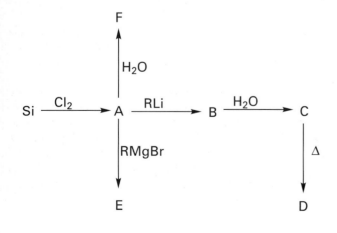

13.5 (a) Summarize the trends in relative stabilities of the oxidation states of the elements of Group 14, and indicate the elements that display the inert-pair effect. (b) With this information in mind, write balanced chemical reactions or NR (for no reaction) for the following combinations, and explain how the answer fits the trends.

(i) $Sn^{2+}(aq) + PbO_2(s)(excess) \rightarrow$ (air excluded)

(ii) $Sn^{2+}(aq) + O_2(air) \rightarrow$

13.6 Use data from *Resource section* 3 to determine the standard potential for each of the reactions in Exercise 13.5b. In each case, comment on the agreement or disagreement with the qualitative assessment you gave for the reactions.

13.7 Give balanced chemical equations and conditions for the recovery of silicon and germanium from their ores.

13.8 (a) Describe the trend in band gap energy, E_g, for the elements carbon (diamond) to tin (grey). (b) Does the electrical conductivity of silicon increase or decrease when its temperature is changed from $20°C$ to $40°C$?

13.9 Preferably without consulting reference material, draw a periodic table and indicate the elements that form saline, metallic, and metalloid carbides.

13.10 Describe the preparation, structure, and classification of (a) KC_8, (b) CaC_2, (c) K_3C_{60}.

13.11 Write balanced chemical equations for the reactions of K_2CO_3 with HCl(aq) and of Na_4SiO_4 with aqueous acid.

13.12 Describe in general terms the nature of the $[SiO_3]_n^{2n-}$ ion in jadeite and the silica–alumina framework in kaolinite.

13.13 (a) How many bridging O atoms are in the framework of a single sodalite cage? (b) Describe the (supercage) polyhedron at the centre of the zeolite A structure in Fig. 13.17.

PROBLEMS

13.1 Correct any inaccuracies in the following descriptions of Group 14 chemistry. (a) None of the elements in this group is a metal. (b) At very high pressures, diamond is a thermodynamically stable phase of carbon. (c) Both CO_2 and CS_2 are weak Lewis acids and the hardness increases from CO_2 to CS_2. (d) Zeolites are layered materials exclusively composed of aluminosilicates. (e) The reaction of calcium carbide with water yields ethyne and this product reflects the presence of a highly basic C_2^{2-} ion in calcium carbide.

13.2 The lightest p-block elements often display different physical and chemical properties from the heavier members. Discuss the similarities and differences by comparison of:

(a) The structures and electrical properties of carbon and silicon.

(b) The physical properties and structures of the oxides of carbon and silicon.

(c) The Lewis acid/base properties of the tetrahalides of carbon and silicon.

13.3 Describe the physical properties of pyrophyllite and muscovite mica and explain how these properties arise from the composition and structures of these closely related aluminosilicates.

13.4 There are major commercial applications for semicrystalline and amorphous solids, many of which are formed by Group 14 elements or their compounds. List four different examples of amorphous or partially crystalline solids described in this chapter and briefly state their useful properties.

13.5 The layered silicate compound $CaAl_2Si_2O_8$ contains a double aluminosilicate layer, with both Si and Al in four-coordinate sites. Sketch an edge-on view of a reasonable structure for the double layer, involving only vertex sharing between the SiO_4 and AlO_4 units. Discuss the likely sites occupied by Ca^{2+} in relation to the silica–alumina double layer.

13.6 Discuss the solid-state chemistry of silicon with reference to silicon dioxide, mica, asbestos, and silicate glasses.

13.7 One of your friends is studying English and is taking a course in science fiction. A common theme in the sources is silicon-based life forms. Your friend wonders why silicon should be chosen and why all life is carbon based. Prepare a short article that presents arguments for and against silicon-based life.

The Group 15 elements

14

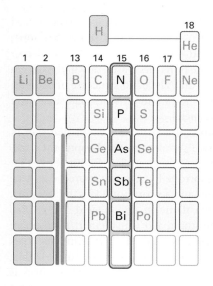

The chemical properties of the Group 15 elements are very diverse. Although the simple trends that we observed for Groups 13 and 14 are still apparent, they are complicated by the fact that the Group 15 elements exhibit a wide range of oxidation states and form many complex compounds with oxygen. Nitrogen makes up a large proportion of the atmosphere and is widely distributed in the biosphere. Phosphorus is essential for both plant and animal life. In stark contrast, arsenic is a well-known poison.

The Group 15 elements—nitrogen, phosphorus, arsenic, antimony, and bismuth—are some of the most important elements for life, geology, and industry. The members of Group 15, the nitrogen group, are sometimes referred to collectively as the **pnictogens** but this name is neither widely used nor officially sanctioned. As in the rest of the *p* block, the element at the head of Group 15, nitrogen, differs significantly from its congeners. Its coordination number is generally lower and it is the only member of the group to exist as a gaseous, diatomic molecule under normal conditions.

The elements

Key points: Nitrogen is gaseous; the heavier elements are all solids that exist in several allotropic forms.

The properties of the Group 15 elements are diverse and more difficult to rationalize in terms of atomic radii and electron configuration than the *p*-block elements we have studied so far. The usual trends of increasing metallic character down a group and stability of low oxidation states at the foot of the group are still evident but they are complicated by the wide range of oxidation states available.

All the members of the group other than nitrogen are solids under normal conditions. However, the trend to increasing metallic character down the group is not clear-cut, because the electrical conductivities of the heavier elements actually decrease from arsenic to bismuth (Table 14.1). The normal increase in conductivity down a group reflects the closer spacing of the atomic energy levels in heavier elements and hence a smaller separation of the valence and conduction bands (Section 3.17). The opposite

Table 14.1 Selected properties of the Group 15 elements

	N	P	As	Sb	Bi
Melting point/°C	−210	44 (white) 590 (red)	613 (sublimes)	630	271
Atomic radius/pm	74	110	121	141	170
First ionization energy/(kJ mol^{-1})	1400	1060	966	833	774
Electrical conductivity/(10^{-10} S m^{-1})		10	3.33	2.50	0.77
Pauling electronegativity	3.0	2.1	2.0	1.9	1.9
Electron affinity/(kJ mol^{-1})	−3.0	−70.0	78	103	91
B(E—H)/(kJ mol^{-1})	390	322	275		

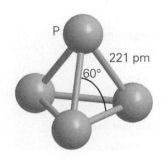

1 P_4, T_d

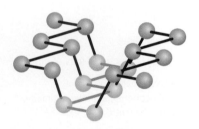

Fig. 14.1 One of the puckered layers of black phosphorus. Note the trigonal pyramidal coordination of the atoms. Dark atoms are closest to the viewer and open circles are furthest away.

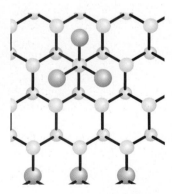

Fig. 14.2 The puckered network structure of bismuth. Each Bi atom has three nearest neighbours; the dark shaded atoms indicate next-nearest neighbours from an adjacent puckered sheet. Lightest atoms are furthest from the viewer.

trend in conductivity in this group suggests that there must be a more pronounced molecular character in the solid state. Indeed, the structures of solid arsenic, antimony, and bismuth have three nearest-neighbour atoms and three more at significantly larger distances. The ratio of these long and short interactions decreases down the group indicating the onset of a polymeric, delocalized structure. The band structure of bismuth suggests a low density of conduction electrons and holes and it is best classified as a metalloid rather than as a semiconductor or a true metal.

The solid elements of Group 15 exist as a number of allotropes.[1] *White phosphorus*, for example, is a waxy solid consisting of tetrahedral P_4 molecules (**1**). Despite the small PPP angle (60°), the molecules persist in the vapour up to about 800°C, but above that temperature the equilibrium concentration of the allotrope P_2 becomes appreciable. Like the gaseous N_2 molecule, P_2 has a formal triple bond and a short bond length (189 pm). White phosphorus is very reactive and bursts into flame in air to yield P_4O_{10}. *Red phosphorus* can be obtained by heating white phosphorus at 300°C in an inert atmosphere for several days. It is normally obtained as an amorphous solid, but crystalline materials can be prepared that have very complex three-dimensional network structures. Unlike white phosphorus, red phosphorus does not ignite readily in air. When phosphorus is heated under high pressure, a series of phases of *black phosphorus* are formed, the most thermodynamically stable form. One of the phases consists of puckered layers composed of pyramidal three-coordinate P atoms (Fig. 14.1). In contrast to the usual practice of choosing the most stable phase of an element as the reference phase for thermodynamic calculations, white phosphorus is adopted because it is more accessible and better characterized than the other forms.

Arsenic exists in two forms; *yellow arsenic* and *metallic arsenic*. Yellow arsenic and gaseous arsenic both consist of tetrahedral As_4 molecules. The most stable structure at room temperature of metallic arsenic, and of antimony and bismuth, is built from puckered hexagonal layers in which each atom has three nearest neighbours. The layers stack in a way that gives three more distant neighbours in the adjacent net as described above (Fig. 14.2).

Bismuth has recently been found to be radioactive, decaying by α emission with a half-life of 1.9×10^{19} years.

14.1 Occurrence and recovery

Key points: Nitrogen is recovered by distillation from liquid air; it is used as an inert gas and in the production of ammonia. Elemental phosphorus is recovered from the minerals fluorapatite and hydroxyapatite by carbon arc reduction; the resulting white phosphorus is a molecular solid, P_4. Treatment of apatite with sulfuric acid yields phosphoric acid, which is converted to fertilizers and other chemicals.

---

[1] The term *allotrope* refers to different structures of the same element; thus ozone and dioxygen are allotropes of oxygen. The word *polymorph* denotes the same substance (an element or a compound) in different crystal forms. For example, phosphorus occurs in several solid polymorphs (including white and black phosphorus).

Nitrogen is readily available as dinitrogen, N_2, as it makes up 78 per cent by mass of the atmosphere and is obtained from it on a massive scale by the distillation of liquid air. Liquid nitrogen is a very convenient way of storing and handling N_2 in the laboratory. Although the bulk price of N_2 is low, the scale on which it is used is an incentive for the development of less expensive processes than liquefaction and distillation. Membrane materials that are more permeable to O_2 than to N_2 are used in laboratory-scale separations of oxygen and nitrogen at room temperature (Fig. 14.3).

Phosphorus was first isolated by Hennig Brandt in 1669. Brandt, misinterpreting their colour, was trying to extract gold from urine and sand and instead extracted a white solid, which glowed in the dark. This element was called phosphorus after the Greek for 'light bearer'. Today, the principal raw material for the production of elemental phosphorus and phosphoric acid is phosphate rock, the insoluble, crushed, and compacted remains of ancient organisms, which consists primarily of the minerals fluorapatite, $Ca_5(PO_4)_3F$, and hydroxyapatite, $Ca_5(PO_4)_3OH$. Phosphoric acid can be produced by the action of concentrated sulfuric acid:

$$Ca_5(PO_4)_3F(s) + 5\ H_2SO_4(l) \rightarrow 3\ H_3PO_4(l) + 5\ CaSO_4(s) + HF(g)$$

The potential pollutant, hydrogen fluoride, from the fluoride component of the rock is scavenged by reaction with silicates to yield the less reactive SiF_6^{2-} complex ion.

The product of the treatment of phosphate rock with acid contains d-metal contaminants that are difficult to remove completely, so its use is largely confined to fertilizers and metal treatment. Most pure phosphoric acid and phosphorus compounds are still produced from the element because it can be purified by sublimation. The production of elemental phosphorus starts with crude calcium phosphate (as calcined phosphate rock), which is reduced with carbon in an electric arc furnace. Silica is added (as sand) to produce a slag of calcium silicate:

$$2\ Ca_3(PO_4)_2(s) + 6\ SiO_2(s) + 10\ C(s) \xrightarrow{1500°C} 6\ CaSiO_3(l) + 10\ CO(g) + P_4(g)$$

The slag is molten at these high temperatures and so can easily be removed from the furnace. The phosphorus vaporizes and is condensed to the solid, which is stored under water to protect it from reaction with air. Most phosphorus produced in this way is burned to form P_4O_{10}, which is then hydrated to yield pure phosphoric acid.

The chemically softer elements arsenic, antimony, and bismuth are often found in sulfide ores. Arsenic is found naturally in the ores realgar, As_4S_4, orpiment, As_2S_3, arsenolite, As_2O_3, and arsenopyrite, FeAsS (Box 14.1). Arsenic is usually extracted from

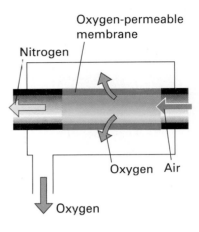

Fig. 14.3 Schematic diagram of a membrane separator for nitrogen and oxygen.

Box 14.1 Arsenic in the environment

The environmental toxicity of arsenic is a problem of groundwater contamination. The worst occurrence of arsenic pollution is in Bangladesh, and the neighbouring Indian province of West Bengal, where hundreds of thousands of people have been diagnosed with arsenicosis. Three major rivers drain into this region, bringing iron-laden sediments from the mountains. The fertile delta is heavily farmed, and organic matter leaches into the shallow aquifer, creating reducing conditions. High levels of arsenic are found in various parts of this aquifer.

Paradoxically, the problem grew out of a United Nations sponsored effort, starting in the 1960s, to provide clean drinking water (replacing contaminated surface water) by sinking inexpensive tube wells into this aquifer. These wells did in fact improve health greatly by reducing the incidence of water-borne diseases, but the high arsenic content went unrecognized for many years. As many as half the four million tube wells exceed the Bangladesh arsenic standard

of 50 ppb (the World Health Organization's guideline is 10 ppb), whereas levels routinely exceed 500 ppb in the more contaminated areas. There are several schemes to treat the well water to remove arsenic, and new wells could be dug into deeper, uncontaminated aquifers. The World Bank is coordinating a mitigation plan, but the massive effort could take years.

Arsenicosis develops over a period of up to 20 years. The first symptoms are keratoses of the skin, which develop into cancers; the liver and kidneys also deteriorate. The early stage is reversible if arsenic ingestion is discontinued, but once cancers develop, effective treatment becomes more difficult. The biochemistry of these effects is uncertain. Arsenate is reduced to As(III) complexes in the body, which probably act by binding sulfhydryl groups. A plausible link to cancer is suggested by the laboratory finding that low levels of arsenic inhibit hormone receptors that turn on cancer suppressor genes.

the flue dust of copper and lead smelters. However, it is also obtained by heating the ores in the absence of oxygen:

$$FeAsS(s) \xrightarrow{700°C} FeS(s) + As(g)$$

Antimony occurs naturally as the minerals stibnite, Sb_2S_3, and ullmanite, NiSbS. It is extracted by heating the ore with scrap iron, which produces the metal and iron sulfide:

$$Sb_2S_3(s) + 3\ Fe(s) \rightarrow 2\ Sb(s) + 3\ FeS(s)$$

Bismuth occurs as bismite, Bi_2O_3, and bismuthinite, Bi_2S_3. It is produced as a by-product of the extraction of copper, tin, lead, and zinc.

14.2 Uses

Key points: Nitrogen is essential for the industrial production of ammonia and nitric acid; the major use of phosphorus is in the manufacture of fertilizers.

The major nonchemical use of nitrogen gas is as an inert atmosphere in metal processing, petroleum refining, and food processing. Nitrogen gas is used to provide an inert atmosphere in the laboratory, and liquid nitrogen (b.p. $-196°C$, 77 K) is a convenient refrigerant in both industry and the laboratory. The major industrial use of nitrogen is in the production of ammonia by the Haber process (Section 14.3) and conversion to nitric acid by the Ostwald process (Section 14.6a). Ammonia provides a route to a wide range of nitrogen compounds, which include fertilizers, plastics, and explosives (Fig. 14.4). Nitrogen plays a crucial role in biology as it is a constituent of amino acids, nucleic acids, and proteins and the nitrogen cycle is one of the most important processes in the ecosystem (Box 14.2 and Section 26.13).

Phosphorus is used in pyrotechnics, smoke bombs, steel making, and alloys. Red phosphorus mixed with sand is used as the striking strip on matchboxes. Sodium phosphate is used as a cleaning agent, a water softener, and to prevent scaling in boilers and pipes. Condensed phosphates are added to detergents as builders that enhance the detergency. In the natural environment, phosphorus is usually present as phosphate ions. Phosphorus (together with nitrogen and potassium) is an essential plant nutrient. However, as a result of the low solubility of many metal phosphates it is often depleted in soil, and hence hydrogenphosphates are important components of balanced fertilizers. Approximately 85 per cent of the phosphoric acid produced goes into fertilizer manufacture. Phosphorus is also an important constituent of bones and teeth (which are predominantly calcium phosphate), cell membranes (phosphate esters of fatty acids), and nucleic acids, including DNA and RNA, and of adenosine triphosphate (ATP), the energy-transfer unit of living organisms. The phosphines, PX_3, are widely used ligands (Section 8.5).

Arsenic is used as a dopant in solid-state devices such as integrated circuits and lasers. Although arsenic is a well-known poison it is also an essential trace element in chickens, rats, goats, and pigs and arsenic deficiency leads to restricted growth (Box 14.3).

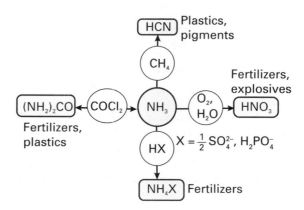

Fig. 14.4 The industrial uses of ammonia.

Box 14.2 The nitrogen cycle

Most of the molecules used by biological systems contain nitrogen, including proteins, nucleic acids, chlorophyll, various enzymes and vitamins, and many other cellular constituents. In all these compounds, nitrogen is in its reduced form with an oxidation state of -3. The principal reservoir of nitrogen is N_2 in the atmosphere. Therefore, a major challenge to biology (and technology) involves the reduction of N_2 for incorporation into essential nitrogen compounds.

The nitrogen cycle is shown in Fig. B14.1. The cycle can be viewed as a set of enzymatically catalysed redox reactions that lead to an accessible supply of reduced nitrogen compounds. Microorganisms are almost entirely responsible for the interconversion of inorganic forms of nitrogen. The enzymes that catalyse these conversions have Fe, Mo, and Cu at their active site. Enzymes of the nitrogen cycle are discussed in Section 6.13.

Although N_2 is the most abundant constituent of the Earth's atmosphere, its usefulness is limited by its unreactivity. Although small amounts of N_2 react with O_2 under extreme conditions (for instance, in lightning discharges and high-temperature combustions), such processes provide only a small percentage of the nitrogen requirements of the biosphere. The remainder comes from the process of nitrogen fixation. The biological reduction of nitrogen to ammonia is carried out exclusively by prokaryotes, including both bacteria and archaea. The enzyme system for nitrogen fixation functions anaerobically and oxygen rapidly and irreversibly destroys the enzyme. Nevertheless, nitrogen fixation also occurs in aerobic bacteria. In addition, important symbioses and associations exist between higher plants and nitrogen-fixing bacteria (such as rhizobia). In these symbioses, bacteria live within controlled environments in the plant, such as root nodules, that have low oxygen levels. The plant provides the bacterium with reduced carbon compounds from photosynthesis, while the bacterium provides fixed nitrogen to the plant.

A reduction potential of about -0.28 V is required for biological nitrogen fixation. Reduced ferredoxins or flavoproteins with reduction potentials of -0.4 to -0.5 V are readily available in biological systems (Chapter 26). Whereas these potentials indicate that nitrogen fixation is thermodynamically feasible, this is not the case kinetically. The kinetic barrier to N_2 reduction apparently arises from the need to form bound intermediates in the conversion of N_2 to ammonia. Organisms invest metabolic energy from adenosine triphosphate (ATP) hydrolysis, for which $\Delta_r G^{\ominus} \approx -130 \text{ kJ mol}^{-1}$, for conversion to adenosine diphosphate (ADP) and inorganic phosphate (P_i) to produce the key intermediates in the N_2 fixation process. The reduction of N_2 consumes 16 molecules of ATP for each molecule of N_2 reduced. Given the opportunity, most organisms capable of nitrogen fixation use available fixed nitrogen sources (ammonia, nitrate, or nitrite) and repress the synthesis of the elaborate nitrogen fixation system.

Once nitrogen is reduced, organisms incorporate the nitrogen into organic molecules, where it enters the biosynthetic pathways of the cell. When organisms die and biomass decays, organonitrogen compounds decompose and release nitrogen to the environment in the form of NH_3 or NH_4^+, depending on the conditions.

The growing human population and its dependence on synthetic fertilizers have had enormous impact on the nitrogen cycle. Ammonia synthesis is carried out by the Haber–Bosch process (Section 25.13), which augments the total fixed nitrogen available to life on Earth. Between a third and a half of all nitrogen fixed on Earth occurs through technological and agricultural, rather than natural, means. In addition to ammonia itself, nitrate salts are produced industrially from ammonia for use in fertilizers. Both ammonia and nitrates enter the nitrogen cycle as fertilizer, which increases all segments of the natural cycle. The natural reservoirs are inadequate sinks for the excess input. Under such conditions, nitrate or nitrite may accumulate as an undesirable component of ground water or produce eutrophication in lakes, wetlands, river deltas, and coastal areas.

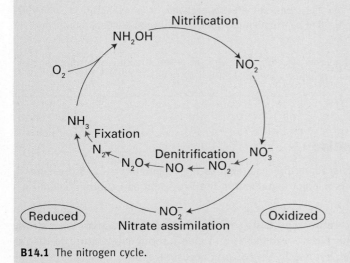

B14.1 The nitrogen cycle.

Antimony is used in semiconductor technologies to produce infrared detectors and light-emitting diodes. It is used in alloys where it leads to stronger and harder products. Antimony oxide is used as a flame retardant.

In keeping with the general trend down the p block, the $+3$ oxidation state becomes more favourable relative to $+5$ on going from phosphorus to bismuth. Consequently, bismuth(V) compounds are useful oxidizing agents. The other major uses of bismuth compounds are in medicine. Bismuth subsalicylate, $HOC_6H_4CO_2BiO$, is used in conjunction with antibiotics and as a treatment of peptic ulcers. Bismuth oxide is used in haemorrhoid creams.

Box 14.3 Arsenicals

'Arsenicals' is the term used to describe chemicals containing arsenic. Arsenic and its compounds are intensely toxic and all the applications of arsenicals are based on this broad spectrum toxicity.

Inorganic arsenicals in the form of the mineral realgar and arsenolite were used in ancient times to treat ulcers, skin diseases, and leprosy. In the early 1900s an organoarsenic compound was found to be an effective treatment for syphilis and led to a rapid increase in research in this area. This initial treatment has now been replaced by penicillin, but organoarsenic compounds are still used today to treat trypanosiosis, or sleeping sickness, which is caused by a parasite in the blood. Arsenoamide, $C_{11}H_{12}AsNO_5S_2$, is used in vetinary medicines to treat heartworm in dogs.

Arsenilic acid, $C_6H_8AsNO_3$, and sodium arsenilate, $NaAsC_6H_8$, are used as antimicrobial agents in animal and poultry feed to prevent the growth of moulds. Another powerful antimicrobial agent is 10,10′-oxybisphenoxarsine (OBPA, **B1**), which is used extensively in the manufacture of plastics.

Arsenicals are also used as insecticides and herbicides. Methylarsonic acid, CH_3AsO_3, is used to control weeds in cotton and

B14.1 10,10′-Oxybisphenoxarsine

turf crops. It is water soluble but is strongly absorbed into the soil and is not easily leached out. The first arsenic-containing insecticide was Paris green, $Cu(CH_3CO_2)_2 \cdot 3Cu(AsO_2)_2$, which was manufactured in 1865 for the treatment of the Colorado potato beetle. Sodium arsenite, $NaAsO_2$, is used in poison baits to control grasshoppers and as a dip to prevent parasites in livestock.

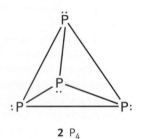

2 P_4

Example 14.1 Examining the electronic structure and chemistry of P₄

Draw the Lewis structure of P_4, and discuss its possible role as a ligand.

Answer In the Lewis structure of P_4 (**2**), there is a lone pair of electrons on each P atom. This structure, together with the fact that the electronegativity of phosphorus is moderate ($\chi_P = 2.06$), suggests that P_4 might be a moderately good donor ligand. Indeed, though rare, P_4 complexes are known.

Self-test 14.1 Consider the Lewis structure of a segment of the structure of bismuth shown in Fig. 14.2. Is this puckered structure consistent with the VSEPR model?

14.3 Nitrogen activation

Key points: The commercial Haber process requires high temperatures and pressures to yield ammonia, which is a major ingredient in fertilizers and an important chemical intermediate.

Nitrogen occurs in many compounds, but N_2 itself is strikingly unreactive. A few strong reducing agents can transfer electrons to the N_2 molecule at room temperature, leading to scission of the N—N bond, but usually the reaction needs extreme conditions. The prime example of this reaction is the slow reaction of lithium metal at room temperature, which yields Li_3N. Similarly, when magnesium (lithium's diagonal neighbour) burns in air it forms the nitride as well as the oxide.

The slowness of the reactions of N_2 appears to be the result of several factors. One is the strength of the N—N bond and hence the high activation energy required for breaking it. (The strength of this bond also accounts for the lack of nitrogen allotropes.) Another factor is the relatively large size of the HOMO–LUMO gap in N_2 (Section 2.8b), which makes the molecule resistant to simple electron-transfer redox processes. A third factor is the low polarizability of N_2, which does not encourage the formation of the highly polar transition states that are often involved in electrophilic and nucleophilic displacement reactions.

Cheap methods of nitrogen activation are highly desirable because they would have a profound effect on the economy, particularly in poorer agricultural economies. In the

Haber process for the production of ammonia, H_2 and N_2 are combined at high temperatures and pressures over an iron catalyst, as we discuss in detail in Section 14.6a. Much of the recent research aimed at more economical conversion of atmospheric N_2 into useful chemicals has focused on the way in which bacteria carry out the transformation at room temperature. The dominant process for the fixation of nitrogen in soils is the catalytic conversion of nitrogen to NH_4^+ by the enzyme nitrogenase, which occurs in the nitrogen-fixing bacteria found in the root nodules of legumes. The identity of this metalloenzyme is the topic of considerable research. In this connection, dinitrogen complexes of metals were discovered in 1965, and at about the same time the partial elucidation of the structure of nitrogenase indicated that the active site contains Fe and Mo atoms (Section 26.13). These developments led to optimism that efficient homogeneous catalysts might be developed in which metal ions would coordinate to N_2 and promote its reduction. Many N_2 complexes have in fact been prepared, and in some cases the preparation is as simple as bubbling N_2 through an aqueous solution of a complex:

$$[Ru(NH_3)_5(H_2O)]^{2+}(aq) + N_2(g) \rightarrow [Ru(NH_3)_5(N_2)]^{2+}(aq) + H_2O(l)$$

As with the isoelectronic CO molecule, end-on bonding is typical of N_2 when it acts as a ligand (**3**, Section 21.11). The N—N bond in the Ru(II) complex is only slightly altered from that in the free molecule. However, when N_2 is coordinated to a more strongly reducing metal centre, this bond is considerably lengthened by back-donation of electron density into the π^* orbitals of N_2.

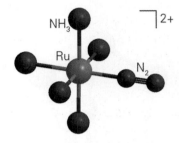

3 $[Ru(NH_3)_5(N_2)]^{2+}$

Simple compounds

The wide variety of possible oxidation states of the Group 15 elements can be understood to a large extent by considering the electron configuration of the elements, which is ns^2np^3. This configuration suggests that the highest oxidation state should be +5, as is indeed the case. According to the inert-pair effect, we should also expect the +3 oxidation state to become progressively more stable from nitrogen to bismuth, as is in fact observed.

Nitrogen has a very high electronegativity (exceeded by only oxygen, fluorine, and chlorine) and in many compounds, for example the nitrides, which contain the N^{3-} ion, and ammonia, NH_3, nitrogen has a negative oxidation state. Nitrogen achieves positive oxidation states only in compounds with more electronegative elements. Nitrogen does achieve the group oxidation state (+5), but only under much stronger oxidizing conditions than are necessary to achieve this state for the other elements of the group.

The distinctive nature of nitrogen is due in large part to its high electronegativity, its small atomic radius, and the absence of accessible d orbitals. Thus, nitrogen seldom has coordination numbers greater than 4 in simple molecular compounds, but the heavier elements frequently reach coordination numbers of 5 and 6, as in PCl_5 and AsF_6^-.

14.4 Binary compounds

The Group 15 elements form binary compounds on direct interaction with many elements.

(a) Nitrides

Key point: Nitrides are classified as saline, covalent, and interstitial.

Nitrogen forms binary compounds, the nitrides, with almost all the elements. They can be prepared by direct interaction of the element with nitrogen or ammonia or by thermal decomposition of an amide:

$$6\,Li(s) + N_2(g) \rightarrow 2\,Li_3N(s)$$
$$3\,Ca(s) + 2\,NH_3(l) \rightarrow Ca_3N_2(s) + 3\,H_2(g)$$
$$3\,Zn(NH_2)_2(s) \rightarrow Zn_3N_2(s) + 4\,NH_3(g)$$

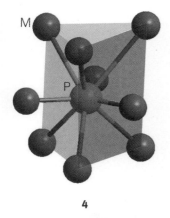

4

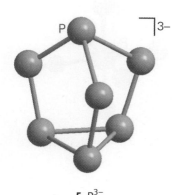

5 P_7^{3-}

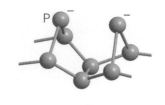

6 $(P_8^{2-})_n$

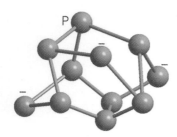

7 P_{11}^{3-}

The nitrides are classified as saline, covalent, and interstitial. The compounds of nitrogen with hydrogen, oxygen, and the halogens are treated separately.

The **saline nitrides** can be regarded as containing the nitride ion, N^{3-}. However, the high charge of this ion means that it is highly polarizing (Section 1.9e) and saline nitrides are likely to have considerable covalent character. Saline nitrides occur for lithium, Li_3N, and the Group 2 elements, M_3N_2.

The **covalent nitrides**, in which the E—N bond is covalent, possess a wide range of properties depending on the element to which nitrogen is bonded. Some examples of covalent nitrides are boron nitride, BN, cyanogen, $(CN)_2$, phosphorus nitride, P_3N_5, tetrasulfur tetranitride, S_4N_4, and disulfur dinitride, S_2N_2. These compounds are discussed in the context of the other element.

The largest category of nitrides consists of the **interstitial nitrides** of the d-block elements with formula MN, M_2N, or M_4N. The N atom occupies some or all of the octahedral sites within the cubic or hexagonal close-packed lattice of metal atoms. The compounds are hard and inert, with a metallic lustre and conductivity. They are widely used as refractory materials and find applications as crucibles, high-temperature reaction vessels, and thermocouple sheaths.

(b) Phosphides

Key point: Phosphides may be metal-rich or phosphorus-rich.

Like nitrogen, phosphorus forms compounds with almost every element in the periodic table. The compounds of phosphorus with hydrogen, oxygen, and the halogens are discussed separately. The phosphides of other elements can be prepared by heating the appropriate element with red phosphorus in an inert atmosphere:

$$n\,M + m\,P \rightarrow M_nP_m$$

There are many varieties of phosphides, with formulas ranging from M_4P to MP_{15}. They include metal-rich phosphides, in which $M:P > 1$, monophosphides, in which $M:P = 1$, and phosphorus-rich phosphides, in which $M:P < 1$. Metal-rich phosphides are usually very inert, hard, brittle refractory materials and resemble the parent metal in having high electrical and thermal conductivities. The structures have a trigonal prismatic arrangement of six, seven, eight, or nine metal ions around a P atom (**4**). Monophosphides adopt a variety of structures depending on the relative size of the other atom. For example, AlP adopts the zinc-blende structure, SnP adopts the rock-salt structure, and VP adopts the nickel-arsenide structure (Section 3.9). Phosphorus-rich phosphides have lower melting points and are less stable than metal-rich phosphides and monophosphides. They are semiconductors rather than conductors. The phosphorus atoms may be arranged in rings, chains, or cages; for example, P_7^{3-} (**5**), $(P_8^{2-})_n$ (**6**), and P_{11}^{3-} (**7**).

(c) Arsenides, antimonides, and bismuthides

Key point: Indium and gallium arsenides and antimonides are semiconductors.

The compounds formed between metals and arsenic, antimony, and bismuth can be prepared by direct reaction of the elements:

$$Ni(s) + As(s) \rightarrow NiAs(s)$$

The arsenides and antimonides of the Group 13 elements indium and gallium are semiconductors. Gallium arsenide (GaAs) is the most important and is used to make devices such as integrated circuits, light-emitting diodes, and laser diodes. Its band gap is similar to that of silicon and larger than those of other Group 13/15 semiconductors (Table 12.6; Section 23.18). Gallium arsenide is superior to silicon for such applications because it has higher electron mobility and the devices produce less electronic noise. Silicon still has major advantages over GaAs in the sense that silicon is cheap and the wafers are stronger than those of GaAs, so processing is easier. Gallium arsenide integrated circuits are commonly used in mobile phones, satellite communications, and some radar systems.

14.5 Azides

Key points: Azides are toxic and unstable; they are used as detonators in explosives.

Azides may be synthesized by the oxidation of sodium amide with either NO_3^- ions or N_2O at elevated temperatures:

$$3\ NH_2^- + NO_3^- \xrightarrow{175°C} N_3^- + 3\ OH^- - + NH_3$$

$$2\ NH_2^- + N_2O \xrightarrow{190°C} N_3^- + OH^- + NH_3$$

The average oxidation number of nitrogen in the azide ion, N_3^-, is $-\frac{1}{3}$. The ion is isoelectronic with both dinitrogen oxide, N_2O, and carbon dioxide and, like these two molecules, is linear. It is a reasonably strong Brønsted base, the pK_a of its conjugate acid, hydrazoic acid, HN_3, being 4.77. It is also a good ligand towards d-block ions. However, heavy-metal complexes or salts, such as $Pb(N_3)_2$ and $Hg(N_3)_2$, are shock-sensitive detonators.

Ionic azides such as NaN_3 are thermodynamically unstable but kinetically inert; they can be handled at room temperature. Sodium azide is toxic and is used as a chemical preservative and in pest control. When alkali metal azides are heated or detonated by impact they explode, liberating N_2; this reaction is used in the inflation of air bags in cars.

Compounds containing the polynitrogen cation, N_5^+, have been synthesized from species containing the N_3^- and N_2F^+ ions. For example, N_5AsF_6 is prepared from N_2FAsF_6 and HN_3 in anhydrous HF solvent:

$$N_2FAsF_6(sol) + HN_3(sol) \rightarrow N_5AsF_6(sol) + HF(l)$$

The compound is a white solid that decomposes explosively above 250°C. It is a powerful oxidizing agent and ignites organic material even at low temperatures. Salts of larger dipositive cations can be prepared from the salts of monopositive cations in anhydrous HF:

$$N_5SbF_6(sol) + Cs_2SnF_6(sol) \rightarrow (N_5)_2SnF_6(sol) + 2\ CsSbF_6(s)$$

The product is a white solid that is friction sensitive and decomposes to N_5SnF_6 above 250°C. This product, N_5SnF_6, is stable up to 500°C.

Example 14.2 The decomposition of sodium azide

A typical airbag contains approximately 50 g of NaN_3. Calculate the volume of nitrogen produced when the azide is detonated at room temperature and pressure (20°C and 1.0 atm).

Answer The decomposition reaction is $2\ NaN_3(s) \rightarrow 2\ Na(s) + 3\ N_2(g)$. The 50 g of sodium azide contains 0.77 mol NaN_3, which liberates 1.2 mol N_2. This amount occupies 26 dm^3 at 20°C and 1.0 atm. As the airbag is restricted in volume, the pressure of nitrogen in the airbag will be high, so providing protection to the driver.

Self-test 14.2 Why does an airbag also contain KNO_3 and SiO_2?

14.6 Hydrides

All the Group 15 elements form simple compounds with hydrogen. Ammonia is of great industrial importance and is produced on a massive scale. All the EH_3 hydrides are toxic. Nitrogen also forms a catenated hydride, hydrazine, N_2H_4.

(a) Ammonia

Key points: Ammonia is produced by the Haber process; it is used to manufacture fertilizers and many other useful nitrogen-containing chemicals.

Ammonia, NH_3 (**8**), is a pungent, colourless gas that can be toxic at high levels of exposure. It is produced in huge quantities worldwide for use as a fertilizer and as a

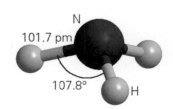

8 Ammonia, NH_3, C_{3v}

primary source of nitrogen in the production of many chemicals. As already mentioned, the Haber process is used for the entire global production. In this process, nitrogen and hydrogen combine at high temperature (450°C) and pressure (100 atm):

$$N_2(g) + 3\ H_2(g) \rightarrow 2\ NH_3(g)$$

This process is the direct combination of N_2 and H_2 carried out over a promoted iron catalyst; the promoters required include SiO_2, MgO, and other oxides (Chapter 25). The reaction is carried out at high temperature and in the presence of a catalyst to overcome the kinetic inertness of N_2, and at high pressure to overcome the thermodynamic effect of an unfavourable equilibrium constant at the operating temperature.

So novel and great were the chemical and engineering problems arising from the then (early twentieth century) uncharted area of large-scale high-pressure technology, that two Nobel Prizes were awarded. One went to Fritz Haber (in 1918), who developed the chemical process. The other went to Carl Bosch (in 1931), the chemical engineer who designed the first plants to realize Haber's process. The **Haber–Bosch process** has had a major impact on civilization because ammonia is the primary source of most nitrogen-containing compounds, including fertilizers and most commercially important compounds of nitrogen. Before the development of the process the main sources of nitrogen for fertilizers were guano (bird droppings) and saltpetre, which had to be mined and transported from South America. In the early twentieth century there were predications of widespread starvation across Europe, predications that were never realized because of the widespread availability of nitrogen-based fertilizers.

The boiling point of ammonia is −33°C, which is higher than that of the hydrides of the other elements in the group and indicates the influence of extensive hydrogen bonding. Liquid ammonia is a useful nonaqueous solvent for solutes such as alcohols, amines, ammonium salts, amides, and cyanides. Reactions in liquid ammonia closely resemble those in aqueous solution, as indicated by the following autoprotolysis equilibria:

$$2\ H_2O(l) \rightleftharpoons H_3O^+(aq) + OH^-(aq)$$
$$2\ NH_3(l) \rightleftharpoons NH_4^+(am) + NH_2^-(am)$$

Many of the reactions are analogous to those carried out in water. For example, simple acid–base neutralization reactions can be carried out:

$$NH_4Cl(am) + NaNH_2(am) \rightleftharpoons NaCl(am) + 2\ NH_3(l)$$

Ammonia is an excellent solvent for the Group 1 metals: for instance, it is possible to dissolve 330 g of caesium in 100 g of liquid ammonia at −50°C. These highly coloured, electrically conducting solutions contain solvated electrons (Section 10.10).

Ammonia is a water-soluble weak base:

$$NH_3(aq) + H_2O(l) \rightleftharpoons NH_4^+(aq) + OH^-(aq) \qquad pK_b = 4.75$$

The chemical properties of ammonium salts are very similar to those of Group 1 salts, especially of K^+ and Rb^+. They are soluble in water and solutions of the salts of strong acids, such as NH_4Cl, are acidic due to the equilibrium:

$$NH_4^+(aq) + H_2O(l) \rightleftharpoons H_3O^+(aq) + NH_3(aq) \qquad pK_a = 9.25$$

Ammonium salts decompose readily on heating and for many salts, such as the halides, carbonate, and sulfates, ammonia is evolved:

$$NH_4Cl(s) \rightarrow NH_3(g) + HCl(g)$$
$$(NH_4)_2SO_4(s) \rightarrow NH_3(g) + H_2SO_4(l)$$

When the anion is oxidizing, as in the case of NO_3^-, ClO_4^-, and $Cr_2O_7^{2-}$, the NH_4^+ is oxidized to N_2 or N_2O.

$$NH_4NO_3(s) \rightarrow N_2O(g) + 2\ H_2O(g)$$

Ammonium nitrate is used in great quantities as a fertilizer. It is also a component of explosives. When it is heated strongly or detonated, the decomposition of 2 mol

$NH_4NO_3(s)$ produces 7 mol of gaseous molecules, corresponding to an increase in volume from about approximately $200\,cm^3$ to about $140\,dm^3$, a factor of 700:

$$2\,NH_4NO_3(s) \rightarrow 2\,N_2(g) + O_2(g) + 4\,H_2O(g)$$

Nitrate fertilizers are often mixed with materials such as calcium carbonate or ammonium sulfate to make them more stable. Ammonium sulfate and the ammonium hydrogenphosphates, $NH_4H_2PO_4$ and $(NH_4)_2HPO_4$, are also used as fertilizers because phosphate is a plant nutrient. Ammonium perchlorate is used as the oxidizing agent in solid-fuel rocket propellants.

(b) Hydrazine and hydroxylamine

Key point: Hydrazine is a weaker base than ammonia and forms two series of salts.

Hydrazine, N_2H_4, is a fuming, colourless liquid with an odour like that of ammonia. It has a liquid range similar to that of water (2–114°C), indicating the presence of hydrogen bonding. In the liquid phase hydrazine adopts a *gauche* conformation around the N—N bond (**9**).

Hydrazine is manufactured by the *Raschig process*, in which ammonia and sodium hypochlorite react in dilute aqueous solution. The reaction proceeds through several steps, which can be simplified to:

$$NH_3(aq) + NaOCl(aq) \rightarrow NH_2Cl(aq) + NaOH(aq)$$
$$2\,NH_3(aq) + NH_2Cl(aq) \rightarrow N_2H_4(aq) + NH_4Cl(aq)$$

There is a competing side reaction that is catalysed by *d*-metal ions:

$$N_2H_4(aq) + 2\,NH_2Cl(aq) \rightarrow N_2(g) + 2\,NH_4Cl(aq)$$

Gelatine is added to the reaction mixture to trap the *d*-metal ions by forming a complex with them. The dilute aqueous solution of hydrazine so produced is converted to a concentrated solution of hydrazine hydrate, $N_2H_4 \cdot H_2O$, by distillation. This product is often preferred commercially as it is cheaper than hydrazine and has a wider liquid range. Hydrazine is produced by distillation of the hydrate in the presence of a drying agent such as solid NaOH or KOH.

Hydrazine is a weaker base than ammonia:

$$N_2H_4(aq) + H_2O(l) \rightleftharpoons N_2H_5^+(aq) + OH^-(aq) \qquad pK_b = 7.93$$
$$N_2H_5^+(aq) + H_2O(l) \rightleftharpoons N_2H_6^{2+}(aq) + OH^-(aq) \quad pK_b = -1.05$$

It reacts with acids to form two series of salts, N_2H_5X and $N_2H_6X_2$.

The major use of hydrazine and its methyl derivatives, CH_3NHNH_2 and $(CH_3)_2NNH_2$, is as a rocket fuel. Hydrazine is also used as a foam blowing agent and a treatment in boiler water to scavenge dissolved oxygen and prevent oxidation of pipes. Both N_2H_4 and $N_2H_5^+$ are reducing agents and are used in the recovery of precious metals.

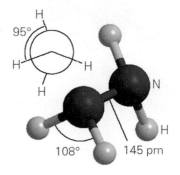

9 Hydrazine, N_2H_4

Exercise 14.3 Evaluating rocket fuels

Both hydrazine, N_2H_4, and dimethylhydrazine, $N_2H_2(CH_3)_2$, are used as rocket fuels. Given the following data, suggest which would be the most efficient fuel thermochemically.

	$\Delta_f H^{\ominus}/(kJ\,mol^{-1})$
$N_2H_4(l)$	+50.6
$N_2H_2(CH_3)_2(l)$	+42.0
$CO_2(g)$	−394
$H_2O(g)$	−242

Answer The combustion reactions are $N_2H_4(l) + O_2(g) \rightarrow N_2(g) + 2\,H_2O(g)$ and $N_2H_2(CH_3)_2(l) + O_2(g) \rightarrow N_2(g) + 4\,H_2O(g) + 2\,CO_2(g)$. The enthalpy of reaction is calculated from $\sum \Delta_f H^{\ominus}$ (products) $- \sum \Delta_f H^{\ominus}$ (reactants), giving values of $-534\,kJ\,mol^{-1}$ for N_2H_4 and $-1796\,kJ\,mol^{-1}$

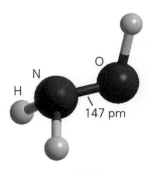

10 Hydroxylamine, NH₂OH

for $N_2H_2(CH_3)_2$. An important factor for selecting rocket fuels is the specific enthalpy, which for these fuels has values -16.7 and $-29.9\,kJ\,g^{-1}$, respectively, indicating that $N_2H_2(CH_3)_2$ is the better fuel when mass is significant.

Self-test 14.3 Refined hydrocarbons and liquid hydrogen are also used as rocket fuel. What are the advantages of dimethylhydrazine over these fuels?

Hydroxylamine, NH_2OH (**10**), is a colourless, hygroscopic solid with a low melting point ($32°C$). It is usually available as one of its salts or in aqueous solution. It is a weaker base than either ammonia or hydrazine:

$$NH_2OH(aq) + H_2O(l) \rightleftharpoons NH_3OH^+(aq) + OH^-(aq) \qquad pK_b = 8.18$$

Anhydrous hydroxylamine can be prepared by adding sodium butoxide, NaC_4H_9O (NaOBu), to a solution of hydroxylamine hydrochloride in 1-butanol. The NaCl produced is filtered off and the hydroxylamine precipitated by the addition of ether.

$$[NH_3OH]Cl(sol) + NaOBu \rightarrow NH_2OH(sol) + NaCl(s) + BuOH(l)$$

The major commercial use of hydroxylamine is in the synthesis of caprolactam, which is an intermediate in the manufacture of nylon.

(c) Phosphine, arsine, and stibine

Key points: Unlike liquid ammonia, liquid phosphine, arsine, and stibine do not associate through hydrogen bonding; their much more stable alkyl and aryl analogues are useful soft ligands.

In contrast to the commanding role that ammonia plays in nitrogen chemistry, the highly poisonous hydrides of the heavier nonmetallic elements of Group 15 (particularly phosphine, PH_3, and arsine, AsH_3) are of minor importance in the chemistry of their respective elements. Both phosphine and arsine are used in the semiconductor industry to dope silicon or to prepare other semiconductor compounds, such as GaAs, by chemical vapour deposition. These thermal decomposition reactions reflect the positive Gibbs energy of formation of these hydrides.

The commercial synthesis of PH_3 uses the disproportionation of white phosphorus in basic solution:

$$P_4(s) + 3\,OH^-(aq) + 3\,H_2O(l) \rightarrow PH_3(g) + 3\,H_2PO_2{}^-(aq)$$

Arsine and stibine may be prepared by the protolysis of compounds that contain an electropositive metal in combination with arsenic or antimony:

$$Zn_3E_2(s) + 6\,H_2O^+(aq) \rightarrow 2\,EH_3(g) + 3Zn^{2+}(aq) + H_2O(l) \quad (E = As, Sb)$$

Phosphine and arsine are poisonous gases that readily ignite in air, but the much more stable organic derivatives PR_3 and AsR_3 ($R = $ alkyl or aryl groups) are widely used as ligands in metal coordination chemistry. In contrast with the hard donor properties of ammonia and alkylamine ligands, the organophosphines and organoarsines, such as $P(C_2H_5)_3$ and $As(C_6H_5)_3$, are soft ligands and are therefore often incorporated into metal complexes having central metal atoms in low oxidation states. The stability of these complexes correlates with the soft acceptor nature of the metals in low oxidation states, and the stability of soft-donor–soft-acceptor combinations (Section 4.15).

It is evident from the boiling points plotted in Fig. 9.4 that PH_3, AsH_3, and SbH_3 are subject to little, if any, hydrogen bonding with themselves; however, PH_3 and AsH_3 can be protonated by strong acids, such as HI, to form phosphonium and stibonium ions, PH_4^+ and SbH_4^+, respectively. All the Group 15 hydrides are pyramidal, but the bond angle decreases down the group:

$$NH_3, 107.8° \qquad PH_3, 93.6° \qquad AsH_3, 91.8° \qquad SbH_3, 91.3°$$

The large change in bond angle has been attributed to a decrease in the extent of sp^3 hybridization from NH_3 to SbH_3 but is more likely to be due to steric effects. The E—H bonding pairs of electrons will repel each other. This repulsion is greatest when the central element, E, is small, as in NH_3 and the H atoms will be as far away from each other as possible in a near-tetrahedral arrangement. As the size of the central atom is increased down the series, the repulsion between bonding pairs decreases and the bond angle is close to 90°.

14.7 Halides

Halogen compounds of phosphorus, arsenic, and antimony are numerous and important in synthetic chemistry. Trihalides are known for all the Group 15 elements. However, whereas pentafluorides are known for all members of the group from P to Bi, pentachlorides are known only for P, As, and Sb, and the pentabromide is known only for P. Nitrogen does not reach its group oxidation state (+5) in neutral binary halogen compounds, but it does achieve it in NF_4^+. Presumably, an N atom is too small for NF_5 to be sterically feasible. The difficulty of oxidizing Bi(III) to Bi(V) by chlorine or bromine is an example of the inert-pair effect (Chapter 12). Bismuth pentafluoride, BiF_5, exists, but $BiCl_5$ and $BiBr_5$ do not.

(a) Nitrogen halides

Key point: Except for NF_3, nitrogen trihalides have limited stability and nitrogen triiodide is dangerously explosive.

Nitrogen trifluoride, NF_3, is the only exoergic binary halogen compound of nitrogen. This pyramidal molecule is not very reactive. Thus, unlike NH_3, it is not a Lewis base because the strongly electronegative F atoms make the lone pair of electrons unavailable: whereas the polarity in of the N—H bond in NH_3 is $^{\delta-}N-H^{\delta+}$, that of the N—F bond in NF_3 is $^{\delta+}N-H^{\delta-}$. Nitrogen trifluoride can be converted into the N(V) species NF_4^+ by the reaction

$$NF_3(l) + 2\,F_2(g) + SbF_3(l) \rightarrow [NF_4^+][SbF_6^-](sol)$$

Nitrogen trichloride, NCl_3, is a highly endoergic, explosive, yellow oil. It is prepared commercially by the electrolysis of an aqueous solution of ammonium chloride and was once used as an oxidizing bleach for flour. The electronegativities of nitrogen and chlorine are the same and the N—Cl bond is almost nonpolar. Nitrogen tribromide, NBr_3, is an explosive, deep red oil. Nitrogen triiodide, NI_3, is known only in the form of the highly explosive ammoniate $NI_3 \cdot NH_3$. Nitrogen is more electronegative than both bromine and iodine, and so the N—X bonds are polar in the sense $^{\delta-}N-H^{\delta+}$. Formally, the oxidation numbers are −3 for the N and +1 for each halogen.

(b) Halides of the heavy elements

Key points: Whereas the halides of nitrogen have limited stability, their heavier congeners form an extensive series of compounds; the trihalides and pentahalides are useful starting materials for the synthesis of derivatives by metathetical replacement of the halide.

The trihalides and pentahalides of Group 15 elements other than nitrogen are used extensively in synthetic chemistry and their simple empirical formulas conceal an interesting and varied structural chemistry.

The trihalides range from gases and volatile liquids, such as PF_3 (b.p. −102°C) and AsF_3 (b.p. 63°C), to solids, such as BiF_3 (m.p. 649°C). A common method of preparation is direct reaction of the element and halogen. For phosphorus, the trifluoride is prepared by metathesis of the trichloride and a fluoride:

$$2\,PCl_3(l) + 3\,ZnF_2(s) \rightarrow 2\,PF_3(g) + 3\,ZnCl_2(s)$$

The trichlorides PCl_3, $AsCl_3$, and $SbCl_3$ are useful starting materials for the preparation of a variety of alkyl, aryl, alkoxy, and amino derivatives because they are susceptible to protolysis and metathesis:

$$ECl_3(sol) + 3\,EtOH(l) \rightarrow E(OEt)_3(sol) + 3\,HCl(sol) \quad (E = P, As, Sb)$$

$$ECl_3(sol) + 6\,Me_2NH(sol) \rightarrow E(NMe_2)_3(sol) + 3\,[Me_2NH_2]Cl(sol) \quad (E = P, As, Sb)$$

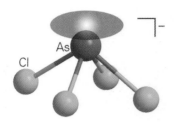

11 AsCl$_4^-$, C_{4v}

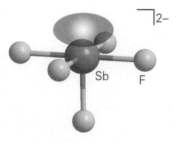

12 SbF$_5^{2-}$, C_{4v}

Phosphorus trifluoride, PF$_3$, is an interesting ligand because in some respects it resembles CO. Like CO, it is a weak σ donor but a strong π acceptor, and complexes of PF$_3$ exist that are the analogues of carbonyls, such as Ni(PF$_3$)$_4$, the analogue of Ni(CO)$_4$ (Section 21.18). The π-acceptor character is attributed to a P—F antibonding LUMO, which has mainly phosphorus p-orbital character. The trihalides also act as mild Lewis acids towards Lewis bases such as trialkylamines and halides. Many halide complexes have been isolated, such as the simple mononuclear species AsCl$_4^-$ (**11**) and SbF$_5^{2-}$ (**12**). More complex dinuclear and polynuclear anions linked by halide bridges, such as the polymeric chain ([BiBr$_3$]$^{2-}$)$_n$ in which Bi(I) is surrounded by a distorted octahedron of Br atoms, are also known.

The pentahalides vary from gases, such as PF$_5$ (b.p. $-85°$C) and AsF$_5$ (b.p. $-53°$C), to solids, such as PCl$_5$ (sublimes at $162°$C) and BiF$_5$ (m.p. $154°$C). The five-coordinate gas-phase molecules are trigonal bipyramidal. In contrast to PF$_5$ and AsF$_5$, SbF$_5$ is a highly viscous liquid in which the molecules are associated through F-atom bridges. In solid SbF$_5$ these bridges result in a cyclic tetramer (**13**), which reflects the tendency of Sb(V) to achieve a coordination number of 6. A related phenomenon occurs with PCl$_5$, which in the solid state exists as [PCl$_4^+$][PCl$_6^-$]. In this case, the ionic contribution to the lattice enthalpy provides the driving force for the transfer of a Cl$^-$ ion from one PCl$_5$ molecule to another. Another contributing factor may be the more efficient packing of the PCl$_4$ and PCl$_6$ units compared to the less efficient stacking of PCl$_5$ units. The pentafluorides of phosphorus, arsenic, antimony, and bismuth are strong Lewis acids (Section 4.11). SbF$_5$ is a very strong Lewis acid; it is much stronger, for example, than the aluminium halides. When SbF$_5$ or AsF$_5$ is added to anhydrous HF, a *superacid* is formed that is more strongly acidic than HF (see Box 4.2):

$$\text{SbF}_5(\text{l}) + \text{HF}(\text{l}) \rightarrow \text{H}_2\text{F}^+(\text{sol}) + \text{SbF}_6^-(\text{sol})$$

Of the pentachlorides, PCl$_5$ and SbCl$_5$ are stable whereas AsCl$_5$ is very unstable. This difference is a manifestation of the alternation effect. The instability of AsCl$_5$ is attributed to the increased effective nuclear charge arising from the poor shielding of the $3d$ electrons, which leads to a 'd-block contraction' and a lowering of the energy of the $4s$ orbitals in As. Consequently, it is more difficult to promote a $4s$ electron to form AsCl$_5$.

The pentahalides of phosphorus and antimony are very useful in syntheses. Phosphorus pentachloride, PCl$_5$, is widely used in the laboratory and in industry as a starting material and some of its characteristic reactions are shown in Fig. 14.5. Note, for example, that reaction of PCl$_5$ with Lewis acids yields PCl$_4^+$ salts, simple Lewis bases like F$^-$ give six-coordinate complexes such as PF$_6^-$. Compounds containing the NH$_2$ group lead to the formation of PN bonds, and the interaction of PCl$_5$ with either H$_2$O or P$_4$O$_{10}$ yields O=PCl$_3$.

14.8 Oxohalides

Key points: Nitrosyl and nitryl halides are useful halogenating agents; phosphoryl halides are important industrially in the synthesis of organophosphorus derivatives.

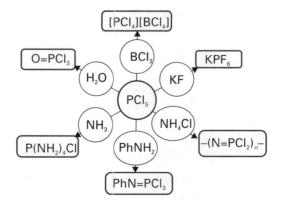

Fig. 14.5 The uses of phosphorus pentachloride.

Nitrogen forms the nitrosyl halides, NOX, and the nitryl halides, NO_2X. The nitrosyl halides and NO_2F are prepared by direct interaction of the halogen with NO or NO_2, respectively:

$$2\,NO(g) + Cl_2(g) \rightarrow 2\,NOCl(g)$$
$$2\,NO_2(g) + F_2(g) \rightarrow 2\,NO_2F(g)$$

They are all reactive gases and the oxofluorides and oxochlorides are useful fluorinating and chlorinating agents.

Phosphorus readily forms the phosphoryl halides $POCl_3$ and $POBr_3$ by the reaction of the trihalides PX_3 with O_2 at room temperature. The fluorine and iodine analogues are prepared by the reaction of $POCl_3$ with a metal fluoride or iodide:

$$POCl_3(l) + 3\,NaF(s) \rightarrow POF_3(g) + 3\,NaCl(s)$$

All the molecules are tetrahedral and contain a P=O bond. POF_3 is gaseous, $POCl_3$ is a colourless liquid, $POBr_3$ is a brown solid, and POI_3 is a violet solid. They are all readily hydrolysed, fume in air, and form adducts with Lewis acids. They provide a route to the synthesis of organophosphorus compounds, which are manufactured on a large scale for use as plasticizers, oil additives, pesticides, and surfactants. For example, reaction with alcohols and phenols gives $(RO)_3PO$ and Grignard reagents (Section 11.10) yield R_nPOCl_{n-1}:

$$3\,ROH(l) + POCl_3(l) \rightarrow (RO)_3PO(sol) + 3\,HCl(sol)$$
$$n\,RMgBr(sol) + POCl_3(sol) \rightarrow R_nPOCl_{n-1}(sol) + n\,MgBrCl(s)$$

14.9 Oxides and oxoanions of nitrogen

Key point: Reactions of nitrogen–oxygen compounds that liberate or consume N_2 are generally very slow at normal temperatures and pH = 7.

We can infer the redox properties of the compounds of the elements in Group 15 in acidic aqueous solution from the Frost diagram shown in Fig. 14.6. The steepness of the slopes of the lines on the far right of the diagram show the thermodynamic tendency for reduction of the +5 oxidation states of the elements. They show, for instance, that Bi_2O_5 is potentially a very strong oxidizing agent, which is consistent with the inert-pair effect and the tendency of Bi(V) to form Bi(III). The next strongest oxidizing agent is NO_3^-. Both As(V) and Sb(V) are milder oxidizing agents, and P(V), in the form of phosphoric acid, is a very weak oxidant.

The redox properties of nitrogen are important because of its widespread occurrence in the atmosphere, the biosphere, industry, and the laboratory. Nitrogen chemistry is quite complex, partly because of the large number of accessible oxidation states but also because reactions that are thermodynamically favourable are often slow or have rates that depend crucially on the identity of the reactants. As the N_2 molecule is kinetically inert, redox reactions that consume N_2 are slow. Moreover, for mechanistic reasons that seem to be different in each case, the formation of N_2 is often slow and may be sidestepped in aqueous solution (Fig. 14.7). As with several other p-block elements, the barriers to reaction of high oxidation state oxoanions, such as NO_3^-, are greater than for low oxidation state oxoanions, such as NO_2^-. We should also remember that low pH enhances the oxidizing power of oxoanions (Section 5.12). Low pH also often accelerates their oxidizing reactions by protonation, and this step is thought to facilitate subsequent NO bond breaking.

Table 14.2 summarizes some of the properties of the nitrogen oxides, and Table 14.3 does the same for the nitrogen oxoanions. Both tables will help us to navigate through the details of their properties.

(a) Nitrogen(V) oxides and oxoanions

Key points: The nitrate ion is a strong but slow oxidizing agent at room temperature; strong acid and heating accelerate the reaction.

The most common source of N(V) is nitric acid, HNO_3, which is a major industrial chemical used in the production of fertilizers, explosives, and a wide variety of

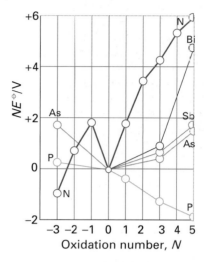

Fig. 14.6 Frost diagram for the elements of the nitrogen group in acidic solution. The species with oxidation number −3 are NH_3, PH_3, and AsH_3, and those with oxidation numbers −2 and −1 are N_2H_4 and NH_2OH, respectively. The positive oxidation states refer to the most stable oxo or hydroxo species in acidic solution, and may be oxides, oxoacids, or oxoanions.

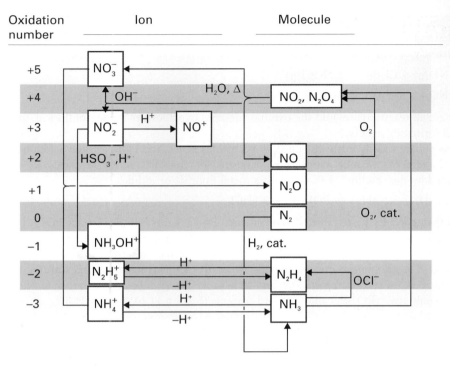

Oxidation number	Ion		Molecule

Fig. 14.7 The interconversion of important nitrogen species.

nitrogen-containing chemicals. It is produced by modern versions of the *Ostwald process*, which make use of an indirect route from N_2 to the highly oxidized compound HNO_3 via the fully reduced compound NH_3. Thus, after nitrogen has been reduced to the -3 state as NH_3 by the Haber process, it is oxidized to the $+4$ state:

$$4\,NH_3(g) + 7\,O_2(g) \rightarrow 6\,H_2O(g) + 4\,NO_2(g) \quad \Delta_r G^{\ominus} = -308.0\,kJ\,(mol\,NO_2)^{-1}$$

The NO_2 then undergoes disproportionation into N(II) and N(V) in water at elevated temperatures:

$$3\,NO_2(aq) + H_2O(l) \rightarrow 2\,HNO_3(aq) + NO(g) \quad \Delta_r G^{\ominus} = -5.0\,kJ\,(mol\,NO_3)^{-1}$$

All the steps are thermodynamically favourable. The by-product NO is oxidized with O_2 to NO_2 and recirculated. Such an indirect route is used because the direct oxidation of N_2 to NO_2 is thermodynamically unfavourable, with $\Delta_f G^{\ominus}(NO_2) = +51\,kJ\,mol^{-1}$. In part, this endoergic character is due to the great strength of the $N\equiv N$ bond.

Standard potential data imply that the NO_3^- ion is a moderately strong oxidizing agent. However, its reactions are generally slow in dilute acid solution. Because protonation of an oxygen atom promotes NO bond breaking, concentrated HNO_3 (in which NO_3^- is protonated) undergoes more rapid reactions than the dilute acid (in which HNO_3 is fully ionized). It is also a thermodynamically more potent oxidizing agent at low pH. A sign of this oxidizing character is the yellow colour of the concentrated acid, which indicates its instability with respect to decomposition into NO_2:

$$4\,HNO_3(aq) \rightarrow 4\,NO_2(aq) + O_2(g) + 2\,H_2O(l)$$

This decomposition is accelerated by light and heat.

The reduction of NO_3^- ions rarely yields a single product, as so many lower oxidation states of nitrogen are available. For example, a strong reducing agent such as zinc can reduce a substantial proportion of dilute HNO_3 as far as oxidation state -3:

$$HNO_3(aq) + 4\,Zn(s) + 9\,H^+(aq) \rightarrow NH_4^+(aq) + 3\,H_2O(l) + 4\,Zn^{2+}(aq)$$

A weaker reducing agent, such as copper, proceeds only as far as oxidation state $+4$ in the concentrated acid:

$$2\,HNO_3(aq) + Cu(s) + 2\,H^+(aq) \rightarrow 2\,NO_2(g) + Cu^{2+}(aq) + 2\,H_2O(l)$$

Table 14.2 Oxides of nitrogen

Oxidation number	Formula	Name	Structure (gas phase)	Comments
−1	N_2O	Nitrous oxide (Dinitrogen oxide)	 119 pm	Colourless gas, not very reactive
+2	NO	Nitric oxide (Nitrogen monoxide)	 115 pm	Colourless, reactive, paramagnetic gas
+3	N_2O_3	Dinitrogen trioxide	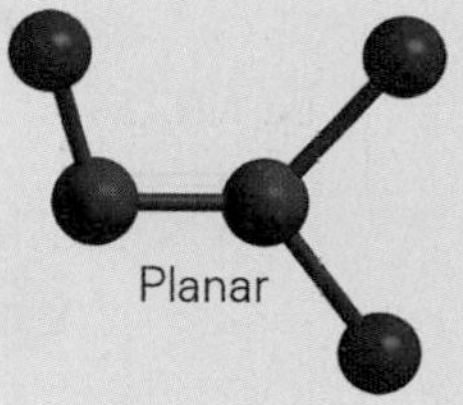 Planar	Blue solid (m.p. −101°C) and dissociates into NO and NO_2 in the gas phase
+4	NO_2	Nitrogen dioxide	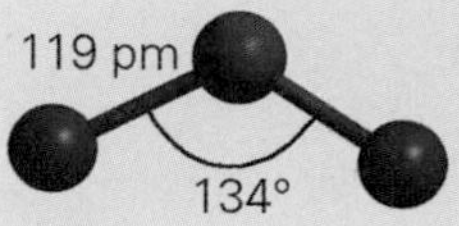 119 pm 134°	Brown, reactive, paramagnetic gas
+4	N_2O_4	Dinitrogen tetroxide	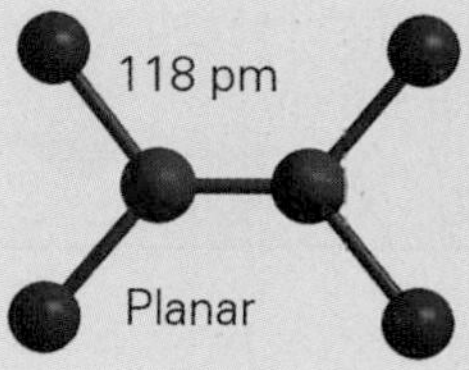 118 pm Planar	Colourless liquid (m.p. −11°C); in equilibrium with NO_2 in the gas phase
+5	N_2O_5	Dinitrogen pentoxide (Dinitrogen oxide)	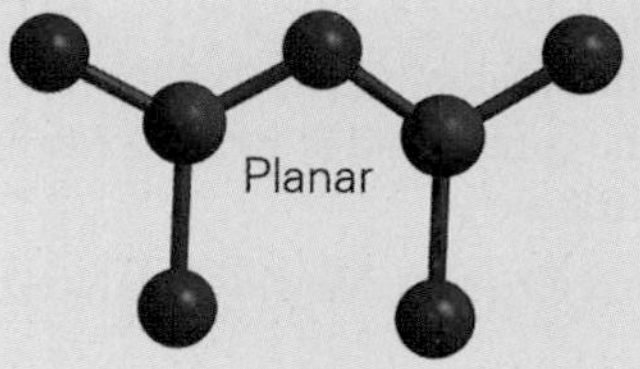 Planar	Colourless, ionic solid $[NO_2][NO_3]$ (m.p. 32°C); unstable

With the dilute acid, the +2 oxidation state is favoured, and NO is formed:

$$2\ NO_3^-(aq) + 3\ Cu(s) + 8\ H^+(aq) \rightarrow 2\ NO(g) + 3\ Cu^{2+}(aq) + 4\ H_2O(l)$$

When concentrated nitric acid is mixed with concentrated hydrochloric acid, the orange, fuming *aqua regia* is formed:

$$HNO_3(aq) + 3\ HCl(aq) \rightarrow NOCl(g) + Cl_2(g) + 2\ H_2O(l)$$

Aqua regia is Latin for 'royal water' and was so called by alchemists because of its ability to dissolve the noble metals gold and platinum. Gold will dissolve to a very small extent in concentrated nitric acid. In aqua regia the chloride ions present react immediately with the Au^{3+} ions formed to produce $[AuCl_4]^-$ and thereby remove Au^{3+} from the product side of the oxidation reaction:

$$Au(s) + NO_3^-(aq) + 4\ Cl^-(aq) + 4\ H^+(aq) \rightarrow [AuCl_4]^-(aq) + NO(g) + 2\ H_2O(l)$$

The anhydride of nitric acid is N_2O_5. It is a crystalline solid with the more accurate formula $[NO_2^+][NO_3^-]$ and can be prepared by dehydration of nitric acid

Table 14.3 Nitrogen oxoanions

Oxidation number	Formula	Common name	Structure	Comments
+1	$N_2O_2^{2-}$	Hyponitrite	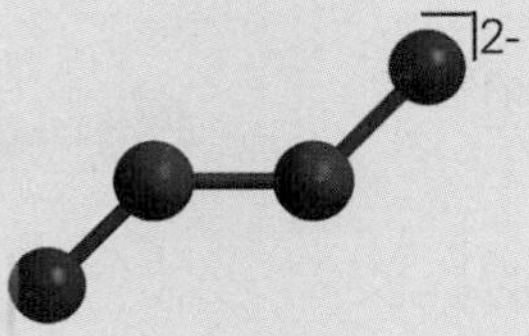	Usually acts as a reducing agent
+3	NO_2^-	Nitrite	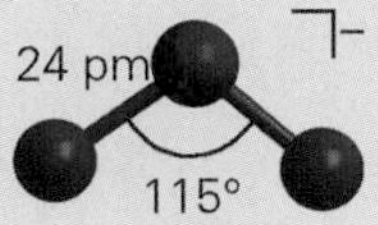 24 pm, 115°	Weak base; acts as an oxidizing agent and a reducing agent
+3	NO^+	Nitrosonium (nitrosyl cation)		Oxidizing agent and Lewis acid; π-acceptor ligand
+5	NO_3^-	Nitrate	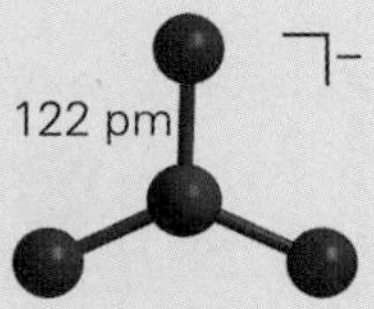 122 pm	Very weak base; an oxidizing agent
+5	NO_2^+	Nitronium (nitryl cation)	 115 pm	Oxidizing agent, nitrating agent, Lewis acid

with P_4O_{10}:

$$HNO_3(l) + P_4O_{10}(s) \rightarrow 2\,N_2O_5(s) + 4\,HPO_3(l)$$

The solid sublimes at 320°C and the gaseous molecules dissociate to give NO_2 and O_2. The compound is a strong oxidizing agent and can be used for the synthesis of anhydrous nitrates:

$$N_2O_5(s) + Na(s) \rightarrow NaNO_3(s) + NO_2(g)$$

Example 14.4 Correlating trends in the stabilities of N(V) and Bi(V)

Compounds of N(V) and Bi(V) are stronger oxidizing agents than the +5 oxidation states of the three intervening elements. Correlate this observation with trends in the periodic table.

Answer The light *p*-block elements are more electronegative than the elements immediately below them in the periodic table; accordingly, these light elements are generally less easily oxidized. Nitrogen is generally a good oxidizing agent in its positive oxidation states. Bismuth is much less electronegative, but favours the +3 oxidation state in preference to the +5 state on account of the inert-pair effect.

Self-test 14.4 From trends in the periodic table, decide whether phosphorus or sulfur is likely to be the stronger oxidizing agent.

(b) **Nitrogen(IV) and nitrogen(III) oxides and oxoanions**

Key point: The intermediate oxidation states of nitrogen are often susceptible to disproportionation.

Nitrogen(IV) oxide, which is commonly called nitrogen dioxide, exists as an equilibrium mixture of the brown NO_2 radical and its colourless dimer, N_2O_4 (dinitrogen tetroxide):

$$N_2O_4(g) \rightleftharpoons 2\ NO_2(g) \quad K = 0.115 \text{ at } 25°C$$

This readiness to dissociate is consistent with the N—N bond in N_2O_4 (**14**) being long and weak and arises because the molecular orbital occupied by the unpaired electron is spread almost equally over all three atoms in NO_2 rather than being concentrated on the N atom.

Nitrogen(IV) oxide is a poisonous oxidizing agent that is present in low concentrations in the atmosphere, especially in photochemical smog. In basic aqueous solution it disproportionates into N(III) and N(V), forming NO_2^- and NO_3^- ions:

$$2\ NO_2(aq) + 2\ OH^-(aq) \rightarrow NO_2^-(aq) + NO_3^-(aq) + H_2O(l)$$

In acidic solution (as in the Ostwald process) the reaction product is N(II) in place of N(III) because nitrous acid itself readily disproportionates:

$$3\ HNO_2(aq) \rightarrow NO_3^-(aq) + 2\ NO(g) + H_3O^+(aq) \quad E^\ominus = +0.05V, K = 50$$

Nitrous acid, HNO_2, is a strong oxidizing agent:

$$HNO_2(aq) + H^+(aq) + e^- \rightarrow NO(g) + H_2O(l) \quad E^\ominus = +1.00\,V$$

and its reactions as an oxidizing agent are often more rapid than its disproportionation.

The rate at which nitrous acid oxidizes is increased by acid as a result of its conversion to the nitrosonium ion, NO^+:

$$HNO_2(aq) + H^+(aq) \rightarrow NO^+(aq) + H_2O(l)$$

The nitrosonium ion is a strong Lewis acid and forms complexes rapidly with anions and other Lewis bases. The resulting species may not themselves be susceptible to oxidation (as in the case of SO_4^{2-} and F^- ions, which form $[SO_3ONO]^-$ (**15**) and ONF (**16**), respectively) but the association is often the initial step in an overall redox reaction. Thus there is good experimental evidence that the reaction of HNO_2 with I^- ions leads to the rapid formation of INO:

$$I^-(aq) + NO^+(aq) \rightarrow INO(aq)$$

followed by the rate-determining second-order reaction between two INO molecules:

$$2\ INO(aq) \rightarrow I_2(aq) + 2\ NO(g)$$

Nitrosonium salts containing poorly coordinating anions, such as $[NO][BF_4]$, are useful reagents in the laboratory as facile oxidizing agents and as a source of NO^+.

Dinitrogen trioxide, N_2O_3, the anhydride of nitrous acid, is a blue solid that melts above $-100°C$ to give a blue liquid that dissociates to give NO and NO_2:

$$N_2O_3(l) \rightarrow NO(g) + NO_2(g)$$

The brown colour of NO_2 means that the liquid becomes progressively more green as the dissociation proceeds.

(c) Nitrogen(II) oxide

Key points: Nitric oxide is a strong π-acceptor ligand, and a troublesome pollutant in urban atmospheres; the molecule acts as a neurotransmitter.

Nitrogen(II) oxide, more commonly nitric oxide, NO, is an odd-electron molecule. However, unlike NO_2 it does not form a stable dimer in the gas phase. This difference reflects the greater delocalization of the odd electron in the π^* orbital in NO than the corresponding HOMO in NO_2 (Section 2.8). Nitric oxide reacts with O_2 to generate NO_2, but in the gas phase the rate law is second-order in NO, because a transient dimer, $(NO)_2$, is produced that subsequently collides with an O_2 molecule. Because the reaction is second-order, atmospheric NO (which is produced in low

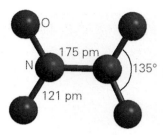

14 N_2O_4, D_{2h}

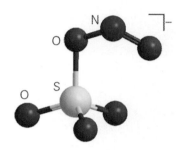

15 $[O_3SONO]^-$

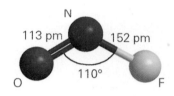

16 ONF

concentrations by coal-fired power plants and by internal combustion engines) is slow to convert to NO_2.

Until the late 1980s, no beneficial biological roles were known for NO. However, since then it has been found that NO is generated *in vivo*, and that it performs functions such as the reduction of blood pressure, neurotransmission, and the destruction of microbes. Thousands of scientific papers have been published on the physiological functions of NO but our fundamental knowledge of its biochemistry is still quite meagre.

Because NO is endoergic, it should be possible to find a catalyst to convert the pollutant NO to the natural atmospheric gases N_2 and O_2 at its source in exhausts. Cu^+ in a zeolite catalyses the decomposition of NO, and a reasonable understanding of the mechanism has been developed; however, this system is not yet sufficiently robust for a practical automobile exhaust catalytic converter.

(d) Low oxidation state nitrogen–oxygen compounds

Key points: Dinitrogen oxide is unreactive, for kinetic reasons; the isoelectronic azide ion, N_3^-, forms many metal complexes.

The average oxidation number of nitrogen in dinitrogen oxide, N_2O (specifically, NNO), which is commonly called nitrous oxide, is +1. N_2O is a colourless, unreactive gas and is produced by the comproportionation of molten ammonium nitrate. Care must be taken to avoid an explosion in this reaction, in which the cation is oxidized by the anion:

$$NH_4NO_3(l) \xrightarrow{250°C} N_2O(g) + 2\,H_2O(g)$$

Standard potential data suggest that N_2O should be a strong oxidizing agent in acidic and basic solutions:

$$N_2O(g) + 2\,H^+(aq) + 2\,e^- \rightarrow N_2(g) + H_2O(l) \quad E^\ominus = +1.77\,V \text{ at pH} = 0$$

$$N_2O(g) + H_2O(l) + 2\,e^- \rightarrow N_2(g) + 2\,OH^-(aq) \quad E = +0.94\,V \text{ at pH} = 14$$

However, kinetic considerations are paramount, and the gas is unreactive towards many reagents at room temperature. One sign of this inertness is that N_2O has been used as the propellant gas for instant whipping cream. Similarly, N_2O was used for many years as a mild anaesthetic; however, this practice has been discontinued because of undesirable physiological side-effects, particularly mild hysteria, indicated by its common name of *laughing gas*. It is still used as a 50:50 mixture with oxygen, commonly known as an analgesic in childbirth and in clinical procedures such as wound suturing.

Example 14.5 Comparing the redox properties of nitrogen oxoanions and oxo compounds

Compare (a) NO_3^- and NO_2^- as oxidizing agents; (b) NO_2, NO, and N_2O with respect to their ease of oxidation in air; (c) N_2H_4 and H_2NOH as reducing agents.

Answer (a) NO_3^- and NO_2^- ions are both strong oxidizing agents. The reactions of the former are often sluggish but are generally faster in acidic solution. The reactions of NO_2^- ions are generally faster and become even faster in acidic solution, where the NO^+ is a common identifiable intermediate. (b) NO_2 is stable with respect to oxidation in air. N_2O and NO are thermodynamically susceptible to oxidation. However, the reaction of N_2O with oxygen is slow, and at low NO concentrations the reaction between NO and O_2 is slow because the rate law is second-order in NO. (c) Hydrazine and hydroxylamine are both good reducing agents. In basic solution hydrazine becomes a stronger reducing agent.

Self-test 14.5 Summarize the reactions that are used for the synthesis of hydrazine and hydroxylamine. Are these reactions best described as electron-transfer processes or nucleophilic displacements?

14.10 Oxygen compounds of phosphorus, arsenic, antimony, and bismuth

Phosphorus, arsenic, antimony, and bismuth form oxides and oxoanions in a range of oxidation states from +5 to +1. The most common oxidation state is +5 but the +3 state becomes increasingly more important for bismuth.

(a) Oxides

Key points: The oxides of phosphorus include P_4O_6 and P_4O_{10}, both of which are cage compounds with T_d symmetry; on progressing from arsenic to bismuth, the +5 oxidation state is more readily reduced to +3.

The complete combustion of phosphorus yields phosphorus(V) oxide, P_4O_{10}. Each P_4O_{10} molecule has a cage structure in which a tetrahedron of P atoms is held together by bridging O atoms, and each P atom has a terminal O atom (**17**). Combustion in a limited supply of oxygen results in the formation of phosphorus(III) oxide, P_4O_6; this molecule has the same O-bridged framework as P_4O_{10}, but lacks the terminal O atoms (**18**). It is also possible to isolate the intermediate compositions having one, two, or three O atoms terminally attached to the P atoms. Both principal oxides can be hydrated to yield the corresponding acids, the P(V) oxide giving phosphoric acid, H_3PO_4, and the P(III) oxide giving phosphonic acid, H_3PO_3. As remarked in Section 4.4, phosphonic acid has one H atom attached directly to the P atom; it is therefore a diprotic acid and better represented as $PHO(OH)_2$.

In contrast to the high stability of phosphorus(V) oxide, arsenic, antimony, and bismuth more readily form oxides with oxidation number +3, specifically As_2O_3, Sb_2O_3, and Bi_2O_3. In the gas phase, the arsenic(III) and antimony(III) oxides have the molecular formula E_4O_6, with the same tetrahedral structure as P_4O_6. Arsenic, antimony, and bismuth do form oxides with oxidation state +5, but Bi(V) oxide is unstable and has not been structurally characterized. This is another example of the consequences of the inert-pair effect. Note the coordination number of 6 for antimony and 4 for the lighter elements phosphorus and arsenic, with their smaller atoms.

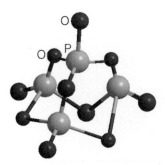

17 P_4O_{10}, T_d

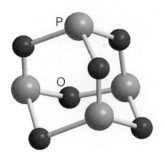

18 P_4O_6, T_d

(b) Oxoanions

Key points: Important oxoanions are the P(I) species hypophosphite, $H_2PO_2^-$, the P(III) species phosphite, HPO_3^{2-}, and the P(V) species phosphate, PO_4^{3-}. The existence of P—H bonds and the highly reducing character of the two lower oxidation states is notable. Phosphorus(V) also forms an extensive series of O-bridged polyphosphates. In contrast to N(V), P(V) species are not strongly oxidizing. As(V) is more easily reduced than P(V).

It can be seen from the Latimer diagram in Table 14.4 that elemental phosphorus and most of its compounds other than P(V) are strong reducing agents. White phosphorus disproportionates into phosphine, PH_3 (oxidation number −3), and hypophosphite ions (oxidation number +1) in basic solution:

$$P_4(s) + 3\,OH^-(aq) + 3\,H_2O(l) \rightarrow PH_3(g) + 3\,H_2PO_2^-(aq)$$

Table 14.4 Latimer diagrams for phosphorus

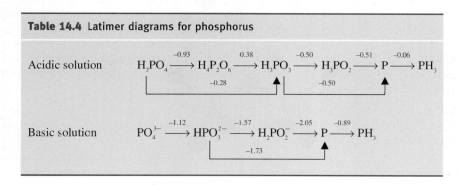

Table 14.5 lists some common phosphorus oxoanions (Box 14.4). The approximately tetrahedral environment of the P atom in their structures should be noted, as should the existence of P—H bonds in the hypophosphite and phosphite anions. The synthesis of various P(III) oxoacids and oxoanions, including HPO_3^{2-} and alkoxophosphanes, is conveniently performed by solvolysis of phosphorus(III) chloride under mild conditions, such as in cold tetrachloromethane solution:

$$PCl_3(l) + 3 H_2O(l) \rightarrow H_3PO_3(aq) + 3 HCl(aq)$$

$$PCl_3(l) + 3 ROH(sol) + 3 N(CH_3)_3(sol) \rightarrow P(OR)_3(sol) + 3 [HN(CH_3)_3]Cl(sol)$$

Table 14.5 Some phosphorus oxoanions

Oxidation number	Formula	Name	Structure	Comments
+1	$H_2PO_2^-$	Hypophosphite (dihydrodioxophosphate)		Facile reducing agent
+3	HPO_3^{2-}	Phosphite		Facile reducing agent
+4	$P_2O_6^{4-}$	Hypophosphate	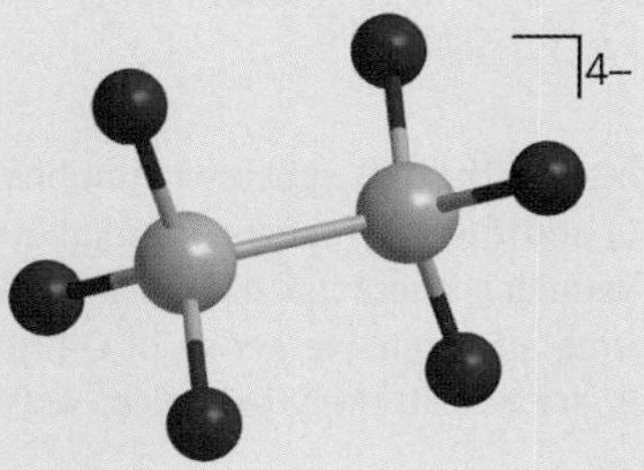	Basic
+5	PO_4^{3-}	Phosphate	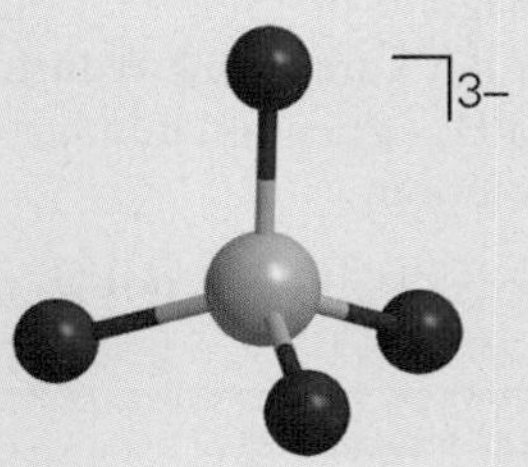	Strongly basic
+5	$P_2O_7^{4-}$	Diphosphate		Basic; longer chain analogues are known

Box 14.4 Phosphates and the food industry

Phosphorus in the form of phosphates is essential to life and phosphate fertilizers in forms such as bone, fish, and guano have been used since ancient times. The phosphate industry started in the mid-nineteenth century when sulfuric acid was used to decompose bones and phosphate minerals to make the phosphate more readily available. The development of more economical routes led to the diversification of the industrial applications of phosphoric acid and phosphate salts.

More than 90 per cent of world production of phosphoric acid is used to make fertilizers but there are several other applications. One of the most important is the food industry. A dilute solution of phosphoric acid is nontoxic and has an acidic taste. It is used extensively in beverages to give a tart taste, as a buffering agent in jams and jellies, and as a purifying agent in sugar refining.

The phosphates and hydrogenphosphates have many applications in the food industries. Sodium dihydrogenphosphate, NaH_2PO_4, is added to animal feeds as a dietary supplement. The disodium salt, Na_2HPO_4, is used as an emulsifier for processing cheese. It interacts with the protein casein and prevents separation of the fat and water. The potassium salts are more soluble and more expensive than the sodium salts. The dipotassium salt, K_2HPO_4, is used as an anti-coagulant in coffee creamer. It interacts with the protein and prevents coagulation by the coffee acids. Calcium dihydrogenphosphate, CaH_2PO_4, is used as a raising agent in bread, cake mixes, and self-raising flour. Together with $NaHCO_3$ it produces CO_2 during the baking process but it also reacts with the protein in the flour to control the elasticity and viscosity of the dough or mixture. The largest use of the dicalcium salt, Ca_2HPO_4, is as a dental polish in toothpaste. Calcium phosphate, Ca_3PO_4, is added to sugar and salt to improve their flow.

Reductions with $H_2PO_2^-$ and HPO_3^{2-} are usually fast. One of the commercial applications of this lability is the use of $H_2PO_2^-$ to reduce $Ni^{2+}(aq)$ ions and so coat surfaces with metallic nickel in the process called 'electrodeless plating'.

$$Ni^{2+}(aq) + 2H_2PO_2^-(aq) + 2H_2O(l) \rightarrow Ni(s) + 2H_2PO_3^-(aq) + H_2(g) + 2H^+(aq)$$

The Frost diagram for the elements shown in Fig. 14.6 reveals similar trends in aqueous solution, with oxidizing character following the order $PO_4^{3-} \approx AsO_4^{3-} < Sb(OH)_6^- \approx Bi(V)$. The thermodynamic tendency and kinetic ease of reducing AsO_4^{3-} is thought to be key to its toxicity towards animals. Thus, As(V) as AsO_4^{3-} readily mimics PO_4^{3-}, and so may be incorporated into cells. There, unlike phosphorus, it is reduced to an As(III) species, which is thought to be the real toxic agent. This toxicity may stem from the affinity of As(III) for sulfur-containing amino acids. The enzyme arsenite oxidase is produced by certain bacteria and is used to reduce the toxicity of As(III) by converting it to As(V).

(c) Condensed phosphates

Key point: Dehydration of phosphoric acid leads to the formation of chain or ring structures that may be based on many PO_4 units.

When phosphoric acid, H_3PO_4, is heated above 200°C, condensation occurs resulting in the formation of P—O—P bridges between two neighbouring PO_4^{3-} units. The extent of this condensation depends upon the temperature and duration of heating.

$$2 H_3PO_4(l) \rightarrow H_4P_2O_7(l) + H_2O(g)$$
$$H_3PO_4(l) + H_4P_2O_7(l) \rightarrow H_5P_3O_{10}(l) + H_2O(g)$$

The simplest condensed phosphate is thus $H_4P_2O_7$. The most commercially important condensed phosphate is the sodium salt of the triacid, $Na_5P_3O_{10}$ (**19**). It is widely used in detergents for laundry and dishwashers, and in other cleaning products and in water treatment (Box 14.5). Polyphosphates are also used in various ceramics and as food additives. Triphosphates such as adenosine triphosphate (ATP) are of vital importance in living organisms (Section 26.8).

A range of condensed phosphates occurs with chain lengths ranging from those based on two PO_4 units to polyphosphates having chain lengths of several thousand units. Di-, tri-, tetra-, and pentapolyphosphates have been isolated but higher members of the series always contain mixtures. However, the average chain length can be determined by the usual methods used in polymer analysis or by titration. Just as the three acidity constants for phosphoric acid differ, so do the acidity constants for the two types of OH group of the polyphosphoric acids. The terminal OH groups, of which there are two per molecule, are weakly acidic. The remaining OH groups, of which there is one per P atom, are

19 $P_3O_{10}^{5-}$

Box 14.5 Polyphosphates

The most widely used polyphosphate is sodium tripolyphosphate, $Na_3P_3O_{10}$. Its major use is as a 'builder' for synthetic detergents used in domestic laundry products, car shampoos, and industrial cleaners. Its role in these applications is to form stable complexes with the calcium and magnesium ions in hard water, effectively making them unavailable for precipitation, the process called 'sequestration'. It also acts as a buffer and prevents the flocculation of dirt and the redeposition of soil particles.

Food-grade sodium tripolyphosphate is used in the curing of hams and bacon. It interacts with the proteins and leads to good moisture retention during curing. It also used to improve the quality of processed chicken and seafood products. The technical-grade product is used as a water softener, by sequestration as above, and

in the paper pulping and textile industries, where it is used to help break down cellulose.

Potassium tripolyphosphate is more soluble and more expensive than the sodium analogue and is used in liquid detergents. For some applications an effective trade-off between solubility and cost can be achieved by using sodium potassium tripolyphosphate, $NaK_2P_3O_{10}$.

The use of polyphosphates in detergents has been implicated in excessive algae growth and eutrophication of some natural waters. This has led to restrictions in their use in many countries and the reduction in their use in home laundry products. However, phosphate is still widely used as an essential fertilizer and more of it enters rivers and lakes by run-off from farm land than from detergents.

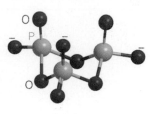

20 $P_3O_9^{3-}$

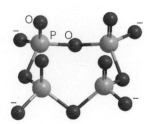

21 $P_4O_{12}^{4-}$

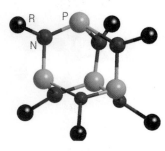

22 $P_4(NR)_6$

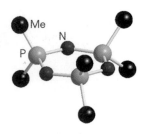

23 (Me_2PN_3)

strongly acidic because they are situated opposite the strongly electron-withdrawing $=O$ groups. The ratio of weakly to strongly acidic protons gives an indication of average chain length. The long-chain polyphosphates are viscous liquids or glasses.

Example 14.6 Determining the chain length of a polyphosphoric acid

A sample of a polyphosphoric acid was dissolved in water and titrated with dilute NaOH(aq). Two stoichiometric points were observed at 16.8 and $28.0\ cm^3$. Determine the chain length of the polyphosphate.

Answer The strongly acidic OH groups are titrated by the first $16.8\ cm^3$. The two terminal OH groups are titrated by the remaining $28.0 - 16.8\ cm^3 = 11.2\ cm^3$. Because the concentrations of analyte and titrant are such that each OH group requires $5.6\ cm^3$ of the titrant, we conclude that there are $(16.8\ cm^3)/(5.6\ cm^3) = 3$ strongly acidic OH groups per molecule. A molecule with two terminal OH groups and three further OH groups is a tripolyphosphate.

Self-test 14.6 When titrated against base a sample of polyphosphate gave end points at 30.4 and $45.6\ cm^3$. Calculate the chain length.

If NaH_2PO_4 is heated and the water vapour allowed to escape, then the tricyclo anion $P_3O_9^{3-}$ (**20**) is formed. If this reaction is carried out in a closed system, the product is 'Maddrell's salt', a crystalline material that contains long chains of PO_4 units. The tetracyclo anion (**21**) is formed when P_4O_{10} is treated with cold aqueous solutions of NaOH or $NaHCO_3$.

14.11 Phosphazenes

Key points: The range of PN compounds is extensive, and includes cyclic and polymeric phosphazenes, $(PX_2N)_n$; phosphazenes form highly flexible elastomers.

Many analogues of phosphorus–oxygen compounds exist in which the O atom is replaced by the isolobal NR or NH group, such as $P_4(NR)_6$ (**22**), the analogue of P_4O_6. Other compounds exist in which OH or OR groups are replaced by the isolobal NH_2 or NR_2 groups. An example is $P(NMe_2)_3$, the analogue of $P(OMe)_3$. Another indication of the scope of PN chemistry, and a useful point to remember, is that PN is isoelectronic with SiO. For example, various phosphazenes, which are chains and rings containing R_2PN units (**23**), are analogous to the siloxanes (Chapter 13) and their R_2SiO units (**24**).

The cyclic phosphazene dichlorides are good starting materials for the preparation of the more elaborate phosphazenes. They are easily synthesized:

$$n\ PCl_5 + n\ NH_4Cl \rightarrow (Cl_2PN)_n + 4n\ HCl \quad n = 3\ or\ 4$$

Box 14.6 Biomedical applications of polyphosphazenes

Biodegradable polymers are attractive biomedical materials because they survive only for a limited time *in vivo*. Polyphosphazenes are proving to be very useful in this respect as they degrade to harmless by-products and their physical properties can be tuned by altering the substituents on the P atoms. They are used as bioinert housing materials for implantation of devices, as structural materials for construction of heart valves and blood vessels, and as biodegradable supports for *in vivo* bone regeneration. The best polyphosphazenes for this last application form fibres in which the P—N backbone consists of alkyoxy groups that form bonds to Ca^{2+} ions. The polymer fibre becomes populated with the patient's

osteoblasts (bone-making cells). The polymer degrades as the osteoblasts multiply and fill the space between the fibres. Polyphosphazenes have been designed that hydrolyse at a specific rate and maintain their strength as the erosion process proceeds.

Polyphosphazenes are also used as drug delivery systems. The bioactive molecule is trapped within the structure of the polymer or incorporated into the P—N backbone and the drug is released when the polymer degrades. The rate of degradation can be controlled by altering the structure of the polymer backbone, thus giving control over the rate of drug delivery. Drugs that can be delivered in this way include cisplatin, dopamine, and steroids.

A chlorocarbon solvent and temperatures near 130°C produce the cyclic trimer (**25**) and tetramer (**26**), and when the trimer is heated to about 290°C it changes to polyphosphazene (Box 14.6). The Cl atoms in the trimer, tetramer, and polymer are readily displaced by other Lewis bases

$$(Cl_2PN)_n + 2n\ CF_3CF_2O^- \rightarrow [(CF_3CF_2O)_2PN]_n + 2n\ Cl^-$$

Like silicone rubber, the polyphosphazenes remain rubbery at low temperatures because, like the isoelectronic SiOSi group, the molecules are helical and the PNP groups are highly flexible.

A phosphorus–nitrogen compound that was particularly useful in the laboratory until its carcinogenic properties were recognized is the aprotic solvent hexamethylphosphoramide, $((CH_3)_2N)_3P{=}O$, which is sometimes designated HMPA. The large bis(triphenylphosphine)iminium cation, $[Ph_3P{=}N{=}PPh_3]^+$, which is commonly abbreviated as PPN^+, is very useful in forming salts of large anions. The salts of this cation are usually soluble in polar aprotic solvents such as HMPA, dimethylformamide, and even dichloromethane.

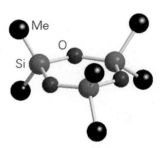

24 $(Me_2SiO)_3$

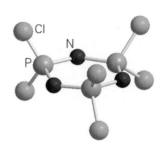

25 $(Cl_2PN)_3$

Example 14.7 Devising a synthetic route to an alkoxy-substituted cyclophosphazene

Suggest syntheses and give balanced equations for the preparation of $[NP(OCH_3)_2]_4$ from PCl_5, NH_4Cl, and $NaOCH_3$.

Answer The cyclic chlorophosphazene can be synthesized first:

$$4\ PCl_5 + 4\ NH_4Cl \xrightarrow{130°C} (Cl_2PN)_4 + 16\ HCl$$

The Cl atoms are readily replaced by strong Lewis bases, such as alkoxides, in the present case to give the desired product

$$(Cl_2PN)_4 + 8\ NaOCH_3 \rightarrow [(CH_3O)_2PN]_4 + 8\ NaCl$$

Self-test 14.7 Suggest a procedure and give the balanced chemical equations for the preparation of a high polymer phosphazene containing a PN backbone with two $N(CH_3)_2$ side groups attached to each P atom.

26 $(Cl_2PN)_4$

Organometallic compounds

Oxidation states +3 and +5 are encountered in many of the organometallic compounds of arsenic, antimony, and bismuth. An example of a compound with an element in the +3 oxidation state is $As(CH_3)_3$ (**27**) and an example of the +5 state is $As(C_6H_5)_5$ (**28**). Organoarsenic compounds were once widely used to treat bacterial infections and as herbicides and fungicides. However, because of their high toxicity they no longer have major commercial applications.

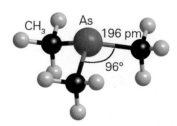

27 $AS(CH_3)_3$

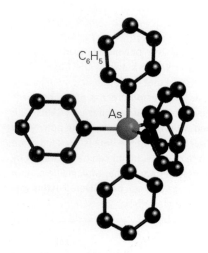

28 As(C$_6$H$_5$)$_5$

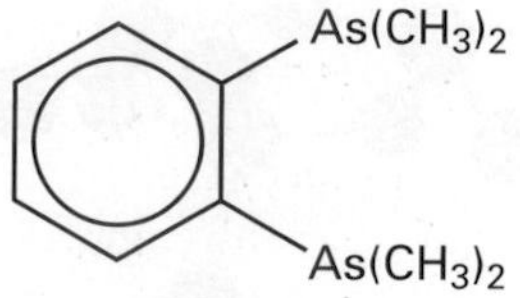

29 C$_6$H$_5$(As(CH$_3$)$_2$)$_2$, diars

14.12 Oxidation state +3

Key points: The stability of the organometallic compounds decreases in the order As > Sb > Bi; the aryl compounds are more stable than the alkyl compounds.

Organometallic compounds of arsenic(III), antimony(III), and bismuth(III) can be prepared in an ether solvent by using a Grignard reagent, an organolithium compound, or an organohalide:

$$\text{AsCl}_3(\text{et}) + 3\,\text{RMgCl}(\text{et}) \rightarrow \text{AsR}_3(\text{et}) + 3\,\text{MgCl}_2(\text{et})$$

$$2\,\text{As}(\text{et}) + 3\,\text{RBr}(\text{et}) \xrightarrow{\text{Cu}/\Delta} \text{AsRBr}_2(\text{et}) + \text{AsR}_2\text{Br}(\text{et})$$

$$\text{AsR}_2\text{Br}(\text{et}) + \text{R}'\text{Li}(\text{et}) \rightarrow \text{AsR}_2\text{R}'(\text{et}) + \text{LiBr}(\text{et})$$

The compounds are all readily oxidized but are stable to water. The M—C bond strength decreases for a given R group in the order As > Sb > Bi. Consequently, the stability of the compounds decreases in the same order. In addition, the aryl compounds, such as (C$_6$H$_5$)$_3$As, are generally more stable than the alkyl compounds. The halogen-substituted compounds R$_n$MX$_{3-n}$ have been prepared and characterized.

All the compounds act as Lewis bases and form complexes with *d*-metal ions. The basicity decreases in the order As > Sb > Bi. Many complexes of alkyl- and arylarsanes have been prepared but fewer stibane complexes are known. A useful ligand, for example, is the bidentate compound known as diars (**29**). Because of their soft-donor character, many aryl- and alkylarsane complexes of the soft species Rh(I), Ir(I), Pd(II), and Pt(II) have been prepared. However, hardness criteria are only approximate, so we should not be surprised to see phosphine and arsane complexes of some metals in higher oxidation states. For example, the unusual +4 oxidation state of palladium is stabilized by the diars ligand (**30**).

The synthesis of diars provides a good illustration of some common reactions in the synthesis of organoarsenic compounds. The starting material is (CH$_3$)$_2$AsI. This compound is not conveniently prepared by metathesis reaction between AsI$_3$ and a Grignard or similar carbanion reagent because that reaction is not selective to partial substitution on the As atom when the organic group is compact. Instead, the compound can be prepared by the direct action of a haloalkane, CH$_3$I, on the metallic allotrope of arsenic:

$$4\,\text{As}(\text{s}) + 6\,\text{CH}_3\text{I}(\text{l}) \rightarrow 3\,(\text{CH}_3)_2\text{AsI}(\text{sol}) + \text{AsI}_3(\text{sol})$$

In the next step, the action of sodium on (CH$_3$)$_2$AsI is used to produce [(CH$_3$)$_2$As]$^-$:

$$(\text{CH}_3)_2\text{AsI}(\text{sol}) + 2\,\text{Na}(\text{sol}) \rightarrow \text{Na}[(\text{CH}_3)_2\text{As}](\text{sol}) + \text{NaI}(\text{s})$$

The resulting powerful nucleophile [(CH$_3$)$_2$As]$^-$ is then used to displace chlorine from 1,2-dichlorobenzene:

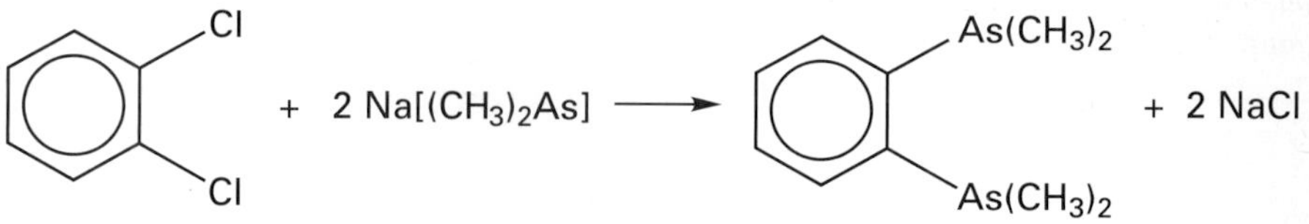

Polyarsine compounds, (RAs)$_n$, can be prepared in ether by reduction of a pentavalent organometallic compound, R$_5$As, or by treating an organohaloarsenic compound with lithium:

$$n\,\text{RAsX}_2(\text{et}) + 2n\,\text{Li}(\text{et}) \rightarrow (\text{RAs})_n(\text{et}) + 2n\,\text{LiX}(\text{et})$$

The compound R$_2$AsAsR$_2$ is very reactive because the As—As bond is readily cleaved. It reacts with oxygen, sulfur, and species containing C=C bonds, and forms complexes

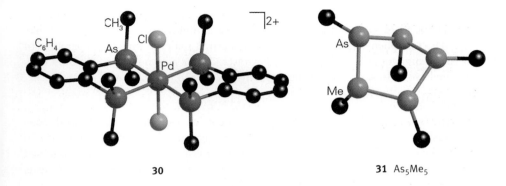

30

31 As₅Me₅

with *d*-metal species in which the As—As bond may be cleaved or left intact:

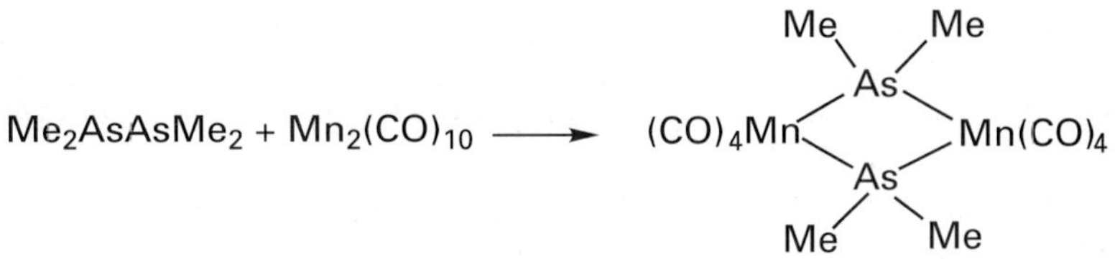

$$Me_2AsAsMe_2 + Mn_2(CO)_{10} \longrightarrow (CO)_4Mn \overset{Me \quad Me}{\underset{Me \quad Me}{\overset{As}{\underset{As}{\diamond}}}} Mn(CO)_4$$

Polyarsines of up to six units have been characterized. Polymethylarsine exists as a yellow, puckered cyclic pentamer (**31**) and as a purple–black ladder-like structure (**32**). The strength of the M—M bond decreases in the order As > Sb > Bi, so although arsenic forms catenated organometallic compounds, only R_2Bi—BiR_2 has been isolated.

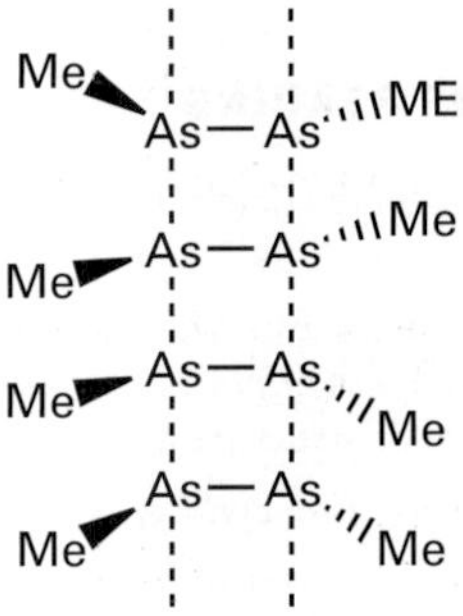

As well as forming single M—C bonds, arsenic, antimony, and bismuth also form M=C bonds. A well-studied group of compounds is the arylometal compounds in which a metal atom forms part of a heterocyclic six-membered benzene-like ring (**33**). Arsabenzene, C_5H_5As, is stable up to 200°C, stibabenzene, C_5H_5Sb, can be isolated but readily polymerizes, and bismabenzene, C_5H_5Bi, is very unstable. These compounds exhibit typical aromatic character although arsabenzene is 1000 times more reactive than benzene. A related group of compounds is arsole, stibole, and bismuthole, C_4H_4M, in which the metal atom forms part of a five-membered ring (**34**).

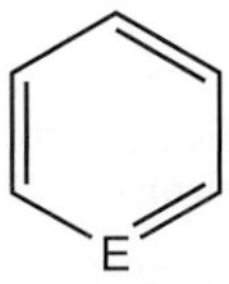

14.13 Oxidation state +5

Key point: The tetraphenylarsonium ion is also a starting material for the preparation of other As(V) organometallic compounds.

The trialkylarsanes act as nucleophiles towards haloalkanes to produce tetraalkylarsonium salts, which contain As(V):

$$As(CH_3)_3(sol) + CH_3Br(sol) \rightarrow [As(CH_3)_4]Br(sol)$$

This type of reaction cannot be used for the preparation of the tetraphenylarsonium ion, $[AsPh_4]^+$, because triphenylarsane is a much weaker nucleophile than trimethylarsane. Instead, a suitable synthetic reaction is:

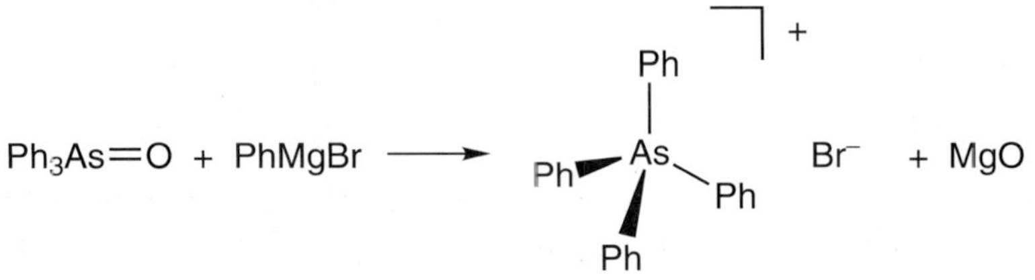

$$Ph_3As{=}O + PhMgBr \longrightarrow \left[\begin{matrix} Ph \\ | \\ Ph{-}As{-}Ph \\ | \\ Ph \end{matrix} \right]^+ Br^- + MgO$$

This reaction may look unfamiliar, but it is simply a metathesis in which the Ph^- anion replaces the formal O^{2-} ion attached to the As atom, resulting in a compound in which the arsenic retains its +5 oxidation state. The formation of the highly exoergic compound MgO also contributes to the Gibbs energy of this reaction, and its formation drives the reaction forward.

32 (AsMe)ₙ

33 C_5H_5E

34 C_4H_6E

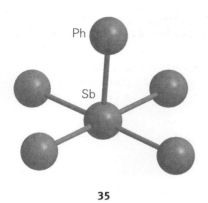

35

The tetraphenylarsonium, tetraalkylammonium, and tetraphenylphosphonium cations are used in synthetic inorganic chemistry as bulky cations to stabilize bulky anions. The tetraphenylarsonium ion is also a starting material for the preparation of other As(V) organometallic compounds. For instance, the action of phenyllithium on a tetraphenylarsonium salt produces pentaphenylarsenic (**28**), a compound of As(V):

$$[AsPh_4]Br(sol) + LiPh(sol) \rightarrow AsPh_5(sol) + LiBr(s)$$

Pentaphenylarsenic, $AsPh_5$, is trigonal bipyramidal, as expected from VSEPR considerations. We have seen (Section 2.3) that a square-pyramidal structure is often close in energy to the trigonal-bipyramidal structure, and the antimony analogue, $SbPh_5$, is in fact square pyramidal (**35**). A similar reaction under carefully controlled conditions yields the unstable compound $As(CH_3)_5$.

FURTHER READING

R.B. King, *Inorganic chemistry of the main group elements.* Wiley (1994).

D.M.P. Mingos, *Essential trends in inorganic chemistry.* Oxford University Press (1998). A survey of inorganic chemistry from the perspective of structure and bonding.

R.B. King (ed.), *Encyclopedia of inorganic chemistry.* Wiley (2005).

H.R. Allcock, *Chemistry and applications of polyphosphazenes.* Wiley (2002).

J. Emsley, *The shocking history of phosphorus: a biography of the devil's element.* Pan (2001).

W.T. Frankenberger, *The environmental chemistry of arsenic.* Marcel Dekker (2001).

G.J. Leigh, *The world's greatest fix: a history of nitrogen and agriculture.* Oxford University Press (2004).

N.N. Greenwood and A. Earnshaw, *Chemistry of the Elements.* Butterworth-Heinemann (1997).

EXERCISES

14.1 List the elements in Groups 15 and indicate the ones that are (a) diatomic gases, (b) nonmetals, (c) metalloids, (d) true metals. Indicate those elements that display the inert-pair effect.

14.2 (a) Give complete and balanced chemical equations for each step in the synthesis of H_3PO_4 from hydroxyapatite to yield (a) high-purity phosphoric acid and (b) fertilizer-grade phosphoric acid. (c) Account for the large difference in costs between these two methods.

14.3 Ammonia can be prepared by (a) the hydrolysis of Li_3N or (b) the high-temperature, high-pressure reduction of N_2 by H_2. Give balanced chemical equations for each method starting with N_2, Li, and H_2, as appropriate. (c) Account for the lower cost of the second method.

14.4 Compare and contrast the formulas and stabilities of the oxidation states of the common nitrogen chlorides with the phosphorus chlorides.

14.5 Use Lewis structures and the VSEPR model to predict the probable structure of (a) PCl_4^+, (b) PCl_4^-, (c) $AsCl_5$.

14.6 Give balanced chemical equations for each of the following reactions: (a) oxidation of P_4 with excess oxygen; (b) reaction of the product from part (a) with excess water; (c) reaction of the product from part (b) with a solution of $CaCl_2$ and name the product.

14.7 Starting with $NH_3(g)$ and other reagents of your choice, give the chemical equations and conditions for the synthesis of (a) HNO_3, (b) NO_2^-, (c) NH_2OH, (d) N_3^-.

14.8 Write the balanced chemical equation corresponding to the standard enthalpy of formation of $P_4O_{10}(s)$. Specify the structure, physical state (s, l, or g), and allotrope of the reactants. Do either of these reactants differ from the usual practice of taking as reference state the most stable form of an element?

14.9 Without reference to the text, sketch the general form of the Frost diagrams for phosphorus (oxidation states 0 to +5) and bismuth (0 to +5) in acidic solution and discuss the relative stabilities of the +3 and +5 oxidation states of both elements.

14.10 Are reactions of NO_2^- as an oxidizing agent generally faster or slower when pH is lowered? Give a mechanistic explanation for the pH dependence of NO_2^- oxidations.

14.11 When equal volumes of nitric oxide (NO) and air are mixed at atmospheric pressure a rapid reaction occurs, to form NO_2 and N_2O_4. However, nitric oxide from an automobile exhaust, which is present in the parts per million concentration range, reacts slowly with air. Give an explanation for this observation in terms of the rate law and the probable mechanism.

14.12 Give balanced chemical equations for the reactions of the following reagents with PCl_5 and indicate the structures of the products: (a) water (1:1), (b) water in excess, (c) $AlCl_3$, (d) NH_4Cl.

14.13 Use standard potentials (*Resource section* 3) to calculate the standard potential of the reaction of H_3PO_2 with Cu^{2+}. Are HPO_3^{2-} and $H_2PO_2^{2-}$ useful as oxidizing or reducing agents?

14.14 Identify the compounds A, B, C, and D.

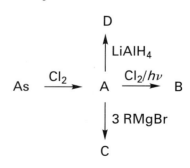

14.15 Identify the nitrogen compounds A, B, C, D, and E.

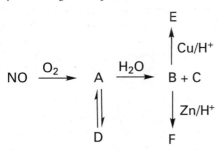

14.16 Sketch the two possible geometric isomers of the octahedral $[AsF_4Cl_2]^-$ and explain how they could be distinguished by ^{19}F NMR.

14.17 Use the Latimer diagrams in *Resource section* 3 to determine which species of N and P disproportionate in acid conditions.

PROBLEMS

14.1 Explain how you could use ^{31}P NMR to distinguish between PF_3 and POF_3.

14.2 The tetrahedral P_4 molecule may be described in terms of localized $2c,2e$ bonds. Determine the number of skeletal valence electrons and from this decide whether P_4 is *closo*, *nido*, or *arachno*. If it is not *closo*, determine the parent *closo* polyhedron from which the structure of P_4 could be formally derived by the removal of one or more vertices.

14.3 Describe sewage treatment methods that result in a decrease in phosphate levels in wastewater. Outline a laboratory method that could be used to monitor phosphate levels in water.

14.4 On account of their slowness at electrodes, the potentials of most redox reactions of nitrogen compounds cannot be measured in an electrochemical cell. Instead, the values must be determined from other thermodynamic data. Illustrate such a calculation by using $\Delta_f G^{\ominus}(NH_3, aq) = -26.5 \, kJ \, mol^{-1}$ to calculate the standard potential of the N_2/NH_3 couple in basic aqueous solution.

14.5 A compound containing pentacoordinate nitrogen has been characterized (A. Frohmann, J. Riede, and H. Schmidbaur, *Nature*, 1990, **345**, 140). Describe (a) the synthesis and (b) the structure of the compound and (c) give a description of the bonding.

14.6 Two articles (A. Lykknes and L. Kvittingen, 'Arsenic: not so evil after all?', *J. Chem. Educ.*, 2003, **80**, 497 and J. Wang and C.M. Chien, 'Arsenic in drinking water—a global environmental problem', *J. Chem. Educ.*, 2004, **81**, 207) present opposing perspectives on the toxic nature of arsenic. Use these references to produce a critical assessment of the beneficial and detrimental affects or arsenic.

14.7 A paper published by N. Tokitoh and co-workers (*Science*, 1997, **277**, 78) gives an account of the synthesis and characterization of a stable bismuthene, containing Bi—Bi double bonds. Give the equations for the synthesis of the compound. Name and sketch the structure of the steric protecting group that was used. Why was the isolation of the product simple? What methods were used to determine the structure of the compound?

15

The Group 16 elements

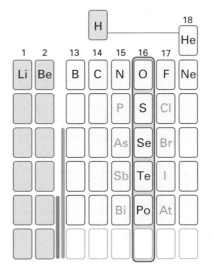

Group 16 contains what is undoubtedly one of the most important elements for life. Elemental oxygen is essential for respiration of most species of organisms and for the existence of water. In stark contrast, other members of the group form some compounds that are highly toxic. The simple trends that we observed towards the left of the periodic table are becoming more difficult to ascertain, as the elements of this group, particularly sulfur, form compounds in a wide variety of oxidations states.

The Group 16 elements, oxygen, sulfur, selenium, tellurium and polonium, are often called the **chalcogens**. The name derives from the Greek word for bronze, and refers to the association of sulfur and its congeners with copper. As in the rest of the *p* block, the element at the head of the group, oxygen, differs significantly from the other members of the group. The coordination numbers of its compounds are generally lower, and it is the only member of the group to exist as diatomic molecules under normal conditions.

The elements

Key points: Oxygen is the most electronegative element in Group 16 and is the only gas; all the elements occur in several allotropic forms.

All the members of Group 16 other than oxygen are solids under normal conditions (Table 15.1) and, as we have seen previously, metallic character generally increases down the group. Thus, oxygen, sulfur, and selenium are nonmetals, tellurium is a metalloid, and polonium is a metal. Allotropy is an important feature of Group 16 and sulfur occurs in more natural allotropes than any other element.

The group electron configuration of ns^2np^4 suggests a group maximum oxidation number of +6. Oxygen never achieves this maximum oxidation state, although the other elements do in some circumstances. The electron configuration also suggests that stability may be achieved with an oxidation number of −2, which is overwhelmingly common for oxygen. The most remarkable feature of sulfur is that it forms stable compounds with oxidation numbers between −2 and +6.

Table 15.1 Selected properties of the elements

	O	S	Se	Te	Po
Covalent radius/pm	74	104	117	137	140
Ionic radius/pm	140	184	198	221	
First ionization energy/(kJ mol^{-1})	1310	1000	941	870	812
Melting point/°C	−218	113 (α)	217	450	254
Boiling point/°C	−183	445	685	990	960
Pauling electronegativity	3.5	2.5	2.4	2.1	2.0
Electron affinity*/(kJ mol^{-1})	141	200	195	190	183
	−844	−532			

* The first value is for $X(g) + e^-(g) \rightarrow X^-(g)$, the second value is for $X^-(g) + e^-(g) \rightarrow X^{2-}(g)$.

In addition to its distinctive physical properties, oxygen is significantly different chemically from the other members of the group. Thus, it is the second most electronegative element in the periodic table and significantly more electronegative than its congeners. This high electronegativity has an enormous influence on the chemical properties of the element. The small atomic radius of oxygen and absence of accessible d orbitals also contribute to its distinctive chemical character. Thus, O seldom has coordination numbers greater than 4 in simple molecular compounds, but its heavier congeners frequently reach coordination numbers of 5 and 6, as in SF_6.

15.1 Occurrence, recovery, and uses of the elements

The principal elements of the group are oxygen and sulfur, and both are found in native form.

(a) Oxygen

Key points: Oxygen has two allotropes, dioxygen and ozone. Dioxygen has a triplet ground state that oxidizes hydrocarbons by a radical chain mechanism. Reaction with an excited state molecule can produce a fairly long-lived singlet state that can react as an electrophile. Ozone is an unstable and highly aggressive oxidizing agent.

Elemental oxygen makes up 21 per cent by mass of the atmosphere (Box 15.1), is the third most abundant element in the Sun, and the most abundant element on the surface of the Moon (46 per cent by mass). It is the most abundant element in the Earth's crust at 46 per cent by mass and is present in all silicate minerals. It comprises 86 per cent by mass of the oceans and 89 per cent of water. The average person is two-thirds oxygen by mass.

Oxygen gas is colourless, odourless, and reactive. It is soluble in water to the extent of 3.08 cm^3 per 100 cm^3 water at 25°C and atmospheric pressure. This solubility falls to below 2.0 cm^3 in seawater but is still sufficient to support marine life. The solubility of oxygen in organic solvents is approximately ten times greater than in water. This high solubility leads to painstaking degassing of solvents required for synthesis of oxygen-sensitive compounds.

Oxygen is readily available as O_2 from the atmosphere and is obtained on a massive scale by the liquefaction and distillation of liquid air. The main commercial motivation is to recover O_2 for steel making, in which it reacts exothermically with coke (carbon) to produce carbon monoxide. The high temperature is necessary to achieve a fast reduction of iron oxides by CO and carbon (Section 5.15). Pure oxygen, rather than air, is advantageous in this process because energy is not wasted in heating the nitrogen. About 1 tonne of oxygen is needed to make 1 tonne of steel. Oxygen is also required by industry in the production of the white pigment TiO_2 by the *chloride process*:

$$TiCl_4(l) + O_2(g) \rightarrow TiO_2(s) + 2\,Cl_2(g)$$

Oxygen is used in many oxidation processes, for example the production of oxirane (ethylene oxide) from ethene. Oxygen is also supplied on a large scale for sewage

Box 15.1 The oxygen atmosphere

In the course of the evolution of the Earth's atmosphere, the proliferation of oxygen-evolving photosynthesis eventually resulted in the presence of O_2 in the atmosphere at the present level of 21 per cent by volume. Photosynthesis also produced organic carbon compounds, which became the materials of cellular biomass (Box 13.3).

Oxygen was a toxic constituent of the atmosphere of the early Earth and led to the extinction of many species. Some species retreated to habitats deeper in the soil or waters, where anaerobic conditions remained and where their descendents remain today. Other organisms adapted differently and learned to use this now abundant and powerful oxidant. These organisms are the *aerobes*, among which were our ancestors. The shift from an anaerobic atmosphere to an oxygenic atmosphere had a profound effect on the composition of the waters. Sulfur, which was present largely in the form of sulfide in anaerobic waters, was oxidized to sulfate. Metal ion concentrations also changed dramatically. Two of the metals most profoundly affected were molybdenum and iron.

In Earth's modern oceans, molybdenum is the most abundant *d* metal (at 0.01 ppm). However, before the oxygenation of the oceans and atmosphere, molybdenum was present as insoluble solids, mainly MoO_2 and MoS_2. Oxidation of these solids produced the molybdate ion:

$$2\,MoS_2(s) + 7\,O_2(g) + 2\,H_2O(l) \rightarrow 2\,MoO_4^{2-}(aq) + 4\,SO_2(g) + 4\,H^+(aq)$$

The highly soluble molybdate ion became available to aquatic organisms and is now transported into cells by methods that differ dramatically from those used to acquire the more widespread cationic *d*-metal species in the marine environment. In addition, competition between molybdate and sulfate, both tetrahedral dinegative ions, remains a challenge for modern organisms.

Iron suffered the opposite fate to molybdenum. In the ancient oceans, the element was present as iron(II). Iron(II) hydroxide and sulfide are essentially soluble and so iron would have been readily available to aquatic organisms. However, upon oxygenation of the atmosphere, the oxidation of Fe^{2+} to Fe^{3+} led to the precipitation of iron(III) hydroxides and oxides. Massive banded formations containing magnetite (Fe_3O_4) and hematite (Fe_2O_3) in Canada and Australia are testimony to the precipitation of the iron from the oceans between 2 and 3 Ga ago.

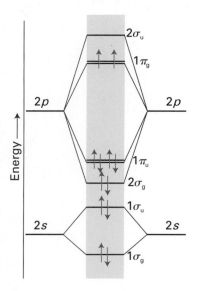

Fig. 15.1 Molecular orbital diagram for O_2.

treatment, renewal of polluted waterways, paper-pulp bleaching, and as an artificial atmosphere in medical and submarine applications.

Liquid oxygen is very pale blue and boils at $-183°$C. Its colour arises from electronic transitions involving pairs of neighbouring molecules: one photon from the red–yellow–green region of the visible spectrum can raise two such O_2 molecules to an excited state. The molecular orbital description of O_2 implies the existence of a double bond; however, as we saw in Section 2.8, the outermost two electrons occupy different antibonding π orbitals with parallel spins; as a result, the molecule is paramagnetic (Fig. 15.1). The term symbol for the ground state is $^3\Sigma_g^-$, and henceforth the molecule will be denoted $O_2(^3\Sigma_g^-)$ when it is appropriate to specify the spin state.[1] The singlet state $^1\Sigma_g^+$, with paired electrons in the same two π^* orbitals as in the ground state, is higher in energy by 1.61 eV (155 kJ mol^{-1}), and another singlet state $^1\Delta_g$ ('singlet delta'), with electrons paired in one orbital, lies between these two terms at 0.95 eV (92.7 kJ mol^{-1}) above the ground state. Of the two singlet states, the latter has much the longer excited state lifetime. This $O_2(^1\Delta_g)$ survives long enough to participate in chemical reactions. When it is needed for reactions, $O_2(^1\Delta_g)$ can be generated in solution by energy transfer from a photoexcited molecule. Thus $[Ru(bipy)_3]^{2+}$ can be excited by absorption of blue light (452 nm) to give an electronically excited state, denoted $^*[Ru(bipy)_3]^{2+}$ (Section 20.14), and this state transfers energy to $O_2(^3\Sigma_g^-)$:

$$^*[Ru(bipy)_3]^{2+} + O_2(^3\Sigma_g^-) \rightarrow [Ru(bipy)_3]^{2+} + O_2(^1\Delta_g)$$

Another efficient way to generate $O_2(^1\Delta_g)$ is through the thermal decomposition of an ozonide:

[1] The symbols Σ, Π, and Δ are used for linear molecules such as dioxygen in place of the symbols S, P, and D used for atoms. The Greek letters represent the magnitude of the total orbital angular momentum around the internuclear axis.

In contrast to the radical character of many $O_2(^3\Sigma_g^-)$ reactions, $O_2(^1\Delta_g)$ reacts as an electrophile. This mode of reaction is feasible because $O_2(^1\Delta_g)$ has a vacant π^* orbital. For example, $O_2(^1\Delta_g)$ adds across a diene, thus mimicking the Diels–Alder reaction of butadiene with an electrophilic alkene:

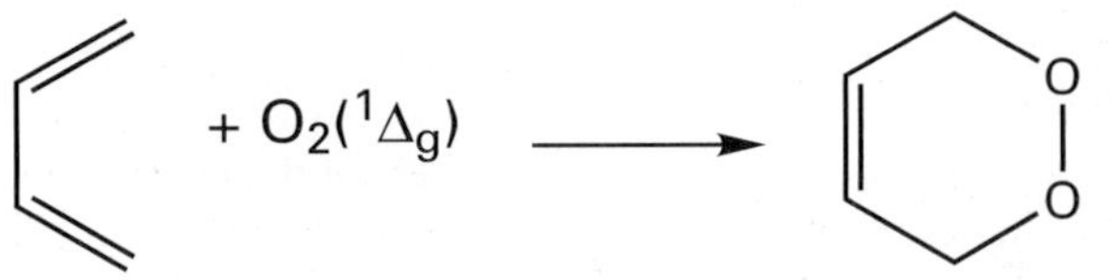

Singlet oxygen is implicated as one of the biologically hazardous products of photo-chemical smog.

Dioxygen oxidizes many elements and reacts with many organic and inorganic compounds under suitable conditions. Only the noble gases He, Ne, and Ar do not form oxides. The oxides of the elements are discussed in each relevant chapter and will not be revisited here. Even though the O=O bond energy of $+494$ kJ mol^{-1} is high, many exothermic combustion reactions occur. One of the most important reactions of dioxygen is the coordination to the oxygen-transport protein haemoglobin (Section 26.2).

The other allotrope of oxygen, *ozone*, O_3, boils at $-112°C$ and is an explosive and highly reactive endoergic blue gas ($\Delta_f G^\circ = +163$ kJ mol^{-1}). It decomposes into dioxygen:

$$2 O_3(g) \rightarrow 3 O_2(g)$$

but this reaction is slow in the absence of a catalyst or ultraviolet radiation.

Ozone has a pungent odour; this property is reflected in its name, which is derived from the Greek *ozein*, to smell. The O_3 molecule is angular, in accord with the VSEPR model (**1**) and has bond angle 117°; it is diamagnetic. Gaseous ozone is blue, liquid ozone is blue–black, and solid ozone is violet–black. Ozone is produced from electrical discharges or ultraviolet radiation acting on O_2. This second method is used to produce low concentrations of ozone for the preservation of foodstuffs. The ability of O_3 to absorb strongly in the 220–290 nm region of the spectrum is vital in preventing the harmful ultraviolet rays of the Sun from reaching the Earth's surface. Ozone reacts with unsaturated polymers, causing undesirable cross-linking and degradation.

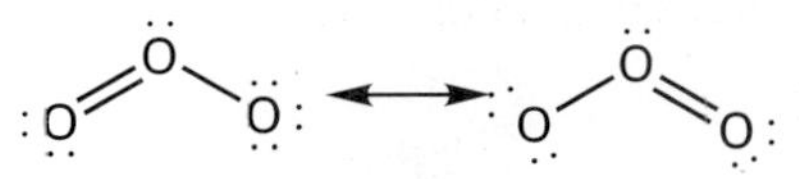

1 O_3, C_{2v}

Reactions of ozone typically involve oxidation and transfer of an O atom. Ozone is very unstable in acidic solution and much more stable in basic conditions:

$$O_3(g) + 2 H^+(aq) + 2 e^- \rightarrow O_2(g) + H_2O(l) \qquad E^\ominus = +2.08 \text{ V}$$
$$O_3(g) + H_2O(l) + 2 e^- \rightarrow O_2(g) + 2 OH^-(aq) \qquad E^\ominus = +1.25 \text{ V}$$

Ozone is exceeded in oxidizing power only by F_2, atomic O, the OH radical, and perxenate ions (Section 17.5). Ozone forms ozonides with Group 1 and 2 elements (Sections 1.5 and 11.5). They are prepared by passing gaseous ozone over the powdered hydroxide, MOH or M(OH)$_2$, at temperatures below $-10°C$. The ozonides are red–brown solids that decompose on warming:

$$MO_3(s) \rightarrow MO_2(s) + \tfrac{1}{2} O_2(g)$$

The ozonide ion, O_3^-, is angular, like O_3, but with the slightly larger bond angle of 119.5°.

(b) Sulfur

Key points: Sulfur is extracted as the element from underground deposits and its most stable form is the cyclic S_8 molecule; it has many allotropic forms, including a metastable polymer.

Sulfur occurs as deposits of the native element, in meteorites, volcanoes, and hot springs, as the ores galena, PbS, and barite, BaSO$_4$, and as Epsom salts, MgSO$_4$. It also occurs as H$_2$S in natural gas and as organosulfur compounds in crude oil. Sulfur can be extracted by the *Frasch process*, in which underground deposits are forced to the surface using superheated water and steam and compressed air. The extracted sulfur is molten and is allowed to cool in large basins. The process is energy intensive and commercial success depends upon access to cheap water and energy. Extraction from natural gas and crude oil by the *Claus process* has become increasingly important. In this process, H_2S is first oxidized in air at 1000–1400°C. This step produces some SO_2 that then reacts with the remaining H_2S at 200–350°C over a catalyst:

$$2 \ H_2S(g) + SO_2(g) \rightarrow 3 \ S(l) + 2 \ H_2O(l)$$

Unlike oxygen, sulfur (and all the heavier members of the group) tends to form single bonds with itself rather than double bonds because of the poor π overlap of its orbitals. As a result, it aggregates into larger molecules or extended structures and hence is a solid at room temperature. Sulfur vapour, which is formed at high temperatures, consists partially of paramagnetic disulfur molecules, S_2, that resemble O_2 in having a triplet ground state and a formal double bond.

That sulfur can exist in a large number of allotropic forms can be explained by the ability of S atoms to catenate because of the high S—S bond energy of 265 kJ mol^{-1}, which is exceeded only by C—C (330 kJ mol^{-1}) and H—H (435 kJ mol^{-1}). All the crystalline forms of sulfur that can be isolated at room temperature consist of S_n rings. The common yellow orthorhombic polymorph, α-S_8, consists of crown-like eight-membered rings (**2**) and all other forms of sulfur eventually revert to this form. Orthorhombic α-sulfur is an electrical and thermal insulator. When it is heated to 93°C, the packing of the S_8 rings is modified and monoclinic β-S_8 forms. When molten sulfur that has been heated above 150°C is cooled slowly, monoclinic γ-sulfur is formed. This allotrope consists of S_8 rings like the α and β forms but the packing of the rings is more efficient, resulting in a higher density.

It is possible to synthesize and crystallize sulfur rings with from six to 20 S atoms (Table 15.2). An additional complexity is that some of these allotropes exist in several crystalline forms. For example, S_7 is known in four crystalline forms and S_{18} is known in two different forms. Orthorhombic sulfur melts at 113°C; the yellow liquid darkens above 160°C and becomes more viscous as the sulfur rings break open and polymerize. The resulting helical S_n polymers (**3**) can be drawn from the melt and quenched to form metastable rubber-like materials that slowly revert to α-S_8 at room temperature. In the gas phase, S_2 and S_3 are observed. S_3 is a cherry red, angular molecule like ozone. The more stable species is the violet S_2 molecule that, like O_2, is doubly bonded with a bond dissociation energy of 421 kJ mol^{-1}.

Sulfur is reactive and interacts directly with many elements at room or elevated temperatures. It ignites in F_2 to form SF_6, reacts rapidly with Cl_2 to form S_2Cl_2, and dissolves in Br_2 to give S_2Br_2, which readily dissociates. It does not react with liquid I_2, which can therefore be used as a low-temperature solvent for sulfur. Atomic sulfur, S, is extremely reactive and triplet and singlet states are possible with different reactivities, as with O.

Most of the sulfur produced is used to manufacture sulfuric acid, H_2SO_4, which is one of the most important manufactured chemicals in the world. Sulfuric acid has many uses, including the synthesis of fertilizers and as the electrolyte in lead–acid batteries. Sulfur is a component of gunpowder (a mixture of potassium nitrate, KNO_3, carbon, and sulfur). It is also used in the vulcanization of natural rubber.

(c) Selenium, tellurium, and polonium

Key points: Selenium and tellurium crystallize in helical chains; polonium crystallizes in a primitive cubic form.

The chemically soft elements selenium and tellurium occur in metal sulfide ores and their principal source is the electrolytic refining of copper. Selenium is also extracted from the waste sludge from sulfuric acid plants. The extraction method depends on the other compounds or elements present. The first step usually involves oxidation in the presence of sodium carbonate:

$$Cu_2Se(aq) + Na_2CO_3(aq) + 2 \ O_2(g) \rightarrow 2 \ CuO(s) + Na_2SeO_3(aq) + CO_2(g)$$

The solution containing Na_2SeO_3 and Na_2TeO_3 is acidified with sulfuric acid. The tellurium precipitates out as the dioxide leaving selenous acid, H_2SeO_3, in solution.

2 S_8

3 S_n

Table 15.2 Properties of selected sulfur allotropes

Allotrope	Melting point*/°C	Appearance
S_3	Gas	Cherry red
S_6	50d	Orange red
S_7	39d	Yellow
α-S_8	113	Yellow
β-S_8	119	Yellow
γ-S_8	107	Pale yellow
S_{10}	0d	Yellow green
S_{12}	148	Pale yellow
S_{18}	128	Lemon yellow
S_{20}	124	Pale yellow
S_∞	104	Yellow

* d, decomposes.

Selenium is recovered by treatment with SO_2

$$H_2SeO_3(aq) + 2\ SO_2(g) + H_2O(l) \rightarrow Se(s) + 2\ H_2SO_4(aq)$$

Tellurium is liberated by dissolving the TeO_2 in aqueous sodium hydroxide followed by electroytic reduction.

$$TeO_2(s) + 2\ NaOH(aq) \rightarrow Na_2TeO_3(aq) + H_2O(l) \rightarrow Te(s) + 2\ NaOH(aq) + O_2(g)$$

Selenium exists as several different polymorphs. As with sulfur, three allotropes of selenium exist that contain Se_8 rings and differ only in the packing of the rings to give red, monoclinic α, β, and γ forms. The most stable form at room temperature is grey selenium, a crystalline material composed of helical chains. Monoclinic crystalline selenium is deep red. The common commercial form of the element is amorphous black selenium; it has a very complex structure comprising rings containing up to 1000 Se atoms. Another amorphous form of selenium, obtained by deposition of the vapour, is used as the photoreceptor in the xerographic photocopying process. Selenium is an essential element for humans, but there is only a narrow range of concentration between the minimum daily requirement and toxicity. An early indication of selenium poisoning is a garlicky smell on the breath, which is due to methylated selenium.

Selenium exhibits both photovoltaic action, where light is converted directly into electricity, and photoconductive action. The photoconductivity of grey selenium arises from the ability of incident light to excite electrons across its reasonably small band gap (2.6 eV in the crystalline material, 1.8 eV in the amorphous material). These properties make selenium useful in the production of photocells and exposure meters for photographic use, as well as solar cells. Selenium is also a p-type semiconductor and is used in electronic and solid-state applications. It is also used in photocopier toner and in the glass industry to make red glasses and enamels.

Tellurium crystallizes in a chain structure like that of grey selenium. Polonium crystallizes in a primitive cubic structure and a closely related higher temperature form above 36°C. We remarked in Section 3.5 that a simple cubic array represents inefficient packing of atoms, and polonium is the only element that adopts this structure under normal conditions. Tellurium and polonium are both highly toxic; the toxicity of polonium is enhanced by its intense radioactivity. Weight for weight it is about 2.5×10^{11} times as toxic as hydrocyanic acid. Polonium has more isotopes than any other element, all 29 of which are radioactive. It has been found in tobacco as a contaminant and in uranium ores. It is produced in small amounts through a nuclear reaction of bismuth. Irradiation of ^{209}Bi (atomic number 83) with neutrons gives ^{210}Po (atomic number 84).

$$^{209}_{83}Bi + {}^1_1n \rightarrow {}^{210}_{84}Po + e^-$$

Metallic polonium can then be separated from the remaining bismuth by fractional distillation or electrodeposited on to a metal surface.

Selenium, tellurium, and polonium combine directly with most elements, although less readily than oxygen or sulfur. The occurrence of multiple bonds is lower than with O or S, as is the tendency towards catenation (compared with S) and the number of allotropes. The unexpected difficulty of oxidizing Se to Se(IV) is due to the lanthanide-like contraction in radius following the $3d$ elements.

Simple compounds

The Group 16 elements are reactive and form compounds by direct reaction with most other elements, with the exception of the noble gases and some d metals.

15.2 Hydrides

The impact of hydrogen bonding is seen clearly in the hydrides of the Group 16 elements. The hydrides of oxygen are water and hydrogen peroxide, which are both liquids. The hydrides of the heavier elements are all toxic, foul-smelling gases.

Table 15.3 Selected properties of the Group 16 hydrides

Property	H_2O	H_2S	H_2Se	H_2Te	H_2Po
Melting point/°C	0.0	−85.6	−65.7	−51	−36
Boiling point/°C	100.0	−60.3	−41.3	−4	37
$\Delta_f H^\ominus$/(kJ mol^{-1})	−285.9	+20.1	+73.0	+99.6	
Bond length/pm	96	134	146	169	
Bond angle/°	104.5	92.1	91	90	
Acidity constants					
pK_{a1}	15.74	6.89	3.89	2.64	
pK_{a2}		14.15	~11	10.80	

(a) Water

Key point: Hydrogen bonding in water results in a high boiling liquid and a highly structured arrangement in the solid, ice.

The most important hydride of any element is that of oxygen, namely *water*. The properties and reactions of water are of paramount importance to inorganic chemists and are discussed throughout this text.

The most remarkable feature of water is its liquid range. Water is the only Group 16 hydride that is not a poisonous, malodorous gas. Its boiling point of 100°C is very high compared to molecules of similar molecular mass and the analogous molecules in Group 16 (Table 15.3). This high boiling point is due to extensive hydrogen bonding between the hydrogen and the highly electronegative oxygen, O—H···O (Section 9.4). Additional evidence for hydrogen bonding comes from the fact that when water freezes the resulting ice is highly structured. At least nine distinct forms of ice have been identified. At 0°C and atmospheric pressure, hexagonal ice I_h forms (Fig. 9.5) but between −120 and −140°C the cubic form, I_c, is produced. At very high pressures several higher density polymorphs are formed, some of which are based on silica-like structures (Section 13.8).

Water is formed by the direct interaction of the elements:

$$H_2(g) + \tfrac{1}{2} O_2(g) \rightarrow H_2O(l) \qquad \Delta_f H^\ominus(H_2O, l) = -286 \text{ kJ mol}^{-1}$$

This reaction is very exothermic and provides the basis for the development of the hydrogen economy and hydrogen fuel cells (Section 9.3a). More important sources of water are the biological cycling of water using indirect electron-transfer mechanisms, and its formation by widespread combustion of organic matter.

Water is the most widely used solvent not only because it is so widely available but also because of its high relative permittivity (dielectric constant), wide liquid range, and solvating ability. Many anhydrous and hydrated compounds dissolve in water to give hydrated cations and anions. Some predominantly covalent compounds, such as ethanol and ethanoic (acetic) acid, are soluble in water or miscible with it because of hydrogen-bonded interactions with the solvent. Many other covalent compounds react with water in hydrolysis reactions and examples of these reactions are discussed in the appropriate chapters. In addition to simple dissolution and hydrolysis reactions, the importance of aqueous solution chemistry can be seen in redox reactions (Chapter 5) and in Brønsted acids and bases (Chapter 4). Water also acts as a ligand in metal complexes (Chapter 8).

(b) Hydrogen peroxide

Key point: Hydrogen peroxide is susceptible to decomposition by disproportionation at elevated temperatures or in the presence of catalysts.

Hydrogen peroxide is a very pale blue, viscous liquid. It has a higher boiling point than water (150°C) and a greater density (1.445 g cm^{-3} at 25°C). It is miscible in water and is usually handled in aqueous solution. The Frost diagram for oxygen (Fig. 15.2)

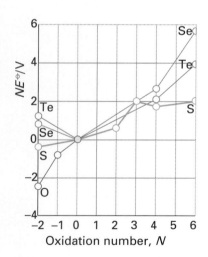

Fig. 15.2 Frost diagram for the elements of Group 16 in acidic solution. The species with oxidation number −2 are H_2E. For oxidation number −1 the compound is H_2O_2. The positive oxidation numbers refer to the oxo acids or oxoanions.

shows that H_2O_2 is a good oxidizing agent, but it is unstable with respect to disproportionation:

$$H_2O_2(aq) \rightarrow H_2O(aq) + \tfrac{1}{2} O_2(g) \qquad \Delta_r G^{\ominus} = -119 \text{ kJ mol}^{-1}$$

This reaction is slow but is explosive when catalysed by a metal surface or alkali dissolved from glass. For this reason, hydrogen peroxide and its solutions are stored in plastic bottles and a stabilizer is added. This reaction can be considered in terms of the reduction half-reactions

$$\tfrac{1}{2} H_2O_2(aq) + H^+(aq) + e^- \rightarrow H_2O(l) \qquad E^{\ominus} = +1.78 \text{ V}$$
$$H^+(aq) + \tfrac{1}{2} O_2(g) + e^- \rightarrow \tfrac{1}{2} H_2O_2(aq) \qquad E^{\ominus} = +0.70 \text{ V}$$

Any substance with a standard potential in the range $0.70 - 1.78$ V that has suitable binding sites will catalyse this reaction. As can be inferred from the standard potentials above, hydrogen peroxide is a very powerful oxidizing agent in acid solution:

$$2 \text{ Ce}^{3+}(aq) + H_2O_2(aq) + 2 \text{ H}^+(aq) \rightarrow 2 \text{ Ce}^{4+}(aq) + 2 \text{ H}_2O(l)$$

However, in basic solution hydrogen peroxide can act as a reducing agent:

$$2 \text{ Ce}^{4+}(aq) + H_2O_2(aq) + 2 \text{ OH}^-(aq) \rightarrow 2 \text{ Ce}^{3+}(aq) + 2 \text{ H}_2O(l) + O_2(g)$$

Hydrogen peroxide is a slightly stronger acid than water:

$$H_2O_2(aq) + H_2O(l) \rightleftharpoons H_3O^+(aq) + OOH^-(aq) \qquad pK_a = 11.65$$

Deprotonation occurs in other basic solvents such as liquid ammonia, and NH_4OOH has been isolated and found to consist of NH_4^+ and OOH^- ions. When solid NH_4OOH melts (at 25°C), the melt contains hydrogen-bonded NH_3 and H_2O_2 molecules.

The oxidizing ability of hydrogen peroxide and the harmless nature of its by-products lead to its many applications. It is used in water treatment to oxidize pollutants, as a mild antiseptic, and as a bleach in the textile, paper, and hair-care industries (Box 15.2).

Example 15.1 Deciding whether an ion can catalyse H_2O_2 disproportionation

Is Fe^{3+} thermodynamically capable of catalysing the decomposition of H_2O_2?

Answer The standard potential of the Fe^{3+}/Fe^{2+} couple is $+0.77$ V. This value falls between the potentials for the reduction of H_2O_2 to H_2O ($+1.76$ V) and for the reduction of O_2 to H_2O_2 ($+0.070$ V), so catalytic decomposition is expected. We can verify that the potentials are favourable. Subtraction of the equations and potentials for reduction of O_2 to H_2O_2 from that for reduction of Fe^{3+} gives:

$$2 \text{ Fe}^{3+}(aq) + H_2O_2(aq) \rightarrow 2 \text{ Fe}^{2+}(aq) + O_2(g) + 2 \text{ H}^+(aq) \qquad E^{\ominus} = +0.07 \text{ V}$$

Because $E^{\ominus} > 0$, the reaction is thermodynamically favourable in the sense $K > 1$. Next we subtract the equation and potential for the reduction of Fe^{3+} from those for the reduction of H_2O_2 to H_2O and obtain

$$2 \text{ Fe}^{2+}(aq) + H_2O_2(aq) + 2 \text{ H}^+(aq) \rightarrow 2 \text{ Fe}^{3+}(aq) + 2 \text{ H}_2O(l) \qquad E^{\ominus} = +0.99 \text{ V}$$

This reaction is also spontaneous (in the sense $K > 1$), so catalytic decomposition is thermodynamically favoured. In fact, the rates also are high, so Fe^{3+} is a highly effective catalyst for the decomposition of H_2O_2, and in its manufacture great pains are taken to minimize contamination by iron.

Self-test 15.1 Determine whether the decomposition of H_2O_2 is spontaneous in the presence of either Br^- or Cl^-.

(c) Hydrides of sulfur, selenium, and tellurium

Key points: The extent of hydrogen bonding is much less for these hydrides than in water; all the hydrides are gases.

Box 15.2 Environmentally friendly bleach

Hydrogen peroxide is rapidly replacing chlorine and hypochlorite bleaches in industrial applications as it is environmentally benign, producing only water and oxygen.

The major consumers of hydrogen peroxide bleach are the paper, textile, and wood pulp industries. A growing market is in de-inking of recycled paper and in the manufacture of kraft paper (strong brown paper). Approximately 85 per cent of all cotton and wool is bleached with hydrogen peroxide. One of its advantages over chorine-based bleaches is that it does not affect many modern dyes. It is also used to decolour oils and waxes.

Hydrogen peroxide is used to treat domestic and industrial effluent and sewage. It minimizes odours by preventing the production of H_2S by anaerobic reactions in sewers and pipes. It also acts as a source of oxygen in sewage sludge treatment plants. Other industrial uses of hydrogen peroxide are the epoxidation of soybean and linseed oils to produce plasticizers and stabilizers for the plastics industry and as a propellant for torpedoes and missiles. It has been speculated that the sinking of the Russian submarine *Kursk* in 2000 was due to an explosion involving the hydrogen peroxide used to fuel its torpedoes.

Hydrogen sulfide is a foul-smelling, poisonous gas. The fact that it is a gas under normal conditions (in contrast to water) demonstrates how the extent of hydrogen bonding is decreased from H—O···H to H—S···H. Its toxicity is made more hazardous by the fact that it tends to anaesthetize the sense of smell, making intensity of smell a dangerously inaccurate guide to concentration. Hydrogen sulfide is produced by volcanoes and by some micro-organisms (Box 15.3). It is an impurity in natural gas and must be removed before the gas is used.

Pure H_2S can be prepared by direct combination of the elements above 600°C:

$$H_2(g) + S(s) \rightarrow H_2S(g)$$

Hydrogen sulfide is easily generated in the laboratory by trickling dilute hydrochloric or phosphoric acid on to FeS:

$$FeS(s) + 2\ HCl(aq) \rightarrow H_2S(g) + FeCl_2(aq)$$

It is readily soluble in water, and is a weak acid:

$$H_2S(aq) + H_2O(l) \rightleftharpoons H_3O^+(aq) + SH^-(aq) \qquad pK_1 = 6.88$$
$$SH^-(aq) + H_2O(l) \rightleftharpoons H_3O^+(aq) + S^{2-}(aq) \qquad pK_2 = 14.15$$

Acidic solutions of H_2S are mild reducing agents and produce elemental sulfur on standing.

Hydrogen selenide, H_2Se, and hydrogen telluride, H_2Te, are both toxic, foul-smelling gases. Hydrogen selenide can be made by direct combination of the elements or by the reaction of FeSe with hydrochloric acid:

$$H_2(g) + Se(s) \rightarrow H_2Se(g)$$
$$FeSe(s) + 2\ HCl(aq) \rightarrow H_2Se(g) + FeCl_2(aq)$$

Hydrogen telluride is made by hydrolysis of Al_2Te_3 or by the action of hydrochloric acid on Mg, Zn, or Al tellurides.

$$Al_2Te_3(s) + 6\ H_2O(l) \rightarrow 3\ H_2Te(g) + 2\ Al(OH)_3(aq)$$
$$MgTe(s) + 2\ HCl(aq) \rightarrow H_2Te(g) + MgCl_2(aq)$$

The solubilities of H_2Se and H_2Te in water are similar to that of H_2S. The acidity of the solutions increases from H_2S to H_2Te (Table 15.3). Aqueous solutions of H_2Se and H_2Te are readily oxidized and produce elemental selenium and tellurium.

15.3 Halides

Key points: The halides of oxygen have limited stability but its heavier congeners form an extensive series of halogen compounds; typical formulas are EX_2, EX_4, and EX_6.

Oxygen forms many halogen oxides and oxoanions (Chapter 16). The oxidation number of oxygen is −2 in all its compounds with the halogens other than fluorine. Oxygen

Box 15.3 The sulfur cycle

Sulfur is essential to all life forms through its presence in the amino acids cysteine and methionine, and in many key active site structures, including the inorganic sulfide in Fe–S proteins, and all molybdenum and tungsten enzymes. Moreover, many organisms obtain energy by the oxidation or reduction of inorganic sulfur compounds. The resultant transformations define the sulfur cycle.

The redox extremes of sulfur chemistry are demonstrated by sulfate, the most oxidized form, and by H_2S and its ionized forms, HS^- or S^{2-}, the most reduced forms. Many classes of organisms occupy ecological niches defined by the sulfur. The sulfate-reducing bacteria (SRBs), such as *Desulforibrio*, reduce sulfate to sulfide under anaerobic conditions, oxidizing organic compounds in the process. Sulfide oxidizing organisms, such as *Thiobacilli*, are generally, but not always, aerobic and use O_2 to oxidize sulfide, polysulfide ions, elemental sulfur, or thiosulfate to sulfate. Figure B15.1 is an incomplete version of the sulfur cycle, highlighting some of the known participating molecules.

SRBs use sulfate as their electron acceptor and generate sulfide under anaerobic conditions. These anaerobic bacteria are found in environments where both SO_4^{2-} and reduced organic matter are found, for example in anoxic marine sediments and in the rumen of sheep and cattle. SRBs are important in sulfide ore formation, bio-corrosion, the souring of petroleum under anaerobic conditions, the Cu–Mo antagonism in ruminants, and many other physiological, ecological, and biogeochemical contexts.

The reduction of sulfate is carried out in two steps:

$$SO_4^{2-} + 8\ e^- + 10\ H^+ \rightarrow H_2S + 4\ H_2O$$

First, the relatively unreactive sulfate must be activated. This step is achieved through reaction with ATP to form adenosine phosphosulfate (APS) and pyrophosphate. The further hydrolysis of pyrophosphate ($\Delta_r G^\ominus = -30.5\ \text{kJ mol}^{-1}$) ensures that the APS formation reaction goes to the right:

$$ATP + SO_4^{2-} \rightarrow APS + P_2O_7^{4-}$$

The enzyme APS reductase carries out the catalytic reduction of the sulfate intermediate to sulfite:

$$APS + 2\ e^- + H^+ \rightarrow AMP + HSO_3^-$$

Conversion of sulfite to sulfide is then catalysed by the enzyme sulfite reductase:

$$HSO_3^- + 6\ e^- + 7\ H^+ \rightarrow H_2S + 3\ H_2O$$

The oxidative part of the sulfur cycle is the province of bacteria that gain energy from various interconversions. Some *Thiobacilli* species can oxidize sulfide in ores; for example, iron sulfides. The oxidation of sulfide to sulfate produces an acidic environment in which some species of *Thiobacilli* thrive and can alter the pH to produce acidic conditions favourable to their own metabolic processes. Acid mine drainage water can have a microbially produced pH as low as 1.5 and *Thiobacilli* are used commercially to mobilize metals from sulfide ores. For example, *Thiobacillus ferrooxidans* not only oxidizes the sulfur in iron sulfide deposits but also oxidizes the iron(II) to the soluble iron(III):

$$4\ FeS_2 + 15\ O_2 + 2\ H_2O \rightarrow 4\ Fe^{3+} + 8\ SO_4^{2-} + 4\ H^+$$

Thiobacilli live solely on inorganic materials: they use energy obtained from sulfide oxidation to drive all their cellular reactions, including the fixation of carbon from CO_2.

B15.1 The sulfur cycle.

---

difluoride, OF_2, is the highest fluoride of oxygen and hence contains oxygen in its highest oxidation state ($+2$).

Sulfur, selenium, tellurium, and polonium have a rich halogen chemistry, and some of the most common halides are summarized in Table 15.4. Sulfur forms very unstable iodides, but tellurium and polonium form more robust compounds with iodine, which is an example of a large anion stabilizing a large cation. Of the halogens, the small, electronegative F atom alone brings out the maximum group oxidation state of the chalcogen elements, but it does not form stable binary compounds of selenium, tellurium, and polonium in low oxidation states ($+1$ and $+2$). A series of catenated subhalides exist for the heavy members of the group. For example, Te_2I and Te_2Br consist of ribbons of edge-shared Te hexagons with halogen bridges (**4**). The inability of the halogens other than fluorine to bring out the higher oxidation states is understandable on the basis that they are less electronegative than fluorine, and their single bond strengths to other elements are generally weaker. The lack of low oxidation state fluorides may be

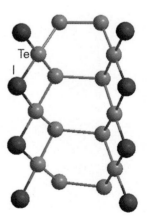

4 Te_2I

Table 15.4 Some halides of sulfur, selenium, and tellurium

Oxidation number	Formula	Structure	Remarks
$+\frac{1}{2}$	Te_2X (X = Br, I)	Halide bridges	Silver-grey
$+1$	S_2F_2	Two isomers: 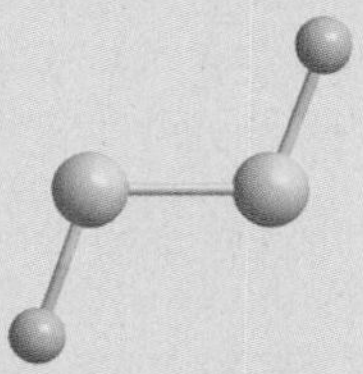 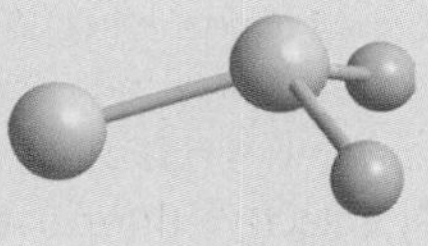	
	S_2Cl_2	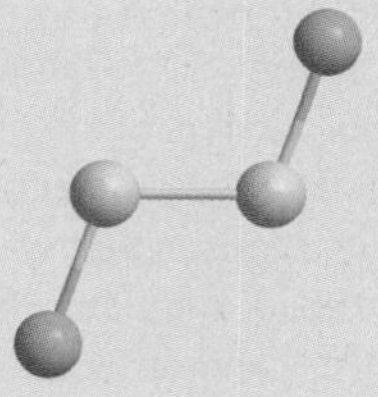	Reactive
$+2$	SCl_2	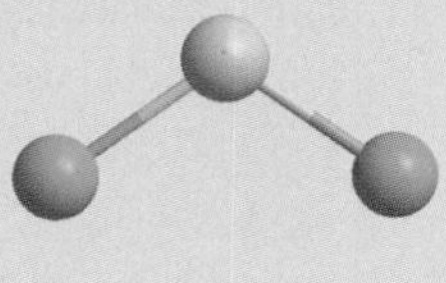	Reactive
$+4$	SF_4		Gas
	SeX_4 (X = F, Cl, Br)		SeF_4 liquid
	TeX_4 (X = F, Cl, Br I)		TeF_4 solid
$+5$	S_2F_{10}		Reactive
	Se_2F_{10}		
$+6$	SF_6, SeF_6		Colourless gases
	TeF_6		Liquid (b.p. 36°C)

a consequence of their instability towards disproportionation into the element and a higher oxidation state fluoride.

The structures of the sulfur halides S_2F_2, SF_4, SF_6, and S_2F_{10} (Table 15.4) are all in line with the VSEPR model. Thus, SF_4 has ten valence electrons around the S atom, two of which form a lone pair in an equatorial position of a trigonal bipyramid. We have already mentioned the theoretical evidence that the molecular orbitals bonding the F atoms to the central atom in SF_6 primarily use the sulfur $4s$ and $4p$ orbitals, with the $3d$ orbitals playing a relatively unimportant role (Section 2.1d). The same seems to be true of SF_4 and S_2F_{10}.

Sulfur hexafluoride is a gas at room temperature. It is very unreactive and its inertness stems from the suppression, presumably by steric protection of the central S atom, of thermodynamically favourable reactions, such as the hydrolysis

$$SF_6(g) + 4 H_2O(l) \rightarrow 6 HF(aq) + H_2SO_4(aq)$$

The less sterically crowded SeF_6 molecule is easily hydrolysed and is generally more reactive than SF_6. Similarly, the sterically less hindered molecule SF_4 is reactive and undergoes rapid partial hydrolysis:

$$SF_4(g) - H_2O(l) \rightarrow OSF_2(aq) + 2 HF(aq)$$

Both SF_4 and SeF_4 are very selective fluorinating agents for the conversion of $-COOH$ into $-CF_3$ and $C=O$ and $P=O$ groups into CF_2 and PF_2 groups:

$$2 R_2CO(l) + SF_4(g) \rightarrow 2 R_2CF_2(sol) + SO_2(g)$$

Sulfur chlorides are commercially important. The reaction of molten sulfur with Cl_2 yields the foul-smelling and toxic substance disulfur dichloride, S_2Cl_2, which is a yellow liquid at room temperature (b.p. 138°C). Disulfur dichloride and its further chlorination product sulfur dichloride, SCl_2, an unstable red liquid, are produced on a large scale for use in the vulcanization of rubber. In this process, S atom bridges are introduced between polymer chains so the rubber object can retain its shape.

15.4 Oxides

Oxygen is a reactive element and forms oxides with most other elements. The oxides of the elements are described within the relevant group chapters. In this section, we concentrate on compounds formed between oxygen and the other Group 16 elements.

(a) Reactivity of oxygen

Key point: The reactions of dioxygen are often thermodynamically favourable but sluggish.

Oxygen is by no means an inert molecule, yet many of its reactions are sluggish (a point first made in connection with overpotentials in Section 5.17). For example, a solution of Fe^{2+} is only slowly oxidized by air even though the reaction is thermodynamically favoured.

Several factors contribute to the appreciable activation energy of many reactions of O_2. One factor is that, with weak reducing agents, single-electron transfer to O_2 is mildly unfavourable thermodynamically:

$$O_2(g) + H^+(aq) + e^- \rightarrow HO_2(g) \qquad E^\ominus = -0.13\,V \text{ at } pH = 0$$
$$O_2(g) + e^- \rightarrow O_2^-(aq) \qquad E^\ominus = -0.33\,V \text{ at } pH = 14$$

A single-electron reducing reagent must exceed these potentials to achieve a significant reaction rate. Second, the ground state of O_2, with both π^* orbitals singly occupied, is neither an effective Lewis acid nor an effective Lewis base, and therefore has little tendency to undergo displacement reactions with p-block Lewis bases or acids. Finally, the high bond energy of O_2 (497 kJ mol^{-1}) results in a high activation energy for reactions that depend on its dissociation. Radical chain mechanisms can provide reaction paths that circumvent some of these activation barriers in combustion processes at elevated temperatures, and radical oxidations also occur in solution.

(b) Sulfur oxides and oxohalides

Key points: Sulfur dioxide is a mild Lewis acid towards p-block bases, and $OSCl_2$ is a useful drying agent.

The molecules of the two common oxides of sulfur, SO_2 (b.p. $-10°C$) and SO_3 (b.p. 44.8°C), are angular (**5**) and trigonal planar (**6**) in the gas phase, respectively. They

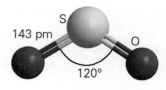

143 pm · 120°

5 SO_2, C_{2v}

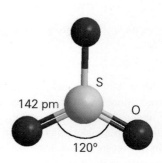

142 pm · 120°

6 SO_3, D_{3h}

7 (SO$_3$)$_3$, C_{3v}

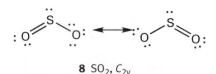

8 SO$_2$, C_{2v}

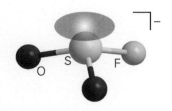

9 SO$_2$F$^-$

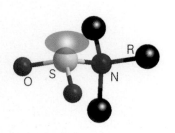

10 SO$_2$NR$_3$

are both Lewis acids, with the S atom the acceptor site, but SO$_3$ is the much the stronger and harder acid. The high Lewis acidity of SO$_3$ accounts for its occurrence as a polymeric O-bridged solid at room temperature and pressure (**7**).

Sulfur dioxide is a poisonous gas with a sharp, choking odour. It is manufactured on a large scale by combustion of sulfur or H$_2$S or by roasting sulfide ores in air:

$$4 \text{ FeS(s)} + 7 \text{ O}_2\text{(g)} \rightarrow 4 \text{ SO}_2\text{(g)} + 2 \text{ Fe}_2\text{O}_3\text{(s)}$$

The major use of SO$_2$ is in the manufacture of sulfuric acid, where it is first oxidized to SO$_3$. It is also used as a bleach, disinfectant, and food preservative. Sulfur dioxide is soluble in water and gives a solution of sulfurous acid, H$_2$SO$_3$, which is in fact a complex mixture of numerous species (Box 15.4). Sulfur dioxide forms weak complexes with simple *p*-block Lewis bases. For example, although it does not form a stable complex with H$_2$O, it does form stable complexes with stronger Lewis bases, such as trimethylamine and F$^-$ ions. Sulfur dioxide is a useful solvent for acidic substances (see Box 15.5).

Example 15.2 Deducing the structures and chemistry of SO$_2$ complexes

Suggest the probable structures of SO$_2$F$^-$ and (CH$_3$)$_3$NSO$_2$, and predict their reactions with OH$^-$.

Answer Although the Lewis structure of SO$_2$ (**8**) has an electron octet around the S atom, that atom can act as a Lewis acid (Section 4.7). Both resulting complexes have a lone pair on S, and the four electron pairs form a tetrahedron around the S atom, yielding trigonal-pyramidal complexes (**9**) and (**10**). The OH$^-$ ion is a stronger Lewis base than either F$^-$ or N(CH$_3$)$_3$, so exposure of either complex to OH$^-$ will yield the hydrogensulfite ion, HSO$_3^-$, which has been found to exist in two isomers, (**11**) and (**12**).

Self-test 15.2 Draw the Lewis structures and identify the point groups of (a) SO$_3$(g) and (b) SO$_3$F$^-$.

Sulfur trioxide, SO$_3$, is made on a huge scale by catalytic oxidation of SO$_2$. It is seldom isolated but immediately converted to sulfuric acid, H$_2$SO$_4$. Because sulfur trioxide is extremely corrosive, anhydrous SO$_3$ is seldom handled in the laboratory. It is available as oleum, or fuming sulfuric acid, H$_2$S$_2$O$_7$, which is solution of 25–65 per cent SO$_3$ by mass in concentrated sulfuric acid. Sulfur trioxide reacts with water to give H$_2$SO$_4$ in a vigorous and very exothermic reaction. It extracts water from organic matter to leave

---

Box 15.4 Acid rain

The main components of acid rain are nitric and sulfuric acids produced by interaction of the oxides with hydroxyl radicals:

$$\text{HO}\cdot + \text{NO}_2 \rightarrow \text{HNO}_3$$
$$\text{HO}\cdot + \text{SO}_2 \rightarrow \text{HSO}_3\cdot$$
$$\text{HSO}_3\cdot + \text{O}_2 + \text{H}_2\text{O} \rightarrow \text{H}_2\text{SO}_4 + \text{HO}_2\cdot$$

The resulting hydroperoxyl radicals produce additional hydroxyl radicals:

$$\text{HO}_2\cdot + \text{X} \rightarrow \text{O} + \text{HO}\cdot \qquad \text{X} = \text{NO or SO}_2$$

Sulfuric and nitric acid molecules form hydrogen bonds and interact strongly with one another, with metal oxides and gases in the atmosphere, and with water, to form particles. These small particles are a major health threat in polluted air. Recent studies have convincingly associated increased concentrations of particulate matter in the size range of 2.5 μm or less with increased mortality from pulmonary, and especially heart, disease. These particles

are small enough to lodge deep in the lungs, and they can carry noxious chemicals on their surface.

In addition to their physiological effects, these particles affect the ecosystem because of the acids they contain. As the acidity of rainfall increases, the protons increasingly wash alkali metal (Na$^+$, K$^+$) and alkaline earth (Ca^{2+}, Mg^{2+}) ions from the soil, where they are held in ion exchange sites in clay and humus or in limestone. Depletion of these nutrients limits plant growth. The same chemistry also erodes marble statues and buildings. Granite-lined lakes (which have low buffer capacity) can be acidified, leading to the disappearance of fish and other aquatic life. Because emissions from combustion sources can travel long distances, acid rain is a regional problem, with large areas at risk, particularly downwind of coal-fired power plants, which have tall smokestacks to disperse the NO and SO$_2$ exhaust gases. These environmental and health effects combine to make NO and SO$_2$ a main focus of air pollution regulatory activity.

Box 15.5 Synthesis in nonaqueous solvents

The nonaqueous solvents liquid ammonia, liquid sulfur dioxide, and sulfuric acid show an interesting set of contrasts as they range from a good Lewis base with a resistance to reduction (ammonia) to a strong Brønsted acid with resistance to oxidation (sulfuric acid).

Liquid ammonia (b.p. $-33°C$) and the solutions it forms may be handled in an open Dewar flask in an efficient hood (Fig. B15.2), in a vacuum line when the liquid is kept below its boiling point, or (with caution) in sealed heavy-walled glass tubes at room temperature (the vapour pressure of ammonia is about 10 atm at room temperature). Liquid ammonia solutions of the s-block metals are excellent reducing agents. One example of their application is in the preparation of a complex of nickel in its unusual $+1$ oxidation state:

$$2\ K(am) + 2\ [Ni(CN)_4]^{2-}(am) \rightarrow [Ni_2(CN)_6]^{4-}(am) + 2\ KCN(am)$$

Liquid sulfur dioxide (b.p. $-10°C$) can be handled like liquid ammonia. An example of its use is the reaction

$$NH_2OH(s) + SO_2(l) \rightarrow H_2NSO_2OH(s)$$

This reaction is performed in a sealed thick-walled borosilicate glass tube, which is sometimes called a 'Carius tube' (Fig. B15.3). The hydroxylamine is introduced into the tube and the sulfur dioxide is condensed into the tube after it has been cooled to about $-45°C$. After the tube has been sealed, it is allowed to warm to room temperature behind a blast shield in a hood. The reaction is complete after several days at room temperature; the tube is then cooled (to reduce the pressure) and broken open.

Sulfuric acid and solutions of SO_3 in sulfuric acid (fuming sulfuric acid, $H_2S_2O_7$) are used as oxidizing acidic media for the preparation of polychalcogen cations:

$$8\ Se(s) + 5\ H_2SO_4(l) \rightarrow Se_8^{2+}(sol) + 2\ H_3O^+(sol)$$
$$+4\ HSO_4^-(sol) + SO_2(g)$$

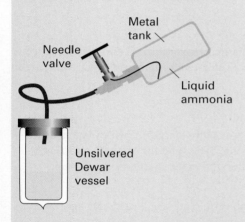

B15.2 The transfer of liquid ammonia from a pressurized tank. (Based on W.L. Jolly, *The synthesis and characterization of inorganic compounds*. Waveland Press, Prospect Heights (1991).)

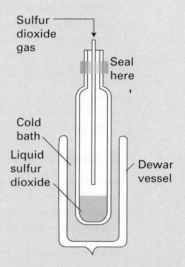

B15.3 The condensation of sulfur dioxide into a Carius tube.

a charred, carbonaceous residue. The reaction with metal oxides to produce sulfates is used to scrub undesirable SO_3 from effluent gases from industrial processes.

(c) Oxides of selenium and tellurium

Key points: Selenium and tellurium dioxides exist in α and β forms; selenium dioxide is less stable than SO_2 or TeO_2 and selenium trioxide, SeO_3, is less stable than SeO_2.

The dioxides of Se, Te, and Po can be prepared by direct reaction of the elements. Selenium dioxide is a white solid that sublimes at 315°C. It has a polymeric structure in the solid state (**13**). It is thermodynamically less stable than SO_2 or TeO_2 and is decomposed to the elements by reaction with NH_3, N_2H_4, or aqueous SO_2. It is used an oxidizing agent in organic chemistry.

Tellurium dioxide occurs naturally as the mineral tellurite, $\beta\text{-}TeO_2$, which has a layer structure in which TeO_4 units form dimers (**14**), Synthetic $\alpha\text{-}TeO_2$ consists of similar TeO_4 units that share all vertices to form a three-dimensional rutile-like structure (**15**). Polonium dioxide exists as the yellow form with the fluorite structure and the red tetragonal form.

Selenium trioxide, unlike SO_3 or TeO_3, is thermodynamically less stable than the dioxide (Table 15.5). It is a white hygroscopic solid that sublimes at 100°C and is

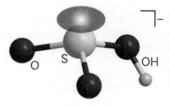

11 HSO_3^-

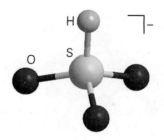

12 HSO_3^-

13 SeO_2

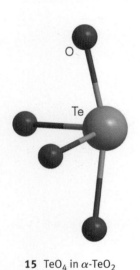

14 $(TeO_4)_2$

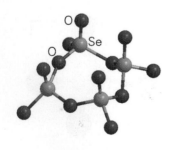

15 TeO_4 in α-TeO_2

Table 15.5 Standard enthalpies of formation, $\Delta_f H^{\circ}$/(kJ mol^{-1}), of sulfur, selenium, and tellurium oxides

SO_2	−297	SO_3	−432
SeO_2	−230	SeO_3	−184
TeO_2	−325	TeO_3	−348

16 $(SeO_3)_4$

decomposed at 165°C. In the solid state the structure is based on Se_4O_{12} tetramers (**16**) but it is monomeric in the vapour phase. Tellurium trioxide exists as the yellow α-TeO_3, which is prepared by dehydration of $Te(OH)_6$, and the more stable β-TeO_3, which is made by heating α-TeO_3 or $Te(OH)_6$ in oxygen.

(d) Chalcogen oxohalides

Key points: The most important oxohalides are those of sulfur; selenium and tellurium oxofluorides are known and the 'teflate' ion is a useful ligand.

Many chalcogen oxohalides are known. The most important are the thionyl dihalides, OSX_2, and the sulfuryl dihalides, O_2SX_2. One laboratory application of thionyl dichloride is the dehydration of metal chlorides:

$$MgCl_2 \cdot 6H_2O(s) + 6\ OSCl_2(l) \rightarrow MgCl_2(s) + 6\ SO_2(g) + 12\ HCl(g)$$

The compound $F_5TeOTeF_5$ and its selenium analogue are known, and the $OTeF_5^-$ ion, which is known informally as 'teflate', is a bulky, electronegative anion. It is a well-established ligand for high oxidation state d-metal and main-group element complexes such as $[W(OTeF_5)_6]$, $[Xe(OTeF_5)_6]$, and $[M(C_5H_5)_2(OTeF_5)_2]$, where $M = $ Ti, Zr, Hf, W, and Mo.

15.5 Oxoacids of sulfur

Sulfur (like nitrogen and phosphorus) forms many oxoacids. These exist in aqueous solution or as the solid salts of the oxoanions (Table 15.6). Many of them are important in the laboratory and in industry.

(a) Redox properties of the oxoanions

Key points: The oxoanions of sulfur include the sulfite ion, SO_3^{2-}, which is a good reducing agent, the rather unreactive sulfate ion, SO_4^{2-}, and the strongly oxidizing peroxodisulfate ion, $O_3SOOSO_3^{2-}$. As with sulfur, the redox reactions of selenium and tellurium oxoanions are often slow.

The peroxodisulfate ion, $O_3SOOSO_3^{2-}$, is a powerful and useful oxidizing agent:

$$[O_3S-O-O-SO_3]^{2-} + 2\ e^- \longrightarrow 2\ [SO_3]^{2-} \qquad E^{\ominus} = +1.96\ V$$

Sulfur's common oxidation numbers are −2, 0, +2, +4, and +6, but there are also many S—S bonded species that are assigned odd and fractional average oxidation numbers. A simple example is the thiosulfate ion, $S_2O_3^{2-}$, in which the average oxidation number of S is +2, but in which the environments of the two S atoms are quite different. The thermodynamic relations between the oxidation states are summarized by the Frost diagram (Fig. 15.2). As with many other p-block oxoanions, many of the thermodynamically favourable reactions are slow when the element is in its maximum oxidation state (+6), as in SO_4^{2-}. Another kinetic factor is suggested by the fact that oxidation numbers of compounds containing a single S atom generally change in steps of 2, which requires an O atom transfer path for the mechanism. In some cases a radical mechanism operates, as in the oxidation of thiols and alcohols by peroxodisulfate, in which O—O bond cleavage produces the transient radical anion SO_4^-.

We saw in Section 5.12 that the pH of a solution has a marked effect on the redox properties of oxoanions. This strong dependence is true for SO_2 and SO_3^{2-}, because the former is easily reduced in acidic solution and is therefore a good oxidizing agent, whereas the latter in basic solution is primarily a reducing agent:

$$SO_2(aq) + 4\ H^+(aq) + 4\ e^- \rightarrow S(s) + 2\ H_2O(l) \qquad E^{\ominus} = +0.50\ V$$

$$SO_4^{2-}(aq) + H_2O(l) + 2\ e^- \rightarrow SO_3^{2-}(aq) + 2\ OH^-(aq) \qquad E^{\ominus} = -0.94\ V$$

Table 15.6 Some sulfur oxoanions

Oxidation number	Formula	Name	Structure	Remarks
One S atom				
+4	SO_3^{2-}	Sulfite	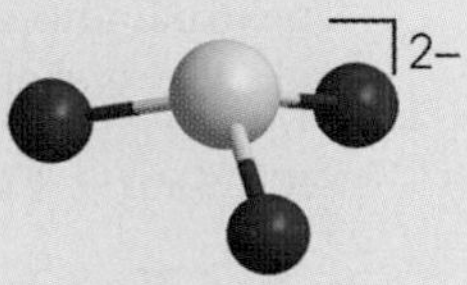	Basic, reducing agent
+6	SO_4^{2-}	Sulfate	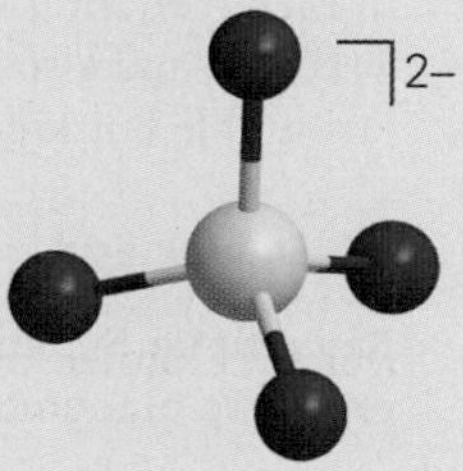	Weakly basic
Two S atoms				
+2	$S_2O_3^{2-}$	Thiosulfate	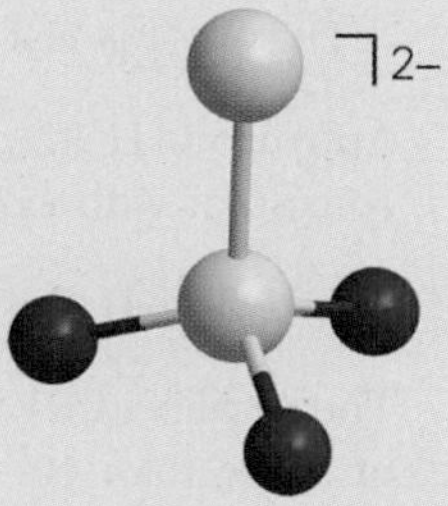	Moderately strong reducing agent
+3	$S_2O_4^{2-}$	Dithionite	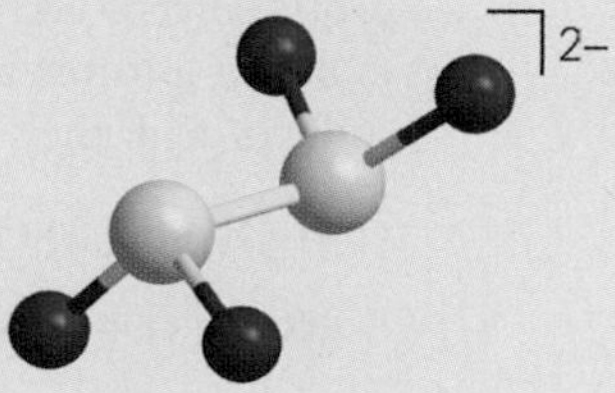	Strong and facile reducing agent
+4	$S_2O_5^{2-}$	Disulfite		
+5	$S_2O_6^{2-}$	Dithionate		Resists oxidation and reduction
Polysulfur oxanions				
Variable	$S_nO_{2n+2}^{2-}$ $3 \leq n \leq 20$	$n = 3$, Trithionate	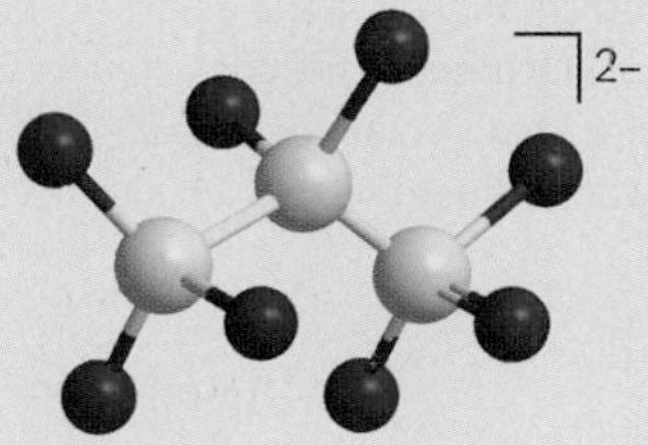	

The principal species present in acidic solution is SO_2, not H_2SO_3, but in more basic solution HSO_3^- exists in equilibrium with H—SO_3^- and H—OSO_2^-. The oxidizing character of SO_2 accounts for its use as a mild disinfectant and bleach for foodstuffs, such as dried fruit and wine.

The oxoanions of selenium and tellurium are a much less diverse and extensive group. Selenic acid is thermodynamically a strong oxidizing acid:

$$SeO_4^{2-}(aq) + 4\ H^+(aq) + 2\ e^- \rightarrow H_2SeO_3(aq) + H_2O(l) \qquad E^{\ominus} = +1.1\ V$$

However, like SO_4^{2-} and in common with the behaviour of oxoanions of other elements in high oxidation states, the reduction of SeO_4^{2-} is generally slow. Telluric acid exists as $Te(OH)_6$ and also as $(HO)_2TeO_2$ in solution. Again, its reduction is thermodynamically favourable but kinetically sluggish.

(b) Sulfuric acid

Key points: Sulfuric acid is a strong acid; it is a useful nonaqueous solvent due its extensive autoprotolysis.

Sulfuric acid is a dense viscous liquid. It dissolves in water in a highly exothermic reaction:

$$H_2SO_4(l) + H_2O(l) \rightarrow H_2SO_4(aq) \qquad \Delta_r H^{\ominus} = -880\ kJ\ mol^{-1}$$

Anhydrous H_2SO_4 has a very high relative permittivity and high electrical conductivity consistent with extensive autoprotolysis:

$$2\ H_2SO_4(l) \rightleftharpoons H_3SO_4^+(sol) + HSO_4^-(sol) \qquad K = 2.7 \times 10^{-4}$$

The equilibrium constant for this autoprotolysis is greater than that of water by a factor of more than 10^{10}. This property leads to the use of sulfuric acid as a nonaqueous, protic solvent (Box 15.5).

Bases (proton acceptors) increase the concentration of HSO_4^- ions in anhydrous sulfuric acid: they include water and salts of weaker acids, such as nitrates:

$$H_2O(l) + H_2SO_4(sol) \rightarrow H_3O^+(sol) + HSO_4^-(sol)$$

$$NO_3^-(s) + H_2SO_4(l) \rightarrow HNO_3(sol) + HSO_4^-(sol)$$

Another example of this kind is the reaction of concentrated sulfuric acid with concentrated nitric acid to produce the nitronium ion, NO_2^+, which is responsible for the nitration of aromatic species:

$$HNO_3(aq) + 2\ H_2SO_4(aq) \rightarrow NO_2^+(aq) + H_3O^+(aq) + 2\ HSO_4^-(aq)$$

Box 15.6 The manufacture of sulfuric acid

Sulfuric acid is one of the most important chemicals produced on an industrial scale. Over 80 per cent of it is used to manufacture fertilizers. It is also used to remove impurities from petroleum, to pickle (clean) iron and steel before electroplating, as the electrolyte in lead–acid batteries (Box 13.5), and in the manufacture of many other bulk chemicals such as hydrochloric and nitric acids.

Concentrated sulfuric acid is manufactured by the *contact process*. The first stage is the oxidation of a sulfur compound to SO_2. Most plants use elemental sulfur but metal sulfides and H_2S are also used:

$$S(s) + O_2(g) \rightarrow SO_2(g)$$

$$4\ FeS(s) + 7\ O_2(g) \rightarrow 2\ Fe_2O_3(s) + 4\ SO_2(g)$$

$$2\ H_2S(g) + 3\ O_2(g) \rightarrow 2\ SO_2(g) + 2\ H_2O(g)$$

The second stage is the oxidation of SO_2 to SO_3. This reaction is carried out at high temperatures and pressures over a V_2O_5 catalyst supported on silica beads:

$$2\ SO_2(g) + O_2(g) \rightarrow 2\ SO_3(g)$$

The SO_3 is then passed into the bottom of a packed column and washed by running oleum, $H_2S_2O_7$, from the top of the column. The gas is then washed in a second column with 98 per cent by mass H_2SO_4. The SO_3 reacts with the 2 per cent of water to produce sulfuric acid, H_2SO_4:

$$SO_3(g) + H_2O(l) \rightarrow H_2SO_4(l)$$

The number of species that are acidic in sulfuric acid is much smaller than in water because the acid is a poor proton acceptor. For example, HSO_3F is a weak acid in sulfuric acid:

$$HSO_3F(sol) + H_2SO_4(l) \rightleftharpoons H_3SO_4^+(sol) + SO_3F^-(sol)$$

As well as undergoing autoprotolysis, H_2SO_4 dissociates into H_2O and SO_3, which react further with H_2SO_4 to give a number of products:

$$H_2O + H_2SO_4 \rightleftharpoons H_3O^+ + HSO_4^-$$
$$SO_3 + H_2SO_4 \rightleftharpoons H_2S_2O_7$$
$$H_2S_2O_7 + H_2SO_4 \rightleftharpoons H_3SO_4 + HS_2O_7^-$$

Consequently, rather than being a single substance, anhydrous sulfuric acid is made up of a complex mixture of at least seven characterized species.

Sulfuric acid forms two series of salts, the sulfates, SO_4^{2-}, and the hydrogensulfates, HSO_4^-. These are often stable compounds and are important minerals of some electropositive elements. They have been discussed in earlier chapters.

(c) Sulfurous acid and disulfurous acid

Key points: Sulfurous and disulfurous acids have never been isolated. However, salts of both acids exist; sulfites are moderately strong reducing agents and are used as bleaches; disulfites rapidly decompose in acidic conditions.

Although an aqueous solution of SO_2 is referred to as sulfurous acid, H_2SO_3 has never been isolated and the predominant species present are the hydrates $SO_2 \cdot nH_2O$. The first and second proton donation are therefore best represented as follows::

$$SO_2 \cdot nH_2O(aq) + H_2O(l) \rightleftharpoons H_3O^+(aq) + HSO_3^-(aq) + nH_2O(l) \qquad pK_a = 1.79$$
$$HSO_3^-(aq) + H_2O(l) \rightleftharpoons H_3O^+(aq) + SO_3^{2-}(aq) \qquad pK_a = 7.00$$

Two series of salts, the sulfites, SO_3^{2-}, and the hydrogensulfites, HSO_3^-, are known and are moderately strong reducing agents, being oxidized to sulfates, SO_4^{2-}, or thionates, $S_2O_6^{2-}$. Anhydrous sodium sulfite, Na_2SO_3, is produced on an industrial scale and used as a bleach in the pulp and paper industry, as a reducing agent in photography, and as an oxygen scavenger in boiler treatments.

Disulfurous acid, $H_2S_2O_5$ (**17**), does not exist in the free state but salts are readily obtained from a concentrated solution of hydrogensulfites:

$$2\ HSO_3^-(aq) \rightleftharpoons S_2O_5^{2-}(aq) + H_2O(l)$$

Acidic solutions of disulfites rapidly decompose to give HSO_3^- and SO_3^{2-}.

Example 15.3 Predicting the reducing power of sulfites

Use the standard potential data in *Resource section 3* to predict which oxidation states of manganese will be reduced by sulfite or hydrogensulfite ions in acidic conditions.

Answer The relevant potential is

$$S_2O_6^{2-} + 2\ e^- \rightarrow 2\ SO_3^{2-} \qquad E^{\ominus} = +0.569\ V$$

Any species with $E^{\ominus} > 0$ will be reduced. Consequently, all the Mn species except Mn^{2+} will be reduced by sulfite ions.

Self-test 15.3 Predict which oxidation states of Mn will be reduced by sulfite ions in basic conditions.

(d) Thiosulfuric acid

Key points: Thiosulfuric acid decomposes but the salts are stable; thiosulfate ion is a moderately strong reducing agent.

Aqueous thiosulfuric acid, $H_2S_2O_3$ (**18**), decomposes rapidly in a complex process that produces a number of products such as SO_2, H_2S, and H_2SO_4. The anhydrous acid

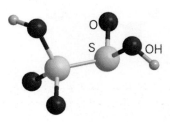

17 Disulfurous acid, $H_2S_2O_5$

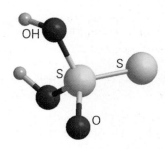

18 Thiosulfuric acid, $H_2S_2O_3$

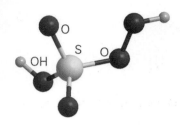

19 Peroxomonosulfuric acid, H_2SO_5

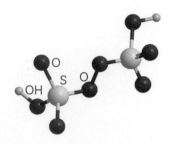

20 Peroxodisulfuric acid, $H_2S_2O_8$

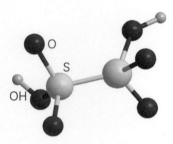

21 Dithionous acid, $H_2S_2O_4$

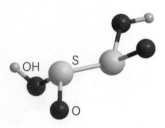

22 Dithionic acid, $H_2S_2O_6$

is more stable and decomposes slowly to H_2S and SO_3. In contrast to the acid, the thiosulfate salts are stable and can be prepared by boiling the sulfites or hydrogensulfites with elemental sulfur or by oxidation of polysulfides:

$$8 K_2SO_3(aq) + S_8(s) \rightarrow 8 K_2S_2O_3(aq)$$

$$2 CaS_2(s) + 3 O_2(g) \rightarrow 2 CaS_2O_3(g)$$

The thiosulfate ion, $S_2O_3^-$, is a moderately strong reducing agent:

$$\tfrac{1}{2} S_4O_6^{2-}(aq) + e^- \rightarrow S_2O_3^{2-}(aq) \qquad E^\ominus = +0.09 \text{ V}$$

The reaction with iodine is the basis of iodometric titrations in analytical chemistry:

$$\tfrac{1}{2} I_2(aq) + e^- \rightarrow I^-(aq) \qquad E^\ominus = +0.54 \text{ V}$$

$$2 S_2O_3^{2-}(aq) + I_2(aq) \rightarrow S_4O_6^{2-}(aq) + 2 I^-(aq)$$

Stronger oxidizing agents, such as chlorine, oxidize thiosulfate to sulfate, which has led to the use of thiosulfate to remove excess chlorine in the bleaching industries.

(e) Peroxosulfuric acids

Key point: Peroxodisulfate salts are strong oxidizing agents.

Peroxomonosulfuric acid, H_2SO_5 (**19**), is a crystalline solid that can be prepared by reacting H_2SO_4 with peroxodisulfates or as a by-product of the synthesis of $H_2S_2O_8$ by electrolysis of H_2SO_4. The salts are unstable and decompose, releasing H_2O_2. Peroxodisulfuric acid, $H_2S_2O_8$ (**20**), is also a crystalline solid. Its ammonium and potassium salts are prepared on an industrial scale by the oxidation of ammonia and potassium sulfates. They are strong oxidizing and bleaching agents.

$$\tfrac{1}{2} S_2O_8^{2-}(aq) + H^+(aq) + e^- \rightarrow HSO_4^- \qquad E^\ominus = +2.12 \text{ V}$$

When $K_2S_2O_8$ is heated, ozone and oxygen are evolved.

(f) Dithionous and dithionic acids

Key point: Dithionate salts may be oxidized or reduced.

Neither anhydrous dithionous acid, $H_2S_2O_4$ (**21**), nor dithionic acid, $H_2S_2O_6$ (**22**), can be isolated. However, the dithionous and dithionate salts are stable crystalline solids. Dithionites, $S_2O_4^{2-}$, can be prepared by the reduction of sulfites with zinc dust or sodium amalgam. Sodium dithionite is an important reducing agent in biochemistry. Neutral and acidic solutions of dithionite disproportionate into HSO_3^{2-} and $S_2O_3^{2-}$:

$$2 S_2O_4^{2-}(aq) + H_2O(l) \rightarrow 2 HSO_3^-(aq) + S_2O_3^{2-}(aq)$$

Dithionates, $S_2O_6^{2-}$, are prepared by oxidation of the corresponding sulfite. Strong oxidizing agents such as MnO_4^- oxidize dithionate to sulfate:

$$SO_4^{2-}(aq) + 2H^+(aq) + e^- \rightarrow \tfrac{1}{2} S_2O_6^{2-}(aq) + H_2O(l) \qquad E^\ominus = -0.22 \text{ V}$$

Strong reducing agents such as sodium amalgam reduce it to SO_3^{2-}:

$$\tfrac{1}{2} S_2O_6^{2-}(aq) + 2 H^+(aq) + e^- \rightarrow H_2SO_3(aq) \qquad E^\ominus = 0.56 \text{ V}$$

Neutral and acidic solutions of dithionate slowly decompose to SO_2 and SO_4^{2-}:

$$S_2O_6^{2-}(aq) \rightarrow SO_2(g) + SO_4^{2-}(aq)$$

(g) Polythionic acids

Key point: Polythionic acids can be prepared with up to six S atoms.

Many polythionic acids, $H_2S_nO_6$, were first identified by the study of 'Wackenroder's solution', which consists of H_2S in aqueous SO_2. Among those first characterized are the terathionate, $S_4O_6^{2-}$ (**23**), and pentathionate, $S_4O_6^{2-}$ (**24**), ions. More recently, a wide variety of preparative routes have been developed, many of which are complicated by numerous redox and catenation reactions. Typical examples are oxidation of thiosulfates with I_2 or H_2O_2, and reaction of polysulfanes, H_2S_n, with SO_3 to yield $H_2S_{n+2}O_6$, where $n = 2-6$.

$$H_2S_n(aq) + 2\,SO_3(aq) \rightarrow H_2S_{n+2}O_6(aq)$$

23 Tetrathionate ion, $O_3S(S_2)SO_3^{2-}$

24 Pentathionate ion, $O_3S(S_3)SO_3^{2-}$

Example 15.4 **Predicting outcomes of reaction of sulfur oxoanions**

Use the reduction potentials given below to predict what will happen when H_2S is bubbled into a sulfurous acid solution.

$$H_2SO_3(aq) + 4\,H^+(aq) + 4\,e^- \rightarrow S(s) + 3\,H_2O(l) \qquad E^{\ominus} = +0.34\ V$$
$$S(s) + 2\,H^+(aq) + 2\,e^- \rightarrow H_2S(g) \qquad E^{\ominus} = +0.14\ V$$

Answer Because the potential for the S/H_2S couple is less positive than that for the H_2SO_3/S couple the reaction will proceed as an oxidation. Therefore, the overall reaction will be

$$H_2SO_3(aq) + 2\,H_2S(g) \rightarrow 3\,S(s) + 3\,H_2O(l)$$

Yellow elemental sulfur will be formed.

Self test 15.4 Predict what will happen when MnO_4^- solution is added to sulfurous acid.

15.6 Metal oxides

Key points: The oxides formed by metals include the basic oxides with high oxygen coordination number that are formed with most M^+ and M^{2+} ions. Oxides of metals in intermediate oxidation states often have more complex structures and are amphoteric. Metal peroxides and superoxides are formed between O_2 and alkali metals and alkaline earth metals. Terminal E=O linkages and E—O—E bridges are common with nonmetals and with metals in high oxidation states.

The O_2 molecule readily removes electrons from metals to form a variety of metal oxides containing the anions O^{2-} (oxide), O_2^- (superoxide), and O_2^{2-} (peroxide). Even though these solids can be expressed in these terms, and the existence of O^{2-} can be rationalized in terms of a closed-shell noble gas electron configuration, the formation of $O^{2-}(g)$ from $O(g)$ is highly endothermic, and the ion is stabilized in the solid state by high lattice energies.

Alkali metals and alkaline earth metals often form peroxides or superoxides (Section 10.5), but peroxides and superoxides of other metals are rare. Among the metals, only some of the noble metals[2] do not form thermodynamically stable oxides. However, even where no bulk oxide phase is formed, an atomically clean metal surface (which can be prepared only in an ultrahigh vacuum) is quickly covered with a surface layer of oxide when it is exposed to traces of oxygen.

An important trend in the chemical properties of metal oxides is the high Brønsted basicity of oxides with metal ions of low charge and large radius (low electrostatic parameter, ξ; and the progression through amphoteric to acidic oxides as the charge-to-radius ratio increases (high ξ). These trends were examined in Section 4.5.

Structural trends in the metal oxides are not so readily summarized but, for oxides in which the metal has oxidation number $+1$, $+2$, or $+3$, the oxide is generally in a site of high coordination number. Thus, oxides of M^{2+} ions usually have the rock-salt structure (6:6 coordination), oxides of M^{3+} ions (of formula M_2O_3) often have 6:4 coordination, and oxides with the formula MO_2 often have a rutile or fluorite structure (6:3 and 8:4 coordination, respectively). At the other extreme, MO_4 compounds are molecular: the tetrahedral compound osmium tetroxide, OsO_4, is an example. The structures of the

---

[2] Noble metals are very resistant to oxidation and corrosion and include silver, gold, platinum, osmium, rhodium, iridium, and rhenium.

oxides with metals in high oxidation states and the oxides of nonmetallic elements often have multiple bond character. Deviations from these simple structures are common with p-block metals, where the less symmetric packing of O^{2-} ions around the metal can often be rationalized in terms of the existence of a stereochemically active lone pair, as in PbO (Section 13.9). Another common structural motif for nonmetals and some metals in high oxidation states is a bridging oxygen atom, as in E—O—E, in angular and linear structures.

15.7 Metal sulfides, selenides, tellurides, and polonides

Sulfur, selenium, and tellurium form compounds with many metals. Sulfur forms sulfides, containing the S^{2-} ion, disulfides, containing S_2^{2-}, and sulfido complexes. Selenium and tellurium form selenides and tellurides by direct reaction with many elements.

(a) Sulfides

Key point: Monatomic and polyatomic sulfide ions are known as discrete anions and as ligands.

Many metals occur naturally as their sulfide ores. The ores are roasted in air to form the oxide or the water-soluble sulfate, from which the metals are extracted. The sulfides can be prepared in the laboratory or industry by a number of routes; direct combination of the elements, reduction of a sulfate, or precipitation of an insoluble sulfide from solution by addition of H_2S.

$$Fe(s) + S(s) \rightarrow FeS(s)$$
$$MgSO_4(s) + 4\ C(s) \rightarrow MgS(s) + 4\ CO(g)$$
$$M^{2+}(aq) + H_2S(g) \rightarrow MS(s) + 2\ H^+(aq)$$

The solubilities of the metal sulfides vary enormously. The Group 1 and 2 sulfides are soluble, whereas the heavy elements of Group 11 and 12 are among the least soluble compounds known. The wide variation enables selective separation of metals to take place on the basis of solubilities of the sulfides.

The Group 1 sulfides, M_2S, adopt the antifluorite structure (Section 3.9). The Group 2 elements and some of the f-block elements form monosulfides, MS, with a rock-salt structure. The first-row d-block metals form monosulfides with the NiAs structure, whereas the heavier elements have a greater tendency to covalency and adopt a layered zinc-blende structure. The d metals form disulfides that have a layered structure or contain discrete S_2^{2-} ions. These compounds are discussed extensively in Chapter 18.

(b) Selenides, tellurides, and polonides

Key points: Selenides and tellurides are readily oxidized; polonides are stable to oxidation.

The selenides and tellurides are the most common naturally occurring sources of the elements. Group 1 and 2 selenides, tellurides, and polonides are prepared by direct interaction of the elements in liquid ammonia. They are water-soluble solids that are rapidly oxidized in air to give the elements, with the exception of the polonides for which they are among the most stable compounds of the element. The selenides and tellurides of Li, Na, and K adopt the antifluorite structure; those of the heavier elements adopt the rock-salt structure. Selenides, tellurides, and polonides of the d metals are also prepared by direct interaction of the elements and are nonstoichiometric. Two examples are $Ti_{\approx 2}Se$ and $Ti_{\approx 3}Se$.

Ring and cluster compounds

Sulfur forms chain, ring, and cluster compounds containing S—S and S—N bonds. Selenium and tellurium also form polymeric chains and rings containing Se—Se and Te—Te bonds.

15.8 Polyanions

Key points: Sulfur forms polyanions with up to six sulfur atoms; polyselenides form chains and rings and polytellurides form chains and bicyclic structures.

Many polysulfides of electropositive elements have been characterized. They all contain the S_n^{2-} ions where $n = 2-6$, as in (**25**), (**26**), and (**27**). Typical examples are Na_2S_2, BaS_2, Na_4S_4, K_2S_4, and Cs_2S_6. They can be prepared by heating stoichiometric amounts of sulfur and the element in a sealed tube.

The polysulfides can also act as ligands. An example is $[Mo_2(S_2)_6]^{2-}$ (**28**), which is formed from ammonium polysulfide and MoO_4^{2-}; it contains side-bonded S_2^{2-} ligands. The larger polysulfides bond to metal atoms forming chelate rings, as in $[MoS(S_4)_2]^{2-}$, which contains chelating S_4 ligands.

The number of polyselenides and polytellurides characterized is more extensive than for the polysulfides. Structurally, the smaller anions resemble the polysulfides. The structures of the larger ones are more complex and depend to some extent on the nature of the cation. The polyselenides up to Se_9^{2-} are chains but larger molecules form rings such as Se_{11}^{2-}, which has a Se atom at the centre of two six-membered rings in a square-planar arrangement (**29**). The polytellurides are structurally more complex and there is a greater occurrence of bicyclic arrangements such as Te_7^{2-} (**30**) and Te_8^{2-} (**31**). d-Metal complexes of larger polyselenides and polytellurides are known, such as $[Ti(Cp)_2Se_5]$ (**32**). It appears that in polysulfides, polyselenides, and polytellurides electron density is concentrated at the ends of an E_n^{2-} chain, which accounts for coordination through the terminal atoms, as shown in (**33**).

15.9 Polycations

Key point: Polyatomic cations of S and Se can be produced by the action of mild oxidizing agents on the elements in strong acid media.

Many cationic chain, ring, and cluster compounds of the p-block elements have been prepared. The majority of them contain sulfur, selenium, or tellurium. Because these cations are oxidizing agents and Lewis acids, the preparative conditions are quite different from those used to synthesize the highly reducing polyanions. For example, S_8 is oxidized by AsF_5 in liquid sulfur dioxide to yield the S_8^{2+} ion:

$$S_8 + 2\,AsF_5 \xrightarrow{SO_2} [S_8][AsF_6]_2 + AsF_3$$

A strong acid solvent, such as fluorosulfuric acid, is used and the strongly oxidizing peroxide compound FO_2SOOSO_2F (that is, $S_2O_6F_2$) oxidizes Se to Se_4^{2+}:

$$4\,Se + S_2O_6F_2 \xrightarrow{HSO_3F} [Se_4][SO_3F]_2$$

The Se_4^{2+} ion has a square-planar structure (**34**). In the molecular orbital model of the bonding, this cation has a closed-shell configuration in which the six electrons fill the a_{2u} and e_g orbitals, leaving the higher energy antibonding b_{2u} vacant. By contrast, most of the larger ring systems can be understood in terms of localized $2c,2e$ bonds. For these larger rings, the removal of two electrons brings about the formation of an additional $2c,2e$

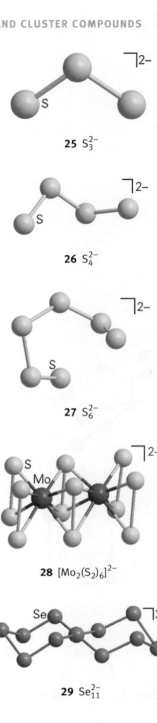

25 S_3^{2-}

26 S_4^{2-}

27 S_6^{2-}

28 $[Mo_2(S_2)_6]^{2-}$

29 Se_{11}^{2-}

30 Te_7^{2-}

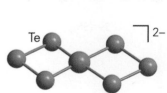

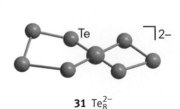

31 Te_8^{2-}

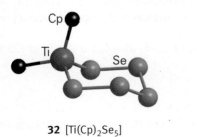

32 $[Ti(Cp)_2Se_5]$

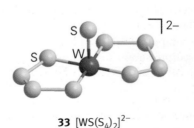

33 $[WS(S_4)_2]^{2-}$

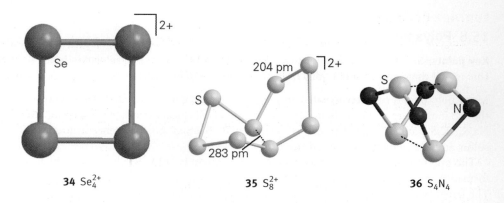

34 Se_4^{2+} **35** S_8^{2+} **36** S_4N_4

bond, thereby preserving the local electron count on each element. This change is readily seen for the oxidation of S_8 to S_8^{2+} (**35**). An X-ray single-crystal structure determination shows that the transannular bonds in S_8^{2+} are long compared with the other bonds. Long transannular bonds are common in these types of compounds.

15.10 Sulfur–nitrogen compounds

Key points: Neutral heteroatomic ring and cluster compounds of the *p*-block elements include P_4S_{10} and cyclic S_4N_4. Disulfurdinitride transforms into a polymer, $(SN)_n$, that is superconducting at very low temperatures.

Sulfur–nitrogen compounds have structures that can be related to the polycations discussed above. The oldest known, and easiest to prepare, is the pale yellow–orange tetrasulfurtetranitride, S_4N_4 (**36**), which is made by passing ammonia through a solution of SCl_2:

$$6\ SCl_2(l) + 16\ NH_3(g) \rightarrow S_4N_4(s) + 2\ S_8(s) + 14\ NH_4Cl(sol)$$

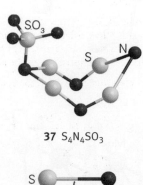

37 $S_4N_4SO_3$

Tetrasulfurtetranitride is endoergic ($\Delta_f G^{\ominus} = +536$ kJ mol^{-1}) and may decompose explosively. The molecule is an eight-membered ring with the four N atoms in a plane and bridged by S atoms that project above and below the plane. The short S—S distance (258 pm) suggests that there is a weak interaction between pairs of S atoms. Lewis acids such as BF_3, SbF_5, and SO_3 form 1:1 complexes with one of the N atoms and in the process the S_4N_4 ring rearranges (**37**).

Disulfurdinitride, S_2N_2 (**38**), is formed (together with Ag_2S and N_2) when S_4N_4 vapour is passed over hot silver wool. It is even more sensitive than its precursor, and it explodes above room temperature. When allowed to stand at 0°C for several days, disulfurdinitride transforms into a bronze-coloured polymer of composition $(SN)_n$ (**39**), which is much more stable than its precursor, not exploding until 240°C. The chains have a zigzag shape and the compound exhibits metallic conductivity along the chain axis. The polymer is superconducting below 0.3 K. The discovery of this superconductivity was important because it was the first example of a superconductor that had no metal constituents. Halogenated derivatives have been synthesized that have even higher conductivity. For example, partial bromination of $(SN)_x$ produces blue–black single crystals of $(SNBr_{0.4})_x$, which has room temperature conductivity an order of magnitude greater than $(SN)_x$. Treatment of S_4N_4 with ICl, IBr, and I_2 produces highly conducting, nonstoichiometric polymers with conductivities greater by 16 orders of magnitude than $(SN)_x$.

The compound S_4N_2 can be prepared by heating S_4N_4 with sulfur in CS_2 at 120°C and increased pressure:

$$S_4N_4 + 4\ S \xrightarrow{\ CS_2/120°C\ } 2\ S_4N_2$$

It forms dark red, needle-like crystals that melt to a dark red liquid at 25°C. It decomposes explosively at 100°C.

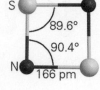

38 S_2N_2

89.6°
90.4°
166 pm

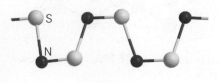

39 $(SN)_n$

FURTHER READING

R.B. King, *Inorganic chemistry of the main group elements*. Wiley (1994).

D.M.P. Mingos, *Essential trends in inorganic chemistry*. Oxford University Press (1998). A survey of inorganic chemistry from the perspective of structure and bonding.

R.B. King (ed.), *Encyclopedia of inorganic chemistry*. Wiley (2005).

N. Saunders, *Oxygen and the elements of group 16*. Heinemann (2003).

P. Ball, *H₂O: a biography of water*. Phoenix (2004). An entertaining look at the chemistry and physics of water.

R. Steudel, *Elemental sulfur and sulfur-rich compounds*. Springer-Verlag (2003).

N.N. Greenwood and A. Earnshaw, *Chemistry of the Elements*. Butterworth-Heinemann (1997).

EXERCISES

15.1 State whether the following oxides are acidic, basic, neutral, or amphoteric: CO_2, P_2O_5, SO_3, MgO, K_2O, Al_2O_3, CO.

15.2 (a) Use standard potentials (*Resource section 3*) to calculate the standard potential of the disproportionation of H_2O_2 in acid solution. (b) Is Cr^{2+} a likely catalyst for the disproportionation of H_2O_2? (c) Given the Latimer diagram

$$O_2 \xrightarrow{-0.13} HO_2^- \xrightarrow{1.51} H_2O_2$$

in acidic solution, calculate $\Delta_r G^{\ominus}$ for the disproportionation of hydrogen superoxide (HO_2^-) into O_2 and H_2O_2, and compare the result with its value for the disproportionation of H_2O_2.

15.3 Which hydrogen bond would be stronger: S—H $\cdots$ O or O—H $\cdots$ S?

15.4 Which of the solvents ethylenediamine (which is basic and reducing) or SO_2 (which is acidic and oxidizing) might not react with (a) Na_2S_4, (b) K_2Te_3?

15.5 Rank the following species from the strongest reducing agent to the strongest oxidizing agent: SO_4^{2-}, SO_3^{2-}, $O_3SO_2SO_3^{2-}$.

15.6 (a) Give the formula for Te(VI) in acidic aqueous solution and contrast it with the formula for S(VI). (b) Offer a plausible explanation for this difference.

15.7 Use the standard potential data in *Resource section 3* to predict which oxoanions of sulfur will disproportionate in acidic conditions.

15.8 Predict whether any of the following will be reduced by thiosulfate ions, $S_2O_3^{2-}$, in acidic conditions: VO^{2+}, Fe^{3+}, Cu^+, Co^{3+}.

15.9 SF_4 reacts with BF_3 to form $[SF_3][BF_4]$. Use VSEPR theory to predict the shapes of the cation and anion.

15.10 Tetramethylammonium fluoride (0.70 g) reacts with SF_4 (0.81 g) to form an ionic product. (a) Write a balanced equation for the reaction and (b) sketch the structure of the anion. (c) How many lines would be observed in the ^{19}F NMR spectrum of the anion?

15.11 Identify the sulfur-containing compounds A, B, C, D, E, and F.

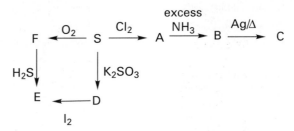

PROBLEMS

15.1 The bond lengths in O_2, O_2^+, and O_2^{2-} are 121, 112, and 149 pm, respectively. Describe the bonding in these molecules in terms of molecular orbital theory and use this description to rationalize the differences in bond lengths.

15.2 Correct any inaccuracies in the following statements and, after correction, provide examples to illustrate each statement. (a) Elements in the middle of Group 16 are easier to oxidize to the group oxidation number than are the lightest and heaviest members. (b) In its ground state, O_2 is a triplet and it undergoes Diels–Alder electrophilic attack on dienes. (c) The diffusion of ozone from the stratosphere into the troposphere poses a major environmental problem.

15.3 Write a comparative account to the properties of sulfuric, selenic, and telluric acids.

15.4 A mechanistic study of reaction between chloramine and sulfite has been reported (B.S. Yiin, D.M. Walker, and D.W. Margerum, *Inorg. Chem.*, 1987, **26**, 3435). Summarize the observed rate law and the proposed mechanism. Accepting the proposed mechanism, why should $SO_2(OH)^-$ and HSO_3^- display different rates of reaction? Explain why it was not possible to distinguish the reactivity of $SO_2(OH)^-$ from that of HSO_3^-.

15.5 Tetramethyltellurium, $Te(CH_3)_4$, was prepared in 1989 (R.W. Gedrige, D.C. Harris, K.R. Higa, and R.A. Nissan, *Organometallics*, 1989, **8**, 2817), and its synthesis was soon followed by the preparation of

the hexamethyl compound (L. Ahmed and J.A. Morrison, *J. Am. Chem. Soc.*, 1990, **112**, 7411). Explain why these compounds are so unusual, give equations for their syntheses, and speculate on why these synthetic procedures are successful. In relation to the last point, speculate on why reaction of TeF_4 with methyllithium does not yield tetramethyltellurium.

15.6 The bonding in the square-planar Se_4^{2+} ion is described in Section 15.9. Explore this proposition in more detail by carrying out computations, using software of your choice, on S_4^{2+} with S—S bond distances of 200 pm (sulfur is recommended because its semiempirical parameters are more reliable than those of Se). From the output (a) draw the molecular orbital energy level diagram, (b) assign the symmetry of each level, and (c) sketch the highest energy molecular orbital. Is a closed-shell molecule predicted?

15.7 The nature of the sulfur cycle in ancient times has been investigated (J. Farquhar, H. Bao, and M. Thiemen, *Science*, 2000, **289**, 756). What three factors influence the modern day cycle? When did the authors establish that a significant change in the cycle had occurred and how were the differences between the ancient and modern cycles explained?

15.8 H. Keppler has investigated the concentration of sulfur in volcano magma (*Science*, 1999, **284**, 1652). In what forms is sulfur erupted from volcanoes? What concentration of sulfur was found in the magma erupted from Mount Pinatubo in 1991? Discuss whether this concentration was expected and how any deviation from the expected value was explained.

16 The Group 17 elements

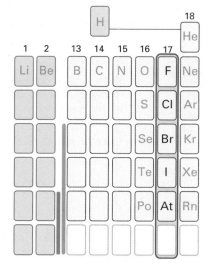

As we discuss the elements of the penultimate group in the *p* block, we shall see that many of the systematic themes that were helpful for discussing preceding groups again prove useful. For instance, the VSEPR model can be used to predict the shapes of the wide range of molecules that the halogens form among themselves, with oxygen, and with the noble gases. As with the elements in Groups 15 and 16, we shall see that the oxoanions of the halogens are oxidizing agents that often react by atom transfer. There is a useful correlation between the oxidation number of the central atom and the rates of redox reactions. A wide range of oxidation states is observed for most of the halogens. The reactions of the dihalogens are often fast.

The Group 17 elements, fluorine, chlorine, bromine, iodine, and astatine, are known as the **halogens** from the Greek for 'salt giver'. Fluorine and chlorine are poisonous gases, bromine is a toxic, volatile liquid, and iodine is a sublimable solid. They are among the most reactive nonmetallic elements. The chemical properties of the halogens are extensive and their compounds have been mentioned many times already. Therefore, in this chapter we highlight their systematic features and discuss the interhalogens and their compounds with oxygen.

The elements

Key points: Except for fluorine and the highly radioactive astatine, the halogens exist with oxidation numbers ranging from −1 to +7; the small and highly electronegative fluorine atom is effective in oxidizing many elements to high oxidation states.

The atomic properties of the halogens are listed in Table 16.1; they all have the electron configuration ns^2np^5. The features to note include their high ionization energies and their high electronegativities and electron affinities. The halogens have high electron affinities because the incoming electron can occupy an orbital of an incomplete valence shell and experience a strong nuclear attraction: recall that Z_{eff} increases progressively across the period (Section 1.7). As a consequence of the high electronegativities and abundances of the lighter halogens, their compounds are important in practically every area of chemistry and their properties are encountered throughout the text.

We have seen when discussing the earlier groups in the *p* block that the element at the head of each group has properties that are distinct from those of its heavier congeners.

Table 16.1 Selected properties of the elements

	F	Cl	Br	I	At
Covalent radius/pm	72	99	114	133	140
Ionic radius/pm	117	167	182	206	
First ionization energy/(kJ mol^{-1})	1680	1250	1140	1008	
Melting point/°C	−220	−101	−7.2	114	302
Boiling point/°C	−188	−34.7	58.8	184	
Pauling electronegativity	4.0	3.0	2.8	2.5	2.2
Electron affinity/(kJ mol^{-1})	334	355	325	295	270
$\frac{1}{2}X_2 + e^- \rightarrow X^-$ $E^{\ominus}$/V	2.87	1.36	1.07	0.54	

The anomalies are much less striking for the halogens, and the most notable difference is that fluorine has a lower electron affinity than chlorine. Intuitively, this feature seems to be at odds with the high electronegativity of fluorine, but it stems from the larger electron–electron repulsion in the compact F atom as compared with the larger Cl atom. This electron–electron repulsion is also responsible for the weakness of the F—F bond in F_2. Despite this difference in electron affinity, the enthalpies of formation of metal fluorides are generally much greater than those of metal chlorides. The explanation of this observation is that the low electron affinity of fluorine is more than offset by the high lattice enthalpies of ionic compounds containing the small F^- ion (Fig. 16.1) and the strengths of bonds in covalent species (for example, the fluorides of metals in high oxidation states).

Because fluorine is the most electronegative of all elements, it is never found in a positive oxidation state (except in the transient gas-phase species F_2^+). With the possible exception of astatine, the other halogens occur with oxidation numbers ranging from −1 to +7. The dearth of chemical information on astatine stems from its lack of any stable isotopes and the relatively short half-life (8.3 h) of its most long-lived isotope. Astatine solutions are intensely radioactive and can be studied only in high dilution. Astatine appears to exist as the anion At^- and as At(I) and At(III) oxoanions; no evidence for At(VII) has yet been obtained.

All the Group 17 elements undergo thermal or photochemical dissociation in the gas phase to form radicals. These radicals take part in chain reactions, such as:

$$X_2 \xrightarrow{\Delta/h\nu} X\cdot + X\cdot$$
$$H_2 + X\cdot \rightarrow HX + H\cdot$$
$$H\cdot + X_2 \rightarrow HX + X\cdot$$

16.1 Occurrence, recovery, and uses

Key points: Fluorine, chlorine, and bromine are prepared by electrochemical oxidation of halide salts; chlorine is used to oxidize Br^- and I^- to the corresponding dihalogen.

Fluorine is a pale yellow reactive gas, reacting with most inorganic and organic molecules and the noble gases krypton, xenon, and radon (Section 17.8). Consequently, it is very difficult to handle as it reacts with most materials but it can be stored in steel or monel metal (a nickel/copper alloy) as these alloys form a passivating metal fluoride surface film. Chlorine is a green–yellow toxic gas. Bromine is the only liquid nonmetallic element and is a dark red, toxic, volatile liquid. Iodine is a purple–grey solid that sublimes to a violet vapour. The violet colour persists when it is dissolved in nonpolar solvents such as CCl_4. However, in polar solvents it dissolves to give red–brown solutions, indicating the presence of polyiodide ions such as I_3^- (Section 16.7).

The halogens are so reactive that they are found naturally only as compounds. They occur mainly as halides, but the most easily oxidized element, iodine, is also found as

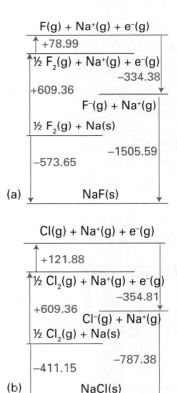

Fig. 16.1 Thermochemical cycles for (a) sodium fluoride and (b) sodium chloride. All values are in kJ mol^{-1}.

Box 16.1 Preparation of anhydrous metal halides

Because anhydrous halides are very common starting materials for inorganic syntheses, their preparation and the removal of water from impure commercial halides is of considerable practical importance. The principal synthetic methods are the direct reaction of a metal and a halogen and the reaction of halogen compounds with metals or metal oxides.

An example of direct reaction is shown in Fig. B16.1, in which the halogen in the gas phase reacts with the metal contained in a heated tube. When the halide is volatile at the temperature of the reaction, it sublimes to the exit end of the tube.

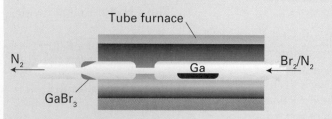

B16.1 A tube furnace used for the synthesis of anhydrous gallium bromide.

It is sometimes advantageous to use a halogen compound for the synthesis of a metal halide. For example, the reaction of ZrO_2 with CCl_4 vapour in a heated tube is a good method for the preparation of $ZrCl_4$. A contribution to the Gibbs energy of this reaction comes from the production of the $C=O$ bond in phosgene, $COCl_2$:

$$ZrO_2(s) + 2\ CCl_4(g) \rightarrow ZrCl_4(s) + 2\ COCl_2(g)$$

Anhydrous halides can often be prepared by the removal of water from hydrated metal chlorides. Simply heating the sample in a dry gas stream is usually not satisfactory for metals with high charge-to-radius ratio because significant hydrolysis may lead to the oxide or oxohalide:

$$2\ CrCl_3 \cdot 6H_2O(s) \rightarrow Cr_2O_3(s) + 6\ HCl(g) + 9\ H_2O(g)$$

This hydrolysis can be suppressed by carrying out the dehydration in a heated tube in a stream of hydrogen halide, or by dehydration with thionyl chloride (b.p. 79°C), which produces the desired anhydrous chloride and volatile HCl and SO_2 by-products:

$$FeCl_3 \cdot 6H_2O(s) + 6\ SOCl_2(l) \rightarrow FeCl_3(s) + 6\ SO_2(g) + 12\ HCl(g)$$

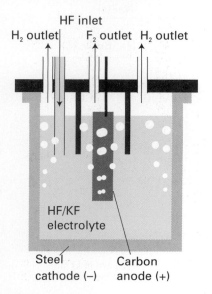

Fig. 16.2 Schematic diagram of an electrolysis cell for the production of fluorine from KF dissolved in liquid HF.

sodium or potassium iodate, KIO_3, in alkali metal nitrate deposits. Because many chlorides, bromides, and iodides are soluble, these anions occur in the oceans and in brines. The primary source of fluorine is calcium fluoride, which has low solubility in water and is often found in sedimentary deposits (as fluorite, CaF_2). Chlorine occurs as sodium chloride in rock salt. Iodine accumulates in seaweed, from which it can be extracted (Box 16.1).

All the dihalogens (except the radioactive At_2) are produced commercially on a large scale, with chlorine production by far the greatest, followed by fluorine. The principal method of production of the elements is by electrolysis of the halides (Section 5.17). The strongly positive standard potentials $E^{\ominus}(F_2,F^-) = +2.87$ V and $E^{\ominus}(Cl_2,Cl^-) = +1.36$ V indicate that the oxidation of F^- and Cl^- ions requires a strong oxidizing agent. Only electrolytic oxidation is commercially feasible. An aqueous electrolyte cannot be used for fluorine production because water is oxidized at a much lower potential ($+1.23$ V) and any fluorine produced would react rapidly with water. The isolation of elemental fluorine is achieved by electrolysis of a 1:2 mixture of molten KF and HF in a cell like that shown in Fig. 16.2.

Most commercial chlorine is produced by the electrolysis of aqueous sodium chloride solution in a *chloralkali cell* (Fig. 16.3). The half-reactions are

Anode half-reaction: $2\ Cl^-(aq) \rightarrow Cl_2(g) + 2\ e^-$

Cathode half-reaction: $2\ H_2O(l) + 2\ e^- \rightarrow 2\ OH^-(aq) + H_2(g)$

The oxidation of water at the anode is suppressed by using an electrode material that has a higher overpotential for O_2 evolution than for Cl_2 evolution. The best anode material seems to be RuO_2 (Section 18.8). This process is the basis of the chloralkali industry, which produces sodium hydroxide on a massive scale (see Box 10.2):

$$2\ NaCl(s) + 2\ H_2O(l) \rightarrow 2\ NaOH(aq) + H_2(g) + Cl_2(g)$$

Bromine is obtained by the chemical oxidation of Br^- ions in seawater. A similar process is used to recover iodine from certain natural brines that are rich in I^-. The more strongly oxidizing halogen, chlorine, is used as the oxidizing agent in both processes, and the resulting Br_2 and I_2 are driven from the solution in a stream of air:

$$Cl_2(g) + 2\ X^-(aq) \xrightarrow{air} 2\ Cl^-(aq) + X_2(g) \qquad (X = Br\ or\ I)$$

Box 16.2 Fluoridation of water and dental health

In the first half of the twentieth century extensive tooth decay was prevalent among most of the population in the developed world. In 1901, F.S. McKay, a Colorado dentist, noticed that many of his patients had a mottled brown stain on their tooth enamel. He also noticed that these patients seemed to be less affected by tooth decay. He suspected that something in the local water supply was responsible. It was not until the 1930s, when analytical science had advanced, that McKay's suspicions were confirmed and high levels (12 ppm) of fluoride ions were identified in the local water supply.

Studies confirmed that there was an inverse relationship between the appearance of the mottled brown stain (termed *fluorosis*) and the occurrence of dental cavities. It was found that fluoride ion levels of up to 1 ppm were sufficient to decrease the occurrence of tooth cavities without causing fluorosis. This led to widespread fluoridation of drinking water supplies in the western world, which has drastically reduced the level of tooth decay in all sectors of the population. Fluoride has also been added to toothpastes, mouthwashes, and even to table salt.

Research suggests that fluoride prevents dental cavities by inhibition of demineralization and inhibition of bacterial activity in dental plaque. Enamel and dentine are composed of hydroxyapatite, calcium phosphate. The hydroxyapatite is dissolved by acids present in food, or produced by bacterial action on food. The fluoride ions form fluoroapatite with the enamel, which is less soluble in acid than the hydroxyapatite. The fluoride ions are also taken up by the bacteria, where they disrupt enzyme activity and reduce acid production.

The fluoridation of public water supplies is controversial. Opponents have claimed that it is linked to increased risk of cancer, Down's syndrome, and heart disease, although, thus far, there has been no convincing evidence to support these claims. Opponents also claim that mass fluoridation infringes civil liberties and that the long-term effects of increased fluoride levels in river habitats is not yet known.

Fluorine is added to some water supplies and toothpaste as F^- ions to prevent tooth decay (Box 16.2). It is used as UF_6 in the nuclear power industry for the separation of the isotopes of uranium. Hydrogen fluoride is used to etch glass and as a nonaqueous solvent. Chlorine is widely used in industry to make chlorinated hydrocarbons and in applications in which a strong and effective oxidizing agent is needed, including disinfectants and bleaches. These applications are in decline, however, because some organic chlorine compounds are carcinogenic and chlorofluorocarbons (CFCs) are implicated in the destruction of ozone in the stratosphere (Box 16.3). Hydrofluorocarbons (HFCs) are now replacing CFCs in applications such as refrigeration and air conditioning. Silver halides are sensitive to light and the bromide and iodide are used in black-and-white photography. Iodine is an essential element and iodine deficiency is a cause of goitre, the enlargement of the thyroid gland. For this reason, small amounts of iodide are added to table salt.

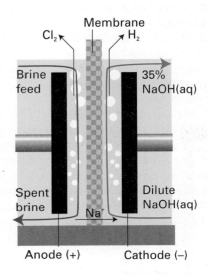

Fig. 16.3 Schematic diagram of a chloralkali cell using a cation-exchange membrane, which has high permeability to Na^+ ions and low permeability to OH^- and Cl^- ions.

Example 16.1 Analysing the recovery of Br_2 from brine

Give the chemical equation and potential for the commercial conversion of Br^- in brine to Br_2. Show that from a thermodynamic standpoint bromide could be oxidized to Br_2 by O_2, and suggest a reason why O_2 is not used for this purpose.

Answer Chlorine is used to oxidize Br^-:

$$Cl_2(g) + 2\,Br^-(aq) \rightarrow 2\,Cl^-(aq) + Br_2(g) \qquad E^\ominus = +0.26\ V$$

and the resulting volatile Br_2 is removed in a steam–air mixture. Oxygen would be thermodynamically capable of carrying out this reaction in acidic solution:

$$O_2(g) + 4\,Br^-(aq) + 4\,H^+(aq) \rightarrow 2\,H_2O(l) + 2\,Br_2(l) \qquad E^\ominus = +0.16\ V$$

but the reaction is not favourable at pH $= 7$, when $E^\ominus = -0.15$ V. Even though the reaction is thermodynamically favourable in acidic solution, it is doubtful that the rate would be adequate, because an overpotential of about 0.6 V is associated with the reactions of O_2 (Section 5.17). Even if the oxidation by O_2 in acidic solution were kinetically favourable, the process would be unattractive because of the cost of acidifying large quantities of brine and then neutralizing the effluent.

Self-test 16.1 One source of iodine is sodium iodate, $NaIO_3$. Which of the reducing agents $SO_2(aq)$ or $Sn^{2+}(aq)$ would seem practical from the standpoints of thermodynamic feasibility and plausible judgements about cost? Standard potentials are given in *Resource section 1*.

Box 16.3 Chlorofluorocarbons and the ozone hole

The ozone layer extends from 10 km to 50 km above the Earth's surface and plays a crucial role in protecting us from the harmful effects of the Sun's ultraviolet rays. Ozone performs this function by absorbing light at wavelengths below 300 nm, thereby attenuating the spectrum of sunlight at ground level. When ozone absorbs a photon, it undergoes a transition to an excited electronic state. The excitation energy is sufficient to break one of the bonds in the molecule:

$$O_3 \rightarrow O_2 + O$$
$$O_3 + O \rightarrow 2\,O_2$$

The stratosphere contains naturally occurring species, such as the hydroxyl radical and nitric oxide, that catalyse this ozone destruction:

$$X + O_3 \rightarrow XO + O_2$$
$$XO + O \rightarrow X + O_2$$

However, the main concern about the loss of ozone centres on Cl and Br atoms, which catalyse O_3 destruction very efficiently. Chlorine and bromine are carried into the stratosphere as part of organohalogen molecules, RHal, which release the halogen atoms when the C—X bond is fragmented by far-UV photons. The ozone-destroying potential of these molecules was pointed out in a 1974 research article by Mario Molina and Sherwood Rowland, who won the 1995 Nobel Prize in chemistry (along with Paul Crutzen) for their work on ozone.

International action followed only 13 years later (in the form of the 1987 Montreal Protocol), and was given added impetus by the discovery of the ozone 'hole' over Antarctica, which provided dramatic evidence of the vulnerability of atmospheric ozone. This hole surprised even the scientists working on the problem; its explanation required additional chemistry, involving the polar stratospheric clouds that form in winter. The ice crystals in these clouds absorb

molecules of chlorine or bromine nitrate, $ClONO_2$ or $BrONO_2$, which form when stratospheric $ClO^{\cdot}$ or $BrO^{\cdot}$ combine with NO_2. Once on the ice surface, these molecules react with water:

$$H_2O + XONO_2 \rightarrow HOX + HNO_3$$

where X = Cl or Br. They also react with co-adsorbed HCl or HBr (formed by attack of $Cl^{\cdot}$ and $Br^{\cdot}$ on the methane escaping from the troposphere)

$$HC + XONO_2 \rightarrow X_2 + HNO_3$$

The nitric acid, being very hygroscopic, enters the ice crystals, while the HOX or X_2 molecules are released during the dark polar winter. When the Sun rises in the spring, these molecules photolyse, releasing high concentrations of ozone-destroying radicals.

$$HOX \xrightarrow{h\nu} HO\cdot + X\cdot$$
$$X_2 \xrightarrow{h\nu} 2\,X\cdot$$

To threaten the ozone layer, organohalogen molecules must survive transit from the Earth's surface. Those that contain H atoms are mostly broken down in the troposphere by reaction with hydroxyl radicals. Even so, they may be a problem if released in sufficient amounts. There is currently a major controversy over the use of methyl bromide, CH_3Br, as an agricultural fumigant. However, the greatest potential for ozone destruction rests with molecules that lack H atoms, the chlorofluorocarbons (CFCs), useful in many industrial applications, and their brominated analogues (halons), used to extinguish fires. These have no tropospheric sink, and eventually reach the stratosphere unaltered. They are the main focus of the international regulatory regime worked out in 1987 (and amended in 1990 and 1992). Most CFCs and halons have been phased out of production, and their atmospheric concentrations are beginning to decline. The CFCs present an additional problem as they are also important greenhouse gases.

16.2 Trends in properties

Unlike the structures of the elements of the preceding groups in the *p* block, the molecular structures of the halogens are strikingly similar. They are all diatomic, and many properties change smoothly down the group.

(a) Molecular structure and properties

Key points: The F—F bond is weak relative to the Cl—Cl bond; in the solid state I_2 molecules cohere to their nearest neighbours more than the other dihalogens do.

Among the most striking physical properties of the halogens are their colours. In the vapour they range from the almost colourless F_2, through yellow–green Cl_2 and red–brown Br_2, to purple I_2. The progression of the maximum absorption to longer wavelengths reflects the decrease in the HOMO–LUMO gap on descending the group. In each case, the optical absorption spectrum arises primarily from transitions in which an electron is promoted from the highest filled σ and $\pi^{\star}$ orbitals into the vacant antibonding $\sigma^{\star}$ orbital (Fig. 16.4).

Except for F_2, the analysis of the UV absorption spectra gives precise values for the dihalogen bond dissociation energies (Fig. 16.5). It is found that bond strengths decrease down the group from Cl_2. The UV spectrum of F_2, however, is a broad continuum that

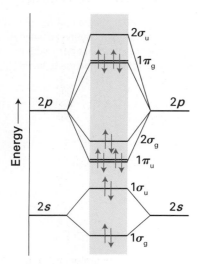

Fig. 16.4 Schematic molecular orbital energy level diagram for Cl_2, Br_2, and I_2. For F_2 the order of the π_u and upper σ_g orbitals is reversed.

Table 16.2 Bonding and shortest nonbonding distances for solid dihalogens

Element	Temperature/°C	Bond length/pm	Nonbonding distance/pm	Ratio
Cl_2	−160	198	332	1.68
Br_2	−106	227	332	1.46
I_2	−163	272	350	129

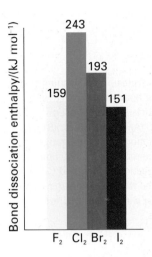

Fig. 16.5 Bond dissociation enthalpies of the halogens (kJ mol^{-1}).

lacks structure because absorption is accompanied by dissociation of the F_2 molecule. The lack of discrete absorption bands makes it difficult to estimate the dissociation energy spectroscopically, and thermochemical methods are complicated by the highly corrosive nature of this reactive halogen. When these corrosion problems were solved, the F—F bond enthalpy was found to be less than that of Br_2 and thus out of line with the trend in the group. However, fluorine's low bond enthalpy is consistent with the low single-bond enthalpies of N—N, O—O, and various combinations of N, F, and O (Fig. 16.6). The simplest explanation (like the explanation of the low electron affinity of fluorine) is that the bond is weakened by the strong repulsions between nonbonding electrons in the small F_2 molecule. In molecular orbital terms, the molecule has numerous electrons in strongly antibonding orbitals.

Chlorine, bromine, and iodine all crystallize in lattices of the same symmetry (Fig. 16.7), so it is possible to make a detailed comparison of distances between bonded and nonbonded adjacent atoms (Table 16.2). The important conclusion is that nonbonded distances do not increase as rapidly as the bond lengths. This observation suggests the presence of weak intermolecular bonding interactions that strengthen on going from Cl_2 to I_2. The interaction also leads to a weakening of the I—I bond within the I_2 molecules, as is demonstrated by the lower I—I stretching frequency and greater I—I bond length in the solid as compared with in the gas phase. Moreover, solid iodine is a semiconductor, and under high pressure exhibits metallic conductivity.

(b) Reactivity trends

Key points: Fluorine is the most oxidizing halogen; the oxidizing power of the halogens decreases down the group.

Fluorine, F_2, is the most reactive nonmetal and is the strongest oxidizing agent among the halogens. The rapidity of many of its reactions with other elements may in part be due to a low kinetic barrier associated with the weak F—F bond. Despite the thermodynamic stability of most metal fluorides, fluorine can be handled in containers made from metals, such as nickel, because a substantial number of them form a passive metal fluoride surface film upon contact with fluorine gas. Fluorocarbon polymers, such as polytetrafluoroethylene, PTFE, are also useful materials for the construction of apparatus to contain fluorine and oxidizing fluorine compounds (Fig. 16.8), as well as for the well-known coating on nonstick pans. Relatively few laboratories have the equipment and expertise for research involving elemental F_2.

The standard potentials for the halides (Table 16.1) indicate that F_2 is a much stronger oxidizing agent than Cl_2. The decrease in oxidizing strength continues in more modest steps from Cl_2 through Br_2 to I_2. Although the half-reaction

$$\tfrac{1}{2} X_2(g) + e^- \rightarrow X^-(aq)$$

is favoured by a high electron affinity (which suggests that fluorine should have a lower reduction potential than chlorine), the process is favoured by the low bond enthalpy of F_2 and by the highly exothermic hydration of the small F^- ion (Fig. 16.9). The net outcome of these three competing effects of size is that fluorine is the most strongly oxidizing element of the group.

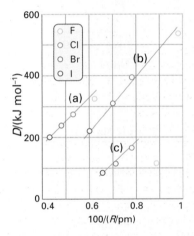

Fig. 16.6 Dissociation enthalpies of (a) carbon–halogen, (b) hydrogen–halogen, and (c) halogen–halogen bonds plotted against the reciprocal of the bond length.

Fig. 16.7 Solid chlorine, bromine, and iodine have similar structures. The closest nonbonded interactions are relatively less compressed in Cl_2 and Br_2 than in I_2.

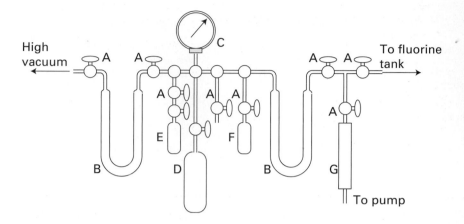

Fig. 16.8 A typical metal vacuum system for handling fluorine and reactive fluorides. Nickel tubing is used throughout. (A) Monel valves, (B) nickel U-traps, (C) monel pressure gauge, (D) nickel container, (E) PTFE reaction tube, (F) nickel reaction vessel, and (G) nickel canister filled with soda lime to neutralize HF and react with F_2 and fluorine compounds.

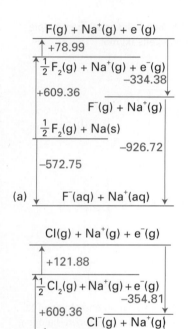

Fig. 16.9 Thermochemical cycles for the enthalpy of formation of (a) aqueous sodium fluoride and (b) aqueous sodium chloride. The hydration is much more exothermic for F^- than for Cl^-. All values are in kJ mol^{-1}.

Table 16.3 Normal boiling points (in °C) of compounds of fluorine and their analogues

F_2	−188.2	H_2	−252.8	Cl_2	−34.0
CF_4	−127.9	CH_4	−161.5	CCl_4	76.7
PF_3	−101.5	PH_3	−87.7	PCl_3	75.5

(c) Special properties of fluorine compounds

Key points: Fluorine substituents promote volatility, increase the strengths of Lewis and Brønsted acids, and stabilize high oxidation states.

The boiling points in Table 16.3 demonstrate that molecular fluorine compounds tend to be highly volatile, in some cases even more volatile than the corresponding hydrogen compounds (compare, for example, PF_3, b.p. −101.5°C, and PH_3, b.p. −87.7°C) and in all cases much more volatile than the chlorine analogues. The volatilities of the compounds are a result of variations in the strength of the dispersion interaction (the interaction between instantaneous transient electric dipole moments), which is strongest for highly polarizable molecules. The electrons in the small F atoms are gripped tightly by the nuclei, and consequently fluorine compounds have low polarizabilities and hence weak dispersion interactions.

Another characteristic of fluorine is the ability of an F atom in a compound to withdraw electrons from the atoms to which it is attached. This withdrawal leads to enhanced Brønsted acidity. An example of this effect is the increase by three orders of magnitude in the acidity of trifluoromethanesulfonic acid, HSO_3CF_3 ($pK_a = 3.0$ in nitromethane) over that of methanesulfonic acid, HSO_3CH_3 ($pK_a = 6.0$ in nitromethane). The presence of F atoms in a molecule also results—for the same reason—in high Lewis acidity. For example, we saw in Sections 4.11 and 14.7b that SbF_5 is one of the strongest Lewis acids of its type (it is a much stronger acid than $SbCl_5$).

The ability of fluorine to stabilize high oxidation states is second only to that of oxygen. Some examples of high oxidation state fluorine compounds are IF_7, PtF_6, BiF_5, and $KAgF_4$. All these compounds are examples of the highest oxidation state attainable for these elements, the rare oxidation state Ag(III) being perhaps the most notable. Another example is the stability of PbF_4, compared to all other Pb(IV) halides. A related phenomenon is the tendency of fluorine to disfavour low oxidation states. Thus solid copper(I) fluoride, CuF, is unknown but CuCl, CuBr, and CuI are well known. Similar trends were discussed in Section 3.11 in terms of a simple ionic model in which the small size of the F^- ion in combination with a small, highly charged cation results in a high lattice enthalpy. As a result, there is a thermodynamic tendency for CuF to disproportionate and form copper metal and CuF_2 (because Cu^{2+} is doubly charged and its ionic radius is smaller than that of Cu^+ and it has a greater lattice enthalpy).

Table 16.4 Selected properties of the hydrogen halides

	HF	HCl	HBr	HI
Melting point/°C	−84	−114	−89	−51
Boiling point/°C	20	−85	−67	−35
Relative permittivity	83.6 (at 0°C)	9.3 (at −95°C)	7.0 (at −85°C)	3.4 (at −50°C)
Electrical conductivity/ (S cm^{-1})	$c.\ 10^{-6}$ (at 0°C)	$c.\ 10^{-9}$ (at −85°C)	$c.\ 10^{-9}$ (at −85°C)	$c.\ 10^{-10}$ (at −50°C)
$\Delta_f G^{\ominus}$/(kJ mol^{-1})	−273.2	−95.3	−54.4	1.72
Bond dissociation energy/(kJ mol^{-1})	574.0	431.6	362.5	294.6

The properties of HF contrast starkly with those of the other hydrogen halides (Table 16.4). Hydrogen fluoride is a volatile liquid whereas HCl, HBr, and HI are diatomic gases at room temperature. The volatility of HF is low for such a light molecule. It has a wide liquid range, a high relative permittivity, and high electrical conductivity. As with water, these properties are attributed to extensive hydrogen bonding in liquid HF. The structure of solid HF is a planar zigzag chain polymer of F—H$\cdots$F. Liquid HF has a lower density and viscosity than water, which suggests the absence of an extensive three-dimensional network of H bonds. In the gas phase HF forms H-bonded oligomers, $(HF)_n$. As with its neighbouring hydrides H_2O and NH_3, the properties of HF make it an excellent nonaqueous solvent. Its autoprotolysis is given by the following equilibrium:

$$2\ HF \rightleftharpoons H_2F^+ + F^- \quad pK_{auto} = 12.3$$

Hydrogen fluoride is a much weaker acid ($pK_a = 2.95$) than the other hydrogen halides. Although in water this difference is sometimes attributed to the formation of a strongly hydrogen-bonded pair ($H_3O^+F^-$), theoretical considerations show that its poor proton donor properties are a direct result of the very strong H–F bond. Carboxylic acids act as bases in anhydrous HF and are protonated:

$$HCOOH(l) + 2\ HF(l) \rightarrow HC(OH)_2^+(sol) + HF_2^-(sol)$$

Compounds that accept F^- ions are Lewis acids and compounds that donate F^- are Lewis bases.

$$SbF_5(s) + HF(l) \rightarrow SbF_6^-(sol) + H^+(sol)$$
$$XeF_6(s) + HF(l) \rightarrow XeF_5^+(sol) + HF_2^-(sol)$$

Ionic fluorides dissolve in HF to give highly conducting solutions. The fact that chlorides, bromides, and iodides react with HF to give the corresponding fluoride and HX provides a preparative route to anhydrous fluorides:

$$TiCl_4(l) + 4\ HF(l) \rightarrow TiF_4(s) + 4\ HCl(g)$$

Although hydrofluoric acid is a weak acid it is one of the most toxic and corrosive substances known. It can attack glass, metals, concrete, and organic matter. It is much more hazardous to handle than other acids: it is very readily absorbed through the skin and even brief contact can lead to severe burning and necrosis of the skin and deep tissue, and damage to bone by decalcification when CaF_2 is formed.

16.3 Pseudohalogens

Key points: Pseudohalogens and pseudohalides mimic halogens and halides, respectively; the pseudohalogens exist as dimers and form molecular compounds with nonmetals and ionic compounds with alkali metals.

A number of compounds have properties so similar to those of the halogens that they are called **pseudohalogens** (Table 16.5). For example, like the dihalogens, cyanogen, $(CN)_2$, undergoes thermal and photochemical dissociation in the gas phase; the resulting CN

Table 16.5 Pseudohalides, pseudohalogens, and corresponding acids

Pseudohalide	Pseudohalogen	$E^{\ominus}/V$	Acid	pK_a
CN^-	NCCN	$+0.27$	HCN	9.2
Cyanide	Cyanogen		Hydrogen cyanide	
SCN^-	NCSSCN	$+0.77$	HNCS	-1.9
Thiocyanate	Dithiocyanogen		Hydrogen thiocyanate	
OCN^-			HNCO	3.5
Cyanate			Isocyanic acid	
CNO^-			HCNO	
Fulminate			Fulminic acid	
NNN^-			HNNN	4.92
Azide			Hydrazoic acid	

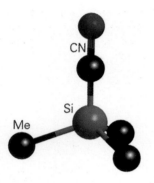

1 Me$_3$SiCN

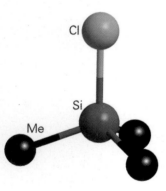

2 Me$_3$SiCl

radicals are isolobal with halogen atoms and undergo similar reactions, such as a chain reaction with hydrogen:

$$NC{-}CN \xrightarrow{\text{heat or light}} 2\ CN\cdot$$

$$H_2 + CN\cdot \rightarrow HCN + H\cdot$$

$$H\cdot + NC{-}CN \rightarrow HCN + CN\cdot$$

Overall: $H_2 + C_2N_2 \rightarrow 2\ HCN$

Another similarity is the reduction of a pseudohalogen:

$$\tfrac{1}{2}(CN)_2(aq) + e^- \rightarrow CN^-(aq)$$

The anion formally derived from a pseudohalogen is called a *pseudohalide ion*. An example is the cyanide anion, CN^-. Covalent pseudohalides similar to the covalent halides of the *p*-block elements are also common. They are often structurally similar to the corresponding covalent halides (compare (**1**) and (**2**), and undergo similar metathesis reactions.

As with all analogies, the concepts of pseudohalogen and pseudohalide have many limitations. For example, pseudohalogen ions are not spherical, so the structures of their ionic compounds often differ: NaCl is fcc but NaSCN is similar to CaC$_2$ (Section 11.6). The pseudohalogens are generally less electronegative than the lighter halogens and some pseudohalides have more versatile donor properties. The thiocyanate ion, SCN^-, for instance, acts as an ambidentate ligand with a soft base site, S, and a hard base site, N (see Section 4.15).

Interhalogens

The halogens form compounds among themselves. These binary **interhalogens** are molecular compounds with formulas XY, XY$_3$, XY$_5$, and XY$_7$, where the heavier, less electronegative halogen X is the central atom. They are of special importance as highly reactive intermediates and for providing useful insights into bonding.

16.4 Physical properties and structures

Key points: Molecular interhalogen compounds include diatomic species and polyatomic species with structures in accord with the VSEPR model; fluorine brings out high oxidation states of the other halogens.

The diatomic interhalogens, XY, have been made for all combinations of the elements, but many of them do not survive for long. All the fluorine interhalogen compounds are exoergic ($\Delta_f G^{\ominus} < 0$). The least labile interhalogen is ClF, but ICl and IBr can also be obtained in pure crystalline form. Their physical properties are intermediate between those of their component molecules. For example, the deep red ICl (m.p. 27°C, b.p. 97°C) is intermediate between yellowish-green Cl_2 (m.p. −101°C, b.p. −35°C) and dark purple I_2 (m.p. 114°C, b.p. 184°C). Photoelectron spectra indicate that the molecular orbital energy levels in the mixed dihalogen molecules lie in the order $3\sigma^2 < 1\pi^4 < 2\pi^4$, which is the same as in the homonuclear dihalogen molecules (Fig. 16.10). An interesting historical note is that ICl was discovered before Br_2 in the early nineteenth century, and, when later the first samples of the dark red–brown Br_2 (m.p. −7°C, b.p. 59°C) were prepared, they were mistaken for ICl.

Most of the higher interhalogens are fluorides (Table 16.6). The only neutral interhalogen with the central atom in a +7 oxidation state is IF_7, but the cation ClF_6^+, a compound of Cl(VII), is known. The absence of a neutral ClF_7 reflects the destabilizing effect of nonbonding repulsions between fluorine atoms (indeed, coordination numbers greater than six are not observed for other p-block central atoms in Period 3). The lack of BrF_7 might be rationalized in a similar way, but in addition we shall see later that bromine is reluctant to achieve its maximum oxidation state. In this respect, it resembles some other Period 4 p-block elements, notably arsenic and selenium. This is another manifestation of the alternation effect (Chapter 12).

The shapes of interhalogen molecules (**3**), (**4**), and (**5**) are largely in accord with the VSEPR model. For example, the XY_3 compounds (such as ClF_3) have five valence electron pairs around the X atom in a trigonal-bipyramidal arrangement. The Y atoms attach to the two axial pairs and one of the three equatorial pairs, and then the two axial bonding pairs move away from the two equatorial lone pairs. As a result, XY_3 molecules have a C_{2v} bent T shape. There are some discrepancies: for example, ICl_3 is a Cl-bridged dimer.

The Lewis structure of XF_5 puts five bonding pairs and one lone pair on the central halogen atom and, as expected from the VSEPR model, the XF_5 molecules are square pyramidal. As already mentioned, the only known XY_7 compound is IF_7, which is predicted to be pentagonal bipyramidal. The experimental evidence for its actual structure is inconclusive; it has been both claimed and denied that electron diffraction data support the prediction. As with other hypervalent molecules, the bonding in IF_7 can be explained without invoking d-orbital participation by adopting a molecular orbital model in which bonding and nonbonding orbitals are occupied but antibonding orbitals are not.

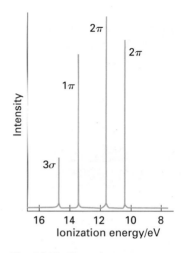

Fig. 16.10 Photoelectron spectrum of ICl. The 2π levels give rise to two peaks because of spin-orbit interaction in the positive ion.

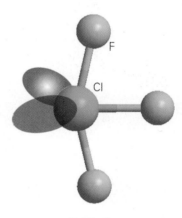

3 ClF_3, C_{2v}

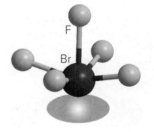

4 BrF_5, C_{4v}

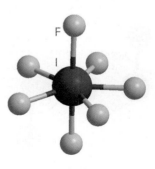

5 IF_5, D_{5h}

Table 16.6 Representative interhalogens			
XY	**XY₃**	**XY₅**	**XY₇**
CF	CF₃	CF₅	
colourless	colourless	colourless	
b.p. −100°C	b.p. 12°C	b.p. −13°C	
BrF*	BrF₃	BrF₅	
light brown	yellow	colourless	
b.p. −20°C	b.p. 126°C	b.p. 41°C	
IF*	(IF₃)ₙ	IF₅	IF₇
	yellow	colourless	colourless
	dec. −28°C	b.p. 105°C	subl. 5°C
BrCl*			
red-brown			
b.p. 5°C			
ICl	I₂Cl₆		
red solid	bright yellow		
	m.p. 101°C (16 atm)		
IBr			
black solid			

* Very unstable. dec., decomposes; subl., sublimes.

6 I_2Cl_6

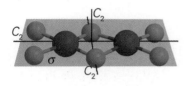

7 D_{2h}

Example 16.2 Predicting the shape of an interhalogen molecule

Predict the shape of the doubly chlorine-bridged I_2Cl_6 molecule by using the VSEPR model, and assign the point group.

Answer Each ICl_3 moiety has five electron pairs, and the bridging Cl atom increases this number to six (**6**). According to the VSEPR model, these pairs adopt an octahedral configuration, with the lone pairs in nonadjacent *trans* positions. Thus the bonds to surrounding Cl atoms take on a square-planar configuration and the molecule should be planar. This shape is verified by X-ray diffraction on single crystals. To assign the point group, we note that there are three perpendicular twofold axes (**7**), so the molecule has D_2 symmetry. The presence of a mirror plane identifies the group as D_{2h}.

Self-test 16.2 Predict the structure and identify the point group of ClO_2F.

16.5 Chemical properties

Key point: Fluorine-containing interhalogens are typically Lewis acids and strong oxidizing agents.

All the interhalogens are oxidizing agents. As with all the known interhalogen fluorides, ClF_3 is an exoergic compound, so thermodynamically it is a weaker fluorinating agent than F_2. However, the rate at which it fluorinates substances generally exceeds that of fluorine, so it is in fact an aggressive fluorinating agent towards many elements and compounds. In general, the rates of oxidation of interhalogens do not have a simple relation to their thermodynamic stabilities. Thus, ClF_3 and BrF_3 are much more aggressive fluorinating agents than BrF_5, IF_5, and IF_7; iodine pentafluoride, for instance, is a convenient mild fluorinating agent that can be handled in glass apparatus. One use of ClF_3 as a fluorinating agent is in the formation of a passivating metal fluoride film on the inside of nickel apparatus used in fluorine chemistry.

Both ClF_3 and BrF_3 react vigorously (often explosively) with organic matter, burn asbestos, and expel oxygen from many metal oxides:

$$2\ Co_3O_4(s) + 6\ ClF_3(g) \rightarrow 6\ CoF_3(s) + 3\ Cl_2(g) + 4\ O_2(g)$$

Bromine trifluoride autoionizes in the liquid state:

$$2\ BrF_3(l) \rightleftharpoons BrF_2^+(sol) + BrF_4^-(sol)$$

This Lewis acid–base behaviour is shown by its ability to dissolve a number of halide salts:

$$CsF(s) + BrF_3(l) \rightarrow Cs^+(sol) + BrF_4^-(sol)$$

Bromine trifluoride is a useful solvent for ionic reactions that must be carried out under highly oxidizing conditions. The Lewis acid character of BrF_3 is shared by other interhalogens, which react with alkali metal fluorides to produce anionic fluoride complexes.

16.6 Cationic interhalogens

Key points: Cationic interhalogen compounds have structures in accord with the VSEPR model; with fluorine substituents they are Lewis acids.

Under special strongly oxidizing conditions, such as in fuming sulfuric acid, I_2 is oxidized to the blue paramagnetic diiodinium cation, I_2^+. The dibrominium cation, Br_2^+, is also known. The bonds of these cations are shorter than those of the corresponding neutral dihalogens, which is the expected result for loss of an electron from a π^* orbital and the accompanying increase in bond order from 1 to 1.5 (see Fig. 16.4). Three higher polyhalogen cations, Br_5^+, I_3^+, and I_5^+, are known, and X-ray diffraction studies of the iodine species have established the structures shown in (**8**) and (**9**). The angular shape of I_3^+ is in line with the VSEPR model because the central I atom has two lone pairs of electrons.

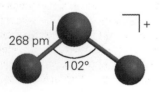

8 I_3^+, C_{2v}

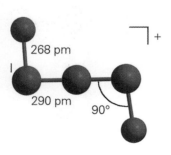

9 I_5^+

Another class of polyhalogen cations of formula XF_n^+ is obtained when a strong Lewis acid, such as SbF_5, abstracts F^- from interhalogen fluorides:

$$ClF_3 + SbF_5 \rightarrow [ClF_2^+][SbF_6^-]$$

This formulation is idealized because X-ray diffraction of solid compounds that contain these cations indicates that the F^- abstraction from the cations is incomplete and that the anions remain weakly associated with the cations by fluorine bridges (**10**). Table 16.7 lists a variety of interhalogen cations that are prepared in the same manner.

16.7 Halogen complexes and polyhalides

The Lewis acidity of diiodine and the other heavy dihalogens towards electron pair donor molecules was mentioned in Section 4.8f. This Lewis acidity is also evident in the interaction of halogen molecules with ion donors to give the range of ions known as **polyhalides**.

(a) Polyiodides

Key points: Polyiodides, such as I_3^-, are formed by adding I_2 to I^-; they are stabilized by large cations.

A deep brown colour develops when I_2 is added to a solution of I^- ions. This colour is characteristic of the homoatomic polyiodides, which include triiodide ions, I_3^-, and pentaiodide ions, I_5^-. These polyiodides are Lewis acid–base complexes in which I^- and I_3^- act as the bases and I_2 acts as the acid (Fig. 16.11). The Lewis structure of I_3^- has three equatorial lone pairs on the central I atom and two axial bonding pairs in a trigonal-bipyramidal arrangement. This hypervalent Lewis structure is consistent with the observed linear structure of I_3^-, which is described in more detail below.

An I_3^- ion can interact with other I_2 molecules to yield larger mononegative polyiodides of composition $[(I_2)_n I^-)]$. The I_3^- ion is the most stable member of this series. In combination with a large cation, such as $[N(CH_3)_4]^+$, it is symmetrical and linear with a longer I—I bond than in I_2. However, the structure of the triiodide ion, like that of the polyiodides in general, is highly sensitive to the identity of the counterion. For example, Cs^+, which is smaller than the tetramethylammonium ion, distorts the I_3^- ion and produces one long and one short I—I bond (**11**). The ease with which the ion responds to its environment is a reflection of the weakness of bonds that just manage to hold the atoms together. An example of sensitivity to the cation is provided by NaI_3, which can be formed in aqueous solution but decomposes when the water is evaporated:

$$Na^+(aq) + I_3^-(aq) \xrightarrow{\text{remove water}} NaI(s) + I_2(s)$$

A more extreme example is $NI_3 \cdot NH_3$, which is a black powder formed when iodine crystals are added to concentrated ammonia solution. The free NI_3 can be prepared from reacting iodine monofluoride with boron nitride:

$$3\ IF(g) + BN(s) \rightarrow NI_3(s) + BF_3(g)$$

Nitrogen triiodide and the ammoniate are extremely unstable and detonate at the slightest touch or vibration:

$$2\ NI_3 \cdot NH_3(s) \rightarrow N_2(g) + 3\ I_2(g) + 2\ NH_3(g)$$

Although the formula of nitrogen triiodide is usually written NI_3, it would be more accurate to write I_3N, as the compound is thought to consist of I^+ and N^{3-} ions and its sensitivity to shock is due to the redox instability of these ions. This behaviour is also another example of the instability of large anions in combination with small cations, which, as we saw in Section 3.15, can be rationalized by the ionic model.

The existence and structures of the higher polyiodides are sensitive to the counterion for similar reasons, and large cations are necessary to stabilize them in the solid state. In fact, entirely different shapes are observed for polyiodide ions in combination with

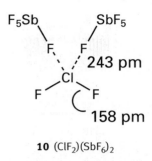

10 $(ClF_2)(SbF_6)_2$

Table 16.7 Representative interhalogen cations

$[XF_2]^+$	$[XF_4]^+$	$[XF_6]^+$
$[ClF_2]^+$	$[ClF_4]^+$	$[ClF_6]^+$
$[BrF_2]^+$	$[BrF_4]^+$	$[BrF_6]^+$
	$[IF_4]^+$	$[IF_6]^+$

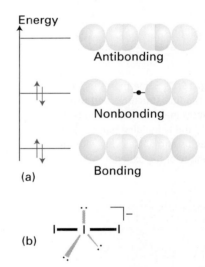

Fig. 16.11 Some representations of the I_3^- polyiodide ion. (a) The σ interaction. (b) Lewis and VSEPR rationalization of the linear structure, where the five electron pairs are arranged around the central atom in a trigonal-pyramidal array.

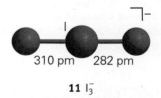

11 I_3^-

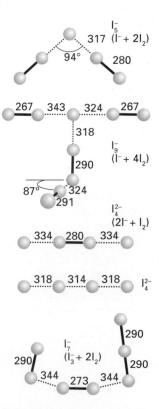

I_5^- (I^- + $2I_2$)

317

94°

280

267 343 324 267

318

I_9^- (I^- + $4I_2$)

290

87° 324

291

I_4^{2-} ($2I^-$ + I_2)

334 280 334

318 314 318 I_4^{2-}

290

I_7^- (I_3^- + $2I_2$)

290 290

344 273 344

Fig. 16.12 Some representative polyiodide structures and their approximate description in terms of I^-, I_3^-, and I_2 building blocks. Bond lengths and angles vary with the identity of the cation.

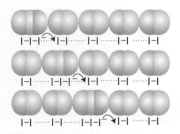

|–|–|–|–|–| |–| |–| |–|

|–| |–|–|–|–| |–| |–|

|–| |–| |–|–|–| |–| |–|

Fig. 16.13 One possible mode of charge transport along a polyiodide chain is the shift of long and short bonds resulting in the effective migration of an I^- ion along a chain. Three successive stages in the migration are shown. Note that the iodide ion from the I_3^- on the left is not the same one emerging on the right.

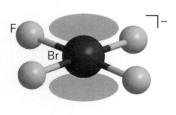

12 BrF_4^-

various large cations, as the structure of the anion is determined in large measure by the manner in which the ions pack together in the crystal. The bond lengths in a polyiodide ion often suggest that it can be regarded as a chain of associated I^-, I_2, I_3^-, and sometimes I_4^{2-} units (Fig. 16.12). Solids containing polyiodides exhibit electrical conductivity, which may arise either from the hopping of electrons (or holes) or by an ion relay along the polyiodide chain (Fig. 16.13).

Some dinegative polyiodides are known. They contain an even number of I atoms and their general formula is $[I^-(I_2)_nI^-]$. They have the same sensitivity to the cation as their mononegative counterparts.

(b) Other polyhalides

Key points: Some of the most stable polyhalogen anions contain fluorine as the substituent; their structures usually conform to the VSEPR model.

Although polyhalide formation is most pronounced for iodine, other polyhalides are also known. They include Cl_3^-, Br_3^-, and BrI_2^-, which are known in solution and (in partnership with large cations) as solids too. Even F_3^- has been detected spectroscopically at low temperatures in an inert matrix. This technique, known as *matrix isolation*, makes use of the co-deposition of the reactants with a large excess of noble gas at very low temperatures (in the region of 4–14 K). The solid noble gas forms an inert matrix within which the F_3^- ion can sit in chemical isolation.

In addition to complex formation between dihalogens and halide ions, some interhalogens can act as Lewis acids towards halide ions. The reaction results in the formation of polyhalides that, in contrast to the chain-like polyiodides, are assembled around a central halogen acceptor atom in a high oxidation state. As mentioned earlier, for instance, BrF_3 reacts with CsF to form $CsBrF_4$, which contains the square-planar BrF_4^- anion (**12**). Many of these interhalogen anions have been synthesized (Table 16.8). Their shapes generally agree with the VSEPR model, but there are some interesting exceptions. Two such exceptions are ClF_6^- and BrF_6^-, in which the central halogen has a lone pair of electrons, but the apparent structure is octahedral. The ions IF_6^- participates in an extended array through $I-F \cdots I$ interactions.

Example 16.3 Proposing a bonding model for I^+ complexes

In some cases the interaction of I_2 with strong donor ligands leads to the formation of cationic complexes such as bis(pyridine)iodine(+1), $[py-I-py]^+$. Propose a bonding model for this linear complex from (a) the standpoint of the VSEPR model and (b) simple molecular orbital considerations.

Answer (a) The Lewis electron structure places 10 electrons around the central I^+ in $[py-I-py]^+$, six from the iodine cation and four from the lone pairs on the two pyridine ligands. According to the VSEPR model, these pairs should form a trigonal bipyramid. The lone pairs will occupy the equatorial positions and consequently the complex should be linear. (b) From a molecular orbital perspective, the orbitals of the N—I—N array can be pictured as being formed from an iodine $5p$ orbital and an orbital of σ symmetry from each of the two ligand atoms. Three orbitals can be constructed: 1σ (bonding), 2σ (nearly nonbonding), and 3σ (antibonding). There are four electrons to accommodate (two from each ligand atom; the iodine $5p$ orbital is empty). The resulting configuration is $1\sigma^2 2\sigma^2$, which is net bonding.

Self-test 16.3 From the perspective of structure and bonding, indicate several polyhalides that are analogous to $[py-I-py]^+$, and describe their bonding.

Compounds with oxygen

In contrast to the rather simple formulas and structures of most halogen fluorides, the molecular compounds of the halogens with oxygen are very diverse. The aqueous chemistry of the halogen oxoanions is the most uniform and important, so we shall concentrate on them after a brief discussion of the neutral oxides.

16.8 Halogen oxides

Key points: The only fluorine oxides are OF_2 and O_2F_2; chlorine oxides are known for Cl oxidation numbers of $+1$, $+4$, $+6$, and $+7$; the strong and facile oxidizing agent ClO_2 is the most commonly used halogen oxide.

Many binary compounds of the halogens and oxygen are known, but most are unstable and not commonly encountered in the laboratory. We shall mention only a few of the most important.

Oxygen difluoride (FOF; m.p. $-224°C$, b.p. $-145°C$), the most stable binary compound of oxygen and fluorine, is prepared by passing fluorine through dilute aqueous hydroxide solution:

$$2\,F_2(g) + 2\,OH^-(aq) \rightarrow OF_2(g) + 2\,F^-(aq) + H_2O(l)$$

The pure difluoride survives in the gas phase above room temperature and does not react with glass. It is a strong fluorinating agent, but less so than fluorine itself. As suggested by the VSEPR model, the OF_2 molecule is angular.

Dioxygen difluoride (FOOF; m.p. $-154°C$, b.p. $-57°C$) can be synthesized by photolysis of a liquid mixture of the two elements. It is unstable in the liquid state and decomposes rapidly above $-100°C$, but can be transferred (with some decomposition) as a low-pressure gas in a metal vacuum line. Dioxygen difluoride is an even more aggressive fluorinating agent than ClF_3. For example, it oxidizes plutonium metal and its compounds to PuF_6, which is an intermediate in reprocessing of nuclear fuels, in a reaction that ClF_3 cannot accomplish:

$$Pu(s) + 3\,O_2F_2(g) \rightarrow PuF_6(g) + 3\,O_2(g)$$

Chlorine occurs with many different oxidation numbers in its oxides (Table 16.9). Some of these oxides are odd-electron species, including ClO_2, in which chlorine has the unusual oxidation number $+4$, and Cl_2O_6, which exists as a mixed oxidation state ionic solid, $[ClO_2][ClO_4]$. All the chlorine oxides are endoergic ($\Delta_f G^\ominus > 0$) and unstable. They all explode when heated.

Chlorine dioxide is the only halogen oxide produced on a large scale. The reaction used is the reduction of ClO_3^- with HCl or SO_2 in strongly acidic solution:

$$2\,ClO_3^-(aq) + SO_2(g) \xrightarrow{\text{acid}} 2\,ClO_2(g) + SO_4^{2-}(aq)$$

Because chlorine dioxide is a strongly endoergic compound ($\Delta_f G^\ominus = +121\,kJ\,mol^{-1}$), it must be kept dilute to avoid explosive decomposition and is therefore used at the site of production. Its major uses are to bleach paper pulp and to disinfect sewage and drinking water. Some controversy surrounds these applications because the action of chlorine (or its product of hydrolysis, HClO) and chlorine dioxide on organic matter produces low concentrations of chlorocarbon compounds, some of which are potential carcinogens. However, the disinfection of water undoubtedly saves many more lives than the carcinogenic by-products may take. Chlorine bleaches are being replaced by oxygen-based bleaches such as hydrogen peroxide (Box 15.1).

Table 16.8 Representative interhalogen anions

Compound	Shape
IF_8^-	Square antiprism
ClF_6^-	Octahedral
BrF_6^-	Octahedral
IF_6^-	Trigonally distorted octahedron

Table 16.9 Selected oxides of chlorine

Oxidation number	+1	+3	+4		+6	+7
Formula	Cl_2C	Cl_2O_3	ClO_2	Cl_2O_4	Cl_2O_6	Cl_2O_7
Colour	Brown–yellow	Dark brown	Yellow	Pale yellow	Dark red	Colourless
State	Gas	Solid	Gas	Liquid	Liquid	Liquid

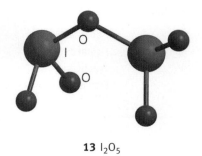

13 I_2O_5

There are fewer oxides of bromine than chlorine. The most well-known compounds are:

Oxidation number	+1	+3	+4
Formula	Br_2O	Br_2O_3	BrO_2
Colour	Dark brown	Orange	Pale yellow
State	Solid	Solid	Solid

The structure of BrO_2 has been found to be a mixed Br(I)/Br(VII) oxide, $BrOBrO_3$. All the bromine oxides are thermally unstable above $-40°C$ and explode on heating.

The most stable halogen oxides are those of formed by iodine. The most important of these is I_2O_5 (**13**). The compound is a white, hygroscopic solid. It dissolves in water to give iodic acid, HIO_3. The less stable iodine oxides I_2O_4 and I_4O_9 are both yellow solids that decompose on heating to give I_2O_5:

$$5\ I_2O_4(s) \rightarrow 4\ I_2O_5(s) + I_2(g)$$
$$4\ I_4O_9(s) \rightarrow 6\ I_2O_5(s) + 2\ I_2(g) + 3\ O_2(g)$$

16.9 Oxoacids and oxoanions

Key points: The halogen oxoanions are thermodynamically strong oxidizing agents; perchlorates of oxidizable cations are unstable and should be avoided.

The wide range of oxoanions and oxoacids of the halogens presents a challenge to those who devise systems of nomenclature. We shall use the common names, such as chlorate for ClO_3^-, rather than the systematic names, such as trioxochlorate(V). Table 16.10 is a brief dictionary for converting between the two systems of nomenclature.

The strengths of the oxoacids vary systematically with the number of O atoms on the central atom (Table 16.11; see Pauling's rules in Section 4.4b). Periodic acid, H_5IO_6, is the I(VII) analogue of perchloric acid. It is a weak acid ($pK_{a1} = 3.29$), which can be explained as soon as we note that its formula is $(HO)_5IO$. The O atoms in the conjugate base $H_4IO_6^-$ are very labile on account of the rapid equilibration:

$$H_4IO_6^-(aq) \rightleftharpoons IO_4^-(aq) + 2\ H_2O(l) \quad K = 40$$

In basic solution IO_4^- is the dominant ion. The tendency to have an expanded coordination shell is shared by the oxoacids of the neighbouring Group 16 element tellurium, which in its maximum oxidation state forms the weak acid $Te(OH)_6$.

The halogen oxoanions, like many oxoanions, form metal complexes, including the metal perchlorates and periodates discussed here. In this connection we note that, because $HClO_4$ is a very strong acid and H_5IO_6 is a weak acid, it follows that ClO_4^- is a very weak base and $H_4IO_6^-$ is a relatively strong base.

In view of the low Brønsted basicity and single negative charge of the perchlorate ion, ClO_4^-, it is not surprising that it is a weak Lewis base with little tendency to form

Table 16.11 Acidities of chlorine oxoacids

Acid	q/p	pK_a
HOCl	0	7.53 (weak)
HOClO	1	2.00
$HOClO_2$	2	−1.2
$HOClO_3$	3	−10 (strong)

Table 16.10 Halogen oxoanions

Oxidation number	Formula	Name*	Point group	Shape	Remarks
+1	ClO^-	Hypochlorite [monoxochlorate(I)]	$C_{\infty v}$	Linear	Good oxidizing agent
+2	ClO_2^-	Chlorite [dioxochlorate(III)]	C_{2v}	Angular	Strong oxidizing agent, disproportionates
+5	ClO_3^-	Chlorate [trioxochlorate(V)]	C_{3v}	Pyramidal	Oxidizing agent
+7	ClO_4^-	Perchlorate [tetraoxochlorate(VII)]	T_d	Tetrahedral	Oxidizing agent, very weak ligand

* IUPAC names in square brackets.

complexes with cations in aqueous solution. Therefore, metal perchlorates are often used to study the properties of hexaaqua ions in solution. The ClO_4^- ion is used as a weakly coordinating ion that can readily be displaced from a complex by other ligands, or as a medium-sized anion that might stabilize solid salts containing large cationic complexes with easily displaced ligands.

However, the ClO_4^- ion is a treacherous ally. Because it is a powerful oxidizing agent, solid compounds of perchlorate should be avoided whenever there are oxidizable ligands or ions present (which is commonly the case). In some cases the danger lies in wait, as the reactions of ClO_4^- are generally slow and it is possible to prepare many metastable perchlorate complexes or salts that may be handled with deceptive ease. However, once reaction has been initiated by mechanical action, heat, or static electricity, these compounds can detonate with disastrous consequences. Such explosions have injured chemists who may have handled a compound many times before it unexpectedly exploded. Some readily available and more docile weakly basic anions may be used in place of ClO_4^-; they include trifluoromethanesulfonate $[SO_3CF_3]^-$, tetrafluoroborate BF_4^-, and hexafluorophosphate $[PF_6^-]$.

In contrast to perchlorate, periodate is a rapid oxidizing agent and a stronger Lewis base than perchlorate. These properties lead to the use of periodate as an oxidizing agent and stabilizing ligand for metal ions in high oxidation states. Some of the high oxidation states it can be used to form are very unusual: they include Cu(III) in a salt containing the $[Cu(HIO_6)_2]^{5-}$ complex and Ni(IV) in an extended complex containing the $[Ni(IO_6)]^-$ unit. The periodate ligand is bidentate in these complexes, and in the last example it forms a bridge between Ni(IV) ions.

16.10 Thermodynamic aspects of redox reactions

Key point: The oxoanions of halogens are strong oxidizing agents, especially in acidic solution.

The thermodynamic tendencies of the halogen oxoanions and oxoacids to participate in redox reactions are well understood. As we shall see, we can summarize their behaviour with a Frost diagram that is quite easy to rationalize. It is a very different story with the rates of the reactions, which vary widely. Their mechanisms are only partly understood despite many years of investigation. Recent progress in the understanding of some of these mechanisms stems from advances in techniques for fast reactions and interest in oscillating reactions (Box 16.4).

Box 16.4 Oscillating reactions

Clock reactions and oscillating reactions are an active topic of research and provide fascinating lecture demonstrations. Most examples of oscillating reactions are based on reactions of halogen oxoanions, apparently because of the variety of oxidation states and their sensitivity to changes in pH. In 1895, H. Landot discovered that a mixture of sulfite, iodate, and starch in acidic solution remains nearly colourless for an initial period and then suddenly switches to the dark purple of the I_2–starch complex. When the concentrations are properly adjusted the reaction oscillates between nearly colourless and opaque blue. The reactions leading to this oscillation are the reduction of periodate to iodide by sulfite, where all reactants are colourless:

$$IO_4^-(aq) + 4\ SO_3^{2-}(aq) \rightarrow I^-(aq) + 4\ SO_4^{2-}(aq)$$

A comproportionation reaction between I^- and IO_3^- then produces I_2, which forms an intensely coloured complex with starch:

$$IO_3^-(aq) + 6\ H^+(aq) + 5\ I^-(aq) \rightarrow 3\ H_2O(l) + 3\ I_2(starch)$$

Under some conditions this is the final state, but adjustment of concentrations may lead to bleaching of the I_2–starch complex by sulfite reduction of iodine to the colourless $I^-(aq)$ ion:

$$3\ I_2(starch) + 3\ SO_3^{2-}(aq) + 3\ H_2O(l) \rightarrow 6\ I^-(aq)$$
$$+ 6\ H^+(aq) + 3\ SO_4^{2-}(aq)$$

The reaction may then oscillate between colourless and blue as the I_2/I^- ratio changes.

The detailed analysis of the kinetic conditions for oscillating reactions is pursued by chemists and chemical engineers. In the former case the challenge is to use kinetic data determined separately for the individual steps to model the observed oscillations with a view to testing the validity of the overall scheme. As oscillating reactions have been observed in commercial catalytic processes, the concern of the chemical engineer is to avoid large fluctuations or even chaotic reactions that might degrade the process.

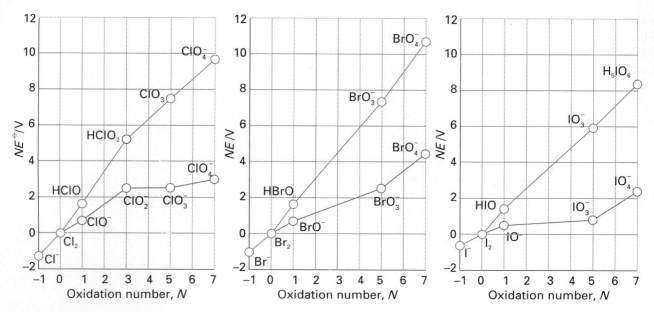

Fig. 16.14 Frost diagrams for chlorine, bromine, and iodine in acidic solution (red line) and in basic solution (blue line).

We saw in Section 5.11 that, if in a Frost diagram a species lies above the line joining its two neighbours of higher and lower oxidation state, then it is unstable with respect to disproportionation into them. From the Frost diagram for the halogen oxoanions and oxoacids in Fig. 16.4 we can see that many of the oxoanions in intermediate oxidation states are susceptible to disproportionation. Chlorous acid, $HClO_2$, for instance, lies above the line joining its two neighbours, and is liable to disproportionation:

$$2\ HClO_2(aq) \rightarrow ClO_3^-(aq) + HClO(aq) + H^+(aq) \qquad E^{\ominus} = +0.52V$$

Although BrO_2^- is well characterized, the corresponding I(III) species is so unstable that it does not exist in solution, except perhaps as a transient intermediate.

We also saw in Section 5.11 that the more positive the slope for the line from a lower to higher oxidation state species in a Frost diagram, the stronger the oxidizing power of the couple. A glance at Fig. 16.14 shows that all three Frost diagrams have steep positively sloping lines, which immediately shows that all the oxidation states except the lowest (Cl^-, Br^-, and I^-) are strongly oxidizing.

Finally, basic conditions decrease reduction potentials for oxoanions as compared with their conjugate acids (Section 5.5). This decrease is evident in the less steep slopes of the lines in the Frost diagrams for the oxoanions in basic solution. The numerical comparison for ClO_4^- ions in 1 M acid compared with 1 M base makes this clear:

At pH = 0: $ClO_4^-(aq) + 2\ H^+(aq) + 2\ e^- \rightarrow ClO_3^-(aq) + H_2O(l)$

$E^{\ominus} = +1.20$ V

At pH = 14: $ClO_4^-(aq) + H_2O(l) + 2\ e^- \rightarrow ClO_3^-(aq) + 2\ OH^-(aq)$

$E_B^{\ominus} = +0.37$ V

The reduction potentials show that perchlorate is thermodynamically a weaker oxidizing agent in basic solution than in acidic solution.

16.11 Trends in rates of redox reactions

Key points: Oxidation by halogen oxoanions is faster for the lower oxidation states; rates and thermodynamics of oxidation are both enhanced by an acidic medium.

Mechanistic studies show that the redox reactions of halogen oxoanions are complex. Nevertheless, despite this complexity there are a few discernible patterns that help to correlate the trends in rates of reaction. These correlations have practical value and give some clues about the mechanisms that may be involved.

The oxidation of many molecules and ions by halogen oxoanions becomes progressively faster as the oxidation number of the halogen decreases. Thus the rates observed are often in the order

$$ClO_4^- < ClO_3^- < ClO_2^- \approx ClO^- \approx Cl_2$$

$$BrO_4^- < BrO_3^- \approx BrO^- \approx Br_2$$

$$IO_4^- < IO_3^- < I_2$$

For example, aqueous solutions containing Fe^{2+} and ClO_4^- are stable for many months in the absence of dissolved oxygen, but an equilibrium mixture of aqueous HClO and Cl_2 rapidly oxidizes Fe^{2+}.

Oxoanions of the heavier halogens tend to react most rapidly, particularly for the elements in their highest oxidation states:

$$ClO_4^- < BrO_4^- < IO_4^-$$

As we have remarked, perchlorates in dilute aqueous solution are usually unreactive, but periodate oxidations are fast enough to be used for titrations. The mechanistic details are often complex, but the existence of both four- and six-coordinate periodate ions shows that the I atom in periodate is accessible to nucleophiles.

We have already seen that the thermodynamic tendency of oxoanions to act as oxidizing agents increases as the pH is lowered. It is also found that their rates are increased too. Thus, kinetics and equilibria unite to bring about otherwise difficult oxidations. The oxidation of halides by BrO_3^- ions, for instance, is second-order in H^+:

$$Rate = k[BrO_3^-][X^-][H^+]^2$$

and so the rate increases as the pH is decreased. The acid is thought to act by protonating the oxo group in the oxoanion, so aiding oxygen–halogen bond scission. Another role of protonation is to increase the electrophilicity of the halogen. An example is HClO, where, as described below, the Cl atom may be viewed as an electrophile towards an incoming reducing agent (**14**). An illustration of the effect of acidity on rate is the use of a mixture of H_2SO_4 and $HClO_4$ in the final stages of the oxidation of organic matter in certain analytical procedures.

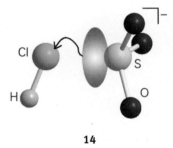

14

16.12 Redox properties of individual oxidation states

With the general redox properties of the halogens now outlined, we can consider the characteristic properties and reactions of specific oxidation states, with emphasis on the mechanism of reaction. Although we are dealing here with halogen oxides, it is convenient to mention, for the sake of completeness, the redox properties of halogen(0) species. Figure 16.15 summarizes some of the reactions that interconvert the oxoanions and oxoacids of chlorine in its various oxidation states. One point to note is the major role of disproportionation and electrochemical reactions in the scheme. For example, the figure includes the production of Cl_2 by the electrochemical oxidation of Cl^-, which was discussed in Section 16.1. Once Cl_2 is available, the figure indicates that two successive disproportionations lead to the important Cl(V) species, ClO_3^-. The highest oxidation state Cl(VII) is achieved by the electrochemical oxidation of ClO_3^- to ClO_4^-.

(a) Halogen(0)

Key point: Disproportionation of dihalogen molecules in aqueous solution occurs with the formation of the +1 (XO^-) and −1 (X^-) species.

One of the most characteristic redox reactions of the halogens is disproportionation, which is thermodynamically favourable for basic solutions of Cl_2, Br_2, and I_2. The equilibria in basic aqueous solution are:

$$X_2(aq) + 2\,OH^-(aq) \rightleftharpoons XO^-(aq) + X^-(aq) + H_2O(l) \qquad K = \frac{[XO^-][X^-]}{[X_2][OH^-]^2}$$

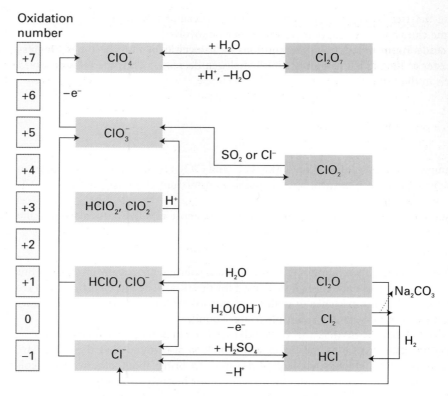

Oxidation number

Fig. 16.15 The interconversion of oxidation states of some important chlorine species.

with $K = 7.5 \times 10^{15}$ for X = Cl, 2×10^8 for Br, and 30 for I.[1]

Disproportionation is much less favourable in acidic solution, as would be expected from the fact that H^+ is a product of the reaction:

$$Cl_2(aq) + H_2O(l) \rightleftharpoons HClO(aq) + H^+(aq) + Cl^-(aq) \quad K = 3.9 \times 10^{-4}$$

The mechanism of this reaction appears to involve attack of H_2O on the Cl_2 molecule:

$$H_2O + Cl_2 \rightarrow [H_2O{-}Cl{-}Cl] \rightarrow H^+ + [HO{-}Cl{-}Cl]^- \rightarrow H^+ + HOCl + Cl^-$$

It is not known if the Lewis complex H_2OCl_2 is in fact formed. According to one interpretation, proton transfer is so rapid that the reaction yields $HOCl_2^-$ directly. The $[HO{-}Cl{-}Cl]^-$ complex is at most transitory, as Cl^- is rapidly displaced by the strong nucleophile OH^-. The important insight gained from this mechanism is that this reaction may be mechanistically a nucleophilic displacement.

Because the redox reactions of HClO and Cl_2 are often fast, Cl_2 in water is widely used as an inexpensive and powerful oxidizing agent. A part of the reason for its usefulness is that the rapid equilibration between ClO^- and Cl_2 gives rise to a variety of possible reaction pathways. For example, the nonlabile complex $[Fe(phen)_3]^{2+}$ reacts more rapidly with Cl_2 than with HClO because the former can more readily undergo an outer-sphere electron transfer, whereas ClO^- or HClO appear to require the detachment of phen prior to an inner-sphere redox process. In other environments HClO may be a more facile oxidizing agent than Cl_2. In general, labile aqua complexes, such as $[Cr(OH_2)_6]^{2+}$, react faster with ClO^- than with Cl_2.

The equilibrium constants for the hydrolysis of Br_2 and I_2 in acid solution are smaller than for Cl_2, and both elements are unchanged when dissolved in slightly acidified water. Because F_2 is a much stronger oxidizing agent than the other halogens, it produces mainly O_2 and H_2O_2 when in contact with water. As a result, hypofluorous acid, HFO, was discovered long after the other hypohalous acids. The first indication of its existence came from IR spectroscopic studies of H_2O and F_2 trapped in a matrix of frozen N_2 at about

---

[1] When writing the associated equilibrium expressions in aqueous solution, we conform to the usual convention that the activity of water is 1, so H_2O does not appear in the equilibrium constant.

Box 16.5 Chlorine-based bleaches

Substances that are used as bleaches are powerful oxidizing agents. As was mentioned in Section 16.2 the oxidizing power of the halogen oxoanions increases as the oxidation number of the halogen decreases. It is not surprising then that the chlorine-based bleaches contain chlorine in a low oxidation state.

Chlorine gas is a widely used bleach for the textiles industry and water treatment. Chlorine disproportionates in water to produce the hypochlorite ion, ClO^-, and H^+ and solutions of hypochlorite salts find widespread use. Solutions of up to 15 percent by mass of sodium hypochlorite are used as industrial bleaches in the paper, textiles, and laundry industries and for disinfecting swimming pools. Household bleach is a more dilute (5 per cent) solution of $NaOCl$.

Other hypochlorite salts are also used. Calcium hypochlorite, $Ca(ClO)_2$, is used as a disinfectant in dairies, breweries, food processing, and bottling plants. It is also used in domestic mildew removers. *Bleaching powder* is a mixture of $Ca(ClO)_2$ and $CaCl_2$ and is used for large-scale applications such as disinfecting seawater, reservoirs, and sewers. It is also used as a decontaminant in areas where chemical weapons, such as mustard gas, have been deployed.

Chlorine dioxide gas is widely used as a bleach in the wood pulp industry where it produces whiter and stronger paper than other bleaches. Unlike oxidizing bleaches such as chlorine, ozone, and hydrogen peroxide, ClO_2 does not attack the cellulose and therefore preserves the mechanical strength of the pulp. Chlorine-based bleaches lead to the production of toxic chlorinated organic compounds. The most toxic polychlorinated phenols, such as dioxins, are mostly produced by ClO_2, but the levels can be drastically reduced by substituting some of the ClO_2 with Cl_2.

20 K. With this encouragement, HFO was finally isolated and chemically characterized in the early 1970s by the controlled reaction of F_2 with ice at $-40°C$, which leads to the highly unstable HFO in approximately 50 per cent yield:

$$F_2(g) + H_2O(s) \rightarrow HFO(g) + HF(g)$$

The product decomposes to O_2 and HF with a half-life of 30 min at $20°C$.

(b) Halogen(I)

Key points: Hypochlorite is a facile oxidizing agent; hypohalite ions undergo disproportionation.

The aqueous Cl(I) species hypochlorous acid, HClO (the angular molecular species H—O—Cl),[2] and hypochlorite ions, ClO^-, are facile oxidizing agents that are used as household bleach and disinfectant and as laboratory oxidizing agents (Box 16.5). The mechanism invoked for oxidation reactions of HClO was thought for many years to involve O atom transfer, but mechanistic studies strongly implicate Cl^+ transfer in the reactions of ClO^- and HClO. For example, kinetic data indicate that the first step in the mechanism of reaction of HOCl with SO_3^{2-} is Cl^+ transfer to produce the chlorosulfate anion:

$$HClO(aq) + SO_3^{2-}(aq) \rightarrow OH^-(aq) + ClSO_3^-(aq)$$

The activated complex is pictured as a Cl-bridged species $[HO—Cl—SO_3]^{2-}$ (as in **14**). To account for the outcome of this reaction, the Cl atom is thought to carry only six electrons with it as it transfers to SO_3^{2-}; so in effect a Cl^+ ion is transferred even though an isolated Cl^+ cation is not involved. This step is followed by the hydrolysis of chlorosulfate, to yield the sulfate and chloride ions:

$$ClSO_3^-(aq) + H_2O(l) \rightarrow SO_4^{2-}(aq) + Cl^-(aq) + 2\,H^+(aq)$$

A similar Cl^+ ion transfer is indicated for the reactions of HClO with other nucleophilic anions and the rate of reaction, which lies in the order

$$Cl^- < Br^- < I^- < SO_3^{2-} < CN^-$$

correlates with the nucleophilicity of the anion. The ready access to the unobstructed, electrophilic Cl atom in HClO appears to be one feature that leads to the very fast redox reactions of this compound. These rates contrast with the much slower redox reactions of perchlorate ions, in which access to the Cl atom is blocked by the surrounding O atoms.

---

[2] Hypohalous acids are widely denoted HOX to emphasize their structure. We have adopted the formula HXO to emphasize their relationship to the other oxoacids, HXO_n.

Hypohalite ions undergo disproportionation. For instance, ClO^- disproportionates into Cl^- and ClO_3^-:

$$3\ ClO^-(aq) \rightleftharpoons 2\ Cl^-(aq) + ClO_3^-(aq) \qquad K = 1.5 \times 10^{27}$$

This reaction (which is used for the commercial production of chlorates) is slow at or below room temperature for ClO^- but is much faster for BrO^-. It is so fast for IO^- that this ion has been detected only as a reaction intermediate.

(c) Halogen(III)

Key point: Halite ions typically undergo disproportionation.

Chlorite ions, ClO_2^-, and bromite ions, BrO_2^-, are both susceptible to disproportionation. However, the rate is strongly dependent on pH and ClO_2^- (and to a lesser extent BrO_2^-) can be handled in basic solution with only slow decomposition. By contrast, chlorous acid, $HClO_2$, and bromous acid, $HBrO_2$, both disproportionate rapidly. As the following standard reaction Gibbs energies indicate, disproportionation is thermodynamically favourable in acidic solution.

$$2\ HClO_2(aq) \rightleftharpoons H^+(aq) + HClO(aq) + ClO_3^-(aq) \quad \Delta_r G^{\ominus} = -49\ kJ\ mol^{-1}$$
$$K = 3.9 \times 10^8$$

$$2\ HBrO_2(aq) \rightleftharpoons H^+(aq) + HBrO(aq) + BrO_3^-(aq) \quad \Delta_r G^{\ominus} = -48\ kJ\ mol^{-1}$$
$$K = 2.6 \times 10^8$$

Iodine(III) is even more elusive, and HIO_2 has only been identified as a transient species in aqueous solution.

These primary reactions are usually accompanied by further reaction. For example, the disproportionation of $HClO_2$ under a broad range of conditions is quite well described by the equation

$$4\ HClO_2(aq) \rightarrow 2\ H^+(aq) + 2\ ClO_2(aq) + ClO_3^-(aq) + Cl^-(aq) + H_2O(l)$$

This complex disproportionation is shown in the centre of Fig. 16.15. Because two oxidized products (ClO_2 and ClO_3^-) are produced, the equation can be balanced differently by changing their ratio. The ratio observed reflects the relative rates of the alternative reactions.

(d) Halogen(V)

Key point: Chlorate ions undergo disproportionation in solution but bromates and iodates do not.

The Frost diagram for chlorine shown in Fig. 16.14 indicates that chlorate ions, ClO_3^-, are unstable with respect to disproportionation in both acidic and basic solution:

$$4\ ClO_3^-(aq) \rightleftharpoons 3\ ClO_4^-(aq) + Cl^-(aq) \quad \Delta_r G^{\ominus} = -24\ kJ\ mol^{-1} \quad K = 1.4 \times 10^{25}$$

Because $HClO_3$ is a strong acid, and this reaction is slow at both low and high pH, ClO_3^- ions can be handled readily in aqueous solution. Bromates and iodates are thermodynamically stable with respect to disproportionation.

Example 16.4 Expressing the decomposition of HFO

It was remarked that HFO is unstable in water. Write plausible equations for the decomposition of HFO(aq).

Answer Oxidation states 0, +1, and +3 are accessible for all the halogens except fluorine. In part, this distinction is simply an artefact of the rule for assigning the most negative oxidation number to the most electronegative element, but the higher electronegativity of fluorine is also reflected in the chemical properties of HFO. We might, for example, expect this thermodynamically unstable compound to yield more stable compounds, such as HF, O_2, H_2O_2, and possibly OF_2. In fact, some of these products are observed:

$$2\ HFO(aq) \rightarrow 2\ HF(aq) + O_2(g)$$
$$HFO(aq) + H_2O(l) \rightarrow H_2O_2(aq) + HF(aq)$$

In the presence of excess fluorine a further reaction occurs:

$$F_2(aq) + HFO(aq) \rightarrow OF_2(aq) + HF(aq)$$

Self-test 16.4 Use standard potentials to decide which dihalogens are thermodynamically capable of oxidizing H_2O to O_2. Which of these reactions are facile?

(e) Halogen(VII)

Key point: The rates of reactions in which ClO_4^- acts as an oxidizing agent are often slow in basic solution but rapid in acidic solution.

The perchlorate ion, ClO_4^-, and the periodate ion, IO_4^-, were known in the nineteenth century, but the perbromate ion, BrO_4^-, was not discovered until the late 1960s. This difference is another example of the instability of the maximum oxidation state of the Period 4 elements arsenic, selenium, and bromine. Although perbromate is not readily available commercially, it may be synthesized by the action of fluorine on bromate ions in basic aqueous solution:

$$BrO_3^-(aq) + F_2(g) + 2\ OH^-(aq) \rightarrow BrO_4^-(aq) + 2\ F^-(aq) + H_2O(l)$$

Of the three XO_4^- ions, BrO_4^- is the strongest oxidizing agent. That perbromate is out of line with its adjacent halogen congeners fits a general pattern for anomalies in the chemistry of p-block elements of Period 4.

The reduction of BrO_4^- is faster than that of ClO_4^- because perbromate ions undergo electron transfer more rapidly (by an outer-sphere mechanism). However, reduction of periodate ions in dilute acid solution is by far the fastest. Periodates are therefore used in analytical chemistry as oxidizing titrants and also in syntheses, such as the oxidative cleavage of diols:

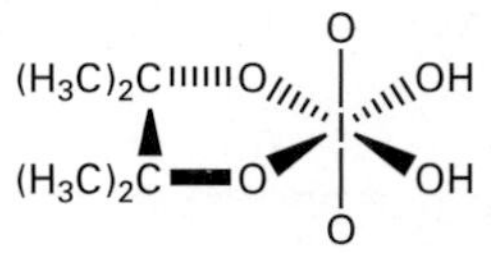

We have already commented on the low Lewis basicity of ClO_4^-, which leads to very weak coordination with metal aqua-ions, and the fact that ClO_4^- in combination with many reducing cations forms metastable compounds that do not react under mild conditions but explode when provoked. With certain reducing agents, ClO_4^-, however, reacts smoothly in aqueous solution. The order of reactivity observed for a series of metal ion reducing agents is

$$Ru^{2+} > Ti^{3+} > Mo^{3+} > V^{2+} \approx V^{3+} > Cr^{2+} > > Fe^{2+}$$

For the most part, this series correlates with the ligand-field splitting parameter Δ_O but not with the values of $E^{\ominus}$ for the individual ions. The decreasing stability across the series due to ligand field effects appears to hinder the reduction of Cl(VII) to Cl(V).

Anhydrous perchloric acid is a particularly aggressive oxidizing medium that must not be allowed to come into accidental contact with organic matter or similar reducing agents. The presence of Cl_2O_7 as a result of the equilibrium

$$3\ HClO_4 \rightleftharpoons Cl_2O_7 + H_3O^+ + ClO_4^-$$

and at elevated temperatures of the radical species $\cdot OH$, $\cdot ClO_3$, and $\cdot ClO_4$ seems to be responsible for its rapid oxidation reactions.

Fluorocarbons

Key point: Fluorocarbon molecules and polymers are resistant to oxidation but are attacked by alkali metals.

The synthesis of fluorocarbon compounds is of great technological importance because they are useful in applications ranging from coatings for nonstick cookware and halogen-resistant laboratory vessels to the volatile fluorocarbons used as refrigerants in air conditioners and refrigerators (Box 16.6). Fluorocarbon derivatives have also been the topic of considerable exploratory synthetic research because their derivatives often have unusual properties.

The direct reaction of an aliphatic hydrocarbon with an oxidizing metal fluoride leads to the formation of strong C—F bonds ($456 \, kJ \, mol^{-1}$) and produces HF as a by-product:

$$RH(l) + 2 \, CoF_3(s) \rightarrow RF(l) + 2 \, CoF_2(s) + HF(l) \quad R = \text{alkyl or aryl}$$

When R is aryl, CoF_3 yields the cyclic saturated fluoride:

$$C_6H_6(l) + 18 \, CoF_3(s) \rightarrow C_6F_{12}(l) + 18 \, CoF_2(s) + 6 \, HF(l)$$

The strongly oxidizing fluorinating agent used in these reactions, CoF_3, is regenerated by the reaction of CoF_2 with fluorine:

$$2 \, CoF_2(s) + F_2(s) \rightarrow 2 \, CoF_3(s)$$

Another important method of CF bond formation is halogen exchange by the reaction of a nonoxidizing fluoride, such as HF, with a chlorocarbon in the presence of a catalyst, such as SbF_3:

$$CCl_4(l) + HF(l) \rightarrow CCl_3F(l) + HCl(g)$$
$$CHCl_3(l) + 2 \, HF(l) \rightarrow CHClF_2(l) + 2 \, HCl(g)$$

These processes used to be performed on a large scale to produce the chlorofluorocarbons (CFCs) and hydrochlorofluorocarbons (HCFCs) that were used as refrigerant fluids, the propellant in spray cans, and in the blowing agent in plastic foam products. These applications have been banned in some countries and are being phased out worldwide because of the role of CFCs and HCFCs in ozone depletion. They are being replaced by

---

Box 16.6 PTFE—a high-performance polymer

Polytetrafluoroethene, PTFE, is a unique product in the plastics industry. It is chemically inert, thermally stable over a wide temperature range (-196 to $260°C$), is an excellent electrical insulator, and has a low coefficient of friction. It is a white solid that is manufactured by the polymerization of tetrafluoro-ethene, TFE:

$$n \, CF_2{=}CF_2 \rightarrow (CF_2CF_2)_n$$

TFE is an expensive polymer because of the cost of synthesizing and purifying the monomer by a multistage process:

$$CaF_2(s) + H_2SO_4(aq) \rightarrow CaSO_4(aq) + 2 \, HF(aq)$$
$$CH_4(g) + 3 \, Cl_2(g) \rightarrow CHCl_3(g) + 3 \, HCl(g)$$
$$CHCl_3(g) + 2 \, HF(g) \rightarrow CHClF_2(g) + 2 \, HCl(g)$$
$$2 \, CHClF_2(g) \xrightarrow{\Delta} CF_2{=}CF_2(g) + 2 \, HCl(g)$$

As the process generates HF and HCl the reactors have to be lined with platinum. Many by-products are produced, which leads to complex purification of the final product.

The TFE is polymerized in two ways: solution polymerization with vigorous agitation produces a resin known as *granular PTFE*; emulsion polymerization with a dispersing agent and gentle agitation produces small particles known as *dispersed PTFE*. The molten polymer does not flow, so the usual methods of processing cannot be used. Instead processes similar to those used for metals are applied. For example, the dispersed form can be cold extruded, which is a method used for processing lead.

The remarkable properties of PTFE arise from the protective sheath that the fluorine atoms form over the carbon polymer backbone. The fluorine atoms are just the right size to form a smooth sheath. This reduces the surface energy, leading to a low coefficient of friction and the familiar nonstick properties. The polymer is used in a wide range of applications. Its low electrical conductivity leads to its use in electrical tapes, wires, and coaxial cable. Its mechanical properties make it an ideal material for seals, piston rings, and bearings. It is used as a packaging material, in hose lines, and as thread sealant tape. Familiar applications are as the nonstick coating on cookware, and as the porous fabric Gore-Tex®.

hydrofluorocarbons (HFCs) after investment by the chemical industry because, in contrast to the simple one-step synthesis of CFCs and HCFCs, HFC production is a complex multistage process. For example, the preferred route to CF_3CH_2F, which is one of the preferred CFC replacements is

$$CCl_2{=}CCl_2 \xrightarrow{HF+Cl_2} CClF_2CCl_2F \xrightarrow{isomerize} CF_3CCl_3 \xrightarrow{HF} CF_3Cl_2F \xrightarrow{H_2} CF_3CH_2F$$

When heated, dichlorofluoromethane is converted to the useful monomer, C_2F_4:

$$2\,CHClF_2 \xrightarrow{600-800°C} C_2F_4 + 2\,HCl$$

The polymerization of tetrafluoroethene is carried out with a radical initiator:

$$n\,C_2F_4 \xrightarrow{ROO\cdot} (-CF_2-CF_2-)_n$$

Polytetrafluoroethylene (PTFE) is sold under many trade names, one of which is Teflon (DuPont). It can be depolymerized at high temperatures and this is the most convenient method of preparing tetrafluoroethene in the laboratory:

$$(-CF_2-CF_2-)_n \xrightarrow{600°C} n\,C_2F_4$$

Although tetrafluoroethylene is not highly toxic, a by-product, perfluoroisobutylene, is toxic and its presence dictates care in handling the crude tetrafluoroethylene.

FURTHER READING

A.G. Massey, *Main group chemistry*. Wiley (2000).

D.M.P. Mingos, *Essential trends in inorganic chemistry*. Oxford University Press (1998).

R.B. King (ed.), *Encyclopedia of inorganic chemistry*. Wiley (2005).

N. N. Greenwood and A. Earnshaw, *Chemistry of the elements*. Butterworth-Heinemann (1997).

P. Schmittinger, *Chlorine: principles and industrial practice*. Wiley-VCH (2000).

M. Howe-Grant, *Fluorine chemistry*. Wiley (1995).

EXERCISES

16.1 Preferably without consulting reference material, write out the halogens and noble gases as they appear in the periodic table, and indicate the trends in (a) physical state (s, l, or g) at room temperature and pressure, (b) electronegativity, (c) hardness of the halide ion, and (d) colour.

16.2 Describe how the halogens are recovered from their naturally occurring halides and rationalize the approach in terms of standard potentials. Give balanced chemical equations and conditions where appropriate.

16.3 Sketch a choralkali cell. Show the half-cell reactions and indicate the direction of diffusion of the ions. Give the chemical equation for the unwanted reaction that would occur if OH^- migrated through the membrane and into the anode compartment.

16.4 Sketch the form of the vacant σ^* orbital of a dihalogen molecule and describe its role in the Lewis acidity of the dihalogens.

16.5 Nitrogen trifluoride, NF_3, boils at $-129°C$ and is devoid of Lewis basicity. By contrast, the lower molar mass compound NH_3 boils at $-33°C$ and is well known as a Lewis base. (a) Describe the origins of this very large difference in volatility. (b) Describe the probable origins of the difference in basicity.

16.6 Based on the analogy between halogens and pseudohalogens write: (a) the balanced equation for the probable reaction of cyanogens, $(CN)_2$, with aqueous sodium hydroxide; (b) the equation for the probable reaction of excess thiocyanate with the oxidizing agent $MnO_2(s)$ in acidic aqueous solution; (c) a plausible structure for trimethylsilyl cyanide.

16.7 Given that 1.84 g of IF_3 reacts with 0.93 g of $[(CH_3)_4N]F$ to form a product X: (a) identify X; (b) use the VSEPR model to predict the shapes of IF_3 and the cation and anion in X; (c) predict how many ^{19}F NMR signals would be observed in IF_3 and X.

16.8 Use the VSEPR model to predict the shapes of $SbCl_5$, $FClO_3$, and $[ClF_6]^+$.

16.9 Indicate the product of the reaction between ClF_5 and SbF_5.

16.10 Sketch all the isomers of the complexes MCl_4F_2 and MCl_3F_3. Indicate how many fluorine environments would be indicated in the ^{19}F NMR spectrum of each isomer.

16.11 (a) Use the VSEPR model to predict the probable shapes of $[IF_6]^+$ and IF_7. (b) Give a plausible chemical equation for the preparation of $[IF_6][SbF_6]$.

16.12 Predict whether each of the following solutes is likely to make liquid BrF_3 a stronger Lewis acid or a stronger Lewis base: (a) SbF_5, (b) SF_6, (c) CsF.

16.13 Predict whether each of the following compounds is likely to be dangerously explosive in contact with BrF_3, and explain your answer: (a) SbF_5, (b) CH_3OH, (c) F_2, (d) S_2Cl_2.

16.14 The formation of Br_3^- from a tetraalkylammonium bromide and Br_2 is only slightly exoergic. Write an equation (or NR for no reaction) for the interaction of $[NR_4][Br_3]$ with I_2 in CH_2Cl_2 solution, and give your reasoning.

16.15 Explain why $CsI_3(s)$ is stable with respect to the elements but $NaI_3(s)$ is not.

16.16 Write plausible Lewis structures for (a) ClO_2 and (b) I_2O_6 and predict their shapes and the associated point group.

16.17 (a) Give the formulas and the probable relative acidities of perbromic acid and periodic acid. (b) Which is the more stable?

16.18 (a) Describe the expected trend in the standard potential of an oxoanion in a solution with decreasing pH. (b) Demonstrate this phenomenon by calculating the reduction potential of ClO_4^- at $pH = 7$ and comparing it with the tabulated value at $pH = 0$.

16.19 With regard to the general influence of pH on the standard potentials of oxoanions, explain why the disproportionation of an oxoanion is often promoted by low pH.

16.20 Which oxidizing agent reacts more readily in dilute aqueous solution, perchloric acid or periodic acid? Give a mechanistic explanation for the difference.

16.21 (a) For which of the following anions is disproportionation thermodynamically favourable in acidic solution: OCl^-, ClO_2^-, ClO_3^-, and ClO_4^-? (If you do not know the properties of these ions, determine them from a table of standard potentials.) (b) For which of the favourable cases is the reaction very slow at room temperature?

16.22 Which if the following compounds present an explosion hazard? (a) NH_4ClO_4, (b) $Mg(ClO_4)_2$, (c) $NaClO_4$, (d) $[Fe(H_2O)_6][ClO_4]_2$. Explain your reasoning.

16.23 Use standard potentials to predict which of the following will be oxidized by ClO^- ions in acidic conditions: (a) Cr^{3+}, (b) V^{3+}, (c) Fe^{2+}, (d) Co^{2+}.

PROBLEMS

16.1 Many of the acids and salts corresponding to the positive oxidation numbers of the halogens are not listed in the catalogue of a major international chemical supplier: (a) $KClO_4$ and KIO_4 are available but $KBrO_4$ is not; (b) $KClO_3$, $KBrO_3$, and KIO_3 are all available; (c) $NaClO_2$ and $NaBrO_2 \cdot 3H_2O$ are available but salts of IO_2^- are not; (d) only ClO^- salts are available but the bromine and iodine analogues are not. Describe the probable reason for the missing salts of the oxoanions.

16.2 Identify the incorrect statements among the following descriptions and provide correct statements. (a) Oxidation of the halides is the only commercial method of preparing the halogens F_2 through I_2. (b) ClF_4^- and I_5^- are isolobal and isostructural. (c) Atom-transfer processes are common in the mechanisms of oxidations by the halogen oxoanions, and an example is the O atom transfer in the oxidation of SO_3^{2-} by ClO^-. (d) Periodate appears to be a more facile oxidizing agent than perchlorate because the former can coordinate to the reducing agent at the I(VII) centre, whereas the Cl(VI) centre in perchlorate is inaccessible to reducing agents.

16.3 The reaction of I^- ions is often used to titrate ClO^-, giving deeply coloured I_3^- ions, along with Cl^- and H_2O. Although never proved, it was thought that the initial reaction proceeds by O atom transfer from Cl to I. However, it is now believed that the reaction proceeds by Cl atom transfer to give ICl as the intermediate (K. Kumar, R.A. Day, and

D.W. Margerum, *Inorg. Chem.*, 1986, **25**, 4344). Summarize the evidence for Cl atom transfer.

16.4 Given the bond lengths and angles in I_5^+ (**9**), describe the bonding in terms of two-centre and three-centre σ bonds and account for the structure in terms of the VSEPR model.

16.5 Until the work of K.O. Christe (*Inorg. Chem.*, 1986, **25**, 3721), F_2 could be prepared only electrochemically. Give chemical equations for Christe's preparation and summarize the reasoning behind it.

16.6 The use of templates to synthesize long-chain polyiodide ions has been described (A.J. Blake *et al.*, *Chem. Soc. Rev.*, 1998, **27**, 195). (a) According to the authors what is the longest polyiodide that has been characterized? (b) How does the nature of the cation influence the structure of the polyanion? (c) What was the templating agent used for the synthesis of I_7^- and I_{12}^{2-}? (c) Which spectroscopic method was used for the characterization of the polyanions in this study?

16.7 Review published studies on the fluoridation of drinking water in your country. Summarize both the reasons for continuing fluoridation and the main concerns expressed by those opposed to it.

16.8 Write a review of the environmental problems associated with the use of chlorine-based bleaches in industry and suggest possible solutions.

The Group 18 elements

17

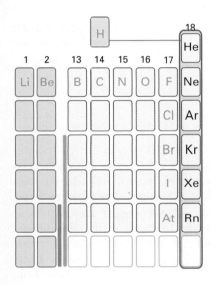

The last group in the *p* block contains six elements that are so unreactive that they form only a very limited range of compounds. The existence of the Group 18 elements was not suspected until late in the nineteenth century and their discovery led to the redrawing of the Periodic Table and played a key role in the development of bonding theories.

The Group 18 elements, helium, neon, argon, krypton, xenon, and radon, are all monatomic gases. They are the least reactive elements and have been given and then lost various collective names over the years as different aspects of their properties have been identified and then disproved. Thus they have been called the *rare gases* and the *inert gases*, and are currently called the **noble gases**. The first name is inappropriate because argon is by no means rare (it is substantially more abundant than CO_2 in the atmosphere). The second has been inappropriate since the discovery of compounds of xenon. The name 'noble gases' is now accepted because it gives the sense of low but significant reactivity.

The elements

Key point: Of the noble gases, only xenon forms a significant range of compounds with fluorine and oxygen.

The atomic properties of the noble gases are listed in Table 17.1. The features to note include their high ionization energies and negative electron affinities. These two features can be understood by considering their electron configuration, ns^2np^6. The electron affinities are negative because an incoming electron occupies an orbital of a new shell. The first ionization energy is high because the effective nuclear charge is high at the far right of the period.

The noble gas with the most extensive chemical properties is xenon. The most important oxidation numbers of xenon in its compounds are +2, +4, and +6 and compounds with Xe—F, Xe—O, Xe—N, Xe—C, and Xe—metal bonds are known.

Table 17.1 Selected properties of the elements

	He	Ne	Ar	Kr	Xe	Rn
Covalent radius/pm	99	160	192	197	217	
Melting point/°C	−270	−249	−189	−157	−112	−71
Boiling point/°C	−269	−246	−186	−152	−108	−62
Electron affinity/(kJ mol^{-1})	−48.2	−115.8	−96.5	−96.5	−77.2	
First ionization energy/(kJ mol^{-1})	2370	2080	1520	1350	1170	1040

Xenon can behave as a ligand. The chemical properties of xenon's lighter neighbour, krypton, are much more limited. The study of radon chemistry, like that of astatine, is inhibited by the high radioactivity of the element.

Argon, krypton, and xenon form clathrates when frozen with water at high pressure (see Box 13.2). The noble gas atoms are guests within the three-dimensional ice structure and have the composition $E \cdot 6H_2O$.

17.1 Occurrence and recovery

Key points: The noble gases are monatomic; radon is radioactive and highly mobile because it is a gas.

Because they are so unreactive and are rarely concentrated in nature, the noble gases eluded recognition until the end of the nineteenth century. Indeed, Mendeleev did not make a place for them in his periodic table because the chemical regularities of the other elements, upon which his table was based, did not suggest their existence. However, in 1868 a new spectral line observed in the spectrum of the Sun did not correspond to a known element. This line was eventually attributed to a new element, helium, and in due course the element itself and its congeners were found on Earth. These new elements were given names that reflect their curious nature; helium from the Greek *helios* for sun, neon from the Greek *neos* for new, argon from *argos* meaning inactive, krypton from *kryptos* meaning hidden, and xenon from *xenos* meaning strange. Radon was named after radium as it is a by-product of the radioactive decay of that element.

Helium makes up 23 per cent by mass of the Universe and the Sun and is the second most abundant element after hydrogen. Helium atoms are too light to be retained by the Earth's gravitational field, so most of the helium on Earth (5 parts per million by volume) is the product of α-emission in the decay of radioactive elements. High helium concentrations of up to 7 per cent by mass are found in certain natural gas deposits (mainly in the USA and eastern Europe) from which it can be recovered by low-temperature distillation. Some helium arrives from the Sun as a solar wind of α particles. All the other noble gases occur in the atmosphere. The crustal abundances of argon (0.94 per cent by volume) and neon (1.5×10^{-3} per cent) make these two elements more plentiful than many familiar elements, such as arsenic and bismuth, in the Earth's crust (Fig. 17.1). Xenon and radon are the rarest elements of the group. Radon is a product of radioactive decay and, as it lies beyond lead, is itself unstable and accounts for around 50 per cent of background radiation. Neon, argon, krypton, and xenon are extracted from liquid air by low-temperature distillation.

The elements are all monatomic gases at room temperature. In the liquid phase they form low concentrations of dimers that are held together by dispersion forces. The low boiling points of the lighter noble gases (Table 17.1) follow from the weakness of these forces between the atoms and the absence of other forces. When helium is cooled below 2.178 K it undergoes a transformation into a second liquid phase known as **helium-II**. This phase is classed as a superfluid, as it flows without viscosity. Solid helium is formed only under pressure.

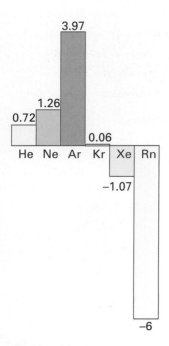

Fig. 17.1 Abundances of the noble gases in the Earth's crust, $\log_{10}$(atmospheric parts per million by volume).

17.2 Uses

Key points: Helium is used as an inert gas and as a light source in lasers and electric discharge lamps; liquid helium is a very low temperature refrigerant.

On account of its low density and nonflammability, helium is used in balloons and lighter-than-air craft. Its very low boiling point leads to its wide use in cryogenics and as a very low temperature refrigerant; it is the coolant for superconducting magnets used for NMR spectroscopy and magnetic resonance imaging. It is also used as an inert atmosphere for growing crystals of semiconducting materials, such as silicon. It is mixed with O_2 in a 4:1 ratio to provide an artificial atmosphere for divers.

The most widespread use of argon is to provide an inert atmosphere for the production of air-sensitive compounds and as an inert gas blanket to suppress the oxidation of metals when they are welded. Argon is also used as a cryogenic refrigerant.

Radon, which is a product of nuclear power plants and of the radioactive decay of naturally occurring uranium and thorium, is a health hazard because of the ionizing nuclear radiation it produces. It is usually a minor contributor to the background radiation arising from cosmic rays and terrestrial sources. However, in regions where the soil, underlying rocks, or building materials contain significant concentrations of uranium, excessive amounts of the gas have been found in buildings.

The noble gases are extensively used in various light sources, including conventional sources (neon signs, fluorescent lamps, xenon flash lamps) and lasers (helium–neon, argon-ion, and krypton-ion lasers). In each case, an electric discharge through the gas ionizes some of the atoms and promotes both ions and neutral atoms into excited states that then emit electromagnetic radiation upon return to a lower state.

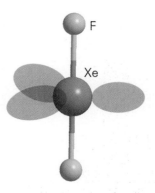

1 XeF_2

Compounds of the elements

The reactivity of the noble gases has been investigated sporadically ever since their discovery, but all early attempts to coerce them into compound formation were unsuccessful. Until the 1960s the only known compounds were the unstable diatomic species such as He_2^+ and Ar_2^+, which were detected only spectroscopically. However, in March 1962, Neil Bartlett, then at the University of British Columbia, observed the reaction of a noble gas. Bartlett's report, and another from Rudolf Hoppe's group in the University of Munster a few weeks later, set off a flurry of activity throughout the world. Within a year, a series of xenon fluorides and oxo compounds had been synthesized and characterized. The field does appear to be somewhat limited, but progress has been made in recent years and compounds with bonds to nitrogen, carbon, and metals have been prepared.

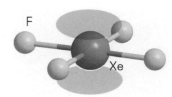

2 XeF_4

17.3 Synthesis and structure of xenon fluorides

Key point: Xenon reacts with fluorine to form XeF_2, XeF_4, and XeF_6.

Bartlett's motivation for studying xenon was based on the observations that PtF_6 can oxidize O_2 to form a solid, and that the ionization energy of xenon is similar to that of molecular oxygen. Indeed, reaction of xenon with PtF_6 did give a solid, but the reaction is complex and the complete formulation of the product (or products) is not clear. The direct reaction of xenon and fluorine leads to a series of compounds with oxidation numbers +2 (XeF_2), +4 (XeF_4), and +6 (XeF_6).

The molecules XeF_2, XeF_4, and XeF_6 are linear (**1**), square-planar (**2**), and distorted octahedral (**3**), respectively. Solid XeF_6 is more complex, and contains fluoride ion-bridged XeF_5^+ cations. In solution, it forms Xe_4F_{24} tetramers. The first two structures are consistent with the VSEPR model for a Xe atom with five and six electron pairs, respectively; they bear a molecular and electronic structural resemblance to the isoelectronic polyhalide anions I_3^- and ClF_4^- (Section 16.7). The structures of XeF_2 and XeF_4 are well established over a wide timescale by diffraction and spectroscopic methods. Similar measurements on XeF_6 in the gas phase, however, led to the conclusion that this molecule

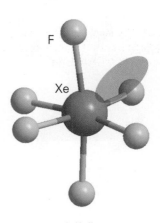

3 XeF_6

is fluxional. Infrared spectra and electron diffraction on XeF_6 show that a distortion occurs about a threefold axis, suggesting that a triangular face of F atoms opens up to accommodate a lone pair of electrons, as in (**3**). One interpretation is that the fluxional process arises from the migration of the lone pair from one triangular face to another.

The xenon fluorides are synthesized by direct reaction of the elements, usually in a nickel reaction vessel that has been passivated by exposure to F_2 to form a thin protective NiF_2 coating. This treatment also removes surface oxide, which would react with the xenon fluorides. The synthetic conditions indicated in the following equations show that formation of the higher halides is favoured by a higher proportion of fluorine and higher total pressure:

$$Xe(g) + F_2(g) \xrightarrow{400°C, 1\ atm} XeF_2(g) \quad (Xe\ in\ excess)$$

$$Xe(g) + 2\,F_2(g) \xrightarrow{600°C, 6\ atm} XeF_4(g) \quad (Xe:F_2 = 1:5)$$

$$Xe(g) + 3\,F_2(g) \xrightarrow{300°C, 60\ atm} XeF_6(g) \quad (Xe:F_2 = 1:20)$$

A simple 'window-sill' synthesis is also possible. Xenon and fluorine are sealed in a glass bulb (previously rigorously dried to prevent the formation of HF and the attendant etching of the glass) and the bulb is exposed to sunlight, whereupon beautiful crystals of XeF_2 slowly form in the bulb. It will be recalled that fluorine undergoes photodissociation (Chapter 16), and in this synthesis the photochemically generated F atoms react with Xe atoms.

17.4 Reactions of xenon fluorides

Key points: Xenon fluorides are strong oxidizing agents and form complexes with F^-, such as XeF_5^-, XeF_7^-, and XeF_8^{2-}; they are used in the preparation of compounds containing Xe—O and Xe—N bonds.

The reactions of the xenon fluorides are similar to those of the high oxidation state interhalogens, and redox and metathesis reactions dominate. One important reaction of XeF_6 is metathesis with oxides:

$$XeF_6(s) + 3\,H_2O(l) \rightarrow XeO_3(aq) + 6\,HF(g)$$

$$2\,XeF_6(s) + 3\,SiO_2(s) \rightarrow 2\,XeO_3(s) + 3\,SiF_4(g)$$

Another important chemical property of the xenon halides is their strong oxidizing power:

$$2\,XeF_2(s) + 2\,H_2O(l) \rightarrow 2\,Xe(g) + 4\,HF(g) + O_2(g)$$

$$XeF_4(s) + Pt(s) \rightarrow Xe(g) + PtF_4(s)$$

As with the interhalogens, the xenon fluorides react with strong Lewis acids to form xenon fluoride cations:

$$XeF_2(s) + SbF_5(l) \rightarrow [XeF]^+[SbF_6]^-(s)$$

These cations are associated with the counterion by F^- bridges.

Another similarity with the interhalogens is the reaction of XeF_4 with the Lewis base F^- in acetonitrile (cyanomethane, CH_3CN) solution to produce the XeF_5^- ion:

$$XeF_4 + [N(CH_3)_4]F \rightarrow [N(CH_3)_4][XeF_5]$$

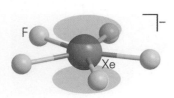

4 XeF_5^-

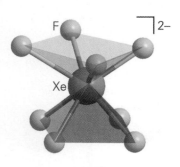

5 XeF_8^{2-}

The XeF_5^- ion is planar-pentagonal (**4**), and in the VSEPR model the two electron pairs on Xe occupy axial positions. Similarly, it has been known for many years that reaction of XeF_6 with an F^- source produces the XeF_7^- or XeF_8^{2-} ions depending on the proportion of fluoride. Only the shape of XeF_8^{2-} is known: it is a square antiprism (**5**), which is difficult to reconcile with the simple VSEPR model because this shape does not provide a site for the lone pair on Xe.

The xenon fluorides are the gateway to the preparation of compounds of the noble gases with elements other than fluorine and oxygen. The reaction of nucleophiles with a xenon fluoride is one useful strategy for the synthesis of such bonds. For instance, the reaction

$$XeF_2 + HN(SO_2F)_2 \rightarrow FXeN(SO_2F)_2 + HF$$

is driven forward by the stability of the product HF and the energy of formation of the Xe—N bond (**6**). A strong Lewis acid such as AsF_5 can extract F^- from the product of this reaction to yield the cation $[XeN(SO_2F)_2]^+$. Another route to XeN bonds is the reaction of one of the fluorides with a strong Lewis acid:

$$XeF_2 + AsF_5 \rightarrow [XeF][AsF_6]$$

followed by the introduction of a Lewis base, such as CH_3CN, to yield $[CH_3CNXeF]$ $[AsF_6]$.

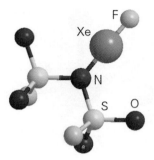

6 $FXeN(SO_2F)_2$

17.5 Xenon–oxygen compounds

Key point: The xenon oxides are unstable and highly explosive.

Xenon oxides are endoergic ($\Delta_f G^{\ominus} > 0$) and cannot be prepared by direct interaction of the elements. The oxides and oxofluorides are prepared by the hydrolysis of xenon fluorides:

$$XeF_6(s) + 3\ H_2O(l) \rightarrow XeO_3(s) + 6\ HF(aq)$$
$$3\ XeF_4(s) + 6\ H_2O(l) \rightarrow XeO_3(s) + 2\ Xe(g) + \tfrac{3}{2}\ O_2(g) + 12\ HF(aq)$$
$$XeF_6(s) + H_2O(l) \rightarrow XeOF_4(s) + 2\ HF(aq)$$

The pyramidal xenon trioxide, XeO_3 (**7**) presents a serious hazard because this endoergic compound is highly explosive. It is a very strong oxidizing agent in acidic solution, with $E^{\ominus}(XeO_3, Xe) = +2.10$ V. In basic aqueous solution the Xe(VI) oxoanion $HXeO_4^-$ slowly decomposes in a coupled disproportionation and water oxidation to yield a Xe(VIII) perxenate ion, XeO_6^{4-}, and xenon:

$$2\ HXeO_4^-(aq) + 2\ OH^-(aq) \rightarrow XeO_6^{4-}(aq) + Xe(g) + O_2(g) + 2\ H_2O(l)$$

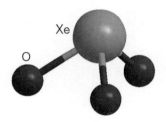

7 XeO_3

The perxenates of several alkali metal ions have been prepared by treatment of XeO_3 with ozone in basic conditions. These compounds are white crystalline solids with octahedral XeO_6^{4-} units (**8**). They are powerful oxidizing agents in acidic aqueous solution:

$$2\ H_2XeO_6 + 2\ H^+(aq) \rightarrow 2\ HXeO_4^- + O_2(g) + 2\ H_2O(l)$$

Treating Ba_2XeO_6 with concentrated sulfuric acid produces the only other known oxide of xenon, XeO_4 (**9**), which is an explosively unstable gas.

The oxofluoride $XeOF_2$ has a T-shape (**10**) and XeO_3F_2 is a trigonal bipyramid (**11**). The square-pyramidal molecule $XeOF_4$ (**12**) is remarkably similar to IF_5 in its physical and chemical properties. When alkali metal fluorides are dissolved in it, solvated fluoride

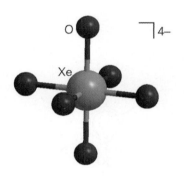

8 XeO_6^{4-}

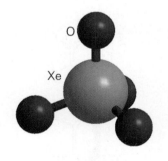

9 XeO_4

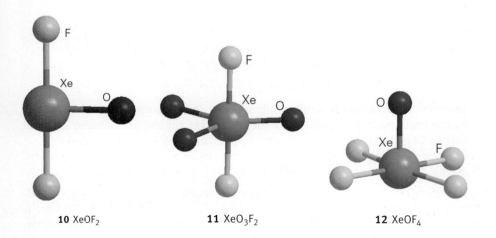

10 $XeOF_2$

11 XeO_3F_2

12 $XeOF_4$

13 $XeOF_5^-$

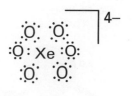

14 XeO_6^{4-}

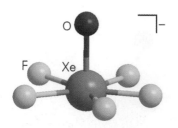

15 $C_6F_6Xe^+$

16 $Xe(C_6F_5)_2$

ions are formed, $F^- \cdot 3XeOF_4$. Attempted removal of $XeOF_4$ from the solvate yields $XeOF_5^-$ (**13**), which is a pentagonal pyramid.

Example 17.1 Describing the synthesis and a structure of a noble gas compound

(a) Describe a procedure for the synthesis of potassium perxenate, starting with xenon and other reagents of your choice. (b) Using VSEPR theory, determine the probable structure of the perxenate ion.

Answer (a) Xe—O compounds are endoergic, so they cannot be prepared by direct reaction of xenon and oxygen. The hydrolysis of XeF_6 yields XeO_3 and, as described in the text, the latter in basic solution undergoes disproportionation yielding perxenate, XeO_6^{4-}. Thus the reaction of xenon and excess F_2 would be carried out at 300°C and 6 MPa in a stout nickel container. The resulting XeF_6 might then be converted to the perxenate in one step by exposing it to aqueous KOH solution. The resulting potassium perxenate (which turns out to be a hydrate) could then be crystallized. (b) A Lewis structure of the perxenate ion is shown in (**14**). With six electron pairs around the Xe atom, the VSEPR model predicts an octahedral arrangement of bonding electron pairs and an octahedral overall structure.

Self-test 17.1 Write a balanced equation for the decomposition of xenate ions in basic solution for the production of perxenate ions, xenon, and oxygen.

17.6 Organoxenon compounds

Key point: Organoxenon compounds can be prepared by xenodeborylation of an organoboron compound.

The first compound containing Xe—C bonds was reported in 1989. Since then a wide variety of organoxenon compounds have been prepared. The most useful routes to organoxenon compounds are via the fluorides XeF_2 and XeF_4.

Organoxenon(II) salts can be prepared from organoboranes by **xenodeborylation**, the substitution of boron by xenon. For example, tris(pentafluorophenyl)borane reacts with XeF_2 in dichloromethane to produce arylxenon(II) fluorobrates (**15**):

$$(C_6F_5)_3B + XeF_2 \xrightarrow{CH_2Cl_2} [C_6F_5Xe]^+ + [(C_6F_5)_nBF_{4-n}]^- \quad n = 1, 2$$

When this reaction is carried out in anhydrous HF, all the C_6F_5 groups are transferred to the Xe:

$$(C_6F_5)_3B + 3\ XeF_2 \xrightarrow{HF} 3\ [C_6F_5Xe]^+ + [BF_4]^- + 2\ [F(HF)_n]^-$$

A general route that can be used to introduce other organic groups uses organodifluoroboranes, RBF_2:

$$RBF_2 + XeF_2 \xrightarrow{CH_2Cl_2} [RXe]^+ + [BF_4]^-$$

In addition to xenodeborylation, organoxenon(II) compounds (**16**) can be prepared from $C_6F_5SiMe_3$:

$$3\ C_6F_5SiMe_3 + 2\ XeF_2 \xrightarrow{CH_2Cl_2} Xe(C_6F_5)_2 + C_6F_5XeF + 3\ Me_3SiF$$

Organoxenon(II) compounds are thermally unstable and decompose above −40°C.

The first organoxenon(IV) compound was prepared by the reaction between XeF_4 and $C_6F_5BF_2$ in CH_2Cl_2:

$$C_6F_5BF_2 + XeF_4 \xrightarrow{CH_2Cl_2} [C_6F_5XeF_2][BF_4]$$

Organoxenon(IV) compounds are less thermally stable than the analogous Xe(II) compounds and decompose above −20°C. In all Xe(II) and Xe(IV) salts the Xe atom is bonded to a C atom that is part of a π system. Extended π systems, such as in aryl groups, increase the stability of the Xe—C bond. This stability is further favoured by the presence of electron-withdrawing substituents (such as fluorine) on the aryl group.

17.7 Coordination compounds

Key points: Argon, xenon, and krypton form coordination compounds that are usually studied by matrix isolation; the stability of the complexes deceases in the order Xe > Kr > Ar.

Coordination compounds of noble gases have been known since the mid-1970s. The first stable noble-gas coordination compound to be synthesized was $[AuXe_4]^{2+}[SbF_{11}]^{2-}$, which consists of a square planar $[AuXe_4]^{2+}$ cation (**17**). The compound is made by reduction of AuF_3 by HF/SbF_5 in elemental xenon to yield dark red crystals that are stable up to $-78°C$. Alternatively, addition of Xe to a solution of Au^{2+} in HF/SbF_5 gives a dark red solution that is stable up to $-40°C$. This solution is stable at room temperature under a xenon pressure of 1 MPa (about 10 atm). During the reduction of Au^{3+} to Au^{2+} the extreme Brønsted acidity of HF/SbF_5 (Section 14.7b) is essential and the overall reaction indicates the role of protons:

$$AuF_3 + 6\ Xe + 3\ H^+ \xrightarrow{HF/SbF_5} [AuXe_4]^{2+} + Xe_2^+ + 3\ HF$$

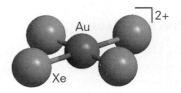

17 $[AuXe_4]^{2+}$

Green crystals of $[Xe_2]^+[Sb_4F_{21}]^-$ are also detected at $-60°C$. The Xe—Xe bond length is 309 pm and is the longest homonuclear bond known for a main group element.

Many complexes of the noble gases are transient species that have been characterized by matrix isolation. The complex $Fe(CO)_4Xe$ is formed when $Fe(CO)_5$ is photolysed in solid xenon at 12 K. Similarly, $M(CO)_5E$ is formed when $M(CO)_6$ (M = Cr, Mo, or W) is photolysed in solid argon, krypton, or xenon at 20 K, where E = Ar, Kr, or Xe, respectively. An alternative method of synthesizing these complexes is to generate them in an argon, krypton, or xenon atmosphere. The Xe complex has also been isolated in liquid xenon. The stabilities of the complexes decrease in the order W > Mo ≈ Cr and Xe > Kr > Ar. The complexes are octahedral (**18**) and the bonding is thought to involve interactions between the p orbitals on the noble gas and orbitals on the equatorial CO groups. Thus, noble gases can be thought of as potential ligands, and indeed have been fully characterized as such (including by NMR; Section 21.17).

When $[Rh(\eta^5\text{-}Cp)(CO)_2]$ or $[Rh(\eta^5\text{-}Cp^*)(CO)_2]^+$ is photolysed in supercritical xenon or krypton at room temperature, the complexes $[Rh(\eta^5\text{-}Cp)(CO)E]$ and $[Rh(\eta^5\text{-}Cp^*)(CO)E]$ are formed, where E = Xe or Kr. The $\eta^5\text{-}Cp^*$ complexes are less stable than the $\eta^5\text{-}Cp$ analogues and the Kr complexes are less stable than those of Xe.

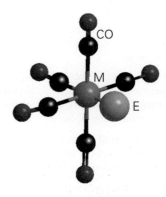

18 $[M(CO)_5E]$, E=Ar, Kr, Xe

17.8 Other compounds of noble gases

Key point: Krypton and radon fluorides are known but their chemical properties are much less extensive than those of xenon.

Radon has a lower ionization energy than xenon, so it can be expected it to form compounds even more readily. Evidence exists for the formation of RnF_2, and cationic compounds, such as $[RnF^+][SbF_6^-]$, but detailed characterization is frustrated by their radioactivity. Krypton has a much higher ionization energy than xenon (Table 17.1) and its ability to form compounds is much more limited. Krypton difluoride, KrF_2, is prepared by passing an electric discharge or ionizing radiation through a fluorine–krypton mixture at low temperatures ($-196°C$). As with XeF_2, the krypton compound is a colourless volatile solid and the molecule is linear. It is an endoergic and highly reactive compound that must be stored at low temperatures.

When monomeric HF is photolysed in solid argon and annealed to 18 K, HArF is formed. This compound is stable up to 27 K and contains the HAr^+ and F^- ions.

Endohedral fullerene complexes, in which the 'guest' atom or ion sits inside a fullerene cage (Section 13.4), have been observed for C_{60}^{n+} and C_{70}^{n+} ($n = 1$, 2, or 3) with He and C_{60}^+ with Ne. Molecular orbital calculations indicate that Ar should be able to penetrate the C_{60}^+ cage, although this complex has not yet been observed.

Other than the fullerene complexes, those identified transiently in high-energy molecular beams, or van der Waals complexes in the gas phase, there are no known compounds of He. However, theoretical calculations predict HeBeO to be exogeric.

FURTHER READING

A.G. Massey, *Main group chemistry*. Wiley (2000).

D.M.P. Mingos, *Essential trends in inorganic chemistry*. Oxford University Press (1998).

R. B. King (ed.), *Encyclopedia of inorganic chemistry*. Wiley (2005).

M. Ozima and F.A. Podosec, *Noble gas geochemistry*. Cambridge University Press (2002)

P. Lazlo and G.J. Schrobilgen, *Angew. Chem., Int. Ed. Engl.*, 1988, **27**, 479. An enjoyable account of the early failure and final success in the quest for compounds of the noble gases.

J. Holloway, Twenty-five years of noble gas chemistry. *Chemistry in Britain*, 1987, 658. A good summary of the development of the field.

H. Frohn and V.V. Bardin, *Organometallics*, 2001, **20**, 4750. A readable review of the organo compounds of the noble gases.

EXERCISES

17.1 Explain why helium is present in low concentration in the atmosphere even though it is the second most abundant element in the universe.

17.2 Which of the noble gases would you choose as (a) the lowest temperature liquid refrigerant, (b) an electric discharge light source requiring a safe gas with the lowest ionization energy, (c) the least expensive inert atmosphere?

17.3 By means of balanced chemical equations and a statement of conditions, describe a suitable synthesis of (a) xenon difluoride, (b) xenon hexafluoride, (c) xenon trioxide.

17.4 Draw the Lewis structures of (a) $XeOF_4$, (b) XeO_2F_2, and (c) XeO_6^{2-}.

17.5 Give the formula and describe the structure of a noble gas species that is isostructural with (a) ICl_4^-, (b) IBr_2^-, (c) BrO_3^-, (d) ClF.

17.6 (a) Give a Lewis structure for XeF_7^-. (b) Speculate on its possible structures by using the VSEPR model and analogy with other xenon fluoride anions.

17.7 Use molecular orbital theory to calculate the bond order of the diatomic species E_2^+ with E = He and Ne.

17.8 Identify the xenon compounds A, B, C, D, and E.

$$D \xleftarrow{H_2O} C \xleftarrow{xs\ F_2} Xe \xrightarrow{F_2} A \xrightarrow{MeBF_2} B$$
$$\downarrow 2F_2$$
$$E$$

PROBLEMS

17.1 The first compound containing an Xe—N bond was reported by R.D. LeBlond and K.K. DesMarteau (*J. Chem. Soc., Chem. Commun.*, 1974, 554). Summarize the method of synthesis and characterization. (The proposed structure was later confirmed in an X-ray crystal structure determination.)

17.2 (a) Use the references in the paper by O.S. Jina, X.Z. Sun, and M.W. George (*J. Chem. Soc., Dalton Trans.*, 2003, 1773) to produce a review of the use of matrix isolation in the characterization of organometallic noble gas complexes. (b) How did the methods used by these authors differ from the usual techniques of matrix isolation? (c) Place the complexes [MnCp(CO)$_2$Xe], [RhCp(CO)Xe], [MnCp(CO)$_2$Kr], [Mn(CO)$_5$Kr], and [Kr(CO)$_5$Kr] in order of increasing stability towards CO substitution.

17.3 In their paper 'Xenon as a complex ligand: the tetra xenon gold(II) cation in AuXe$_4^{2+}$(Sb$_2$F$_{11}^-$)$_2$,', S. Seidel and K. Seppelt describe the first

synthesis of a stable noble gas coordination compound. Give details of the synthesis and characterization of the compound.

17.4 The synthesis and characterization of the $XeOF_5^-$ anion has been described by A. Ellern and K. Seppelt (*Angew. Chem., Int. Ed. Engl*, 1995, **34**, 1586). (a) Summarize the similarities between $XeOF_4$ and IF_5. (b) Give possible reasons for the differences in structure between $XeOF_5^-$ and IF_6^-. (c) Summarize how $XeOF_5^-$ was prepared.

17.5 In their paper 'Helium chemistry: theoretical predictions and experimental challenge', (*J. Am. Chem. Soc.*, 1987, **109**, 5917) W. Koch and co-workers used quantum mechanical calculations to demonstrate that helium can form strong bonds with carbon in cations. (a) Give the range of bond lengths calculated for these He—C cations. (b) What are the requirements for an element to form a strong bond with He? (c) To which branch of science do the authors suggest that this work would be particularly relevant?

The *d*-block metals

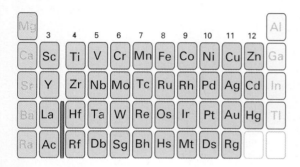

The metallic elements are the most numerous of the elements, and the thirty-nine *d*-block elements are the most important: their chemical properties are central to both industry and contemporary research. We shall see systematic trends across the block and trends in their chemical properties within each group. One of the most important trends is the variation in stability of the oxidation states, because these stabilities are closely related to the ease of recovery of metals from their ores and to the ways in which the various metals and their compounds are handled in the laboratory. It will be seen that simple binary compounds, such as metal halides and oxides, often follow systematic trends; however, when the metal is in a low oxidation state, interesting variations such as metal–metal bonding may be encountered.

The two terms **d-block metal** and **transition metal** are often used interchangeably; however, they do not mean the same thing. The IUPAC definition of a **transition element** is that it is an element that has an incomplete *d* subshell in either the neutral atom or its ions. Thus the Group 12 elements (Zn, Cd, Hg) are members of the *d* block but are not transition elements. In the following discussion, it will be convenient to refer to each row of the *d* block as a series, with the 3*d* series the first row of the block (Period 4), the 4*d* series the second row (Period 5), and so on. It will prove important to note the intrusion of the *f* block, the **inner transition elements**, the lanthanoids, into the 5*d* series. Elements towards the left of the *d* block are often referred to as *early* and those towards the right are referred to as *late*.

The elements

The chemical properties of the *d*-block elements (or *d* metals as we shall commonly call them) will occupy this and the next three chapters. It is therefore appropriate to begin with a survey of their occurrence and properties. The trends in their properties and the correlation with their electronic structures and therefore location in the periodic table should be kept in mind throughout these chapters.

18.1 Occurrence and recovery

Key point: Chemically soft members of the block occur as sulfide minerals and are partially oxidized to obtain the metal; the more electropositive 'hard' metals occur as oxides and are extracted by reduction.

The elements on the left of the 3*d* series occur in nature primarily as metal oxides or as metal cations in combination with oxoanions (Table 18.1). Of these elements, titanium

Table 18.1 Mineral sources and methods of recovery of some commercially important *d* metals

Metal	Principal minerals	Method of recovery	Note
Titanium	Ilmenite, $FeTiO_3$	$TiO_2 + 2\,C + 2\,Cl_2 \rightarrow TiCl_4 + 2\,CO$	
	Rutile, TiO_2	followed by reduction of $TiCl_4$ with Na or Mg	
Chromium	Chromite, $FeCr_2O_4$	$FeCr_2O_4 + 4\,C \rightarrow Fe + 2\,Cr + 4\,CO$	(a)
Molybdenum	Molybdenite, MoS_2	$2\,MoS_2 + 7\,O_2 \rightarrow 2\,MoO_3 + 4\,SO_2$ followed by either $MoO_3 + 2\,Fe \rightarrow Mo + Fe_2O_3$ or $MoO_3 + 3\,H_2 \rightarrow Mo + 3\,H_2O$	
Tungsten	Scheelite, $CaWO_4$	$CaWO_4 + 2\,HCl \rightarrow WO_3 + CaCl_2 + H_2O$	
	Wolframite, $FeMn(WO_4)_2$	followed by $WO_3 + 3\,H_2 \rightarrow W + 3\,H_2O$	
Manganese	Pyrolusite, MnO_2	$MnO_2 + 2\,C \rightarrow Mn + 2\,CO$	(b)
Iron	Haematite, Fe_2O_3	$Fe_2O_3 + 3\,CO \rightarrow 2\,Fe + 3\,CO_2$	
	Magnetite, Fe_3O_4		
	Limonite, $FeO(OH)$		
Cobalt	CoAsS	By-product of copper and nickel production	
	Smaltite, $CoAs_2$		
	Linnaeite, Co_3S_4		
Nickel	Pentlandite, $(Fe,Ni)_6S_8$	$NiS + O_2 \rightarrow Ni + SO_2$	(c)
Copper	Chalcopyrite, $CuFeS_2$	$2\,CuFeS_2 + 2\,SiO_2 + 5\,O_2 \rightarrow 2\,Cu + 2\,FeSiO_3 + 4\,SO_2$	
	Chalocite, Cu_2S		

(a) The iron–chromium alloy is used directly for stainless steel.
(b) The reaction is carried out in a blast furnace with Fe_2O_3 to produce alloys.
(c) NiS is formed by melting the mineral and separated by physical processes. NiO is used in a blast furnace with iron oxides to produce steel. Nickel is purified by electrolysis or the Mond process via $Ni(CO)_4$.

ores are the most difficult to reduce, and the element is widely produced by heating TiO_2 with chlorine and carbon to produce $TiCl_4$, which is then reduced by molten magnesium at about 1000°C in an inert-gas atmosphere. The oxides of chromium, manganese, and iron are reduced with carbon (Section 5.15), a much cheaper reagent. To the right of iron in the 3*d* series, cobalt, nickel, copper, and zinc occur mainly as sulfides and arsenides, which is consistent with the increasingly soft Lewis acid character of their dipositive ions. Sulfide ores are usually roasted in air either to the metal directly (for example, nickel) or to an oxide that is subsequently reduced (for example, zinc). Copper is used in large quantities for electrical conductors; electrolysis is used to refine crude copper to achieve the high purity needed for high electrical conductivity.

The difficulty of reducing the early 4*d* and 5*d* metals molybdenum and tungsten is apparent from Table 18.1. It reflects the tendency of these elements to have stable high oxidation states, as discussed later. The platinum metals (Ru and Os, Rh and Ir, and Pd and Pt), which are found at the lower right of the *d* block, occur as sulfide and arsenide ores, usually in association with larger quantities of copper, nickel, and cobalt. They are collected from the sludge that forms during the electrolytic refinement of copper and nickel. Gold (and to some extent silver) is found in its elemental form.

The role that certain *d* metals play in the environment is considered in Box 18.1.

18.2 Physical properties

Key points: The properties of the *d* metals are largely derived from their electronic structure, with the strength of metallic bonding peaking at Group 6; the lanthanide contraction is responsible for some of the anomalous behaviour of the metals in the 5*d* series.

Box 18.1 Toxic metals in the environment

The biosphere evolved in close association with all the elements in the periodic table, and has had a long time to adapt to them. Metals are released from rocks by weathering, and are processed by a variety of mechanisms, including biological ones. Indeed, many metals have been harnessed for essential biochemical functions (Chapter 26), and other biochemical systems have evolved to sequester metals and keep them out of harm's way. However, the natural biogeochemical cycles have been greatly perturbed by mining: circulating levels of most metals have increased dramatically since pre-industrial times.

Many metals, including beryllium, manganese, chromium, nickel, cadmium, mercury, lead, selenium, and arsenic, are hazardous in occupational or environmental settings. Exposure to these elements varies widely, and depends on the pattern of their industrial use, and on their environmental chemistry. The metals of greatest environmental concern are heavy metals, such as mercury, which, being chemically very soft, bind tightly to thiol groups in proteins.

The effect of a substance on a living organism is often plotted as a *dose–response curve*, such as those in the illustration. The shape of the dose–response curves for various metals varies widely, depending on physiological variables, and on the chemistry of the metals. For example, iron and copper are both essential elements but have very different dose–response curves: toxicity occurs at a much lower concentration for copper than for iron. It is reasonable to associate this toxicity with the higher affinity of copper ions for sulfur and nitrogen ligands, which makes copper more likely to interfere with crucial sites in proteins. Nevertheless, iron is harmful at higher doses, because it can catalyse the production of oxygen radicals, and also because excess iron can stimulate the growth of bacteria.

The ability of elements to exist in alternative redox states can be of crucial importance to the hazard they present. For example, the solubilities of different oxidation states of a particular metal are often very different. Many metals are found as insoluble sulfides in the Earth's crust and are thus immobile. However, when they are disturbed and brought into contact with air, which oxidizes them, they can become very mobile. An example is mercury(II)

sulfide, which is oxidized to the mobile mercury(II) sulfate on exposure to air.

Another effect of oxidation is exemplified by iron, which occurs widely as the mineral pyrite, FeS_2. Mining operations often lead to acid drainage due to the reaction

$$FeS_2 + \tfrac{15}{4} O_2 + \tfrac{7}{2} H_2O \rightarrow Fe(OH)_3 + 2\,SO_4^{2-} + 4\,H^+$$

In this process (which is generally catalysed by certain bacteria), S_2^{2-} is oxidized to sulfuric acid and Fe^{2+} is oxidized to Fe^{3+}, which hydrolyses to $Fe(OH)_3$, yielding additional H^+ ions. Streams emanating from iron mines, many of them abandoned, are thus often highly acidic.

Redox status can have opposite effects on solubility for different metals. Thus, whereas manganese is mobile under reducing conditions, other metals are mobilized under oxidizing conditions. For instance, Cr(III) is immobile, forming an insoluble oxide or kinetically inert complexes with soil constituents; in addition, there is no mechanism for transporting free Cr(III) ions across biological membranes. However, Cr(III) is solubilized upon oxidation to the chromate(VI) ion, CrO_4^{2-}. Chromate(VI) is highly toxic, and is implicated in cancer. The CrO_4^{2-} ion has the same shape and charge as SO_4^{2-}, and can be ferried across biological membranes by sulfate transport proteins. Once inside a cell, chromate is converted to Cr(III) complexes by endogenous reductants, such as ascorbate. In the process, reactive oxygen species are generated, which can damage DNA; in addition, the Cr(III) reduction product can complex and cross-link DNA. Cr(III) accumulates inside the chromate-exposed cells, as it does not have a mechanism for penetrating biological membranes.

For other metals, toxicity is controlled by the interplay of membrane transport and intracellular binding with the redox state. For example, mercury salts are relatively harmless because membranes, which present a hydrocarbon barrier to ionic species, are impermeable to Hg^{2+}. Likewise, metallic mercury is not absorbed through the gut, so it is not toxic when swallowed. However, mercury vapour is highly toxic (that is why all mercury spills must be rigorously cleaned up) because the neutral atoms readily pass through the lungs' membranes, and also across the blood–brain barrier. Once in the brain, the Hg(0) is oxidized to Hg(II) by the vigorous oxidizing activity of brain cell mitochondria, and the Hg(II) binds tightly to crucial thiolate groups of neuronal proteins. Mercury is a powerful neurotoxin, but only if it gets inside nerve cells. Even more hazardous than mercury vapour are organomercurials, particularly methylmercury. Thus, CH_3Hg^+ is taken up through the gut because it is complexed by the stomach's chloride, forming CH_3HgCl, which, being electrically neutral, can pass through a membrane. Once inside cells, CH_3Hg^+ binds to thiolate groups and accumulates.

The environmental toxicity of mercury is associated almost entirely with eating fish. Methylmercury is produced by the action of sulfate-reducing bacteria on Hg^{2+} in sediments, and accumulates as little fish are eaten by bigger fish further up the aquatic food chain. Fish everywhere have some level of mercury present. Mercury levels can increase markedly if sediments are contaminated by additional mercury. The worst known case of environmental mercury

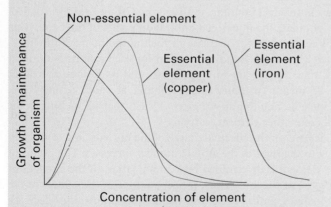

B18.1 The idealized dose–response curves for a non-essential element and two essential elements.

poisoning occurred in the 1950s in the Japanese fishing village of Minamata. A polyvinyl chloride plant, using Hg^{2+} as a catalyst, discharged mercury-laden residues into the bay, where fish accumulated methylmercury to levels approaching 100 ppm. Thousands of people were poisoned by eating the fish, and a number of infants suffered mental disabilities and motor disturbance from exposure *in utero*. This disaster led to strict standards for fish consumption. Limited consumption is advised for fish at the top of the food chain, such as pike and bass in fresh waters, and swordfish and tuna in the oceans.

Regulatory action has been aimed at reducing mercury discharges and emissions, and industrial point sources have been largely controlled. Chloralkali plants, producing the large-volume industrial chemicals Cl_2 and NaOH by electrolysis of NaCl, were a major source as a mercury pool electrode was used to transfer metallic sodium to a separate hydroxide-generating compartment. However, this can be now be accomplished by separating the two electrode compartments with a cation-exchange membrane, which prevents migration of the anions. Combustion can vent mercury to the atmosphere if the fuel contains mercury compounds. Whereas municipal waste and hospital incinerators have been equipped with filters to reduce mercury emissions to the air, coal contains small amounts of mercury minerals. Because of the huge quantities burned, coal is a major contributor to environmental mercury.

Mercury is a global problem, because mercury vapour and volatile organomercurial compounds can travel long distances in the atmosphere. Eventually elemental mercury is oxidized and organomercurials are decomposed, both to Hg^{2+}, by reaction with ozone or by atmospheric hydroxyl or halogen radicals. The Hg^{2+} ions are solvated by water molecules and deposited in rainfall. Thus mercury deposition can occur far from the emission source, and is distributed fairly uniformly around the globe. For example, it is estimated that only a third of North American mercury emissions are in fact deposited in the USA, and this accounts for only half of the mercury deposition in the USA. Even the gold fields of Brazil contribute, because miners use mercury to extract the gold, which is recovered by heating the resultant amalgam to drive off the mercury. This practice is estimated to account for 2 per cent of global mercury emissions (half of South American emissions). The picture is further complicated by the fact that much of the deposited mercury is recirculated through processes that produce volatile compounds or mercury vapour. For example, much of the biomethylation activity of the sulfate-reducing bacteria produces dimethylmercury, $(CH_3)_2Hg$, which, being volatile, is vented to the atmosphere. Other bacteria have an enzyme (methylmercury lyase) that breaks the methyl–mercury bond of CH_3Hg^+, and another enzyme (methylmercury reductase) that reduces the resulting Hg(II) to Hg(0); this is a protective mechanism for the microorganisms, ridding them of mercury as volatile Hg(0).

The *d* block of the periodic table contains the metals most important to modern society. It contains the immensely strong and light titanium, the major components of most steels (iron, chromium, manganese, molybdenum), the highly electrically conducting copper, the malleable gold and platinum, and the very dense osmium and iridium. To a large extent these properties derive from the nature of the metallic bonding that binds the atoms together.

Metallic bonding was covered in Chapter 3, which introduced the concept of band structure. Generally speaking, the same band structure is present for all the *d*-block metals and arises from the overlap of the $(n+1)s$ orbitals to give an *s* band and the *nd* orbitals to give a *d* band. The principal differences between the metals is the number of electrons available to occupy these bands: titanium ($3d^2 4s^2$) has four bonding electrons, vanadium ($3d^3 4s^2$) five, chromium ($3d^5 4s^1$) six, and so on. The lower, net bonding region of the valence band is therefore progressively filled with electrons on going to the right across the block, which results in stronger bonding, until around Group 7 (at Mn, Tc, Re) when the electrons begin to populate the upper, net antibonding part of the band. This trend in bonding strength is reflected in the increase in melting point from the low-melting alkali metals (effectively only one bonding electron for each atom, resulting in melting points typically less than 100°C) up to manganese, and its decline thereafter to the low-melting Group 12 metals (mercury being a liquid at room temperature, Fig. 18.1). The strength of metallic bonding in tungsten is such that its melting point (3410°C) is exceeded by only one other element, carbon.

The radii of *d*-metal ions depend on the effective charge of the nucleus, and ionic radii generally decrease on moving to the right as the atomic number increases. The radius of the metal atoms in the solid element is determined by a combination of the strength of the metallic bonding and the size of the ions. Thus, the separations of the centres of the atoms in the solid generally follow a similar pattern to the melting points: they decrease to the middle of the *d* block, followed by an increase back up to Group 12, with the smallest separations occurring in and near Groups 7 and 8.

The atomic radii of the elements in the 5*d* series (Hf, Ta, W, . . .) are not much bigger than those of their 4*d*-series congeners (Zr, Nb, Mo, . . .). In fact, the atomic radius of

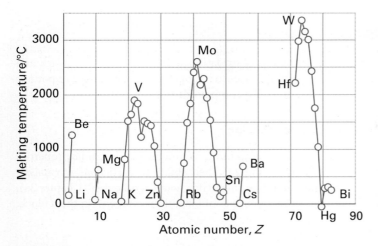

Fig. 18.1 The melting points of the metals in Groups 1–12.

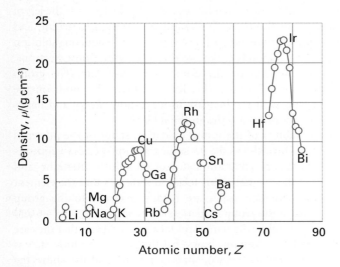

Fig. 18.2 The densities of the elements in Groups 1–12.

Hf is smaller than that of Zr even though it appears in a later period. To understand this anomaly, we need to consider the effect of the lanthanoids (the first row of the *f* block).

The atomic radii of the members of Group 3 (Sc, Y, La) increase down the group, as expected from the increase in principal quantum number of the outermost electrons. However, the 14 lanthanoid elements intervene between lanthanum and hafnium, with the corresponding additional electrons occupying the poorly shielding 4*f* orbitals. Because the atomic number has increased by 32 between zirconium and hafnium without a corresponding increase in shielding, the overall effect is that the atomic radii of the 5*d*-series elements are much smaller than expected. This reduction in radius is the lanthanide contraction introduced in Section 1.9a. The lanthanide contraction also affects the ionization energies of the 5*d*-series elements, making them higher than expected on the basis of a straightforward extrapolation. Some of the metals—specifically Au, Pt, Ir, and Os—have such high ionization energies that they are unreactive under normal conditions.

Atomic mass increases with atomic number, and the combination of this increase with the changes in the radii of the metal atoms in the metal lattice means that the mass densities of the elements reach a peak with iridium (density 22.65 g cm^{-3}). Figure 18.2 illustrates this trend.

Trends in chemical properties

Many of the *d* metals display a wide range of oxidation states, which leads to a rich and fascinating chemistry. They also form an extensive range of coordination compounds (Chapters 8, 19, and 20) and organometallic compounds (Chapter 21).

18.3 Oxidation states across a series

The range of oxidation states of the *d* metals accounts for interesting electronic properties of many solid compounds (Chapters 23 and 24), their ability to participate in catalysis (Chapter 25), and their subtle and interesting role in biochemical processes (Chapter 26). In this section we focus on trends in the stabilities of oxidation states within the *d* block.

(a) High oxidation states

Key point: Oxygen is usually more effective than fluorine at bringing out the highest oxidation states because less crowding is involved.

The **group oxidation number** of an element is its group number (in the 1–18 notation). The corresponding oxidation state can be achieved by elements that lie towards the left of the *d* block but not by elements on the right. For example, scandium, yttrium, and lanthanum in Group 3, with configurations $nd^1(n+1)s^2$, are found in aqueous solution only in oxidation state +3, which corresponds to the loss of all their outermost electrons, and the majority of their complexes contain the elements in this state. The group oxidation state is never achieved after Group 8 (Fe, Ru, and Os). This limit on the maximum oxidation state correlates with the increase in ionization energy and hence noble character from left to right across each series in the *d* block.

The trend in thermodynamic stability of the group oxidation states of the 3*d*-series elements is illustrated in Fig. 18.3, which shows the Frost diagram for species in aqueous acidic solution. We see that the group oxidation states of scandium, titanium, and vanadium fall in the lower part of the diagram. This location indicates that the element and any species in intermediate oxidation states are readily oxidized to the group oxidation state. By contrast, species in the group oxidation state for Cr and Mn (+6 and +7, respectively) lie in the upper part of the diagram. This location indicates that they are very susceptible to reduction. The Frost diagram shows that the group oxidation state is not achieved in Groups 8–12 of the 3*d* series (Fe, Co, Ni, Cu, and Zn), and also shows the oxidation states that are most stable under acid conditions; namely Ti^{3+}, V^{3+}, Cr^{3+}, Mn^{2+}, Fe^{2+}, Co^{2+}, and Ni^{2+}.

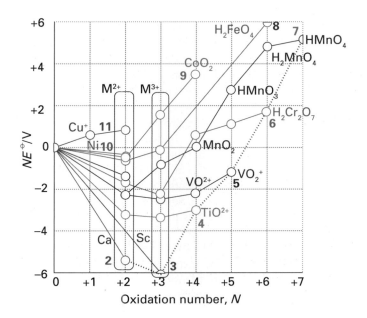

Fig. 18.3 A Frost diagram for the first series of the *d*-block elements in acidic solution (pH = 0). The bold numbers designate the group numbers and the broken line connects species in their group oxidation states.

The binary compounds of the 3d-series elements with the halogens and with oxygen also illustrate the trend in stabilities of the group oxidation states. The earliest metals can achieve their group oxidation states in compounds with chlorine (for example, $ScCl_3$ and $TiCl_4$), but the more strongly oxidizing halogen fluorine is necessary to achieve the group oxidation state of V (Group 5) and Cr (Group 6), which form VF_5 and CrF_6, respectively. Beyond Group 6 in the 3d series, even fluorine cannot produce the group oxidation state, and MnF_7 and FeF_8 have not been prepared. Oxygen brings out the group oxidation state for many elements more readily than does fluorine because fewer O atoms than F atoms are needed to achieve the same oxidation number, thus decreasing steric crowding. For example, the group oxidation state of +7 for Mn is achieved in manganate(VII) salts, such as potassium permanganate, $KMnO_4$.

As can be inferred from the Frost diagram in Fig. 18.3, chromate(VI) (CrO_4^{2-}), manganate(VII) (MnO_4^-), and ferrate(VI) (FeO_4^{2-}), are strong oxidizing agents and become stronger from CrO_4^{2-} to FeO_4^{2-}. This trend is another illustration of the decreasing stability of the maximum attainable oxidation state for Groups 6, 7, and 8. Yet another example of the greater difficulty of oxidizing an element to the right of Cr to its group oxidation state is that the air oxidation of MnO_2 in molten potassium hydroxide does not take Mn to its group oxidation state but instead yields the deep green compound potassium manganate(VI), K_2MnO_4. The disproportionation of MnO_4^{2-} in acidic aqueous solution yields manganese(IV) oxide (MnO_2) and the deep purple manganate(VII) ion (MnO_4^-):

$$3\ MnO_4^{2-}(aq) + 4\ H^+(aq) \rightarrow 2\ MnO_4^-(aq) + MnO_2(s) + 2\ H_2O(l)$$

(b) Intermediate oxidation states in the 3d series

Key point: Oxidation state +3 is common to the left of the 3d series and +2 is common for metals from the middle to the right of the block.

Most +1 d-metal cations (M^+) disproportionate (to M and M^{2+}) because the bonds in the solid metal are so strong, and for the 3d metals, the +2 oxidation state is generally the lowest one to consider in aqueous solution and in combination with hard ligands. Exceptions are mainly confined to metal–metal bonded and organometallic compounds, such as the metal carbonyls $Ni(CO)_4$ and $Mo(CO)_6$ discussed in Chapter 21, in which the metal is in oxidation state 0.

The dipositive aqua ions, $M^{2+}(aq)$ (specifically, the octahedral complexes $[M(OH_2)_6]^{2+}$), play an important role in the chemistry of the 3d-series metals. Many of these ions are coloured as a result of d–d transitions in the visible region of the spectrum (Section 19.4). For example, $Cr^{2+}(aq)$ is blue, $Fe^{2+}(aq)$ is green, $Co^{2+}(aq)$ is pink, $Ni^{2+}(aq)$ is green, and $Cu^{2+}(aq)$ is blue.

The +3 oxidation state is common at the left of the period, and is the only oxidation state normally encountered for scandium. Titanium, vanadium, and chromium all form a wide range of compounds in oxidation state +3. In line with increasing ionization energy, the stability of M(III) decreases across the period. Manganese(II) is especially stable due to its half-filled d shell, and relatively few Mn(III) compounds are known. Beyond manganese, iron(III) complexes are known, but are often oxidizing. In acid solution, $Co^{3+}(aq)$ is powerfully oxidizing and O_2 is evolved:

$$Co^{3+}(aq) + e^- \rightarrow Co^{2+}(aq) \qquad E^{\ominus} = +1.81\ V$$

Species such as Co(III) are stabilized as oxo complexes (for instance, $CoO(OH)$) when formed under basic conditions (Section 18.6) or when complexed by other good donor ligands. Aqua ions of Ni^{3+} and Cu^{3+} have not been prepared.

Conversely, M(II) becomes increasingly common from left to right across the period. For example, among the early members of the 3d series, $Sc^{2+}(aq)$ (Group 3) is unknown and $Ti^{2+}(aq)$ (Group 4) is formed only upon bombarding solutions of Ti^{3+} with electrons in the technique known as **pulse radiolysis**. For Groups 5 and 6, $V^{2+}(aq)$ and $Cr^{2+}(aq)$ are thermodynamically unstable with respect to oxidation by H^+:

$$2\ V^{2+}(aq) + 2\ H^+(aq) \rightarrow 2\ V^{3+}(aq) + H_2(g) \qquad E^{\ominus} = +0.26\ V$$

Table 18.2 Structures of the *d*-metal dihalides*

	Ti	V	Cr	Mn	Fe	Co	Ni	Cu
F	R	R†	R†	R	R	R	R†	
Cl	L	L	R†	L	L	L	L	L
Br	L	L	L	L	L	L	L	L
I	L	L	L	L	L	L	L	L

* R = rutile, L = layered (CdI₂, CdCl₂ or related structures).
† The structure is distorted from ideal.
Adapted from A.F. Wells, *Structural inorganic chemistry*. Oxford University Press (1984).

Despite this instability, the slowness of the evolution of H_2 makes it possible to work with aqueous solutions of these dipositive ions in the absence of air, and as a result they are useful reducing agents. Beyond chromium (for Mn^{2+}, Fe^{2+}, Co^{2+}, Ni^{2+}, and Cu^{2+}), M(II) is stable with respect to reaction with water, and only Fe^{2+} is oxidized by air.

Water is incompatible with many metal ions because it can serve as an oxidizing agent. A wider range of M^{2+} ions is therefore known in solids than in aqueous solution. For example, $TiCl_2$ can be prepared by the reduction of $TiCl_4$ with hexamethyldisilane:

$$(CH_3)_3Si—Si(CH_3)_3(l) + TiCl_4(l) \rightarrow TiCl_2(s) + 2\ (CH_3)_3SiCl(l)$$

but it is oxidized by both water and air. With the exception of $ScCl_2$, these dihalides are well described by structures containing isolated M^{2+} ions (Table 18.2). The fluorides have a rutile structure (Section 3.9a) and most of the heavier halides are layered; in both cases the metal is in an octahedral site. This shift in structure is understandable because the rutile structure is associated with ionic bonding and the layered structures are associated with more covalent bonding.

Example 18.1 Judging trends in redox stability in the *d* block

On the basis of trends in the properties of the 3*d*-series elements, suggest possible M^{2+} aqua ions for use as reducing agents, and write a balanced chemical equation for the reaction of one of these ions with O_2 in acidic solution.

Answer Because oxidation state +2 is most stable for the late 3*d*-series elements, strong reducing agents include ions of the metals on the left of the series: such ions include $V^{2+}(aq)$ and $Cr^{2+}(aq)$. The $Fe^{2+}(aq)$ ion is only weakly reducing. The $Co^{2+}(aq)$, $Ni^{2+}(aq)$, and $Cu^{2+}(aq)$ ions are not oxidized in water. The Latimer diagram for iron indicates that Fe^{3+} is the only accessible higher oxidation state of iron in acidic solution:

$$Fe^{3+} \xrightarrow{+0.77} Fe^{2+} \xrightarrow{-0.44} Fe$$

The chemical equation for the oxidation is then

$$4\ Fe^{2+}(aq) + O_2(g) + 4\ H^+(aq) \rightarrow 4\ Fe^{3+}(aq) + 2\ H_2O(l)$$

Self-test 18.1 Refer to the appropriate Latimer diagram in *Resource section 3* and identify the oxidation state and formula of the species that is thermodynamically favoured when an acidic aqueous solution of V^{2+} is exposed to oxygen.

(c) Intermediate oxidation states in the 4*d* and 5*d* series

Key points: Complexes of M(II) with σ-donor ligands are common for the 3*d*-series metals but complexes of M(II) of the 4*d*- and 5*d*-series metals are less common; they generally contain π-acceptor ligands.

In contrast to the 3*d*-series, the 4*d*- and 5*d*-series metals only rarely form simple $M^{2+}(aq)$ ions. A few examples have been characterized, including $[Ru(OH_2)_6]^{2+}$, $[Pd(OH_2)_4]^{2+}$, and $[Pt(OH_2)_4]^{2+}$. However, the 4d- and 5d-series metals do form many M(II) complexes with ligands other than H_2O; they include the very stable d^6 octahedral complexes, such as (1), and the much rarer square-pyramidal d^6 complexes, such as (2), which form with bulky ligands. Palladium(II) and platinum(II) form many square-planar d^8 complexes, such as $[PtCl_4]^{2-}$. Much of the rationalization of this behaviour relates to the bonding of the ligands, and cannot be fully understood without reference to the concepts introduced in Chapter 19; however, examples of complexes are included here for completeness.

The Ru(II) complex in (1) is obtained by reducing $RuCl_3 \cdot 3H_2O$ with zinc in the presence of ammonia; it is a useful starting material for the synthesis of a range of ruthenium(II) pentaammine complexes that have π-acceptor ligands, such as CO, as the sixth ligand:

$$[Ru(NH_3)_5(OH_2)]^{2+}(aq) + L(aq) \rightarrow [Ru(NH_3)_5L]^{2+}(aq) + H_2O(l)$$
$$(L = CO, N_2, N_2O)$$

These ruthenium and the related osmium pentaammine species are strong π donors, and consequently the complexes they form with the π acceptors CO and N_2 are stable.

18.4 Oxidation states down a group

Key points: In Groups 4–10, the highest oxidation state of an element becomes more stable on descending a group, with the greatest change in stability occurring between the first two rows of the d block; ease of oxidation of the metal does not correlate with the highest available oxidation state.

The stability of an oxidation state of an element changes on descending a group. Figure 18.4 is the Frost diagram for the chromium group: note that Mo(VI) and W(VI) lie below Cr(VI) in $H_2Cr_2O_7$, indicating that the maximum oxidation state is more stable for molybdenum and tungsten, with the greatest change in stability occurring between Cr and Mo. This pattern is repeated across the d block: for Groups 4–10 the highest oxidation state of an element becomes more stable on descending a group, with the greatest change in stability occurring between the first two series of the d block.

The increasing stability of high oxidation states for the heavier d metals can be seen in the formulas of their halides (Table 18.3), and the limiting formulas MnF_4, TcF_6, and ReF_7 show the greater ease of oxidizing the 4d- and 5d-series metals than the 3d-series. The hexafluorides of the heavier d metals (as in PtF_6) have been prepared from Group 6 through to Group 10 except for palladium. In keeping with the stability of high oxidation

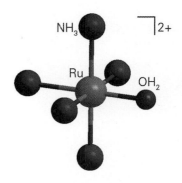

1 $[Ru(OH_2)(NH_3)_5]^{2+}$

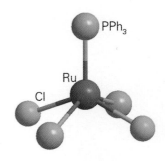

2 $[Ru(Cl)_2(PPh_3)_3]$

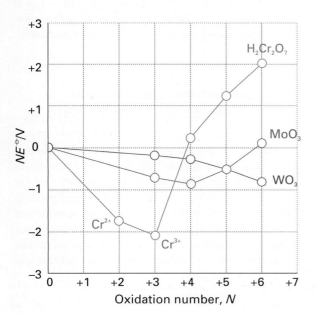

Fig. 18.4 A Frost diagram for the chromium group in the d block (Group 6) in acidic solution (pH = 0).

Table 18.3 Highest oxidation states of the *d*-block binary halides*

Group							
4	**5**	**6**	**7**	**8**	**9**	**10**	**11**
TiI_4	VF_5	CrF_5†	MnF_4	$FeBr_3$	CoF_3	NiF_4	$CuBr_2$
ZrI_4	NbI_5	MoF_6	$TcCl_6$	RuF_6	RhF_6	PdF_4	AgF_3
HfI_4	TaI_5	WF_6	ReF_7	OsF_6	IrF_6	PtF_6	$AuCl_6$

*The formulas show the least electronegative halide that brings out the highest oxidation state of the *d* metal.
† CrF_6 exists for several days at room temperature in a passivated Monel chamber.

states for the heavier metals, WF_6 is not a significant oxidizing agent. However, the oxidizing character of the hexafluorides increases to the right, and PtF_6 is so potent that it can oxidize O_2 to O_2^+:

$$O_2(g) + PtF_6(s) \rightarrow (O_2)PtF_6(s)$$

PtF_6 can even oxidize Xe, but the identity of the products is not entirely clear (Section 17.3).

The ability to achieve the highest oxidation state does not correlate with the ease of oxidation of the bulk metal to an intermediate oxidation state. For example, although elemental iron is susceptible to oxidation by $H^+(aq)$ under standard conditions,

$$Fe(s) + 2\,H^+(aq) \rightarrow Fe^{2+}(aq) + H_2(g) \qquad E^\ominus = +0.44\ V$$

no chemical oxidizing agent has been found that will take it to its group oxidation state of $+8$ in solution. By contrast, although the two heavier metals in Group 8 (ruthenium and osmium) are not oxidized by H^+ in acidic aqueous solution:

$$Os(s) + 2\,H_2O(aq) \rightarrow OsO_2(s) + H_2(g) \qquad E^\ominus = -0.65\ V$$

they can be oxidized by oxygen to the $+8$ state, as RuO_4 and OsO_4:

$$Os(s) + 2\,O_2(g) \rightarrow OsO_4(s)$$

We see from Fig. 18.4 that the Frost diagrams for the positive oxidation states of molybdenum and (especially) tungsten are quite flat. This flatness indicates that neither element exhibits the marked tendency to form M(III) that is so characteristic of chromium. Mononuclear complexes of molybdenum and tungsten with oxidation states $+3$, $+4$, $+5$, and $+6$ are common. We shall see later that M—M bonded dinuclear or polynuclear complexes of molybdenum and tungsten with oxidation states $+2$ and $+3$ are numerous.

18.5 Structural trends

Key points: The 4*d*- and 5*d*-series elements often exhibit higher coordination numbers than their 3*d*-series congeners; high oxidation state compounds tend to have covalent structures.

As may be anticipated from considerations of atomic and ionic radii, the 4*d*- and 5*d*-series elements often have higher coordination numbers than their smaller 3*d*-series congeners. Table 18.4 illustrates this trend for the fluoro and cyano complexes of the early *d*-metals. Note that with the small F^- ligand, these 3*d* elements tend to form six-coordinate complexes but that the larger 4*d*- and 5*d*-series metals in the same oxidation state tend to form seven-, eight-, and nine-coordinate complexes. The octacyano-molybdate complex, $[Mo(CN)_8]^{3-}$, illustrates the tendency towards high coordination numbers with compact ligands. The same complex is readily reduced electrochemically or chemically without change of coordination number:

$$[Mo(CN)_8]^{3-}(aq) + e^- \rightarrow [Mo(CN)_8]^{4-}(aq) \qquad E^\ominus = +0.73\ V$$

Table 18.4 Coordination numbers of some early *d*-block fluoro and cyano complexes*

	Group		
	3	**4**	**5**
3*d*	$(NH_4)_3[ScF_6]$ (6)	$Na_2[TiF_6]$ (6)	$K[VF_6]$ (6); $K_2[V(CN)_7]\cdot 2H_2O$ (7)
4*d*	$Na_6[YF_9]$ (9)	$Na_3[ZrF_7]$ (7)	$K_2[NbF_7]$ (7); $K_2[Nb(CN)_8]$ (8)
5*d*	$Na_6[LaF_9]$ (9)	$Na_3[HfF_7]$ (7)	$K_3[TaF_8]$ (8)

*The number in parentheses is the coordination number of the *d*-metal atom in the complex anion enclosed in square brackets.

Structural changes also result from changes in oxidation state. Low oxidation state compounds often exist as ionic solids, whereas high oxidation state compounds tend to take on covalent character: compare OsO_2, which is an ionic solid with the rutile structure, and OsO_4, which is a covalent molecular species (Section 18.6). We discussed the effect in Section 1.9e.

18.6 Noble character

Key point: Metals on the right of the *d* block tend to exist in low oxidation states and form compounds with soft ligands.

With the exception of Group 12, metals on the bottom right of the *d* block are resistant to oxidation. This resistance is largely due to strong intermetallic bonding and high ionization energies. It is most evident for silver, gold, and the 4*d*- and 5*d*-series metals in Groups 8–10 (Fig. 18.5). The latter are referred to as the **platinum metals** because they occur together in platinum-bearing ores. In recognition of their traditional use, copper, silver, and gold are referred to as the **coinage metals**. Gold occurs as the metal; silver, gold, and the platinum metals are also recovered in the electrolytic refining of copper. The prices of the individual platinum metals vary widely because they are recovered together but their consumption is not proportional to their abundance. Rhodium is by far the most expensive metal in this group because it is widely used in industrial catalytic processes and in automotive catalytic converters (Section 25.2). Rhodium is about twenty times more costly than the less catalytically useful metal palladium even though they occur in similar abundance.

Copper, silver, and gold are not susceptible to oxidation by H^+ under standard conditions, and this noble character accounts for their use, together with platinum, in jewellery and ornaments. *Aqua regia*, a 3:1 mixture of concentrated hydrochloric and nitric acids, is an old but effective reagent for the oxidation of gold and platinum. Its function is twofold: the NO_3^- ions provide the oxidizing power and the Cl^- ions act as complexing agents. The overall reaction is

$$Au(s) + 4\,H^+(aq) + NO_3^-(aq) + 4\,Cl^-(aq) \rightarrow [AuCl_4]^-(aq) + NO(g) + 2\,H_2O(l)$$

The active species in solution are thought to be Cl_2 and $NOCl$, which are generated in the reaction

$$3\,HCl(aq) + HNO_3(aq) \rightarrow Cl_2(aq) + NOCl(aq) + 2\,H_2O(l)$$

Oxidation state preferences are erratic in Group 11. For copper, the +1 and +2 states are most common, but for silver +1 is typical, and for gold +1 and +3 are common. The simple aqua ions $Cu^+(aq)$ and $Au^+(aq)$ undergo disproportionation in aqueous solution:

$$2\,Cu^+(aq) \rightarrow Cu(s) + Cu^{2+}(aq)$$

$$3\,Au^+(aq) \rightarrow 2\,Au(s) + Au^{3+}(aq)$$

Complexes of Cu(I), Ag(I), and Au(I) are often linear. For example, $[H_3NAgNH_3]^+$ forms in aqueous solution and linear $[XAgX]^-$ complexes have been identified by X-ray

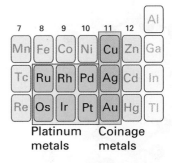

Fig. 18.5 The location of the platinum and coinage metals in the periodic table.

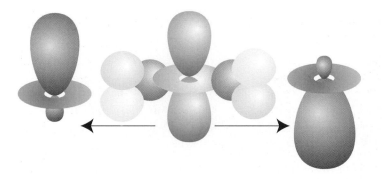

Fig. 18.6 The hybridization of *s*, p_z, and d_{z^2} with the choice of phases shown here produces a pair of collinear orbitals that can be used to form strong σ bonds.

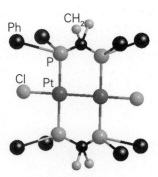

3 $[Pt_2(Cl)_2(Ph_2PCH_2PPh_2)_2]$

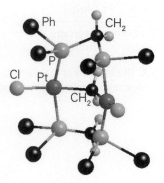

4 $[Pt_2(Cl)_2(\mu\text{-}CH_2)(Ph_2PCH_2PPh_2)_2]$

crystallography. The currently preferred explanation for the tendency towards linear coordination is the similarity in energy of the outer *nd* and $(n+1)s$ and $(n+1)p$ orbitals, which permits the formation of collinear *spd* hybrids (Fig. 18.6).

The soft character of Cu^+, Ag^+, and Au^+ is illustrated by their affinity order, which is $I^- > Br^- > Cl^-$. Complex formation, as in the formation of $[Cu(NH_3)_2]^+$ and $[AuI_2]^-$, provides a means of stabilizing the +1 oxidation state of these metals in aqueous solution. Many tetrahedral complexes are also known for Cu(I), Ag(I), and Au(I) (Section 8.2).

Square-planar complexes are common for the platinum metals and gold in oxidation states that yield the d^8 electronic configuration, which include Rh(I), Ir(I), Pd(II), Pt(II), and Au(III). An example is $[Pt(NH_3)_4]^{2+}$. Characteristic reactions for these complexes are ligand substitution (Section 20.4) and, except for gold(III) complexes, oxidative addition (Section 21.22).

The properties of platinum complexes with organic ligands are discussed in Chapter 21. An interesting series of Pt(I) and Pd(I) complexes has been found in which two M—M bonded metal atoms are bridged by phosphine ligands (**3**). These compounds undergo reactions in which a CH_2 group is inserted into the M—M bond:

$$
\begin{array}{ccc}
\begin{array}{c}
Ph_2P \overbrace{\qquad} PPh_2 \\
| \qquad\qquad | \\
Cl\!-\!Pt\!-\!-\!-\!-\!Pt\!-\!Cl \\
| \qquad\qquad | \\
Ph_2P \underbrace{\qquad} PPh_2
\end{array}
&
\xrightarrow[-N_2]{CH_2N_2}
&
\begin{array}{c}
Ph_2P \; H_2 \; PPh_2 \\
| \quad C \quad | \\
Cl\!-\!Pt\!\diagdown\!\diagup\!Pt\!-\!Cl \\
| \qquad\qquad | \\
Ph_2P \underbrace{\qquad} PPh_2
\end{array}
\end{array}
$$

In this example, the CH_2 group inserts into the M—M bond and the resulting compound (**4**), which is informally called an 'A-frame complex', now contains two square-planar Pt(II) groups bridged by CH_2 as well as the phosphine ligands.

The noble character that these elements developed across the *d* block is suddenly lost at Group 12, where the metals once again become susceptible to atmospheric oxidation. The greater ease of oxidation of the Group 12 metals is due to a reduction in the extent of intermetallic bonding and an abrupt lowering of *d*-orbital energies at the end of the *d* block, with the higher energy $(n+1)s$ electrons participating in reactions.

Representative compounds

The traditional view of the chemical properties of the *d* metals is from the perspective of ionic solids and complexes that contain a single central metal ion surrounded by a set of ligands (Chapter 8). However, with the development of improved techniques for the determination of structure, it has been recognized that there are also many *d*-metal compounds that have metal–metal (M—M) bond distances comparable to or shorter than those in the elemental metal. These findings have stimulated additional X-ray structural investigations of compounds that might contain M—M bonds and have inspired research into syntheses designed to broaden the range of such compounds. Examples of these metal (and nonmetal) cluster compounds are now known in every block of the periodic table (Fig. 18.7), but they are most numerous in the *d* block.

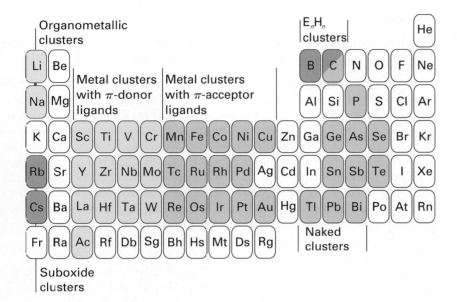

Fig. 18.7 The major classes of cluster-forming elements. Note that carbon is in two classes (for example it occurs as C_8H_8 and as C_{60}). (Adapted from D.M.P. Mingos and D.J. Wales, *Introduction to cluster chemistry*. Prentice Hall, Englewood Cliffs (1990).)

As a result of the convention of assigning a negative oxidation number to any non-metal in combination with a metal, high formal oxidation states for *d* metals may be encountered in their compounds, such as Re(VII) in $[ReH_9]^{2-}$ and W(VI) in $W(CH_3)_6$. However, these compounds are not oxidizing agents in the usual sense, and are best discussed together with other organometallic compounds (Chapter 21). This discussion is confined to compounds containing electronegative ligands such as the halogens, oxygen, nitrogen and sulfur.

18.7 Metal halides

Key points: Binary halides of the *d*-block elements span all metals and most oxidation states; dihalides are typically ionic solids, with higher halides taking on covalent character.

Binary metal halides of the *d*-block elements occur for all the elements with nearly all oxidation states represented. As we should expect, the more strongly oxidizing halogens bring out the higher oxidation states, with the corollary that the low oxidation state binary halides are more stable as iodides and bromides.

Of all the groups, only the members of Group 11 (Cu, Ag, Au) have simple mono-halides. For Cu, these salts are highly insoluble in water and dissolve only when complexed by other ligands (which includes halides). Silver(I) halides are sparingly soluble and photosensitive, decomposing to the metal (a process that is exploited in photography). The only monohalide of Au that exists is the chloride; it is oxidized by water.

More common are the dihalides which, as noted in Section 18.3b (Table 18.2), are typically ionic solids that dissolve in water to give M^{2+} ions; some of the earlier dihalides are reducing. Many of the intermediate halides exhibit metal–metal bonds and form clusters (Section 18.11).

Higher halides exist for most of the *d* block, and covalent character becomes more prevalent with high oxidation state, especially for the lower halogens. For instance, in Group 4, whereas TiF_4 is a solid with melting point 284°C, $TiCl_4$ melts at −24°C and boils at 136°C. In Group 6, not even the fluoride has ionic character and both MoF_6 and WF_6 are liquids at room temperature (Table 18.3).

18.8 Metal oxides and oxo complexes

Because oxygen is readily available in an aqueous environment, in the atmosphere, and as the donor atom in many organic molecules, it is not surprising that oxides and oxygen-containing ligands play a major role in the chemistry of the metallic elements.

(a) Metal oxides

Key point: Many different oxides of the *d*-block elements exist, with a wide variety of structures, varying from ionic lattices to covalent molecules.

Many different oxides are known for the *d*-block elements, with a number of different structures (Table 18.5). We have already noted the ability of oxygen to bring out the highest oxidation state for some elements, but oxides exist for some elements in very low oxidation state: in Cu_2O, copper is present as Cu(I). Monoxides are known for many of the 3*d*-series metals. They have the rock-salt structure characteristic of ionic solids but their properties, which are discussed in more detail in Chapter 23, indicate significant deviations from the simple ionic $M^{2+}O^{2-}$ model. For example, TiO has metallic conductivity and FeO is always deficient in iron. The early *d*-block monoxides are strong reducing agents. Thus, TiO is easily oxidized by water or oxygen, and MnO is a convenient oxygen scavenger that is used in the laboratory to remove oxygen impurity in inert gases down to the parts-per-billion range.

As we have already noted, very high oxidation state oxides can show covalent structures. For example ruthenium tetroxide and osmium tetroxide are low melting, highly volatile, toxic, molecular compounds that are used as selective oxidizing agents. Indeed, osmium tetroxide is used as the standard reagent to oxidize alkenes to *cis*-diols:

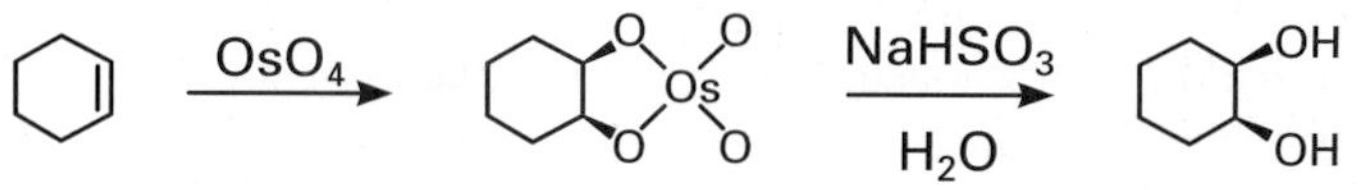

(b) Mononuclear oxo complexes

Key points: The conversion of an aqua ligand to an oxo ligand is favoured by a high pH and by a high oxidation state of the central metal atom; in vanadium complexes in oxidation state +4 or +5, the site *trans* to the oxo ligand may be vacant or occupied by a weak ligand.

Our focus in this section is on the oxo ligand (O^{2-}) and its ability to bring out the high oxidation states of the chemically hard metals on the left of the *d* block. Another important issue is the relation between oxo and aqua (OH_2) ligands resulting from proton transfer equilibria.

Elements in high oxidation states typically occur as oxoanions in aqueous solution, such as permanganate (MnO_4^-), which contains Mn(VII), and chromate (CrO_4^{2-}), which contains Cr(VI). The existence of these oxoanions contrasts with the existence of simple aqua ions for the same metals in lower oxidation states, such as $[Mn(OH_2)_6]^{2+}$ for manganese(II) and $[Cr(OH_2)_6]^{2+}$ for chromium(II).

The formation of an oxo complex rather than an aqua complex is favoured by high pH because the OH^- ions in basic solution tend to remove protons from the aqua ligands. The occurrence of aqua complexes for low oxidation state metal cations is explained by noting the relatively small electron-withdrawing effect exerted by these cations on the O atoms of the OH_2 ligand. As a result, the aqua ligands are only weak proton donors.

Table 18.5 Structures of *d*-block MO compounds

	Group						
	4	5	6	7	8	9	10
Rock-salt	Ti	V		Mn	Fe	Co	Ni
structure	Zr						Pd*
(shaded)							

*PtS structure (square-planar, four-coordinate metal atom).
Adapted from A.F. Wells, *Structural inorganic chemistry*. Oxford University Press (1984).

A metal ion in a high oxidation state, however, depletes the electron density on the O atoms attached to it and thereby increases the Brønsted acidity of OH_2 and OH^- ligands.

The influence of pH on the stabilities of different oxidation states is best summarized by a Pourbaix diagram (Section 5.13). The example shown in Fig. 18.8 is for a complex (**5**) containing a tetradentate ligand L, which results in a similar coordination geometry for a wide range of conditions. The aqua complex *cis*-$[RuLCl(OH_2)]^+$ in solution at pH = 2 is stable up to +0.40 V. Just above this potential, simple oxidation (that is, electron removal) occurs to give *cis*-$[RuLCl(OH_2)]^{2+}$. Under even more strongly oxidizing conditions (at +0.95 V) and pH = 2, both oxidation and deprotonation occur and result in the formation of a Ru(IV) oxo species, *cis*-$[RuLCl(O)]^+$. At an even higher potential (about +1.4 V), further oxidation yields *cis*-$[RuLCl(O)]^{2+}$. As already remarked, the influence of more basic conditions is to deprotonate an OH_2 ligand and thus to favour hydroxo or oxo complexes. For example, at pH = 8, the Ru(III) species is deprotonated to the hydroxo complex *cis*-$[RuLCl(OH)]^+$, which in turn is converted to the oxo-Ru(IV) complex at a lower potential than for the Ru(III)—Ru(IV) transformation at pH = 2.

Table 18.6 gives a list of simple complexes containing oxo ligands. Many complexes are known that contain the vanadium(IV) moiety, VO^{2+} (sometimes referred to as vanadyl), with vanadium in its penultimate oxidation state (+4). Vanadyl complexes generally contain four additional ligands and are square pyramidal (**6**). Many of these d^1 complexes are blue (as a result of a *d–d* transition) and may take on a weakly bound sixth ligand *trans* to the oxo ligand. The vanadyl VO bond length (158 pm) in $[VO(acac)_2]$ is short compared with the four VO bond lengths to the acac ligand (197 pm). The short bond lengths in vanadyl complexes, together with high VO stretching wavenumbers (940–1000 cm^{-1}) provide strong evidence for VO multiple bonding in which lone pairs on the oxo ligand are donated to the central vanadium atom. The V3*d*—O2*p* π-orbital overlap involved in a multiple bond is illustrated in Fig. 18.9; thus it can be seen that the O^{2-} ligand is not only a σ donor but also a π donor capable of making two π bonds and resulting in a bond order of up to three. This strong multiple bonding with the oxygen atom appears to be responsible for the *trans* influence of the oxo ligand, which disfavours attachment of the ligand *trans* to O (Section 20.5).

Vanadium in its highest oxidation state, +5, forms an extensive series of oxo compounds, many of which are the polyoxo species discussed later. The simplest oxo complex $[V(O)_2(OH_2)_4]^+$ exists in the acidic solution formed when the sparingly soluble vanadium(V) oxide, V_2O_5, dissolves in water. This pale yellow complex has a *cis* geometry (**7**). Here again we see the *trans* influence of an oxo ligand: the *cis* geometry minimizes competition between the oxo ligands for π bonding to the vanadium atom. As shown in Fig. 18.10, the two O^{2-} ligands in the *trans* configuration can π-bond with the same two *d* orbitals; in the *cis* configuration the metal atom has only one *d* orbital in common with the π orbitals on the O atoms.

In contrast to the *cis* structure of the d^0 complex $[V(O)_2(OH_2)_4]^+$, many *trans*-dioxo complexes are known for the d^2 metal centres Os(VI) and Re(V) (**8**); Table 18.6 gives some examples. This geometry is probably favoured because the *trans* configuration leaves a vacant low-energy orbital that can be occupied by the two *d* electrons. According to this explanation, the avoidance of the destabilizing effect of the d^2 electrons more

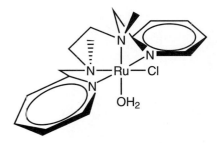

5 *cis*-$[RuLCl(OH_2)]^{2+}$

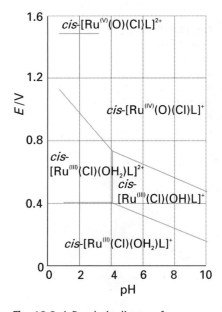

Fig. 18.8 A Pourbaix diagram for *cis*-$[RuLCl(OH_2)]^{2+}$ (**5**) and related species. (Adapted from C.-K. Li, W.-T. Wang, C.-M. Chi, K.-Y. Wong, R.-J. Wang, and T.C.W. Mak, *J. Chem. Soc., Dalton Trans.*, 1991, 1909.)

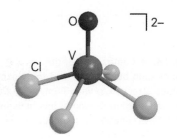

6 $[VOCl_4]^{2-}$

Table 18.6 Some common monoxo and dioxo complexes

Group	Element, configuration	Structure	Formula
5	V(IV), d^1	Square pyramidal	$[V(O)(acac)_2]$, $[V(O)Cl_4]^{2-}$
	V(V), d^0	*cis*-Octahedral	$[V(O)_2(OH_2)_4]^+$
6	Mo(VI), d^0; W(VI), d^0	Tetrahedral	$[M(O)_2(Cl)_2]$
7, 8	Re(V), d^2; Os(VI), d^2	*trans*-Octahedral	$[Re(O)_2(CN)_4]^{3-}$
			$[Os(O)_2(Cl)_2]$
			$[Os(O)_2(Cl)_4]^{2-}$

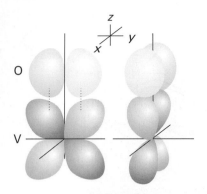

Fig. 18.9 The *d–p* π bonding between oxygen and vanadium in the vanadyl moiety VO²⁺. It is convenient to think of the oxygen atom as having oxidation number −2 and vanadium +4. Thus both electron pairs for the two π bonds are donated from the oxide into the metal d_{yz} and d_{zx} orbitals.

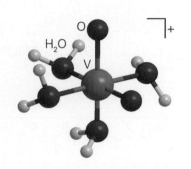

7 $[V(O)_2(OH_2)_4]^+$

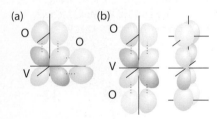

Fig. 18.10 Comparison of the competition between *cis* and *trans* oxo ligands attached to vanadium. (a) In the *cis* configuration the metal has only one *d* orbital is common with the π orbitals on the O atoms. (b) In the *trans* configuration the metal atom has two *d* orbitals in common with the π orbitals on the two O atoms.

than offsets the disadvantage of the ligands having to compete for the same *d* orbital in the *trans*-oxo geometry.

(c) Polyoxometallates

Key points: In their highest oxidation states, metals in Groups 5 and 6 readily form polyoxometallates and heteropolyoxometallates; pH is crucial in determining which compounds form.

A **polyoxometallate** is an oxoanion containing more than one metal atom. The OH_2 ligand created by protonation of an oxo ligand at low pH can be eliminated from the central metal atom and thus lead to the condensation of mononuclear oxometallates. A familiar example is the reaction of a basic chromate solution, which is yellow, with excess acid to form the oxo-bridged dichromate ion, which is orange:

$$2 CrO_4^{2-}(aq) + 2 H^+(aq) \rightarrow Cr_2O_7^{2-}(aq) + H_2O(l)$$

In highly acidic solution, oxo-bridged Cr(VI) species with longer chains are formed. The tendency for Cr(VI) to form polyoxo species is limited by the fact that the O tetrahedra link only through vertices: edge and face bridging would result in too close an approach of the metal atoms. By contrast, it is found that five- and six-coordinate metal oxo complexes, which are common with the larger 4*d*- and 5*d*-series metal atoms, can share oxo ligands between either vertices or edges. These structural possibilities lead to a richer variety of polyoxometallates than found with the 3*d*-series metals.

Chromium's neighbours in Groups 5 and 6 form six-coordinate polyoxo complexes (Fig. 18.11). In Group 5, the polyoxometallates are most numerous for vanadium, which forms many V(V) complexes and a few V(IV) or mixed oxidation state V(IV)—V(V) polyoxo complexes. Polyoxometallate formation is most pronounced in Groups 5 and 6 for vanadium(V), molybdenum(VI), and tungsten(VI).

It is often convenient to represent the structures of the polyoxometallate ions by polyhedra, with the metal atom understood to be in the centre and O atoms at the vertices. For example, the sharing of O atom vertices in the dichromate ion, $Cr_2O_7^{2-}$, may be depicted in either the traditional way (**9**) or in the polyhedral representation (**10**). Similarly, the important M_6O_{19} structure of $[Nb_6O_{19}]^{8-}$, $[Ta_6O_{19}]^{8-}$, $[Mo_6O_{19}]^{2-}$, and $[W_6O_{19}]^{2-}$ is depicted by the conventional or polyhedral structures shown in Fig. 18.12. The structures for this series of polyoxometallates contain terminal O atoms (those projecting out from a single metal atom) and two types of bridging O atoms: two-metal bridges, M—O—M, and one hypercoordinated O atom in the centre of the structure that is common to all six metal atoms. The structure consists of six MO_6 octahedra, each sharing an edge with four neighbours. The overall symmetry of the M_6O_{19} array is O_h. Another example of a polyoxometallate is $[W_{12}O_{40}(OH)_2]^{10-}$ (**11**). As shown by this formula, the polyoxoanion is protonated.

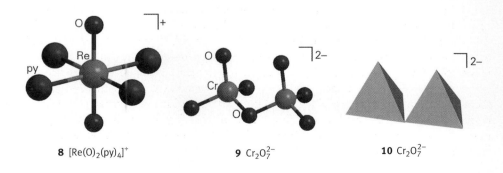

8 $[Re(O)_2(py)_4]^+$ **9** $Cr_2O_7^{2-}$ **10** $Cr_2O_7^{2-}$

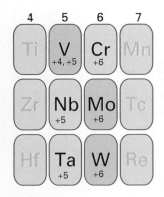

4	5	6	7
Ti	V +4, +5	Cr +6	Mn
Zr	Nb +5	Mo +6	Tc
Hf	Ta +5	W +6	Re

Fig. 18.11 Elements in the *d* block that form polyoxometallates. The elements coloured yellow form the greater variety of polyoxometallates.

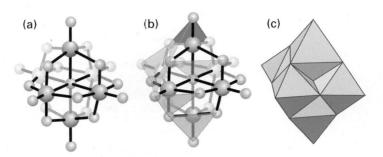

Fig. 18.12 (a) Conventional and (c) polyhedral representation of the six edge-shared octahedra as found in $[M_6O_{19}]^{2-}$. The intermediate structure (b) shows the process of constructing the polyhedral representation in progress.

Proton transfer equilibria are common for the polyoxometallates, and may occur together with condensation and fragmentation reactions. Polyoxometallate anions can be prepared by carefully adjusting pH and concentrations: for example, poly-oxomolybdates and polyoxotungstates are formed by acidification of solutions of the simple molybdate or tungstate:

$$6\,[MoO_4]^{2-}(aq) + 10\,H^+(aq) \rightarrow [Mo_6O_{19}]^{2-}(aq) + 5\,H_2O(l)$$

$$8\,[MoO_4]^{2-}(aq) + 12\,H^+(aq) \rightarrow [Mo_8O_{26}]^{4-}(aq) + 6\,H_2O(l)$$

Mixed metal polyoxometallates are also common, as in $MoV_9O_{28}^{5-}$, and there is a large class of heteropolyoxometallates, such as the molybdates and tungstates, that also incorporate phosphorus, arsenic, and other heteroatoms. For example, $[PMo_{12}O_{40}]^{3-}$ contains a PO_4^{3-} tetrahedron that shares O atoms with surrounding octahedral MoO_6 groups (**12**). Many different heteroatoms can be incorporated into this structure, and the general formulation is $[X(+N)Mo_{12}O_{40}]^{(8-N)-}$ where $X(+N)$ represents the oxidation state of the heteroatom X, which may be As(V), Si(IV), Ge(IV), or Ti(IV). An even broader range of heteroatoms is observed with the analogous tungsten hetero-polyoxoanions. Heteropolyoxomolybdates and tungstates can undergo one-electron reduction with no change in structure but with the formation of a deep blue colour. The colour seems to arise from the excitation of the added electron from Mo(V) or W(V) to an adjacent Mo(VI) or W(VI) site.

11 $[W_{12}O_{40}(OH)_2]^{10-}$

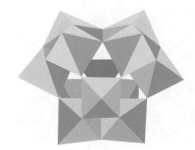

12 $[PMo_{12}O_{40}]^{3-}$

18.9 Metal sulfides and sulfide complexes

Key points: The 4*d*- and 5*d*-series metals often form disulfides with alternating layers of metal ions and sulfide ions; binary disulfides of the early *d* metals often have a layered structure, whereas Fe^{2+} and many of the later *d*-metal disulfides contain discrete S_2^{2-} ions; chelating polysulfide ligands are common in metal–sulfur coordination compounds of the 4*d*- and 5*d*-series metals; sulfur prefers metal centres that are not highly oxidizing.

Sulfur is softer and less electronegative than oxygen; it therefore has a broader range of oxidation states and a stronger affinity for the softer metals on the right of the *d* block. For example, zinc(II) sulfide is readily precipitated when $Zn^{2+}(aq)$ is added to an aqueous solution containing H_2S and NH_4OH (to adjust the pH) whereas the harder $Sc^{3+}(aq)$ gives $Sc(OH)_3(s)$ instead. The covalent contribution to the lattice enthalpy of all the softer metal sulfides cannot be overcome in water, and in general the sulfides are only very sparingly soluble. Another striking feature of sulfur is its tendency to form catenated sulfide ions. A broad set of compounds containing S_n^{2-} ions can be prepared, and the smaller polysulfides can serve as chelating ligands towards *d*-block metal ions.

Table 18.7 Structures of *d*-block MS compounds*

	Group						
	4	5	6	7	8	9	10
Nickel-arsenide structure (shaded)	Ti	V		Mn	Fe	Co	Ni
Rock-salt structure (unshaded)	Zr	Nb					

* Metal monosulfides of Group 6 are not shown; some of the heavier metals have more complex structures.
MnS has two polymorphs; one has a rock-salt structure, the other has a wurtzite structure.

Table 18.8 Structures of *d*-block MS$_2$ compounds*

	Group							
	4	5	6	7	8	9	10	11
Layered (shaded)	Ti			Mn	Fe	Co	Ni	Cu
Pyrite or marcasite	Zr	Nb	Mo		Ru	Rh		
(unshaded)	Hf	Ta	W	Re	Os	Ir	Pt	

* Metals not shown do not form disulfides or have disulfides with complex structures.
Adapted from A.F. Wells, *Structural inorganic chemistry*. Oxford University Press (1984).

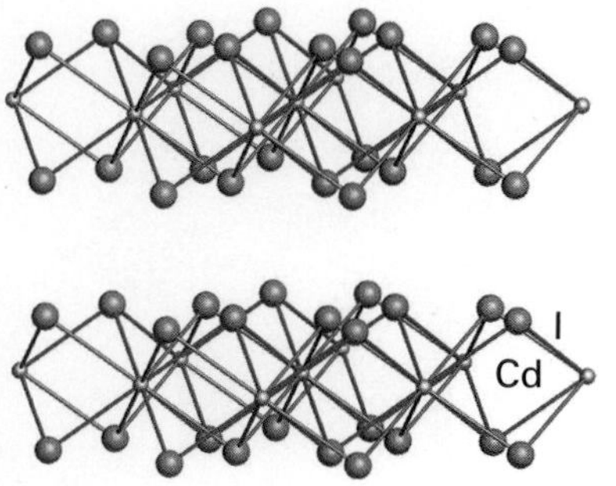

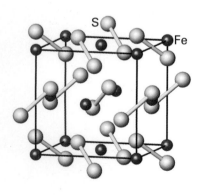

Fig. 18.13 The CdI$_2$ structure adopted by many disulfides.

Fig. 18.14 The structure of pyrite, FeS$_2$.

(a) Monosulfides

As with the *d*-metal monoxides, the monosulfides are most common in the 3*d* series (Table 18.7). In contrast to the monoxides, however, most of the monosulfides have the nickel-arsenide structure (Fig. 3.35). The different structures are consistent with the rock-salt structure of the monoxides being favoured by the ionic (harder) cation–anion combination and the nickel-arsenide structure being favoured by more covalent (softer) combinations. The nickel-arsenide structure is expected when there is appreciable metal–metal bonding, resulting in the shorter metal–metal distances found for the hexagonal packing. Monosulfides are formed by most 3*d*-series metal ions.

(b) Disulfides

The disulfides of the *d* metals fall into two broad classes (Table 18.8). One class consists of layered compounds with the CdI$_2$ or MoS$_2$ structure; the other consists of compounds containing discrete S$_2^{2-}$ groups.

The layered disulfides are built from a sulfide layer, a metal layer, and then another sulfide layer (Fig. 18.13). These sandwiches stack together in the crystal with sulfide layers in one slab adjacent to a sulfide layer in the next. Clearly, this crystal structure is not in harmony with a simple ionic model and its formation is a sign of covalence in the bonds between the soft sulfide ion and *d*-metal cations. The metal ion in these layered structures is surrounded by six S atoms. Its coordination environment is octahedral in some cases (PtS$_2$, for instance) and trigonal prismatic in others (MoS$_2$). The layered MoS$_2$ structure is favoured by S—S bonding as indicated by short S—S distances within each of the MoS$_2$ slabs. The common occurrence of the trigonal-prismatic structure in many of these compounds is in striking contrast to the isolated metal complexes, where the octahedral arrangement of ligands is by far the most common. Some of the layered metal sulfides readily undergo intercalation reactions in which ions or molecules penetrate between adjacent sulfide layers (Section 23.11).

Compounds containing discrete S$_2^{2-}$ ions adopt the pyrite or marcasite structure (Fig. 18.14). The stability of the formal S$_2^{2-}$ ion in metal sulfides is much greater than that of the O_2^{2-} ion in peroxides, and there are many more metal sulfides in which the anion is S_2^{2-} than there are peroxides.

Example 18.2 Contrasting the structures of two different *d*-block disulfides

Compare the structures of MoS_2 and FeS_2 and explain their existence in terms of the oxidation states of the metal ions.

Answer The point to be decided is whether the metals are likely to be present as M(IV), in which case the two S atoms would be present as S^{2-} ions. If M(IV) is not readily accessible, the metal might be present as M(II), and the S atoms present as the S—S bonded species S_2^{2-}. Because it is readily oxidized, S^{2-} will be found only with a metal ion in an oxidation state that is not easily reduced. As with many of the 4*d*- and 5*d*-series metals, molybdenum is easily oxidized to Mo(IV); therefore, Mo(IV) can coexist with S^{2-}. Molybdenum(IV) sulfide (MoS_2) has the layered structure typical of metal disulfides. Iron is readily oxidized to Fe(II) but not to Fe(IV). Therefore, Fe(IV) cannot coexist with S^{2-}. The compound is therefore likely to contain Fe(II) and S_2^{2-}. The mineral name for FeS_2 is pyrite; its common name, 'fools' gold', indicates its misleading colour.

Self-test 18.2 Molybdenum(IV) sulfide is a very effective lubricant. Present a plausible reason for this property.

(c) Sulfido complexes

The coordination chemistry of sulfur is quite different from that of oxygen. Much of this difference is connected with the ability of sulfur to catenate (form chains), the availability of low-lying empty 3*d* orbitals that allow sulfur to act as a π acceptor, and the preference of sulfur for metal centres that are not highly oxidizing.

Simple thiometallate complexes such as $[MoS_4]^{2-}$ can be synthesized easily by passing H_2S gas through a strongly basic aqueous solution of molybdate or tungstate ions:

$$[MoO_4]^{2-}(aq) + 4\,H_2S(g) \rightarrow [MoS_4]^{2-}(aq) + 4\,H_2O(l)$$

These tetrathiometallate anions are building blocks for the synthesis of complexes containing more metal atoms. For example, they will coordinate to many dipositive metal ions, such as Co^{2+} and Zn^{2+}:

$$Co_2^+(aq) + 2\,[MoS_4]^{2-}(aq) \rightarrow [S_2MoS_2CoS_2MoS_2]^{2-}(aq)$$

The polysulfide ions, such as S_2^{2-} and S_3^{2-}, which are formed by addition of elemental sulfur to a solution of ammonium sulfide, can also act as ligands. An example is $[Mo_2(S_2)_6]^{2-}$ (**13**), which is formed from ammonium polysulfide and MoO_4^{2-}; it contains side-bonded S_2^{2-} ligands. The larger polysulfides bond to metal atoms forming chelate rings, as in $[MoS(S_4)_2]^{2-}$ (**14**), which contains chelating S_4 ligands.

Clusters of Fe and S atoms (Fe—S clusters) are cofactors in many enzymes, including nitrogenase, which converts N_2 to NH_3 under ambient conditions (Chapter 14). Synthetic analogues such as the cubane shown in (**15**) have been prepared and characterized. The cubane contains Fe and S atoms on alternate corners, so each S atom bridges three Fe atoms; each of the Fe atoms has a thiolate group, RS^-, occupying a terminal position. These compounds can be obtained from simple starting materials in the absence of air using a polar aprotic solvent such as dimethylsulfoxide:

$$4\,FeCl_3 + 4\,NaSH + 6\,RSH + 10\,NaOCH_3 \rightarrow [Fe_4S_4(SR)_4]^{2-} + RS\text{—}SR$$
$$+ 14\,Na^+ + 12\,Cl^- + 10\,CH_3OH$$

The HS^- ion provides the sulfide ligands for the cage, the RS^- ion serves as both a ligand and a reducing agent, and the CH_3O^- ion acts as a base. Addition of $(Et_4N)Cl$ with its bulky cation results in the formation of black crystals of $(Et_4N)_2[Fe_4S_4(SR)_4]$. The observation that this reaction is successful with many different R groups, and its good yield, indicate that the Fe_4S_4 cage is thermodynamically more stable than other

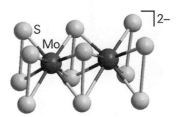

13 $[Mo_2(S_2)_6]^{2-}$

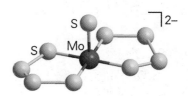

14 $[MoS(S_4)_2]^{2-}$

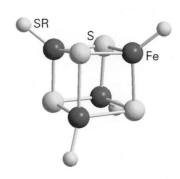

15

possibilities. The cage undergoes reversible one-electron reduction to $[Fe_4S_4(SR)_4]^{3-}$, a reaction that is important in biological oxidation and reduction.

18.10 Nitrido and alkylidyne complexes

Key points: Multiple M—L bonds are common with high oxidation state metals of the early *d*-metal series; these bonds weaken the bonds to the *trans* ligands.

The nitrido ligand (N≡) and alkylidyne ligand (RC≡, previously known as carbido) are isolobal and are present formally as N^{3-} and RC^{3-}. Thus, complexes containing them are usually of very high oxidation state. The highly reactive compound $(Me_3CO)_3W\equiv W(OCMe_3)_3$ cleaves the C≡N bond in benzonitrile to give both a nitrido and an alkylidyne ligand:

$$(Me_3CO)_3W\equiv W(OCMe_3)_3 + PhC\equiv N \rightarrow (Me_3CO)_3W\equiv CPh + (Me_3CO)_3W\equiv N$$

The remarkable feature of this reaction is that the C≡N bond is broken despite its considerable strength ($890\ kJ\ mol^{-1}$), so the W≡C and W≡N bond enthalpies must be substantial.

Both nitrido and alkylidyne complexes are known to have short M—N and M—C bond lengths, which confirms the existence of multiple metal–ligand bonds. Like the formally isolobal O^{2-}, these ligands weaken the *trans* metal–ligand bond. The order of this influence is RC≡ > N≡ > O≡; this weakening is attributed to competition for π donation into a common set of metal *d* orbitals.

Many nitrido complexes are known for the elements in Groups 5–8. They are most numerous for molybdenum and tungsten in Group 6, rhenium in Group 7, and ruthenium and osmium in Group 8. Square-pyramidal nitrido complexes of formula $[MNX_4]^-$ are known for M(VI), with M = Mo, Re, Ru, and Os, and X = F, Cl, Br, and (in some cases) I. These square-pyramidal structures (**16**) have short M≡N bonds (157 to 166 pm). In a few cases, a ligand is attached in the sixth site, but the resulting bond is long and weak, as with the oxo complexes discussed earlier. Examples of M=N=M species are known, such as $[Ta_2NBr_8]^{3-}$ (**17**), but nitrogen analogues of the polyoxometallates described earlier are unknown.

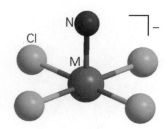

16 $[M(N)Cl_4]^-$

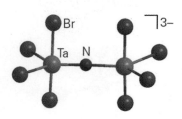

17 $[(Br)_4Ta(\mu\text{-}N)Ta(Br)_4]^{3-}$

18.11 Metal–metal bonded compounds and clusters

Key points: Metal–metal bonds with bond orders up to 4 are formed by many 4*d*- and 5*d*-series metals in low oxidation states, and metal clusters may be formed; for early *d*-block elements this is favoured by good donor ligands, whereas for later *d*-block elements π-acceptor ligands are needed.

A general feature of the low oxidation state metals from the early *d* block is their ability to form metal–metal bonds. Figure 18.7 showed how cluster compounds may be classified according to ligand type. Although this classification is not appropriate in detail for all cluster compounds, it provides information on the principal ligand types and some insight into the bonding. For example, alkyllithium compounds often have alkyl groups attached to a metal cluster by two-electron multicentre bonds. Clusters of the early *d*-block elements and the lanthanoids generally contain donor ligands such as Br^-. These ligands can fill some of the low-lying orbitals of these electron-poor metal atoms by σ- and π-electron donation. By contrast, electron-rich metal clusters on the right of the *d* block generally contain π-acceptor ligands such as CO, which remove some of the electron density from the metal. Many *p*-block elements do not require ligands to complete the valence shells of the individual atoms in clusters and can exist as **naked clusters**, which are clusters of the form E_n with no accompanying ligands. The ionic clusters Pb_5^{2-} and Sn_9^{4-} are two examples of naked clusters formed by metals.

The first *d*-block metal–metal bonded species to be identified was the Hg_2^{2+} ion of mercury(I) compounds, as occurs in Hg_2Cl_2, and examples of metal–metal bonded compounds and clusters are now known for most of the *d* metals. These clusters may involve M—M bonds within a discrete molecular cluster or in extended solid-state compounds. When no bridging ligands are present in a cluster, as in $[Re_2Cl_8]^{2-}$ (**18**), the presence of a metal–metal bond is unambiguous. Although this compound may be prepared by the reduction of ReO_4^- by a conventional reducing agent (such as zinc and a dilute acid), it is best prepared by the reduction of ReO_4^- with benzyl chloride. When bridging ligands are present, careful observations and measurement (typically of bond lengths and magnetic properties) are needed to identify direct metal–metal bonding. We shall see in Section 21.20 that there is an extensive range of organometallic metal cluster compounds of the middle-to-late d-block elements that are stabilized by π-acceptor ligands, particularly CO.

The bonding patterns in most metal cluster compounds are so intricate that metal–metal bond strengths cannot be determined with great precision. Some evidence, however, such as the stability of compounds and the magnitudes of M—M force constants, indicates that there is an increase in M—M bond strength down a group, perhaps on account of the greater spatial extension of d orbitals in heavier atoms. This trend may be the reason why there are so many more metal–metal bonded compounds for the 4d- and 5d-series metals than for their 3d-series counterparts. Figure 18.1 indicates that for the bulk metals, the metal–metal bonds in the d block are strongest in the 4d and 5d series, and this feature carries over into their compounds. By contrast, element–element bonds weaken down a group in the p block.

Not all the early d-block metal–metal bonded compounds are discrete clusters. Many extended metal–metal bonded compounds exist, such as the multiple-chain scandium subhalides, Sc_7Cl_{10} and Sc_5Cl_6 (Fig. 18.15) and layered compounds such as ZrCl (Fig. 18.16). In ZrCl, the Zr atoms are within bonding distance in two adjacent layers of metal atoms sandwiched between Cl^- layers. We have already seen that Zr and Sc adopt their group oxidation states ($+3$ and $+4$, respectively) when exposed to air and moisture, so these metal–metal bonded compounds are prepared out of contact with air and moisture. For example, ZrCl is prepared by reducing zirconium tetrachloride with zirconium metal in a sealed tantalum tube at high temperatures:

$$3 \ Zr(s) + ZrCl_4(g) \xrightarrow{600-800\,^\circ C} 4 \ ZrCl(s)$$

Discrete clusters—as distinct from the extended M—M bonded solid-state compounds just described—are often soluble, and can be manipulated in solution. The π-donor ligands in these clusters typically occur in one of several locations, namely, in a terminal position (**19**), bridging two metal atoms (**20**), or bridging three metal atoms (**21**). Clusters

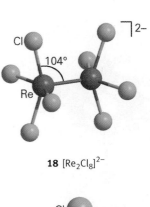

18 $[Re_2Cl_8]^{2-}$

19

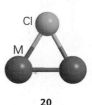

20

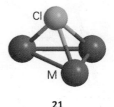

21

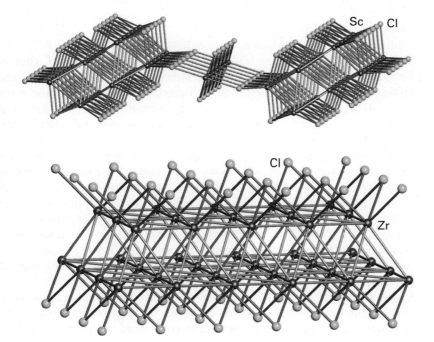

Fig. 18.15 The structure of Sc_7Cl_{10} showing a single chain of Sc atoms in the centre and multiple chains on either side.

Fig. 18.16 The structure of ZrCl consists of layers of metal atoms in graphite-like hexagonal nets.

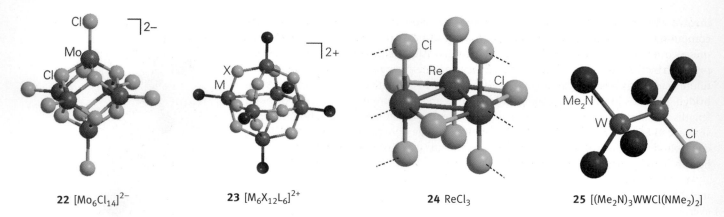

22 [Mo₆Cl₁₄]²⁻   **23** [M₆X₁₂L₆]²⁺   **24** ReCl₃   **25** [(Me₂N)₃WWCl(NMe₂)₂]

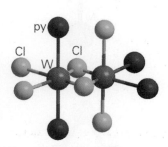

26 [(Cl)₂(py)₂W(μ-Cl)₂W(Cl)₂(py)₂]

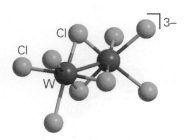

27 [(Cl)₃W(μ-Cl)₃W(Cl)₃]³⁻

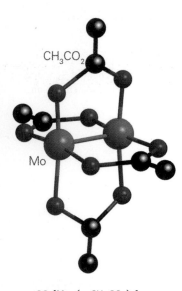

28 [Mo₂(μ-CH₃CO₂)₄]

may also be linked by bridging halides or chalcogenides in the solid state. For example, the solid compound of formula MoCl₂ consists of octahedral Mo clusters linked by Cl bridges. The bridged structure of MoCl₂ is in sharp contrast to CrCl₂, which has discrete Cr atoms arranged in a distorted rutile structure. Samples of MoCl₂ may be prepared in a sealed glass tube by the reaction of MoCl₅ in a mixture of molten NaAlCl₄ and AlCl₃, together with aluminium metal as the reducing agent:

$$MoCl_5(s) + Al(s) \xrightarrow{\text{NaAlCl}_4/\text{AlCl}_3(l),\ 200°C} MoCl_2(s) + AlCl_3(l)$$

The compound can withstand oxidation under mild conditions. When treated with hydrochloric acid it forms the anionic cluster [Mo₆Cl₁₄]²⁻. This cluster contains an octahedral array of Mo atoms with a Cl atom bridging each triangular face and a terminal Cl atom on each Mo vertex (**22**). These terminal Cl atoms can be replaced by other halogens, alkoxides, and phosphines. An analogous series of tungsten cluster compounds is known. Once formed, these molybdenum and tungsten compounds can be handled in air and water at room temperature on account of their kinetic barriers to decomposition. The oxidation number of the metal is +2, so the metal valence electron count is $(6-2) \times 6 = 24$. One-electron oxidation and reduction of the clusters is possible.

Similar octahedral clusters are known for Nb and Ta in Group 5, and for Zr in Group 4. The cluster [Nb₆Cl₁₂L₆]²⁺ and its tantalum analogue have an octahedral framework with edge-bridging Cl atoms and six terminal ligands (**23**). The metal valence electron count in this case is 16, but these clusters can be oxidized in several steps. One example from Group 4 is [Zr₆Cl₁₈C]⁴⁻, which has the same metal and Cl atom array together with a C atom in the centre of the octahedron.

Rhenium trichloride, which consists of Re₃Cl₉ clusters linked by weak halide bridges (**24**), provides the starting material for the preparation of a series of M₃ clusters that have been studied thoroughly. As with molybdenum dichloride, the intercluster bridges in the solid state can be broken by reaction with potential ligands. For example, treatment of Re₃Cl₉ with Cl⁻ ions produces the discrete complex [Re₃Cl₁₂]³⁻. Neutral ligands, such as trialkylphosphines, can also occupy these coordination sites and result in clusters of the general formula Re₃Cl₉L₃.

Many compounds containing bonds between two metals are known. Some of their common structural motifs are an ethane-like structure (**25**), an edge-shared biocta-hedron (**26**), a face-shared bioctahedron (**27**), and a tetragonal prism, which we have already encountered with [Re₂Cl₈]²⁻. We shall focus on bonding in the last of these structural types, in which the bond order may be considered as lying between 1 and 4.

Figure 18.17 shows that a σ bond between two metal atoms can arise from the overlap of a d_{z^2} orbital from each atom, π bonds can arise from the overlap of d_{xz} or d_{yz} orbitals (two such π bonds are possible), and a δ bond can be formed from the overlap of two face-to-face d_{xy} orbitals (the remaining $d_{x^2-y^2}$ orbital is used in the M—L σ bonds). A quadruple M≡M bond results when all the bonding orbitals are occupied and the configuration is $\sigma^2\pi^4\delta^2$ (Fig. 18.18). Eight bonding electrons are required for this configuration, implying that the metal atoms concerned should have a d^4 configuration.

Evidence for quadruple bonding comes from the observation that $[Re_2Cl_8]^{2-}$ has an eclipsed array of Cl ligands, which is sterically unfavourable. It is argued that the δ bond, which is formed only when the d_{xy} orbitals are confacial, locks the complex in the eclipsed configuration. Another well-known quadruply bonded compound is molybdenum(II) acetate (**28**), which is prepared by heating $Mo(CO)_6$ with acetic acid:

$$2\ Mo(CO)_6 + 4\ CH_3COOH \rightarrow Mo_2(O_2CCH_3)_4 + 2\ H_2 + 12\ CO$$

The quadruply bonded dimolybdenum acetato complex is an excellent starting material for other Mo—Mo compounds. For example, the quadruply bonded chloro complex is obtained when the acetato complex is treated with concentrated hydrochloric acid at below room temperature:

$$Mo_2(OCCH_3)_4 + 4\ H^+(aq) + 8\ Cl^-(aq) \rightarrow [Mo_2Cl_8]^{4-}(aq) + 4\ CH_3COOH(aq)$$

As shown in Table 18.9, triply bonded M≡M systems arise for tetragonal-prismatic complexes when δ and δ^* orbitals are both occupied. These complexes are more

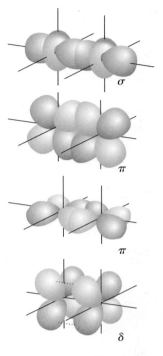

Fig. 18.17 The origin of σ, π, and δ interactions between the d orbitals of two d-metal atoms situated along the z-axis. Only bonding combinations are shown.

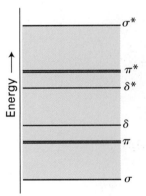

Fig. 18.18 Approximate molecular orbital energy level scheme for the M—M interactions in a quadruply bonded system.

Table 18.9 Examples of metal–metal bonded tetragonal prismatic complexes[†]

Complex	Configuration	Bond order	M—M bond length/pm
(Mo₂ dithio-acetato)⁴⁻	$\sigma^2\pi^4\delta^2$	4	211
(Mo₂ dithio-acetato)³⁻	$\sigma^2\pi^4\delta^1$	3.5	217
(Mo₂ phospho-acetato)²⁻	$\sigma^2\pi^4$	3	222
(Ru₂ chloro-acetato)⁻	$\sigma^2\pi^4\delta^2\delta^{\star 1}\pi^{\star 2}$	2.5	227

Table 18.9 (*Continued*)

Complex	Configuration	Bond order	M—M bond length/pm
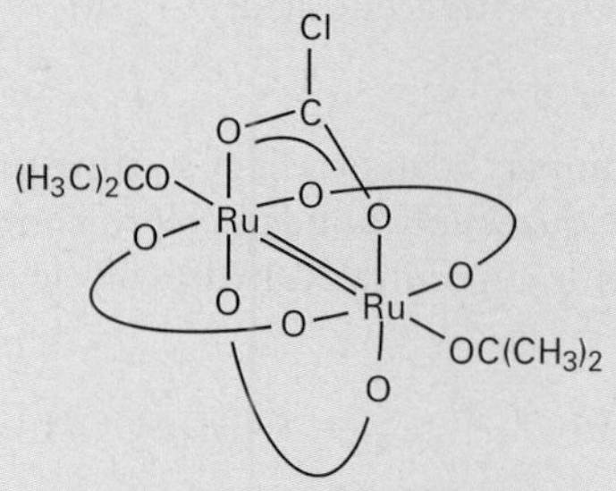	$\sigma^2\pi^4\delta^2\delta^{\star 2}\pi^{\star 2}$	2	226
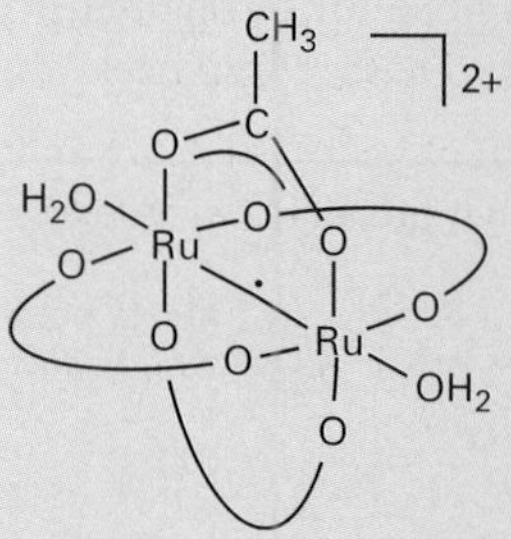	$\sigma^2\pi^4\delta^2\delta^{\star 1}\pi^{\star 4}$	1.5	232
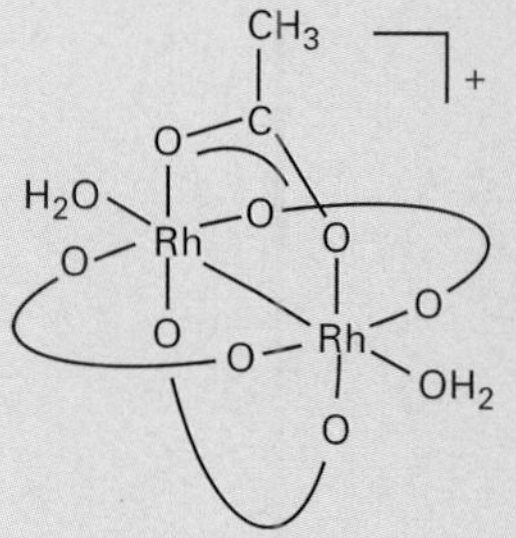	$\sigma^2\pi^4\delta^2\delta^{\star 2}\pi^{\star 4}$	1	239

[†] When multiple bridging ligands are present, only one is shown in detail.
Source: F.A. Cotton, *Chem. Soc. Rev.*, 1983, **12**, 35.

numerous than the quadruply bonded complexes and, because δ bonds are weak, M≡M bond lengths are often similar to those of quadruply bonded systems. Many triply bonded systems also include bridging ligands (Table 18.9). Table 18.9 shows that singly occupied δ or $\delta^{\star}$ orbitals lead to a formal bond order of 3.5. Once the δ and $\delta^{\star}$ orbitals are fully occupied, successive occupation of the two higher lying $\pi^{\star}$ orbitals leads to further reduction in the bond order from 2.5 to 1.

As for carbon–carbon multiple bonds, metal–metal multiple bonds are centres of reaction. However, the variety of structures resulting from the reactions of metal–metal multiply bonded compounds is more diverse than for organic compounds. For example,

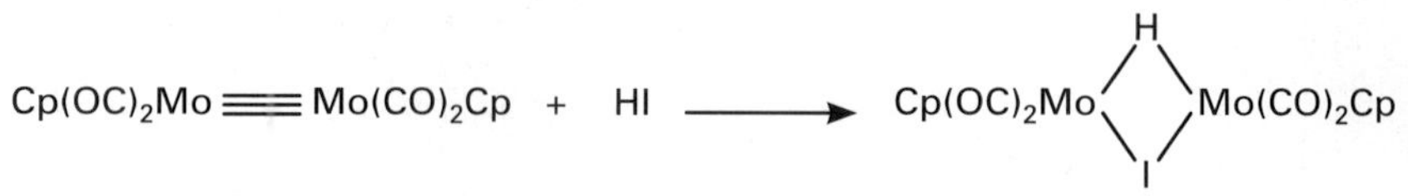

where Cp denotes the cyclopentadienyl group, $C_5H_5^-$. In this reaction, HI adds across a triple bond but both the H and I bridge the metal atoms; the outcome is quite unlike the addition of HX to an alkyne, which results in a substituted alkene. The reaction product can be regarded as containing a $3c,2e$ MHM bridge and an I atom bonded by two conventional $2c,2e$ bonds, one to each Mo atom.

Larger metal clusters can be synthesized by addition to a metal–metal multiple bond. For example, $Pt(PPh_3)_4$ loses two triphenylphosphine ligands when it adds to the Mo—Mo triple bond, resulting in a three-metal cluster:

$$Cp(OC)_2Mo \equiv Mo(CO)_2Cp \; + \; Pt(PPh_3)_4 \longrightarrow Cp(OC)_2Mo = Mo(CO)_2Cp \; + \; 2\,PPh_3$$

Example 18.3 Predicting probable structures of metal halides

List the major structural classes of d-metal halides and decide the most likely class for (a) MnF_2, (b) WCl_2, (c) RuF_6, and (d) FeI_2.

Answer The difluorides of the $3d$ metals (MnF_2) have the simple rutile structure characteristic of ionic AB_2 compounds; heavier halides (FeI_2) are commonly layered. The chlorides, bromides, and iodides of the low oxidation state early $4d$- and $5d$-series metals have metal–metal bonds (WCl_2 is in this class; it contains a W_6 cluster). The hexahalides are molecular (RuF_6).

Self-test 18.3 Describe the probable structure of the compound formed when Re_3Cl_9 is dissolved in a solvent containing PPh_3.

FURTHER READING

D.M.P. Mingos, *Essential trends in inorganic chemistry*. Oxford University Press (1998). A survey of inorganic chemistry from the perspective of structure and bonding.

R.B. King (ed.), *Encyclopedia of inorganic chemistry*. Wiley (2005).

M.T. Pope, Polyoxoanions: synthesis and structure. In *Comprehensive coordination chemistry II*, Vol. 4 (ed. J.A. McCleverty and T.J. Meyer),

Chapter 10. Elsevier (2004). A discussion of metal oxides in solution, focusing on their tendency to form polyoxoanions.

M.H. Chisholm (ed.) *Early transition metal clusters with π-donor ligands*. VCH, Weinheim (1995).

EXERCISES

18.1 Without reference to a periodic table, sketch the first series of the d block, including the symbols of the elements. Indicate those elements for which the group oxidation number is common by C; those for which the group oxidation number can be reached but is a powerful oxidizing agent by O; and those for which the group oxidation number is not achieved by N.

18.2 Explain why the enthalpy of sublimation of Re(s) is $704\ kJ\ mol^{-1}$, whereas that of Mn(s) is $221\ kJ\ mol^{-1}$.

18.3 State the trend in the stability of the group oxidation number on descending a group of metallic elements in the d block. Illustrate the trend using standard potentials in acidic solution for Groups 5 and 6.

18.4 For each part, give balanced chemical equations or NR (for no reaction) and rationalize your answer in terms of trends in oxidation states.

(a) $Cr^{2+}(aq) + Fe^{3+}(aq) \rightarrow$

(b) $CrO_4^{2-}(aq) + MoO_2(s) \rightarrow$

(c) $MnO_4^-(aq) + Cr^{3+}(aq) \rightarrow$

18.5 (a) Which ion, $Ni^{2+}(aq)$ or $Mn^{2+}(aq)$, is more likely to form a sulfide in the presence of H_2S? (b) Rationalize your answer with the trends in hard and soft character across Period 4. (c) Give a balanced chemical equation for the reaction.

18.6 Preferably without reference to the text (a) write out the d block of the periodic table, (b) indicate the metals that form difluorides with the rutile or fluorite structures, and (c) indicate the region of the periodic table in which metal–metal bonded halide compounds are formed, and give one example.

18.7 Write a balanced chemical equation for the reaction that occurs when cis-$[RuLCl(OH_2)]^+$ (see Fig. 18.8) in acidic solution at 0.2 V is made strongly basic at the same potential. Write a balanced equation for each of the successive reactions when this same complex at $pH = 6$ and 0.2 V is exposed to progressively more oxidizing environments up to 1.0 V. Give other examples and a reason for the redox state of the metal centre affecting the extent of protonation of coordinated oxygen.

18.8 Give plausible balanced chemical reactions (or NR for no reaction) for the following combinations, and state the basis for your answer: (a) $MoO_4^-(aq)$ plus $Fe^{2+}(aq)$ in acidic solution; (b) the preparation of $[Mo_6O_{19}]^{2-}(aq)$ from $K_2MoO_4(s)$; (c) $ReCl_5(s)$ plus aqueous $KMnO_4$; (d) $MoCl_2$ plus warm HBr(aq); (e) TiO with aqueous HCl under an inert atmosphere; (f) $Hg^{2+}(aq)$ added to Cd(s).

18.9 Speculate on the structures of the following species and present bonding models to justify your answers: (a) $[Re(O)_2(py)_4]^+$, (b) $[V(O)_2(ox)_2]^{3-}$, (c) $[Mo(O)_2(CN)_4]^{4-}$, (d) $[VOCl_4]^{2-}$.

18.10 Which of the following are likely to have structures that are typical of (a) predominantly ionic, (b) significantly covalent, (c) metal–metal bonded compounds: NiI_2, $NbCl_4$, FeF_2, PtS, and WCl_2? Rationalize the differences and speculate on the structures.

18.11 Indicate the probable occupancy of σ, π, and δ bonding and antibonding orbitals and the bond order for the following tetragonal prismatic complexes: (a) $[Mo_2(O_2C(CH_3)_4]$, (b) $[Cr_2(O_2CC_2H_5)_4]$, (c) $[Cu_2(O_2CCH_3)_4]$.

18.12 Explain the differences in the following redox couples, measured at 25°C:

$$MnO_4^-/MnO_2 \quad +1.69 \text{ V}$$
$$TcO_4^-/TcO_2 \quad +0.74 \text{ V}$$
$$ReO_4^-/ReO_2 \quad +0.53 \text{ V}$$

18.13 Addition of sodium ethanoate to aqueous solutions of Cr(II) gives a red diamagnetic product. Draw the structure of the product, noting any features of interest.

PROBLEMS

18.1 An amateur chemist claimed the existence of a new metallic element, grubium (Gr), which has the following characteristics. Metallic Gr reacts with 1 M H^+(aq) in the absence of air to produce Gr^{3+}(aq) and H_2(g). In the absence of air $GrCl_2$(s) dissolves in 1 M H^+(aq) and very slowly yields H_2(g) plus Gr^{3+}(aq). When Gr^{3+}(aq) is exposed to air GrO^{2+}(aq) is produced. From this information, estimate the range of potentials for (a) the reduction of Gr^{3+}(aq) to Gr(s), (b) the reduction of Gr^{3+}(aq) to $GrCl_2$(s), and (c) the reduction of GrO^{2+}(aq) to Gr^{3+}(aq). Suggest a known element that fits the description of grubium.

18.2 Compared with the *p* block, the variation in the properties of the elements of the *d* block across each period is rather modest. Provide evidence for this statement by reference to isostructural compounds. Speculate on reasons why this statement is true.

18.3 Think about the mechanism of electrical conduction and then suggest how electrical conductivity might vary across the *d* block.

18.4 Many metal salts can be vaporized to a small extent at high temperatures and their structures studied in the vapour phase by electron diffraction. Speculate on the structures that you might expect to find for the following gas-phase species and present your reasoning: (a) TaF_5; (b) MoF_6.

18.5 Discuss ways in which the triamidoamine ligands $[(RNCH_2CH_2)_3N]^{3-}$ (where R is a bulky substituent) can be used to stabilize nitrido, phosphido, and arsenido groups (see R.R. Schrock, *Acc. Chem. Res.*, 1997, **30**, 9).

d-Metal complexes: electronic structure and spectra

d-Metal complexes play an important role in inorganic chemistry. In this chapter, we discuss the nature of the ligand–metal bonding in terms of two theoretical models. We start with the simple but useful crystal-field theory, which is based on an electrostatic model of the bonding, and then progress to the more sophisticated ligand-field theory. Both theories invoke a parameter, the ligand-field splitting parameter, to correlate spectroscopic and magnetic properties. We then examine the electronic spectra of complexes and see how ligand-field theory allows us to interpret the energies and intensities of electronic transitions.

We now examine in detail the bonding, electronic structure, and electronic spectra of the *d*-metal complexes introduced in Chapter 8. The striking colours of many *d*-metal complexes were a mystery to Werner, and their origin was clarified only when the description of electronic structure in terms of orbitals was applied to the problem in the period from 1930 to 1960. Tetrahedral and octahedral complexes are the most important, and the discussion begins with them.

Electronic structure

There are two widely used models of the electronic structure of *d*-metal complexes. One ('crystal-field theory') emerged from an analysis of the spectra of *d*-metal ions in solids; the other ('ligand-field theory') arose from an application of molecular orbital theory. Crystal-field theory is more primitive, but it captures the essence of the electronic structure of complexes in a straightforward manner. Ligand-field theory builds on crystal-field theory, and accounts for a wider range of properties.

19.1 Crystal-field theory

In **crystal-field theory**, a ligand lone pair is modelled as a point negative charge (or as the partial negative charge of an electric dipole) that repels electrons in the *d* orbitals of the central metal ion. The theory concentrates on the resulting splitting of the *d* orbitals into groups with different energies and uses that splitting to rationalize and correlate the optical spectra, thermodynamic stability, and magnetic properties of complexes.

(a) Octahedral complexes

Key points: In the presence of an octahedral crystal field, *d* orbitals are split into a lower-energy triply degenerate set (t_{2g}) and a higher-energy doubly degenerate set (e_g) separated by an energy Δ_O; the ligand-field splitting parameter increases along a spectrochemical series of ligands and varies with the identity and charge of the metal atom.

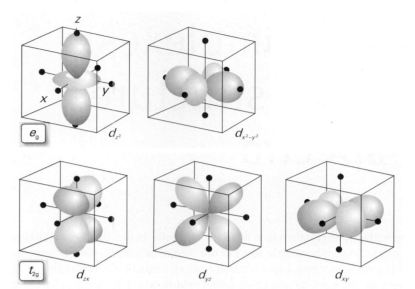

Fig. 19.1 The orientation of the five *d* orbitals with respect to the ligands of an octahedral complex: the degenerate (a) e_g and (b) t_{2g} orbitals.

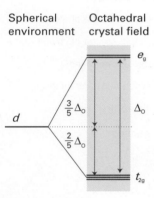

Fig. 19.2 The energies of the *d* orbitals in an octahedral crystal field. Note that the mean energy remains unchanged relative to the energy of the *d* orbitals in a spherically symmetric environment (such as in a free atom).

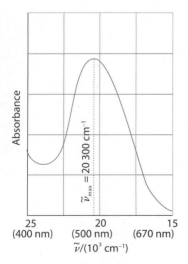

Fig. 19.3 The optical absorption spectrum of $[Ti(OH_2)_6]^{3+}$.

In the model of an octahedral complex used in crystal-field theory, six point negative charges representing the ligands are placed in an octahedral array around the central metal ion. These charges (which henceforth we shall refer to as the 'ligands') interact strongly with the central metal ion, and the stability of the complex stems in large part from this attractive interaction between opposite charges. However, there is a much smaller but very important secondary effect arising from the fact that electrons in different *d* orbitals interact with the ligands to different extents. Although this differential interaction is little more than about 10 per cent of the overall metal–ligand interaction energy, it has major consequences for the properties of the complex and is the principal focus of this section.

Electrons in d_{z^2} and $d_{x^2-y^2}$ orbitals (which are of symmetry type e_g in O_h; Section 7.5) are concentrated close to the ligands, along the axes, whereas electrons in d_{xy}, d_{yz}, and d_{zx} orbitals (which are of symmetry type t_{2g}) are concentrated in regions that lie between the ligands (Fig. 19.1). As a result, the former are repelled more strongly by the negative charge on the ligands than the latter and lie at a higher energy. Group theory shows that the two e_g orbitals have the same energy (although this is not readily apparent from drawings), and that the three t_{2g} orbitals also have the same energy. This simple model leads to an energy-level diagram in which the three degenerate t_{2g} orbitals lie below the two degenerate e_g orbitals (Fig. 19.2). The separation of the two sets of orbitals is called the **ligand-field splitting parameter**, Δ_O (where the subscript O signifies an octahedral crystal field).[1] The energy level that corresponds to the spherical symmetric environment (in which the negative charge due to the ligands is evenly distributed over a sphere instead of being localized at six points) defines the **barycentre** of the array of levels, with the two e_g orbitals lying at $\frac{3}{5}\Delta_O$ above the barycentre and the three t_{2g} orbitals lying at $\frac{2}{5}\Delta_O$ below it. As in the representation of the configurations of atoms, a superscript is used to indicate the number of electrons in each set.

The simplest property that can be interpreted by crystal-field theory is the absorption spectrum of a one-electron complex. Figure 19.3 shows the optical absorption spectrum of the d^1 hexaaquatitanium(III) ion, $[Ti(OH_2)_6]^{3+}$. Crystal-field theory assigns the first absorption maximum at $20\,300$ cm^{-1} (493 nm) to the transition $e_g \leftarrow t_{2g}$ and identifies $20\,300$ cm^{-1} with Δ_O for the complex.[2] It is more difficult to obtain values of Δ_O for complexes with more than one *d* electron because the energy of a transition then depends not only on orbital energies but also on the electron–electron repulsion energies. This

---

[1] In the context of crystal-field theory, the ligand-field splitting parameter should be called the *crystal-field splitting parameter*, but we use ligand-field splitting parameter to avoid a proliferation of names.

[2] Throughout this discussion we use the convention of indicating spectroscopic transitions with the lower energy state on the right and the upper energy state on the left of the arrow.

Table 19.1 Ligand-field splitting parameters Δ_O of ML_6 complexes*

	Ions	Ligands				
		Cl^-	H_2O	NH_3	en	CN^-
d^3	Cr^{3+}	13 700	17 400	21 500	21 900	26 600
d^5	Mn^{2+}	7500	8500		10 100	30 000
	Fe^{3+}	11 000	14 300			(35 000)
d^6	Fe^{2+}		10 400			(32 800)
	Co^{3+}		(20 700)	(22 900)	(23 200)	(34 800)
	Rh^{3+}	(20 400)	(27 000)	(34 000)	(34 600)	(45 500)
d^8	Ni^{2+}	7500	8500	10 800	11 500	

* Values are in cm^{-1}; entries in parentheses are for low-spin complexes.

Source: H.B. Gray, *Electrons and chemical bonding*. Benjamin, Menlo Park (1965).

aspect is treated more fully in Section 19.4 and the results from the analyses described there have been used to obtain the values of Δ_O in Table 19.1.

The ligand-field splitting parameter varies systematically with the identity of the ligand. For instance, in the series of complexes $[CoX(NH_3)_5]^{n+}$ with $X = I^-$, Br^-, Cl^-, H_2O, and NH_3, the colours range from purple (for $X = I^-$) through pink (for Cl^-) to yellow (with NH_3). This sequence indicates that the energy of the lowest energy electronic transition (and therefore Δ_O) increases as the ligands are varied along the series. The same order is followed regardless of the identity of the metal ion. On the basis of observations like these, Ryutaro Tsuchida proposed that ligands could be arranged in a **spectrochemical series**, in which the members are arranged in order of increasing energy of transitions that occur when they are present in a complex:

$$I^- < Br^- < S^{2-} < \underline{S}CN^- < Cl^- < N\underline{O}_2^- < N^{3-} < F^- < OH^- < C_2O_4^{2-} < H_2O < \underline{N}CS^-$$
$$< CH_3CN < py < NH_3 < en < bipy < phen < \underline{N}O_2^- < PPh_3 < \underline{C}N^- < CO$$

(The donor atom in an ambidentate ligand is underlined.) Thus, the series indicates that, for the same metal, the optical absorption of the cyano complex will occur at much higher-energy than that of the corresponding chloro complex. A ligand that gives rise to a high-energy transition (such as CO) is referred to as a **strong-field ligand**, whereas one that gives rise to a low-energy transition (such as Br^-) is referred to as a **weak-field ligand**. Crystal-field theory alone cannot explain these strengths, but ligand-field theory can, as we shall see in Section 19.2.

The ligand-field strength also depends on the identity of the central metal ion, the order being approximately:

$$Mn^{2+} < Ni^{2+} < Co^{2+} < Fe^{2+} < V^{2+} < Fe^{3-} < Co^{3+} < Mo^{3+} < Rh^{3+} < Ru^{3+} < Pd^{4+}$$
$$< Ir^{3+} < Pt^{4+}$$

The value of Δ_O increases with increasing oxidation number of the central metal ion (compare the two entries for Fe and Co) and also increases down a group (compare, for instance, the locations of Ni, Pd, and Pt). The variation with oxidation number reflects the smaller size of more highly charged ions and the consequently shorter metal–ligand distances and stronger interaction energies. The increase down a group reflects the larger size of the 4d and 5d orbitals compared with the compact 3d orbitals and the consequent stronger interactions with the ligands.

(b) Ligand-field stabilization energies

Key point: The ligand-field stabilization energy is a measure of the net energy of occupation of the d orbitals relative to their mean energy.

In an octahedral ion, relative to the barycentre, the energy of a t_{2g} orbital is $-0.4\Delta_O$ and that of an e_g orbital is $+0.6\Delta_O$. It follows that the net energy of a $t_{2g}^x e_g^y$

Table 19.2 Ligand-field stabilization energies*

d^n	Example	Octahedral				Tetrahedral	
		N	LFSE			N	LFSE
d^0		0	0			0	0
d^1	Ti^{3+}	1	0.4			1	0.6
d^2	V^{3+}	2	0.8			2	1.2
d^3	Cr^{3+}, V^{2+}	3	1.2			3	0.8
		Strong-field			**Weak-field**		
d^4	Cr^{2+}, Mn^{3+}	2	1.6	4	0.6	4	0.4
d^5	Mn^{2+}, Fe^{3+}	1	2.0	5	0	5	0
d^6	Fe^{2+}, Co^{3+}	0	2.4	4	0.4	4	0.6
d^7	Co^{2+}	1	1.8	3	0.8	3	1.2
d^8	Ni^{2+}	2	1.2			2	0.8
d^9	Cu^{2+}	1	0.6			1	0.4
d^{10}	Cu^+, Zn^{2+}	0	0			0	0

* *N* is the number of unpaired electrons; LFSE is in units of Δ_O for octahedra or Δ_T for tetrahedra; the calculated relation is $\Delta_T \approx 0.45\Delta_O$.

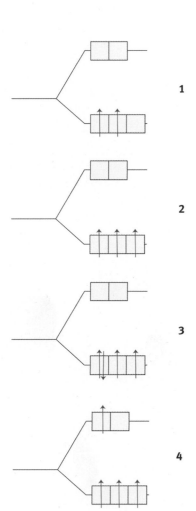

1

2

3

4

configuration relative to the barycentre, which is called the **ligand-field stabilization energy** (LFSE),[3] is

$$LFSE = (0.4x - 0.6y)\Delta_O \tag{19.1}$$

Table 19.2 lists the LFSE values for various configurations. The LFSE is generally only a small fraction of the overall interaction between the metal atom and the ligands, which increases from left to right across a period on account of the decrease in radius of the M^{2+} ions along that series.

(c) Weak-field and strong-field limits

Key points: The ground-state configuration of a complex reflects the relative values of the ligand-field splitting parameter and the pairing energy; for $3d^n$ species with $n = 4-7$, high-spin and low-spin complexes occur in the weak-field and strong-field cases, respectively. Complexes of 4*d*- and 5*d*-series metals are typically low-spin.

To obtain the ground-state electron configurations of *d*-metal complexes, we use the *d*-orbital energy level diagram shown in Fig. 19.2 as a basis for applying the building-up principle. We look for the lowest energy configuration subject to the Pauli exclusion principle (a maximum of two electrons in an orbital) and (if more than one degenerate orbital is available) to the requirement that electrons first occupy separate orbitals and do so with parallel spins. We begin by considering complexes formed by the 3*d*-series elements.

In an octahedral complex, the first three *d* electrons of a $3d^n$ complex occupy separate t_{2g} nonbonding orbitals, and do so with parallel spins. For example, the ions Ti^{2+} and V^{2+} have electron configurations $3d^2$ and $3d^3$, respectively. The *d* electrons occupy the lower t_{2g} orbitals as shown in (**1**) and (**2**), respectively, and the complexes are stabilized by $2 \times (0.4\Delta_O) = 0.8\Delta_O$ (for Ti^{2+}) and $3 \times (0.4\Delta_O) = 1.2\Delta_O$ (for V^{2+}).

The next electron needed for the $3d^4$ ion Cr^{2+} may enter one of the t_{2g} orbitals and pair with the electron already there (**3**). However, if it does so, it experiences a strong Coulombic repulsion, which is called the **pairing energy**, *P*. Alternatively, the electron may occupy one of the e_g orbitals (**4**). Although the pairing penalty is now avoided, the orbital energy is higher by Δ_O. In the first case (t_{2g}^4), the LFSE is $1.6\Delta_O$, the pairing energy is *P*, and the net stabilization is $1.6\Delta_O - P$. In the second case ($t_{2g}^3 e_g^1$), the LFSE is

---

[3] The term *crystal-field stabilization energy* (CFSE) is widely used in place of LFSE, but strictly speaking the term is appropriate only for ions in crystals.

$3 \times (0.4\Delta_O) - 0.6\Delta_O = 0.6\Delta_O$, and there is no pairing energy to consider. Which configuration is adopted depends on which of $1.60\Delta_O - P$ and $0.60\Delta_O$ is the larger.

If $\Delta_O < P$, which is called the **weak-field case**, a lower energy is achieved if the upper orbital is occupied to give the configuration $t_{2g}^3 e_g^1$. If $\Delta_O > P$, which is called the **strong-field case**, a lower energy is achieved by occupying only the lower orbitals despite the cost of the pairing energy. The resulting configuration is now t_{2g}^4. For example, $[Cr(OH_2)_6]^{2+}$ has the ground-state configuration $t_{2g}^3 e_g^1$ whereas $[Cr(CN)_6]^{4-}$, with relatively strong-field ligands (as given by the spectrochemical series), has the configuration t_{2g}^4.

The ground-state electron configurations of $3d^1$, $3d^2$, and $3d^3$ complexes are unambiguous because there is no competition between the LFSE and the pairing energy: the configurations are t_{2g}^1, t_{2g}^2, and t_{2g}^3, respectively. There are two possible configurations for $3d^n$ complexes in which $n = 4$ or 5: in the strong-field case the lower orbitals will be occupied, giving rise to t_{2g}^n configurations, whereas in the weak-field case electrons will avoid the pairing energy by occupying the upper orbitals, when the configurations will be $t_{2g}^3 e_g^1$ and $t_{2g}^3 e_g^2$. Because in the latter case all the electrons occupy different orbitals, they will have parallel spins.

When alternative configurations are possible, the species with the smaller number of parallel electron spins is called a **low-spin complex**, and the species with the greater number of parallel electron spins is called a **high-spin complex**. An octahedral $3d^4$ complex is likely to be low-spin if the ligand field is strong but high-spin if the field is weak (Fig. 19.4). The same applies to $3d^5$ complexes (see Table 19.2). High- and low-spin configurations are also found for $3d^6$ and $3d^7$ complexes. In these cases, a strong crystal field results in the low-spin configurations t_{2g}^6 (no unpaired electrons) and $t_{2g}^6 e_g^1$ (one unpaired electron), respectively, and a weak crystal field results in the high-spin configurations $t_{2g}^4 e_g^2$ (four unpaired electrons) and $t_{2g}^5 e_g^2$ (three unpaired electrons), respectively.

The strength of the crystal field (as measured by the value of Δ_O) and the spin-pairing energy (as measured by P) depend on the identity of both the metal and the ligand, so it is not possible to specify a universal point in the spectrochemical series at which a complex changes from high spin to low spin. For $3d$-metal ions, low-spin complexes commonly occur for ligands that are high in the spectrochemical series (such as CN^-) and high-spin complexes are common for ligands that are low in the series (such as F^-). For octahedral d^n complexes with $n = 1-3$ and $8-10$ there is no ambiguity about the configuration (see Table 19.2), and the designations high-spin and low-spin are not used.

As we have seen, the values of Δ_O for complexes of $4d$- and $5d$-series metals are typically higher than for the $3d$-series metals. Consequently, complexes of these metals generally have electron configurations that are characteristic of strong crystal fields and typically have low spin. An example is the $4d^4$ complex $[RuCl_6]^{2-}$, which has a t_{2g}^4 configuration, corresponding to a strong crystal field, despite Cl^- being low in the spectrochemical series. Likewise, $[Ru(ox)_3]^{3-}$ has the low-spin configuration t_{2g}^5 whereas $[Fe(ox)_3]^{3-}$ has the high-spin configuration $t_{2g}^3 e_g^2$.

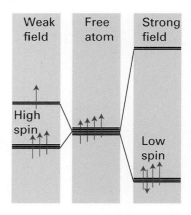

Fig. 19.4 The effect of weak and strong ligand fields on the occupation of electrons for a d^4 complex. The former results in a high-spin configuration and the latter in a low-spin configuration.

Example 19.1 Calculating the LFSE

What is the LFSE for octahedral ions of the following configurations: (a) d^3, (b) high-spin d^5 (c) d^9?

Answer Using the notation $t_{2g}^x e_g^y$ and eqn 19.1, we can assign the following values: (a) $x = 3$, $y = 0$, therefore LFSE $= 1.2\Delta_O$, (b) $x = 3$, $y = 2$, therefore LFSE $= 0$, (c) $x = 6$, $y = 3$, therefore LFSE $= 0.6\Delta_O$.

Self test 19.1 What is the LFSE for both high- and low-spin d^6 configurations?

(d) Magnetic measurements

Key points: Magnetic measurements can be used to determine the number of unpaired spins in a complex and hence to identify its ground-state configuration. A spin-only calculation may fail for low-spin d^5 and for high-spin $3d^6$ and $3d^7$ complexes.

The experimental distinction between high-spin and low-spin octahedral complexes is based on the determination of their magnetic properties. Compounds are classified as **diamagnetic** if they are repelled by a magnetic field and **paramagnetic** if they attracted by a magnetic field. The two can be distinguished experimentally as explained in Chapter 6, Magnetometry. The magnitude of the paramagnetism of a complex is commonly reported in terms of the magnetic dipole moment it possesses: the higher the magnetic dipole moment of the complex, the greater the paramagnetism of the sample.

In a free atom or ion, both the orbital and the spin angular momenta give rise to a magnetic moment and contribute to the paramagnetism. When the atom or ion is part of a complex, any orbital angular momentum is normally suppressed—the technical term is **quenched**—as a result of the interactions of the electrons with their nonspherical environment. However, the electron spin angular momentum survives, and gives rise to **spin-only paramagnetism**, which is characteristic of many *d*-metal complexes. The spin-only magnetic moment, μ, of a complex with total spin quantum number S is

$$\mu = 2\{S(S+1)\}^{1/2}\mu_B \tag{19.2}$$

where μ_B is the collection of fundamental constants known as the **Bohr magneton**,

$$\mu_B = \frac{e\hbar}{2m_e}$$

with the value $9.274 \times 10^{-24}\,\mathrm{J\,T^{-1}}$.

A measurement of the magnetic moment of a *d*-block complex can usually be interpreted in terms of the number of unpaired electrons it contains, and hence the measurement can be used to distinguish between high-spin and low-spin complexes. As shown later, we can also use magnetic data to determine coordination geometry. For example, magnetic measurements on a d^6 complex distinguishes between a high-spin $t_{2g}^4 e_g^2$ ($S=2$, $\mu = 4.90\mu_B$) configuration and a low-spin t_{2g}^6 ($S=0$, $\mu=0$) configuration.

The spin-only magnetic moments for the configurations $t_{2g}^x e_g^y$ are listed in Table 19.3 and compared there with experimental values for a number of 3*d* complexes. For most 3*d* complexes (and some 4*d* complexes), experimental values lie reasonably close to spin-only predictions, so it becomes possible to identify correctly the number of unpaired electrons and assign the ground-state configuration. For instance, $[\mathrm{Fe(OH_2)_6}]^{3+}$ is paramagnetic with a magnetic moment of $5.9\mu_B$. As shown in Table 19.3, this value is consistent with there being five unpaired electrons (and $S=\frac{5}{2}$), which implies a high-spin $t_{2g}^3 e_g^2$ configuration.

The interpretation of magnetic susceptibility measurements is usually less straightforward than this example might suggest. For example, the potassium salt of $[\mathrm{Fe(CN)_6}]^{3-}$ has $\mu = 2.3\mu_B$, which is between the spin-only values for one and two unpaired electrons ($1.7\mu_B$ and $2.8\mu_B$, respectively). In this case, the spin-only assumption has failed because the orbital magnetic contribution is substantial.

For orbital angular momentum to contribute, and hence for the paramagnetism to differ significantly from the spin-only value, there must be an unfilled or half-filled orbital similar in energy to that of the orbitals occupied by the unpaired spins and of the appropriate symmetry (one that is related to the occupied orbital by rotation

Table 19.3 Calculated spin-only magnetic moments

Ion	N	S	μ/μ_B	
			Calculated	Experimental
Ti^{3+}	1	$\frac{1}{2}$	1.73	1.7–1.8
V^{3+}	2	1	2.83	2.7–2.9
Cr^{3+}	3	$\frac{3}{2}$	3.87	3.8
Mn^{3+}	4	2	4.90	4.8–4.9
Fe^{3+}	5	$\frac{5}{2}$	5.92	5.9

round the direction of the applied field). If that is so, then the electrons can make use of the available orbital to circulate around the metal ion, hence generating orbital angular momentum and an orbital contribution to the total magnetic moment (Fig. 19.5). Departure from spin-only values is generally large for low-spin d^5 and for high-spin $3d^6$ and $3d^7$ complexes.

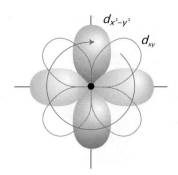

Example 19.2 Inferring an electron configuration from a magnetic moment

The magnetic moment of a certain octahedral Co(II) complex is $4.0\mu_B$. What is its d-electron configuration?

Answer A Co(II) complex is d^7. The two possible configurations are $t_{2g}^5 e_g^2$ (high spin, $S = \frac{3}{2}$) with three unpaired electrons or $t_{2g}^6 e_g^1$ (low spin, $S = \frac{1}{2}$) with one unpaired electron. The spin-only magnetic moments are $3.87\mu_B$ and $1.73\mu_B$, respectively (see Table 19.3). Therefore, the only consistent assignment is the high-spin configuration $t_{2g}^5 e_g^2$.

Self-test 19.2 The magnetic moment of the complex $[Mn(NCS)_6]^{4-}$ is $6.06\mu_B$. What is its electron configuration?

Fig. 19.5 If there is a low-lying orbital of the correct symmetry, the applied field may induce the circulation of the electrons in a complex and hence generate orbital angular momentum. This diagram shows the way in which circulation may arise when the field is applied perpendicular to the xy-plane (perpendicular to this page).

(e) Thermochemical correlations

Key point: The experimental variation in hydration enthalpies reflects a combination of the variation in radii of the ions (the linear trend) and the variation in LFSE (the saw-tooth variation).

The concept of ligand-field stabilization energy helps to explain the double-humped variation in the hydration enthalpies of octahedral $3d$-metal M^{2+} ions (Fig. 19.6). The nearly linear increase across a period shown by the filled circles represents the increasing strength of the bonding between H_2O ligands and the central metal ion as the ionic radii decrease from left to right across the period. The deviation of hydration enthalpies from a straight line reflects the variation in the ligand-field stabilization energies. As Table 19.2 shows, the LFSE increases from d^1 to d^3, decreases again to d^5, then rises to d^8. The filled circles in Fig. 19.6 were calculated by subtracting the high-spin LFSE from $\Delta_{hyd}H$ by using the spectroscopic values of Δ_O in Table 19.1. We see that the LFSE calculated from spectroscopic data accounts for the additional ligand binding energy for the complexes shown in the illustration.

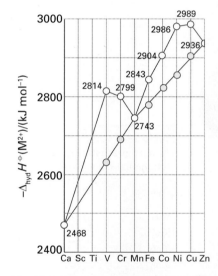

Fig. 19.6 The hydration enthalpy of M^{2+} ions of the first row of the d block. The straight line shows the trend when the ligand-field stabilization energy has been subtracted from the observed values. Note the general trend to greater hydration enthalpy (more exothermic hydration) on crossing the period from left to right.

Example 19.3 Using the LFSE to account for thermochemical properties

The oxides of formula MO, which all have octahedral coordination of the metal ions in a rock-salt structure, have the following lattice enthalpies:

CaO	TiO	VO	MnO
3460	3878	3913	3810 kJ mol^{-1}

Account for the trends in terms of the LFSE.

Answer The general trend across the d block is the increase in lattice enthalpy from CaO (d^0) to MnO (d^5) as the ionic radii of the metal decrease (recall that lattice enthalpy is proportional to $1/(r_+ + r_-)$, Section 3.12). Both Ca^{2+} and Mn^{2+} have an LFSE of zero. TiO (d^2) has an LFSE of $0.8\Delta_O$ and VO (d^3) has an LFSE of $1.2\Delta_O$. It follows that the greater lattice enthalpies of TiO and VO arise from the ligand-field stabilization energy.

Self-test 19.3 Account for the variation in lattice enthalpy of the solid fluorides in which each metal ion is surrounded by an octahedral array of F^- ions: MnF_2 (2780 kJ mol^{-1}), FeF_2 (2926 kJ mol^{-1}), CoF_2 (2976 kJ mol^{-1}), NiF_2 (3060 kJ mol^{-1}), and ZnF_2 (2985 kJ mol^{-1}).

(f) Tetrahedral complexes

Key points: In a tetrahedral complex, the e orbitals lie below the t_2 orbitals; only the high-spin case need be considered.

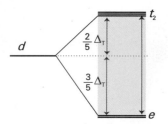

Spherical environment   Tetrahedral crystal field

Fig. 19.7 The orbital energy level diagram used in the application of the building-up principle in a crystal-field analysis of a tetrahedral complex.

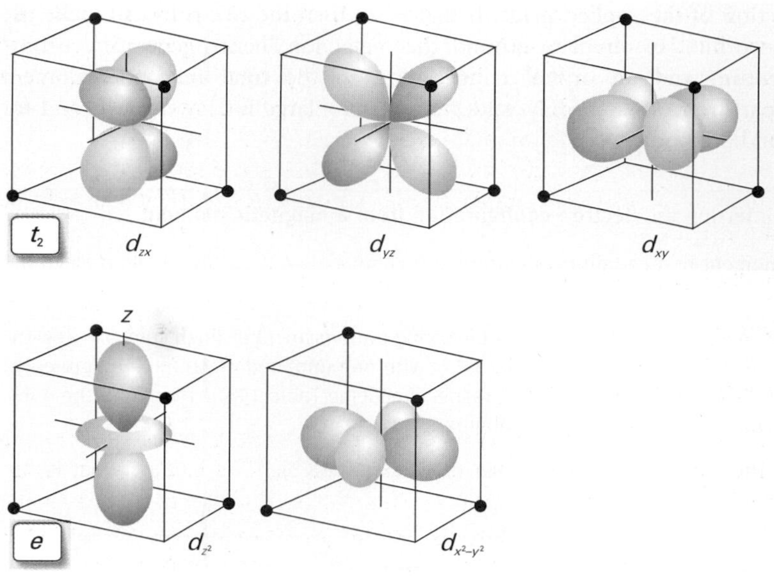

Fig. 19.8 The effect of a tetrahedral crystal field on a set of d orbitals is to split them into two sets; the e pair (which point less directly at the ligands) lie lower in energy than the t_2 triplet.

Second only in abundance to octahedral complexes are the four-coordinate tetrahedral complexes. The same kind of arguments based on crystal-field theory can be applied to these species as we used for octahedral complexes.

A tetrahedral crystal field splits d orbitals into two sets but with the two e orbitals lower in energy than the three t_2 orbitals (Fig. 19.7).[4] The fact that the e orbitals lie below the t_2 orbitals can be understood from a detailed analysis of the spatial arrangement of the orbitals: the e orbitals point between the positions of the ligands and their partial negative charges whereas the t_2 orbitals point more directly towards the ligands (Fig. 19.8). A second difference is that the ligand-field splitting parameter in a tetrahedral complex, Δ_T, is less than Δ_O, as should be expected for complexes with fewer ligands, none of which is oriented directly at the d orbitals (in fact, $\Delta_T \approx \frac{4}{9}\Delta_O$). The pairing energy is invariably more unfavourable than Δ_T, and only high-spin tetrahedral complexes are commonly encountered.

Ligand-field stabilization energies can be calculated in exactly the same way as for octahedral complexes, the only differences being the order of occupation (e before t_2) and the contribution of each orbital to the total energy ($\frac{3}{5}\Delta_T$ for an e orbital and $-\frac{2}{5}\Delta_T$ for a t_2 orbital). Table 19.2 lists some calculated values of the LFSE and Table 19.4 lists some experimental values of Δ_T for a number of complexes. As can be inferred from Table 19.2, the configurations of d^n complexes are

$$e^1 \quad e^2 \quad e^2t_2^1 \quad e^2t_2^2 \quad e^2t_2^3 \quad e^3t_2^3 \quad e^4t_2^3 \quad e^4t_2^4 \quad e^4t_2^5 \quad e^4t_2^6$$

as n increases from 1 to 10.

(g) Square-planar complexes

Key point: A d^8 configuration, coupled with a strong ligand field, favours the formation of square-planar complexes.

The four ligands in a square-planar arrangement gives the d-orbital splitting shown in Fig 19.9, with $d_{x^2-y^2}$ raised above all the others. This arrangement becomes energetically favourable when there are eight d electrons and the crystal field is strong enough to favour the low-spin arrangement $d_{yz}^2 d_{zx}^2 d_{z^2}^2 d_{xy}^2$. Thus, many square-planar complexes are found for the $4d^8$ and $5d^8$ complexes Rh(I), Ir(I), Pt(II), Pd(II), and Au(III), in which

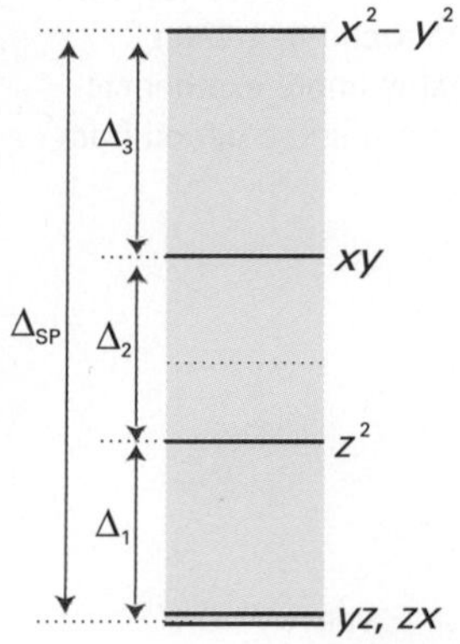

Table 19.4 Values of Δ_T for representative tetrahedral complexes

Complex	Δ_T/cm^{-1}
VCl_4	9010
$[CoCl_4]^{2-}$	3300
$[CoBr_4]^{2-}$	2900
$[CoI_4]^{2-}$	2700
$[Co(NCS)_4]^{2-}$	4700

Fig. 19.9 The orbital splitting parameters for a square-planar complex.

---

[4] Because there is no centre of inversion in a tetrahedral complex, the orbital designation does not include the parity label g or u.

there is a large ligand-field splitting associated with the $4d$- and $5d$-series metals. By contrast, $3d$-series metal complexes such as $[NiX_4]^{2-}$, with X a halogen, are generally tetrahedral because the ligand-field splitting parameter is generally quite small. Only when the ligand is high in the spectrochemical series is the LFSE large enough to result in the formation of a square-planar complex, as for example with $[Ni(CN)_4]^{2-}$.

The sum of the three distinct orbital splittings in Fig. 19.9 is denoted Δ_{SP}. Simple theory predicts that $\Delta_{SP} = 1.3\Delta_O$ for complexes of the same metal and ligands with the same M—L bond lengths.

(h) Tetragonally distorted complexes: the Jahn–Teller effect

Key points: A tetragonal distortion can be expected when the ground electronic configuration of a complex is orbitally degenerate; the complex will distort so as to remove the degeneracy and achieve a lower energy

Six-coordinate d^9 complexes of copper(II) usually depart considerably from octahedral geometry and show pronounced tetragonal distortions. High-spin d^4 (for instance, Mn^{3+}) and low-spin d^7 six-coordinate complexes (for instance, Ni^{3+}) may show a similar distortion, but they are less common. These distortions are manifestations of the **Jahn–Teller effect**: if the ground electronic configuration of a nonlinear complex is orbitally degenerate, and asymmetrically filled, then the complex distorts so as to remove the degeneracy and achieve a lower energy.

The physical origin of the effect is quite easy to identify. Thus, a tetragonal distortion of a regular octahedron, corresponding to extension along the z-axis and compression on the x-and y-axes, lowers the energy of the $e_g(d_{z^2})$ orbital and increases the energy of the $e_g(d_{x^2-y^2})$ orbital (Fig. 19.10). Therefore, if one or three electrons occupy the e_g orbitals (as in high-spin d^4, low-spin d^7, and d^9 complexes) a tetragonal distortion may be energetically advantageous. For example, in a d^9 complex (with configuration that would be $t_{2g}^6 e_g^3$ in O_h), such a distortion leaves two electrons in the d_{z^2} orbital with a lower energy and one in the $d_{x^2-y^2}$ orbital with a higher energy.

The Jahn–Teller effect identifies an unstable geometry (a nonlinear complex with an orbitally degenerate ground state); it does not predict the preferred distortion. For instance, instead of elongation of two axial bonds and the compression of the four equatorial bonds, axial compression and elongation in a plane also removes the degeneracy. Which distortion occurs in practice is a matter of energetics, not symmetry. However, because axial elongation weakens two bonds but equatorial elongation weakens four, axial elongation is more common than axial compression.

A Jahn–Teller distortion can hop from one orientation to another and give rise to the **dynamic Jahn–Teller effect**. For example, below 20 K the electron paramagnetic resonance (EPR) spectrum of $[Cu(OH_2)_6]^{2+}$ shows a static distortion (more precisely, one effectively stationary on the timescale of the resonance experiment). However, above 20 K the distortion disappears because it hops more rapidly than the timescale of the EPR observation.

A Jahn–Teller effect is possible for other electron configurations (for an octahedral complex the d^1, d^2, low-spin d^4 and d^5, high-spin d^6, and d^7 configurations, for a tetrahedral complex the d^1, d^3, d^4, d^6, d^8, and d^9 configurations). However, as neither the e_g orbitals in an octahedral complex nor any of the d orbitals in a tetrahedral complex point directly at the ligands, the effect is too small to induce a measurable distortion in the structure.

(i) Octahedral versus tetrahedral coordination

Key points: Consideration of LFSE allows us to predict that d^3 and d^8 ions will strongly prefer an octahedral geometry over a tetrahedral one; for other configurations the preference is less pronounced, and LFSE has no bearing on the geometry of d^0, d^5, and d^{10} ions.

An octahedral complex has six M—L bonding interactions and, in the absence of significant steric and electronic effects, this arrangement will have a lower energy than a tetrahedral complex with just four M—L bonding interactions. We have already

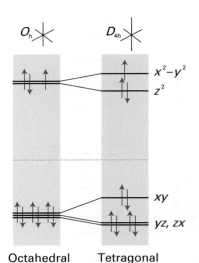

Fig. 19.10 The effect of tetragonal distortions (compression along x and y and extension along z) on the energies of d orbitals. The electron occupation is for a d^9 complex.

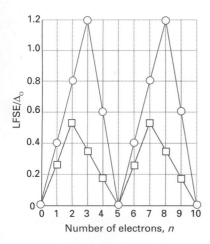

Fig. 19.11 The ligand-field stabilization energy for d^n complexes in octahedral (high-spin, circles) and tetrahedral (squares) complexes. The LFSE is shown in terms of Δ_O, by applying the relation $\Delta_T = \frac{4}{9}\Delta_O$

discussed the effects of steric bulk on a complex (Section 8.1), and have just seen the electronic reasons that favour a square-planar complex. We can now complete the discussion by considering the electronic effects that favour an octahedral complex over a tetrahedral one.

Figure 19.11 illustrates the variation of the LFSE for tetrahedral and high-spin octahedral complexes for all electronic configurations. It is apparent that, in terms of LFSE, octahedral geometries are strongly preferred over tetrahedral for d^3 and d^8 complexes: chromium(III) (d^3) and nickel(II) (d^8) do indeed show an exceptional preference for octahedral geometries. Similarly, d^4 and d^9 configurations show a preference for octahedral complexes (as for Mn(III) and Cu(II); note that the Jahn–Teller effect enhances this preference), whereas tetrahedral complexes of d^1, d^2, d^6, and d^7 ions will not be too disfavoured; thus V(II) (d^2) forms tetrahedral VX_4 complexes with the halides, and Co(II) (d^7) often forms tetrahedral complexes. The geometry of complexes of ions with d^0, d^5, and d^{10} configurations will not be affected by the number of d electrons, as there is no LFSE for these species.

Because the size of the d-orbital splitting, and hence the LFSE, depends on the ligand, it follows that a preference for octahedral coordination will be least pronounced for weak-field ligands. For low-spin complexes, the situation is complicated by the pairing energy, but the LFSE will be greater, with a corresponding greater preference for octahedral over tetrahedral coordination.

The LFSE plays an important role in the solid state by influencing the structures that are adopted by d-metal compounds. This influence is demonstrated by the ways in which the different metal ions A and B in spinels (of formula AB_2O_4, Section 3.9b) occupy the octahedral or tetrahedral sites. Thus, Co_3O_4 is a normal spinel because the d^6 Co(III) ion strongly favours octahedral coordination, resulting in $(Co^{2+})_t(2Co^{3+})_oO_4$, whereas Fe_3O_4 (magnetite) is an inverse spinel because Fe(II), but not Fe(III), can acquire greater LFSE by occupying an octahedral site. Thus magnetite is formulated as $(Fe^{3+})_t$ $(Fe^{2+}Fe^{3+})_oO_4$.

Finally, the LFSE has an important influence on the rates of substitution of ligands in d-metal complexes. This role is described in Section 20.2.

19.2 Ligand-field theory

Crystal-field theory provides a simple conceptual model that can be used to interpret magnetic, spectroscopic, and thermochemical data by using empirical values of Δ_O. However, the theory is defective because it treats ligands as point charges or dipoles and does not take into account the overlap of ligand and metal atom orbitals. One consequence of this oversimplification is that crystal-field theory cannot account for the ligand spectrochemical series. **Ligand-field theory**, which is an application of molecular orbital theory that concentrates on the d orbitals of the central metal atom, provides a more substantial framework for understanding the origins of Δ_O.

The strategy for describing the molecular orbitals of a d-metal complex follows procedures similar to those described in Chapter 2 for bonding in polyatomic molecules: the valence orbitals on the metal and ligand are used to form symmetry-adapted linear combinations (SALCs; Section 7.6), and then estimating the relative energies of the molecular orbitals by using empirical energy and overlap considerations. These relative energies can be verified and positioned more precisely by comparison with experimental data (particularly optical absorption and photoelectron spectroscopy).[5]

The structure of this section is as follows. We start with octahedral complexes, initially considering only σ bonding between the metal atom and the ligands. Then we consider the effect of π bonding, and see that it is essential for understanding Δ_O (which is one reason why crystal-field theory cannot explain the spectrochemical series). Finally, we consider complexes with different symmetries, and see that similar arguments apply to

[5] High-resolution photoelectron spectroscopy (PES) data are available only for complexes in the gas phase, and are therefore restricted to uncharged complexes.

them. In the next part of the chapter, we see how information from optical spectroscopy is used to refine the discussion and provide quantitative data on the ligand-field splitting parameter and electron–electron repulsion energies.

(a) σ Bonding

Key point: In ligand-field theory, the building-up principle is used in conjunction with a molecular orbital energy level diagram constructed from metal orbitals and linear combinations of ligand orbitals.

We begin by considering an octahedral complex in which each ligand (L) has a single valence orbital directed towards the central metal atom (M); each of these orbitals has local σ symmetry with respect to the M–L axis. Examples of such ligands include the NH_3 molecule and the F^- ion.

In an octahedral (O_h) environment, the orbitals of the central metal atom divide by symmetry into four sets (Fig. 19.12 and *Resource section 4*):

Metal orbital	Symmetry label	Degeneracy
s	a_{1g}	1
p_x, p_y, p_z	t_{1u}	3
$d_{x^2-y^2}, d_{z^2}$	e_g	2
d_{xy}, d_{yz}, d_{zx}	t_{2g}	3

Six symmetry-adapted linear combinations of the six ligand σ orbitals can also be formed. These combinations can be taken from *Resource section 5* and are also shown in Fig. 19.12. One (unnormalized) SALC has symmetry a_{1g}:

$$a_{1g}: \quad \sigma_1 + \sigma_2 + \sigma_3 + \sigma_4 + \sigma_5 + \sigma_6$$

where σ_i denotes a σ orbital on ligand i. There are three SALCs of symmetry t_{1u}:

$$t_{1u}: \quad \sigma_1 - \sigma_3, \quad \sigma_2 - \sigma_4, \quad \sigma_5 - \sigma_6$$

and two SALCs of symmetry e_g:

$$e_g: \quad \sigma_1 - \sigma_2 + \sigma_3 - \sigma_4, \quad 2\sigma_6 + 2\sigma_5 - \sigma_1 - \sigma_2 - \sigma_3 - \sigma_4$$

These six SALCs account for all the ligand orbitals of σ symmetry: there is no combination of ligand σ orbitals that has the symmetry of the metal t_{2g} orbitals, so the latter do not participate in σ bonding.[6]

Molecular orbitals are formed by combining SALCs and metal atomic orbitals of the same symmetry. For example, the (unnormalized) form of an a_{1g} molecular orbital is $c_M\psi_{Ms} + c_L\psi_{La_{1g}}$, where ψ_{Ms} is the s orbital on the metal atom M and $\psi_{La_{1g}}$ is the ligand SALC of symmetry a_{1g}. The metal s orbital and and ligand a_{1g} SALC overlap to give two molecular orbitals, one bonding and one antibonding. Similarly, the doubly degenerate metal e_g orbitals and the ligand e_g SALC overlap to give four molecular orbitals (two degenerate bonding, two degenerate antibonding), and the triply degenerate metal t_{1u} orbitals and the three t_{1u} SALCs overlap to give six molecular orbitals (three degenerate bonding, three degenerate antibonding). There are therefore six bonding combinations in all and six antibonding combinations. The three triply degenerate metal t_{2g} orbitals remain nonbonding and fully localized on the metal atom. Calculations of the resulting energies (adjusted to agree with a variety of spectroscopic data of the kind to be discussed in Section 19.4) result in the molecular orbital energy level diagram shown in Fig. 19.13.

The greatest contribution to the molecular orbital of lowest energy is from atomic orbitals of lowest energy (Section 2.9). For NH_3, F^-, and most other ligands, the ligand σ orbitals are derived from atomic orbitals with energies that lie well below those of the metal d orbitals. As a result, the six bonding molecular orbitals of the complex are mainly ligand-orbital in character (that is, $c_L^2 > c_M^2$). These six bonding orbitals can accommodate the 12 electrons provided by the six ligand lone pairs. The electrons that we can

---

[6] The normalization constants (with overlap neglected) are $N(a_{1g}) = (\frac{1}{6})^{1/2}$, $N(t_{1u}) = (\frac{1}{2})^{1/2}$, for all three orbitals, and $N(e_g) = \frac{1}{2}$ and $(\frac{1}{12})^{1/2}$, respectively.

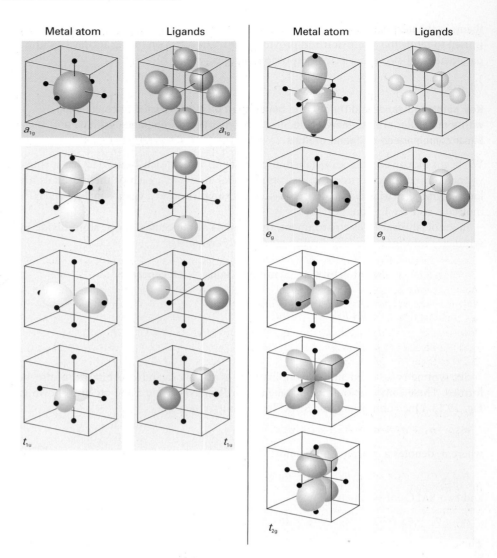

Fig. 19.12 Symmetry-adapted combinations of ligand σ orbitals (represented here by spheres) in an octahedral complex. For symmetry adapted orbitals in other point groups, see *Resource section 5*.

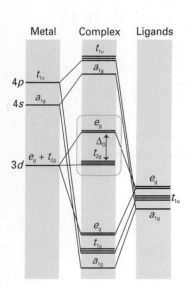

Fig. 19.13 Molecular orbital energy levels of a typical octahedral complex. The frontier orbitals are inside the tinted box.

regard as provided by the ligands are therefore largely confined to the ligands in the complex, just as the crystal-field theory presumes. However, because the coefficients c_M are nonzero, the bonding molecular orbitals do have some *d*-orbital character and the 'ligand electrons' are partly delocalized on to the central metal atom.

The total number of electrons to accommodate, in addition to those supplied by the ligands, now depends on the number of *d* electrons, *n*, supplied by the metal atom. These additional electrons enter the orbitals next in line for occupation, which are the nonbonding *d* orbitals (the t_{2g} orbitals) and the antibonding combination (the upper e_g orbitals) of the *d* orbitals and ligand orbitals. The t_{2g} orbitals are wholly confined (in the present approximation) to the metal atom and the antibonding e_g orbitals are largely metal-atom in character too, so the *n* electrons supplied by the central atom remain largely on that atom. The frontier orbitals of the complex are therefore the nonbonding entirely metal t_{2g} orbitals and the antibonding mainly metal e_g orbitals. Thus, we have arrived at an arrangement that is qualitatively the same as in crystal-field theory. In the ligand-field approach the octahedral ligand-field splitting parameter, Δ_O, is the separation between the molecular orbitals largely, but not completely, confined to the metal atom.

With the molecular orbital energy level diagram established, we use the building-up principle to construct the ground-state electron configuration of the complex. For a d^n complex, there are $12 + n$ electrons to accommodate. The six bonding molecular orbitals accommodate the 12 electrons supplied by the ligands. The remaining *n* electrons are accommodated in the nonbonding t_{2g} orbitals and the antibonding e_g orbitals. Now the story is essentially the same as for crystal-field theory, the types of complexes that are obtained (high-spin or low-spin, for instance) depending on the relative values of Δ_O and the pairing energy P. The principal difference from the crystal-field discussion is that ligand-field theory gives deeper insight into the origin of the ligand-field splitting, and we can begin to understand why some ligands are strong and others are weak. For instance, a good σ-donor ligand should result in strong metal–ligand overlap, hence a more strongly antibonding e_g set and consequently a larger value of Δ_O. However, before drawing further conclusions, we must go on to consider what crystal-field theory ignores completely, the role of π bonding.

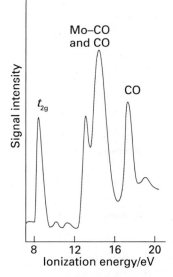

Fig. 19.14 The He(II) (30.4 nm) photoelectron spectrum of $Mo(CO)_6$. With six electrons from Mo and twelve from :CO, the ground-state configuration of the complex is $a_{1g}^2 t_{1u}^6 e_g^4 t_{2g}^6$. (From B.R. Higginson, D.R. Lloyd, P. Burroughs, D.M. Gibson, and A.F. Orchard, *J. Chem. Soc., Faraday II*, 1973, **69**, 1659.)

Example 19.4 Using a photoelectron spectrum to obtain information about a complex

The photoelectron spectrum of gas-phase $[Mo(CO)_6]$ is shown in Fig. 19.14. Use the spectrum to infer the energies of the molecular orbitals of the complex.

Answer Twelve electrons are provided by the six CO ligands (treated as :CO); they enter the bonding orbitals and result in the configuration $a_{1g}^2 t_{1u}^6 e_g^4$. The oxidation number of molybdenum, Group 6, is 0, so Mo provides a further six valence electrons. The ligand and metal valence electrons are distributed over the orbitals shown in the box in Fig. 19.13 and, as CO is a strong-field ligand, the ground-state electron configuration of the complex is expected to be low-spin: $a_{1g}^2 t_{1u}^6 e_g^4 t_{2g}^6$. The HOMOs are the three t_{2g} orbitals largely confined to the Mo atom, and their energy can be identified by ascribing the peak of lowest ionization energy (close to 8 eV) to them. The group of ionization energies around 14 eV are probably due to the Mo–CO σ-bonding orbitals. The value of 14 eV is close to the ionization energy of CO itself, so the variety of peaks at that energy also arise from bonding orbitals in CO.

Self-test 19.4 Suggest an interpretation of the photoelectron spectra of $[Fe(C_5H_5)_2]$ and $[Mg(C_5H_5)_2]$ shown in Fig. 19.15.

(b) π Bonding

Key points: π-Donor ligands decrease Δ_O whereas π-acceptor ligands increase Δ_O; the spectrochemical series is largely a consequence of the effects of π bonding when such bonding is feasible.

If the ligands in a complex have orbitals with local π symmetry with respect to the M–L axis (as two of the p orbitals of a halide ligand have), they may form bonding and antibonding π orbitals with the metal orbitals (Fig. 19.16). For an octahedral complex the combinations that can be formed from the ligand π orbitals include SALCs of t_{2g} symmetry. These ligand combinations have net overlap with the metal t_{2g} orbitals, which are therefore no longer purely nonbonding on the metal atom. Depending on the relative energies of the ligand and metal orbitals, the energies of the now molecular t_{2g} orbitals lie above or below the energies they had as nonbonding atomic orbitals, so Δ_O is decreased or increased, respectively.

To explore the role of π bonding in more detail, we need two of the general principles described in Chapter 2. First, we shall make use of the idea that, when atomic orbitals overlap strongly, they mix strongly: the resulting bonding molecular orbitals are significantly lower in energy and the antibonding molecular orbitals are significantly higher in energy than the atomic orbitals. Second, we note that atomic orbitals with similar energies interact strongly, whereas those of very different energies mix only slightly even if their overlap is large.

A **π-donor ligand** is a ligand that, before any bonding is considered, has filled orbitals of π symmetry around the M–L axis. Such ligands include Cl^-, Br^-, OH^-, and H_2O. In Lewis acid–base terminology, π-donor ligands are **π bases** (Section 4.8). The energies of the full π orbitals on the ligands will not normally be higher than their σ-donor orbitals (HOMO) and must therefore also be lower in energy that the metal d orbitals. Because the full π orbitals of π donor ligands lie lower in energy than the partially filled d orbitals of the metal, when they form molecular orbitals with the metal t_{2g} orbitals, the bonding

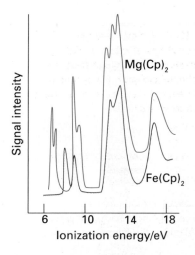

Fig. 19.15 Photoelectron spectra of ferrocene and magnesocene.

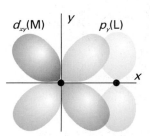

Fig. 19.16 The π overlap that may occur between a ligand p orbital perpendicular to the M–L axis and a metal d_{xy} orbital.

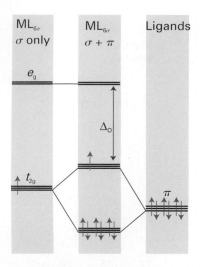

Fig. 19.17 The effect of π bonding on the ligand-field splitting parameter. Ligands that act as π donors decrease Δ_O. Only the π orbitals of the ligand are shown.

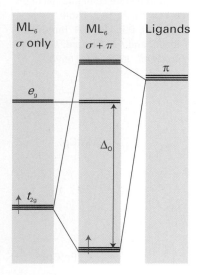

Fig. 19.18 Ligands that act as π acceptors increase Δ_O. Only the π orbitals of the ligand are shown.

combination lies lower than the ligand orbitals and the antibonding combination lies above the energy of the *d* orbitals of the free metal atom (Fig. 19.17). The electrons supplied by the ligand π orbitals occupy and fill the bonding combinations, leaving the electrons originally in the *d* orbitals of the central metal atom to occupy the antibonding t_{2g} orbitals. The net effect is that the previously nonbonding metal t_{2g} orbitals become antibonding and hence are raised closer in energy to the antibonding e_g orbitals. It follows that *π-donor ligands decrease* Δ_O.

A **π-acceptor ligand** is a ligand that has empty π orbitals that are available for occupation. In Lewis acid–base terminology, a π-acceptor ligand is a **π acid**. Typically, the π-acceptor orbitals are vacant antibonding orbitals on the ligand (usually the LUMO), as in CO and N_2, which are higher in energy than the metal *d* orbitals. The $π^\star$ orbital of CO, for instance, has its largest amplitude on the C atom and has the correct symmetry for overlap with the metal t_{2g} orbitals, so CO can act as a π-acceptor ligand (Section 21.5). Phosphines (PR_3) are also able accept π-electron density and also act as π acceptors (Section 21.6).

Because the π-acceptor orbitals on most ligands are higher in energy than the metal *d* orbitals, they form molecular orbitals in which the bonding t_{2g} combinations are largely of metal *d*-orbital character (Fig. 19.18). These bonding combinations lie lower in energy than the *d* orbitals themselves. The net result is that *π-acceptors increase* Δ_O.

We can now put the role of π bonding in perspective. The order of ligands in the spectrochemical series is partly that of the strengths with which they can participate in M–L σ bonding. For example, both CH_3^- and H^- are very high in the spectrochemical series because they are very strong σ donors. However, when π bonding is significant, it has a strong influence on Δ_O: π-donor ligands decrease Δ_O and π-acceptor ligands increase Δ_O. This effect is responsible for CO (a strong π acceptor) being high on the spectrochemical series and for OH^- (a strong π donor) being low in the series. The overall order of the spectrochemical series may be interpreted in broad terms as dominated by π effects (with a few important exceptions), and in general the series can be interpreted as follows:

$$- \text{ increasing } \Delta_O \rightarrow$$
$$\text{π donor} < \text{weak π donor} < \text{no π effects} < \text{π acceptor}$$

Representative ligands that match these classes are

π donor	weak π donor	no π effects	π acceptor
I^-, Br^-, Cl^-, F^-	H_2O	NH_3	PR_3, CO

Notable examples of where the effect of σ bonding dominates include amines (NR_3), CH_3^-, and H^-, none of which has orbitals of π symmetry of an appropriate energy and thus are neither π-donor nor π-acceptor ligands. For *d* metals, the spectrochemical series of ligands largely replicates the hard/soft classification of ligands outlined in Section 4.10; thus we can see that hard ligands are either strong σ or π donors, and soft ligands are π acceptors. It is important to note that the classification of a ligand as strong field or weak-field does not give any guide as to the strength of the M—L bond.

Electronic spectra

Now that we have considered the electronic structure of *d*-metal complexes, we are in a position to understand their electronic spectra and to use the data they provide to refine the discussion of structure. The magnitudes of ligand-field splittings are such that the energy of electronic transitions corresponds to an absorption of ultraviolet radiation and visible light. However, the presence of electron–electron repulsions within the metal orbitals mean that the absorption frequencies are not in general a direct portrayal of the ligand-field splitting. The role of electron–electron repulsion was originally determined by the analysis of atoms and ions in the gas phase, and much of that information can be used in the analysis of the spectra of metal complexes provided we take into account the lower symmetry of a complex.

Keep in mind that the purpose of the following development is to find a way to extract the value of the ligand-field splitting parameter from the electronic absorption spectrum of a complex with more than one d electron, when electron–electron repulsions are important. First, we discuss the spectra of free atoms, and see how to take electron–electron repulsions into account. Then we see what energy states atoms adopt when they are embedded in an octahedral ligand field. Finally, we see how to represent the energies of these states for various field strengths and electron–electron repulsion energies (in the Tanabe–Sugano diagrams of Section 19.4e), and how to use these diagrams to extract the value of the ligand-field splitting parameter.

19.3 Electronic spectra of atoms

Key point: Electron–electron repulsions result in multiple absorptions in the electronic spectrum.

Figure 19.19 sets the stage for our discussion by showing the electronic absorption spectrum of the d^3 complex $[Cr(NH_3)_6]^{3+}$ in aqueous solution. The band at lowest energy (longest wavelength) is very weak; later we shall see that it is an example of a 'spin-forbidden' transition. Next are two bands with intermediate intensities; these are 'spin-allowed' transitions between the t_{2g} and e_g orbitals of the complex, which are mainly derived from the metal d orbitals. The third feature in the spectrum is an intense charge-transfer band at short wavelength (labelled CT, denoting 'charge transfer'), of which only the low-energy tail is evident in the illustration.

One problem that immediately confronts us is why two absorptions can be ascribed to the apparently single transition $t_{2g}^2 e_g^1 \leftarrow t_{2g}^3$. This splitting of a single transition into two bands is in fact an outcome of the electron–electron repulsions mentioned above. To understand how it arises, and to extract the information it contains, we need to consider the spectra of free atoms.

(a) Spectroscopic terms

Key points: Different microstates exist for the same electronic configuration; for light atoms, Russell–Saunders coupling is used to describe the terms, which are specified by symbols in which the value of L is indicated by one of the letters S, P, D, ..., and the value of $2S + 1$ is given as a left superscript.

In Chapter 1 we expressed the electronic structures of atoms by giving their electronic configurations, the designation of the number of electrons in each orbital (as in $1s^2 2s^1$ for Li). However, a configuration is an incomplete description of the arrangement of electrons in atoms. In the configuration $2p^2$, for instance, the two electrons might

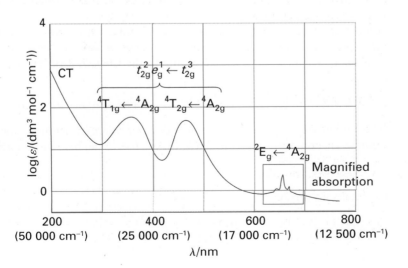

Fig. 19.19 The spectrum of the d^3 complex $[Cr(NH_3)_6]^{3+}$, which illustrates the features studied in this section, and the assignments of the transitions as explained in the text.

occupy orbitals with different orientations of their orbital angular momenta (that is, with different values of m_l from among the possibilities $+1$, 0, and -1 that are available when $l = 1$). Similarly, the designation $2p^2$ tells us nothing about the spin orientations of the two electrons $m_s = +\frac{1}{2}$ or $-\frac{1}{2}$. The atom may in fact have several different states of total orbital and spin angular momenta, each one corresponding to the occupation of orbitals with different values of m_l by electrons with different values of m_s. The different ways in which the electrons can occupy the orbitals specified in the configuration are called the **microstates** of the configuration. For example, one micro-state of a $2p^2$ configuration is $(1^+, 1^-)$; this notation signifies that both electrons occupy an orbital with $m_l = +1$ but do so with opposite spins, the superscript $+$ indicating $m_s = +\frac{1}{2}$ and $-$ indicating $m_s = -\frac{1}{2}$. Another microstate of the same configuration is $(-1^+, 0^+)$. In this microstate, both electrons have $m_s = +\frac{1}{2}$ but one occupies the $2p$ orbital with $m_l = -1$ and the other occupies the orbital with $m_l = 0$.

The microstates of a given configuration have the same energy only if elecron–electron repulsions on the atom are negligible. However, because atoms and most molecules are compact, interelectronic repulsions are strong and cannot always be ignored. As a result, microstates that correspond to different relative spatial distributions of electrons have different energies. If we group together the microstates that have the same energy when electron–electron repulsions are taken into account, we obtain the spectroscopically distinguishable energy levels called **terms.**

For light atoms and the $3d$ series, it turns out that the most important property of a microstate for helping us to decide its energy is the relative orientation of the spins of the electrons. Next in importance is the relative orientation of the orbital angular momenta of the electrons. It follows that we can identify the terms of light atoms and put them in order of energy by sorting the microstates according to their total spin quantum number S (which is determined by the relative orientation of the individual spins) and then according to their total orbital angular momentum quantum number L (which is determined by the relative orientation of the individual orbital angular momenta of the electrons). For heavy atoms, such as those of the $4d$ and $5d$ series, the relative orientations of orbital momenta or of spin momenta are less important. In these atoms the spin and orbital angular momenta of individual electrons are strongly coupled together by **spin–orbit coupling,** so the relative orientation of the spin and orbital angular momenta of each electron is the most important feature for deter-mining the energy. The terms of heavy atoms are therefore sorted on the basis of the values of the total angular momentum quantum number j for an electron in each microstate.

The process of combining electron angular momenta by summing first the spins, then the orbital momenta, and finally combining the two resultants is called **Russell–Saunders coupling.** This coupling scheme is used to identify the terms of light atoms (i.e. the $3d$ metals), and we consider it in detail here. The coupling scheme most appropriate to heavy atoms (that is, atoms of the $4d$ and $5d$ series of elements) is called *jj*-**coupling,** but we shall not consider it further.

Our first task is to identify the values of L and S that can arise from the orbital and spin angular momenta of individual electrons. Suppose we have two electrons with quantum numbers l_1, s_1 and l_2, s_2. Then, according to the **Clebsch–Gordan series,** the possible values of L and S are

$$L = l_1 + l_2, l_1 + l_2 - 1, \ldots, |l_1 - l_2| \quad S = s_1 + s_2, s_1 + s_2 - 1, \ldots, |s_1 - s_2| \quad (19.3)$$

(The modulus signs appear because neither L nor S can, by definition, be negative.) For example, an atom with configuration d^2 ($l_1 = 2$, $l_2 = 2$) can have the following values of L:

$$L = 2 + 2, 2 + 2 - 1, \ldots, |2 - 2| = 4, 3, 2, 1, 0$$

The total spin (because $s_1 = \frac{1}{2}$, $s_2 = \frac{1}{2}$) can be

$$S = \frac{1}{2} + \frac{1}{2}, \frac{1}{2} + \frac{1}{2} - 1, \ldots, \left|\frac{1}{2} - \frac{1}{2}\right| = 1, 0$$

To find the values of L and S for atoms with three electrons, we continue the process by combining l_3 with the value of L just obtained, and likewise for s_3.

Once L and S have been found, we can write down the allowed values of the quantum numbers M_L and M_S,

$$M_L = L, L-1\ldots, -L \qquad M_S = S, S-1, \ldots, -S$$

These quantum numbers give the orientation of the angular momentum relative to an arbitrary axis: there are $2L+1$ values of M_L for a given value of L and $2S+1$ values of M_S for a given value of S. The values of M_L and M_S for a given microstate can be found very easily by adding together the values of m_l or m_s for the individual electrons. Therefore, if one electron has the quantum number m_{l_1} and the other has m_{l_2}, then

$$M_L = m_{l1} + m_{l2}$$

A similar expresson applies to the total spin:

$$M_S = m_{s1} + m_{s2}$$

Thus, for example, $(0^+, -1^-)$ is a microstate with $M_L = 0-1 = -1$ and $M_S = \frac{1}{2} + \left(-\frac{1}{2}\right) = 0$ and may contribute to any term for which these two quantum numbers apply.

By analogy with the notation $s, p, d, \ldots$ for orbitals with $l = 0, 1, 2, \ldots$, the total orbital angular momentum of an atomic term is denoted by the equivalent uppercase letter:

$$L = 0 \quad 1 \quad 2 \quad 3 \quad 4 \quad \ldots$$
$$ S \quad P \quad D \quad F \quad G \quad \text{then alphabetical (omitting J)}$$

The total spin is normally reported as the value of $2S+1$, which is called the **multiplicity** of the term:

$$S = 0 \quad \frac{1}{2} \quad 1 \quad \frac{3}{2} \quad 2 \quad \ldots$$
$$2S + 1 = 1 \quad 2 \quad 3 \quad 4 \quad 5 \quad \ldots$$

The multiplicity is written as a left superscript on the letter representing the value of L, and the entire label of a term is called a **term symbol**. Thus, the term symbol ^3P denotes a term (a collection of nearly degenerate states) with $L = 1$ and $S = 1$, and is called a *triplet term*.

Example 19.5 Deriving term symbols

Give the term symbols for an atom with the configurations (a) s^1, (b) p^1, and (c) s^1p^1.

Answer (a) The single s electron has $l = 0$ and $s = \frac{1}{2}$. Because there is only one electron, $L = 0$ (an S term), $S = \frac{1}{2}$, and $2S + 1 = 2$ (a doublet term). The term symbol is therefore ^2S. (b) For a single p electron, $l = 1$ so $L = 1$ and the term is ^2P. (These terms arise in the spectrum of an alkali metal atom, such as Na.) (c) With one s and one p electron, $L = 0 + 1 = 1$, a P term. The electrons may be paired ($S = 0$) or parallel ($S = 1$). Hence both ^1P and ^3P terms are possible.

Self-test 19.5 What terms arise from a p^1d^1 configuration?

(b) The classification of microstates

Key point: The allowed terms of a configuration are found by identifying the values of L and S to which the microstates of an atom can contribute.

The Pauli principle restricts the microstates that can occur in a configuration and consequently affects the terms that can occur. For example, two electrons cannot both have the same spin and be in a d orbital with $m_l = +2$. Therefore, the microstate $(2^+, 2^+)$ is forbidden and so are the values of L and S to which such a microstate might contribute. We shall illustrate how to determine what terms are allowed by considering a d^2

Table 19.5 Microstates of the d^2 configuration

M_L	M_S		
	-1	0	$+1$
$+4$		$(2^+,2^-)$	
$+3$	$(2^-,1^-)$	$(2^+,1^-)(2^-,1^+)$	$(2^+,1^+)$
$+2$	$(2^-,0^-)$	$(2^+,0^-)(2^-,0^+)$ $(1^+,1^-)$	$(2^+,0^+)$
$+1$	$(2^-,-1^-)(1^-,0^-)$	$(2^+,-1^-)(2^-,-1^+)$ $(1^+,0^-)$ $(1^-,0^+)$	$(2^+,-1^+)(1^+,0^+)$
0	$(1^-,-1^-)(2^-,-2^-)$	$(1^+,-1^-)(1^-,-1^+)$ $(2^+,-2^-)(2^-,-2^+)$ $(0^+,0^-)$	$(1^+,-1^+)(2^+,-2^+)$
-1 to -4^*			

* The lower half of the diagram is a reflection of the upper half.

configuration, as the outcome will be useful in the discussion of the complexes encountered later in the chapter. An example of a species with a d^2 configuration is a Ti^{2+} ion.

We start the analysis by setting up a table of microstates of the d^2 configuration (Table 19.5); only the microstates allowed by the Pauli principle have been included. We then use a process of elimination to classify all the microstates. First, we note the largest value of M_L, which for a d^2 configuration is $+4$. This state must belong to a term with $L=4$ (a G term). Table 19.5 shows that the only value of M_S that occurs for this term is $M_S=0$, so the G term must be a singlet. Moreover, as there are nine values of M_L when $L=4$, one of the microstates in each of the boxes in the column below $(2^+,2^-)$ must belong to this term.[7] We can therefore strike out one microstate from each row in the central column of Table 19.5, which leaves 36 microstates.

The next largest value is $M_L=+3$, which must stem from $L=3$ and hence belong to an F term. That row contains one microstate in each column (that is, each box contains one unassigned combination for $M_S=-1$, 0, and $+1$), which signifies $S=1$ and a triplet term. Hence the microstates belong to ^3F. The same is true for one microstate in each of the rows down to $M_L=-3$, which accounts for a further $3\times 7=21$ microstates. If we strike out one state in each of the 21 boxes, we are left with 15 to be assigned.

There is one unassigned microstate in the row with $M_L=+2$ (which must arise from $L=2$) and the column under $M_S=0$ ($S=0$), which must therefore belong to a ^1D term. This term has five values of M_L, which removes one microstate from each row in the column headed $M_S=0$ down to $M_L=-2$, leaving 10 microstates unassigned. Because these unassigned microstates include one with $M_L=+1$ and $M_S=+1$, nine of these microstates must belong to a ^3P term. There now remains only one microstate in the central box of the table, with $M_L=0$ and $M_S=0$. This microstate must be the one and only state of a ^1S term (which has $L=0$ and $S=0$).

At this point we can conclude that the terms of a $3d^2$ configuration are ^1G, ^3F, ^1D, ^3P, and ^1S. These terms account for all 45 permitted states (see table in the margin).

(c) **The energies of the terms**

Key point: Hund's rules indicate the likely ground term of a gas-phase atom or ion.

Once the values of L and S that can arise from a given configuration are known, it is possible to identify the term of lowest energy by using Hund's rules. The first of these

Term	Number of states
^1G	$9\times 1=9$
^3F	$7\times 3=21$
^1D	$5\times 1=5$
^3P	$3\times 3=9$
^1S	$1\times 1=1$
Total:	45

---

[7] In fact, it is unlikely that one of the microstates itself will correspond to one of these states: in general, a state is a linear combination of microstates. However, as N linear combinations can be formed from N microstates, each time we cross off one microstate, we are taking one linear combination into account, so the bookkeeping is correct even though the detail may be wrong.

empirical rules was introduced in Section 1.8, where it was expressed as 'the lowest energy configuration is achieved if the electron spins are parallel'. Because a high value of S stems from parallel electron spins, an alternative statement is

1 For a given configuration, the term with the greatest multiplicity lies lowest in energy.

The rule implies that a triplet term of a configuration (if one is permitted) has a lower energy than a singlet term of the same configuration. For the d^2 configuration, this rule predicts that the ground state will be either ^3F or ^3P.

By inspecting spectroscopic data, Hund also identified a second rule for the relative energies of the terms of a given multiplicity:

2 For a term of given multiplicity, the term with the greatest value of L lies lowest in energy.

The physical justification for this rule is that when L is high, the electrons can stay clear of one another and hence experience a lower repulsion. If L is low, the electrons are more likely to be closer to each other, and hence repel one another more strongly. The second rule implies that, of the two triplet terms of a d^2 configuration, the ^3F term is lower in energy than the ^3P term. It follows that the ground term of a d^2 species such as Ti^{2+} is expected to be ^3F.

The spin multiplicity rule is fairly reliable for predicting the ordering of terms, but the 'greatest L' rule is reliable only for predicting the ground term, the term of lowest energy; there is generally little correlation of L with the order of the higher terms. Thus, for d^2 the rules predict the order

$$^3F < {}^3P < {}^1G < {}^1D < {}^1S$$

but the order observed for Ti^{2+} from spectroscopy is

$$^3F < {}^1D < {}^3P < {}^1G < {}^1S$$

Normally, all we want to know is the identity of the ground term of an atom or ion. The procedure may then be simplified and summarized as follows:

1 Identify the microstate that has the highest value of M_S.

This step tells us the highest multiplicity of the configuration.

2 Identify the highest permitted value of M_L for that multiplicity.

This step tells us the highest value of L consistent with the highest multiplicity.

Example 19.6 Identifying the ground term of a configuration

What is the ground term of the configurations (a) $3d^5$ of Mn^{2+} and (b) $3d^3$ of Cr^{3+}?

Answer (a) The d^5 configuration permits occupation of each d orbital singly, so the maximum value of S is $\frac{5}{2}$ and the maximum multiplicity is $2 \times \frac{5}{2} + 1 = 6$, a sextet term. If each of the electrons is to have the same spin quantum number, all must occupy different orbitals and hence have different M_L values. Thus, the M_L values of the occupied orbitals will be $+2, +1, 0, -1$, and -2. The sum of these values is 0, so $L = 0$ and the term is ^6S. (b) For the configuration d^3, the maximum multiplicity corresponds to all three electrons having the same spin quantum number, so $S = \frac{3}{2}$. The multiplicity is therefore $2 \times \frac{3}{2} + 1 = 4$, a quartet. The three M_L values must be different if all spins are the same, which allows a maximum value of $M_L = 2 + 1 + 0 = +3$, indicating $L = 3$, an F term. Hence, the ground term of d^3 is ^4F.

Self-test 19.6 Identify the ground terms of (a) $2p^2$ and (b) $3d^9$. (*Hint*: Because d^9 is one electron short of a closed shell with $L = 0$ and $S = 0$, treat it on the same footing as a d^1 configuration.)

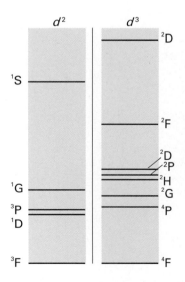

Fig. 19.20 The relative energies of the terms arising from d^2 (left) and d^3 (right) configurations of a free atom.

Figure 19.20 shows the relative energies of the terms for the d^2 and d^3 configurations of free atoms. Later, we shall see how to extend these diagrams to include the effect of a crystal field (Section 19.4).

(d) Racah parameters

Key points: The Racah parameters summarize the effects of electron–electron repulsion on the energies of the terms that arise from a single configuration; the parameters are the quantitative expression of the ideas underlying Hund's rules and account for deviations from them.

Different terms of a configuration have different energies on account of the repulsion between electrons. To calculate the energies of the terms we must evaluate these electron–electron repulsion energies as complicated integrals over the orbitals occupied by the electrons. Mercifully, though, all the integrals for a given configuration can be collected together in three specific combinations and the repulsion energy of any term of a configuration can be expressed as a sum of these three quantities. The three combinations of integrals are called the **Racah parameters** and denoted A, B, and C. The parameter A corresponds to an average of the total interelectron repulsion and B and C relate to the repulsion energies between individual d electrons. We do not even need to know the theoretical values of the parameters or the theoretical expressions for them, because it is more reliable to use A, B, and C as empirical quantities obtained from gas-phase atomic spectroscopy.

Each term stemming from a given configuration has an energy that may be expressed as a linear combination of all three Racah parameters. For a d^2 configuration a detailed analysis shows that

$$E(^1S) = A + 14B + 7C \quad E(^1G) = A + 4B + 2C \quad E(^1D) = A - 3B + 2C$$
$$E(^3P) = A + 7B \qquad\qquad E(^3F) = A - 8B$$

The values of A, B, and C can be determined by fitting these expressions to the observed energies of the terms. Note that A is common to all the terms (as remarked above it is the average of the total interelectron repulsion energy); therefore, if we are interested only in their relative energies, we do not need to know its value. Likewise, if we are interested only in the relative energies of the two triplet terms, we also do not need to know the value of C.

All three Racah parameters are positive as they represent electron–electron repulsions. Therefore, provided $C > 5B$, the energies of the terms of the d^2 configuration lie in the order

$$^3F < {}^3P < {}^1D < {}^1G < {}^1S$$

This order is nearly the same as obtained by using Hund's rules. However, if $C < 5B$, the advantage of having an occupation of orbitals that corresponds to a high orbital angular momentum is greater than the advantage of having a high multiplicity, and the ^3P term lies above ^1D (as is in fact the case for Ti^{2+}). Table 19.6 shows some experimental values of B and C. The values in parentheses indicate that $C \approx 4B$, so the ions listed there are in

Table 19.6 Racah parameters for some *d*-block ions*

	1+	2+	3+	4+
Ti		720 (3.7)		
V		765 (3.9)	860 (4.8)	
Cr		830 (4.1)	1030 (3.7)	1040 (4.1)
Mn		960 (3.5)	1130 (3.2)	
Fe		1060 (4.1)		
Co		1120 (3.9)		
Ni		1080 (4.5)		
Cu	1220 (4.0)	1240 (3.8)		

* The table gives the B parameter in cm^{-1} with the value of C/B in parentheses.

the region where Hund's rules are not reliable for predicting anything more than the ground term of a configuration.

If, as is usual, we are interested only in spin-allowed transitions, we do not need to consider C, as it only appears in the expression for states that differ in multiplicity from the ground state.[8] The parameter B is of the most interest, and we return to factors that affect its value in Section 19.4f.

19.4 Electronic spectra of complexes

The preceding discussion related only to free atoms, and we will now expand our discussion to encompass complex ions. The spectrum of $[Cr(NH_3)_6]^{3+}$ in Fig. 19.19 has two central bands with intermediate intensities and with energies that differ on account of the electron–electron repulsions (as we explain soon). Because both the transitions are between orbitals that are predominantly metal d orbital in character, with a separation characterized by the strength of the ligand-field splitting parameter Δ_O, these two transitions are called **d–d transitions** or **ligand-field transitions**.

(a) Ligand-field transitions

Key point: Electron–electron repulsion splits ligand-field transitions into components with different energies.

According to the discussion in Section 19.1, we expect the octahedral d^3 complex $[Cr(NH_3)_6]^{3+}$ to have the ground-state configuration t_{2g}^3. The absorption near $25\,000\,\mathrm{cm}^{-1}$ can be identified as arising from the excitation $t_{2g}^2 e_g^1 \leftarrow t_{2g}^3$ because the corresponding energy (close to 3 eV) is typical of ligand-field splittings in complexes.

Before we embark on a Racah-like analysis of the transition, it will be helpful to see qualitatively from the viewpoint of molecular orbital theory why the transition gives rise to two bands. First, note that a $d_{z^2} \leftarrow d_{xy}$ transition, which is one way of achieving $e_g \leftarrow t_{2g}$, promotes an electron from the xy-plane into the already electron-rich z-direction: that axis is electron-rich because both d_{yz} and d_{zx} are occupied (Fig. 19.21). However, a $d_{z^2} \leftarrow d_{zx}$ transition, which is another way of achieving $e_g \leftarrow t_{2g}$, merely relocates an electron that is already largely concentrated along the z-axis. In the former case, but not in the latter, there is a distinct increase in electron repulsion and, as a result, the two $e_g \leftarrow t_{2g}$ transitions lie at different energies. There are six possible $t_{2g}^2 e_g^1 \leftarrow t_{2g}^3$ transitions, and all resemble one or other of these two cases: three of them fall into one group and the other three fall into the second group.

(b) The spectroscopic terms

Key points: The terms of an octahedral complex are labelled by the symmetry species of the overall orbital state; a superscript prefix shows the multiplicity of the term.

The two bands we are discussing in Fig. 19.19 are labelled $^4T_{2g} \leftarrow {}^4A_{2g}$ (at $21\,550\,\mathrm{cm}^{-1}$) and $^4T_{1g} \leftarrow {}^4A_{2g}$ (at $28\,500\,\mathrm{cm}^{-1}$). The labels are **molecular term symbols** and serve a purpose similar to that of atomic term symbols. The left superscript denotes the multiplicity, so the superscript 4 denotes a quartet state with $S = \frac{3}{2}$, as expected when there are three unpaired electrons. The rest of the term symbol is the symmetry label of the overall electronic orbital state of the complex. For example, the nearly totally symmetric ground state of a d^3 complex (with an electron in each of the three t_{2g} orbitals) is denoted A_{2g}. We say *nearly* totally symmetric, because close inspection of the behaviour of the three occupied t_{2g} orbitals shows that the C_3 rotation of the O_h point group transforms the product $t_{2g} \times t_{2g} \times t_{2g}$ into itself, which identifies the complex as an A symmetry species (see the character table in *Resource section 4*). Moreover, because each orbital has even parity (g), the overall parity is also g. However, each C_4 rotation transforms one t_{2g} orbital into the negative of itself and the other two t_{2g} orbitals into each other (Fig. 19.22),

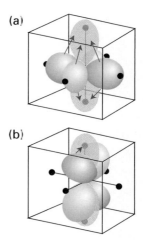

Fig. 19.21 The shifts in electron density that accompany the two transitions discussed in the text. There is a considerable relocation of electron density towards the ligands on the z-axis in (a), but a much less substantial relocation in (b).

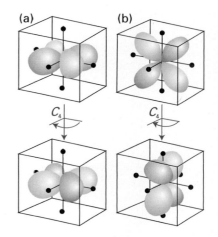

Fig. 19.22 The changes in sign that occur under C_4 rotations about the z-axis: (a) a d_{xy} orbital is rotated into the negative of itself; (b) a d_{yz} orbital is rotated into a d_{zx} orbital.

---

[8] We deal with the selection rules that tell us whether a transition is allowed or forbidden in Section 19.6.

Table 19.7 The correlation of spectroscopic terms for *d* electrons in O_h complexes

Atomic term	Number of states	Terms in O_h symmetry
S	1	A_{1g}
P	3	T_{1g}
D	5	$T_{2g} + E_g$
F	7	$T_{1g} + T_{2g} + A_{2g}$
G	9	$A_{1g} + E_g + T_{1g} + T_{2g}$

so overall there is a change of sign under this operation and its character is -1. The term is therefore A_{2g} rather than the totally symmetrical A_{1g} of a closed shell.

It is more difficult to establish that the term symbols that can arise from the quartet $t_{2g}^2 e_g^1$ excited configuration are $^4T_{2g}$ and $^4T_{1g}$ and we shall not consider this aspect here. The superscript 4 implies that the upper configuration continues to have the same number of unpaired spins as in the ground state, and the subscript g stems from the even parity of all the contributing orbitals.

(c) Correlating the terms

Key point: The free atom terms split in the ligand field of an octahedral complex and are then labelled by their symmetry species as enumerated in Table 19.7.

In a free atom, where all five *d* orbitals in a shell are degenerate, we needed to consider only the electron–electron repulsions to arrive at the relative ordering of the terms of a given d^n configuration. In a complex, the *d* orbitals are not all degenerate and it is necessary to take into account the difference in energy between the t_{2g} and e_g orbitals as well as the electron–electron repulsions.

Consider the simplest case of an atom or ion with a single valence electron. Because a totally symmetrical orbital in one environment becomes a totally symmetrical orbital in another environment, an *s* orbital in a free atom becomes an a_{1g} orbital in an octahedral field. We express the change by saying that the *s* orbital of the atom 'correlates' with the a_{1g} orbital of the complex. Similarly, the five *d* orbitals of a free atom correlate with the triply degenerate t_{2g} and doubly degenerate e_g sets in an octahedral complex.

Now consider a many-electron atom. In exactly the same way as for a single electron, the totally symmetrical overall S term of a many-electron atom correlates with the totally symmetrical A_{1g} term of an octahedral complex. Likewise, an atomic D term splits into a T_{2g} term and an E_g term in O_h symmetry. The same kind of analysis can be applied to other states, and Table 19.7 summarizes the correlations between free atom terms and terms in an octahedral complex.

Example 19.7 Identifying correlations between terms

What terms in a complex with O_h symmetry correlate with the ^3P term of a free atom with a d^2 configuration?

Answer We argue by analogy: if we know how *p* orbitals correlate with orbitals in a complex, then we can use that information to express how the overall states correlate, simply by changing to uppercase letters. The three *p* orbitals of a free atom become the triply degenerate t_{1u} orbitals of an octahedral complex. Therefore, if we disregard parity for the moment, a P term of a many-electron atom becomes a T_1 term in the point group O_h. Because *d* orbitals have even parity, the term overall must be g, and specifically T_{1g}. The multiplicity is unchanged in the correlation, so the ^3P term becomes a $^3T_{1g}$ term.

Self-test 19.7 What terms in a d^2 complex of O_h symmetry correlate with the ^3F and ^1D terms of a free atom?

(d) The energies of the terms: weak- and strong-field limits

Key points: For a given metal ion, the energies of the individual terms respond differently to ligands of increasing field strength and the correlation between free atom terms and terms of a complex can be displayed on an Orgel diagram.

Electron–electron repulsions are difficult to take into account, but the discussion is simplified by considering two extreme cases. In the weak-field limit the ligand field, as measured by Δ_O, is so weak that only electron–electron repulsions are important. As the Racah parameters B and C fully describe the interelectron repulsions, these are the only parameters we need at this limit. In the strong-field limit the ligand field is so strong that electron–electron repulsions can be ignored and the energies of the terms can be expressed solely in terms of Δ_O. Then, with the two extremes established, we can consider intermediate cases by drawing a correlation diagram between the two. We shall illustrate what is involved by considering two simple cases, namely, d^1 and d^2. Then we show how the same ideas are used to treat more complicated cases.

The only term arising from the d^1 configuration of a free atom is ^2D. In an octahedral complex the configuration is either t_{2g}^1, which gives rise to a $^2T_{2g}$ term, or e_g^1, which gives rise to a ^2E_g term. Because there is only one electron, there are no electron–electron repulsions to worry about, and the separation of the $^2T_{2g}$ and ^2E_g terms is the same as the separation of the t_{2g} and e_g orbitals, which is Δ_O. The correlation diagram for the d^1 configuration will therefore resemble that shown in Fig. 19.23.

We saw earlier that for a d^2 configuration the lowest energy term in the free atom is the triplet ^3F. We need consider only electronic transitions that start from the ground state, and, in this section, will discuss only those in which there is no change in spin. There is an additional triplet term (^3P); relative to the lower term (^3F), the energies of the terms are $E(^3F) = 0$ and $E(^3P) = 15B$. These two energies are marked on the left of Fig. 19.24. Now consider the very strong field limit. A d^2 atom has the configurations

$$t_{2g}^2 < t_{2g}^1 e_g^1 < e_g^2$$

In an octahedral field, these configurations have different energies; that is, as we noted earlier, the ^3F term splits into three terms. From the information in Fig. 19.2, we can write their energies as

$$E(t_{2g}^2) = 2\left(-\frac{2}{5}\Delta_O\right) = -0.8\Delta_O$$

$$E(t_{2g}^1 e_g^1) = \left(-\frac{2}{5} + \frac{3}{5}\right)\Delta_O = +0.2\Delta_O$$

$$E(e_g^2) = 2\left(\frac{3}{5}\right)\Delta_O = +1.2\Delta_O$$

Therefore, relative to the energy of the lowest term, their energies are

$$E(t_{2g}^2, T_{1g}) = 0 \qquad E(t_{2g}^1 e_g^1, T_{2g}) = \Delta_O \qquad E(e_g^2, A_{2g}) = 2\Delta_O$$

These energies are marked on the right in Fig. 19.24.

Our problem now is to account for the energies when neither the ligand-field nor the electron repulsion terms is dominant. To do so, we correlate the terms in the two extreme cases. The triplet t_{2g}^2 configuration gives rise to a $^3T_{1g}$ term, and this correlates with the ^3F term of the free atom. The remaining correlations can be established similarly, and we see that the $t_{2g}^1 e_g^1$ configuration gives rise to a T_{2g} term and that the e_g^2 configuration gives rise to a A_{2g} term; both terms correlate with the ^3F term of the free atom. Note that some terms, such as the $^3T_{1g}$ term that correlates with ^3P, are independent of the ligand-field strength. All the correlations are shown in Fig. 19.24, which is a simplified version of an **Orgel diagram**. An Orgel diagram can be constructed for any d-electron configuration, and several electronic configurations can be combined on the same diagram. Orgel diagrams are of considerable value for simple discussions of the electronic spectra of complexes; however, they consider only some of the possible transitions (the

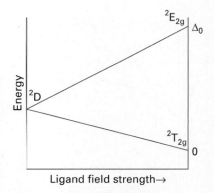

Fig. 19.23 Correlation diagram for a free ion (left) and the strong-field terms (right) of a d^1 configuration.

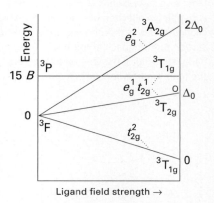

Fig. 19.24 Correlation diagram for a free ion (left) and the strong-field terms (right) of a d^2 configuration.

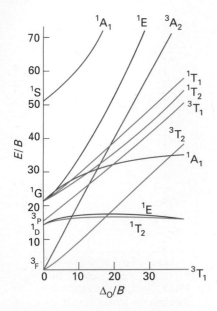

Fig. 19.25 The Tanabe–Sugano diagram for the d^2 configuration. Note that the left-hand axis corresponds to Fig. 19.20(left). A complete collection of diagrams for d^n configurations is given in *Resource section 6* at the back of the book. The parity subscript g has been omitted from the term symbols for clarity.

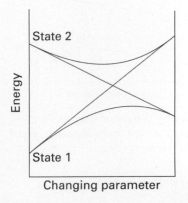

Fig. 19.26 The noncrossing rule states that if two states of the same symmetry are likely to cross as a parameter is changed (as shown by the blue lines), they will in fact mix together and avoid the crossing (as shown by the brown lines).

spin-allowed transitions, which is why we considered only the triplet terms) and cannot be used to extract a value for the ligand-field splitting parameter, Δ_O.

(e) Tanabe–Sugano diagrams

Key point: Tanabe–Sugano diagrams are correlation diagrams that depict the energies of electronic states of complexes as a function of the strength of the ligand field.

Diagrams showing the correlation of all terms can be constructed for any electron configuration and strength of ligand field. The most widely used versions are called **Tanabe–Sugano diagrams**, after the scientists who devised them. Figure 19.25 shows the diagram for d^2 and we can see splittings for all the atomic terms that split; thus the ^3F splits into 3, the ^1D into 2, and the ^1G into 4. In these diagrams the term energies, E, are expressed as E/B and plotted against Δ_O/B, where B is the Racah parameter. The relative energies of the terms arising from a given configuration are independent of A, and by choosing a value of C (typically setting $C \approx 4B$), terms of all energies can be plotted on the same diagrams. Some lines in Tanabe–Sugano diagrams are curved because of the mixing of terms of the same symmetry type. Terms of the same symmetry obey the **noncrossing rule**, which states that, if the increasing ligand field causes two weak field terms of the same symmetry to approach, then they do not cross but bend apart from each other (Fig. 19.26). The effect of the noncrossing rule can be seen for the two ^1E terms, the two $^1T_2$ terms, and the two $^1A_1$ terms in Fig. 19.25.

Tanabe–Sugano diagrams for O_h complexes with configurations d^2 to d^8 are given in *Resource section 5*. The zero of energy in a Tanabe–Sugano diagram is always taken as that of the lowest term. Hence the lines in the diagrams have abrupt changes of slope when there is a change in the identity of the ground term brought about by the change from high-spin to low-spin with increasing field strength (see that for d^4, for instance).

The purpose of the preceding discussion has been to find a way to extract the value of the ligand-field splitting parameter from the electronic absorption spectrum of a complex with more than one *d* electron, when electron–electron repulsions are important, such as that in Fig. 19.19. The strategy involves fitting the observed transitions to the correlation lines in a Tanabe–Sugano diagram and identifying the values of Δ_O and B where the observed transition energies match the pattern. This procedure is illustrated in Example 19.8.

As we shall see in Section 19.6, certain transitions are allowed and certain transitions are forbidden. In particular, those that correspond to a change of spin state are forbidden, whereas those that do not are allowed. In general, spin-allowed transitions will dominate the UV/visible absorption spectrum, thus we might only expect to see three transitions for a d^2 ion, specifically $^3T_{2g} \leftarrow {}^3T_{1g}$, $^3T_{1g} \leftarrow {}^3T_{1g}$, and $^3A_{2g} \leftarrow {}^3T_{1g}$. However, some complexes (for instance high-spin d^5 ions such as Mn^{2+}) do not have any spin-allowed transitions and none of the eleven possible transitions dominates.

Example 19.8 Calculating Δ_O and B using a Tanabe–Sugano diagram

Deduce the values of Δ_O and B for $[Cr(NH_3)_6]^{3+}$ from the spectrum in Fig. 19.19 and a Tanabe–Sugano diagram.

Answer The relevant diagram (for d^3) is shown in Fig. 19.27. We need concern ourselves only with the spin-allowed transitions, of which there are three for a d^3 ion (a $^4T_{2g} \leftarrow {}^4A_{2g}$ and two $^4T_{1g} \leftarrow {}^4A_{2g}$ transitions). We have seen that the spectrum in Fig. 19.19 exhibits two low-energy ligand-field transitions at 21 550 and 28 500 cm^{-1}, which correspond to the two lowest energy transitions ($^4T_{2g} \leftarrow {}^4A_{2g}$ and $^4T_{1g} \leftarrow {}^4A_{2g}$). The ratio of the energies of the transitions is 1.32, and the only point in Fig. 19.27 where this energy ratio is satisfied is on the far right. Hence, we can read off the value of $\Delta_O/B = 33.0$ from the location of this point. The tip of the arrow representing the lower energy transition lies at $32.8B$ vertically, so equating $32.8B$ and 21 550 cm^{-1} gives $B = 657$ cm^{-1} and therefore $\Delta_O = 21\,700$ cm^{-1}.

Self-test 19.8 Use the same Tanabe–Sugano diagram to predict the energy of the first two spin-allowed quartet bands in the spectrum of $[Cr(OH_2)_6]^{3+}$ for which $\Delta_O = 17\,600$ cm^{-1} and $B = 700$ cm^{-1}.

A Tanabe–Sugano diagram also provides some understanding of the widths of some absorption lines. Consider Fig. 19.25 and the $^3T_{2g} \leftarrow {}^3T_{1g}$ and the $^3A_{2g} \leftarrow {}^3T_{1g}$ transitions. The line representing $^3T_{2g}$ is not parallel to that representing $^3T_{1g}$, and so any variation in the value of Δ_O (such as those caused by molecular vibration) results in a change in the energy of the transition and thus a broadening of the absorption band. The line representing $^3A_{2g}$ is even less parallel to the line representing $^3T_{1g}$; the energy of this transition will be affected even more by variations in Δ_O and will thus be even broader. By contrast, the line representing the lower $^1T_{2g}$ term is almost parallel to that representing $^3T_{1g}$ and thus the energy of this transition is largely unaffected by variations in Δ_O and the absorption is consequently very sharp (albeit weak, because it is forbidden).

(f) The nephelauxetic series

Key points: Electron–electron repulsions are lower in complexes than in free ions because of electron delocalization; the nephelauxetic parameter is a measure of the extent of d-electron delocalization on to the ligands of a complex; the softer the ligand, the smaller the nephelauxetic parameter.

In Example 19.8 we found that $B = 657$ cm^{-1} for $[Cr(NH_3)_6]^{3+}$, which is only 64 per cent of the value for a Cr^{3+} ion in the gas phase. This reduction is a general observation and indicates that electron repulsions are weaker in complexes than in the free atoms and ions. The weakening occurs because the occupied molecular orbitals are delocalized over the ligands and away from the metal. The delocalization increases the average separation of the electrons and hence reduces their mutual repulsion.

The reduction of B from its free ion value is normally reported in terms of the **nephelauxetic parameter**, β:[9]

$$\beta = \frac{B(\text{complex})}{B(\text{free ion})} \tag{19.4}$$

The values of β depend on the identity of the metal ion and the ligand, and a list of ligands ordered by the value of β gives the **nephelauxetic series**:

$$Br^- < CN^-, Cl^- < NH_3 < H_2O < F^-$$

A small value of β indicates a large measure of d-electron delocalization on to the ligands and hence a significant covalent character in the complex. Thus the series shows that a Br^- ligand results in a greater reduction in electron repulsions in the ion than an F^- ion, which is consistent with a greater covalent character in bromo complexes than in analogous fluoro complexes. As an example, compare $[NiF_6]^{4-}$, for which $B = 843$ cm^{-1}, with $[NiBr_4]^{2-}$, for which $B = 600$ cm^{-1}. Another way of expressing the trend represented by the nephelauxetic series is: *the softer the ligand, the smaller the nephelauxetic parameter.*

19.5 Charge-transfer bands

Key points: Charge-transfer bands arise from the movement of electrons between orbitals that are predominantly ligand in character and orbitals that are predominantly metal in character; such transitions are identified by their high intensity and the sensitivity of their energies to solvent polarity.

Another feature in the spectrum of $[Cr(NH_3)_6]^{3+}$ in Fig. 19.19 that remains to be explained is the very intense shoulder of an absorption that appears to have a maximum at well above $50\,000$ cm^{-1}. The high intensity suggests that this transition is not a simple ligand-field transition, but is consistent with a **charge-transfer transition** (CT transition). In a CT transition, an electron migrates between orbitals that are predominantly ligand in character and orbitals that are predominantly metal in character. The transition is classified as a **ligand-to-metal charge-transfer transition** (LMCT transition) if the migration of the electron is from the ligand to the metal, and as a **metal-to-ligand charge-transfer transition** (MLCT transition) if the charge migration occurs in the opposite direction. An example of an MLCT transition is the one responsible for the red colour of

[9] The name is from the Greek words for 'cloud expanding'.

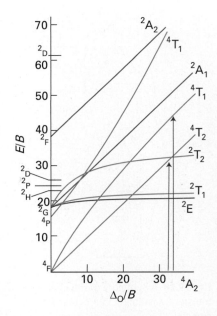

Fig. 19.27 The Tanabe–Sugano diagram for the d^3 configuration. Note that the left-hand axis corresponds to Fig. 19.20(right). The parity subscript g has been omitted from the term symbols for clarity.

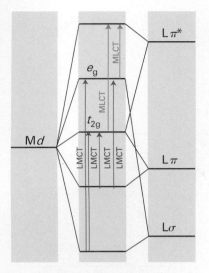

Fig. 19.28 A summary of the charge-transfer transitions in an octahedral complex.

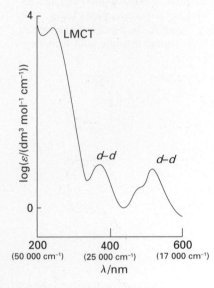

Fig. 19.29 The absorption spectrum of $[CrCl(NH_3)_5]^{2+}$ in water in the visible and ultraviolet regions. The peak corresponding to the transition $^2E \leftarrow {}^4A$ is not visible on this magnification.

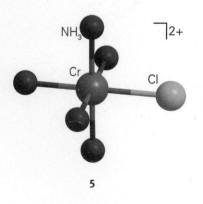

5

tris(bipyridyl)iron(II), the complex used for the colorimetric analysis of Fe(II). In this case, an electron makes a transition from a *d* orbital of the central metal into a π^* orbital of the ligand. Figure 19.28 summarizes the transitions we classify as charge transfer.

Several lines of evidence are used to identify a band as due to a CT transition. The high intensity of the band, which is evident in Fig. 19.19, is one strong indication. Another indication is if such a band appears following the replacement of one ligand with another, as this implies that the band is strongly dependent on the ligand. The CT character is most often identified (and distinguished from $\pi^* \leftarrow \pi$ transitions on ligands) by demonstrating **solvatochromism**, the variation of the transition frequency with changes in solvent permittivity. Solvatochromism indicates that there is a large shift in electron density as a result of the transition, which is more consistent with a metal–ligand transition than a ligand–ligand or metal–metal transition.

Figure 19.29 shows another example of a CT transition in the visible and ultraviolet spectrum of $[CrCl(NH_3)_5]^{2+}$ (**5**). If we compare this spectrum with that of $[Cr(NH_3)_6]^{3+}$ in Fig. 19.19, then we can recognize the two ligand-field bands in the visible region. The replacement of one NH_3 ligand by a weaker field Cl^- ligand moves the lowest energy ligand-field bands to lower energy than those of $[Cr(NH_3)_6]^{3+}$. Also, a shoulder appears on the high-energy side of one of the ligand-field bands, indicating an additional transition that is the result of the reduction in symmetry from O_h to C_{4v}. The major new feature in the spectrum is the strong absorption maximum in the ultraviolet, near $42\,000\ cm^{-1}$. This band is at lower energy than the corresponding band in the spectrum of $[Cr(NH_3)_6]^{3+}$ and is due to an LMCT transition from the Cl^- ligand to the metal. The LMCT character of similar bands in $[CoX(NH_3)_5]^{2+}$ is confirmed by the decrease in energy in steps of about $8000\ cm^{-1}$ as X is varied from Cl to Br to I. In this LMCT transition, a lone-pair electron of the halide ligand is promoted into a predominantly metal orbital.

(a) LMCT transitions

Key points: Ligand-to-metal charge-transfer transitions are observed in the visible region of the spectrum when the metal is in a high oxidation state and ligands contain nonbonding electrons; the variation in the position of LMCT bands can be parameterized in terms of optical electronegativities.

Charge-transfer bands in the visible region of the spectrum (and hence contributing to the intense colours of many complexes) may occur if the ligands have lone pairs of relatively high energy (as in sulfur and selenium) or if the metal atom has low-lying empty orbitals.

The tetraoxoanions of metals with high oxidation numbers (such as MnO_4^-) provide what are probably the most familiar examples of LMCT bands. In them, an O lone-pair electron is promoted into a low-lying empty metal *e* orbital. High metal oxidation numbers correspond to a low *d*-orbital population (many are formally d^0), so the acceptor level is available and low in energy. The trend in LMCT energies is:

Oxidation number	
+7	$MnO_4^- < TcO_4^- < ReO_4^-$
+6	$CrO_4^{2-} < MoO_4^{2-} < WO_4^{2-}$
+5	$VO_4^{3-} < NbO_4^{3-} < TaO_4^{3-}$

The UV/visible spectra of the tetraoxo anions of the Group 6 metals, CrO_4^{2-}, MoO_4^{2-}, and WO_4^{2-}, are shown in Fig. 19.30. The energies of the transitions correlate with the order of the electrochemical series, with the lowest energy transitions taking place to the most easily reduced metal ions. This correlation is consistent with the transition being the transfer of an electron from the ligands to the metal ion, corresponding, in effect, to the reduction of the metal ion by the ligands. Polymeric and monomeric oxoanions follow the same trends, with the oxidation state of the metal the determining factor. The similarity suggests that these LMCT transitions are localized processes that take place on discrete molecular fragments.

Table 19.8 Optical electronegativities

Metal	O_h	T_d	Ligand	π	σ
Cr(III)	1.8–1.9		F^-	3.9	4.4
Co(III)*	2.3		Cl^-	3.0	3.4
Ni(II)		2.0–2.1	Br^-	2.8	3.3
Co(II)		1.8–1.9	I^-	2.5	3.0
Rh(III)*	2.3		H_2O	3.5	
Mo(VI)	2.1		NH_3	3.3	

* Low-spin complexes.

The variation in the position of LMCT bands can be expressed in terms of the **optical electronegativities** of the metal, χ_{metal}, and the ligands, χ_{ligand}. The energy of the transition is then written as the difference between the two electronegativities:

$$\tilde{v} = |\chi_{ligand} - \chi_{metal}|\tilde{v}_0$$

Optical electronegativities have values comparable to Pauling electronegativities if we set $\tilde{v}_0 = 3.0 \times 10^4\,cm^{-1}$ (Table 19.8), and can be used in a similar fashion. If the LMCT transition terminates in an e_g orbital, Δ_O must be added to the energy predicted by this equation. Electron pairing energies must also be taken into account if the transition results in the population of an orbital that already contains an electron. The values for metals are different in complexes of different symmetry, and the ligand values are different if the transition originates from a π orbital rather than a σ orbital.

(b) MLCT transitions

Key point: Charge-transfer transitions from metal to ligand are observed when the metal is in a low oxidation state and the ligands have low-lying acceptor orbitals.

Charge-transfer transitions from metal to ligand are most commonly observed in complexes with ligands that have low-lying π^* orbitals, especially aromatic ligands. The transition occurs at low energy and appears in the visible spectrum if the metal ion is in a low oxidation state, as its d orbitals are then relatively close in energy to the empty ligand orbitals.

The family of ligands most commonly involved in MLCT transitions are the diimines, which have two N donor atoms: two important examples are 2,2′-bipyridine (bipy, **6**) and 1,10-phenanthroline (phen, **7**). Complexes of diimines with strong MLCT bands include tris(diimine) species such as tris(2,2′-bipyridyl)ruthenium(II) (**8**), which is orange. A diimine ligand may also be easily substituted into a complex with other ligands that favour a low oxidation state. Two examples are $[W(CO)_4(phen)]$ and $[Fe(CO)_3(bipy)]$. However, the occurrence of MLCT transitions is by no means limited to diimine ligands. Another important ligand type that shows typical MLCT transitions is dithiolene, $S_2C_2R_2^{2-}$ (**9**). Resonance Raman spectroscopy (Section 6.4) is a powerful technique for the study of MLCT transitions.

The MLCT excitation of tris(2,2′-bipyridyl)ruthenium(II) has been the subject of intense research efforts because the excited state that results from the charge transfer has a lifetime of microseconds, and the complex is a versatile photochemical redox reagent. The photochemical behaviour of a number of related complexes has also been studied on account of their relatively long excited-state lifetimes.

19.6 Selection rules and intensities

Key point: The strength of an electronic transition is determined by the transition dipole moment.

The contrast in intensity between typical charge-transfer bands and typical ligand-field bands raises the question of the factors that control the intensities of absorption bands.

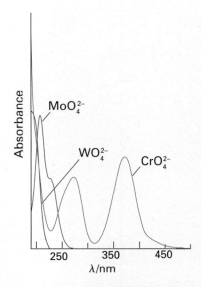

Fig. 19.30 Optical absorption spectra of the ions CrO_4^{2-}, WO_4^{2-}, and MoO_4^{2-}. On descending the group, the absorption maximum moves to shorter wavelengths, indicating an increase in the energy of the LMCT band.

In an octahedral, nearly octahedral, or square-planar complex, the maximum molar absorption coefficient ε_{max} (which measures the strength of the absorption)[10] is typically less than or close to 100 $dm^3\,mol^{-1}\,cm^{-1}$ for ligand-field transitions. In tetrahedral complexes, which have no centre of symmetry, ε_{max} for ligand-field transitions might exceed 250 $dm^3\,mol^{-1}\,cm^{-1}$. By contrast, charge-transfer bands usually have an ε_{max} in the range 1000–50 000 $dm^3\,mol^{-1}\,cm^{-1}$.

To understand the intensities of transitions in complexes we have to explore the strength with which the complex couples with the electromagnetic field. Intense transitions indicate strong coupling; weak transitions indicate feeble coupling. The strength of coupling when an electron makes a transition from a state with wavefunction ψ_i to one with wavefunction ψ_f is measured by the **transition dipole moment**, which is defined as the integral

$$\mu_{fi} = \int \psi_f^* \mu \psi_i d\tau \tag{19.5}$$

where $\boldsymbol{\mu}$ is the electric dipole moment operator, $-e\boldsymbol{r}$. The transition dipole moment can be regarded as a measure of the impulse that a transition imparts to the electromagnetic field: a large impulse corresponds to an intense transition; zero impulse corresponds to a forbidden transition. The intensity of a transition is proportional to the square of its transition dipole moment.

A spectroscopic **selection rule** is a statement about which transitions are allowed and which are forbidden. An **allowed transition** is a transition with a nonzero transition dipole moment, and hence nonzero intensity. A **forbidden transition** is a transition for which the transition dipole moment is calculated as zero. Formally forbidden transitions may occur in a spectrum if the assumptions on which the transition dipole moment were calculated are invalid, such as the complex having a lower symmetry than assumed.

(a) Spin selection rules

Key points: Electronic transitions with a change of multiplicity are forbidden; intensities of spin-forbidden transitions are greater for *4d*- or *5d*-series metal complexes than for comparable *3d*-series complexes.

The electromagnetic field of the incident radiation cannot change the relative orientations of the spins of the electrons in a complex. For example, an initially antiparallel pair of electrons cannot be converted to a parallel pair, so a singlet $(S=0)$ cannot undergo a transition to a triplet $(S=1)$. This restriction is summarized by the rule $\Delta S = 0$ for **spin-allowed transitions**.

The coupling of spin and orbital angular momenta can relax the spin selection rule, but such **spin-forbidden**, $\Delta S \neq 0$, transitions are generally much weaker than spin-allowed transitions. The intensity of spin-forbidden bands increases as the atomic number increases because the strength of the spin–orbit coupling is greater for heavy atoms than for light atoms. The breakdown of the spin selection rule by spin–orbit coupling is often called the **heavy-atom effect**. In the *3d* series, in which spin–orbit coupling is weak, spin-forbidden bands have ε_{max} less than about 1 $dm^3\,mol^{-1}\,cm^{-1}$; however, spin-forbidden bands are a significant feature in the spectra of heavy *d*-metal complexes.

The very weak transition labelled $^2E_g \leftarrow {}^4A_{2g}$ in Fig. 19.19 is an example of a spin-forbidden transition. Some metal ions, such as the d^5 Mn^{2+} ion, have no spin-allowed transitions, and hence are only weakly coloured.

(b) The Laporte selection rule

Key points: Transitions between *d* orbitals are forbidden in octahedral complexes; asymmetric vibrations relax this restriction.

The **Laporte selection rule** states that *in a centrosymmetric molecule or ion, the only allowed transitions are those accompanied by a change in parity.* That is, transitions

---

[10] The molar absorption coefficient is the constant in the Beer–Lambert law for the transmittance $\tau = I_f/I_i$ when light passes through a length l of solution of molar concentration [X] and is attenuated from an intensity I_i to an intensity I_f: $\log \tau = -\varepsilon l[X]$ (the logarithm is a common logarithm, to the base 10). Its older but still widely used name is the 'extinction coefficient'.

between g and u terms are permitted, but a g term cannot undergo a transition to another g term and a u term cannot undergo a transition to another u term:

$$g \leftrightarrow u \quad g \nleftrightarrow g \quad u \nleftrightarrow u$$

In many cases it is enough to note that in a centrosymmetric complex, if there is no change in quantum number l, then there is no change in parity. Thus, s–s, p–p, d–d, and f–f transitions are forbidden whereas s–p, p–d, and d–f transitions are allowed. A more formal treatment of the Laporte selection rule is based on the properties of the transition dipole moment r. Because r changes sign under inversion (and is therefore u), the entire integral in eqn 19.5 will also change sign under inversion if ψ_i and ψ_f have the same parity, because

$$g \times u \times g = u \quad \text{and} \quad u \times u \times u = u$$

Therefore, because the value of an integral cannot depend on the choice of coordinates used to evaluate it,[11] it vanishes if ψ_i and ψ_f have the same parity. However, if they have opposite parity, the integral does not change sign under inversion of the coordinates because

$$g \times u \times u = g$$

and therefore need not vanish.

In a centrosymmetric complex, d–d ligand-field transitions are $g \leftrightarrow g$ and are therefore forbidden. Their forbidden character accounts for the relative weakness of these transitions in octahedral complexes (which are centrosymmetric) compared with those in tetrahedral complexes, on which the Laporte rule is silent (they are noncentrosymmetric, and have no g or u as a subscript).

The question remains why d–d ligand-field transitions in octahedral complexes occur at all, even weakly. The Laporte selection rule may be relaxed in two ways. First, a complex may depart slightly from perfect centrosymmetry in its ground state, perhaps on account of the intrinsic asymmetry in the structure of polyatomic ligands or a distortion imposed by the environment of a complex packed into a crystal. Alternatively, the complex might undergo an asymmetrical vibration, which also destroys its centre of inversion. In either case, a Laporte-forbidden d–d ligand-field band tends to be much more intense than a spin-forbidden transition.

Table 19.9 summarizes typical intensities of electronic transitions of complexes of the $3d$-series elements. The width of spectroscopic absorption bands is due principally to the simultaneous excitation of vibration when the electron is promoted from one distribution to another. According to the **Franck–Condon principle**, the electronic transition takes place within a stationary nuclear framework. As a result, after the transition has occurred, the nuclei experience a new force field and the molecule begins to vibrate anew.

Table 19.9 Intensities of spectroscopic bands in $3d$ complexes

Band type	$\varepsilon_{max}/$ $(dm^3\ mol^{-1}\ cm^{-1})$
Spin-forbidden	< 1
Laporte-forbidden d–d	20–100
Laporte-allowed d–d	ca 250
Symmetry-allowed (e.g. CT)	1000–50 000

Example 19.9 Assigning a spectrum using selection rules

Assign the bands in the spectrum in Fig. 19.29 by considering their intensities.

Answer If we assume that the complex is approximately octahedral, examination of the Tanabe–Sugano diagram reveals that the ground term is $^4A_{2g}$. Transitions to the higher terms ^2E_g, $^2T_{1g}$, and $^2T_{2g}$ are spin-forbidden and will have $\varepsilon_{max} < 1\ dm^3\ mol^{-1}\ cm^{-1}$. Thus, very weak bands for these transitions are predicted, and will be difficult to distinguish. The next two higher terms of the same multiplicity are $^4T_{2g}$ and $^4T_{1g}$. These terms are reached by spin-allowed but Laporte-forbidden ligand-field transitions, and have $\varepsilon_{max} \approx 100\ dm^3\ mol^{-1}\ cm^{-1}$: these are the two bands at 360 and 510 nm. In the near UV, the band with $\varepsilon_{max} \approx 10\ 000$ $dm^3\ mol^{-1}\ cm^{-1}$ corresponds to the LMCT transitions in which an electron from a chlorine π lone pair is promoted into a molecular orbital that is principally metal d orbital in character.

Self-test 19.9 The spectrum of $[Cr(NCS)_6]^{3-}$ has a very weak band near $16\ 000\ cm^{-1}$, a band at $17\ 700\ cm^{-1}$ with $\varepsilon_{max} = 160\ dm^3\ mol^{-1}\ cm^{-1}$, a band at $23\ 800\ cm^{-1}$ with $\varepsilon_{max} = 130$ $dm^3\ mol^{-1}\ cm^{-1}$, and a very strong band at $32\ 400\ cm^{-1}$. Assign these transitions using the d^3 Tanabe–Sugano diagram and selection rule considerations. (*Hint*: NCS$^-$ has low-lying π^* orbitals.)

---

[11] An integral is an area, and areas are independent of the coordinates used for their evaluation.

19.7 Luminescence

Key points: A luminescent complex is one that re-emits radiation after it has been electronically excited. Fluorescence occurs when there is no change in multiplicity, whereas phosphorescence occurs when an excited state undergoes intersystem crossing to a state of different multiplicity and then undergoes radiative decay.

A complex is **luminescent** if it emits radiation after it has been electronically excited by the absorption of radiation. Luminescence competes with nonradiative decay by thermal degradation of energy to the surroundings. Relatively fast radiative decay is not especially common at room temperature for *d*-metal complexes, so strongly luminescent systems are comparatively rare. Nevertheless, they do occur, and we can distinguish two types of process. Traditionally, rapidly decaying luminescence was called 'fluorescence' and luminescence that persists after the exciting illumination is extinguished was called 'phosphorescence'. However, because the lifetime criterion is not reliable, the modern definitions of the two kinds of luminescence are based on the distinctive mechanisms of the processes. **Fluorescence** is radiative decay from an excited state of the same multiplicity as the ground state. The transition is spin-allowed and is fast; fluorescence half-lives are a matter of nanoseconds. **Phosphorescence** is radiative decay from a state of different multiplicity from the ground state. It is a spin-forbidden process, and hence is often slow.

The initial excitation of a phosphorescent complex usually populates a state by a spin-allowed transition, so the mechanism of phosphorescence involves **intersystem crossing**, the nonradiative conversion of the initial excited state into another excited state of different multiplicity. This second state acts as an energy reservoir because radiative decay to the ground state is spin-forbidden. However, just as spin–orbit coupling allows the intersystem crossing to occur, it also breaks down the spin selection rule, so the radiative decay can occur. Radiative decay back to the ground state is slow, so a phosphorescent state of a *d*-metal complex may survive for microseconds or even longer.

An important example of phosphorescence is provided by ruby, which consists of a low concentration of Cr^{3+} ions in place of Al^{3+} in alumina. Each Cr^{3+} ion is surrounded octahedrally by six O^{2-} ions and the initial excitations are the spin-allowed processes

$$t_{2g}^2 e_g^1 \leftarrow t_{2g}^3 : \quad {}^4T_{2g} \leftarrow {}^4A_{2g} \text{ and } {}^4T_{1g} \leftarrow {}^4A_{2g}$$

These absorptions occur in the green and violet regions of the spectrum and are responsible for the red colour of the gem (Fig. 19.31). Intersystem crossing to a ${}^2E$ term of the t_{2g}^3 configuration occurs in a few picoseconds or less, and red 627 nm phosphorescence occurs as this doublet decays back into the quartet ground state. This red emission adds to the red perceived by the subtraction of green and violet light from white light, and adds lustre to the gem's appearance. When the optical arrangement is such that the emission is stimulated by 627 nm photons reflecting back and forth between two mirrors, it grows strongly in intensity. This stimulated emission in a resonant cavity is the process used by Theodore Maiman in the first successful laser (in 1960) based on ruby.

A similar ${}^2E \rightarrow {}^4A$ phosphorescence can be observed from a number of Cr(III) complexes in solution. The ${}^2E$ term belongs to the t_{2g}^3 configuration, which is the same as the ground state, and thus the strength of the ligand field is not important. Hence the emission is always in the red (and close to the wavelength of ruby emission). If the ligands are rigid, as in $[Cr(bipy)_3]^{3+}$, the ${}^2E$ term may live for several microseconds in solution.

Another interesting example of a phosphorescent state is found in $[Ru(bipy)_3]^{2+}$. The excited singlet term produced by a spin-allowed MLCT transition of this d^6 complex undergoes intersystem crossing to the lower energy triplet term of the same configuration, $t_{2g}^5 \pi^{*1}$. Bright orange emission then occurs with a lifetime of about 1 μs (Fig. 19.32). The effects of other molecules (quenchers) on the lifetime of the emission may be used to monitor the rate of electron transfer from the excited state.

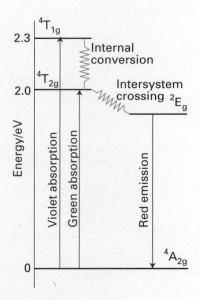

Fig. 19.31 The transitions responsible for the absorption and luminescence of Cr^{3+} ions in ruby.

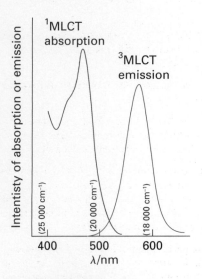

Fig. 19.32 The absorption and phosphorescence spectra of $[Ru(bipy)_3]^{2+}$.

FURTHER READING

F.A. Cotton, *Chemical applications of group theory*. Wiley, Chichester (1996).

S.F.A. Kettle, *Physical inorganic chemistry: a co-ordination chemistry approach*. Oxford University Press (1998).

B.N. Figgis and M.A. Hitchman, *Ligand field theory and its applications*. Wiley, Chichester (2000).

J.S. Griffith, *The theory of transition-metal ions*. Cambridge University Press (1964). A thorough and authoritative account of the material covered in this chapter.

E.U. Condon and G.H. Shortley, *The theory of atomic spectra*. Cambridge University Press (1935); revised as E.U. Condon and H. Odabaşi, *Atomic structure*. Cambridge University Press (1980). The standard reference text on atomic spectra.

A.B.P. Lever, *Inorganic electronic spectroscopy*. Elsevier, Amsterdam (1984). A useful discussion of solvatochromism.

EXERCISES

19.1 Determine the configuration (in the form $t_{2g}^x e_g^y$ or $e^x t_2^y$, as appropriate), the number of unpaired electrons, and the ligand-field stabilization energy as a multiple of Δ_O or Δ_T for each of the following complexes using the spectrochemical series to decide, where relevant, which are likely to be strong-field and which weak-field. (a) $[Co(NH_3)_6]^{3+}$, (b) $[Fe(OH_2)_6]^{2+}$, (c) $[Fe(CN)_6]^{3-}$, (d) $[Cr(NH_3)_6]^{3+}$, (e) $[W(CO)_6]$, (f) tetrahedral $[FeCl_4]^{2-}$, and (g) tetrahedral $[Ni(CO)_4]$.

19.2 Both H^- and $P(C_6H_5)_3$ are ligands of similar field strength, high in the spectrochemical series. Recalling that phosphines act as π acceptors, is π-acceptor character required for strong-field behaviour? What orbital factors account for the strength of each ligand?

19.3 Estimate the spin-only contribution to the magnetic moment for each complex in Exercise 19.1.

19.4 Solutions of the complexes $[Co(NH_3)_6]^{2+}$, $[Co(H_2O)_6]^{2+}$ (both O_h), and $[CoCl_4]^{2-}$ are coloured. One is pink, another is yellow, and the third is blue. Considering the spectrochemical series and the relative magnitudes of Δ_T and Δ_O, assign each colour to one of the complexes.

19.5 For each of the following pairs of complexes, identify the one that has the larger LFSE:

(a) $[Cr(OH_2)_6]^{2+}$ or $[Mn(OH_2)_6]^{2+}$

(b) $[Mn(OH_2)_6]^{2+}$ or $[Fe(OH_2)_6]^{3+}$

(c) $[Fe(OH_2)_6]^{2+}$ or $[Fe(CN)_6]^{3-}$

(d) $[Fe(CN)_6]^{3-}$ or $[Ru(CN)_6]^{3-}$

(e) tetrahedral $[FeCl_4]^{2-}$ or tetrahedral $[CoCl_4]^{2-}$

19.6 Interpret the variation, including the overall trend across the $3d$ series, of the following values of oxide lattice enthalpies (in $kJ\ mol^{-1}$). All the compounds have the rock-salt structure. CaO (3460); TiO (3878); VO (3913); MnO (3810); FeO (3921); CoO (3988); NiO (4071).

19.7 A neutral macrocyclic ligand with four donor atoms produces a red diamagnetic low-spin d^8 complex of Ni(II) if the anion is the weakly coordinating perchlorate ion. When perchlorate is replaced by two thiocyanate ions, SCN^-, the complex turns violet and is high-spin with two unpaired electrons. Interpret the change in terms of structure.

19.8 Bearing in mind the Jahn–Teller theorem, predict the structure of $[Cr(OH_2)_6]^{2+}$.

19.9 The spectrum of d^1 $Ti^{3+}(aq)$ is attributed to a single electronic transition $e_g \leftarrow t_{2g}$. The band shown in Fig. 19.3 is not symmetrical and suggests that more than one state is involved. Suggest how to explain this observation using the Jahn–Teller theorem.

19.10 Write the Russell–Saunders term symbols for states with the angular momentum quantum numbers (L,S): (a) $(0, \frac{5}{2})$, (b) $(3, \frac{3}{2})$, (c) $(2, \frac{1}{2})$, (d) $(1,1)$.

19.11 Identify the ground term from each set of terms: (a) ^1P, ^3P, ^3F, ^1G; (b) ^3P, ^5D, ^3H, ^1I, ^1G; (c) ^6S, ^4P, ^4G, ^2I.

19.12 Give the Russell–Saunders terms of the configurations: (a) $4s^1$, (b) $3p^2$. Identify the ground term.

19.13 Identify the ground terms of the gas-phase species B^+, Na, Ti^{2+}, and Ag^+.

19.14 The gas-phase ion V^{3+} has a ^3F ground term. The ^1D and ^3P terms lie, respectively, 10 642 and 12 920 cm^{-1} above it. The energies of the terms are given in terms of Racah parameters as $E(^3F) = A - 8B$, $E(^3P) = A + 7B$, $E(^1D) = A - 3B + 2C$. Calculate the values of B and C for V^{3+}.

19.15 Write the d-orbital configurations and use the Tanabe–Sugano diagrams (*Resource section 6*) to identify the ground term of (a) low-spin $[Rh(NH_3)_6]^{3+}$, (b) $[Ti(OH_2)_6]^{3+}$, (c) high-spin $[Fe(OH_2)_6]^{3+}$.

19.16 Using the Tanabe–Sugano diagrams in Appendix 4, estimate Δ_O and B for (a) $[Ni(OH_2)_6]^{2+}$ (absorptions at 8500, 15 400, and 26 000 cm^{-1}) and (b) $[Ni(NH_3)_6]^{2+}$ (absorptions at 10 750, 17 500, and 28 200 cm^{-1}).

19.17 If an octahedral Fe(II) complex has a large paramagnetic susceptibility, what is the ground-state label according to the Tanabe–Sugano diagram?

19.18 The spectrum of $[Co(NH_3)_6]^{3+}$ has a very weak band in the red and two moderate intensity bands in the visible to near-UV. How should these transitions be assigned?

19.19 Explain why $[FeF_6]^{3-}$ is colourless whereas $[CoF_6]^{3-}$ is coloured but exhibits only a single band in the visible.

19.20 The Racah parameter B is 460 cm^{-1} in $[Co(CN)_6]^{3-}$ and 615 cm^{-1} in $[Co(NH_3)_6]^{3+}$. Consider the nature of bonding with the two ligands and explain the difference in nephelauxetic effect.

19.21 An approximately 'octahedral' complex of Co(III) with ammine and chloro ligands gives two bands with ε_{max} between 60 and 80 $dm^3\ mol^{-1}\ cm^{-1}$, one weak peak with $\varepsilon_{max} = 2\ dm^3\ mol^{-1}\ cm^{-1}$ and a strong band at higher energy with $\varepsilon_{max} = 2 \times 10^4\ dm^3\ mol^{-1}\ cm^{-1}$. What do you suggest for the origins of these transitions?

19.22 Ordinary bottle glass appears nearly colourless when viewed through the wall of the bottle but green when viewed from the end so that the light has a long path through the glass. The colour is associated with the presence of Fe^{3+} in the silicate matrix. Describe the transitions.

19.23 Solutions of $[Cr(OH_2)_6]^{3+}$ ions are pale blue–green but the chromate ion CrO_4^{2-} is an intense yellow. Characterize the origins of the transitions and explain the relative intensities.

19.24 Classify the symmetry type of the *d* orbital in a tetragonal C_{4v} symmetry complex, such as $[CoCl(NH_3)_5]^+$, where the Cl lies on the *z*-axis. (a) Which orbitals will be displaced from their position in the octahedral molecular orbital diagram by π interactions with the lone pairs of the Cl^- ligand? (b) Which orbital will move because the Cl^- ligand is not as strong a σ base as NH_3? (c) Sketch the qualitative molecular orbital diagram for the C_{4v} complex.

19.25 Consider the molecular orbital diagram for a tetrahedral complex (based on Fig. 19.7) and the relevant *d*-orbital configuration.

Show that the purple colour of MnO_4^- ions cannot arise from a ligand-field transition.

19.26 Given that the wavenumbers of the two transitions in MnO_4^- are 18 500 and 32 200 cm^{-1}, explain how to estimate Δ_T from an assignment of the two charge-transfer transitions, even though Δ_T cannot be observed directly.

19.27 Prussian blue was one of the first synthetic pigments. It has the overall composition $K[Fe_2(CN)_6]$ in which each Fe(II) is surrounded by six C-coordinated CN^- ligands that bridge to Fe(III) bound to the nitrogen. Assign the intense absorption bands at 15 000 and 25 000 cm^{-1}.

PROBLEMS

19.1 In a fused magma liquid from which silicate minerals crystallize, the metal ions can be four-coordinate. In olivine crystals, the M(II) coordination sites are octahedral. Partition coefficients, which are defined as $K_p = [M(II)]_{olivine}/[M(II)]_{melt}$, follow the order Ni(II) > Co(II) > Fe(II) > Mn(II). Account for this in terms of ligand-field theory. (See I.M. Dale and P. Henderson, *24th Int. Geol. Congress, Sect.* 1972, **10**, 105.)

19.2 By considering the splitting of the octahedral orbitals as the symmetry is lowered, draw the symmetry-adapted linear combinations and the molecular orbital energy level diagram for σ bonding in a *trans*-$[ML_4X_2]$ complex. Assume that the ligand X is lower in the spectrochemical series than L.

19.3 By referring to *Resource section* 3, draw the appropriate symmetry-adapted linear combinations and the molecular orbital diagram for σ bonding in a square-planar complex. The point group is D_{4h}. Take note of the small overlap of the ligand with the d_{z^2} orbital. What is the effect of π bonding?

19.4 Figures 19.9 and 19.10 show the relation between the frontier orbitals of octahedral and square-planar complexes. Construct similar diagrams for two-coordinate linear complexes.

19.5 Consider the trigonal prismatic six-coordinate ML_6 complex with D_{3h} symmetry. Use the D_{3h} character table to divide the *d* orbitals of the metal into sets of defined symmetry type. Assume that the ligands are at the same angle relative to the *xy*-plane as in a tetrahedral complex.

19.6 Vanadium(IV) species that have the V=O group have quite distinct spectra. What is the *d*-electron configuration of V(IV)? The most symmetrical of such complexes are VOL_5 with C_{4v} symmetry with the O atom on the *z*-axis. What are the symmetry species of the five *d* orbitals in VOL_5 complexes? How many *d–d* bands are expected in the spectra of these complexes? A band near 24 000 cm^{-1} in these complexes shows vibrational progressions of the V=O vibration, implicating an orbital involving V=O bonding. Which *d–d* transition is a candidate? (See C.J. Ballhausen and H.B. Gray, *Inorg. Chem.*, 1962, **1**, 111.)

19.7 The compound $Ph_3Sn\!-\!Re(CO)_3(^tBu\text{-}DAB)$ ($^tBu\text{-}DAB = {}^tBu\!-\!N\!=\!CH\!=\!CH\!=\!N\!-\!{}^tBu$) has a metal-to-diimine ligand MLCT band as the lowest energy transition in its spectrum. Irradiation of this band gives an $Re(CO)_3(^tBu\text{-}DAB)$ radical as a photochemical product. The EPR spectrum shows extensive hyperfine splitting by ^{14}N and ^1H. The radical is assigned as a complex of the DAB radical anion with Re(I). Explain the argument (see D.J. Stufkens, *Coord. Chem. Rev.*, 1990, **104**, 39).

19.8 MLCT bands can be recognized by the fact that the energy is a sensitive function of the polarity of the solvent (because the excited state is more polar than the ground state). Two simplified molecular orbital diagrams are shown in Fig. 19.33. In (a) is a case with a ligand π level higher than the metal *d* orbital. In (b) is a case in which the metal *d* orbital and the ligand level are at the same energy. Which of the two MLCT bands should be more solvent sensitive? These two cases are realized by $[W(CO)_4(phen)]$ and $[W(CO)_4(^iPr\text{-}DAB)]$, where DAB = 1,4-diaza-1,3-butadiene, respectively. (See P.C. Servas, H.K. van Dijk, T.L. Snoeck, D.J. Stufkens, and A. Oskam, *Inorg. Chem.*, 1985, **24**, 4494.) Comment on the CT character of the transition as a function of the extent of back-donation by the metal atom.

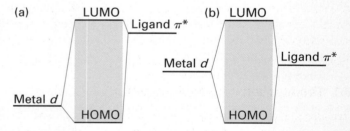

Fig. 19.33 Representation of the orbitals involved in MLCT transitions for cases in which the energy of the ligand π^* orbital varies with respect to the energy of the metal *d* orbital. See Problem 19.8.

Coordination chemistry: reactions of complexes

We now look at the evidence and experiments that are used in the analysis of the reaction pathways of metal complexes and so develop a deeper understanding of their mechanisms. Because the mechanism is rarely known definitively, the nature of the evidence for a mechanism should always be kept in mind in order to recognize that there might be other possibilities consistent with it. In the first part of this chapter we consider ligand exchange reactions and describe how reaction mechanisms are classified. We consider the steps by which the reactions take place and the details of the formation of the transition state. These concepts are then used to describe the mechanisms of the redox reactions of complexes.

Coordination chemistry is not the sole preserve of *d* metals. Whereas Chapter 19 dealt exclusively with *d* metals, this chapter builds on the introduction to coordination chemistry in Chapter 8 and applies it to all metals regardless of the block to which they belong. However, there are special features of each block, and we shall point them out.

Ligand substitution reactions

The most fundamental reaction a complex can undergo is **ligand substitution**, a reaction in which one Lewis base displaces another from a Lewis acid:

$$Y + M—X \rightarrow M—Y + X$$

This class of reaction includes complex formation reactions, in which the **leaving group**, the displaced base X, is a solvent molecule and the **entering group**, the displacing base Y, is some other ligand. An example is the replacement of a water ligand by Cl^-:

$$[Co(OH_2)_6]^{2+}(aq) + Cl^-(aq) \rightarrow [CoCl(OH_2)_5]^+(aq) + H_2O(l)$$

20.1 Thermodynamic considerations

When assessing chemical reactions we need to consider both thermodynamic and kinetic aspects because although a reaction may be thermodynamically feasible, there might be kinetic constraints.

(a) Formation constants

Key points: A formation constant expresses the strength of a ligand relative to the strength of the solvent molecules (usually H_2O) as a ligand; a stepwise formation constant is the formation constant for each individual solvent replacement in the synthesis of the complex; an overall formation constant is the product of the stepwise formation constants.

An example of a **coordination equilibrium** is the reaction of Fe(III) with SCN^- to give $[Fe(SCN)(OH_2)_5]^{2+}$, a red complex used to detect either iron(III) or the thiocyanate ion:[1]

$$[Fe(OH_2)_6]^{3+}(aq) + SCN^-(aq) \rightleftharpoons [Fe(SCN)(OH_2)_5]^{2+}(aq) + H_2O(l)$$

---

[1] In expressions for equilibrium constants and rate equations, we omit the brackets that are part of the chemical formula of the complex; the surviving square brackets denote molar concentration of a species.

Table 20.1 Formation constants for the reaction $[M(H_2O)_n]^{m+} + L \rightleftharpoons [M(L)(OH_2)_{n-1}]^{m+} + H_2O$

Ion	Ligand	K_f	log K_f	Ion	Ligand	K_f	log K_f
Mg^{2+}	NH_3	1.7	0.23	Pd^{2+}	Cl^-	1.25×10^5	6.1
Ca^{2+}	NH_3	0.64	-0.2	Na^+	SCN^-	1.2×10^4	4.08
Ni^{2+}	NH_3	525	2.72	Cr^{3+}	SCN^-	1.2×10^3	3.08
Cu^+	NH_3	8.50×10^5	5.93	Fe^{3+}	SCN^-	234	2.37
Cu^{2+}	NH_3	2.0×10^4	4.31	Co^{2+}	SCN^-	11.5	1.06
Hg^{2+}	NH_3	6.3×10^8	8.8	Fe^{2+}	pyridine	5.13	0.71
Rb^+	Cl^-	0.17	-0.77	Zn^{2+}	pyridine	8.91	0.95
Mg^{2+}	Cl^-	4.17	0.62	Cu^{2+}	pyridine	331	2.52
Cr^{3+}	Cl^-	7.24	0.86	Ag^+	pyridine	93	1.97
Co^{2+}	Cl^-	4.90	0.69				

$$K_f = \frac{[Fe(SCN)(OH_2)_5^{2+}]}{[Fe(OH_2)_6^{3+}][SCN^-]}$$

The equilibrium constant K_f is the **formation constant** of the complex. The concentration of solvent (normally H_2O) does not appear because it is taken to be constant in dilute solution and is absorbed into the numerical value of K_f. The value of K_f indicates the strength of binding of the ligand relative to H_2O: if K_f is large, the incoming ligand binds more strongly than the solvent, H_2O; if K_f is small, the incoming ligand binds more weakly than H_2O. Because the values of K_f can vary over a huge range (Table 20.1), they are often expressed as logarithms, log K_f.

The discussion of stabilities is more involved when more than one ligand may be replaced. For instance, in the reaction of $[Ni(OH_2)_6]^{2+}$ to give $[Ni(NH_3)_6]^{2+}$,

$$[Ni(OH_2)_6]^{2+}(aq) + 6\,NH_3(aq) \rightarrow [Ni(NH_3)_6]^{2+}(aq) + 6\,H_2O(l)$$

there are at least six steps, even if *cis–trans* isomerization is ignored. For the general case of the complex ML_n, for which the overall reaction is $M + n\,L \rightarrow ML_n$, the **stepwise formation constants** are

$$M + L \rightleftharpoons ML \qquad K_{f1} = \frac{[ML]}{[M][L]}$$

$$ML + L \rightleftharpoons ML_2 \qquad K_{f2} = \frac{[ML_2]}{[ML][L]}$$

and so on, and in general

$$ML_{n-1} + L \rightleftharpoons ML_n \qquad K_{fn} = \frac{[ML_n]}{[ML_{n-1}][L]}$$

These stepwise constants are the ones to consider when seeking to understand the relations between structure and reactivity.

When we want to calculate the concentration of the final product (the complex ML_n) we use the **overall formation constant**, β_n:

$$M + n\,L \rightleftharpoons ML_n \qquad \beta_n = \frac{[ML_n]}{[M][L]^n}$$

As may be verified by multiplying together the individual stepwise constants, the overall formation constant is the product of the stepwise constants:

$$\beta_n = K_{f1}K_{f2} \ldots K_{fn}$$

The inverse of each K_f, the **dissociation constant**, K_d, is also sometimes useful, and is often preferred when we are interested in the concentration of ligand that is required to give a certain concentration of complex:

$$ML \rightleftharpoons M + L \qquad K_{d1} = \frac{[M][L]}{[ML]} = \frac{1}{K_{f1}}$$

For a 1:1 reaction, like the one above, when half the metal ion is complexed and half is not, so that $[M] = [ML]$, then $K_{d1} = [L]$. In practice, if initially $[L] \gg [M]$, so that there is an insignificant change in the concentration of L when M is added and undergoes complexation, K_d is the ligand concentration required to obtain 50 per cent complexation.

Because K_d has the same form as K_a for acids, with L taking the place of H^+, its use facilitates comparisons between metal complexes and Brønsted acids. The values of K_d and K_a can be tabulated together if the proton is considered to be simply another cation. For instance, HF can be considered as the complex formed from the Lewis acid H^+ with the Lewis base F^- playing the role of a ligand.

(b) Trends in successive formation constants

Key points: Stepwise formation constants typically lie in the order $K_n > K_{n+1}$, as expected statistically; deviations from this order indicate a major change in structure.

The magnitude of the formation constant is a direct reflection of the sign and magnitude of the Gibbs energy of formation (because $\Delta_r G^{\ominus} = -RT \ln K_f$). It is commonly observed that stepwise formation constants lie in the order $K_{f1} > K_{f2} > \ldots > K_{fn}$. This general trend can be explained quite simply by considering the decrease in the number of the ligand H_2O molecules available for replacement in the formation step, as in

$$M(OH_2)_5L(aq) + L(aq) \rightarrow M(OH_2)_4L_2(aq) + H_2O(l)$$

compared with

$$M(OH_2)_4L_2(aq) + L(aq) \rightarrow M(OH_2)_3L_3(aq) + H_2O(l)$$

The decrease in the stepwise formation constants reflects the diminishing statistical factor as successive ligands are replaced. That such a simple explanation is more or less correct is illustrated by data for the successive complexes in the series from $[Ni(OH_2)_6]^{2+}$ to $[Ni(NH_3)_6]^{2+}$ (Table 20.2). The reaction enthalpies for the six successive steps are known to vary by less than 2 kJ mol^{-1}.

A reversal of the relation $K_{fn} > K_{fn+1}$ is usually an indication of a major change in the electronic structure of the complex as more ligands are added. An example is the observation that the tris(bipyridine) complex of Fe(II), $[Fe(bipy)_3]^{2+}$, is strikingly stable compared with the bis complex, $[Fe(bipy)_2(OH_2)_2]^{2+}$. This observation can be correlated with the change from a high-spin (weak-field) $t_{2g}^4 e_g^2$ configuration in the bis complex (note the presence of weak-field H_2O ligands) to a low-spin (strong-field) t_{2g}^6

Table 20.2 Formation constants of Ni(II) ammines, $[Ni(NH_3)_n(OH_2)_{6-n}]^{2+}$

n	K_f	$\log K_f$	K_n/K_{n-1} Experimental	Statistical*
1	525	2.72		
2	148	2.17	0.28	0.42
3	45.7	1.66	0.31	0.53
4	13.2	1.12	0.29	0.56
5	4.7	0.63	0.35	0.53
6	1.1	0.03	0.2	0.42

*Based on ratios of numbers of ligands available for replacement, with the reaction enthalpy assumed constant.

configuration in the tris complex, where there is a considerable increase in the LFSE.

$$[Fe(OH_2)_6]^{2+}(aq) + bipy(aq) \rightleftharpoons [Fe(bipy)(OH_2)_4]^{2+}(aq) + 2\ H_2O(l) \quad \log K_{f1} = 4.2$$

$$[Fe(bipy)(OH_2)_4]^{2+}(aq) + bipy(aq) \rightleftharpoons [Fe(bipy)_2(OH_2)_2]^{2+}(aq) + 2\ H_2O(l)$$
$$\log K_{f2} = 3.7$$

$$[Fe(bipy)_2(OH_2)_2]^{2+}(aq) + bipy(aq) \rightleftharpoons [Fe(bipy)_3]^{2+}(aq) + 2\ H_2O(l) \quad \log K_{f3} = 9.3$$

A contrasting example is the halogeno complexes of Hg(II), where K_{f3} is anomalously low compared with K_{f2}:

$$[Hg(OH_2)_6]^{2+}(aq) + Cl^-(aq) \rightleftharpoons [HgCl(OH_2)_5]^+(aq) + H_2O(l) \quad \log K_{f1} = 6.74$$

$$[HgCl(OH_2)_5]^+(aq) + Cl^-(aq) \rightleftharpoons [HgCl_2(OH_2)_4](aq) + H_2O(l) \quad \log K_{f2} = 6.48$$

$$[HgCl_2(OH_2)_4](aq) + Cl^-(aq) \rightleftharpoons [HgCl_3(OH_2)]^-(aq) + 3\ H_2O(l) \quad \log K_{f3} = 0.85$$

The decrease between the second and third values is too large to be explained statistically and suggests a major change in the nature of the complex, such as the onset of four-coordination:

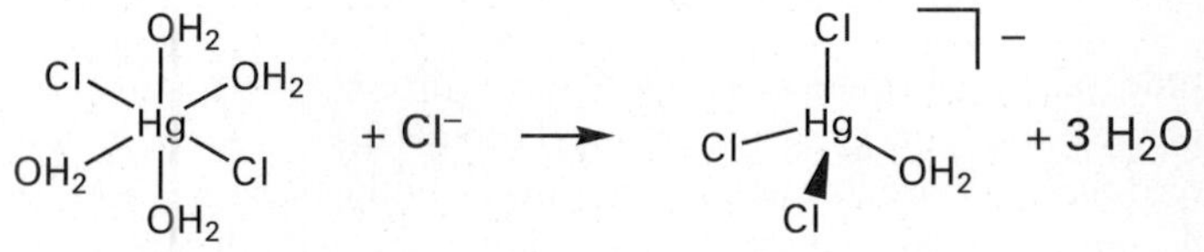

Example 20.1 Interpreting irregular successive formation constants

The successive formation constants for complexes of cadmium with Br^- are $K_{f1} = 36.3$, $K_{f2} = 3.47$, $K_{f3} = 1.15$, $K_{f4} = 2.34$. Suggest an explanation of why $K_{f4} > K_{f3}$.

Answer The anomaly suggests a structural change. Aqua complexes are usually six-coordinate whereas halo complexes of M^{2+} ions are commonly tetrahedral. The reaction of the complex with three Br^- groups to add the fourth is

$$[CdBr_3(OH_2)_3]^-(aq) + Br^-(aq) \rightarrow [CdBr_4]^{2-}(aq) + 3\ H_2O(l)$$

This step is favoured by the release of three H_2O molecules from the relatively restricted coordination sphere environment. The result is an increase in K_f.

Self-test 20.1 A square-planar four-coordinate Fe(II) porphyrin (P) complex may add two further ligands (L) axially. The maximally coordinated complex [FePL₂] is low-spin, whereas [FePL] is high-spin. Account for an increase of the second formation constant with respect to the first.

(c) The chelate effect

Key points: The chelate effect is the greater stability of a complex containing a coordinated polydentate ligand compared with a complex containing the equivalent number of analogous monodentate ligands; it is largely an entropic effect.

When K_{f1} for the formation of a complex with a bidentate chelate ligand, such as ethylenediamine (en), is compared with the value of β_2 for the corresponding bis(ammine) complex, it is found that the former is generally larger:

$$[Cd(OH_2)_6]^{2+}(aq) + en(aq) \rightleftharpoons [Cd(en)(OH_2)_4]^{2+}(aq) + 2\ H_2O(l)$$
$$\log K_{f1} = 5.84 \qquad \Delta_r H^\ominus = -29.4\ kJ\,mol^{-1} \qquad \Delta_r S^\ominus = +13.0\ J\,K^{-1}mol^{-1}$$

$$[Cd(OH_2)_6]^{2+}(aq) + 2\ NH_3(aq) \rightleftharpoons [Cd(NH_3)_2(OH_2)_4]^{2+}(aq) + 2\ H_2O(l)$$
$$\log \beta_2 = 4.95 \qquad \Delta_r H^\ominus = -29.8\ kJ\,mol^{-1} \qquad \Delta_r S^\ominus = -5.2\ J\,K^{-1}mol^{-1}$$

Two similar Cd—N bonds are formed in each case, yet the formation of the chelate-containing complex is distinctly more favourable. This greater stability of chelated complexes compared with their nonchelated analogues is called the **chelate effect**.

The chelate effect can be traced primarily to differences in reaction entropy between chelated and nonchelated complexes in dilute solutions. The chelation reaction results in an increase in the number of independent molecules in solution. By contrast, the nonchelating reaction produces no net change (compare the two chemical equations above). The former therefore has the more positive reaction entropy and hence is the more favourable process. The reaction entropies measured in dilute solution support this interpretation.

The entropy advantage of chelation extends beyond bidentate ligands, and applies, in principle, to any polydentate ligand. In fact, the greater the number of donor sites the multidentate ligand has, the greater is the entropic advantage of displacing monodentate ligands. The chelate effect accounts in part for the great stability of complexes containing the tetradentate porphyrin ligand and the hexadentate edta^{4-} ligand.

The chelate effect is of great practical importance. The majority of reagents used in complexometric titrations in analytical chemistry are polydentate chelates like edta^{4-}, and most biochemical metal binding sites are chelating ligands. A formation constant as high as 10^{12} to 10^{25} is generally a sign that the chelate effect is in operation.

In addition to the thermodynamic rationalization for the chelate effect described above, there is an additional role in the chelate effect for kinetics. Once one ligating group of a polydentate ligand has bound to a metal ion, it becomes more likely that its other ligating groups will bind, as they are now constrained to be in close proximity to the metal ion; thus chelate complexes are favoured kinetically too.

(d) Steric effects and electron delocalization

Key point: The stability of chelate complexes of d metals involving diimine ligands is a result of the chelate effect in conjunction with the ability of the ligands to act as π acceptors as well as σ donors.

Steric effects have an important influence on formation constants. They are particularly important in chelate formation because ring completion may be difficult geometrically. Chelate rings with five members are generally very stable, as their bond angles are near ideal in the sense of there being no ring strain. Six-membered rings are reasonably stable and may be favoured if their formation results in electron delocalization. Three-, four-, and seven-membered (and larger) chelate rings are found only rarely because they normally result in distortions of bond angles and unfavourable steric interactions.

Complexes containing chelating ligands with delocalized electronic structures may be stabilized by electronic effects in addition to the entropy advantages of chelation. For example, diimine ligands (**1**), such as bipyridine (**2**) and phenanthroline (**3**), are constrained to form five-membered rings with the metal atom. The great stability of their complexes with d metals is probably a result of their ability to act as π acceptors as well as σ donors and to form π bonds by overlap of the full metal d orbitals and the empty ring π^* orbitals. This bond formation is favoured by electron population in the metal t_{2g} orbitals, which allows the metal atom to act as a π donor and transfer electron density to the ligand rings. An example is the complex $[Ru(bipy)_3]^{2+}$ (**4**). In some cases the chelate ring that forms can have appreciable aromatic character, which stabilizes the chelate ring even more.

(e) The Irving–Williams series

Key point: The Irving–Williams series summarizes the relative stabilities of complexes formed by M^{2+} ions, and reflects a combination of electrostatic effects and LFSE.

Figure 20.1 shows $\log K_f$ values for complexes of the M^{2+} ions of the $3d$ series. The variation in formation constants shown there is summarized by the **Irving–Williams series**:

$$Ba^{2+} < Sr^{2+} < Ca^{2+} < Mg^{2+} < Mn^{2+} < Fe^{2+} < Co^{2+} < Ni^{2+} < Cu^{2+} > Zn^{2+}$$

The order is relatively insensitive to the choice of ligands.

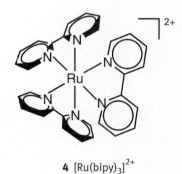

1

2 bipy

3 phen

4 $[Ru(bipy)_3]^{2+}$

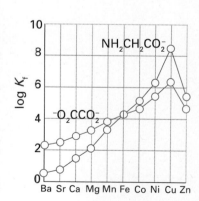

Fig. 20.1 The variation of formation constants for the M^{2+} ions of the Irving–Williams series.

In general, the increase in stability correlates with ionic radius, which suggests that the Irving–Williams series reflects electrostatic effects. However, beyond Mn^{2+} there is a sharp increase in the value of K_f for d^6 Fe(II), d^7 Co(II), d^8 Ni(II), and d^9 Cu(II), with strong-field ligands. These ions experience an additional stabilization proportional to the ligand-field stabilization energies (Table 19.2). There is one important exception: the stability of Cu(II) complexes is greater than that of Ni(II) even though Cu(II) has an additional antibonding e_g electron. This anomaly is a consequence of the stabilizing influence of the Jahn–Teller effect (Section 19.1h), which results in strong binding of four of the ligands in the plane of the tetragonally distorted Cu(II) complex, and that stabilization enhances the value of K_f.

20.2 Rates of ligand substitution

Key points: The rates of substitution reactions span a very wide range and correlate with the structures of the complexes; complexes that react quickly are called labile, those that react slowly are called inert or nonlabile.

Rates of reaction are as important as equilibria in coordination chemistry. The numerous isomers of the ammines of Co(III) and Pt(II), which were so important to the development of the subject, could not have been isolated if ligand substitutions and interconversion of the isomers had been fast. But what determines whether one complex will survive for long periods whereas another will undergo rapid reaction?

The rate at which one complex converts into another is governed by the height of the activation energy barrier that lies between them. Thermodynamically unstable complexes that survive for long periods (by convention, at least a minute) are commonly called 'inert', but **nonlabile** is more appropriate and is the term we shall use. Complexes that undergo more rapid equilibration are called **labile**. An example of each type is the labile complex $[Ni(OH_2)_6]^{2+}$, which has a half-life of the order of milliseconds before the H_2O is replaced by a stronger base, and the nonlabile complex $[Co(NH_3)_5(OH_2)]^{3+}$, in which H_2O survives for several minutes as a ligand before it is replaced by a stronger base.

Figure 20.2 shows the characteristic lifetimes of the important aqua metal ion complexes. We see a range of lifetimes starting at about 1 ns, which is approximately the time it takes for a molecule to diffuse one molecular diameter in solution. At the other end of the scale are lifetimes in years. Even so, the illustration does not show the longest times that could be considered, which are comparable to geological eras.

We shall examine the lability of complexes in greater detail when we discuss the mechanism of reactions later in this section, but we can make two broad generalizations now. The first is that complexes of metals that have no additional factor to provide extra stability (for instance, the LFSE and chelate effects) are among the most labile. Any additional stability of a complex results in an increase in activation energy for a ligand replacement reaction and hence decreases the lability of the complex. A second

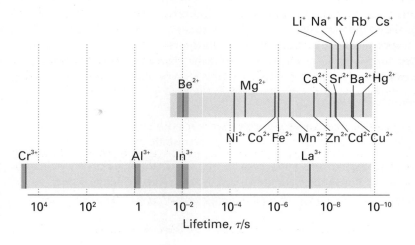

Fig. 20.2 Characteristic lifetimes for exchange of water molecules in aqua complexes.

Table 20.3 Representative timescales of chemical and physical processes

Timescale*	Process	Example
10^8 s	Ligand exchange (inert complex)	$[Cr(OH_2)_6]^{3+} - H_2O$ (c. 32 years)
60 s	Ligand exchange (nonlabile complex)	$[V(OH_2)_6]^{3+} - H_2O$ (50 s)
1 ms	Ligand exchange (labile complex)	$[Pt(OH_2)_4]^{2+} - H_2O$ (0.4 ms)
1 μs	Intervalence charge transfer	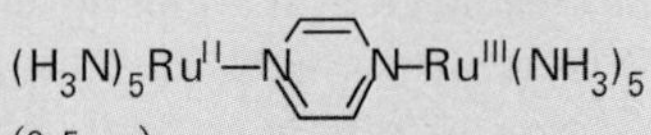 $(H_3N)_5Ru^{II}-N\bigcirc N-Ru^{III}(NH_3)_5$ (0.5 μs)
1 ns	Ligand exchange (labile complex)	$[Ni(OH_2)_5(py)]^{2+} - H_2O$ (1 ns)
10 ps	Ligand association	$Cr(CO)_5 + THF$ (10 ps)
1 ps	Rotation time in liquid	CH_3CN (1 ps)
1 fs	Molecular vibration	Sn—Cl stretch (300 fs)

* Approximate time at room temperature.

generalization is that very small ions are often less labile because they have greater M—L bond strengths and it is sterically very difficult for incoming ligands to approach the metal atom closely.

Some further generalizations are as follows:

1 All complexes of *s*-block ions except the smallest (Be^{2+} and Mg^{2+}) are very labile.

2 Complexes of the M(III) ions of the *f*-block are all very labile.

3 Complexes of the d^{10} ions (Zn^{2+}, Cd^{2+}, and Hg^{2+}) are normally very labile.

4 Across the 3*d* series, complexes of *d*-block M(II) ions are generally moderately labile, with distorted Cu(II) complexes among the most labile.

5 Complexes of M(III) ions are distinctly less labile than M(II) ions.

6 *d*-Metal complexes with d^3 and low-spin d^6 configurations (for example Cr(III), Fe(II), and Co(III)) are generally nonlabile as they have large LFSEs. Chelate complexes with the same configuration, such as $[Fe(phen)_3]^{2+}$, are particularly inert.

7 Nonlability is common among the complexes of the 4*d* and 5*d* series, which reflects the high LFSE and strength of the metal–ligand bonding.

Table 20.3 illustrates the range of timescales for a number of reactions.

The identities of ligands also affect the rates of reactions. The identity of the incoming ligand has the greatest effect, and equilibrium constants of displacement reactions can be used to rank ligands in order of their strength as Lewis bases. However, a different order may be found if bases are ranked according to the rates at which they displace a ligand from the central metal ion. Therefore, for kinetic considerations, we replace the equilibrium concept of basicity by the kinetic concept of **nucleophilicity**, the rate of attack on a complex by a given Lewis base relative to the rate of attack by a reference Lewis base. The shift from equilibrium to kinetic considerations is emphasized by referring to ligand displacement as **nucleophilic substitution**.

Ligands other than the entering and leaving groups may play a significant role in controlling the rates of reactions; these ligands are referred to as **spectator ligands**. For instance, it is observed for square-planar complexes that the ligand *trans* to the leaving group X has a great effect on the rate of substitution of X by the entering group Y.

20.3 The classification of mechanisms

The **mechanism** of a reaction is the sequence of elementary steps by which the reaction takes place. Once the mechanism has been unravelled, attention turns to the details of the activation process of the rate-determining step.[2] In some cases the overall mechanism is not fully resolved, and the only information available is the rate-determining step.

---

[2] The latter is sometimes called the 'intimate mechanism' of the reaction.

(a) Association, dissociation, and interchange

Key points: The mechanism of a nucleophilic substitution reaction is the sequence of elementary steps by which the reaction takes place and is classified as associative, dissociative, or interchange; an associative mechanism is distinguished from an interchange mechanism by demonstrating that the intermediate has a relatively long life.

The first stage in the kinetic analysis of a reaction is to study how its rate changes as the concentrations of reactants are varied. This type of investigation leads to the identification of **rate laws**, the differential equations governing the rate of change of the concentrations of reactants and products. For example, the observation that the rate of formation of $[Ni(OH_2)_5(NH_3)]^{2+}$ from $[Ni(OH_2)_6]^{2+}$ is proportional to the concentration of both NH_3 and $[Ni(OH_2)_6]^{2+}$ implies that the reaction is first-order in each of these two reactants, and that the overall rate law is[3]:

$$\text{rate} = k[Ni(OH_2)_6^{2+}][NH_3] \tag{20.1}$$

In simple reaction schemes, the slowest elementary step of the reaction dominates the overall reaction rate and the overall rate law, and is called the **rate-determining step**. However, in general, all the steps in the reaction may contribute to the rate law and affect its rate. Therefore, in conjunction with stereochemical and isotopic labelling studies, the determination of the rate law is the route to the elucidation of the mechanism of the reaction.

Three main classes of reaction mechanism have been identified. A **dissociative mechanism**, denoted D, is a reaction sequence in which an intermediate of reduced coordination number is formed by the departure of the leaving group:

$$ML_nX \rightarrow ML_n + X$$
$$ML_n + Y \rightarrow ML_nY$$

Here ML_n (the metal atom and any spectator ligands) is a true intermediate that can, in principle, be detected (or even isolated). For example, the substitution of hexacarbonyltungsten by phosphine takes place by dissociation of CO from the complex

$$W(CO)_6 \rightarrow W(CO)_5 + CO$$

followed by coordination of phosphine:

$$W(CO)_5 + PPh_3 \rightarrow W(CO)_5PPh_3$$

Under the conditions in which this reaction is usually performed in the laboratory, the intermediate $W(CO)_5$ is rapidly captured by the solvent, such as tetrahydrofuran, to form $[W(CO)_5(thf)]$. This complex in turn is converted to the phosphine product, presumably by a second dissociative process. The typical form of the reaction profile is shown in Fig. 20.3.

An **associative mechanism**, denoted A, involves a step in which an intermediate is formed with a higher coordination number than the original complex:

$$ML_nX + Y \rightarrow ML_nXY$$
$$ML_nXY \rightarrow ML_nY + X$$

Once again, the intermediate ML_nXY can, in principle at least, be detected. This mechanism plays a role in many reactions of square-planar Pt(II), Pd(II), Ni(II), and Ir(I) d^8 complexes. A specific example is the exchange of $^{14}CN^-$ with the ligands in the square-planar complex $[Ni(CN)_4]^{2-}$. The first step is the coordination of a ligand to the complex

$$[Ni(CN)_4]^{2-} + {}^{14}CN^- \rightarrow [Ni(CN)_4({}^{14}CN)]^{3-}$$

Subsequently, a ligand is discarded:

$$[Ni(CN)_4({}^{14}CN)]^{3-} \rightarrow [Ni(CN)_3({}^{14}CN)]^{2-} + CN^-$$

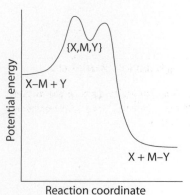

Fig. 20.3 The typical form of the reaction profile of a reaction with a dissociative mechanism.

---

[3] Recall that in rate equations, as in expressions for equilibrium constants, we omit the brackets that are part of the chemical formula of the complex; the surviving brackets denote molar concentration.

The radioactivity of carbon-14 provides a means of monitoring this reaction, and the intermediate $[Ni(CN)_5]^{3-}$ has been detected and isolated. The typical form of the reaction profile is similar to that of the dissociative mechanism, and is shown in Fig. 20.4.

An **interchange mechanism**, denoted I, takes place in one step:

$$ML_nX + Y \rightarrow X \cdots ML_n \cdots Y \rightarrow ML_nY + X$$

The leaving and entering groups exchange in a single step by forming a transition state[4] but not a true intermediate. The interchange mechanism is common for many reactions of six-coordinate complexes. The typical form of the reaction profile is shown in Fig. 20.5.

The distinction between the A and I mechanisms hinges on whether or not the intermediate persists long enough to be detectable. One type of evidence is the isolation of an intermediate in another related reaction or under different conditions. If an argument by extrapolation to the actual reaction conditions suggests that a moderately long-lived intermediate might exist during the reaction in question, then the A path is indicated. For example, the synthesis of the first trigonal-bipyramidal Pt(II) complex, $[Pt(SnCl_3)_5]^{3-}$, indicates that a five-coordinate platinum complex may be plausible in substitution reactions of square-planar Pt(II) ammine complexes. Similarly, the fact that $[Ni(CN)_5]^{3-}$ is observed spectroscopically in solution, and that it has been isolated in the crystalline state, provides support for the view that it is involved when CN^- exchanges with the square-planar tetracyanonickelate(II) ion.

A second indication of the persistence of an intermediate is the observation of a stereochemical change, which implies that the intermediate has lived long enough to undergo rearrangement. *Cis* to *trans* isomerization is observed in the substitution reactions of certain square-planar phosphine Pt(II) complexes, which is in contrast to the retention of configuration usually observed. This difference implies that the trigonal-bipyramidal intermediate lives long enough for an exchange between the axial and equatorial ligand positions to occur.

Direct spectroscopic detection of the intermediate, and hence an indication of A rather than I, may be possible if a sufficient amount accumulates. Such direct evidence, however, requires an unusually stable intermediate with favourable spectroscopic characteristics.

(b) The rate-determining step

Key point: The rate-determining step is classified as associative or dissociative according to the dependence of its rate on the identity of the entering group.

Now we consider the rate-determining step of a reaction and the details of its formation. The step is called **associative** and denoted a if its rate depends strongly on the identity of the incoming group. Examples are found among reactions of the d^8 square-planar complexes of Pt(II), Pd(II), and Au(III), including

$$[PtCl(dien)]^+(aq) + I^-(aq) \rightarrow [PtI(dien)]^+(aq) + Cl^-(aq)$$

where dien is diethylenetriamine $(NH_2CH_2CH_2NHCH_2CH_2NH_2)$. It is found, for instance, that use of I^- instead of Br^- decreases the rate constant by an order of magnitude. Experimental observations on the substitution reactions of square-planar complexes support the view that the rate-determining step is associative.

The strong dependence on entering group Y of an associative rate-determining step indicates that the transition state must involve significant bonding to Y. A reaction with an associative mechanism (A) will be associatively activated (a) if the attachment of Y to the initial reactant ML_nX is the rate-determining step; such a reaction is designated A_a; in this case, the intermediate ML_nXY would not be detected. A reaction with a dissociative mechanism (D) is associatively activated (a) if the attachment of Y to the intermediate ML_n is the rate-determining step; such a reaction is designated D_a. An interchange mechanism (I) is associatively activated if the rate of formation of the transition state depends on the rate at which the new $Y \cdots M$ bond forms. Figure 20.6 shows the reaction profiles for associatively activated A and D mechanisms. For the reactions to proceed, it is

[4] In this text we shall not distinguish between the terms 'transition state' and 'activated complex'.

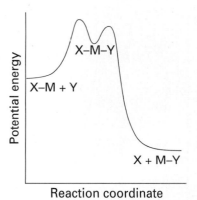

Fig. 20.4 The typical form of the reaction profile of a reaction with an associative mechanism.

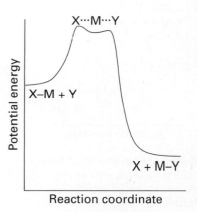

Fig. 20.5 The typical form of the reaction profile of a reaction with an interchange mechanism.

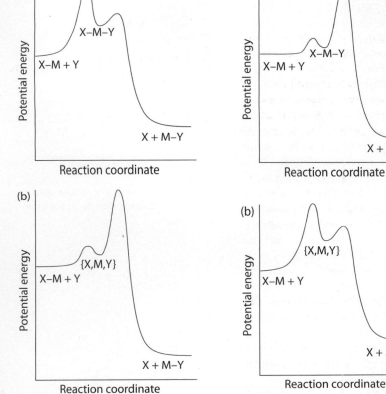

Fig. 20.6 The typical form of the reaction profile of reactions with an associatively activated steps: (a) associative mechanism, A_a; (b) a dissociative mechanism, D_a.

Fig. 20.7 The typical form of the reaction profile of reactions with a dissociatively activated steps: (a) an associative mechanism, A_d; (b) a dissociative mechanism, D_d.

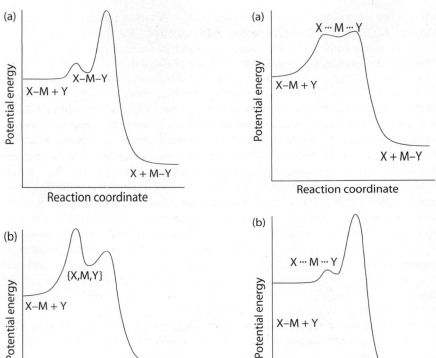

Fig. 20.8 The typical form of the reaction profile of reactions with an interchange mechanism: (a) associatively activated, I_a; (b) dissociatively activated, I_d.

necessary to have established a population of an encounter complex X–M, Y in a pre-equilibrium step.

The rate-determining step is called **dissociative** and denoted d if its rate is largely independent of the identity of Y. This category includes some of the classic examples of ligand substitution in octahedral d-metal complexes, including

$$[\text{Ni(OH}_2)_6]^{2+}(\text{aq}) + \text{NH}_3(\text{aq}) \rightarrow [\text{Ni(OH}_2)_5(\text{NH}_3)]^{2+}(\text{aq}) + \text{H}_2\text{O(l)}$$

It is found that replacement of NH_3 by pyridine in this reaction changes the rate by at most a few per cent.

The weak dependence on Y of a dissociatively activated process indicates that the rate of formation of the transition state is determined largely by the rate at which the bond to the leaving group X can break. A reaction with an associative mechanism (A) will be dissociatively activated (d) provided the loss of X from the intermediate YML_nX is the rate-determining step; such a reaction is designated A_d. A reaction with a dissociative mechanism (D) is dissociatively activated (d) if the initial loss of X from the reactant ML_nX is the rate-determining step, such a reaction is designated D_d. In this case, the intermediate ML_n would not be detected. Figure 20.7 shows the reaction profiles for dissociatively activated A and D mechanisms.

A reaction that has an interchange mechanism (I) can be either associatively or dissociatively activated, and is designated either I_a or I_d, respectively. In an I_a mechanism, the rate of reaction depends on the rate at which the M$\cdots$Y bond forms, whereas in an I_d reaction the rate of reaction depends on the rate at which the M$\cdots$X bond breaks (Fig. 20.8).

Ligand substitution in square-planar complexes

The mechanism of ligand exchange in square-planar complexes of platinum has been studied extensively, largely because the reactions occur on a timescale that is very amenable to study. We might expect an associative mechanism of ligand exchange because square-planar complexes are sterically uncrowded—they can be considered as octahedral complexes with two ligands missing—but it is rarely that simple. The elucidation of the mechanism of the substitution of square-planar complexes is often complicated by the occurrence of alternative pathways. For instance, if a reaction such as

$$PtCl(dien)]^+(aq) + I^-(aq) \rightarrow [PtI(dien)]^+(aq) + Cl^-(aq)$$

is first-order in the complex and independent of the concentration of I^-, then the rate of reaction will be equal to $k_1[PtCl(dien)^+]$. However, if there is a pathway in which the rate law is overall second-order, first-order in the complex, and first-order in the incoming group, then the rate would be given by $k_2[PtCl(dien)^+][I^-]$. If both reaction pathways occur at comparable rates, the rate law has the form

$$rate = (k_1 + k_2[I^-])[PtCl(dien)^+] \tag{20.2}$$

A reaction like this is usually studied under the conditions $[I^-] \gg [complex]$ so that $[I^-]$ does not change significantly during the reaction. This simplifies the treatment of the data as $k_1 + k_2[I^-]$ is effectively constant and the rate law is now pseudo first-order:

$$rate = k_{obs}[PtCl(dien)^+] \qquad k_{obs} = k_1 + k_2[I^-] \tag{20.3}$$

A plot of the observed pseudo-first-order rate constant against $[I^-]$ gives k_2 as the slope and k_1 as the intercept.

In the following sections, we examine the factors that affect the second-order reaction and then consider the first-order process.

20.4 The nucleophilicity of the entering group

Key points: The nucleophilicity of an entering group is expressed in terms of the nucleophilicity parameter defined in terms of the substitution reactions of a specific square-planar platinum complex; the sensitivity of other platinum complexes to changes in the entering group is expressed in terms of the nucleophilic discrimination factor.

We start by considering the variation of the rate of the reaction as the entering group Y is varied. The reactivity of Y (for instance, I^- in the reaction above) can be expressed in terms of a **nucleophilicity parameter**, n_{Pt}:

$$n_{Pt} = \log \frac{k_2(Y)}{k_2^\circ} \tag{20.4}$$

where $k_2(Y)$ is the second-order rate constant for the reaction

$$trans\text{-}[PtCl_2(py)_2] + Y \rightarrow trans\text{-}[PtClY(py)_2]^+ + Cl^-$$

and k_2° is the rate constant for the same reaction with the reference nucleophile methanol. The entering group is highly nucleophilic, or has a high nucleophilicity, if n_{Pt} is large.

Table 20.4 gives some values of n_{Pt}. One striking feature of the data is that, although the entering groups in the table are all quite simple, the rate constants span nearly nine orders of magnitude. Another feature is that the nucleophilicity of the entering group towards Pt appears to correlate with soft Lewis basicity, with $Cl^- < I^-$, $O < S$, and $NH_3 < PR_3$.

The nucleophilicity parameter is defined in terms of the reaction rates of a specific platinum complex. When the complex itself is varied we find that the reaction rates show a range of different sensitivities towards changes in the entering group. To express this range of sensitivities we rearrange eqn 20.4 into

$$\log k_2(Y) = n_{Pt}(Y) + C \tag{20.5}$$

Table 20.4 A selection of n_{Pt} values for a range of nucleophiles

Nucleophile	Donor atom	n_{Pt}
Cl^-	Cl	3.04
I^-	I	5.42
CN^-	C	7.00
CH_3OH	O	0
C_6H_5SH	S	4.15
NH_3	N	3.06
$(C_6H_5)_3P$	P	8.79

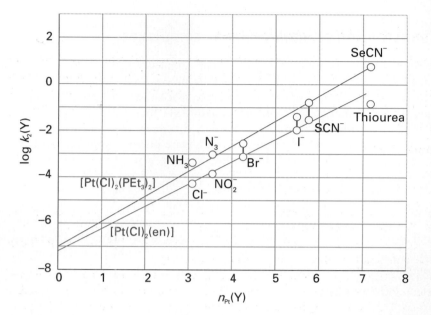

Fig. 20.9 The slope of the straight line obtained by plotting log k_2(Y) against the nucleophilicity parameter n_{Pt}(Y) for a series of ligands is a measure of the responsiveness of the complex to the nucleophilicity of the entering group.

Table 20.5 Nucleophilic discrimination factors

	S
trans-[PtCl₂(PEt₃)₂]	1.43
trans-[PtCl₂(py)₂]	1.00
[PtCl₂(en)]	0.64
trans-[PtCl(dien)]⁺	0.65

where $C = \log k_2^{\circ}$. Now consider the analogous substitution reactions for the general complex [PtL₃X]:

$$[PtL_3X] + Y \rightarrow [PtL_3Y] + X$$

The relative rates of these reactions can be expressed in terms of the same nucleophilicity parameter n_{Pt} provided we replace eqn 20.5 by

$$\log k_2(Y) = S n_{Pt}(Y) + C \tag{20.6}$$

The parameter S, which characterizes the sensitivity of the rate constant to the nucleophilicity parameter, is called the **nucleophilic discrimination factor**. We see that the straight line obtained by plotting log k_2(Y) against n_{Pt} for reactions of Y with *trans*-[PtCl₂(PEt₃)₂] is steeper than that for reactions with *cis*-[PtCl₂(en)] (Fig. 20.9). Hence, S is larger for the former reaction, which indicates that rate of the reaction is more sensitive to changes in the nucleophilicity of the entering group.

Some values of S are given in Table 20.5. Note that S is close to 1 in all cases, so all the complexes are quite sensitive to n_{Pt}. This sensitivity is what we expect for associatively activated reactions. Another feature to note is that larger values of S are found for complexes of platinum with softer base ligands.

Example 20.2 Using the nucleophilicity parameter

The second-order rate constant for the reaction of I⁻ with *trans*-[PtCl(CH₃)(PEt₃)₂] in methanol at 30°C is 40 dm³ mol⁻¹ s⁻¹. The corresponding reaction with N₃⁻ has $k_2 = 7.0$ dm³ mol⁻¹ s⁻¹. Estimate S and C for the reaction given the n_{Pt} values of 5.42 and 3.58, respectively, for the two nucleophiles.

Answer Substituting the two values of n_{Pt} into eqn 20.6 gives

   $1.60 = 5.42 S + C$ (for I⁻)
   $0.85 = 3.58 S + C$ (for N₃⁻)

Solving these two simultaneous equations gives $S = 0.41$ and $C = -0.61$. The value of S is fairly small, showing that the discrimination of this complex among different nucleophiles is not great. This lack of sensitivity is related to the fairly large value of C, which corresponds to the rate constant being large and hence to the complex being reactive. It is commonly found that high reactivity correlates with low selectivity.

Self-test 20.2 Calculate the second-order rate constant for the reaction of the same complex with NO₂⁻, for which $n_{Pt} = 3.22$.

20.5 The shape of the transition state

Careful studies of the variation of the reaction rates of square-planar complexes with changes in the composition of the reactant complex and the conditions of the reaction shed light on the general shape of the transition state. They also confirm that substitution almost invariably has an associative rate-determining stage; hence intermediates are rarely detected.

(a) The *trans* effect

Key point: A strong σ-donor ligand or π-acceptor ligand greatly accelerates substitution of a ligand that lies in the *trans* position.

The spectator ligands T that are *trans* to the leaving group in square-planar complexes influence the rate of substitution. This phenomenon is called the **trans effect**. It is generally accepted that the *trans* effect arises from two separate influences: one arising in the ground state and the other in the transition state itself.

The **trans influence** is the extent to which the ligand T weakens the bond *trans* to itself in the ground state of the complex. The *trans* influence correlates with the σ-donor ability of the ligand T because, broadly speaking, ligands *trans* to each other use the same orbitals on the metal for bonding. Thus if one ligand is a strong σ donor, then the ligand *trans* to it cannot donate electrons to the metal so well, and thus has a weaker interaction with the metal. The *trans* influence is assessed quantitatively by measuring bond lengths, stretching frequencies, and metal-to-ligand NMR coupling constants (Section 6.5c). The **transition state effect** correlates with the π-acceptor ability of the ligand. Its origin is thought to be the increase in electron density on the metal atom arising due to the incoming ligand: any ligand that can accept this increased electron density will stabilize the transition state (**5**). The *trans* effect is the combination of both effects; it should be noted that the same factors contribute to a large ligand-field splitting. *Trans* effects are listed in Table 20.6 and follow the order:

For a T σ-donor : $OH^- < NH_3 < Cl^- < Br^- < CN^-, CH_3^- < I^- < SCN^- < PR_3, H^-$

For a T π-acceptor : $Br^- < I^- < NCS^- < NO_2^- < CN^- < CO, C_2H_4$

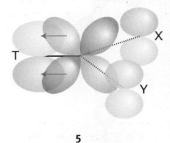

5

Table 20.6 The effect of the *trans* ligand in reactions of *trans*-[PtCl(PEt$_3$)$_2$L]

L	k_1/s^{-1}	$k_2/(dm^3\ mol^{-1}\ s^{-1})$
CH$_3^-$	1.7×10^{-4}	6.7×10^{-2}
C$_6$H$_5^-$	3.3×10^{-5}	1.6×10^{-2}
Cl$^-$	1.0×10^{-6}	4.0×10^{-4}
H$^-$	1.8×10^{-2}	4.2
PEt$_3$	1.7×10^{-2}	3.8

Example 20.3 Using the *trans* effect synthetically

Use the *trans* effect series to suggest synthetic routes to *cis*- and *trans*-[PtCl$_2$(NH$_3$)$_2$] from [Pt(NH$_3$)$_4$]$^{2+}$ and [PtCl$_4$]$^{2-}$.

Answer Reaction of [Pt(NH$_3$)$_4$]$^{2+}$ with HCl leads to [PtCl(NH$_3$)$_3$]$^+$. Because the *trans* effect of Cl$^-$ is greater than that of NH$_3$, substitution reactions will occur preferentially *trans* to Cl$^-$, and further action of HCl gives *trans*-[PtCl$_2$(NH$_3$)$_2$]:

$$[Pt(NH_3)_4]^{2+} + Cl^- \rightarrow [PtCl(NH_3)_3]^+ \rightarrow trans\text{-}[PtCl_2(NH_3)_2]$$

When the starting complex is [PtCl$_4$]$^{2-}$, reaction with NH$_3$ leads first to [PtCl$_3$(NH$_3$)]$^-$. A second step should substitute one of the two mutually *trans* Cl$^-$ ligands with NH$_3$ to give *cis*-[PtCl$_2$(NH$_3$)$_2$].

$$[PtCl_4]^{2-} + NH_3 \rightarrow [PtCl_3(NH_3)]^- \rightarrow cis\text{-}[PtCl_2(NH_3)_2]$$

Self-test 20.3 Given the reactants PPh$_3$, NH$_3$, and [PtCl$_4$]$^{2-}$, propose an efficient route to both *cis*- and *trans*-[PtCl$_2$(NH$_3$)(PPh$_3$)].

(b) Steric effects

Key point: Steric crowding at the reaction centre usually inhibits associative reactions and facilitates dissociative reactions.

Steric crowding at the reaction centre by bulky groups that can block the approach of attacking nucleophiles will inhibit associative reactions. The rate constants for

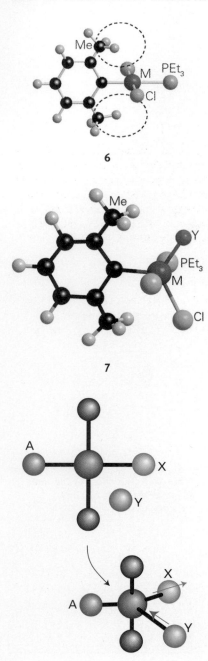

6

7

8

the replacement of Cl^- by H_2O in *cis*-$[PtClL(PEt_3)_2]$ complexes at 25°C illustrate the point:

L =	pyridine	2-methylpyridine	2, 6-dimethylpyridine
k/s^{-1}	8×10^{-2}	2.0×10^{-4}	1.0×10^{-6}

The methyl groups adjacent to the N donor atom greatly decrease the rate. In the 2-methylpyridine complex they block positions either above or below the plane. In the 2,6-dimethylpyridine complex they block positions both above and below the plane (**6**). Thus, along the series, the methyl groups increasingly hinder attack by H_2O. The effect is smaller if L is *trans* to Cl^-. This difference is explained by the methyl groups then being further from the entering and leaving groups in the trigonal-bipyramidal transition state if the pyridine ligand is in the trigonal plane (**7**). Conversely, the reduction of coordination number that occurs in a dissociative reaction can relieve the steric overcrowding, and thus increase the rate of the dissociative reaction.

(c) Stereochemistry

Key point: Substitution of a square-planar complex preserves the original geometry, which suggests a trigonal-pyramidal transition state.

Further insight into the nature of the transition state is obtained from the observation that substitution of a square-planar complex preserves the original geometry. That is, a *cis* complex gives a *cis* product and a *trans* complex gives a *trans* product. This behaviour is explained by the formation of an approximately trigonal-bipyramidal transition state with the entering, leaving, and *trans* groups in the trigonal plane (**8**).[5] Trigonal-bipyramidal intermediates of this type account for the relatively small influence that the two *cis* spectator ligands have on the rate of substitution, as their bonding orbitals will be largely unaffected by the course of the reaction.

The steric course of the reaction is shown in Fig. 20.10. We can expect a *cis* ligand to exchange places with the T ligand in the trigonal plane only if the intermediate lives long enough to be stereomobile. That is, it must be a long-lived associative (*A*) intermediate, with release of the ligand from the five-coordinate intermediate being the rate-determining step.

(d) Temperature and pressure dependence

Key point: Negative volumes and entropies of activation support the view that the rate-determining step of square-planar Pt(II) complexes is associative.

Another clue to the nature of the transition state comes from the entropies and volumes of activation for reactions of Pt(II) and Au(III) complexes (Table 20.7). The entropy of activation is obtained from the temperature dependence of the rate constant, and indicates the change in disorder (of reactants and solvent) when the transition state forms. Likewise, the volume of activation, which is obtained (with considerable difficulty) from the pressure dependence of the rate constant, is the change in volume that occurs on formation of the transition state. The limiting cases for the volume of activation in ligand substitution reactions correspond to the increase in molar volume of the outgoing ligand (for a dissociative reaction) and the decrease in molar volume of the incoming ligand (for an associative reaction).

The two striking aspects of the data in the table are the consistently strongly negative values of both quantities. The simplest explanation of the decrease in disorder and the reduction in volume is that the entering ligand is being incorporated into the transition state without release of the leaving group. That is, we can conclude that the rate-determining step is associative.

(e) The first-order pathway

Key point: The first-order contribution to the rate law in eqn 20.7 is in fact a pseudo-first-order process in which the solvent participates.

---

[5] Note that the stereochemistry is quite different from that of *p*-block central atoms, such as Si(IV) and P(V), where the leaving group departs from an axial position.

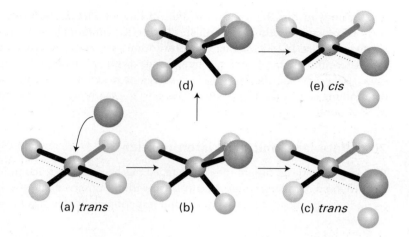

Fig. 20.10 The stereochemistry of substitution in a square-planar complex. The normal path (resulting in retention) is from (a) to (c). However, if intermediate (b) is sufficiently long-lived, it can undergo pseudorotation to (d), which leads to isomer (e).

Table 20.7 Activation parameters for substitution in square-planar complexes (in methanol)

Reaction	k_1			k_2		
	$\Delta^{\ddagger}H$	$\Delta^{\ddagger}S$	$\Delta^{\ddagger}V$	$\Delta^{\ddagger}H$	$\Delta^{\ddagger}S$	$\Delta^{\ddagger}V$
trans-$[PtCl(NO_2)(py)_2] + py$				50	-100	-38
trans-$[PtBrP_2(mes)] + SC(NH_2)_2$	71	-84	-46	46	-138	-54
cis-$[PtBrP_2(mes)] + I^-$	84	-59	-67	63	-121	-63
cis-$[PtBrP_2(mes)] + SC(NH_2)_2$	79	-71	-71	59	-121	-54
$[AuCl(dien)]^{2+} + Br^-$				54	-17	

$[PtBrP_2(mes)]$ is $[PtBr(PEt_3)_2(2,4,6-Me_3C_6H_2)]$.
Enthalpy in kilojoules per mole (kJ mol^{-1}); entropy in J K^{-1} mol^{-1} and volume in cm^3 mol^{-1}.

Having considered factors that affect the second-order pathway, we can now consider the first-order pathway for the substitution of square-planar complexes. The first issue we must address is the first-order pathway in the rate equation and decide whether k_1 in the rate law in eqn 20.2 and its generalization

$$\text{rate} = (k_1 + k_2[\text{Y}])[\text{PtL}_4] \tag{20.7}$$

does indeed represent the operation of an entirely different reaction mechanism. It turns out that it does not, and k_1 represents an associative reaction involving the solvent. In this pathway the substitution of Cl$^-$ by pyridine in methanol as solvent proceeds in two steps, with the first rate-determining:

$$[PtCl(dien)]^+ + CH_3OH \rightarrow [Pt(dien)(CH_3OH)]^{2+} + Cl^- \quad \text{(slow)}$$

$$[Pt(dien)(CH_3OH)]^{2+} + py \rightarrow [Pt(dien)(py)]^{2+} + CH_3OH \quad \text{(fast)}$$

The evidence for this two-step mechanism comes from a correlation of the rates of these reactions with the nucleophilicity parameters of the solvent molecules and the observation that reactions of entering groups with solvent complexes are rapid when compared to the step in which the solvent displaces a ligand. Thus, the substitution of ligands at a square-planar platinum complex results from two competing associative reactions.

Ligand substitution in octahedral complexes

Octahedral complexes occur for a wide variety of metals in a wide range of oxidation states and with a great diversity of bonding modes. We might therefore expect a wide variety of mechanisms of substitution; however, almost all octahedral complexes react by the interchange mechanism. The only real question is whether the rate-determining step is associative or dissociative. The analysis of rate laws for reactions that take place by such a mechanism helps to formulate the precise conditions for distinguishing these two possibilities and identifying the substitution as I_a (interchange with an associative rate-determining stage) or I_d (interchange with a dissociative rate-determining stage). The difference between the two classes of reaction hinges on whether the rate-determining step is the formation of the new $Y \cdots M$ bond or the breaking of the old $M \cdots X$ bond.

20.6 Rate laws and their interpretation

Rate laws provide an insight into the detailed mechanism of a reaction in the sense that any proposed mechanism must be consistent with the observed rate law. In the following section we see how the rate laws found experimentally for ligand substitution are interpreted.

(a) The Eigen–Wilkins mechanism

Key points: In the Eigen–Wilkins mechanism, an encounter complex is formed in a pre-equilibrium step and the encounter complex forms products in a subsequent rate-determining step.

As an example of a ligand substitution reaction, we consider

$$[Ni(OH_2)_6]^{2+} + NH_3 \rightarrow [Ni(OH_2)_5(NH_3)]^{2+} + H_2O$$

The first step in the **Eigen–Wilkins mechanism** is an encounter in which the complex ML_6, in this case $[Ni(OH_2)_6]^{2+}$, and the entering group Y, in this case NH_3, diffuse together and come into contact:

$$[Ni(OH_2)_6]^{2+} + NH_3 \rightarrow \{[Ni(OH_2)_6]^{2+}, NH_3\}$$

The two components of the encounter pair, the entity $\{A,B\}$, may also separate at a rate governed by their ability to migrate by diffusion through the solvent:

$$\{[Ni(OH_2)_6]^{2+}, NH_3\} \rightarrow [Ni(OH_2)_6]^{2+} + NH_3$$

Because in aqueous solution the lifetime of an encounter pair is approximately 1 ns, the formation of the pair can be treated as a pre-equilibrium in all reactions that take longer than a few nanoseconds. Consequently, we can express the concentrations in terms of a pre-equilibrium constant K_E:

$$ML_6 + Y \rightleftharpoons \{ML_6, Y\} \qquad K_E = \frac{[\{ML_6, Y\}]}{[ML_6][Y]}$$

The second step in the mechanism is the rate-determining reaction of the encounter complex to give products:

$$\{[Ni(OH_2)_6]^{2+}, NH_3\} \rightarrow [Ni(OH_2)_5(NH_3)]^{2+} + H_2O$$

and in general

$$\{ML_6, Y\} \rightarrow ML_5Y + L \qquad \text{rate} = k[\{ML_6, Y\}]$$

We cannot simply substitute $[\{ML_6,Y\}] = K_E[ML_6][Y]$ into this expression because the concentration of ML_6 must take into account the fact that some of it is present as the encounter pair; that is, $[M]_{tot} = [\{ML_6,Y\}] + [ML_6]$, the total concentration of the complex. It follows that

$$\text{rate} = \frac{kK_E[M]_{tot}[Y]}{1 + K_E[Y]} \tag{20.8}$$

It is rarely possible to conduct experiments over a range of concentrations wide enough to test eqn 20.8 exhaustively. However, at such low concentrations of the entering group that $K_E[Y] \ll 1$, the rate law reduces to

$$\text{rate} = k_{obs}[M]_{tot}[Y] \qquad k_{obs} = kK_E \tag{20.9}$$

Table 20.8 Complex formation by the $[Ni(OH_2)_6]^{2+}$ ion

Ligand	$k_{obs}/(dm^3\,mol^{-1}\,s^{-1})$	$K_E/(dm^3\,mol^{-1})$	$(k_{obs}/K_E)/s^{-1}$
$CH_3CO_2^-$	1×10^5	3	3×10^4
F^-	8×10^3	1	8×10^3
HF	3×10^3	0.15	2×10^4
H_2O^*			3×10^3
NH_3	5×10^3	0.15	3×10^4
$[NH_2(CH_2)_2NH_3]^+$	4×10^2	0.02	2×10^4
SCN^-	6×10^3	1	6×10^3

*The solvent is always in encounter with the ion so that K_E is undefined and all rates are inherently first-order.

Because k_{obs} can be measured and K_E can be either measured or estimated as we describe below, the rate constant k can be found from k_{obs}/K_E. The results for reactions of Ni(II) hexaaqua complexes with various nucleophiles are shown in Table 20.8. The very small variation in k indicates a model I_d reaction with very slight sensitivity to the nucleophilicity of the entering group.

When Y is a solvent molecule the encounter equilibrium is 'saturated' in the sense that, because the complex is always surrounded by solvent, a solvent molecule is always available to take the place of one that leaves the complex. In such a case $K_E[Y] \gg 1$ and $k_{obs} = k$. Thus, reactions with the solvent can be directly compared to reactions with other entering ligands without needing to estimate the value of K_E.

(b) The Fuoss–Eigen equation

Key point: The Fuoss–Eigen equation provides an estimate of the pre-equilibrium constant based on the strength of the Coulombic interaction between the reactants and their distance of closest approach.

The equilibrium constant K_E for the encounter pair can be estimated by using a simple equation proposed independently by R.M. Fuoss and M. Eigen. Both sought to take the complex size and charge into account, expecting larger, oppositely charged ions to meet more frequently than small ions of the same charge. Fuoss used an approach based on statistical thermodynamics and Eigen used one based on kinetics. Their result, which is called the **Fuoss–Eigen equation**, is

$$K_E = \tfrac{4}{3}\pi a^3 N_A e^{-V/RT} \tag{20.10}$$

In this expression a is the distance of closest approach of ions of charge numbers z_1 and z_2 in a medium of permittivity ε, V is the Coulombic potential energy ($z_1 z_2 e^2/4\pi\varepsilon a$) of the ions at that distance, and N_A is the Avogadro constant. Although the value predicted by this equation depends strongly on the details of the charges and radii of the ions, typically it favours the encounter if the reactants are large (so a is large) or oppositely charged (V negative). If one of the reactants is uncharged (as in the case of substitution by NH_3), then $V = 0$ and $K_E = \tfrac{4}{3}\pi a^3 N_A$. For an encounter distance of 200 pm for neutral species, we find $K_E = 20.2\ dm^3\,mol^{-1}$.

20.7 The activation of octahedral complexes

Many studies of substitution in octahedral complexes support the view that the rate-determining step is dissociative, and we summarize these studies first. However, the reactions of octahedral complexes can acquire a distinct associative character in the case of large central ions (as in the 4d and 5d series) or where the d-electron population at the metal is low (the early members of the d block). More room for attack or lower π^* electron density appears to facilitate nucleophilic attack and hence permit association.

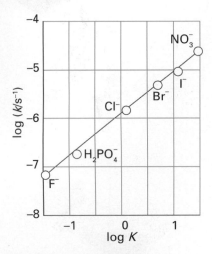

Fig. 20.11 The straight line obtained when the logarithm of a rate constant is plotted against the logarithm of an equilibrium constant shows the existence of a linear free energy relation. The graph is for the reaction $[CoX(NH_3)_5]^{2+} + H_2O \rightarrow [Co(NH_3)_5(OH_2)]^{3+} + X^-$ with different leaving groups X.

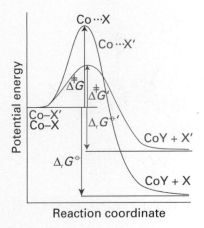

Fig. 20.12 The existence of a linear free energy relation with unit slope shows that changing X has the same effect on $\Delta^{\ddagger}G$ for the conversion of M—X to the transition state as it has on $\Delta_r G^{\ominus}$ for the complete elimination of X^-. The reaction profile shows the effect of changing the leaving group from X to X′.

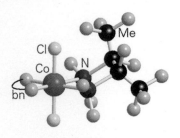

9 $[Co(Cl)_2(bn)_2]^+$

(a) Leaving-group effects

Key points: A large effect of the leaving group X is expected in I_d reactions; a linear relation is found between the logarithms of the rate constants and equilibrium constants.

We can expect the identity of the leaving group X to have a large effect in dissociatively activated reactions because their rates depend on the scission of the $M \cdots X$ bond. When X is the only variable, as in the reaction

$$[CoX(NH_3)_5]^{2+} + H_2O \rightarrow [Co(NH_3)_5(OH_2)]^{3+} + X^-$$

it is found that the rate constant and equilibrium constant of the reaction are related by

$$\ln k = \ln K + c \tag{20.11}$$

This correlation is illustrated in Fig. 20.11. Because both logarithms are proportional to Gibbs energies ($\ln k$ is approximately proportional to the activation Gibbs energy, $\Delta^{\ddagger}G$, and $\ln K$ is proportional to the standard reaction Gibbs energy, $\Delta_r G^{\ominus}$, we can write the following **linear free energy relation** (LFER):

$$\Delta^{\ddagger}G = p\Delta_r G^{\ominus} + b \tag{20.12}$$

with p and b constants (and $p \approx 1$).

The existence of an LFER of unit slope, as for the reaction of $[CoX(NH_3)_5]^{2+}$, shows that changing X has the same effect on $\Delta^{\ddagger}G$ for the conversion of Co—X to the transition state as it has on $\Delta_r G^{\ominus}$ for the complete elimination of X^- (Fig. 20.12). This observation in turn suggests that in a reaction with an interchange mechanism and a dissociative rate-determining step (I_d), the leaving group (an anionic ligand) has become a solvated ion in the transition state. An LFER with a slope of less than 1, indicating some associative character, is observed for the corresponding complexes of Rh(III). For Co(III), the reaction rates are in the order $I^- > Br^- > Cl^-$ whereas for Rh(III) they are reversed, and are in the order $I^- < Br^- < Cl^-$. This difference should be expected as the softer Rh(III) centre forms more stable complexes with I^-, compared with Br^- and Cl^-, whereas the harder Co(III) centre forms more stable complexes with Cl^-.

(b) The effects of spectator ligands

Key points: In octahedral complexes, spectator ligands affect rates of substitution; the effect is related to the strength of the metal–ligand interaction, with stronger donor ligands increasing the reaction rate by stabilizing the transition state.

In Co(III), Cr(III), and related octahedral complexes, both *cis* and *trans* ligands affect rates of substitution in proportion to the strength of the bonds they form with the metal atom. There is no important *trans* effect. For instance, hydrolysis reactions such as

$$[NiXL_5]^+ + H_2O \rightarrow [NiL_5(OH_2)] + X^-$$

are much faster when L is NH_3 than when it is H_2O. This difference can be explained on the grounds that, as NH_3 is a stronger σ donor than H_2O, it increases the electron density at the metal atom and hence facilitates the scission of the M—X bond and the formation of X^-. In the transition state, the stronger donor stabilizes the reduced coordination number.

(c) Steric effects

Key point: Steric crowding favours dissociative activation because formation of the transition state can relieve strain.

Steric effects on reactions with dissociative rate-determining steps can be illustrated by considering the rate of hydrolysis of the first Cl^- ligand in two complexes of the type $[CoCl_2(bn)_2]^+$:

$$[CoCl_2(bn)_2]^+ + H_2O \rightarrow [CoCl(bn)_2(OH_2)]^{2+} + Cl^-$$

The ligand bn is 2,3-butanediamine, and may be either chiral (**9**) or achiral (**10**). The important observation is that the complex formed with the chiral form of the ligand

Table 20.9 Tolman cone angles for various ligands

Ligand	$\theta/°$	Ligand	$\theta/°$
CH_3	90	$P(OC_6H_5)_3$	127
CO	95	PBu_3	130
Cl, Et	102	PEt_3	132
PF_3	104	$\eta^5\text{-}C_5H_5(Cp)$	136
Br, Ph	105	PPh_3	145
I, $P(OCH_3)_3$	107	$\eta^5\text{-}C_5Me_5 \ (Cp^*)$	165
PMe_3	118	$2,4\text{-}Me_2C_5H_3$	180
t-Butyl	126	$P(t\text{-}Bu)_3$	182

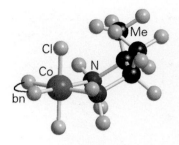

10 $[Co(Cl)_2(bn)_2]^+$

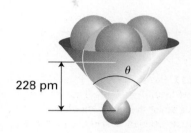

Fig. 20.13 The determination of the ligand cone angles from space-filling molecular models of the ligand and an assumed M—P bond length of 228 pm.

hydrolyses 30 times more slowly than the complex of the achiral form. The two ligands have very similar electronic effects, but the CH_3 groups are on the opposite sides of the chelate ring in (**9**) but adjacent and crowded in (**10**). The latter arrangement is more reactive because the strain is relieved in the dissociative transition state with its reduced coordination number. In general, steric crowding favours an I_d process because the five-coordinate transition state can relieve strain.

Quantitative treatments of the steric effects of ligands have been developed using molecular modelling computer software that takes into account van der Waals interactions. However, a more pictorial semiquantitative approach was introduced by C.A. Tolman. In this approach, the extent to which various ligands (especially phosphines) crowd each other is assessed by approximating the ligand by a cone with an angle determined from a space-filling model and, for phosphine ligands, an M—P bond length of 228 pm (Fig. 20.13 and Table 20.9).[6] The ligand CO is small in the sense of having a small cone angle; P(^{t}Bu)$_3$ is regarded as bulky because it has a large cone angle. Bulky ligands have considerable steric repulsion with each other when packed around a metal centre. They favour dissociative activation and inhibit associative activation.

As an illustration, the rate of the reaction of $[Ru(SiCl_3)_2(CO)_3(PR_3)]$ (**11**) with Y to give $[Ru(SiCl_3)_2(CO)_2Y(PR_3)]$ is independent of the identity of Y, which suggests that the rate-determining step is dissociative. Furthermore, it has been found that there is only a small variation in rate for the substituents Y with similar cone angles but significantly different values of pK_a. This observation supports the assignment of the rate changes to steric effects because changes in pK_a should correlate with changes in electron distributions in the ligands.

(d) Associative activation

Key point: Negative volumes of activation indicate association of the entering group into the transition state.

As we have seen, activation volumes reflect the changes in compactness (including that of the surrounding solvent) when the transition state forms from the reactants. The last column in Table 20.10 gives $\Delta^{\ddagger}V$ for some H_2O ligand exchange reactions. We see that $\Delta^{\ddagger}V$ increases from -4.1 cm^3 mol^{-1} for V^{2+} to $+7.2$ cm^3 mol^{-1} for Ni^{2+}. Because the negative volume of activation can be interpreted as the result of shrinkage when an H_2O molecule becomes part of the transition state, we can infer that the activation has significant associative character.[7] The increase in $\Delta^{\ddagger}V$ follows the increase in the number of nonbonding d electrons from d^3 to d^8 across the 3d series. In the earlier part of the d block, associative reaction appears to be favoured by a low population of d electrons.

Negative volumes of activation are also observed in the 4d and 5d series, such as for Rh(III), and indicate an associative interaction of the entering group in the transition

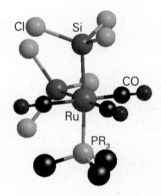

11 $[Ru(SiCl_3)_2(CO)_3(PR_3)]^+$

---

[6] Tolman's studies were on nickel complexes, so strictly speaking we are talking about a Ni—P distance of 228 pm.

[7] The limiting value for $\Delta^{\ddagger}V$ is approximately ± 18 cm^3 mol^{-1}, the molar volume of water, with A reactions having negative and D reactions positive values.

Table 20.10 Activation parameters for the H_2O exchange reactions
$[M(OH_2)_6]^{2+} + H_2{}^{17}O \rightarrow [M(OH_2)_5({}^{17}OH_2)]^{2+} + H_2O$

	$\Delta^{\ddagger}H/(\text{kJ mol}^{-1})$	LFSE*/Δ_o	LFSE†/Δ_o	LFAE/Δ_o	$\Delta^{\ddagger}V/(\text{cm}^3\,\text{mol}^{-1})$
$Ti^{2+}(d^2)$		0.8	0.91	−0.11	
$V^{2+}(d^3)$	68.6	1.2	1	0.2	−4.1
$Cr^{2+}(d^4, \text{hs})$		0.6	0.91	−0.31	
$Mn^{2+}(d^5, \text{hs})$	33.9	0	0	0	−5.4
$Fe^{2+}(d^6, \text{hs})$	31.2	0.4	0.46	−0.06	+3.8
$Co^{2+}(d^7, \text{hs})$	43.5	0.8	0.91	−0.11	+6.1
$Ni^{2+}(d^8)$	58.1	1.2	1	0.2	+7.2

* Octahedral.
† Square pyramidal.
hs, high spin.

Table 20.11 Kinetic parameters for anion attack on Cr(III)*

X	L = H_2O				L = NH_3
	$k/$ $(10^{-8}\,\text{mol}^{-1}\,\text{s}^{-1})$	$\Delta^{\ddagger}H/$ (kJ mol^{-1})	$\Delta^{\ddagger}S/$ $(\text{J K}^{-1}\,\text{mol}^{-1})$		$k/$ $(10^{-8}\,\text{dm}^3\,\text{mol}^{-1}\,\text{s}^{-1})$
Br^-	0.46	122	8		3.7
Cl^-	1.15	126	38		0.7
NCS^-	48.7	105	4		4.2

state of the reaction. Associative activation begins to dominate when the metal centre is more accessible to nucleophilic attack, either because it is large or has a low (nonbonding or $\pi^\star$) d-electron population, and the mechanism shifts from I_d towards I_a. Table 20.11 shows some data for the formation of Br^-, Cl^-, and NCS^- complexes from $[Cr(NH_3)_5(OH_2)]^{3+}$ and $[Cr(OH_2)_6]^{3+}$. In contrast to the strong dependence of the hexaaqua complex, the pentaammine complex shows only a weak dependence on the identity of the nucleophile. The two complexes probably mark a transition from I_d to I_a. The rate constants for the replacement of H_2O in $[Cr(OH_2)_6]^{3+}$ by Cl^-, Br^-, or NCS^- are smaller by a factor of about 10^4 than those for the analogous reactions of $[Cr(NH_3)_5(OH_2)]^{3+}$ (see Table 20.11). This difference suggests that the NH_3 ligands, which are stronger σ donors than H_2O, promote dissociation of the sixth ligand more effectively. As we saw above, this behaviour is to be expected in dissociatively activated reactions.

Example 20.4 Interpreting kinetic data in terms of a mechanism

The second-order rate constants for formation of $[VX(OH_2)_5]^+$ from $[V(OH_2)_6]^{2+}$ and X^- for $X^- = Cl^-$, NCS^-, and N_3^- are in the ratio 1:2:10. What do the data suggest about the rate-determining step for the substitution reaction?

Answer Because all three ligands are singly charged anions of similar size, the encounter equilibrium constants are similar. Therefore, the second-order rate constants are proportional to first-order rate constants for substitution in the encounter complex. The second-order rate constant is equal to $K_E k_2$, where K_E is the pre-equilibrium constant and k_2 is the first-order rate constant for substitution of the encounter complex. The greater rate constants for NCS^- than for Cl^-, and especially the fivefold difference of NCS^- from its close structural analogue N_3^-, suggest some contribution from nucleophilic attack and an associative reaction. By contrast, there is no such systematic pattern for the same anions reacting with Ni(II), for which the reaction is believed to be dissociative.

Self-test 20.4 Use the data in Table 20.10 to estimate an appropriate value for K_E and calculate k_2 for the reactions of V(II) with Cl^- if the observed second-order rate constant is $1.2 \times 10^2 \ dm^3 \ mol^{-1} s^{-1}$.

20.8 Base hydrolysis

Key points: Octahedral substitution can be greatly accelerated by OH^- ions when ligands with acidic hydrogens are present as a result of the decrease in charge of the reactive species and the increased ability of the deprotonated ligand to stabilize the transition state.

Consider a substitution reaction in which the ligands possess acidic protons, such as

$$[CoCl(NH_3)_5]^{2+} + OH^- \rightarrow [Co(OH)(NH_3)_5]^{2+} + Cl^-$$

An extended series of studies has shown that whereas the rate law is overall second-order, with rate = $k[CoCl(NH_3)_5^{2+}][OH^-]$, the mechanism is not a simple bimolecular attack by OH^- on the complex. For instance, whereas the replacement of Cl^- by OH^- is fast, the replacement of Cl^- by F^- is slow, even though F^- resembles OH^- more closely in terms of size and nucleophilicity. There is a considerable body of indirect evidence relating to the problem but one elegant experiment makes the essential point. This conclusive evidence comes from a study of $^{18}O/^{16}O$ isotope distribution in the product $[Co(OH)(NH_3)_5]^{2+}$, where it is found that the $^{18}O/^{16}O$ ratio differs between H_2O and OH^- at equilibrium. The isotope ratio in the product matches that for H_2O, not that for the OH^- ions. Therefore, an H_2O molecule and not an OH^- ion appears to be the entering group.

The mechanism that takes these observations into account supposes that the role of OH^- is to act as a Brønsted base, not an entering group:

$$[CoCl(NH_3)_5]^{2+} + OH^- \rightleftharpoons [CoCl(NH_2)(NH_3)_4]^+ + H_2O$$

$$[CoCl(NH_2)(NH_3)_4]^+ \rightarrow [Co(NH_2)(NH_3)_4]^{2+} + Cl^- \ (slow)$$

$$[Co(NH_2)(NH_3)_4]^{2+} + H_2O \rightarrow [Co(OH)(NH_3)_5]^{2+} \ (fast)$$

In the first step, an NH_3 ligand acts as a Brønsted acid, resulting in the formation of its conjugate base, the NH_2^- ion, as a ligand. Because the deprotonated form of the complex has a lower charge, it will be able to lose a Cl^- ion more readily than the protonated form, thus accelerating the reaction. In addition, according to the **conjugate base mechanism**, loss of a proton from an NH_3 ligand changes it from a pure σ-donor ligand to a strong σ and π donor (as NH_2^-) and so helps to stabilize the five-coordinate transition state, greatly accelerating the loss of the Cl^- ion (see Example 20.5).

20.9 Stereochemistry

Key point: Reaction through a square-pyramidal intermediate results in retention of the original geometry but reaction through a trigonal-bipyramidal intermediate can lead to isomerization.

Classic examples of octahedral substitution stereochemistry are provided by Co(III) complexes. Table 20.12 shows some data for the hydrolysis of *cis*- and *trans*-$[CoAX(en)_2]^{2+}$ where X is the leaving group (either Cl^- or Br^-) and A is OH^-, NCS^-, or Cl^-. The stereochemical consequences of substitution of octahedral complexes are much more intricate than those of square-planar complexes. The *cis* complexes do not undergo isomerization when substitution occurs, whereas the *trans* forms show a tendency to isomerize in the order $A = NO_2^- < Cl^- < NCS^- < OH^-$.

The data can be understood in terms of an I_d mechanism and by recognizing that the five-coordinate metal centre in the transition state may resemble either of the two stable geometries for five-coordination, namely square-pyramidal or trigonal-bipyramidal. As can

Table 20.12 Stereochemical course of hydrolysis reactions of $[CoAX(en)_2]$

	A	X	Percentage *cis* in product
cis	OH^-	Cl^-	100
	Cl^-	Cl^-	100
	NCS^-	Cl^-	100
	Cl^-	Br^-	100
trans	NO_2^-	Cl^-	0
	NCS^-	Cl^-	50–70
	Cl^-	Cl^-	35
	OH^-	Cl^-	75

X is the leaving group.

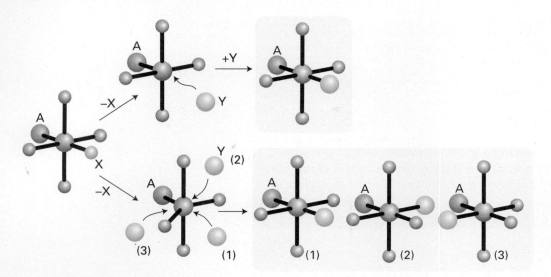

Fig. 20.14 Reaction through a square-pyramidal complex (top path) results in retention of the original geometry but reaction through a trigonal-bipyramidal complex (bottom path) can lead to isomerization.

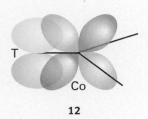

12

be seen from Fig. 20.14, reaction through the square-pyramidal complex results in retention of the original geometry but reaction through the trigonal-bipyramidal complex can lead to isomerization. The *cis* complex gives rise to a square-pyramidal intermediate but the *trans* isomer gives a trigonal-bipyramidal intermediate. For *d* metals, trigonal-bipyramidal complexes are favoured when the ligands in the equatorial positions are good π donors, and a good π-donor ligand *trans* to the leaving group Cl⁻ favours isomerization (**12**).

20.10 Isomerization reactions

Key points: Isomerization of a complex can take place by mechanisms that involve substitution, bond cleavage, and reformation, or twisting.

Isomerization reactions are closely related to substitution reactions; indeed, a major pathway for isomerization is often via substitution. The square-planar Pt(II) and octahedral Co(III) complexes we have discussed can form five-coordinate trigonal-bipyramidal transition states. The interchange of the axial and equatorial ligands in a trigonal-bipyramidal complex can be pictured as occurring by a Berry pseudorotation through a square-pyramidal conformation (Section 8.2b and Fig. 20.15). As we have seen, when a trigonal-bipyramidal complex adds a ligand to produce a six-coordinate complex, a new direction of attack of the entering group can result in isomerization.

If a chelate ligand is present, isomerization can occur as a consequence of metal–ligand bond breaking, and substitution need not occur. An example is the exchange of the 'outer' CD₃ group with the 'inner' CH₃ group during the isomerization of tris (acetylacetonato)cobalt(III), (**13**) → (**14**). An octahedral complex can also undergo isomerization by an intramolecular twist without loss of a ligand or breaking of a bond. There is evidence, for example, that racemization of [Ni(en)₃]²⁺ occurs by such an internal twist. Two possible paths are the **Bailar twist** and the **Ray–Dutt twist** (Fig. 20.16).

Fig. 20.15 The exchange of axial and equatorial ligands by a twist through a square-pyramidal conformation of the complex.

13

14

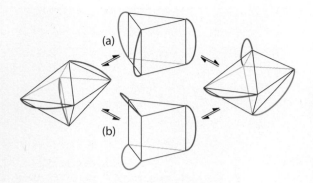

Fig. 20.16 (a) The Bailar twist and (b) the Ray–Dutt twist by which an octahedral complex can undergo isomerization without losing a ligand or breaking a bond.

Example 20.5 Interpreting the stereochemical course of a reaction

Substitution of Co(III) complexes of the type $[CoAX(en)_2]^+$ results in *trans* to *cis* isomerization, but only when the reaction is catalysed by a base. Show that this observation is consistent with the model in Fig. 20.14.

Answer In a base hydrolysis reaction, one of the ligands becomes its conjugate base, the strong π-donor ligand $:NHR^-$. A trigonal bipyramid of the type shown in Fig. 20.14 will be favoured, and it may be attacked in the way shown there.

Self-test 20.5 If the directions of attack on the trigonal-bipyramidal complex are random, what *cis–trans* isomer distribution does the analogy with Fig. 20.14 suggest for products of the type $[CoA(NH_3)(OH_2)]^{2+}$?

The reaction of ligands with other molecules to give other, more complex, ligands is discussed in Box 20.1.

Redox reactions

As remarked in Chapter 5, redox reactions can occur by the direct transfer of electrons (as in some electrochemical cells and in many solution reactions) or they may occur by the transfer of atoms and ions, as in the transfer of O atoms in reactions of oxoanions. Because redox reactions in solution involve both an oxidizing and a reducing agent, they are usually bimolecular in character. The rare exceptions are reactions in which one molecule has both oxidizing and reducing centres.

20.11 The classification of redox reactions

Key points: In an inner-sphere redox reaction a ligand is shared to form a transition state; in an outer-sphere redox reaction there is no bridging ligand between the reacting species.

In the 1950s, Henry Taube identified two mechanisms of redox reactions. One is the **inner-sphere mechanism**, which includes atom transfer processes. In an inner-sphere mechanism, the coordination spheres of the reactants share a ligand transitorily and form a bridged transition state. The other is an **outer-sphere mechanism**, which includes many simple electron transfers. In an outer-sphere mechanism, the complexes come into contact without sharing a bridging ligand and the electron tunnels from one metal atom to the other.

The mechanisms of some redox reactions are definitively assigned as inner- or outer-sphere. However, the mechanisms of a vast number of reactions are unknown because it is difficult to make unambiguous assignments when complexes are labile. Much of the study of well-defined examples is directed towards the identification of the parameters that differentiate the two paths with the aim of being able to making correct assignments in more difficult cases.

Box 20.1 Template synthesis

A metal ion such as Ni(II) can be used to assemble a group of ligands that then undergo a reaction among themselves to form a *macrocyclic ligand*, a cyclic molecule with several donor atoms. A simple example is

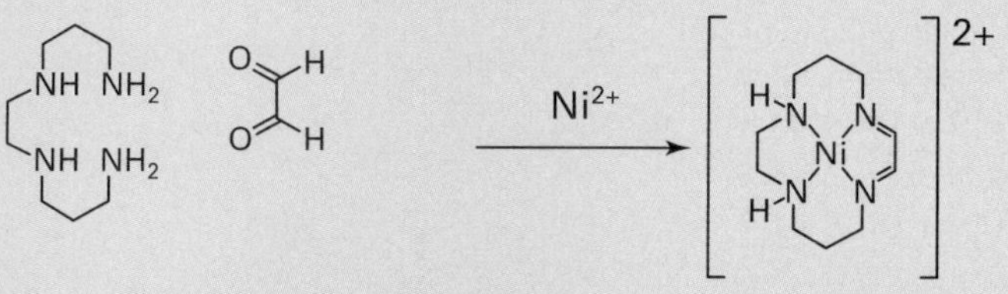

This phenomenon, which is called the *template effect*, can be applied to produce a surprising variety of macrocyclic ligands. The reaction shown above is an example of a *condensation reaction*, a reaction in which a bond is formed between two molecules, and a small molecule (in this case H_2O) is eliminated. If the metal ion had not been present, the condensation reaction of the component ligands would have been an ill-defined polymeric mixture, not a macrocycle. Once the macrocycle has been formed, it is normally stable on its own, and the metal ion may be removed to leave a multidentate ligand that can be used to complex other metal ions.

A wide variety of macrocyclic ligands can be synthesized by the template approach – two more complicated ligands are shown below.

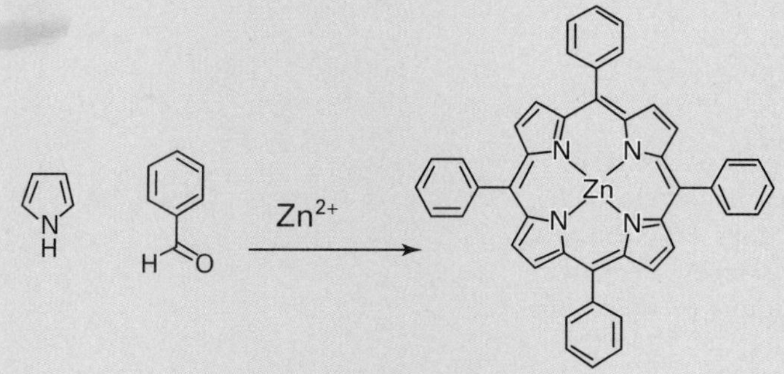

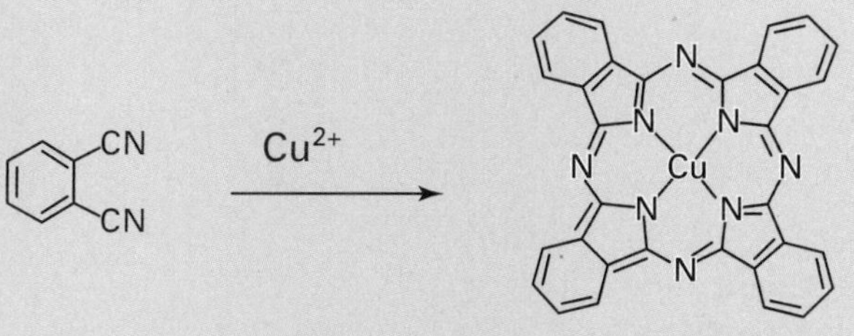

The origin of the template effect may be either kinetic or thermodynamic. For example, the condensation may stem either from the increase in the rate of the reaction between coordinated ligands (on account of their proximity or electronic effects) or from the added stability of the chelated ring product.

More complicated template syntheses can be used to construct topologically complex molecules, such as the chain-like *catenanes*, molecules that consist of interlinked rings. An example of the synthesis of a catenane containing two rings is shown below.

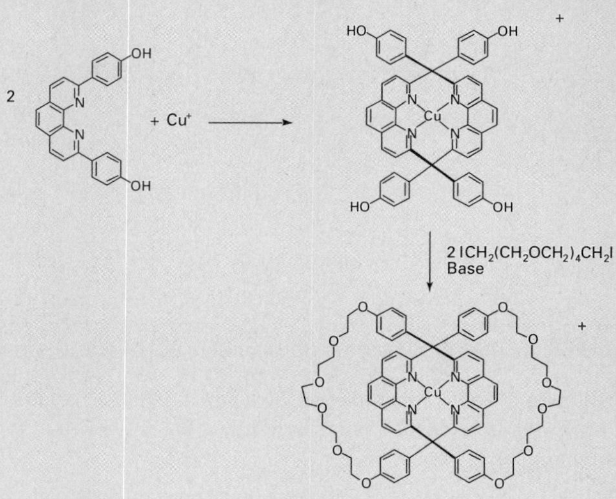

Here, two bipyridine-based ligands are coordinated to a copper ion, and then the ends of each ligand are joined by a flexible linkage. The metal ion can then be removed to give a *catenand* (*catenane ligand*), which can be used to complex other metal ions.

Even more complicated systems, equivalent to knots and links,[1] can be constructed with multiple metals. The following synthesis gives rise to a single molecular strand tied in a trefoil knot:

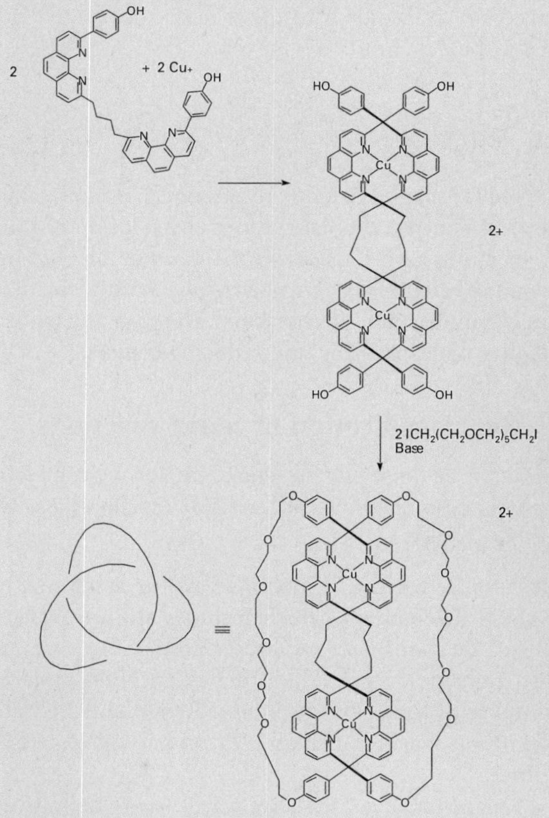

[1] Knotted and linked systems are far from being purely of academic interest and many proteins exist in these forms: see C. Liang and K. Mislow, *J. Am. Chem. Soc.,* 1994, **116**, 3588 and 1995, **117**, 4201.

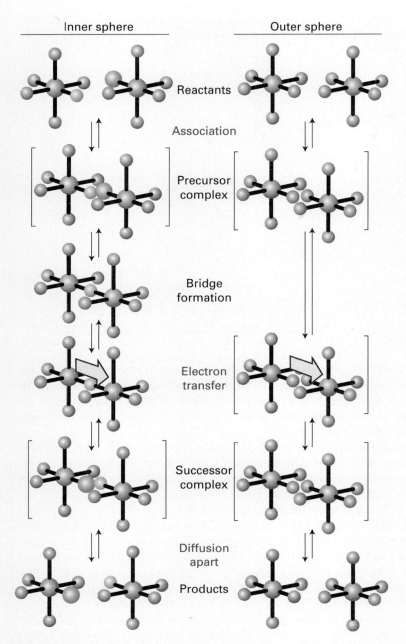

Inner sphere   Outer sphere

Reactants

Association

Precursor
complex

Bridge
formation

Electron
transfer

Successor
complex

Diffusion
apart

Products

Fig. 20.17 The different pathways followed by inner- and outer-sphere mechanisms.

20.12 The inner-sphere mechanism

Key point: The rate-determining step of an inner-sphere redox reaction may be any one of the component processes, but a common one is electron transfer.

The inner-sphere mechanism was first confirmed for the reduction of the nonlabile complex $[CoCl(NH_3)_5]^{2+}$ by $Cr^{2+}(aq)$. The products of the reaction included both Co^{2+} and $[CrCl(OH_2)_5]^{2+}$, and addition of $^{36}Cl^-$ to the solution did not lead to the incorporation of any of the isotope into the Cr(III) product. Furthermore, the reaction is faster than reactions that remove Cl^- from nonlabile Co(III) or introduce Cl^- into the non-labile $[Cr(OH_2)_6]^{3+}$ complex. These observations suggest that Cl has moved directly from the coordination sphere of one complex to that of the other during the reaction. The Cl^- attached to Co(III) can easily enter into the labile coordination sphere of $[Cr(OH_2)_6]^{2+}$ to produce a bridged intermediate (**15**).

Inner-sphere reactions, though involving more steps than outer-sphere reactions, can be fast. Figure 20.17 summarizes the steps necessary for such a reaction to occur. The first

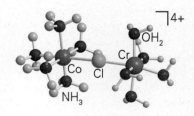

15

Table 20.13 Second-order rate constants for selected inner-sphere reactions with variable bridging ligands

Oxidant	Reductant	Bridging ligand	$k/(dm^3\ mol^{-1}\ s^{-1})$
$[Co(NH_3)_6]^{3+}$	$[Cr(OH_2)_6]^{2+}$		8×10^{-5}
$[CoF(NH_3)_5]^{2+}$	$[Cr(OH_2)_6]^{2+}$	F^-	2.5×10^5
$[CoCl(NH_3)_5]^{2+}$	$[Cr(OH_2)_6]^{2+}$	Cl^-	6.0×10^5
$[CoI(NH_3)_5]^{2+}$	$[Cr(OH_2)_6]^{2+}$	I^-	3.0×10^6
$[Co(NCS)(NH_3)_5]^{2+}$	$[Cr(OH_2)_6]^{2+}$	NCS^-	1.9×10^1
$[Co(SCN)(NH_3)_5]^{2+}$	$[Cr(OH_2)_6]^{2+}$	SCN^-	1.9×10^5
$[Co(OH_2)(NH_3)_5]^{2+}$	$[Cr(OH_2)_6]^{2+}$	H_2O	1.0×10^{-1}
$[CrF(OH_2)_5]^{2+}$	$[Cr(OH_2)_6]^{2+}$	F^-	7.4×10^{-3}
$Co(NH_3)_5O_2C-\langle\ \rangle-N-Ru(NH_3)_4(OH_2)$		$^-O_2C-\langle\ \rangle-N$	1.0×10^2
$Co(NH_3)_5O_2C-\ \ N-Ru(NH_3)_4(OH_2)$		$^-O_2C-\ \ N$	1.6×10^{-3}

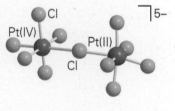

16

17 Pyrazine

18 4,4′-Bipyridine

19 Dimethylaminopyridine

two steps of an inner-sphere reaction are the formation of a precursor complex and the formation of the bridged binuclear intermediate. These two steps are identical to the first two steps in the Eigen–Wilkins mechanism (Section 20.6). The final steps are electron transfer through the bridging ligand to give the successor complex, followed by dissociation to give the products.

The rate-determining step of the overall reaction may be any one of these processes, but the most common one is the electron-transfer step. However, if both metal ions have a nonlabile electron configuration after electron transfer, then the break-up of the bridged complex is rate determining. An example is the reduction of $[RuCl(NH_3)_5]^{2+}$ by $[Cr(OH_2)_6]^{2+}$ in which the rate-determining step is the dissociation of the chlorine-bridged complex $[Ru^{II}(NH_3)_5(\mu\text{-}Cl)Cr^{III}(OH_2)_5]^{4+}$. Reactions in which the formation of the bridged complex is rate determining tend to have similar rate constants for a series of partners of a given species. For example, the oxidation of $V^{2+}(aq)$ has similar rate constants for a long series of Co(III) oxidants with different bridging ligands. The explanation is that the rate-determining step is the substitution of an H_2O molecule from the coordination sphere of V(II), which is quite slow (Table 20.10).

The numerous reactions in which electron transfer is rate determining do not display such simple regularities. Rates vary over a wide range as metal ions and bridging ligands are varied.[8] The data in Table 20.13 show some typical variations as bridging ligand, oxidizing metal, and reducing metal are changed.

All the reactions in Table 20.13 result in the change of oxidation number by ±1. Such reactions are still often called **one-equivalent processes**, the name reflecting the largely outmoded term 'chemical equivalent'. Similarly, reactions that result in the change of oxidation number by ±2 are often called **two-equivalent processes** and may resemble nucleophilic substitutions. This resemblance can be seen by considering the reaction

$$[Pt^{II}Cl_4]^{2-} + [Pt^{IV}Cl_6]^{2-} \rightarrow [Pt^{IV}Cl_6]^{2-} + [Pt^{II}Cl_4]^{2-}$$

which occurs through a Cl^- bridge (**16**). The reaction depends on the transfer of a Cl^- ion in the break-up of the successor complex.

There is no difficulty in assigning an inner-sphere mechanism when the reaction involves ligand transfer from an initially nonlabile reactant to a nonlabile product. With more labile complexes, inner-sphere reactions should always be suspected when ligand transfer occurs as well as electron transfer, and if good bridging groups such as Cl^-, Br^-,

---

[8] Some bridged intermediates have been isolated with the electron clearly located on the bridge, but we shall not consider these here.

I^-, N_3^-, CN^-, SCN^-, pyrazine (**17**), 4,4′-bipyridine (**18**), and 4-dimethylaminopyridine (**19**) are present. Although all these ligands have lone pairs to form the bridge, this may not be an essential requirement. For instance, just as the carbon atom of a methyl group can act as a bridge between OH^- and I^- in the hydrolysis of iodomethane, so it can act as a bridge between Cr(II) and Co(III) in the reduction of methylcobalt species by Cr(II).

20.13 The outer-sphere mechanism

Key points: An outer-sphere redox reaction involves electron tunnelling between two reactants without any major disturbance of their covalent bonding or inner coordination spheres; the rate constant depends on the electronic and geometrical structures of the reacting species and on the Gibbs energy of reaction.

A good conceptual starting point for understanding the principles of outer-sphere electron transfer is the deceptively simple reaction called **electron self-exchange**. A typical example is the exchange of an electron between $[Fe(OH_2)_6]^{3+}$ and $[Fe(OH_2)_6]^{2+}$ ions in water.

$$[Fe(OH_2)_6]^{3+} + [Fe(OH_2)_6]^{2+} \rightarrow [Fe(OH_2)_6]^{2+} + [Fe(OH_2)_6]^{3+}$$

Self-exchange reactions can be studied over a wide dynamic range with techniques ranging from isotopic labelling to NMR, with EPR being useful for even faster reactions. The rate constant of the Fe^{3+}/Fe^{2+} reaction is about 1 $dm^3\ mol^{-1}s^{-1}$ at 25°C.

To set up a mechanism, we suppose that Fe^{3+} and Fe^{2+} come together to form a weak outer-sphere complex (Fig. 20.18). We need to consider, by assuming that the overlap of their respective acceptor and donor orbitals is sufficient to give a reasonable tunnelling probability,[9] how rapidly an electron transfers between the two metal ions. To explore this problem, we invoke the Franck–Condon principle, which states that electronic transitions are so fast that they take place in a stationary nuclear framework. In Fig. 20.19 the nuclear motions associated with the 'reactant' Fe^{3+} and its 'conjugate product' Fe^{2+} are represented as displacements along a reaction coordinate. If $[Fe(OH_2)_6]^{3+}$ lies at its energy minimum, then an instantaneous electron transfer would give a compressed state of $[Fe(OH_2)_6]^{2+}$. Likewise, the removal of an electron from Fe^{2+} at its energy minimum would give an expanded state of $[Fe(OH_2)_6]^{3+}$. The only instant at which the electron can transfer within the precursor complex is when both $[Fe(OH_2)_6]^{3+}$ and $[Fe(OH_2)_6]^{2+}$ have achieved the same nuclear configuration by thermally induced fluctuations. That configuration corresponds to the point of intersection of the two curves, and the energy required to reach this position is the Gibbs energy of activation, $\Delta^\ddagger G$. If $[Fe(OH_2)_6]^{3+}$ and $[Fe(OH_2)_6]^{2+}$ differ in their nuclear configurations, $\Delta^\ddagger G$ is larger and electron exchange is slower. The difference in rates is expressed quantitatively by the **Marcus equation**

$$k_{ET} = \nu_N \kappa_e e^{-\Delta^\ddagger G/RT} \tag{20.13}$$

in which k_{ET} is the rate constant for electron transfer and $\Delta^\ddagger G$ is given by

$$\Delta^\ddagger G = \tfrac{1}{4}\lambda \left(1 + \frac{\Delta_r G^\ominus}{\lambda}\right)^2 \tag{20.14}$$

with $\Delta_r G^\ominus$ the standard reaction Gibbs energy (which is obtained from the difference in standard potentials of the redox partners) and λ the **reorganization energy**, the energy required to move the nuclei associated with the reactant to the positions they adopt in the product immediately before the transfer of the electron. This energy depends on the changes in metal–ligand bond lengths (the so-called *inner-sphere reorganization energy*) and alterations in solvent polarization, principally the orientation of the solvent molecules around the complex (the *outer-sphere reorganization energy*).

The pre-exponential factor in eqn 20.14 has two components, the **nuclear frequency factor** ν_N and the **electronic factor** κ_e. The former is the frequency at which the two complexes, having already encountered each other in the solution, attain the transition

[9] Tunnelling refers to a process where, according to classical physics, the electrons do not have sufficient energy to overcome the barrier.

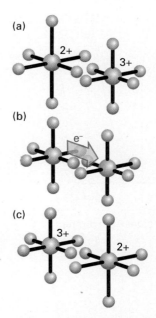

Fig. 20.18 Electron transfer between two metal ions in a precursor complex is not productive until their coordination shells have reorganized to be of equal size. (a) Reactants, (b) reactant complexes having distorted into the same geometry, (c) products.

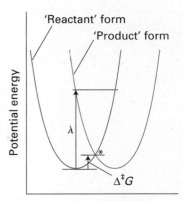

Fig. 20.19 The potential energy curves for electron self-exchange. The nuclear motions of both the oxidized and reduced species (shown displaced along the reaction coordinate) and the surrounding solvent are represented by potential wells. Electron transfer to oxidized metal ion (left) occurs once fluctuations of its inner and outer coordination shell bring it to a point (denoted by an asterisk) on its energy surface that coincides with the energy surface of its reduced state (right). This point is at the intersection of the two curves. The activation energy depends on the horizontal displacement of the two curves (representing the difference in sizes of the oxidized and reduced forms).

(a)

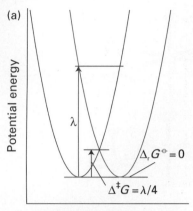

(b)

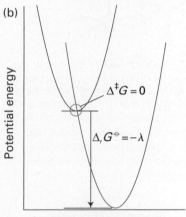

(c)

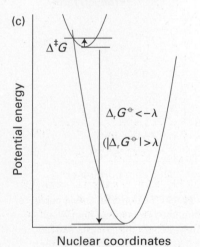

Fig. 20.20 Variation of the activation Gibbs energy ($\Delta^{\ddagger}G$) with reaction Gibbs energy ($\Delta_r G^{\ominus}$). (a) In a self-exchange reaction, $\Delta_r G^{\ominus} = 0$ and $\Delta^{\ddagger}G = \lambda/4$. (b) A reaction is 'activationless' when $\Delta_r G^{\ominus} = -\lambda$. (c) $\Delta^{\ddagger}G$ increases (the rate diminishes) as $\Delta_r G^{\ominus}$ becomes more negative beyond $\Delta_r G^{\ominus} = -\lambda$.

state. The electronic factor gives the probability on a scale from 0 to 1 that an electron will transfer when the transition state is reached; its precise value depends on the extent of overlap of the donor and acceptor orbitals.

A small reorganization energy and a value of κ_e close to 1 corresponds to a redox couple capable of fast electron self-exchange. The first requirement is achieved if the transferred electron is removed from or added to a nonbonding orbital, as the change in metal–ligand bond length is then least. It is also likely if the metal ion is shielded from the solvent, in the sense of it being sterically difficult for solvent molecules to approach close to the metal ion, because the polarization of the solvent is normally a major component of the reorganization energy. Simple metal ions such as aqua species typically have λ well in excess of 1 eV, whereas buried redox centres in enzymes, which are very well shielded from the solvent, can have values as low as 0.25 eV. A value of κ_e close to 1 is achieved if there is good orbital overlap between the two components of the precursor complex.

For a self-exchange reaction, $\Delta_r G^{\ominus} = 0$ and therefore, from eqn 20.14, $\Delta^{\ddagger}G = \frac{1}{4}\lambda$ and the rate of electron transfer is controlled by the reorganization energy (Fig. 20.20a). To a considerable extent, the rates of self-exchange can be interpreted in terms of the types of orbitals involved in the transfer (Table 20.14). In the $[Cr(OH_2)_6]^{3+/2+}$ self-exchange reaction, an electron is transferred between antibonding $\sigma^{\star}$ orbitals, and the consequent extensive change in metal–ligand bond lengths results in a large inner-sphere reorganization energy and therefore a slow reaction. The $[Co(NH_3)_6]^{3+/2+}$ couple has an even greater reorganization energy because two electrons are moved into the $\sigma^{\star}$ orbital as rearrangement occurs and the reaction is even slower. With the other hexaaqua and hexaammine complexes in the table, the electron is transferred between weakly antibonding or nonbonding π orbitals, the inner-sphere reorganization is less extensive, and the reactions are faster. The bulky, hydrophobic chelating ligand bipyridyl acts as a solvent shield, thus decreasing the outer-sphere reorganization energy.

Bipyridyl and other π-acceptor ligands allow electrons in an orbital with π symmetry on the metal ion to delocalize on to the ligand. This delocalization effectively lowers the reorganization energy when the electron is transferred between π orbitals, as occurs with Fe and Ru, where the electron transfer is between t_{2g} orbitals (which, as explained in Section 19.2, can participate in π bonding), but not with Ni, where the electron transfer is between e_g orbitals. Delocalization can also increase the electronic factor.

Self-exchange reactions are helpful for pointing out the concepts that are involved in electron transfer, but chemically useful redox reactions occur between different species and involve net electron transfer. For the latter reactions, $\Delta_r G^{\ominus}$ is nonzero and contributes to the rate through eqns 20.13 and 20.14. Provided $|\Delta_r G^{\ominus}| \ll |\lambda|$ eqn 20.14 becomes

$$\Delta^{\ddagger}G = \tfrac{1}{4}\lambda(1 + \Delta_r G^{\ominus}/\lambda)^2 \approx \tfrac{1}{4}\lambda(1 + 2\Delta_r G^{\ominus}/\lambda) = \tfrac{1}{4}(\lambda + 2\Delta_r G^{\ominus})$$

Table 20.14 Correlations between rate constants for electron self-exchange reactions

Reaction	Electron configuration	Δd/pm	k_{11}/(dm³ mol⁻¹ s⁻¹)
$[Cr(OH_2)_6]^{3+/2+}$	$t_{2g}^3/t_{2g}^3 e_g^1$	20	1×10^{-5}
$[V(OH_2)_6]^{3+/2+}$	t_{2g}^2/t_{2g}^3	13	1×10^{-2}
$[Fe(OH_2)_6]^{3+/2+}$	$t_{2g}^3 e_g^2/t_{2g}^4 e_g^2$	13	1.1
$[Ru(OH_2)_6]^{3+/2+}$	t_{2g}^5/t_{2g}^6	9	20
$[Ru(NH_3)_6]^{3+/2+}$	t_{2g}^5/t_{2g}^6	4	6.6×10^3
$[Co(NH_3)_6]^{3+/2+}$	$t_{2g}^6/t_{2g}^5 e_g^2$	22	2×10^{-8}
$[Fe(bipy)_3]^{3+/2+}$	t_{2g}^5/t_{2g}^6	0	3×10^8
$[Ru(bipy)_3]^{3+/2+}$	t_{2g}^5/t_{2g}^6	0	4×10^8
$[Ni(bipy)_3]^{3+/2+}$	$t_{2g}^6 e_g^1/t_{2g}^6 e_g^2$	12	1.5×10^3

and then according to eqn 20.13

$$k_{ET} \approx \nu_N \kappa_e e^{-(\lambda + 2\Delta_r G^{\ominus})/4RT}$$

Because $\lambda > 0$ and $\Delta_r G^{\ominus} < 0$ for thermodynamically feasible reactions, provided $|\Delta_r G^{\ominus}| \ll |\lambda|$ the rate constant increases exponentially as $\Delta_r G^{\ominus}$ becomes increasingly favourable (that is, more negative). However, as $|\Delta_r G^{\ominus}|$ becomes comparable to $|\lambda|$, this equation breaks down and we see that the reaction rate peaks before declining as $|\Delta_r G^{\ominus}| > |\lambda|$.

Equation 20.14 shows that $\Delta^{\ddagger}G = 0$ when $\Delta_r G^{\ominus} = -\lambda$. That is, the reaction becomes 'activationless' when the standard reaction Gibbs energy and the reorganization energy cancel (Fig. 20.20b). The activation energy now *increases* as $\Delta_r G^{\ominus}$ becomes more negative and the reaction rate decreases. This slowing of the reaction as the standard Gibbs energy of the reaction becomes more exergonic is called **inverted behaviour** (Fig. 20.20c). Inverted behaviour has important consequences, a notable one relating to the long-range electron transfer involved in photosynthesis. Photosystems are complex proteins containing light-excitable pigments such as chlorophyll and a chain of redox centres having low reorganization energies. In this chain, one highly exergonic recombination of the photoelectron with oxidized chlorophyll is sufficiently retarded (to 30 ns) to allow the electron to escape (in 200 ps) and proceed down the photosynthetic electron-transport chain, ultimately to produce reduced carbon compounds (Chapter 26). The theoretical dependence of the reaction rate of a reaction on the standard Gibbs energy is plotted in Fig. 20.21, and Fig. 20.22 shows the observed variation of reaction rate with $\Delta_r G^{\ominus}$ for the iridium complex (**20**). These results represent the first unambiguous experimental observation of an inverted region for a synthetic complex.

The Marcus equation can be used to predict the rate constants for outer-sphere electron transfer reactions between different species. Consider the electron-transfer reaction between an oxidant Ox_1 and reductant Red_2:

$$Ox_1 + Red_2 \rightarrow Red_1 + Ox_2$$

If we suppose that the reorganization energy for this reaction is the average of the values for the two self-exchange processes, we can write $\lambda_{12} = \frac{1}{2}(\lambda_{11} + \lambda_{22})$; then manipulation of eqns 20.13 and 20.14 gives the **Marcus cross-relation**

$$k_{12} = (k_{11}k_{22}K_{12}f_{12})^{1/2} \qquad\qquad 20.15$$

in which k_{12} is the rate constant, K_{12} is the equilibrium constant obtained from $\Delta_r G^{\ominus}$, and k_{11} and k_{22} are the respective self-exchange rate constants for the two reaction partners. For reactions between simple ions in solution, when the standard Gibbs reaction energy is not too large, an LFER of the kind expressed by eqn 20.12 exists between $\Delta^{\ddagger}G$ and $\Delta_r G^{\ominus}$ and f_{12} can normally be set to 1. However, for reactions that are highly favourable thermodynamically (that is, $\Delta_r G^{\ominus}$ large and negative) the LFER breaks down. The term f_{12} takes into consideration the nonlinearity of the relation between $\Delta^{\ddagger}G$ and $\Delta_r G^{\ominus}$ and is given by

$$\log f_{12} = \frac{(\log K_{12})^2}{4 \log(K_{11}k_{22}/Z^2)} \qquad\qquad (20.16)$$

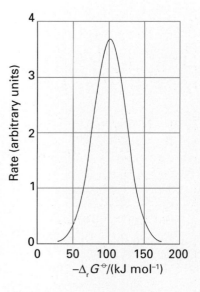

Fig. 20.21 The theoretical dependence of the reaction rate (in arbitrary units) on $\Delta_r G^{\ominus}$ for a redox reaction with $\lambda = 1.0$ eV (100 kJ mol^{-1}).

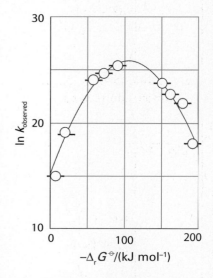

Fig. 20.22 Plot of ln k against $-\Delta_r G^{\ominus}$ for the iridium complex (**20**) in acetonitrile solution at room temperature (data from L.S. Fox, M. Kozik, J.R. Winkler, and H.B. Gray, *Science*, 1990, **297**, 1069).

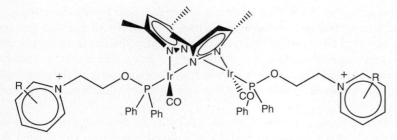

20

where Z is the constant of proportionality between the encounter density in solution (in moles of encounters per cubic decimetre per second) and the molar concentrations of the reactants; it is often taken to be 10^{11} mol^{-1} dm^3 s^{-1}.

Example 20.6 Using the Marcus cross-relation

Calculate the rate constant at 0°C for the outer-sphere reduction of $[Co(bipy)_3]^{3+}$ by $[Co(terpy)_2]^{2+}$ (where bipy is bipyridyl and terpy is tripyridyl).

Answer The required self-exchange reactions are

$$[Co(bipy)_3]^{2+} + [Co(bipy)_3]^{3+} \xrightarrow{k_{11}} [Co(bipy)_3]^{3+} + [Co(bipy)_3]^{2+}$$

$$[Co(terpy)_2]^{2+} + [Co(terpy)_2]^{3+} \xrightarrow{k_{22}} [Co(terpy)_2]^{3+} + [Co(terpy)_2]^{2+}$$

where $k_{11} = 9.0$ dm^3 mol^{-1} s^{-1} and $k_{22} = 48$ dm^3 mol^{-1} s^{-1}, $K_{12} = 3.57$ and setting $f_{12} = 1$ (as noted above) gives

$$k_{12} = (9.0 \times 48 \times 3.57)^{1/2} \ dm^3 \ mol^{-1} \ s^{-1} = 39 \ dm^3 \ mol^{-1} \ s^{-1}$$

This result compares reasonably well with the experimental value, which is 64 dm^3 mol^{-1} s^{-1}.

Self-test 20.6 Calculate the rate constants for electron transfer in the oxidation of $[Cr(OH_2)_6]^{2+}$ ($E^{\ominus}(Cr^{3+}, Cr^{2+}) = -0.41$ V) and each of the oxidants $[Ru(NH_3)_6]^{3+}$ ($E^{\ominus}(Ru^{3+}, Ru^{2+}) = +0.07$ V), $[Fe(OH_2)_6]^{3+}$ ($E^{\ominus}(Fe^{3+}/Fe^{2+}) = +0.77$ V), and $[Ru(bipy)_3]^{3+}$ ($E^{\ominus}(Ru^{3+}, Ru^{2+}) = +1.26$ V).

Photochemical reactions

The absorption of a photon of ultraviolet radiation or visible light increases the energy of a complex by between 170 and 600 kJ mol^{-1}. Because these energies are larger than typical activation energies it should not be surprising that new reaction channels are opened. However, when the high energy of a photon is used to provide the energy of the primary forward reaction, the back reaction is almost always very favourable, and much of the design of efficient photochemical systems lies in trying to avoid the back reaction.

20.14 Prompt and delayed reactions

Key point: Reactions of electronically excited species are classified as prompt or delayed.

In some cases, the excited state formed after absorption of a photon dissociates almost immediately after it is formed. Examples include formation of the pentacarbonyl intermediates that initiate ligand substitution in metal carbonyl compounds.

$$Cr(CO)_6 \xrightarrow{h\nu} Cr(CO)_5 + CO$$

and the scission of Co—Cl bonds:

$$[Co^{III}Cl(NH_3)_5]^{2+} \xrightarrow{h\nu(\lambda < 350 \text{ nm})} [Co^{II}(NH_3)_5]^{2+} + Cl\cdot$$

Both processes occur in less than 10 ps and hence are called **prompt reactions.**

In the second reaction, the **quantum yield**, the amount of reaction per mole of photons absorbed, increases as the wavelength of the radiation is decreased (and the photon energy correspondingly increased, $E_{\text{photon}} = hc/\lambda$). The energy in excess of the bond energy is available to the newly formed fragments and increases the probability that they will escape from each other through the solution before they have an opportunity to recombine.

Some excited states have long lifetimes. They may be regarded as energetic isomers of the ground state that can participate in **delayed reactions.** The excited state of $[Ru^{II}(bipy)_3]^{2+}$ created by photon absorption in the metal-to-ligand charge-transfer band (Section 19.5) may be regarded as a Ru(III) cation complexed to a radical anion of the ligand. Its redox reactions can be explained by adding the excitation energy (expressed as a potential by using $-FE = \Delta_r G$ and equating $\Delta_r G$ to the molar excitation energy) to the ground-state reduction potential (Fig. 20.23).

20.15 d–d and charge-transfer reactions

Key point: A useful first approximation is to associate photosubstitution and photoisomerization with *d–d* transitions and photoredox reactions with charge-transfer transitions, but the rule is not absolute.

We have seen (Sections 19.4 and 19.5) that there are two main types of spectroscopically observable electron promotion in *d*-metal complexes, namely *d–d* transitions and charge-transfer transitions. A *d–d* transition corresponds to the essentially *angular* redistribution of electrons within a *d* shell. In octahedral complexes, this redistribution often corresponds to the occupation of M—L antibonding e_g orbitals. An example is the $^4T_{1g} \leftarrow {}^4A_{2g}$ ($t_{2g}^2 e_g^1 \leftarrow t_{2g}^3$) transition in $[Cr(NH_3)_6]^{3+}$. The occupation of the antibonding e_g orbital results in a quantum yield close to 1 (specifically 0.6) for the photosubstitution

$$[Cr(NH_3)_6]^{3+} + H_2O \xrightarrow{h\nu} [Cr(NH_3)_5(OH_2)]^{3+} + NH_3$$

This is a prompt reaction, as it occurs in less than 5 ps.

Charge-transfer transitions correspond to the *radial* redistribution of electron density. They correspond to the promotion of electrons into predominantly ligand orbitals if the transition is metal-to-ligand or into orbitals of predominantly metal character if the transition is ligand-to-metal. The former process corresponds to oxidation of the metal centre and the latter to its reduction. These excitations commonly initiate photoredox reactions of the kind already mentioned in connection with Co(III) and Ru(II).

Although a useful first approximation is to associate photosubstitution and photoisomerization with *d–d* transitions and photoredox with charge-transfer transitions, the rule is not absolute. For example, it is not uncommon for a charge-transfer transition to result in photosubstitution by an indirect path:

$$[Co^{III}Cl(NH_3)_5]^{2+} + H_2O \xrightarrow{h\nu} [Co^{II}(NH_3)_5(OH_2)]^{2+} + Cl\cdot$$
$$[Co^{II}(NH_3)_5(OH_2)]^{2+} + Cl\cdot \rightarrow [Co^{III}(NH_3)_5(OH_2)]^{3+} + Cl^-$$

In this case, the aqua complex formed after the homolytic fission of the Co—Cl bond is reoxidized by the Cl atom. The net result leaves the Co substituted. Conversely, some excited states show no differences in substitutional reactivity compared with the ground state: the long-lived excited ^2E state of $[Cr(bipy)_3]^{3+}$ results from a pure *d–d* transition and its lifetime of several microseconds allows the excess energy to enhance its redox reactions. The standard potential ($+1.3$ V), calculated by adding the excitation energy to the ground-state value, accounts for its function as a good oxidizing agent, in which it undergoes reduction to $[Cr(bipy)_3]^{2+}$.

20.16 Transitions in metal–metal bonded systems

Key points: Population of a metal–metal antibonding orbital can sometimes initiate photodissociation; such excited states have been shown to initiate multielectron redox photochemistry.

We might expect the δ–δ^* transition in metal–metal bonded systems to initiate photodissociation as it results in the population of an antibonding orbital of the metal–metal system. It is more interesting that such excited states have also been shown to initiate multielectron redox photochemistry.

One of the best characterized systems is the dinuclear platinum complex $[Pt_2(\mu\text{-}P_2O_5H_2)_4]^{4-}$, called informally 'PtPOP' (**21**). There is no metal–metal bonding in the ground state of this Pt(II)—Pt(II) d^8–d^8 species. The HOMO–LUMO pattern indicates that excitation populates a bonding orbital between the two metal atoms (Fig. 20.24).

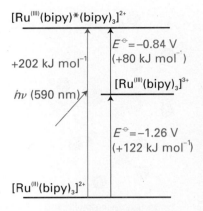

Fig. 20.23 The photoexcitation of $[Ru^{II}(bipy)_3]^{2+}$ can be treated as if the excited state is a Ru(III) cation complexed to a radical anion of the ligand.

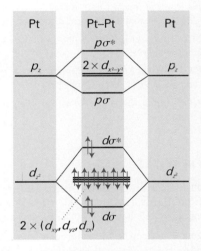

Fig. 20.24 The dinuclear complex $[Pt_2(\mu\text{-}P_2O_5H_2)_4]^{4-}$ consists of two face-to-face square-planar complexes held together by a bridging pyrophosphito ligand. The metal p_z and d_{z^2} orbitals interact along the Pt–Pt axis. The other *p* and *d* orbitals are considered to be nonbonding. Photoexcitation results in an electron in the antibonding σ^* orbital moving into the bonding σ orbital.

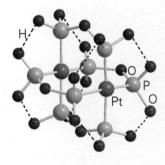

21 $[Pt_2(\mu\text{-}P_2O_5H_2)_4]^{4-}$, PtPOP

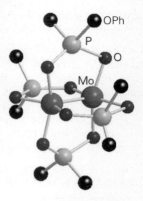

22 [Mo$_2$(O$_2$P(OPh)$_2$)$_4$]

The lowest-lying excited state has a lifetime of 9 µs and is a powerful reducing agent, reacting by both electron and halogen-atom transfer. The most interesting oxidation products are Pt(III)—Pt(III), which contain X$^-$ ligands (where X is halogen or pseudohalogen) at both ends and a metal–metal single bond. Irradiation in the presence of (Bu)$_3$SnH gives a dihydrido product that can eliminate H$_2$.

Irradiation of the quadruply bonded dinuclear cluster [Mo$_2$(O$_2$P(OC$_6$H$_5$)$_2$)$_4$] (**22**) at 500 nm in the presence of ClCH$_2$CH$_2$Cl results in production of ethene and the addition of two Cl atoms to the two Mo atoms, with a two-electron oxidation. The reaction proceeds in one-electron steps, and requires a complex with the metal atoms shielded by sterically crowding ligands. If smaller ligands are present, the reaction that occurs instead is a photochemical oxidative addition of the organic molecule.

FURTHER READING

G.J. Leigh and N. Winterbottom (ed.), *Modern coordination chemistry: the legacy of Joseph Chatt*. Royal Society of Chemistry, Cambridge (2002). A readable historical discussion of this area.

M.L. Tobe and J. Burgess, *Inorganic reaction mechanisms*. Longman, Harlow (1999).

R.G. Wilkins, *Kinetics and mechanism of reactions of transition metal complexes*. VCH, Weinheim (1991).

A.G. Lappin, *Redox mechanisms in inorganic chemistry*. Ellis Horwood, London (1994).

A special issue of *Coordination Chemistry Reviews* has been dedicated to the work of Henry Taube. See *Coord. Chem. Rev.*, 2005, **249**.

Two readable discussions of redox processes are to be found in Taube's 1983 Nobel Prize lecture reprinted in *Science*, 1984, **226**, 1028 and in Marcus's 1992 Nobel Prize lecture published in *Nobel Lectures: Chemistry 1991–1995*, World Scientific, Singapore (1997).

EXERCISES

20.1 The stepwise formation constants for complexes of NH$_3$ with [Cu(OH$_2$)$_6$]$^{2+}$(aq) are log K_{f1} = 4.15, log K_{f2} = 3.50, log K_{f3} = 2.89, log K_{f4} = 2.13, log K_{f5} = −0.52. Suggest a reason why K_{f5} is so different.

20.2 The stepwise formation constants for complexes of NH$_2$CH$_2$CH$_2$NH$_2$ (en) with [Cu(OH$_2$)$_6$]$^{2+}$(aq) are log K_{f1} = 10.72, log K_{f2} = 9.31. Compare these values with those of ammonia given in Exercise 20.1 and suggest why they are different.

20.3 The rate constants for the formation of [CoX(NH$_3$)$_5$]$^{2+}$ from [Co(NH$_3$)$_5$OH$_2$]$^{3+}$ for X = Cl$^-$, Br$^-$, N$_3^-$, and SCN$^-$ differ by no more than a factor of two. What is the mechanism of the substitution?

20.4 If a substitution process is associative, why may it be difficult to characterize an aqua ion as labile or inert?

20.5 The reactions of Ni(CO)$_4$ in which phosphines or phosphites replace CO to give the family Ni(CO)$_3$L occur at the same rate for different phosphines or phosphites. Is the reaction *d* or *a*?

20.6 Write the rate law for formation of [MnX(OH$_2$)$_5$]$^+$ from the aqua ion and X$^-$. How would you undertake to determine if the reaction is *d* or *a*?

20.7 Octahedral complexes of metal centres with high oxidation numbers or of *d* metals of the second and third series are less labile than those of low oxidation number and *d* metals of the first series of the block. Account for this observation on the basis of a dissociative rate-determining step.

20.8 A Pt(II) complex of tetramethyldiethylenetriamine is attacked by Cl$^-$ 10^5 times less rapidly than the diethylenetriamine analogue. Explain this observation in terms of an associative rate-determining step.

20.9 The rate of loss of chlorobenzene, PhCl, from [W(CO)$_4$L(PhCl)] increases with increase in the cone angle of L. What does this observation suggest about the mechanism?

20.10 Does the fact that [Ni(CN)$_5$]$^{3-}$ can be isolated help to explain why substitution reactions of [Ni(CN)$_4$]$^{2-}$ are very rapid?

20.11 Design two-step syntheses of *cis*- and *trans*-[PtCl$_2$(NO$_2$)(NH$_3$)]$^-$ starting from [PtCl$_4$]$^{2-}$.

20.12 How does each of the following modifications affect the rate of a square-planar complex substitution reaction? (a) Changing a *trans* ligand from H to Cl. (b) Changing the leaving group from Cl to I. (c) Adding a bulky substituent to a *cis* ligand. (d) Increasing the positive charge on the complex.

20.13 The rate of attack on Co(III) by an entering group Y is nearly independent of Y with the spectacular exception of the rapid reaction with OH$^-$. Explain the anomaly. What is the implication of your explanation for the behaviour of a complex lacking Brønsted acidity on the ligands?

20.14 Predict the products of the following reactions:

(a) [Pt(PR$_3$)$_4$]$^{2+}$ + 2 Cl$^-$

(b) [PtCl$_4$]$^{2-}$ + 2 PR$_3$

(c) *cis*-[Pt(NH$_3$)$_2$(py)$_2$]$^{2+}$ + 2 Cl$^-$

20.15 Put in order of increasing rate of substitution by H$_2$O the complexes (a) [Co(NH$_3$)$_6$]$^{3+}$, (b) [Rh(NH$_3$)$_6$]$^{3+}$, (c) [Ir(NH$_3$)$_6$]$^{3+}$, (d) [Mn(OH$_2$)$_6$]$^{2+}$, (e) [Ni(OH$_2$)$_6$]$^{2+}$.

20.16 State the effect on the rate of dissociatively activated reactions of Rh(III) complexes of (a) an increase in the overall charge on the complex; (b) changing the leaving group from NO_3^- to Cl^-; (c) changing the entering group from Cl^- to I^-; (d) changing the *cis* ligands from NH_3 to H_2O.

20.17 The pressure dependence of the replacement of chlorobenzene (PhCl) by piperidine in the complex $[W(CO)_4(PPh_3)(PhCl)]$ has been studied. The volume of activation is found to be $+11.3\ cm^3\ mol^{-1}$. What does this value suggest about the mechanism?

20.18 Write out the inner- and outer-sphere pathways for reduction of azidopentaamminecobalt(III) ion with $V^{2+}(aq)$. What experimental data might be used to distinguish between the two pathways?

20.19 The compound $[Fe(SCN)(OH_2)_5]^{2+}$ can be detected in the reaction of $[Co(NCS)(NH_3)_5]^{2+}$ with $Fe^{2+}(aq)$ to give $Fe^{3+}(aq)$ and $Co^{2+}(aq)$. What does this observation suggest about the mechanism?

20.20 Calculate the rate constants for electron transfer in the oxidation of $[V(OH_2)_6]^{2+}$ ($E^{\ominus}(V^{3+}, V^{2+}) = -0.255\ V$) and the oxidants (a) $[Ru(NH_3)_6]^{3+}$ ($E^{\ominus}(Ru^{3+}, Ru^{2+}) = +0.07\ V$), (b) $[Co(NH_3)_6]^{3+}$

$(E^{\ominus}(Co^{3+}, Co^{2+}) = +0.10\ V)$. Comment on the relative sizes of the rate constants.

20.21 The photochemical substitution of $[W(CO)_5(py)]$ (py = pyridine) with triphenylphosphine gives $W(CO)_5(P(C_6H_5)_3)$. In the presence of excess phosphine, the quantum yield is approximately 0.4. A flash photolysis study reveals a spectrum that can be assigned to the intermediate $W(CO)_5$. What product and quantum yield do you predict for substitution of $[W(CO)_5(py)]$ in the presence of excess triethylamine? Is this reaction expected to be initiated from the ligand field or MLCT excited state of the complex?

20.22 From the spectrum of $[CrCl(NH_3)_5]^{2+}$ shown in Fig. 19.29, propose a wavelength for photoinitiation of reduction of Cr(III) to Cr(II) accompanied by oxidation of a ligand.

20.23 Reactions of $[Pt(Ph)_2(SMe_2)_2]$ with the bidentate ligand 1,10-phenanthroline (phen) give $[Pt(Ph)_2phen]$. There is a kinetic pathway with activation parameters $\Delta^{\ddagger}H = +101\ kJ\ mol^{-1}$ and $\Delta^{\ddagger}S = +42\ J\ K^{-1}\ mol^{-1}$. Propose a mechanism.

PROBLEMS

20.1 The equilibrium constants for the sucessive reactions of ethylenediamine with Co^{2+}, Ni^{2+}, and Cu^{2+} are as follows.

$$[M(OH_2)_6]^{2+} + en \rightleftharpoons [M(en)(OH_2)_4]^{2+} + 2\ H_2O \qquad K_1$$

$$[M(en)(OH_2)_4]^{2+} + en \rightleftharpoons [M(en)_2(OH_2)_2]^{2+} + 2\ H_2O \quad K_2$$

$$[M(en)_2(OH_2)_2]^{2+} + en \rightleftharpoons [M(en)_3]^{2+} + 2\ H_2O \qquad K_3$$

Ion	$\log K_1$	$\log K_2$	$\log K_3$
Co^{2+}	5.89	4.83	3.10
Ni^{2+}	7.52	6.28	4.26
Cu^{2+}	10.72	9.31	−1.0

Discuss whether these data support the generalizations in the text about sucesssive formation constants and the Irving–Williams series. How do you account for the very low value of K_3 for Cu^{2+}?

20.2 How may the aromatic character of a chelate ring provide additional stabilization of a complex? See A. Crispini and M. Ghedini, *J. Chem. Soc., Dalton Trans.* 1997, 75.

20.3 Given the following mechanism for the formation of a chelate complex

$$[Ni(OH_2)_6]^{2+} + L\!-\!L \rightleftharpoons [Ni(OH_2)_6]^{2+}, L\!-\!L \qquad K_E, \text{ rapid}$$

$$[Ni(OH_2)_6]^{2+}, L\!-\!L \rightleftharpoons [Ni(OH_2)_5L\!-\!L]^{2+} + H_2O \qquad k_a, k_a'$$

$$[Ni(OH_2)_5L\!-\!L]^{2+} \rightleftharpoons [Ni(OH_2)_4L\!-\!L]^{2+} + H_2O \qquad k_b, k_b'$$

derive the rate law for the formation of the chelate. Discuss the step that is different from that for two monodentate ligands. The formation of chelates with strongly bound ligands occurs at the rate of formation of the analogous monodentate complex but the formation of chelates of weakly bound ligands is often significantly slower. Assuming an I_d mechanism, explain this observation. (See R.G. Wilkins, *Acc. Chem. Res.*, 1970, **3**, 408.)

20.4 The complex $[PtH(PEt_3)_3]^+$ was studied in deuterated acetone in the presence of excess PEt_3. In the absence of excess ligand the ^1H NMR spectrum in the hydride region exhibits a doublet of triplets. As excess PEt_3 ligand is added the hydride signal begins to change, the line shape depending on the ligand concentration. Suggest a mechanism to account for the effects of excess PEt_3.

20.5 Solutions of $[PtH_2(PMe_3)_2]$ exist as a mixture of *cis* and *trans*

isomers. Addition of excess PMe_3 led to formation of $[PtH_2(PMe_3)_3]$ at a concentration that could be detected using NMR. This complex exchanged phosphine ligands rapidly with the *trans* isomer but not the *cis*. Propose a pathway. What are the implications for the *trans* effect of H versus PMe_3? (See D.L. Packett and W.G. Trogler, *Inorg. Chem.*, 1988, **27**, 1768.)

20.6 Figure 20.25 (which is based on J.B. Goddard and F. Basolo, *Inorg. Chem.*, 1968, **7**, 936) shows the observed first-order rate constants for the reaction of $[PdBrL]^+$ with various Y^- to give $[PdYL]^+$, where L is $Et_2NCH_2CH_2NHCH_2CH_2NEt_2$. Note the steep slope for $S_2O_3^{2-}$ and zero slopes for $Y^- = N_3^-, I^-, NO_2^-$, and SCN^-. Propose a mechanism.

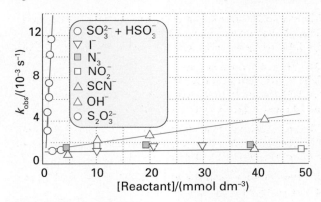

Fig. 20.25 The data required for Problem 20.6.

20.7 The substitution reactions of the bridged dinuclear Rh(II) complex $[Rh_2(\mu\text{-}O_2CCH_3)_4XY]$ (**23**) have been studied by M.A.S. Aquino and

D.H. Macartney (*Inorg. Chem.*, 1987, **26**, 2696). Reaction rates show little dependence on the choice of the entering group. The table below shows the dependence on the leaving group, X, and the ligand on the opposite Rh, (*trans*) Y, at 298 K. What conclusions can you draw concerning the mechanism?

X	Y	$k/(\text{dm}^3 \text{ mol}^{-1} \text{ s}^{-1})$
H_2O	H_2O	10^5–10^7
CH_3OH	CH_3OH	2×10^6
CH_3CN	CH_3CN	1.1×10^5
PPh_3	PPh_3	1.5×10^5
CH_3CN	PR_3	10^7–10^9
PR_3	CH_3CN	10^{-1}–10^2
N-donor	H_2O	10^2–10^3

Note that the complex has a d^7 configuration at each Rh and a single Rh—Rh bond.

20.8 The activation enthalpy for the reduction of *cis*-$[CoCl_2(en)_2]^+$ by $Cr^{2+}(aq)$ is -24 kJ mol^{-1}. Explain the negative value. (See R.C. Patel, R.E. Ball, J.F. Endicott, and R.G. Hughes, *Inorg. Chem.*, 1970, **9**, 23.)

20.9 The rate of reduction of $[Co(NH_3)_5(OH_2)]^{3+}$ by Cr(II) is seven orders of magnitude slower than reduction of its conjugate base, $[Co(NH_3)_5(OH)]^{2+}$, by Cr(II). For the corresponding reductions with $[Ru(NH_3)_6]^{2+}$, the two differ by less than a factor of 10. What do these observations suggest about mechanisms? Comment on H_2O and OH^- as bridging ligands.

20.10 Calculate the rate constants for outer-sphere reactions from the following data. Compare your results to the measured values in the last column.

Reaction	$k_{11}/(\text{dm}^3 \text{ mol}^{-1} \text{ s}^{-1})$	$k_{22}/(\text{dm}^3 \text{ mol}^{-1} \text{ s}^{-1})$	$E^{\ominus}$/V	$k_{\text{obs}}/(\text{dm}^3 \text{ mol}^{-1} \text{ s}^{-1})$
$Cr^{2+} + Fe^{2+}$	2×10^{-5}	4.0	$+1.18$	2.3×10^3
$W(CN)_8^{4-} + Ce(IV)$	$>4 \times 10^4$	4.4	$+0.90$	$>10^8$
$[Fe(CN)_6]^{4-} + MnO_4^-$	7.4×10^2	3×10^3	$+0.20$	1.7×10^5
$[Fe(phen)_3]^{2+} + Ce(IV)$	$>3 \times 10^7$	4.4	$+0.36$	1.4×10^5

20.11 In the presence of catalytic amounts of $[Pt(P_2O_5H_2)_4]^{4-}$ (**21**) and light, 2-propanol produces H_2 and acetone (E.L. Harley, A.E. Stiegman, A. Vlcek, Jr., and H.B. Gray, *J. Am. Chem. Soc.*, 1987, **109**, 5233; D.C. Smith and H.B. Gray, *Coord. Chem. Rev.*, 1990, **100**, 169). (a) Give the equation for the overall reaction. (b) Give a plausible molecular orbital scheme for the metal–metal bonding in this tetragonal-prismatic complex and indicate the nature of the excited state that is thought to be responsible for the photochemistry. (c) Indicate the metal complex intermediates and the evidence for their existence.

d-Metal organometallic chemistry

21

Organometallic chemistry is the chemistry of compounds containing metal–carbon bonds. Much of the basic organometallic chemistry of the s- and p-block metals was understood by the early part of the twentieth century and we have discussed it in Chapters 10–14. By contrast, the organometallic chemistry of the d and f blocks has been developed much more recently. Since the mid-1950s this field has grown into a thriving area that spans new types of reactions, unusual structures, and practical applications in organic synthesis and industrial catalysis. We discuss the organometallic chemistry of the d and f blocks separately, covering d metals in this chapter and f metals in the next. The widespread use of organometallic compounds in synthesis is covered in Chapter 25 (on catalysis).

A few d-block organometallic compounds were synthesized and partially characterized in the nineteenth century. The first of them (1), an ethene complex of platinum(II), was prepared by W.C. Zeise in 1827, with the first metal carbonyls, $[PtCl_2(CO)_2]$ and $[PtCl_2(CO)]_2$, being reported by P. Schützenberger in 1868. The next major discovery was the metal carbonyl, tetracarbonylnickel (2), which was synthesized by L. Mond, C. Langer, and F. Quinke in 1890. Beginning in the 1930s, W. Hieber synthesized a wide variety of metal carbonyl cluster compounds, many of which are anionic, including $[Fe_4(CO)_{13}]^{2-}$ (3). It was clear from this work that metal carbonyl chemistry was potentially a very rich field. However, as the structures of these and other d- and f-block organometallic compounds are difficult or impossible to deduce by chemical means alone, fundamental advances had to await the development of X-ray diffraction for precise structural data on solid samples and of IR and NMR spectroscopy for structural information in solution. The discovery of the remarkably stable organometallic compound ferrocene, $Fe(C_5H_5)_2$ (4), occurred at a time (in 1951) when these techniques were becoming widely available. The 'sandwich' structure of ferrocene was soon correctly inferred from its IR spectrum and then determined in detail by X-ray crystallography.

The stability, structure, and bonding of ferrocene defied the classical Lewis description and therefore captured the imagination of chemists. This puzzle in turn set off a train of synthesizing, characterizing, and theorizing that led to the rapid development of d-block organometallic chemistry. Two highly productive research workers in the formative stage of the subject, Ernst-Otto Fischer in Munich and Geoffrey Wilkinson in London, were awarded the Nobel Prize in 1973 for their contributions. Similarly, f-block organometallic chemistry blossomed soon after the discovery in the late 1970s that the pentamethylcyclopentadienyl ligand, C_5Me_5, forms stable f-block compounds (5).

We adhere to the convention that an organometallic compound contains at least one metal–carbon (M—C) bond. Thus, compounds (1) to (5) clearly qualify as organometallic, whereas a complex such as $[Co(en)_3]^{3+}$, which contains carbon but has no M—C bonds, does not. Cyano complexes, such as hexacyanoferrate(II) ions, do have M—C bonds, but as their properties are more akin to those of conventional coordination complexes they are generally not considered as organometallic. By contrast, complexes of the isoelectronic ligand CO are considered to be organometallic. The justification for this somewhat arbitrary distinction is that many metal carbonyls are significantly different from coordination complexes both chemically and physically.

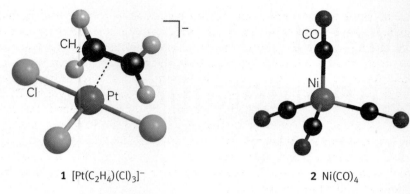

1 $[Pt(C_2H_4)(Cl)_3]^-$

2 $Ni(CO)_4$

In general, the distinctions between the two classes of compounds are clear: coordination complexes normally are charged, with variable *d*-electron count, and are soluble in water; organometallic compounds are often neutral, with fixed *d*-electron count, and are soluble in organic solvents such as tetrahydrofuran. Most organometallic compounds have properties that are much closer to organic compounds than inorganic salts, with many of them having low melting points (some are liquid at room temperature).

Bonding

Although there are many organometallic compounds of the *s* and *p* blocks, the bonding is often relatively simple, and normally adequately described solely by σ bonds. The *d* metals, by contrast, exhibit a large number of organometallic compounds with many different bonding modes. For instance, to describe fully the bonding of a cyclopentadienyl group in ferrocene (and in general to any *d* metal), we need to invoke σ, π, and δ bonds.

Unlike coordination compounds, *d*-metal organometallic compounds normally have relatively few stable electron configurations and often have a total of 16 or 18 valence electrons around the metal atom. This restriction to a limited number of electronic configurations is due to the strength of the π (and δ, where appropriate) bonding interactions between the metal atom and the carbon-containing ligands.

21.1 Stable electron configurations

We start by examining the bonding patterns so that we can appreciate the importance of π bonds and understand the origin of the restriction of the *d*-metal organometallics to certain electron configurations.

(a) 18-electron compounds

Key points: Six σ bonding interactions are possible in an octahedral complex and, when π-acceptor ligands are present, bonding combinations can be made with the three orbitals of the t_{2g} set, leading to nine bonding MOs, and space for a total of 18 electrons.

In the 1920s, N.V. Sidgwick recognized that the metal atom in a simple metal carbonyl, such as $Ni(CO)_4$, has the same valence electron count (18) as the noble gas that terminates the long period to which the metal belongs. Sidgwick coined the term 'inert gas rule' for this indication of stability, but it is now usually referred to as the **18-electron rule**.[1] It becomes readily apparent, however, that the 18-electron rule is not as uniformly obeyed for *d*-block organometallic compounds as the octet rule is obeyed for compounds of Period 2 elements, and we need to look more closely at the bonding to establish the reasons for the stability of both the compounds that have the 18-electron configurations and those that do not.

Recall the picture of bonding in octahedral complexes discussed in Section 19.2. Figure 21.1 shows the energy levels that arise when a strong-field ligand such as carbon

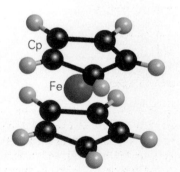

3 $[Fe_4(CO)_{13}]^{2-}$

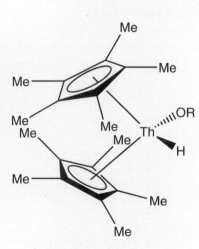

4 $FeCp_2$, $Cp = C_5H_5$

5

---

[1] The 18-electron rule is sometimes referred to as the *effective atomic number* or *EAN* rule.

monoxide bonds to a *d*-metal atom. Carbon monoxide is a strong-field ligand, even though it is a poor σ donor, because it can use its empty π* orbitals to act as a good π acceptor. In this picture of the bonding, the t_{2g} set of orbitals on the metal atom are no longer simply nonbonding, as they would be in the absence of π interactions, but are bonding. The energy level diagram shows six bonding MOs that result from the ligand–metal σ interactions, and three bonding MOs that result from π interactions. Thus there is room for a total of 18 electrons in the nine bonding MOs. Compounds that have this configuration are remarkably stable: for instance, the 18-electron $Cr(CO)_6$ is a colourless air-stable compound. An indication of the size of the HOMO–LUMO gap (Δ_O) can be gained from a consideration of its lack of colour, which results from a lack of any electronic transitions in the visible region of the spectrum; that is, Δ_O is so large that such transitions are shifted to the UV.

The only way to accommodate more than 18 valence electrons in an octahedral complex with strong-field ligands is to partially occupy an antibonding orbital. As a result, such complexes are unstable, being particularly prone to electron loss (acting as reducing agents). Compounds with fewer than 18 electrons will not necessarily be very unstable, but such complexes will find it energetically favourable to acquire extra electrons by reaction and so populate their bonding MOs fully. As we shall see later, compounds with fewer than 18 electrons often occur as intermediates in reaction pathways.

When we examine the mode of bonding of the many diverse organic ligands to the metal atom, the situation with the carbonyl ligand is replicated: the ligands we are considering are often poor σ donors but good π acceptors. Hence, octahedral organometallic compounds are most stable when they have a total of 18 valence electrons around their central metal ion.

Similar arguments can be used to rationalize the stability of the 18-electron configuration for other geometries, such as tetrahedral and trigonal-bipyramidal, although in practice relatively few tetrahedral organometallic compounds are known. The steric requirements of most ligands normally preclude coordination numbers of greater than 6 for *d*-metal organometallic compounds.

(b) 16-electron square-planar compounds

Key point: With strong-field ligands, a square-planar complex has only eight bonding MOs, thus a 16-electron configuration is the most energetically favourable configuration.

One other geometry already discussed in the context of coordination chemistry is the square-planar arrangement of four ligands (Section 19.1), where we noted that it occurred only for strong-field ligands and a d^8 metal ion. Because organometallic ligands often produce a strong field, many square-planar organometallic compounds exist. Stable square-planar complexes are normally found only with a total of 16 valence electrons, which results in the population of all the bonding and none of the antibonding MOs (Fig. 21.2).

The ligands in square-planar complexes can normally provide only two electrons each, for a total of eight electrons. Therefore, to reach 16 electrons, the metal ion must provide an additional eight electrons. As a result, organometallic compounds with 16 valence electrons are common only on the right of the *d* block, particularly in Groups 9 and 10 (Table 21.1). Examples of such complexes include $[IrCl(CO)(PPh_3)_2]$ (**6**) and the anion of Zeise's salt, $[PtCl_3(C_2H_4)]^-$ (**1**). Square-planar 16-electron complexes are particularly common for the heavier elements in Groups 9 and 10, especially for Rh(I), Ir(I), Pd(II),

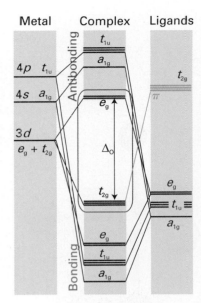

Fig. 21.1 The energy levels of the *d* orbitals of an octahedral complex with strong-field ligands.

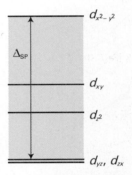

Fig. 21.2 The energy levels of the molecular orbitals of a square-planar complex with strong-field ligands. The four lowest MOs correspond to bonding interactions, and the highest corresponds to an antibonding interaction; the MOs are labelled with the *d* orbitals from which they are derived.

Table 21.1 Validity of the 16/18-electron rule for *d*-metal organometallic compounds							
Usually less than 18 eletrons			**Usually 18 electrons**			**16 or 18 electrons**	
Sc	Ti	V	Cr	Mn	Fe	Co	Ni
Y	Zr	Nb	Mo	Tc	Ru	Rh	Pd
La	Hf	Ta	W	Re	Os	Ir	Pt

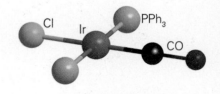

6 *trans*-[IrCl(CO)(PPh₃)₂]

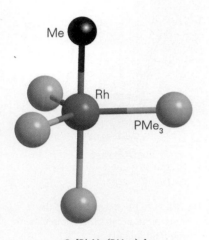

7

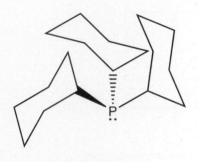

8 [RhMe(PMe₃)₄]

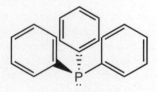

9 PCy₃, Cy=*cyclo*-C₆H₁₁

10 PPh₃

and Pt(II), because the ligand-field splitting is large and the ligand-field stabilization energy of these complexes favours the square-planar configuration.

21.2 Electron count preference

Key point: On the left of the *d* block, steric requirements can mean that it is not possible to assemble enough ligands around a metal atom to achieve a total of 16 or 18 valence electrons.

The preference of a metal atom for one particular geometry and electron count is not normally so strong as to preclude formation of others. For instance, although the chemistry of both palladium(II) and rhodium(I) is dominated by 16-electron square-planar complexes, the cyclopentadienyl palladium complex (**7**) and the rhodium complex (**8**) are 18-electron trigonal bipyramidal.

Steric factors may restrict the number of ligands that can bond to a metal atom and stabilize compounds with lower electron counts than might have been expected. For instance, the tricyclohexylphosphine ligands (**9**) in the trigonal platinum(0) compound $Pt(PCy_3)_3$ are so large that only three can be fitted around the metal atom, which consequently has only 16 valence electrons. Steric stabilization of the metal centre is partially a kinetic effect, in which the large groups protect the metal centre from reaction. Many complexes readily lose or gain ligands, forming other configurations transitorily during reactions—indeed, the accessibility of these other configurations is precisely why the organometallic chemistry of the *d* metals is so interesting.

Unusual electron configurations are common on the left of the *d* block, where the metal atoms have fewer electrons, and it is often not possible to crowd enough ligands around the atom to satisfy an electron count of 16 or 18 electrons. For example, the simplest carbonyl in Group 5, $[V(CO)_6]$, is a 17-electron complex. Other examples include $[W(CH_3)_6]$, which has 12 valence electrons, and $[Cr(\eta^5\text{-}Cp)(CO)_2(PPh_3)]$ with 17. The latter compound provides a good example of the role of steric crowding. When the compact CO ligand is present in place of bulky triphenylphosphine (**10**), a dimeric compound with a long but definite Cr—Cr bond is observed in the solid state and in solution. The formation of the Cr—Cr bond in $[Cr(\eta^5\text{-}Cp)(CO)_3]_2$ raises the electron count on each metal to 18.

21.3 Electron counting and oxidation states

The dominance of 16- and 18-electron configurations in organometallic chemistry makes it imperative that we are able to count the number of valence electrons on a central metal atom, because knowing that number allows us to predict the stabilities of compounds and to suggest patterns of reactivity. Although the concept of 'oxidation state' for organometallic compounds is regarded by many as tenuous at best, the vast majority of the research community uses it as a convenient shorthand for describing electron configurations. Oxidation states (and the corresponding oxidation number) help to systematize reactions such as oxidative addition (Section 21.23), and also bring out analogies between the chemical properties of organometallic and coordination complexes. Fortunately, the business of counting electrons and assigning oxidation numbers can be combined.

Two models are routinely used to count electrons, the so-called **neutral-ligand method** (sometimes called the *covalent method*) and the **donor-pair method** (sometimes known as the *ionic method*). We introduce both briefly—they give identical results for electron counting—but in what follows we use the donor-pair method as it may easily be used to assign oxidation numbers too.

(a) Neutral-ligand method

Key point: All ligands are treated as neutral and are categorized according to how many electrons they are considered to donate.

For the sake of counting electrons, each metal atom and ligand is treated as neutral. If the complex is charged, we simply add or subtract the appropriate number of electrons to the total. We must include in the count all valence electrons of the metal atom and all the electrons donated by the ligands. Ligands are defined as **L type** if they are neutral two-electron donors (like CO, PMe$_3$), and **X type** if, when they are considered to be neutral, they are one-electron radical donors (like halogen atoms, H, CH$_3$). For example, Fe(CO)$_5$ acquires 18 electrons from the eight valence electrons on the Fe atom and the 10 electrons donated by the five CO ligands. Some ligands are considered combinations of these types; for instance, cyclopentadienyl is considered as a five-electron L$_2$X donor; see Table 21.2.

Table 21.2 Typical ligands and their electron counts

(a) Neutral-ligand method

Ligand	Formula	Designation*	Electrons donated
Carbonyl	CO	L	2
Phosphine	PR$_3$	L	2
Hydride	H	X	1
Dihydrogen	H$_2$	L	2
η^1-Alkyl, -alkenyl, -alkynyl, and -aryl groups	R	X	1
η^2-Alkene	CH$_2$=CH$_2$	L	2
η^2-Alkyne	RCCR	L	2
Dinitrogen	N$_2$	L	2
Butadiene	CH$_2$=CH—CH=CH$_2$	L$_2$	4
Benzene	C$_6$H$_6$	L$_3$	6
η^3-Allyl	CH$_2$CHCH$_2$	LX	3
η^5-Cyclopentadienyl	C$_5$H$_5$	L$_2$X	5

*Ligands are defined as **L type** if they are neutral two-electron donors (such as CO, PMe$_3$), and **X type** if, when they are considered to be neutral, they are one-electron radical donors (such as halogen atoms, H, CH$_3$).

*(b) Donor-pair method**

Ligand	Formula	Electrons donated
Carbonyl	CO	2
Phosphine	PR$_3$	2
Hydride	H$^-$	2
Dihydrogen	H$_2$	2
η^1-Alkyl, -alkenyl, -alkynyl, and -aryl groups	R$^-$	2
η^2-Alkene	CH$_2$=CH$_2$	2
η^2-Alkyne	RCCR	2
Dinitrogen	N$_2$	2
Butadiene	CH$_2$=CH—CH=CH$_2$	4
Benzene	C$_6$H$_6$	6
η^3-Allyl	CH$_2$CHCH$_2^-$	4
η^5-Cyclopentadienyl	C$_5$H$_5^-$	6

*We use this method throughout this book.

Example 21.1 Counting electrons using the neutral-ligand method

Do (a) $[IrBr_2(CH_3)(CO)(PPh_3)_2]$ and (b) $[Cr(\eta^5\text{-}C_5H_5)(\eta^6\text{-}C_6H_6)]$ obey the 18-electron rule?

Answer (a) An Ir atom (Group 9) has nine valence electrons; the two Br atoms and the CH_3 group are one-electron donors; CO and PPh_3 are two-electron donors. Thus, the number of valence electrons on the metal atom is $9 + (3 \times 1) + (3 \times 2) = 18$, in accord with the common occurrence of 18-electron complexes. (b) A Cr atom (Group 6) has six valence electrons, the $\eta^5\text{-}C_5H_5$ ligand donates five electrons, and the $\eta^6\text{-}C_6H_6$ ligand donates six; so the number of metal valence electrons is $6 + 5 + 6 = 17$. This complex does not obey the 18-electron rule and is not stable. A related but stable 18-electron compound is $[Cr(\eta^6\text{-}C_6H_6)_2]$.

Self-test 21.1 Is $[Mo(CO)_7]$ likely to be stable?

The advantage of the neutral-ligand method is that, given the information in Table 21.2, it is trivial to establish the electron count. The disadvantage, however, is that the method overestimates the degree of covalence and thus underestimates the charge at the metal. Moreover, it becomes confusing to assign an oxidation number to a metal, and meaningful information on some ligands is lost.

(b) Donor-pair method

Key point: Ligands are considered to donate electrons in pairs, resulting in the need to treat some ligands as neutral and others as charged.

The donor-pair method requires a calculation of the oxidation number. The rules for calculating the oxidation number of an element in an organometallic compound are the same as for conventional coordination compounds. Neutral ligands, such as CO and phosphine, are considered to be two-electron donors and are formally assigned an oxidation number of 0. Ligands such as halides, H, and CH_3 are formally considered to take an electron from the metal atom, and are treated as Cl^-, H^-, and CH_3^- (and hence are assigned oxidation number -1); in this anionic state they are considered to be two-electron donors. The cyclopentadienyl ligand, C_5H_5 (Cp), is treated as $C_5H_5^-$ (it is assigned an oxidation number of -1); in this anionic state it is considered to be a six-electron donor. Then:

> The *oxidation number* of the metal atom is the total charge of the complex minus the charges of any ligands.
> The *number of electrons* the metal provides is its group number minus its oxidation number.
> The *total electron count* is the sum of the number of electrons on the metal atom and the number of electrons provided by the ligands.

The main advantage of this method is that with a little practice both the electron count and the oxidation number may be determined in a straightforward manner. The main disadvantage is that it overestimates the charge on the metal atom and can suggest reactivity that might be incorrect (see Section 21.17 on hydrides). Table 21.2 lists the maximum number of electrons available for donation to a metal for most common ligands.

Example 21.2 Assigning oxidation numbers and counting the electrons using the donor-pair method

Assign the oxidation number and count the valence electrons on the metal atom in (a) $[IrBr_2(CH_3)(CO)(PPh_3)_2]$, (b) $[Cr(\eta^5\text{-}C_5H_5)(\eta^6\text{-}C_6H_6)]$, and (c) $[Mn(CO)_5^-]$.

Answer (a) The two Br groups and the CH_3 are treated as three singly negatively charged two-electron donors and the CO and the two PPh_3 ligands are treated as three two-electron donors, providing 12 electrons in all. Because the complex is neutral overall, the Group 9 Ir atom must have a charge of $+3$ (that is, have oxidation number $+3$) to balance the charge of the three anionic ligands, and thus contributes $9 - 3 = 6$ electrons. This analysis gives a total of 18 electrons for the Ir(III) complex. (b) The η^5-C_5H_5 ligand is treated as $C_5H_5^-$ and donates six electrons, with the η^6-C_6H_6 ligand donating a further six. To maintain neutrality, the Group 6 Cr atom must have a charge of $+1$ (and an oxidation number of $+1$) and contributes $6 - 1 = 5$ electrons. The total number of metal electrons is $12 + 5 = 17$ for a Cr(I) complex. As noted before, this complex does not obey the 18-electron rule and is unlikely to be stable. (c) Each CO ligand is neutral and contributes two electrons, giving 10 electrons in all. The overall charge of the complex is -1; because all the ligands are neutral, we consider this charge to reside formally on the metal atom, giving it an oxidation number of -1. The Group 7 Mn atom thus contributes $7 + 1$ electrons, giving a total of 18 for a Mn(-1) complex.

Self-test 21.2 What is the electron count for and oxidation number of platinum in the anion of Zeise's salt, $[PtCl_3(CH_2{=}CH_2)]^-$? (Treat $CH_2{=}CH_2$ as a neutral two-electron donor.)

21.4 Nomenclature

Key point: The naming of organometallic compounds is similar to the naming of coordination compounds, but certain ligands have multiple bonding modes, which is reported as the hapticity.

According to the recommended convention, we use the same system of nomenclature for organometallic compounds as set out for coordination complexes in Section 8.6. Thus, ligands are listed in alphabetical order followed by the name of the metal, all of which is written as one word. The name of the metal should be followed by its oxidation number in parentheses. The nomenclature used in research journals, however, does not always obey these rules, and it is common to find the name of the metal buried in the middle of the name of the compound and the oxidation number omitted. For example, (**11**) is sometimes referred to as benzenemolybdenumtricarbonyl, rather than the preferred name benzene(tricarbonyl)molybdenum(0).

The IUPAC recommendation for the formula of an organometallic compound is to write it in the same form as for a coordination complex: the symbol for the metal is written first, followed by formally anionic ligands, listed in alphabetical order. The neutral ligands are then listed in alphabetical order based on their chemical symbol. We shall follow these conventions unless a different order of ligands helps to clarify a particular point.

Often a ligand with carbon donor atoms can exhibit multiple bonding modes—for instance, the cyclopentadienyl group can commonly bond to a d-metal atom in three different ways—thus, we need some additional nomenclature. Without going into the intimate details of the bonding of the various ligands (we do that later in this chapter), the extra information we need to determine a bonding mode is normally conveniently described by the number of points of attachment. This procedure gives rise to the notion of **hapticity**, η (eta), the number of ligand atoms that are considered formally to be bonded to the metal atom. The number of atoms is denoted by a superscript. For example, a CH_3 group attached by a single M—C bond is monohapto, η^1, and if the two C atoms of an ethene ligand are both within bonding distance of the metal, the ligand is dihapto, η^2. Thus, the three cyclopentadienyl complexes might be described as having η^1 (**12**), η^3 (**13**), or η^5 (**14**) cyclopentadienyl groups.

Some ligands (including the simplest of them all, the hydride ligand, H^-) can bond to more than one metal atom in the same complex, and are then referred to as **bridging ligands**. We do not require any new concepts to understand bridging ligands other than those introduced in Section 2.12. Recall from Section 8.6 that the Greek letter μ (mu) is used to indicate how many atoms the ligand bridges. Thus a μ_2-CO is a carbonyl group that bridges two metal atoms and a μ_3-CO bridges three.

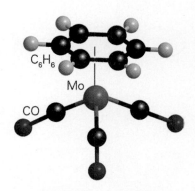

11 $[Mo(\eta^6\text{-}C_6H_6)(CO)_3]$

12 η^1-Cyclopentadienyl

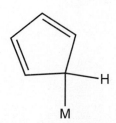

13 η^3-Cyclopentadienyl

14 η^5-Cyclopentadienyl

Ligands

A large number of ligands are found in organometallic complexes, with many different bonding modes. Because the reactivity of the metal atom and the ligands is affected by the M—L bonding, it is important to look at each ligand in some detail.

21.5 Carbon monoxide

Key point: The 3σ orbital of CO serves as a very weak donor and the π* orbitals act as acceptors.

Carbon monoxide is a very common ligand in organometallic chemistry, where it is known as the *carbonyl group*. Carbon monoxide is particularly good at stabilizing very low oxidation states, with many compounds (such as Fe(CO)$_5$) having the metal in its zero oxidation state. We looked at the molecular orbital structure of CO in Section 2.9b, and it would be sensible to review that section.

A simple picture of the bonding of CO to a metal atom is to treat the lone pair on the carbon atom as a Lewis σ base (an electron-pair donor) and the empty CO antibonding orbital as a Lewis π acid (an electron-pair acceptor), which accepts π-electron density from the filled *d* orbitals on the metal atom. In this picture, the bonding can be considered to be made up of two parts: a σ bond from the ligand to the metal atom (**15**), and a π bond from the metal atom to the ligand (**16**). This type of π bonding is sometimes referred to as **π backbonding**.

Carbon monoxide is not appreciably nucleophilic, which suggests that σ bonding to a *d*-metal atom is weak. As many *d*-metal carbonyl compounds are very stable, we can also infer that π back-bonding is strong, and the stability of carbonyl complexes arises mainly from the π-acceptor properties of CO. Further evidence for this view comes from the observation that stable carbonyl complexes exist only for metals that have filled *d* orbitals of an energy suitable for donation to the CO antibonding orbital (thus, elements in the *s* and *p* blocks do not form stable carbonyl complexes). However, the bonding of CO to a *d* metal is best regarded as a synergistic (that is, mutually enhancing) outcome of both σ and π bonding: the π backbonding from the metal to the CO increases the electron density on the CO, which in turn increases the ability of the CO to form a σ bond to the metal atom.

A more formal description of the bonding can be derived from the molecular orbital scheme for CO (Fig. 21.3), which shows that the HOMO has σ symmetry and is essentially a lobe that projects away from the C atom. When CO acts as a ligand, this 3σ orbital serves as a very weak donor to a metal atom, and forms a σ bond with the central metal atom. The LUMOs of CO are the π* orbitals. These two orbitals play a crucial role because they can overlap with metal *d* orbitals that have local π symmetry (such as the t_{2g} orbitals in an O$_h$ complex). The π interaction leads to the delocalization of electrons from filled *d* orbitals on the metal atom into the empty π* orbitals on the CO ligands, so the ligand also acts as a π acceptor.

One important consequence of this bonding scheme is the effect on the strength of the CO triple bond: the stronger the metal–carbon bond becomes through pushing electron

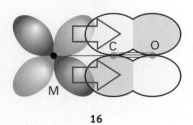

15

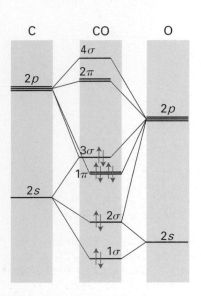

16

Fig. 21.3 The molecular orbital scheme for CO shows that the HOMO has σ symmetry and is essentially a lobe that projects away from the C atom. The LUMO has π symmetry.

density from the metal atom into the π bond, the weaker the CO bond becomes, as this electron density enters a CO antibonding orbital. In the extreme case, when two electrons are fully donated by the metal atom, a formal metal–carbon double bond is formed; because the two electrons occupy a CO antibonding orbital, this donation results in a decrease in the bond order of the CO to 2.

In practice, the bonding is somewhere between M—C≡O, with no backbonding, and M=C=O, with backbonding. Infrared spectroscopy is a very convenient method of assessing the extent of π bonding; the CO stretch is clearly identifiable as it is both strong and normally clear of all other absorptions. In CO gas, the absorption for the triple bond is at 2143 cm^{-1}, whereas a typical metal carbonyl complex has a stretching mode in the range 2100–1700 cm^{-1} (Table 21.3). The number of IR absorptions that occur in a particular carbonyl compound is discussed in Section 21.18h.

Carbonyl stretching frequencies are often used to determine the order of acceptor or donor strengths for the other ligands present in a complex. The basis of the approach is that the CO stretching frequency is decreased when it serves as a π acceptor. However, as other π acceptors in the same complex compete for the d electrons of the metal atom, they cause the CO frequency to increase. This behaviour is opposite to that observed with donor ligands, which cause the CO stretching frequency to decrease as they supply electrons to the metal atom and hence, indirectly, to the CO π* orbitals. Thus, strong σ-donor ligands attached to a metal carbonyl and a formal negative charge on a metal carbonyl anion both result in slightly greater CO bond lengths and significantly lower CO stretching frequencies.

Carbon monoxide is versatile as a ligand because, as well as the bonding mode described above (often referred to as 'terminal'), it can bridge two (**17**) or three (**18**) metal atoms. Although the description of the bonding is now more complicated, the concepts of σ-donor and π-acceptor ligands remain useful. The CO stretching frequencies generally follow the order MCO > M₂CO > M₃CO, which suggests an increasing occupation of the π* orbital as the CO molecule bonds to more metal atoms and more electron density from the metals enters the CO π* orbitals. As a rule of thumb, carbonyls bridging two metal atoms typically have stretching bands between 1900 and 1750 cm^{-1}, and those that bridge three atoms have stretching bands between 1800 and 1600 cm^{-1} (Fig. 21.4). We consider carbon monoxide to be a two-electron neutral ligand when terminal or bridging (thus a carbonyl bridging two metals can be considered to give one electron to each).

A further bonding mode for CO that is sometimes observed is when the CO is terminally bound to one metal atom and the CO triple bond binds side-on to another metal (**19**). This description is best considered as two separate bonding interactions, with the terminal interaction being the same as that described above and the side-on bonding being essentially identical to that of other side-on π donors such as alkynes and N₂, which are discussed later.

The synthesis, properties, and reactivities of compounds containing the carbonyl ligand are discussed in more detail in Section 21.18.

21.6 Phosphines

Key point: Phosphines bond to metals by a combination of σ donation from the P atom and π back-bonding from the metal atom.

Although phosphines are not organometallic, because they do not bond to metals through a carbon atom, phosphine ligands are best discussed here as their bonding has many similarities to that of carbon monoxide.

Phosphine, PH₃ (formally, phosphane), is a reactive, noxious, poisonous, and flammable gas (Section 14.6). Like ammonia, phosphine can behave as a Lewis base and use its lone pair to donate electron density to a Lewis acid, and so act as a ligand. However, given the problems associated with handling phosphine, it is rarely used as a ligand. By contrast, substituted phosphines, such as trialkylphosphines (for example, PMe₃, PEt₃), or triarylphosphines (for example, PPh₃) (**10**), or trialkyl- or triarylphosphites (for example, P(OMe)₃, P(OPh)₃ (**20**)) and a whole host of bridged multidentate di- and triphosphines (for example, Ph₂PCH₂CH₂PPh₂ = dppe (**21**)) are easy to handle (indeed some are

Table 21.3 The influence of coordination and charge on CO stretching wavenumbers	
Compound	$\tilde{\nu}/\mathrm{cm}^{-1}$
CO	2143
[Mn(CO)₆]⁺	2090
Cr(CO)₆	2000
[V(CO)₆]⁻	1860
[Ti(CO)₆]²⁻	1750

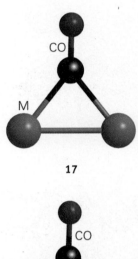

17

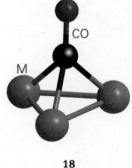

18

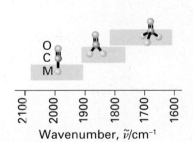

Fig. 21.4 Approximate ranges for CO stretching bands in neutral metal carbonyls. Note that high wavenumbers (and hence high frequencies) are on the left, in keeping with the way infrared spectra are generally plotted.

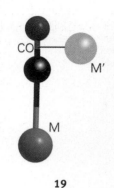

19

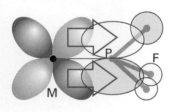

20 P(OPh)$_3$

21 Ph$_2$PCH$_2$CH$_2$PPh$_2$, dppe

22

23 2,2′-bis(diphenylphosphino)-
1,1′-binaphthyl, BINAP

air-stable odourless solids with no appreciable toxicity) and are widely used as ligands; all are colloquially referred to as 'phosphines'.

Phosphines have a lone pair on the P atom that is appreciably basic and nucleophilic and can serve as a σ donor. Phosphines also have empty orbitals on the P atom that can overlap with filled d orbitals on $3d$-metal ions and behave as π acceptors (**22**). The bonding of phosphines to a d-metal atom, made up of a σ bond from the ligand to the metal and a π bond from the metal back to the ligand, is completely analogous to the bonding of CO to a d-metal atom.

As mentioned above, a huge variety of phosphines are both possible and widely available, including chiral systems such as 2,2′-bis(diphenylphosphino)-1,1′-binaphthyl (BINAP) (**23**), in which steric constraints result in compounds that can be resolved into diastereomers. Generally there are two properties of phosphine ligands that are considered important in discussions of the reactivity of their complexes: their steric bulk, and their electron donating (and accepting) ability.

As we saw in Section 20.7c, the steric bulk of phosphines can be expressed in terms of the notional cone occupied by the bonded ligand. Table 21.4 lists some of the derived cone angles. The bonding of phosphines to d-metal atoms is, as we have seen, a composite of σ bonding from the ligand to the metal atom, and π backbonding from the metal atom to the ligand. The σ-donating ability and π-acceptor ability of phosphines are inversely correlated in the sense that electron-rich phosphines, such as PMe$_3$, are good σ donors and poor π acceptors, whereas electron-poor phosphines, such as PF$_3$, are poor σ donors and good π acceptors. Thus Lewis basicity can normally be used as a single scale to indicate their donor/acceptor ability. The generally accepted order of basicity of phosphines is:

$$PCy_3 > PEt_3 > PMe_3 > PPh_3 > P(OMe)_3 > P(OPh)_3 > PCl_3 > PF_3$$

and is easily understood when the electronegativity of the substituents on the P atom is considered. The basicity of a phosphine is not simply related to the strength of the M—P bond in a complex: for instance, an electron-poor metal atom forms a stronger bond with an electron-rich (basic) phosphine, whereas an electron-rich metal atom will form a stronger bond with an electron-poor phosphine.

If there are carbonyl ligands present in a metal–phosphine complex, then the carbonyl stretching frequency can be used to give an insight into the basicity of the phosphine ligand: this method allows us to conclude that PF$_3$ is as good a π acceptor as CO.

The vast range of phosphines that are commonly used in organometallic chemistry is a testament to their versatility as ligands: judicious choice allows control over both steric and electronic properties of the metal atom in a complex. Phosphorus-31 (which occurs in 100 per cent natural abundance) is easy to observe by NMR and both the ^{31}P chemical shift and the coupling constant to the metal atom (where appropriate) give considerable insight into the bonding and reactivity of a complex. Like carbonyls, phosphines can bridge either two or three metals atoms, providing additional variety in bonding modes.

Example 21.4 Interpreting carbonyl stretching frequencies and phosphine complexes

(a) Which of the two isoelectronic compounds Cr(CO)$_6$ and [V(CO)$_6$]$^-$ will have the higher CO stretching frequency? (b) Which of the two chromium compounds [Cr(CO)$_5$(PEt$_3$)] and [Cr(CO)$_5$(PPh$_3$)] will have the lower CO stretching frequency? Which will have the shorter M—C bond?

Answer (a) The negative charge on the V complex will result in greater π backbonding to the CO π* orbitals, compared with the Cr complex. This backbonding results in a weakening of the CO bond, with a corresponding decrease in stretching frequency. Thus the Cr complex has the higher CO stretching frequency. (b) PEt$_3$ is more basic than PPh$_3$ and thus the PEt$_3$ complex will have greater electron density on the metal atom than the PPh$_3$ complex. The greater electron density will result in greater backbonding and thus both a lower CO stretching frequency and a shorter M—C bond.

Self-test 21.4 Which of the two iron compounds Fe(CO)$_5$ and [Fe(CO)$_4$(PEt$_3$)] will have the higher CO stretching frequency? Which will have the longer M—C bond?

21.7 Hydrides and dihydrogen complexes

Key points: The bonding of a hydrogen atom to a metal atom is a σ interaction, whereas the bonding of a dihydrogen ligand involves π backbonding.

A hydrogen atom directly bonded to a metal is commonly found in organometallic complexes, and is referred to as a **hydride ligand**. The name 'hydride' can be misleading as it implies a H^- ligand. Whereas the formulation H^- might be appropriate for most hydrides, such as $[CoH(PMe_3)_4]$, some hydrides are appreciably acidic and behave as H^+, for example $[CoH(CO)_4]$ has $pK_a = 8.3$. The acidity of organometallic carbonyls is described in Section 21.18e.

The bonding of a hydrogen atom to a metal atom is simple because the only orbital of appropriate energy for bonding on the hydrogen is H1s and the M—H bond can be considered solely on the basis of a σ interaction between the two atoms. Hydrides are readily identified by NMR spectroscopy as their chemical shift is rather unusual, typically occurring in the range $-50 < \delta < 0$. Infrared spectroscopy can also be useful in identifying metal hydrides as they normally have a stretching band in the range $2850 - 2250\ cm^{-1}$. X-ray diffraction, normally so valuable for identifying the structure of crystalline materials, is of little use in identifying hydrides because the diffraction is related to electron density, and the hydride ligand will have at most two electrons around it, compared with, for instance, 78 for a platinum. Neutron diffraction is of more use in locating hydride ligands, especially if the hydrogen atom is replaced by a deuterium atom, because deuterium has a large neutron scattering cross-section.

An M—H bond can sometimes be produced by protonation of an organometallic compound, such as neutral and anionic metal carbonyls (Section 21.18e). For example, ferrocene can be protonated in strong acid to produce an Fe—H bond:

$$Cp_2Fe + HBF_4 \rightarrow [Cp_2FeH]^+[BF_4]^-$$

Bridging hydrides exist, where an H atom bridges either two or three metal atoms: here the bonding can be treated in exactly the same way we considered bridging hydrides in diborane, B_2H_6 (Section 2.12).

In the donor-pair method of electron counting, we consider the hydride ligand to contribute two electrons and to have a single negative charge (that is, to be H^-).

Although the first organometallic metal hydride was reported in 1931, complexes of hydrogen gas, H_2, were identified only in 1984. In such compounds, the dihydrogen molecule, H_2, bonds side-on to the metal atom (in the older literature, such compounds were sometimes called *non-classical hydrides*). The bonding of dihydrogen to the metal atom is considered to be made up of two components: a σ donation of an electron in the H_2 bond to the metal atom (**24**) and a π backdonation from the metal to the σ^* H—H antibonding orbital (**25**). This picture of the bonding raises a number of interesting issues. In particular, as the π backbonding from the metal increases, the strength of the H—H bond decreases and the structure tends to that of a dihydride:

It is now recognized that complexes exist with structures at all points between these two extremes, and in some cases an equilibrium can be identified between the two. Work by G. Kubas on tungsten complexes used the H—D coupling constant to show that it is possible to detect both the dihydrogen complex (**26**, $^1J_{HD} = 34\ Hz$) and the dihydride (**27**, $^2J_{HD} < 2\ Hz$), and to follow the conversion from one to the other.

The reaction of dihydrogen with a metal to give a dihydride is an example of an *oxidative addition* reaction (Section 21.22). Certain microbes contain enzymes known as *hydrogenases* that use Fe and Ni at their catalytic centres to catalyse the rapid oxidation of H_2 and reduction of H^+, via intermediate metal–dihydrogen and hydride species (Section 26.14).

Table 21.4 Tolman cone angles (in degrees) for selected phosphines

PF_3	104
$P(OMe)_3$	107
PMe_3	118
PCl_3	125
$P(OPh)_3$	127
PEt_3	132
PPh_3	145
PCy_3	169
P^tBu_3	182
$P(o\text{-}tol)_3$	193

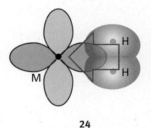

24

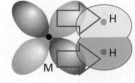

25

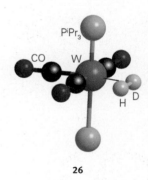

26

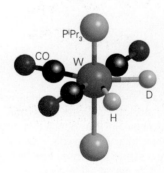

27

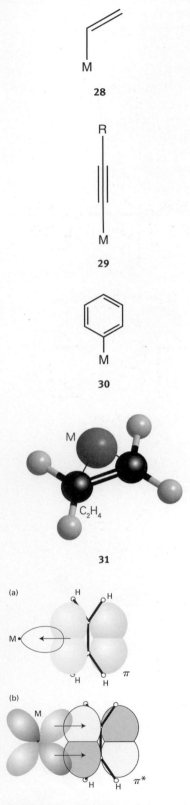

28

29

30

C_2H_4

31

(a)

(b)

Fig. 21.5 The interaction of ethene with a metal atom. (a) Donation of electron density from the filled π molecular orbital of ethene to a vacant metal σ orbital. (b) Acceptance of electron density from a filled $d\pi$ orbital into the vacant π^* orbital of ethene.

21.8 η^1-Alkyl, -alkenyl, -alkynyl, and -aryl ligands

Key point: The metal–ligand bonding of η^1-hydrocarbon ligands is a σ interaction.

Alkyl groups are often found as ligands in *d*-metal organometallic chemistry, and their bonding presents no new features: it is best considered a simple covalent σ interaction between the metal atom and the carbon atom of the organic fragment. Alkyl groups with a hydrogen atom on a carbon atom adjacent to the one that bonds to the metal are prone to decompose by a process known as β-*hydrogen elimination* (Section 21.25) and hence those alkyl groups that cannot react in this fashion, such as methyl, benzyl, $(CH_2C_6H_5)$, *t*-butyl, and trimethylsilylmethyl (CH_2SiMe_3), are more stable than those that can, such as ethyl.

Alkenyl (**28**), alkynyl (**29**), and aryl (**30**) groups can bond to a metal atom in a similar fashion, binding to the metal atom through a single carbon atom, and hence are described as monohapto (η^1). Although there is potential for each of these three groups to accept π electron density into antibonding orbitals, there is little evidence that this happens. For instance, even though an η^1-alkynyl group might be considered analogous to a CO group, the stretching frequency of the triple bond in alkynyl complexes changes little on attachment to a metal. Bridging alkyl and aryl groups also exist, and the bonding can be considered in the same way as we have considered other bridging ligands, with 3*c*,2*e* bonds.

Alkyl, alkenyl, alkynyl, and aryl groups are commonly introduced into organometallic complexes by the displacement of a halide at a metal centre with a lithium or Grignard reagent. For example:

$$\text{Cl}_{\textit{////}}\text{Pd}^{\textbackslash\textbackslash\textbackslash}\text{PPh}_3 \quad \xrightarrow{\text{2 PhLi}} \quad \text{Ph}_{\textit{////}}\text{Pd}^{\textbackslash\textbackslash\textbackslash}\text{PPh}_3 \quad + \ 2\ \text{LiCl}$$

We consider alkyl, alkenyl, alkynyl, and aryl ligands to be two-electron donors with a single negative charge (for example, Me^-, Ph^-) in the donor-pair scheme of electron counting.

21.9 η^2-Alkene and -alkyne ligands

Key point: The bonding of an alkene or an alkyne to a metal atom is best described as a σ interaction from the multiple bond to the metal atom, with a π backbonding interaction from the metal atom to the π^* antibonding orbital on the alkene or alkyne.

Alkenes are routinely found bound to metal centres: the first organometallic compound isolated, Zeise's salt (**1**), was a complex of ethene. Alkenes normally bond side-on to a metal atom with both carbon atoms of the double bond equidistant from the metal with the other groups on the alkene approximately perpendicular to the plane of the metal atom and the two carbon atoms (**31**). In this arrangement, the electron density of the C=C π bond can be donated to an empty orbital on the metal atom to form a σ bond. In parallel with this interaction, a filled metal *d* orbital can donate electron density back to the empty π^* orbitals of the alkene to form a π bond. This description is called the **Dewar–Chatt–Duncanson model** (Fig. 21.5) and we consider η^2-alkenes to be two-electron neutral ligands.

Electron donor and acceptor character appear to be fairly evenly balanced in most ethene complexes of the *d* metals, but the degree of donation and backdonation can be altered by substituents on the metal atom and on the alkene. When the π backbonding from the metal atom increases, the strength of the C=C bond decreases as the electron density is located in the C=C antibonding orbital and the structure tends to that of a C—C singly bonded structure, known as a metallocylopropane:

Dihaptoalkenes with only a small degree of electron donation from the metal have their substituents bent slightly away from the metal atom, and the C=C bond length is only slightly greater than in the free alkene. When the degree of backdonation is greater, substituents on the alkene are bent away more from the metal atom and the C—C bond length approaches that of a single C—C bond. Steric constraints can also force the other groups on the alkene to bend away from the metal atom.

Alkynes have two π bonds and hence the potential to be four-electron donors. When side-on to a single metal atom, the η^2-carbon–carbon triple bond is best considered as a two-electron donor, with the π* orbitals accepting electron density from a metal atom in the same way as for alkenes. When strongly electron-withdrawing groups are attached to an alkyne, the ligand can become an excellent π acceptor and displace other ligands such as phosphines; the compound commonly known as dimethylacetylenedicarboxylate, $CH_3OCOC≡CCO_2CH_3$, is a good example.

Substituted alkynes can form very stable polymetallic complexes in which the alkyne can be regarded as a four-electron donor. An example is η^2-diphenylethyne(hexacarbonyl)dicobalt(0), in which we can view one π bond as donating to one of the Co atoms and the second π bond as overlapping with the other Co atom (**32**). In this example, the alkyl or aryl groups present on the alkyne impart stability by lowering the tendency towards secondary reactions of the coordinated ethyne, such as loss of the slightly acidic ethynic H atom to the metal atom.

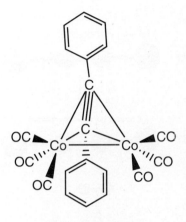

32 $[Co_2(PhC≡CPh)(CO)_6]$

21.10 Nonconjugated diene and polyene ligands

Key point: The bonding of nonconjugated alkenes to a metal atom is best described as independent multiple alkenes bonding to a metal centre.

Nonconjugated diene (—C=C—X—C=C—) and polyene ligands can also bond to metal atoms. It is simplest to consider them as linked alkenes, and hence they present no new bonding concepts. As with the chelate effect in coordination complexes (Section 20.1c), the resulting polyene complexes are usually more stable than the equivalent complex with individual ligands because the entropy of dissociation of the complex is much smaller than when the liberated ligands can move independently. For example, bis(η^4-cycloocta-1,5-diene)nickel(0) (**33**) is more stable than the corresponding complex containing four ethene ligands. Cycloocta-1,5-diene (**34**) is a fairly common ligand in organometallic chemistry, where it is referred to engagingly as 'cod', and is normally introduced into the metal coordination sphere by simple ligand displacement reactions. An example is:

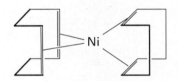

33 Ni(cod)$_2$

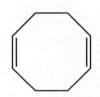

34 Cycloocta-1,5-diene, cod

Metal complexes of cod are commonly used as starting materials because they often have intermediate stability. Many of them are sufficiently stable to be isolated and handled, but cod can be displaced by many stronger ligands. For example, if the highly toxic $Ni(CO)_4$ molecule is needed in a reaction, then it may be generated from Ni(cod)$_2$ directly in the reaction flask:

$$[Ni(cod)_2](soln) + 4\ CO(g) \rightarrow Ni(CO)_4(soln) + 2\ cod(soln)$$

21.11 Dinitrogen and nitrogen monoxide

Key points: The bonding of dinitrogen to a metal atom is weak but contains both a σ-donating and a π-accepting component; nitrogen monoxide can bind to a metal in two different ways—either bent or straight.

Neither dinitrogen, N_2, nor nitrogen monoxide, NO, is strictly an organometallic ligand, although they are sometimes found in organometallic compounds. Dinitrogen is a much

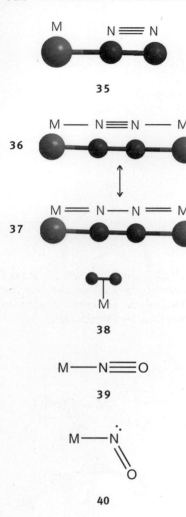

35

36

37

38

M—N≡≡O

39

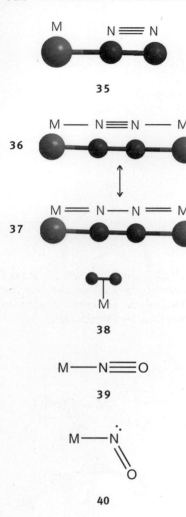

40

sought after ligand as complexes can potentially take part in a catalytic reduction of nitrogen to more useful species. Dinitrogen can bind to metals in a number of different ways. The majority of complexes have a terminal monohapto link, η^1-N_2, in which the bonding can be considered to be like that of the isoelectronic CO ligand (**35**). Dinitrogen is both a weaker σ donor and a weaker π acceptor than CO, and hence is bound less strongly; in fact only good π-donor metal atoms bind N_2. Like CO, the N_2 ligand has a distinctive IR stretching band lying between 2150 and 1900 cm^{-1}.

A dinitrogen molecule can participate in two bonding interactions and bridge two metal atoms (**36**). If the backdonation to the nitrogen is extensive in this kind of complex, it can formally be considered to have been reduced to a hydrazine (**37**). Occasionally, dinitrogen ligands are found bound in a dihapto (η^2) side-on fashion (**38**). In these complexes the ligand is best considered analogous to an η^2-alkyne. The side-on bonding mode seems to be particularly common in complexes of the *f* metals.

Nitrogen monoxide (nitric oxide) is a radical with 11 valence electrons. When bound, NO is referred to as the *nitrosyl ligand* and can bond in one of two modes to *d*-metal atoms, in either a bent or a straight fashion. In the straight arrangement (**39**), the ligand is considered to be the NO$^+$ cation. The NO$^+$ cation is isoelectronic with CO, and the bonding can be considered in a similar fashion (a two-electron σ donor, with strong π-acceptor ability). In the bent arrangement (**40**), NO is considered to behave as NO$^-$, again donating two electrons. In many complexes, NO can change its coordination mode; in effect, moving from the straight to the bent mode reduces the number of electrons on the metal by 2.

21.12 Butadiene, cyclobutadiene, and cyclooctatetraene

Key points: Some insight into the bonding of butadiene and cyclobutadiene can be gained by treating them as containing two alkene units, but a full understanding needs a consideration of the molecular orbitals. Cyclooctatetraene bonds in a number of different ways; the most common mode in *d*-metal chemistry is as an η^4-donor, analogous to butadiene.

The temptation with both butadiene and cyclobutadiene is to treat them like two isolated double bonds. However, a proper molecular orbital approach is necessary to understand the bonding fully because the ligand–metal atom interactions are different in the two cases.

Figure 21.6 shows the molecular orbitals for the π system in butadiene. The two occupied lower energy MOs can behave as donors to the metal, the lowest a σ donor and the next a π donor. The next higher unoccupied MO, the LUMO, can act as a π acceptor from the metal atom. Thus, attachment of a butadiene molecule to a metal atom results

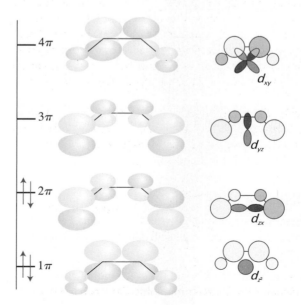

Fig. 21.6 The molecular orbitals of the π system in butadiene; also shown are metal *d* orbitals of appropriate symmetry to form bonding interactions.

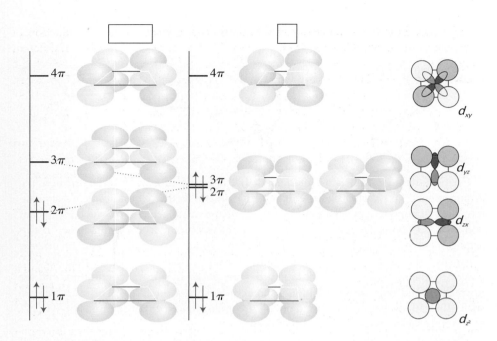

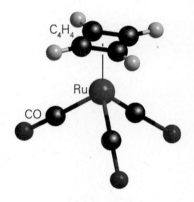

Fig. 21.7 The molecular orbitals of the π system in cyclobutadiene; also shown are metal d orbitals of appropriate symmetry to form bonding interactions.

in population of an MO that is bonding between the two central C atoms (which are already nominally singly bonded) and antibonding between the nominally doubly bonded C atoms. The resulting modification of electron density results in a shortening of the central C—C bond and a lengthening of the double C—C bonds. In theory, a δ-bonding interaction is possible between the d_{xy}-metal atomic orbital and the most antibonding of the butadiene MOs, but there is no definitive evidence that it occurs. Butadiene is therefore considered to be a four-electron neutral ligand in our electron counting scheme.

Cyclobutadiene is rectangular (D_{2h}) and unstable as a free molecule, as a result of bond angle constraints and its four-electron anti-aromatic configuration. However, stable complexes are known, including $[Ru(CO)_3(\eta^4\text{-}C_4H_4)]$ (**41**). This species is one of many in which coordination to a metal atom stabilizes an otherwise unstable molecule.

As a result of a distortion from the square arrangement,[2] cyclobutadiene has a similar MO diagram to butadiene (Fig. 21.7). Population of the LUMO by backbonding now leads to greater bonding on the long sides of the rectangular cyclobutadiene molecule, tending towards a square (D_{4h}) arrangement. If the cyclobutadiene formally accepts two electrons from the metal atom, it will have six π electrons with three MOs occupied and no incentive to distort from square; two of the MOs are degenerate in this aromatic configuration. Because all cyclobutadiene metal complexes are found to be square, this six-electron aromatic configuration must represent the bonding. This view of the bonding has led some to regard cyclobutadiene complexes as complexes of the $R_4C_4^{2-}$ dianion (a six-electron donor), although most regard cyclobutadiene as a four-electron neutral ligand. Once again, a δ-bonding interaction is possible between the d_{xy}-metal orbital and the most antibonding of the butadiene MOs, but again there seems to be no definitive evidence that it occurs.

Because cyclobutadiene is unstable, the ligand must be generated in the presence of the metal to which it is to be coordinated. This synthesis can be accomplished in a variety of ways. One method is the dehalogenation of a halogenated cyclobutene:

41 $[Ru(C_4H_4)](CO)_3$

---

[2] We can regard this distortion as an organic example of the Jahn–Teller effect (Section 19.1h).

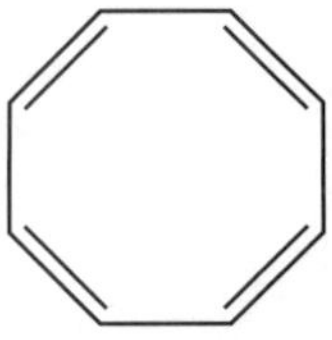

42 Cyclooctatetraene

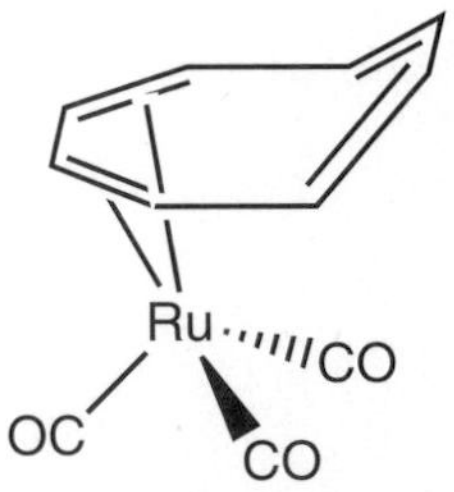

43 [Ru(η⁴-C₈H₈)(CO)₃]

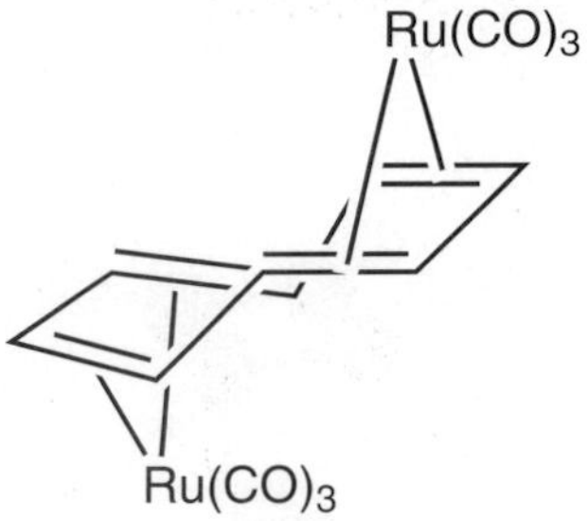

Ru(CO)₃

Ru(CO)₃

44

Cp

Co

Co

Cp

45

C₆H₆

Cr

46

Another procedure is the dimerization of a substituted ethyne:

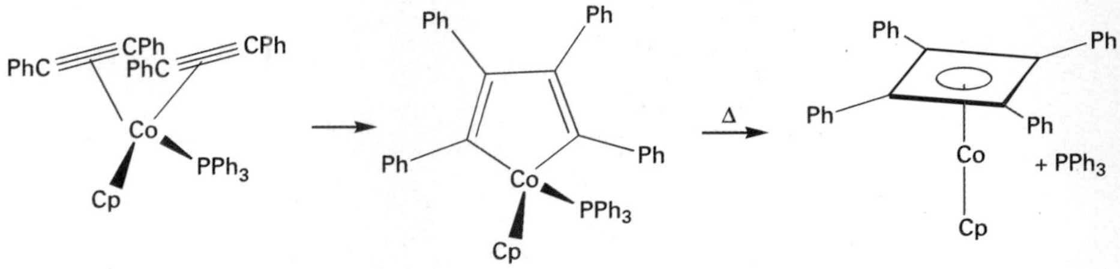

Cyclooctatetraene (**42**) is a large ligand that is found in a wide variety of bonding arrangements. Like cyclobutadiene, it is anti-aromatic as the free molecule. Cyclooctatetraene can bond to a metal atom in an η⁸-octahapto arrangement, in which it is planar and all C—C bond lengths are equal. In a similar fashion to cyclobutadiene, in this arrangement cyclooctatetraene is considered to extract two electrons and to become (formally) the aromatic dinegative ligand $[C_8H_8]^{2-}$ (that is, a 10-electron donor). Cyclooctetraenes bound in this fashion are rarely found in *d*-metal compounds and are normally found only with lanthanoids and actinoids (Section 22.10). For example, two such dinegative ligands are found in bis(η⁸-cyclooctatetraenyl)uranium(IV), [U(η⁸-C₈H₈)₂], commonly referred to as uranocene.

The more usual bonding modes for cyclooctetraenes in *d*-metal compounds are as puckered η⁴-C₈H₈ ligands such as (**43**), where the bonding part of the ligand can be treated as a butadiene. Bridging modes, (**44**) and (**45**), are also possible.

21.13 Benzene and other arenes

Key point: A consideration of the MOs of benzene leads to a picture of bonding of benzene to a metal atom that includes a significant δ backbonding interaction.

If benzene is considered to have three localized double bonds, each double bond can behave as a ligand and the molecule could behave as a tridentate η⁶-ligand. A compound such as bis(η⁶-benzene)chromium (**46**) could then be considered to be made up of six coordinated double bonds, each donating two electrons, bound to a d^6 metal atom, giving a total of 18 valence electrons for the octahedral complex. Bis(η⁶-benzene)chromium does exist, and is remarkably stable (it can be handled in air, and sublimes with no decomposition). Although this simplistic view of the bonding is a first step towards understanding its structure, the true picture needs a deeper consideration of the molecular orbitals involved.

In the molecular orbital picture of the π bonding in benzene there are three bonding and three antibonding orbitals. If we consider a single benzene molecule bonding to a single metal, and consider only the *d* orbitals, the strongest interaction is a σ interaction between the most strongly bonding a_1 benzene MO and the d_{z^2} metal orbital; π bonds are possible between the other bonding MOs on benzene and the d_{zx} and d_{yz} orbitals. Backbonding from the metal atom to the benzene is possible only as a δ interaction between both of the $d_{x^2-y^2}$ and d_{xy} metal orbitals and the antibonding e_2 orbitals on benzene. Figure 21.8 shows these interactions.

Hexahapto (η⁶) arene complexes are very easy to make, often simply by refluxing a compound that has three replaceable ligands in an arene:

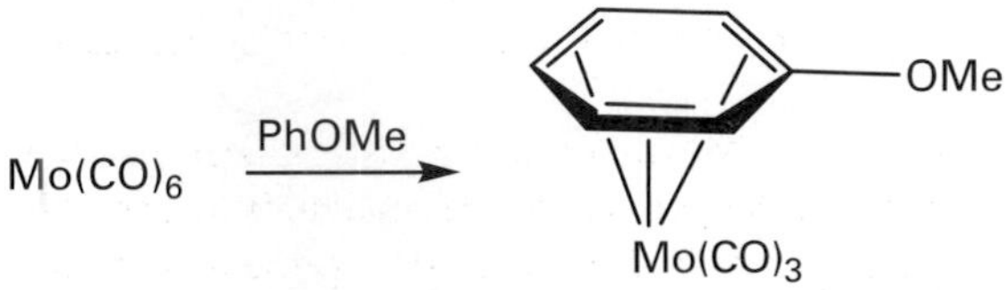

$$Mo(CO)_6 \xrightarrow{PhOMe} \text{(arene)Mo(CO)}_3$$

η⁶-Arenes are considered to be neutral ligands that donate six electrons and, sterically, they are normally considered to take up three coordination sites at a metal.

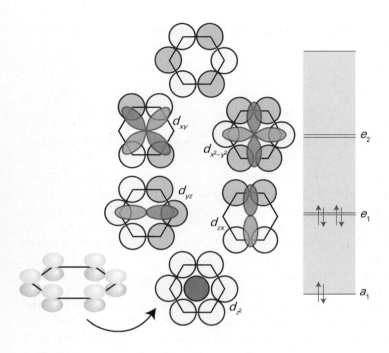

Fig. 21.8 The molecular orbitals of the π system in benzene; also shown are metal d orbitals of appropriate symmetry to form bonding interactions.

One commonly invoked reaction intermediate of η^6-arene complexes is a 'slipping' to an η^4 complex, which donates only four electrons to the metal, and therefore allows a substitution reaction to proceed without an initial ligand loss:

$$\eta^6\text{-arene} \quad \xrightarrow{+\,PR_3} \quad \eta^4\text{-arene} \quad \xrightarrow{-\,CO} \quad \eta^6\text{-arene}$$

η^2-Arenes are also known and are analogous to η^2-alkenes; they have an important role in the activation of arenes by metal complexes.

21.14 The allyl ligand

Key points: A consideration of the MOs of η^3-allyl complexes leads to a picture that gives two identical C—C bond lengths; because the type of bonding of the allyl ligand is so variable, η^3-allyl complexes are often highly reactive.

The allyl ligand, $CH_2{=}CH{-}CH_2^-$, can bind to a metal atom in either of two con-figurations. As an η^1-ligand (**47**) it should be considered just like an η^1-alkyl group (that is, as a two-electron donor with a single negative charge). However, the allyl ligand can also use its double bond as an additional two-electron donor and act as a η^3-ligand (**48**) (a four-electron donor with a single negative charge). The η^3-allyl ligand can be thought of as a resonance between two forms (**49**), and because all evidence points towards a symmetric structure, it is often depicted with a dotted line representing all the bonding electrons (**50**).

As for benzene, a more detailed understanding of the bonding of an allyl group needs a consideration of the molecular orbitals of the organic fragment (Fig. 21.9), whereupon it becomes apparent that the totally symmetric arrangement is the correct form.

The terminal substituents of an η^3-allyl group are bent slightly out of the plane of the three-carbon backbone and are either *syn* (**51**) or *ante* (**52**) relative to the central hydrogen. It is common to observe *anti* and *syn* groups exchange, which in some cases

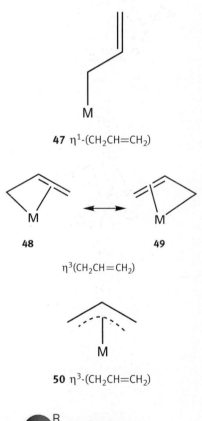

47 η^1-$(CH_2CH{=}CH_2)$

48　　**49**

$\eta^3(CH_2CH{=}CH_2)$

50 η^3-$(CH_2CH{=}CH_2)$

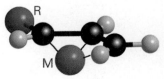

51 *syn*

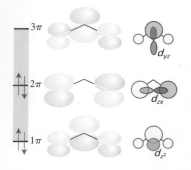

Fig. 21.9 The molecular orbitals of the π system of the allyl (CH$_2$CHCH$_2^-$) group; also shown are metal d orbitals of appropriate symmetry to form bonding interactions.

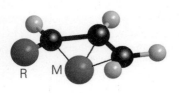

52 *anti*

is fast on an NMR timescale. A mechanism that involves the transformation η^3 to η^1 to η^3 is often invoked to explain this exchange.

Because of this flexibility in the bonding, η^3-allyl complexes are often highly reactive as they can readily bind another ligand in the η^1-form.

There are many routes to allyl complexes. One is the nucleophilic attack of an allyl Grignard reagent on a metal halide:

$$2\ C_3H_5MgBr + NiCl_2 \xrightarrow{\text{Ether}} [Ni(\eta^3\text{-}C_3H_5)_2] + 2\ MgBrCl$$

Nucleophilic attack on a haloalkane by a metal atom in a low oxidation state also yields allyl complexes:

$$[Mn(CO)_5]^- + CH_2{=}CHCH_2Cl \longrightarrow$$

Assuming the metal centre is not protonated directly, the protonation of a butadiene complex leads to an η^3-allyl complex:

21.15 Cyclopentadiene and cycloheptatriene

Key points: The common η^5-bonding mode of a cyclopentadienyl ligand can be understood on the basis of both σ and π donation from the organic fragment to the metal in conjunction with δ backbonding; cycloheptatriene commonly forms either η^6-complexes or η^7-complexes of the aromatic tropilium cation (C$_7$H$_7$)$^+$.

Cyclopentadiene, C_5H_6, is a mildly acidic hydrocarbon that can be deprotonated to form the cyclopentadienyl anion, $C_5H_5^-$. The stability of the cyclopentadienyl anion can be understood when it is realized that the six electrons in its π system make it aromatic. The delocalization of these six electrons results in a ring structure with five equal bond lengths. As a ligand, the cyclopentadienyl group, commonly denoted Cp, has played a major role in the development of organometallic chemistry and continues to be the archetype of cyclic polyene ligands. We have already alluded to the role ferrocene (**4**) had in the development of organometallic chemistry. A huge number of metal cyclopentadienyl and substituted cyclopentadienyl compounds are known. Some compounds have $C_5H_5^-$ as a monohapto ligand, in which case it is treated like an η^1-alkyl; others contain $C_5H_5^-$ as a trihapto ligand, in which case it is treated like an η^3-allyl group. Usually, though, $C_5H_5^-$ is present as a pentahapto ligand, bound through all five carbons of its ring.

We treat the η^5-$C_5H_5^-$ group as a six-electron donor. Formally, the electron donation to the metal now comes from the filled a_1 (σ bonding) and e_1 (π bonding) MOs (Fig. 21.10) with δ backbonding from the d_{xy} and $d_{x^2-y^2}$ e_2 orbitals on the metal atom. As we shall see in Section 21.19, coordinated Cp ligands behave as though they maintain their six-electron aromatic structure.

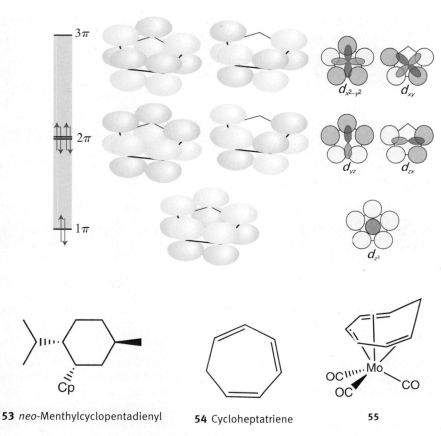

53 *neo*-Menthylcyclopentadienyl   **54** Cycloheptatriene   **55**   **56**

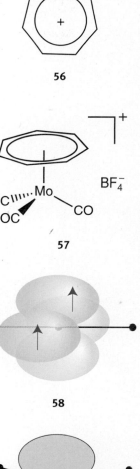

The electronic and steric properties of cyclopentadiene can easily be tuned: electron withdrawing and donating groups can be attached to the five-membered ring, and steric bulk can be enhanced by additional substitution. The pentamethylcyclopentadienyl ligand (Cp*) is commonly used to provide greater electron density on, and greater steric protection of, a metal atom. Chiral groups are often added to Cp groups so that complexes can be used in stereoselective reactions: the neo-menthyl group (**53**) is commonly used.

The synthesis, properties, and reactivities of compounds containing the cyclopentadienyl ligand are discussed in more detail in Section 21.19.

Cycloheptatriene, C_7H_8 (**54**), can form η^6-complexes such as (**55**), which may be treated as having three η^2-alkene molecules bound to the metal atom. Hydride abstraction from these complexes results in formation of η^7-complexes of the six-electron aromatic tropilium cation $(C_7H_7)^+$ (**56**), for example (**57**). In η^7-tropilium complexes, all carbon–carbon bond lengths are equal; bonding to the metal atom and backbonding from the metal are similar to those found in arene and cyclopentadienyl complexes.

21.16 Carbenes

Key points: Fischer- and Schrock-type carbene complexes are considered to have a metal–carbon double bond and substantial π backbonding from the metal, whereas *N*-heterocyclic carbenes are considered to have a metal–carbon single bond and minimal π backbonding.

Carbene, CH_2, has only six electrons around its C atom and is consequently highly reactive. Other substituted carbenes exist and are substantially less reactive and can behave as ligands towards metals.

In principle, carbenes can exist in one of two electronic configurations: with a linear arrangement of the two groups bound to the carbon atom and the two remaining electrons unpaired in two p orbitals (**58**), or with the two groups bent and the two remaining electrons paired and an empty p orbital (**59**). Carbenes with a linear

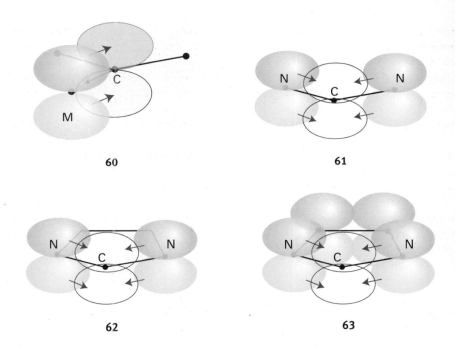

60   61

62   63

arrangement of groups are referred to as 'triplet carbenes' (because the two unpaired electrons are unpaired and $S = 1$) and are favoured when sterically very bulky groups are attached to the carbene carbon. Carbenes with the bent arrangement are known as 'singlet carbenes' (the two electrons are paired and $S = 0$) and are the normal form for carbenes. The electron pair on the carbon of a singlet carbene is suitable to bond to a metal atom, resulting in a ligand-to-metal bond. The empty *p* orbital on the carbon atom can then accept electron density from the metal atom, thus stabilizing the electron-poor carbon (**60**). For historical reasons, carbenes bonded to metal atoms in this fashion are known as **Fischer carbenes**, and are represented by a metal–carbon double bond. Fischer carbenes are electron deficient at the C atom and consequently are easily attacked by nucleophiles. When the backbonding to the C atom is very strong, the carbene can become electron rich and is thus prone to attack by electrophiles. Carbenes of this type are known as **Schrock carbenes**, after their discoverer. The term **alkylidene** technically refers only to carbenes with alkyl substituents (CR_2), but is sometimes used to mean both Fischer and Schrock carbenes.

More recently, a large number of derivatives of what are known as **N-heterocyclic carbenes** (NHCs) have reportedly been used as ligands. In an NHC, two nitrogen atoms are adjacent to the carbene carbon atom, and if the lone pair on the nitrogen is considered to be largely *p*-orbital based, then strong π-donor interactions from the two nitrogen atoms can stabilize the carbene (**61**). Tying the carbene carbon and the two nitrogens into a ring helps to stabilize the carbene, and five-membered rings are common (**62**). Additional stability can be achieved by a double bond in the ring, which provides an additional two electrons that may be considered part of a six-electron aromatic resonance structure (**63**). The bonding of an NHC ligand to a metal is currently considered to be largely a two-electron σ donation from the carbon, with minimal π backbonding from the metal atom.

21.17 Alkanes, agostic hydrogens, and noble gases

Key point: Alkanes can donate the electron density from C—H single bonds to a metal atom and, in the absence of other donors, even the electron density of a noble gas atom can enable it to behave as a ligand.

Highly reactive metal intermediates can be generated by photolysis and, in the absence of any other ligands, alkanes and noble gases have been observed to coordinate to the metal atom. Such species were first identified in the 1970s in solid methane or noble gas matrices and were initially regarded as mere curiosities. However, both species have recently been fully characterized in solution and are now accepted as important intermediates in many reactions.

Alkanes are considered to donate electron density from a C—H σ bond to the metal atom (**64**), and accept π-electron density back from the metal atom into the σ* orbital, just like dihydrogen (**65**). Although most alkane complexes are short-lived, with the alkane being readily displaced, in 1998 the cyclopentane complex (**66**) was unambiguously identified in solution by NMR.

Interactions between the C—H bond of an already coordinated ligand and the metal atom have also been observed. These species are referred to as having **agostic** C—H interactions, from the Greek for 'to hold on to oneself', and are thought to have additional stability due to the chelate effect. Many examples of compounds with agostic interactions are now known; an example is (**67**). Though weakly bonding, each C—H to metal atom interaction, whether it be agostic or not, is considered formally to donate two electrons to the metal.

Unlikely as it might seem, noble gas atoms can behave as ligands towards metal centres, and a number of complexes of Kr and Xe have been identified by IR spectroscopy, with the relatively long-lived Xe complex (**68**) characterized in solution by NMR. These complexes are stable only in the absence of better ligands (such as alkanes). The noble gas is formally considered to be neutral and donate two electrons.

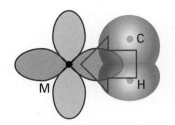

64

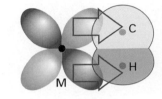

65

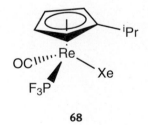

66

Example 21.5 Counting electrons in complexes

Which of the following compounds have 18 electrons: (a) the agostic Pd compound shown as (**67**), (b) [^{i}Pr-CpRe(CO)(PF$_3$)Xe] (**68**)?

Answer (a) Both the η^1-alkenyl and the bromide ligands are considered to be two-electron singly negatively charged donors, with PPh$_3$ ligands considered as two-electron neutral donors. Thus the complex is of Pd(II), which provides eight further electrons. The total number of electrons, before we consider the agostic interaction, is therefore $(4 \times 2) + 8 = 16$. The agostic interaction can be considered to donate a further two electrons, leading to an 18-electron compound. (b) The ^{i}Pr-Cp ligand is considered a six-electron donor with a single negative charge, the CO, PF$_3$, and Xe ligands are each considered to be two-electron neutral donors, implying that the Re must have oxidation number +1, so providing a further six electrons. The total electron count is therefore $6 + 2 + 2 + 2 + 6 = 18$.

Self-test 21.5 Show that both (a) [(η^6-C$_7$H$_8$)Mo(CO)$_3$] (**55**) and (b) [(η^7-C$_7$H$_7$)Mo(CO)$_3$]$^+$ (**57**) are 18-electron species.

Compounds

The preceding discussion of ligands and their bonding modes suggests that there are likely to be a large number of organometallic compounds that have either 16 or 18 valence electrons. A detailed discussion of all these compounds is well beyond the scope of this book. However, we shall examine a number of different classes of compounds because they provide insight into the structures and properties of many other compounds that can be derived from them. Thus, we shall now consider the structures, bonding, and reactions of metal carbonyls, which historically formed the foundation of much of *d*-block organometallic chemistry. Then we consider some sandwich compounds, before describing the structures and reactions of metal cluster compounds.

21.18 *d*-Block carbonyls

d-Block carbonyls have been studied extensively since the first discovery of nickel tetracarbonyl in 1890. Interest in carbonyl compounds has not waned, with many important industrial processes relying on carbonyl intermediates.

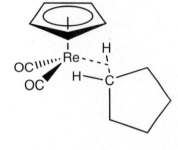

67

68

(a) Homoleptic carbonyls

Key points: The carbonyls of the Period 4 elements of Groups 6 to 10 obey the 18-electron rule; they have alternately one and two metal atoms and a decreasing number of CO ligands.

A **homoleptic complex** is a complex with only one kind of ligand. Simple homoleptic metal carbonyls can be prepared for most of the *d* metals, but those of palladium and platinum are so unstable that they exist only at low temperatures. No simple neutral metal carbonyls are known for copper, silver, and gold or for the members of Group 12. The metal carbonyls are useful synthetic precursors for other organometallic compounds and are used in organic syntheses and as industrial catalysts.

The 18-electron rule helps to systematize the formulas of metal carbonyls. As shown in Table 21.5, the carbonyls of the Period 4 elements of Groups 6 to 10 have alternately one and two metal atoms and a decreasing number of CO ligands. The binuclear carbonyls are formed by elements of the odd-numbered groups, which have an odd number of valence electrons and therefore dimerize by forming metal–metal (M—M) bonds. The decrease in the number of CO ligands from left to right across a period matches the need for fewer CO ligands to achieve 18 valence electrons. The simple vanadium carbonyl $V(CO)_6$ is an exception as it only has 17 valence electrons and is too sterically crowded to dimerize; it is, however, readily reduced to the 18-electron $V(CO)_6^-$ anion.

Simple metal carbonyl molecules often have well-defined, simple, symmetrical shapes that correspond to the CO ligands taking up the most-distant locations, like electron pairs in the VSEPR model. Thus, the Group 6 hexacarbonyls are octahedral, pentacarbonyliron(0) is trigonal bipyramidal, tetracarbonylnickel(0) is tetrahedral, and decacarbonyldimanganese(0) consists of two square-pyramidal $Mn(CO)_5$ groups joined by a metal–metal bond. Bridging carbonyls are also found; for example, one isomer of octacarbonyldicobalt(0) has its metal–metal bond bridged by CO.

Table 21.5 Formulas and electron count for some 3*d*-series carbonyls

Group	Formula	Valence electrons		Structure
6	$Cr(CO)_6$	Cr	6	
		6(CO)	12	
			18	
7	$Mn_2(CO)_{10}$	Mn	7	
		5(CO)	10	
		M—M	1	
			18	
8	$Fe(CO)_5$	Fe	8	
		5(CO)	10	
			18	
9	$Co_2(CO)_8$	Co	9	
		4(CO)	8	
		M—M	1	
			18	
8	$Ni(CO)_4$	Ni	10	
		4(CO)	8	
			18	

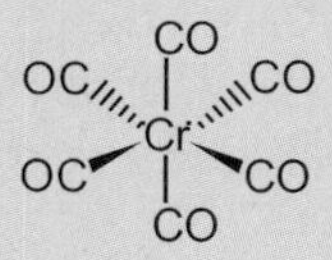

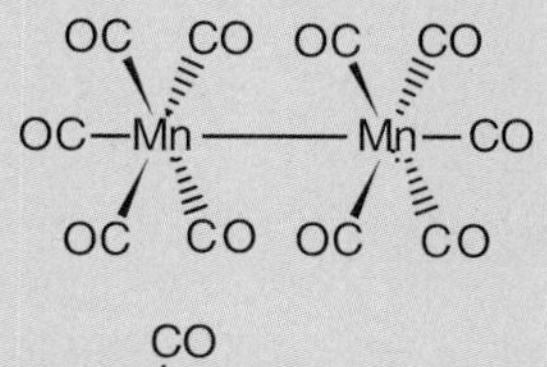

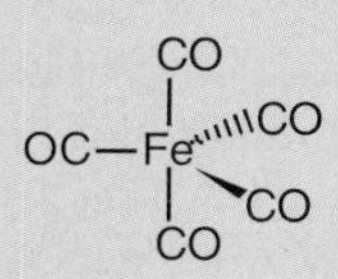

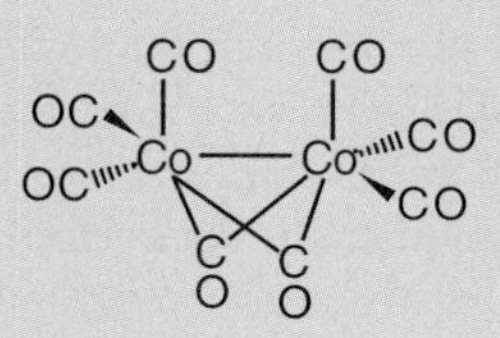

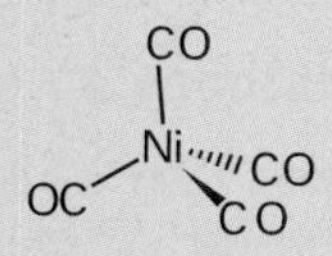

(b) Synthesis of homoleptic carbonyls

Key points: Some metal carbonyls are formed by direct reaction, but of those that can be formed in this way most require high pressures and temperatures; metal carbonyls are commonly formed by reductive carbonylation.

The two principal methods for the synthesis of monometallic metal carbonyls are direct combination of carbon monoxide with a finely divided metal and the reduction of a metal salt in the presence of carbon monoxide under pressure. Many polymetallic carbonyls are synthesized from monometallic carbonyls.

Mond, Langer, and Quinke discovered that the direct combination of nickel and carbon monoxide produced nickel carbonyl, $Ni(CO)_4$:[3]

$$Ni(s) + 4\ CO(g) \xrightarrow{30^\circ C,\, 1\,atm\ CO} Ni(CO)_4(l)$$

Tetracarbonylnickel(0) is in fact the metal carbonyl that is most readily synthesized in this way, with other metal carbonyls, such as $Fe(CO)_5$, being formed more slowly. They are therefore synthesized at high pressures and temperatures (Fig. 21.11):

$$Fe(s) + 5\ CO(g) \xrightarrow{200^\circ C,\, 200\,atm} Fe(CO)_5(l)$$

$$2\ Co(s) + 8\ CO(g) \xrightarrow{150^\circ C,\, 35\,atm} Co_2(CO)_8(s)$$

Direct reaction is impractical for most of the remaining *d* metals, and **reductive carbonylation**, the reduction of a salt or metal complex in the presence of CO, is normally used instead. Reducing agents vary from active metals such as aluminium and sodium, to alkylaluminium compounds, H_2, and CO itself:

$$CrCl_3(s) + Al(s) + 6\ CO(g) \xrightarrow{AlCl_3,\, benzene} AlCl_3(soln) + Cr(CO)_6(soln)$$

$$3\ Ru(acac)_3(soln) + H_2(g) + 12\ CO(g) \xrightarrow{150^\circ C,\, 200\,atm,\, CH_3OH} Ru_3(CO)_{12}(soln) + \cdots$$

$$Re_2O_7(s) + 17\ CO(g) \xrightarrow{250^\circ C,\, 350\,atm} Re_2(CO)_{10}(s) + 7\ CO_2(g)$$

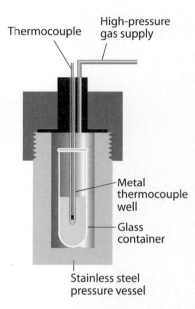

Fig. 21.11 A high-pressure reaction vessel. The reaction mixture is in the glass container.

(c) Properties of homoleptic carbonyls

Key points: All the mononuclear carbonyls are volatile; all the mononuclear and many of the polynuclear carbonyls are soluble in hydrocarbon solvents; polynuclear carbonyls are coloured.

Iron and nickel carbonyls are liquids at room temperature and pressure but all other common carbonyls are solids. All the mononuclear carbonyls are volatile; their vapour pressures at room temperature range from approximately 50 kPa for tetracarbonylnickel(0) to approximately 10 Pa for hexacarbonyltungsten(0). The high volatility of $Ni(CO)_4$, coupled with its extremely high toxicity, means that unusual care is required in its handling. Although the other carbonyls appear to be less toxic, they too must not be inhaled or allowed to touch the skin.

Because they are nonpolar, all the mononuclear and many of the polynuclear carbonyls are soluble in hydrocarbon solvents. The most striking exception among the common carbonyls is nonacarbonyldiiron(0), $[Fe_2(CO)_9]$, which has a very low vapour pressure and is insoluble in solvents with which it does not react. By contrast, $[Mn_2(CO)_{10}]$ and $[Co_2(CO)_8]$ are soluble in hydrocarbon solvents and sublime readily.

Most of the mononuclear carbonyls are colourless or lightly coloured. Polynuclear carbonyls are coloured, the intensity of the colour increasing with the number of metal atoms. For example, pentacarbonyliron(0) is a light straw-coloured liquid, nonacarbonyldiiron(0) forms golden-yellow flakes, and dodecacarbonyltriiron(0) is a deep green compound that looks black in the solid state. The colours of polynuclear carbonyls

---

[3] L. Mond, C. Langer, and F. Quinke discovered $Ni(CO)_4$ in the course of studying the corrosion of nickel valves in process gas containing CO. They quickly applied their discovery to develop a new industrial process for the separation of nickel from cobalt. The centennial of this important discovery is celebrated in a special volume of *J. Organomet. Chem.*, 1990, **383**, which is devoted to metal carbonyl chemistry.

arise from electronic transitions between orbitals that are largely localized on the metal framework.

The principal reactions of the metal centre of simple metal carbonyls are substitution (Section 21.21), oxidation, reduction, and condensation into clusters (Section 21.20d). In certain cases, the CO ligand itself is also subject to attack by nucleophiles or electrophiles.

(d) Oxidation and reduction of carbonyls

Key points: Most metal carbonyls can be reduced to metal carbonylates; some metal carbonyls disproportionate in the presence of a strongly basic ligand, producing the ligated cation and a carbonylate anion; metal carbonyls are susceptible to oxidation by air; metal–metal bonds undergo oxidative cleavage.

Most neutral metal carbonyl complexes can be reduced to an anionic form known as a **metal carbonylate**. In monometallic carbonyls, two-electron reduction is generally accompanied by loss of the two-electron donor CO ligand, thus preserving the electron count at 18:

$$2\,Na + Fe(CO)_5 \xrightarrow{\text{THF}} (Na^+)_2[Fe(CO)_4]^{2-} + CO$$

The metal carbonylate contains Fe with oxidation number -2, and it is rapidly oxidized by air. That much of the negative charge is delocalized over the CO ligands is confirmed by the observation of a low CO stretching band in the IR spectrum at about $1730\,cm^{-1}$. Polynuclear carbonyls, which obey the 18-electron rule through the formation of M—M bonds, are generally cleaved by strong reducing agents. The 18-electron rule is obeyed in the product and a mononegative mononuclear carbonylate results:

$$2\,Na + (OC)_5Mn-Mn(CO)_5 \xrightarrow{\text{THF}} 2\,Na[Mn(CO)_5]$$

Some metal carbonyls disproportionate in the presence of a strongly basic ligand, producing the ligated cation and a carbonylate. Much of the driving force for this reaction is the stability of the metal cation when it is surrounded by strongly basic ligands. Octacarbonyldicobalt(0) is highly susceptible to this type of reaction when exposed to a good Lewis base such as pyridine (py):

$$3\,[Co_2^{(0)}(CO)_8] + 12\,py \longrightarrow 2\,[Co^{(+2)}(py)_6][Co^{(-1)}(CO)_4]_2 + 8\,CO$$

It is also possible for the CO ligand to be oxidized in the presence of the strongly basic ligand OH^-, the net outcome being the reduction of a metal centre:

$$3\,[Fe^{(0)}(CO)_5] + 4\,OH^- \rightarrow [Fe_3^{(-2/3)}(CO)_{11}]^{2-} + CO_3^{2-} + 2\,H_2O + 3\,CO$$

Carbonyl compounds that have only 17 electrons are particularly prone to reduction to give 18-electron carbonylates.

Metal carbonyls are susceptible to oxidation by air. Although uncontrolled oxidation produces the metal oxide and CO or CO_2, of more interest in organometallic chemistry are the controlled reactions that give rise to organometallic halides. One of the simplest of these is the oxidative cleavage of an M—M bond:

$$[(OC)_5Mn^{(0)}-Mn^{(0)}(CO)_5] + Br_2 \rightarrow 2\,[Mn^{(+1)}Br(CO)_5]$$

In keeping with the loss of electron density from the metal when a halogen atom is attached, the CO stretching frequencies of the product are significantly higher than those of $[Mn_2(CO)_{10}]$.

(e) Metal carbonyl basicity

Key points: Most organometallic metal compounds can be protonated at the metal centre; the acidity of the protonated form depends on the other ligands on the metal.

Many organometallic compounds can be protonated at the metal centre. Metal carbonylates provide many examples of this basicity:

$$[Mn(CO)_5]^- + H^+ \rightarrow [MnH(CO)_5](s)$$

The affinity of metal carbonylates for the proton varies widely (Table 21.6). It is observed that, the more negative the anion, the higher its Brønsted basicity and hence the lower the acidity of its conjugate acid (the metal carbonyl hydride).

As we noted in Section 21.7, d-block M—H complexes are commonly referred to as 'hydrides', which reflects the assignment of oxidation number -1 to an H atom attached to a metal atom. Nevertheless, most of the carbonyl hydrides of metals to the right of the d block are Brønsted acids. The Brønsted acidity of a metal carbonyl hydride is a reflection of the π-acceptor strength of a CO ligand, which stabilizes the conjugate base. Thus, $[CoH(CO)_4]$ is acidic whereas $[CoH(PMe_3)_4]$ is strongly hydridic. In striking contrast to p-block hydrogen compounds, the Brønsted acidity of d-block M—H compounds decreases on descending a group.

Neutral metal carbonyls (such as pentacarbonyliron, $[Fe(CO)_5]$) can be protonated in air-free concentrated acid; the Brønsted basicity of a metal atom with oxidation number zero is associated with the presence of nonbonding d electrons. Compounds having metal–metal bonds, such as clusters (Section 21.20), are even more easily protonated; here the Brønsted basicity is associated with the ready protonation of M—M bonds to produce a formal $3c,2e$ bond like that in diborane:

$$[Fe_3(CO)_{11}]^{2-} + H^+ \rightarrow [Fe_3H(CO)_{11}]^-$$

The M—H—M bridge is by far the most common bonding mode of hydrogen in clusters.

Metal basicity is turned to good use in the synthesis of a wide variety of organometallic compounds. For example, alkyl and acyl groups can be attached to metal atoms by the reaction of an alkyl or acyl halide with an anionic metal carbonyl:

$$[Mn(CO)_5]^- + CH_3I \rightarrow [Mn(CH_3)(CO)_5] + I^-$$
$$[Co(CO)_4]^- + CH_3COI \rightarrow [Co(COCH_3)(CO)_4] + I^-$$

A similar reaction with organometallic halides may be used to form M—M bonds:

$$[Mn(CO)_5]^- + [ReBr(CO)_5] \rightarrow [(OC)_5Mn-Re(CO)_5] + Br^-$$

(f) Reactions of the CO ligand

Key points: The C atom of CO is susceptible to attack by nucleophiles if it is attached to a metal atom that is electron poor; the O atom of CO is susceptible to attack by electrophiles in electron-rich carbonyls.

The C atom of CO is susceptible to attack by nucleophiles if it is attached to a metal atom that is not electron rich. Thus, terminal carbonyls with high CO stretching frequencies are liable to attack by nucleophiles. The d electrons in these neutral or cationic metal carbonyls are not extensively delocalized on to the carbonyl C atom and so that atom can be attacked by electron-rich reagents. For example, strong nucleophiles (such as methyllithium, Section 10.11) attack the CO in many neutral metal carbonyl compounds:

$$\tfrac{1}{4}[Li_4(CH_3)_4] + [Mo(CO)_6] \rightarrow Li[Mo(COCH_3)(CO)_5]$$

The resulting anionic acyl compound reacts with carbocation reagents to produce a stable and easily handled neutral product:

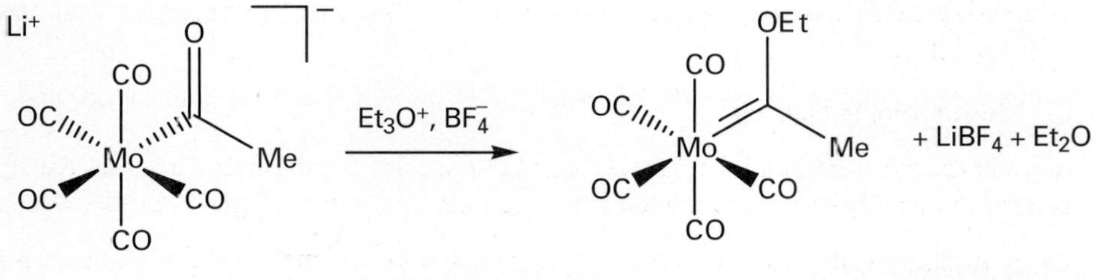

The product of this reaction, with a direct M=C bond, is a Fischer carbene (Section 21.16). The attack of a nucleophile on the C atom is also important for the mechanism

Table 21.6 Acidity constants of d-metal hydrides in acetonitrile at 25°C

Hydride	pK_a
$[CoH(CO)_4]$	8.3
$[CoH(CO)_3P(OPh)_3]$	11.3
$[Fe(H)_2(CO)_4]$	11.4
$[CrH(Cp)(CO)_3]$	13.3
$[MoH(Cp)(CO)_3]$	13.9
$[MnH(CO)_5]$	15.1
$[CoH(CO)_3PPh_3]$	15.4
$[WH(Cp)(CO)_3]$	16.1
$[MoH(Cp^*)(CO)_3]$	17.1
$[Ru(H)_2(CO)_4]$	18.7
$[FeH(Cp)(CO)_2]$	19.4
$[RuH(Cp)(CO)_2]$	20.2
$[Os(H)_2(CO)_4]$	20.8
$[ReH(CO)_5]$	21.1
$[FeH(Cp^*)(CO)_2]$	26.3
$[WH(Cp)(CO)_2PMe_3]$	26.6

of the hydroxide-induced dissociation of metal carbonyls:

$$[(OC)_nM(CO)] + OH^- \rightarrow [(OC)_nM(COOH)]^-$$

$$[(OC)_nM(COOH)]^- + 3\,OH^- \rightarrow [M(CO)_n]^{2-} + CO_3^{2-} + 2\,H_2O$$

In electron-rich metal carbonyls, considerable electron density is delocalized on to the CO ligand. As a result, in some cases the O atom of a CO ligand is susceptible to attack by electrophiles. Once again, IR data provide an indication of when this type of reaction should be expected, as a low CO stretching frequency indicates significant backdonation to the CO ligand and hence appreciable electron density on the O atom. Thus a bridging carbonyl is particularly susceptible to attack at the O atom:

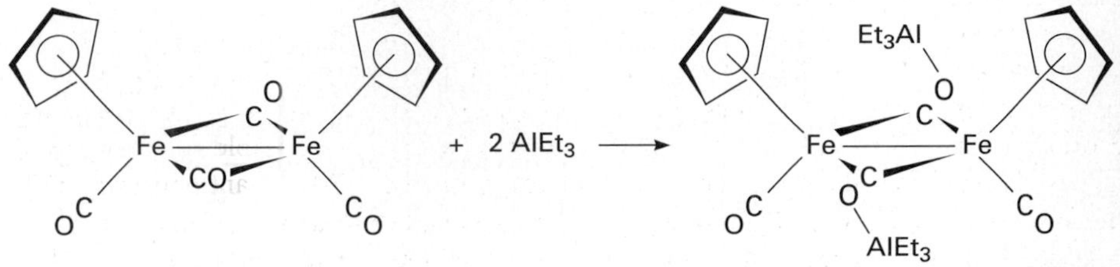

The attachment of an electrophile to the oxygen of a CO ligand, as in the structure on the right, promotes migratory insertion reactions and C—O cleavage reactions.

The ability of some alkyl-substituted metal carbonyls to undergo a migratory insertion reaction to give acyl ligands, —(CO)R, is discussed in detail in Section 21.24.

Example 21.6 Converting CO to carbene and acyl ligands

Propose a set of reactions for the formation of [W(C(OCH₃)Ph)(CO)₅] starting with hexa-carbonyltungsten(0) and other reagents of your choice.

Answer The CO ligands in hexacarbonyltungsten(0) are susceptible to attack by nucleophiles, and the reaction with phenyllithium should give a C-phenyl intermediate:

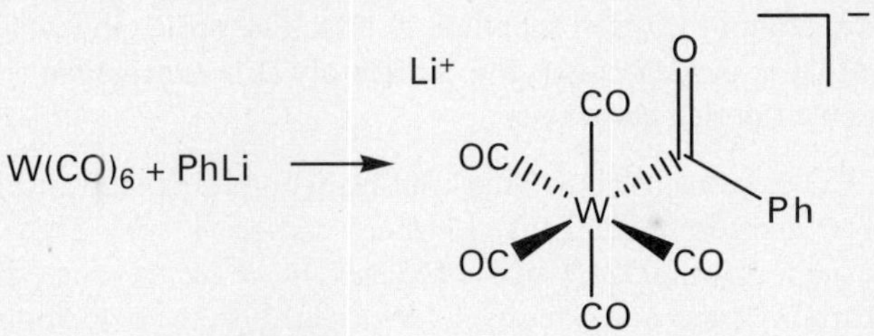

The anion can then react with a carbon electrophile to attach an alkyl group to the O atom of the CO ligand:

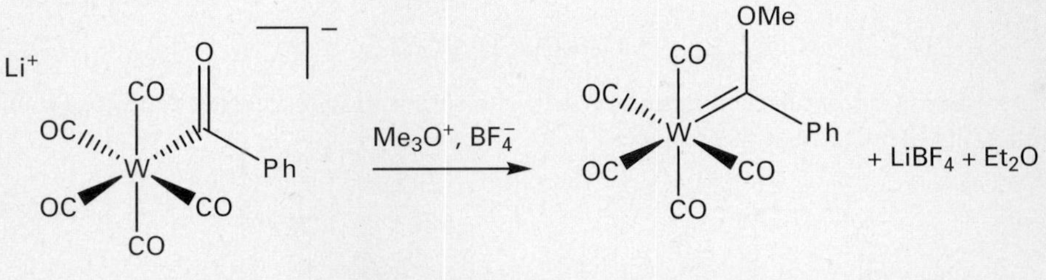

Self-test 21.6 Propose a synthesis for [Mn(CO)₄(PPh₃)(COCH₃)] starting with [Mn₂(CO)₁₀], PPh₃, Na, and CH₃I.

(g) Spectroscopic properties of carbonyl compounds

Key points: The CO stretching frequency is decreased when it serves as a π acceptor; donor ligands cause the CO stretching frequency to decrease as they supply electrons

(a) Synthesis and reactivity of cyclopentadienyl compounds

Key points: Deprotonation of cyclopentadiene gives a convenient precursor to many metal cyclopentadienyl compounds; bound cyclopentadienyl rings behave as aromatic compounds and will undergo Friedel–Crafts electrophilic reactions.

Sodium cyclopentadienide is a common starting material for the preparation of cyclopentadienyl compounds. It can conveniently be prepared by the action of metallic sodium on cyclopentadiene in tetrahydrofuran solution:

$$2\,Na + 2\,C_5H_6 \xrightarrow{\text{THF}} 2\,Na[C_5H_5] + H_2$$

Sodium cyclopentadienide can then be used to react with d-metal halides to produce metallocenes. Cyclopentadiene itself is acidic enough that potassium hydroxide will deprotonate it in solution and, for example, ferrocene can be prepared with a minimum of fuss:

$$2\,KOH + 2\,C_5H_6 + FeCl_2 \xrightarrow{\text{dmso}} Fe(C_5H_5)_2 + 2\,H_2O + 2\,KCl$$

Because of their great stability, the 18-electron Group 8 compounds ferrocene, ruthenocene, and osmocene maintain their ligand–metal bonds under rather harsh conditions, and it is possible to carry out a variety of transformations on the cyclopentadienyl ligands. For example, they undergo reactions similar to those of simple aromatic hydrocarbons, such as Friedel–Crafts acylation:

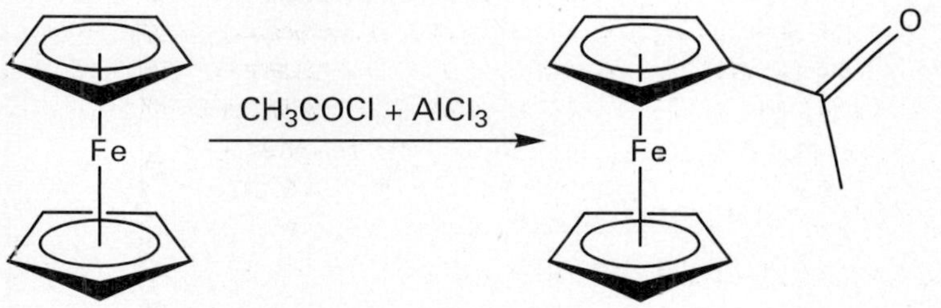

It also is possible to replace H on a C_5H_5 ring by Li:

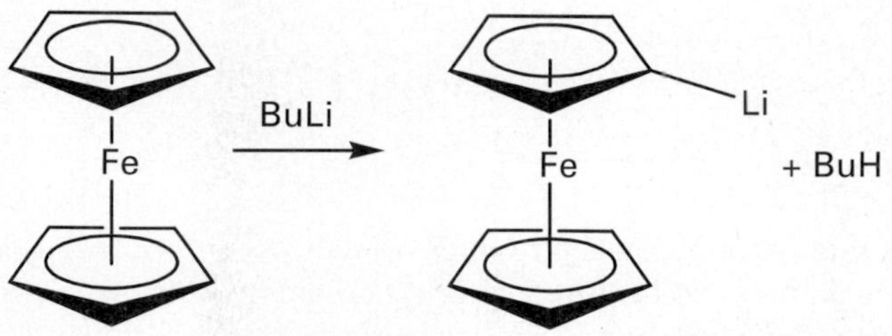

As might be imagined, the lithiated product is an excellent starting material for the synthesis of a wide variety of ring-substituted products and in this respect resembles simple organolithium compounds (Section 10.11). Most Cp complexes of other metals undergo reactions similar to these two types where the five-membered ring behaves as an aromatic system.

(b) Bonding in bis(cyclopentadienyl)metal complexes

Key points: The MO picture of bonding in bis(Cp) metal complexes shows that the frontier orbitals are neither strongly bonding nor strongly antibonding; thus complexes that do not obey the 18-electron rule are possible.

We start by looking at ferrocene, where, although details of the bonding are not settled, the molecular orbital energy level diagram shown in Fig. 21.13 accounts for a number of experimental observations. This diagram refers to the eclipsed (D_{5h}) form of the complex, which in the gas phase is about $4\,kJ\,mol^{-1}$ lower in energy than the staggered conformation; almost all metallocenes have low staggered–eclipsed conversion barriers.[4]

[4] The steric bulk of substituents changes the barrier and can make the staggered the preferred conformation with the result that in 1,1'-disubstituted ferrocenes the barrier to rotation can sometimes be significant. In

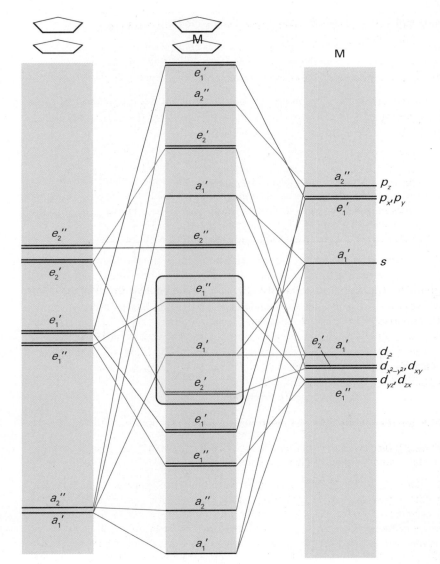

Fig. 21.13 Molecular orbital energy diagram for M(Cp)$_2$ with D_{5h} symmetry. The energies of the symmetry-adapted π orbitals of the C$_5$H$_5$ ligands are shown on the left, relevant *d* orbitals of the metal are on the right, and the resulting molecular orbital energies are in the centre. Eighteen electron can be accommodated by filling the molecular orbitals up to and including the a_1' orbital in the box. The box denotes the orbitals typically regarded as frontier orbitals in these molecules.

We shall focus our attention on the frontier orbitals. As shown in Fig. 21.13, the e_1'' symmetry-adapted linear combinations of ligand orbitals have the same symmetry as the d_{zx} and d_{yz} orbitals of the metal atom. The lower energy frontier orbital (a_1') is composed of d_{z^2} and the corresponding SALC of ligand orbitals. However, there is little interaction between the ligands and the metal orbitals because the ligand π orbitals happen to lie, by accident, in the conical nodal surface of d_{z^2} orbital of the metal atom. In ferrocene and the other 18-electron bis(cyclopentadienyl) complexes, the a_1' frontier orbital and all lower orbitals are full but the e_1'' frontier orbital and all higher orbitals are empty.

The frontier orbitals are neither strongly bonding nor strongly antibonding. This characteristic permits the possibility of the existence of metallocenes that diverge from the 18-electron rule, such as the 17-electron complex [Fe(η^5-Cp)$_2$]$^+$ and the 20-electron complex [Ni(η^5-Cp)$_2$]. Deviations from the 18-electron rule, however, do lead to significant changes in M—C bond lengths that correlate fairly well with the molecular orbital scheme (Table 21.8). Similarly, the redox properties of the complexes can be understood in terms of the electronic structure.

We have already mentioned that ferrocene is fairly readily oxidized to the ferrocinium ion, [Fe(η^5-Cp)$_2$]$^+$. From an orbital viewpoint, this oxidation corresponds to the removal of an electron from the nonbonding a_1' orbital. The 19-electron complex

practice, in solution, both rings normally behave as though there is totally free rotation. The rings are often drawn in a staggered conformation simply because there is then a little more space to illustrate substitutions.

Table 21.8 Electronic configuration and M—C bond length in $[M(\eta^5\text{-Cp})_2]$ complexes

Complex	Valence electrons	Electron configuration	M—C bond length/pm
$[V(\eta^5\text{-Cp})_2]$	15	$e_2'^2 a_1'^1$	228
$[Cr(\eta^5\text{-Cp})_2]$	16	$e_2'^3 a_1'^1$	217
$[Mn(\eta^5\text{-Me-C}_5\text{H}_4)_2]^*$	17	$e_2'^3 a_1'^2$	211
$[Fe(\eta^5\text{-Cp})_2]$	18	$e_2'^4 a_1'^2$	206
$[Co(\eta^5\text{-Cp})_2]$	19	$e_2'^4 e_1''^1 a_1'^2$	212
$[Ni(\eta^5\text{-Cp})_2]$	20	$e_2'^4 e_1''^2 a_1'^2$	220

*Data are quoted for this complex because $[Mn(\eta^5\text{-Cp})_2]$ has a high-spin configuration and hence an anomalously long M—C bond (238 pm).

$[Co(\eta^5\text{-Cp})_2]$ is much more readily oxidized than ferrocene because the electron is lost from the antibonding e_1'' orbital to give the 18-electron $[Co(\eta^5\text{-Cp})_2]^+$ ion.

A useful comparison can be made with octahedral complexes. The e_1'' frontier orbital of a metallocene is the analogue of the e_g orbital in an octahedral complex, and the a' orbital plus the e_2' pair of orbitals are analogous to the t_{2g} orbitals of an octahedral complex. This formal similarity extends to the existence of high- and low-spin bis(cyclopentadienyl) complexes.

Example 21.8 Identifying metallocene electronic structure and stability

Refer to Fig. 21.13. Discuss the occupancy and nature of the HOMO in $[Co(\eta^5\text{-Cp})_2]^+$ and the change in metal–ligand bonding relative to neutral cobaltocene.

Answer The $[Co(\eta^5\text{-Cp})_2]^+$ ion contains 18 valence electrons (six from Co(III), 12 from the two Cp ligands). Assuming that the molecular orbital energy level diagram for ferrocene is applicable, the 18-electron count leads to double occupancy of the orbitals up to a_1'. The 19-electron cobaltocene molecule has an additional electron in the e_1'' orbital, which is antibonding with respect to the metal and ligand. Therefore, the metal–ligand bond should be stronger and shorter in $[Co(\eta^5\text{-Cp})_2]^+$ than in $[Co(\eta^5\text{-Cp})_2]$. This conclusion is borne out by structural data.

Self-test 21.8 By using the same molecular orbital diagram, comment on whether the removal of an electron from $[Fe(\eta^5\text{-Cp})_2]$ to produce $[Fe(\eta^5\text{-Cp})_2]^+$ should produce a substantial change in M—C bond length relative to neutral ferrocene.

(c) Fluxional behaviour of metallocenes

Key point: Many metallocenes exhibit fluxionality and undergo internal rotation because the barrier to the interconversion of the various forms is low.

One of the most remarkable aspects of many cyclic polyene complexes is their stereochemical nonrigidity (their fluxionality). For example, at room temperature the two rings in ferrocene rotate rapidly relative to each other. This type of fluxional process is called **internal rotation**, and is similar to the process by which the two CH$_3$ groups rotate relative to each other in ethane.

Of greater interest is the stereochemical nonrigidity that is often seen when a conjugated cyclic polyene is attached to a metal atom through some, but not all, of its C atoms. In such complexes the metal–ligand bonding may hop around the ring; in the informal jargon of organometallic chemists this internal rotation is called 'ring whizzing'. A simple example is found in $[Ge(\eta^1\text{-Cp})(CH_3)_3]$, in which the single site of attachment of the Ge atom to the cyclopentadiene ring hops around the ring in a series of **1,2-shifts**,

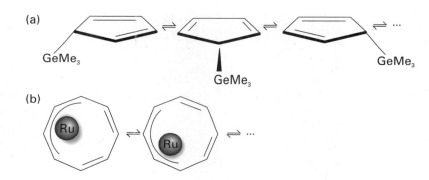

Fig. 21.14 (a) The fluxional process in $[Ge(\eta^1\text{-Cp})(Me)_3]$ occurs by a series of 1,2-shifts. (b) the fluxionality of $[Ru(\eta^4\text{-}C_8H_8)(CO)_3]$ can be described similarly. We need to imagine that the Ru atom is out of the plane of the page with the CO ligands omitted for clarity.

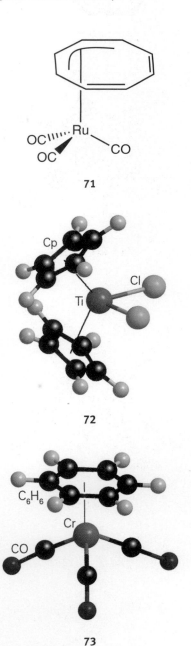

a motion in which a C—M bond is replaced by a C—M bond to the next C atom around the ring; this motion is known as a 1,2-shift, because the bond starts on atom 1 and ends up on the adjacent atom 2 (Fig. 21.14). The great majority of fluxional conjugated polyene complexes that have been investigated migrate by 1,2-shifts, but it is not known whether these shifts are controlled by a principle of least motion or by some aspect of orbital symmetry.

Nuclear magnetic resonance provides the primary evidence for the existence and mechanism of these fluxional processes, as they occur on a timescale of 10^{-2} to 10^{-4} s, and can be studied by ^1H and ^{13}C NMR. The compound $[Ru(\eta^4\text{-}C_8H_8)(CO)_3]$ (**71**) provides a good illustration of the approach. At room temperature, its ^1H NMR spectrum consists of a single, sharp line that could be interpreted as arising from a symmetrical $\eta^8\text{-}C_8H_8$ ligand. However, X-ray diffraction studies of single crystals show unambiguously that the ligand is tetrahapto. This conflict is resolved by ^1H NMR spectra at lower temperatures, because as the sample is cooled the signal broadens and then separates into four peaks. These peaks are expected for the four pairs of protons of an $\eta^4\text{-}C_8H_8$ ligand. The interpretation is that at room temperature the ring is 'whizzing' around the metal atom rapidly compared with the timescale of the NMR experiment, so an averaged signal is observed. At lower temperatures the motion of the ring is slower, and the distinct conformations exist long enough to be resolved. A detailed analysis of the line shape of the NMR spectra can be used to measure the activation energy of the migration.

(d) Bent metallocene complexes

Key point: The structures of bent sandwich compounds can be systematized in terms of a model in which three metal atom orbitals project towards the open face of the bent Cp_2M fragment.

In addition to the simple bis(cyclopentadienyl) and bis(arene) complexes with parallel rings, there are many related structures. In the jargon of this area, these species are referred to as 'bent sandwich compounds' (**72**), 'half-sandwich' or 'piano stool' compounds (**73**), and, inevitably, 'triple deckers' (**74**). Bent sandwich compounds play a major role in the organometallic chemistry of the early and middle *d*-block elements, and examples include $[TiCl_2(\eta^5\text{-Cp})_2]$, $[Re(\eta^5\text{-Cp})_2(Cl)]$, $[W(\eta^5\text{-Cp})_2(H)_2]$, and $[Nb(\eta^5\text{-Cp})_2(Cl)_3]$.

As shown in Fig. 21.15, bent sandwich compounds occur with a variety of electron counts and stereochemistries. Their structures can be systematized in terms of a model in which three metal atom orbitals project out of the open face of the bent $M(Cp)_2$ fragment. According to this model, the metal atom often satisfies its electron deficiency when the electron count is less than 18 by interaction with lone pairs or agostic C—H groups on the ligands.

21.20 Metal–metal bonding and metal clusters

Organic chemists must go to heroic efforts to synthesize their cage-like molecules, such as cubane (**75**). By contrast, one of the distinctive characteristics of inorganic chemistry is the large number of closed polyhedral molecules, such as the tetrahedral P_4 molecule (Chapter 14), the octahedral halide-bridged early *d*-block clusters (Section 18.11), the

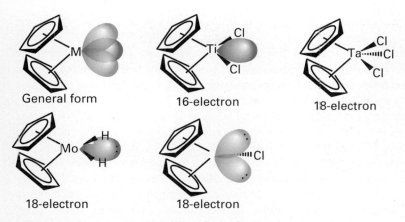

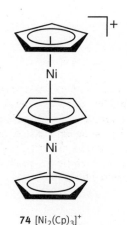

74 $[Ni_2(Cp)_3]^+$

Fig. 21.15 Bent sandwich compounds with their electron counts.

polyhedral carboranes (Section 12.10), and the organometallic cluster compounds that we discuss here. The structures of clusters often resemble the close-packed structures of the metal itself, and this similarity provides the main rationale for studying them—the idea that the chemical properties of the ligands of a cluster reflect the behaviour on a metal surface. In more recent years, electronic properties of clusters that depend on how large the cluster is have emerged, such as quantum dots and nanoparticles (Section 24.2).

(a) Structure of clusters

Key points: A cluster includes all compounds with metal–metal bonds that form triangular or larger cyclic structures.

A rigorous definition of **metal clusters** restricts them to molecular complexes with metal–metal bonds that form triangular or larger cyclic structures. This definition excludes linear M—M compounds and cage compounds, in which several metal atoms are held together exclusively by ligand bridges. However, this rigorous definition is normally relaxed, and we shall consider any M—M bonded system as a cluster. The distinction between cage and cluster compounds can seem arbitrary as the presence of bridging ligands in a cluster such as (**76**) raises the possibility that the atoms are held together by M—L—M interactions rather than M—M bonds. Bond lengths are of some help in resolving this issue. If the M—M distance is much greater than twice the metallic radius, then it is reasonable to conclude that the M—M bond is either very weak or absent. However, if the metal atoms are within a reasonable bonding distance, the proportion of the bonding that is attributable to direct M—M interaction is ambiguous. For example, there has been much debate about the extent of Fe—Fe bonding in $[Fe_2(CO)_9]$ (**77**).

Metal–metal bond strengths in metal complexes cannot be determined with great precision, but a variety of pieces of evidence—such as the stability of compounds and M—M force constants—indicate that there is an increase in M—M bond strengths down a group in the d block. This trend contrasts with that in the p block, where element–element bonds are usually weaker for the heavier members of a group. As a consequence of this trend, metal–metal bonded systems are most numerous for the $4d$ and $5d$ metals.

(b) Electron counting in clusters

Key points: The EAN rule is suitable for identifying the correct number of electrons for clusters with fewer than six metal atoms; the Wade–Mingos–Lauher rules identify a correlation between the valence electron count and the structures of larger organometallic complexes.

Organometallic cluster compounds are rare for the early d metals and unknown for the f metals, but a large number of metal carbonyl clusters exist for the elements of Groups 6 to 10. The bonding in the smaller clusters can be readily explained in terms of a local M—M and M—L electron pair bonding and the 18-electron rule.

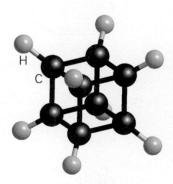

75 Cubane, C_8H_8

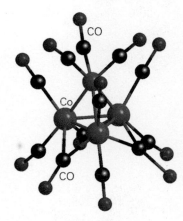

76 $[Co_4(CO)_{12}]$

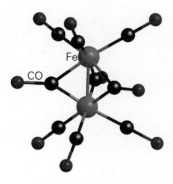

77 $[Fe_2(CO)_9]$

If we take $Mn_2(CO)_{10}$ and $Os_3(CO)_{12}$ as examples we can arrive at a simple, yet illuminating picture. In $Mn_2(CO)_{10}$ each Mn atom is considered to have 17 electrons (seven from Mn and ten from the five CO ligands) before we take into account the Mn—Mn bond. This Mn—Mn bond consists of two electrons shared between the two metal atoms, and hence raises the electron count of each by 1, resulting in two 18-electron metals, but with a total electron count of only 34, not 36. In $Os_3(CO)_{12}$, each $Os(CO)_4$ fragment has 16 electrons before metal–metal bonding is taken into consideration and each metal shares two further electrons with adjacent metals, so increasing the number of electrons around each metal to 18, but with a total of only 48 and not 54 electrons. The bonding electrons we are dealing with are referred to as the **cluster valence electrons** (CVEs), and it quickly becomes apparent that a cluster of x metal atoms with y metal–metal bonds needs $18x - 2y$ electrons. Octahedral M_6 and larger clusters do not conform to this pattern, and the polyhedral skeletal electron pair rules, known as 'Wade's rules' (Section 12.8), have been refined by D.M.P. Mingos and J. Lauher to apply to metal clusters. These **Wade–Mingos–Lauher rules** are summarized in Table 21.9; they apply most reliably to metal clusters in Groups 6 to 9. In general, and as in the boron hydrides where similar considerations apply, more open structures (which have fewer metal–metal bonds) occur when there is a higher CVE count.

Table 21.9 Correlation of cluster valence electron (CVE) count and structure

Number of metal atoms	Structure of metal framework	CVE count	Example
1	Single metal ○	18	$Ni(CO)_4$ **(2)**
2	Linear ○——○	34	$Mn_2(CO)_{10}$
3	Closed triangle 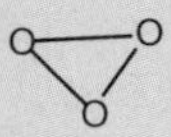	48	$[Co_3(CH)(CO)_9]$ **(78)**
4	Tetrahedron 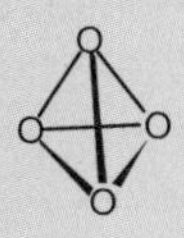	60	$Co_4(CO)_{12}$ **(76)**
	Butterfly 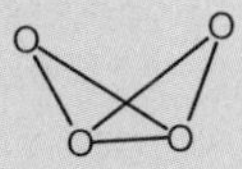	62	$[Fe_4(CO)_{12}C]^{2-}$
	Square 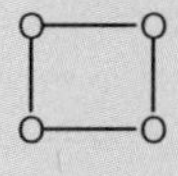	64	$Os_4(CO)_{16}$
5	Trigonal bipyramid 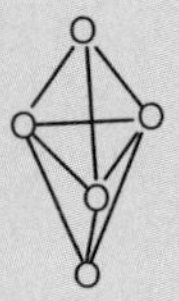	72	$Os_5(CO)_{16}$
	Square pyramid 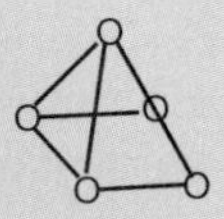	74	$Fe_5C(CO)_{15}$
6	Octahedron 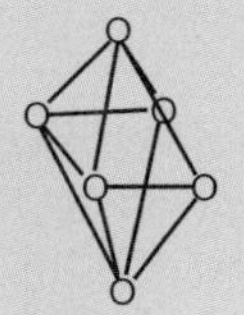	86	$Ru_6C(CO)_{17}$
	Trigonal prism 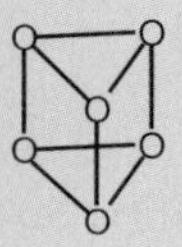	90	$[Rh_6C(CO)_{15}]^{2-}$

Example 21.9 Correlating spectroscopic data, cluster valence electron count, and structure

The reaction of chloroform (trichloromethane) with $[Co_2(CO)_8]$ yields a compound of formula $[Co_3(CH)(CO)_9]$. Both NMR and IR data indicate the presence of only terminal CO ligands and the presence of a CH group. Propose a structure consistent with the spectra and the correlation of CVE with structure.

Answer We assume that the CH ligand is simply C-bonded: one C electron is used for the C—H bond, so three are available for bonding in the cluster. Electrons available for the cluster are then 27 for three Co atoms, 18 for nine CO ligands, and 3 for CH. The resulting total CVE of 48 indicates a triangular cluster (see Table 21.9). A structure consistent with this conclusion and the presence of only terminal CO ligands and a capping CH ligand is (**78**)

Self-test 21.9 The compound $[Fe_4(Cp)_4(CO)_4]$ is a dark-green solid. Its IR spectrum shows a single CO stretch at $1640\ cm^{-1}$. The ^1H NMR spectrum is a single line even at low temperatures. From this spectroscopic information and the CVE, propose a structure for $[Fe_4(Cp)_4(CO)_4]$.

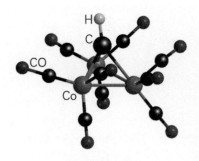

78 $[Co_3(CH)(CO)_9]$

(c) Isolobal analogies

Key point: Isolobal analogies allow correlations between seemingly unrelated fragments, allowing the rationalization of diverse structures.

A good way to picture the incorporation of hetero atoms into a metal cluster is to use isolobal analogies (Section 2.6c). These analogies let us draw a parallel between $[Co_3(CO)_9(CH)]$ (**78**) and $[Co_4(CO)_{12}]$ (**76**), both of which can be regarded as triangular $Co_3(CO)_9$ fragments capped on one side either by $Co(CO)_3$ or by CH. The CH and $Co(CO)_3$ groups are isolobal because each one has three orbitals and three electrons available for participation in framework bonding. A minor complication in this comparison is the occurrence of $Co(CO)_2$ groups together with bridging CO ligands in $[Co_4(CO)_{12}]$ because bridging and terminal ligands often have similar energies.

Isolobal analogies allow us to draw many parallels between metal–metal bonded systems containing only d metals, mixed systems containing d-block and p-block metals, and organic fragments. Examples of isolobal fragments are given in Table 21.10, where it will be noted that a P atom is isolobal with CH; accordingly, a cluster similar to $[Co_3(CH)(CO)_9]$ is known, but with a capping P atom. Similarly, the ligands CR_2 and $Fe(CO)_4$ are both capable of bonding to two metal atoms in a cluster; CH_3 and $Mn(CO)_5$ can bond to one metal atom. As a final example of isolobal analogies, consider the mixed manganese and platinum complex (**79**): the three-membered

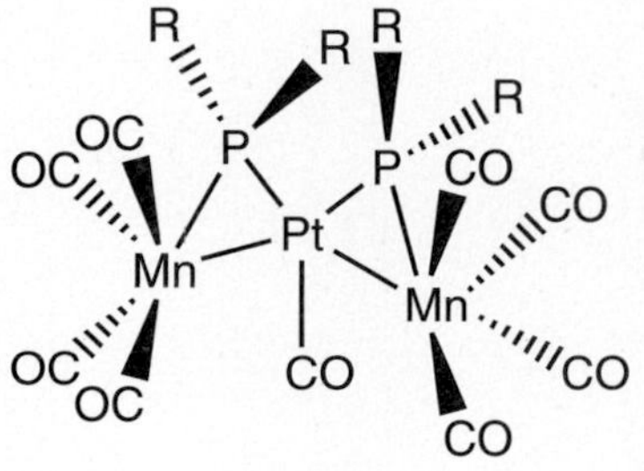

79

Table 21.10 Selected isolobal fragments

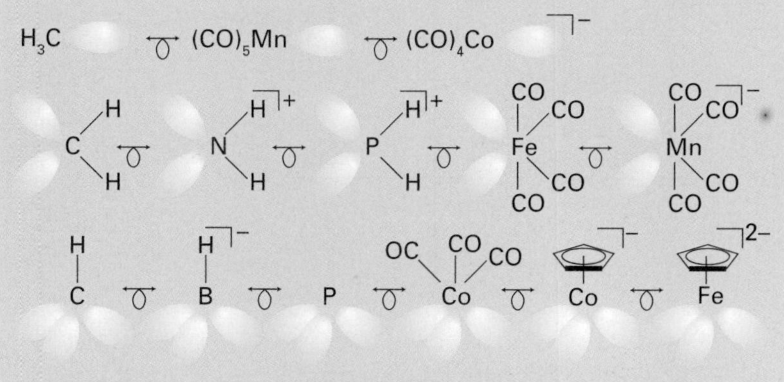

H_3C $\longrightarrow$ $(CO)_5Mn$ $\longrightarrow$ $(CO)_4Co$ ⌉⁻

Note that electrons can be added to or subtracted from each member of the isolobal group and still maintain isolobality. For example, CH_3^+ ⟷ $Mn(CO)_5^+$ ⟷ $Co(CO)_4$.

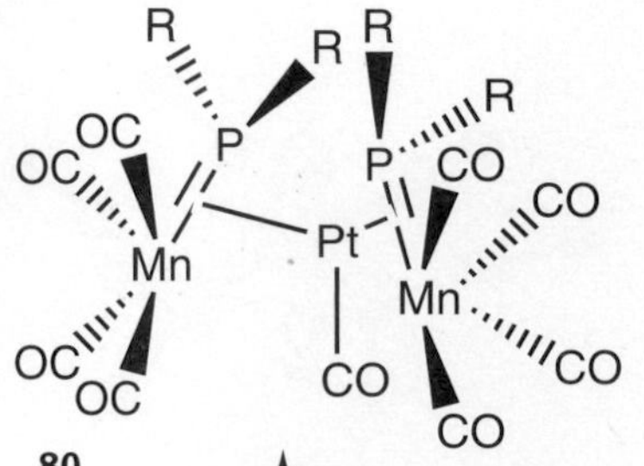

80

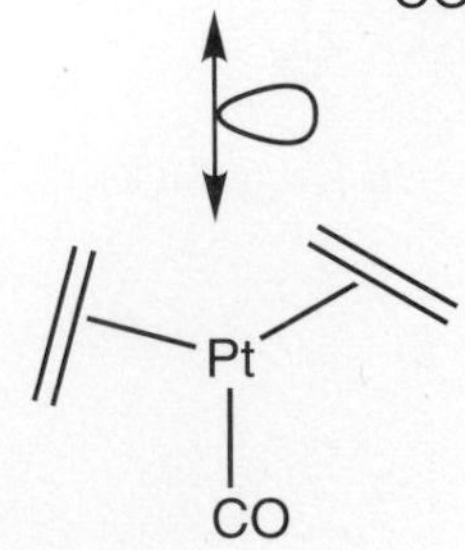

81

{Mn,Pt,P} rings can be considered as the metallocyclopropane form of a coordinated double bond (**80**). Both halves of the Mn=P fragments are isolobal with CH_2 if we treat them as PR_2^+ and $Mn(CO)_4^-$; the complete fragment can then be treated as analogous to an ethene molecule. This treatment means that (**79**) can be considered analogous to bis-(ethene)carbonylplatinum(0) (**81**), a known simple 16-electron organometallic compound.

(d) Synthesis of clusters

Key point: Three methods are commonly used to prepare metal clusters: thermal expulsion of CO from a metal carbonyl, the condensation of a carbonyl anion and a neutral organometallic complex, and the condensation of an organometallic complex with an unsaturated organometallic compound.

One of the oldest methods for the synthesis of metal clusters is the thermal expulsion of CO from a metal carbonyl. The pyrolytic formation of metal cluster compounds can be viewed from the standpoint of electron count: a decrease in valence electrons around the metal resulting from loss of CO is compensated by the formation of M—M bonds. One example is the synthesis of $[Co_4(CO)_{12}]$ by heating $[Co_2(CO)_8]$:

$$2\ [Co_2(CO)_8] \xrightarrow{\Delta} [Co_4(CO)_{12}] + 4\ CO$$

This reaction proceeds slowly at room temperature, so samples of octacarbonyldi-cobalt(0) are usually contaminated with dodecacarbonyltetracobalt(0).

A widely used and more controllable reaction is based on the condensation of a carbonyl anion and a neutral organometallic complex:

$$[Ni_5(CO)_{12}]^{2-} + [Ni(CO)_4] \rightarrow [Ni_6(CO)_{12}]^{2-} + 4\ CO$$

The Ni_5 complex has a CVE of 76 whereas the Ni_6 complex has a count of 86. The descriptive name **redox condensation** is often given to reactions of this type, which are very useful for the preparation of anionic metal carbonyl clusters. In this example, a trigonal-bipyramidal cluster containing Ni with formal oxidation number $-\frac{2}{5}$ and $Ni(CO)_4$ containing Ni(0) is converted into an octahedral cluster having Ni with oxidation number $-\frac{1}{3}$. The $[Ni_5(CO)_{12}]^{2-}$ cluster, which has four electrons in excess of the 72 expected for a trigonal bipyramid, illustrates a fairly common tendency for the Group 10 metal clusters to have an electron count in excess of that expected from the Wade–Mingos–Lauher rules.

A third method, pioneered by F.G.A. Stone, is based on the condensation of an organometallic complex containing displaceable ligands with an unsaturated organo-metallic compound. The unsaturated complex may be a metal alkylidene, $L_nM=CR_2$, a metal alkylidyne, $L_nM\equiv CR$, or a compound with multiple metal–metal bonds:

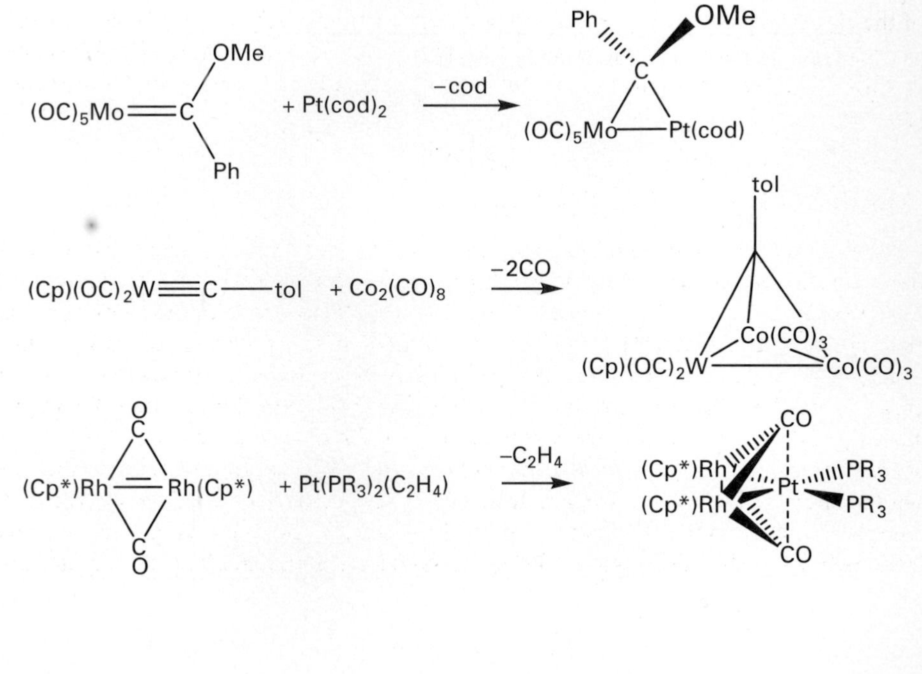

$(Cp)(OC)_2Mo\equiv\!\!\equiv\!\!\equiv Mo(CO)_2(Cp)$ + $Pt(PPh_3)_4$ $\xrightarrow{-2\ PPh_3}$

PPh$_3$ PPh$_3$

Pt

$(Cp)(OC)_2Mo\!\!-\!\!-\!\!Mo(CO)_2(Cp)$

Reactions

The fact that most organometallic compounds can be made to react in a variety of ways is responsible for their use as catalysts. In the previous sections we looked at ligands and how to introduce them into a metal centre, and in this section we look at how ligands might react further or with each other. Implicit in the following discussion is the fact that coordinatively saturated complexes are less reactive that unsaturated ones.

21.21 Ligand substitution

Key points: The substitution of ligands in organometallic complexes is very similar to the substitution of ligands in coordination complexes, with the additional constraint that the valence electron count at the metal atom does not increase above 18; steric crowding of ligands increases the rate of dissociative processes and decreases the rate of associative processes.

Extensive studies of CO substitution reactions of simple carbonyl complexes have revealed systematic trends in mechanisms and rates and much that has been established for these compounds is applicable to all organometallic complexes. The simple replacement of one ligand by another in organometallic complexes is very similar to that observed with coordination compounds, where reactions sometimes go by an associative, a dissociative, or an interchange pathway, with the reaction being either associatively or dissociatively activated (Section 20.3).

The 18-electron metal carbonyls generally undergo dissociatively activated substitution reactions; this class of reaction is characterized by insensitivity of its rate to the identity of the entering group. Organometallic reaction intermediates with more than 18 valence electrons are rarely observed, as their formation corresponds to populating high-energy antibonding MOs, so associative activation is not normally expected for 18-electron organometallic compounds.

The simplest examples of substitution reactions involve the replacement of CO by another electron-pair donor, such as a phosphine. Studies of the rates at which trialkylphosphines and other ligands replace CO in $Ni(CO)_4$, $Fe(CO)_5$, and the hexacarbonyls of the chromium group do indeed indicate that a dissociatively activated mechanism is in operation. In some cases a solvated intermediate such as $[Cr(CO)_5(THF)]$ has been detected. This intermediate then combines with the entering group in a bimolecular process:

$$Cr(CO)_6 + sol \rightleftharpoons Cr(CO)_5(sol) + CO$$
$$Cr(CO)_5(sol) + L \rightarrow Cr(CO)_5L + sol$$

Whereas the loss of the first CO group from $Ni(CO)_4$ occurs easily, and substitution is fast at room temperature, the CO ligands are much more tightly bound in the Group 6 carbonyls, and loss of CO often needs to be promoted thermally or photochemically. For example, the substitution of CO by CH_3CN is carried out in refluxing acetonitrile, using a stream of nitrogen to sweep away the carbon monoxide and hence drive the reaction to completion. To achieve photolysis, mononuclear carbonyls (which do not absorb strongly in the visible region) are exposed to near-UV radiation in an apparatus like that shown in Fig. 21.16. As with the thermal process, there is strong evidence that the photoassisted substitution reaction leads to the formation of a labile intermediate complex with the solvent, which is then displaced by the entering group. Solvated intermediates in the photolysis of metal carbonyls have been detected, not only in polar

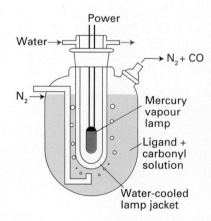

Fig. 21.16 Apparatus for photochemical ligand substitution of metal carbonyls.

solvents such as THF but also in every solvent that has been tried, even in alkanes and noble gases.[5]

The rates of substitution of 16-electron complexes are sensitive to the identity and concentration of the entering group, which indicates associative activation. For example, the reactions of $[IrCl(CO)(PPh_3)_2]$ with triethylphosphine are associatively activated. Sixteen-electron organometallic compounds appear to undergo associatively activated substitution reactions because the 18-electron activated complex is energetically more favourable than the 14-electron activated complex that would occur in dissociative activation.

As in the reactions of coordination complexes, we can expect steric crowding between ligands to accelerate dissociative processes and to decrease the rates of associative processes (Section 20.7). The extent to which various ligands crowd each other is approximated by the Tolman cone angle and we can see how it influences the equilibrium constant for ligand binding by examining the dissociation constants of $[Ni(PR_3)_4]$ complexes (Table 21.11). These complexes are slightly dissociated in solution if the phosphine ligands are compact, such as PMe_3, with a cone angle of 118°. However, a complex such as $[Ni(P^tBu_3)_4]$, where the cone angle is huge (182°), is highly dissociated. The ligand P^tBu_3 is so bulky that the 14-electron complex $[Pt(P^tBu_3)_2]$ can be identified.

The rate of CO substitution in six-coordinate metal carbonyls often decreases as more strongly basic ligands replace CO, and two or three alkylphosphine ligands often represent the limit of substitution. With bulky phosphine ligands, further substitution may be thermodynamically unfavourable on account of ligand crowding, but increased electron density on the metal centre, which arises when a π-acceptor ligand is replaced by a net donor ligand, appears to bind the remaining CO ligands more tightly and therefore reduce the rate of CO dissociative substitution. The explanation of the influence of σ-donor ligands on CO bonding is that the increased electron density contributed by the phosphine leads to stronger π backbonding to the remaining CO ligands and therefore strengthens the M—CO bond. This stronger M—C bond decreases the tendency of CO to leave the metal atom and therefore decreases the rate of dissociative substitution. It is also observed that the second carbonyl that is replaced is normally *cis* to the site of the first and that replacement of a third carbonyl results in a *fac* complex. The reason for this regiochemistry is that CO ligands have very high *trans* effects (Section 20.5).

Table 21.11 Cone angles and dissociation constants for some Ni complexes

L	$\theta/°$	K_d
PMe_3	118	$< 10^{-9}$
PEt_3	137	1.2×10^{-5}
$PMePh_2$	136	5.0×10^{-2}
PPh_3	145	Large
P^tBu_3	182	Large

Data are for $NiL_4 \rightleftharpoons NiL_3 + L$ in benzene at 25°C.

Example 21.10 Preparing substituted metal carbonyls

Starting with MoO_3 as a source of Mo, and CO and PPh_3 as the ligand sources, plus other reagents of your choice, give equations and conditions for the synthesis of $[Mo(CO)_5PPh_3]$.

Answer A typical procedure is to synthesize $Mo(CO)_6$ first and then carry out a ligand substitution. Reductive carbonylation of MoO_3 might be performed using $Al(CH_2CH_3)_3$ as a reducing agent in the presence of carbon monoxide under pressure. The temperature and pressure required for this reaction are less than those for the direct combination of molybdenum and carbon monoxide.

$$MoO_3 + Al(CH_2CH_3)_3 + 6\ CO \xrightarrow{\text{50 atm, 150°C, heptane}} Mo(CO)_6$$
$$+ \text{oxidation products of } Al(CH_2CH_3)_3$$

The subsequent substitution could be carried out photochemically by using the apparatus illustrated in Fig. 21.16:

$$Mo(CO)_6 + PPh_3 \xrightarrow{\text{THF}, h\nu} Mo(CO)_5PPh_3 + CO$$

The progress of the reaction can be followed by IR spectroscopy in the CO stretching region using small samples that are removed periodically from the reaction vessel.

Self-test 21.10 If the highly substituted complex $[Mo(CO)_3L_3]$ is desired, which of the ligands $P(CH_3)_3$ or $P(^tBu)_3$ would be preferred? Give reasons for your choice.

---

[5] Infrared spectroscopy has been used to demonstrate that, as quickly as can be measured (that is, in less than one picosecond) the photolysis of $W(CO)_6$ results in the formation of $W(CO)_5(sol)$.

Although the generalizations above apply to a wide range of reactions, some exceptions are observed, especially if cyclopentadienyl or nitrosyl ligands are present. In these cases it is common to find evidence of associatively activated substitution even for 18-electron complexes. The common explanation is that NO may switch from being linear (as in **39**) to being angular (as in **40**), whereupon it donates two fewer electrons (Section 21.11). Similarly, the η^5-Cp six-electron donor can slip relative to the metal and become an η^3-Cp four-electron donor. In this case, the $C_5H_5^-$ ligand is regarded as having a three-carbon interaction with the metal while the remaining two electrons form a simple $C=C$ bond that is not engaged with the metal (**82**), and the relatively electron-depleted central metal atom becomes susceptible to substitution.

It has been found that, for some metal carbonyls, the displacement of CO can be catalysed by electron-transfer processes that create anion or cation radicals. These radicals do not have 18 electrons, and a typical process of this type is illustrated in Fig. 21.17. As can be seen, the key features are the lability of CO in the 19-electron anion radical and the more negative reduction potential of the phosphine-substituted product compared to the metal carbonyl starting material. Similarly, the less common 19- and 17-electron metal compounds are labile with respect to substitution.

The substitution of ligands at a cluster is often not a straightforward process because fragmentation is common. Fragmentation occurs because the M—M bonds in a cluster are generally comparable in strength to the M—L bonds, and so the breaking of the M—M bonds provides a reaction pathway with a low activation energy. For example, dodecacarbonyltriiron(0) reacts with triphenylphosphine under mild conditions to yield simple mono- and disubstituted products as well as some cluster fragmentation products:

$$[Fe_3(CO)_{12}] + PPh_3 \rightarrow [Fe_3(CO)_{11}PPh_3]$$
$$+ [Fe_3(CO)_{10}(PPh_3)_2] + [Fe(CO)_5]$$
$$+ [Fe(CO)_4PPh_3] + [Fe(CO)_3(PPh_3)_2] + CO$$

However, for somewhat longer reaction times or elevated temperatures, only monoiron cleavage products are obtained. Because the strength of M—M bonds increases down a group, substitution products of the heavier clusters, such as $[Ru_3(CO)_{10}(PPh_3)_2]$ or $[Os_3(CO)_{10}(PPh_3)_2]$, can be prepared without significant fragmentation into mononuclear complexes.

Example 21.11 Assessing substitutional reactivity

Which of the compounds (**83**) or (**84**) will undergo substitution of a CO ligand for a phosphine more readily?

Answer The 18-electron compound (**84**) contains the indenyl ligand, which can ring slip to a 16-electron η^3 bound form (**85**) more readily that the normal Cp compound (**83**), as the double bond that forms becomes part of a six-membered aromatic ring. This ring slipping provides a low-energy route to coordinative unsaturation and thus the indenyl compound reacts with an incoming ligand much more rapidly than the Cp compound.

Self-test 21.11 Assess the relative substitutional reactivities of indenyl and fluorenyl (**86**) compounds.

21.22 Oxidative addition and reductive elimination

Key points: Oxidative addition occurs when a molecule X—Y adds to a metal atom to form new M—X and M—Y bonds with cleavage of the X—Y bond; oxidative addition results in increasing the coordination number of the metal atom by 2 and increasing the oxidation number by 2; reductive elimination is the reverse of oxidative addition.

When we discussed the bonding of dihydrogen to a metal atom in Section 21.7, we noted that the oxidation number on the metal atom increased by 2 when the dihydrogen reacted

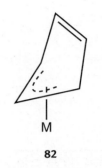

82

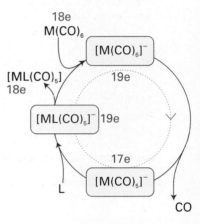

Fig. 21.17 Schematic diagram of an electron transfer catalysed CO substitution. After addition of a small amount of a reducing initiator, the cycle continues until the limiting reagent $M(CO)_6$ or L has been consumed.

83

84

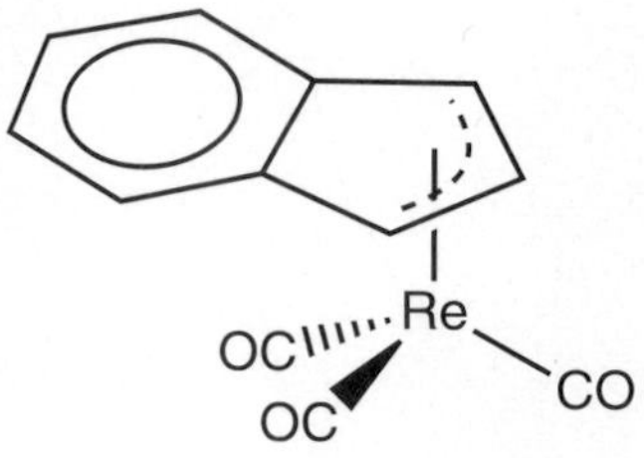

85

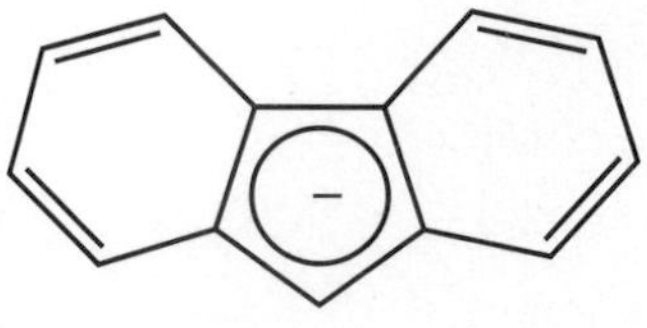

86 Fluorenyl⁻

to give a dihydride:

$$M^{(n)} + H_2 \rightarrow [M^{(n+2)}(H)_2]$$

The increase in oxidation number of the metal by 2 arises because dihydrogen is treated as a neutral ligand, whereas the hydride ligands are treated as H⁻: thus the formation of two M—H bonds from a H₂ molecule corresponds to a formal increase in the charge on the metal charge by 2.[6] This type of reaction is quite general and is known as **oxidative addition**. A large number of molecules add oxidatively to a *d*-metal, including the alkyl and aryl halides, dihydrogen, and simple hydrocarbons. In general, the addition of any molecule X—Y to a metal atom, where both X and Y are more electronegative than the metal, can be classed as oxidative addition. Thus the reaction of metal with an acid such as HCl is an oxidative addition reaction. Oxidative addition reactions are not restricted to *d*-block metals: the reaction of magnesium to form Grignard reagents (Section 11.10) is an oxidative addition reaction.

Oxidative addition reactions result in two more ligands bound to the metal with an increase in the total electron count at the metal of 2. Thus oxidative addition reactions normally require a coordinatively unsaturated metal centre, and are particular common for 16-electron square-planar metal complexes:

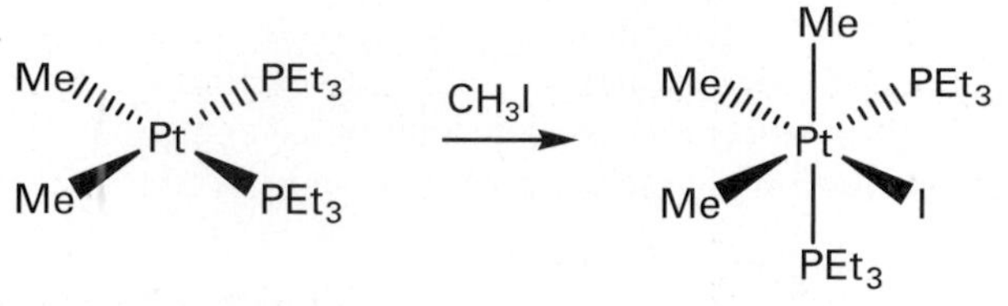

16e Pt(II)   18e Pt(IV)

The oxidative addition of hydrogen is a concerted reaction: dihydrogen coordinates to form a σ-bonded H₂ ligand, and then backbonding from the metal results in cleavage of the H—H bond and the formation of *cis* dihydrides:

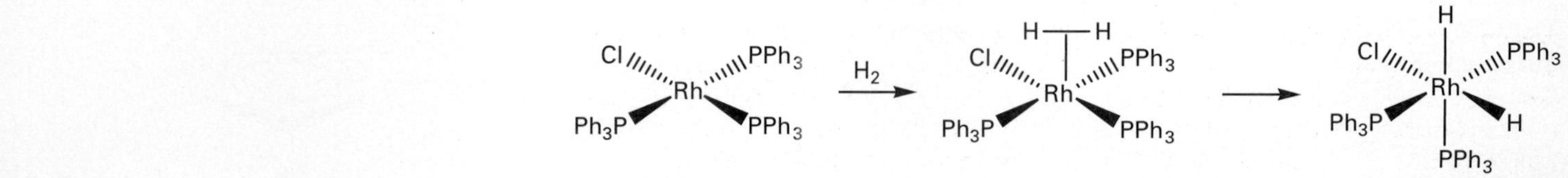

Four-coordinate   Five-coordinate   Six-coordinate
16e Rh(I)   18e Rh(I)   18e Rh(III)

Other molecules, such as alkanes and aryl halides, are known to react in a concerted fashion, and in all these cases the two incoming ligands end up *cis* to each other.

Some oxidative addition reactions are not concerted and either go through radical intermediates or are best thought of as S_N2 displacement reactions. Radical oxidative addition reactions are rare and will not be discussed further here. In an S_N2 oxidative addition reaction, a lone pair on the metal attacks the X—Y molecule displacing Y⁻, which subsequently bonds to the metal.

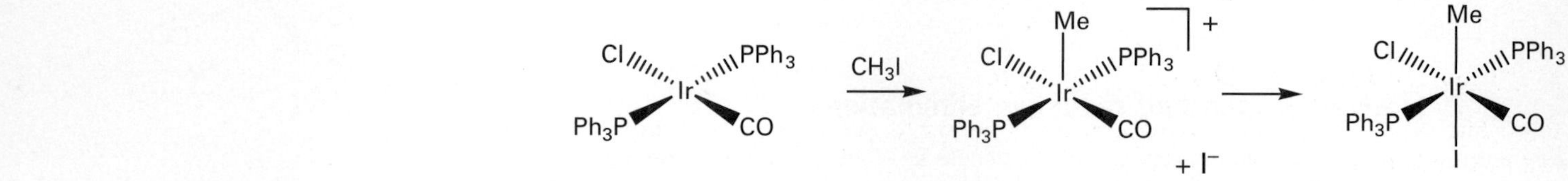

There are two stereochemical consequences of this reaction. First, the two incoming ligands need not end up *cis* to each other and, second, any chirality at the X group is

---

[6] Another way to view the increase in oxidation state of the metal is to consider the metal as using two more electrons when it bonds to two hydrides compared with a single dihydrogen molecule.

inverted. An S_N2-type oxidative addition is common for polar molecules such as alkyl halides.

The opposite of oxidative addition, where two ligands couple and eliminate from a metal centre, is known as **reductive elimination**.

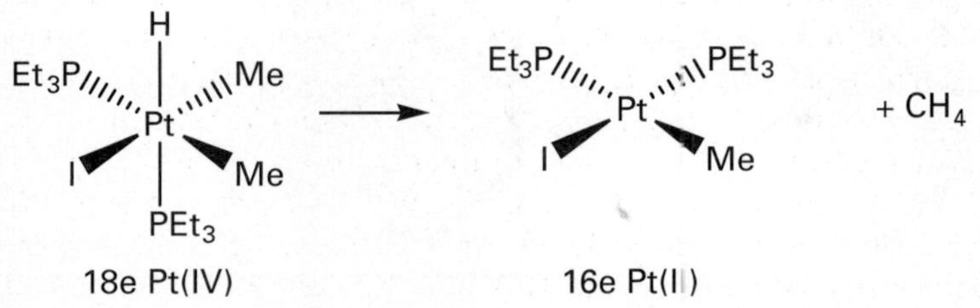

18e Pt(IV)   16e Pt(II)

Reductive elimination reactions require both eliminating fragments to be *cis* to each other, and are best thought of as the reverse of the concerted form of oxidative addition.

Oxidative addition and reductive elimination reactions are, in principle, reversible. However, in practice, one direction is normally thermodynamically favoured over the other.

Example 21.12 Identifying oxidative addition and reductive elimination

Show how the reaction:

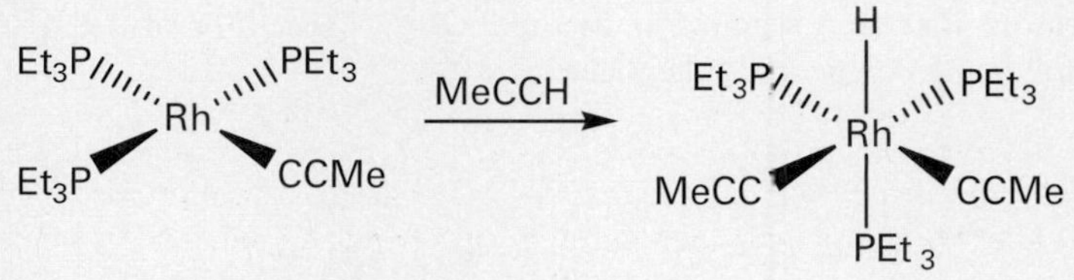

is an example of an oxidative addition reaction.

Answer The four-coordinate square-planar rhodium starting material contains an η^1-alkynyl as well as three neutral phosphine ligands; it is therefore a 16-electron Rh(I) species. The six-coordinate octahedral product contains two η^1-alkynyls, a hydride ligand, and three neutral phosphine ligands; it is therefore an 18-electron Rh(III) species. The increase in both coordination number and oxidation number by 2 identifies it as an oxidative addition.

Self-test 21.12 Show that the reaction:

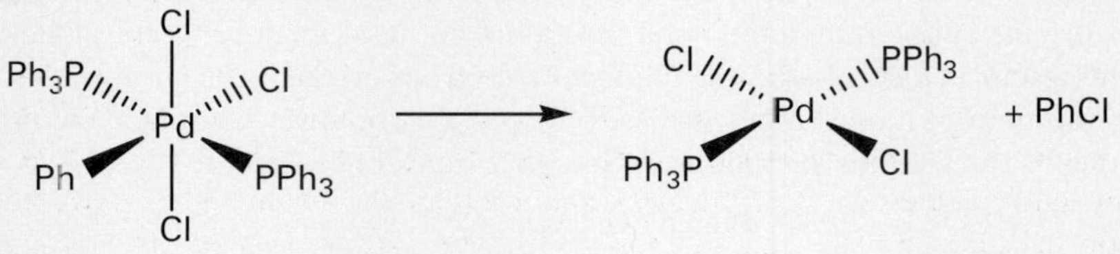

is an example of reductive elimination.

21.23 σ-Bond metathesis

Key point: A σ-bond metathesis reaction is a concerted process that occurs when oxidative addition cannot take place.

A reaction sequence that appears to be a oxidative addition followed by a reductive elimination may in fact be the exchange of two species via a process known as **σ-bond metathesis**. σ-Bond metathesis reactions are common for early *d*-metal complexes where there are not enough electrons on the metal atom for it to participate in oxidative addition. For instance, the 16-electron compound $[Zr(Cp)_2HMe]$ cannot react with H_2

to give a trihydride as all its electrons are involved in bonding to the existing ligands. A four-membered transition state is proposed in such cases, and a concerted bond-making and bond-breaking step results in elimination of methane.

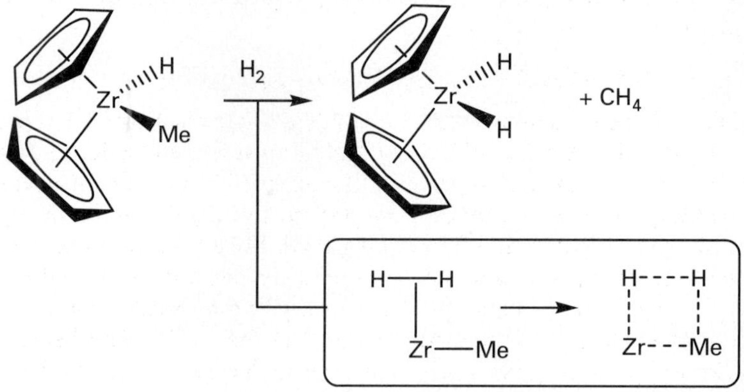

21.24 1,1-Migratory insertion reactions

Key point: 1,1-Migratory insertion reactions result from the migration of a species such as a hydride or alkyl group to an adjacent ligand such as carbonyl to give a metal complex with two fewer electrons on the metal atom.

A **1,1-migratory insertion reaction** is exemplified by a reactions of the η^1-CO ligand, where the following change can take place:

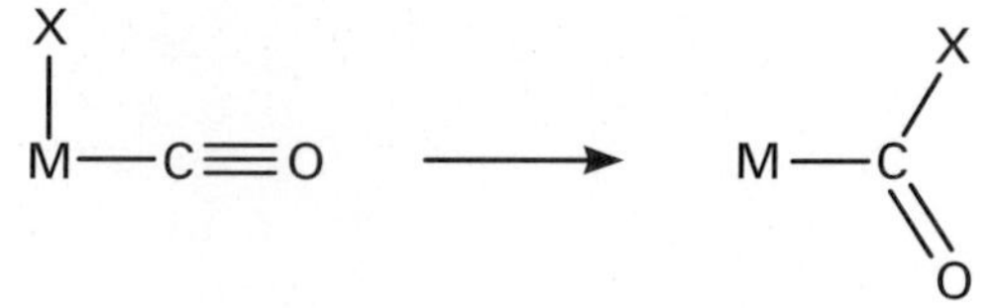

The reaction is called a '1,1-reaction' because the X group that was one bond away from the metal atom ends up on an atom that is one bond away from the metal atom. Typically the X group is an alkyl or aryl species and then the product contains an acyl group. In principle, the reaction could go via a migration of the X group, or an insertion of the CO into the M—X bond. An uncertainty about the actual mechanism has led to the apparently contradictory name **migratory insertion**. Colloquially, however, the terms 'migratory insertion', 'migration', and 'insertion' are used interchangeably. The overall reaction results in a decrease in the number of electrons on the metal atom by 2, with no change in the oxidation state. It is therefore possible to induce 1,1-migratory insertion reactions by the addition of another species that can act as a ligand:

$$[Mn(CH_3)(CO)_5] + PPh_3 \rightarrow [Mn(CH_3CO)(CO)_4PPh_3]$$

The classic study of the migratory insertion of CO with $[CH_3Mn(CO)_5]$ illustrates a number of key features of reactions of this type.[7] First, in the reaction

$$[Mn(CH_3)(CO)_5] + {}^{13}CO \rightarrow [Mn(CH_3CO)(CO)_4({}^{13}CO)]$$

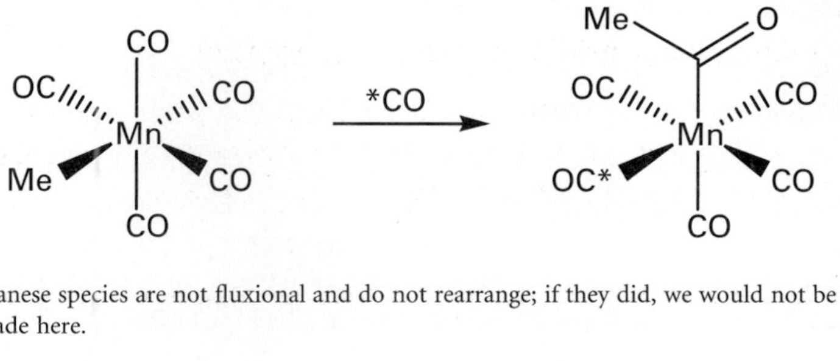

---

[7] The manganese species are not fluxional and do not rearrange; if they did, we would not be able draw the conclusions made here.

the product has only one labelled CO, and that group is *cis* to the newly formed acyl group. This stereochemistry demonstrates that the incoming CO group does not insert into the Mn—CH$_3$ bond, and that either the methyl group migrates to an adjacent CO ligand, or a CO ligand adjacent to the methyl group inserts into the Mn—CH$_3$ bond. Second, in the reverse reaction

$$cis\text{-}[Mn(CH_3CO)(CO)_4(^{13}CO)] \rightarrow [MnCH_3(CO)_5] + CO$$

it is possible to distinguish between the migration of the methyl group and the insertion of the CO ligand: cis-[Mn(CH$_3$CO)(CO)$_4$(^{13}CO)] must lose a CO ligand *cis* to the acyl group in order for the reaction to proceed. The scheme below summarizes the potential reaction pathways. In one quarter of the instances, this ligand will be the labelled CO and there will be no significant information gained. In half the instances, a nonlabelled CO will be lost, leaving a vacant site *cis* to both the labelled CO ligand and the acyl group. In this case, either migration of the methyl group back to the metal atom or extrusion of CO will lead to the methyl group and the ^{13}CO ligand being *cis* to each other. However, in the remaining one quarter of the instances, the CO that is *trans* to the labelled CO ligand will be lost, and in this case it is possible to distinguish methyl group migration from CO ligand extrusion. If the methyl group migrates, it ends up *trans* to the labelled CO, whereas if the CO is extruded, the methyl group ends up *cis* to the CO.

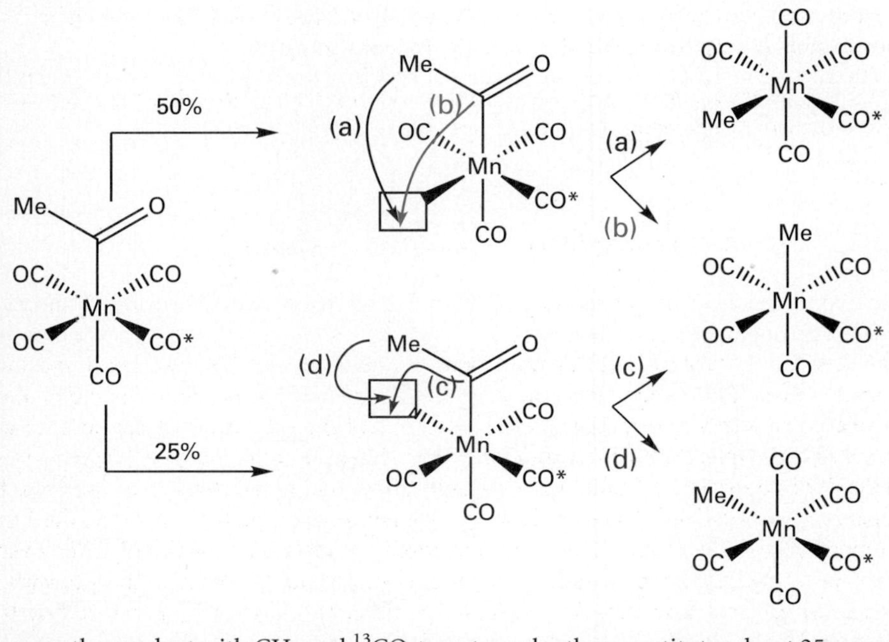

Because the product with CH$_3$ and ^{13}CO *trans* to each other constitutes about 25 per cent of the product, we can conclude that the CH$_3$ group does indeed migrate. Applying the principle of microscopic reversibility[8] allows us to conclude that the forward reaction proceeds by methyl-group migration. All 1,1-migratory insertions are now thought to proceed by migration of the X group. An important consequence of this pathway is that the relative positions of the other groups on the migrating atom are left unchanged, so the stereochemistry at the X group is preserved.

21.25 1,2-Insertions and β-hydride elimination

Key points: 1,2-Insertion reactions are observed with η^2 ligands such as alkenes and result in the formation of an η^1 ligand with no change in oxidation state of the metal; β-hydride elimination is the reverse of 1,2-insertion.

---

[8] The principle of microscopic reversibility states that both forward and reverse reactions proceed via the same mechanism.

1,2-Insertion reactions are commonly observed with η^2-ligands, such as alkenes and alkynes, and are exemplified by the reaction

The reaction is a **1,2-insertion** because the X group that was one bond away from the metal atom ends up on an atom that is two bonds away from the metal. Typically, the X group is a hydride, alkyl, or aryl species, in which case the product contains a (substituted) alkyl group. Like 1,1-insertion reactions, the overall reaction results in a decrease in the number of electrons on the metal atom by 2, with no change in the oxidation state.

If, in the above reaction with X=H, another ethene molecule were to coordinate, the resultant ethyl group could migrate to give a butyl group:

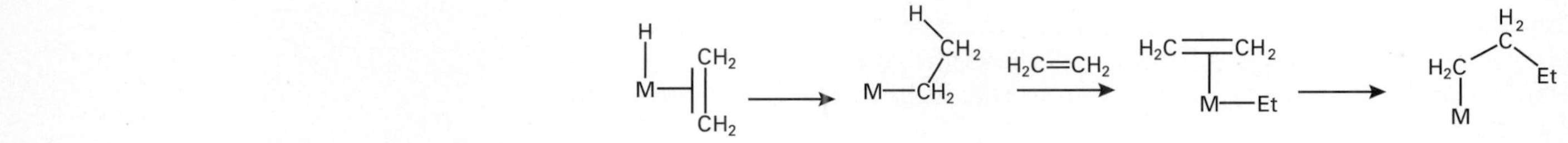

Repetition of this process gives polyethene. Catalytic reactions of this kind are of very considerable industrial importance, and are discussed in Chapter 25.

The reverse of 1,2-insertions can occur but this is rare except when X=H, when the reaction is known as β-**hydride elimination**:[9]

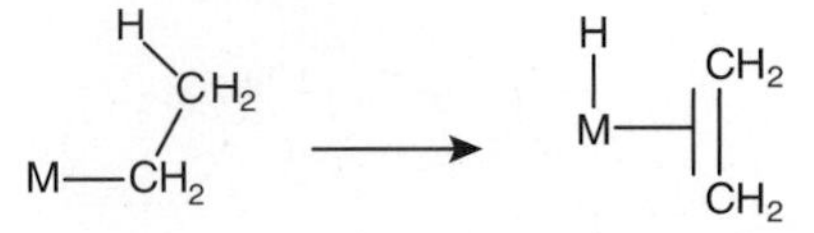

The experimental evidence shows that both 1,2-insertion and β-hydride elimination proceed through a syn intermediate:

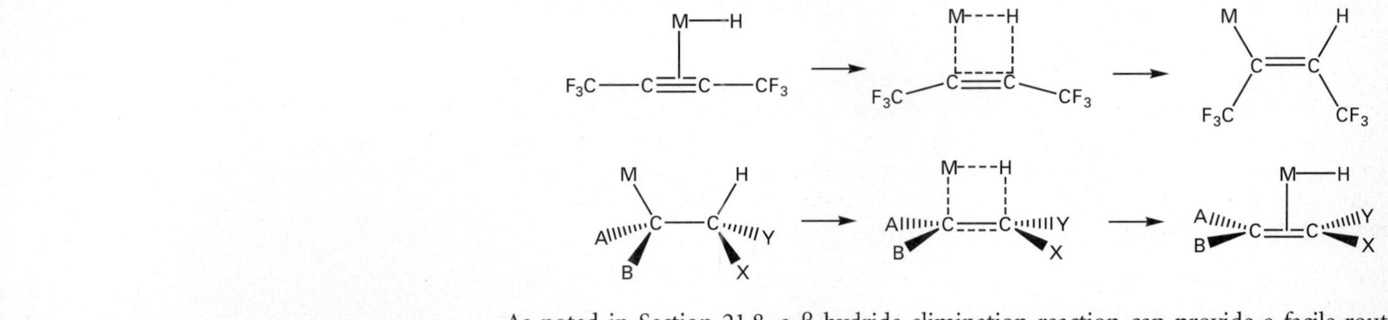

As noted in Section 21.8, a β-hydride elimination reaction can provide a facile route for decomposition of alkyl-containing compounds. The 1,2-insertion reaction, coupled with the β-hydride elimination, can also provide a low-energy route to alkene isomerization:

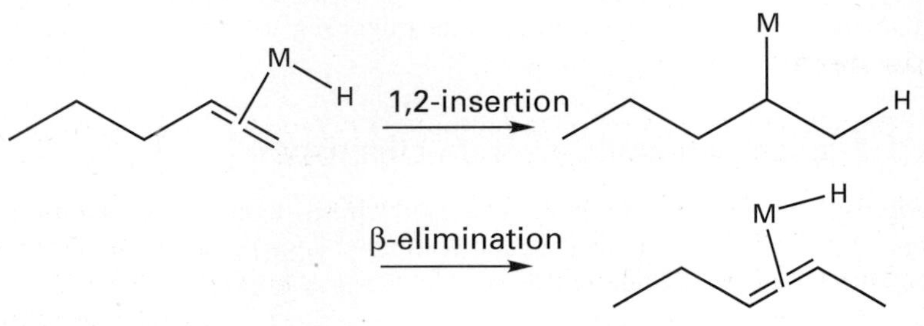

---

[9] The reaction is known as 'β-hydride elimination' because the H atom that is eliminated is on the second carbon atom from the metal atom (the carbon atom bonded to the metal atom is the α carbon, the third one the γ carbon, etc.).

21.26 α-, γ-, and δ-Hydride eliminations and cyclometallations

Key point: Cyclometallation reactions, in which a metal inserts into a remote C—H bond, are equivalent to hydride elimination reactions.

α-Hydride eliminations are occasionally found for complexes that have no β-hydrogens and the reaction gives rise to a carbene that is often highly reactive:

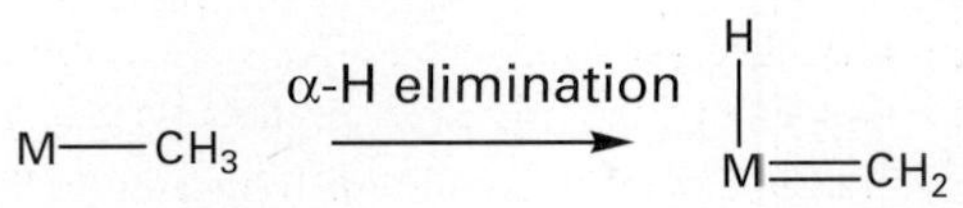

γ-Hydride and δ-hydride eliminations are more commonly observed. Because the product contains a **metallocycle**, a cyclic structure incorporating a metal atom, these reactions are normally described as **cyclometallation** reactions:

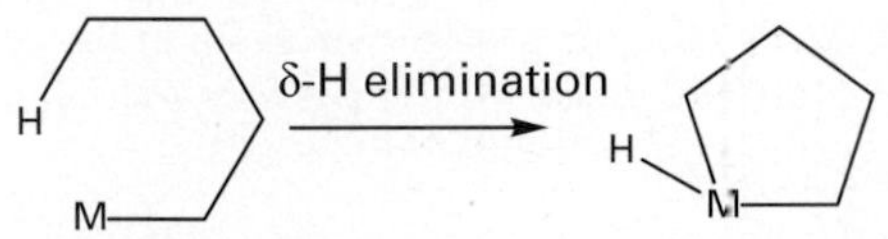

A cyclometallation reaction is often also thought of as the oxidative addition to an adjacent C—H bond. Both α- and β-hydride eliminations can also be considered as cyclometallation reactions. This identification is more obvious for the β-hydride elimination if we consider an alkene in its metallacyclopropane form:

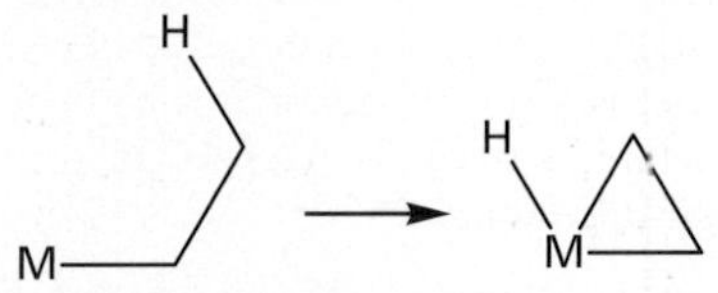

Example 21.13 Predicting the outcomes of insertion and elimination reactions

What product, including its stereochemistry, would you expect from the reaction between [MnMe(CO)$_5$] and PPh$_3$?

Answer The reaction between [MnMe(CO)$_5$] and PPh$_3$ is unlikely to be the simple replacement of a carbonyl ligand by the phosphine as this reaction would require a strongly bound carbonyl ligand to dissociate. A more likely reaction would be the migration of the methyl group on to an adjacent CO ligand to give an acyl group, with the phosphine filling the vacated coordination site. This reaction has a low activation barrier, and the product would therefore be expected to be *cis*-[Mn(MeCO)(PPh$_3$)(CO)$_4$].

Self-test 21.13 Explain why [Pt(PEt$_3$)$_2$(Et)(Cl)] readily decomposes, whereas [Pt(PEt$_3$)$_2$(Me)(Cl)] does not.

FURTHER READING

R.H. Crabtree, *The organometallic chemistry of the transition metals.* Wiley, New York (2001). The best single volume book on the subject.

G.O. Spessard and G.L. Miessler, *Organometallic Chemistry.* Prentice-Hall, London (1996).

C. Elschenbroich and A Salzer, *Organometallics.* Wiley-VCH, Weinheim (2003).

D.M.P. Mingos and R.H. Crabtree (eds). *Comprehensive Organometallic Chemistry III.* Elsevier, Oxford, 2006. The definitive reference work that builds on the two earlier editions.

See *J. Organomet. Chem.*, 1975, **100**, 273 for Wilkinson's personal account of the development of metallocene chemistry.

G.J. Kubas, *Acc. Chem. Res.*, 1988, **21**, 120. An historical perspective and a full account of the discovery of dihydrogen complexes.

W. Scherer and G.S. McGrady, *Angew. Chem., Int. Ed. Engl.*, 2004, **43**, 1782. A review of agostic interactions that gives a good historical perspective.

D.C. Grills and M.W. George, *Adv. Inorg. Chem.*, 2001, **52**, 113. A review of the topic of noble gas complexes.

D.M.P. Mingos and D.J. Wales, *Introduction to cluster chemistry*. Prentice Hall, Englewood Cliffs (1990); J.W. Lauher, *J. Am. Chem. Soc.*, 1978, **100**, 5305. Descriptions of some of the concepts of cluster bonding.

R. Hoffmann, *Angew. Chem., Int. Ed. Engl.*, 1982, **21**, 711. The application of isolobal analogies to metal cluster compounds (Hoffmann's Nobel Prize lecture).

J. Ellis, *J. Chem. Educ.*, 1976, **53**, 2. Structures of organometallics and clusters systematized by isoelectronic relations.

EXERCISES

21.1 Name the species, draw the structures of, and give valence electron counts to the metal atoms in: (a) $Fe(CO)_5$, (b) $Mn_2(CO)_{10}$, (c) $V(CO)_6$, (d) $[Fe(CO)_4]^{2-}$, (e) $La(\eta^5\text{-}Cp^*)_3$, (f) $Fe(\eta^3\text{-allyl})(CO)_3Cl$, (g) $Fe(CO)_4(PPh_3)$, (h) $Rh(CO)_2(Me)(PPh_3)$, (i) $Pd(Cl)(Me)(PPh_3)_2$, (j) $Co(\eta^5\text{-}C_5H_5)(\eta^4\text{-}C_4Ph_4)$, (k) $[Fe(\eta^5\text{-}C_5H_5)(CO)_2]^-$, (l) $Cr(\eta^6\text{-}C_6H_6)(\eta^6\text{-}C_7H_8)$, (m) $Ta(\eta^5\text{-}C_5H_5)_2Cl_3$, (n) $Ni(\eta^5\text{-}C_5H_5)NO$. Do any of the complexes deviate from the 18-electron rule? If so, how is this reflected in their structure or chemical properties?

21.2 (a) Sketch an η^2 interaction of 1,4-butadiene with a metal atom, and (b) do the same for an η^4 interaction.

21.3 What hapticities are possible for the interaction of each of the following ligands with a single d-block metal atom such as cobalt? (a) C_2H_4, (b) cyclopentadienyl, (c) C_6H_6, (d) cyclooctadiene, (e) cyclooctatetraene.

21.4 Draw plausible structures and give the electron count of (a) $Ni(\eta^3\text{-}C_3H_5)_2$, (b) η^4-cyclobutadiene-η^5-cyclopentadienylcobalt, (c) $(\eta^3\text{-}C_3H_5)Co(CO)_2$. If the electron count deviates from 18, is the deviation explicable in terms of periodic trends?

21.5 State the two common methods for the preparation of simple metal carbonyls and illustrate your answer with chemical equations. Is the selection of method based on thermodynamic or kinetic considerations?

21.6 Suggest a sequence of reactions for the preparation of $Fe(dppe)(CO)_3$, given iron metal, CO, dppe $(Ph_2PCH_2CH_2PPh_2)$, and other reagents of your choice.

21.7 Suppose that you are given a series of metal tricarbonyl compounds having the respective symmetries C_{2v}, D_{3h}, and C_s. Without consulting reference material, which of these should display the greatest number of CO stretching bands in the IR spectrum? Check your answer and give the number of expected bands for each by consulting Table 21.7.

21.8 Provide plausible reasons for the differences in IR wavenumbers between each of the following pairs: (a) $Mo(PF_3)_3(CO)_3$ 2040, 1991 cm^{-1} versus $Mo(PMe_3)_3(CO)_3$ 1945, 1851 cm^{-1}, (b) $MnCp(CO)_3$ 2023, 1939 cm^{-1} versus $MnCp^*(CO)_3$ 2017, 1928 cm^{-1}.

21.9 The compound $Ni_3(C_5H_5)_3(CO)_2$ has a single CO stretching absorption at 1761 cm^{-1}. The IR data indicate that all C_5H_5 ligands are pentahapto and probably in identical environments. (a) On the basis of these data, propose a structure. (b) Does the electron count for each metal in your structure agree with the 18-electron rule? If not, is nickel in a region of the periodic table where deviations from the 18-electron rule are common?

21.10 Decide which of the two complexes (a) $W(CO)_6$ or (b) $IrCl(PPh_3)_2(CO)$ should undergo the fastest exchange with ^{13}CO. Justify your answer.

21.11 Which metal carbonyl in each of (a) $[Fe(CO)_4]^{2-}$ or $[Co(CO)_4]^-$, (b) $[Mn(CO)_5]^-$ or $[Re(CO)_5]^-$ should be the most basic towards a proton? What are the trends upon which your answer is based?

21.12 Using the 18-electron rule as a guide, indicate the probable number of carbonyl ligands in (a) $W(\eta^6\text{-}C_6H_6)(CO)_n$, (b) $Rh(\eta^5\text{-}C_5H_5)(CO)_n$, and (c) $Ru_3(CO)_n$.

21.13 Propose two syntheses for $MnMe(CO)_5$, both starting with $Mn_2(CO)_{10}$, with one using Na and one using Br_2. You may use other reagents of your choice.

21.14 Give the probable structure of the product obtained when $Mo(CO)_6$ is allowed to react first with LiPh and then with the strong carbocation reagent, $CH_3OSO_2CF_3$.

21.15 $Na[(\eta^5\text{-}C_5H_5)W(CO)_3]$ reacts with 3-chloroprop-1-ene to give a solid, A, which has the molecular formula $(C_3H_5)(C_5H_5)(CO)_3W$. Compound A loses carbon monoxide on exposure to light and forms compound B, which has the formula $(C_3H_5)(C_5H_5)(CO)_2W$. Treating compound A with hydrogen chloride and then potassium hexafluorophosphate, $K^+PF_6^-$, results in the formation of a salt, C. Compound C has the molecular formula $[(C_3H_6)(C_5H_5)(CO)_3W]PF_6$. Use this information and the 18-electron rule to identify the compounds A, B, and C. Sketch a structure for each, paying particular attention to the hapticity of the hydrocarbon.

21.16 Treatment of $TiCl_4$ at low temperature with EtMgBr gives a compound that is unstable above $-70°C$. However, treatment of $TiCl_4$ at low temperature with MeLi or $LiCH_2SiMe_3$ gives compounds that are stable at room temperature. Rationalize these observations.

21.17 Suggest syntheses of (a) $[Mo(\eta^7\text{-}C_7H_7)(CO)_3]BF_4$ from $Mo(CO)_6$ and (b) $[IrCl_2(COMe)(CO)(PPh_3)_2]$ from $[IrCl(CO)(PPh_3)_2]$.

21.18 Which compound would you expect to be more stable, $Rh(\eta^5\text{-}C_5H_5)_2$ or $Ru(\eta^5\text{-}C_5H_5)_2$? Give a plausible explanation for the difference in terms of simple bonding concepts.

21.19 Give the equation for a workable reaction that will convert $Fe(\eta^5\text{-}C_5H_5)_2$ into (a) $Fe(\eta^5\text{-}C_5H_5)(\eta^5\text{-}C_5H_4COCH_3)$ and (b) $Fe(\eta^5\text{-}C_5H_5)(\eta^5\text{-}C_5H_4CO_2H)$.

21.20 Sketch the a_1' symmetry-adapted orbitals for the two eclipsed C_5H_5 ligands stacked together with D_{5h} symmetry. Identify the s, p, and d orbitals of a metal atom lying between the rings that may have nonzero overlap, and state how many a_1' molecular orbitals may be formed.

21.21 The compound $Ni(\eta^5\text{-}C_5H_5)_2$ readily adds one molecule of HF to yield $[Ni(\eta^5\text{-}C_5H_5)(\eta^4\text{-}C_5H_6)]^+$ whereas $Fe(\eta^5\text{-}C_5H_5)_2$ reacts with strong acid to yield $[Fe(\eta^5\text{-}C_5H_5)_2H]^+$. In the latter compound the H atom is attached to the Fe atom. Provide a reasonable explanation for this difference.

21.22 Write a plausible mechanism, giving your reasoning, for the reactions

(a) $[Mn(CO)_5(CF_2)]^+ + H_2O \rightarrow [Mn(CO)_6]^+ + 2\,HF$

(b) $Rh(CO)(C_2H_5)(PR_3)_2 \rightarrow RhH(CO)(PR_3)_2 + C_2H_4$

21.23 The mechanism of CO insertion is thought to be

$$RMn(CO)_5 \rightleftharpoons (RCO)Mn(CO)_4$$
$$(RCO)Mn(CO)_4 + L \rightleftharpoons (RCO)MnL(CO)_4$$

In the first step, the acyl intermediate is formed, leaving a vacant coordination position. In the second step, a ligand enters. At high ligand concentrations, the rate is independent of [L]. At this limit, what rate constant can be extracted from rate data?

21.24 (a) What cluster valence electron (CVE) count is characteristic of octahedral and trigonal prismatic complexes? (b) Can these CVE values be derived from the 18-electron rule? (c) Determine the probable geometry (octahedral or trigonal prismatic) of $[Fe_6(C)(CO)_{16}]^{2-}$ and $[Co_6(C)(CO)_{16}]^{2-}$. (The C atom in both cases resides in the centre of the cluster and can be considered to be a four-electron donor.)

21.25 Based on isolobal analogies, choose the groups that might replace the group in boldface in

(a) $Co_2(CO)_9\mathbf{CH}$ $OCH_3, N(CH_3)_2$, or $SiCH_3$

(b) $(OC)_5Mn\mathbf{Mn(CO)}_5$ I, CH_2, or CCH_3

21.26 Ligand substitution reactions on metal clusters are often found to occur by associative mechanisms, and it is postulated that these occur by initial breaking of an M—M bond, thereby providing an open coordination site for the incoming ligand. If the proposed mechanism is applicable, which would you expect to undergo the fastest exchange with added ^{13}CO? $Co_4(CO)_{12}$ or $Ir_4(CO)_{12}$? Suggest an explanation.

PROBLEMS

21.1 Propose the structure of the product obtained by the reaction of $[Re(CO)(\eta^5-C_5H_5)(PPh_3)(NO)]^+$ with $Li[HBEt_3]$. The latter contains a strongly nucleophilic hydride. (For full details see: W. Tam, G. Y. Lin, W.K. Wong, W.A. Kiel, V. Wong, and J.A. Gladysz, *J. Am. Chem. Soc.*, 1982, **104**, 141.)

21.2 Treatment of $TiCl_4$ with 4 equivalents of NaCp gives a single organometallic compound (together with NaCl by-product). At room temperature the ^1H NMR spectrum shows a single sharp singlet; on cooling to $-40°C$ this singlet separates into two singlets of equal intensity; further cooling results in one of the singlets separating into three signals with intensity ratios 1:2:2. Explain these results.

21.3 When several CO ligands are present in a metal carbonyl, an indication of the individual bond strengths can be determined by means of force constants derived from the experimental IR frequencies. In $Cr(CO)_5(Ph_3P)$ the *cis*-CO ligands have the higher force constants, whereas in $Ph_3SnCo(CO)_4$ the force constants are higher for the *trans*-CO. Suggest why, and explain which carbonyl C atoms should be susceptible to nucleophiles to attack in these two cases. (For details see D.J. Darensbourg and M.Y. Darensbourg, *Inorg. Chem.*, 1970, **9**, 1691.)

21.4 It is often possible to assign different tautomeric structures to an organometallic compound; for example, the structure below can be thought of as existing as either the carbene structure on the left, or the charge separated aromatic form on the right.

Suggest ways of distinguishing the two forms. (For details see: G.W.V. Cave, A.J. Hallett, W. Errington, and J.P. Rourke, *Angew. Chem., Int. Ed. Engl.*, 1998, **37**, 3270.)

21.5 Describe how NMR has been used to identify alkane complexes of *d* metals unambiguously. See S. Geftakis and G.E. Ball, *J. Am. Chem. Soc.*, 1998, **120**, 9953 and D.J. Lawes, S. Geftakis, and G.E. Ball, *J. Am. Chem. Soc.*, 2005, **127**, 4134.

21.6 The dinitrogen complex $Zr_2(\eta^5-Cp^*)_4(N_2)_3$ has been isolated and its structure determined by single-crystal X-ray diffraction. Each Zr atom is bonded to two Cp* and one terminal N_2. The third N_2 bridges between the Zr atoms in a nearly linear ZrNNZr array. Before consulting the reference, write a plausible structure for this compound that accounts for the ^1H NMR spectrum obtained on a sample held at $-7°C$. This spectrum shows two singlets indicating that the Cp* rings are in two different environments. At somewhat above room temperature these rings become equivalent on the NMR timescale and ^{15}N NMR indicates that N_2 exchange between the terminal ligands and dissolved N_2 is correlated with the process that interconverts the Cp* ligand sites. Propose a way in which this equilibration could interconvert the sites of the Cp* ligands. (For further details see J.M. Manriquez, D.R. McAlister, E. Rosenberg, H.M. Shiller, K.L. Willamson, S.I. Chan, and J.E. Bercaw, *J. Am. Chem. Soc.*, 1978, **100**, 3078.)

21.7 How might you unambiguously identify an organometallic complex containing a noble gas atom as a ligand? See G.E. Ball, T.A. Darwish, S. Geftakis, M.W. George, D.J. Lawes, P. Portius, and J.P. Rourke, *Proc. Natl Acad. Sci.*, 2005, **102**, 1853.

21.8 If any of the following statements are incorrect, provide a corrected counterpart and an example that bears our your correction. (a) A high-field NMR signal is characteristic of the hydride ligand in Groups 8 and 9 organometallic complexes. (b) The appearance of a single feature in the ^1H NMR spectrum of a cyclooctatetraene metal complex is proof that the ligand is bound in a symmetric η^8 manner. (c) Square-planar 16-electron compounds generally undergo ligand substitution by a dissociative mechanism. (d) Metal–metal bonds in cluster compounds generally become weaker going down a group in the *d* block.

21.9 What conclusions can you draw from the bonding and reactivity of dihydrogen bound to a low oxidation state *d*-block metal that might be applicable to the bonding of an alkane to a metal? What implications might this have for the oxidative addition of a hydrocarbon to a metal atom? (See, for example, R.H. Crabtree, *J. Organomet. Chem.*, 2004, **689**, 4083.)

21.10 Compare and contrast Fischer and Schrock carbenes. See E.O. Fischer, *Adv. Organomet. Chem.*, 1976, **14**, 1 and R.R. Schrock, *Acc. Chem. Res.*, 1984, **12**, 98.

22

The *f*-block metals

In this chapter we discuss the chemistry of the *f*-block elements. There are two series, each of 14 elements, in the *f* block, and relatively little diverse chemistry has been reported for such a large number of elements. For the 4*f* elements (the lanthanoids) this lack of diversity is normally ascribed to the buried nature of the 4*f* electrons. The chemical properties of the lanthanoids are similar to those of other electropositive metals. The fact that only two of the 5*f* elements (the actinoids) occur naturally—all other actinoids are synthetic and often intensely radioactive—has hindered their study and little is known about the chemical properties of many of them.

The two series of elements in the *f* block correspond to the filling of the seven 4*f* and 5*f* orbitals, respectively. This occupation of *f* orbitals from f^1 to f^{14} corresponds to the elements cerium to lutetium in Period 6 and thorium to lawrencium in Period 7; however, given the similarity of their chemical properties, the elements lanthanum and actinium are normally included in discussion of the *f* block, as we shall do here. The 4*f* elements are normally referred to as the **lanthanoids** (formerly and still commonly 'the lanthanides') and the 5*f* elements as the **actinoids** (the 'actinides') There is a striking uniformity in the properties of the 4*f* elements and greater diversity in the chemistry of the 5*f* elements. A general lanthanoid is represented by the symbol Ln and an actinoid by An. The lanthanoids are sometimes referred to as the 'rare earth elements'; however, that name is inappropriate because they are not particularly rare, except for promethium, which has no stable isotopes.

The elements

We begin our study of the *f*-block elements by considering the properties and the extraction of the elements. The chemical properties of the lanthanoids are quite different from those of the actinoids and we discuss them separately.

22.1 Occurrence and recovery

Key points: The principal sources of the lanthanoids are phosphate minerals; the most important actinoid, uranium, is recovered from its oxide.

Other than promethium, the lanthanoids are reasonably common in the Earth's crust; indeed, even the 'rarest' lanthanoid, thulium, has a crustal abundance that is similar to that of iodine. The principal mineral source for the early lanthanoids is monazite, $(Ln,Th)PO_4$, which contains mixtures of lanthanoids and thorium. Another phosphate mineral, xenotime (of similar composition), is the principal source of the heavier lanthanoids (and yttrium, which has similar atomic radius and chemical properties). The common oxidation state for the lanthanoids is Ln(III), which makes separation difficult, although cerium, which can be oxidized to Ce(IV), and europium, which can be reduced to Eu(II), are separable from the other lanthanoids by exploiting their redox chemistry. Separation of the remaining Ln^{3+} ions is accomplished on a large scale by multistep liquid–liquid extraction in which the ions distribute themselves between an aqueous phase and an organic phase containing complexing agents. Ion-exchange chromatography is used to separate the individual lanthanoid ions when high purity is required. Pure and mixed lanthanoid metals are prepared by the electrolysis of molten lanthanoid halides.

Beyond lead ($Z=82$) no element has a stable isotope, but two of the actinoids, thorium (Th, $Z=90$) and uranium (U, $Z=92$), have isotopes that are sufficiently long-lived that significant quantities have persisted from their formation in the supernova of the star that preceded the formation of the Sun (Section 1.2). Table 22.1 gives the half-lives of the most stable actinoid isotopes.

Thorium is extracted from either monazite or thorite (essentially $ThSiO_4$); the major ores for uranium, uranite and pitchblende, have the approximate formulas UO_2 and U_3O_8, respectively. The primary source of the other actinoids is synthesis by nuclear reactions; all of them are more radioactive than thorium and uranium.

22.2 Physical properties and uses

Key points: Lanthanoids are reactive metals that find small-scale uses in specialist applications; the primary uses of actinoids are restricted by their radioactivity.

The lanthanoids are soft white metals that have densities comparable to those of the $3d$ metals (6–10 g cm^{-3}). For metals, they are relatively poor conductors of both heat and electricity, with thermal and electrical conductivities 25 and 50 times less than that of copper, respectively. The metals react with steam and dilute acids but are somewhat passivated by an oxide coating.

A mixture of the early lanthanoid metals, including cerium, is referred to in commerce as *mischmetal*. It is used in steel making to remove impurities such as oxygen, hydrogen, sulfur, and arsenic, which reduce the mechanical strength and ductility of steel. Compounds of the lanthanoids find a wide range of applications, many of which depend on the optical properties associated with their *f–f* electronic transitions (Section 22.7): europium oxide and europium orthovanadate are used as red phosphors in cathode-ray

Table 22.1 Half-lives of the most stable actinoid isotopes

Z	Name	Symbol	Mass number	$t_{\frac{1}{2}}$
89	Actinium	Ac	227	21.8 years
90	Thorium	Th	232	1.41×10^{10} years
91	Protactinium	Pa	231	3.28×10^4 years
92	Uranium	U	238	4.47×10^9 years
93	Neptunium	Np	237	2.14×10^6 years
94	Plutonium	Pu	244	8.1×10^7 years
95	Americium	Am	243	7.38×10^3 years
96	Curium	Cm	247	1.6×10^7 years
97	Berkelium	Bk	247	1.38×10^3 years
98	Californium	Cf	251	900 years
99	Einsteinium	Es	252	460 days
100	Fermium	Fm	257	100 days
101	Mendelevium	Md	258	55 days
102	Nobelium	No	259	1.0 h
103	Lawrencium	Lr	260	3 min

Box 22.1 Nuclear fission

The fission of heavy elements, such as ^{235}U, can be induced by bombardment by neutrons. Thermal neutrons (neutrons with low velocities) bring about the fission of ^{235}U to produce two nuclides of medium mass, and a large amount of energy is released because the binding energy per nucleon decreases steadily for

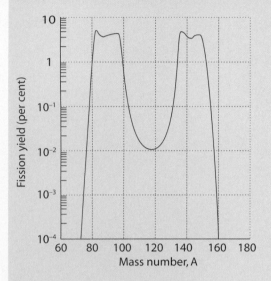

atomic numbers beyond about 26 (Fe; see Fig. 1.1). That unsymmetrical fission of the uranium nucleus has a high probability of occurring is shown by the double-humped distribution of fission products (see the illustration), which has maxima close to mass numbers 95 (isotopes of Mo) and 135 (isotopes of Ba). Almost all the fission products are unstable nuclides. The most troublesome are those with half-lives in the range of years to centuries: these nuclides decay fast enough to be highly radioactive but not sufficiently fast to disappear in a convenient time.

The first nuclear power plants relied on the fission of uranium to generate heat. The heat was used to produce steam, which can drive turbines in much the same way as conventional power plants use the burning of fuels to produce heat. However, the energy produced by the fission of a heavy element is huge in comparison with the burning of conventional fuels: for instance, the complete combustion of 1 kg of octane produces approximately 50 MJ, whereas the energy liberated by the fission of 1 kg of ^{235}U is approximately 2 TJ (1 TJ $= 10^{12}$ J), 40 000 times as much. Later designs of nuclear power plants used plutonium, normally mixed with uranium. Although nuclear power offers the potential of enormous quantities of energy at low cost, no satisfactory method of disposal of the radioactive waste it produces has yet been found.

tube displays, and neodymium (Nd^{3+}), samarium (Sm^{3+}), and holmium (Ho^{3+}) are used in solid-state lasers.

The density of the actinoids increases from 10.1 g cm^{-3} for actinium to 20.4 g cm^{-3} for neptunium, before decreasing for the remainder of the series. Plutonium metal has at least six phases at atmospheric pressure with density differences of more than 20 per cent. Many of the physical and chemical properties of the actinoids are unknown because only miniscule quantities have ever been isolated; in addition to being radioactive, many actinoids are known to be chemically toxic and they are all treated as hazardous. The current primary use for actinoids is in nuclear reactors (Box 22.1) and nuclear weapons.

Lanthanoid chemistry

The lanthanoids are all electropositive metals with a remarkable uniformity of chemical properties. Frequently the only significant difference between two lanthanoids is their size; the ability to choose a lanthanoid of a particular size often lets us 'tune' the properties of their compounds.

22.3 General trends

Key points: The lanthanoids are electropositive metals that commonly occur as Ln(III); other oxidation states are stable only when an empty, half-filled, or full *f* subshell is present.

The elements La through to Lu are all highly electropositive, with electronegativities lying between those of Li and Mg. The elements favour the oxidation state Ln(III) with a uniformity that is unprecedented in the periodic table. Other properties of the elements vary significantly. For example, the radii of the Ln^{3+} ions contract steadily from 116 pm for La^{3+} to 98 pm for Lu^{3+} (Table 22.2). The decrease in ionic radius is attributed in part to the increase in Z_{eff} as electrons are added to the 4*f* subshell (Section 1.8), but detailed calculations indicate that subtle relativistic effects also make a substantial contribution.

Table 22.2 Names, symbols, and selected properties of the lanthanoids

Z	Name	Symbol	Configuration (M^{3+})	$E^{\ominus}$/V	r(M^{3+})/pm*	O.N.†
57	Lanthanum	La	[Xe]	−2.38	116	2(n), **3**, 4
58	Cerium	Ce	[Xe]f^1	−2.34	114	2(n), **3**, 4
59	Praseodymium	Pr	[Xe]f^2	−2.35	113	2(n), **3**, 4
60	Neodymium	Nd	[Xe]f^3	−2.32	111	2(n), **3**
61	Promethium	Pm	[Xe]f^4	−2.29	109	**3**
62	Samarium	Sm	[Xe]f^5	−2.30	108	2(n), **3**
63	Europium	Eu	[Xe]f^6	−1.99	107	2, **3**
64	Gadolinium	Gd	[Xe]f^7	−2.28	105	**3**
65	Terbium	Tb	[Xe]f^8	−2.31	104	**3**, 4
66	Dysprosium	Dy	[Xe]f^9	−2.29	103	2(n), **3**
67	Holmium	Ho	[Xe]f^{10}	−2.33	102	**3**
68	Erbium	Er	[Xe]f^{11}	−2.32	100	**3**
69	Thulium	Tm	[Xe]f^{12}	−2.32	99	2(n), **3**
70	Ytterbium	Yb	[Xe]f^{13}	−2.22	99	2, **3**
71	Lutetium	Lu	[Xe]f^{14}	−2.30	98	**3**

*Ionic radii for coordination number 8 from R.D. Shannon, *Acta Cryst.*, 1976, **A32**, 751.
†Oxidation numbers in bold type indicate the most stable states; (n) indicates that the state is stable only in nonaqueous conditions.

This 18 per cent decrease in radius leads to an increase in the hydration enthalpy across the series. The reduction potentials of the lanthanoids are all very similar, with the standard potential of the La^{3+}/La couple, −2.38 V, similar to that for Lu^{3+}/Lu, −2.30 V, at the other end of the series.

The common occurrence of the lanthanoids as Ln(III) is normally ascribed to the fact that once the valence *s* and *d* electrons have been removed, the *f* electrons are held tightly by the nucleus and do not extend beyond the xenon-like core of the atom. Thus, a Ln^{3+} ion has no frontier orbitals with directional preference, and their bonding is best considered as due to the electrostatic attraction of ions. Ligand-field stabilization plays no part in the chemical properties of the lanthanoid complexes. Superimposed on this uniformity of formation of Ln^{3+} there are some atypical oxidation states that are most prevalent when the ion can attain an empty (f^0), half-filled (f^7), or filled (f^{14}) subshell (Table 22.2). Thus, Ce^{3+}, which is an f^1 ion, can be oxidized to the f^0 ion Ce^{4+}, a strong and useful oxidizing agent. The next most common of the atypical oxidation states is Eu^{2+}, which is an f^7 ion that reduces water.

A Ln^{3+} ion is a hard Lewis acid, as indicated by its preference for F$^-$ and oxygen-containing ligands and its occurrence with PO$_4^{3-}$ in minerals.

22.4 Binary ionic compounds

Key points: The structures of ionic lanthanoid compounds are determined by the size of the lanthanoid ion; binary oxides, halides, hydrides, and nitrides are all known.

Lanthanoid(III) ions have radii that vary between 116 and 98 pm; by comparison, Fe^{3+} has an ionic radius of 64 pm. Thus the volume occupied by a Ln^{3+} ion is typically four to five times that occupied by a 3*d*-metal ion. Unlike the 3*d*-metal atoms, which rarely exceed a coordination number of 6 (with 4 being common too), compounds of lanthanoids often have high coordination numbers and a wide variety of coordination environments.

The binary lanthanoid(III) oxides, Ln$_2$O$_3$, have moderately complex structures with the coordination number of the lanthanoid ions typically 7 (or a mixture of 6 and 7). Several related structure types termed A-, B-, C-Ln$_2$O$_3$ are known and many of the oxides are polymorphic with transitions between the structures occurring as the temperature is changed. The coordination geometries are determined by the radius of the lanthanoid ion, with the average cation coordination number in the structures decreasing with decreasing ionic radius: for example, the La^{3+} ion in La_2O_3 has coordination number 7, whereas the Lu^{3+} ion in Lu_2O_3 has coordination number 6. In cases where Ln^{4+} ions can be obtained (for example, with cerium, praseodymium, and terbium), the LnO_2 adopts the fluorite structure as expected from radius ratio rules (Section 3.10).

The lanthanoid(III) trihalides have complex structural characteristics as a result of large coordination numbers for these relatively large ions. So, for example, in LaF_3 the La^{3+} is in an irregular 11-coordinate environment and in $LaCl_3$ it is in a nine-coordinate tricapped trigonal prismatic environment (Fig. 22.1). Towards the end of the lanthanoid series the smaller lanthanoids have different structure types for the trihalides, with distorted tricapped trigonal prisms for LnF_3 (Fig. 22.2) and layer structures based on six-coordinate cubic close packing for $LnCl_3$. Cerium is the only lanthanoid to form a tetrahalide (CeF_4); it crystallizes with a structure formed from vertex-sharing CeF_8 polyhedra (Fig. 22.3).

All the lanthanoids form hydrides of stoichiometry LnH_2. They all adopt the fluorite structure (Section 3.9a) based on cubic close packed hydride ions with lanthanoid ions in all the tetrahedral holes. These compounds show metallic properties as the remaining valence electrons partially fill a conduction band. Cerium hydride may be further oxidized by hydrogen to form a series of nonstoichiometric phases of formula CeH_{2+x}, with additional hydride ions incorporated into the fluorite lattice. Some of the smaller lanthanoids (for instance, dysprosium, ytterbium, and lutetium) form stoichiometric trihydrides, LnH_3. Complex metal hydrides based on lanthanum, such as $LaNi_5H_6$, have been studied intensively as possible hydrogen storage materials. Binary lanthanoid nitrides, LnN, exist for all the lanthanoids and adopt the expected rock-salt structure with alternating Ln^{3+} and N^{3-} ions.

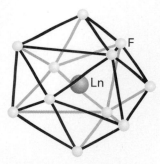

Fig. 22.2 The distorted tricapped trigonal prismatic structure of LnF_3.

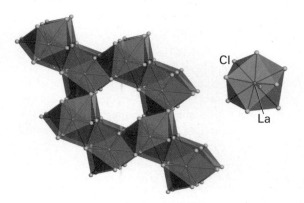

Fig. 22.1 The structure of $LaCl_3$, shown as vertex-linked $LaCl_9$ capped antiprisms.

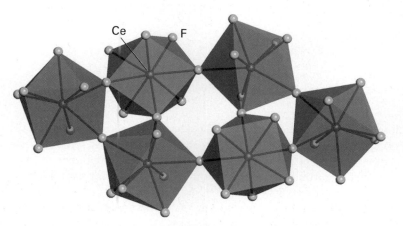

Fig. 22.3 The structure of CeF_4 contains vertex-sharing CeF_8 antiprisms.

22.5 Ternary and complex oxides

Key point: Lanthanoid ions are often found in perovskites and garnets, where the ability to change the size of the ion allows the properties of the materials to be modified.

The lanthanoids are a good source of stable, large, trivalent cations with a reasonable range of ionic radii: they can thus take one or more of the cation positions in ternary and more complex oxides. For example, perovskites of the type ABO_3 can readily be prepared with lanthanum on the A cation site; an example is $LaFeO_3$. Indeed, some distorted structure types are named after lanthanoids; an example is the structural type $GdFeO_3$, which has vertex-linked FeO_6 octahedra around the Gd^{3+} ion (as in the parent perovskite structure, Fig. 3.38) but the octahedra are tilted relative to each other. This tilting allows better coordination to the central Gd^{3+} ion. The ability to change the size of the B^{3+} ion in a series of compounds $LnBO_3$ allows the physical properties of the complex oxide to be modified in a controlled manner. For example, in the series of compounds $LnNiO_3$ for Ln = Pr to Eu, the insulating–metallic transition temperature T_{IM} increases with decreasing lanthanoid ionic radius:

	$PrNiO_3$	$NdNiO_3$	$EuNiO_3$
$r(Ln^{3+})$/pm	113	111	107
T_{IM}/K	135	200	480

As the perovskite unit cell is a structural building block often found in more complex oxide structures, lanthanoids are frequently used in such materials. One famous example is the high-temperature cuprate '1-2-3' superconductors, $LnBa_2Cu_3O_7$. The best known of these HTSCs have Ln = Y, but they are also found for all the lanthanoids (Section 23.8).

The spinel structure (Fig. 3.40) has only small tetrahedral and octahedral holes in the close-packed O^{2-} ion array and thus cannot accommodate lanthanoid ions. However, the garnet structure adopted by materials of stoichiometry $M_3M'_2(XO_4)_3$, where M and M′ are normally di- and tripositive cations and X is Si, Al, Ga, Ge, etc., has eight-coordinate sites that can be occupied by lanthanoid ions. Yttrium aluminium garnet (YAG) is the host material for neodymium ions in the lasing material Nd-YAG, and yttrium iron garnet (YIG) is an important ferrimagnetic material that is used in microwave and optical communication devices (Fig. 22.4).

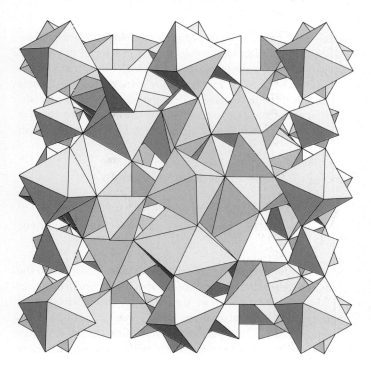

Fig. 22.4 The garnet structure shown in the form of linked AO_8, BO_4, and MO_6 polyhedra. The eight-coordinate sites occupied by yttrium can be occupied by other lanthanoids.

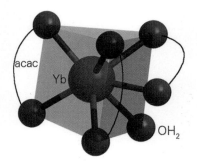

1 [Yb(acac)$_3$(OH$_2$)]

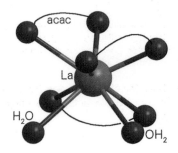

2 [La(acac)$_3$(OH$_2$)$_2$]

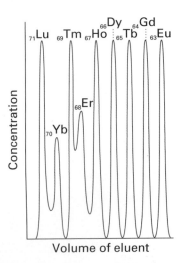

Fig. 22.5 Elution of heavy lanthanoid ions from a cation exchange column using ammonium 2-hydroxyisobutyrate as the eluent. Note that the higher atomic number lanthanoids elute first because they have smaller radii and are more strongly complexed by the eluent.

22.6 Coordination compounds

Key point: High coordination numbers with ligands adopting geometries that minimize interligand repulsions are the norm for the lanthanoids.

The adoption by the relatively large Ln^{3+} ions of structures with high coordination numbers and with a variety of coordination environments in the solid state is repeated in solution. The variation in structure adopted is in harmony with the view that the spatially buried *f* electrons have no significant stereochemical influence, and consequently ligands adopt positions that minimize ligand–ligand repulsions. In addition, polydentate ligands must satisfy their own stereochemical constraints, much as for the *s*-block and Al^{3+} complexes. For example, many lanthanoid complexes have been formed with crown ether and β-diketonate ligands. The coordination numbers for [Ln(OH$_2$)$_n$]$^{3+}$ in aqueous solution are thought to be 9 for the early lanthanoids and 8 for the later, smaller members of the series, but these ions are highly labile and the measurements are subject to considerable uncertainty. Similarly, a striking variation is observed for the coordination numbers and structures of lanthanoid salts and complexes. For example, the small ytterbium cation, Yb^{3+}, forms the seven-coordinate complex [Yb(acac)$_3$(OH$_2$)], and the larger La^{3+} is eight-coordinate in [La(acac)$_3$(OH$_2$)$_2$]. The structures of these two complexes are approximately a capped trigonal prism (**1**) and a square antiprism (**2**), respectively.

The partially fluorinated β-diketonate ligand CF$_3$COCHCOCF$_3$, nicknamed fod, produces complexes with Ln^{3+} that are volatile and soluble in organic solvents. On account of their volatility, these complexes are used as precursors for the synthesis of lanthanoid-containing superconductors by vapour deposition (Section 23.6).

Charged ligands generally have the highest affinity for the smallest Ln^{3+} ion, and the resulting increase in formation constants from large, lighter Ln^{3+} (on the left of the series) to small, heavier Ln^{3+} (on the right of the series) provides a convenient method for the chromatographic separation of these ions (Fig. 22.5). In the early days of lanthanoid chemistry, before ion-exchange chromatography was developed, tedious repetitive crystallizations were used to separate the elements.

22.7 Optical properties and spectra of lanthanoid complexes

Key point: Lanthanoid ions typically display weak but sharp absorption spectra because the *f* orbitals are shielded from the ligands.

Most lanthanoid ions are weakly coloured. The spectra of their complexes generally show much narrower and more distinct absorption bands than those for *d*-metal complexes. These spectra are associated with weak *f–f* electronic transitions. Both the narrowness of the spectral features and their insensitivity to the nature of coordinated ligands indicate that the *f* orbitals have a smaller radial extension than the filled 5*s* and 5*p* orbitals. Similarly, the magnetic properties of the ions can be explained on the assumption that the *f* electrons in Ln^{3+} ions are only slightly perturbed by ligands because they are buried so deeply.

One systematizing feature is that the colours of the aqua ions from La^{3+} (f^0) to Gd^{3+} (f^7) tend to repeat themselves in the reverse sequence Lu^{3+} (f^{14}) to Gd^{3+} (f^7):

f electrons:	0	1	2	3	4	5	6	7
Colour:	colourless	colourless	green	lilac	pink	yellow	pink	colourless
f electrons:	14	13	12	11	10	9	8	

This sequence suggests that the absorption maxima are related in a simple manner to the number of unpaired *f* electrons and largely independent of the specific stereochemistry of the complex. Unfortunately, however, the apparent simplicity of this correlation is deceptive and is not fully borne out by detailed analysis.

We shall not go into as deep an analysis of *f–f* electronic transitions as for *d–d* electronic transitions (Sections 19.3–19.7) as they are very complex (for instance, there are

91 microstates for an f^2 configuration). However, the situation is simplified somewhat by the fact that f orbitals are relatively deep inside the atom and overlap only weakly with ligand orbitals. Hence, as a first approximation, their spectra can be discussed in the free-ion limit.

For a slightly more in-depth discussion of the visible spectra of the lanthanoids, we find that the Russell–Saunders coupling scheme remains a reasonable approximation despite the elements having high atomic numbers, because the f electrons penetrate only slightly through the inner shells. As a result, they do not sample the high electric field at the nucleus and hence their spin–orbit coupling is weak.

The large number of microstates for each electronic configuration means a correspondingly large number of terms, and hence of transitions. As the f orbitals interact only weakly with the ligands bound to the metal ion there is little coupling of the electronic transitions, with molecular vibrations leading to narrow bands. Hence the visible spectra of lanthanoids usually have a large number of sharp peaks, in contrast to d metals, which normally show one or two broad bands. Figure 22.6 shows the spectrum of f^2 Pr^{3+}(aq), which has a ^3H ground state, from the infrared to the ultraviolet regions, with the six bands labelled with the free-ion term symbols.

All the lanthanoid ions except La^{3+} (f^0) and Lu^{3+} (f^{14}) show luminescence, with Eu^{3+} (f^6) and Tb^{3+} (f^8) showing particularly strong emissions. In part, this is due to the large number of excited states that exist, increasing the probability of intersystem crossing and the population of excited states of different spin multiplicity from the ground state. In part, also, the luminescence derives from the excited electron interacting only weakly with its environment leading to a nonradiative lifetime of the excited state that is quite long. The luminescence of lanthanoid complexes is the reason they are used as phosphors on TV screens.

22.8 Organometallic compounds

Key points: The organometallic compounds of the lanthanoids are dominated by good donor ligands, with complexes of acceptor ligands being rare; the bonding in complexes is best treated on the basis of ionic interactions; there are similarities between the organometallic compounds of the lanthanoids and those of the early d metals.

In line with the picture of the lanthanoid ions as having no directional frontier orbitals, it should not be surprising that they do not exhibit a rich organometallic chemistry. In particular, the lack of any orbitals that can backbond to organic fragments (because the $5d$ orbitals are empty and the $4f$ orbitals are buried) restricts the number of bonding modes that are available to lanthanoid ions. Thus most of the ligands discussed in Chapter 21 cannot bond in the fashion described there. In addition, the strongly electropositive nature of the lanthanoids means that they need good donor, not good acceptor, ligands. Thus alkoxide, amide, and halide ligands, which are both σ and π donors, are common, whereas CO and phosphine ligands, which are both σ donors and π acceptors, and which do such a good job in stabilizing low oxidation state d-metal complexes, are rarely seen in lanthanoid chemistry. Indeed, under normal laboratory conditions, neutral metal carbonyl compounds are unknown for the f-block elements.

As an example of the contrasting behaviour of organometallic complexes of the d and f blocks it should be noted that the first η^2 alkene complex of a lanthanoid was characterized only in 1987, a century and a half after the first d-metal alkene complex, Zeise's salt, was isolated. This lanthanoid alkene complex, $[(Cp^*)_2Yb(C_2H_4)Pt(PPh_3)_3]$ (**3**), has a particularly electron-rich alkene as it is already bound to an electron-rich Pt(0) centre, and it is thought that the alkene therefore needs no backbonding from the ytterbium to stabilize it.

The bonding in the organometallic lanthanoid complexes that do form is predominantly ionic and is governed by electrostatic factors and steric requirements. Consequently there is no need for the 18-electron rule to be obeyed. Although there is a chemical uniformity across the f block, there is also a gradation of change with steric factors tending to dominate: a small change in ligand size can make a significant

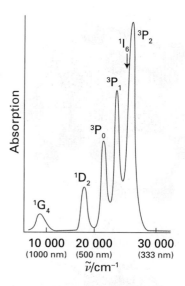

Fig. 22.6 The absorption spectrum of the f^2 (^3H) Pr^{3+}(aq) ion in the visible region.

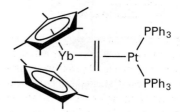

3 $[(Cp^*)_2Yb(C_2H_4)Pt(PPh_3)_2]$

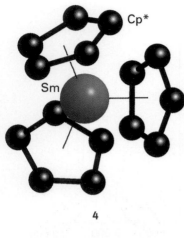

4

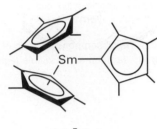

5

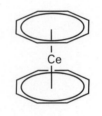

6 Ce(C₈H₈)₂

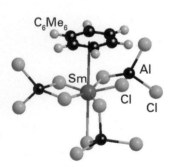

7 [(C₆Me₆)Sm(AlCl₄)₃]

difference to reactivity. All complexes are strong Lewis acids and are very sensitive to air and moisture.

The first organometallic compounds of the lanthanoids were cyclopentadienyl compounds: Geoffrey Wilkinson made a large number of $Ln(Cp)_3$ compounds with a wide variety of electron counts in 1954. The large lanthanoid ions can easily accommodate three cyclopentadienyl ligands, and they even tend to oligomerize, indicating that there is yet more space for additional ligands. Thus, compounds of substituted cyclopentadienyl ligands are possible and have been developed in more recent years. The limit seems to have been reached with the very sterically demanding pentamethylcyclopentadienyl (Cp^*) ligand. It was, however, nearly forty years after the original $Ln(Cp)_3$ compounds were isolated that $Ln(Cp^*)_3$ compounds were obtained, and even then they exist in equilibrium:

$$Sm(\eta^5\text{-}Cp^*)_3 (\mathbf{4}) \rightleftharpoons Sm(\eta^1\text{-}Cp^*)(\eta^5\text{-}Cp^*)_2 (\mathbf{5})$$

The great majority of lanthanoid organometallic compounds formally contain the Ln^{3+} ion with a limited number of Ln^{2+} compounds: no other oxidation states are known. σ-Bonded alkyl groups are common, with compounds containing the cyclopentadienyl ligands tending to dominate. It is best to consider cyclopentadienyl compounds as containing Cp^- groups electrostatically bound to a central cation. Compounds containing η^8-cyclooctatetraene ligands are known, such as $Ce(C_8H_8)_2$ (**6**), and as noted in Section 21.12, it is best to consider these to be complexes of the electron-rich $C_8H_8^{2-}$ ligand.

Current research into lanthanoid organometallic compounds typically involves compounds of the type $[(Cp)_2LnR]_2$, $[(Cp)_2LnR(sol)]$, $[(Cp^*)LnRX(sol)]$, and $[(Cp^*)LnR_2(sol)]$. A number of arene complexes are also known, such as $[(C_6Me_6)Sm(AlCl_4)_3]$ (**7**), where it is thought that the bonding is largely the result of an electrostatically induced dipole between the Sm^{3+} and the electron-rich ring.

No lanthanoid has two stable states with oxidation numbers differing by 2, so there is no possibility of oxidative addition or reductive elimination reactions: σ-bond metathesis-type reactivity dominates (Section 21.23).

In addition to the handling problems caused by the extreme sensitivity to air and moisture of the lanthanoid organometallic compounds, their study by NMR has been inhibited by the fact that they are all paramagnetic. Useful comparisons can be drawn through the study of compounds of the $4d$ element yttrium, as Y^{3+} has the same charge and a similar size to a typical Ln^{3+} ion but is diamagnetic. In addition, yttrium is 100 per cent ^{89}Y with $I = \frac{1}{2}$, so it is possible to measure Y—C and Y—H coupling constants quite easily and thus get additional structural information.

There are strong similarities between the chemical properties of the early d-block organometallic compounds (those of Groups 3 to 5) and those of the f block. These similarities are to be expected because the early d metals are also strongly electropositive, have a limited number of d electrons to backbond with ligands, and have a limited number of accessible oxidation states.

Although the organometallic chemistry of the lanthanoids is less rich than that of the d metals, there are some striking examples of unusual reactions. For instance, a lanthanoid organometallic compound can be used for the C—H activation of methane. This discovery was based on the observation that $^{13}CH_4$ exchanges ^{13}C with the CH_3 group attached to Lu:

$$Lu(Cp^*)_2(CH_3) + {}^{13}CH_4 \rightarrow Lu(Cp^*)_2({}^{13}CH_3) + CH_4$$

This reaction can be carried out in deuterated cyclohexane with no evidence for activation of the cyclohexane C—D bond, presumably because cyclohexane is too bulky to gain access to the metal centre. Methane activation has also been observed with electrophilic organoactinoid complexes; a four-centre σ-bond metathesis-type intermediate has been proposed (Section 21.23):

$$
\begin{array}{ccccc}
\text{Me}\!-\!\!-\!\text{H} & & \text{Me}\cdots\text{H} & & \text{Me} \quad \text{H} \\
| & \longrightarrow & \vdots \quad \vdots & \longrightarrow & | \quad \quad | \\
\text{Lu}\!-\!\!-\!\text{Me} & & \text{Lu}\cdots\text{Me} & & \text{Lu} \quad \text{Me}
\end{array}
$$

Lanthanoid complexes are used in the Ziegler–Natta polymerization of alkenes (Section 25.17). Dinitrogen complexes of lanthanoids were first reported in 1988.

Example 22.1 Accounting for the organometallic reactivity of a lanthanoid

Suggest a likely reaction pathway for the following transformation:

$$Ln-Bu + H_2 \longrightarrow Ln-H + BuH$$

Answer The reaction with dihydrogen cannot proceed via a dihydrogen complex, oxidative addition to a dihydride, and then reductive elimination of butane because the lanthanoid is incapable of oxidative addition reactions. The only possible scenario involves a σ-bond metathesis reaction via an intermediate such as

$$\begin{array}{ccc} H & \cdots\cdots & H \\ | & & | \\ Li & \text{—} & Bu \end{array}$$

Self-test 22.1 The product of the reaction above is in fact a hydride bridged dimer (**8**). Suggest a strategy to ensure that the hydride is monomeric.

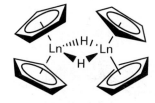

8 $[(Cp)_2Ln(\mu-H)_2Ln(Cp)_2]$

Actinoid chemistry

The chemical properties of the actinoids show less uniformity across the series than those of the lanthanoids. However, the radioactivity associated with most of the actinoids has hindered their study. Because the later actinoids are available in such tiny amounts, little is known about their reactions.

22.9 General trends

Key point: The early actinoids do not exhibit the chemical uniformity of the lanthanoids and exist in diverse oxidation states with An(III) progressively more stable for the later actinoids.

The fifteen elements from actinium (Ac, $Z = 89$) to lawrencium (Lr, $Z = 103$) involve the filling of the $5f$ subshell, and in this sense are analogues of the lanthanoids. However, the actinoids do not exhibit the chemical uniformity of the lanthanoids. Whereas a common oxidation state of the actinoids is An(III), unlike the lanthanoids the early members of the series occur in a rich variety of oxidation states. The Frost diagrams in Fig. 22.7 show that oxidation numbers higher than +3 are preferred for the early elements of the block (Th, Pa, U, and Np): linear or nearly linear AnO^{2+} and AnO_2^{2+} ions are the dominant aqua species for oxidation numbers +5 and +6. Unlike in the lanthanoids, f orbitals extend into the bonding region for the early actinoids, so the spectra of actinoid complexes are strongly affected by ligands.

Like the lanthanoids, the actinoids have large atomic and ionic radii (the radius of an An^{3+} ion is typically only 5 pm larger than its Ln^{3+} congener) and as a result often have high coordination numbers. For example, uranium in solid UCl_4 is eight-coordinate and in solid UBr_4 it is seven-coordinate in a pentagonal-bipyramidal array. Solid-state structures with coordination numbers up to 12 have been observed (Section 8.3). The reason for the contrasting behaviour of the lanthanoid and early actinoids is normally attributed to extension of the $5f$ electrons beyond the [Rn] core for the early actinoids. For americium (Am, $Z = 95$) and beyond, the properties of the actinoids begin to converge with those of the lanthanoids. With increasing atomic number, An(III) becomes progressively more stable relative to higher oxidation states and is dominant for curium

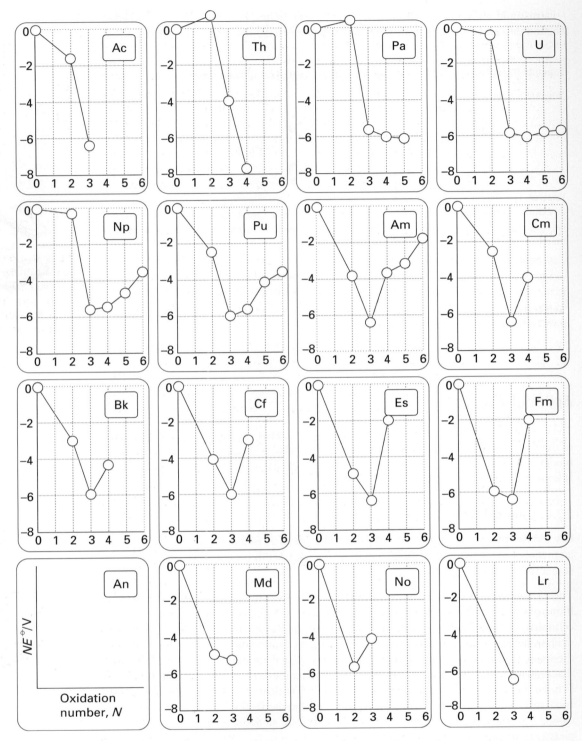

Fig. 22.7 Frost diagrams for the actinoids in acidic solution. (Based on J.J. Katz, G.T. Seaborg, and L. Morss, *Chemistry of the actinide elements*. Chapman and Hall, London (1986).)

(Cm), berkelium (Bk), californium (Cf), and einsteinium (Es); these elements therefore resemble the lanthanoids.

The striking differences between the chemical properties of the lanthanoids and early actinoids led to controversy about the most appropriate placement of the actinoids in the periodic table. For example, before 1945, periodic tables usually showed uranium below tungsten because both elements have a maximum oxidation number of +6. The emergence of oxidation state An(III) for the later actinoids was a key point in the way that chemists thought about the location of the actinoids in the periodic table. The similarity of the heavy actinoids and the lanthanoids is illustrated by their similar elution behaviour in ion-exchange separation (compare Figs 22.5 and 22.8).

On account of the small quantities of material available in most cases and their intense radioactivity, most of the chemical properties of the transamericium elements (the elements following americium, $Z = 95$) have been established by experiments carried out on a microgram scale or even a few hundreds of atoms. For example, the actinoid ion complexes have been adsorbed on and eluted from a single bead of ion-exchange material of about 0.2 mm in diameter. For the heaviest and most unstable post-actinoids, such as hassium ($Z = 108$), the lifetimes are too short for chemical separation and the identification of the element is based exclusively on the properties of the radiation it emits.

22.10 Thorium and uranium

Key points: The common nuclides of thorium and uranium exhibit only low levels of radioactivity, so their chemical properties have been extensively developed; the uranyl cation, with a linear OUO^{2+} array, is found in complexes with many different ligand donor atoms; the organometallic compounds of Th and U are dominated by pentamethylcyclopentadienyl complexes.

Because of their ready availability and low level of radioactivity, the chemical manipulation of thorium and uranium can be carried out with ordinary laboratory techniques. As indicated in Fig. 22.7, the most stable oxidation state of thorium in aqueous solution is Th(IV). This oxidation state also dominates the solid-state chemistry of the element. Eight-coordination is common in simple thorium(IV) compounds. For example, ThO_2 has the fluorite structure (in which a Th atom is surrounded by a cubic array of O^{2-} ions) and in $ThCl_4$ the coordination number is also 8 with dodecahedral symmetry. The coordination number of Th in $[Th(NO_3)_4(OPPh_3)_2]$ is 10, with the NO_3^- ions and triphenylphosphine oxide groups arranged in a capped cubic array around the metal atom.

The chemical properties of uranium are more varied than those of thorium because the element has access to oxidation states from U(III) to U(VI), with U(IV) and U(VI) the most common. Uranium halides are known for the full range of oxidation states U(III) to U(VI), with a trend towards decreasing coordination number with increasing oxidation number. The uranium atom is nine-coordinate in solid UCl_3, eight-coordinate in UCl_4, and six-coordinate for the U(V) and U(VI) chlorides U_2Cl_{10} and UCl_6, both of which are molecular compounds. The high volatility of UF_6 (it sublimes at 57°C) together with the occurrence of fluorine in a single isotopic form account for the use of this compound in the separation of the uranium isotopes by gaseous diffusion or centrifugation.

Uranium metal does not form a passivating oxide coating, and so it is corroded on prolonged exposure to air to give a complex mixture of oxides. The most important oxide is UO_3, which dissolves in acid to give the uranyl ion, UO_2^{2+}. In water, this bright fluorescent yellow ion forms complexes with many anions, such as NO_3^- and SO_4^{2-}. In contrast to the angular shape of the VO_2^+ ion and similar d^0 complexes, the AnO_2^{2+} group, with An = U, Np, Pu, and Am, maintains its linearity in all complexes. Both f-orbital bonding and relativistic effects have been invoked to explain this linearity.

The separation of uranium from most other metals is accomplished by the extraction of the neutral uranyl nitrato complex $[UO_2(NO_3)_2(OH_2)_4]$ from the aqueous phase into a polar organic phase, such as a solution of tributylphosphate dissolved in a hydrocarbon solvent. This kind of solvent-extraction process is used to separate actinoids from other fission products in spent nuclear fuel.

The organometallic chemistry of thorium and uranium is reasonably well developed and shows a lot of similarities to that of the lanthanoids, except that Th and U occur in a number of oxidation states and are larger than the typical Ln ion. Thus, compounds are dominated by those containing good donor ligands, such as σ-bonded alkyl and cyclopentadienyl groups. The increased size of Th and U compared with typical lanthanoids means that the tetrahedral Th(Cp)$_4$ (**9**) and U(Cp)$_4$ (**10**) can be isolated as monomers,

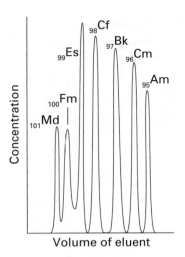

Fig. 22.8 Elution of heavy actinoid ions from a cation exchange column using ammonium 2-hydroxyisobutyrate as the eluent. Note the similarity in elution sequence with Fig. 22.5: the heavy (smaller) An^{3+} ions elute first.

9 Th(Cp)$_4$

10 U(Cp)$_4$

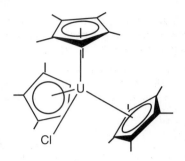

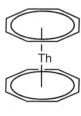

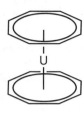

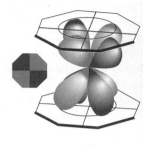

11 U(Cp*)₃Cl

12 Th(C₈H₈)₂

13 U(C₈H₈)₂

14

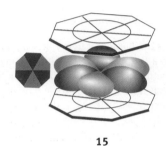

15

and not only can $U(Cp^*)_3$ be isolated but so too can $U(Cp^*)_3Cl$ (**11**). As with lanthanoid organometallics, actinoid organometallics do not obey the 18-electron rule.

Sandwich compounds are possible with the η^8-cyclooctatetraene ligand, and both thorocene, $Th(C_8H_8)_2$ (**12**), and uranocene, $U(C_8H_8)_2$ (**13**), are known (indeed, they are stable in water). Compared with lanthanoid complexes of cyclooctatetraene, the bonding in uranocene and thorocene is complicated by the extension of the f orbitals beyond the core of the atom. In theory, therefore, not only is δ bonding possible (**14**), but so is ϕ bonding (**15**). Quite how much of a role ϕ bonds play in the bonding in uranocene or thorocene is the subject of considerable debate, but it is interesting to note that actinoid cyclooctatetraene compounds are the only 'real' compounds (as opposed to metal dimers in the gas phase) that might contain any contribution from ϕ bonding.

Example 22.2 Assessing the redox stability of actinoid ions

Use the Frost diagram for thorium (Fig. 22.7) to describe the relative stability of Th(II) and Th(III).

Answer The initial slope in the Frost diagrams indicates that the Th^{2+} ion might be readily attained with a mild oxidant. However, Th^{2+} is above lines connecting Th(0) with the higher oxidation states, so it is susceptible to disproportionation. Thorium(III) is readily oxidized to Th(IV) and the steep negative slope indicates that it would be oxidized by water:

$$Th^{3+}(aq) + H^+(aq) \rightarrow Th^{4+}(aq) + \tfrac{1}{2} H_2(g)$$

We can confirm from *Resource section* 2 that because $E^\circ = +3.8\,V$, this reaction is highly favoured. Thus, Th(IV) will dominate in aqueous solution.

Self-test 22.2 Use the Frost diagrams and data in *Resource section* 2 to determine the most stable uranium ion in acid aqueous solution in the presence of air and give its formula.

FURTHER READING

S.A. Cotton, *Lanthanides and actinides*. Macmillan, London (1991).

N. Kaltsoyannis and P. Scott, *The f elements*. Oxford University Press (1999).

G. Seaborg, J. Katz, and L.R. Morss, *Chemistry of the actinide elements*. Chapman and Hall, London (1986).

D.M.P. Mingos and R.H. Crabtree (eds). *Comprehensive organometallic chemistry III*. Elsevier, Oxford (2006). Volume 4 (ed. M. Bochmann) deals with Groups 3 and 4 and the lanthanoids and actinoids.

W.J. Evans, *Adv. Organomet. Chem.*, 1985, **24**, 131, C.J. Schaverien, *Adv. Organomet. Chem.*, 1994, **36**, 283, and W.J. Evans and B.L. Davis, *Chem. Rev.*, 2002, **102**, 2119. Reviews of organolanthanoid chemistry.

T.J. Marks and A. Streitwieser, p. 1547, and T.J. Marks, p. 1588, in G. Seaborg, J. Katz, and L.R. Morss, *Chemistry of the actinide elements*, Vol. 2. Chapman and Hall, London (1986). For a survey of organoactinoid chemistry.

EXERCISES

22.1 (a) Give a balanced equation for the reaction of any of the lanthanoid metals with aqueous acid. (b) Justify your answer with redox potentials and with a generalization on the most stable positive oxidation states for the lanthanoids. (c) Name two lanthanoids that have the greatest tendency to deviate from the usual positive oxidation state and correlate this deviation with electronic structure.

22.2 From a knowledge of their chemical properties, speculate on why cerium and europium were the easiest lanthanoids to isolate before the development of ion-exchange chromatography.

22.3 How would you expect the first and second ionization energies of the lanthanoids to vary across the series? Sketch the graph that you would get if you plotted the third ionization energy of the lanthanoids versus atomic number. Identify elements at any peaks or troughs.

22.4 Explain why stable and readily isolable carbonyl complexes are unknown for the lanthanoids.

22.5 Suggest a synthesis of uranocene from UCl_4.

22.6 Account for the similar electronic spectra of Eu^{3+} complexes with various ligands and the variation of the electronic spectra of Am^{3+} complexes as the ligand is varied.

22.7 The M(II) oxidation state is not common among the lanthanoids but a 'normal' M(II) chemistry does exist for Sm^{2+}, Eu^{2+} and Yb^{2+}. Write the f-electron configurations for these species and identify the ground terms.

22.8 Describe the general nature of the distribution of the elements formed in the thermal neutron fission of ^{235}U, and decide which of the following highly radioactive nuclides are likely to present the greatest radiation hazard in the spent fuel from nuclear power reactors: (a) ^{39}Ar, (b) ^{228}Th, (c) ^{90}Sr, (d) ^{144}Ce.

PROBLEMS

22.1 Lanthanoid coordination compounds rarely exhibit isomerism in solution. Suggest two factors that might cause this phenomenon, explaining your reasoning. (See D. Parker, R.S. Dickins, H. Puschmann, C. Crossland, and J.A.K. Howard, *Chem. Rev.*, 2002, **102**, 1977.)

22.2 Neither lanthanoid nor actinoid organometallic compounds obey the 18-electron rule. Discuss the reasons, using the structures of the tris(Cp) and tris(Cp*) Ln and An complexes as examples. (See W.J. Evans and B.L. Davis, *Chem. Rev.*, 2002, **102**, 2119.)

22.3 The bonding in the linear uranyl ion, OUO^{2+}, is often explained in terms of significant π bonding using $5f$ orbitals on the metal. Using the

f orbitals illustrated in Fig. 1.10, construct a reasonable molecular orbital diagram for π bonding with the appropriate oxygen p orbitals.

22.4 The existence of a maximum oxidation number of +6 for both uranium ($Z = 92$) and tungsten ($Z = 74$) prompted the placement of uranium under tungsten in early periodic tables. When the element after uranium, neptunium ($Z = 93$), was discovered in 1940 its properties did not correspond to those of rhenium ($Z = 75$), and this cast doubt on the original placement of uranium. (See G.T. Seaborg and W.D. Loveland, *The elements beyond uranium.* Wiley-Interscience, New York (1990), p. 9 *et seq.*) Using standard potential data from *Resource section* 2, discuss the differences in oxidation state stability between Np and Re.

PART 3
Frontiers

Inorganic chemistry is advancing rapidly at its frontiers, especially where research impinges on other disciplines such as the life sciences, condensed-matter physics, materials science, and environmental chemistry. These swiftly developing fields also represent many areas of inorganic chemistry where novel types of compounds are used in catalysis, electronics, and pharmaceuticals. The aim of this section of the book is to demonstrate the vigorous nature of contemporary inorganic chemistry by building upon the introductory and descriptive material in Parts 1 and 2.

These *Frontiers* chapters open with a discussion in Chapter 23 of materials chemistry, focusing on solid-state compounds, their synthesis, structure, and their electronic, magnetic, and optical properties. One area that has developed enormously in the past decade has been that of nanomaterials, and in Chapter 24 we provided a comprehensive introduction to inorganic nanochemistry. Chapter 25 covers catalysis involving inorganic compounds and discusses the basic concepts behind catalytic reactions at metal centres.

Finally, we turn to the frontier of inorganic chemistry with life. Chapter 26 discusses the function of different elements in cells and intracellular compartments and the various and extraordinarily subtle ways in which they are exploited. It also describes the structures and functions of complexes and materials that are formed in the biological environment and how inorganic elements are used in medical treatments.

23

Solid-state and materials chemistry

The area of 'materials chemistry' is developing rapidly and there has been a great increase in interest in the synthesis and properties of novel inorganic solids. In this chapter we discuss a number of important areas of current investigation. Initially, we describe the role of defects, nonstoichiometry, and ion migration in solids, all of which are important for understanding structures composed of infinite arrays of interacting atoms. We then extend these concepts to key classes of inorganic materials, with sections on intercalation compounds, complex electronic oxides (such as high-temperature superconductors), magnetic compounds (including materials exhibiting giant magnetoresistance), framework structures (such as zeolites and their analogues), and molecular materials. Throughout the chapter we shall see how the synthesis and properties of these solids correlate with their crystal and electronic structures. This chapter concentrates on the inorganic chemistry of various bulk solid-state compounds with infinite arrays of atoms and ions; nanochemistry is covered in the next chapter and heterogeneous catalysis in Chapter 25.

Solid-state inorganic chemistry is concerned with the synthesis, structure, and properties of solids. Many solids exhibit new phenomena or possess desirable properties, such as high-temperature superconductivity, ferromagnetism, molecular porosity, or intense colour. The chemistry of the solid state is a vigorous and exciting area of inorganic research, partly on account of the technological applications of these novel materials but also because they are challenging to understand. We shall draw on some of the concepts developed in earlier chapters for discussing solids, such as lattice energetics and band structure, and introduce some additional concepts that are needed to discuss the dynamics of events that occur in the interior of materials. We also need to go beyond the view that solids have a well-defined stoichiometry, for interesting phenomena arise from nonstoichiometry and ion mobility. The fact that solid materials have continuous structures, in which atoms and ions can interact in a cooperative manner, gives rise to many of the fascinating and useful aspects of their chemical properties.

General principles

Chapter 3 provided a summary of prototypical structures adopted by many metals and ionic solids; it also explored some general rules for the stability of ionic solids. Most of that discussion centred on the structures of stoichiometric compounds, with well-defined composition. In this chapter, we describe a wider range of solids, including 'nonstoichiometric compounds', in which atomic-scale defects and deviations from simple whole-number ratios of the constituent elements influence their physical and chemical properties.

Much of the current research in solid-state chemistry is motivated by the search for commercially useful materials, such as components of batteries and fuel cells, catalysts for hydrocarbon interconversion, and improved electronic and photonic devices for information processing and storage. The scope for the synthesis of new inorganic solids is enormous; for example, although it is known that 100 structural types account for

95 per cent of the known binary or ternary intermetallic compounds, there are plenty of opportunities for extending these studies to the synthesis and characterization of four-, five-, and six-component systems.

Among the many other areas of materials chemistry that are currently being explored are new microporous solids for use in molecular separations and heterogeneous catalysis; these materials were discussed in Section 13.13 but recent advances in this area are developed here and in Chapter 25. Furthermore, because the properties and uses of a material are very dependent on its physical size and form, there has been an explosion of work on nanomaterials, in which the focus of work is on inorganic solids with controlled sub-micrometre dimensions; this topic is treated in Chapter 24.

Computer modelling is used to develop numerical models of the structures and properties of materials (Chapter 6). It has been applied to a wide range of inorganic materials to understand and predict how atomic and electronic structure control their physical properties. Some of the most successful applications have been in areas related to defect structures, ionic mobility, and catalysis on surfaces. This chapter and Section 25.10 describe some recent applications of these methods.

23.1 Defects

All solids contain **defects**, or imperfections of structure or composition. Defects are important because they influence properties such as mechanical strength, electrical conductivity, and chemical reactivity. We need to consider both **intrinsic defects**, which are defects that occur in the pure substance, and **extrinsic defects**, which stem from the presence of impurities. It is also common to distinguish **point defects**, which occur at single sites, from **extended defects**, which are ordered in one, two, and three dimensions. Point defects are random errors in a periodic lattice, such as the absence of an atom at its usual site or the presence of an atom at a site that is not normally occupied. Extended defects involve various irregularities in the stacking of the planes of atoms.

(a) The origin of defects

Key point: All solids have a thermodynamic tendency to acquire defects.

Solids contain defects because they introduce disorder into an otherwise perfect structure and hence increase its entropy. The Gibbs energy, $G = H - TS$, of a solid with defects has contributions from the enthalpy and the entropy of the sample. The formation of defects is normally endothermic because, as the lattice is disrupted, the enthalpy of the solid rises. However, the term $-TS$ becomes more negative as defects are formed because they introduce disorder into the lattice and the entropy rises. Provided $T > 0$, therefore, the Gibbs energy will have a minimum at a nonzero concentration of defects and their formation will be spontaneous (Fig. 23.1a). Moreover, as the temperature is raised, the minimum in G shifts to higher defect concentrations (Fig. 23.1b), so solids have a greater number of defects as their melting points are approached. A further factor than can help to stabilize the incorporation of defects is **clustering**, the segregation of defects to certain parts of the structure, so reducing the overall enthalpy penalty associated with the disruption of the array.

(b) Intrinsic point defects

Key points: Schottky defects are site vacancies, formed in cation/anion pairs and Frenkel defects are displaced, interstitial atoms; the structure of a solid influences the type of defect that occurs, with Frenkel defects forming in solids with lower coordination numbers and more covalency and Schottky defects in more ionic materials.

The solid-state physicists W. Schottky and J. Frenkel identified two specific types of point defect. A **Schottky defect** (Fig. 23.2) is a vacancy in an otherwise perfect arrangement of atoms on a lattice. That is, it is a point defect in which an atom or ion is missing from its normal site in the structure. The overall stoichiometry of a solid is not affected by the presence of Schottky defects because, to ensure charge balance, the defects occur in pairs in a compound of stoichiometry MX and there are equal numbers of vacancies at cation

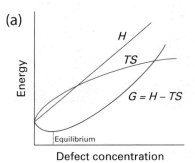

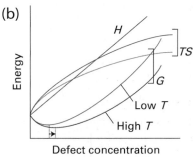

Fig. 23.1 (a) The variation of the enthalpy and entropy of a crystal as the number of defects increases. The resulting Gibbs energy $G = H - TS$ has a minimum at a nonzero concentration, and hence defect formation is spontaneous. (b) As the temperature is increased, the minimum in the Gibbs energy moves to higher defect concentrations, so more defects are present at equilibrium at higher temperatures than at low.

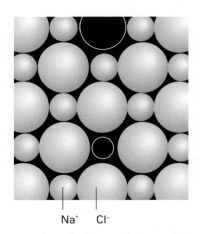

Fig. 23.2 A Schottky defect is the absence of ions on normally occupied sites; for charge neutrality here must be equal numbers of cation and anions vacancies in a 1:1 compound.

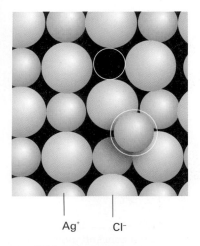

Ag⁺ | Cl⁻

Fig. 23.3 A Frenkel defect forms when an ion moves to an interstitial site.

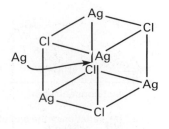

1 Interstitial Ag⁺

and anion sites. In solids of different composition, for example MX$_2$, the defects must occur with balanced charges, so two anion vacancies must be created for each cation lost. Schottky defects occur at low concentrations in purely ionic solids, such as NaCl; they occur most commonly in structures with high coordination numbers, such a close-packed metals, where the enthalpy penalty of reducing the average coordination number of the remaining atoms (from 12 to 11, for instance) is relatively low.

A **Frenkel defect** (Fig. 23.3) is a point defect in which an atom or ion has been displaced onto an interstitial site. For example, in silver chloride, which has the rock-salt structure, a small number of Ag$^+$ ions reside in tetrahedral sites (**1**). The stoichiometry of the compound is unchanged when a Frenkel defect forms and it is possible to have Frenkel defects involving either one (M or X displaced) or both (some M and some X interstitials) of the ion types in a binary compound, MX. Thus the Frenkel defects that occur in, for example, PbF$_2$ involve the displacement of a small number of F$^-$ ions from their normal sites in the fluorite structure, on the tetrahedral holes in the close-packed Pb^{2+} ion array, to sites that correspond to the octahedral holes. A useful generalization is that Frenkel defects are most often encountered in structures such as wurtzite and sphalerite in which coordination numbers are low (six or less) and the more open structure provides sites that can accommodate the interstitial atoms. This is not to say that Frenkel defects are exclusive to such structures; as we have seen, the 8:4 coordination fluorite structure can accommodate such interstitials although some local repositioning of adjacent anions is required to allow for the presence of the displaced anion.

The concentration of Schottky defects varies considerably from one type of compound to the next. The concentration of vacancies is very low in the alkali metal halides, being of the order of 10^6 cm^{-3} at 130°C. That concentration corresponds to about one defect per 10^{14} formula units. On the contrary, some *d*-metal oxides, sulfides, and hydrides have very high concentrations of vacancies. An extreme example is the high-temperature form of TiO, which has vacancies on both the cation and anion sites at a concentration corresponding to about one defect per 10 formula units.

Example 23.1 Inferring the type of defect from density measurements

Titanium monoxide has a rock-salt structure. X-ray diffraction data show that the cubic unit cell parameter for TiO with a 1:1 ratio of Ti to O is 418 pm, and the density is determined as 4.92 g cm^{-3}. Do the data indicate that defects are present? If so, are they vacancy or interstitial defects?

Answer The presence of vacancies (Schottky defects) at the Ti and O sites should be reflected in a lower measured density than that calculated from the size of the unit cell and the assumption that every Ti and O site is occupied. Interstitial (Frenkel) defects would result in little, if any, difference between the measured and theoretical densities. There are four TiO formula units per unit cell (Fig. 3.28). The molar mass of TiO is 63.88 g mol^{-1}, so the corresponding theoretical mass of one unit cell is (4 × 63.88 g mol^{-1})/N_A = 4.24 × 10^{-22} g. The corresponding theoretical density is

$$\rho = \frac{4.24 \times 10^{-22}\text{g}}{(4.18 \times 10^{-8}\text{cm})^3} = 5.81\text{g cm}^{-3}$$

which is significantly greater than the measured density. Therefore, the crystal must contain numerous vacancies. Because the overall composition of the solid is TiO, there must be equal numbers of vacancies on the cation and anion sites provided the metal oxidation state is unchanged.

Self-test 23.1 The measured density of VO (1:1 stoichiometry) is 5.92 g cm^{-3} and the theoretical density is 6.49 g cm^{-3}. Do the data indicate the presence of vacancies or interstitials?

Schottky and Frenkel defects are only two of the many possible types of defect. Another type is an **atom-interchange** or **anti-site defect**, which consists of an interchanged pair of atoms. This type of defect is common in metal alloys with exchange of neutral atoms. It is expected to be very unfavourable for binary ionic compounds on account of the introduction of strongly repulsive interactions between neighbouring similarly charged ions. For example, a copper–gold alloy of exact overall composition CuAu has extensive disorder at high temperatures, with a significant fraction of Cu and Au atoms interchanged (Fig. 23.4). The interchange of similarly charged species on different sites in ternary and compositionally more complex compounds is common; thus in spinels (Section 23.8) the partial swapping of the metal ions between tetrahedral and octahedral sites is often observed.

Both Schottky and Frenkel defects are stoichiometry defects in that they do not change the overall composition of the material because the vacancies occur in charge-balanced pairs (Schottky) or each interstitial is derived from one displaced atom or ion (Frenkel, a vacancy interstitial pair). Similar types of defects, vacancies and interstitials, occur in many inorganic materials and may be balanced by changes in the oxidation number of one component in the system rather than by their creation as charge-balanced pairs. This behaviour, as seen for example in $La_2CuO_{4.1}$ with extra O^{2-} ion interstitials, is discussed more fully in Section 23.2.

(c) Extrinsic point defects

Key point: Extrinsic defects are defects introduced into a solid as a result of doping with an impurity atom.

Extrinsic defects (those resulting from the presence of impurities) are inevitable because perfect purity is unattainable in practice in crystals of any significant size. In Section 3.18 we discussed impurities that had been introduced intentionally by doping one material with another. An example is the introduction of As atoms into silicon to modify its semiconducting properties. We can begin to see how the electronic structure of the solid influences the type of defect it is likely to form. Thus, when an As atom replaces an Si atom, the additional electron from each As atom enters the conduction band. In the more ionic substance ZrO_2, the introduction of Ca^{2+} impurities in place of Zr^{4+} ions is accompanied by the formation of an O^{2-} ion vacancy to maintain charge neutrality (Fig. 23.5). In many cases, a change in oxidation state (if that is possible) may be induced by an impurity. For example, the introduction of Li_2O into NiO, to give $Li_x(Ni^{2+})_{1-x}(Ni^{3+})_xO$, places Li^+ in Ni^{2+} sites, and charge balance is achieved by the oxidation of one Ni^{2+} ion to Ni^{3+} for each Li^+ ion present. This process introduces holes into the valence band of NiO and the conductivity increases greatly.

Another example of an extrinsic point defect is a **colour centre**, a generic term for defects responsible for modifications to the IR, visible, and UV absorption characteristics of solids that have been irradiated or exposed to chemical treatment. One type of colour centre is produced by heating an alkali metal halide crystal in the vapour of the alkali metal and gives a material with a colour characteristic of the system: NaCl becomes orange, KCl violet, and KBr blue–green. The process results in the introduction of an alkali metal cation at a normal cation site and the associated electron from the metal atom occupies a halide ion vacancy. A colour centre consisting of an electron in a halide ion vacancy (Fig. 23.6) is called an **F-centre**.[1] The colour results from the excitation of the electron in the localized environment of its surrounding ions. An alternative method of producing F-centres involves exposing a material to an X-ray beam that ionizes electrons into anion vacancies.

(d) Extended defects

Key point: Wadsley defects are shear planes that collect defects along certain crystallographic directions.

All the point defects discussed so far entail a significant local distortion of the structure and in some instances localized charge imbalances too. Therefore, it should not be surprising that defects may cluster together and sometimes form lines and planes.

Tungsten oxides illustrate the formation of planes of defects. As illustrated in Fig. 23.7, the idealized structure of WO_3 (which is usually referred to as the 'ReO₃ structure'; see

---

[1] The name comes from the German word for colour centre, *Farbenzentrum*.

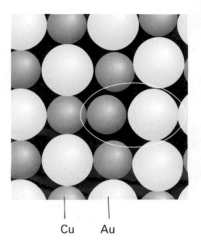

Fig. 23.4 Atom exchange can also give rise to a point defect as in CuAu.

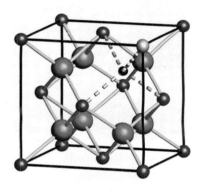

Fig. 23.5 Introduction of a calcium ion (grey sphere) into the ZrO_2 lattice produces a vacancy (black sphere) on the oxide sublattice. This substitution helps to stabilize the cubic fluorite structure of ZrO_2.

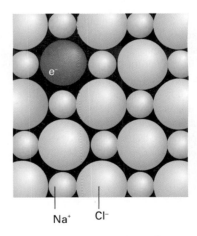

Fig. 23.6 An F-centre is an electron that occupies an anion vacancy. The energy levels of the electron resemble those of a particle in a three-dimensional square well.

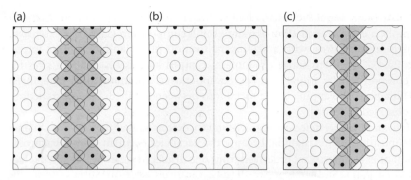

Fig. 23.7 The concept of a crystallographic shear plane illustrated by the (100) plane of the ReO$_3$ structure. (a) A plane of metal, •, and oxygen, ○, atoms. The octahedron around each metal atom is completed by a plane of oxygen atoms above and below the plane illustrated here. Some of the octahedra are shaded to clarify the processes that follow. (b) Oxygen atoms in the plane perpendicular to the page are removed, leaving two planes of metal atoms that lack their sixth oxygen ligand. (c) The octahedral coordination of the two planes of metal atoms is restored by translating the right slab as shown. This creates a plane (shear plane) vertical to the paper in which the MO$_6$ octahedra share edges.

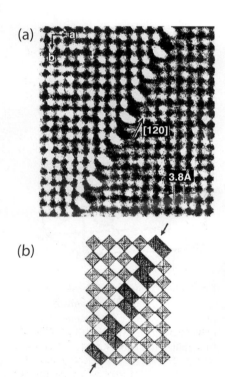

Fig. 23.8 (a) High-resolution electron micrograph lattice image of a crystallographic shear plane in WO$_{3-x}$. (b) Drawing of the oxygen octahedral polyhedra that surround the W atoms imaged in the electron micrograph. Note the edge-shared octahedra along the crystallographic shear plane. (Reproduced by permission from S. Iijima, *J. Solid State Chem.*, 1975, **14**, 52.)

below) consists of WO$_6$ octahedra sharing all vertices. To picture the formation of the defect plane, we imagine the removal of shared O atoms along a diagonal. Then adjacent slabs slip past each other in a motion that results in the completion of the vacant coordination sites around each W atom. This shearing motion creates edge-shared octahedra along a diagonal direction. The resulting structure was named a **crystallographic shear plane** by A.D. Wadsley, who first devised this way of describing extended planar defects.

Crystallographic shear planes randomly distributed in the solid are called **Wadsley defects**. Such defects lead to a continuous range of compositions, as in tungsten oxide, which ranges from WO$_3$ to WO$_{2.93}$ (made by heating and reducing WO$_3$ with tungsten metal). If, however, the crystallographic shear planes are distributed in a nonrandom, periodic manner, so giving rise to a new unit cell, then we should regard the material as a new stoichiometric phase. Thus, when even more O^{2-} ions are removed from tungsten oxide, a series of discrete phases having ordered crystallographic shear planes and compositions W$_n$O$_{3n-2}$ ($n = 20$, 24, 25, and 40) are observed. Closely spaced compositions containing shear planes are known for oxides of tungsten, molybdenum, titanium, and vanadium. Electron microscopy provides an excellent method of observing these defects experimentally because it reveals both ordered and random arrays of shear planes (Fig. 23.8).

23.2 Nonstoichiometric compounds and solid solutions

The statement that the stoichiometry of a molecular compound is fixed by the molecular formula and a crystalline aggregate of such molecules must have the same precise overall stoichiometry is not always true for solids with extended structures. In such cases the overall formula of the compound denotes the ratio of atoms present in a unit cell. For example, the composition of a unit cell of the perovskite lanthanum iron(III) oxide is LaFeO$_3$ and that of sodium chloride is NaCl (with exactly four Na$^+$ and Cl$^-$ ions in the unit cell). Thus, the chemical formula of an ionic solid is based on the presumption that every unit cell has identical contents. In practice, however, differences in the composition of unit cells can occur throughout a solid, perhaps because there are defects at one or more atom sites, interstitial atoms are present, or substitutions have occurred at one position. This variability results in nonstoichiometric compounds and the formation of solid solutions.

(a) Nonstoichiometry

Key point: Deviations from ideal stoichiometry are common in the solid-state compounds of the *d*-, *f*-, and heavy *p*-block elements.

A **nonstoichiometric compound** is a substance that exhibits variable composition but retains the same structure type. For example, at 1000°C the composition of 'iron monoxide', which is sometimes referred to as wüstite, $Fe_{1-x}O$, varies from $Fe_{0.89}O$ to $Fe_{0.96}O$. Gradual changes in the size of the unit cell occur as the composition is varied, but all the features of the rock-salt structure are retained throughout this composition range. The fact that the lattice parameter of the compound varies smoothly with composition is a defining criterion of a nonstoichiometric compound, because a discontinuity in the value of the lattice parameter indicates the formation of a new crystal phase. Moreover, the thermodynamic properties of nonstoichiometric compounds also vary continuously as the composition changes. For example, as the partial pressure of oxygen above a metal oxide is varied, both the lattice parameter and the equilibrium composition of the oxide change continuously (Figs 23.9 and 23.10). The gradual change in the lattice parameter of a solid as a function of its composition is known as **Vegard's rule**.

Table 23.1 lists some representative nonstoichiometric hydrides, oxides, and sulfides. Note that as the formation of a nonstoichiometric compound requires overall changes in composition, it also requires at least one element to exist in more than one oxidation state. Thus in wüstite, $Fe_{1-x}O$, as x increases some iron(II) must be oxidized to iron(III) in the structure. Hence deviations from stoichiometry are common only for d- and f-block elements, which commonly adopt two or more oxidation states, and for some heavy p-block metals that have two accessible oxidation states.

(b) Solid solutions

Key point: A solid solution occurs where there is a continuous variation in compound stoichiometry without a change in structural type.

Because many substances adopt the same structural type, it is often energetically feasible to replace one type of atom or ion with another. Such behaviour is seen in many simple metal alloys such as those discussed in Section 3.8. Thus zinc/copper brasses exist for the complete range of compositions $Cu_{1-x}Zn_x$ with $0 < x < 0.38$, where Cu atoms in the structure are gradually replaced by Zn atoms. This replacement occurs randomly throughout the solid and individual unit cells contain an arbitrary number of Cu and Zn atoms (but such that the sum of their contents gives the overall brass stoichiometry).

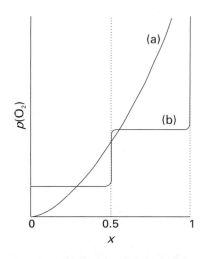

Fig. 23.9 Schematic representation of the variation of the partial pressure of oxygen with composition at constant pressure for (a) a nonstoichiometric oxide MO_{1+x} and (b) a stoichiometric pair of metal oxides MO and MO_2. The x axis is the atom ratio in MO_{1+x}.

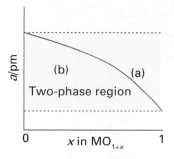

Fig. 23.10 Schematic representation of the variation of one lattice parameter with composition for (a) a nonstoichiometric oxide MO_{1+x} and (b) a stoichiometric pair of metal oxides MO and MO_2 with no intermediate stoichiometric phases (which would produce a two-phase mixture for $0 < x < 1$), each phase in the mixture having the lattice parameter of the end member.

Table 23.1 Representative composition ranges* of nonstoichiometric binary hydrides, oxides, and sufides

d block			f block		
Hydrides					
				Fluorite type	Hexagonal
TiH_x	1–2				
ZrH_x	1.5–1.6		GdH_x	1.9–2.3	2.85–3.0
HfH_x	1.7–1.8		ErH_x	1.95–2.31	2.82–3.0
NbH_x	0.64–1.0		LuH_x	1.85–2.23	1.74–3.0
Oxides					
	Rock-salt type	Rutile type			
TiO_x	0.7–1.25	1.9–2.0			
VO_x	0.9–1.20	1.8–2.0			
NbO_x	0.9–1.04				
Sulfides					
ZrS_x	0.9–1.0				
YS_x	0.9–1.0				

* Expressed as the range of values x may take.

Another good example is the perovskite structure adopted by many compounds of stoichiometry ABX_3 (Section 3.9), in which the composition can be varied continuously by varying the ions that occupy the A, B, and X sites. For instance, both $LaFeO_3$ and $SrFeO_3$ adopt the perovskite structure and we can consider a perovskite crystal that has, randomly distributed, half $SrFeO_3$ unit cells (with Sr on the A-type cation site) and half $LaFeO_3$ unit cells (with La on the A-site). The overall compound stoichiometry is $LaSrFe_2O_6$, which is better written $(La_{0.5}Sr_{0.5})FeO_3$, to reflect the normal ABO_3 perovskite stoichiometry. Other proportions of these unit cells are possible and the series of compounds $La_{1-x}Sr_xFeO_3$ for $0 \leq x \leq 1$ can be prepared. This system is called a **solid solution**, because all the phases formed as x is varied have the same perovskite structure. A solid solution occurs when there is a single structural type for a range of compositions and there is a smooth variation in lattice parameter over that range.

Solid solutions occur most frequently for d-metal compounds because the change in one component might require a change in the oxidation state of another component to preserve the charge balance. Thus, as x increases in $La_{1-x}Sr_xFeO_3$ and La(III) is replaced by Sr(II), the oxidation state of iron must change from Fe(III) to Fe(IV). This change can occur through a gradual replacement of one exact oxidation state, here Fe(III), by another, Fe(IV), on a proportion of the cation sites within the structure. Alternatively, if the material is metallic and has delocalized electrons, then the change can be accommodated by altering the number of electrons in a conduction band, which corresponds to the delocalization of the change in oxidation state rather than its identification with individual atoms. Some other solid solutions include the high-temperature superconductors of composition $La_{2-x}Ba_xCuO_4$ ($0 \leq x \leq 0.4$), which are superconducting for $0.12 \leq x \leq 0.25$, and the spinels $Mn_{1-x}Fe_{2+x}O_4$ ($0 \leq x \leq 1$). It is also possible to combine solid-solution behaviour on a cation site with nonstoichiometry caused by defects on a different ion site. An example is the system $La_{1-x}Sr_xFeO_{3-y}$, with $0 \leq x \leq 1.0$ and $0.5 < y \leq 0$, which has vacancies on the O^{2-} ion sites.

Example 23.2 Identifying solid solutions

Use the ionic radii data in *Resource section 1* and a knowledge of available oxidation states to decide whether (a) $La_{2-x}Sr_xCuO_4$ ($0 \leq x \leq 0.4$), (b) $La_{2-x}Mg_xCuO_4$ ($0 \leq x \leq 0.4$), (c) $Ca_{1-x}Na_xCl_2$ ($0 \leq x \leq 0.4$) might form as solid solutions.

Answer (a) The ionic radius of Sr^{2+} is similar to that of La^{3+} and copper can access the Cu(III) oxidation state so this solid solution does exist. (b) Mg^{2+} is too small to substitute for La^{3+}. (c) This system cannot be synthesized as it would require the formation of the chemically nonexistent Ca^{3+}.

Self-test 23.2 Would the system $Pr_{1-x}Sr_xMnO_3$ be expected to form a solid solution over the range $0 \leq x \leq 1$?

23.3 Atom and ion diffusion

Key point: The diffusion of ions in solids is markedly dependent on the presence of defects.

One reason why diffusion in solids is much less familiar than diffusion in gases and liquids is that at room temperature it is generally very much slower. This slowness is why most solid-state reactions are undertaken at high temperatures (see Section 23.5), when diffusion is faster. However, there are some striking exceptions to this generalization. Diffusion of atoms or ions in solids is in fact very important in many areas of solid-state technology, such as semiconductor manufacture, the synthesis of new solids, fuel cells, sensors, metallurgy, and heterogeneous catalysis.

The rates at which ions move through a solid can often be understood in terms of the mechanism for their migration and the activation barriers the ions encounter as they move. The lowest energy pathway generally involves defect sites with the roles summarized in Fig. 23.11. Materials that show high rates of diffusion at moderate temperatures thus have the following characteristics:

- Low-energy barriers; so temperatures at or a little above room temperature are sufficient to permit ions to jump from site to site.

- Low charges and small radii; so, for example, the most mobile cation (other than the proton) and anion are Li^+ and F^-, respectively. Reasonable mobilities are also found for Na^+ and O^{2-}.

- High concentrations of intrinsic or extrinsic defects; defects typically provide a low-energy pathway for diffusion through a structure that does not involve the energy penalties associated with continuously displacing ions from normal, favourable ion sites.

- Mobile ions are present as a significant proportion of the total number of ions.

Figure 23.12 shows the temperature dependence of the diffusion coefficients, which are a good measure of the ion mobility, for specified ions in a selection of solids at high temperature. The slopes of the lines are proportional to the activation energy for that ion transport. Thus Na^+ is highly mobile and has a low activation energy for motion through β-alumina, whereas Ca^{2+} in CaO is much less mobile and has a high activation energy for hopping through the rock-salt structure.

As with the mechanisms of most chemical reactions, the evidence for the operation of a particular diffusion process is circumstantial. The individual events are never directly observed but are inferred from the influence of experimental conditions on atom or ion diffusion rates. For example, a detailed analysis of the thermal motion and distributions of ions in crystals based on X-ray and neutron diffraction provides strong hints about the ability of ions to move through the crystal, including their most likely paths. Computer modelling, which is often an elaboration of the ionic model (Chapter 3), also gives very useful guidance to the feasibility of these migration mechanisms.

23.4 Solid electrolytes

Any electrochemical cell, such as a battery, fuel cell, electrochromic display, or electrochemical sensor, requires an electrolyte. In many applications, an ionic solution (for example, dilute sulfuric acid in a lead–acid battery) is an acceptable electrolyte, but because it is often desirable to avoid a liquid phase, there is considerable interest in the development of solid electrolytes. Two important and well-studied solid electrolytes with mobile cations are silver tetraiodomercurate(II), Ag_2HgI_4, and sodium β-alumina with the composition $Na_{1+x}Al_{11}O_{17+x/2}$. Other recently developed fast-cationic conductors include NASICON (a name formed from the letters in sodium, Na, SuperIonic CONductor) of composition $Na_{1+x}Zr_2P_{3-x}Si_xO_{12}$ and a number of proton conductors, such as $CsHSO_4$, above 160°C.

Solids exhibiting high anion mobility are rarer than cationic conductors and generally show high conductivity only at elevated temperatures: anions typically have larger ionic radii than cations and so the energy barrier for diffusion through the solid is high. As a consequence, fast anion conduction in solids is limited to F^- and O^{2-} (with ionic radii 133 pm and 140 pm, respectively). Despite these limitations, anionic conductors play an important role in sensors and fuel cells, where a typical material is 'yttrium-stabilized zirconia' (YSZ), of composition $Y_xZr_{1-x}O_{2-x/2}$). Table 23.2 summarizes some typical conductivity values of these solid electrolytes in comparison with other ionically conducting media.

(a) Solid cationic electrolytes

Key points: Solid inorganic electrolytes often have a low-temperature form in which the ions are ordered on a subset of sites in the structure; at higher temperatures the ions become disordered over the sites and the ionic conductivity increases.

Below 50°C, Ag_2HgI_4 has an ordered crystal structure in which Ag^+ and Hg^{2+} ions are tetrahedrally coordinated by I^- ions and there are unoccupied tetrahedral holes

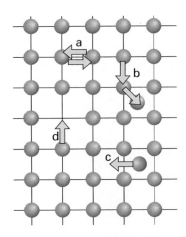

Fig. 23.11 Some diffusion mechanisms for ions or atoms in a solid: (a) exchange, (b) interstitialcy, (c) interstitial, and (d) vacancy.

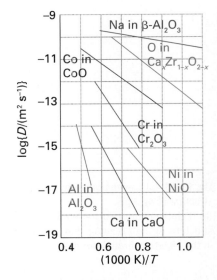

Fig. 23.12 The diffusion coefficients (on a logarithmic scale) as a function of inverse temperature for the mobile ion in selected solids.

Table 23.2 Comparative values of electrical conductivity of solid electrolytes

Material	Conductivity/(S m^{-1})
Ionic conductors	
Ionic crystals	$<10^{-16}$–10^{-2}
Example: LiI at 298°C	10^{-4}
Solid electrolytes	10^{-1}–10^{3}
Example: YSZ at 600°C	1
AgI at 500°C	10^{2}
Strong (liquid) electrolytes	10^{-1}–10^{3}
Example: 1 M NaCl(aq)	10^{2}
Electronic conductors	
Metals	10^{3}–10^{6}
Semiconductors	10^{-3}–10^{2}
Insulators	$<10^{-7}$

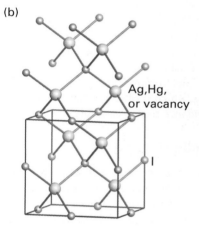

Fig. 23.13 (a) Low-temperature ordered structure of Ag_2HgI_4.
(b) High-temperature disordered structure. Ag_2HgI_4 is an Ag^+ ion conductor in the high-temperature form.

(Fig. 23.13a). At this temperature its electrical conductivity is low. Above 50°C, however, the Ag^+ and Hg^{2+} ions are randomly distributed over the tetrahedral sites (Fig. 23.13b). At this temperature the material is a good electrical conductor, largely on account of the mobility of the Ag^+ ions. The close-packed array of polarizable I^- ions is easily deformed and this results in a low activation energy for the migration of an Ag^+ ion from one ion site to the next. There are many related solid electrolytes having similar structures containing soft anions, such as AgI and $RbAg_4I_5$, both of which have highly mobile Ag^+ ions so that the conductivity of $RbAg_4I_5$ at room temperature is greater than that of aqueous sodium chloride.

Sodium β-alumina is an example of a mechanically hard material that is a good ionic conductor. In this case, the rigid and dense Al_2O_3 slabs are bridged by a sparse array of O^{2-} ions (Fig. 23.14). The plane containing these bridging ions also contains Na^+ ions, which can move from site to site because there are no major bottlenecks to hinder their motion. Many similar rigid materials having planes or channels through which ions can move are known and are called **framework electrolytes**. Another closely related material, sodium β″-alumina, has even less restricted motion of ions than β-alumina, and it has been found possible to substitute dipositive cations such as Mg^{2+} or Ni^{2+} for Na^+. Even the large lanthanoid cation Eu^{2+} can be introduced into β″-alumina, although the diffusion of such ions is slower than that of their smaller counterparts. The material NASICON, mentioned earlier, is a nonstoichiometric, solid-solution system with a framework constructed from ZrO_6 octahedra and PO_4 tetrahedra, corresponding to the parent phase of composition $NaZr_2P_3O_{12}$ (see Fig. 23.14). A solid solution can be obtained by partially replacing P by Si to give $Na_{1+x}Zr_2P_{3-x}Si_xO_{12}$ with an increase in the number of Na^+ ions for charge balance. In this material, the full set of possible Na sites is only partially filled and these sites lie within a three-dimensional network of channels that allow rapid migration of the remaining Na^+ ions through the structure. Other classes of materials currently being investigated as fast cation conductors include Li_4GeO_4 doped with Zn or V on the Ge sites (a lithium-ion conductor), the perovskite $La_{0.6}Li_{0.2}TiO_3$, and sodium yttrium silicate, $Na_5YSi_4O_{12}$ (a sodium-ion conductor).

Example 23.3 Correlating conductivity and ion size in a framework electrolyte

Conductivity data on β-alumina containing monopositive ions of various radii show that Ag^+ and Na^+ ions, both of which have radii close to 100 pm, have activation energies for conductivity close to 17 kJ mol^{-1} whereas that for Tl^+ (radius 149 pm) is about 35 kJ mol^{-1}. Suggest an explanation of the difference.

Answer In sodium β-alumina and related β-aluminas, a fairly rigid framework provides a two-dimensional network of passages that permit ion migration. Judging from the experimental results, the bottlenecks for ion motion appear to be large enough to allow Na^+ or Ag^+ (ionic radii close to 100 pm) to pass quite readily (with a low activation energy) but too small to let the larger Tl^+ (ionic radius 149 pm) pass through as readily.

Self-test 23.3 Why does increased pressure reduce the conductivity of K^+ in β-alumina more than that of Na^+ in β-alumina?

Because the reactivity of a bulk material is related to the presence of crystal defects and to the processes of atom and ion diffusion, by modelling an ion diffusion process we can obtain information on both ion conduction and the reaction mechanisms of a solid. Such a study has been undertaken on Li_3N with the aim of determining the energy barriers for Li^+ ions moving through the structure. Comparison of the values for the various barriers (between all the different possible sites that Li^+ could occupy as it migrates through the solid) in turn allows proposals to be made for the conduction pathway and a value for ionic conductivity to be calculated. Diffusion of Li^+ ions in the Li_2N plane (Fig. 23.15) was found to have a very low energy barrier in comparison to that for ion motion perpendicular to the Li_2N plane. The effect of replacing the Li^+ ions between the layers, as in Li_2MN (where M is a $3d$-series metal) and thereby reducing the energy barrier can also be studied; this kind of investigation is important for understanding potential applications of these materials as the electrolyte in rechargeable lithium-ion batteries.

(b) Solid anionic electrolytes

Key point: Anion mobility can occur at high temperatures in certain structures that contain high levels of anion vacancies.

Michael Faraday reported in 1834 that red-hot solid PbF_2 is a good conductor of electricity. Much later it was recognized that the conductivity arises from the mobility of F^- ions through the solid. The property of anion conductivity is shared by other crystals having the fluorite structure. Ion transport in these solids is thought to be by an 'interstitialcy mechanism' in which an F^- ion first migrates from its normal position into an interstitial site and then moves to a vacant F^- site.

Structures that have large numbers of vacant sites generally show the highest ionic conductivities because they provide a path for ion motion (although at very high levels of defects their clustering can cause a reduction in conductivity). These vacancies, which are equivalent to extrinsic defects, can be introduced in fairly high numbers into many simple oxides and fluorides by doping with appropriately chosen metal ions in different oxidation states. Zirconia, ZrO_2, at high temperature has a fluorite structure, but on cooling the pure material to room temperature it distorts to a monoclinic polymorph. The cubic fluorite structure may be stabilized at room temperature by doping a small level of various cations for Zr(IV), such as the similarly sized Ca^{2+} and Y^{3+}. Doping with

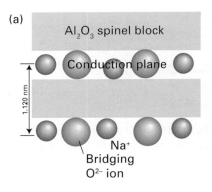

(a)

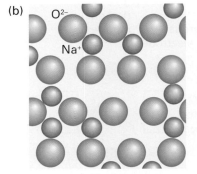

(b)

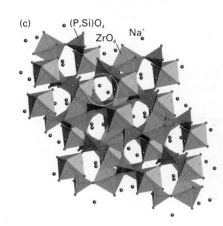

(c)

Fig. 23.14 Schematic side-view of β-alumina showing the Na_2O conduction planes between Al_2O_3 slabs. The O atoms in these planes bridge the two slabs. (b) A view of the conduction plane. Note the abundance of mobile ions and vacancies to which they can move. (c) The $Na_{1+x}Zr_2P_{3-x}Si_xO_{12}$ (NASICON) structure shown as linked $(P,Si)O_4$ tetrahedra (purple) and ZrO_6 octahedra (green) circumscribing channels (one highlighted by a yellow circle) containing the mobile sodium cations (red spheres).

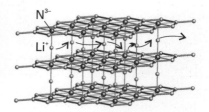

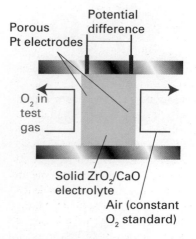

Fig. 23.15 The Li_3N structure showing possible pathways for Li^+ ion diffusion between the Li_2N planes. Li^+ ions are shown in blue and N^{3-} ions in red.

Porous Pt electrodes

Potential difference

O_2 in test gas

Solid ZrO_2/CaO electrolyte

Air (constant O_2 standard)

Fig. 23.16 An oxygen sensor based on the solid electrolyte $Zr_{1-x}Ca_xO_{2-x}$.

these ions with lower oxidation numbers results in the introduction of vacancies on the anion sites to give, for example, $Y_xZr_{1-x}O_{2-x/2}$, the material mentioned previously as 'yttrium stabilized zirconia' (YSZ). This material has completely occupied cation sites in the fluorite structure but high levels of anion vacancies, with $0 \leq x \leq 0.15$. These vacant sites provide a path for oxide-ion diffusion through the structure so that a typical electrical conductivity, in, for example, $Ca_{0.15}Zr_{0.85}O_{1.85}$ is 50 mS cm^{-1} at 1000°C;[2] note that this conductivity is much lower than typical solid-state cation conductivities—even at these very high temperatures—due to the large anion size. Behaviour similar to that of YSZ is encountered for other compounds that adopt the fluorite structure, such as PbF_2 as discovered by Faraday. As noted previously, the anionic conductivities remain low even at high temperatures, so many other complex metal oxides are currently being investigated with the aim of generating good mobilities at low temperatures. Some compounds that show promising behaviour include $LaLnMo_2O_9$, barium indate ($Ba_2In_2O_5$), BIMEVOX (a d-MEtal doped BIsmuth Vanadium OXide), the apatite structure of $La_{9.33}Si_6O_{26}$, and strontium- and magnesium-doped lanthanum gallate (Sr,Mg-doped $LaGaO_3$, or LSGM).

Calcium-oxide doped zirconia exploits its high oxide ion conductivity in a solid-state electrochemical sensor for measuring the partial pressure of oxygen in automobile exhaust systems (Fig. 23.16).[3] The platinum electrodes in this cell adsorb O atoms and, if the partial pressures of oxygen are different between the sample and reference side, there is a thermodynamic tendency for oxygen to migrate through the electrolyte as the O^{2-} ion. The thermodynamically favoured processes are:

High $p(O_2)$ side:

$$O_2(g) + Pt(s) \rightarrow 2\, O(Pt, surface)$$
$$O(Pt, surface) + 2\, e^- \rightarrow O^{2-}(ZrO_2)$$

Low $p(O_2)$ side:

$$O^{2-}(ZrO_2) \rightarrow O(Pt, surface) + 2\, e^-$$
$$O(Pt, surface) \rightarrow O_2(g) + Pt(s)$$

The cell potential is related to the two oxygen partial pressures (p_1 and p_2) by the Nernst equation (Section 5.5), for the half-cell reaction $O_2 + 4\, e^- \rightarrow 2\, O^{2-}$, which occurs at both electrodes

$$E = \frac{RT}{F} \ln \frac{p_1}{p_2}$$

so a simple measurement of the potential difference provides a measure of the oxygen partial pressure in the exhaust gases. According to this equation, the potential difference produced by an oxygen sensor operating at 1000 K in an exhaust system with air on one side ($p_{O_2} = 0.2$ atm) and a burnt fuel/air mixture ($p_{O_2} = 0.001$ atm) on the other is about 0.4 V.

(c) Mixed ionic–electronic conductors

Key point: Solid materials can exhibit both ionic and electronic conductivity.

Most ionic conductors, such as sodium β''-alumina and YSZ, have low electronic conductivity (that is, conduction by electron rather ion motion). Their application as solid electrolytes, in sensors for instance, requires this feature to avoid shorting-out the cell. In some cases a combination of electronic and ionic conductivity is desirable, and this type of behaviour can be found in some d-metal compounds where defects allow O^{2-} conduction and the metal d orbitals provide an electronic conduction band. Many such materials are perovskite-based structures with mixed oxidation states at the B cation sites. Two examples are $La_{1-x}Sr_xMnO_{3-y}$ and $La_{1-x}Sr_xFeO_{3-y}$. These oxide systems are good electronic conductors with partially filled bands as a result of the nonintegral

---

[2] The symbol S denotes siemens; $1\, S = 1\, \Omega^{-1}$, where $1\, \Omega$ (ohm) $= 1\, V\, A^{-1}$.

[3] The signal from this sensor is used to adjust the air/fuel ratio and thereby the composition of the exhaust gas being fed to the catalytic converter.

Box 23.1 Solid oxide fuel cells

A fuel cell consists of two electrodes sandwiched around an electrolyte; oxygen passes over one electrode and the fuel over the other, generating electricity, water, and heat. The general operation and construction of a fuel cell that converts a fuel, such as hydrogen, methane, or methanol, by reaction with oxygen into electrical energy (and combustion products H_2O and CO_2) was described in Box 9.1. A variety of materials can be used as the electrolyte in such cells including phosphoric acid, proton-exchange membranes, and, in *solid oxide fuel cells* (SOFCs), oxide ion conductors.

SOFCs operate at high temperatures and use an oxide ion conductor as the electrolyte. The design of a typical SOFC is shown in Fig. B23.1. Each cell generates a limited voltage but, just as with the cells of a battery, a connected stack may be constructed to increase the potential difference and power supplied. Cells are connected electrically through an 'interconnect' that can also be used to isolate the fuel and air supplies for each cell.

The attraction of SOFCs is based on several aspects including the clean conversion of chemical energy to electricity, low levels of

noise pollution, the ability to cope with different fuels, but most significantly a high efficiency. Their high efficiency is a result of their high operating temperatures, which in some designs may exceed 1000°C but is typically between 500 and 1000°C. In high-temperature SOFCs the interconnect may be a ceramic such as lanthanum chromite (the perovskite $LaCrO_3$) or, if the temperature is below 1000°C, an alloy such as Y/Cr may be used. The oxide ion conductor used as the electrolyte in these very high-temperature SOFCs is normally yttrium-doped zirconia (YSZ).

Intermediate-temperature SOFCs, which typically operate between 500 and 700°C, have a number of advantages over very high temperature devices in that there is reduced corrosion, a simpler design, and the time to heat the system to the operating temperature is much reduced. However, such devices require a material with excellent oxide ion conductivities at lower temperatures to act as the electrolyte. The best developed intermediate-temperature SOFC consists of an anode of CeO_2/Ni, an electrolyte of gadolinium-doped cerium oxide (CGO), and a cathode of the perovskite LSCF $(La,Sr)(Fe,Co)O_3$. The electrolyte material CGO possesses a much higher ionic conductivity than YSZ at these lower temperatures. Unfortunately, however, its electrical conductivity is also higher and the use of a CGO electrolyte results in reduced efficiency as energy is wasted due to electrical resistive effects. For this reason new and better oxide ion conductors are being sought, such as those mentioned in the text.

One of the main advantages of SOFCs over other fuel cell types is their ability to handle more convenient hydrocarbon fuels: other types of fuel cells have to rely on a clean supply of hydrogen for their operation. Because SOFCs operate at high temperature there is the opportunity to convert hydrocarbons catalytically (see Section 25.16) to hydrogen and carbon oxides within the system. Because of their size and the requirement to be heated to, and operate at, high temperatures, the applications of SOFCs focus on large-scale static systems.

Oxygen or air

$$O_2 + 4\,e^- \rightarrow 2\,O^{2-}$$

(La,Sr)FeO$_{3-x}$ cathode

YSZ electrolyte

Power

Ni–YSZ anode

$$2\,H_2 + 2\,O^{2-} \rightarrow 2\,H_2O + 4\,e^-$$

Hydrogen gas

B23.1 The structure of a solid-oxide fuel cell.

d-metal oxidation number and can conduct by O^{2-} migration through the perovskite O^{2-} ion sites. This type of material is of use in Solid Oxide Fuel Cells (SOFCs, Box 23.1), one type of fuel cell mentioned in Box 9.1, in which one electrode has to allow diffusion of ions through a conducting electrode

Synthesis of materials

Much of synthetic inorganic chemistry, including the coordination chemistry of the metals and organometallic chemistry, involves the conversion of molecules by, for example, the replacement of one ligand by another. These reactions generally involve molecular rearrangements or substitution of one group or ligand by another. Processes of this type generally have relatively small activation energies and can be undertaken at low temperatures, typically between 0 and 150°C, and in solvents that aid the diffusion of reacting species. This rapid molecular diffusion in solvents also results in fairly short reaction times. Formation of solid materials, however, involves rather different reactions as the high lattice energies of their extended structures need to be overcome and ion diffusion in the solid state is normally slow except at very elevated temperatures; we consider this synthetic chemistry here.

As well as the requirement to produce inorganic solids as bulk phases for many applications, there is a need to deposit thin films of the material on a surface or substrate.

For this type of deposited product, an inorganic molecular species may be used as a precursor in the synthesis. The synthesis of inorganic materials with controlled particle size and morphology is discussed in Chapter 24.

23.5 The formation of extended structures

New materials can be obtained by two main methods: the breaking-up of one or more continuous, or so-called extended-lattice, structures followed by the slow diffusion of ions and crystallization of a new structure, and the linking of polyhedral building units from solution and deposition of the newly formed solid.

(a) Methods of direct synthesis

Key point: Many complex solids can be obtained by direct reaction of the components at high temperatures.

The most widely used method for the synthesis of inorganic materials involves heating the components together at a high temperature for an extended period. Normally a complex oxide may be obtained by heating a mixture of all the oxides of the various metals present; alternatively, simple compounds that decompose to give the oxides may be used instead of the oxide. Thus ternary phases, such as $BaTiO_3$, and quaternary oxides, such as $YBa_2Cu_3O_7$, are synthesized by heating together the following mixtures for several days:

$$BaCO_3(s) + TiO_2(s) \xrightarrow{1000°C} BaTiO_3(s) + CO_2(g)$$

$$\tfrac{1}{2}Y_2O_3(s) + 2\,BaCO_3(s) + 3\,CuO(s) \xrightarrow{930°C/air\ and\ 450°C/O_2} YBa_2Cu_3O_7(s)$$
$$+2\,CO_2(g)$$

The need for the high temperatures in these syntheses is a result of the slow diffusion of ions in most solids at lower temperatures and the requirement to overcome the high Coulombic attractions between the ions. The direct method is applicable to many other inorganic material types, such as the syntheses of a complex chloride and a dense, nonhydrous metal aluminosilicate:

$$3\,CsCl(s) + 2\,ScCl_3(s) \rightarrow Cs_3Sc_2Cl_9(s)$$

$$NaAlO_2(s) + SiO_2(s) \rightarrow NaAlSiO_4(s)$$

Most simple binary oxides are available commercially as pure, polycrystalline powders with typical particle dimensions of a few micrometres. Alternatively, the decomposition of a simple metal salt precursor, either prior to or during reaction, leads to a finely divided oxide. Such precursors include metal carbonates, hydroxides, oxalates, and nitrates. An additional advantage of precursors is that they are normally air stable whereas many oxides are hygroscopic and pick up carbon dioxide from the air. Thus, in the synthesis of $BaTiO_3$, barium carbonate, $BaCO_3$, which starts to decompose to BaO above 900°C, would be ground together with TiO_2 in the correct stoichiometric proportions using a pestle and mortar. The mixture is then transferred to a crucible, normally constructed of an inert material such as vitreous silica, recrystallized alumina, or platinum, and placed in a furnace. The reaction, even at high temperatures, is slow and typically takes several days as the energies involved in breaking down the structures by overcoming lattice enthalpies and diffusing the ions through the solid structures over significant distances are very high. A variety of methods can be used to improve reaction rates, including pelletizing the reaction mixture under high pressure to increase interfacial contact between the reactant particles, regrinding the mixture periodically to introduce virgin reactant interfaces, and the use of fluxes, low melting solids that aid the ion diffusion processes.

The reaction environment may need to be controlled if a particular oxidation state is required or one of the reactants is volatile. Solid-state reactions can be carried out in a controlled atmosphere, using a tube furnace, where a gas can be passed over the reaction mixture during heating. An example of such a reaction is the use of an inert gas to prevent oxidation, as in the preparation of $TlTaO_3$

$$Tl_2O(s) + Ta_2O_5(s) \xrightarrow{N_2/600°C} 2\ TlTaO_3(s)$$

Another example is the partial reduction in hydrogen of V_2O_5 to V_2O_3

$$V_2O_5(s) \xrightarrow{H_2/1000°C} V_2O_3(s)$$

High gas pressures may also be used to control the composition of the reaction product. For example, the use of oxygen under pressures of several hundred atmospheres leads to the formation of Sr_2FeO_4, containing Fe(IV), from mixtures of SrO and Fe_2O_3. This outcome is in contrast to the formation of compounds containing Fe(III) under normal pressures. For volatile reactants the reaction mixture is normally sealed in a glass tube, under vacuum, prior to heating. Examples of such reactions are

$$Ta(s) + S_2(l) \xrightarrow{500°C} TaS_2(s)$$

$$Tl_2O_3(l) + 2\ BaO(s) + 3\ CaO(s) + 4\ CuO(s) \xrightarrow{860°C} Tl_2Ba_2Ca_3Cu_4O_{12}(s)$$

where the sulfur and thallium(III) oxide, respectively, are volatile at the reaction temperature and would be lost from the reaction mixture in an open vessel, leading to products of the incorrect stoichiometry (Section 23.8).

High pressures can also be used to affect the outcome of a solid-state chemical reaction. Apparatus exists, typically based on large presses, that allows reactions between solids to be undertaken at pressures up to about 100 GPa at temperatures close to 1500°C. Reactions carried out under such conditions promote the formation of denser, higher coordination number structures. Examples include the synthesis of diamond from graphite and of $MgSiO_3$, with a perovskite-like structure having six-coordinate Si rather than the normal tetrahedral SiO_4 unit. Small-scale reactions can be undertaken at very high pressures in **diamond anvil cells** in which the faces of two opposed diamonds are pushed together in a vice-like apparatus to generate pressures of up to 100 GPa.

(b) Solution methods

Key point: Frameworks formed from polyhedral species can often be obtained by condensation reactions in solution.

Many inorganic materials, especially framework structures, can be synthesized by crystallization from solution. Although the methods used are very diverse, the following are typical reactions that occur in water:

$$ZrO_2(s) + 2\ H_3PO_4(l) \rightarrow Zr(HPO_4)_2 \cdot H_2O(s) + H_2O(l)$$

$$12\ NaAlO_2(s) + 12\ Na_2SiO_3 \cdot 9H_2O(s)$$

$$\xrightarrow{90°C} Na_{12}[Si_{12}Al_{12}O_{48}] \cdot nH_2O \ (\text{Zeolite LTA})(s) + 24\ NaOH(aq)$$

The application of solution methods is extended by using hydrothermal techniques, in which the reacting solution is heated above its boiling point (at 1 atm) in a sealed vessel. Such reactions are important in the synthesis of open-structure aluminosilicates (zeolites) and analogous porous structures based on linked oxo-polyhedra (Section 23.13). These porous aluminosilicate structures are thermodynamically metastable, with respect to conversion to dense aluminosilicates, and thus not obtainable through high-temperature reactions. More recently, other solvents such as liquid ammonia, supercritical CO_2, and organic amines have been used in these 'solvothermal' reactions.

A reaction in solution can also be used to as an initial stage in the synthesis of many materials, particularly oxides, normally obtained through direct high-temperature reaction. The advantages of starting with solutions is that the reactants are mixed at the atomic level, so overcoming the problems associated with the direct reaction of two or more solid phases consisting of micrometre-sized particles. In the simplest reaction of this type, a solution of metal ions, for example as their nitrates, is converted to a solid form through a variety of methods such as evaporation of the solvent, precipitation as a simple mixed metal salt, or formation of a gel. This solid is then heated, and thus decomposed, to produce the target phase. So, for example,

$$2\ La^{3+}(aq) + Cu^{2+}(aq) \xrightarrow{OH^-(aq)} 2\ La(OH)_3 \cdot Cu(OH)_2(s)$$
$$\xrightarrow{600°C} La_2CuO_4(s) + 4\ H_2O(g)$$

or

$$Zn^{2+}(aq) + 2\ Fe^{2+}(aq) + 3\ C_2O_4{}^{2-}(aq) \rightarrow ZnFe_2(C_2O_4)_3(s)$$
$$\xrightarrow{700°C} ZnFe_2O_4(s) + 4\ CO(g) + 2\ CO_2(g)$$

As well as the advantages of reduced reaction times, a result of the intimate mixing of the reactants, the final decomposition temperature is somewhat lower than that needed for the direct reaction of the oxides. The use of a lower temperture can also have the effect of reducing the size of the particles formed in the reaction. Further discussion of routes involving gel formation, or so-called 'sol–gel processes', are included in Sections 23.9 and 24.5.

Although high-temperature, direct-combination methods and solvothermal techniques are the most commonly used synthesis methods in materials chemistry, some reactions involving solids can occur at low temperatures if there is no major change in structure. These so-called 'intercalation reactions' are discussed in Section 23.11.

23.6 Chemical deposition

Key point: Thermal decomposition of volatile inorganic compounds can be used to deposit electronic materials on substrates.

The technological applications of the materials described in this chapter often require the inorganic material to be generated as a thin layer or film, for example on a silicon substrate, and it has become necessary to develop the chemistry to deposit films of many inorganic materials. The main method by which this is done is **chemical vapour deposition** (CVD), in which a volatile inorganic compound is decomposed above the substrate. When the compound is a metallo-organic complex (that is, a complex of a metal atom with organic ligands), this route is known as **metallo-organic chemical vapour deposition** (MOCVD). A large area of research has grown up around the design and synthesis of volatile inorganic compounds that can be used for CVD.

For many electronic compounds, simple metal alkyls provide a route to depositing the metal or, through reaction with other gas molecules, its compounds. Thus Me_2Zn, which has a vapour pressure of 0.3 bar at room temperature, can react with H_2S above a substrate to generate the Group 12/16 (II/VI) semiconductor ZnS. For others, such as indium, the metal alkyl is a solid with a low vapour pressure at room temperature (for Me_3In, less than 2 mbar) and it cannot be easily used. In such cases other volatile compounds must be found and this work centres on other organometallics such as carbonyls and cyclopentadienyl complexes, acetoacetonates, and thiocarbamates. For complex oxides, such as the superconducting cuprates, $YBa_2Cu_3O_7$, suitable precursors are required for each of the metals and the trick is to find highly volatile molecules of the electropositive elements barium and yttrium, which normally form ionic compounds. Metal β-diketonates have been found to be useful in this respect, so compounds such as the 2,2,6,6-tetramethyl-3,5-heptanedione complex of yttrium sublime at about 150°C.

Another approach to molecules that can be used for CVD involves incorporating into a single-molecule precursor more than one of the atom types to be deposited. This procedure has the potential advantage of improving the control of the product stoichiometry. Thus, zinc sulfide can be deposited from a variety of zinc thiocomplexes, such as $Zn(S_2PMe_2)_2$. A future goal is to make complex volatile molecules containing, for example, several different metal atoms that can be deposited simultaneously to make a complex oxide.

Other methods of deposition of thin films, at the nanometre scale, of inorganic materials on substrates include sputtering and laser ablation: these techniques are dealt with in greater detail in Section 24.7.

Example 23.4 Synthesizing complex oxides

How would you synthesize a sample of the high-temperature superconductor $NdBa_2Cu_3O_7$?

Answer The same method as used for preparing $YBa_2Cu_3O_7$ can be used but with the appropriate lanthanoid oxide. That is, reaction of neodymium oxide, barium carbonate, and copper(II) oxide at 940°C following by annealing under pure oxygen at 450°C.

Self-test 23.4 How would you prepare a sample of Sr_2MoO_4?

Metal oxides, nitrides, and fluorides

In this section, we explore the binary compounds of oxygen, nitrogen, and fluorine with metals. These compounds, particularly oxides, are central to much solid-state chemistry on account of their stability, ease of synthesis, and variety in composition and structure. These attributes lead to the vast number of compounds that have been synthesized and the ability to tune the properties of a compound for a specific application based on their electronic or magnetic characteristics. As we shall see, a discussion of the chemical properties of these compounds also provides insight into defects, nonstoichiometry, and ion diffusion, and into the influence of these characteristics on physical properties. This section also describes the chemical properties of some classes of solids that are of interest for the production of magnetic and superconducting materials. The chemical properties of metal fluorides parallel much of that of metal oxides, but the lower charge of the F^- ion means that equivalent stoichiometries are produced with cations having lower charges, as in $KMn(II)F_3$ compared with $SrMn(IV)O_3$. Compounds containing the nitride ion, N^{3-}, in combination with one or more metal ions have only recently been developed to a significant extent; the area of metal nitride chemistry is one where rapid advances are being made in terms of new materials and structure types.

23.7 Monoxides of the 3*d* metals

The monoxides of most of the 3*d* metals (Section 18.8) adopt the rock-salt structure, so it might at first appear that there is little to say about them and that their properties should be simple. In fact, the actual stoichiometries formed around the ideal composition MO and the structures and properties of these oxides are rather more interesting, and as a result they have been repeatedly investigated (Table 23.3). In particular, the compounds provide examples of how mixed oxidation states and defects lead to nonstoichiometry for the compositions TiO, VO, FeO, CoO, and NiO. All these compounds are usually obtained with significant deviations from the nominal MO stoichiometry.

Table 23.3 Monoxides of the Period 4 (3*d*-series) metals

Compound	Structure	Composition, x	Electrical character
CaO_x	Rock-salt	1	Insulator
TiO_x	Rock-salt	0.65–1.25	Metallic
VO_x	Rock-salt	0.79–1.29	Metallic
MnO_x	Rock-salt	1–1.15	Semiconductor
FeO_x	Rock-salt	1.04–1.17	Semiconductor
CoO_x	Rock-salt	1–1.01	Semiconductor
NiO_x	Rock-salt	1–1.001	Semiconductor
CuO_x	PtS (linked CuO_4 square-planes)	1	Semiconductor
ZnO_x	Wurtzite	Slight Zn excess	Wide band gap n-type semiconductor

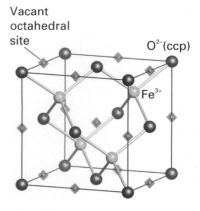

Vacant octahedral site

O^{2-}(ccp)

Fe^{3+}

Fig. 23.17 Defect sites proposed for $Fe_{1-x}O$. Note that the tetrahedral Fe^{3+} interstitials (green spheres) and octahedral Fe^{2+} vacancies (grey diamonds) have clustered together.

(a) Defects and nonstoichiometry

Key point: The nonstoichiometry of $Fe_{1-x}O$ arises from the creation of vacancies on the Fe^{2+} octahedral sites, with each vacancy charge-compensated by the conversion of two Fe^{2+} ions to two Fe^{3+} ions.

The origin of nonstoichiometry in FeO has been studied in more detail than that in most other compounds. In fact, it is found that stoichiometric FeO does not exist but rather a range of iron-deficient compounds in the range $Fe_{1-x}O$, $0.13 < x < 0.04$, can be obtained by quenching (cooling very rapidly) iron(II) oxide from high temperatures. $Fe_{1-x}O$ is in fact metastable at room temperature: it is thermodynamically unstable with respect to disproportionation into iron metal and Fe_3O_4 but does not convert for kinetic reasons. The general consensus is that the structure of $Fe_{1-x}O$ is derived from the rock-salt FeO structure by the presence of vacancies on the Fe^{2+} octahedral sites, and that each vacancy is charge-compensated by the conversion of two adjacent Fe^{2+} ions to two Fe^{3+} ions. The relative ease of oxidizing Fe(II) to Fe(III) accounts for the fairly broad range of compositions of $Fe_{1-x}O$. At high temperatures, the interstitial Fe^{3+} ions associate with the Fe^{2+} vacancies to form clusters distributed throughout the structure (Fig. 23.17).

Similar defects and the clustering of defects appear to occur with all other 3d-metal monoxides, with the possible exception of NiO. The range of nonstoichiometry in $Ni_{1-x}O$ is extremely narrow, but conductivity and the rate of atom diffusion vary with oxygen partial pressure in a manner that suggests the presence of isolated point defects. Both CoO and NiO occur in a metal-deficient state, although their range of compositions is again not as broad as that of FeO. As indicated by standard potentials in aqueous solution, Fe(II) is more easily oxidized than either Co(II) or Ni(II); this solution redox chemistry correlates well with the much smaller range of oxygen deficiency in NiO and CoO. Chromium(II) oxide, like FeO, spontaneously disproportionates:

$$3\,Cr(II)O(s) \rightarrow Cr(III)_2O_3(s) + Cr(0)(s)$$

However, the material can be stabilized by crystallization in a copper(II) oxide matrix. Both CrO and TiO have structures that show high levels of defects on both the cation and anion sites forming metal-rich or metal-deficient stoichiometries ($Ti_{1-x}O$ and TiO_{1-x}); in fact, TiO has large numbers of vacancies, in equal amounts, on both cation and anion sublattices rather than the expected perfect, defect-free structure. Note that significant deviations from the stoichiometry MO is out of the question for Group 2 metal oxides, such as CaO, because for these elements M^{3+} ions are chemically inaccessible.

(b) Electronic and magnetic properties

Key points: The 3d-metal monoxides MnO, FeO, CoO, and NiO are semiconductors; TiO and VO are metallic conductors.

The 3d-metal monoxides MnO, $Fe_{1-x}O$, CoO, and NiO have low electrical conductivities that increase with temperature (corresponding to semiconducting behaviour) or have such large band gaps that they are insulators. The electron or hole migration in these oxide semiconductors is attributed to a hopping mechanism. In this model, the electron or hole hops from one localized metal atom site to the next. When it lands on a new site it causes the surrounding ions to adjust their locations and the electron or hole is trapped temporarily in the potential well produced by this atomic polarization. The electron resides at its new site until it is thermally activated to migrate into another nearby site. Another aspect of this charge-hopping mechanism is that the electron or hole tends to associate with local defects, so the activation energy for charge transport may also include the energy of freeing the hole from its position next to a defect.

Hopping contrasts with the band model for semiconductivity discussed in Section 3.15, where the conduction and valence electrons occupy orbitals that spread through the whole crystal. The difference stems from the less diffuse d orbitals in the monoxides of the mid-to-late 3d metals, which are too compact to form the broad bands necessary for metallic conduction. We have already mentioned that when NiO is doped with Li_2O in an O_2 atmosphere a solid solution $Li_x(Ni^{2+})_{1-2x}(Ni^{3+})_xO$ is obtained, which has greatly increased conductivity for reasons similar to the increase in conductivity of silicon when doped with boron. The characteristic pronounced increase in electronic conductivity with increasing temperature of metal oxide semiconductors is used in 'thermistors' to measure temperature.

In contrast to the semiconductivity of the monoxides in the centre and right of the $3d$ series, TiO and VO have high electronic conductivities that decrease with increasing temperature. This metallic conductivity persists over a broad composition range from highly oxygen-rich $Ti_{1-x}O$ to metal-rich TiO_{1-x}. In these compounds, a conduction band is formed by the overlap of the t_{2g} orbitals of metal ions in neighbouring octahedral sites that are oriented towards each other (Fig. 23.18). The radial extension of the d orbitals of these early d-block elements is greater than for elements later in the period, and a band results from their overlap (Fig. 23.19); this band is only partly filled. The widely varying compositions of these monoxides also appears to be associated with the electronic delocalization: the conduction band serves as a rapidly accessible source and sink of electrons that can readily compensate for the formation of vacancies.

As well as having electronic properties that are the result of interactions between the d-orbital electrons, the transition metal monoxides have magnetic behaviours that derive from cooperative interaction of the individual atomic magnetic moments. These cooperative magnetic effects and the resulting magnetic properties of the oxides MO, $M = Mn$, Fe, Co, Ni, are discussed fully in Box 23.2.

23.8 Higher oxides and complex oxides

Binary metal oxides that do not have a 1:1 metal:oxygen ratio are known as **higher oxides**. Compounds containing ions of more than one metal are often termed **complex oxides** and include compounds containing three elements (for instance, $LaFeO_3$) and four or more elements (for instance, $YBa_2Cu_3O_7$). This section describes the structures and properties of some of the more important higher and complex oxides.

(a) M₂O₃ corundum structure

Key point: The corundum structure is adopted by many oxides of the stoichiometry M_2O_3 including chromium-doped aluminium oxide (ruby).

α-Aluminium oxide (the mineral corundum) adopts a structure that can be modelled as a hexagonal close-packed array of O^{2-} ions with the cations in two-thirds of the octahedral holes (Fig. 23.20). The corundum structure is also adopted by the oxides of titanium, vanadium, chromium, rhodium, iron, and gallium in their +3 oxidation states. Two of these oxides, Ti_2O_3 and V_2O_3, exhibit metallic to semiconducting transitions below 410 and 150 K, respectively (Fig. 23.21). In V_2O_3, the transition is accompanied by antiferromagnetic ordering of the spins. The two insulators Cr_2O_3 and Fe_2O_3 also display antiferromagnetic ordering.

Another interesting aspect of the M_2O_3 compounds is the formation of solid solutions of dark-green Cr_2O_3 and colourless Al_2O_3 to form brilliant red ruby. As we remarked in Section 19.1, this shift in the ligand-field transitions of Cr^{3+} stems from the compression of the O^{2-} ions around Cr^{3+} in the Al_2O_3 host structure. (In Al_2O_3, $a = 475$ pm and $c = 1300$ pm; in Cr_2O_3 the lattice constants are 493 pm and 1356 pm, respectively.) The compression shifts the absorption towards the blue as the ligand field increases, and the solid appears red in white light. The responsiveness of the absorption (and fluorescence) spectrum of Cr^{3+} ions to compression is sometimes used to measure pressure in high-pressure experiments. In this application, a tiny crystal of ruby in one segment of the sample can be interrogated by visible light and the shift in its fluorescence spectrum provides an indication of the pressure inside the cell.

(b) Rhenium trioxide

Key point: The rhenium trioxide structure can be constructed from ReO_6 octahedra sharing all vertices in three dimensions.

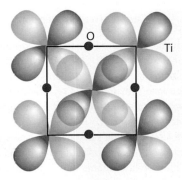

Fig. 23.18 Overlap of the d_{zx} orbitals in TiO to give a t_{2g} band.

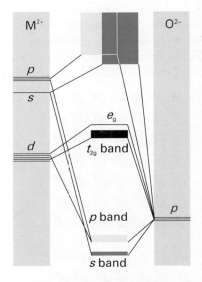

Fig. 23.19 Molecular orbital energy level diagram for early d-metal monoxides. The s and p bands are full. The t_{2g} band is only partly filled and metallic conduction results.

Box 23.2 Cooperative magnetism

The diamagnetic and paramagnetic properties of compounds were covered in Section 19.1; they are characteristic of individual atoms or complexes. By contrast, properties such as ferromagnetism and antiferromagnetism depend on interactions between electron spins on many atoms and arise from the cooperative behaviour of many unit cells in a crystal.

In a **ferromagnetic** substance the spins on different metal centres are coupled into a parallel alignment (Fig. B23.2a) that is sustained over thousands of atoms in a magnetic domain. The net magnetic moment may be very large because the magnetic moments of individual spins augment each other. Moreover, once established and with the temperature maintained below the **Curie temperature** (T_C), the magnetization persists because the spins are locked together. Ferromagnetism is exhibited by materials containing unpaired electrons in d or f orbitals that couple with unpaired electrons in similar orbitals on surrounding atoms. The key feature is that this interaction is strong enough to align spins but not so strong as to form covalent bonds, in which the electrons would be paired.

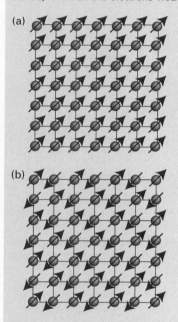

(a)

(b)

B23.2 (a) The parallel alignment of individual magnetic moments in a ferromagnetic material and (b) The antiparallel arrangement of individual magnetic moments in an antiferromagnetic material.

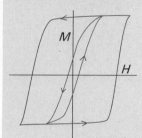

B23.3 Magnetization curves for ferromagnetic materials. A hysteresis loop results because the magnetization of the sample with increasing field ($\rightarrow$) is not retraced as the field is decreased ($\leftarrow$). Blue line: hard ferromagnet; red line: a soft ferromagnet.

The magnetization, M, of a ferromagnet is not proportional to the applied field strength H. Instead, a hysteresis loop (Fig. B23.3) is observed. For **hard ferromagnets** the loop is broad and M is large when the applied field has been reduced to zero. Hard ferromagnets are used for permanent magnets. A **soft ferromagnet** has a narrower hysteresis loop and is therefore much more responsive to the applied field. Soft ferromagnets are used in transformers, where they must respond to a rapidly oscillating field.

In an **antiferromagnetic** substance, neighbouring spins are locked into an antiparallel alignment (Fig. B23.2b) and the sample has a low magnetic moment. Antiferromagnetism is often observed when a paramagnetic material is cooled to a low temperature and is signalled by a decrease in magnetic susceptibility with decreasing temperature (Fig. B23.4). The critical temperature for the onset of antiferromagnetism is called the **Néel temperature**, T_N. The spin coupling responsible for antiferromagnetism generally occurs through intervening ligands by a mechanism called **superexchange**. As indicated in Fig. B23.5, the spin on one metal atom induces a small spin polarization on an occupied orbital of a ligand, and this spin polarization results in an antiparallel alignment of the spin on the adjacent metal atom. Many compounds exhibit antiferromagnetic behaviour in the appropriate temperature range; for example, MnO is antiferromagnetic below 122 K, and Cr_2O_3 is antiferromagnetic below 310 K. The overall magnetic structure of MnO and the other $3d$-series metal monoxides is shown in Fig. B23.6. The Néel temperatures

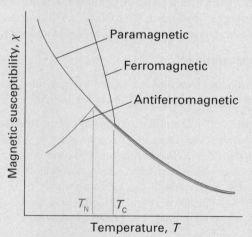

B23.4 The temperature dependence of the susceptibilities of paramagnetic, ferromagnetic, and antiferromagnetic substances.

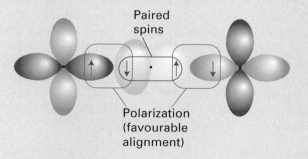

B23.5 Antiferromagnetic coupling between two metal centres created by spin polarization of a bridging ligand.

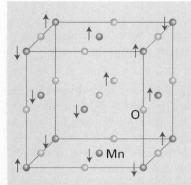

B23.6 The arrangement of electron spins on the Mn^{2+} ions (blue) in the antiferromagnetic state of MnO.

(in parentheses) of the series of d-metal oxides MnO (122 K), FeO (198 K), CoO (271 K), and NiO (523 K) reflect the strength of the superexchange spin interactions along the M—O—M directions. As the size of the M^{2+} ion decreases from Mn to Ni, the superexchange mechanism becomes stronger due to the increased metal–oxygen orbital overlap and T_N increases. Coupling of spins through intervening ligands is frequently observed in molecular complexes containing two ligand-bridged metal ions but this is weaker than with a simple oxide ion link between metal sites and thus ordering temperatures are much lower, typically below 100 K.

In a third type of collective magnetic interaction, **ferrimagnetism**, net magnetic ordering is observed below the Curie temperature. However, ferrimagnets differ from ferromagnets because ions with different local moments are present. These ions order with opposed spins, as in antiferromagnetism, but because of the different magnitude of the individual spin moments there is incomplete cancellation and the sample has a net overall moment. As with antiferromagnetism, these interactions are generally transmitted by the ligand.

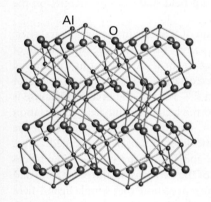

Fig. 23.20 The corundum structure, as adopted by Al_2O_3, with cations (red) occupying two-thirds of the octahedral holes between layers of closed-packed oxide ions (blue).

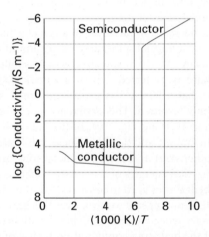

Fig. 23.21 The temperature dependence of the electrical conductivity of V_2O_3 showing the metal to semiconductor transition.

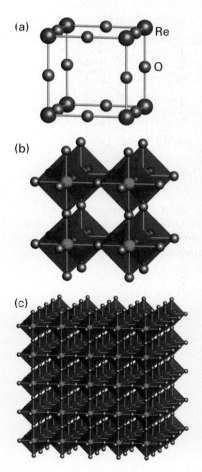

Fig. 23.22 The ReO_3 structure shown as (a) the unit cell, (b) the ReO_6 octahedra forming the unit cell and (c) an extended, three-dimensional array formed from the vertex-linked ReO_6 octahedra.

Materials adopting the rhenium trioxide structure are relatively rare. This rarity is in part due to the requirement of the oxidation state M(VI) when M is in combination with oxygen. Rhenium(VI) oxide, ReO_3, itself and one form of UO_3 (δ-UO_3) have this structure type and WO_3 exists in a slightly distorted version of it. The structure type is very simple, consisting of a cubic unit cell with Re atoms at the corners and O atoms at the mid-point of each edge (Fig. 23.22). Alternatively, the structure can be considered to be derived from ReO_6 octahedra sharing all vertices. In WO_3, the WO_6 octahedra are slightly distorted and tilted relative to each other so that the W—O—W bond angle is not 180°. The rhenium trioxide structure is also closely related to a perovskite structure (Section 3.9) in which the A-type cation has been removed from the unit cell.

Rhenium trioxide itself is a bright red lustrous solid. Its electrical conductivity at room temperature is similar to that of copper metal. The band structure for this compound contains a band derived from the rhenium t_{2g} orbitals and the O2p orbitals (Fig. 23.23). This band can contain up to six electrons per Re atom but is only partially filled for the Re^{6+} d^1 configuration, so producing the observed metallic properties.

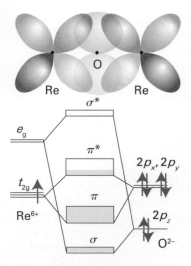

Fig. 23.23 The band structure of ReO$_3$.

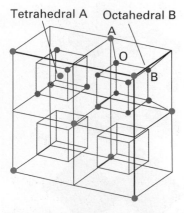

Fig. 23.24 A segment of the spinel (AB$_2$O$_4$) unit cell showing the tetrahedral environment of A ions and the octahedral environments of B ions. (Compare with Fig. 3.40.)

(c) Spinels

Key point: The observation that many *d*-metal spinels do not have the normal spinel structure is related to the effect of ligand-field stabilization energies on the site preferences of the ions.

The *d*-block higher oxides Fe$_3$O$_4$, Co$_3$O$_4$, and Mn$_3$O$_4$, and many related mixed-metal compounds, such as ZnFe$_2$O$_4$, have very useful magnetic properties. They all adopt the structural type of the mineral spinel, MgAl$_2$O$_4$, and may be given the general formula AB$_2$O$_4$. Most oxide spinels are formed with a combination of A^{2+} and B^{3+} cations (that is, as A^{2+}B$_2^{3+}$O$_4$), although there are a number of spinels that can be formulated with A^{4+} and B^{2+} cations (as A^{4+}B$_2^{2+}$O$_4$). The spinel structure was described briefly in Section 3.9, where we saw that it consists of an fcc array of O^{2-} ions in which the A ions reside in one-eighth of the tetrahedral holes and the B ions inhabit half the octahedral holes (Fig. 23.24); this structure is commonly denoted A[B$_2$]O$_4$, where the atom type in the square bracket represents that occupying the octahedral sites. In the inverse spinel structure, the cation distribution is B[AB]O$_4$, with the more abundant B-type cation distributed over both coordination geometries. Lattice enthalpy calculations based on a simple ionic model indicate that, for A^{2+} and B^{3+}, the normal spinel structure, A[B$_2$]O$_4$, should be the more stable. The observation that many *d*-metal spinels do not conform to this expectation has been traced to the effect of ligand-field stabilization energies on the site preferences of the ions.

The **occupation factor**, λ, of a spinel is the fraction of B atoms in the tetrahedral sites: $\lambda = 0$ for a normal spinel and $\lambda = 0.5$ for an inverse spinel, B[AB]O$_4$. The distribution of cations in (A^{2+},B^{3+}) spinels (Table 23.4) illustrates that for d^0 A and B ions the normal structure is preferred ($\lambda = 0$) as predicted by electrostatic considerations. Table 23.4 shows that, when A^{2+} is a d^6, d^7, d^8, or d^9 ion and B^{3+} is Fe^{3+}, the inverse structure is generally favoured. This preference can be traced to the lack of ligand-field stabilization (Section 19.1 and Fig. 19.9) of the high-spin d^5 Fe^{3+} ion in either the octahedral or the tetrahedral site and the ligand-field stabilization of the other d^n ions in the octahedral site. For other combinations of *d*-metal ions on the A and B sites the relative ligand-field stabilization energies of the different arrangements of the two ions on the octahedral and tetrahedral sites need to be calculated. It is also important to note that simple ligand-field stabilization appears to work over this limited range of cations. More detailed analysis is necessary when cations of different radii are present or any ions that are present do not adopt the high-spin configuration typical of most metals in spinels (for instance, Co^{3+} in Co$_3$O$_4$, which is low-spin d^6). Moreover, because λ is often found to depend on the temperature, care has to be taken in the synthesis of a spinel with a specific distribution of cations because slow cooling or quenching a sample from a high reaction temperature can produce quite different cation distributions.

The inverse spinels of formula AFe$_2$O$_4$ are sometimes classified as **ferrites** (the term ferrite also applies in different circumstances to other iron oxides). When $RT > J$, where J is the energy of interaction of the spins on different ions, ferrites are paramagnetic. However, when $RT < J$, a ferrite may be either ferrimagnetic or antiferromagnetic

Table 23.4 Occupation factor, λ, in some spinels*

	A	Mg^{2+}	Mn$^+$	Fe^{2+}	Co^{2+}	Ni^{2+}	Cu^{2+}	Zn^{2+}
B		d^0	d^5	d^6	d^7	d^8	d^9	d^{10}
Al^{3+}	d^0	0	0	0	0	0.38	0	
Cr^{3+}	d^3	0	0	0	0	0	0	0
Mn^{3+}	d^4	0						0
Fe^{3+}	d^5	0.45	0.1	0.5	0.5	0.5	0.5	0
Co^{3+}	d^6					0		0

* $\lambda = 0$ corresponds to a normal spinel; $\lambda = 0.5$ corresponds to an inverse spinel.

(Box 23.2). The antiparallel alignment of spins characteristic of antiferromagnetism is illustrated by $ZnFe_2O_4$, which has the cation distribution $Fe[ZnFe]O_4$. In this compound the Fe^{3+} ions (with $S = \frac{5}{2}$) in the tetrahedral and octahedral sites are antiferromagnetically coupled below 9.5 K to give nearly zero net magnetic moment to the solid as a whole; note that Zn^{2+} as a d^{10} ion makes no contribution to the magnetic moment of the material.

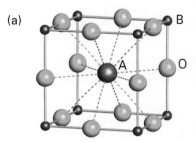

Example 23.5 Predicting the structures of spinel compounds

Is $MnCr_2O_4$ likely to have a normal or inverse spinel structure?

Answer A normal spinel structure is expected because Cr^{3+} will have a ligand-field stabilization energy in the octahedral site whereas the d^5 Mn^{2+} ion will not. Table 23.4 shows that this prediction is verified experimentally.

Self-test 23.5 Table 23.4 indicates that $FeCr_2O_4$ is a normal spinel. Rationalize this observation.

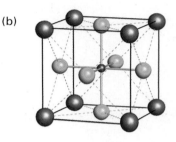

$CoAl_2O_4$ is among the normal spinels in Table 23.4 with $\lambda = 10$ and thus has the Co ions at the tetrahedral sites. The colour of $CoAl_2O_4$ (an intense blue) is that expected of tetrahedral Co^{2+} and this property coupled with the ease of synthesis and stability of the spinel structure has led to cobalt aluminate being used as a pigment ('cobalt blue'). Other mixed d-metal spinels that exhibit strong colours, for example $CoCr_2O_4$ (green), $CuCr_2O_4$ (black), and $(Zn,Fe)Fe_2O_4$ (orange/brown), are also used as pigments, with applications that include colouring certain construction materials, such as concrete.

(d) Perovskites and related phases

Key points: The perovskites have the general formula ABX_3, in which the 12-coordinate hole of a ReO_3-type BX_3 structure is occupied by a large A ion; the perovskite barium titanate, $BaTiO_3$, exhibits ferroelectric and piezoelectric properties associated with cooperative displacements of the ions.

The perovskites have the general formula ABX_3, in which the 12-coordinate hole of BX_3 (as in ReO_3) is occupied by a large A ion (Fig. 23.25; a different view of this structure was given in Fig. 3.38). The X ion is most frequently O^{2-} or F^- (as in $NaFeF_3$) although nitride- and hydride-containing perovskites can also be synthesized, such as in $LiSrH_3$. Perovskite itself is named after the naturally occurring oxide mineral $CaTiO_3$ and the largest class of perovskites are those with the anion as oxide. This breadth of perovskites is widened by the observation that solid solutions and nonstoichiometry are also common features of the perovskite structure, as in $Ba_{1-x}Sr_xTiO_3$ and $SrFeO_{3-y}$. Some metal-rich materials adopt the perovskite structure with the normal distribution of cations and anions partially inverted; for instance, $SnNCo_3$.

The perovskite structure is often observed to be distorted in such a manner that the unit cell is no longer centrosymmetric and the crystal develops an overall permanent electric polarization as a result of ion displacements. Some polar crystals are **ferroelectric** in the sense that they resemble ferromagnets, but instead of the electron spins being aligned over a region of the crystal, the electric polarizations of many unit cells are aligned. As a result, the relative permittivity, which reflects the polarity of a compound, for a ferroelectric material often exceeds 1×10^3 and can be as high as 1.5×10^4; for comparison, the relative permittivity of liquid water is about 80 at room temperature. Barium titanate, $BaTiO_3$, is the most extensively studied example of such a material. At high temperatures, above 120°C, this compound has the perfect cubic perovskite structure. At room temperature, it adopts a lower symmetry, tetragonal unit cell in which the various ions can be considered as having been displaced from their normal high symmetry sites (Fig. 23.26). This displacement results in a spontaneous polarization of the material; the application and then removal of an electric field results in a material that is polarized in a particular direction. The temperature below which this spontaneous

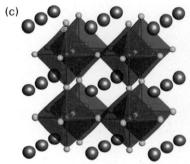

Fig. 23.25 Views of the perovskite (ABO_3) structure (a) emphasizing the 12-fold coordination of the larger A cation (blue) and showing the relationship with the ReO_3 structure of Fig. 23.22b, (b) highlighting the octahedral coordination of the B cation, and (c) a polyhedral representation accentuating the BO_6 octahedra.

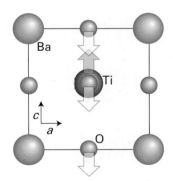

Fig. 23.26 The tetragonal $BaTiO_3$ structure showing the local ion displacements that lead to the ferroelectric behaviour of this material.

polarization can occur is known as the **Curie temperature**, denoted T_C, and is 120°C for $BaTiO_3$. The high relative permittivity of barium titanate (about 100 times that of water) leads to its use in capacitors, where its presence allows up to 1000 times the charge to be stored in comparison with a simple capacitor with air between the plates. The introduction of dopants into the barium titanate structure, forming solid solutions, allows various properties of the compound to be tuned. For example, the replacement of Ba by Sr or of Ti by Zr causes a sharp lowering of T_C.

Another characteristic of many crystals, including a number of perovskites, that lack a centre of symmetry is **piezoelectricity**, the ability to generate an electrical field when the crystal is under stress or to change dimensions when an electrical field is applied. Piezoelectric materials are used for a variety of applications, such as pressure transducers, ultramicromanipulators (where very small movements can be controlled), sound detectors, and as the probe support in scanning tunnelling microscopy. Some important examples are $BaTiO_3$, $NaNbO_3$, $NaTaO_3$, and $KTaO_3$.

Although a noncentrosymmetric structure is required for both ferroelectric and piezoelectric behaviour, the two phenomena do not necessarily occur for the same crystal. For example, quartz is piezoelectric but not ferroelectric. Quartz is widely used to set the clock rate of microprocessors and watches because a thin sliver oscillates at a specific frequency to produce a small oscillating electrical field. This frequency is very insensitive to temperature.

Another prototypical structure, that of potassium tetrafluoronickelate(II), K_2NiF_4 (Fig. 23.27), is related to perovskite. The compound can be thought of as containing individual slices from the perovskite structure that share the four F atoms from the octahedra within the layer and have terminal fluorine atoms above and below the layer. These layers are displaced relative to each other and separated by the potassium ions (which are nine-coordinate, to eight F atoms of one layer and one terminal F atom from the next).

Compounds with the K_2NiF_4 structure have come under renewed investigation because some high-temperature superconductors, such as $La_{1.85}Sr_{0.15}CuO_4$, crystallize with this structure. Apart from their importance in superconductivity, compounds with the K_2NiF_4 structure also provide an opportunity to investigate two-dimensional magnetic domains as coupling between electron spins is much stronger within the layers of linked octahedra than between these layers.

The K_2NiF_4 structure has been introduced as being derived from a single slice of the perovskite structure; other related structures are possible where two or more perovskite layers are displaced horizontally relative to each other. Structures with K_2NiF_4 at one end of the range (a single perovskite layer) and perovskite itself at the other (an infinite number of such layers) are known as **Ruddlesden–Popper phases**. They include $Sr_3Fe_2O_7$, with double layers, and $Ca_4Mn_3O_{10}$, with triple layers (Fig. 23.28).

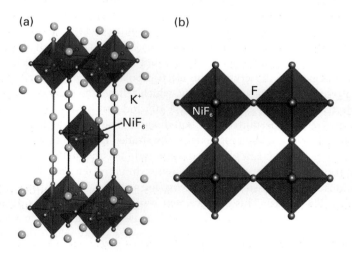

Fig. 23.27 The K_2NiF_4 structure. (a) The displaced layers of NiF_6 octahedra interspersed with potassium ions and (b) a view of one layer of composition NiF_4 showing the corner-sharing octahedra linked through F (blue).

Table 23.5 Some materials that exhibit superconductivity below the critical temperature, T_c

Element	T_c/K	Compound	T_c/K
Zn	0.88	Nb_3Ge	23.2
Cd	0.56	Nb_3Sn	18.0
Hg	4.15	$LiTi_2O_4$	13
Pb	7.19	$K_{0.4}Na_{0.6}BiO_3$	29.8
Nb	9.50	$YBa_2Cu_3O_{10}$	93
		$Tl_2Ba_3Ca_3Cu_4O_{12}$	134
		MgB_2	40
		K_3C_{60}	39

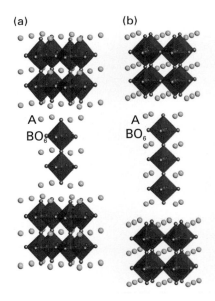

Fig. 23.28 The Ruddlesden–Popper phases of stoichiometry $A_3B_2O_7$ and $A_4B_3O_{10}$ formed, respectively, from two and three perovskite layers of linked BO_6 octahedra separated by A-type cations.

(e) High-temperature superconductors

Key point: High-temperature cuprate superconductors have structures related to perovskite.

The versatility of the perovskites extends to superconductivity, because most of the high-temperature superconductors first reported in 1986 can be viewed as variants of the perovskite structure. Superconductors have two striking characteristics. Below a critical temperature, T_c (as distinct from T_C, the Curie temperature), they enter the superconducting state and have zero electrical resistance. In this superconducting state they also exhibit the **Meissner effect**, the exclusion of a magnetic field.

Following the discovery in 1911 that mercury is a superconductor below 4.2 K, physicists and chemists made slow but steady progress in the discovery of superconductors with higher values of T_c at a rate of about 3 K per decade. After 75 years, T_c had been edged up to 23 K in Nb_3Ge. Most of these superconducting materials were metal alloys, although superconductivity had been found in some oxides and sulfides (Table 23.5); magnesium diboride is superconducting below 39 K (see Box 12.3). Then, in 1986, the first **high-temperature superconductor** (HTSC) was discovered. Several materials are now known with T_c well above 77 K, the boiling point of the relatively inexpensive refrigerant liquid nitrogen, and in a few years the maximum T_c was increased by more than a factor of five to around 134 K.

Two types of superconductors are known. Those classed as **Type I** show abrupt loss of superconductivity when an applied magnetic field exceeds a value characteristic of the material. **Type II** superconductors, which include high-temperature materials, show a gradual loss of superconductivity above a critical field denoted H_c.[4] Figure 23.29 shows that there is a degree of periodicity in the elements that exhibit superconductivity. Note in particular that the ferromagnetic metals iron, cobalt, and nickel do not display superconductivity; nor do the alkali metals and the coinage metals copper, silver, and gold. For simple metals, ferromagnetism and superconductivity never coexist, but in some of the oxide superconductors ferromagnetism and superconductivity appear to coexist on different portions of the structure of the same solid.

The first HTSC compound reported was $La_{1.8}Ba_{0.2}CuO_4$ ($T_c = 35$ K), which is a member of the solid-solution series $La_{2-x}Ba_xCuO_4$ in whch Ba replaces a proportion of the La sites in La_2CuO_4. This material has the K_2NiF_4 structure type with layers of edge-sharing CuO_6 octahedra separated by the La^{3+} and Ba^{2+} cations, although the octahedra are axially elongated by a Jahn–Teller distortion (Section 19.18). A similar compound with Sr replacing Ba in this structure type, as in $La_{1.8}Sr_{0.2}CuO_4$ ($T_c = 38$ K), is also known.

One of the most widely studied HTSC oxide materials, $YBa_2Cu_3O_{7-x}$ ($T_c = 93$ K; informally this compound is called '123', from the proportions of metal atoms in the

---

[4] The Chevrel phases discussed in Section 23.12 have the highest observed values of H_c.

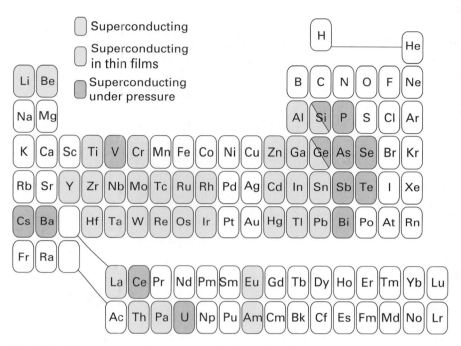

Fig. 23.29 Elements that show superconductivity under the specified conditions.

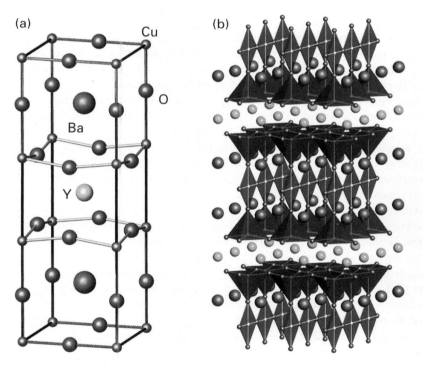

Fig. 23.30 Structure of the $YBa_2Cu_3O_7$ superconductor: (a) the unit cell; (b) oxygen polyhedra around the copper ions showing the layers formed from linked CuO_5 square pyramids and chains formed from corner-linked CuO_4 square planes.

compound, or YBCO, pronounced 'ib-co'), has a structure similar to perovskite but with missing O atoms. In terms of the structure shown in Fig. 23.30, the stoichiometric $YBa_2Cu_3O_7$ unit cell consists of three simple perovskite cubes stacked vertically with Y and Ba in the A sites of the original perovskite and Cu atoms in the B sites. However, unlike in a true perovskite structure, the B sites are not surrounded by an octahedron of O atoms: the 123 structure has a large number of sites that would normally be occupied by O but are in fact vacant. As a result, some Cu atoms have five O atom neighbours in a square-pyramidal arrangement and others have only four, as square-planar CuO_4 units. Similarly, the Y and Ba in the A sites have less than 12-coordination. $YBa_2Cu_3O_7$ readily loses oxygen from some sites within the CuO_4 square planes forming $YBa_2Cu_3O_{7-x}$ ($0 < x < 1$), but as x increases above 0.1 the critical temperature drops rapidly from 93 K. Typically, a sample of the 123 material made in the laboratory and heated under pure oxygen at 450°C as the final stage of its preparation is oxygen deficient with x about 0.1.

If we assign the usual oxidation numbers (Y +3; Ba +2; O −2), then the average oxidation number of copper turns out to be +2.33, so it is inferred that $YBa_2Cu_3O_7$ is a mixed oxidation state material that contains Cu^{2+} and Cu^{3+}. Note that in $YBa_2Cu_3O_{7-x}$ the material formally contains some Cu^{3+} until x increases above 0.5. An alternative view is that the number of electrons in $YBa_2Cu_3O_7$ is such that a partially filled band is present: this view is consistent with the high electrical conductivity and metallic behaviour of this oxide at room temperature (Section 3.7). If this band can be considered as being constructed from $Cu3d$ orbitals, then the partial filling is a result of holes in this level (corresponding to Cu^{3+}); another possible description is that the band also involves $O2p$ orbitals, suggesting that the material can be considered to contain Cu^{2+} and O^-.

The square-planar CuO_4 units in $YBa_2Cu_3O_7$ are arranged in chains and the CuO_5 units link together to form infinite sheets. The stoichiometry of an infinite sheet of vertex-sharing CuO_4 square planes is CuO_2. The addition of one or two additional apical O atoms in the cases where the layers are constructed from, respectively, linked square-based pyramids or octahedra maintains the CuO_2 sheet. This structural feature is also seen in all other oxocuprate high-temperature superconductors. It is thought that it is an important component of the mechanism of superconduction.

Some HTSC materials are listed in Table 23.5 together with other superconducting materials. All these HTSC materials may be considered to have at least part of their structure derived from that of perovskite, as a layer of linked CuO_n ($n = 4, 5, 6$) polyhedra is a section of that structural type. Lying between these cuprate layers (which may include up to six such perovskite-derived CuO_2 sheets) can be a variety of other simple structural units, containing s- and p-block metals in combination with oxygen, such as rock-salt and fluorite structures. Thus $Tl_2Ba_2Ca_2Cu_3O_{10}$ can be considered as having three perovskite layers based on Cu, O, and Ca separated by double layers of a rock-salt structure built from Tl and O; the Ba ions lie between the rock-salt and perovskite layers (Fig. 23.31).

The synthesis of high-temperature superconductors has been guided by a variety of qualitative considerations, such as the demonstrated success of the layered structures and of mixed oxidation state copper in combination with heavy p-block elements. Additional considerations are the radii of ions and their preference for certain coordination environments. Many of these materials are prepared simply by heating an intimate mixture of the metal oxides to 800–900°C in an open alumina crucible. Others, such as mercury- and thallium-containing complex copper oxides, require reactions involving the volatile and toxic oxides Tl_2O and HgO; in such cases the reactions are normally carried out in sealed gold or silver tubes.

Thin films are needed if superconductors are to be used in electronic devices. Their preparation is an active area of research and a promising strategy is chemical vapour deposition, mentioned in Section 23.6. The general strategy is to form a thin film by decomposing a thermally unstable compound on a hot solid substrate material (Fig. 23.32). Fluorinated acetylacetonato complexes of the metals (using ligands such as fod, **2**) are sometimes used because they are more volatile than simple acetylacetonato complexes. These complexes, such as $Cu(acac)_2$, $Y(dmpm)_3$ (dmpm is **3**), and $Ba(fod)_2$, are swept into the reaction chamber by slightly moist oxygen gas. When conditions are properly controlled, this gaseous mixture reacts on the hot substrate to produce the desired $YBa_2Cu_3O_{7-x}$ film. The film may be amorphous and require subsequent heating to form a crystalline product.

There is, as yet, no settled explanation of high-temperature superconductivity. It is believed that the Cooper pairs responsible for conventional superconductivity are important in the high-temperature materials but the mechanism for pairing is hotly debated.

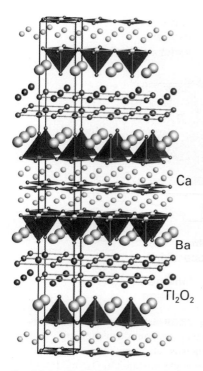

Fig. 23.31 $Tl_2Ba_3Ca_2Cu_3O_{10}$ structure formed from three oxygen-deficient perovskite layers produced from linked CuO_4 square planes separated by Ca on the A-type cation position. Double layers of stoichiometry Tl_2O_2 with rock-salt type arrangements of the Tl and O atoms are interleaved between the multiple perovskite layers.

2 fod

3 (Dimethylphosphino)methane

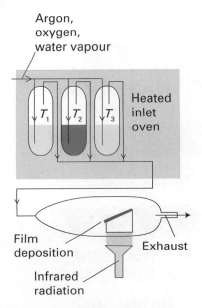

Fig. 23.32 Schematic diagram of the chemical vapour deposition of superconductor thin films. A reactive carrier gas mixture (Ar, O_2, and H_2O) is passed through traps of the volatile metal precursors held at temperatures T_1, T_2, and T_3, which provide the desired vapour pressure of each reactant. Deposition occurs on the wedge-shaped block, which is heated using an IR lamp. After the deposition, the resulting film is annealed at high temperature to improve the crystallinity of the film.

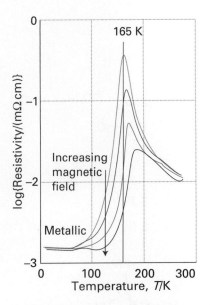

Fig. 23.33 The resistivity as a function of temperature at various magnetic fields of a material displaying colossal magnetoresistance. At 165 K, application of a magnetic field will cause a change in resistivity of about 2 orders of magnitude.

(f) Colossal magnetoresistance

Key point: Perovskites with manganese on the B cation sites can show very large changes in resistance on application of a magnetic field, known as colossal magnetoresistance.

Manganites, which are Mn(III) and Mn(IV) complex oxides, with the generic solid solution formulation $Ln_{1-x}A_xMnO_3$ (A = Ca, Sr, Pb, Ba; Ln = lanthanoid, typically La, Pr, or Nd), order ferromagnetically on cooling below room temperature, with Curie temperatures typically between 100 and 250 K, and simultaneously transform from insulators (at the higher temperature) to poor metallic conductors. These materials also exhibit **magnetoresistance**, a marked decrease of their resistance on the application of a magnetic field near, but just above, their Curie temperatures (Fig. 23.33). Recent investigations have shown that, for these manganites, the decrease in resistance can be by as much as 11 orders of magnitude; and for this reason, these compounds have been named **colossal magnetoresistance manganites**.

Colossal magnetoresistance (CMR) manganites have perovskite-type structures with the A cation sites occupied by a mixture of Ln^{3+} and A^{2+} cations and the B site occupied by Mn. The oxidation state of manganese in these solid solutions varies between +3 and +4 as the proportion of A^{2+} is changed. Pure $LaMnO_3$ orders antiferromagnetically below its Néel temperature, $T_N = 150$ K, but with an increase in x in $Ln_{1-x}A_xMnO_3$, corresponding to an increase in Mn^{4+} content, the manganites order ferromagnetically on cooling.

The observation that the transition to conducting electron behaviour and ferromagnetism occur simultaneously on cooling the $Ln_{1-x}A_xMnO_3$ manganites and the origin of the CMR effect are not completely understood, but it is known that they are based on a so-called **double exchange** (DE) mechanism between the Mn(III) and Mn(IV) species present in these materials. The basic process of this mechanism is the hopping of an electron from $Mn^{3+}(t_{2g}^3 e_g^1)$ to $Mn^{4+}(t_{2g}^3)$ via the O atom, so that the Mn(III) and Mn(IV) positions change places. In manganites at high temperatures, the electrons in these systems effectively become trapped on a specific site leading to ordering of the Mn(III) and Mn(IV) species; that is, the trapping results in charge ordering. The charge-ordered state is generally associated with insulating and paramagnetic behaviour whereas the charge-disordered state, in which the electron can hop between sites, is associated with metallic behaviour and ferromagnetism. The high-temperature charge-ordered state can be transformed into a metallic, spin-ordered (ferromagnetic) state just by the application of a magnetic field; therefore, application of a magnetic field to a manganite just above a critical temperature, T_c, causes the transformation of charge ordering into electron delocalization and hence a massive decrease in resistance.

The CMR effect is being developed for magnetic data-storage devices, such as computer hard drives. Further work using these compounds is aimed at **spintronics**, where, rather than use electron movements to transmit information, as is the basis of electronics and the functioning of the silicon chip, the movement of spin through materials could be used in a similar way.

(g) Rechargeable battery materials

Key point: The redox chemistry associated with the extraction and insertion of metal ions into oxide structures is exploited in rechargeable batteries.

The existence of complex oxide phases that demonstrate good ionic conductivity associated with the ability to vary the oxidation state of a d-metal ion has led to the development of materials for use as the cathode in rechargeable batteries (see Box 10.1). Examples include $LiCoO_2$, with a layer-type structure based on sheets of edge-linked CoO_6 octahedra separated by Li^+ ions (Fig. 23.34), and various lithium manganese spinels, such as $LiMn_2O_4$. In each of these compounds the battery is charged by removing the mobile Li^+ ions from the complex metal oxide, as in

$$LiCoO_2 \rightarrow CoO_2 + Li^+ + e^-$$

The battery is discharged, while in use, through the reverse electrochemical reaction. Lithium cobalt oxide, which is used in many commercial lithium-ion batteries, has many

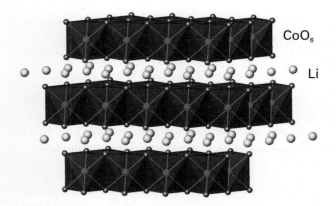

Fig. 23.34 The structure of $LiCoO_2$ shown as layers of linked CoO_6 octahedra separated by Li^+ ions; lithium may be de-intercalated electrochemically from between the layers.

of the characteristics required for this type of application. The specific energy (the stored energy divided by the mass) of $LiCoO_2$ ($140\,W\,h\,kg^{-1}$) is maximized by using light elements such as lithium and cobalt; the $3d$ metals are almost invariably used in such applications because they are the lowest-density elements with variable oxidation states. The high mobility of the Li^+ ion and good reversibility of electrochemical charging and discharging stem from lithium ion's small ionic radius and the layer-like structure of $LiCoO_2$, which allows the Li^+ to be extracted without major disruption of the structure. High capacities are obtained from the large amount of Li (one Li^+ ion per $LiCoO_2$ formula unit) that may be reversibly extracted (about 500 discharge/recharge cycles) from the compound, and the current is delivered at a constant and high voltage (of between 3.5 and 4 V). The high potential difference characteristic of the system is partly due to the high oxidation states of cobalt (+3 and +4) that are involved.

Because cobalt is expensive and fairly toxic, the search continues for even better oxide materials than $LiCoO_2$. New materials will be required to demonstrate the high levels of reversibility found for lithium cobaltate and considerable effort is being directed at doped forms of $LiCoO_2$ and of $LiMn_2O_4$ spinels and at nanostructured complex oxides (Section 24.13), which, because of their small particle size, can offer excellent reversibility. There is also a high level of interest in $LiFePO_4$, which shows good characteristics for a cathode material and contains cheap and nontoxic iron.

The other electrode in a rechargeable lithium ion battery can simply be lithium metal, which completes the overall cell reaction through the process

$$Li(s) \rightarrow Li^+ + e^-$$

Lithium ions then migrate to the cathode through an electrolyte, which is typically an anhydrous lithium salt, such as $LiPF_4$ or $LiC(SO_2CF_3)_3$ dissolved in a polymer, such as propylene carbonate. However, the use of lithium metal has a number of problems associated with its reactivity and volume changes that occur in the cell. Therefore, an alternative anode material that is frequently used in rechargeable batteries is graphitic carbon, which can intercalate, electrochemically, large quantities of Li to form C_6Li. As the cell is discharged, Li is transferred from between the carbon layers at the anode and intercalated into the metal oxide at the cathode (and vice versa on charging) with the following overall processes (Fig. 23.34):

$$Li_yC_6 + Li_{1-x}CoO_2 \underset{\text{discharge}}{\overset{\text{charge}}{\rightleftharpoons}} C_6 + Li_{1-x+y}CoO_2$$

23.9 Oxide glasses

The term **ceramic** is often applied to all inorganic nonmetallic, nonmolecular materials, including both amorphous and crystalline materials, but the term is commonly reserved for compounds or mixtures that have undergone heat treatment. The term **glass** is used in a variety of contexts, but for our present purposes it implies an amorphous ceramic with a viscosity so high that it can be considered rigid. A substance in its glassy form is said to be in its **vitreous state**. Although ceramics and glasses have been utilized since antiquity, their development is currently an area of rapid scientific and technological

(a)   (b)

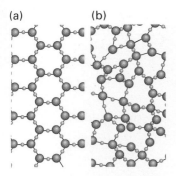

Fig. 23.35 Schematic representation of (a) a two-dimensional crystal and (b) a two-dimensional glass.

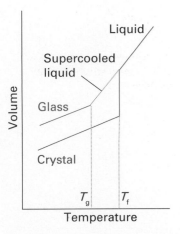

Fig. 23.36 Comparison of the volume change for supercooled liquids and glasses with that for a crystalline material. The glass transition temperature is T_g.

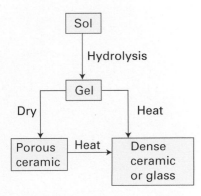

Fig. 23.37 Schematic diagram of the sol–gel process. When the gel is dried at high temperatures, dense ceramics or glasses are formed. Drying at low temperatures above the critical pressure of water produces porous solids known as xerogels or aerogels.

progress. This enthusiasm stems from interest in the scientific basis of their properties and the development of novel synthetic routes to new high-performance materials. We confine our attention here to glasses. The most familiar glasses are alkali-metal or alkaline-earth-metal silicates and borosilicates.

(a) Glass formation

Key points: Silicon dioxide readily forms a glass because the three-dimensional network of strong covalent Si—O bonds in the melt does not readily break and reform upon cooling; the Zachariasen rules summarize the properties likely to lead to glass formation.

A glass is prepared by cooling a melt more quickly than it can crystallize. Thus, cooling molten silica gives vitreous quartz. Under these conditions the solid mass has no long-range order as judged by the lack of X-ray diffraction peaks, but spectroscopic and other data indicate that each Si atom is surrounded by a tetrahedral array of O atoms. The lack of long-range order results from variations of the Si—O—Si angles. Figure 23.35 illustrates in two dimensions how a local coordination environment can be preserved but long-range order lost by variation of the bond angles around oxygen. Silicon dioxide readily forms a glass because the three-dimensional network of strong covalent Si—O bonds in the melt does not readily break and reform upon cooling. The lack of strong directional bonds in metals and simple ionic substances makes it much more difficult to form glasses from these materials. Recently, however, techniques have been developed for ultrafast cooling and, as a result, a wide variety of metals and simple inorganic materials can now be frozen into a vitreous state.

The concept that the local coordination sphere of the glass-forming element is preserved but that bond angles around O are variable was originally proposed by W.H. Zachariasen in 1932. He reasoned that these conditions would lead to similar molar Gibbs energies and molar volumes for the glass and its crystalline counterpart. Zachariasen also proposed that the vitreous state is favoured by corner-shared O atoms, rather than edge- or face-shared, which would enforce greater order. These and other **Zachariasen rules** hold for common glass-forming oxides, but exceptions are known.

An instructive comparison between vitreous and crystalline materials is seen in their change in volume with temperature (Fig. 23.36). When a molten material crystallizes, an abrupt change in volume (usually a decrease) occurs. By contrast, a glass-forming material that is cooled sufficiently rapidly persists as a metastable supercooled liquid. When cooled below the **glass transition temperature**, T_g, the supercooled liquid becomes rigid, and this change is accompanied by only an inflection in the cooling curve rather than an abrupt change in slope. The rates of crystallization are very slow for many complex metal silicates, phosphates, and borates and it is these compounds that often form glasses. Another route to glasses is the **sol–gel process**, which is described schematically in Fig. 23.37 and discussed more fully in Section 24.5. As described there, the sol–gel process is also used to produce crystalline ceramic materials and high-surface-area compounds such as silica gel. A typical process involves the addition of a metal alkoxide precursor to an alcohol, followed by the addition of water to hydrolyse the reactants. This hydrolysis leads to a thick gel that can be dehydrated and **sintered** (heated below its melting point to produce a compact solid). For example, ceramics containing TiO_2 and Al_2O_3 can be prepared in this way at much lower temperatures than required to produce the ceramic from the simple oxides. Often a special shape can be fashioned at the gel stage. Thus, the gel may be shaped into a fibre and then heated to expel water; this process produces a glass or ceramic fibre at much lower temperatures than would be necessary if the fibre were made from the melt of the components.

(b) Glass composition, production, and application

Key points: Low-valence metal oxides, such as Na_2O and CaO, are often added to silica to reduce its softening temperature by disrupting the silicon–oxygen framework. Other cations may be incorporated into glasses giving applications as diverse as lasers and nuclear waste containment.

Although vitreous silica is a strong glass that can withstand rapid cooling or heating without cracking, it has a high glass-transition temperature and therefore must be worked at inconveniently high temperatures. Therefore, a **modifier**, such as Na_2O or CaO, is commonly added to SiO_2. A modifier disrupts some of the Si—O—Si linkages and replaces them with terminal Si—O$^-$ links that associate with the cation (Fig. 23.38). The consequent partial disruption of the Si—O network leads to glasses that have lower softening points. The common glass used in bottles and windows is called 'sodalime glass' and contains Na_2O and CaO as modifiers. When B_2O_3 is used as a modifier, the resulting 'borosilicate glasses' have lower thermal expansion coefficients than sodalime glass and are less likely to crack when heated. Borosilicate glass (such as Pyrex®) is therefore widely used for ovenware and laboratory glassware.

Glass formation is a property of many oxides, and practical glasses have been made from sulfides, fluorides, and other anionic constituents. Some of the best glass formers are the oxides of elements near silicon in the periodic table (B_2O_3, GeO_2, and P_2O_5), but the solubility in water of most borate and phosphate glasses and the high cost of germanium limit their usefulness.

The development of transparent crystalline and vitreous materials for light transmission and processing has led to a revolution in signal transmission. For example, optical fibres for light transmission are currently being produced with a composition gradient from the interior to the surface. This composition gradient modifies the refractive index and thereby decreases light loss. Fluoride glasses are also being investigated as possible substitutes for oxide glasses, because oxide glasses contain small numbers of OH groups, which absorb near-infrared radiation and attenuate the signal. Doping lanthanoid ions into the glass fibres produces materials that can be used to amplify the signals by using lasing effects. Optical circuit elements are being developed that may eventually replace all the components in an electronic integrated circuit and lead to very fast optical computers.

As modifier cations effectively become trapped in a glass, which is chemically inert and thermodynamically very stable with respect to transformation to a soluble, crystalline phase, such glassy materials offer a potential method for containing and storing nuclear waste. Thus vitrification of metal oxides containing a radioactive species with glass-forming oxides produces a stable glass that can be stored for long periods, allowing the radionucleide to decay. Examples of such materials are borosilicate glasses and combinations of such glasses with crystalline oxide phases ('Synroc').

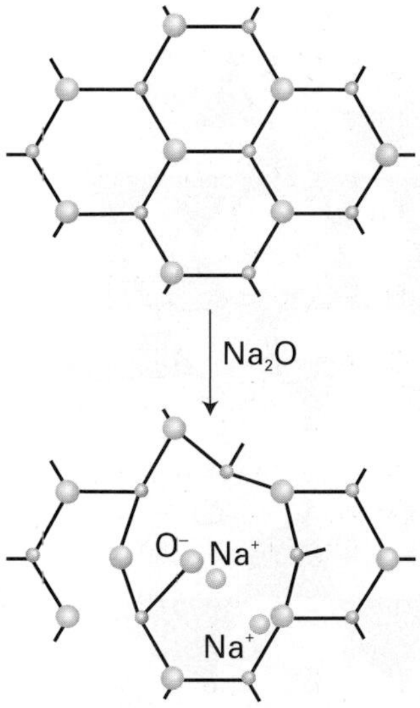

Fig. 23.38 The role of a modifier is to introduce O^{2-} ions and cations that disrupt the lattice.

23.10 Nitrides and fluorides

The solid-state chemistry of the metals in combination with anions other than oxide is not as highly developed or extensive as that of the complex oxides described above. However, complex nitrides and fluorides and mixed anion compounds are of growing importance.

(a) Nitrides

Key points: Complex metal nitrides and oxide nitrides are materials containing the N^{3-} anion; many new compounds of this type have recently been synthesized.

Simple metal nitrides of main-group elements, such AlN, GaN, and Li_3N, have been known for decades and have well-defined applications. Many of the recent advances in nitride chemistry have centred on d-metal compounds and complex nitrides. That nitrides are less common than oxides stems, in part, from the high enthalpy of formation of N^{3-} compared with that of O^{2-}. Furthermore, because many nitrides are sensitive to oxygen and water, their synthesis and handling are problematic. Some simple metal nitrides can be obtained by the direct reaction of the elements; for example, Li_3N is obtained by heating lithium in a stream of nitrogen at 400°C. The instability of sodium nitride allows sodium azide to be used as a nitriding agent:

$$2\,NaN_3(s) + 9\,Sr(s) + 6\,Ge(s) \xrightarrow{\;750°C,\ \text{sealed Nb tube}\;} 3\,Sr_3Ge_2N_2(s) + 2\,Na(g)$$

(a)

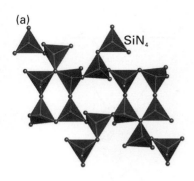

(b)

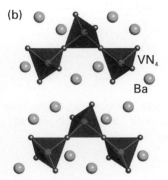

Fig. 23.39 The structure of two nitrides: (a) Si_3N_4 shown as linked SiN_4 tetrahedra; (b) Ba_2VN_3, containing one-dimensional chains of corner-sharing VN_4 tetrahedra separated by Ba^{2+} ions

The ammonolysis of oxides (the deprotonation of NH_3 by oxides with the formation of water as a by-product) provides a convenient route to some nitrides. For instance, tantalum nitride can be obtained by heating tantalum pentoxide in a fast-flowing stream of ammonia:

$$3\ Ta_2O_5(s) + 10\ NH_3(l) \xrightarrow{700°C} 2\ Ta_3N_5(s) + 15\ H_2O(g)$$

In such reactions, the equilibrium is driven towards the products by removal of the steam in the gas flow. Similar reactions may be used for the preparation of complex nitrides from complex oxides, although competing reactions involving partial reduction of the metal oxide by ammonia can also occur. In all these reactions the complete elimination of O^{2-} ions from the product can be troublesome, so the reactions give products that contain both the oxide and nitride ions:

$$Ca_2Ta_2O_5(s) + 2\ NH_3(g) \xrightarrow{800°C} 2\ CaTaO_2N(s) + 3\ H_2O(g)$$

In comparison with the O^{2-} ion, the higher charge of the N^{3-} ion results in a greater degree of covalence in its bonding. There is also a tendency with nitrides for the formation of compounds in which the metallic element is in a lower oxidation state because nitrogen, on account of its high bond energy, is not as potent an oxidant as oxygen or fluorine. Thus, whereas heating titanium in oxygen readily produces TiO_2, Ti_2N and TiN are known but Ti_3N_4 is difficult to prepare and poorly characterized. Likewise, V_3N_5 is unknown whereas V_2O_5 is readily obtained from the decomposition of many vanadium salts in air.

Many of the early d-metal nitrides are interstitial compounds and are used as high-temperature refractory ceramics. Similarly, the nitrides of silicon and aluminium, such as Si_3N_4 (Fig. 23.39a), are stable at very high temperatures, particularly under nonoxidizing conditions, and are used for crucibles and furnace elements. Recently, GaN, which can exist in both wurtzite and sphalerite structural types, has been the focus of considerable research on account of its semiconducting properties. Li_3N has an unusual structure based on hexagonal Li_2N^- layers separated by Li^+ ions (Fig. 23.15). These Li^+ ions are highly mobile, as is expected from the existence of free space between the layers, and this compound and other structurally related materials are being studied for possible use in rechargeable batteries. Among the many complex nitrides that have been synthesized are materials of stoichiometry AMN_2, such as $SrZrN_2$ and $CaTaN_2$, with structures based on sheets formed from MN_6 octahedra sharing edges (see the discussion of $LiCoO_2$ in Section 23.8) and A_2MN_3, such as Ba_2VN_3, containing one-dimensional chains of corner-sharing MN_4 tetrahedra (Fig. 23.39). Analogues of the zeolites containing nitride bridges rather than oxide bridges have also been prepared.

(b) Fluorides and other halides

Key point: Because fluorine and oxygen have similar ionic radii, fluoride solid-state chemistry parallels much of oxide chemistry.

The ionic radii of F and O are very similar (at between 130 and 140 pm), and as a result metal fluorides show many stoichiometric and structural analogies with the complex oxides but with lower charge on the metal ion to reflect the lower charge on the F^- ion. Many binary metal fluorides adopt the simple structural types expected on the basis of the radius-ratio rule (Section 3.10). For example, FeF_2 and PdF_2 have a rutile structure and AgF has a rock-salt structure; similarly, NbF_3 adopts the ReO_3 structure. For complex fluorides, analogues of typical oxide structural types are well known, including perovskites (such as $KMnF_3$), Ruddlesden–Popper phases (for example $K_3Co_2F_7$), and spinels (Li_2NiF_4). Synthetic routes to complex fluorides also parallel those for oxides. For instance, the direct reaction of two metal fluorides yields the complex fluoride, as in

$$2\ LiF(s) + NiF_2(s) \rightarrow Li_2NiF_4(s)$$

Like some complex oxides, some complex fluorides may be precipitated from solution

$$MnBr_2(aq) + 3\ KF(aq) \rightarrow KMnF_3(s) + 2\ KBr(aq)$$

As with the ammonolysis of oxides to produce nitrides and oxide-nitrides, the formation of oxide-fluorides is possible by the appropriate treatment of a complex oxide, as in

$$Sr_2CuO_3(s) \xrightarrow{F_2, \ 200°C} Sr_2CuO_2F_{2+x}(s)$$

The product is a superconductor with $T_c = 45$ K. Fluoride analogues of the silicate glasses, which are based on linked SiO_4 tetrahedra, exist for small cations that form tetrahedral units in combination with F^-; an example is $LiBF_4$, which contains linked BF_4 tetrahedra. Lithium borofluoride glasses are used to contain samples for X-ray work because they are highly transparent to X-rays due to their low electron densities. Framework and layer structures based on linked MF_4 (M = Li, Be) tetrahedra have also been described. Some metal fluorides are used as fluorination agents in organic chemistry. However, few of the solid complex metal fluorides are technologically important in comparison with the wealth of applications associated with analogous complex oxides.

Metal chloride structures reflect the greater covalence associated with bonding to chloride in comparison with fluoride: the chlorides are less ionic and have structures with lower coordination numbers than the corresponding fluorides. Thus, simple metal chlorides, bromides, and iodides normally adopt the cadmium-chloride or cadmium-iodide structures based on sheets formed from edge-sharing MX_6 octahedra. Complex chlorides often contain the same structural unit: for example, $CsNiCl_3$ has chains of edge-sharing $NiCl_6$ octahedra separated by Cs^+ ions. Many analogues of oxide structures also occur among the complex chlorides, such as $KMnCl_3$, K_2MnCl_4, and Li_2MnCl_4, which have the perovskite, K_2NiF_4, and spinel structures, respectively.

Chalcogenides, intercalation chemistry, and metal-rich phases

The soft chalcogens sulfur, selenium, and tellurium form binary compounds with metals that commonly have quite different structures from the corresponding oxides, nitrides, and fluorides. As we saw in Section 2.2, this difference is consistent with the greater covalence of the compounds of sulfur and its heavier congeners. For example, we noted there that MO compounds generally adopt the rock-salt structure whereas ZnS and CdS can crystallize with either of the sphalerite or the wurtzite structures in which the lower coordination numbers indicate the presence of directional bonding. Similarly, the d-block monosulfides generally adopt the more characteristically covalent nickel-arsenide structure rather than the rock-salt structure of alkaline-earth oxides such as MgO. Even more striking are the layered MS_2 compounds formed by many d-block elements in contrast to the fluorite or rutile structures of many d-block dioxides.

Before we discuss these compounds, we should note that there are many metal-rich compounds that disobey simple valence rules and do not conform to an ionic model. Some examples are Ti_2S, Pd_4S, V_2O, and Fe_3N, and even the alkali metal suboxides, such as Cs_3O (Section 10.5). The occurrence of metal-rich phases is generally associated with M—M interactions. Many other intermetallic compounds display stoichiometries that cannot be understood in terms of conventional valence rules. Included among them are the important permanent-magnet materials $Nd_2Fe_{17}B$ and $SmCo_5$.

23.11 Layered MS_2 compounds and intercalation

We introduced the layered metal sulfides and their intercalation compounds in Section 18.9. Here we develop a broader picture of their structures and properties.

(a) Synthesis and crystal growth

Key point: d-Metal disulfides can be synthesized by the direct reaction of the elements in a sealed tube and purified by using chemical vapour transport with iodine.

Compounds of the chalcogens with d metals are prepared by heating mixtures in a sealed tube (to prevent the loss of the volatile elements). The products obtained in this

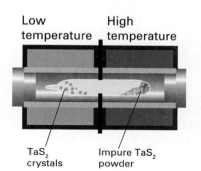

Low temperature	High temperature

TaS₂ crystals   Impure TaS₂ powder

Fig. 23.40 Vapour transport crystal growth and purification of TaS_2. A small quantity of I_2 is present to serve as a transport agent.

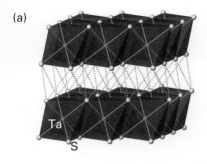

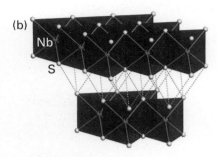

Fig. 23.41 (a) The structure of TaS_2 (CdI_2-type). The Ta atoms reside in octahedral sites between the AB layers of S atoms. (b) The NbS_2 structure; the Nb atoms reside in trigonal-prismatic sites, between the sulfide layers.

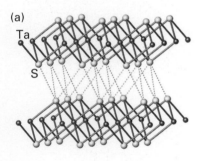

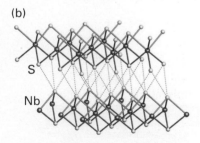

Fig. 23.42 The metal disulfide structures of Fig. 23.41 drawn as layers of MS_6 polyhedra sharing edges: (a) octahedra in TaS_2; (b) trigonal prisms in NbS_2.

manner can have a variety of stoichiometries. The preparation of crystalline dichalcogenides suitable for chemical and structural studies is often performed by **chemical vapour transport** (CVT), as described below. It is possible in some cases simply to sublime a compound, but the CVT technique can also be applied to a wide variety of nonvolatile compounds in solid-state chemistry.

In a typical procedure, the crude material is loaded into one end of a borosilicate or fused quartz tube. After evacuation, a small amount of a CVT agent is introduced, the tube is sealed and placed in a furnace with a temperature gradient. The polycrystalline and possibly impure metal chalcogenide is vaporized at one end and redeposited as pure crystals at the other (Fig. 23.40). The technique is called chemical vapour transport rather than sublimation because the CVT agent, which is often a halogen, produces an intermediate volatile species, such as a metal halide. Generally, only a small amount of transport agent is needed because upon crystal formation it is released and diffuses back to pick up more reactant. For example, TaS_2 can be transported with I_2 in a thermal gradient. The reaction with I_2 to produce gaseous products

$$TaS_2(s) + 2\ I_2(g) \rightarrow TaI_4(g) + S_2(g)$$

is endothermic, so the equilibrium lies further to the right at 850°C than at 750°C. Consequently, although TaI_4 is formed at 850°C, at 750°C the mixture deposits TaS_2. If, as occasionally is the case, the transport reaction is exothermic, the solid is carried from the cooler to the hotter end of the tube.

(b) Structure

Key points: Elements on the left of the *d* block form sulfides consisting of sandwich-like layers of the metal coordinated to six sulfur ions; the bonding between the layers is very weak.

As we saw in Section 18.9, the *d*-block disulfides fall into two classes: layered materials are formed by metals on the left of the *d* block, and compounds containing formal S_2^{2-} ions are formed by metals in the middle and towards the right of the block (such as pyrite, FeS_2). We concentrate here on the layered materials.

In TaS_2 and many other layered disulfides, the *d*-metal ions are located in octahedral holes between close-packed AB layers (Fig. 23.41a). The Ta ions form a close-packed layer denoted *x*, so the metal and adjoining sulfide layers can be portrayed as an A*x*B sandwich. These sandwich-like slabs form a three-dimensional crystal by stacking in sequences such as A*x*BA*x*BA*x*B . . . , where the strongly bound A*x*B slabs are held to their neighbours by weak dispersion forces. An alternative view of these MS_2 structures, which have the metal ions in octahedral holes, is as MS_6 octahedra sharing edges (Fig. 23.42), which reinforces the idea of the greater degree of covalent bonding that occurs in these materials than in, for instance, Li_2S, which has the anti-fluorite structure.

The Nb atoms in NbS_2 reside in the trigonal-prismatic holes between sulfide layers that are in register with one another (AA, Fig. 23.41). The Nb atoms, which are strongly bonded to the adjacent sulfide layers, form a close-packed array denoted *m*, so we can represent each slab as A*m*A or C*m*C. These slabs form a three-dimensional crystal by stacking in a pattern such as A*m*AC*m*CA*m*AC*m*C . . . . Weak dispersion forces also contribute to holding these A*m*A and C*m*C slabs together. Polytypes (versions that differ only in the stacking arrangement along a direction perpendicular to the plane of the slabs) can occur. Thus, NbS_2 and MoS_2 form several polytypes including one with the sequence C*m*CA*m*AB*m*B.

It has been convenient to describe—somewhat simplistically—the layered structures in terms of cations and anions. However, to account for the significant covalence, the electrical conductivities, and the chemical properties of these layers we need to invoke the more sophisticated band model of their electronic structure. Some approximate band structures derived primarily from molecular orbital calculations and photoelectron spectra are shown in Fig. 23.43. They show that the dichalcogenides with octahedral and trigonal-prismatic metal sites have low-lying bands composed primarily of chalcogen *s* and *p* orbitals, higher-energy bands derived primarily from metal *d* orbitals, and still higher in energy a variety of metal and chalcogen bands.

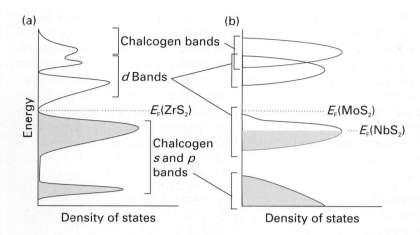

Fig. 23.43 Approximate band structures of dichalcogenides. (a) Octahedral MS_2 compounds; for ZrS_2 (d^0) only the sulfur s and p bands are full and the compound is a semiconductor. (b) Trigonal-prismatic MS_2 compounds. NbS_2 (d^1) is metallic whereas MoS_2 (d^2) is a semiconductor.

When the metal atom occupies an octahedral site, its d orbitals are split into a lower t_{2g} set and a higher e_g set, just as in localized complexes. These atomic orbitals combine to give a t_{2g} band and an e_g band. The broad t_{2g} band may accommodate up to six electrons per metal atom. Thus TaS_2, with one d electron per metal atom, has an only partly filled t_{2g} band, and is a metallic conductor. A trigonal-prismatic ligand field leads to a low-energy a_1' level and two doubly-degenerate higher energy orbitals designated e' and e'', respectively. In this case, the lowest band needs only two electrons per atom to fill it. Thus MoS_2, with trigonal-prismatic Mo sites and two d electrons, has a filled a_1' band and is an insulator (more precisely, a large band-gap semiconductor).

(c) Intercalation and insertion

Key points: Insertion compounds can be formed from the d-metal disulfides either by direct reaction or electrochemically; insertion compounds can also be formed with molecular guests.

We have already introduced the idea that alkali metal ions may insert between graphite sheets (Section 13.11), metal disulfide slabs (Section 18.11), and metal oxide layers (as in Li_xCoO_2, Section 23.8), to form intercalation compounds. For a reaction to qualify as an intercalation, or as an **insertion reaction**, the basic structure of the host should not be altered when it occurs.[5]

The π conduction and valence bands of graphite are contiguous in energy (we have seen in fact that graphite is formally a semimetal, Section 3.14) and the favourable Gibbs energy for intercalation arises from the transfer of an electron from the alkali metal atom to the graphite conduction band. The insertion of an alkali metal atom into a dichalcogenide involves a similar process: the electron is accepted into the d band and the charge-compensating alkali metal ion diffuses to positions between the slabs. Some representative alkali metal insertion compounds are listed in Table 23.6.

The insertion of alkali metal ions into host structures can be achieved by direct combination of the alkali metal and the disulfide:

$$TaS_2(s) + x\,Na(g) \xrightarrow{800°C} Na_xTaS_2(s)$$

with $0.4 < x < 0.7$. Insertion may also be achieved by using a highly reducing alkali-metal compound, such as butyllithium, or the electrochemical technique of **electrointercalation** (Fig. 23.44). One advantage of electrointercalation is that it is possible to measure the amount of alkali metal incorporated by monitoring the current passed during the synthesis (and using $n_{e^-} = It/F$). It also is possible to distinguish solid-solution formation from discrete-phase formation. As illustrated in Fig. 23.45, the formation of a solid solution is characterized by a gradual change in potential as intercalation proceeds, whereas the formation of a new discrete phase yields a steady potential over

Table 23.6 Some alkali metal intercalation compounds of chalcogenides

Compound	Δ/pm*
$K_{1.0}ZrS_2$	160
$Na_{1.0}TaS_2$	117
$K_{1.0}TiS_2$	192
$Na_{0.6}MoS_2$	135
$K_{0.4}MoS_2$	214
$Rb_{0.3}MoS_2$	245
$Cs_{0.3}MoS_2$	366

* The change in interlayer spacing as compared with the parent MS_2 phase.

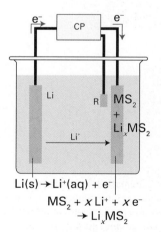

$Li(s) \rightarrow Li^+(aq) + e^-$

$MS_2 + x\,Li^+ + x\,e^-$
$\rightarrow Li_xMS_2$

Fig. 23.44 Schematic experimental arrangement for electrointercalation. A polar organic solvent (such as polypropylene carbonate) containing an anhydrous lithium salt is used as an electrolyte. R is a reference electrode and CP is a coulometer (to measure the charge passed) and a potential controller.

---

[5] Reactions in which the structure of one of the solid starting materials is not radically altered are called 'topotactic' reactions. They are not limited to the type of insertion chemistry we are discussing here. For example, hydration, dehydration, and ion exchange reactions may be topotactic.

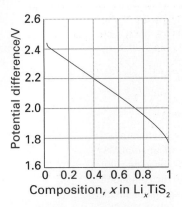

Fig. 23.45 Potential versus composition diagram for the electrointercalation of lithium into titanium disulfide. The composition, x, in Li_xTiS_2 is calculated from the charge passed in the course of electrointercalation.

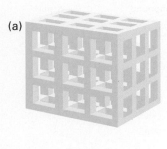

(a)

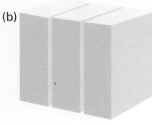

(b)

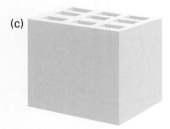

(c)

Fig. 23.46 Schematic representation of host materials for intercalation reactions. (a) A three-dimensional host with intersecting channels, (b) a two-dimensional layered compound, and (c) a host containing one-dimensional channels.

the range in which one solid phase is being converted into the other, followed by an abrupt change in potential when that reaction is complete.

Insertion compounds are examples of mixed ionic and electronic conductors. In general, the insertion process can be reversed either chemically or electrochemically. This reversibility makes it possible to recharge a lithium cell by removal of Li from the compound. In a clever synthetic application of these concepts, the previously unknown layered disulfide VS_2 can be prepared by first making the known layered compound $LiVS_2$ in a high-temperature process. The Li is then removed by reaction with I_2 to produce the metastable layered VS_2, which has the TiS_2 structure:

$$2 \, LiVS_2(s) + I_2(s) \rightarrow 2 \, LiI(s) + 2 \, VS_2(s)$$

Insertion compounds also can be formed with molecular guests. Perhaps the most interesting guest is the metallocene $Co(\eta^5\text{-}Cp)_2$, where $Cp = C_5H_5$ (Section 21.19), which can be incorporated into a variety of hosts with layered structures, such as TiS_2, $TiSe_2$, and TaS_2, to the extent of about 0.25 $Co(\eta^5\text{-}Cp)_2$ per MS_2 or MSe_2. This limit appears to correspond to the space available for forming a complete layer of $Co(\eta^5\text{-}Cp)_2^+$. The organometallic compound appears to undergo oxidation upon intercalation, so the favourable Gibbs energy in these reactions arises in the same way as in alkali metal intercalation. In agreement with this interpretation, $Fe(\eta^5\text{-}Cp)_2$, which is more difficult to oxidize than its cobalt analogue, does not intercalate.

We can imagine the insertion of ions into one-dimensional channels, between two-dimensional planes of the type we have been discussing, or into channels that intersect to form three-dimensional networks (Fig. 23.46). Aside from the availability of a site for a guest to enter, the host must provide a conduction band of suitable energy to take up electrons reversibly (or, in some cases, be able to donate electrons to the host). Table 23.7 illustrates that a wide variety of hosts are possible, including metal oxides and various ternary and quaternary compounds. We see that intercalation chemistry is by no means limited to graphite and layered disulfides.

Example 23.6 Explaining the structure of a staged intercalation compound

The lattice parameter of TaS_2 perpendicular to the layers is 569 pm and that of the intercalation compound $LiTaS_2$ is 621 pm. The lattice parameter of the intermediate composition $Li_{0.5}TaS_2$ is 1190 pm. Explain these observations.

Answer $Li_{0.5}TaS_2$ is a second-stage intercalation compound in which Li^+ ions occur only between alternate TaS_2 sheets. Thus, the repeat distance perpendicular to the TaS_2 layers consist of the sequence TaS_2–TaS_2·Li·TaS_2 and is equal to 569 pm + 621 pm = 1190 pm.

Self-test 23.6 Estimate the lattice parameter of $Li_{0.25}TaS_2$ a third-stage intercalation compound with lithium incorporated between every third pair of TaS_2 layers.

23.12 Chevrel phases

Key point: A Chevrel phase has a formula such as Mo_6X_8 or $M_xMo_6S_8$, where Se or Te may take the place of S and the intercalated M atom may be a variety of metals such as Li, Mn, Fe, Cd, and Pb.

We close this section on sulfide materials with a brief discussion of an interesting class of ternary compounds first reported by R. Chevrel in 1971. These compounds, which illustrate three-dimensional intercalation, have formulas such as Mo_6X_8 and $A_xMo_6S_8$; Se or Te may take the place of S and the intercalated A atom may be a variety of metals such as Li, Mn, Fe, Cd, or Pb. The parent compounds Mo_6Se_8 and Mo_6Te_8 are prepared by heating the elements at about 1000°C. A structural unit common to this series is M_6S_8, which may be viewed as an octahedron of M atoms face-bridged by S atoms, or alternatively as an octahedron of M atoms in a cube of S atoms (Fig. 23.47). This type cf cluster is also observed for some halides of the Period 4 and 5 early *d*-block elements, such as the $[M_6X_8]^{4-}$ cluster found in Mo and W dichlorides, bromides, and iodides.

Figure 23.47 shows that in the three-dimensional solid the Mo_6S_8 clusters are tilted relative to each other and relative to the sites occupied by intercalated ions. This tilting allows a secondary donor–acceptor interaction between vacant Mo $4d_{z^2}$ orbitals (which project outward from the faces of the Mo_6S_8 cube) and a filled donor orbital on the S atoms of adjacent clusters.

One of the physical properties that has drawn attention to the Chevrel phases is their superconductivity. Superconductivity persists up to 14 K in $PbMo_6S_8$, and it also persists to very high magnetic fields, which is of considerable practical interest because many applications involve high fields. In this respect the Chevrel phases appear to be significantly superior to the newer oxocuprate high-temperature superconductors. Further possible application of Chevrel phases is in the thermoelectric devices used to convert heat into electrical energy or in devices that use electrical energy directly for cooling purposes. The ideal thermoelectric materials have good electrical conductivity associated with low thermal conductivities. In Chevrel phases the ability to incorporate various cations between the M_6X_8 blocks that 'rattle' around on their sites reduces the material's thermal conductivity but not at the expense of the electronic conductivity.

Framework structures

Much of this chapter has concerned structures derived from close-packed anions accompanying *d*-metal ions typically with coordination number 6. Many of these structures (for example, ReO_3, perovskites, and MS_2) can also be described in terms of linked polyhedra in which MX_6 octahedra connect through their vertices or edges to form a variety of arrays. For most *d* metals in their typical oxidation states the coordination number 6 is preferred but for smaller metal species (for example, the later 3*d*-series and the lighter *p*-block metals and metalloids such as Al and Si), fourfold tetrahedral coordination to oxygen is commonplace and leads to structures that are best described as based on linked MO_4 tetrahedra. These tetrahedral units may be the only building block present, as in zeolites, or may link together with metal–oxygen octahedra to produce new structural types (Fig. 23.48). Many of these linked polyhedral structures are known as **framework structures**.

23.13 Structures based on tetrahedral oxoanions

As remarked above, the elements able to form very stable tetrahedral MO_4 species that can link together into framework structures are the later 3*d*-series and the lighter *p*-block metals and metalloids. These ions are so small that they coordinate strongly to four O atoms in preference to higher coordination numbers; the principal examples are SiO_4, AlO_4, and PO_4, although GaO_4, GeO_4, AsO_4, BO_4, BeO_4, LiO_4, $Co(II)O_4$, and ZnO_4 are all well known in these structural types. Other tetrahedral units that have been found only rarely in framework structures include $Ni(II)O_4$, $Cu(II)O_4$, and InO_4. We concentrate on

Table 23.7 Some three-dimensional intercalation compounds

Phase	Composition, x
$Li_x[Mo_6S_8]$	0.65–2.4
$Na_x[Mo_6S_8]$	3.6
$Ni_x[Mo_6Se_8]$	1.8
H_xWO_3	0–0.6
H_xReO_3	0–1.36

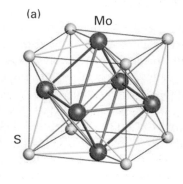
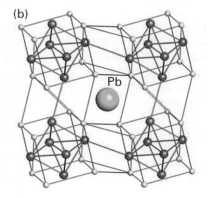

Fig. 23.47 (a) The Mo_6S_8 unit present in a Chevrel phase $Pb_xMo_6S_8$ and (b) the structure of a Chevrel phase showing the canted Mo_6S_8 units forming a slightly distorted cube around a Pb atom. An Mo atom in one cube can act as the acceptor for an electron pair donated by an S atom in a neighbouring cage.

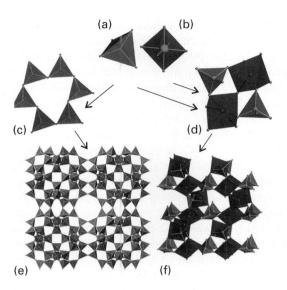

Fig. 23.48 Tetrahedral (a) and octahedral units (b) are linked together through their vertices to form larger units, called secondary building units, such as (c) and (d), which in turn bridge through their vertices to form framework structures such as (e) and (f).

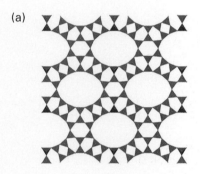

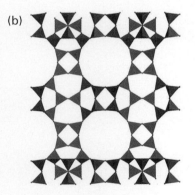

Fig. 23.49 Representations as linked tetrahedra of synthesized zeolites showing the main channels in (a) UTD-1 and (b) ITQ-7.

the structures derived from the most commonly found units in the zeolites, alumino-phosphates, and phosphates.

(a) Contemporary zeolite chemistry

Key points: New zeolite framework structures continue to be synthesized by using complex template molecules; important applications of zeolites include gas adsorption and ion exchange.

The role of structure-directing agents in the synthesis of zeolitic materials was described in Section 13.13. Following the discovery of a large number of new microporous structures from the 1950s onwards, many of which were prepared in the laboratory by using organic templates, more recent work on zeolites has been directed towards the systematic study of template–framework relationships. These studies can be divided into two major categories: understanding the interaction between the framework and template through computer modelling and experiment (Section 25.14), and designing templates with specific geometries to direct the formation of zeolites with particular pore sizes and connectivity.

One particular area that has become the focus of much attention is the use of bulky organic and organometallic molecules as templates in the quest for new, very large pore structures. This approach has been used to make the first zeolitic materials containing 14-ring channels.[6] Thus, the microporous silica UTD-1 (Fig. 23.49) has been prepared by using a permethylated bis-cyclopentadienyl cobalt metallocene and the siliceous CIT-5 (CFI structure type) was prepared by using a polycyclic amine and lithium. Other more complex amines have been synthesized with the aim of using them as templates in zeolite synthesis and to obtain desirable pore geometries. The addition of fluorides into the zeolite precursor gel improves reaction rates and acts as a template for some of the smaller cage units, for example where linked TO_4 tetrahedra arranged at the corners of a cube surround a central F^- ion.

As part of the overall growth in the number of known synthetic zeolites, there has been a significant increase in the proportion that can be made in (essentially) pure silica form so that over 20 structural types of zeolitic silica polymorphs are now known. The use of low H_2O/SiO_2 ratios is a key factor for producing these materials and the new 'silica' phases so produced are of unusually low density; for example, a purely siliceous framework with the same topology as the naturally occurring mineral chabazite, $Ca_{1.85}(Al_{3.7}Si_{8.3}O_{24})$, is the least dense silica polymorph known, with—according to the normal atomic radii—only 46 per cent of the unit cell volume occupied.

---

[6] A n-ring channel (in this case $n = 14$) refers to the number of tetrahedral units (MO_4) linked together to define the circumference of the channel: the larger n is, the greater is the diameter of the channel.

The experimental determination of the location of template molecules by X-ray and neutron diffraction has proved important for establishing template–framework relationships. Examples of such work include establishing the location of the templating ions in fluoride–silicalite at the channel intersections (Fig. 23.50). These approaches to the determination of structure are often used in association with computer modelling (Section 6.13).

Much of the attention directed at new zeolites focuses on obtaining ever larger pore sizes or a particular pore geometry, with the eventual aim of improving their catalytic properties (Section 25.14). For example, larger pore sizes would allow larger, more complex organic molecules to undergo transformations inside the zeolite cavity. However, we should also consider the other main applications of zeolites, which are as adsorbents for small molecules and as ion exchangers. Zeolites are excellent adsorbents for most small molecules such as H_2O, NH_3, H_2S, NO_2, SO_2, and CO_2, linear and branched hydrocarbons, aromatic hydrocarbons, alcohols, and ketones in the gas or liquid phase. Zeolites with different sized pores may be used to separate mixtures of molecules based on size, and this application has led to their description as **molecular sieves**. Through the correct selection of zeolite pore, it is possible to control the rates of diffusion of various molecules with different effective diameters, leading to separation and purification. Figure 23.51 illustrates this application schematically.

Industrial applications of zeolites for separation and purification include petroleum refining processes, where they are used to remove water, CO_2, chlorides, and mercury, the

(a)

(b)

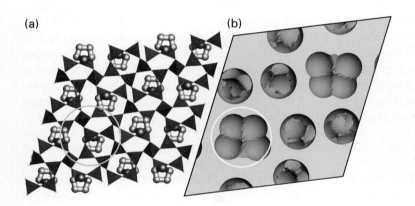

Fig. 23.50 (a) Calculated and experimental position of a pyrrolidinium cation in zeolite in the cavities; (b) framework space-filling diagram of the same structure showing that the shape of the cavities replicate the shape of the templating amine cation.

(a)

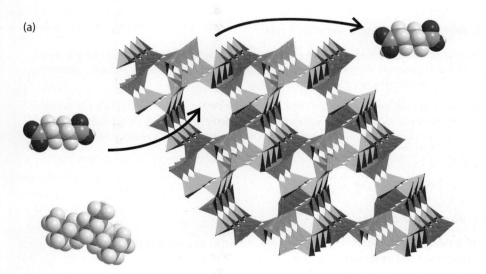

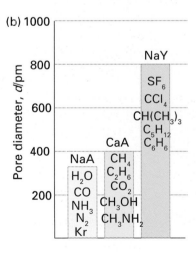

Fig. 23.51 The use of porous zeolite structures for the separation of molecules of various sizes. (a) Only the smaller molecule can diffuse into and then out of the zeolite pore so a membrane of this material may be used to separate this mixture. (b) The maximum molecular size that can be adsorbed into various zeolite channel diameters. NaY is foujacite, similar to molecular sieve X in Table 13.7.

desulfurization of natural gas (where the removal of H_2S and other sulfurous compounds protects transmission pipelines and removes the undesirable smell from home supplies), the removal of H_2O and CO_2 from air before liquefaction and separation by cryogenic distillation, and the drying and the removal of odours from pharmaceutical products. One area of growing importance is the separation of air into its main components other than by cryogenic means. Many dehydrated zeolites adsorb N_2 more strongly into their pores than O_2, and by passing air over a bed of zeolite at controlled pressures it is possible to produce oxygen of over 95 per cent purity. Zeolites such as NaX (the sodium-exchanged form of an aluminium-rich zeolite of structure type X[7] and CaA (a calcium-exchanged form of zeolite framework structure type A) were originally developed for this purpose. The selectivity of the process can be much improved by exchanging selected cations into the zeolite pores and thus changing the dimensions of the sites onto which the nitrogen and oxygen molecules adsorb. Thus lithium-, calcium-, strontium-, and magnesium-exchanged forms of the faujasite (FAU, type X) and Linde Type A (LTA) structures are now used very effectively in this process.

The excellent ion-exchange properties of zeolites result from their open structures and ability to trap significant quantities of cations selectivity within these pores. High capacities for ion exchange are derived from the large numbers of exchangeable cations. This is especially true of those zeolites with a high proportion of Al in the framework, which includes the zeolites with the LTA and gismondine (GIS) topologies that have the highest attainable Si:Al ratio of 1:1. The ion-exchange selectivity has led to a major application of zeolites as a **builder** in laundry detergents, where they are used to remove the hard ions Ca^{2+} and Mg^{2+} and to replace them with soft ions such as Na^+. Phosphates, which may also be used as detergent builders, have been the subject of environmental concerns linked to rapid algal growth and eutrophication of natural waters (Section 5.14). Spherical Na-form LTA zeolite particles, a few micrometres in diameter, are small enough to pass through the openings in the weave of clothing and so may be added to detergents to remove the hard cations in natural waters, and are then washed away harmlessly into the environment. A different zeolite framework, zeolite P with the GIS framework topology, has been developed as an alternative builder, with improved properties associated with its extremely high selectively for Ca^{2+} ions. Another important area where the ion-exchange properties of zeolites are exploited is in the trapping and removal of radionuclides from nuclear waste. Several zeolites, including the widely used clinoptilolite, have high selectivities for the larger alkali metal and alkaline earth metal cations, which in nuclear waste include ^{137}Cs and ^{90}Sr (Fig. 23.52). These zeolites may be vitrified by further reaction with glass-forming oxides as discussed in Section 23.9.

Fig. 23.52 The clinoptilolite structure highlighting the relationship between the trapped Cs^+ ions, shown as spheres with ionic radius 180 pm, and the framework.

Example 23.7 Applying Lowenstein's rule

In zeolites, where all the vertices of the SiO_4 and AlO_4 building units are shared between two tetrahedra, Lowenstein's rule states that no O atom is shared between two AlO_4 tetrahedra. Determine the maximum Al:Si ratio in a zeolite.

Answer The maximum aluminium content is reached when alternate tetrahedra in the structure are those of AlO_4 and each vertex is linked to four SiO_4 units (and, similarly, each SiO_4 tetrahedra is surrounded by four AlO_4 units. Thus the highest ratio achievable is 1:1.

Self-test 23.7 Write the formula of such a zeolite when the nonframework cations are Ca^{2+}.

(b) Aluminophosphates

Key point: The structures and physical properties of aluminophosphates parallel much of those of zeolites.

The structural and electronic equivalence of two silicate tetrahedra (SiO_4) and the aluminophosphate unit (AlO_4PO_4) can be recognized in the simple compounds SiO_2 and

---

[7] The designations of some zeolites are specified in Table 13.7.

$AlPO_4$, both of which adopt a similar range of dense polymorphs, including the quartz structure. The development of zeolites with very high Si content, which are effectively silica polymorphs, in turn led to the discovery of the aluminophosphate (ALPO) framework structures based on a 1:1 mixture of AlO_4 and PO_4 tetrahedra; $AlPO_4$ itself has the same structure as quartz (SiO_2) but with an arrangement of alternating Al and Si atoms at the centres of the tetrahedra. A wide range of ALPOs has been developed that parallel the zeolites in, for example, their synthesis under hydrothermal conditions (although in acid conditions rather than the basic ones used for zeolites) and their adsorption and catalytic properties. Again, organic template molecules have been designed and used to prepare many different ALPO structures, such as those with the structure codes VPI (with a large channel formed from 18 AlO_4/PO_4 tetrahedra), DAF, CIT, and STA (Fig. 23.53). Although aluminophosphate frameworks are neutral, substitution of either Al^{3+} or P^{5+} with metal ions of lower charge leads to the formation of 'solid acid catalysts' that can, for example, convert methanol into hydrocarbons selectively. The incorporation of cobalt and manganese, both of which have redox properties, into ALPO frameworks yields materials that can be used for the oxidation of alkanes.

(c) Phosphates and silicates

Key point: Calcium hydrogenphosphates are inorganic materials used in bone formation.

The complexities of silicate chemistry and the myriad ways in which SiO_4 tetrahedra can link together in crystalline silicates and some glasses have been discussed in Sections 23.9 and 13.13. In combination with other tetrahedral units, such as AlO_4, a vast range of other structures such as the naturally occurring and synthetic aluminosilicates, including the feldspars, clays, and zeolites, becomes possible. Another tetrahedral oxoanion that is frequently incorporated into framework materials is the phosphate group, PO_4^{3-}, although, as we shall see in the next section, many other tetrahedral units also form such structures.

Simple phosphate structures were described in Section 14.10 and are generally formed from linked PO_4 tetrahedra in chains, cross-linked chains, and cyclic units. We consider just one metal phosphate material in more detail here, namely calcium hydrogenphosphate, and closely related materials. The principal mineral present in bone and teeth is hydroxyapatite, $Ca_5(OH)(PO_4)_3$, the structure of which consists of Ca^{2+} ions coordinated by PO_4^{3-} and OH^- groups to produce a rigid three-dimensional structure (Fig. 23.54). The mineral apatite is the partially fluoride-substituted $Ca_5(OH,F)(PO_4)_3$. Related biominerals are octacalcium hydrogenphosphate, $Ca_8H_2(PO_4)_6$, and amorphous forms of calcium phosphate itself. Biominerals are discussed in more detail in Section 24.15 and Chapter 26.

23.14 Structures based on octahedra and tetrahedra

Many metals adopt MO_6 octahedral coordination in their oxo compounds. This polyhedral structural unit can be regarded as being formed by locating metal ions in

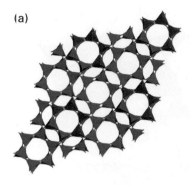

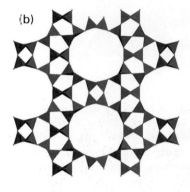

Fig. 23.53 The frameworks, formed from linked oxotetrahedra, of (a) DAF and (b) CIT. In each case the main channels present are emphasized by viewing the structure along them.

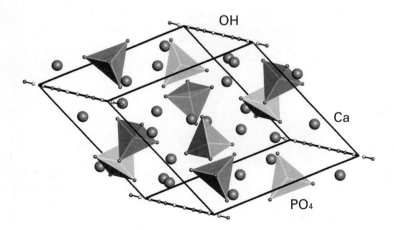

Fig. 23.54 The structure of hydroxyapatite, $Ca_5(OH)(PO_4)_3$, shown as Ca^{2+} ions coordinated by phosphate (shown as tetrahedra) and OH^- ions into a strong three-dimensional structure.

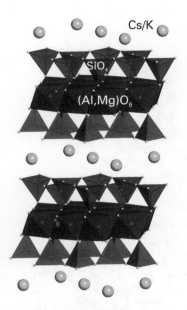

Fig. 23.55 Sheet-like structures of the clay hectorite, consisting of layers of linked octahedra and tetrahedra centred on, typically, Al, Si, or Mg and separated by cations such as K^+ or Cs^+.

the octahedral holes of a close-packed oxide ion array in, for example, MgO with the rock-salt structure. The MO_6 polyhedral building unit may also be incorporated into framework-type structures often in combination with tetrahedral oxo species.

(a) Clays, pillared clays, and layered double hydroxides

Key point: Sheet-like structures, found in many metal hydroxides and clays, can be constructed from linked metal oxo tetrahedra and octahedra.

The diameters of the largest pores found in synthetic zeolites are of the order of 1.2 nm. In an attempt to increase this diameter and allow larger molecules to be absorbed into inorganic structures, chemists have turned to mesoporous materials (Section 24.13), and structures produced by 'pillaring' (that is, stacking and connecting together two-dimensional materials). The two-dimensional nature of many *d*-metal disulfides and some of their intercalation compounds were discussed in Sections 8.9 and 23.11. Similar intercalation reactions when applied to aluminosilicates from the clay family allow the synthesis of large-pore materials.

Clays such as hectorite and montmorillonite have layer structures like those shown in Fig. 23.55. The layers are constructed from vertex- and edge-sharing octahedra, MO_6, and tetrahedra, TO_4. The metal atoms, M and T, contained within the layers, which have an overall negative charge, are typically silicon and aluminium. Small singly and doubly charged ions, such as Li^+ and Mg^{2+}, occupy sites between the layers. These interlayer cations are often hydrated and can readily be replaced by ion exchange. Other materials with similar structures are the *layered double hydroxides* with structures similar to that of $Mg(OH)_2$, the naturally occurring mineral brucite.

In the pillaring of clays, the species exchanged into the interlayer region is selected for size. Ions such as alkylammonium ions and polynuclear hydroxometal ions may replace the alkali metal as shown schematically in Fig. 23.56. The most widely used pillaring species are of the polynuclear hydroxide type and include $Al_{13}O_4(OH)_{28}^{3+}$, $Zr_4(OH)_{16-n}^{n+}$, and $Si_8O_{12}(OH)_8$; the first consists of a central AlO_4 tetrahedron surrounded by octahedrally coordinated aluminium ions as $Al(O,OH)_6$ species. The pillaring process can be followed by powder X-ray diffraction because it leads to expansion of the interlayer spacing, corresponding to an increase in the *c* lattice parameter.

Once an ion such as $Al_{13}O_4(OH)_{28}^{3+}$ has been incorporated between the layers, heating the modified clay results in its dehydration and the linking of the ion to the layers (Fig. 23.56). The resulting product is a pillared clay with excellent thermal stability

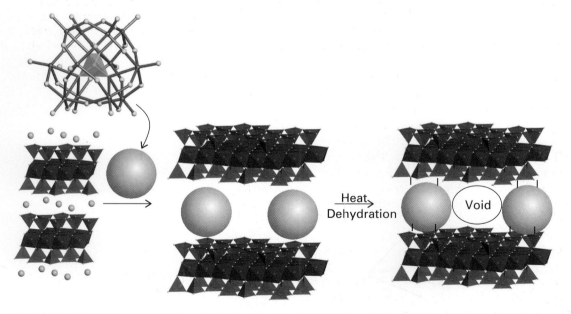

Fig. 23.56 Schematic representation of pillaring of a clay by ion exchange of a simple monatomic interlayer cation with a large polynuclear hydroxometallate followed by dehydration and cross-linking of the layers to form the cavities.

to at least 500°C. The expanded interlayer region can now absorb large molecules in the same way as zeolites. However, because the distribution of pillaring ions between the layers is difficult to control, the pillared clay structures are less regular than zeolites. Despite this lack of uniformity, pillared clays have been widely studied for their potential as catalysts because they act in a similar way to zeolites, as acid catalysts promoting isomerization and dehydration.

(b) Advances in inorganic framework chemistry

Key point: Enormous structural diversity can be obtained from linked polyhedra and, with the use of templates, can lead to remarkable porous frameworks.

The extensive development of aluminosilicates and zeolites has motivated synthetic inorganic chemists to seek similar structural types built from other tetrahedral and octahedral polyhedra for use in ion exchange, absorption, and catalysis. The use of different and larger polyhedra provides more flexibility in the framework topologies, and there is also the potential to incorporate d-metal ions with their associated properties of colour, redox properties, and magnetism. Tetrahedral species that have been incorporated into such frameworks, as well as the aluminate, silicate, and phosphate groups mentioned previously, include ZnO_4, AsO_4, CoO_4, GaO_4, and GeO_4. Octahedral units are mainly based on d metals and the heavier and larger metals from Groups 13 and 14. Other polyhedral units such as five-coordinate square pyramids also occur, but less commonly.

Zeolite analogues are called **zeotypes**. Ring sizes larger than 12 tetrahedrally coordinated atoms were first seen in metallophosphate systems, and structures with 20 and 24 linked tetrahedra have been made. Aluminophosphates were the first microporous frameworks synthesized that contain polyhedra with coordination numbers greater than four. Other so-called **hypertetrahedral frameworks** are now well-established, such as the titanosilicate families (which have four-coordinate Si and five- and six-coordinate Ti sites) and a series of octahedral molecular sieves based on linked MnO_6 units.

Good examples of this structural family are the titanosilicate zeotypes built from SiO_4 tetrahedra and various TiO_n polyhedra with $n = 4$–6. These compounds are made under the hydrothermal conditions similar to those used for synthesizing many zeolites but by using a source of titanium, such as $TiCl_4$ or $Ti(OC_2H_5)_4$, that hydrolyses under the basic conditions in the autoclave. Templates may also be used in such media and act as structure-directing units giving rise to particular pore sizes and geometries. In a typical reaction the titanosilicate ETS-10 (Engelhard TitanoSilicate 10) is prepared by the reaction of $TiCl_4$, sodium silicate, sodium hydroxide, and sometimes a template such as tetraethylammonium bromide, in a sealed polytetrafluorethylene-lined autoclave at between 150 and 230°C. The number of titanosilicates continues to grow, but two materials, ETS-10 and $Na_2Ti_2O_3SiO_4\cdot2H_2O$, are worthy of further consideration here as specific examples of this type of material.

ETS-10 is a microporous material built from TiO_6 octahedra and SiO_4 tetrahedra, with the TiO_6 groups linked together in chains (Fig. 23.57). The structure has 12-membered rings (that is, pores formed from 12-unit polyhedra) in all three directions. $Na_2Ti_2O_3SiO_4\cdot2H_2O$ also has TiO_6 octahedra, but in this structure these octahedra form clusters with four TiO_6 units linked by the tetrahedral SiO_4 units (see Fig. 23.48). This connectivity gives rise to large octagonal pores containing hydrated Na^+ ions. This compound, in common with several other titanosilicates (such as potassium hydrogen titanosilicate, $K_3H(TiO)_4(SiO_4)_3\cdot4H_2O$, a synthetic analogue of a natural mineral pharmacosiderite, Fig. 23.57), shows excellent ion-exchange properties, particularly with large cations. These large ions replace Na^+ with very high selectivity, so that, for example, Cs^+ and Sr^{2+} ions may be extracted into the titanosilicate structure from their dilute solutions. This ability has led to the development of these materials for removal of radionuclides from nuclear waste, in which ^{137}Cs and ^{90}Sr are highly active.

A wealth of framework structures have been reported in recent years. One of the systems studied most systematically is that of the numerous gallophosphate fluoride phases, where the role of various synthesis parameters on the outcome of syntheses have been identified. pH, for instance, has been found to have a major and complex

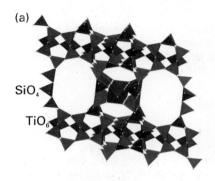

(a)

SiO_4

TiO_6

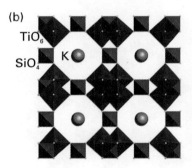

(b)

TiO_6

SiO_4 K

Fig. 23.57 The structures of (a) the titanosilicate ETS-10, which is built from chains of TiO_6 octahedra (red) linked by SiO_4 tetrahedra (blue), and (b) $K_3H(TiO)_4(SiO_4)_3\cdot4H_2O$, a synthetic analogue of a natural mineral pharmacosiderite; the K^+ ions are shown as magenta spheres.

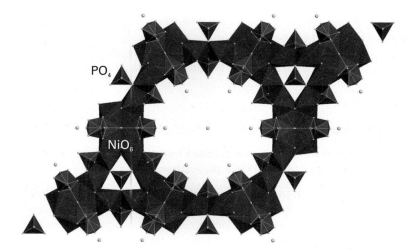

Fig. 23.58 The framework structure of VSB-1 consists of linked NiO_6 octahedra and PO_4 tetrahedra. The main channel is bounded by 24 such units and has a diameter such that molecules up to 0.88 nm in diameter may pass through it.

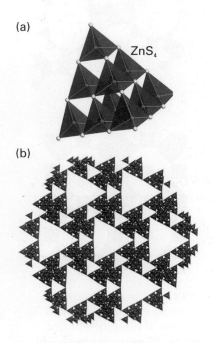

Fig. 23.59 (a) An isolated supertetrahedral unit of the stoichiometry $Zn_{10}S_{20}$ formed from individual ZnS_4 tetrahedral. (b) These supertetrahedral units may be linked together as building blocks to form large porous three-dimensional structures.

role, as it influences the nature of the phases, the hydration of the templating amine, and the nature and the kind of linkage of the polyhedra around the metal.

The aim of much of the recent work on porous framework structures has been to incorporate *d*-metal ions. Materials that have been synthesized include antiferromagnetic porous iron(III) fluorophosphates with Néel temperatures in the range 10–40 K. These temperatures are relatively high for iron clusters linked by phosphate groups and indicate the presence of moderately strong magnetic interactions. Zeotypic cobalt(II), vanadium(V), titanium(IV), and nickel(II) phosphates have been prepared, including the material known as VSB-1, which was the first microporous solid with 24-membered ring tunnels, and is simultaneously porous, magnetic, and an ion exchanger (Fig. 23.58).

Two other rapidly growing areas of research into framework structures are materials based on sulfides and on the use of small organic molecules to link the metal atoms, so giving **metal–organic frameworks**. Tetrahedral coordination is common in simple metal sulfide chemistry, as we saw in the discussion of ZnS in wurtzite and sphalerite (Section 3.9), and some compounds may be considered as containing linked fragments of these structures known as **supertetrahedral clusters**. For example, $[N(CH_3)_4]_4[Zn_{10}S_4(SPh)_{16}]$ contains the isolated supertetrahedral unit $Zn_{10}S_{20}$ terminated by phenyl groups (Fig. 23.59).

Metal–organic frameworks have structures that are based on bidentate or polydentate organic ligands lying between the metal atoms. Examples of simple ligands used to build these often porous frameworks are CN^-, nitriles, amines, and carboxylates. Many hundreds of these compounds have now been made, and include positively charged frameworks with balancing anions in the cavities, for instance, $Ag(4,4'-bipy)\cdot NO_3$, and electrically neutral frameworks, such as $Zn_2(1,3,5-benzenetricarboxylate)\cdot NO_3 \cdot H_2O\cdot C_2H_5OH$, from which the H_2O molecules may be removed reversibly. The compounds have properties and applications analogous to those of zeolites (Fig. 23.60).

Inorganic pigments

We saw in Section 23.7 that the adoption of a tetrahedral site by Co(II) in the spinel $CoAl_2O_4$ and the resulting deep blue colour led to its use as a pigment. Many inorganic solids are intensely coloured and are used as pigments in colouring inks, plastics, glasses, and glazes. Whereas many insoluble organic compounds (for example the C.I. Pigment Red 48, which is calcium 4-((5-chloro-4-methyl-2-sulfophenyl)azo)-3-hydroxy-2-naphthalenecarboxylic acid) are also used as pigments, inorganic materials often have advantages in terms of applications associated with their chemical, light, and thermal stability. Pigments were originally developed from naturally occurring compounds such as hydrated iron oxides, manganese oxides, lead carbonate, vermilion (HgS), orpiment (As_2S_3), and copper carbonates. These compounds were used even in cave paintings. Synthetic pigments, which are often analogues of naturally occurring compounds, were developed by some of the earliest chemist and alchemists, and the first synthetic chemists were probably those involved in making pigments. Thus, the pigment Egyptian blue ($CaCuSi_4O_{10}$) was made from sand, calcium carbonate, and copper ores as long as 3000 years ago. This compound and a structural analogue, Chinese blue ($BaCuSi_4O_{10}$), which was first made about 2500 years ago, have a structure containing square-planar copper(II) ions surrounded by Si_4O_{10} groups (Fig. 23.61). Inorganic pigments continue to be important commercial materials and this section summarizes some of the recent advances in this field.

As well as producing the colours of inorganic pigments as a result of the absorption and reflection of visible light, some solids are able to absorb energy of other wavelengths (or types, for example electron beams) and emit light in the visible region. This **luminescence** is responsible for the properties of inorganic phosphors as discussed in Box 23.3.

23.15 Coloured pigments

Key point: Intense colour in inorganic solids can arise through *d–d* transitions, charge transfer (and the analogous interband electron transfer), or intervalence charge transfer.

The blue colour of $CoAl_2O_4$ and $CaCuSi_4O_{10}$ stems from the presence of *d–d* transitions in the visible portion of the electromagnetic spectrum. The characteristic, intense colour of cobalt aluminate is a result of having a tetrahedral site for the metal ion, which removes the constraint of the Laporte selection rule of octahedral symmetries (Section 19.6). The chemical and thermal stabilities are due to the location of the Co^{2+} ion in the close-packed oxide arrangement. Other inorganic pigments with colours based on *d–d* transitions include nickel-doped TiO_2 (yellow) and $Cr_2O_3 \cdot nH_2O$ (green). Colour also arises in many inorganic compounds from charge transfer (Section 19.5) or what is often electronically an equivalent process in solids, the promotion of an electron from a valence band (derived mainly from anion orbitals) into a conduction band (originating from metal orbitals). Charge-transfer pigments include compounds such as lead chromate ($PbCrO_4$), containing the yellow–orange chromate(VI) anion, and $BiVO_4$ with the yellow vanadate(V) anion. CdS (yellow) and CdSe (red) both adopt the wurtzite structure and colour arises from the filled valence band (which is mainly derived from chalcogenide *p* orbitals) to orbitals based mainly on cadmium. For a material with a band gap of 2.4 eV (as for CdS at 300 K) these transitions occur as a broad absorption corresponding to wavelengths shorter than 515 nm; thus CdS is bright yellow as the blue part of the visible spectrum is fully absorbed. For CdSe the band gap is smaller on account of the higher energies of the Se 4*p* orbitals and the absorption edge shifts to lower energies. As a result, only red light is not absorbed by the material. In some mixed-valence compounds, electron transfer between differently charged metal centres can also occur in the visible region, and as these transfers are often fully allowed they give rise to intense colour. Prussian blue, $Fe(III)_4[Fe(II)(CN)_6]_3$ (Fig. 23.62), is one such compound and its dark blue colour has resulted in its widespread use in inks.

Inorganic radicals often have fairly low-energy electronic transitions that can occur in the visible region. Two examples are NO_2 (brown) and ClO_2 (yellow). One inorganic pigment is based on an inorganic radical, but because of the high reactivity normally associated with main-group compounds containing unpaired electrons, this species is trapped inside a zeolite cage. Thus the royal-blue pigment ultramarine, a synthetic analogue of the naturally occurring semi-precious stone lapis lazuli, has the idealized formula $Na_8[SiAlO_4]_6 \cdot (S_3)_2$ and contains the S_3^- polysulfide radical anion occupying a sodalite cage formed by the aluminosilicate framework (Fig. 23.63).

Current developments in inorganic pigment chemistry are centred on finding replacements for some of the yellow and red materials that contain heavy metals, such as cadmium and lead. Although these materials themselves are not toxic, as the compounds are very stable and the metal is difficult to leach into the environment, their synthesis and disposal can be problematic. Compounds that have been investigated to replace cadmium

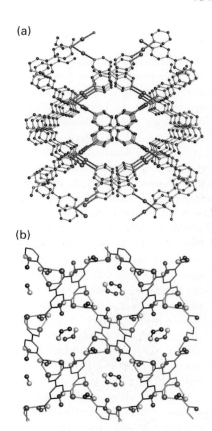

(a)

(b)

Fig. 23.60 Framework structures made from metal ions linked by organic ligands. (a) Ag(bipy)·NO_3 (the NO_3 is not shown) and (b) Zn_2(1,3,5-benzenetricarboxylate)·$NO_3 \cdot H_2O \cdot C_2H_5OH$ with ethanol molecules (C_2O backbone only shown) adsorbed within the pores.

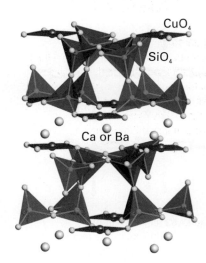

Fig. 23.61 The square-planar copper/oxygen environment formed by Si_4O_{10} groups in Egyptian blue ($CaCuSi_4O_{10}$) and Chinese blue ($BaCuSi_4O_{10}$).

Box 23.3 Inorganic phosphors

Luminescence is the emission of light by materials that have absorbed energy in some form. Photoluminescence occurs when photons, usually in the ultraviolet region of the electromagnetic spectrum, are the source of energy and the output usually is visible light. **Cathodoluminescence** uses electron beams as a source of energy and **electroluminescence** uses electrical energy as the energy source. Two types of photoluminescence can be distinguished: **fluorescence**, which has a period of less than 10^{-8} s between photon absorption and emission, and **phosphorescence**, for which there are much longer delay times.

Photoluminescent materials, frequently referred to as **phosphors**, generally consist of a host structure, such as ZnS (wurtzite structure; Section 3.9), CaWO$_4$ (scheelite structure type with discrete WO$_4^{2-}$ tetrahedra separated by Ca^{2+} ions), or Zn$_2$SiO$_4$, into which an **activator ion** is introduced by doping. These activator ions are certain d-metal or lanthanoid ions, such as Mn^{2+}, Cu^{2+}, and Eu^{2+}, that have the ability to absorb and emit light of the desired wavelengths. In some cases a second dopant is added as a **sensitizer** to aid the absorption of light of the desired wavelength. Well-known applications of such materials include fluorescent lamps and television screens, where there is a need for materials that fluoresce in specific regions of the visible spectrum.

Many host–activator combinations have been studied as potential phosphor materials with the aim of producing a material that efficiently converts UV radiation or cathode rays (electron beams) to pure emission of a desirable colour. The modification of the host structure and the nature and environment of the activator ion allows these properties to be tuned (Table B23.1).

In many fluorescent lamps, a mercury discharge produces UV radiation at 254 and 185 nm. Then a coating of ZnS that has been doped with various activators produces fluorescences at several wavelengths that combine together to give an effective white light.

Table B23.1

Phosphor host	Activator	Colour
Zn$_2$SiO$_4$	Mn^{2+}	Green
Y$_2$O$_3$	Eu^{3+}	Red
CaMg(SiO$_3$)$_2$ diopside	Ti	Blue
CaSiO$_3$	Mn	Yellow/orange
Ca$_5$(PO$_4$)$_3$(F,Cl)	Mn	Orange
ZnS	Ag$^+$, Cu^{2+}, Mn^{2+}	Blue, green, yellow

Colour television requires three primary-colour, cathodoluminescent materials to produce a picture. These phosphors are typically ZnS:Ag$^+$ (blue), ZnS:Cu$^+$ (green), and YVO$_4$:Eu^{3+} (red). In YVO$_4$:Eu the vanadate group absorbs the incident electron energy and the activator is Eu^{3+}. The emission mechanism involves electron transfer between the vanadate group and Eu^{3+} and the efficiency of this process depends on the M—O—M bond angle in a similar way to the superexchange process in antiferromagnets (Box 23.2). The nearer this angle is to 180°, the faster and more efficient is the electron transfer process. In YVO$_4$:Eu this angle is 170° and this material is a highly efficient phosphor.

Anti-Stokes phosphors convert two or more photons of lower energy to one of higher energy (such as infrared radiation to visible light). They act by absorbing two or more photons in the excitation process before emitting one photon. The best anti-Stokes phosphors have ionic host structures such as YF$_3$, NaLa(WO$_4$)$_2$, and NaYF$_4$ doped with Yb^{3+} as a sensitizer ion (to absorb the IR radiation) and Er^{3+} as an activator (emitting visible light). Applications include night-vision binoculars.

Fig. 23.62 In Prussian blue, [Fe(III)]$_4$[Fe(II)(CN)$_6$]$_3$, the iron ions are linked through cyanide, forming a cubic unit cell.

chalcogenides and lead-based pigments are the lanthanoid sulfides, such as Ce$_2$S$_3$ (red), and early d-metal oxide-nitrides, such as Ca$_{0.5}$La$_{0.5}$Ta(O$_{1.5}$N$_{1.5}$) (orange). In both cases, the replacement of O^{2-} by S^{2-} or N^{3-} ion has the effect of narrowing the band gap in these solids (compared to the colourless solids CeO$_2$ and Ca$_2$Ta$_2$O$_5$, which have large band gaps and absorb only in the UV region of the spectrum) and bringing the electron excitation energy into the visible. However, neither of these materials has the stability of the cadmium- and lead-based pigments.

23.16 White and black inorganic materials

Some of the most important compounds used to modify the visual characteristics of polymers and paints have visible-region absorption spectra that result in them appearing either white (ideally no absorption in the visible region) or black (complete absorption between 380 and 800 nm).

(a) White pigments

Key point: TiO$_2$ is used almost universally as a white pigment.

White inorganic materials can also be classified as pigments and vast quantities of these compounds are synthesized for applications such as the production of white plastics and paints. Important commercial compounds of this class that have been used extensively historically are TiO$_2$, ZnO, and ZnS, lead(II) carbonate, and lithopone (a mixture of ZnO and $BaSO_4$); note that none of the metals in these materials has an incomplete d-electron shell that might otherwise induce colour through d–d transitions. TiO_2, in either its rutile or anatase forms (Fig. 23.64), is produced from titanium ores, often ilmenite $FeTiO_3$, by the *sulfate process* (which involves dissolution in concentrated H_2SO_4 and subsequent precipitation through hydrolysis) or the *chloride process* (which is based on the reaction of mixed complex titanium oxides with chlorine to produce $TiCl_4$, which is then combusted with oxygen at over 1000°C). These routes produce very high quality TiO_2 free from impurities (which is essential for a bright white pigment) of controlled particle size. The desirable qualities of TiO_2 as a white pigment derive from its excellent light scattering power, which in turn is a result of its high refractive index ($n_r = 2.70$), the ability to produce very pure materials of a desired particle size, and its good light-fastness and weather resistance. Uses of titanium dioxide, which nowadays dominates the white-pigment market, include paints, coatings, and printing ink (where it is often used in combination with coloured pigments to increase their brightness and hiding power), plastics, fibres, paper, and white cements.

(b) Black, absorbing, and specialist pigments

Key points: Special colour, light absorbing, and interference effects can be induced in inorganic materials used as pigments.

The most important black pigment is carbon black, which is a better defined, industrially manufactured form of soot. Carbon black is obtained by partial combustion or pyrolysis (heating in the absence of air) of hydrocarbons. The material has excellent absorption properties right across the visible region of the spectrum and applications include printing inks, paints, plastics, and rubber. Copper(II) chromite, $CuCr_2O_4$, with the spinel structure (Fig. 23.24), is used less frequently as a black pigment. These black pigments also absorb light outside the visible region, including the infrared, which means that they heat up readily on exposure to sunlight. Because this heating can have drawbacks in a number of applications, there is interest in the development of new materials that absorb in the visible region but reflect infrared wavelengths; $Bi_2Mn_4O_{10}$ is one compound that exhibits these properties.

Examples of more specialist inorganic pigments are magnetic pigments based on coloured ferromagnetic compounds such as Fe_3O_4 and CrO_2 and anticorrosive pigments such as zinc phosphates. The deposition of inorganic pigments as thin layers on to surfaces can produce additional optical effects beyond light absorption. Thus deposition of TiO_2 or Fe_3O_4, as thin layers a few hundred nanometres thick, on to flakes of mica produces lustrous or pearlescent pigments where interference effects between light reflected from the various surfaces and layers produces shimmering and iridescent colours.

Semiconductor chemistry

The basic inorganic chemistry of semiconducting materials, particularly their electronic band structures, was covered in Chapter 3. The aim of this section is to discuss the inorganic semiconducting compounds themselves in more detail and to describe some of the applications that result from using chemistry to control their electronic properties.

Semiconductors are classified on the basis of their composition. To produce materials with band gaps typical of a semiconductor (a few electronvolts corresponding to 100–200 kJ mol^{-1}), compounds usually contain the p-block metals and Group 13/14 (III/IV) metalloids often in combination with heavier chalcogenides and pnictides. For these combinations, the atomic orbitals form into bands with energies such that the valence and conduction band separation is in the desired range of 0.2–4 eV. Materials based on more electropositive and electronegative elements (for example, alkaline earth metal oxides) are much more ionic in character with much wider separations of the valence and conduction bands, and are insulators. A further factor that influences the band gap in semiconductors is particle size (Section 24.2).

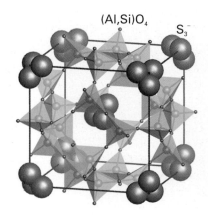

Fig. 23.63 One of the sodalite cages present in ultramarine, $Na_8[SiAlO_4]_6\cdot(S_3)_2$; the framework consists of linked SiO_4 and AlO_4 tetrahedra surrounding a cavity containing the polysulfide radical ion S_3^- and Na^+ ions (the latter omitted for clarity).

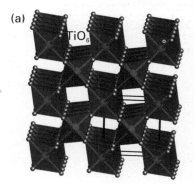

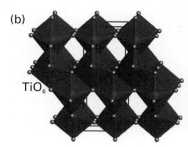

Fig. 23.64 TiO_2 exists as several polymorphs that can be described in terms of linked TiO_6 octahedra including the (a) rutile and (b) anatase forms.

23.17 Group 14 semiconductors

Key point: Crystalline and amorphous silicon are cheap semiconducting materials and are widely used in electronic devices.

The most important semiconducting material is silicon, which in its pure crystalline form (with a diamond structure) has a band gap of 1.1 eV. As would be expected from considerations of atomic radii, orbital energies, and the extent of orbital overlap, germanium has a smaller band gap, 0.66 eV, and carbon in diamond form has a band gap of 5.47 eV. When doped with a Group 13 or 15 element, silicon—and indeed diamond and germanium—are extrinsic semiconductors. The conductivity of pure silicon, an intrinsic semiconductor, is around 10^{-2} S cm^{-1} at room temperature but increases by several orders of magnitude on doping with either a Group 13 element (to give a p-type semiconductor) or a Group 15 element (to give an n-type semiconductor) and thus the properties of doped silicon can be tuned for a particular semiconductor application.

Amorphous silicon can be obtained by chemical vapour deposition or by heavy-ion bombardment of crystalline silicon. The deposited material, which is often obtained by thermal decomposition of SiH$_4$, contains a small proportion of hydrogen that is present in Si—H groups in a three-dimensional glass-like structure with many Si—Si links. The lack of regular structure in this material and the presence of Si—H groups alters the semiconducting properties of the material considerably. One of the main applications of amorphous silicon is in silicon solar cells (Fig. 23.65). Thin films of p- and n-type amorphous silicon forming a p–n junction generate current when illuminated (Box 23.4). The electron and hole pairs produced from the energy supplied by the incident photon separate rather than recombine because of the normal p–n junction bias, with the tendency for the electrons to travel towards the p-type silicon and holes towards the n-type. If a load is connected across the junction then a current can flow and electrical energy is generated from the electromagnetic illumination. The efficiency of such devices depends on a number of factors. For example, amorphous silicon absorbs solar radiation 40 times more efficiently than does single-crystal silicon, so a film only about 1 μm thick can absorb 90 per cent of the usable solar energy. Also, the lifetimes and mobility of the electrons and holes are longer in amorphous silicon, which results in high photoelectric efficiencies (the proportion of radiant energy converted into electrical energy) of the order of 10 per cent. Other economic advantages are that amorphous silicon can be produced at a lower temperature and can be deposited on low-cost substrates. Amorphous silicon solar cells are widely used in pocket calculators but, as production costs diminish, are likely to find much wider applications as renewable energy sources.

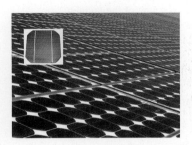

Fig. 23.65 Photograph of a solar cell.

23.18 Semiconductor systems isoelectronic with silicon

Key points: Semiconductors formed from equal amounts of Group 13/15 or Group 12/16 elements are isoelectronic with silicon and can have enhanced properties based on changes in the electronic structure and electron motion.

Gallium arsenide, GaAs, is one of a number of so-called Group 13/15 or, still more commonly, III/V semiconductors, which also include GaP, InP, AlAs, and GaN, formed by combination of equal amounts of a Group 13 (old name, Group III) and a Group 15 (old name, Group V) element. Ternary and quaternary Group 13/15 compounds, such as Al$_x$Ga$_{1-x}$As, InAs$_{1-y}$P$_y$, and In$_x$Ga$_{1-x}$As$_{1-y}$P$_y$, can also be formed and many of them also have valuable semiconducting properties. Note that these compositions are isoelectronic with pure Group 14 elements, but the changes in the element electronegativity and thus bonding type (for instance, pure Si silicon can be considered as having purely covalent bonding whereas GaAs has a small degree of ionic character due to the difference in the electronegativities of Ga and As) leads to changes in the band structures and also fundamental properties associated with electron motion through the structures.

One of the advantageous properties of GaAs is that semiconductor devices based on it respond more rapidly to electrical signals than those based on silicon. This responsiveness

Box 23.4 p–n Junctions and LEDs

A p–n junction can be constructed from two pieces of silicon, one of which is n-type and the other p-type. Figure B23.7 shows the band structure of the junction. The Fermi levels in the differently doped materials are at different energies and electrons flow from the n-type (high potential) to the p-type (low potential) region across the junction. This process continues for as long as a potential difference exists across the junction. Thus a current may flow across a p–n junction, but will do so only in one direction, from n-type to p–type. The p–n junction thus forms the basis of a rectifier, allowing current to pass in only one direction.

A light-emitting diode (LED) is a p–n junction semiconductor diode that emits light when current is passed. Effectively the transfer of the electron from the conduction band of the n-type material to the valence band of the p-type semiconductor is associated with emission of light. LEDs are highly monochromatic, emitting a pure colour in a narrow frequency range. The colour emitted from an LED is controlled by the band gap, with small band gaps producing radiation in the infrared and red regions of the electromagnetic spectrum and larger band gaps resulting in emission in the blue and ultraviolet regions (Table B23.2). It is possible to produce white light with a single LED by using a phosphor layer (yttrium aluminium garnet) on the surface of a blue, gallium nitride, LED.

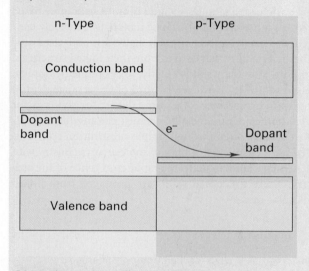

B23.7 The structure of a p–n junction.

Table B23.2 LED colours

LED colour	Chip material	
	Low brightness	High brightness
Red	GaAsP/GaP	AlInGaP
Orange	GaAsP/GaP	AlInGaP
Amber	GaAsP/GaP	AlInGaP
Yellow	GaP	–
Green	GaP	GaN
Turquoise	–	GaN
Blue	–	GaN

makes GaAs better than silicon for a number of tasks, such as amplifying the high-frequency (1–10 GHz) signals of satellite TVs. Gallium arsenide can be used with signal frequencies up to about 100 GHz. At even higher frequencies materials such as indium phosphide (InP) may be used. At present, frequencies above about 50 GHz are rarely used commercially, so most of the electronics in the world tend to be based on silicon, with some GaAs, and there are only a few InP devices. Gallium arsenide is also far more expensive than Si, both in terms of the cost of raw materials and the chemical processes required to produce the pure material.

In some Group 13/15 semiconductors, such as GaN, the cubic, diamond-like, sphalerite structure is only metastable, the stable polymorph being the hexagonal, wurtzite structure. Both structures can be grown by altering the synthesis routes and conditions. Because of their large and direct energy band gaps, these semiconductors allow the fabrication of luminescence devices that produce blue light at high intensity (Box 23.4), and their stabilities to high temperatures and good thermal conductivities also make them valuable for the fabrication of high-power transistors.

The Group 12/16 or II/VI semiconductors (as before, the name is derived from the former group notation) comprise the compounds containing Zn, Cd, and Hg as cations and O, S, Se, and Te as anions. These semiconductor materials can crystallize in either the cubic sphalerite phase or the hexagonal wurtzite phase and the form synthesized has its characteristic semiconducting properties. For example, the band gap is 3.64 eV in cubic ZnS but 3.74 eV in hexagonal ZnS. These Group 12/16 compounds are more ionic in nature than the Group 13/15 semiconductors and Group 14 elements, particularly for the lighter elements, and the band gap is around 3–4 eV for ZnO and ZnS but 1.475 eV for CdTe. Although amorphous silicon is the leading thin-film photovoltaic (PV) material, cadmium telluride (CdTe) is also being studied for similar applications.

Some other semiconducting oxides and sulfides were mentioned in Section 3.18, and research continues to seek other complex metal oxides and chalcogenides with improved characteristics. For example, the current world record thin-film solar cell efficiency of 17.7 per cent is held by a device based on copper indium diselenide ($CuInSe_2$; CIS), which has a similar structure to cubic ZnS but with an ordered distribution of Cu and In atoms in the tetrahedral holes.

Molecular materials and fullerides

The majority of compounds discussed so far in this chapter have been materials with extended structures in which ionic or covalent interactions link all the atoms and ions together into a three-dimensional structure. Examples include infinite structures based on ionic, as in NaCl, or covalent interactions, as in SiO_2. These materials are widely used in applications such as heterogeneous catalysis, rechargeable batteries, and electronic devices due to the chemical and thermal stability that derives from their linked structures. It is often possible to tune the properties of many solids exactly, for example by doping, introducing defects or forming solid solutions. However, control of the arrangement of atoms into a particular structure is not achievable to the same degree for solids as for molecular systems. Thus a coordination or organometallic chemist can modify a molecule by introducing a wide range of ligands, often through simple substitution reactions. The desire to combine the synthetic and chemical flexiblity of molecular chemistry with the properties of classical solid-state materials has led to the rapid emergence of the area of **molecular materials chemistry**, where functional solids are produced from linked and interacting molecules or molecular ions.

23.19 Fullerides

Key points: Solid C_{60} can be considered as a close-packed array of fullerene molecules interacting only weakly through van der Waals forces; holes in arrays of C_{60} molecules may be filled by simple and solvated cations and small inorganic molecules.

The chemical properties of C_{60} span many of the conventional borders of chemistry and include the chemistry of C_{60} as a ligand (Section 13.4). In this section we describe the solid-state chemistry of solid fullerene, $C_{60}(s)$, and the M_nC_{60} fulleride derivatives that contain discrete C_{60}^{n-} molecular anions. The synthesis and chemistry of the more complex carbon nanotubes are discussed in Section 24.13.

Crystals of C_{60} grown from solution may contain included solvent molecules, but with the correct crystallization and purification methods—for example, using sublimation to eliminate the solvent molecules—pure C_{60} crystals may be grown. The solid structure has a face-centred cubic array of the C_{60} molecules as shown in Fig. 23.66, as would be expected on the basis of efficient packing of these almost spherical molecules. At room temperature the molecules can rotate freely on their lattice positions and powder X-ray diffraction data collected from crystalline C_{60} are typical of an fcc lattice with a lattice parameter of 1417 pm. The molecules are separated by a distance of 296 pm, which is similar to the value found for the interlayer separation in graphite (335 pm). On cooling the solid, the rotation halts and adjacent molecules align relative to each other such that an electron-rich region of one C_{60} molecule is close to an electron-poor region in its neighbour.

Exposure of solid C_{60} to alkali metal vapour results in the formation of a series of compounds of formula M_xC_{60}, the precise stoichiometry of the product depending on the composition of the reactant mixture. With excess alkali metal, compounds of composition M_6C_{60}, M = K, Rb, Cs, and sometimes Na and Li, are formed. The structure of K_6C_{60} is body-centred cubic; C_{60}^{6-} molecular ions occupy sites at the cell corners and body centre and the K^+ ions fill a portion of the sites, with approximately tetrahedral coordination to four C_{60} molecular ions, near the centres of each of the faces (Fig. 23.67). Of most interest are the compounds with the stoichiometries M_3C_{60}, which become superconducting in the temperature range 10–40 K depending on the type of metal, M.

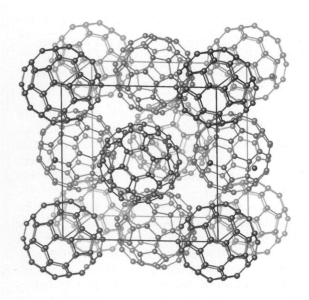

Fig. 23.66 The arrangement of C_{60} molecules in a face-centred cubic lattice in the crystalline material.

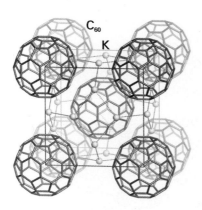

Fig. 23.67 The structure of K_6C_{60} with a body-centred cubic unit cell with C_{60}^{6-} molecular ions at the cell corners and body centre and K^+ ions (yellow) occupying half of the sites in the faces of the cell, which have approximate tetrahedral coordination to four C_{60}

The stoichiometry K_3C_{60} is obtained by filling all the tetrahedral and all the octahedral holes in the cubic close-packed C_{60}^{3-} arrangement (Fig. 23.68). K_3C_{60} becomes superconducting on cooling to 18 K, although gradual replacement of potassium by the larger alkali-metal ions raises T_c, so that for Rb_3C_{60} $T_c = 29$ K and for $CsRb_2C_{60}$ $T_c = 33$ K. Note that Cs_3C_{60} does not form the same fcc structure as the other M_3C_{60} phases (in fact it has a structure based on a body-centred arrangement of C_{60}^{3-} anions) and is not superconducting at normal pressures; however, it can be made superconducting, with a critical temperature of 40 K at 12 kbar.

Other more complex species can be incorporated into a matrix of C_{60} units. Molecular species such as iodine (I_2) or phosphorus molecules (P_4 tetrahedra) can fill spaces between the C_{60} molecules in close-packed arrays. Solvated cations can also occupy the tetrahedral and octahedral holes in a similar way to the simple alkali metal cations. Thus, $Na(NH_3)_4CsNaC_{60}$, which is obtained by the ammoniation of Na_2CsC_{60}, contains Na^+ ions solvated with ammonia molecules on the octahedral site and uncoordinated Cs^+ and Na^+ ions on the (twice as abundant) tetrahedral sites in an fcc arrangement of C_{60}^{3-} molecular ions.

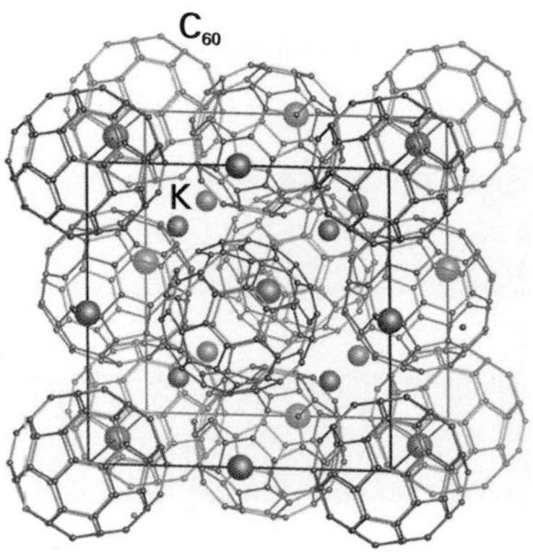

Fig. 23.68 The structure of K_3C_{60} is obtained by filling all the tetrahedral and all the octahedral holes in the close-packed C_{60}^{3-} lattice with K^+ ions.

Example 23.8 Synthesizing stoichiometric K_nC_{60} phases

The reaction between potassium and fullerene is difficult to control to a specific K_nC_{60} level although reaction with excess potassium leads to K_6C_{60} as the only fulleride product. How might a pure sample of K_3C_{60} be synthesized ?

Answer Prepare a sample of K_6C_{60} and then reduce this with a stoichiometric portion of C_{60} heated together in a sealed tube.

Self-test 23.8 How could a sample of Na_2CsC_{60} be synthesized ?

23.20 **Molecular materials chemistry**

The ability to modify the shapes, and thus the packing and arrangements of inorganic molecules in the solid state, is one valuable aspect of molecular materials chemistry. This capability, when associated with some of the specific properties of inorganic compounds such as the unpaired d electrons of transition metals, can allow control of magnetic and electronic properties. This section considers a number of such inorganic molecular materials being developed at the frontiers of the subject.

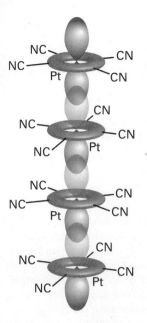

Fig. 23.69 A representation of the infinite chain structure of KCP, $K_2Pt(CN)_4Br_{0.3}\cdot3H_2O$, and a schematic illustration of its d band.

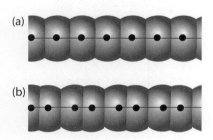

Fig. 23.70 The formation of a Peierls distortion: the energy of the line of atoms with alternating bond lengths (b) is lower than that of the uniformly spaced atoms (a).

(a) One-dimensional metals

Key points: A stack of molecules that interact with each other along one dimension, as occurs in a number of crystalline platinum complexes, can show conductivity in that direction; a Peierls distortion ensures that no one-dimensional solid is a metallic conductor below a critical temperature.

A one-dimensional (1D) metal is a material that exhibits metallic properties along one direction in the crystal and nonmetallic properties orthogonal to that direction. Such properties arise when the orbital overlap occurs along a single direction in the crystal (as in VO_2). Several classes of 1D metals are known and include $(SN)_x$ and organic polymers, such as doped polyacetylenes $[(CH)I_{0.25}]_n$, but this section is concerned specifically with chains of interacting d metals, particularly platinum.

In such materials, the structural requirements for a 1D metal are satisfied by the presence of square-planar complexes that stack one above another (Fig. 23.69). The ligands surrounding the metal atom ensure large interchain separations, of at least 900 pm, while the average intrachain metal–metal distance is less than 300 pm. Square-planar complexes are commonly found for metal ions with d^8 configurations, and the overlap of orbitals between d^8 species is greatest for the heavy d metals of Period 6 (which use $5d$ orbitals). Hence the compounds of interest are mainly associated with Pt(II) and Ir(I), where a band is formed from overlapping d_{z^2} and p_z orbitals. The d_{z^2} band is full for Pt(II), and a partially full level is achieved by oxidation of the platinum. Many d^8 integral oxidation number tetracyanoplatinate(II) complexes are semiconductors with $d_{Pt—Pt} > 310$ pm, and the partial oxidation of these salts results in Pt—Pt distances of less than 290 pm and metallic behaviour. Typical ways of oxidizing the chain are through the incorporation of extra anions into the structure or by removing cations. The first 1D metal Pt complex was made in 1846 by oxidation of a solution of $K_2Pt(CN)_4\cdot3H_2O$ with bromine, which on evaporation gave crystals of $K_2Pt(CN)_4Br_{0.3}\cdot3H_2O$, known as KCP.

The electronic properties of 1D solids are not quite as simple as has been implied by the discussion so far, as a theorem due to Rudolph Peierls states that, at $T = 0$, *no one-dimensional solid is a metal!* The origin of Peierls' theorem can be traced to a hidden assumption in the discussion so far: we have supposed that the atoms lie in a line with a regular separation. However, the actual spacing in a one-dimensional solid (and any solid) is determined by the distribution of the electrons, not vice versa, and there is no guarantee that the state of lowest energy is a solid with a regular lattice spacing. In fact, in a one-dimensional solid at $T = 0$, there always exists a distortion, a **Peierls distortion**, which leads to a lower energy than in the perfectly regular solid.

An idea of the origin and effect of a Peierls distortion can be obtained by considering a one-dimensional solid of N atoms and N valence electrons (Fig. 23.70). Such a line of atoms distorts to one that has long and short alternating bonds. Although the longer bond is energetically unfavourable, the strength of the short bond more than compensates for the weakness of the long bond and the net effect is a lowering of energy below that of the regular solid. Now, instead of the electrons near the Fermi surface being free to move through the solid, they are trapped between the longer-bonded atoms (these electrons have antibonding character, and so are found outside the internuclear region between strongly bonded atoms). The Peierls distortion introduces a band gap in the centre of the original conduction band, and the filled orbitals are separated from the empty orbitals. Hence, the distortion results in a semiconductor or insulator, not a metallic conductor.

The conduction band in KCP is a d band formed principally by overlap of $Pt5d_{z^2}$ orbitals. The small proportion of Br in the compound, which is present as Br^-, removes a small number of electrons from this otherwise full d band, so turning it into a conduction band. Indeed, at room temperature, doped KCP is a lustrous bronze colour with its highest conductivity along the axis of the Pt chain. However, below 150 K the conductivity drops sharply on account of the onset of a Peierls distortion. At higher temperatures, the motion of the atoms averages the distortion to zero, the separation is regular (on average), the gap is absent, and the solid is metallic. Mixed-valence platinum chain complexes, such as $[Pt(en)_2][PtCl_2(en)_2](ClO_4)_4$, may be isolated as insulated, one-dimensional wires by surrounding the platinum chain with a sheath of electronically inert molecules such as anionic lipids. Nanowires of this type are discussed more fully in Section 24.13.

The observation of metallic behaviour in 1D solids formed from interacting molecules has in turn led to a search for superconductivity in this class of materials. One type of material in which superconductivity has been found is a series of metal complexes derived from the organic metal and superconductors based on stacked molecules with interacting π systems. Thus salts of TCNQ (7,7,8,8-tetracyano-*p*-quinodimethane, (**4**) with TTF (tetrathiafulvalene, **5**) show metallic properties and salts of tetramethyltetraselenafulvalene (TMTSF, **6**) such as $(TMTSF)_2ClO_4$ show superconductivity, albeit at less than 10 K and often only under pressure. Molecular metal complexes involving types of sulfur-containing ligand, such as dmit ($dmit^{2-}$ = 1,3-dithiol-2-thione-4,5-dithiolato), can also show superconductivity. The compound $[TTF][Ni(dmit)_2]_2$, which consists of stacks of both TTF and $Ni(dmit)_2$, shows superconductivity at 10 kbar and below 2 K.

(b) Molecular magnets

Key point: Molecular solids containing individual molecules, clusters, or linked chains of molecules can show three-dimensional magnetic effects such as ferromagnetism.

Molecular inorganic magnetic materials, in which individual molecules, or units constructed from such molecules, contain *d*-metal atoms with unpaired electrons are a class of compounds of growing interest. Generally the phenomena associated with long-range interaction of electron spins, such as ferromagnetism and antiferromagnetism, are much weaker as the short, superexchange-type pathways that are found in metal oxides do not exist. However, as with all molecular systems the opportunity exists to tune interactions between metal centres by tailoring the ligand properties.

Examples of ferromagnetic molecular inorganic compounds are decamethylferrocene tetracyanoethenide (TCNE), $[Fe(\eta^5\text{-}Cp^*)_2(C_2(CN)_4)]$ (**7**), with $Cp^* = C_5Me_5$, and the analogous manganese compound. These materials, which have structures based on chains of alternating $M(\eta^5\text{-}Cp^*)_2^+$ and $TCNE^-$ ions show ferromagnetism along the chain direction below $T_C = 4.8$ K (for M = Fe) and 6.2 K (for M = Mn). An alternative approach to molecular-based compounds exhibiting magnetic ordering consists of assembling chains of magnetically interacting centres. For example, MnCu(2-hydroxy-1,3-propyle-nebisoxamato)·3H$_2$O (**8**) consists of chains of alternating manganese(II) and copper(II) ions bridged by the ligand. The magnetic moments on the metal ions in the chains order ferromagnetically at and below 115 K. This ferromagnetic ordering occurs initially only along the chains, that is in one dimension, due to the stronger interactions through the ligands and shorter Mn—Cu distances, but this material orders fully in three dimensions at 4.6 K once the individual chains interact with each other magnetically.

The incorporation of several *d*-metal ions into a single complex provides an opportunity to produce a molecule that acts as a tiny magnet. Such compounds have been termed **single-molecule magnets** (SMMs). One example is the complex manganese acetate $Mn_{12}O_{12}(O_2CMe)_{16}(H_2O)_4\cdot2MeCO_2H\cdot4H_2O$, which contains a cluster of 12 Mn(III) and Mn(IV) ions linked through O atoms with the metal oxide unit terminated by the acetate groups, Fig. 23.71. Another is $[Mn_{84}O_{72}(O_2CMe)_{78}(OMe)_{24}(MeOH)_{12}(H_2O)_{42}(OH)_6]\cdot xH_2O\cdot yCHCl_3$, which contains 84 Mn(III) ions in a large doughnut-shaped molecule 4 nm in diameter. The ability to magnetize such individual SMMs potentially provides a route to storing information at extremely high densities because their dimensions, a few nanometres, are much less than that of the typical domain of magnetic material used in a conventional magnetic data storage medium.

(c) Inorganic liquid crystals

Key point: Inorganic metal complexes with disc- or rod-like geometries can show liquid crystalline properties.

Liquid crystalline, or **mesogenic**, compounds possess properties that lie between those of solids and liquids and include both. For instance, they are fluid, but with positional order in at least one dimension. Such materials have become widely used in displays.

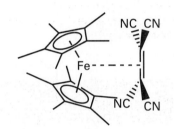

4 TCNQ

5 TTF

6 TMTSF

7 $[(\eta^5\text{-}Cp^*)_2Fe(TCNE)_2]$

8

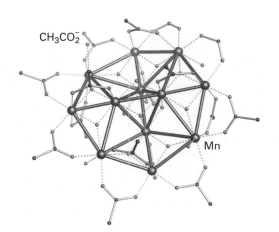

$CH_3CO_2^-$

Mn

Fig. 23.71 The core of the single molecular magnetic $Mn_{12}O_{12}(O_2CMe)_{16}(H_2O)_4 \cdot 2MeCO_2H \cdot 4H_2O$ that contains 12 Mn(III) and Mn(IV) ions.

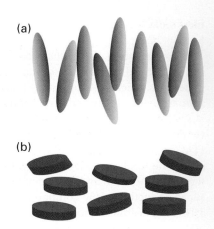

(a)

(b)

Fig. 23.72 Schematic diagram of liquid crystalline materials based on (a) calamitic (rod-like) or (b) discotic (disc-like) molecules.

The molecules that form liquid crystalline materials are generally **calamitic** (rod-like) or **discotic** (disc-like), and these shapes lead to the ordered liquid-type structures in which the molecules align in a particular direction (Fig. 23.72). Although most liquid crystalline materials are totally organic there is a growing number of inorganic liquid crystals based on the coordination compounds of metals and on organometallic compounds. These metal-containing liquid crystals show similar properties to the purely organic systems but offer additional properties associated with a *d*-metal centre, such as redox and magnetic effects.

As the requirement for liquid crystalline behaviour is a rod- or disc-shaped molecule, many of the metal-containing systems are based around the low coordination geometries of the later *d* metals, particularly the square-planar complexes found for Groups 10 and 11. Thus the β-diketone complex (**9**) has a square-planar Cu^{2+} ion coordinated to four O atoms from two β-diketones with long pendant alkyl groups. This copper(II) material is paramagnetic but also forms a **nematic phase** in which the rod-shaped molecules align predominantly in one direction (see Fig. 23.72).

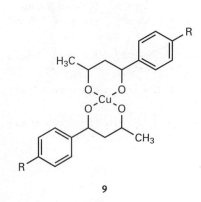

9

FURTHER READING

A.R. West, *Basic solid state chemistry*, 2nd edn. Wiley, New York (1999). A good comprehensive guide to the fundamentals of the solid state from an inorganic chemist's perspective.

A.K. Cheetham and P. Day (ed.), *Solid state chemistry: compounds*. Oxford University Press (1992). A useful collection of chapters covering the key compound types in materials chemistry.

R.M. Hazen, *The breakthrough: the race for the superconductor*. Summit Books, New York (1988). A readable narrative of the discovery of high-temperature superconductors.

R.C. Mehrotra, Present status and future potential of the sol–gel process. *Struct. Bonding*, 1992, **77**, 1. A good review of sol–gel chemistry.

A.K. Cheetham, G. Férey, and T. Loiseau, Open-framework inorganic materials. *Angew. Chem., Int. Ed. Engl.*, 1999, **38**, 3268. An excellent review of progress in, and the structural chemistry of, framework solids.

R.L. Carlin, *Magnetochemistry*. Springer-Verlag, Berlin (1986). A more advanced text covering the magnetic behaviour of solids.

D.W. Bruce and D. O'Hare, *Inorganic materials*. Wiley, New York (1992). Collection of reviews on various topics of materials chemistry, highlighting molecular systems.

M.T. Weller, *Inorganic materials chemistry*. Oxford Chemistry Primers 23. Oxford University Press (1994). Introductory text covering some aspects of solid-state chemistry and materials characterization.

C.N.R. Rao and J. Golalakrishnan, *New directions in solid state chemistry*. Cambridge University Press (1997). Review of current hot topics and research in solid-state chemistry.

L.V. Interrante, L.A. Casper, and A.B. Ellis (ed.), *Materials chemistry: an emerging discipline*. Advances in Chemistry Series no. 245. American Chemical Society, Washington, DC (1995). A series of chapters on a broad range of inorganic and organic solids.

S.E. Dann, *Reactions and characterization of solids*. Royal Society of Chemistry, Cambridge (2000). A good introductory text on solid-state chemistry.

A.F. Wells, *Structural inorganic chemistry*, 5th edn. Oxford University Press (1985). A comprehensive and systematic volume on structural solid-state chemistry.

U. Müller, *Inorganic structural chemistry*. Wiley, New York (1993). A useful text on structural solid-state chemistry with numerous illustrations.

EXERCISES

23.1 From each of the pairs of compounds (a) NaCl and NiO, (b) CaF_2 and PbF_2, and (c) Al_2O_3 and Fe_2O_3, pick the one most likely to have a high concentration of defects. State the nature of the possible defects.

23.2 Bearing in mind that the redox chemistry of metal oxides can often be correlated with the redox behaviour of the ions in solution, explain which of FeO, Fe_2O_3, and NiO is (a) the most difficult to oxidize with air, (b) the easiest to reduce.

23.3 When NiO is doped with small quantities of Li_2O the electronic conductivity of the solid increases. Provide a plausible chemical explanation for this observation. (*Hint:* Li^+ occurs on Ni^{2+} sites.)

23.4 Is a crystallographic shear plane a defect, a way of describing a new structure, or both? Explain your answer.

23.5 Sketch the rock-salt crystal structure and locate in the sketch the sites the interstitial sites to which a cation might migrate. What is the nature of the bottleneck between the normal and interstitial sites?

23.6 How might you experimentally distinguish the existence of a solid solution from a series of crystallographic shear plane structures for a material that appears to have variable composition?

23.7 Outline how you could prepare samples of (a) $SrTiO_3$, (b) Sr_2FeO_3Cl, and (c) TiN.

23.8 Identify the likely products of the reactions
(a) $Li_2CO_3 + NiO \xrightarrow{800°C, O_2}$?
(b) $2 Sr(OH)_2 + WO_3 + MnO \xrightarrow{900°C, O_2}$?

23.9 Draw one unit cell in the ReO_3 structure showing the M and O atoms. Does this structure appear to be sufficiently open to undergo Na^+ ion intercalation? If so, where might the Na^+ ions reside?

23.10 Magnetic measurements on the ferrite $CoFe_2O_4$ indicate 3.4 spins per formula unit. Suggest a distribution of cations between octahedral and tetrahedral sites that would satisfy this observation.

23.11 Given the ligand-field Δ_O and Δ_T values of 1400 and 620 cm^{-1}, respectively, for Fe^{3+} and 860 and 360 cm^{-1}, respectively, for Ni^{2+}, determine the site preference for $A = Ni^{2+}$ and $B = Fe^{3+}$ in normal compared with inverse spinel, assuming that ligand-field stabilization is dominant.

23.12 Which of the materials (a) $YBa_2Cu_4O_8$, (b) $Ca_{1.8}Na_{0.2}CuO_2Cl_2$, (c) $Gd_2Ba_2Ti_2Cu_2O_{11}$, and (d) $SrCuO_{2.12}$, which all contain layers of the stoichiometry CuO_2, might be expected to be high-temperature superconductors?

23.13 Contrast the mechanism of interaction between unpaired spins in a ferromagnetic material to that for an antiferromagnet.

23.14 Classify the oxides (a) BeO, (b) TiO_2, (c) La_2O_3, (d) B_2O_3, and (e) GeO_2 into glass forming and nonglass forming.

23.15 Write possible formulas for a sulfide and a fluoride that might adopt the spinel structure.

23.16 Describe two methods that could be used to prepare the intercalation compound $LiTaS_2$.

23.17 Which of the elements Be, Ca, Ga, Zn, P, and Cl might form structures in which the element is incorporated into a framework as an oxotetrahedral species?

23.18 Propose formulas for structures that would be isomorphous with SiO_2 and zeolites of the same stoichiometry involving Al, P, B, and Zn replacing Si.

23.19 Egyptian blue, $CaCuSi_4O_{10}$, is pale blue and the spinel $CuAl_2O_4$ is an intense blue–green. Explain the difference.

23.20 Place the semiconductors AlP, BN, InSb, and C(diamond) in order of the expected band gap.

23.21 Describe the structures of Na_2C_{60} and Na_3C_{60} in terms of hole filling in a close-packed array of fulleride molecular ions.

PROBLEMS

23.1 Describe the nature of Frenkel and Schottky defects, and some experimental tests for their occurrence. Does either of these defects alone give rise to nonstoichiometry?

23.2 TiO adopts the rock-salt structure, but it contains a high concentration of Schottky defects at both the cation and anion sites. LiCl also has the same structure, but the concentration of defects is very low. Describe the probable influence of redox behaviour of the cations on this difference in defect density.

23.3 Fe_xO generally has $x < 1$; describe the probable metal ion defect that leads to x being less than 1.

23.4 Ag_2HgI_4 is a good electrical conductor above 50°C. Speculate on why Ag^+ rather than Hg^{2+} is the mobile species.

23.5 Solid PbF_2 and ZrO_2 at elevated temperatures are good anion conductors. Describe their crystal structure and why it may be conducive to ion transport.

23.6 To obtain high oxidation states for d metals in complex oxides, compounds are normally prepared at as low temperature as possible commensurate with the reaction. Discuss the thermodynamic reasons for the use of these conditions and explain why the optimum temperature for producing YBCO involves a final annealing stage at about 450°C.

23.7 Heating a complex oxide under ammonia can lead to nitridation or reduction. Describe possible products that might be formed when Sr_2WO_4 is heated in ammonia.

23.8 Describe the issue of occupation factors in Fe_3O_4, Cr_3O_4, and Mn_3O_4 and how the occupation factor is related to ligand-field stablization factors.

23.9 The perovskites, such as ABO_3, often have the dipositive ion in the A site and an ion of higher oxidation state in the B site. What are the factors that lead to this site preference?

23.10 Superconductors are often classified as type I or II. Describe the physical characteristic that determines the classification of a superconductor into one or the other of these two types.

23.11 The superconducting compound $YBa_2Cu_3O_7$ is described as having a perovskite-like structure. Ignoring problems of charge on the ions, describe the difference between this structure and that of perovskite.

23.12 State Zachariasen's two generalizations that favour glass formation and apply them to the observation that cooling molten CaF_2 leads to a crystalline solid whereas cooling molten SiO_2 at a similar rate produces a glass.

23.13 Starting with $TiCl_4$ and $AlCl_3$, give a plausible procedure for the formation of a ceramic containing polycrystalline Al_2O_3 and TiO_2.

23.14 Reaction of ZrS_2 (*c* lattice parameter 583 pm) with $Co(\eta^5\text{-}C_5H_5)_2$ gives a compound with a *c* lattice parameter of 1164 pm and reaction with $Co(\eta^5\text{-}C_5Me_5)_2$ gives a product with a corresponding lattice parameter of 1161 pm: ZrS_2 does not react with $Fe(\eta^5\text{-}C_5H_5)_2$. Explain these observations.

23.15 Describe the interactions between Mo_6S_8 units in a Chevrel phase.

23.16 Discuss differences in the properties of zeolites, zeotypes (frameworks built from oxotetrahedral species other than AlO_4 and SiO_4), and metal organic frameworks.

23.17 What are the properties of an ideal inorganic pigment?

23.18 Compare and contrast the chemistries of graphite and C_{60} with respect to their compounds in association with the alkali metals.

23.19 'The increased reactivity that allows the design and synthesis of materials based on molecular units also means that these compounds are unsuitable for many applications that currently use inorganic materials.' Discuss this remark.

Nanomaterials, nanoscience, and nanotechnology

24

Inorganic chemistry often focuses on entities with atomic and molecular dimensions ranging from 0.1 to 10 nm, whereas solid-state chemistry and materials chemistry have traditionally been concerned with solid materials with dimensions greater than 100 nm. There is currently a great deal of interest in materials that exist between these two scales, because they can exhibit unique properties. This chapter focuses on materials that have a critical dimension between 1 and 100 nm. It introduces the fundamental physical and chemical principles that illustrate why these nanomaterials have generated such intense and broad interest and describes the technology used to generate and exploit them. In the first part of this chapter we introduce nanomaterials, nanoscience, and nanotechnology from a historical perspective, together with definitions and examples. In the second part we discuss the advances in characterization and fabrication methods that allow us to make high-quality nanomaterials. The bulk of the chapter is the third part, where we use examples of nanomaterials to introduce how chemical principles and materials properties are different or superior in the nanoscale regime.

All materials exhibit a hierarchy of structural levels. Chemists have long been familiar with the atomic, molecular, unit-cell, and electronic structural levels because atom-scale interactions play a primary role in many of the bulk properties exhibited by real materials. Moreover, deviations on an atomic level from the basic structural motif are of great interest because they can affect the bulk properties of materials, as we have seen in connection with the electrical conductivities of semiconductors and solids that contain point defects (Chapter 23).

Materials scientists have long been interested in the structure of materials from the micro (10^{-6} m, 1 µm, 'micron') to macro (bulk) scales, as well as in the properties of materials that are controlled by structures that exist on and interact over these scales. They have studied how the physical properties at these scales deviate from those predicted by extrapolation from longer scales and deviations from the basic structural motif, such as the presence of line, plane, and volume defects. For example, crystals often contain an enormous number of regions where one part of the crystal is translated relative to the other part. These dislocations affect strength, crystal growth, electrical conductivity, and other physical properties.

This chapter is concerned with **nanomaterials**, which are materials that have some critical dimension between 1 and 100 nm. Over this intermediate scale, effects from the sub-nanometre- and micrometre-length scales can play equal roles and quantum effects can intervene to give rise to fascinating properties.

Fundamentals

It will prove helpful to understand nanomaterials from historical, empirical, and chemical perspectives.

24.1 Terminology and history

Key points: A nanomaterial is any material that has a critical dimension on the scale of 1 to 100 nm; a more exclusive definition is that a nanomaterial is a substance that exhibits properties absent in both the molecular and bulk solid state on account of it having a critical dimension in this range.

The prefix 'nano' will occur in many combinations throughout this chapter. **Nanoscience** is the study of the properties of matter that have length scales between 1 and 100 nm. **Nanotechnology** is the collection of procedures for manipulating matter on this scale in order to build nanosized entities for useful purposes. Some definitions, however, are more restrictive. Thus, a 'nanomaterial' is taken to be a solid material that exists over the scale of 1 to 100 nm *and* exhibits novel properties that are related to its scale. Likewise, 'nanoscience' is also sometimes restricted to the study of *new* effects that arise *only* in materials that exist on the nanoscale; and 'nanotechnology' is similarly restricted to the procedures for creating *new* functionalities that are possible *only* by manipulating matter on the nanometre scale.

The original version of nanotechnology occurred in nature, where organisms developed an ability to manipulate light and matter on an atomic scale to build devices that perform specific functions, such as storing information, reproducing themselves, and moving about. In this sense, DNA is the ultimate nanomaterial, as it stores information as the sequence of base pairs that are spaced about 0.3 nm apart. Folded DNA molecules have an information density of more than about 1 Tb cm^{-2} (1 Tb $= 10^{12}$ bits), which is much greater than achieved in most current data storage systems. Photosynthesis is another example of biological nanotechnology in which nanostructures are exploited to absorb light, separate electric charge, shuttle protons around, and ultimately convert solar energy into biologically useful chemical energy. Our quest to harness solar energy has also turned to the nanoscale, where nanocatalysts are improving our ability to convert light to electrical energy by using photovoltaic materials.

Humans have practised nanotechnology for centuries, although it was more of an art than a science or engineering discipline. For example, gold and silver salts have long been used to colour glass: gold has been used to produce red stained glass and silver used to produce yellow. In stained glass, the metal atoms form nanoparticles (known previously as 'colloidal particles') with optical properties that depend strongly on their size. Metallic nanopigments are now becoming a focal point of biomedical nanotechnology because they can be used to tag DNA and other nanoparticles. Other traditional examples of nanotechnology include the photosensitive nanosized particles in silver halide emulsions used in photography and the nanosized carbon granules in the 'carbon black' used for reinforcing tyres and in printer's ink.

The science and engineering of nanotechnology began to take shape in the latter half of the twentieth century. One significant leap forward in nanotechnology occurred when Gerd Binnig and Heinrich Rohrer developed the scanning tunnelling microscope, which was the forerunner to all other forms of scanning probe microscopy (Section 24.3). Later, a scanning probe tip was used to rearrange atoms on a surface to spell out words, so demonstrating an ability to manipulate and characterize nanoscale structures. Work in the areas of nanoscience and nanotechnology has thus been multidisciplinary as well as diverse in scope. This chapter aims to introduce this broad field, highlighting the importance of inorganic chemistry in the multidisciplinary research directed towards understanding nanomaterials.

24.2 Novel optical properties of nanomaterials

Confinement effects lead to some of the most fundamental manifestations of nanoscale phenomena in materials and are frequently used as a point of departure for the study of nanoscience. Novel optical properties appear in nanoparticles as a result of such effects and are being exploited for information, biological, sensing, and energy technologies.

(a) Semiconducting nanoparticles

Key points: The colour of quantum dots is dictated by quantum confinement phenomena and particle localization; quantization of the HOMO–LUMO bands leads to novel optical effects involving interband and intraband transitions.

Semiconducting nanoparticles have been investigated intensively for their optical properties. These particles are often called **quantum dots**, because quantum effects become important in these three-dimensionally confined particles (dots). Two important effects occur in semiconductors when electrons are confined to tiny regions. First, the HOMO–LUMO energy gap increases from the value observed in bulk crystals. Second, the energy levels of electrons in the LUMOs (and holes—the absence of electrons—in the HOMOs) are quantized, like those of a particle in a box. Both effects play an important role in determining the optical properties of quantum dots.

Quantum confinement, the trapping of electrons and holes in tiny regions, provides a method of tailoring or engineering the band gap of materials. Transitions of electrons between states in the valence band (the so-called HOMO states) and the conduction band (the LUMO states) are called **interband transitions**, and the minimum energy for these transitions is increased in quantum dots relative to those in bulk semiconductors. Interband transitions control the luminescent properties of quantum dots, and because the band gap is a function of the size of the dots, it is possible to tailor their luminescence simply by changing their size. In fact, the band gap can be tailored through the entire visible spectrum. One of the exciting uses of quantum dots is as chromophores for **biotags**, where dots of different sizes are functionalized to detect different biological analytes. It turns out (as we shall see) that, although they emit at specific tunable wavelengths, quantum dots exhibit broadband absorption for energies above the band gap. The intriguing point for bioapplications is that it is possible to excite an array of distinct quantum dot chromophores with a single broadband excitation and to detect simultaneously multiple analytes by their distinct optical emissions.

The second manifestation of quantum confinement is that the available quantized energy levels inside a quantum dot have no net linear momentum (they are standing waves) and therefore transitions between them do not require any momentum transfer. As a result, the transition probabilities between any two states are high. This lack of momentum dependence also explains the broadband absorption nature of quantum dots, because probabilities are high for most transitions from the occupied valence band states to unoccupied conduction band states. Probabilities are also high for **intraband transitions**, transitions of electrons between states in the LUMO band or of holes in the HOMO band. These relatively intense intraband transitions are typically in the infrared region of the spectrum and are currently being exploited to make devices such as infrared photodetectors, sensors, and lasers.

(b) Metallic nanoparticles

Key points: The colours of metallic nanoparticles dispersed in a dielectric medium are dominated by localized surface plasmon absorption, the collective oscillation of electrons at the metal–dielectric interface.

The optical properties of metallic nanoparticles arise from a complex electrodynamic effect that is strongly influenced by the surrounding dielectric medium. Light impinging on metallic particles causes optical excitations of their electrons. The principal type of optical excitation that occurs is the collective oscillation of electrons in the valence band of the metal. Such coherent oscillations occur at the interface of a metal with a dielectric medium and are called **surface plasmons**.

In bulk particles, the surface plasmons are travelling waves and are characterized by a linear momentum. To excite plasmons using photons in bulk metals, the momenta of the plasmon and the photon must match. This matching is possible only for very specific geometries of the interaction between light and matter and is a weak contributor to the optical properties of the metal. In nanoparticles, however, the surface plasmons are localized and have no characteristic momentum. As a result, the momenta of the plasmon and the photon do not need to match and plasmon excitation occurs with a greater intensity. The peak intensity of the surface plasmon absorption for gold and silver occurs in the optical region of the spectrum, and so these metallic nanoparticles are useful as pigments.

The characteristics of plasmon absorption depend strongly on the metal and the dielectric surroundings as well as on the size and shape of the nanoparticle. To control the dielectric surroundings, so-called core–shell composite nanoparticles have been designed in which metallic shells of nanometre-scale thickness encapsulate a dielectric nanoparticle. Metallic nanoparticles and metallic nanoshells are used as dielectric sensors because their optical properties change when they come into contact with different dielectric materials. In particular, biological sensing is of interest because biological analytes can bind to the surface of the nanoparticle, causing a detectable shift in the plasmon absorption band.

Characterization and fabrication

The burgeoning activities during the late twentieth century in the areas of nanomaterials, nanoscience, and nanotechnology were intimately tied to advances in characterization and fabrication. This section introduces the basic concepts of nanoscale characterization and fabrication methods that allow us to design nanomaterials and develop nanotechnologies.

24.3 Characterization methods

The great advances made in nanosciences and nanotechnology would not have occurred without the ability to characterize the nanoscale structural, chemical, and physical properties of materials. Moreover, the direct observation of nanostructure allows meaningful relationships between processing and properties to be made.

(a) Scanning probe microscopy

Key points: Scanning tunnelling microscopy uses tunnelling current from a sharp tip to image and characterize a surface; atomic force microscopy responds to intermolecular forces to image the surface.

Scanning tunnelling microscopy (STM) was the first of a series of **scanning probe microscopies**, which are techniques that allow three-dimensional imaging of the surface of materials by using sharp and sensitive physical probes. All these techniques use a sharply pointed probe brought into close proximity (or in contact) with the specimen and construct an image by scanning the probe over the surface of the specimen and monitoring the spatial variation in the value of a physical parameter, such as potential difference, electric current, magnetic field, and mechanical force. The best known examples of scanning probe microscopies are STM itself and **atomic force microscopy** (AFM).

In STM, an atomically sharp conductive tip is scanned at about 0.3–10 nm above the surface of the sample. Electrons in the conductive tip, which is held either at a constant potential or at a constant height above the sample, can tunnel through the gap with a probability that is related exponentially to the distance from the surface. As a result, the tunnelling electron current represents the distance between the sample and the tip. If the tip is scanned over the surface at a constant height, then the current represents the change in the topography of the surface. The movement of the tip at a constant height with great precision is accomplished by using piezoelectric ceramics for the displacements of the tip in different directions.

Typically, STM probes are manufactured to have only a few atoms at their tip. The simplest method of creating such sharp tips is through electrolytic or electrochemical etching. In one technique, a thin tungsten wire is placed in a solution of potassium hydroxide and an electric current is passed through it.

In AFM, the atoms at the tip of the probe interact with the surface atoms of the sample through intermolecular forces (such as van der Waals interactions). The cantilever holding the probe bends up and down in response to the forces and the extent of deflection is monitored with a reflected laser beam. Variations on AFM include the **frictional force microscope**, which measures variations in the lateral forces on the tip that are based on chemical variations on the surface, the **magnetic force microscope**, which uses a magnetic tip to image magnetic structure, the **electrostatic force microscope**, which uses tips that can sense electric fields, and the **scanning capacitance microscope**, in which the tip is used as an electrode in a capacitor. **Molecular recognition AFM** is also now being carried out in which the tip is functionalized with specific ligands and the interaction between the functionalized tip and surface is measured. Such microscopes can provide nanometre resolution of the chemical properties on the surface. AFM can also act as a patterning tool. In **dip-pen nanolithography** (DPN, Fig. 24.1) the tip of an AFM is used as an ink pen. By using an ink containing molecular entities, self-assembled monolayers can be formed through molecular transport of the nanoink to the solid substrate surface.

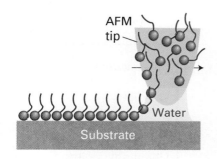

Fig. 24.1 The process of dip-pen nanolithography. (Adopted from C. Mirkin.)

Scanning near-field optical microscopy (SNFOM) combines the local interaction of a scanning probe and a specimen with well-established methods of optical spectroscopy. Conventional optical spectroscopy is limited in resolution by the wavelength of the light source, but when the light source is brought to within a few atomic diameters of the surface and a nanometre-scale aperture is used, the local confinement of the electromagnetic field allows for much improved spatial resolution (of about 20 nm) to investigate local chemical properties of nanoscale structures.

(b) Electron microscopy techniques

Key point: Transmission and scanning electron microscopes use electrons to image the sample in a similar fashion to optical microscopes.

An imaging electron microscope operates like a conventional optical microscope but instead of imaging photons, as in the visible microscope, electron microscopes use electrons. In these instruments, electron beams are accelerated through 1–200 kV and electric and magnetic fields are used to focus the electrons. In **transmission electron microscopy** (TEM) the electron beam passes through the thin sample being examined and is imaged on a phosphorescent screen. In **scanning electron microscopy** (SEM) the beam is scanned over the object and the reflected (scattered) beam is then imaged by the detector. The ultimate resolution of SEM depends on how sharply the incident beam is focused on the sample, how it is moved over the sample, and how much the beam spreads out into the sample before reflecting. In both microscopes, the electron probes cause the production of X-rays with energies characteristic of the elemental composition of the material. As such, **energy-dispersive spectroscopy** of these characteristic X-rays is used to quantify the chemical make-up of materials by using electron microscopes.

The primary advantage of SEM over TEM is that it can form images of electron-opaque samples without the need for carrying out difficult specimen preparations. It is therefore the electron microscopy method of choice for the straightforward characterization of materials. However, SEM samples need to be conductive because otherwise electrons can collect on the sample and interact with the electron beam itself, resulting in blurring. Nonconductive samples must therefore be coated with a thin layer of metal, usually gold or aluminium. TEM is often used for imaging electron-transparent biological samples and, because it offers atomic resolution, for high-resolution materials studies. Great care and considerable time and effort are required in preparing high-resolution specimens, which limits the number of specimens that can be observed in TEM.

24.4 Top-down and bottom-up fabrication

Key points: Top-down fabrication methods carve out or add on nanoscale features to a bulk material by using physical methods; bottom-up fabrication methods assemble atoms or molecules in a controlled manner to build nanomaterials piece-by-piece.

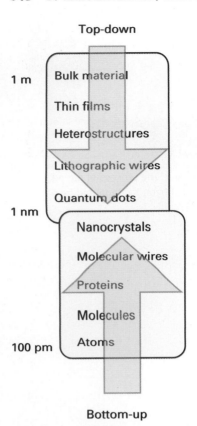

Top-down

1 m

Bulk material

Thin films

Heterostructures

Lithographic wires

1 nm

Quantum dots

Nanocrystals

Molecular wires

Proteins

Molecules

100 pm

Atoms

Bottom-up

Fig. 24.2 The two techniques for making nanoscale structures. The top-down technique starts with larger objects that are whittled down into nanoscale objects; the bottom-up technique starts with smaller objects that are combined into nanoscale objects.

There are two basic techniques for the fabrication of nanoscale entities. The first is to take a macroscale (or microscale) object and to carve out nanoscale patterns. Methods of this sort are called **top-down approaches**. In top-down approaches, patterns are first designed on a large scale, their lateral dimensions are reduced, and then used to transfer the nanoscaled features into or onto the bulk material. Physical interactions are used in top-down fabrication approaches, such as lithography, mechanical stamping, or nanoscale printing. The most common and well-known approach is photolithography, the technique used to fabricate very large-scale integrated circuits having feature dimensions on the 100 nm scale (Fig. 24.2). The second technique is to build larger objects by controlling the arrangement of their component smaller-scale objects. Methods of this sort are called **bottom-up approaches** and start with control over the arrangements of atoms and molecules. This chapter emphasizes the bottom-up approach to nanoscale fabrication because of its focus on the interactions of atoms and molecules and their arrangement into larger functional structures.

The two basic approaches most widely used to prepare nanomaterials, solution methods and vapour-phase methods, are bottom-up methods, because control over the arrangement of individual atoms is exerted to achieve larger-scale structures.

24.5 Solution-based synthesis of nanoparticles

Key points: Solution-based synthetic methods are the main techniques for nanoparticle synthesis because they have atomically mixed and highly mobile reagents, allow for the incorporation of stabilizing molecules, and have been widely successful in practice; the two stages of crystallization from solution are nucleation and growth.

As discussed in Chapter 23, the two basic techniques to generate inorganic solids are direct-combination methods and solution-based methods. The former does not lend itself well to the synthesis of nanoparticles because the reactants tend to be larger than nanoparticles and therefore require long times to reach equilibrium. Moreover, the use of elevated temperatures leads to particle growth and coarsening during the reaction period, resulting in large crystallite sizes. There are, however, some examples of using mechanical ball-milling at low temperature to break macroscopic powders into nanoparticles. However, solution-based methods permit excellent control over the crystallization of inorganic materials and are widely used. The techniques used to make nanoparticles are similar to those described in Chapter 23 to make solid-state compounds. Special care must be taken to control the size and shape of the particles during nanoparticle synthesis, as well as their uniformity in size and shape. By fine-tuning the crystallization process from solution, highly monodisperse, uniformly shaped nanoparticles of a wide range of compositions can be prepared from combinations of elements from throughout the periodic table.

Because the reactants in solution-based methods are mixed on an atomic scale and solvated in a liquid medium, diffusion is fast and diffusion distances are typically small. Therefore, reactions can be carried out at low temperature, which minimizes the thermally driven particle growth that is problematic in direct combination methods. Although the specifics of each reaction differ greatly, the basic stages in solution chemistry are:

1 Solvate the reactant species and additives.

2 Form stable solid nuclei from solution.

3 Grow the solid particles by addition of material until the reactant species are consumed.

Example 24.1 Controlling nucleation during nanoparticle synthesis

Use thermodynamic arguments related to the crystallization of a phase to describe how to generate a short burst of homogeneous nucleation in a solution.

Answer The formation of stable solid nuclei is controlled through the thermodynamic driving force of the reaction, which can be controlled by adjusting the composition of the different solution species and the temperature (as well as the pressure for hydrothermal techniques). As with most chemical reactions, there is an activation barrier to the crystallization reaction, which prevents the instantaneous formation of many stable nuclei. To overcome this barrier in a short single burst, we move the solution quickly to a highly nonequilibrium condition in which the driving force for nucleation becomes very large and the barrier becomes small; widespread homogeneous nucleation of many particles then occurs. This nucleation quickly moves the system back towards equilibrium and suppresses further nucleation.

Self-test 24.1 Use a similar argument to describe the synthesis of core–shell nanoparticles.

The basic aim in solution synthesis is to generate in a controlled manner the simultaneous formation of large numbers of stable nuclei that undergo little further growth. If growth is to occur, it should occur independently of the nucleation step, because then all particles have a chance to grow to similar sizes. If performed successfully, the particles will be monodisperse and in the nanometre range. The drawback to the solution method is that the particles can undergo **Ostwald ripening**, in which smaller particles in the distribution redissolve and their solvated species reprecipitate onto larger particles, so increasing the size distribution and decreasing the total particle count. To prevent this unwanted ripening, **stabilizers**, surfactant molecules that help to stabilize the particles against growth and dissolution, are added.

As there are many methods to synthesize nanoparticles, we limit this discussion to a few well-known examples. In 1857, Michael Faraday found that reduction of an aqueous solution of $[AuCl_4]^-$ with phosphorus in CS_2 produced a deep-red suspension that contained nanoparticles of gold. Because sulfur forms bonds to gold, sulfur-containing species are good stabilizing agents and the most widely used stabilizer molecules for gold nanoparticles contain a thiol group (—SH). An approach has been developed in the same spirit as Faraday's to control the size and dispersity of gold nanoparticles by using $[AuCl_4]^-$ and thiol stabilizers. The reaction is relatively simple and the procedure produces air-stable gold nanoparticles with diameters between 1.5 and 5.2 nm with narrow dispersion in the size. In this **Brust–Schiffrin approach**, $[AuCl_4]^-$ is first transferred from water to methylbenzene (toluene) by using tetraoctylammonium bromide as a phase-transfer agent. The methylbenzene contains dodecanethiol as a stabilizer and, after transfer, $NaBH_4$ is used as a reducing agent to precipitate Au nanoparticles with dodecanethiol surface groups:

Transfer:

$$[AuCl_4]^-(aq) + N(C_8H_{17})_4^+(sol) \rightarrow N(C_8H_{17})_4^+ + AuCl_4^-(sol)$$

Precipitation (of the $[Au_m(C_{12}H_{25}SH)_n]$ nanoparticle):

$$m\,[AuCl_4]^-(sol) + n\,C_{12}H_{25}SH(sol) + 3m\,e^- \rightarrow 4m\,Cl^-(sol)$$
$$+ [Au_m(C_{12}H_{25}SH)_n](sol)$$

where sol is methylbenzene. The ratio of stabilizer ($C_{12}H_{25}SH$) to metal (Au) controls the particle size in the sense that higher stabilizer:metal ratios lead to the smaller metal core sizes. By adding the $NaBH_4$ reductant quickly and cooling the system quickly after the reaction terminates, smaller and more monodisperse nanoparticles are formed. The rapid addition of reductant improves the probability of simultaneous formation of all nuclei. By cooling the solution quickly, both post-nucleation growth and dissolution of particles are minimized. Similar approaches can be used for other metal nanoparticles.

Quantum dots of materials such as GaN, GaP, GaAs, InP, InAs, ZnO, ZnS, ZnSe, CdS, and CdSe have been investigated for their optical properties (Section 24.2), because their interband absorption and fluorescence occur in the visible spectrum. As an early example of their preparation, dimethylcadmium is dissolved in mixture of trioctylphosphine (TOP) and trioctylphosphine oxide (TOPO) and the selenium source, which is often selenium dissolved in TOP or TOPO, is added at room temperature. The solution is injected into a reactant vessel containing vigorously stirred, hot TOPO, which permits widespread nucleation of TOPO-stabilized CdSe quantum dots. The addition of the room-temperature liquid lowers the temperature of the solution and prevents further nucleation or growth (because the activation barriers are so high). The solution is then reheated to a temperature that allows for slow growth but no further nucleation. This step leads to nanoparticles with narrow size distributions and sizes in the range 2–12 nm. Alternative, less hazardous synthetic methods have been developed and are being pursued.

Cadmium sulfide can be grown in pH-controlled aqueous solutions of Cd(II) salts with polyphosphate stabilizers by the addition of a sulfur source. For example, at pH = 10.3 the addition of Na_2S causes the precipitation of CdS nanoparticles from aqueous solutions containing $Cd(NO_3)_2$ and sodium polyphosphate. These quantum dots range in size from 1 to 10 nm, and the size is controllable through the reactant concentrations and the rate of addition of the reactant.

Oxide nanoparticles can also be grown by solution methods. Many applications use colloidal particles of oxides such as SiO_2 and TiO_2 for food, ink, paints, coatings, and so on. Many of the efforts to achieve controlled oxide nanoparticle growth stem from earlier work in traditional ceramic and colloidal applications, where particle sizes from 1 nm to 1 μm are used. Silica, SiO_2, and titania (titanium dioxide), TiO_2, are probably the best known oxides grown from solution. Typical schemes involve the controlled hydrolysis of metal alkoxides. All successful hydrolysis reactions of metal alkoxides aim to follow the same basic rules as described above: the controlled nucleation stage and slow growth stage are performed independently. In all cases, strict monitoring of the pH, precursor chemistry, reactant concentration, rate of addition of reactant, and temperature is required to control the final size and shape of the particles.

An important example of nanoparticle oxide use is in the photoelectrochemical solar cell known as the **Graetzel cell**. This cell uses nanocrystalline TiO_2 as a medium to transfer electrons from an organoruthenium dye to a conductor. Nucleation occurs in the hydrolysis of titanium isopropoxide that is added dropwise into vigorously stirred 0.1 M nitric acid. The filtered nanoparticles are then placed in an autoclave and allowed to grow by the hydrothermal addition of material. The size, shape, and state of agglomeration are controlled by adjusting the conditions of either the nucleation or the growth stage.

24.6 Vapour-phase synthesis of nanoparticles

Key points: Vapour-phase synthetic methods are alternative techniques for nanoparticle synthesis because they have atomically mixed and highly mobile reagents, can be controlled by varying the conditions, and have been widely successful in practice.

The same fundamentals concerning nucleation and growth that are relevant to solution synthesis apply to vapour-phase synthesis. The vapour needs to be supersaturated to the point where a high density of homogeneous nucleation events produce solid particles in one short burst, and growth must be limited and controlled in a subsequent step, if it is to occur at all. Commercially, vapour-phase synthesis is carried out to produce nanoscale carbon black and fumed silica in large quantities. Metals, oxides, nitrides, carbides, and chalcogenides can also easily be formed by using vapour-phase techniques.

There are, however, several differences between vapour-phase and solution-based techniques. In the latter, stabilizers can be added in a straightforward and controllable fashion, and particles remain dispersed and independent of one another. In vapour-phase techniques, however, surfactants or stabilizers are not easily added (although solid-state stabilizers are available). Without surface stabilizers, nanoparticles tend to agglomerate into larger particles. If temperatures during the process are high enough, the particles sinter and coalesce into one larger particle. Although the 'soft agglomerates' formed under the influence of weak intermolecular forces can usually be redispersed, the 'hard agglomerates' (partially sintered-formed at intermediate temperature) and coalesced particles (formed at high temperatures) cannot be. Therefore, techniques that lead to soft agglomerates are generally best suited to nanoparticle production. Because of these differences, the size dispersion of nanoparticles tends to be better for solution-based techniques than for vapour-phase techniques.

Vapour-phase techniques are attractive synthesis methods for particles when continuous operation is required, or when solution methods do not produce good quality nanoparticles. They can also produce complex materials in a straightforward manner, such as doped binary and ternary compounds or composite particles. Vapour-phase techniques are classified by the physical state of the precursor used as a reagent and by the reaction method, such as plasma synthesis or flame pyrolysis. In each case, the reagent is converted to a supersaturated or superheated vapour that is allowed to react or to cool to force nucleation. Solid reagents are vaporized and then recondensed in gas condensation methods (thermal evaporation to produce vapour), laser ablation, sputtering, and spark discharge methods. Liquid or vapour precursors are used in spray pyrolysis, flame synthesis, laser pyrolysis, plasma synthesis, and chemical vapour deposition. In chemical vapour deposition, the vapour is transported to a substrate where reaction and solid nucleation occurs (Fig. 24.3). In a variation of this procedure, gaseous precursors are delivered into a hot-walled reactor and allowed to react homogeneously in the vapour phase to nucleate solids, and particles that are collected are collected downstream. Particle sizes are controlled by flow rates, precursor chemistry, concentrations, and residence times in the reactor. Examples of materials synthesized by this technique include metals, oxides, nitrides, and carbides such as SiC, SiO_2, Si_3N_4, $SiC_xO_yN_z$, TiO_2, TiN, ZrO_2, and ZrN.

Solution and vapour-phase methods can both be used to make composite nanoparticles, such as core–shell nanocomposites. In both techniques the approach is to grow a second phase on an initial nucleus or nanoparticle. Reaction design to produce core–shell nanoparticles by solution is straightforward provided the solution characteristics of both materials are similar. In practice, however, it is difficult to find materials where the synthesis conditions overlap. Vapour-phase techniques offer another approach to the design of core–shell particles, by injecting a second vapour into a reactor at the growth stage.

Another straightforward approach to vapour-phase nanoparticle production is **plasma synthesis**, which can be used to synthesize elemental solids, alloys, and oxides, as well as core–shell nanoparticles. In this method, gas or solid particles are fed into a plasma where they vaporize and ionize to highly energetic charged species. Inside a plasma, the temperatures can exceed 10 kK. Upon leaving the plasma, the temperature falls rapidly and crystallization occurs under conditions far from equilibrium. The nanoparticles are collected downstream from the plasma zone. Depending on the carrier gas (that is, the oxidation conditions), elemental solids, core–shell, or compound particles can be formed. Figure 24.4 shows two examples of nickel/zinc ferrite particles.

24.7 Synthesis using frameworks, supports, and substrates

Heterogeneous nucleation, nucleation that occurs on an existing surface, can be used to generate important kinds of nanostructures, including zero-, one-, and two-dimensional materials. The methods are similar to those already described: they are either physical or chemical and involve crystallization from either liquids or vapours. The principal difference is that an outside agent is also involved and permits the direct control of the nanoparticle formation. The outside agent can be a framework or a support structure that limits the size of the reaction volume, as described in Section 24.7a for nanoparticle synthesis using inverse micelle frameworks and embossed supports. During thin film growth, the outside agent is a substrate. Because of its importance in nanomaterials synthesis, thin film growth on substrates is described in Sections 24.7b and c.

(a) Nanosized reaction vessels

Key points: By carrying out reactions in nanoscale reaction vessels, the ultimate dimensions of solid products are confined to the vessel size; a reverse micelle has an aqueous core in which reactions can occur.

When the synthesis of a particle is carried out in a nanosized vessel, the ultimate particle size is limited by the vessel size. One popular route is the **inverse micelle synthesis** approach. An inverse micelle consists of a two-phase dispersion of immiscible liquids,

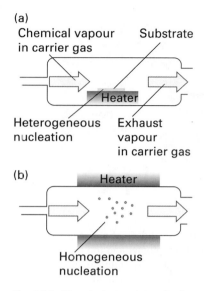

Fig. 24.3 Chemical vapour methods to achieve (a) thin film growth and (b) nanoparticle production.

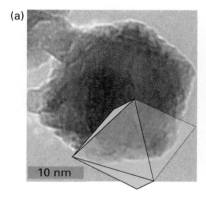

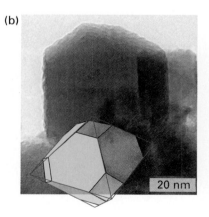

Fig. 24.4 Transmission electron micrograph images of nanoparticles of $Ni_{0.5}Zn_{0.5}Fe_2O_4$ synthesized by plasma torch synthesis. The particle shapes are superimposed: (a) an octahedron and (b) a truncated octahedron. (Based on images provided by M. McHenry and R. Swaminatham.)

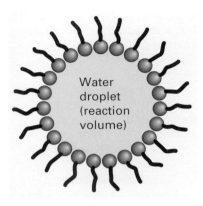

Fig. 24.5 An inverse micelle. The hydrophilic heads of molecules surround the water droplet and the hydrophobic tails contact the nonpolar solvent, which acts as the dispersed medium. Reactants are solvated in the water droplet and then caused to react in these spatially confined vessels.

such as water and a nonpolar oil. By including amphipathic surfactant molecules (molecules having a polar and a nonpolar end), the aqueous phase can be stabilized as dispersed spheres with a size dictated by the water:surfactant ratio (Fig. 24.5). The size of the crystalline particles is limited by the micelle volume, which can be controlled on the nanoscale. Example of nanoparticles formed by this technique are Cu, Fe, Au, Co, CdS, CdSe, ZrO_2, ferrites, and core–shell particles such as Fe/Au.

An alternative approach to nanopatterning is **laser-assisted embossing**, which has been used to generate close-packed arrays of hemispherical nanowells having volumes of the order of zeptolitres ($1 \, zL = 10^{-21} \, L = 10^{-24} \, m^3 = 10^3 \, nm^3$). The volumes of the nanowells are manipulated by changing the intensity of the laser, the pressure between the mould and the substrate, and growing an amorphous oxide on the patterned silicon prior to nanocrystallization. Nanowells with diameters as small as 50 nm have been used as reaction chambers for the preparation of semiconducting nanocrystals. Under most growth conditions, independent nanocrystals are formed in individual nanowells (Fig. 24.6).

Example 24.2 Assessing the role of the volume of nanosized reaction vessels

The use of nanosized reaction vessels combines solution-based synthesis techniques directly with size-limited growth. Assume that a surface is embossed with 100 nm wide hemispherical imprints that are then used as reaction vessels. (a) What is the volume of the reaction vessel? (b) What will be the size of an NaCl particle that is crystallized in this vessel to produce a single spherical particle? Assume a solution of 1.00 M NaCl(aq), and that the crystallization goes to completion. The density of NaCl is 2170 kg m^{-3} and its molar mass is 58.442 g mol^{-1}.

Answer (a) The volume of each hemispherical imprint is half the volume of a sphere of radius 50 nm:

$$V_{solution} = \frac{1}{2}\left(\frac{4}{3}\pi r^3\right) = \frac{2}{3}\pi(5.0 \times 10^{-8} \, m)^3 = 2.62 \times 10^{-22} \, m^3$$

(b) Because the concentration of the solution is 1.00 mol dm^{-3}, the amount of NaCl in the hemispherical imprint is

$$n = (1.00 \times 10^3 \, mol \, m^{-3}) \times (2.62 \times 10^{-22} \, m^3) = 2.62 \times 10^{-19} \, mol$$

The mass of this amount of NaCl is

$$m = (2.62 \times 10^{-19} \, mol) \times (58.442 \, g \, mol^{-1}) = 1.53 \times 10^{-17} g$$

or 1.53×10^{-20} kg of NaCl in the solution. The volume of this NaCl is therefore

$$V_{NaCl} = \frac{1.53 \times 10^{-20} \, kg}{2170 \, kg \, m^{-3}} = 7.06 \times 10^{-24} \, m^3$$

To determine the size of the particle, we use the definition of volume and solve for the diameter (twice the radius):

$$d = 2r = 2\left(\frac{3}{4\pi}V\right)^{1/3} = 2\left(\frac{3}{4\pi} \times 7.06 \times 10^{-24} \, m^3\right)^{1/3} = 2.38 \times 10^{-8} \, m$$

Hence, the diameter of the particle is 23.8 nm.

Self-test 24.2 For the same solution, how small would the hemispherical imprint need to be to make a particle 2 nm in diameter?

As discussed earlier, high-temperature (150–250°C) solution synthesis is one of the most successful methods for achieving size and shape control of nanocrystalline materials through nucleation and growth in the presence of selective surfactant molecules, including TOPO. However, inhomogeneity in the injection of precursors, mixing of the reactants, and temperature gradients in the reaction flask can lead to problems with control over the size distribution of nanoparticles. Small reaction vessels, such as reverse micelles and laser-embossed surfaces, provide an appealing alternative for controlling chemical and thermal homogeneity. The use of nanofabricated structures made from

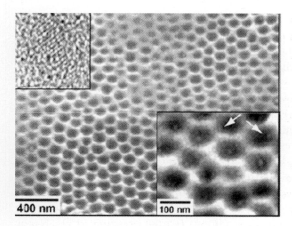

Fig. 24.6 An SEM image of an array of nanoscale wells that contain one CdS nanocrystal per well. The lower inset magnifies a small area of the array; the white arrows indicate crystals formed at the edges of the wells. The upper inset is a TEM image of an individual CdS nanocrystal removed from the array of nanowells by sonication. (J.E. Barton and T.W. Odom, *Nano Lett.*, 2004, **4**, 1525.)

these materials enables the growth and assembly of nanocrystals that are potentially more interesting than inorganic salts and organic molecular crystals. This approach to the growth of nanocrystals has several advantages, including facile syntheses based on bulk reactions, flexible size control achieved by using different concentrations of precursor materials, realization of isolated nanocrystals in ordered arrays, and reusable templates.

(b) Physical vapour deposition

Key points: In the physical vapour deposition methods, a vapour of atoms, ions, or clusters physically adsorb to the surface and combine with other species to create a solid; molecular beam epitaxy is a technique where evaporated species from elemental charges are directed as a beam at a substrate where growth occurs.

In **physical vapour deposition** (PVD) methods, vapours are delivered from their source to a solid substrate on which they crystallize. The arriving gaseous species are typically atoms, ions, or clusters of elements. There are several general forms of PVD that are widely used: **molecular beam epitaxy** (MBE), sputtering, and **pulsed laser deposition** (PLD). The gas-phase species can have either relatively low kinetic energies on arrival (as in MBE) or relatively high kinetic energies (as in sputtering and PLD). The most important feature that all PVD methods have in common is the ability to achieve complex film stoichiometries, either through the transfer of the different species between the source and the film or by using multisource systems to compensate for deviations from the ideal film composition. Because vapour deposition methods allow single atomic layers to be deposited in a controlled fashion on a support or substrate, nanoscale architectures can be built from the bottom-up. In this section we describe some fundamental principles for these processes in systems that allow for real-time, *in situ* control over the nanostructure of materials.

Molecular beam epitaxy is an ultrahigh vacuum technique for growing thin **epitaxial films**, films that have a definite crystallographic relationship with the underlying substrate. In MBE, molecular beams are formed by heating elemental sources until atoms evaporate and are transported ballistically to the substrate surface with relatively low kinetic energies (of the order of 1 eV). The overall pressure is kept very low so that collisions within the beam are avoided. To prevent species from bouncing off other surfaces of the system and back towards the substrate, much of the system is cooled with liquid nitrogen. In that way, stray elemental species adsorb onto surfaces and remain there. This cooling also helps to maintain the vacuum. Film stoichiometry and film growth rates are highly dependent on the beam fluxes, which can be controlled by adjusting the temperatures of the elemental sources.

Homoepitaxy is the epitaxial growth of a thin film of a material on a substrate of the same material. **Heteroepitaxy** is the epitaxial growth of a thin film of a material on a substrate of a different material. Heteroepitaxy introduces a strain between the growing material and the substrate caused by their crystallographic mismatch. This strain can lead to a self-assembly process where nanosized islands form in ordered arrays on the surface to minimize strain energies. By alternating the deposition of two materials by MBE,

(a)

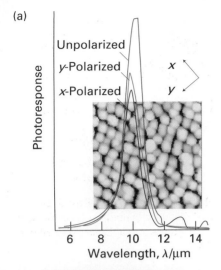

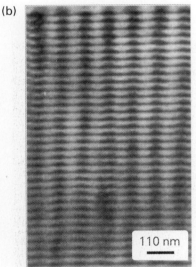

(b)

110 nm

Fig. 24.7 (a) Photoresponse (in arbitrary units) of $(In_{0.2}Ga_{0.8})As$ quantum dots formed using MBE. The IR absorption band arises from intraband absorption. The inset is an AFM image of the quantum dots. The quantum dots vary in lateral dimensions between 15 and 20 nm and in vertical height in the range 3–7 nm. (b) A cross-sectional TEM micrograph of an ordered superlattice of $(In_{0.2}Ga_{0.8})As$ quantum dots in a GaAs matrix. (Courtesy of E. Towe.)

a multilayer structure of ordered quantum dots can be created (Fig. 24.7). By tailoring the structure of the substrate appropriately, nanowires and nanowells can also be created as well as the superlattices and artificially layered crystals we describe later.

Pulsed laser deposition is a versatile PVD technique that can be used to synthesize a wide variety of high-quality thin films (Fig. 24.8). In PLD, a pulsed laser is used to ablate a target, which releases a plume of atomized and ionized particles from its surface that condenses onto a nearby target. The PLD process usually produces films of composition identical to that of the substrate, which is a major simplification compared with techniques that require fine-tuning or expensive control equipment to achieve a specific stoichiometry. This simplicity makes PLD attractive for exploratory materials chemistry, because the cation stoichiometry is controlled externally and independently of the deposition parameters. The technique also allows control over a number of deposition conditions that alter the thermodynamics and kinetics of growth, including substrate temperature, laser energy, and atmospheric conditions (partial pressures of O_2, N_2, or Ar); it can be carried out at pressures at or below 500 mTorr (about 100 Pa). Pulsed laser deposition can achieve average growth rates similar to those of MBE (for example, 100 pm s^{-1}) but the instantaneous growth is much higher because growth occurs only in the short intervals (of about 1 μs) immediately after the 10–20 ns laser pulse ablates the target. Therefore, for a very short period there is a very high arrival rate and a very considerable supersaturation at the substrate surface, leading to very high nucleation rates. An additional difference from MBE is that in PLD the energies of the particles arriving at the substrate are usually 10–100 eV, which is large enough that there is the potential that arriving particles could damage the growing film. Nevertheless, PLD has been used to grow a variety of high-quality superlattices, such as the $SrMnO_3$ and $PrMnO_3$ superlattice film shown in Fig. 24.9 and described later.

(c) Chemical vapour deposition

Key point: In chemical vapour deposition methods, a vapour of molecules chemically interact or decompose at or near the substrate, where they adsorb to the surface and combine with other species to create a solid and residual gaseous products.

The control over complex stoichiometries, the ability to achieve monolayer-by-monolayer growth, and the attainment of high-quality films is not limited to physical vapour techniques. Chemical techniques, such as **metal/organic chemical vapour deposition** (MOCVD) and **atomic layer deposition** (ALD), also provide these levels of control. In contrast to physical methods, in which physical species condense directly onto a substrate and react with one another, chemical techniques require that a precursor decomposes chemically on or near the substrate in order to deliver the reactant species to the growing film. Therefore, in chemical vapour methods, the decomposition thermodynamics of the selected precursors must be considered because the vapour often contains elements that must not be incorporated into the growing films. Typically, the kinetic energy of the arriving species is relatively low in chemical techniques, and the temperature of the substrate can be controlled externally.

The layout of a typical chemical vapour deposition (CVD) system was shown Fig. 24.3. Such systems normally operate at moderate vacuum or even at atmospheric pressure (in the region of 0.1–100 kPa). Their growth rates can be quite high, more than 10 times that of MBE or PLD. In CVD techniques, chemical decomposition of the feed molecules proceeds upstream of a substrate surface upon which the desired product will be grown. The decomposition of the gaseous reactants is activated by high temperatures, lasers, or plasmas. Large numbers of different materials have been grown using CVD methods. For Group 13/15 (III/V) semiconductors (such as GaAs), the typical sources are organo-metallic precursors for the Group 13 element (such as $Ga(CH_3)_3$), and hydrides or chlorides for the Group 15 element (such as AsH_3). A typical reaction is

$$Ga(CH_3)_3(g) + AsH_3(g) \xrightarrow{550-650°C} GaAs(s) + 3\ CH_4(g)$$

carried out in a hydrogen environment. The CVD technique can be fine-tuned to produce very high-quality films.

The drawbacks of CVD include the use of toxic chemicals, turbulent flow in the reaction chamber, the incorporation of unwanted chemical species owing to incomplete decomposition or exhaust, and the use of high pressure (which prevents the use of *in situ* diagnostics). This same technique, however, can be used without a substrate to create nanoparticles by pyrolysis of the vapour species. Furthermore, MOCVD and MBE have been combined to allow for *in situ* monitoring of growth. This technique has been called **chemical beam epitaxy** (CBE). These low pressures decrease the interactions in the reactant stream and lead to more ballistic transport of the chemical vapour species to the substrate where they decompose and react to form a film.

A final chemical approach aims to control the precise chemical interactions that occur at a surface. In the process called **atomic layer deposition** (ALD), chemical species are delivered sequentially to a substrate upon which a single monolayer deposits. The excess reactant is removed. Repetition of this monolayer coverage, subsequent reaction, and removal of excess reactants allows for precise control over the growth of complex materials. In this process, it is necessary to control the chemical species and their interactions to ensure monolayer coverage only and facile subsequent reaction. To do so means that the vapour of each reagent must interact in the proper manner with the film layer deposited previously. The technique produces flat, homogeneously coated layers. Some examples of ALD-grown nanomaterials are Al_2O_3, ZrO_2, HfO_2, CuS, and $BaTiO_3$.

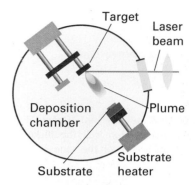

Fig. 24.8 A pulsed laser deposition chamber used to produce nanostructured superlattices and artificially layered thin films.

Artificially layered materials

In this section, we describe the design of so-called two-dimensional materials, which are materials that have macroscopic length scales in two dimensions and a nanoscale length scale along the third dimension.

Thin-film processing methods permit the deposition of films only one atomic (or unit cell) layer thick. By varying sequentially the types of atomic (or unit cell) layers being deposited, it is possible to control the material architecture along the growth direction at a sub-nanometre scale, thereby allowing the bottom-up development of artificially layered nanostructures. A **quantum well** (QW) is a thin layer of one material sandwiched between two thick layers of another material and is the two-dimensional equivalent of a zero-dimensional quantum dot. In a **superlattice** and an **artificial crystal structure**, two (or more) materials alternate with an artificially induced periodicity along the growth direction. Superlattices often have repeat periods of about 1.5–20 nm or greater, and sublayer thicknesses that range from two unit cells to many tens of unit cells. Artificial crystal structures usually have repeat distances similar to bulk crystals (about 0.3–2.0 nm) and have sublayer thicknesses that range from an atomic layer to two unit cells (about 1 nm).

All three types of artificially layered structures are important in materials design. To control a physical property such as absorption and emission, a quantum well can be used in a similar fashion to the quantum dots described earlier. The optical properties of quantum wells can be enhanced in quantum well superlattices (often called multiple quantum well structures), which are used in lasers. To control a physical property such as hardness, which is tied to dislocation formation and motion, it is known that many interfaces between dissimilar materials are required. Hardness can be increased substantially in such superlattices, as compared to monolithic films, ranking them among the hardest known materials and making them successful coatings in the tooling industry. To control a property such as ferroelectricity or superconductivity, which are intimately tied to the crystal structures, it is important to be able to engineer artificial crystal structures.

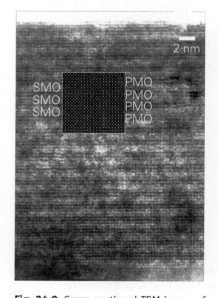

Fig. 24.9 Cross-sectional TEM image of a 2 × 2 superlattice of $SrMnO_3$ and $PrMnO_3$ prepared using laser-MBE. The inset shows a calculated image confirming the $(SrMnO_3)_2(PrMnO_3)_2$ structure. (B. Mercey, A.M. Haghiri-Gosnet, M. Hervieu, Ch. Simon, D. Chippaux, Ph. LeCoeur, *et al.*, *J. Appl. Phys.*, 2003, **94**, 2716.)

24.8 Quantum wells and multiple quantum wells

Key points: Quantum wells consist of a thin small band-gap material sandwiched between thick layers of large-gap materials; multiple quantum wells can enhance the effects of quantum wells when the wells do not interact.

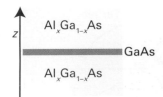

Fig. 24.10 An ($Al_xGa_{1-x}As$)–(GaAs)–($Al_xGa_{1-x}As$) quantum well; the thickness of the GaAs layer is on the nanoscale.

A quantum well is typically composed of two semiconductor materials with different band gaps, such as $Al_{1-x}Ga_xAs$ and GaAs. The smaller band-gap material (GaAs) is sandwiched between layers of the larger band-gap material ($Al_{1-x}Ga_xAs$), and the thickness of the layer of the small band-gap material is confined to the nanometre scale (Fig. 24.10). Electrons are confined in the layer of the small band-gap material along the growth direction; the barrier height is the difference in the energy levels of the band edges for the two materials. Therefore, the energy states in the smaller gap material are quantized along the growth direction, similar to the quantization of the energy in a quantum dot. The electronic wave functions are spatially extended in the direction parallel to the layer, and therefore the energy levels are not quantized.

As for a quantum dot, the optical properties of quantum wells can be tailored and both interband (between the valence and conduction band) and intraband (between the quantized sub-bands) absorption and emission can be controlled. Both $In_{1-x}Ga_xAs$/GaAs and $Al_{1-x}Ga_xAs$/GaAs quantum wells have been widely studied, and the optical transition has been observed to move to higher energies compared to the bulk when the thickness drops down to about 20 nm. The main use of quantum wells is in semiconductor lasers, where the small-gap quantum well is the active layer in the device. Typical QW lasers use the interband transitions, but intraband lasers are also being developed and are discussed below.

Many of the effects that occur in quantum wells can be enhanced by using **superlattice structures**, the periodic repetition of quantum wells along the special direction (the z-axis). In the context of semiconductors these superlattices are called **multiple quantum well** (MQW) structures. If the active layers (the small band-gap layers) do not interact, then the electrons are confined to a given layer and are unable to tunnel between them. The use of MQW structures in this case increases the absorption or emission from a given device, as there are multiple levels. For example, an MQW laser has much higher power output than the corresponding single quantum-well laser.

If the wide band-gap layers are thin enough, one QW interacts with the adjacent QW and electrons can tunnel between them. This phenomenon is used in **quantum cascade** (QC) lasers, which operate at high powers in the IR region. The laser characteristics of these materials are fundamentally different from those of semiconductor diodes and MQW lasers in a variety of ways. In a QC laser, only one type of carrier (electrons) is necessary to achieve laser action; in the other two, both electrons and holes are required. In addition, in the QC laser, the transitions are intraband transitions that arise from the quantization of the valence band.

The QC lasers are constructed from 13/15 (III/V) semiconducting materials such as GaAs, InAs, and AlAs. All these superlattice systems have been grown by solid-source molecular beam epitaxy on single-crystal substrates. The heterostructures that comprise an individual layer in such a superlattice are far more complex than the simple unit-cell periodicities found in most artificially layered structures or superlattices discussed below. For example, a single functional unit of an AlAs/GaAs superlattice QC laser has a layer sequence (in nanometres) of

2.4 **4.3** 1.8 **4.3** 1.4 **4.4** 1.3 4.5 1.2 4.5 1.1 **4.8** 0.8 **5.2** 0.6 **5.8** 0.6 **6.2** 0.5 **6.6** 0.5 **7.2** 0.5 **8.0**

with GaAs layers in red, AlAs in blue, and Si-doped layers in green. This sequence leads to a functional layer thickness of 72.8 nm, which is much larger than most superlattice periods. The superlattice itself consists of 40 such layers.

24.9 Solid-state superlattices

Key points: Artificially layered materials have a periodic repeat along the growth direction of a thin film; the periodic repeat is controlled by the number and type of sublayers deposited in sequence.

The periodic repeat along the growth direction of a thin film in a superlattice is controlled by the number and type of sublayers deposited in sequence, whereas the lateral periodicity is determined by the coherency—the matching of lattice characteristics—between the sublayers (Figs 24.11 and 24.12). It should be kept in mind that the superlattice is

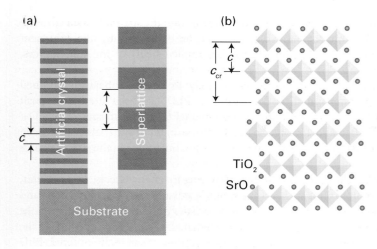

Fig. 24.11 (a) The structure of an AB artificial crystal and an AB superlattice; c and λ represent the repeat period during growth of the artificial structure and the superlattice period, respectively. (b) The structure of Sr_2TiO_4 as an artificially layered oxide. The polyhedra represent Ti-centred, corner-sharing TiO_6 octahedra and the spheres represent Sr^{2+} cations. The lateral coherency between the SrO and TiO_2 sublayers is excellent in this layered structure and the crystallographic repeat parameter, c_{cr}, is twice the growth repeat period, c.

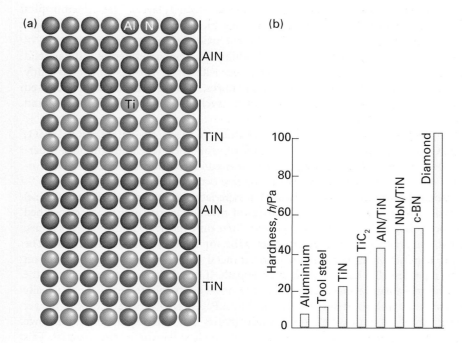

Fig. 24.12 (a) The structure of an ultrahard AlN/TiN superlattice and (b) the hardness of a nitride superlattice and commonly used hard materials. (Adapted from S.A. Barnett and A. Madan, *Physics World*, 1998, **11**, 45.)

built bottom-up with periods in the nanometre range and total thicknesses in the micrometre range. The goal of materials chemists and engineers is to use these synthesis skills to control the properties and functionality of materials of this kind.

Superlattice nitrides are ranked among the hardest known materials. The superlattice period and the chemical composition play important roles in determining the mechanical properties of such compounds, as the interfaces between the two nitride layers are responsible for the enhanced hardness. Figure 24.13 shows, for instance, that the maximum hardness of a typical superlattice occurs for periodicities in the range 5–10 nm. These superlattices have been successfully deposited using sputtering, pulsed laser deposition, and molecular beam epitaxy. Such superlattices are used on cutting tools and are made economically by using sputtering to form them.

Perovskite oxides (Section 23.8) are used in numerous applications on account of their ferroelectric, acoustic, microwave, electronic, magnetic, and optical properties. The most widely used perovskite is $BaTiO_3$, which is a very important dielectric material. It has been discovered that layering ferroelectric $BaTiO_3$ with the isostructural perovskite $SrTiO_3$ can lead to enhanced dielectric properties arising from lattice strain. PLD, MBE, and CVD have all been used to create $SrTiO_3/BaTiO_3$ superlattice thin films (Fig. 24.14). Although the misfit between the two crystal structures is small enough to allow for

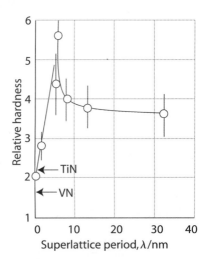

Fig. 24.13 The dependence of hardness on the superlattice period for $(TiN)_m(VN)_m$ superlattices. (Adapted from U. Helmersson, S. Todovora, S.A. Barnett, J.S. Sundgren, H. Markett, and J.E. Greene, *J. Appl. Phys.*, 1987, **62**, 481.)

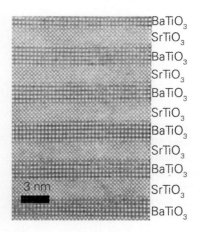

Fig. 24.14 TEM image of a $SrTiO_3/BaTiO_3$ superlattice. (D.G. Schlom, J.H. Haeni, J. Lettieri, C.D. Theis, W. Tian, J.C. Jiang, et al., *Mater. Sci. Eng. B*, 2001, **87**, 282.)

epitaxial growth and coherent interfaces between each bilayer, the stress introduced at the interface is sufficient to improve the dielectric response, specifically the remanent polarization (the polarization in the absence of an applied field) of the superlattices, compared to undoped $BaTiO_3$.

It is important to control layer thickness precisely to ensure coherently strained interfaces that lead to improved dielectric response. PLD has been used to deposit alternating blocks of $SrTiO_3$ and $BaTiO_3$ having unequal layer thicknesses of $SrTiO_3/BaTiO_3 = 15/3$, $10/3$, $3/10$, and $3/15$, and the net dipole moment of superlattices has been reported to be on the order of 46 $\mu C\ cm^{-2}$, whereas that of bulk films of the parent phases are about 14 $\mu C\ cm^{-2}$.

Perovskite-based structures can also exhibit interesting magnetic effects. In particular, manganese-based perovskite films such as $(La,Sr)MnO_3$ possess useful ferromagnetic and magnetoresistive properties. In general, these characteristics arise from electron exchange interactions occurring between Mn and O that are mediated by the various cations in the structure, and can be altered significantly when these cations are spatially ordered. PLD has been used to deposit superlattices of $LaMnO_3/SrMnO_3$; the Mn^{3+} cations are found in the $LaMnO_3$ layers and the Mn^{4+} are found in the $SrMnO_3$ layers. Thus, the thin-film superlattice technique allows for precise ordering of A-site cations (La, Sr), which in turn causes ordering of the Mn in its different oxidation states. In superlattices of $(LaMnO_3)_m(SrMnO_3)_m$, samples with $m \leq 4$ exhibited magnetic properties just like solid solutions of $La_{0.5}Sr_{0.5}MnO_3$, whereas superlattices with larger periods had significantly higher resistivities and lower Curie temperatures. Similar effects were observed in superlattices with unequal layer thicknesses, such as $(LaMnO_3)_m(SrMnO_3)_n$ and $(PrMnO_3)_m(SrMnO_3)_m$ systems (see Fig. 24.9).

The magnetic properties of thin-film superlattices in the family $La(Sr)MnO_3/LaMO_3$ (M = Fe, Cr, Co, and Ni) exhibit a wide range of properties depending on both super-lattice repeat period and the identity of M. For instance, an increase in the Curie temperature, T_C, occurs for Co and Ni, but a decrease occurs for Fe and Cr.

Interesting electronic effects are observed in superlattice thin films consisting of spinel Fe_3O_4 and rock-salt NiO layers. Superlattices of these two structures are well matched because both have the same O sublattice and differ only in the locations of the cations. Fe_3O_4/NiO superlattices have been grown by MBE for a variety of layer thicknesses. The electronic conduction mechanism in many d-metal oxides such as these is **electron hopping**, a process that has a short mean free path. If this distance is approximately the same as the crystal lattice spacings it can lead to novel electrical properties for artificially layered materials with short lattice spacings. Thus, Fe_3O_4/NiO superlattices with bilayer repeat periods of 6.8 nm have electrical conductivities that differ by factors in the range 10^6-10^8 in different directions, which is among the highest known for any material. This result effectively demonstrates both the high degree of chemical modulation available by superlattice deposition and the extent of electrical conduction localization that can be achieved as a consequence.

24.10 Artificially layered crystal structures

Key point: Monolithic films of artificially layered crystals can be developed to be superconducting, dielectric, ferroelectric, electro-optic, and magnetic.

The distinction between artificial crystal structures (like those shown in Fig. 24.11) and superlattices is somewhat arbitrary because both involve the layer-by-layer growth of thin films with artificially induced periodicities along the growth direction. Artificial crystal structures usually have repeat distances similar to bulk crystals (about 0.3–2.0 nm) and have sublayers that range in thickness from an atomic layer to two unit cells. Furthermore, artificial structures most often have excellent lateral coherency between the sublayers. Because of these traits, they often resemble and behave as layered crystals.

Superconducting copper oxides all share a common structural basis, consisting of planes of corner-sharing square-planar CuO_2 nets separated by various other layers that act as charge reservoirs and spacers. The simplest compounds containing the CuO_2 nets as structural elements are the infinite-layer $AeCuO_2$ materials (where Ae denotes an

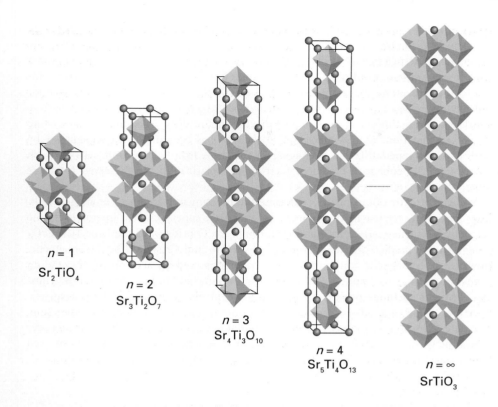

$n = 1$
Sr_2TiO_4

$n = 2$
$Sr_3Ti_2O_7$

$n = 3$
$Sr_4Ti_3O_{10}$

$n = 4$
$Sr_5Ti_4O_{13}$

$n = \infty$
$SrTiO_3$

Fig. 24.15 Structural representations $n = 1–4$ members of the $Sr_{n+1}Ti_nO_{3n+1}$ Ruddlesden–Popper phases.

alkaline earth element). Synthetic strategies have been adapted to develop compounds that incorporate elements of the infinite-layer structure with charge reservoir layers. One of the simplest strategies for using thin-film depositions is to design artificially layered structures closely related to infinite-layer $AeCuO_2$ phases by depositing blocks of $CaCuO_2$, $SrCuO_2$, and $BaCuO_2$ sequentially. A variety of artificially layered $SrCuO_2/CaCuO_2/BaCuO_2$ materials have been deposited, mostly through PLD. The Ae cation size probably plays a significant role in the properties of these artificial structures. The larger Ba^{2+} cations in the $BaCuO_2$ layers allow for the inclusion of extra oxygen in the barium-containing layers, which can then act as a charge reservoir for the CuO_2 superconducting planes (presumably the layers containing Ca and Sr). The successful deposition of high-T_c artificial structures of this sort has led to the discovery of a number of other superconducting families through thin-film methods, including $Sr_{n+1}Cu_nO_{2n+1+\delta}$, $Cu_zBa_2Ca_{n-1}Cu_nO_y$, and oxocarbonate superconductors, such as $(Ba,Sr)_2Cu(CO_3)O_{2+\delta}$.

Another class of artificially layered structures corresponds to the artificial versions of the naturally occurring Ruddlesden–Popper phases mentioned in Section 23.8 and having the general formula $A_{n+1}B_nO_{3n+1}$. Their structures consist of n layers of ABO_3 perovskite with a layer of rock-salt AO interrupting the perovskite stacking along the c axis. One example is the homologous series $Sr_{n+1}Ti_nO_{3n+1}$ (Fig. 24.15): $SrTiO_3$ and related compounds are well known for their dielectric and optical properties and Sr_2TiO_4 and $Sr_3Ti_2O_7$ were the first materials found to adopt this naturally layered crystal structure. Only the members of the series with $n = 1$, 2, and 3 can be prepared by conventional synthesis techniques as polycrystals or single crystals. It is straightforward to grow thin films of $Sr_{n+1}Ti_nO_{3n+1}$ by using MBE and compounds of $n = 1–5$ have been synthesized. The perovskite layer can be grown by alternating one SrO and one TiO_2 layer for each perovskite unit cell, and the rock-salt layer can be grown by depositing an extra layer of SrO after every n perovskite unit cells. As with superconductors, phase-pure Ruddlesden–Popper compounds are attained only when correct cation ratios and total amounts are deposited.

Following the work described in Section 24.9 on $SrTiO_3/BaTiO_3$ superlattices, it was suggested that compositionally ordered structures based on similar superlattices but with broken inversion symmetry along the growth direction might lead to further enhancement of their dielectric properties. Such materials are termed **asymmetric dielectrics** to reflect their key structure–property relationship. An asymmetry is introduced along the growth direction by stacking components in such a manner as to remove any mirror plane perpendicular to the growth direction. Removal of this mirror symmetry leads to asymmetric strain fields in the film that, in turn, lead to built-in electric fields and high polarizations and nonlinear optical coefficients, with applications for technologically desirable capacitive or optical materials. A second, more elegant approach is based on stacking three layers in a manner that automatically removes the inversion along the special direction. Thus, $CaTiO_3/SrTiO_3/BaTiO_3$ has been deposited sequentially by MBE to grow superlattice layers ranging in thickness from 2 to 6 unit cells for each component. An enhanced remanent (zero-field) polarization was observed for these superlattices.

Artificial structures can help to elucidate the fundamental interactions in materials. One example is related to the Fe—O—Cr magnetic superexchange interactions (the magnetic interaction between Fe and Cr through the O bridge). In the $LaFeO_3/LaCrO_3$ system, theory predicts that the coupling between Fe and Cr should be ferromagnetic. However, ferromagnetic coupling is not typically observed on account of the random positioning of the Fe^{3+} and Cr^{3+} cations over the same site in bulk compounds. Thin-film synthesis methods offer the possibility of controlling the order along a specific direction in the unit cell and therefore the bonding interaction along that direction. Alternate deposition of monolayers of $LaFeO_3$ and $LaCrO_3$ would lead to the ideal order to test the theoretical predictions. Increased ferromagnetic ordering is indeed observed for artificial crystal structures grown by sequential $LaFeO_3$ and $LaCrO_3$ deposition.

Example 24.3 Devising syntheses of artificially layered crystal structures

(a) Describe how the well-known perovskite $CaTiO_3$ might be synthesized from the bottom-up using layer-by-layer film growth methods on a $SrTiO_3(100)$ substrate. (b) Describe how a novel $(CaTiO_3)_1(SrTiO_3)_1(BaTiO_3)_1$ artificial structure might be synthesized using these methods. (c) Discuss the loss of all mirror symmetry along the growth direction by depositing this trilayer artificial structure.

Answer (a) The CaO and TiO_2 crystalline planes alternate along the (100) direction of the perovskite structure (Section 23.8). Therefore, by using an atomic layer epitaxial approach to deposit alternating CaO and TiO_2 layers, $CaTiO_3$ can be built from the bottom-up. The $SrTiO_3(100)$ surface is used as a site for heterogeneous nucleation of the initial heteroepitaxial layers into the $CaTiO_3(100)$ layers. (b) In bulk synthesis methods, $CaTiO_3$, $SrTiO_3$, and $BaTiO_3$ can form solid solutions in which the Ca, Sr, and Ba statistically occupy the same crystallographic site. Using atomic layer deposition methods, we can force the Ca, Sr, and Ba to occupy distinct AO planes in the structure. The deposition sequence would be . . . $CaO/TiO_2/SrO/TiO_2/BaO/TiO_2$ . . . . (c) Looking at the repeat period given in (b), we see that there are no mirror planes along the growth direction. Hence by bottom-up growth we have directly removed the centre of symmetry in the structure, so allowing interesting ferroelectric and nonlinear optical properties.

Self-test 24.3 $LaCrO_3$ and $LaFeO_3$ form solid solutions in which Cr and Fe occupy the same crystallographic site. Ordered Fe—Cr solids should exhibit different magnetic character. Describe a layer-by-layer approach to yield (100) and (111) ordered artificial structures.

Self-assembled nanostructures

One of the true frontiers of nanoscience lies at methods that bridge the bottom-up and top-down approaches to achieve nanoscale structures. This frontier is the focus of the section that follows. The specific examples that are discussed reflect the key advantages of and challenges facing bottom-up methods. They include self-assembly, supramolecular chemistry and morphosyntheses, and controlled dimensionality in self-assembled nanostructures.

24.11 Self-assembly and bottom-up fabrication

Key point: Because components that can self-assemble fall between the sizes that can be controlled chemically and that can be manipulated by conventional manufacturing, self-assembly may offer the crucial technique to bridge top-down and bottom-up methods.

Various definitions of self-assembly have been proposed. They include the noncovalent interaction of two or more molecular subunits to form an aggregate with novel structure and properties that are determined by the nature and positioning of the components, the spontaneous assembly of molecules into structured, stable, noncovalently joined aggregates, the spontaneous formation of higher-ordered structures, and the process by which specific components spontaneously assemble in a highly selective fashion into a well-defined, discrete supramolecular architecture. Because the components that assemble fall between the sizes that can be controlled chemically and those that can be manipulated by conventional manufacturing, self-assembly may offer the crucial technique to bridge top-down and bottom-up methods. An example of the application of methods that bridge top-down and bottom-up assembly has been the development of microdevices capable of heterogeneous catalysis, acting as lasers, and as the basis of gas sensing.

Potential self-assembly components must be mobile. Therefore, self-assembly usually takes place in fluid phases or on smooth surfaces. **Static self-assembly** occurs when the system is at a global or local equilibrium, such as in liquid crystals. **Dynamic self-assembly** occurs when the system is dissipating energy, such as in an oscillating chemical reaction. **Templated self-assembly** occurs when systems organize based on interactions between the components and regular features in the environment. The quantum dot self-assembly that occurs during strained heteroepitaxy in MBE and the use of laser-embossed surfaces for small reaction vessel synthesis in ordered arrays (Section 24.7) are examples of templated self-assembly. Finally, **biological self-assembly** occurs in systems involving life, such as cells and tissues: entire organisms are elaborate examples of biological self-assembly.

24.12 Supramolecular chemistry and morphosynthesis

Key point: Morphosynthesis is the control of architecture and morphology and the patterning of inorganic materials with nanoscale to macroscopic-scale dimensions through changes in synthesis parameters.

A general concept for generating ordered structures is based on the recognition-derived spontaneous assembly of complementary subunits. In the field of supramolecular chemistry, one of the key challenges is the characterization of the supramolecules themselves, in particular the characterization of discrete, highly symmetrical, large, self-assembled entities.

A great number of biological systems are formed as a result of the formation of many weak hydrogen bonds and van der Waals interactions. Synthetic systems can take advantage not only of these bonds but also of metal–ligand bonding interactions, the latter being much stronger and generally highly directional. (There are some important examples of biological systems using coordination chemistry for self-assembly, such as the metal-ion activated proteins, including Ca^{2+} signalling proteins and Zn fingers – Sections 26.4 and 26.5). As a result of various d-metal coordination geometries, metal complexes provide a pool of different acceptor subunits that can be linked together by donor building blocks to form rigid frameworks. The final shape of the self-assembled entity is defined by both the metal atom coordination geometry and the orientation of the interaction sites in a given ligand. This methodology has been successfully used to synthesize the cuboctahedron shown in Fig. 24.16.

Many self-assembled nanostructures use similar approaches to take advantage of encoded information within the components of the nanostructure building block. Molecular simulations have been performed to study the self-assembly of nanoparticles functionalized with oligomeric tethers attached to specific locations on the nanoparticle

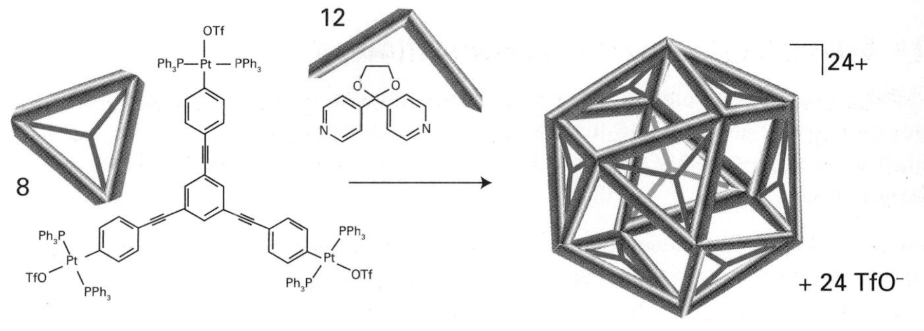

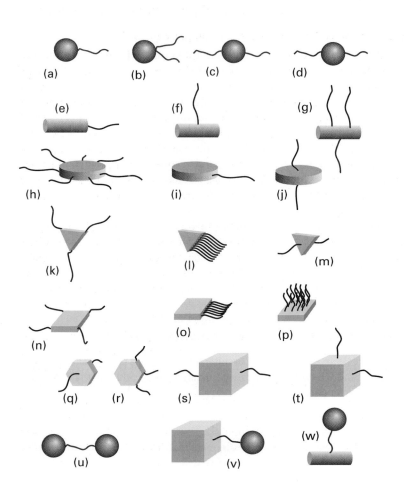

Fig. 24.16 A cuboctahedron synthesized through self-assembly of two distinct building blocks. (Adapted from S. Leininger, B. Olenyuk, and P.J. Stang, *Chem. Rev.*, 2000, **100**, 853.)

Fig. 24.17 Representative tethered nanobuilding blocks: (a–d) tethered nanospheres; (e–g) tethered nanorods; (h–p) tethered nanoplates, including circular, triangular, and rectangular plates; (q, r) tethered nanowheels; (s, t) tethered nanocubes; (u–w) triblock nanobuilding blocks created by joining two nano-objects with a polymer tether. (Adapted from Z. Zhang, M.A. Horsch, M.H. Lamm, and S.C. Glotzer, *Nano Lett.*, 2003, **3**, 1341.)

surface (Fig. 24.17). These results suggest that tethered nanobuilding blocks constitute a unique class of macromolecule that can be used to assemble novel nanoscale materials. Specific tethering groups influence the nature of the self-assembly of the nanobuilding blocks into larger ordered nanostructures.

Morphosynthesis is the control of architecture and morphology and the patterning of inorganic materials with nanoscale to macroscale dimensions through changes in synthesis parameters. As an example, sodium polyacrylate has been used as a structure-directing or templating agent that, when coupled with controls over temperature, pH, and the concentration of reagents, can be used to tune the nanoarchitectures of $BaSO_4$ (Fig. 24.18). Another intriguing example of morphosynthesis is the use of small molecules, such as ethylenediamine, to control crystal face growth to achieve ZnS nanowires. Zinc sulfide is a large band-gap semiconductor that has been widely used commercially as a phosphor in luminescent devices because of its emission in the visible range. Ethylenediamine molecules can act as templates in a solvothermal route to direct the formation of nanorods.

24.13 Dimensional control in nanostructures

Key points: Self-assembly is regarded as the most promising method for designing and controlling the bottom-up assembly of nanometre-scale objects with rationally designed dimensionality; dimensionality plays a crucial role in determining the properties of materials.

An example of how control over dimensionality in materials can yield unique control over their physical properties is the effect that dimensionality has on the density of electronic states (Fig. 24.19). In the next few sections we examine specific examples of how that control has been achieved and the novel properties that have been reported.

(a) One-dimensonal control: carbon nanotubes and inorganic nanowires

The elongated one-dimensional (1D) morphology of nanorods, nanowires, nanofibres, nanowhiskers, nanobelts, and nanotubes has been studied extensively because 1D systems are the lowest dimensional structures that can be used for efficient transport of electrons and optical excitation. They are therefore expected to be crucial to the function and integration of nanoscale devices. Not much is known, however, about the nature of localization that could preclude transport through 1D systems. Such systems should possess discrete molecular-like states extending over large distances and exhibit some exotic phenomena, such as the effective separation of the spin and charge of an electron. There are also many applications where 1D nanostructures could be exploited, including nanoelectronics, very strong and tough composites, functional nanostructured materials, and novel probe microscopy tips.

To address these fascinating fundamental scientific issues and potential applications, two important questions pose key challenges to the fields of condensed matter chemistry and physics research. First, how can atoms or other building blocks be assembled rationally into structures with nanometre diameters but much greater lengths? Second, what are the intrinsic properties of these quantum wires and how do these properties depend, for example, on their diameter and their structure? A key class of nanomaterials that offers potential answers to these questions is **carbon nanotubes** (CNTs). Carbon nanotubes are perhaps the best example of novel nanostructures fabricated through bottom-up chemical synthesis approaches. They have very simple chemical composition and atomic bonding configuration but exhibit remarkably diverse structures and unparalleled physical properties. These novel materials have found application as chemical sensors, fuel cells, field-effect transistors, electrical interconnects, and mechanical reinforcers.

The structure and properties of buckminsterfullerene, C_{60}, and related species have produced much activity in the chemistry and materials sciences communities. The now familiar soccer-ball structure of C_{60} itself was discussed in Chapter 13. Carbon nanotubes

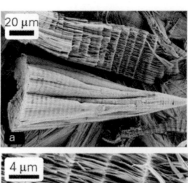

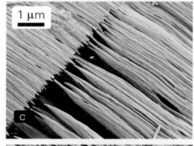

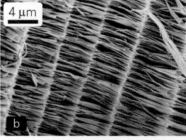

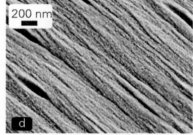

Fig. 24.18 Scanning electron micrographs illustrating the multiple levels of order in $BaSO_4$ nanostructures controlled through morphosynthesis: (a) funnel-like structure with well-aligned cones; (b) magnified section showing the alignment in the superstructure; (c) magnified section showing the alignment of the nanostructured bundles; (d) magnified section showing the surface of the bundles. (Reproduced by permission from S.-H. Yu, M. Antonietti, H. Cölfen, and J. Hartmann, *Nano Lett.*, 2003, **3**, 379.)

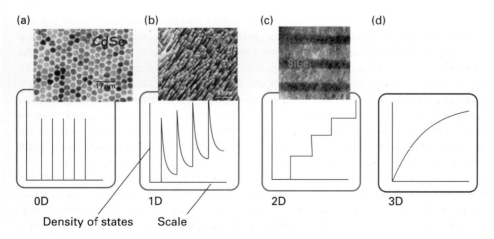

Fig. 24.19 Effects of confinement of charge carriers on the density of states for nanomaterials with specific dimensionality. (Adapted from A. Mrumjadar.)

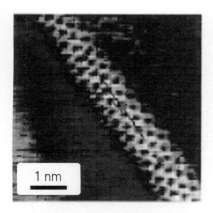

Fig. 24.20 STM image of a single-walled carbon nanotube. (J. Hu, T.W. Odom, and C.M. Lieber, *Acc. Chem. Res.*, 1999, **32**, 435.)

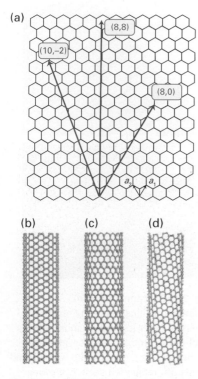

Fig. 24.21 (a) The honeycomb structure of a graphene sheet. Single-walled carbon nanotubes can be formed by folding the sheet along lattice vectors, two of which are shown as a_1 and a_2. Folding the (8,8), (8,0), and (10,−2) vectors leads to armchair (b), zigzag (c), and chiral (d) tubes, respectively. (Based on H. Dai, *Acc. Chem. Res.*, 2002, **35**, 1035.)

were discovered in the early 1990s by electron microscopy. The bonding, local coordination, and general structure of CNTs are similar to those of buckminsterfullerene, but CNTs can have a greatly extended length, leading to a tube rather than a ball structure (Fig. 24.20). Carbon nanotubes are cylindrical shells formed conceptually by rolling graphene (graphite-like) sheets into closed tubular nanostructures with diameters matching that of C_{60} (0.5 nm) but lengths up to micrometres. A single-walled nanotube (SWNT) is formed by rolling a sheet of graphene into a cylinder along an (*m,n*) lattice vector in the graphene plane (Fig. 24.21). The (*m,n*) indices determine the diameter and chirality of the CNT, which in turn control its physical properties. Most CNTs have closed ends where hemispherical units cap the hollow tubes.

Carbon nanotubes self-assemble in two distinct classes, SWNTs and multiwalled carbon nanotubes (MWNTs), in which the tube wall is composed of multiple graphene sheets. **Morphosynthetic control** provides routes to tune the details of these self-assembled nanostructures. This self-assembly occurs during synthesis using several of the fabrication techniques we have already described, including pulsed laser ablation, laser-assisted catalytic growth, chemical vapour deposition (CVD) based on hydrocarbon gases, and carbon arc discharge. All of these strategies rely on vaporizing carbon and condensing some fraction into extended nanostructures. **Laser vaporization methods** typically make relatively small amounts of nanocarbons; specialized CVD techniques have been developed to synthesize CNTs in quantities in excess of a few milligrams. In the **CVD approach**, a hydrocarbon gas such as methane is decomposed at elevated temperatures and carbon atoms are condensed on to a cooled substrate that may contain various catalysts, such as iron. This CVD method is attractive because it produces **open-ended tubes** (which are not produced in the other methods), allows continuous fabrication, and can easily be scaled up to large-scale production. Because the tubes are open, the method also allows for the use of the nanotube as a templating agent.

In the **arc discharge method**, extremely high temperatures are obtained by shorting two carbon rods together, which causes a plasma discharge. Such plasmas easily achieve temperatures in excess of where carbon vaporizes (at about 4500 K). Low voltages and moderately high currents are needed to produce this arc. The typical CNT formed by either the arc method or CVD is multiwalled. To encourage SWNT formation, it is necessary to add a metal catalyst such as Co, Fe, or Ni to the carbon source. These metal catalyst particles block the end-cap of each carbon hemisphere and thus promote SWNT growth. In addition, the growth directions of the nanotubes can be controlled by van der Waals forces, applied electric fields, and patterning of the metal catalyst onto different substrates. The patterned growth approach is feasible with discrete catalytic nanoparticles and scalable on large wafers to achieve large arrays of nanowires (Fig. 24.22).

The repeating axial hexagonal patterns of CNTs are graphitic structures; however, the electrical properties of nanotubes depend on the relative orientation of the repeating hexagons. The nanotubes can be either semiconductors or metallic conductors. When oriented in the chair configuration, CNTs exhibit remarkably high electrical conductivity. Electrons can travel through the micrometre-length nanowire with zero scattering and zero heat dissipation. CNTs also have extremely high thermal conductivity, comparable to the best thermal conductors known (such as their counterparts, diamond and graphite). Therefore, they are being heralded as the ideal nanomaterial for the development of nanoscale interconnects for integrated circuits. They may solve two key challenges in the computer industry: heat dissipation and increased processing speeds.

An additional useful property of SWNTs is their high tensile strength, which is about 100 times that of steel, whereas their density is only about one-sixth that of steel. The incorporation of CNTs into polymer matrices is used to enhance the mechanical strength of the composite material (Section 24.18).

In addition to CNTs, parallel methods have been discovered to make nanotubes out of other materials that share bonding characteristics with carbon, including semiconductors and metal oxides. More specifically, BN, ZnO, ZnSe, ZnS, InP, GaAs, InAs, and GaN have all been made into nanotubes. The novel electronic properties and small sizes of these nanotubes make them appealing as inorganic nanowires.

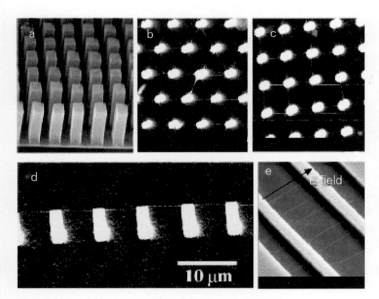

Fig. 24.22 Ordered carbon nanotube structures obtained by direct chemical vapour deposition synthesis. (a) An SEM image of self-oriented MWNT arrays. Each tower-like structure is formed by many closely packed multiwalled nanotubes. Nanotubes in each tower are oriented perpendicular to the substrate. (b) SEM top view of a hexagonal network of SWNTs (the line-like structures) suspended on top of silicon posts (the bright dots). (c) SEM top view of a square network of suspended SWNTs. (d) Side view of a suspended SWNT power line on silicon posts (the bright lines). (e) SWNTs suspended by silicon structures (the bright regions). The nanotubes are aligned along the electric field direction. (H. Dai, *Acc. Chem. Res.*, 2002, **35**, 1035.)

Core–sheath wires similar to macroscopic coaxial wires are of great interest. One method to prepare core–sheath particles uses a colloid-templating strategy that involves a layer-by-layer assembly of polyelectrolytes and inorganic nanoparticles on submicrometre- and micrometre-sized polystyrene (PS) latex particles. A number of methods, including laser ablation, carbothermal reduction, and several solution-based methods, have been developed to generate one-dimensional nanostructures with coaxial structures. For instance, by using a novel nanowire-templating technique based on the layer-by-layer approach and calcining, ordered Au/TiO$_2$ core-sheath nanowire arrays have been fabricated. A template-grown gold nanowire array is used as a positive template, and then a cationic polyelectrolyte and an inorganic precursor are assembled on gold nanowires by the layer-by-layer technique. Calcination then converts the inorganic precursor to titanium dioxide (Fig. 24.23).

An advantage of the layer-by-layer nanowire-templating approach over other methods is that nanoscale control can be exerted over the sheath thickness by varying the number of deposition cycles. Titanium dioxide has been one of the most investigated oxide materials because of its technological importance. It has been widely used for photocatalysis and environmental clean-up applications, on account of its strong oxidizing power, chemical inertness, and nontoxicity. These core–sheath nanowires have well-defined diameters and lengths, largely determined by the dimensions of the templates, and porous sheaths with thicknesses controlled by the number of layers deposited. This method opens a door to the use of a wide variety of nanowires as the positive templates for the fabrication of core–sheath nanostructured materials for chemical sensors, photocatalysis, light energy conversion devices, and nanoscale electronic and optoelectronic devices.

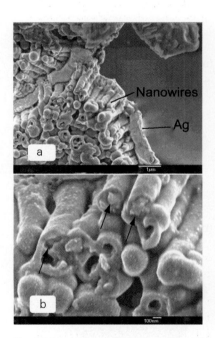

Fig. 24.23 (a) Low-magnification and (b) high-magnification SEM images of Au/TiO$_2$ core–sheath nanowire arrays. (Y.-G. Guo, L.-J. Wan, and C.-L. Bai, *J. Phys. Chem. B*, 2003, **107**, 5441.)

(b) Two-dimensional control: nanoplates

As discussed earlier, nanometre metal and semiconductor materials are being intensively studied in materials chemistry because of their size- and shape-dependent optical, electrical, and magnetic properties. Although the synthesis of monodispersed spherical particles and one-dimensional nanotubes has been successful in many cases, systematic

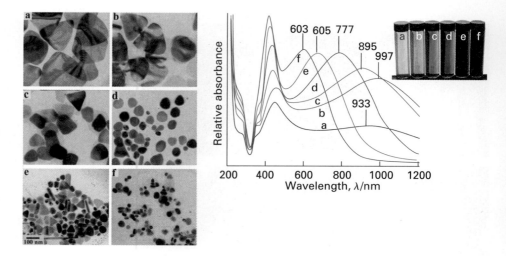

Fig. 24.24 (Left) TEM images of the silver nanoplates obtained with different seed amounts: (a) 0.1 cm³; (b) 0.25 cm³; (c) 0.5 cm³; (d) 1.0 cm³; (e) 2.0 cm³; and (f) 3.0 cm³. (Right) Absorption spectra of the corresponding silver nanoplate solutions. (S. Chen and D.L. Carroll, *J. Phys. Chem. B*, 2004, **108**, 5500.)

control of the size of crystal growth in two dimensions has been more difficult. Synthesis of two-dimensional (2D) nanomaterials has included thermal evaporation of semiconductor oxides of zinc, tin, indium, cadmium, lead, and gallium nanobelts and nanoribbons, use of graphite layers to fabricate platinum nanosheets, exfoliation (removal of a thin coating from its substrate) of titanate films into nanosheets, and of small molecules such as ethylenediamine or polymers as crystal-plane controllers to form *d*-metal nanoplates. However, in the majority of these systems, well-defined, systematic control of size in two dimensions has rarely been achieved.

Some success has been achieved by using a modified solution approach to nanoparticle synthesis. Silver nanoplates with sizes in the range 40–300 nm have been synthesized by a simple room-temperature solution-phase chemical reduction method in the presence of diluted cetyltrimethylammonium bromide (CTAB, $(C_{16}H_{33})(CH_3)_3NBr$). These plates are single crystals having as their basal plane the (111) plane of face-centred cubic silver. The stronger adsorption of CTAB molecules on the (111) basal plane than on the (100) side plane of these plates may account for the anisotropic growth of nanoplates. As discussed earlier, metal nanoparticles have interesting optical properties related to their surface plasmon excitations. The optical (in-plane dipole) plasmon resonance peaks can be shifted to wavelengths of 1000 nm in the near-IR region when the aspect ratio (the ratio of long-axis to short-axis, or width to thickness) of the nanoplates reaches 9. Such control over the optical properties of simple metals opens new possibilities for various near-IR-related applications (Fig. 24.24).

(c) Three-dimensional control: mesoporous materials and metal–organic frameworks

Key points: Mesoporous materials and metal–organic frameworks have ordered pore structures that are defined over the nanometre scale; the synthesis of these materials relies upon control of self-assembly; guests can be incorporated into the inorganic host framework for novel applications.

The design and synthesis of three-dimensional (3D) supramolecular architectures with tunable, nanoporous, open channel structures have attracted considerable attention because of their potential applications as molecular sieves, sensors, size-selective separators, and catalysts. An important class of three-dimensionally ordered nanomaterials is **mesostructured nanomaterials**. Mesoporous materials are well known in heterogeneous catalysis (Chapter 25) and are of great interest because of our ability to tune their pore sizes from 1.5 to 10 nm (Fig. 24.25).

Mesoporous inorganic nanomaterials are synthesized in a multistep process based on the initial self-assembly of surfactant molecules and block copolymers that self-organize into supramolecular structures (which are liquid crystalline assemblies of cylindrical, spherical, or lamellar micelles; Fig. 24.26). These supramolecular frameworks serve as structure-directing templates for the growth of mesostructured inorganic materials (often silica or titania). During a solvothermal reaction step, oxide particles (silica, in the

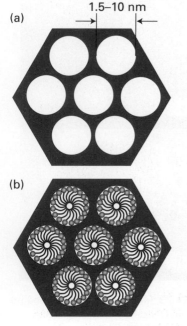

Fig. 24.25 A hexagonal mesoporous structure with (a) controlled nanoporosity and (b) functionalized pores.

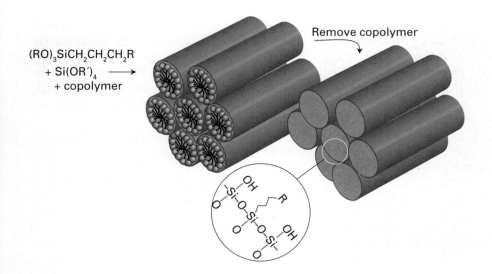

$(RO)_3SiCH_2CH_2CH_2R$
$+ Si(OR')_4 \longrightarrow$
$+$ copolymer

Remove copolymer

Fig. 24.26 Block-copolymer structure-directing agents self-assemble into micellar rods that form hexagonal arrays that can be removed to yield nanoporous silica matrices for inclusion chemistry and catalysis. (Adapted from M.E. Davis, *Chem. Rev.*, 2002, **102**, 3601.)

case depicted in Fig. 24.26) form at the surfaces of the hexagonal rods, assembling around the supramolecular structures. The templating agent can be removed through an acid wash or by calcination (heating in air) to yield inorganic materials having hexagonal pores with uniform and controllable dimensions. This porosity offers unique capabilities for both catalysis and inclusion chemistry.

A wide variety of mesostructured and mesoporous inorganic materials have been obtained by varying the choice of templating agent and reaction conditions (Fig. 24.27). For instance, the family known as M41S has silica or alumina–silica inorganic phases and various cationic surfactants leading to three distinct types of structures: hexagonal lamellar (MCM-50), cubic (MCM-48), and hexagonal (MCM-41) (Fig. 24.27). The shape of the surfactant and the water content used during synthesis controls the resulting nanoarchitecture, as can be seen from the illustration. Surfactants can also be used to tailor the structure. For example, surfactant cetyltrimethylammonium cations ($C_{16}TMAC$) have been used to fabricate silica nanofibres with hexagonal pores.

Mesoporous nanomaterials also offer routes to functionalize pores to increase catalytic activity and selectivity. They have also received much attention as host materials for the inclusion of numerous guests such as organometallic complexes, polymers, *d*-metal complexes, macromolecules, and optical laser dyes. The silica nanofibres shown in Fig. 24.28 can even be used as hosts to grow nanowires of various other oxide materials.

Example 24.4 Controlling nanoporosity in molecular sieves

(a) Compare the nanoarchitecture of ZSM-5 zeolite and MCM-41; include in your comparison descriptions of the dimensionality of tunnels and relative pore sizes. (b) Cetyltrimethylammonium $[C_{16}N(CH_3)_3]^+$ cation and tetrapropylammonium $[N(CH_2CH_2CH_3)_4]^+$ cation are surfactants used in the syntheses of these materials. Which surfactant is a better choice for the synthesis of MCM-41?

Answer (a) ZSM-5 is a microporous zeolite with pore sizes of approximately 0.5 nm. It has three-dimensional intersecting pores similar to the cubic mesoporous phases but with smaller dimensions (Section 13.8). MCM-41 is a mesoporous solid with hexagonal, one-dimensional pores with tunable dimensions of 2 nm up to 10 nm. (b) The choice of surfactant must match the pore size of the solid. The longer chain surfactant (cetyltrimethylammonium cation) is a better choice for the self-assembly of the MCM-41 material owing to its longer hydrocarbon tails and larger spontaneous curvature. This increased tail size encourages larger pore dimensions that are a direct consequence of the packing of the surfactants within the micellar rods that lead to the hexagonal mesophase.

Self-test 24.4 Which of these materials, MCM-41 or ZSM-5, is more likely to be used as a host material for the entrapment of quantum dots?

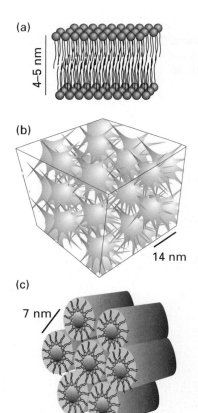

(a)

4–5 nm

(b)

14 nm

(c)

7 nm

Fig. 24.27 Representations of three types of ordered mesoporous solids. (a) Lamellar (layered materials), (b) cubic (complex arrangements), (c) hexagonal (honeycomb). (Adapted from A. Mueller and D.F. O'Brien, *Chem. Rev.*, 2002, **102**, 729.)

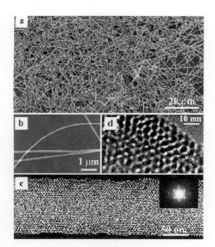

Fig. 24.28 (a) SEM image of mesoporous silica nanofibres. (b) Low-magnification TEM image of the nanofibres. (c) High-magnification TEM image of one nanofibre. The inset is a selected-area electron diffraction pattern of the nanofibre. (d) High-resolution TEM image recorded at the edge of one nanofibre. (J. Wang, C.-K. Tsung, W. Hong, Y. Wu, J. Tang, and G.D. Stucky, *Chem. Mater.*, 2004, **16**, 5169.)

A second class of three-dimensionally controlled nanostructures are **metal–organic frameworks** (MOFs). These frameworks are self-assembled from careful choice of metal ions and bridging organic ligands. They offer an alternative approach to the synthesis of open porous materials using **supramolecular assembly** and are being called the 'new zeolites'. These macroscopic materials have some of the highest reported surface areas and, as such, have found practical application for methane and hydrogen storage and heterogeneous catalysis.

As discussed earlier, supramolecular chemistry relies upon bridging together nano-building blocks (NBBs) to assemble macroscopic materials. One new strategy in design is the incorporation of **nanoporosity** (pores with nanometre dimension) within the NBB. The synthesis of a nanoporous cuboctahedron NBB (Fig. 24.29) was achieved by linking rigid square molecules of $Cu_2(CO_2)_4$ with m-BDC (1,3-benzenedicarboxylate). The overall NBB formula is: $Cu_{24}(m\text{-BDC})_{24}(DMF)_{14}(H_2O)_{10}$ $(H_2O)_{50}(DMF)_6(C_2H_5OH)_6$ and is termed more conveniently MOP-1. MOP-1 is constructed from 12 paddle-wheel units (themselves NBBs) bridged by m-BDC to yield a large metal–carboxylate polyhedron. The simplest way to view the overall structure is by considering its relationship to the cuboctahedron (shown in Fig. 24.16), where each square and link have been replaced by the paddle-wheel NBB and the m-BDC (two-connector) units, respectively, to give an *expanded-augmented* cuboctahedron (or a truncated cuboctahedron). This illustrates the feasibility of obtaining crystals of large porous metal–organic polyhedra in which rigid NBBs are an integral part of a well-defined structure.

Some of the key challenges in the area of mesostructured materials are achieving control over pore size and shape, the presence of counterions and solvents within the channels, the interpenetration of different networks, the framework structural instability in the absence of guest molecules, and the often low thermal stability of the host framework. The assembly of NBBs having polytopic carboxylate linkers may generate rigid porous frameworks with open metal sites and functionalized pores in a controlled fashion, leading to important advances in the design of mesoporous materials that meet these key materials challenges.

Bioinorganic nanomaterials

Biological phenomena, such as DNA condensation, intercellular transport, tissue assembly, respiration, photosynthesis, and reproduction, originate and operate at the nanoscale. Because biological processes use and manipulate materials at this length scale, it is not surprising that scientists and engineers are looking to natural systems for inspiration and to reach a better understanding of the design of robust and useful nanomaterials. This interest has spawned an extensive research effort termed **biomimetics**, the mimicking of biological systems. Of particular interest to us are materials that bridge solid-state inorganic materials and living cells, the so-called **bioinorganic materials**.

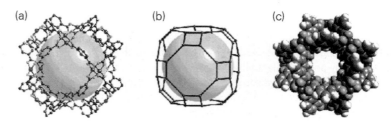

Fig. 24.29 The crystal structure of MOP-1 (drawn using coordinates obtained for c-MOP-1) showing (a) 12 paddle-wheel units (Cu, red; O, blue; C, grey) linked by m-BDC to form (b) a large truncated cuboctahedron of 1.5 nm diameter void (yellow sphere); the grey spheres (carbon atoms) represent the polyhedron constructed by linking together only the carboxylate C atoms in (a) to form linked square NBBs, an arrangement that provides for (c) a large porous polyhedron with triangular and square windows. Hydrogen atoms in light grey, otherwise same colouring scheme as in (a). (M. Eddaoudi, J.-H. Kim, J.B. Wachter, H.K. Chae, M. O'Keeffe, and O.M. Yaghi, *J. Am. Chem. Soc.*, 2001, **123**, 4368.)

Their nano counterparts, the bioinorganic nanomaterials, promise to have a profound impact on the way we live; key applications include drug delivery, medical diagnostics, cancer therapy, and environmental pollution control, as well as chemical and biological sensing.

Many biological structures (people, for instance) are produced from self-assembly of functional building blocks, which have complex morphological architectures themselves. Despite success in understanding the basic principles of the biological assembly process, as well as in making inorganic materials through biological templating, it remains a key challenge to mimic natural pathways as efficient routes for fabricating artificial bionanomaterials. In this section we discuss some of these recent advances and offer a taste of the complexity, diversity, and architectures of some exquisite bionanomaterials by using key examples of these types of materials.

24.14 DNA and nanomaterials

Key point: Interactions between gold nanoparticles and DNA can cause the self-assembly of DNA into condensates and the ordering of nanoparticles into regular arrays that can be used for sensor applications.

DNA condensates are self-assembled nanostructures. The architecture of these condensates is driven by the components that become woven into the nanoassembly. Electrostatic interactions drive the formation of DNA condensates. DNA is a negatively charged polyelectrolyte that interacts with positively charged ions, molecules, or modified nanoparticles to form ordered structures and convert from their random coils to a more compact form. This compact form, called a **condensate**, is essential for the organization of DNA into chromatin, a component of genes. Mimicking or controlling the DNA condensation process is the basis of the design of nonviral gene delivery vectors that have practical application in the treatment of cancer, drug delivery, and biosensing.

Studies of DNA condensation *in vitro* have focused on the use of complex cations including hexaamminecobalt(III) or polyamines as the positive charge centre in the condensate. Complex formation in these systems can be expected to mimic DNA wrapping around charged particles in living cells, such as histone proteins. Because the cationic particles can be fine-tuned with respect to size and charge density, the effect of those parameters on the complex formation patterns with DNA have begun to be understood. Gold nanoparticles have been suggested as an effective transfection agent (a device for introducing exogeneous DNA into a cell) and, as such, it is important to understand the interactions of nanoparticles with DNA.

Figure 24.30 shows AFM images of complexes of lysine-modified gold nanoparticles and DNA at different nanoparticle/DNA ratios. These results indicate the possibility of developing functionalized gold nanoparticles–DNA into a model system to study DNA condensation *in vivo*. These smart nanomaterials have far-reaching consequence for the design of nanoscale self-assembled materials that can be used as building blocks for active nanodevices.

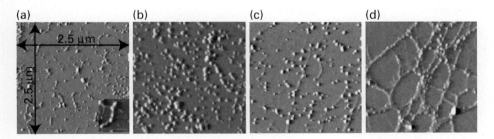

(a)   (b)   (c)   (d)

2.5 μm

2.5 μm

Fig. 24.30 AFM images of complexes of lysine-modified gold nanoparticles and DNA at different nanoparticle/DNA ratios (a) 10:1, (b) 20:1, (c) 50:1, and (d) 100:1. Several wrapped molecules are seen along with free nanoparticles in (a) and (b). Network formation is seen in (c) and (d). The inset in (a) shows a single DNA molecule wrapping around six nanoparticles. (Reproduced by permission from M. Ganguli, J.V. Babu, and S. Maiti, *Langmuir*, 2004, **20**, 5165.)

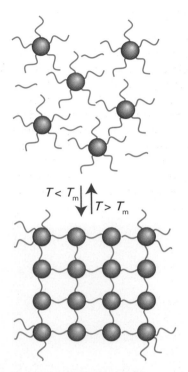

$T < T_m$ ↕ $T > T_m$

Fig. 24.31 DNA templating of inorganic semiconducting nanoparticles. (N.L. Rosi and C.A. Mirkin, *Chem. Rev.*, 2005, **105**, 1547.)

Another exciting accomplishment in the synthesis of bionanomaterials is the use of DNA to drive the assembly of inorganic nanoparticles (Fig. 24.31). First, nanoparticles are functionalized with single strands of DNA. Then, the complementary strand of DNA is introduced from an analyte or bioagent, leading to dimerization with the DNA on the nanoparticles and the self-assembly of the nanoparticles into ordered arrays. Detectable differences in optical properties occur between the disordered and ordered state and result in colour changes. This work has helped to make sophisticated optical-based biosensor arrays with remarkable sensitivity and selectivity for different analytes and bioagents, including the potential to sense anthrax.

24.15 Natural and artificial nanomaterials: biomimetics

Key point: Biological materials can be used as templates in the design of nanoinorganics having specific architectures that mimic the structure of the natural material.

The formation mechanisms in fossilization, including those for siliceous woods, offer efficient methods to reproduce morphological hierarchies of original plant matter through the replacement of the organic components by silica. Artificial fossilization processes can be realized by carefully lining the morphologically complex surfaces of the biological structure with inorganic layers and then removing the organic template. Natural materials such as wood and eggshell membrane have been used as templates for the preparation of macroporous silica, zeolites, and titanium dioxide from both precursor sol–gel solutions and suspensions of nanocrystals. Unfortunately, morphological replication has been achieved only over the micrometre scale and the nanoscale details of the biological templates have not yet been reproduced.

An artificial fossilization process has been developed by taking advantage of a surface sol–gel process that can replicate features in biological templates on a nanometre scale (Box 24.1). The surface sol–gel process is a two-step process where adsorption of metal alkoxides occurs from the solution onto hydroxylated substrate surfaces and then hydrolysis of the adsorbed species occurs to yield nanometre-thick oxide films. Natural cellulose fibres possess surface −OH groups and offer a remarkable template for using the surface sol–gel process. The outer diameter of the TiO_2 nanotube varies from 30 to 100 nm, and the thickness of the tube is uniform along its length, with a wall thickness of about 10 nm. The nanotube assembly exhibits the original morphology of interwoven cellulose fibres. The 'titania paper' produced in this way records the morphological information of the original paper at the nanometre scale and offers a remarkable example of successful biotemplating of metal oxide nanomaterials.

24.16 Bionanocomposites

Key points: Bionanomaterials synthesis offers new routes to realize smart materials with advanced mechanical strengths and improved performance over natural material; protein engineering offers an opportunity to insert species that attach selectively to inorganic materials.

Millions of bone fractures resulting in hospitalization occur worldwide every year. Modern treatments for severe bone injuries replace a non-uniform defect with a permanent biomaterial that may corrode, wear, and ultimately cause severe infection. In addition, these biomaterials typically exhibit mechanical strength far greater than that of bone, resulting in stress shielding and eventual bone resorption around the implant. There is a social need for a bone tissue engineering scaffold that has mechanical properties similar to bone and that will facilitate bone growth into the defect, degrade slowly and naturally, and eventually be replaced by natural bone tissue without threat of infection. The primary obstacle towards this goal is matching the unique mechanical properties of natural bone tissue with a degradable, synthetic nanomaterial.

The versatile and important structural properties of bone are derived from the nanoscale interactions between its inorganic and organic components. There are two types of bone, *trabecular bone*, the central part which is weaker and porous, and *cortical*

Box 24.1 Artificial fossils

An artificial 'titania fossil' of paper has been created. A titania gel film was deposited on the morphologically complex surface of paper, and the resultant paper/titania composite was calcined to remove the original filter paper. In a typical procedure, a piece of commercial filter paper was placed in a suction filtering unit, washed by suction filtration with ethanol, and then dried by air flow. Then 10 cm^3 of titanium butoxide solution was passed through the filter paper during 2 min. Two 20 cm^3 portions of ethanol were filtered immediately to remove the unreacted metal alkoxide and 20 cm^3 of water was passed through to promote hydrolysis and condensation. Finally, the filter paper was dried in flowing air. By repeating this filtration/deposition cycle, thin titania gel layers covered the surface of the cellulose fibres. The resultant paper/titania composite was calcined in air at 723 K for 6 h to remove the filter paper.

The resulting titania fossils possessed, except for some shrinkage in size owing to the calcination step, the morphological characteristics of the original filter paper, as shown in Fig. B24.1. The titania sheet is self-supporting and 0.22 mm thick. In this case, deposition of titania thin films was repeated 20 times. Besides paper, other morphologies have been produced using a similar approach (Fig. B24.1d,e). The sheet size and thickness depend on the original filter paper used. The original morphology of the filter paper was found to be replicated by titania films, and the cellulose fibres were copied precisely as irregular titania nanotubes, which are clearly seen in the illustrations.

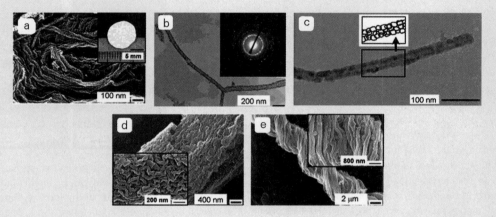

B24.1 (a) A field emission scanning electron micrograph (FE-SEM) of titania paper. The inset shows the photograph of a sheet of titania paper. (b,c) Transmission electron micrographs of individual titania nanotubes, isolated from the assembly. The inset in (c) gives a schematic illustration of the boxed area, showing that the titania nanotube wall is composed of fine anatase particles. (d) An SEM image of titania cloth and (e) an SEM image of titania cotton. (J. Huang and T. Kunitake, *J. Am. Chem. Soc.*, 2003, **125**, 11834.)

bone, the outer shell of bone, which is strong and denser. Hydroxyapatite nanocrystals provide compressive strength as they precipitate onto bundles of collagen fibres that impart tensile strength. Degradable biomaterials composed of inorganic and organic components can have mechanical properties in the range of natural bone tissue, and they may be suitable for load-bearing bone tissue engineering applications. Here we describe one illustrative example of these approaches that combine polymeric components with inorganic components.

Poly(propene fumarate) (PPF) cross-linked with poly(propene fumarate)-diacrylate (PPF-DA) is an injectable, biodegradable biomaterial. Although PPF shows great potential as a material for bone tissue engineering where mechanical properties are not crucial, such as the trabecular bone, its mechanical properties are inferior to those required for the more demanding human bone applications, such as cortical bone. A promising strategy to improve the mechanical properties of a polymer is to mimic the architecture of bone, or to incorporate inorganic particles into the polymer matrix. For example, significant increases have been observed in the compressive mechanical properties of PPF by the incorporation of calcium phosphate. The extent to which inorganic particles modify the polymer properties is associated with their size, shape, and dispersion uniformity, as well as with the degree of interaction between the inorganic and organic components (similar to the red abalone shell, Box 24.2). Chemically modified

Box 24.2 A natural nanocomposite: the shell of red abalone

Nacre (mother-of-pearl) is an outstanding example of a natural nanocomposite that achieves enhanced structural integrity by combining (on the nanoscale) individual components that are either brittle or structurally weak. Red abalone shell is composed of about 95 per cent by mass inorganic aragonite (a polymorph of $CaCO_3$), with only a small percentage of an organic biopolymer. This ceramic/polymer composite material exhibits natural beauty in conjunction with a nanostructure that affords high-quality mechanical properties. Compared to its constituent materials, the laminated structure of red abalone achieves approximately a twofold increase in strength and a thousandfold increase in toughness, owing to the interfaces between the ceramic matrix and the biopolymer assemblies. These remarkable increases in material strength and toughness have inspired chemists and materials scientists to develop synthetic, biomimetic nanocomposites that attempt to reproduce Nature's composite systems.

The abalone shell is a highly ordered nanocomposite material (see Fig. B24.2) that is extraordinarily tough, hard, and strong. Nanoscale asperities (spikes) exist on the ceramic platelet surfaces that serve to lock together neighbouring platelets, thus resisting fracture and enhancing material strength. The formation mechanisms of these nanoscale asperities are still unknown but they play an essential role in stabilizing the material against fracture. In the nacreous layer, cobble-like polygonal nanograins are the basic building blocks that construct individual aragonite platelets. The nanograin-structured aragonite platelets are somewhat ductile and the organic biopolymer serves as an adhesive to hold the aragonite platelets together. The nacreous region of the abalone shell has such a great increase in toughness over its constituent materials because this nanocomposite has excellent crack-tip deflection properties (or the ability to prevent catastrophic failure by crack propagation), the ability to deform by nanograin or platelet slip, and its possession of a well-dispersed organic adhesive holding everything together. By contrast, the outer prismatic layer does not show these crack diversion mechanisms and serves mainly as a brittle outer shield for the abalone.

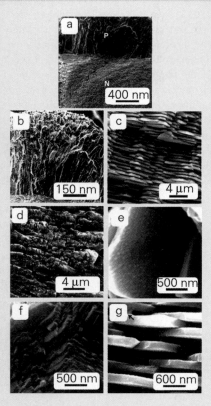

B24.2 SEM images of the fracture shell surface. (a) Low magnification image showing the prismatic and nacreous regions. (b), (d), and (f) SEM images of the fracture surface of the prismatic section taken at increasing magnification. The prismatic section exhibits columnar architecture with cleavage fracture characteristics. (c), (e), and (g) SEM images of the nacreous section taken at increasing magnification. The brick-and-mortar architecture is shown in (c). Nanoscale asperities on the aragonite platelet surface are readily observed in (e). (X. Li, X., W.-C. Chang, Y.-J. Chao, R. Wang, and M. Chang, *Nano Lett.*, 2004, **4**, 613.)

inorganic materials that allow for optimal dispersion and interaction with the polymer matrix therefore seem to be ideal candidates for use in bionanocomposites. A hybrid alumoxane nanoparticle, or a partially hydrolysed aluminium oxyhydroxide with surrounding organic material (Fig. 24.32), with a long carbon chain and a reactive double bond dispersed in PPF/PPF-DA, shows over a threefold increase in strength over polymer resin alone.

Several approaches have been successful in attaching nucleic acids to nanoparticles, which is then followed by DNA hybridization like that mentioned in Section 24.14. Various functional inorganic nanomaterials, such as semiconductors, fullerenes, metals, and oxides, have been considered for use in bottom-up approaches to the design of systems that bridge inorganic and biological materials. One approach is to functionalize the surface of nanoparticles for selective molecular attachment; such materials are used in the development of biosensors, nanoprobes, drug transporters, and other smart devices.

Proteins have been bound to nanoparticles by using a single site on the protein, which yields hybrid nanomaterials in which oriented proteins are attached to the surface of an

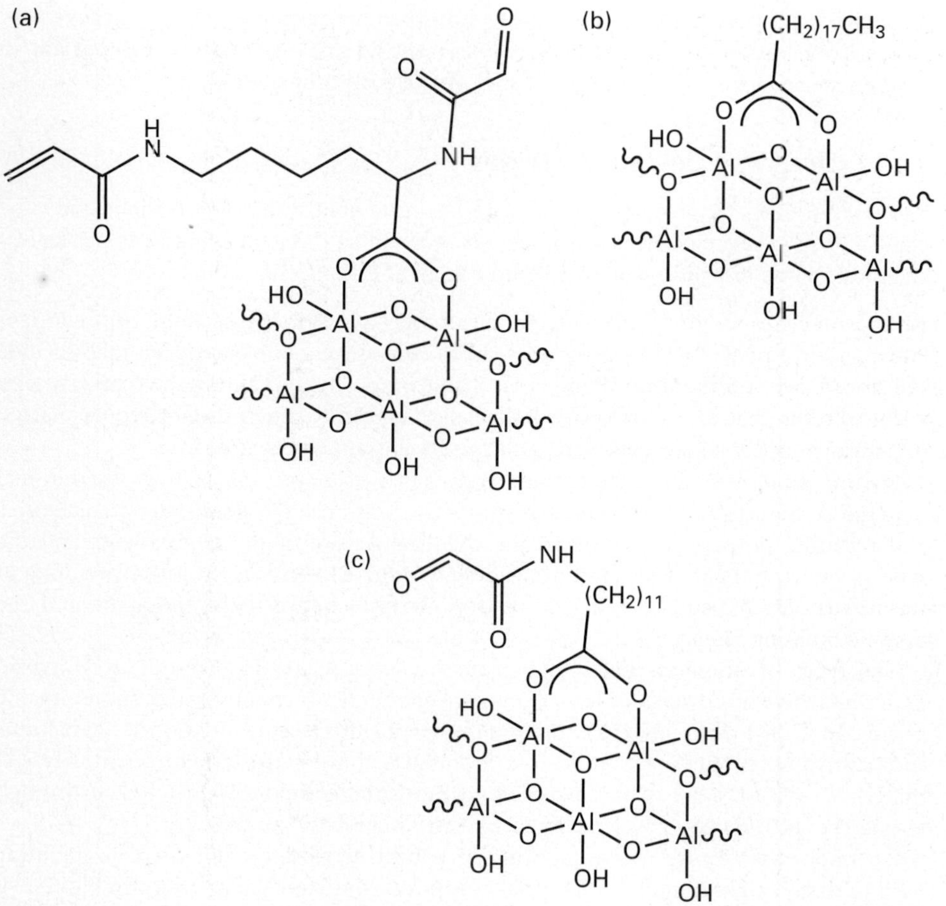

(a)

(b)

(c)

Fig. 24.32 Chemical structures of modified alumoxanes: (a) diacryloyl lysine–alumoxane (activated), (b) stearic acid–alumoxane (surfactant), and (c) acryloyl undecanoic amino acid–alumoxane (hybrid). (R.A. Horch, N. Shahid, A.S. Mistry, M.D. Timmer, A.G. Mikos, and A.R. Barron, *Biomacromolecules*, 2004, **5**, 1990.)

inorganic material. An important class of composite material is that of proteins on fullerenes, which take advantage of the latter's rich chemical and electronic properties, small dimensions, and potential applications. Alterations to the amino acid sequence of a protein are carried out be site-directed mutagenesis (SDM). By using SDM, a single cysteine residue, which selectively attaches to the fullerene, has been introduced in the protein sequence away from the active site to avoid deactivation upon binding to the nanoparticles. Figure 24.33 illustrates one approach to create a covalent linkage between nanosized fullerenes and an active protein by using a functional tether (a bridging molecule). Many of these functionalized fullerenes have found biological applications as inhibitors of HIV protease, as light-sensitive biochemical probes, and in photo-dynamic tumour therapies. Engineering covalent linkages between nanomaterials and proteins has outstanding potential for future applications in drug delivery and enhanced drug efficacy.

Inorganic–organic nanocomposites

Inorganic–organic nanocomposites possess chemical and physical properties that can be tuned by using the synergistic association of organic and inorganic components at nanometre scales. These hybrid materials originated in the paint and polymer industries where inorganic fillers and pigments were dispersed in organic materials (including solvents, surfactants, and polymers) to fabricate commercial products with improved materials performance. Hybrid nanomaterials are also of interest because their mechanical properties occupy a niche between glasses and polymers, so achieving enhanced strength and robustness. Hybrid nanomaterials have been reported with excellent laser

(a)

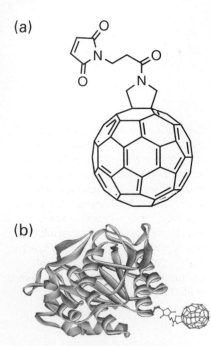

(b)

Fig. 24.33 (a) Structure of *N*-(3-maleimidopropionyl)-3,4-fulleropyrrolidine. (b) The protein conjugated to *N*-(3-maleimidopropionyl)-3,4-fulleropyrrolidine (shown to scale). (P. Nednoor, M. Capaccio, V. Gavalas, V. Meier, J.E. Anthony, and L.G. Bachas, *Biocon J. Chem.*, 2004, **15**, 12.)

efficiencies, photostability, and ultrafast photochromic response. They also have very large second-order and third-order nonlinear optical response, which is important in frequency conversion and optical switching for telecommunications.

24.17 Uses and design strategies

Key points: Class I inorganic–organic materials have noncovalent interactions and class II have some covalent interactions; sol–gel and self-assembly methods are key chemical routes to hybrid nanocomposite design and synthesis.

Nanocomposites have already entered the market-place in sunscreens, fire-retardant fabrics, stain-resistant clothing, thermoplastics, and automobile parts. Examples include the use of nylon-6/montmorillonite clay nanocomposites for timing-belt covers, television screens that are coated with indigo dyes embedded in a silica–zirconia matrix, organically doped sol–gel glassware, and sol–gel-entrapped enzymes.

Hybrid nanomaterials offer the materials chemist novel routes that use rational materials design to optimize structure–property relationships. Their nanoarchitectures and resulting properties depend on the chemical nature of the components and the synergy between them. A key part of the design of these hybrids is the selective tuning of the nature, extent, and accessibility of the interfaces between the inorganic and the organic building blocks.

The types of interfaces fall into two main classes. **Class I** corresponds to hybrid materials where no covalent or ionic bonds are present between the organic and inorganic phases. In Class I materials, the various components interface through weak interactions such as hydrogen bonding, van der Waals contacts, and electrostatic forces. In **Class II** materials, at least some of the organic and inorganic components are linked through strong chemical bonds (covalent, ionic, or Lewis acid–base bonds).

An important feature in the tailoring of hybrid networks is the chemical pathway used to design a given material. The main chemical routes are summarized in Fig. 24.34. As can be seen, it is possible to control the nature of the inorganic precursors (by using the methods described in Section 24.5), the nature of the organic material, and the method of assembling the composite material. The templated-growth fabrication processes often rely upon organic molecules and macromolecules as structure-directing agents to achieve the construction of complex hierarchical architectures. It turns out that molecular and supramolecular interactions between template molecules (surfactants, amphiphilic block copolymers, organogelators, etc.) and hybrid or metal-oxo-based networks and NBBs allow a rich variety of nanocomposite materials to be prepared. Figure 24.35 illustrates two approaches to the assembly of NBBs into larger, ordered structures.

24.18 Polymer nanocomposites

Key points: The properties of polymer nanocomposites are controlled through the nature of the polymeric and inorganic phases, as well as through their dispersions and interactions; their mechanical properties can be controlled using nanocarbons and metal oxides as the dispersed nanophase.

Polymer nanocomposites (PNCs) are composed of inorganic nanoparticles dispersed in a polymeric matrix. Early commercial PNCs used 2D ordered or lamellar clays, such as sodium montmorillonite (Na-MMT), dispersed in the polymer matrix. Such dispersants (or fillers) have sandwich-type structures (with channels between layers), a total thickness of 0.3–1 nm, and a length of 50–100 nm for each layer; these sandwich structures typically agglomerate into micrometre-sized agglomerates. The dispersion of the inorganic phase within the organic polymeric matrix in a PNC is obtained by intercalation (the removal of a thin coating from its substrate). Intercalation involves the expansion of the inter-lamellar spacing as a consequence of ion exchange using organic amines or quaternary ammonium salts. Exfoliation is achieved through reactive chemical compounding or by intensive melt mixing of the clay and polymer phases. Without proper dispersion, the

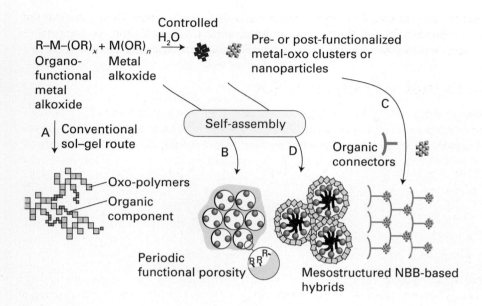

R–M–(OR)$_x$ + M(OR)$_n$
Organo-functional metal alkoxide   Metal alkoxide

Controlled H$_2$O →

Pre- or post-functionalized metal-oxo clusters or nanoparticles

A | Conventional sol–gel route

Self-assembly

B   D   C

Organic connectors

Oxo-polymers
Organic component

Periodic functional porosity

Mesostructured NBB-based hybrids

Fig. 24.34 Different routes to obtain hybrid nanomaterials. Route A is a conventional sol–gel route that leads to 'ordinary' hybrids. Routes B and D involve the use of templates capable of self-assembly, giving rise to meso-organized phases. Routes C and D involve the assembly of nanobuilding blocks (NBBs). (Adapted from C. Sanchez, G.J.A.A. Soler-Illia, F. Ribot, T. Lalot, C.R. Mayer, and V. Cabuil, *Chem. Mater.*, 2001, **13**, 3061.)

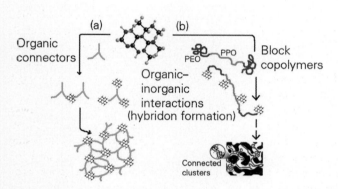

(a)   (b)

Organic connectors

Organic–inorganic interactions (hybridon formation)

PEO   PPO   Block copolymers

Connected clusters

Fig. 24.35 Different approaches to the assembly of nanobuilding blocks (NBBs) involving clusters and different connectors. (a) Ordered cluster dispersions based on the covalent linkage of complementary organic and inorganic NBBs. (b) Creation of a bicontinuous mesostructured hybrid arrangement. Interactions between the template and the inorganic NBB include covalent bonding and van der Waals forces (hydrophobic contacts). (Adapted from C. Sanchez, G.J.A.A. Soler-Illia, F. Ribot, T. Lalot, C.R. Mayer, and V. Cabuil, *Chem. Mater.*, 2001, **13**, 3061.)

nanosized filler particles aggregate into larger clusters causing degradation in the properties of the composite.

Many techniques can be used to control the state of dispersion as well as the nature of the bond between nanoparticle and matrix, including the use of silanes, grafting, and CVD. Silanes are being produced that perform both tasks simultaneously: they both tailor the surface properties of the filler, which promotes the coupling of the nanoparticle to the polymer matrix, and act as a surfactant, reducing the surface energy of the filler to prevent agglomeration and promote dispersion. Radiation copolymer grafting using γ-rays, electron beams, or X-rays is very effective in modifying polymer surfaces. In most cases the polymer chains that are created strengthen the polymer–filler bonds. This is also the case with CVD, and the gaseous molecules that form a film on the substrate typically increase the bond strength between filler and polymer.

The nature of the filler (dispersant) and any cavities (voids) present in a PNC greatly influences the mechanical properties of the composites because they can control the distribution of stress through the composite matrix. Both the yield strength (a measure of

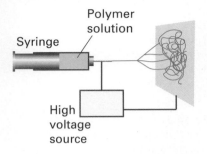

Fig. 24.36 The electrospinning apparatus. (Adapted from R. Sen, B. Zhao, D. Perea, M.E. Itkis, H. Hu, J. Love, *et al.*, *Nano Lett.*, 2004, **4**, 459.)

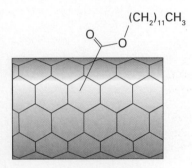

Fig. 24.37 Ester-functionalized SWNTs (SWNT-COO(CH₂)₁₁CH₃, EST-SWNTs). (Based on R. Sen *et al.*, *Nano Lett.*, 2004, **4**, 459.)

the material's ability to resist permanent deformation) and toughness (a measure of the energy absorbed prior to fracture) of nanocomposites are controlled by the nanoparticle size and dispersion and nanoparticle-to-polymer contact interactions. Therefore, control of the nanoparticle dispersion and alignment of these fillers offers a way to tailor the mechanical properties of the nanocomposites.

One important class of PNCs uses nanocarbons as the dispersants. Single-walled nanotubes (SWNTs) have exceptional mechanical properties; their very high Young's modulus (their resistance to stress), low densities, and high aspect ratios make them very attractive as fillers to enhance polymer strength. SWNT-reinforced composites have been developed by the physical mixing of SWNTs in solutions of preformed polymers, *in situ* polymerization in the presence of SWNTs, surfactant-assisted processing of SWNT–polymer composites, and chemical modification of the incorporated SWNTs. Melt processing can be carried out easily by compression moulding at high temperatures and pressures followed by rapid quenching. The so-called **electrospinning methods** use electrostatic forces to distort a droplet of polymer solution into a fine filament, which is then deposited on a substrate (Fig. 24.36). The nanofibres created from the electro-spinning can be configured into a variety of forms including membranes, coatings, and films, and can be deposited on targets of different shapes. Fibres can be prepared with diameters smaller than 3 nm, with high surface-to-volume and length-to-diameter ratios, and with controlled pore sizes. Electrospun polymer nanocomposites have been made with homogeneously dispersed SWNTs and have exhibited significantly improved mechanical strength.

Different organo-functional groups covalently attached to the nanotubes through oxidation reactions have been used to improve their chemical compatibility with specific polymers (Fig. 24.37). The use of SWNTs with different functionalities allows the study of the importance of the interfacial interactions between the filler and the polymer matrix. This chemical functionalization is an effective approach towards improving the processability of the nanocomposite and the chemical compatibility of the components. As described before, specific functionalization can be used to exfoliate the SWNT bundles and prevent their agglomeration.

Two polymers that have been investigated for their use in PNCs with ester-functionalized SWNTs are polystyrene and polyurethane. Polystyrene is a thermoplastic, which softens and melts when heated, and is both tough and brittle; polyurethane (PU) is an elastomer with applications such as coatings, sealants, transdermal patches, and catheters. From stress–strain analysis (Fig. 24.38) it is clear that the ester-functionalized SWNTs (EST-SWNT-PU) perform better than the 'as-prepared' SWNTs (AP-SWNT-PU), and that both composites show improved Young's moduli (the ratio of stress to strain) over the polymer without nanotube fillers.

Metal oxide fillers have also been used to optimize polymer strengths and thermal properties. A particularly interesting example involves the control over the mechanical properties of alumina/polymethylmethacrylate (PMMA) nanocomposites by engineering the distribution of weak particle-to-polymer interactions. The filler nanoparticles (alumina) are dispersed in the composite matrix in the form of 'net-like' structures as opposed to 'isolated island' structures formed when microparticles are dispersed in the polymer matrix. In contrast to microparticle–polymer composites, these net-like structures inhibit crazing, the propagation of microscopic cracks during tensional loading. Alumina/PMMA composites also display higher yield strengths and a shift from brittle to ductile behaviour when nanoparticulate alumina is used. Ductility is advantageous as the composite can then be drawn into thin wires and can withstand sudden impact, thus making the composite more resilient mechanically. The addition of nano-alumina with the anti-agglomerant methacrylic acid lowers the glass transition temperature enough to change the stress–strain curve of the composite and shift from brittle to ductile under tensile loading (Fig. 24.39).

There is a new world of inorganic chemistry, as we have sought to show, waiting in the wings of our subject: this intermediate domain between the atomic and the bulk is waiting for discovery and exploitation. To develop it, it is essential to have a reliable and

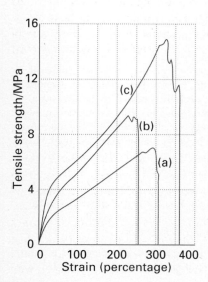

Fig. 24.38 Stress–strain curve for various membranes: (a) PU, (b) AP-SWNT-PU, and (c) EST-SWNT-PU. The SWNT-to-PU mass ratio is 1:100 in the composite membranes; (c) has a higher strength (related to the first transition in slopes, a higher ductility (related to the overall strain), and a higher toughness (related to the area under the curve). (Adapted from R. Sen *et al.*, *Nano Lett.*, 2004, **4**, 459.)

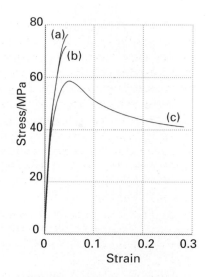

Fig. 24.39 Typical stress–strain curves for (a) neat PMMA, (b) 2 per cent by mass as-received micrometre-sized alumina-filled PMMA composite, and (c) 2.2 per cent by mass 38 nm (MAA) alumina/PMMA nanocomposite. Although the strength decreases slightly for (c) (related to the decreased stress at the curve maximum), the overall ductility (related to the total strain) and toughness (related to the area under the curve) are greatly improved. (Adapted from B.J. Ash, R.W. Siegel, and L.S. Schadler, *Macromolecules*, 2004, **37**, 1358.)

extensive grounding in the conventions of inorganic chemistry, to note what Nature has already achieved, and—above all—to build on it with imagination.

FURTHER READING

R.P. Feynman, There's plenty of room at the bottom. *Eng. Sci.*, 1960, **23**, 22. This lecture is heralded as the beginning of the age of nanotechnology.

C.P. Poole and F.J. Owens, *Introduction to nanotechnology*. Wiley–Interscience, Hoboken (2003). This book provides key chapters on a variety of nanomaterials systems including quantum structures, magnetic nanomaterials, nanoelectromechanical systems (NEMS), carbon nanotubes, and nanocomposites with emphasis on characterization and synthesis strategies for each system.

Understanding nanotechnology. Reprinted by Scientific American, 2002. This collection of easy to understand articles that originally appeared in *Scientific American* in 2001 offers an outstanding overview of the key issues of nanoscience, including coverage of top-down versus bottom-up methods of synthesis and applications of nanotechnology in medicine, computers, and telecommunications.

M. Ratner and D. Ratner, *Nanotechnology: the next big idea*. Prentice Hall, Upper Saddle River (2003). A wonderful general approach to nanoscience with broad coverage of key issues in the field.

M. Wilson, K. Kannangara, G. Smith, M. Simmons, and B. Raguse (eds), *Nanotechnology: basic science and emerging technologies*. CRC Press, Boca Raton (2002). This is another good introductory book on many areas of nanotechnology.

E. Bhushan (ed.), *Handbook of nanotechnology*. Springer, Berlin (2004). An overarching technical introduction to nanotechnology.

P.N. Prasad, *Nanophotonics*. Wiley–Interscience, Hoboken (2003). A good introduction to optical properties and processes in nanostructured materials.

Supramolecular chemistry and self-assembly. Special issue of *Science* (March 2002). This collection of articles describes key research on the overlap between supramolecular chemistry and self-assembled nanostructures; it includes surveys of chemical approaches to link nanobuilding blocks to generate extended framework structures.

Inorganic–organic nanocomposites. Special issue of *Chemistry of Materials* (October 2001). This collection of articles describes nanocomposites with broad applications in photonics, electronics, and chemical sensing.

J. Hu, T.W. Odom, and C.M. Leiber, Chemistry and physics in one dimension: synthesis and properties of nanowires and nanotubes. *Acc. Chem. Res.*, 1999, **32**, 435. An excellent review of issues of controlled dimensionality in nanomaterials including important pioneering work on transport properties of carbon nanotubes.

T. Masciangioli and W.-X. Zhang, Environmental technologies at the nanoscale. *Environ. Sci. Technol., A-Pages*, 2003, **37**, 102A. A short survey of the future impacts of nanoscience on our environment.

C.J. Murphy, Optical sensing with quantum dots. *Anal. Chem.*, 2002, **74**, 520A. A well constructed survey of the key advantages and limitations of using QDs for optical sensing.

EXERCISES

24.1 (a) Compare the surface area for two spherical objects: one has a diameter of 10 nm and the other has a diameter of 1000 nm. (b) Describe whether these two objects are considered nanoparticles using the size-based definition of a nanomaterial. (c) Using a surface area-related property, describe what must hold true for either of these objects to be considered nanoparticles using the size/properties-based definition of a nanomaterial.

24.2 Quantum confinement bestows unique properties on semiconductor nanocrystals compared to the bulk semiconductor. For this to happen, some characteristic length that describes an electron in a crystal and the size of the crystal must be similar. Describe a characteristic length for an electron and how it becomes comparable to the particle size in quantum confinement.

24.3 Explain why quantum dots might be superior to organic fluorophores for bioimaging applications.

24.4 Compare and contrast the band energies for a QD nanocrystal and a bulk semiconductor.

24.5 (a) Explain the difference between the top-down and bottom-up methods of fabrication of materials. Be specific and provide one example of each. (b) Give one advantage and one disadvantage for each synthesis method.

24.6 (a) Give a definition for scanning probe microscopy that makes clear both what it is and why it is called scanning probe microscopy. (b) Using a material of interest to you, pick any scanning probe microscopy method and describe how you might use it to characterize an important aspect of your material.

24.7 Distinguish between SEM and TEM. What is the principal difference in sample preparation and detection?

24.8 (a) Describe the three basic steps in nanoparticle formation from solution. (b) Explain why the latter two steps should occur independently to achieve a uniform size distribution. (c) What are stabilizer molecules used for in nanoparticle synthesis?

24.9 Explain whether vapour-phase or solution-based techniques typically lead to (a) larger size distributions in nanoparticle synthesis, (b) agglomerated particles that are strongly bonded to one another in so-called hard agglomerates.

24.10 (a) Draw a schematic diagram of a core–shell nanoparticle. (b) Briefly describe how core–shell nanoparticles could be made using either vapour-phase or solution-based techniques. (c) For what purpose would core–shell nanoparticles be used?

24.11 (a) Discuss the difference between homogeneous and heterogeneous nucleation from the vapour phase. (b) Which type of nucleation is preferred for the growth of a thin film in this process? (c) Which type of nucleation is preferred for the growth of nanoparticles in this process?

24.12 Describe the difference between a physical vapour and a chemical vapour with respect to the type and stability of the vapour species.

24.13 (a) Superlattices of ordered quantum dots (QDs) and an overlaying semiconductor can be considered core–shell QD arrays. What is the purpose of building multiple layers on the QD?

(b) What are the limitations on the order and types of semiconductors that can be put together?

24.14 In Fig. 24.9, a TEM micrograph is shown of an artificial structure of $(PrMnO_3)_2(SrMnO_3)_2$ deposited on $SrTiO_3$ epitaxially. (a) What is meant by 'epitaxy'? (b) Describe whether the diagram illustrates that there is atomic order between all of the layers (the contrast represents columns of atoms). (c) The material was grown layer-by-layer by using a physical vapour method. Of the following combinations of growth, which refer to homoepitaxial situations and which refer to heteroepitaxial combinations: $PrMnO_3$ on $SrTiO_3$; $PrMnO_3$ on $PrMnO_3$; $SrMnO_3$ on $PrMnO_3$?

24.15 (a) Give two examples of applications of quantum wells. (b) Describe why quantum wells are used and if either molecular materials or traditional solid-state materials can exhibit similar properties. (c) How are quantum wells made?

24.16 (a) Use any of the solid-state superlattices described in the chapter to explain why the use of nanostructured materials led to enhanced properties.

24.17 (a) Use examples of artificially layered crystal structures to explain how the techniques of nanotechnology allow for the growth of materials described in Chapter 23. (b) Use Fig. 24.15 to describe how the techniques of nanotechnology allow for the growth of new materials that cannot be synthesized using the techniques described in that chapter. Bear in mind that only the structures a–c are known in bulk form.

24.18 (a) What is the relevance of self-assembly to the fabrication of nanomaterials? (b) What role will it play in nanotechnology?

24.19 Discuss briefly the features common to self-assembly processes.

24.20 Distinguish between static and dynamic self-assembly. Give an example of each type.

24.21 How is a self-assembled monolayer (SAM) comprising gold–organothiol linkages related to the cell membrane within our bodies? Use a sketch to make your comparison, define the term surfactant, and distinguish between the hydrophilic and hydrophobic regions.

24.22 Define morphosynthesis. Give an example of how this approach can be used to control nanoarchitecture.

24.23 Give an example of a bionanomaterial and its application in nanotechnology.

24.24 (a) Define biomimetics. (b) Describe biomimetics with respect to how artificial fossilization is used to create titania paper.

24.25 Describe why mechanical properties are a key metric of the quality of artificial bone materials. Use the example described in the text to explain how chemistry plays a major role in enhancing the mechanical properties of bionanocomposite bone material.

24.26 (a) Describe the two classes of inorganic–organic nanocomposites based on their bonding types. (b) Give one example of a nanocomposite in each class.

24.27 (a) Explain why the state of dispersion of inorganic nanoparticles is important in inorganic–organic nanocomposites. (b) Use the concept of oil and water dispersions to explain why highly dispersed inorganic nanoparticles might be difficult to achieve in these nanocomposites.

PROBLEMS

24.1 The synthesis method described in the chapter to generate CdSe quantum dots involves the use of rather toxic compounds. Explore the chemical literature to find a more recent example that uses less toxic substances. Describe the solvation step, the nucleation step, and the growth step in each case. Comment on the size dispersions in each case. [See the following articles as a start: G.C. Lisensky and E.M. Boatman, *J. Chem. Educ.*, 2005, **82**, 1360; W. William Yu and X.-G. Peng, *Angew. Chem., Int. Ed. Engl.*, **2002**, 41, 2368.]

24.2 The Graetzel cell has been described as a useful photoelectrochemical cell. Describe how the photoelectrochemical cell differs from the photovoltaic cell. Describe why the nanostructure TiO_2 is important to the improved performance. What other inorganic chemical compound plays an important role in functionalizing the Graetzel cell? (See M. Graetzel, *Nature*, 2001, **414**, 338.)

24.3 Compare and contrast the atomic-layer deposition (based on chemical interactions) and the MBE deposition (based on physical interactions of atomic beams) with respect to system cost, speed of process, ambient atmospheric conditions, *in situ* monitoring abilities, and precision ability to design artificial structures and superlattices. Describe which has been used to produce more materials and describe the general quality of the materials.

24.4 One of the key challenges of nanotechnology is achieving mechanical devices that function on the nanoscale. One such device is a Brownian ratchet. Describe how the Brownian ratchet mechanism has been used to sort DNA. (See J.S. Bader, R.W. Hammond, S.A. Henck, M.W. Deem, G.A. McDermott, J.M. Bustillo, *et al.*, *Proc. Natl. Acad. Sci.*, 1999, **96**, 13165.)

24.5 Carbon nanotubes have been suggested for their use as wires in molecular electronics. Describe the challenges of using nanotubes as wires in terms of connecting two functional electronic devices. Describe a possible technique to overcome some of the problems.

24.6 Show that folded DNA has an information density (density of base-pairs) equal to $1\,Tb\,cm^{-2}$. IBM has proposed a scanning probe data storage device called the Millipede, in which an array of scanning probe tips reads and writes nanoscale marks on a substrate. What would the footprint (the mark plus the space between it and one of its neighbours) of a square mark need to be to attain an information density of $1\,Tb\,cm^{-2}$? Describe the method proposed to make such marks and to read them.

24.7 Describe an example of where a chemical reaction was carried out in an inverse micelle 'beaker' to produce nanoparticles. What was the reaction, the volume, the product, and the size of the product? Is it possible to control the size of the product by using simple modifications to basic set-up? If so, how?

24.8 Nanotubes are widely known for carbon. Find an example of an inorganic nanotube not based on carbon and describe its synthesis and properties as compared to the corresponding bulk material. Describe the structure compared to the carbon nanotubes discussed in this chapter.

25 Catalysis

In this chapter we apply the concepts of organometallic chemistry, coordination chemistry, and materials chemistry to catalysis. We emphasize general principles, such as the nature of catalytic cycles, in which a catalytic species or surface is regenerated in a reaction, and the delicate balance of reactions required for a successful cycle. We see that there are numerous requirements for a successful catalytic process: the reaction being catalysed must be thermodynamically favourable and fast enough when catalysed; the catalyst must have an appropriate selectivity towards the desired product and a lifetime long enough to be economical. We then survey homogeneously catalysed reactions and show how proposals about mechanisms are invoked. The final part of the chapter develops a similar theme in heterogeneous catalysis, and we shall see that many parallels exist between homogeneous and heterogeneous catalysis. In neither type of catalysis are mechanisms necessarily finally settled and there is still considerable scope for making new discoveries.

As will be familiar from introductory chemistry, a **catalyst** is a substance that increases the rate of a reaction but is not itself consumed. Catalysts are widely used in nature, in industry, and in the laboratory, and it is estimated that they contribute to one-sixth of the value of all manufactured goods in industrialized countries. As shown in Table 25.1, 15 of the top 20 synthetic chemicals in the USA are produced directly or indirectly by catalysis. For example, a key step in the production of a dominant industrial chemical, sulfuric acid, is the catalytic oxidation of SO_2 to SO_3. Ammonia, another chemical essential for industry and agriculture, is produced by the catalytic reduction of N_2 by H_2. Inorganic catalysts are also used for the production of the major organic chemicals and petroleum products, such as fuels, petrochemicals, and polyalkene plastics. Catalysts play a steadily increasing role in achieving a cleaner environment, both through the destruction of pollutants (as with the catalytic converters found on the exhaust systems of vehicles) and through the development of cleaner industrial processes with fewer unwanted by-products. Enzymes, a class of elaborate biochemical catalysts, are discussed in Chapter 26.

In addition to their economic importance and contribution to the quality of life, catalysts are interesting in their own right: the subtle influence a catalyst has on reagents can completely change the outcome of a reaction. The understanding of the mechanisms of catalytic reactions has improved considerably in recent years with the greater availability of isotopically labelled molecules, improved methods for determining reaction rates, improved spectroscopic and diffraction techniques, and much more reliable molecular orbital calculations.

General principles

A catalysed reaction is faster than an uncatalysed version of the same reaction because the catalyst provides a different reaction pathway with a lower activation energy. The term *negative catalyst* is sometimes applied to substances that retard reactions. We shall not use the term because these substances are best considered to be **catalyst poisons** that block one or more elementary steps in a catalytic reaction.

Table 25.1 The top 20 synthetic chemicals in the USA in 2004 (based on mass)

Rank	Chemical	Catalytic process	Rank	Chemical	Catalytic process
1	Sulfuric acid	SO_2 oxidation, heterogeneous	11	Urea	NH_3 precursor catalytic
2	Ethene	Hydrocarbon cracking, heterogeneous	12	Ethylbenzene	Alkylation of benzene, homogeneous
3	Propene	Hydrocarbon cracking, heterogeneous	13	Styrene	Dehydrogenation of ethylbenzene, heterogeneous
4	Chlorine	Electrolysis, not catalytic	14	HCl	Precursors catalytic
5	1,2-Dichloroethane	Ethene + Cl_2, heterogeneous	15	Ethylene oxide	Ethene + O_2, heterogeneous
6	Phosphoric acid	Not catalytic	16	Cumene	Alkylation of benzene, heterogeneous
7	Ammonia	$N_2 + H_2$, heterogeneous	17	Ammonium sulfate	Precursors catalytic
8	Sodium hydroxide	Electrolysis, not catalytic	18	Sodium carbonate	Not catalytic
9	Nitric acid	$NH_3 + O_2$, heterogeneous	19	Butadiene	Dehydrogenation of butane, heterogeneous
10	Ammonium nitrate	Precursors catalytic	20	Titanium dioxide	Not catalytic

Source: Facts & Figures for the Chemical Industry, *Chem. Eng. News*, 2005, **83**, 67.

25.1 The language of catalysis

Before we discuss the mechanism of catalytic reactions, we need to introduce some of the terminology used to describe the rate of a catalytic reaction and its mechanism.

(a) Energetics

Key point: A catalyst increases the rates of processes by introducing new pathways with lower Gibbs energies of activation; the reaction profile contains no high peaks and no deep troughs.

A catalyst increases the rates of processes by introducing new pathways with lower Gibbs energies of activation, $\Delta^{\ddagger}G$. It is important to focus on the Gibbs energy profile of a catalytic reaction, not just the enthalpy or energy profile, because the new elementary steps that occur in the catalysed process are likely to have quite different entropies of activation. A catalyst does not affect the Gibbs energy of the overall reaction, $\Delta_r G^{\ominus}$, because G is a state function.[1] The difference is illustrated in Fig. 25.1, where the overall reaction Gibbs energy is the same in both energy profiles. Reactions that are thermodynamically unfavourable cannot be made favourable by a catalyst.

Figure 25.1 also shows that the Gibbs energy profile of a catalysed reaction contains no high peaks and no deep troughs. The new pathway introduced by the catalyst changes the mechanism of the reaction to one with a very different shape and with lower maxima. However, an equally important point is that stable or nonlabile catalytic intermediates do not occur in the cycle. Similarly, the product must be released in a thermodynamically favourable step. If, as shown by the blue line in Fig. 25.1, a stable complex is formed with the catalyst, it would turn out to be the product of the reaction and the cycle would terminate. Similarly, impurities may suppress catalysis by coordinating strongly to catalytically active sites and act as catalyst poisons.

(b) Catalytic cycles

Key point: A catalytic cycle is a sequence of reactions that consumes the reactants and forms products, with the catalytic species being regenerated after the cycle.

The essence of catalysis is a cycle of reactions that consumes the reactants, forms products, and regenerates the catalytic species. A simple example of a catalytic cycle involving a homogeneous catalyst is the isomerization of 3-hydroxypropene (allyl alcohol $(CH_2{=}CHCH_2OH)$) to propanal (CH_3CH_2CHO) with the catalyst $[CoH(CO)_3]$. The

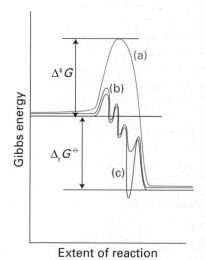

Fig. 25.1 Schematic representation of the energetics of a catalytic cycle. The uncatalysed reaction (a) has a higher $\Delta^{\ddagger}G$ than a step in the catalysed reaction (b). The Gibbs energy of the overall reaction, $\Delta_r G^{\ominus}$, is the same for routes (a) and (b). The curve (c) shows the profile for a reaction mechanism with a stable intermediate.

[1] That is, G depends only on the current state of the system and not on the path that led to the state.

first step is the coordination of the reactant to the catalyst. That complex isomerizes in the coordination sphere of the catalyst and goes on to release the product and reform the catalyst (Fig. 25.2). As with all mechanisms, this cycle has been proposed on the basis of a range of information like that summarized in Fig. 25.3. Many of the components shown in the diagram were encountered in Chapter 20 in connection with the determination of mechanisms of substitution reactions. However, the elucidation of catalytic mechanisms is complicated by the occurrence of several delicately balanced reactions, which often cannot be studied in isolation.

Two stringent tests of any proposed mechanism are the determination of rate laws and the elucidation of stereochemistry. If intermediates are postulated, their detection by magnetic resonance and IR spectroscopy also provides support. If specific atom-transfer steps are proposed, then isotopic tracer studies may serve as a test. The influences of different ligands and different substrates are also sometimes informative. Although rate

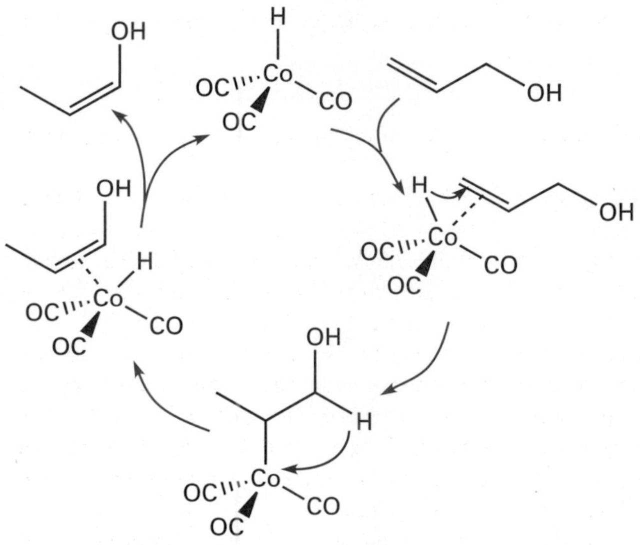

Fig. 25.2 The catalytic cycle for the isomerization of propene-3-ol to propanal.

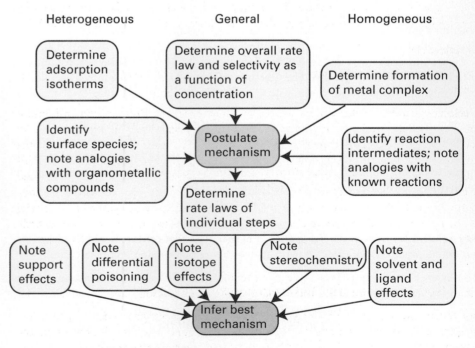

Fig. 25.3 The determination of catalytic mechanisms.

data and the corresponding laws have been determined for many overall catalytic cycles, it is also necessary to determine rate laws for the individual steps in order to have reasonable confidence in the mechanism. However, because of experimental complications, it is rare that catalytic cycles are studied in this detail.

(c) Catalytic efficiency and lifetime

Key points: A highly active catalyst, one that results in a fast reaction even in low concentrations, has a large turnover frequency. A catalyst must be able to survive a large number of catalytic cycles if it is to be of use.

The **turnover frequency**, N (formerly 'turnover number'), is often used to express the efficiency of a catalyst. For the conversion of A to B catalysed by Q and with a rate v,

$$A \xrightarrow{Q} B \quad v = \frac{d[B]}{dt} \tag{1}$$

then provided the rate of the uncatalysed reaction is negligible, the turnover frequency is

$$N = \frac{v}{[Q]} \tag{2}$$

A highly active catalyst, one that results in a fast reaction even in low concentrations, has a high turnover frequency.

In heterogeneous catalysis, the reaction rate is expressed in terms of the rate of change in the amount of product (in place of concentration) and the concentration of catalyst is replaced by the amount present. The determination of the number of active sites in a heterogeneous catalyst is particularly challenging, and often the denominator [Q] in eqn 25.2 is replaced by the surface area of the catalyst.

A catalyst must survive through a large number of cycles if it is to be economically viable. However, it may be destroyed by side reactions to the main catalytic cycle or by the presence of small amounts of impurities in the starting materials (the feedstock). For example, many alkene polymerization catalysts are destroyed by O_2, so in the synthesis of polyethylene and polypropylene the concentration of O_2 in the ethene or propene feedstock should be no more than a few parts per billion.

Some catalysts can be regenerated quite readily. For example, the supported metal catalysts used in the reforming reactions that convert hydrocarbons to high-octane gasoline become covered with carbon because the catalytic reaction is accompanied by a small amount of dehydrogenation. These supported metal particles can be cleaned by interrupting the catalytic process periodically and burning off the accumulated carbon.

(d) Selectivity

Key point: A selective catalyst yields a high proportion of the desired product with minimum amounts of side products.

A **selective catalyst** yields a high proportion of the desired product with minimum amounts of side products. In industry, there is considerable economic incentive to develop selective catalysts. For example, when metallic silver is used to catalyse the oxidation of ethene with oxygen to produce ethylene oxide (**1**), the reaction is accompanied by the more thermodynamically favoured but undesirable formation of CO_2 and H_2O. This lack of selectivity increases the consumption of ethene, so chemists are constantly trying to devise a more selective catalyst for ethylene oxide synthesis. Selectivity can be ignored in only a very few simple inorganic reactions, where there is essentially only one thermodynamically favourable product, as in the formation of NH_3 from H_2 and N_2.

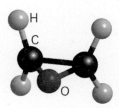

1 Ethylene oxide

25.2 Homogeneous and heterogeneous catalysts

Key points: Homogeneous catalysts are present in the same phase as the reagents, and are often well defined; heterogeneous catalysts are present in a different phase from the reagents with separation of products from the catalyst often being trivial.

Catalysts are classified as **homogeneous** if they are present in the same phase as the reagents; this normally means that they are present as solutes in liquid reaction mixtures. Catalysts are **heterogeneous** if they are present in a different phase from that of the reactants; this normally means that they are present as solids with the reactants present either as gases or in solution. Both types of catalysis are discussed in this chapter, and it will be seen that they are fundamentally similar.

From a practical standpoint, homogeneous catalysis is attractive because it is often highly selective towards the formation of a desired product. In large-scale industrial processes, homogeneous catalysts are preferred for exothermic reactions because it is easier to dissipate heat from a solution than from the solid bed of a heterogeneous catalyst. In principle, every homogeneous catalyst molecule in solution is accessible to reagents, potentially leading to very high activities. It should also be borne in mind that the mechanism of homogeneous catalysis is more accessible to detailed investigation than that of heterogeneous catalysis as species in solution are often easier to characterize than those on a surface and because the interpretation of rate data is frequently easier. The major disadvantage of homogeneous catalysts is that a separation step is required.

Heterogeneous catalysts are used very extensively in industry and have a much greater economic impact than homogeneous catalysts. One attractive feature is that many of these solid catalysts are robust at high temperatures and therefore tolerate a wide range of operating conditions. Another reason for their widespread use is that extra steps are not needed to separate the product from the catalyst, resulting in efficient and more environmentally friendly processes. Typically, gaseous or liquid reactants enter a tubular reactor at one end, pass over a bed of the catalyst, and products are collected at the other end. This same simplicity of design applies to the catalytic converter used to oxidize CO and hydrocarbons and reduce nitrogen oxides in automobile exhausts (Fig. 25.4).

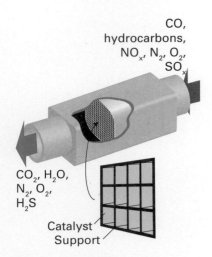

CO, hydrocarbons, NO_x, N_2, O_2, SO_x

CO_2, H_2O, N_2, O_2, H_2S

Catalyst Support

Fig. 25.4 A heterogeneous catalyst in action. The automobile catalytic converter oxidizes CO and hydrocarbons and reduces nitrogen and sulfur oxides. The particles of a metal catalyst are supported on a robust, ceramic honeycomb.

Homogeneous catalysis

In this section we concentrate on some important homogeneous catalytic reactions based on organometallic and coordination complexes. We describe their currently favoured mechanisms, but it should be noted that, as with nearly all mechanistic proposals, catalytic mechanisms are subject to refinement or change as more detailed experimental information becomes available.

Chapters 20 and 21 described reactions that take place at metal centres: these reactions lie at the heart of the catalytic processes we now consider. In general, we need to invoke ligand substitution reactions (Sections 20.1–20.10 and 21.21), redox reactions (Sections 20.11–20.13), oxidative addition and reductive elimination reactions (Section 21.22), migratory insertion reactions (Section 21.25), and 1,2-insertions and β-hydride eliminations (Section 21.26). Together with the direct attack on coordinated ligands, these types of reaction (and in some cases their reverse), in combination, account for the mechanisms of most of the homogeneous catalytic cycles that have been proposed for organic transformations. Other reactions yet to be fully investigated will undoubtedly use other reaction steps.

We shall now see how several individual elementary reactions combine forces and contribute to catalytic cycles. The following examples illustrate our current understanding of the mechanisms of some important types of catalytic reactions and give some insight into the manner in which catalytic activity and selectivity might be altered. As we have already remarked, there is usually even more uncertainty associated with catalytic mechanisms than with mechanisms of simpler kinds of reactions. Unlike simple reactions, a catalytic process frequently contains many steps over which the experimentalist has little control. Moreover, highly reactive intermediates are often present in concentrations too low to be detected spectroscopically. The best attitude to adopt towards these catalytic mechanisms is to learn the pattern of transformations and appreciate their implications but to be prepared to accept new mechanisms that might be indicated by future work.

The scope of homogeneous catalysis ranges across hydrogenation, oxidation, and a host of other processes. Often the complexes of all metal atoms in a group will exhibit catalytic activity in a particular reaction, but the $4d$-metal complexes are often superior as catalysts to their lighter and heavier congeners. In some cases the difference may be associated with the greater substitutional lability of $4d$ organometallic compounds in comparison with their $3d$ and $5d$ analogues. It is often the case that the complexes of costly metals must be used on account of their superior performance compared with the complexes of cheaper metals.

25.3 Hydrogenation of alkenes

Key points: Wilkinson's catalyst, [RhCl(PPh₃)₃], and related complexes are used for the hydrogenation of a wide variety of alkenes at pressures of hydrogen close to 1 atm or less; suitable chiral ligands can lead to enantioselective hydrogenations.

The addition of hydrogen to an alkene to form an alkane is favoured thermodynamically ($\Delta_r G^\ominus = -101\ \mathrm{kJ\ mol^{-1}}$ for the conversion of ethene to ethane). However, the reaction rate is negligible at ordinary conditions in the absence of a catalyst. Efficient homogeneous and heterogeneous catalysts are known for the hydrogenation of alkenes and are used in such diverse areas as the manufacture of margarine, pharmaceuticals, and petrochemicals.

One of the most studied catalytic systems is the Rh(I) complex [RhCl(PPh₃)₃], which is often referred to as **Wilkinson's catalyst**. This useful catalyst hydrogenates a wide variety of alkenes and alkynes at pressures of hydrogen close to 1 atm or less. The dominant cycle (Fig. 25.5) for the hydrogenation of terminal alkenes by Wilkinson's catalyst involves the oxidative addition of H₂ to the 16-electron complex [RhCl(PPh₃)₃] (A), to form the 18-electron dihydrido complex (B). The dissociation of the phosphine ligands from (B) results in the formation of the coordinatively unsaturated complex (C), which then forms the alkene complex (D). Hydrogen transfer from the Rh atom in (D) to the coordinated alkene yields a transient 16-electron alkyl complex (E). This complex takes on a phosphine ligand to produce (F), and hydrogen migration to carbon results in the reductive elimination of the alkane and the formation of (A), which is set to repeat the cycle.

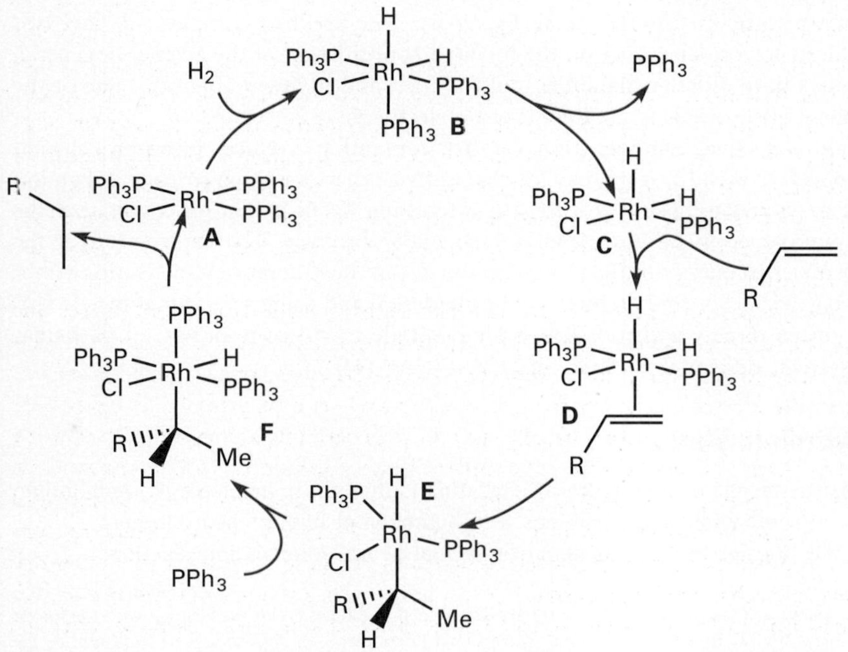

Fig. 25.5 The catalytic cycle for the hydrogenation of terminal alkenes by Wilkinson's catalyst.

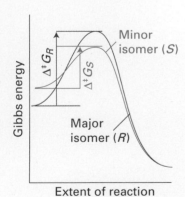

Fig. 25.6 The diasteromeric complexes that may form from a complex with chiral phosphine ligands and a prochiral alkene.

2 DiPAMP

3 L-Dopa

4 Ru(BiNAP)Br$_2$, X=PPh$_2$

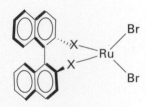

Fig. 25.7 Kinetically controlled stereoselectivity. Note that $\Delta^\ddagger G_S < \Delta^\ddagger G_R$, so the minor isomer reacts faster than the major isomer.

A parallel but slower cycle (which is not shown) is known in which the order of H$_2$ and alkene addition is reversed. Another cycle is known, based around the 14-electron intermediate [RhCl(PPh$_3$)$_2$]. Even though there is very little of this species present, it reacts much faster with hydrogen than [RhCl(PPh$_3$)$_3$] and makes a significant contribution to the catalytic cycle. In this cycle, (E) would eliminate alkane directly, regenerating [RhCl(PPh$_3$)$_2$], which rapidly adds H$_2$ to give (C).

Wilkinson's catalyst is highly sensitive to the nature of the phosphine ligand and the alkene substrate. Analogous complexes with alkylphosphine ligands are inactive, presumably because they are more strongly bound to the metal atom and do not readily dissociate. Similarly, the alkene must be just the right size: highly hindered alkenes or the sterically unencumbered ethene are not hydrogenated by the catalyst, presumably because the sterically crowded alkenes do not coordinate and ethene forms a strong complex that does not react further. These observations emphasize the point made earlier that a catalytic cycle is usually a delicately poised sequence of reactions, and anything that upsets their flow may block catalysis or alter the mechanism.

Wilkinson's catalyst is used in laboratory-scale organic synthesis and in the production of fine chemicals. Related Rh(I) phosphine catalysts that contain a chiral phosphine ligand have been developed to synthesize optically active products in **enantioselective reactions** (reactions that produce a particular chiral product). The alkene to be hydrogenated must be **prochiral**, which means that it must have a structure that leads to R or S chirality when complexed to the metal (Fig. 25.6).[2] The resulting complex will have two diastereomeric forms depending on the mode of coordination of the alkene. In general, diastereomers have different stabilities and labilities, and in favourable cases one or the other of these effects leads to product enantioselectivity.

An enantioselective hydrogenation catalyst containing a chiral phosphine ligand referred to as DiPAMP (**2**) is used by Monsanto to synthesize L-dopa (**3**), a chiral amino acid used to treat Parkinson's disease. An interesting detail of the process is that the minor diastereomer in solution leads to the major product. The explanation of the greater turnover frequency of the minor isomer lies in the difference in activation Gibbs energies (Fig. 25.7). Spurred on by clever ligand design and using a variety of metals, this field has grown rapidly and provides many clinically useful compounds; of particular note are systems derived from ruthenium(II) BiNAP (**4**).

25.4 Hydroformylation

Key point: The mechanism of hydrocarbonylation is thought to involve a pre-equilibrium in which octacarbonyldicobalt combines with hydrogen at high pressure to give a monometallic species that brings about the actual hydrocarbonylation reaction.

---

[2] The designations R and S for a chiral centre are determined as follows. With the element with the lowest atomic number (Z) away from the viewer, the centre is R if the sequence of Z from highest to lowest for the remaining three atoms decreases in a clockwise manner. The centre is S if the decrease in Z occurs in a anticlockwise manner. There are additional rules in the event of atoms having identical atomic numbers.

In a **hydroformylation reaction**, an alkene, CO, and H_2 react to form an aldehyde containing one more C atom than in the original alkene:

$$RCH=CH_2 + CO + H_2 \rightarrow RCH_2CH_2CHO$$

The term 'hydroformylation' derived from the idea that the product resulted from the addition of methanal (formaldehyde) to the alkene, and the name has stuck even though experimental data indicate a different mechanism. A less common but more appropriate name is **hydrocarbonylation**. Both cobalt and rhodium complexes are used as catalysts. Aldehydes produced by hydroformylation are normally reduced to alcohols that are used as solvents and plasticizers, and in the synthesis of detergents. The scale of production is enormous, amounting to millions of tonnes annually.

The general mechanism of cobalt-carbonyl-catalysed hydroformylation was proposed in 1961 by Heck and Breslow by analogy with reactions familiar from organometallic chemistry (Fig. 25.8). Their general mechanism is still invoked, but has proved difficult to verify in detail. In the proposed mechanism, a pre-equilibrium is established in which octacarbonyldicobalt combines with hydrogen at high pressure to yield the known tetra-carbonylhydridocobalt complex:

$$[Co_2(CO)_8] + H_2 \rightarrow 2\,[CoH(CO)_4]$$

This complex, it is proposed, loses CO to produce the coordinatively unsaturated complex $[CoH(CO)_3]$ (B):

$$[CoH(CO)_4] \rightarrow [CoH(CO)_3] + CO$$

It is thought that $[CoH(CO)_3]$ then coordinates an alkene, producing (C) in Fig. 25.8, which undergoes an insertion reaction with the coordinated hydrido ligand, and recoordinates CO. The product at this stage is a normal alkyl complex (D). In the presence of CO at high pressure, (D) undergoes migratory insertion and coordinates another CO, yielding the acyl complex (E), which has been observed by IR spectroscopy under catalytic reaction conditions. The formation of the aldehyde product is thought to occur by attack of either H_2 (as depicted in Fig. 25.8) or the strongly acidic complex $[CoH(CO)_4]$ to yield an aldehyde and generate $[CoH(CO)_4]$ or $[Co_2(CO)_8]$, respectively. Either of these complexes will regenerate the coordinatively unsaturated $[CoH(CO)_3]$.

A significant portion of branched aldehyde is also formed in the cobalt-catalysed hydroformylation. This product may result from a 2-alkylcobalt intermediate formed when reaction of (C) leads to an isomer of (D), with hydrogenation then yielding a

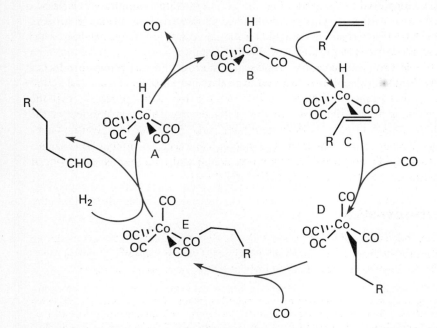

Fig. 25.8 The catalytic cycle for the hydroformylation of alkenes by a cobalt carbonyl catalyst.

Fig. 25.9 The formation of branched aldehydes in hydroformylation reactions occurs when the alkyl group is not terminally bound.

branched aldehyde as set out in Fig. 25.9. When the linear aldehyde is required, such as for the synthesis of biodegradable detergents, the isomerization can be suppressed by the addition of an alkylphosphine to the reaction mixture. One plausible explanation is that the replacement of CO by a bulky ligand disfavours the formation of complexes of sterically crowded 2-alkenes:

$$K \ll 1$$

Here again we see an example of the powerful influence of ancillary ligands on catalysis.

Another effective hydroformylation catalyst precursor is $[RhH(CO)(PPh_3)_3]$ (**5**), which loses a phosphine ligand to form the coordinatively unsaturated 16-electron complex $[RhH(CO)(PPh_3)_2]$, which promotes hydroformylation at moderate temperatures and 1 atm. This behaviour contrasts with the cobalt carbonyl catalyst, which typically requires 150°C and 250 atm. The rhodium catalyst is useful in the laboratory as it is effective under convenient conditions. Because it favours linear aldehyde products, it competes with the phosphine-modified cobalt catalyst in industry.

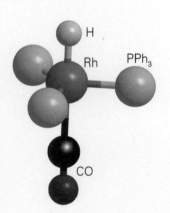

5 $[RhH(CO)(PPh_3)_3]$

Example 25.1 Interpreting the influence of chemical variables on a catalytic cycle

An increase in CO partial pressure above a certain threshold decreases the rate of the cobalt-catalysed hydroformylation of 1-pentene. Suggest an interpretation of this observation.

Answer The decrease in rate with increasing partial pressure suggests that CO suppresses the concentration of one of the catalytic species. An increase in CO pressure will lower the concentration of $CoH(CO)_3$ in the equilibrium

$$[CoH(CO)_4] \rightleftharpoons [CoH(CO)_3] + CO$$

This type of evidence was used as the basis for postulating the existence of $[CoH(CO)_3]$, which is not detected spectroscopically in the reaction mixture.

Self-test 25.1 Predict the influence of added triphenylphosphine on the rate of hydroformylation catalysed by $[RhH(CO)(PPh_3)_3]$.

25.5 Methanol carbonylation: ethanoic acid synthesis

Key point: Both rhodium and iridium complexes are highly active and selective in the carbonylation of methanol to form acetic acid.

The time-honoured method for synthesizing ethanoic (acetic) acid is by aerobic bacterial action on dilute aqueous ethanol, which produces vinegar. However, this process is uneconomical as a source of concentrated ethanoic acid for industry. A highly successful commercial process is based on the carbonylation of methanol:

$$CH_3OH + CO \rightarrow CH_3COOH$$

The reaction is catalysed by all three members of Group 9 (Co, Rh, and Ir). Originally a cobalt complex was used, but then a rhodium catalyst developed at Monsanto greatly reduced the cost of the process by allowing lower pressures to be used. As a result, the rhodium-based **Monsanto process** was used throughout the world. Subsequently, British Petroleum developed the **Cativa process**, which uses a promoted iridium catalyst. Both processes are highly selective and generate ethanoic acid of sufficient purity that it can be used in human food.

The Monsanto and Cativa processes follow essentially the same reaction sequence, so the rhodium-based cycle described here captures the principal features of of the iridium-based process too (Fig. 25.10). Under the conditions used, I^- ions that are present react with methanol to set up an appreciable concentration of iodomethane in the first step of the reaction. Starting with the four-coordinate, 16-electron complex $[RhI_2(CO)_2]^-$ (A), the first step is the oxidative addition of iodomethane to produce the six-coordinate 18-electron complex $[(H_3C)RhI_3(CO)_2]^-$ (B). This step is followed by methyl migration, yielding a 16-electron acyl complex (C). Coordination of CO restores an 18-electron complex (D), which is then set to undergo reductive elimination of acetyl iodide with the regeneration of $[RhI_2(CO)_2]^-$. Water then hydrolyses the acetyl iodide to acetic acid and regenerates HI. Under normal operating conditions, the rate-determining step for the rhodium-based system is the oxidative addition of iodomethane, whereas for the iridium-based system it is the migration of the methyl group. An important feature is that methyl migration on iridium is favoured by formation of a neutral intermediate, and iodide-accepting promoters help facilitate substitution of an iodide ligand by CO in the Ir analogue of complex (B).

25.6 Wacker oxidation of alkenes

Key points: The Wacker process is used to produce acetaldehyde from ethene and oxygen; the most successful system uses a palladium catalyst to oxidize the alkene, with the palladium being reoxidized via a secondary copper catalyst.

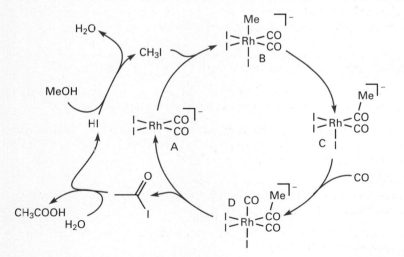

Fig. 25.10 The catalytic cycle for the formation of ethanoic (acetic) acid with a rhodium-based catalyst. The oxidative addition step (A → B) is rate determining.

The **Wacker process** is used primarily to produce ethanal (acetaldehyde) from ethene and oxygen:

$$C_2H_4 + \tfrac{1}{2}O_2 \rightarrow CH_3CHO \qquad \Delta_r G^{\ominus} = -197 \, kJ \, mol^{-1}$$

Its invention at the Wacker Consortium für Elektrochemische Industrie in the late 1950s marked the beginning of an era of production of chemicals from petroleum feedstock. Although the Wacker process is no longer of major industrial concern, it has some interesting mechanistic features that are worth noting.

The actual oxidation of ethene is known to be caused by a palladium(II) salt:

$$C_2H_4 + PdCl_2 + H_2O \rightarrow CH_3CHO + Pd(0) + 2 \, HCl$$

The exact nature of the Pd(0) species is unknown, but it probably is present as a mixture of compounds. The slow oxidation of Pd(0) back to Pd(II) by oxygen is catalysed by the addition of Cu(II), which shuttles back and forth to Cu(I):

$$Pd(0) + 2 \, [CuCl_4]^{2-} \rightarrow Pd^{2+} + 2 \, [CuCl_2]^- + 4 \, Cl^-$$

$$2 \, [CuCl_2]^- + \tfrac{1}{2}O_2 + 2 \, H^+ + 4 \, Cl^- \rightarrow 2 \, [CuCl_4]^{2-} + H_2O$$

The overall catalytic cycle is shown in Fig. 25.11. Detailed stereochemical studies on related systems indicate that the hydration of the alkene–Pd(II) complex (B) occurs by the attack of H_2O from the solution on the coordinated ethene rather than the insertion of coordinated OH. Hydration, to form (C), is followed by two steps that isomerize the coordinated alcohol. First, β-hydrogen elimination occurs with the formation of (D), and then migration of a hydride results in the formation of (E). Elimination of the ethanal and an H^+ ion then leaves Pd(0), which is converted back to Pd(II) by the auxiliary copper(II)-catalysed air oxidation cycle.

One important observation that the mechanism must account for is that, when the reaction is carried out in the presence of D_2O, no deuterium is incorporated into the final product. This observation suggests that either intermediate (D) is very short lived and does not exchange the Pd—H for a Pd—D, or that intermediate (C) rearranges directly to (E).

Alkene ligands coordinated to Pt(II) are also susceptible to nucleophilic attack, but only palladium leads to a successful catalytic system. The principal reason for palladium's unique behaviour appears to be the greater lability of the $4d$ Pd(II) complexes in comparison with their $5d$ Pt(II) counterparts. Furthermore, the potential for the oxidation of Pd(0) to Pd(II) is more favourable than for the corresponding platinum couple.

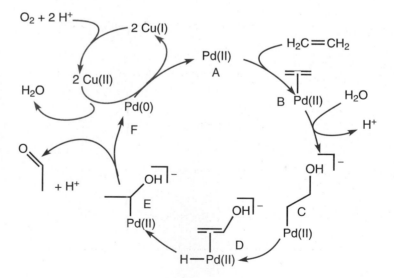

Fig. 25.11 The catalytic cycle for the palladium-catalysed oxidation of alkenes to alkanals (aldehydes).

25.7 Alkene metathesis

Key points: Alkene metathesis reactions are catalysed by homogeneous organometallic complexes that allow considerable control over product distribution; a key step in the reaction mechanism is the dissociation of a ligand from a metal centre to allow an alkene to coordinate.

Alkene metathesis, the redistribution of carbon–carbon double bonds, was first reported in the 1950s. In early studies, poorly defined mixtures of reagents, such as WCl_6/Bu_4Sn and MoO_3/SiO_2, were used to bring about a number of different reactions (Table 25.2). In recent years, a number of newer catalysts have been introduced, and the development of the well-defined alkylidene compound (**6**) by Grubbs in 1992, and now known as **Grubbs' catalyst**, was of seminal importance.[3]

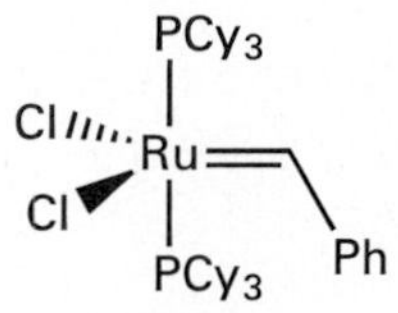

All alkene metathesis reactions were originally thought to proceed via a metallacyclobutane:

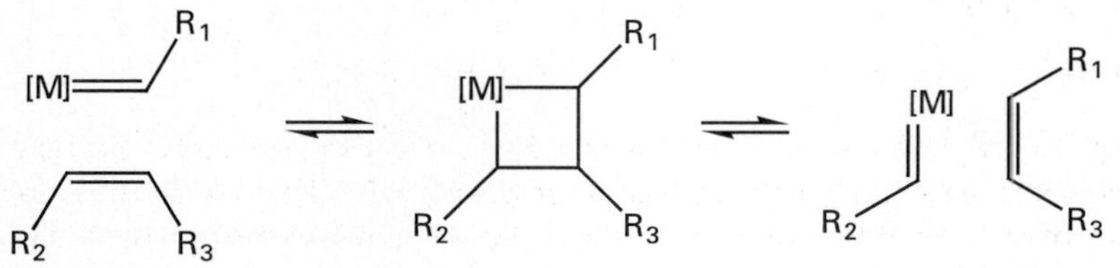

It is now known (at least in the case of Grubbs' catalyst) that the dissociation of phosphine from the metal atom is crucial in allowing the alkene molecule to coordinate, prior to metallacyclobutane formation. The identification of this mechanism led Grubbs to replace one of the phosphine ligands by an N-heterocyclic carbene (NHC) ligand, reasoning that the stronger σ-donor and poorer π-acceptor ability of an NHC ligand would both encourage phosphine ligand dissociation and stabilize the alkene complex. In a triumph of rational design, the so-called **second-generation Grubbs catalyst** (**7**) proved to be more active than the original bisphosphine complex. The second generation Grubbs' catalyst is active in the presence of a large number of different functional groups on substrates and can be used in many solvent systems. It is commercially available and has been widely used, including in the total synthesis of a number of natural products.

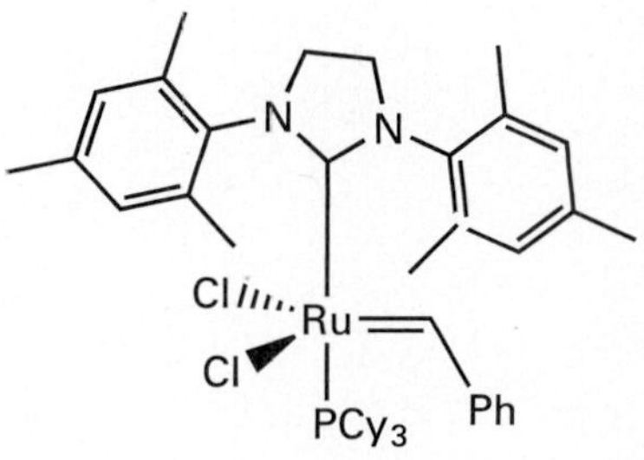

6

7

Table 25.2 The scope of the alkene metathesis reaction

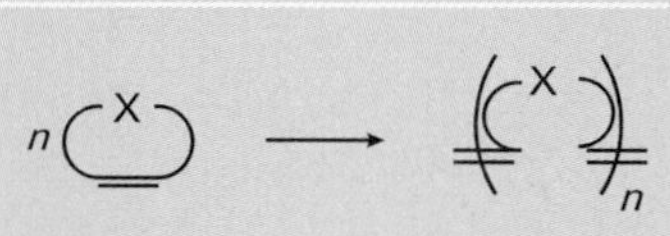

Ring-opening metathesis polymerization (ROMP)

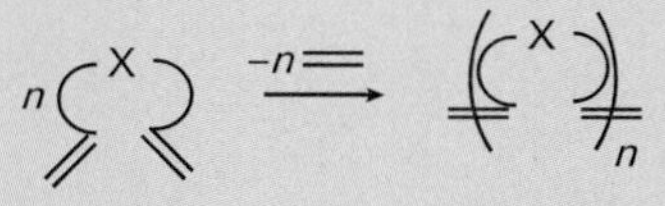

Acyclic diene metathesis polymerization (ADMET)

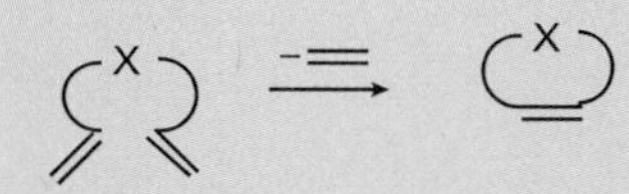

Ring-closing metathesis (RCM)

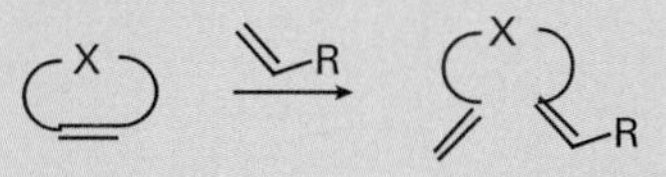

Ring-opening metathesis (ROM)

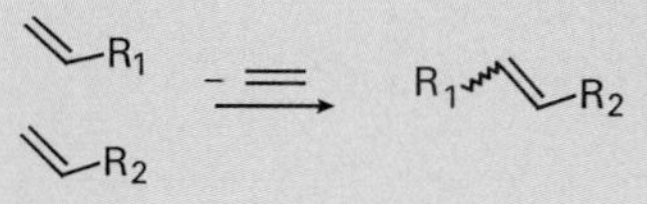

Cross-metathesis (CM or XMET)

---

[3] Yves Chauvin, Robert Grubbs, and Richard Schrock shared the 2005 Nobel Prize for chemistry 'for the development of the metathesis method in organic chemistry'.

25.8 **Palladium-catalysed C—C bond forming reactions**

Key points: A number of palladium-catalysed coupling reactions are known; they all proceed through oxidative addition of reagents at the metal centre followed by the reductive elimination of the two fragments.

A large number of palladium-catalysed carbon–carbon bond forming ('coupling') reactions are known. They include the coupling of a Grignard with an aryl halide and the Heck, Stille, and Suzuki coupling reactions:

Normally, either a palladium(II) complex, such as $Pd(PPh_3)_2Cl_2$, in the presence of additional phosphine or a palladium(0) compound, such as $Pd(PPh_3)_4$, is used as the catalyst, although many other palladium/ligand combinations are active. The precise reaction pathway is unclear (and probably differs with each palladium/ligand/substrate combination) but it is apparent that all these reactions follow the same general sequence. Figure 25.12 shows an idealized catalytic cycle for the coupling of an ethenyl group with an aryl halide. An initial oxidative addition of an aryl–halogen bond to an unsaturated Pd(0) complex (A) results in a Pd(II) species (B). Coordination of an alkene results in complex (C); 1,2-insertion results in a alkyl complex (D), which can be deprotonated with the loss of the halide to give the organic product attached to the palladium atom (E).

In other palladium-catalysed coupling reactions, such as that of a Grignard reagent with an aryl halide, initial oxidative addition proceeds as in Fig. 25.12. The second organic group is thought to be introduced with the Grignard reagent behaving as the nucleophilic R^- group displacing the halide at the metal centre in (B), to give two organic

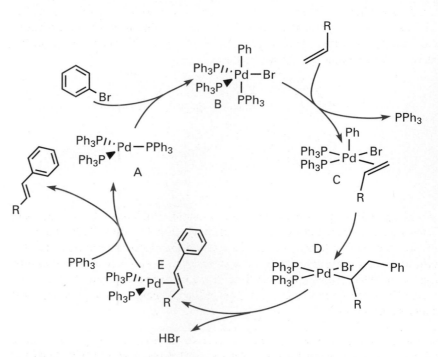

Fig. 25.12 An idealized catalytic cycle for the coupling of a substituted prop-1-ene to an aryl halide in the Heck reaction.

fragments attached to the palladium atom, as indicated in Fig. 25.13. These two adjacent fragments can then couple and reductively eliminate to regenerate the starting Pd(0) species (A).

In all palladium-catalysed coupling reactions, it is necessary for the two fragments that are coupling to be *cis* to each other at the metal centre, before insertion or reductive elimination can take place, and this requirement has led to the use of chelating diphosphines such as dppe (**8**) and the ferrocene derivative (**9**). Palladium-catalysed coupling reactions are tolerant to a wide range of substitution on both fragments and, with suitable substrates and appropriate chiral ligands, asymmetric reactions are possible too; the reaction

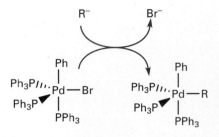

Fig. 25.13 The exchange of a halide for an organic fragment at a Pd centre can be thought of as nucleophilic displacement.

ee = 85%

Ligand:

was the first published example of an **asymmetric Suzuki reaction** with the ligand shown.

25.9 Asymmetric oxidations

Key point: Appropriate chiral ligands can be used in conjunction with *d*-metal catalysts to induce chirality into oxidation products of organic substrates.

In addition to catalysing reductions, *d*-metal complexes are also active in oxidations. For example, in the **Sharpless epoxidation**, 3-hydroxypropene (allyl alcohol) or a derivative is oxidized with *tert*-butylhydroperoxide, in the presence of a Ti catalyst with diethyl tartrate as a chiral ligand, producing an epoxide:

^{t}BuOOH + → Ti(OR)$_4$ → ^{t}BuOH +

Ligand:

The reaction is thought to go through a transition state in which both the peroxide and the allyl alcohol are coordinated to a Ti atom through their O atoms. Each Ti atom is known to have one diethyl tartrate attached to it, and the chiral environment that the diethyl tartrate produces around the Ti atom is sufficient to differentiate the two prochiral faces of the allyl alcohol. Additional experimental evidence points to a dimeric intermediate (**10**). Enantiomeric excesses of greater than 98 per cent have been reported for Sharpless epoxidations.[4]

8 dppe

9

10

---

[4] An enantiomeric excess (ee) is defined as the percentage yield of the major enantiomeric product minus the percentage yield of the minor enantiomeric product. Thus a reaction that gives 51 per cent of one enantiomer and 49 per cent of another would be described as having an enantiomeric excess of 2 per cent; a reaction that gave 99 per cent of one enantiomer and 1 per cent of another would have ee = 98 per cent.

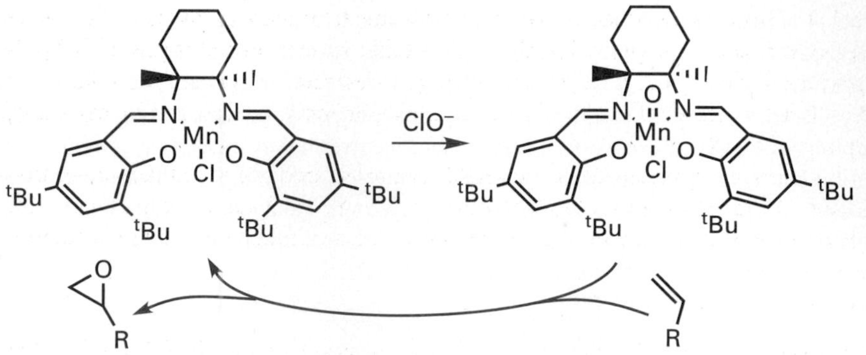

Fig. 25.14 The Jacobsen epoxidation relies on a manganese complex of a salen-based ligand and chlorate(I).

A **Jacobsen oxidation** is a reaction in which the catalyst is a manganese complex of the mixed N_2,O_2 donor ligand known as salen. Hypochlorite ions (ClO^-) are used to oxidize the Mn(III) complex to a Mn(V) oxide, which is then able to deliver its O atoms to an alkene to generate an epoxide. The cycle is depicted in Fig. 25.14. This oxidation has been used with a wide variety of substrates and routinely delivers enantiomeric excesses greater than 95 per cent. The mechanism of the reaction has not been defined precisely, but proposals include the existence of a dimeric form of the catalyst or a radical oxygen transfer step.

Heterogeneous catalysis

Numerous industrial processes are facilitated by heterogeneous catalysis. Practical heterogeneous catalysts are high-surface-area materials that may contain several different phases and operate at pressures of 1 atm and higher. In some cases the bulk of a high-surface-area material serves as the catalyst, and such a material is called a **uniform catalyst.** One example is the catalytic zeolite ZSM-5, which contains channels through which reacting molecules diffuse. More often, **multiphasic catalysts** are used. These consist of a high-surface-area material that serves as a support on to which an active catalyst is deposited (Fig. 25.15). Heterogeneous catalysts generally fall into two categories in terms of the location of the active surfaces. Many heterogeneous catalysts are finely divided solids where the active sites lie on the particle surfaces; others, particularly the microporous zeolite family and mesoporous materials, have pore-like structures and the active sites are the internal surfaces, such as pores and cavities, within the individual crystallites. We shall discuss examples that illustrate some of the range of heterogeneous catalysts, but first we need to describe some of the unique mechanistic features they exhibit.

25.10 The nature of heterogeneous catalysts

There are many parallels between the individual reaction steps encountered in heterogeneous and homogeneous catalysis, but we need to consider some additional points.

(a) Surface area and porosity

Key point: Heterogeneous catalysts are high-surface-area materials formed as either finely divided substrates or crystallites with accessible internal pores.

An ordinary dense solid is unsuitable as a catalyst because its surface area is quite low. Thus α-alumina, which is a dense material with a low specific surface area, is used much less as a catalyst support than the microcrystalline solid γ-alumina, which can be prepared with small particle size, and therefore a high specific surface area (the surface area divided by the mass of the sample). The high surface area results from the many small but connected particles like those shown in Fig. 25.15, and a gram or so of a typical catalyst

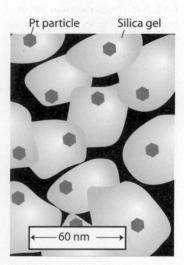

Pt particle   Silica gel

← 60 nm →

Fig. 25.15 Schematic diagram of metal particles supported on a finely divided silica such as silica gel.

support has a surface area equal to that of a tennis court. Similarly, polycrystalline quartz is not used as a catalyst support but the high specific surface area versions of SiO_2, the silica gels, are widely used. In a typical heterogeneous catalyst this substrate surface is coated with active sites or particles, such as metals or metal oxides, producing a large number of active sites.

Both γ-alumina and silica gel are metastable materials, but under ordinary conditions they do not convert to their more stable phases (α-alumina and quartz, respectively). The preparation of γ-alumina involves the dehydration of an aluminium oxide hydroxide:

$$2 \; AlO(OH) \xrightarrow{\Delta} \gamma\text{-}Al_2O_3 + H_2O$$

Similarly, silica gel is prepared from the acidification of silicates to produce $Si(OH)_4$, which rapidly forms a hydrated silica gel from which much of the adsorbed water can be removed by gentle heating. When viewed with an electron microscope, the texture of silica gel or alumina appears to be that of a rough gravel bed with irregularly shaped voids between the interconnecting particles (as in Fig. 25.15). Other high-surface-area materials used as supports in heterogeneous catalysts include TiO_2, Cr_2O_3, ZnO, MgO, and carbon.

Zeolites (Section 23.13) are examples of uniform catalysts. They are prepared as very fine crystals that contain large regular channels and cages defined by the crystal structure (Fig. 25.16; see also Chapter 23). The openings in these channels vary from one crystalline form of the zeolite to the next, but are typically between 0.3 and 2 nm. The zeolite absorbs molecules small enough to enter the channels and excludes larger molecules. This selectivity, in combination with catalytic sites inside the cages, provides a degree of control over catalytic reactions that is unattainable with silica gel or γ-alumina. The synthesis of new zeolites and similar shape-selective solids and the introduction of catalytic sites into them is a vigorous area of research (Section 23.13).

(b) Surface acidic and basic sites

Key point: Surface acids and bases are highly active for catalytic reactions such as the dehydration of alcohols and isomerization of alkenes.

When exposed to atmospheric moisture, the surface of γ-alumina is covered with adsorbed water molecules. Dehydration at 100 to 150°C leads to the desorption of water, but surface OH groups remain and act as weak Brønsted acids:

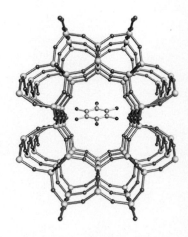

Fig. 25.16 A view into the channels of zeolite theta-1 with an absorbed benzene molecule in the large central channel. (Based on A. Dyer, *An introduction to molecular sieves*. Wiley, Chichester (1988).)

At even higher temperatures, adjacent OH groups condense to liberate more H_2O and generate exposed Al^{3+} Lewis acid sites as well as O^{2-} Lewis base sites (**11**). The rigidity of the surface permits the coexistence of these strong Lewis acid and base sites, which would otherwise immediately combine to form Lewis acid–base complexes. Surface acids and bases are highly active for catalytic reactions such as the dehydration of alcohols and isomerization of alkenes. Similar Brønsted and Lewis acid sites exist on the interior of certain zeolites.

11

Example 25.2 Using IR spectra to probe molecular interaction with surfaces

Infrared spectra of hydrogen-bonded and Lewis acid complexes of pyridine (py), such as $Cl_3Al(py)$, show that bands near $1540 \; cm^{-1}$ can be ascribed to hydrogen-bonded pyridine and that bands near $1465 \; cm^{-1}$ are due to Al–py Lewis acid–base interactions. A sample of γ-alumina that had been pretreated by heating to 200°C and then cooled and exposed to pyridine vapour had absorption bands near $1540 \; cm^{-1}$ and none near $1465 \; cm^{-1}$. Another sample that was heated to 500°C, cooled, and then exposed to pyridine had bands near 1540 and $1465 \; cm^{-1}$. Correlate these results with the statements made in the text concerning the effect of heating γ-alumina. (Much of the evidence for the chemical nature of the γ-alumina surface comes from experiments like these.)

Answer At about 200°C, surface H_2O is lost but OH^- bound to Al^{3+} remains. These groups appear to be mildly acidic as judged by colour indicators and bands at about 1540 cm^{-1} that indicate the presence of hydrogen-bonded pyridine generated in the reaction

$$\begin{array}{ccc} OH + py & & O\text{---}H\text{----}py \\ | & \longrightarrow & | \\ Al & & Al \\ /////// & & /////// \end{array}$$

When heated to 500°C, much but not all the OH is lost as H_2O, leaving behind O^{2-} and exposed Al^{3+}. The evidence is the appearance of the 1465 cm^{-1} bands, which are indicative of Al^{3+}–NC_5H_5, as well as the 1540 cm^{-1} band.

Self-test 25.2 If the statements in this section about the effect of heating γ-alumina are correct, what would be the intensities of the diagnostic IR bands if the γ-alumina sample were pretreated at 900°C, cooled in the absence of water, exposed to pyridine vapour, and a spectrum determined?

(c) Surface metal sites

Key points: Very small metal particles on ceramic oxide substrates are very active catalysts particularly for reactions involving CO; terminal and bridging CO can be identified on surfaces by IR spectroscopy.

Metal particles are often deposited on supports to provide a catalyst. For example, finely divided platinum–rhenium alloys distributed on the surface of γ-alumina particles are used to interconvert hydrocarbons, and finely divided platinum–rhodium alloy particles supported on γ-alumina are used in the catalytic converters of vehicles to promote the combination of O_2 with CO and hydrocarbons to form CO_2 and reduction of nitrogen oxides to nitrogen. A supported metal particle about 2.5 nm in diameter has about 40 per cent of its atoms on the surface, and the particles are protected from fusing together into bulk metal by their separation. The high proportion of exposed atoms is a great advantage for these small supported particles, particularly for metals such as platinum and the even more expensive rhodium.

The metal atoms on the surface of metal clusters are capable of forming bonds such as M—CO, M—CH_2R, M—H, and M—O (Table 25.3). Often the nature of surface ligands is inferred by comparison of IR spectra with those of organometallic or inorganic complexes. Thus, both terminal and bridging CO groups can be identified on surfaces by IR spectroscopy, and the IR spectra of many hydrocarbon ligands on surfaces are similar to those of discrete organometallic complexes. The case of the N_2 ligand is an interesting contrast, because coordinated N_2 was identified by IR spectroscopy on metal surfaces before dinitrogen complexes had been prepared.

The development of new techniques for studying single-crystal surfaces has greatly expanded our knowledge of the surface species that may be present in catalysis. For example, desorption of molecules from surfaces (thermally or by ion or atom impact) combined with mass spectrometric analysis of the desorbed substance provides insight into the chemical identity of surface species. Similarly, Auger and X-ray photoelectron spectroscopy (XPS) provide information on the elemental composition of surfaces. Low-energy electron diffraction (LEED) provides information about the structure of single-crystal surfaces and, when adsorbate molecules are present, their arrangement on the surface. One important finding from LEED is that the adsorption of small molecules on a surface may bring about a structural modification of the surface. This surface reconstruction is often observed to reverse when desorption occurs. Scanning tunnelling microscopy (STM, Section 24.3) provides an unrivalled method for locating adsorbates on surfaces. This striking technique provides a contour map of single-crystal surfaces at or close to atomic resolution.

Although most of these modern surface techniques cannot be applied to the study of supported multiphasic catalysts, they are very helpful for revealing the range of probable surface species and circumscribing the structures that may plausibly be invoked in a

Table 25.3 Chemisorbed ligands on surfaces

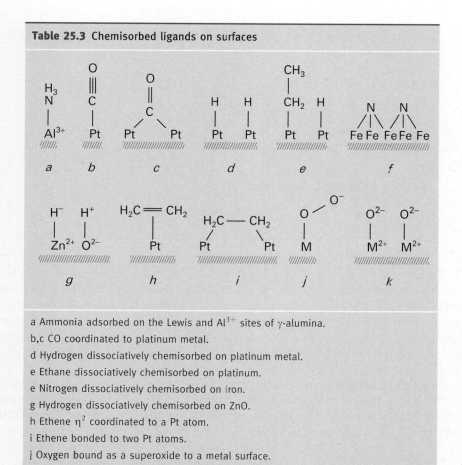

a Ammonia adsorbed on the Lewis and Al^{3+} sites of γ-alumina.

b,c CO coordinated to platinum metal.

d Hydrogen dissociatively chemisorbed on platinum metal.

e Ethane dissociatively chemisorbed on platinum.

e Nitrogen dissociatively chemisorbed on iron.

g Hydrogen dissociatively chemisorbed on ZnO.

h Ethene $η^2$ coordinated to a Pt atom.

i Ethene bonded to two Pt atoms.

j Oxygen bound as a superoxide to a metal surface.

k Oxygen dissociatively chemisorbed on a metal surface.

Adapted from R.L. Burwell, Jr., Heterogeneous catalysis. *Surv. Prog. Chem.*, 1977 **8**, 2.

mechanism of heterogeneous catalysis. The application of these techniques to heterogeneous catalysis is similar to the use of X-ray diffraction and spectroscopy for the characterization of organometallic homogeneous catalyst precursors and model compounds.

(d) Chemisorption and desorption

Key point: Adsorption is essential for heterogeneous catalysis to occur but must not be so strong that it blocks the catalytic sites and prevents further reaction.

The adsorption of molecules on surfaces often activates molecules just as coordination activates molecules in complexes. The desorption of product molecules that is necessary to refresh the active sites in heterogeneous catalysis is analogous to the dissociation of a complex in homogeneous catalysis.

Before a heterogeneous catalyst is used it is usually 'activated'. Activation is a catch-all term; in some instances it refers to the desorption of adsorbed molecules such as water from the surface, as in the dehydration of γ-alumina. In other cases it refers to the preparation of the active site by a chemical reaction, such as by reduction of metal oxide particles to produce active metal particles.

An activated surface can be characterized by the adsorption of various inert and reactive gases. The adsorption may be either **physisorption**, when no new chemical bond is formed, or **chemisorption**, when surface–adsorbate bonds are formed (Fig. 25.17). Low-temperature physisorption of a gas such as nitrogen is useful for the determination of the total surface area of a solid, whereas chemisorption is used to determine the number of exposed reactive sites. For example, the dissociative chemisorption of H_2 on supported platinum particles reveals the number of exposed surface Pt atoms.

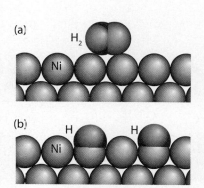

Fig. 25.17 Schematic representation of (a) physisorption and (b) chemisorption of hydrogen and a nickel metal surface.

Table 25.4 The ability of metals to chemisorb simple gas molecules

	O_2	C_2H_2	C_2H_4	CO	H_2	CO_2	N_2
Ti, Zr, Hf, V, Ta, Cr, Mo, W, Fe, Ru, Os	+	+	+	+	+	+	+
Ni, Co	+	+	+	+	+	+	+
Rh, Pd, Pt, Ir	+	+	+	+	+	+	+
Mn, Cu	+	+	+	+	±	+	+
Al, Au	+	+	+	+	−	−	−
Na, K	+	+	−	−	−	−	−
Ag, Zn, Cd, In, Si, Ge, Sn, Pb, As, Sb, Bi	+	−	−	−	−	−	−

+ Strong chemisorption, ± weak, − unobservable.
Adapted from G.C. Bond, *Heterogeneous catalysis*, p. 29. Oxford University Press (1987).

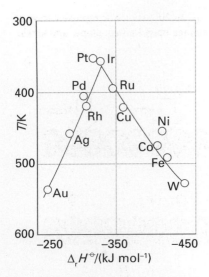

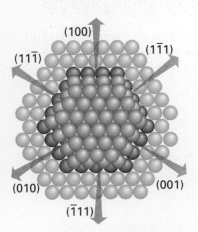

Fig. 25.18 A volcano diagram. In this case the reaction temperature for a set rate of methanoic (formic) acid decomposition is plotted against the stability of the corresponding metal methanoate as judged by its enthalpy of formation (Based on W.J.M. Rootsaert and W.M.H. Sachtler, *Z. Phyzik. Chem.*, 1960, **26**, 16.)

Fig. 25.19 Some possible metal crystal planes that might be exposed on a metal surface to a reactive gas. The planes labelled $(\bar{1}11)$, $(1\bar{1}1)$, etc. are hexagonally close packed and the planes represented by (100), (010), etc. have square arrays of atoms.

The interaction of small molecules with metal surfaces is similar to their interaction with low-oxidation-state metal complexes. Table 25.4 shows that a wide range of metals chemisorb CO, and that fewer are capable of chemisorbing N_2, just as there is a much wider variety of metals that form carbonyls than form dinitrogen complexes. Furthermore, just as with metal carbonyl complexes, both bridging and terminal CO surface species have been identified by IR spectroscopy. The dissociative chemisorption of H_2 is analogous to the oxidative addition of H_2 to metal complexes.

Although adsorption is essential for heterogeneous catalysis to occur, it must not be so strong as to block the catalytic sites and prevent further reaction. This factor is in part responsible for the limited number of metals that are effective catalysts. The catalytic decomposition of methanoic (formic) acid on metal surfaces

$$HCOOH \xrightarrow{M} CO + H_2O$$

provides a good example of this balance between adsorption and catalytic activity. It is observed that the catalysis is most effective using metals for which the metal methanoate is of intermediate stability (Fig. 25.18). The plot in Fig. 25.18 is an example of a 'volcano diagram', and is typical of many catalytic reactions. The implication is that the earlier *d*-block metals form very stable surface compounds whereas the later noble metals such as silver and gold form very weak surface compounds, both of which are detrimental to a catalytic process. Between these extremes the metals in Groups 8 to 10 have high catalytic activity, especially the platinum metals (Group 10). In Section 25.4 we saw a similar high activity of these metal complexes in the homogeneous catalysis of hydrocarbon transformations.

The active sites of heterogeneous catalysts are not uniform and many diverse sites are exposed on the surface of a poorly crystalline solid such as γ-alumina or a noncrystalline solid such as silica gel. However, even highly crystalline metal particles are not uniform. A crystalline solid has typically more than one type of exposed plane, each with its characteristic pattern of surface atoms (Fig. 25.19). In addition, single-crystal metal surfaces have irregularities such as steps that expose metal atoms with low coordination numbers (Fig. 25.20). These highly exposed, coordinatively unsaturated sites appear to be particularly reactive. As a result, the different sites on the surface may serve different functions in catalytic reactions. The variety of sites also accounts for the lower selectivity of many heterogeneous catalysts in comparison with their homogeneous analogues.

(e) Surface migration

Key point: Adsorbed atoms and molecules migrate over metal surfaces.

The surface analogue of fluxional mobility in clusters is diffusion, and there is abundant evidence for the diffusion of chemisorbed molecules or atoms on metal surfaces. For example, adsorbed H atoms and CO molecules are known to move over the surface of a metal particle. These diffusion pathways generally involve the adsorbed molecules moving through a variety of different coordination sites to the metal surface. So, for example, CO migration can result from a molecule moving between sites interacting with one (terminal CO) and between two and four (bridging CO) metal atoms on the surface. The energy barrier to this process is relatively low, a few tens of kilojoules per mole, and thus migration rates are very high under typical catalytic reaction conditions. This mobility is important in catalytic reactions as it allows atoms or molecules to find and approach one another rapidly.

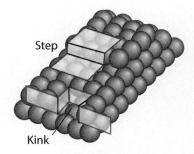

Fig. 25.20 Schematic representation of surface irregularities, steps, and kinks.

25.11 Hydrogenation of alkenes

Key point: Alkenes are hydrogenated on supported metal particles by a process that involves H_2 dissociation and migration of H· to an adsorbed ethene molecule.

A milestone in heterogeneous catalysis was Paul Sabatier's observation in 1890 that nickel catalyses the hydrogenation of alkenes. He was in fact attempting to synthesize $Ni(C_2H_4)_4$ in response to Mond, Langer, and Quinke's synthesis of $Ni(CO)_4$ (Section 21.18). However, when he passed ethene over heated nickel he detected ethane. His curiosity was sparked, so he included hydrogen with the ethene, whereupon he observed a good yield of ethane.

The hydrogenation of alkenes on supported metal particles is thought to proceed in a manner very similar to that in metal complexes. As pictured in Fig. 25.21, H_2, which is dissociatively chemisorbed on the surface, is thought to migrate to an adsorbed ethene molecule, giving first a surface alkyl and then the saturated hydrocarbon. When ethene is hydrogenated with D_2 over platinum, the simple mechanism depicted in Fig. 25.21 indicates that CH_2DCH_2D should be the product. In fact, a complete range of $C_2H_nD_{6-n}$ ethane molecules is observed. It is for this reason that the central step is written as reversible; the rate of the reverse reaction must be greater than the rate at which the ethane molecule is formed and desorbed in the final step.

One of the most important classes of heterogeneous hydrogenation catalysts are the 'Raney nickels', which are used for various processes such as conversion of alkanals (aldehydes) to alkanols (alcohols), as in

$$CH_3CH_2CH_2CHO + H_2 \xrightarrow{85°C,\ 25\ atm\ H_2} CH_3CH_2CH_2CH_2OH$$

and reduction of alkylchloronitroanilines to the corresponding amines. Raney nickels and similar catalytically active metal alloys are produced by preparing a metal alloy such as NiAl at high temperature and then selectively dissolving most of the aluminium by treatment with sodium hydroxide. Other metals, such as molybdenum and chromium, can be added to the original alloy and may act as promoters that affect the catalyst's reactivity and selectivity for certain reactions. The resulting spongy or porous metals are rich in nickel (>90 per cent) and their high surface areas lead to very high catalytic activities. One further application of these catalysts is to the conversion of naturally occurring polyunsaturated fats, which are liquids, to solid polyhydrogenated fats, as in margarine.

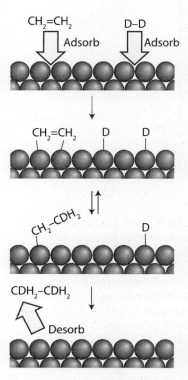

Fig. 25.21 Schematic diagram of the stages involved in the hydrogenation of ethene by deuterium on a metal surface.

25.12 Ammonia synthesis

Key point: Catalysts based on iron metal are used for the synthesis of ammonia from nitrogen and hydrogen.

The synthesis of ammonia has already been discussed from several different viewpoints (Section 14.6). Here we concentrate on details of the catalytic steps. The formation of ammonia is exoergic and exothermic at 25°C, the relevant thermodynamic data being $\Delta_f G^\ominus = -16.5$ kJ mol^{-1}, $\Delta_f H^\ominus = -46.1$ kJ mol^{-1}, and $\Delta_f S^\ominus = -99.4$ J K^{-1}mol^{-1}. The negative entropy of formation reflects the fact that two NH_3 molecules form in place of four reactant molecules.

The great inertness of N_2 (and to a lesser extent H_2) requires that a catalyst be used for the reaction. Iron metal, together with small quantities of alumina and potassium salts and other promoters, is used as the catalyst. Extensive studies on the mechanism of ammonia synthesis indicate that the rate-determining step under normal operating conditions is the dissociation of N_2 coordinated to the catalyst surface. The other reactant, H_2, undergoes much more facile dissociation on the metal surface and a series of insertion reactions between adsorbed species leads to the production of NH_3:

$$N_2(g) \longrightarrow \underset{/////}{N_2(g)} \longrightarrow \underset{/////}{N} \quad \underset{/////}{N}$$

$$H_2(g) \longrightarrow \underset{/////}{H_2(g)} \longrightarrow \underset{/////}{H} \quad \underset{/////}{H}$$

$$\underset{/////}{N} + \underset{/////}{H} \longrightarrow \underset{/////}{NH} \overset{}{\underset{\underset{/////}{H}}{\Big\uparrow}} \underset{/////}{NH_2} \overset{}{\underset{\underset{/////}{H}}{\Big\uparrow}} \underset{/////}{NH_3} \longrightarrow NH_3(g)$$

Because of the slowness of the N_2 dissociation, it is necessary to run the ammonia synthesis at high temperatures, typically 400°C. However, because the reaction is exothermic, high temperature reduces the equilibrium constant of the reaction. To recover some of this reduced yield, pressures on the order of 100 atm are used to favour the formation products. A catalyst operating at room temperature could give good equilibrium yields of NH_3, such as the enzyme nitrogenase (Section 26.13), has long been sought.

In the course of developing the original ammonia synthesis process, Haber, Bosch, and their co-workers investigated the catalytic activity of most of the metals in the periodic table, and found that the best are iron, ruthenium, and uranium promoted by small amounts of alumina and potassium salts. Cost and toxicity considerations led to the choice of iron as the basis of the commercial catalyst.

25.13 Sulfur dioxide oxidation

Key point: The most widely used catalyst for the oxidation of SO_2 to SO_3 is molten potassium vanadate supported on a high-surface-area silica.

The oxidation of SO_2 to SO_3 is a key step in the production of sulfuric acid (Section 15.5). The reaction of sulfur with oxygen to produce SO_3 gas is exoergic ($\Delta_r G^{\ominus} = -371$ kJ mol^{-1}) but very slow, and the principal product of combustion is SO_2:

$$S(s) + O_2(g) \rightarrow SO_2(g)$$

The combustion is followed by the catalytic oxidation of SO_2:

$$SO_2(g) + \tfrac{1}{2}O_2(g) \rightarrow SO_3(g)$$

This step is also exothermic and thus, as with ammonia synthesis, has a less favourable equilibrium constant at elevated temperatures. The process is therefore generally run in stages. In the first stage, the combustion of sulfur raises the temperature to about 600°C, but by cooling and pressurizing before the catalytic stage the equilibrium is driven to the right and high conversion of SO_2 to SO_3 is achieved.

Several quite different catalytic systems have been used to catalyse the combination of SO_2 with O_2. The most widely used catalyst at present is potassium or caesium vanadate molten salt covering a high-surface-area silica. The current view of the mechanism of the reaction is that the rate-determining step is the oxidation of V(IV) to V(V) by O_2 (Fig. 25.22). In the melt, the vanadium and oxide ions are part of a polyvanadate complex (Section 18.8), but little is known about the evolution of the oxo species.

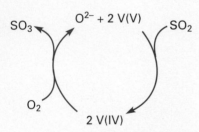

Fig. 25.22 Cycle showing the key elements involved in the oxidation of SO_2 by V(V) compounds.

25.14 Interconversion of aromatics by zeolites

Key points: Zeolite catalysts have strongly acidic sites that promote reactions such as isomerization via carbonium ions; shape selectivity may arise at various stages of the reaction due to the relative dimensions of the zeolite channels and reactant, intermediate, and product molecules.

Fig. 25.23 Diagram of the zeolite ZSM-5 structure highlighting the channels along which small molecules may diffuse.

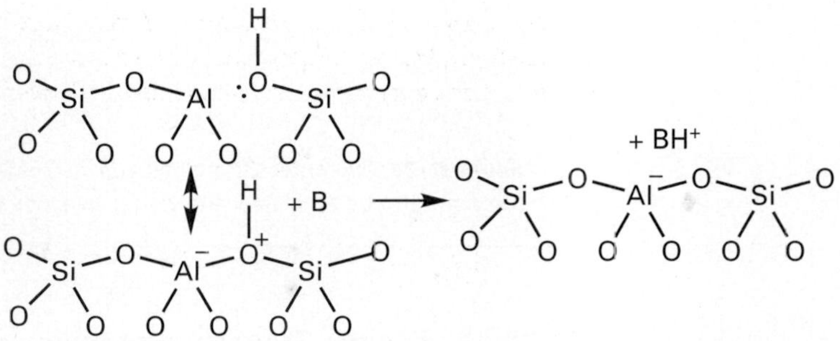

Fig. 25.24 The Brønsted acid site in H-ZSM-5 and its interaction with a base, typically an organic molecule. (Based on W.O. Haag, R.M. Lago, and P.B. Weisz, *Nature*, 1984, **309**, 589.)

The zeolite-based heterogeneous catalysts play an important role in the interconversion of hydrocarbons and the alkylation of aromatics as well as in oxidation and reduction.

The synthetic zeolite ZSM-5 is widely used by the petroleum industry for a variety of hydrocarbon interconversions.[5] ZSM-5 is an aluminosilicate zeolite with a higher silica content and a much lower Al content than the zeolites discussed in Section 13.13. Its channels consist of a three-dimensional maze of intersecting tunnels (Fig. 25.23). As with other aluminosilicate catalysts, the aluminium sites are strongly acidic. The charge imbalance of Al^{3+} in place of tetrahedrally coordinated Si(IV) requires the presence of an added positive ion. When this ion is H^+ (Fig. 25.24) the Brønsted acidity of the aluminosilicate can be higher than that of concentrated H_2SO_4 and the turnover frequency for hydrocarbon reactions at these sites can be very high.

Reactions such as dimethylbenzene (xylene) isomerization and methylbenzene (toluene) disproportionation illustrate the selectivity that can be achieved with acidic zeolite catalysis. The shape selectivity of these catalysts has been attributed to faster diffusion of product molecules that have dimensions compatible with the channels. According to this idea, molecules that do not fit the channels diffuse slowly and, on account of their long residence in the zeolite, have ample opportunity to be converted to the more mobile isomers that can escape rapidly. A currently more favoured view, however, is that the orientation of reactive intermediates within the zeolite channels favours specific products, such as 1,4-dialkylbenzenes.

Aside from their important shape selectivity, acidic zeolite catalysts appear to promote reactions by standard carbonium ion mechanisms. For example, the isomerization of 1,3-dimethylbenzene to 1,4-dimethylbenzene may occur by the following steps:

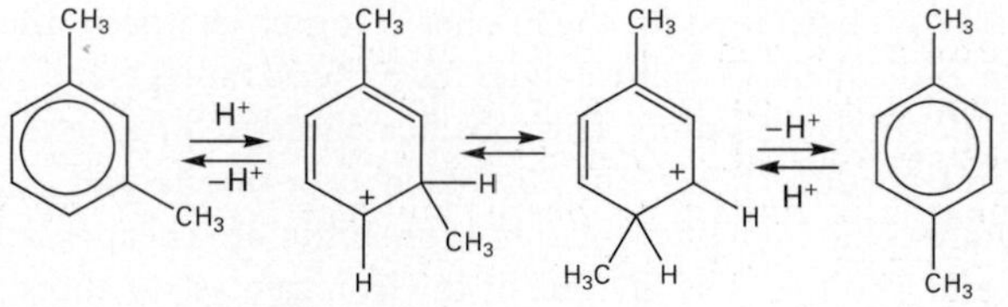

Another common reaction in zeolites is the alkylation of aromatics with alkenes.

Example 25.3 Proposing a mechanism for the alkylation of benzene

In its protonated form, ZSM-5 catalyses the reaction of ethene with benzene to produce ethylbenzene. Write a plausible mechanism for the reaction.

Answer The acidic form of ZSM-5 is strong enough to generate carbocations:

$$CH_2CH_2 + H^+ \rightarrow CH_2CH_3^+$$

---

[5] The catalyst was developed in the research laboratories of Mobil Oil; the initials stand for Zeolite Socony-Mobil.

As we saw in Section 4.10, the carbocation can attack benzene. Subsequent deprotonation of the product yields ethylbenzene:

$$CH_2CH_3^+ + C_6H_6 \rightarrow C_6H_5CH_2CH_3 + H^+$$

Self-test 25.3 A pure silica analogue of ZSM-5 can be prepared. Would you expect this compound to be an active catalyst for benzene alkylation? Explain your reasoning.

25.15 Fischer–Tropsch synthesis

Key point: Hydrogen and carbon monoxide can be converted to hydrocarbons and water by reaction over iron or cobalt catalysts.

The conversion of **syngas**, a mixture of H_2 and CO, to hydrocarbons over metal catalysts was first discovered by Franz Fischer and Hans Tropsch at the Kaiser Wilhelm Institute for Coal Research in Müllheim in 1923. In the Fischer–Tropsch reaction, CO reacts with H_2 to produce hydrocarbons, which can be written symbolically as the formation of the chain extender ($-CH_2-$), and water:

$$CO + 2\,H_2 \rightarrow -CH_2- + H_2O$$

The process is exothermic, with $\Delta_r H^\ominus = -165\ kJ\ mol^{-1}$. The product range consists of aliphatic straight-chain hydrocarbons that include methane (CH_4) and ethane, LPG (C_3-C_4), gasoline (C_5-C_{12}), diesel ($C_{13}-C_{22}$), and light and heavy waxes ($C_{23}-C_{32}$ and $>C_{33}$, respectively). Side reactions include the formation of alcohols and other oxygenated products. The distribution of the products depends on the catalyst and the temperature, pressure, and residence time. Typical conditions for the Fischer–Tropsch synthesis are a temperature range of 200–350°C and pressures of 15–40 atm.

There is general agreement that the first stages of the hydrocarbon synthesis involve the adsorption of CO on the metal, followed by its cleavage to give a surface carbide (and water), and the successive hydrogenation of such species to surface methyne (CH), methylene (CH_2), and methyl (CH_3) species, but there is still debate on what happens next and how chain growth occurs. One proposal suggests a polymerization of bridging surface methylene groups initiated by a surface methyl group. However, the fact that many such species have been isolated and are stable as metal complexes suggests that the mechanism is probably not so simple. Other mechanistic investigations of these processes have suggested an alternative possibility for chain growth in the hydrocarbon synthesis, namely that it proceeds by combination of surface methylene groups and alkenyl chains ($M-CH=CHR$), rather than by the combination of alkyl chains ($M-CH_2CH_2R$) with methylene groups in the surface.

Several catalysts have been used for the Fischer–Tropsch synthesis; the most important are based on iron and cobalt. Cobalt catalysts have the advantage of a higher conversion rate and a longer life (of over 5 years). The cobalt catalysts are in general more reactive for hydrogenation and produce fewer unsaturated hydrocarbons and alcohols than iron catalysts. Iron catalysts have a higher tolerance for sulfur, are cheaper, and produce more alkene products and alcohols. The lifetime of the iron catalysts is short, however, and in commercial installations generally limited to about 8 weeks.

25.16 Alkene polymerization

Key points: Heterogeneous Ziegler–Natta catalysts are used in alkene polymerization; the Cossee–Arlman mechanism describes their function; low molecular weight homogeneous catalysts also catalyse the alkene polymerization reaction; considerable control over polymer tacticity is possible with judicious ligand design.

Polyalkenes, which are among the most common and useful class of synthetic polymers, are most often prepared by use of organometallic catalysts, either in solution or supported on a solid surface. The development of alkene polymerization catalysts in the second half of the twentieth century, producing polymers, such as polypropylene and

Fig. 25.25 The Cossee–Arlman mechanism for the catalytic polymerization of ethene.

polystyrene, ushered in a revolution in packaging materials, fabrics, and constructional materials.

In the 1950s J.P. Hogan and R.L. Banks discovered that chromium oxides supported on silica, a so-called **Philips catalyst**, polymerized alkenes to polyenes. Also in the 1950s K. Ziegler, working in Germany, developed a catalyst for ethene polymerization based on a catalyst formed from $TiCl_4$ and $Al(C_2H_5)_3$, and soon thereafter G. Natta in Italy used this type of catalyst for the stereospecific polymerization of propene (see below). Both the **Ziegler–Natta catalysts** and the chromium-based polymerization catalysts are widely used today.

The full details of the mechanism of Ziegler–Natta catalysts are still uncertain, but the **Cossee–Arlman mechanism** is regarded as highly plausible (Fig. 25.25). The catalyst is prepared from $TiCl_4$ and the $Al(C_2H_5)_3$, which react to give polymeric $TiCl_3$ in the form of a fine powder. The alkylaluminium alkylates a Ti atom on the surface of the solid and an ethene molecule coordinates to the neighbouring vacant site (represented by the open circle). In the propagation steps for the polymerization, the coordinated alkene undergoes a migratory insertion reaction. This migration opens up another neighbouring vacancy, and so the reaction can continue and the polymer chain can grow. The release of the polymer from the metal atom occurs by β-hydrogen elimination, and the chain is terminated. Some catalyst remains in the polymer, but the process is so efficient that the amount is negligible.

The proposed mechanism of alkene polymerization on a Philips catalyst involves the initial coordination of one or more alkene molecules to a surface Cr(II) site followed by rearrangement to metallocycloalkanes on a formally Cr(IV) site. Unlike Ziegler–Natta catalysts, the solid-phase catalyst does not need an alkylating agent to initiate the polymerization reaction; instead, this species is thought to be generated by the metallocycloalkane directly or by formation of an ethenylhydride by cleavage of a C—H bond at the chromium site.

Homogeneous catalysts related to the Philips and Ziegler–Natta catalysts provide additional insight into the course of the reaction and are of considerable industrial significance in their own right, being used commercially for the synthesis of specialized polymers. Most examples are from Group 4 (Ti, Zr, Hf) and are based on a bis(cyclopentadienyl)metal system: the tilted ring complex $[Zr(\eta^5\text{-}Cp)_2(CH_3)L]^+$ (**12**) is a good example. These Group 4 metallocene complexes catalyse alkene polymerization by successive insertion steps that involve prior coordination of the alkene to the electrophilic metal centre. Catalysts of this type are used in the presence of the so-called methyl aluminoxane (MAO), a poorly defined compound of approximate formula $(MeAlO)_n$, which, among other functions, serves to methylate a starting chloride complex.

Additional complications arise with alkenes other than ethene. We discuss only terminal alkenes such as propene and styrene, as these are relatively simple. The first complication to consider arises because the two ends of the alkene molecule are different.

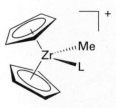

12 $[Zr(\eta^5\text{-}Cp)_2(CH_3)L]^+$

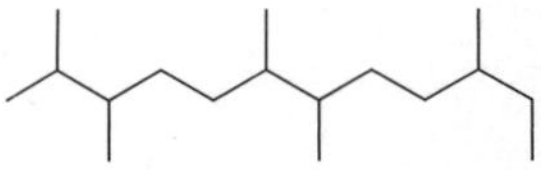

13

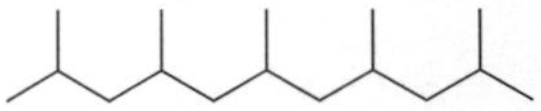

14

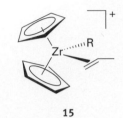

15

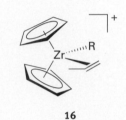

16

17 Isotactic

18 Syndiotactic

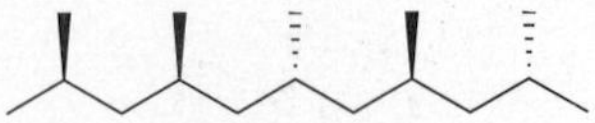

19 Atactic

In principle, it is possible for the polymer to form with the different ends either head-to-head (**13**), head-to-tail (**14**), or randomly. Studies on catalysts such as (**12**) show that the growing chain migrates preferentially to the more highly substituted C atom of the alkene, thus giving a polymer chain that contains only head-to-tail orientations:

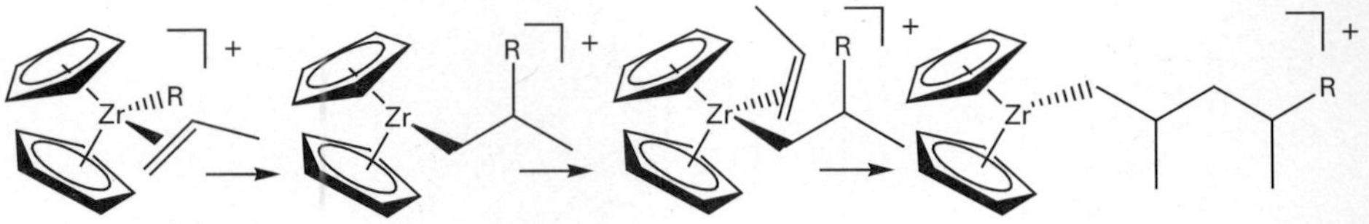

The normal explanation for this regularity relates to the steric requirements of the propene: a coordinated propene molecule induces less steric strain if its smaller CH_2 end is pointing into the cleft of the $(Cp)_2Zr$ catalyst (**15**) rather than (**16**). The migrating polymer chain is thus adjacent to the methyl-substituted end of the propene molecule, and it is to this methyl-substituted C atom that the chain attaches, giving a head-to-tail sequence to the whole polymer chain.

The second structural modification of polypropene is its **tacticity**, the relative orientations of neighbouring groups in the polymer. In a regular **isotactic** polypropene, all the methyl groups are on the same side of the polymer backbone (**17**). In regular **syndiotactic** polypropylene, the orientation of the methyl groups alternates along the polymer chain (**18**). In an **atactic** polypropylene, the orientation of neighbouring methyl groups is random (**19**). Control of the tacticity of a polymer is equivalent to controlling the stereospecificity of the reaction steps. The orientation of neighbouring groups is not simply of academic interest because the orientation has a significant effect on the properties of the bulk polymer. For example, the melting points of isotactic, syndiotactic, and atactic polypropene are 165°C, 130°C and below 0°C, respectively.

With a zirconium catalyst such as (**15**), it is not possible to control the tacticity of polypropylene and an atactic polymer results. However, with other catalysts it is possible to control the tacticity. The type of catalyst normally used to control the tacticity has a metal atom bonded to two indenyl groups that are linked by an ethene spacer. Reaction of the bis(indenyl) fragment with a metal salt gives rise to three compounds: two enantiomers (**20**) and (**21**), which have C_2 symmetry, and a nonchiral compound (**22**). It is possible to separate the two enantiomers from the nonchiral compounds and both enantiomers of these so-called *ansa*-metallocenes can catalyse the stereoregular polymerization of propene.

If we now consider the coordination of propene to one of the enantiomeric compounds (**20**) or (**21**) in addition to the steric factor identified above (CH_2 group pointing towards the cleft), we have another constraint. Of the two potential arrangements of the methyl group shown in (**23**) and (**24**), the latter is disfavoured by a steric interaction with

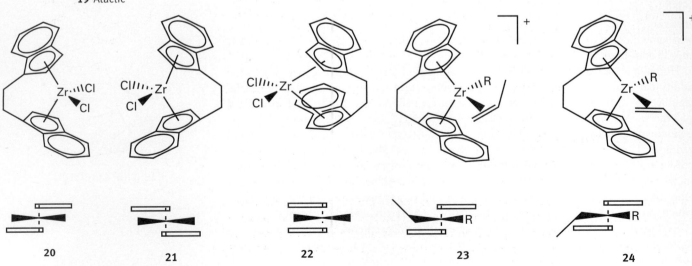

20   **21**   **22**   **23**   **24**

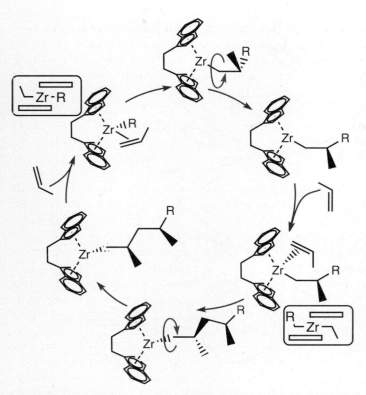

Fig. 25.26 When propene is polymerized with an indenyl metallocene catalyst, isotactic polypropene results. The zirconium species all have a single positive charge; this has been omitted for clarity.

the phenyl ring of the indenyl group. During the polymerization reaction, the R group will migrate preferentially to one side of the propene molecule; coordination of another alkene is then followed by migration, and so on. Figure 25.26 shows how an isotactic polypropene then results.

Example 25.4 Controlling the tacticity of polypropene

Show that polymerization of propene with a catalyst containing ethene-linked fluorenyl and cyclopentadienyl groups (**25**) should result in highly syndiotactic polypropene.

Answer In complex (**25**), as illustrated, a coordinated propene will always coordinate with the methyl group pointing away from the fluorenyl and towards the cyclopentadienyl ring. Following the sequential alkene insertions leads to the conclusion that the product should be syndiotactic. This reaction is outlined in Fig. 25.27.

Self-test 25.4 Demonstrate that polymerization of propene with a simple $Zr(Cp)_2Cl_2$ catalyst would give rise to atactic polypropene.

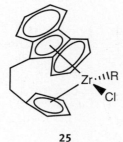

25

25.17 Electrocatalysis

Key points: Overpotentials represent the kinetic barrier for electrochemical reactions and electrocatalysts may be used to increase current densities in such processes.

Kinetic barriers are quite common for electrochemical reactions at the interface between a solution and an electrode and, as mentioned in Section 5.17, it is common to express these barriers as overpotentials η, the potential in addition to the zero-current cell potential (the emf) that must be applied to bring about an otherwise slow reaction within

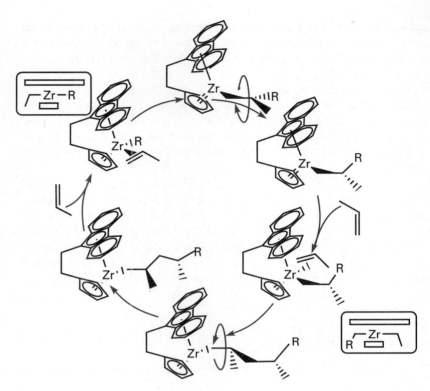

Fig. 25.27 When propene is polymerized with a fluorenyl metallocene catalyst, syndiotactic polypropene results. The zirconium species all have a single positive charge; this has been omitted for clarity.

the cell. The overpotential is related to the current density, j (the current divided by the area of the electrode), that passes through the cell by[6]

$$j = j_0 \, e^{a\eta} \tag{3}$$

where j_0 and a are best regarded for our purposes as empirical constants. The constant j_0, the **exchange current density**, is a measure of the rates of the forward and reverse electrode reactions at dynamic equilibrium. For systems obeying these relations, the reaction rate (as measured by the current density) increases rapidly with increasing applied potential difference when $a\eta > 1$. If the exchange current density is high, an appreciable reaction rate may be achieved with only a small overpotential. If the exchange current density is low, a high overpotential is necessary. There is therefore considerable interest in controlling the exchange current density. In an industrial process an overpotential in a synthetic step is very costly because it represents wasted energy. However, a low but significant exchange current density may block undesirable pathways.

A catalytic electrode surface can increase the exchange current density and hence dramatically decrease the overpotential required for sluggish electrochemical reactions, such as H_2, O_2, or Cl_2 evolution and consumption. For example, 'platinum black', a finely divided form of platinum, is very effective at increasing the exchange current density and hence decreasing the overpotential of reactions involving the consumption or evolution of H_2. The role of platinum is to dissociate the strong H—H bond and thereby reduce the large barrier that this strength imposes on reactions involving H_2. Palladium also has a high exchange current density (and hence requires only a low overpotential) for H_2 evolution or consumption.

The effectiveness of metals can be judged from Fig. 25.28, which also gives insight into the process. The volcano-like plot of exchange current density against M—H bond

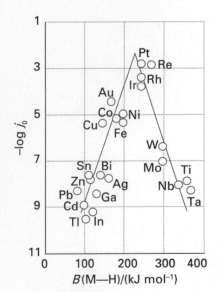

Fig. 25.28 The rate of H_2 evolution expressed as the logarithm of the exchange current density versus M—H bond energy.

[6] The exponential relation between the current and the overpotential is accounted for by the Butler–Volmer equation, which is derived by applying activated complex theory to dynamical processes at electrodes. See Chapter 25 of P. Atkins and J. de Paula, *Physical chemistry*. Oxford University Press and W.H. Freeman & Co (2006).

enthalpy suggests that M—H bond formation and cleavage are both important in the catalytic process. It appears that an intermediate M—H bond energy leads to the proper balance for the existence of a catalytic cycle and the most effective metals for electro-catalysis are clustered around Group 10.

Ruthenium dioxide is an effective catalyst for both O_2 and Cl_2 evolution and it also is a good electrical conductor. It turns out that at high current densities RuO_2 is more effective for the catalysis of Cl_2 evolution than for O_2 evolution. RuO_2 is therefore extensively used as an electrode material in the commercial production of chorine. The electrode processes that contribute to this subtle catalytic effect do not appear to be well understood.

There is great interest in devising new catalytic electrodes, particularly ones that decrease the O_2 overpotential on surfaces such as graphite. Thus, tetrakis(4-*N*-methyl-pyridyl) tetraphenylporphyriniron(II), [Fe(TMPyP)] (**26**), has been deposited on the exposed edges of graphite electrodes (on which O_2 reduction requires a high over-potential) and the resulting electrode surface was found to catalyse the electrochemical reduction of O_2. A plausible explanation for this catalysis is that [Fe^{III}(TMPyP)] attached to the electrode (indicated below by an asterisk) is first reduced electrochemically:

$$[Fe^{III}(TMPyP)]^* + e^- \rightarrow [Fe^{II}(TMPyP)]^*$$

The resulting [Fe^{II}(TMPyP)] forms an O_2 complex:

$$[Fe^{II}(TMPyP)]^* + O_2 \rightarrow [Fe^{II}(O_2)(TMPyP)]^*$$

This iron(II) porphyrin oxygen complex then undergoes reduction to water and hydrogen peroxide:

$$[Fe^{II}(O_2)(TMPyP)]^* + n\,e^- + n\,H^+ \rightarrow [Fe^{II}(TMPyP)]^* + (H_2O_2, H_2O)_n$$

Although details of the mechanism are still elusive, this general set of reactions is in harmony with electrochemical measurements and the known properties of iron porphyrins. The investigation of porphyriniron(III) complex as a catalyst was motivated by Nature's use of metalloporphyrins for oxygen activation (section 26.10).

26

25.18 **New directions in heterogeneous catalysis**

Key point: Continuing developments in heterogeneous catalysis include research on new compositions for the controlled partial oxidation of hydrocarbons.

The development of solid-phase catalysts is very much a frontier subject for inorganic chemistry with the continuing discovery of compositions for promoting reactions, particularly for petrochemicals. One very active area is the investigation of selective heterogeneous oxidation catalysts, which allow partial oxidation of hydrocarbons to useful intermediates in, for example, the polymer and pharmaceutical industries. Examples of such reactions include alkene epoxidation, aromatic hydroxylation, and ammoxidation of alkanes, alkenes, and alkyl aromatics. In all these cases it is desirable to produce the products without complete oxidation of the hydrocarbon to carbon dioxide.

One example where new catalysts are being investigated is in the partial oxidation of benzene to phenol. At present, the three-step **cumene process** produces about 95 per cent of the phenol used in the world and gives propanone (acetone) as a by-product, although the market for acetone is oversupplied from other industrial processes. The cumene process involves three stages, namely alkylation of benzene with propene to form cumene, (a process catalysed by phosphoric acid or aluminium chloride), direct oxida-tion of cumene to cumene hydroperoxide using molecular oxygen, and finally cleavage of cumene hydroperoxide to phenol and acetone, which is catalysed by sulfuric acid. Far better would be a single-stage process that accomplishes the reaction

$$C_6H_6 + \tfrac{1}{2}O_2 \rightarrow C_6H_5OH$$

Examples of catalysts investigated for this and similar processes include iron-containing zeolites with the silicalite framework type (Section 23.13), mixtures of $FeCl_3/SiO_2$, photocatalysts based on $Pt/H_2SO_4/TiO_2$, and various vanadium salts.

Hybrid catalysis

Chemists have begun to look at catalytic systems that cannot simply be defined as either homogeneous or heterogeneous. Research has centred on trying to achieve the best of both systems: the high selectivity of homogeneous catalysts coupled with the ease of separation of heterogeneous catalysts. This approach is sometimes referred to as 'heterogenizing' homogeneous catalysts.

25.19 Tethered catalysts

One popular technique has been the **tethering** of a homogeneous catalyst to a solid support. Thus, a hydrogenation catalyst such as Wilkinson's catalyst can be attached to a silica surface by means of a long hydrocarbon chain:

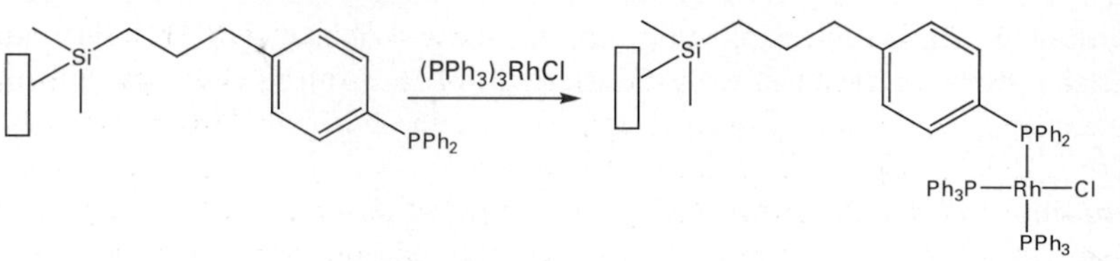

When the silica monolith is immersed in a solvent, the rhodium-based catalytic site behaves as though it is in solution, and reactivity is largely unaffected. Separation of the products from the catalyst simply requires the decanting of the solvent. Commercially available functionalized silica precursors include amino-, acrylate-, allyl-, benzyl-, bromo-, chloro-, cyano-, hydroxy-, iodo-, phenyl-, styryl-, and vinyl-substituted reagents. Relatively simple reactions can result in the synthesis of a whole host of further reagents (such as the phosphine compound in the scheme above). In addition to silica, polystyrene, polyethene, polypropene, and various clays have been used as the solid support and have led to reports of the successful heterogenization of most reactions that rely on soluble metal complexes.

In some cases the activity of a supported catalyst is greater than that of its unsupported analogue. This improvement normally takes the form of enhanced selectivity brought about by the steric demands of approaching a catalyst constrained to a surface or an increase in catalyst turnover frequencies brought about by protection from the support. Often, however, supported catalysts suffer from catalyst leaching and reduced activity.

25.20 Biphasic systems

Another popular method of attempting to combine the best of both homogeneous and heterogeneous catalysts has been the use of two liquid phases that are not miscible at room temperature but are miscible at higher temperatures. The existence of immiscible aqueous/organic phases together with a compound that facilitates the transfer of reagents between the two phases will be familiar; two other systems worthy of note are **ionic liquids** and the **fluorous biphase system**.

Ionic liquids are typically derived from 1,3-dialkylimidazolium cations (**27**) with counterions such as $[PF_6]^-$, $[BF_4]^-$, and $[CF_3SO_3]^-$. These systems have melting points of less than (and often much less than) 100°C, a very low viscosity, and an effectively zero vapour pressure. If it can be arranged that the catalysts are preferentially soluble in the ionic phase (for instance, by making them ionic), immiscible organic solvents can be used to extract organic products. As an example, consider the hydroformylation of alkenes by a Rh catalyst with phosphine ligands. When the ligand used is triphenylphosphine, the catalyst is extracted from the ionic liquid together with the products. However, when an ionic sulfonated triphenylphosphine is used as the ligand, the catalyst remains in the ionic liquid and product separation from the catalyst is complete. It should be borne in mind that the ionic liquid phase is not always unreactive, and may induce alternative reactions.

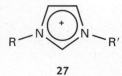

27

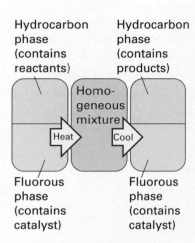

28

29

30

31

Fig. 25.29 The sequence of processes that occurs during the use of a fluorous biphase catalyst.

Hydrocarbon phase (contains reactants)

Hydrocarbon phase (contains products)

Homo-geneous mixture

Heat   Cool

Fluorous phase (contains catalyst)

Fluorous phase (contains catalyst)

The fluorous biphase system offers two principal advantages over aqueous/organic phases: the fluorous phase is unreactive, so sensitive groups (such as hydrolytically unstable groups) are stable in it, and polyfluorinated and hydrocarbon solvents, which are not miscible at room temperatures, become so on heating, giving rise to a genuinely homogeneous system. A catalyst that contains polyfluorinated groups is preferentially retained in the fluorous solvent, with the reactants (and products) preferentially soluble in the hydrocarbon phase. Figure 25.29 indicates the type of sequence that is used. Separation of the products from the catalyst becomes a trivial matter of decanting one liquid from another. A number of ligand systems that confer fluorous solubility on a catalyst have been developed; typically they are phosphine based such as (**28**), (**29**), and (**30**). Rhodium- and palladium-based catalysts of these ligands have then been prepared. With a fluorous solvent such as perfluoro-1,3-dimethylcyclohexane (**31**), miscibility with an organic phase can be achieved at 70°C, and catalysts have been used in hydrogenation (Section 25.3), hydroformylation (Section 25.4), and hydroboration reactions.

FURTHER READING

R. Whyman, *Applied organometallic chemistry and catalysis.* Oxford University Press (2001).

G.W. Parshall and S.D. Ittle, *Homogeneous catalysis.* Wiley, New York (1992).

See H.H. Brintzinger, D. Fischer, R. Mülhaupt, B. Rieger, and R.M. Waymouth, *Angew. Chem., Int. Ed. Engl.,* 1995, **34**, 1143 for a review of the area of control of polymer tacticity.

For a good review of the Stille reaction that touches upon the mechanism of all palladium-catalysed coupling reactions, see P. Espinet and A.M. Echavarren, *Angew. Chem., Int. Ed. Engl.,* 2004, **43**, 4704.

A whole issue of a major journal has been devoted to the subject of supported homogeneous catalysts. See *Chem. Rev.,* 2002, **102**, 3215.

Fluorous biphase systems are discussed in greater detail by E.G. Hope and A.M Stuart, *J. Fluorine Chem.,* 1999, **100**, 75.

For a review of the development of alkene metathesis catalysts, see T.M. Trnka and R.H. Grubbs, *Acc. Chem. Res.,* 2001, **34**, 18.

V. Ponec and G.C. Bond, *Catalysis by metal and alloys.* Elsevier, Amsterdam (1995). Comprehensive discussion of the basis of chemisorption and catalysis by metal.

R.D. Srivtava, *Heterogeneous catalytic science.* CRC Press, Boca Raton (1988). A survey of experimental methods and several major heterogeneous catalytic processes.

M. Bowker, *The basis and applications of heterogeneous catalysis.* Oxford Chemistry Primers 53. Oxford University Press (1998). Concise coverage of heterogeneous catalysis.

J.M. Thomas and W.J. Thomas, *Principles and practice of heterogeneous catalysis.* VCH, Weinheim (1997). Readable introduction to the fundamental principles of heterogeneous catalysis written by world-renowned experts.

K.M. Neyman and F. Illas, Theoretical aspects of heterogeneous catalysis: applications of density functional methods. *Catalysis Today,* 2005, **105**, 15. Modelling methods applied to heterogeneous catalysis.

M.A. Keane, Ceramics for catalysis. *J. Mater. Sci.,* 2003, **38**, 4661. Overview of heterogeneous catalysis illustrated with three established methods: (i) catalysis using zeolites; (ii) catalytic converters; (iii) solid oxide fuel cells.

F.S. Stone, Research perspectives during 40 years of the *Journal of Catalysis. J. Catal.,* 2003, **216**, 2. A historical perspective on developments in catalysis.

EXERCISES

25.1 Which of the following constitute genuine examples of catalysis and which do not? Present your reasoning. (a) The addition of H_2 to C_2H_4 when the mixture is brought into contact with finely divided platinum. (b) The reaction of an H_2/O_2 gas mixture when an electrical arc is struck. (c) The combination of N_2 gas with lithium metal to produce Li_3N, which then reacts with H_2O to produce NH_3 and LiOH.

25.2 Define the terms (a) turnover frequency, (b) selectivity, (c) catalyst, (d) catalytic cycle, and (e) catalyst support.

25.3 Classify the following as homogeneous or heterogeneous catalysis and present your reasoning. (a) The increased rate in the presence of $NO(g)$ of $SO_2(g)$ oxidation by $O_2(g)$ to $SO_3(g)$. (b) The hydrogenation of liquid vegetable oil using a finely divided nickel catalyst. (c) The conversion of an aqueous solution of D-glucose to a D, L mixture catalysed by $HCl(aq)$.

25.4 You are approached by an industrialist with the proposition that you develop catalysts for the following processes at 80°C with no input of electrical energy or electromagnetic radiation:

(a) The splitting of water into H_2 and O_2.

(b) The decomposition of CO_2 into C and O_2.

(c) The combination of N_2 with H_2 to produce NH_3.

(d) The hydrogenation of the double bonds in vegetable oil.

The industrialist's company will build the plant to carry out the process and the two of you will share equally in the profits. Which of these would be easy to do, which are plausible candidates for investigation, and which are unreasonable? Describe the chemical basis for the decision in each case.

25.5 Addition of PPh_3 to a solution of Wilkinson's catalyst, $[RhCl(PPh_3)_3]$, reduces the turnover frequency for the hydrogenation of propene. Give a plausible mechanistic explanation for this observation.

25.6 The rates of H_2 gas absorption (in $dm^3\ mol^{-1}\ s^{-1}$) by alkenes catalysed by $[RhCl(PPh_3)_3]$ in benzene at 25°C are: hexene, 2910; cis-4-methyl-2-pentene, 990; cyclohexene, 3160; 1-methylcyclo-hexene, 60. Suggest the origin of the trends and identify the affected reaction step in the proposed mechanism (Fig. 25.5).

25.7 Infrared spectroscopic investigation of a mixture of CO, H_2, and 1-butene under conditions that bring about hydroformylation indicate the presence of compound (E) in Fig. 25.8 in the reaction mixture. The same reacting mixture in the presence of added tributylphosphine was studied by infrared spectroscopy and neither (E) nor an analogous phosphine-substituted complex was observed. What does the first observation suggest as the rate-limiting reaction in the absence of phosphine? Assuming the sequence of reactions remains unchanged, what are the possible rate-limiting reactions in the presence of tributylphosphine?

25.8 Show how reaction of MeCOOMe with CO under conditions of the Monsanto ethanoic acid process can lead to ethanoic anhydride.

25.9 Suggest reasons why (a) ring-opening alkene metathesis polymerization and (b) ring-closing metathesis reactions proceed.

25.10 (a) Starting with the alkene complex shown in Fig. 25.11 with trans-DHC=CHD in place of C_2H_4, assume dissolved OH^- attacks from the side opposite the metal. Give a stereochemical drawing of the resulting compound. (b) Assume attack on the coordinated trans-DHC=CHD by an OH^- ligand coordinated to Pd, and draw the stereochemistry of the resulting compound. (c) Does the sterochemistry differentiate these proposed steps in the Wacker process?

25.11 Aluminosilicate surfaces in zeolites act as strong Brønsted acids, whereas silica gel is a very weak acid. (a) Give an explanation for the enhancement of acidity by the presence of Al^{3+} in a silica lattice. (b) Name three other ions that might enhance the acidity of silica.

25.12 Why is the platinum/rhodium catalyst in automobile catalytic converters dispersed on the surface of a ceramic rather than used in the form of a thin metal foil?

25.13 Alkanes are observed to exchange hydrogen atoms with deuterium gas over some platinum metal catalysts. When 3,3-dimethylpentane in the presence of D_2 is exposed to a platinum catalyst and the gases are observed before the reaction has proceeded very far, the main product is $CH_3CH_2C(CH_3)_2CD_2CD_3$ plus unreacted 3,3-dimethylpentane. Devise a plausible mechanism to explain this observation.

25.14 The effectiveness of platinum in catalysing the reaction $2\ H^+(aq) + 2\ e^- \rightarrow H_2(g)$ is greatly decreased in the presence of CO. Suggest an explanation.

PROBLEMS

25.1 Consider the truth of each of the following statements and provide corrections where required. (a) A catalyst introduces a new reaction pathway with lower enthalpy of activation. (b) As the Gibbs energy is more favourable for a catalytic reaction, yields of the product are increased by catalysis. (c) An example of a homogeneous catalyst is the Ziegler–Natta catalyst made from $TiCl_4(l)$ and $Al(C_2H_5)_3(l)$. (d) Highly favourable Gibbs energies for the attachment of reactants and products to a homogeneous or heterogeneous catalyst are the key to high catalytic activity.

25.2 A catalyst might not just lower the enthalpy of activation, but might make a significant change to the entropy of activation. Discuss this phenomenon. (See A. Haim, J. Chem., Educ., 1989, **66**, 731.)

25.3 The addition of promoters can further enhance the rate of a catalysed reaction. Describe how the promoters allowed the iridium-based Cativa process to compete with the rhodium-based process in the carbonylation of methanol. (See A. Haynes, P.M. Maitlis, G.E. Morris, G.J. Sunley, H. Adams, P.W. Badger, J. Am. Chem. Soc., 2004, **126**, 2847.)

25.4 When direct evidence for a mechanism is not available, chemists frequently invoke analogies with similar systems. Describe how J.E. Bäckvall, B. Åkermark, and S.O. Ljunggren (J. Am. Chem. Soc., 1979, **101**, 2411) inferred the attack of uncoordinated water on η^2-C_2H_4 in the Wacker process.

25.5 Whereas many enantioselective catalysts require the precoordination of a substrate, this is not always the case. Use the example of asymmetric epoxidation to demonstrate the validity of this statement, and indicate the advantages that such catalysts might have over catalysts that require precoordination of a substrate. (See M. Palucki, N.S. Finney, P.J. Pospisil, M.L. Güler, T. Ishida, and E.N. Jacobsen, J. Am. Chem. Soc., 1998, **120**, 948.)

25.6 Discuss shape selectivity with respect to catalytic processes involving zeolites.

25.7 Discuss the applications and mechanisms of oxidation and ammoxidation catalysts such as bismuth molybdate. (See, for example, R.K. Grasselli, J. Chem. Educ., 1986, **63**, 216.)

25.8 Discuss the advantages of a solid support in catalysis by reference to the use of $(Ni(POEt)_3)_4$ in alkene isomerization. (See A.J. Seen, J. Chem. Educ., 2004, **81**, 383 and K.R. Birdwhistell and J. Lanza, J. Chem. Educ., 1997, **74**, 579.)

Biological inorganic chemistry

26

Organisms have exploited the chemical properties of the elements in remarkable ways, providing examples of coordination specificities that are far higher than observed in simple compounds. This chapter describes how different elements are taken up selectively by different cells and intracellular compartments and the various ways they are exploited. We discuss the structures and functions of complexes and materials that are formed in the biological environment in the context of the chemistry covered earlier in the text.

Biological inorganic chemistry ('bioinorganic chemistry') is the study of the 'inorganic' elements as they are utilized in biology. The main focus is on metal ions, for which we are interested in their interaction with biological ligands (structure and dynamics) and the important chemical properties they are able to exhibit and impart to an organism. These properties include ligand binding, catalysis, signalling, regulation, sensing, defence, and structural support.

The organization of cells

To appreciate the role of the elements (other than C, H, O, and N) in the structure and functioning of organisms we need to know a little about the organization of the 'atom' of biology, the cell, and its 'fundamental particles', a cell's constituent organelles.

26.1 The physical structure of cells

Key points: Living cells and organelles are enclosed by membranes; the concentrations of specific elements may vary greatly between different compartments due to the actions of ion pumps and gated channels.

Cells, the basic unit of any living organism, range in complexity from the simplest types found in prokaryotes (bacteria and bacteria-like organisms now classified as archaea) and the much larger and more complex examples found in eukaryotes (which include animals and plants). The main features of these cells are illustrated in the generic model shown in Fig. 26.1. Crucial to all cells are membranes, which act as barriers to water and ions and make possible the management of all mobile species and of electrical currents. Membranes are lipid bilayers, approximately 4 nm thick, into which are embedded protein molecules and other components. Bilayer membranes have great lateral strength but they are easy to bend. The long hydrocarbon chains of lipids make the membrane interior very hydrophobic and impermeable to ions, which must instead travel through specific channels, pumps, and other receptors provided by special membrane proteins. The structure of a cell also depends on osmotic pressure, which is maintained by high concentrations of solutes, including ions, imported during active transport by pumps.

Prokaryotic cells consist of an enclosed aqueous phase, the **cytoplasm**, which contains the DNA and most of the materials used and transformed in the biochemical reactions. Bacteria are classified according to whether they are enclosed by a single membrane or have an additional intermediate aqueous space, the **periplasm**, between the outer

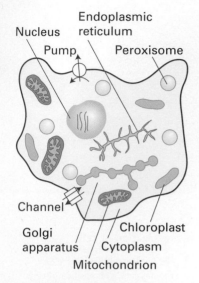

Fig. 26.1 The layout of a generic eukaryotic cell showing the cell membrane, various kinds of compartments (organelles), and the membrane-bound pumps and channels that control the flow of ions between compartments.

membrane and the cytoplasmic membrane, and are known as **Gram-positive** or **Gram-negative**, respectively, depending on their response to a staining test with the dye crystal violet. The much more extensive cytoplasm of eukaryotic cells contains subcompartments (also enclosed within lipid bilayers) known as **organelles**, which have highly specialized functions. Organelles include the **nucleus** (which houses DNA), **mitochondria** (the 'fuel cells' that carry out oxidative respiration), **chloroplasts** (the 'photocells' that harness light energy), the **endoplasmic reticulum** (for protein synthesis), **Golgi** (vesicles containing proteins for export), **lysosomes** (which contain degradative enzymes and help rid the cell of waste), **peroxisomes** (which remove harmful hydrogen peroxide), and other specialized processing zones.

26.2 The inorganic composition of cells

Key points: The major biological elements are oxygen, hydrogen, carbon, nitrogen, phosphorus, sulfur, sodium, magnesium, calcium, and potassium. The trace elements include many *d* metals, as well as selenium, iodine, silicon, and boron.

Table 26.1 lists many of the elements known to be used in living systems, although not necessarily by higher life forms. All the 2*p* and 3*p* elements except Be, Al, and the noble gases are used, as are most of the 3*d* elements, whereas Br, I, Mo, and W are the only heavier elements so far confirmed to have a biological function. Several others, such as Li, Tc, Pt, and Au, have important and increasingly well-understood applications in medicine.

The biologically essential elements can be divided broadly into two classes, **major** and **trace**. Although a good idea of the biological abundances of different elements is given in Table 26.1, the levels vary considerably among organisms and different components of organisms. For example, Ca has little role in microorganisms but is abundant in higher life forms, whereas the use of Co by higher organisms depends upon it being incorporated into a special cofactor (cobalamin) by microorganisms. There is probably a universal requirement for K, Mg, Fe, and Mo. Vanadium is used by lower animals and plants as well as some bacteria. Nickel is essential for most microorganisms, and is used by plants,

Table 26.1 The approximate concentrations (mol dm^{-3}), where known, of elements (apart from C, H, O, N, P, S, Se, Br, I, B, Si and W) in different biological zones

Element	External fluids (sea water)	Free ions in external fluids (blood plasma)	Cytoplasm (free ions)	Comments on status in cell
Na	$> 10^{-1}$	10^{-1}	$< 10^{-2}$	Not bound
K	10^{-2}	4×10^{-3}	$\leq 3 \times 10^{-1}$	Not bound
Mg	$> 10^{-2}$	10^{-3}	$\approx 10^{-3}$	Weakly bound as ATP complex
Ca	$> 10^{-3}$	10^{-3}	$\approx 10^{-7}$	Concentrated in some vesicles
Cl	10^{-1}	10^{-1}	10^{-2}	Not bound
Fe	10^{-17} (Fe(III))	10^{-16} (Fe(III))	$< 10^{-7}$ (Fe (II))	Too much unbound Fe is toxic (Fenton chemistry) in and out of cells
Zn	$< 10^{-8}$	10^{-9}	$< 10^{-11}$	Totally bound, but may be exchangeable
Cu	$< 10^{-10}$ (Cu(II))	10^{-12}	$< 10^{-15}$ (Cu(I))	Totally bound, not mobile. Mostly outside cytoplasm
Mn	10^{-9}		$\approx 10^{-6}$	Higher in chloroplasts and vesicles
Co	10^{-11}		$< 10^{-9}$	Totally bound (cobalamin)
Ni	10^{-9}		$< 10^{-10}$	Totally bound
Mo	10^{-7}		$< 10^{-7}$	Mostly bound

but there is no evidence for any direct role in animals. Biology's use of different elements is largely based on their availability, for example Zn has widespread use (and, along with Fe, ranks among the most abundant biological trace elements) whereas Co (a comparatively rare element) is essentially restricted to cobalamin. The early atmosphere (over 2.3 Ga ago[1]), being highly reducing, enabled Fe to be freely available as soluble Fe(II) salts, whereas Cu was trapped as insoluble sulfides (as was Zn). Indeed, Cu is not found in the archaea (which are believed to have evolved in pre-oxygenic times), including the hyperthermophiles, organisms that are able to survive at temperatures in excess of 100°C. These organisms are found in deep sea hydrothermal vents and terrestrial hot springs and are good sources of enzymes that contain W, the heaviest element known to be essential to life. The finding that W, Co, and for the most part Ni are used only by more primitive life forms probably reflects their special role in the early stages of evolution.

(a) Compartmentalization

Key point: Different elements are strongly segregated inside and outside a cell and among different internal compartments.

Compartmentalization is the distribution of elements inside and outside a cell and between different internal compartments. The maintenance of constant ion levels in different biological zones is an example of 'homeostasis' and it is achieved as a result of membranes being barriers to passive ion flow. An example is the large difference in concentration of K^+ and Na^+ ions across cell membranes. In the cytoplasm, the K^+ concentration may be as high as 0.3 M whereas outside it is usually less than 5×10^{-3} M. By contrast, Na^+ is abundant outside a cell but scarce inside; indeed, the low intracellular concentration of Na^+, which has characteristically weak binding to ligands, means that it has few specific roles in biochemistry. Another important example is Ca^{2+}, which is almost absent from the cytoplasm (its concentration is below 1×10^{-7} M) yet is a common cation in the extracellular environment and is concentrated in certain organelles, such as mitochondria. That pH may also vary greatly between different compartments has particularly important implications because sustaining a transmembrane proton gradient is a key feature in photosynthesis and respiration.

The distributions of Cu and Fe provide another example: Cu enzymes are often **extracellular**, that is they are synthesized in the cell and then secreted outside the cell, where they catalyse reactions involving O_2. By contrast, Fe enzymes are contained inside the cell. This difference can be rationalized on the basis that the inactive trapped states of these elements are Fe(III) and Cu(I) (or even metallic Cu) and organisms have stumbled upon the expediency of keeping Fe in a relatively reducing environment and Cu in a relatively oxidizing environment.

The selective uptake of metal ions has potential industrial applications, for many organisms and organs are known to concentrate particular elements. Thus, liver cells are a good source of cobalamin[2] (Co), and milk is rich in Ca. Certain bacteria accumulate Au and thus provide an unusual way for procuring this precious metal. Compartmentalization is an important factor in the design of metal complexes that are used in medicine (Sections 26.17–26.20).

The very small size of bacteria and organelles raises an interesting point about scale, as species present at very low concentrations in very small regions may be represented by only a few individual atoms or molecules. For example, the cytoplasm in a bacterial cell of volume 10^{-15} dm^3 at pH $= 6$ will contain less than 1000 'free' H^+ ions. Indeed, any element nominally present at less than 1 nmol dm^{-3} may be completely absent in individual cases. The word 'free' is significant, particularly for metal ions such as Zn^{2+} that are high in the Irving–Williams series; even a eukaryotic cell with a total Zn concentration of 0.1 mmol dm^{-3} may contain very few uncomplexed Zn^{2+} ions.

---

[1] Current geological and geochemical evidence date the advent of atmospheric O_2 at between 2.2 and 2.4 Ga ago. It is likely that this gas arose by the earliest catalytic actions of the photosynthetic Mn cluster described in Section 26.10

[2] In nutrition, the common complexes of cobalamin that are ingested are known as vitamin B_{12}.

Two important issues arise in the context of compartmentalization. First, the process requires energy because ions must be pumped against a gradient in chemical potential. However, once a concentration difference has been established, there is a difference in electrical potential across the membrane dividing the two regions. For instance, if the concentrations of K^+ ions on either side of a membrane are $[K^+]_{in}$ and $[K^+]_{out}$, then the contribution to the potential difference across the membrane is

$$\Delta\phi = \frac{RT}{F}\ln\frac{[K^+]_{in}}{[K^+]_{out}} \tag{1}$$

This difference in electrical potential is a way of storing energy, which is released when the ions flood back to their natural concentrations. Second, the selective transport of ions must occur through ion channels built from membrane-spanning proteins, some of which release ions upon receipt of an electrical or chemical signal whereas others, the **transporters** and **pumps**, transfer ions against the concentration gradient by using energy provided by adenosine triphosphate (ATP) hydrolysis. The selectivity of these channels is exemplified by the highly discriminatory transport of K^+ as distinct from Na^+ (Section 26.3).

It is important to note that proteins, the most important sites for metal ion coordination, are not permanent species but are ceaselessly degraded by enzymes (proteases), releasing both amino acids and metal ions to provide materials for new molecules.

Example 26.1 Assessing the role of phosphate ions

Phosphate is the most abundant small anion in the cytoplasm. What implications does this abundance have for the biochemistry of Ca^{2+}?

Answer The solubility product of $Ca_3(PO_4)_2$ is very low and it precipitates inside the cell if the Ca^{2+} concentration rises above a critical value.

Self-test 26.1 Is Fe(II) expected to be present in the cell as uncomplexed ions?

(b) Biological metal-coordination sites

Key points: The major binding sites for metal ions are provided by the amino acids that make up protein molecules; the ligation sites range from backbone peptide carbonyls to the side chains that provide more specific complexation; nucleic acids and lipid head groups are usually coordinated to major metal ions.

Metal ions coordinate to proteins, nucleic acids, lipids, and a variety of other molecules. For instance, ATP is a tetraprotic acid and is always found as its Mg^{2+} complex (**1**), and DNA is stabilized by weak coordination of K^+ and Mg^{2+} to its phosphate groups but destabilized by binding of soft metal ions such as Cu(I) to the bases. Ribozymes may represent an important stage in the early evolution of life forms and are catalytic

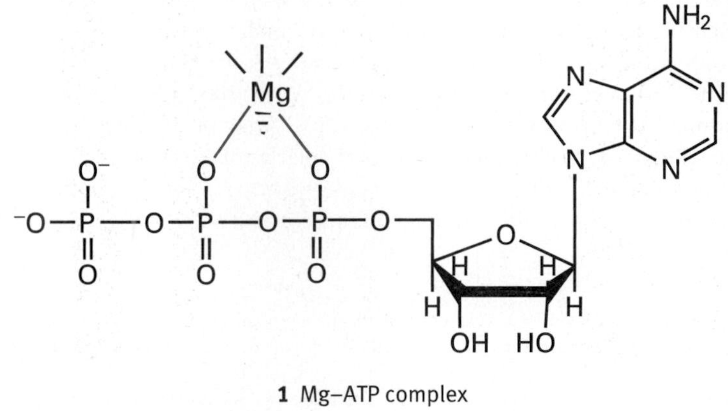

1 Mg–ATP complex

molecules composed of RNA and Mg^{2+}. The binding of Mg^{2+} to phospholipid head groups is important for stabilizing membranes. There are a number of important small ligands, apart from water and free amino acids, which include sulfide, sulfate, carbonate, cyanide, carbon monoxide, and nitrogen monoxide, and as well as organic acids such as citrate that form reasonably strong polydentate complexes with Fe(III).

As will be familiar from introductory chemistry, a 'protein' is a polymer with a specific sequence of amino acids linked by peptide bonds (**2**). A 'small' protein is generally regarded as one with molar mass below 20 kDa, whereas a 'large' protein is one having a molar mass above 100 kDa. The principal amino acids are listed in Table 26.2. Proteins are synthesized (a process called **translation**) on a special assembly called a ribosome and may then be processed further by **post-translational modification**, a change made to the protein structure, which includes the binding of **cofactors** such as metal ions.

Metalloproteins, proteins containing one or more metal ions, perform a wide range of specific functions. These functions include oxidation and reduction (for which the most important elements are Fe, Mn, Cu, and Mo), radical-based rearrangement reactions and methyl-group transfer (Co), hydrolysis (Zn, Fe, Mg, Mn and Ni), and DNA processing (Zn). Special proteins are required for transporting and storing different metal atoms. The action of Ca^{2+} is to alter the conformation of a protein (its shape) as a step in cell signalling (a term used to describe the transfer of information between and within cells). Such proteins are often known as **metal ion-activated proteins**. Hydrogen bonding between main-chain —NH and CO groups of different amino acids results in **secondary structure** (Fig. 26.2). The **α-helix** regions of a polypeptide provide mobility (like springs or screws) and are important in converting processes that occur at the metal site into conformational changes; by contrast, a **β-sheet** region confers rigidity and can provide strain to modify the properties of the metal ion. The secondary structure is largely determined by the sequence of amino acids: thus the α helix is favoured by chains containing alanine and lysine but is destabilized by glycine and proline. A protein that lacks its cofactor (such as the metal ions required for normal activity) is called an **apoprotein**; an enzyme with a complete complement of cofactors is known as a **holoenzyme**.

An important factor influencing metal-ion coordination in proteins is the energy required to locate an electrical charge inside a medium of low permittivity. To a first approximation, protein molecules may be regarded as oil drops in which the interior has a much lower relative permittivity (about 4) than water (about 78). This difference leads to a strong tendency to preserve electrical neutrality at the metal site, and hence influence the redox chemistry and Brønsted acidity of its ligands.

2 Peptide bond

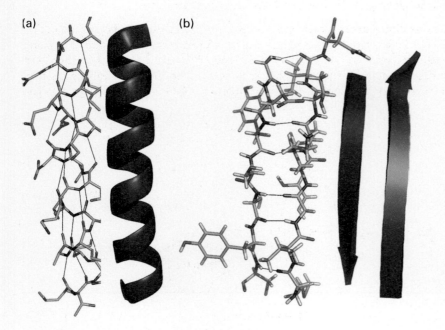

Fig. 26.2 The most important regions of secondary structure, (a) α helix, (b) β sheet, showing hydrogen bonding between main-chain amide and carbonyl groups and their corresponding representations.

Table 26.2 The amino acids and their codes

Amino acid, R—CH(NH$_2$)COOH	R—	Three-letter abbreviation	One-letter abbreviation
Alanine	CH$_3$—	Ala	A
Arginine	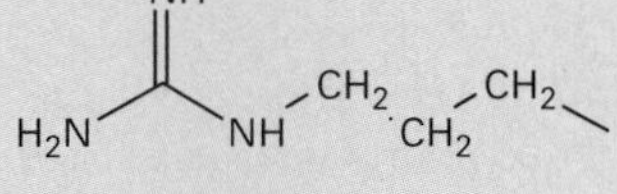	Arg	R
Asparagine	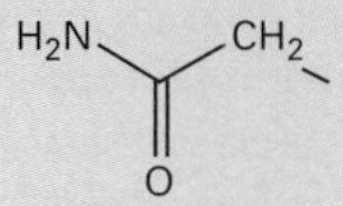	Asn	N
Aspartic acid	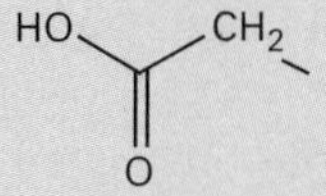	Asp	D
Cysteine	HSCH$_2$—	Cys	C
Glutamic acid	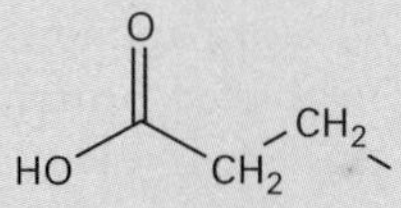	Glu	E
Glutamine	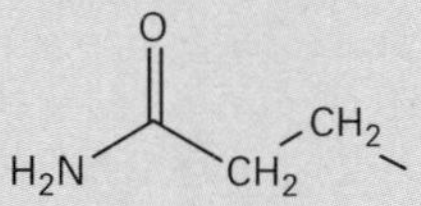	Gln	Q
Glycine	H—	Gly	G
Histidine	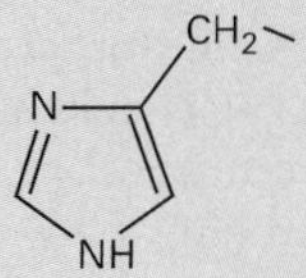	His	H
Isoleucine	CH$_3$CH$_2$CH(CH$_3$)—	Ile	I
Leucine	(CH$_3$)$_2$CHCH$_2$—	Leu	L
Lysine	NH$_2$(CH$_2$)$_4$—	Lys	K
Methionine	CH$_3$SCH$_2$CH$_2$—	Met	M
Phenylalanine	C$_6$H$_5$CH$_2$—	Phe	F
Proline	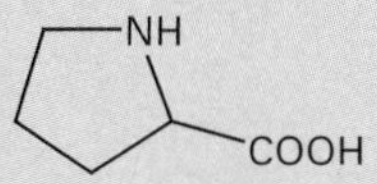 (Complete acid)	Pro	P
Serine	HOCH$_2$—	Ser	S
Threonine	HOCH(CH$_3$)—	Thr	T
Tryptophan	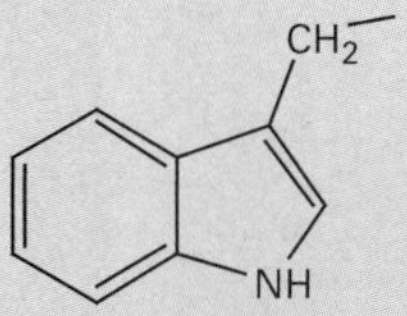	Trp	W
Tyrosine	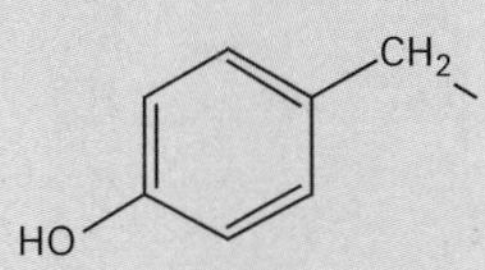	Tyr	Y
Valine	(CH$_3$)$_2$CH—	Val	V

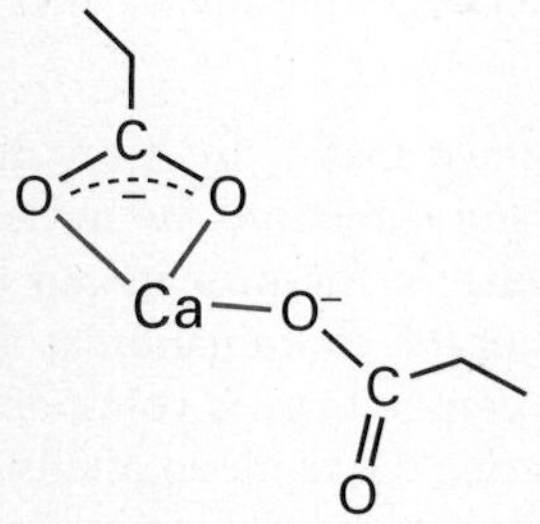

3 Ca^{2+} coordination 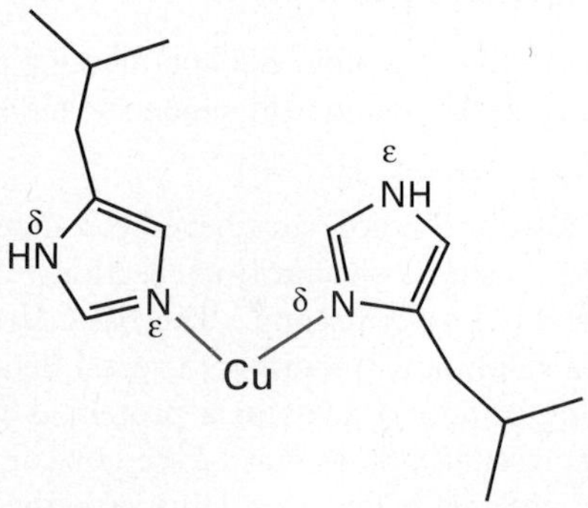 **4** Cu–imidazole coordination 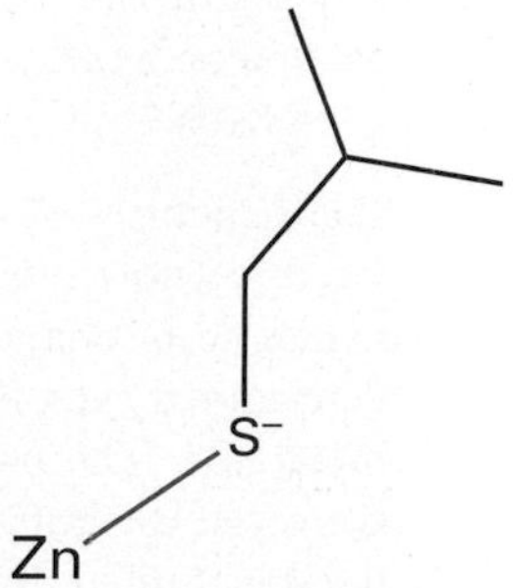 **5** Zn–cysteine coordination 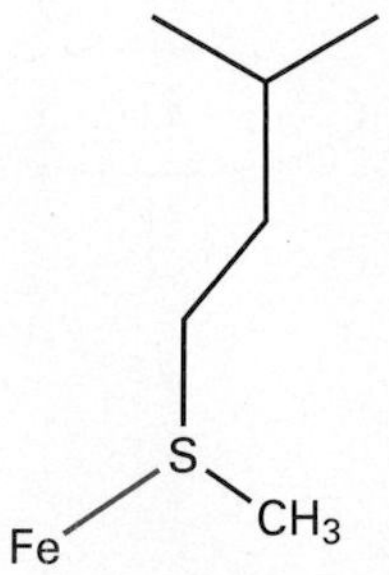 **6** Fe–methionine coordination

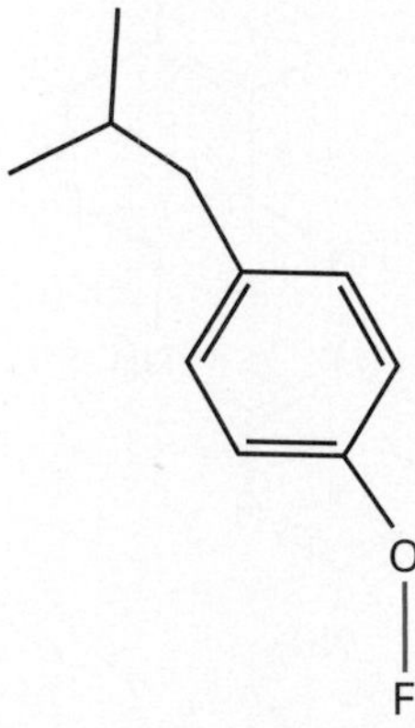

All amino acid residues can use their peptide carbonyl (or amide-N) as a donor group, but it is the side chain that usually provides more selective coordination. By referring to Table 26.2 and from the discussion in Section 4.10, we can recognize donor groups that are either chemically hard or soft and that therefore confer a particular affinity for specific metal ions. Aspartate and glutamate each provide a hard carboxylate group, and may use one or both O atoms as donors (**3**). The ability of Ca^{2+} to have a high coordination number and its preference for hard donors are such that certain Ca^{2+}-binding proteins also contain the unusual amino acids γ-carboxyglutamate and hydroxyaspartate (generated by post-translational modification), which provide additional functionalities to enhance binding. Histidine, which has an imidazole group with two coordination sites, the ε-N atom (more common) and the δ-N atom, is an important ligand for Fe, Cu, and Zn (**4**). Cysteine has a thiol S atom that is expected to be unprotonated (thiolate) when involved in metal coordination. It is a good ligand for Fe, Cu, and Zn (**5**), as well as for toxic metals such as Cd and Hg. Methionine contains a soft thioether S donor that stabilizes Fe(II) and Cu(I) (**6**). Tyrosine can be deprotonated to provide a phenolate O donor atom that is a good ligand for Fe(III) (**7**). Selenocysteine (a specially coded amino acid in which Se replaces S) has also been identified as a ligand. A modified form of lysine, in which the side-chain —NH$_2$ has reacted with a molecule of CO$_2$ to produce a carbamate, is found as a ligand to Mg in the crucial photosynthetic enzyme known as rubisco (Section 26.9).

The primary and secondary structures of a polypeptide molecule can enforce unusual metal coordination geometries that are rarely encountered in small complexes. Protein-induced strain is important; for example, the protein may impose a coordination geometry on the metal ion that resembles the transition state for the particular process being executed.

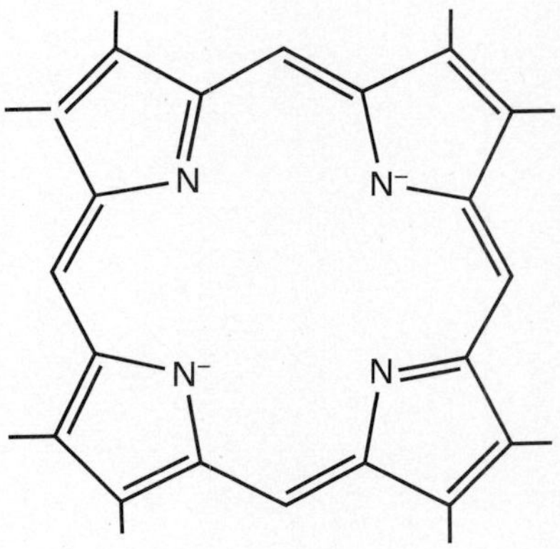

(c) Special ligands

Key point: Metal ions may be bound in proteins by special organic ligands such as porphyrins and pterin-dithiolenes.

The porphyrin group (**8**) was first identified in haemoglobin (Fe) and a similar macrocycle is found in chlorophyll (Mg). There are several classes of this hydrophobic macrocycle, each differing in the nature of the side chains. The corrin ligand (**9**) has a slightly smaller ring size and coordinates Co in cobalamin (Section 26.10). Rather than show these macrocycles in full, we shall use shorthand symbols such as (**10**) to show the complexes they form with metals. Almost all Mo and W enzymes have the metal coordinated by a special ligand known as molybdopterin (**11**). The donors to the metal are a pair of S atoms from a dithiolene group that is covalently attached to a pterin. The phosphate group is often joined to a nucleoside base X, such as guanosine 5′-diphosphate (GDP). Why Mo and W are coordinated by this complex ligand is unknown, but the pterin group could provide a good electron conduit and facilitate redox reactions.

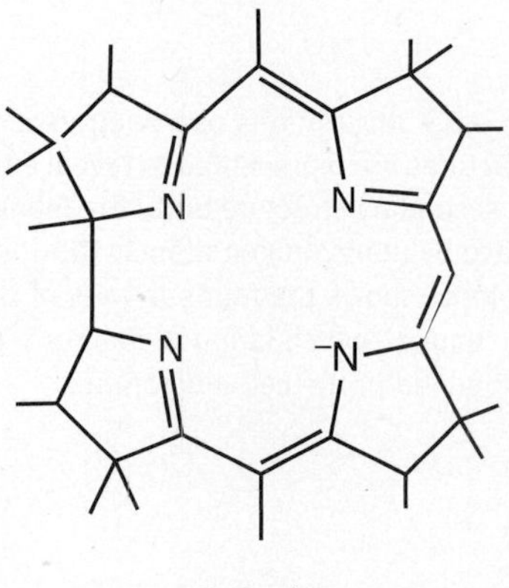

7 Fe–tyrosine coordination

8 Porphyrin^{2-}

9 Corrin^{-}

10

11 Molybdopterin as ligand

(d) The structures of metal coordination sites

Key point: The likelihood that a protein will coordinate a particular kind of metal centre can be inferred from the amino acid sequence and ultimately from the gene itself.

The structures of metal coordination sites have been determined mainly by X-ray diffraction (now mostly by using a synchrotron, Section 6.1) and sometimes by nuclear magnetic resonance (NMR) spectroscopy.[3] The basic structure of the protein can be determined even if the resolution is too low to reveal details of the coordination at the metal site. The packing of amino acids in a protein is far denser than is commonly conveyed by simple representations, as may be seen by comparing the representations of the structure of the K^+ channel in Fig. 26.3. Thus, even the substitution of an amino acid that is far from a metal centre may result in significant structural changes to its coordination shell and immediate environment. Of special interest are channels or clefts that allow a substrate *selective* access to the active site, pathways for long-range electron transfer (metal centres positioned less than 1.5 nm apart), pathways for long-range proton transfer (comprising chains of basic groups such as carboxylates and water molecules in close proximity, usually less than 0.3 nm), and channels for small gaseous molecules (which can be revealed by placing the crystal under Xe, an electron-rich gas).

Example 26.2 Interpreting the coordination environments of metal ions

Simple Cu(II) complexes have four to six ligands with trigonal-bipyramidal or tetragonal geometries, whereas simple complexes of Cu(I) have four or fewer ligands, and geometries that range between tetrahedral and linear. Predict how a Cu-binding protein will have evolved so that the Cu can act as an efficient electron-transfer site.

Answer A fast and efficient electron transfer site is one for which the reorganization energy is small. The protein will enforce upon the Cu atom a coordination geometry that is unable to alter much between Cu(II) and Cu(I) states (see Section 26.8).

Self-test 26.2 In certain chlorophyll cofactors the Mg is axially coordinated by a methionine-S ligand. Why is this unusual coordination choice achieved in a protein ?

Other physical methods described in Chapter 6 provide less information on the overall structure but are useful for identifying ligands. Thus, electron paramagnetic resonance (EPR) spectroscopy is very important for studying *d* metals, especially those engaged in

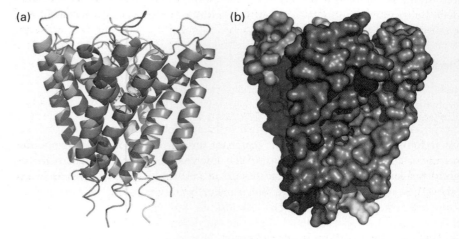

(a)   (b)

Fig. 26.3 Illustrations of how protein structures are represented to reveal either (a) secondary structure or (b) the filling of space by nonhydrogen atoms. The example shows the four subunits of the K^+ channel, which is found mainly embedded in the cell membrane.

---

[3] The atomic coordinates of proteins and other large biological molecules are stored in a public repository known as the Protein Data Bank. Each set of coordinates corresponding to a particular structure determination is identified by its 'pdb code'. A variety of software packages are available to construct and examine protein structures generated from these coordinates.

redox chemistry, because at least one oxidation state usually has an unpaired electron. The use of NMR is restricted to proteins smaller than 20–30 kDa because tumbling rates for larger proteins are too slow and ^{1}H resonances are too broad to observe unless shifted away from the normal region (1–10 ppm) by a paramagnetic metal centre. Extended X-ray absorption fine-structure spectroscopy (EXAFS) can provide structural information on metal sites in amorphous solid samples, including frozen solutions. Infrared spectroscopy is largely restricted to ligands such as CO and CN, and resonance Raman spectroscopy is very helpful when the metal centre has strong electronic transitions, such as occur with Fe porphyrins. Mössbauer spectroscopy plays a special role in studies of Fe sites. Perhaps the greatest challenge is presented by Zn^{2+}, which has a d^{10} configuration that provides no useful magnetic or electronic signatures.

Metal ion binding sites can often be predicted from a gene sequence. The development of **bioinformatics** is significant because many proteins that bind metal ions or have a metal-containing cofactor occur at cellular levels below that normally detectable by analysis and isolation. A particularly common sequence of the human genome encodes the so-called **Zn finger domain**, thereby identifying proteins that are involved in DNA binding (Section 26.5). Likewise, it can be predicted whether the protein that is encoded is likely to bind Cu, Ca, an Fe-porphyrin, or different types of Fe–S clusters. The gene can be cloned and the protein for which it encodes can be produced in sufficiently large quantities by 'overexpression' in suitable hosts, such as the common gut bacterium *Escherichia coli* or yeast, to enable it to be characterized. Furthermore, the use of genetic engineering to alter the amino acids in a protein is a powerful principle in biological inorganic chemistry. This technique often permits identification of the ligands to particular metal ions and the participation of other residues essential to functions such as substrate binding or proton transfer.

Although structural and spectroscopic studies give a good idea of the basic coordination environment of a metal centre, it is by no means certain that the same structure is retained in key stages of a catalytic cycle, in which unstable states are formed as intermediates. The most stable state of an enzyme, in which form it is usually isolated, is called the 'resting state'. Many enzymes are catalytically inactive upon isolation and must be subjected to an activation procedure that may involve reinsertion of a metal ion or other cofactor or removal of an inhibitory ligand.

Intense efforts have been made to model the active sites of metalloproteins by synthesizing analogues. The models may be divided into two classes: those designed to mimic the structure and spectroscopic properties of the real site, and those synthesized with the intention of mimicking a functional activity, most obviously catalysis. As we shall see throughout this chapter, the difficulty is that an enzyme not only enforces strain upon a metal atom's coordination sphere (even a porphyrin ring is puckered in most cases) but also provides, at fixed distances, functional groups that provide additional coulombic and hydrogen-bonding interactions essential for binding and activating substrates. Indeed, the active site of a metalloenzyme is arguably the ultimate example of supramolecular chemistry.

Transport, transfer, and transcription

In this section we turn to three related aspects of the function of biological molecules containing metal ions, and see their role in the transport of ions through membranes, the transport and distribution of molecules through organisms, and the transfer of electrons. Metal ions also play an important role in the transcription of genes.

26.3 Sodium and potassium transport

Key points: Transport across a membrane is active (energized) or passive (spontaneous); the flow of ions is achieved by proteins known as ion pumps (active) and channels (passive).

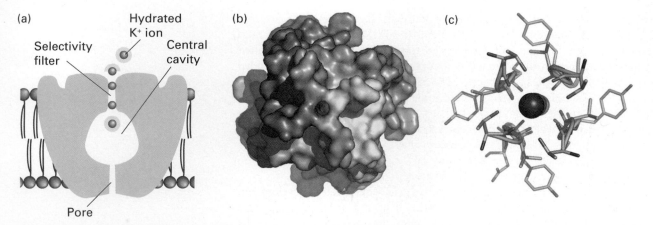

Fig. 26.4 (a) Schematic structure of the K^+ channel showing the different components and the transport of K^+ ions: the blue halo represents hydration. (b) View of the enzyme from inside the cell showing the entrance pore that admits hydrated ions. (c) View looking up the selectivity filter showing how mobile dehydrated K^+ ions are coordinated by peptide carbonyl O atoms provided by each of the four subunits. Note the almost fourfold symmetry axis.

In Chapter 10, we discussed the challenge of differentiating between Na^+ and K^+, two ions that are very similar except for their radii (116 pm and 152 pm, respectively). We saw that selective complexation is achieved through special ligands, such as crown ethers and cryptands, with dimensions appropriate for coordination to one particular kind of ion. Organisms use this principle in the molecules known as **ionophores**, which have hydrophobic exteriors that make them soluble in lipids. The antibiotic valinomycin (Section 10.11) is an ionophore that has a high selectivity for K^+, which is coordinated by six carbonyl groups. It enables K^+ to pass through a bacterial cell membrane and thereby dissipate the electrical potential difference, so causing the bacterium's death.

At the higher end of the complexity scale are the K^+ channels, which are large membrane-spanning proteins that allow selective transport of K^+ and are responsible for electrical conduction in nervous systems.[4] Figure 26.4 shows the important structural aspects of the potential-gated K^+ channel. Moving from the inside surface of the membrane, the enzyme has a pore (which can open and close on receipt of a signal) leading into a central cavity about 1 nm in diameter; up to this stage K^+ ions can remain hydrated. Polypeptide helices pointing at this cavity have their partial charges directed in such a way as to favour population by cations, resulting in a local K^+ concentration of approximately 2 M. Above the central cavity the tunnel contracts into a **selectivity filter** that requires K^+ ions to dehydrate before they enter. The filter has four K^+ binding sites and consists of helical ladders of closely spaced peptide carbonyl-O donors providing eightfold coordination to each K^+. Whereas Na^+ cannot bind effectively and is thus excluded, the larger Rb^+ and Cs^+ ions are taken up, so providing a useful tool for X-ray structural refinement because they are more electron-dense than K^+.

The channel acts by the mechanism illustrated in Fig. 26.5. Charged groups on the molecule move in response to a change in the membrane potential and cause the intracellular pore to open, so allowing entry of hydrated K^+ ions. Selective binding of dehydrated K^+ ions occurs in the filter region, a potential drop across the membrane is sensed, and the cavity closes. At this point the filter opens up to the external surface, where the K^+ concentration is low and the K^+ ions are hydrated and released. This release causes the protein to switch back to the original conformation and K^+ ions again enter the filter. There is a 10^4-fold selectivity for K^+ over Na^+. The binding is weak and fast because it is important to convey but not trap K^+; indeed, transport across the membrane approaches diffusion control.

The Na^+/K^+ pump (Na^+/K^+-ATPase), the enzyme that maintains the concentration differential of Na^+ and K^+ inside and outside a cell, is another example of the high

---

[4] Roderick MacKinnon shared the 2003 Nobel Prize for Chemistry for his elucidation of the structures and mechanisms of ion channels.

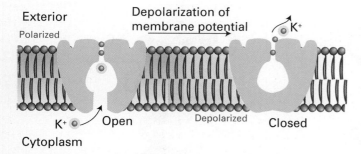

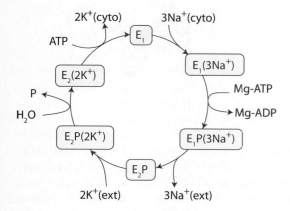

Fig. 26.5 Proposed mechanism of action of the K^+ channel. The potential difference across the membrane is sensed by the protein, which causes the pore to open, allowing hydrated ions to enter the cavity. After shedding their hydration sphere, K^+ ions pass up the selectivity filter at rates close to diffusion control.

Fig. 26.6 General principle of the Na^+, K^+-ATPase (the Na pump). Release of two K^+ ions into the cytoplasm is accompanied by binding of ATP (from the cytoplasm) and conversion of the enzyme into state 1, which binds three Na^+ ions from the cytoplasm. A phosphate group (P) is transferred to the enzyme, which opens to the external side, expels three Na^+ ions, then binds two K^+ ions. Release of the phosphate group causes release of K^+ into the cytoplasm and the cycle begins again.

discrimination between alkali metal ions that has evolved with biological ligands. The ions are pumped against their concentration gradients by coupling the process to ATP hydrolysis. The mechanism, which is outlined in Fig. 26.6, involves conformational changes induced by ATP-driven protein phosphorylation.

Example 26.3 Assessing the role of ions in active and passive transport

The toxic species Tl^+ (radius 164 pm) is used as a probe for K^+ binding in proteins. Explain why Tl^+ is suited for this purpose and account for its high toxicity.

Answer As noted in Section 6.4 $Tl(I = \frac{1}{2})$ can be studied by NMR. The similarity with K^+ also allows Tl^+, a toxic element, free entry into a cell via Na^+/K^+-ATPase, but once inside, differences in chemical properties, such as the tendency of Tl to form more stable complexes with soft ligands, are manifested and become lethal.

Self-test 26.3 Explain why the saline fluid used in hospital procedures contains NaCl.

26.4 Calcium signalling proteins

Key point: Calcium ions are suitable for signalling because they exhibit fast ligand exchange and a large, flexible coordination geometry.

Calcium ions play a crucial role in higher organisms as an intracellular messenger, providing a remarkable demonstration of how organisms have exploited the otherwise rather dull chemistry of this element. Fluxes of Ca^{2+} trigger enzyme action in cells in response to receiving a hormonal or electrical signal from elsewhere in the organism. Calcium is specially suited for signalling because it has fast ligand-exchange rates, intermediate binding constants, and a large, flexible coordination sphere.

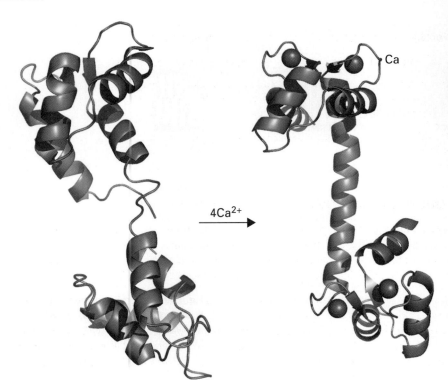

Fig. 26.7 The binding of four Ca^{2+} to apocalmodulin causes a change in the protein conformation, converting it to a form that is recognized by many enzymes. The high proportion of α helix is typical of proteins that are activated by metal ion binding.

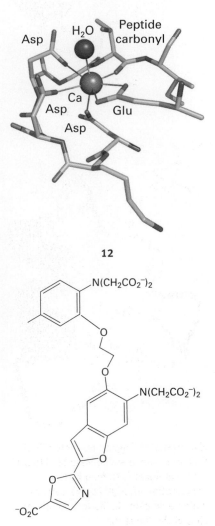

12

13 FURA-2

Calcium signalling proteins are small proteins that change their conformation depending upon the binding of Ca^{2+} at one or more sites; they are thus examples of metal ion-activated proteins mentioned earlier. Every muscle movement we make is stimulated by Ca^{2+} binding to a protein known as troponin C. The best studied Ca^{2+}-regulatory protein is calmodulin (17 kDa, Fig. 26.7): its roles include activating protein kinases that catalyse phosphorylation of proteins and activating NO-synthase, an Fe-containing enzyme responsible for generating the hormone nitric oxide. Calmodulin has four Ca^{2+}-binding sites (one is shown as **12**) with dissociation constants lying in the micromolar range.[5] The binding of Ca^{2+} to the four sites alters the protein conformation and it is then recognized by a target enzyme.

Calcium signalling requires special Ca^{2+} pumps, which are large, membrane-spanning enzymes that pump Ca^{2+} out of the cytoplasm, either out of the cell altogether or into Ca-storing organelles such as the endoplasmic reticulum or the mitochondria. As with Na^+/K^+-ATPases, the energy for Ca^{2+} pumping comes from ATP hydrolysis. Hormones or electrical stimuli open specific channels (analogous to K^+ channels) that release Ca^{2+} into the cell. Because the level in the cytoplasm before the pulse is low, the influx easily raises the Ca^{2+} concentration above that needed for Ca^{2+}-binding proteins such as calmodulin (or troponin C in muscle). The action can be short-lived, so that after a pulse of Ca^{2+} the cell is quickly evacuated by the calcium pump.

Although Ca^{2+} is invisible to most spectroscopic methods, some Ca proteins, such as calmodulin or troponin C, are small enough to be studied by NMR. Because of their preference for large multicarboxylate ligands, the lanthanoid ions (Section 22.6) have been used as probes for Ca binding, exploiting their properties of paramagnetism (in NMR spectroscopy) and fluorescence. Intracellular concentrations of Ca are monitored using special fluorescent polycarboxylate ligands (**13**) that are introduced to the cell as their esters, which are hydrophobic and able to cross the membrane lipid barrier. Once in the cell, enzymes known as esterases hydrolyse the esters and release the ligands, which respond to changes in Ca^{2+} concentration in the range $10^{-7}-10^{-9}$ M.

---

[5] Dissociation constants (the reciprocal of association constants) indicate the concentration of species required to cause 50 per cent occupation of binding sites.

Example 26.4 Explaining why calcium is suitable for signalling

Why is Ca^{2+} more suitable than Mg^{2+} for fast signalling processes in cells?

Answer The exchange of coordinated water molecules is 10^3–10^4 times slower for Mg^{2+} than for Ca^{2+}. This slowness retards events in which speed is important, such as the initiation of muscle contraction (troponin C).

Self-test 26.4 The Ca^{2+} pump is activated by calmodulin. Explain the significance of this observation.

26.5 Zinc in transcription

Key points: Zinc fingers are protein structural features produced by coordination of Zn to specific histidine and cysteine residues; a sequence of these fingers enables the protein to recognize and bind to precise sequences of DNA base pairs and plays a crucial role in transferring information from the gene.

The roles of Zn are either catalytic, which we will deal with later, or structural and regulatory. Unlike Ca and Mg, Zn forms more stable complexes with softer donors, so it is not surprising that we usually find it coordinated in proteins through histidine and cysteine residues. Typical catalytic sites (**14**) usually have three permanent protein ligands and an exchangeable ligand (H_2O) whereas structural Zn sites (**15**) are coordinated by four 'permanent' protein ligands.

Transcription factors are proteins that recognize certain regions of DNA and control how the genetic code is converted into RNA. It has been known since the mid-1980s that many DNA-binding proteins contain repeating domains that are folded in place by the binding of Zn and form characteristic folds known as 'zinc fingers' (Fig. 26.8). In a typical case, one side of the finger provides two cysteine-S donors and the other side provides two histidine-N donors and folds as an α helix. Each 'fingertip' makes recognitory contacts with specific DNA bases. As shown in Fig. 26.9, the zinc fingers wrap around sequences of DNA that they are able to recognize by acting collectively. The high fidelity of transcription factors is the result of a number of such contacts being made along the DNA chain at the beginning of the sequence that is transcribed.

The characteristic residue sequence for a Zn-finger motif is

—(Tyr, Phe)—X—Cys—X_{2-4}—Cys—X_3—Phe—X_5—Leu—X_2—His—X_{3-5}—His—

where the amino acid X is variable. Aside from the 'classical' $(Cys)_2(His)_2$ zinc finger, others have been discovered that have $(Cys)_3His$ or $(Cys)_4$ coordination, together with more elaborate examples having 'Zn-thiolate clusters', such as the so-called GAL4 transcription factor in which two Zn atoms are linked by bridging cysteine-S ligands (**16**). Various protein folds are produced, with faintly jocular names, such as 'zinc knuckles',

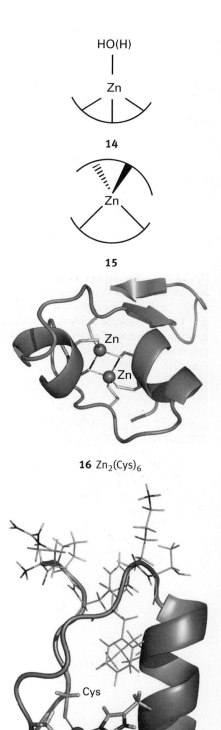

14

15

16 $Zn_2(Cys)_6$

Fig. 26.8 Zinc fingers are protein folds that form a sequence able to bind to DNA. A typical finger is formed by the coordination of Zn(II) to two pairs of amino acid side chains located either side of the 'fingertip'.

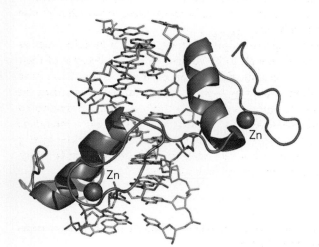

Fig. 26.9 A pair of zinc fingers interacting with a section of DNA.

joining an increasingly large family. Higher order Zn-thiolate clusters are found in proteins known as metallothioneins and some Zn-sensor proteins (see Section 26.16).

Zinc is particularly suited for binding to proteins to hold them in a particular conformation: Zn^{2+} is high in the Irving–Williams series (Section 20.1) and thus forms stable complexes, particularly to S and N donors. It is also redox inactive, which is an important factor because it is crucial to avoid oxidative damage to DNA. Other examples of structural zinc include insulin and alcohol dehydrogenase. The lack of good spectroscopic probes for Zn, however, has meant that even though it has tightly bound in a protein, it has been difficult to confirm its binding or deduce its coordination geometry in the absence of direct structural information from X-ray diffraction or NMR. However, some elegant measurements have exploited the ability of Co(II), which is coloured and paramagnetic, to substitute for Zn (Section 26.9).

26.6 Selective transport and storage of iron

Key points: The uptake of Fe into organisms involves special ligands known as siderophores; transport in the circulating fluids of higher organisms requires a protein called transferrin; Fe is stored as ferritin.

Iron is essential for almost all life forms; however, Fe is also difficult to obtain, yet any excess presents a serious toxic risk. Nature has at least two problems in dealing with this element. The first is the insolubility of Fe(III), which is the stable oxidation state found in most minerals. As the pH increases, hydrolysis, polymerization, and precipitation of hydrated forms of the oxide occur. Polymeric oxide-bridged Fe(III) is the thermodynamic sink of aerobic Fe chemistry (as seen in a Pourbaix diagram, Section 5.13). The insolubility of rust renders the straightforward uptake by a cell very difficult. The second problem is the toxicity of 'free-Fe' species, particularly through the generation of OH radicals. To prevent Fe from reacting with oxygen species in an uncontrolled manner, a protective coordination environment is required. Nature has thus evolved sophisticated chemical systems to execute and regulate all aspects, from the primary acquisition of Fe, to its subsequent transport, storage, and utilization in tissue. The 'Fe cycle' as it affects a human is summarized in Fig. 26.10.

(a) Siderophores

Siderophores are small, polydentate ligands that have a very high affinity for Fe(III). They are secreted from many bacterial cells into the external medium where they sequester Fe to give a soluble complex that re-enters the organism at a specific receptor. Once inside the cell, the Fe is released.

Aside from citrate (the Fe(III) citrate complex is the simplest Fe transport species in biology) there are two main types of siderophore. The first type is based on phenolate or

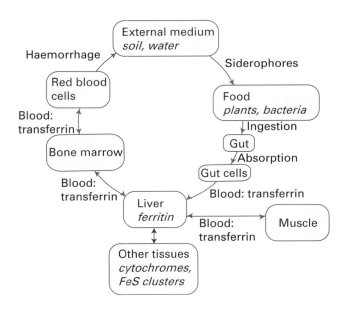

Fig. 26.10 The biological Fe cycle showing how Fe is taken up from the external medium and guarded carefully in its travels through organisms.

catecholate ligands, and is exemplified by enterobactin (**17**) for which the association constant for Fe(III) is $K_A = 10^{52}$ M, an affinity so great that enterobactin enables bacteria to erode steel bridges! The second type of siderophore is based on hydroxamate ligands, and is exemplified by ferrichrome (**18**), a cyclic hexapeptide consisting of three glycine and three *N*-hydroxyl-L-ornithines.

All Fe(III) siderophore complexes are octahedral and high spin. Because the donor atoms are hard O or N atoms and negatively charged, they have a relatively low affinity for Fe(II). Synthetic siderophores are proving to be very useful agents for the control of 'iron overload', a serious condition affecting large populations of the world, particularly South-East Asia (Section 26.17).

(b) Iron-transport proteins in higher organisms

There are several important, structurally similar Fe-transport proteins known collectively as **transferrins**. The best characterized examples are serum transferrin (in blood plasma), ovotransferrin (in egg-white), and lactoferrin (in milk). The apoproteins are potent antibacterial agents as they deprive microbes of their iron. Transferrins are also present in tears, serving to cleanse eyes after irritation. All these transferrins are glycoproteins with molar masses of about 80 kDa and containing two separated and essentially equivalent binding sites for Fe. Complexation of Fe(III) at each site involves simultaneous binding of HCO_3^- or CO_3^{2-} and release of H^+:

$$\text{apo-TF} + \text{Fe(III)} + HCO_3^- \rightarrow \text{Fe —TF} + H^+$$

where TF denotes transferrin. For each site, the association constant K_A under physiological conditions (pH = 7) is in the range 10^{22}–10^{26} M. However, K_A is very pH dependent and pH is the main factor controlling Fe uptake and release.

Transferrin consists of two very similar parts, termed the **N-lobe** and the **C-lobe** (Fig. 26.11). The protein is a product of gene duplication, because the structure of the first half of the molecule can almost be overlaid on the second half. Each half consists of two domains, 1 and 2, which together form a cleft with a binding site for Fe(III). There is a considerable proportion of α helix, resulting in flexibility. Complexation with Fe(III) causes a conformational change consisting of a hinge motion involving domains 1 and 2 at each lobe. Binding of Fe(III) causes the domains to come together.

In each active site (**19**), a single Fe atom is coordinated by widely dispersed amino acid side chains from both domains and the connecting region, hence the change in conformation that occurs. The protein ligands are carboxylate-O (Asp), two phenolate-O (Tyr) and an imidazole-N (His). Only one of the aspartate carboxylate-O atoms is coordinated. The protein ligands form part of a distorted octahedral coordination sphere. The coordination is completed by bidentate binding to the exogenous carbonate, which is

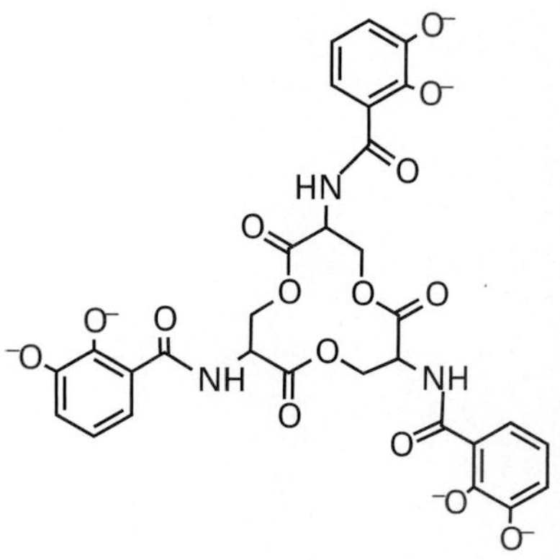

17 Enterobactin

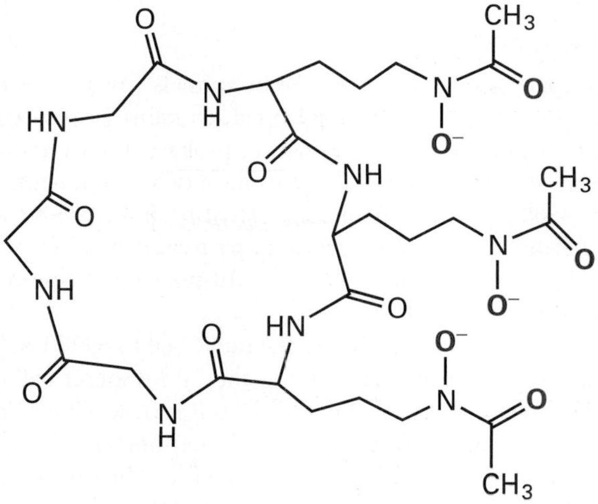

18 Ferrichrome

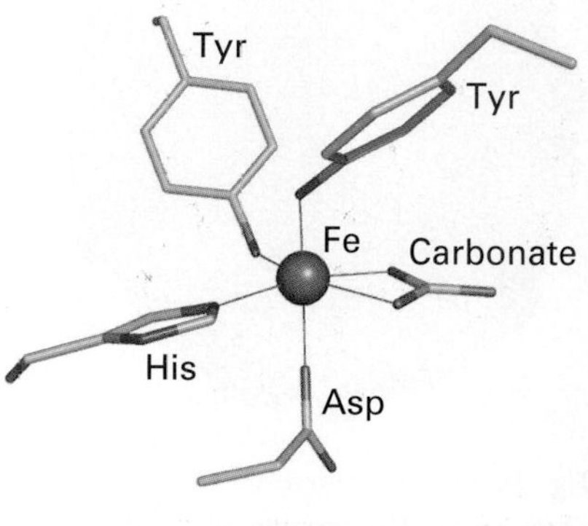

19

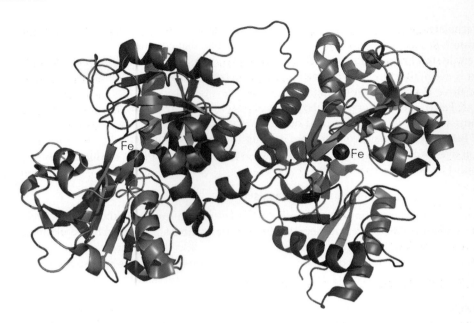

Fig. 26.11 Structure of the Fe-transport protein transferrin: the identical halves of the molecule each coordinates to a single Fe(III) atom (the black spheres) between two lobes. This coordination causes a conformational change that allows transferrin to be recognized by the transferrin receptor.

referred to as a **synergistic ligand** because Fe binding depends on its presence. In certain cases phosphate is bound instead of carbonate. As expected from the predominantly anionic ligand set, Fe(III) binds much more tightly than Fe(II). However, ions similar to Fe(III), particularly Ga(III) and Al(III), also bind tightly, so that these metals too can use the same transport system to gain access to tissues.

(c) Release of iron from transferrin

Cells in need of Fe produce large amounts of a protein called the **transferrin receptor** (180 kDa), which is incorporated within their plasma membrane. This protein binds Fe-loaded transferrin. The most favoured mechanism for Fe uptake involves the Fe-loaded transferrin receptor complex entering the cell by a process known as **endocytosis**. In endocytosis, a section of the cell membrane is engulfed by the wall along with its component membrane-bound proteins, to form a vesicle. The pH within this vesicle is then lowered by a membrane-bound H^+-pumping enzyme that is also swallowed by the cell. The subsequent release of Fe(III) is probably linked to the coordination of carbonate, which is synergistic in the sense that it is necessary for the binding of Fe but is unstable at low pH. Indeed, from *in vitro* studies it is known that Fe is released by lowering the pH to about 5 for serum transferrin and to 2–3 for lactoferrin. The vesicle then splits and the TF-receptor complex is returned to the plasma membrane by **exocytosis**, and Fe(III), probably now complexed by citrate, is released to the cytoplasm.

(d) Ferritin, the cellular Fe store

Ferritin is the principal store of non-haem Fe in animals (most Fe is occupied in haemoglobin and myoglobin) and, when fully loaded, contains 20 per cent Fe by mass! It occurs in all types of organism, from mammals to prokaryotes. In mammals, it is found particularly in the spleen and in blood. Ferritins have two components, a 'mineral' core that contains up to 4500 Fe atoms (mammalian ferritin) and a protein shell. Apoferritin (the protein shell devoid of Fe) can be prepared by treatment of ferritin with reducing agents and an Fe(II) chelating ligand (such as 1,10-phenanthroline or 2,2′-bipyridyl). Dialysis then yields the intact shell.

Apoferritins have average molar masses in the range 460 to 550 kDa. The protein shell (Fig. 26.12) consists of 24 subunits that link together like a soccer ball to form a hollow sphere with twofold, threefold (as shown in the illustration), and fourfold symmetry axes. Each subunit consists of a bundle of four long and one short α helices, with a loop that forms an antiparallel β sheet with a neighbouring subunit. The mineral core is composed of hydrated Fe(III) oxide with varying amounts of phosphate. The structure as revealed by X-ray or electron diffraction resembles that of the mineral ferrihydrite, $5Fe_2O_3 \cdot 9H_2O$,

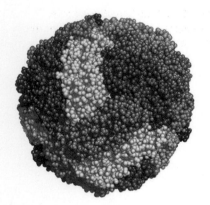

Fig. 26.12 The structure of ferritin, showing the arrangement of subunits that make up the protein shell.

which is based on an hcp array of O^{2-} and OH^- ions, with Fe(III) layered in both the octahedral and tetrahedral sites (**20**).

The threefold and fourfold symmetry axes of apoferritin are, respectively, hydrophilic and hydrophobic pores. The threefold-axis pores are suited for the passage of ions. However, the ferrihydrite core is insoluble and Fe must be mobilized. The most feasible mechanism so far proposed for the reversible incorporation of Fe in ferritin involves its transport in and out as Fe(II), perhaps as the Fe^{2+} ion, which is soluble at neutral pH, but more likely some type of 'chaperone' complex. Oxidation to Fe(III) is thought to occur at specific di-iron binding sites known as **ferroxidase centres**, present in each of the subunits. Oxidation to Fe(III) involves the coordination of O_2 and inner-sphere electron transfer:

$$2\,Fe(II) + O_2 + 2\,H^+ \rightarrow 2\,Fe(III) + H_2O_2$$

The mechanism by which Fe is released almost certainly involves its reduction back to the more mobile Fe(II).

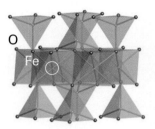

20 Ferrihydrite

26.7 Oxygen transport and storage

Dioxygen, O_2, is a special molecule that has not always been available to biology; in fact, to many life forms it is highly toxic. However, as we shall see in Section 26.10, the great thermodynamic advantage of using such a powerful oxidant undoubtedly led to the evolution of higher organisms. Indeed, the requirement for O_2 became so important as to necessitate special systems for transporting and storing it. Apart from the difficulty in supplying O_2 to buried tissue, there is the problem of achieving a sufficiently high concentration in water. This problem is overcome by special metalloproteins known as **O_2 carriers**. In mammals and most other animals and plants, these special proteins (myoglobin and haemoglobin) contain an Fe porphyrin cofactor. Animals such as molluscs and arthropods use a Cu enzyme called haemocyanin, and some lower invertebrates use the binuclear Fe complex haemerythrin.

(a) Myoglobin

Key points: The deoxy form containing high-spin five-coordinate Fe(II) reacts rapidly and reversibly with O_2 to produce low-spin six-coordinate Fe(II); a slow autooxidation reaction releases superoxide and produces Fe(III), which is inactive in binding O_2.

Myoglobin is an Fe protein (17 kDa, Fig. 26.13) that coordinates O_2 reversibly and controls its concentration in tissue. The molecule contains several regions of α helix, implying mobility, with the single Fe porphyrin group located in a cleft between helices E and F. Two propionate substituents on the porphyrin interact with solvent water molecules on the surface of the protein. The fifth ligand to the Fe is provided by a histidine-N from helix F, and the sixth position is the site at which O_2 is coordinated. In common terminology, the side of the haem plane at which exchangeable ligands are bound is known as the **distal region**, while that below the haem plane is known as the **proximal region**. The histidine on helix F is one of two that are present in all species. Such 'highly conserved' amino acids are a strong indication that evolution has determined that they are essential for function. The other conserved histidine is located on helix E.

Deoxymyoglobin (Mb) is bluish red and contains Fe(II); this is the oxidation state that binds O_2 to give the familiar bright red oxymyoglobin (oxyMb). In some instances deoxymyoglobin becomes oxidized to Fe(III), which is called metmyoglobin (metMb) and is unable to bind O_2. This oxidation may occur by a ligand substitution-induced redox reaction in which Cl^- ions displace bound O_2 as superoxide:

$$Fe(II)O_2 + Cl^- \rightarrow Fe(III)Cl + O_2^-$$

In healthy tissue, an enzyme (methaemoglobin reductase) is available to reduce the met form back to the Fe(II) form.

Deoxymyoglobin is five-coordinate and high-spin. When O_2 binds it is coordinated by a monohapto link to the Fe atom, the electronic structure of which is tuned by the F helix

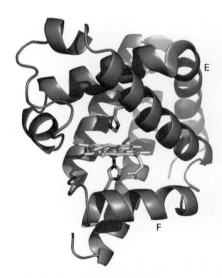

Fig. 26.13 Structure of myoglobin, showing the Fe porphyrin group located between helices E and F.

histidine ligand (Fig. 26.14). The unbound end of the O_2 molecule is fastened by a hydrogen bond to the imidazole-NH of the histidine in helix E. The coordination of O_2 (a strong-field π-acceptor ligand) causes the Fe(II) to switch from high-spin to low-spin, and with all the d electrons in t_{2g} orbitals, to shrink slightly and move into the plane of the ring. The bonding is often expressed in terms of Fe(II) coordination by singlet O_2, in which the doubly occupied antibonding $2π_g$ orbital of O_2 acts as a σ donor and the empty $2π_g$ orbital of O_2 accepts an electron pair from the Fe (Fig. 26.15).

(b) Haemoglobin

Key point: Haemoglobin consists of a tetramer of myoglobin-like subunits, with four Fe sites that bind O_2 cooperatively.

Haemoglobin (Hb, 68 kDa, Fig. 26.16) is the O_2 transport protein found in special cells known as **erythrocytes** (red blood cells): a litre of human blood contains about 150 g of Hb. Simplistically, Hb can be thought of as a tetramer of myoglobin-like units with a cavity in the middle. There are in fact two types of Mb-like subunits, which differ slightly in their structures, and Hb is referred to as an **$α_2β_2$ tetramer**.

The O_2 binding curves for Mb and Hb are shown in Fig. 26.17: it is highly significant that the curve for Hb is sigmoidal, which indicates that uptake and release of successive O_2 molecules is cooperative. At low O_2 partial pressure and greater acidity (as in venous

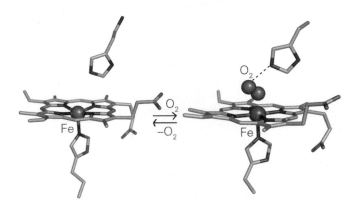

Fig. 26.14 Reversible binding of O_2 to myoglobin: coordination by O_2 causes the Fe to become low-spin and move into the plane of the porphyrin ring.

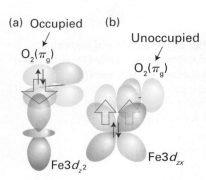

Fig. 26.15 The orbitals used to form the Fe—O_2 adduct of myoglobin and haemoglobin. This model considers the O_2 ligand to be in a singlet state, in which the full $2π_g$ orbital donates an electron pair and the other $2π_g$ orbital acts as a π-electron pair acceptor.

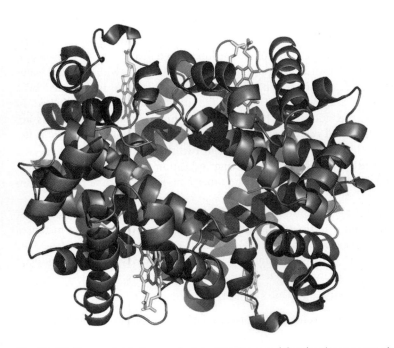

Fig. 26.16 Haemoglobin is an $α_2β_2$ tetramer. Its α and β subunits are very similar to myoglobin. Haem groups are shown in yellow.

blood and muscle tissue following aggressive exercise) Hb has low affinity for O_2. This low affinity enables Hb to transfer its O_2 to Mb. As the pressure increases, so does the affinity of Hb for O_2 and as a result Hb can pick up O_2 in the lungs. This change in affinity can be attributed to there being two conformations. The **tensed state** (T) has a low affinity and the **relaxed state** (R) has a high affinity. Deoxy-Hb is T and fully loaded oxy-Hb is R.

To understand the molecular basis of cooperativity we need to refer to Fig. 26.14. Binding of the first O_2 molecule to the T-state molecule is weak, but the decrease in the size of the Fe allows it to move into the plane of the porphyrin ring. This motion is particularly important for Hb because it pulls on the proximal histidine ligand and helix F moves. This movement is transmitted to the other O_2 binding sites, the effect being to move the other Fe atoms closer to their respective ring planes and thereby convert the protein into the R state. The way is thereby opened for them to bind O_2, which they now do with greater ease.

(c) Other oxygen transport systems

Key point: Arthropods and molluscs use haemocyanin and certain marine worms use haemerythrin.

In many organisms, such as arthropods and molluscs, O_2 is transported by the Cu protein haemocyanin, which, unlike haemoglobin, is extracellular, as is common for Cu proteins. Haemocyanin is oligomeric, with each monomer containing a pair of Cu atoms in close proximity. Deoxyhaemocyanin is colourless but it becomes bright blue when O_2 binds.

The active site is shown in Fig. 26.18. In the deoxy state, each Cu atom is three-coordinate and bound in a pyramidal array by three histidine residues. The two Cu atoms are so far apart (460 pm) that there is no direct interaction between them. The low coordination number is typical of Cu(I), which normally is two- to four-coordinate. Rapid and reversible coordination of O_2 occurs between the two Cu atoms in a bridging dihapto manner (μ-$\eta^2\eta^2$) and the low vibrational wavenumber of the coordinated O_2 molecule ($750\ cm^{-1}$) shows it has been reduced to peroxide (O_2^{2-}), with a concomitant lowering of the bond order from 2 to 1. To accommodate the binding of O_2, the protein adjusts its conformation to bring the two Cu atoms closer together. The Cu sites become five-coordinate, which is typical of Cu(II).

Haemerythrin is an example of a special class of binuclear Fe centres that are found in a number of proteins with diverse functions, such as methane monooxygenase, and some ribonucleotide reductases and acid phosphatases. The two Fe atoms in the active site of haemerythrin (**21**) are each coordinated by amino acid side chains but are also linked by two bridging carboxylate groups and a small ligand. In the reduced form, which binds O_2 reversibly, this small ligand is an OH^- ion. Coordination of O_2 occurs at only one of the Fe atoms and the distal O atom forms a hydrogen bond to the H atom of the bridging hydroxide.

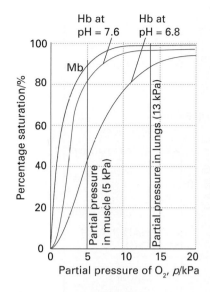

Fig. 26.17 Oxygen binding curves for myoglobin and haemoglobin showing how cooperativity between the four sites in haemoglobin gives rise to a sigmoidal curve. The binding of the first O_2 molecule to haemoglobin is unfavourable, but it results in a greatly enhanced affinity for subsequent O_2 molecules.

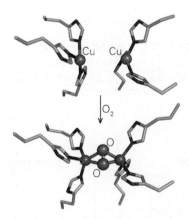

Fig. 26.18 Binding of O_2 at the active site of haemocyanin causes the two Cu atoms to be brought closer together. The O_2 complex is regarded as a binuclear Cu(II) centre in which the two Cu atoms are bridged by an η^2, η^2-peroxide.

Example 26.5 Identifying how biology compensates for strong competition by CO

Carbon monoxide is well known to be a strong inhibitor of O_2 binding by myoglobin and haemoglobin, yet relative to O_2 its binding is much weaker in the protein compared to a simple Fe-porphyrin complex. This suppression of CO binding is important as even trace levels of CO would otherwise have serious consequences for aerobes. Suggest an explanation.

Answer The binding of O_2 is nonlinear (see Figs 26.14 and 26.15) and the distal O atom is well positioned to form a hydrogen bond to the distal imidazole. By contrast, CO adopts a linear FeCO arrangement and does not participate in the additional bonding.

Self-test 26.5 Why are small Fe-porphyrin complexes unable to bind O_2 reversibly?

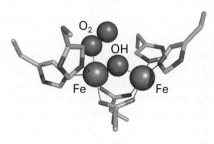

21

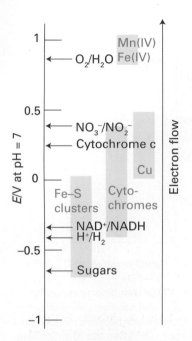

Fig. 26.19 The 'redox spectrum' of life.

26.8 Electron transfer

In all but a few interesting cases, the energy for life stems ultimately from the Sun, either directly in photosynthesis or indirectly by acquiring energy-rich compounds (fuel) from photosynthesizing organisms. Energy can be considered in electrochemical terms as a flow of electrons from fuel to oxidant. Important fuels include fats, sugars, and H_2, and important biological oxidants include O_2, nitrate, and even H^+. As seen from Fig. 26.19, oxidation of sugars by O_2 provides a lot of energy (over 4 eV per O_2 molecule), and is the reason for the success of aerobic organisms over the anaerobic ones that once dominated the Earth.

(a) General considerations

Key points: Electron flow along electron-transport chains is coupled to chemical processes such as ion (particularly H^+) transfer; the simplest electron-transfer centres have evolved to optimize fast electron transfer.

In organisms, electrons are abstracted from food (fuel) and flow to an oxidant, down the potential gradient formed by the sequence of acceptors and donors known as a **respiratory chain** (Fig. 26.20).[6] Apart from flavins and quinones, which are redox-active organic cofactors, these acceptors and donors are metal-containing electron transfer (ET) centres, which fall into three main classes, namely FeS clusters, cytochromes, and Cu sites. These enzymes are generally bound in a membrane, across which the energy from ET is used to sustain a transmembrane proton gradient: this is the basis of the famous **chemiosmotic theory**. The counterflow of H^+, through a rotating enzyme known as ATP synthase, drives the phosphorylation of ADP to ATP. Many membrane-bound redox enzymes are **electrogenic proton pumps**, which means they directly couple long-range ET to proton transfer through specific internal channels.

We shall examine the properties of the three main types of ET centre. The same rules concerning outer-sphere electron-transfer that were discussed in Section 20.13 apply to

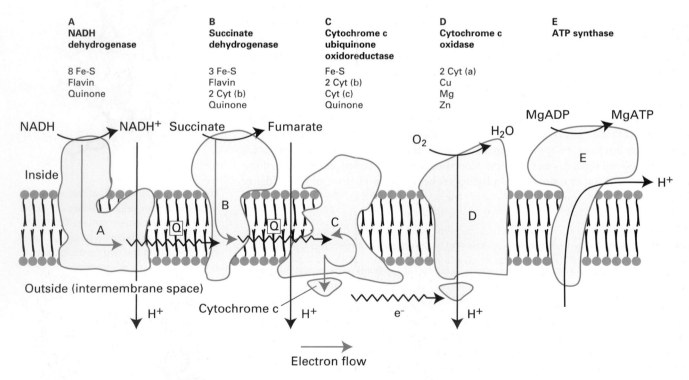

Fig. 26.20 The mitochondrial respiratory electron transfer (ET) chain consists of several metalloenzyme molecules that use the energy of electron transport to transport protons across a membrane. The proton gradient is used to drive ATP synthesis.

---

[6] An overall current of about 80 A flows through the mitochondrial respiratory chains in an average human!

metal centres in proteins, and we should note that organisms have optimized the structures and properties of these centres to achieve efficient long-range electron transfer.

In the following discussion it will be useful to keep in mind that reduction potentials depend on several factors. Besides ionization energy and ligand environment (strong donors stabilize high oxidation states and lower the reduction potential; weak donors, π acceptors, and protons stabilize low oxidation states and raise the reduction potential), an active site in a protein is also influenced by the relative permittivity (which stabilizes centres with low overall charge), the presence of neighbouring charges, including those provided by other bound metal ions, and the availability of hydrogen-bonding interactions that will also stabilize reduced states.

When we consider the kinetics of electron transfer, it will similarly be useful to keep in mind that 'efficiency' means that electron transfer is fast even when the reaction Gibbs energy is low and therefore that the reorganization energy λ of Marcus theory (Section 20.13) is low and intersite distances short (less than 1.4 nm). This requirement is met by providing a ligand environment that does not alter significantly when an electron is added and by burying the site so that water molecules are excluded. It is still debated whether electron transfer in proteins depends mainly on distance alone or whether the protein can provide special pathways.

(b) Electron-transfer cytochromes

Key points: Electron-transfer cytochromes operate in the potential region -0.3 to $+0.4$ V; they have a combination of low reorganization energy and extended electron coupling.

Cytochromes were identified many years ago as cell pigments (hence the name). They contain an Fe porphyrin group and the term 'cytochrome' can refer to both an individual protein and a subunit of a larger enzyme that contains the cofactor. Cytochromes that are involved in electron transfer use the Fe^{3+}/Fe^{2+} couple and are generally six-coordinate (**22**), with two stable axial bonds to amino acid donors, and the Fe is usually low spin in both oxidation states. This contrasts with ligand-binding cytochromes such as haemoglobin, for which the sixth coordination site is either empty or occupied by an H_2O molecule.

A good way to consider the capability of cytochromes for fast electron transfer is to treat the d orbitals of Fe(III) and Fe(II) in terms of an octahedral ligand field and to consider the overlap between the electron-rich but nearly nonbonding t_{2g} orbitals (the configurations are t_{2g}^5 and t_{2g}^6 in Fe(III) and Fe(II), respectively) and the orbitals of the porphyrin. The electron enters or leaves an orbital having π overlap with the π^* antibonding molecular orbital on the ring system. This arrangement provides enhanced electron transfer because the d orbitals of the Fe atom are effectively extended out to the edge of the porphyrin ring, so decreasing the distance over which an electron must transfer between redox partners (Fig. 26.21).

The paradigm of ET cytochromes is **mitochondrial cytochrome c** (12 kDa, Fig. 26.22). This cytochrome is found in the mitochondrial intramembrane space, where it supplies electrons to cytochrome c oxidase, the enzyme responsible for reducing O_2 to H_2O at the end of the energy-transducing respiratory chain (Section 26.10). The fifth and sixth ligands to Fe in cytochrome c are histidine (imidazole-N) and methionine (thioether-S, **22**). Methionine is not a common ligand in metalloproteins but because it is a neutral, soft donor it is expected to stabilize Fe(II) rather than Fe(III). The reduction potential of cytochrome c is $+0.26$ V, at the higher end of values for cytochromes in general. Cytochromes vary in the identity of the axial ligands as well as the structure of the porphyrin ligand (the notation a, b, c, d, etc. defines positions of absorption maxima in the visible region, but also refer to variations in the substituents on the porphyrin ring). Many ET cytochromes, in particular those sandwiched between membrane-spanning helices, have bis(histidine) axial ligation (**23**).

In cytochrome c, the edge of the porphyrin ring is exposed to solvent and is the most likely site for electrons to enter or leave. Specific protein–protein interactions are important for obtaining efficient electron transfer and the region around the exposed

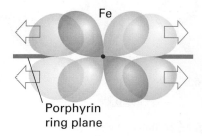

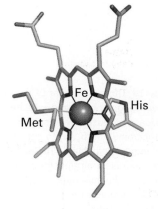

Fig. 26.21 Overlap between the t_{2g} orbitals of the Fe and low-lying empty π^* orbitals on the porphyrin effectively extends the Fe orbitals out to the periphery of the ring.

22

23

(a) 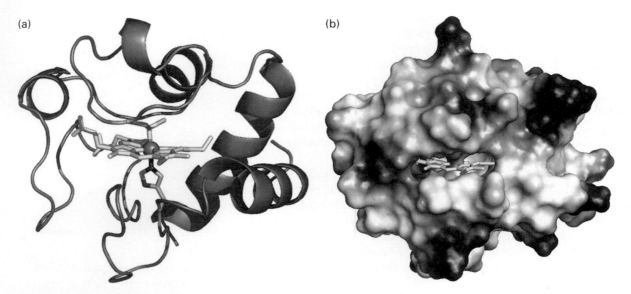 (b)

Fig. 26.22 Different views (but from the same viewpoint) of mitochondrial cytochrome *c*. (a) The secondary structure and the position of the haem cofactor. (b) The surface charge distribution that guides the docking with its natural redox partners (red and blue areas represent patches of negative and positive charge, respectively).

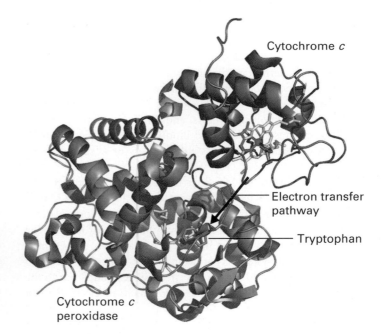

Cytochrome *c*

Electron transfer pathway

Tryptophan

Cytochrome *c* peroxidase

Fig. 26.23 The bimolecular ET complex between cytochrome *c* and cytochrome *c* peroxidase produced by co-crystallization of cytochrome *c* with the Zn derivative of cytochrome *c* peroxidase. The orientation suggests an electron transfer pathway between the haem groups of cytochrome *c* and cytochrome *c* peroxidase that includes tryptophan.

edge of the porphyrin ring in cytochrome *c* provides a pattern of charges that are recognized by cytochrome *c* oxidase and other redox partners. One example in particular has been well studied, that of cytochrome *c* with yeast cytochrome *c* peroxidase. The driving force for electron transfer from either of the two catalytic intermediates of peroxidase (Section 26.9) to each reduced cytochrome *c* is approximately 0.5 V. Figure 26.23 shows the structure of a bimolecular complex formed between cytochrome *c* and cytochrome *c* peroxidase. Electrostatic interactions guide the two proteins together within a distance that is favourable for fast electron tunnelling between cytochrome *c* and two redox centres on the peroxidase, the haem cofactor and tryptophan-191 (Section 26.10).

(c) FeS clusters

Key points: Fe-S clusters generally operate at more negative potentials than cytochromes; they are composed of high-spin Fe(III) or Fe(II) with sulfur ligands in a mainly tetrahedral environment.

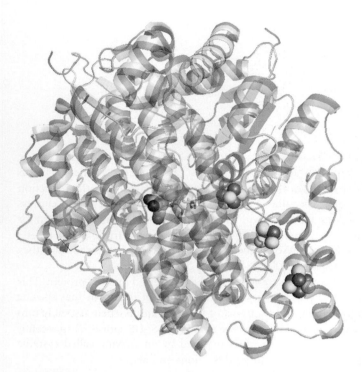

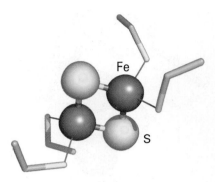

24 [2Fe–2S]

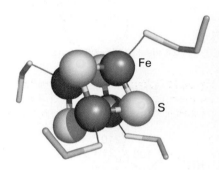

25 [4Fe–4S]

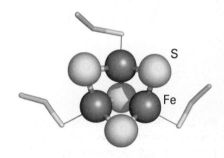

26 [3Fe–4S]

Fig. 26.24 A series of three Fe–S clusters provides a long-range electron transfer pathway to the buried active site in hydrogenases.

Iron–sulfur clusters are widespread in biology, although their importance was not established as early as cytochromes on account of their lack of distinctive optical characteristics. By convention, FeS clusters are represented by square brackets showing how many Fe and S atoms are present, as in [2Fe–2S] (**24**), [4Fe–4S] (**25**), and [3Fe–4S] (**26**). The efficacy of FeS clusters as fast ET centres is largely due to their being able to delocalize the added electron to varying degrees, which minimizes bond length changes and decreases the reorganization energy. The presence of sulfur ligands to provide good lead-in groups is also important. Small ET proteins containing FeS clusters are known as **ferredoxins**, whereas in many large enzymes, FeS clusters are arranged in a relay, less than 1.5 nm apart, to link remote redox sites in the same molecule. The relay concept is illustrated in Fig. 26.24 with a class of enzymes known as **hydrogenases**, which we discuss further in Section 26.14.

In nearly all cases, the Fe atoms are tetrahedrally coordinated by cysteine thiolate (RS^-) groups as the protein ligands. The overall assembly including the protein ligands is known as an 'FeS centre'. Examples are known in which one or more of the Fe atoms is coordinated by non-thiolate amino acid ligands, such as carboxylate, imidazole, and alkoxyl (serine), or by an exogenous ligand such as H_2O or OH^-, and the coordination number about the Fe subsite may be increased to six. The cubane [4Fe–4S] (**25**) and cuboidal [3Fe–4S] (**26**) clusters are obviously closely related, and may even interconvert within a protein by the addition or removal of Fe from one subsite. Larger clusters also occur, such as the 'super clusters' [8Fe–7S] and [Mo–7Fe–8S–X] found in nitrogenase (Section 26.13).

Despite the presence of more than one Fe atom, FeS clusters generally carry out single electron transfers and are good examples of mixed-valence systems comprising Fe(III) and Fe(II). They have negative reduction potentials (usually more negative than -0.2 V) so the reduced forms are good reducing agents:

$$[2Fe\text{–}2S]^{2+} + e^- \rightarrow [2Fe\text{–}2S]^+ \qquad E^{\ominus} = 0 \text{ to } -0.4 \text{ V}$$

2Fe(III)   {Fe(III):Fe(II)}

$S = 0$ $S = \frac{1}{2}$

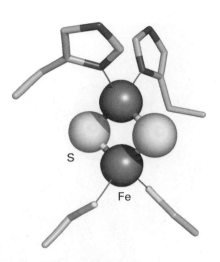

27 Rieske [2Fe–2S] centre

Oxidized plastocyanin

Reduced plastocyanin

28 Cu-ligand bond lengths (pm) in plastocyanin

$$[3Fe–4S]^+ + e^- \rightarrow [3Fe–4S] \qquad E^\ominus = +0.1 \text{ to } -0.4\,\text{V}$$

$$3Fe(III) \qquad\qquad \{2Fe(III){:}Fe(II)\}$$

$$S = \tfrac{1}{2} \qquad\qquad S = 2$$

$$[4Fe–4S]^{2+} + e^- \rightarrow [4Fe–4S]^+ \qquad E^\ominus = -0.2 \text{ to } -0.7\,\text{V}$$

$$\{2Fe(III){:}2Fe(II)\} \quad \{Fe(III){:}3Fe(II)\}$$

$$S = 0 \qquad\qquad S = \tfrac{1}{2}$$

An exception is the so-called **Rieske centre**, which is a [2Fe–2S] cluster having one Fe subsite coordinated by two neutral imidazole ligands rather than cysteine (**27**); this coordination environment stabilizes the iron as Fe(II) and usually raises the reduction potential to a much more positive value (above +0.2 V). The spin states of FeS clusters are shown in the half-reactions above: individual Fe atoms are high spin, as expected for tetrahedral coordination by S^{2-}, and different magnetic states arise from ferromagnetic and antiferromagnetic coupling. These magnetic properties are very important, as they allow the centres to be investigated by EPR (Section 6.5).

A major question is how FeS centres are synthesized and inserted into the protein. This process has been studied mostly in prokaryotes, from which it is known that specific proteins are involved in the supply and transport of Fe and S atoms, their assembly into clusters, and their transfer to target proteins. Free sulfide (H_2S, HS^-, or S^{2-}) in a cell is highly poisonous, so it is produced only when required by an enzyme called cysteine desulfurase, which breaks down cysteine to yield S^{2-} ions and alanine.

(d) Copper electron-transfer centres

Key point: The protein overcomes the large inherent difference in preferred geometries for Cu(II) and Cu(I) by constraining Cu in a geometry that does not alter upon electron transfer.

The so-called 'blue' Cu centre is the active site of a number of small electron-transfer proteins as well as larger enzymes (the blue Cu oxidases) that contain, in addition, other Cu sites. Blue Cu centres have reduction potentials for the Cu(II)/Cu(I) redox couple that lie in the range 0.15–0.7 V and so they are generally more oxidizing than cytochromes. The name stems from the intense blue colour of pure samples in the oxidized state, which arises from ligand(thiolate)-to-metal charge transfer. In all cases, the Cu is shielded from solvent water and coordinated by a minimum of two imidazole-N and one cysteine-S in a nearly trigonal planar manner, with one or two longer bonds to axial ligands. The most studied examples are plastocyanin (Fig. 26.25), a small electron carrier protein in chloroplasts, and azurin, a bacterial electron carrier. These small proteins have a β-barrel structure, which holds the Cu coordination sphere in a very rigid geometry. Indeed, the crystal structures of oxidized, reduced (**28**), and apo forms reveal that the ligands remain in essentially the same position in all cases. As a result, the blue Cu centre is well suited to undergo fast and efficient electron transfer because the reorganization energy is small.

The binuclear Cu centre known as Cu_A is present in cytochrome c oxidase and N_2O reductase. The two Cu atoms (**29**) are each coordinated by two imidazole groups and a pair of cysteine thiolate ligands act as bridging ligands. In the reduced form, both Cu atoms are Cu(I). This form undergoes one-electron oxidation to give a purple, para-magnetic species in which the unpaired electron is shared between the two Cu atoms. Once again, we see how delocalization assists electron transfer because the reorganization energy is lowered.

Example 26.6 Explaining the function of ET centres

The reduction potential of Rieske FeS centres is very pH dependent, unlike the standard FeS centres that have only thiolate ligation. Suggest an explanation.

Answer Each of the two imidazole ligands that coordinate one of the Fe atoms in the Rieske [2Fe–2S] cluster is electrically neutral at pH = 7 and the proton located on the noncoordinating N atom is easily removed. The pK depends on the oxidation level of the cluster, and there is a large region of pH in which the imidazole ligands are protonated in the reduced form but not in the oxidized form. As a result, the reduction potential depends on pH.

Self-test 26.6 Simple Cu(II) compounds show a large EPR hyperfine coupling to the Cu nucleus ($I = \frac{3}{2}$ for ^{65}Cu and ^{63}Cu), whereas the EPR spectra of blue Cu proteins show a much smaller hyperfine coupling. What does this suggest about the nature of the ligand coordination at blue Cu centres?

Fig. 26.25 The plastocyanin molecule.

Catalytic processes

The classic role of enzymes is as highly selective catalysts for the myriad of chemical reactions that take place in organisms and sustain the activities of life. In this section we view some of the most important examples in terms of the suitability of certain elements for their roles.

26.9 Acid–base catalysis

Biological systems rarely have the extreme pH conditions under which catalysis by free H^+ or OH^- can occur; indeed, the result would be indiscriminate because all hydrolysable bonds would be targets. One way that organisms have solved this problem has been to harness properties of certain metal ions and build them into protein structures designed to accomplish specific (Brønsted) acid–base reactions (Section 4.1).

Organisms make extensive use of Zn for achieving acid–base catalysis, but not to the exclusion of other metals. For example, numerous enzymes feature Fe(II) and Fe(III), Mg(II) serves as the catalyst in pyruvate kinase and ribulose bisphosphate carboxylase, Mn is the catalyst in arginase (for the hydrolysis of arginine, yielding urea and L-ornithine), and Ni(II) is the active metal in urease (for the hydrolysis of urea, yielding ammonia and carbon dioxide). Because of their importance to industry and medicine, many of these enzymes have been studied in great detail and model systems have been synthesized in efforts to reproduce catalytic properties and understand the mode of action of inhibitors. Many of these sites (including arginase-[Mn,Mn] and urease-[Ni,Ni]) contain two or more metal ions in an arrangement unique to the protein and difficult to model with simple ligands.

(a) Zinc enzymes

Key point: Zinc is well suited for catalysing acid–base reactions as it is abundant, redox inactive, forms strong bonds to donor groups of amino acid residues, and exogenous ligands such as H_2O are exchanged rapidly.

A Zn^{2+} ion has high rates of ligand exchange and its polarizing power means that the pK of a coordinated water molecule is quite low. Combined with strong binding to protein ligands, rapid ligand exchange (coordinated H_2O or substrate molecules), good electron affinity, flexibility of coordination geometry, and no complicating redox chemistry, Zn is well suited to its role in catalysing specific acid–base reactions. The large family of Zn enzymes include carbonic anhydrase, carboxypeptidases, alkaline phosphatase, β-lactamase (responsible for penicillin resistance in bacteria), and alcohol dehydrogenase. Typically, the Zn is coordinated by three amino acid ligands (in contrast to zinc fingers, which have four) and one exchangeable water molecule (**14**).

The mechanisms of Zn enzymes are normally discussed in terms of two limiting cases. In the **Zn-hydroxide mechanism**, the Zn functions by promoting deprotonation of a bound water molecule, so creating an OH^- nucleophile that can go on to attack the carbonyl C atom:

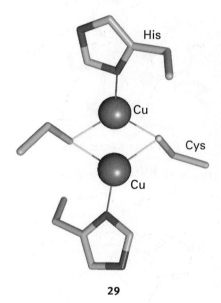

29

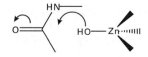

In the **Zn-carbonyl mechanism**, the Zn ion acts directly as a Lewis acid to accept an electron pair from the carbonyl O atom, and its role is therefore analogous to H^+ in acid catalysis:

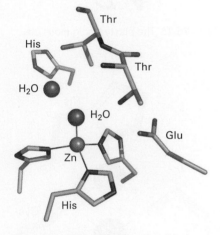

Similar reactions occur with other X=O groups, particularly P=O as occurs in phosphate esters. There is an obvious advantage of achieving such catalysis at Zn or some other acid species that is anchored in a stereoselective environment.

The formation and transport of CO_2 is a fundamental process in biology. The solubility of CO_2 in water depends upon its hydration and deprotonation to form HCO_3^-. However, the uncatalysed reaction at pH = 7 is very slow, the forward process occurring with a rate constant of less than 10^{-3} s^{-1}. Because turnover of CO_2 by biological systems is very high, such a rate is far too slow to sustain vigorous aerobic life in a complex organism. The CO_2/HCO_3^- equilibrium is also important (in addition to its role in CO_2 transport) because it provides a way of regulating tissue pH.

In 1932 an enzyme, carbonic anhydrase (CA, or carbon dioxide dehydratase, was identified in red blood cells that catalyses this reaction with a remarkable rate enhancement, and it was found to contain one Zn atom per molecule. It is now known that there are several forms of CA, all of which are monomers with molar mass close to 30 kDa. The best studied enzyme is CA II from red blood cells, which has a turnover frequency for CO_2 hydration of about 10^6 s^{-1}, making it one of the most active of all enzymes. The crystal structure of human CA II shows that the Zn atom is located in a conical cavity about 1.6 nm deep, which is lined with several histidine residues. The Zn is coordinated by three His-N ligands and one H_2O molecule in a pseudo-tetrahedral arrangement (Fig. 26.26). The neutral N-ligands lower the pK of the bound water molecule by about 3 units (compared to the aqua ion), so creating a high local concentration of OH^- as attacking nucleophile. Other groups in the active site pocket, including noncoordinating histidines and ordered water molecules, are important for mediating proton transfer (which is the rate-limiting factor) and for binding the CO_2 substrate (which does not coordinate to Zn).

The mechanism of action of CA (Fig. 26.27) is best described in terms of a Zn-hydroxide mechanism. The key feature is the acidity of the H_2O molecule coordinated to

Fig. 26.26 The active site of carbonic anhydrase.

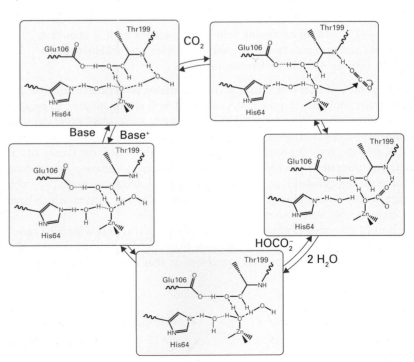

Fig. 26.27 The mechanism of action of carbonic anhydrase indicating the importance of proton transfer in this very fast reaction.

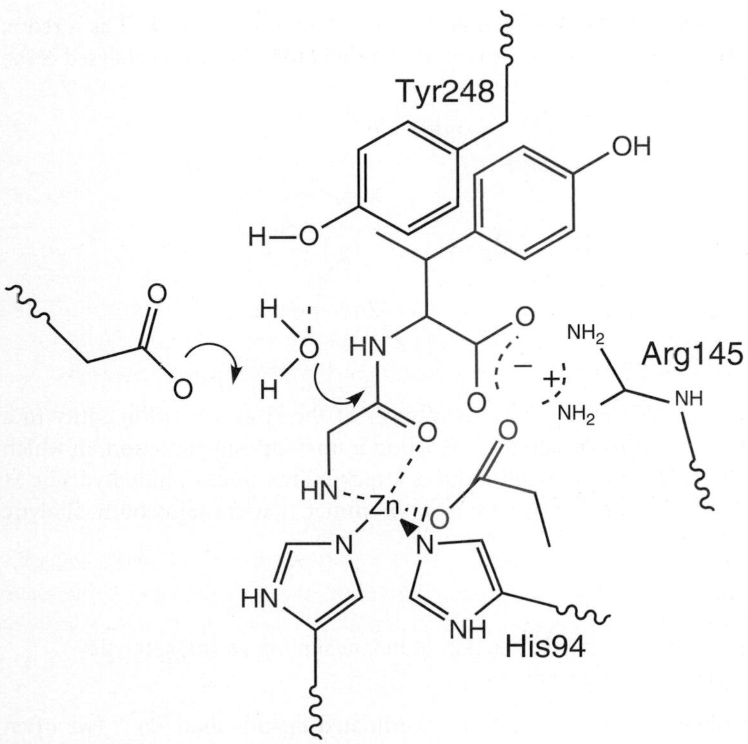

Fig. 26.28 Structure of the active site of carboxypeptidase with a peptide inhibitor (red) bound to it.

Zn, as the coordinated OH$^-$ ion that is produced after deprotonation is sufficiently nucleophilic to attack a nearby CO_2 molecule bound noncovalently. This attack results in a coordinated HCO_3^-, which is then released.

Carboxypeptidase (CPD, 34.6 kDa) is an exopeptidase, an enzyme that catalyses the hydrolysis of C-terminal amino acids containing an aromatic or bulky aliphatic side chain. There are two types of Zn-containing enzyme, and both are synthesized as inactive precursors in the pancreas for secretion into the digestive tract. The better studied is CPD A, which acts upon terminal aromatic residues; whereas CPD B acts upon basic residues.

The X-ray structure of CPD shows that the Zn is located to one side of a groove in which the substrate is bound. It is coordinated by two histidine-N ligands, one glutamate-CO_2^- (bidentate), and one weakly bound H_2O molecule (Fig. 26.28). Structures of the enzyme obtained in the presence of glycyl inhibitors show that the H_2O molecule has moved away from the Zn atom, which has become coordinated instead to the carbonyl-O of the glycine, suggesting a Zn-carbonyl mechanism. The guanidinium group of a nearby arginine binds the terminal carboxyl group, while the tyrosine provides aromatic/hydrophobic recognition.

Alkaline phosphatase (AP) introduces us to catalytic Zn centres that contain more than one metal atom. The enzyme catalyses the hydrolysis of phosphate monoesters:

$$R-OPO_3^{2-} + H_2O \rightarrow ROH + HOPO_3^{2-}$$

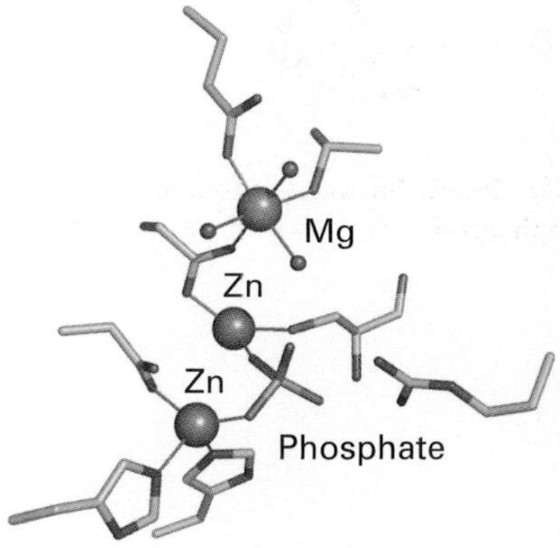

It is found in tissues as diverse as the intestine and bone, where it is found in the membranes of osteoblasts, cells that form the sites of nucleation of hydroxyapatite crystals. AP catalyses the general breakdown of organic phosphates, including ATP, to provide the phosphate required for bone growth. As its name implies, its optimum pH is in the mild alkali region. The active site of AP contains two Zn atoms located only about 0.4 nm apart with a Mg ion nearby. The crystal structure of the enzyme–phosphate complex (**30**) reveals that the phosphate ion (the product of the normal reaction) bridges the two Zn atoms.

30

Alcohol dehydrogenase (ADH) is discussed here even though it is classed as a redox enzyme, because the role of Zn once again is as a Lewis acid. The reaction catalysed is the reduction of NAD^+ by alcohol:

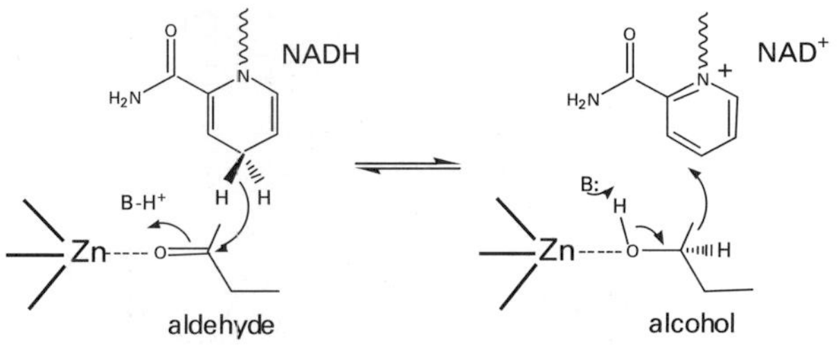

The Zn activates the C—OH group towards transfer of the H as a hydride entity to a molecule of NAD^+. It is easy to visualize this reaction in the opposite direction, in which the Zn polarizes the carbonyl group and induces attack by the nucleophilic hydridic H atom from NADH. Alcohol dehydrogenase is an α_2 dimer that contains both catalytic and structural Zn sites.

(b) Magnesium enzymes

Key points: The major direct catalytic function of magnesium is as the catalytic centre of ribulose bisphosphate carboxylase.

The Mg^{2+} cation confers less polarization of coordinated ligands than Zn^{2+} (we often refer to Mg^{2+} being a 'weaker acid' than Zn^{2+}); however, compared to Zn it is much more mobile and cells contain high concentrations of uncomplexed Mg^{2+} ions. Its major role in enzyme catalysis is as the Mg–ATP complex (**1**), which is the substrate in **kinases**, the enzymes that transfer phosphate groups thereby activating the target compound or causing it to change its conformation. Kinases are controlled by calmodulin (Section 26.4) and other proteins, so they are part of the signalling mechanism in higher organisms.

An important example of an Mg enzyme in which Mg acts separately from ATP is ribulose 1,5-bisphosphate carboxylase, commonly known as *rubisco*. This enzyme, the most abundant in the biosphere, is responsible for the production of biomass by oxygenic photosynthetic organisms and removal of CO_2 from the atmosphere (to the extent globally, of over 10^{11} t of CO_2 per year). Rubisco is an enzyme of the **Calvin cycle**, the stages of photosynthesis that can occur in the dark, in which it catalyses the incorporation of CO_2 into a molecule of ribulose 1,5-bisphosphate (Fig. 26.29). The Mg^{2+} ion is octahedrally coordinated by carboxylate groups from glutamate and aspartate residues, three coordinated H_2O molecules, and a carbamate derived from a lysine residue. The carbamate is formed by a reaction between CO_2 and the terminal —NH_2, in an activation process that is necessary for Mg^{2+} to bind. In the catalytic cycle, the binding of ribulose 1,5-bisphosphate displaces two H_2O molecules, and proton abstraction assisted by the carbamate results in a coordinated enolate. This intermediate reacts with CO_2, forming a new C—C bond, then the product is cleaved to yield two new three-carbon species and the cycle continues. The reactive enolate will also react with O_2, in which case the result is an oxidative degradation of substrate: for this reason the enzyme is often called ribulose 1,5-bisphosphate carboxylase-oxygenase. We note the difference with Zn, which would favour ligation by softer ligands and a lower coordination number. Rubisco requires a metal ion that combines good Lewis acidity with weak binding and high abundance.

(c) Iron enzymes

Key points: Acid phosphatases contain a binuclear metal site containing Fe(III) in conjunction with either Fe, Zn, or Mn; aconitase contains a [4Fe–4S] cluster, one subsite of which is modified to manipulate the substrates.

Acid phosphatases, sometimes known as 'purple' acid phosphatases (PAPs) on account of their intense colour, occur in various mammalian organs, particularly the bovine spleen

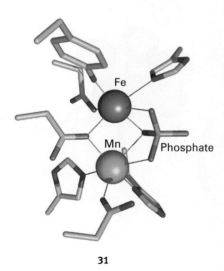

Fig. 26.29 Mechanism of action of ribulose 1,5-bisphosphate carboxylase, the enzyme responsible for removing CO_2 from the atmosphere and 'fixing' it in organic molecules in plants.

and porcine uterus. Acid phosphatases catalyse hydrolysis of phosphate esters, with optimal activity under mild acid conditions. They are involved in bone maintenance and hydrolysis of phosphorylated proteins (therefore they are important in signalling). They may also have other functions such as Fe transport. The pink or purple colours of acid phosphatases are due to a tyrosinate $\rightarrow$ Fe(III) charge-transfer transition at 510–550 nm ($\varepsilon = 4000\ dm^3\ mol^{-1}\ cm^{-1}$). The active site contains two Fe atoms linked by ligands, similar to haemerythrin (**21**). Acid phosphatases are inactive in the oxidized {Fe(III)Fe(III)} state in which they are often isolated. In the active state, one Fe is reduced to Fe(II). Both Fe atoms are high spin and remain so throughout the various stages of reactions.

Acid phosphatases also occur in plants, and in these enzymes the reducible Fe is replaced by Zn or Mn. The active site of an acid phosphatase from kidney bean (**31**) shows how phosphate becomes coordinated to both Fe(III) and Mn(II) ions. In the mechanism shown in Fig. 26.30, rapid binding of the phosphate group of the ester occurs to the divalent metal subsite (M(II)), then this is attacked by an OH^- ion that is formed at the more acidic Fe(III) subsite. The FeZn centre is also found in an important enzyme called *calcineurin*, which catalyses the phosphorylation of serine or threonine residues on certain protein surfaces, in particular a transcription factor involved in controlling the immune response. Calcineurin is activated by Ca^{2+} binding, both directly and through calmodulin.

Aconitase is an essential enzyme of the **tricarboxylic acid cycle**, the main source of energy production in higher organisms, where it catalyses the interconversion of citrate and isocitrate in a reaction that formally involves dehydration and rehydration, and

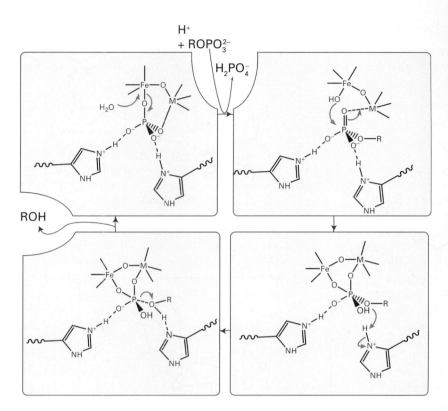

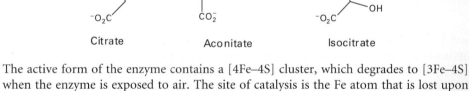

Fig. 26.30 Proposed mechanism of action of acid phosphatase. The metal site M(II) is occupied by Fe (most common in animals) or by Mn or Zn (in plants).

proceeds through an intermediate, aconitate, which is released in small amounts:

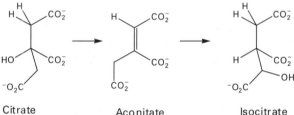

The active form of the enzyme contains a [4Fe–4S] cluster, which degrades to [3Fe–4S] when the enzyme is exposed to air. The site of catalysis is the Fe atom that is lost upon oxidation. The structure shows that this unique subsite is not coordinated by a cysteine-S but by a water molecule, which explains why this Fe is most labile.

A plausible mechanism for the action of aconitase, based on structural, kinetic, and spectroscopic evidence, involves the binding of citrate to the active Fe subsite, which increases its coordination number to six (**32**). The Fe atom polarizes a C—O bond and OH is abstracted, while a nearby base accepts a proton. The substrate now swings round, and the OH and H are reinserted onto a different position. A form of aconitase that is found in cytoplasm has another intriguing role, that of an Fe sensor, and this will be discussed in Section 26.15.

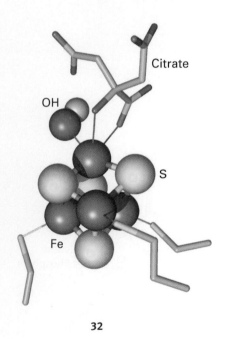

32

26.10 Enzymes dealing with H_2O_2 and O_2

In Section 26.7 we saw how organisms have evolved systems that will transport O_2 reversibly and deliver it unchanged to where it is required. In this section we describe how O_2 is reduced catalytically, either for production of energy or synthesis of oxygenated organic molecules. We start by considering a simpler case, that of the reduction of hydrogen peroxide, as this discussion introduces Fe(IV) as a key intermediate in so many biological processes. We end by completing a remarkable cycle, the production of O_2 from H_2O, catalysed by a unique manganese–calcium cluster.

(a) Peroxidases

Key points: Peroxidases catalyse reduction of harmful hydrogen peroxide; they provide important examples of Fe(IV) intermediates that can be isolated and characterized.

Haem-containing peroxidases, as exemplified by horseradish peroxidase (HRP) and cytochrome c peroxidase (CcP), catalyse the removal of harmful hydrogen peroxide:

$$H_2O_2(aq) + 2\,e^- + 2\,H^+(aq) \rightarrow 2\,H_2O(l)$$

The intense chemical interest in these enzymes lies in the fact that they are the best examples of Fe(IV) in chemistry. Fe(IV) is an important catalytic intermediate in numerous biological processes involving oxygen. Catalase, which catalyses the thermodynamically favourable disproportionation of H_2O_2 and is one of the most active enzymes known, is also a peroxidase. The active site of yeast cytochrome c peroxidase shown in Fig. 26.31 indicates how the substrate is manipulated during the catalytic cycle. The proximal ligand is the imidazole side chain of a histidine and the distal pocket, like myoglobin, also contains an imidazole side chain, but there is also a guanidinium group from arginine.

The catalytic cycle shown in Fig. 26.32 starts from the Fe(III) form. A molecule of H_2O_2 coordinates to Fe(III) and the distal histidine mediates proton transfer so that both H atoms are placed on the remote O atom. The simultaneous bond polarization by the guanidinium side chain results in heterolytic cleavage of the O—O bond: one half leaves as H_2O and the other remains bound to the Fe to produce a highly oxidizing intermediate. Although it is instructive to regard this system as a trapped O atom (or an O^{2-} ion bound to Fe(V)), detailed measurements by EPR and Mössbauer spectroscopy show that this highly oxidizing intermediate (which is known historically as 'Compound I') is in fact Fe(IV) and an organic cation radical. In HRP the radical is located on the porphyrin ring whereas in cytochrome c peroxidase it is located on nearby peptide residue tryptophan-191. Descriptions of the Fe–O bonding range from Fe(IV)=O ('ferryl') to Fe(IV)–O...H, in which the O atom is either protonated or linked by a hydrogen bond to a donor group. Compound I is reduced back to the resting Fe(III) state by two one-electron transfers either from organic substrates or cytochrome c (Fig. 26.23).

(b) Oxidases

Key points: Oxidases are enzymes that catalyse the reduction of O_2 to water or hydrogen peroxide without incorporation of O atoms into the oxidizable substrate;

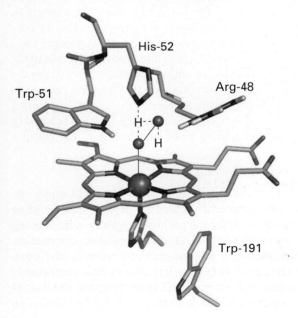

Fig. 26.31 The active site of yeast cytochrome c peroxidase showing amino acids essential for activity and indicating how peroxide is bound in the distal pocket.

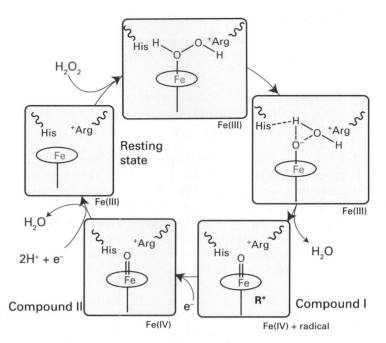

Fig. 26.32 The catalytic cycle of haem-containing peroxidases.

they include cytochrome *c* oxidase, the enzyme that is a basis for all higher life forms.

Cytochrome *c* oxidase is a membrane-bound enzyme that catalyses the four-electron reduction of O_2 to water, using cytochrome *c* as the electron donor. The potential difference between the two half-cell reactions is over 0.5 V but this does not reflect the true thermodynamics because the actual reaction catalysed by cytochrome *c* oxidase is:

$$O_2(g) + 4\ e^- + 8\ H^+(inside) \rightarrow 2\ H_2O(l) + 4\ H^+(outside)$$

This reaction includes four H^+ that are not consumed chemically but are 'pumped' across the membrane against a concentration gradient. Such an enzyme is called an **electrogenic ion pump** (or proton pump). In eukaryotes, cytochrome *c* oxidase is located in the inner membrane of mitochondria and has many subunits (Fig. 26.33), although a simpler enzyme is produced by some bacteria. It contains three Cu atoms and two haem-Fe atoms, as well as an Mg atom and a Zn atom. The Cu and Fe atoms are arranged in three main sites. The active site for O_2 reduction consists of a myoglobin-like Fe-porphyrin (haem-a_3) that is situated close to a 'semi-haemocyanin-like' Cu (known as Cu_B) coordinated by three histidine ligands (**33**). One of the histidine imidazole ligands to the Cu is modified by formation of a covalent bond to an adjacent tyrosine. Electrons to the binuclear site are provided by a second Fe porphyrin (haem-a) that is six-coordinate, as expected for an electron-transfer centre. These centres are located in subunit 1. Subunit 2 contains the binuclear Cu_A centre that was described in Section 26.8, which is believed to be the immediate acceptor of the electron arriving from cytochrome *c*. The electron transfer sequence is therefore

Cytochrome $c \rightarrow Cu_A \rightarrow$ haem $a \rightarrow$ binuclear site

The enzyme contains two proton-transfer channels, one of which is used to supply the protons needed for H_2O production while the other is used for protons that are being pumped across the membrane. Figure 26.34 shows the proposed catalytic cycle. Starting from the state in which the active site is Fe(II)—Cu(I), O_2 binds to give an intermediate (oxy) that resembles oxymyoglobin. However, this intermediate takes up the other electron that is immediately available, producing a peroxy species that immediately breaks down to give an intermediate known as P. Species P has been trapped and studied by optical and EPR spectroscopy, which show that it contains a ferryl Fe(IV) and an

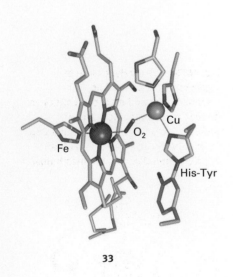

33

Reacts with cytochrome G

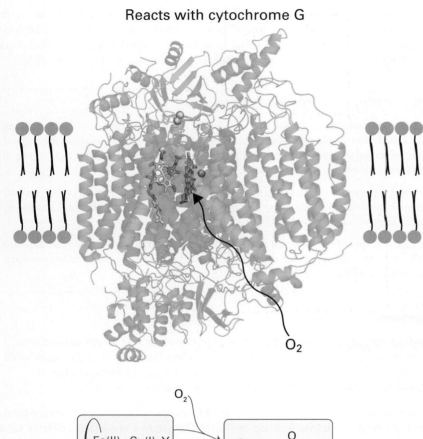

Fig. 26.33 The structure of cytochrome c oxidase as it occurs in the membrane, showing the locations of the redox centres and the sites for reaction with O_2 and cytochrome c.

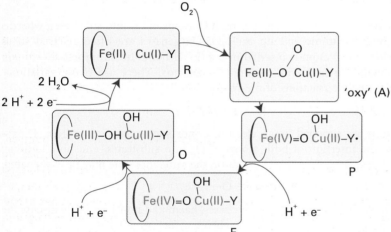

Fig. 26.34 The catalytic cycle of cytochrome c oxidase. The intermediates are labelled according to current convention. Electrons are provided from the other haem and Cu_A. During the cycle, an additional four H^+ are pumped across the membrane.

organic radical that may be located on the unusual His–Tyr pair. The ferryl group is formed by heterolytic cleavage of O_2, producing a water molecule. The role of a cation radical is again noted: without this radical, the Fe would have to be assigned as Fe(V).

The blue Cu oxidases contain a blue Cu centre that removes an electron from a substrate and passes it to a trinuclear Cu site that catalyses the reduction of O_2 to H_2O. Two examples, ascorbate oxidase and a larger class known as laccases, are well characterized, whereas another protein, ceruloplasmin, occurs in mammalian tissue and is the least well understood. Ascorbate oxidase occurs in the skins of fruit such as squash. Its role may be twofold: to protect the flesh of the fruit from O_2 and to oxidize phenolic substrates to intermediates that will form the skin of the fruit. Laccases are widely distributed, particularly in plants and fungi from which they are secreted to catalyse the oxidation of phenolic substrates. The active site at which O_2 is reduced (**34**) is well buried. It contains a pair of Cu atoms linked in the oxidized form by a bridging O atom, with a third Cu atom situated very close by, completing an almost triangular arrangement.

Amine oxidases catalyse oxidation of amines to aldehydes by using just a single Cu atom that shuttles between Cu(II) and Cu(I), yet the enzyme carries out a two-electron

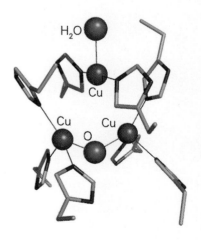

34

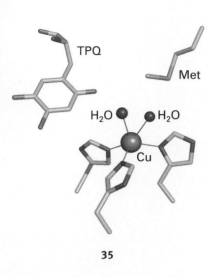

TPQ

Met

H_2O H_2O

Cu

35

reduction of O_2, producing a molecule of H_2O_2. The problem is overcome because, like cytochrome *c* peroxidase and cytochrome *c* oxidase, amine oxidases have an additional oxidizing source located near to the metal, in this case a special cofactor called topa-quinone (TPQ), which is formed by post-translational oxidation of tyrosine (**35**).

Example 26.7 Interpreting reduction potentials

The four-electron reduction potential for O_2 is 0.82 V at pH $= 7$. Cytochrome *c*, the electron donor to cytochrome *c* oxidase, has a reduction potential of 0.26 V, whereas the organic substrates of fungal laccases often have values as high as 0.7 V. What is the significance of these data in terms of energy conservation?

Answer Laccases are efficient catalysts of phenol oxidation, the driving force being small. Cytochrome *c* oxidase is a proton pump and approximately 2 eV (4×0.56 eV) of Gibbs energy is available to drive proton transfer.

Self-test 26.7 Before the discovery of the unusual active site structures in amine oxidase and another Cu enzyme called galactose oxidase, Cu(III) was proposed as a catalytic intermediate. What properties would be expected of this state ?

(c) Oxygenases

Key points: Oxygenases catalyse the insertion of one or both O atoms derived from O_2 into an organic substrate; monooxygenases catalyse insertion of one O atom while the other O atom is reduced to H_2O; dioxygenases catalyse the incorporation of both O atoms.

Oxygenases catalyse the insertion of one or both O atoms of O_2 into substrates, whereas with oxidases both O atoms end up as H_2O. Oxygenases are often referred to as hydroxylases when the O atom is inserted into a C—H bond. Most oxygenases contain Fe, the rest contain Cu or flavin, an organic cofactor. There are many variations. Monooxygenases catalyse reactions of the type

$$R—H + O_2 + 2\ H^+ + 2\ e^- \rightarrow R—O—H + H_2O$$

in which electrons are supplied by an electron donor such as an FeS protein. Dioxy-genases catalyse the insertion of both atoms of O_2 into substrates, and no additional electron donor is required. Two C—H bonds on the same molecule may be oxygenated:

$$H—R—R'—H + O_2 \rightarrow H—O—R—R'—O—H$$

The Fe enzymes are divided into two main classes, haem and non-haem. We discuss the haem enzymes first, the most important type being cytochrome P450.

Cytochrome P450 (or just 'P450') refers to an important and widely distributed group of haem-containing monooxygenases. In eukaryotes, they are localized particularly in mitochondria, and in higher animals they are concentrated in liver tissue. They play an essential role in biosynthesis (for example steroid transformations), such as the production of progesterone. The designation P450 arises from the intense absorption band that appears at 450 nm when solutions containing the enzyme (this can even be crude tissue extracts) are treated with a reducing agent and carbon monoxide, which produces the Fe(II)–CO complex. Most P450s are complex membrane-bound enzymes that are difficult to isolate. Much of what we know about them stems from studies carried out with an enzyme P450cam, which is isolated from the bacterium *Pseudomonas putida*. This organism uses camphor as its sole source of carbon, and the first stage is oxygenation of the 5-position:

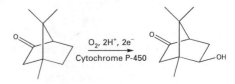

$$\xrightarrow[\text{Cytochrome P-450}]{O_2,\ 2H^+,\ 2e^-}$$

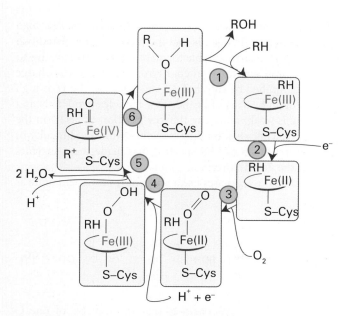

Fig. 26.35 The catalytic cycle of cytochrome P450.

The catalytic cycle has been studied using a combination of kinetic and spectroscopic methods Fig. 26.35. Starting from the resting enzyme, which is Fe(III), the binding of the substrate in the active site pocket (1) induces release of the coordinated H_2O molecule. This step is detected as a change in spin state from low spin ($S = \frac{1}{2}$) to high spin ($S = \frac{5}{2}$) and the reduction potential increases, causing an electron to be transferred (2) from a small [2Fe–2S]-containing protein known as putidaredoxin. The five-coordinate Fe(II) that is formed resembles deoxymyoglobin and binds O_2 (3). Unlike in myoglobin, addition of a second electron is both thermodynamically and kinetically favourable. The subsequent reactions (4–6) are very fast, but it is thought that an Fe(III) peroxide intermediate is formed that undergoes rapid heterolytic O—O cleavage to produce a species similar to Compound I of peroxidases. In what is known as the **oxygen rebound mechanism**, the Fe(IV)=O group abstracts an H atom from the substrate and then inserts it back as an OH radical.

This process is remarkable, as it amounts to the 'taming' of an O atom or OH radical by its attachment to Fe. Other P450s are thought to operate by similar mechanisms but differ in the architecture of the active site pocket. That site, unlike the site in peroxidases, is predominantly hydrophobic, with specific polar groups present to orient the organic substrate so the correct R—H bond is brought close to the Fe=O entity.

Non-haem oxygenases are widely distributed and are usually dioxygenases. Most contain a single Fe atom at the active site and are classified according to whether the active species in the protein is Fe(III) or Fe(II). In the Fe(III) enzymes, which are also (historically) known as **intradiol oxygenases**, the Fe atom functions as a Lewis acid catalyst and activates the organic substrate towards attack by noncoordinating O_2:

By contrast, in the Fe(II) enzymes, which are known historically as **extradiol oxygenases**, the Fe binds O_2 directly and activates it to attack the organic substrate:

The Fe(III) enzymes are exemplified by protocatechuate 3,4-dioxygenase: the Fe is high spin and tightly coordinated by a set of protein ligands that includes two His-N and two Tyr-O, the latter hard donors being particularly suitable for stabilizing Fe(III) relative to Fe(II). The Fe(III) enzymes are deep red due to an intense tyrosinate-to-Fe(III) charge transfer. The Fe(II) enzymes are exemplified by catechol 2,3-dioxygenase: the Fe is high spin and coordinated within the protein by a set of ligands that includes two His-N and one carboxylate group. The binding is weak, reflecting the low position of Fe(II) in the Irving–Williams series (Section 20.1). This weak binding, together with the difficulty of observing useful spectroscopic features (such as EPR spectra), has made these enzymes much more difficult to study than the Fe(III) enzymes.

A particularly important class of Fe(II) oxygenases use a molecule of 2-oxoglutarate as a second substrate:

$$RH + O_2 + \qquad \longrightarrow \qquad ROH + \qquad + CO_2$$

The principle of **oxo-glutarate-dependent oxygenases** is that the transfer of one O atom of O_2 to 2-oxoglutarate results in its irreversible decarboxylation, thus driving insertion of the other O atom into the primary substrate. Examples include enzymes that serve in cell signalling by modifying an amino acid in certain transcription factors (Section 26.15).

Oxygenases play a crucial role in the metabolism of methane, a greenhouse gas. Methane-metabolizing bacteria produce two types of enzyme that catalyse the conversion of methane to methanol, and thus attract much industrial interest. One is a membrane-bound enzyme that contains Cu atoms. This enzyme, known as 'particulate' methane monooxygenase (p-mmo), is expressed when high levels of Cu are available. The other enzyme, soluble methane monooxygenase (s-mmo), contains a binuclear Fe active site (**36**) that is related to haemerythrin (**21**). Figure 26.36 shows a proposed catalytic cycle. The intermediate Fe(IV) species that is proposed differs from those we have encountered up to now, as the O_2-derived O atoms are bridging rather than terminal.

Despite the importance of Fe(IV) as an enzyme intermediate, small Fe(IV) complexes that could provide important models for understanding the enzymes have been elusive. The easiest to prepare are haem analogues in which Fe is equatorially ligated by porphyrin

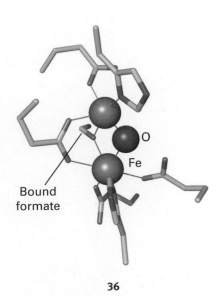

Bound formate

36

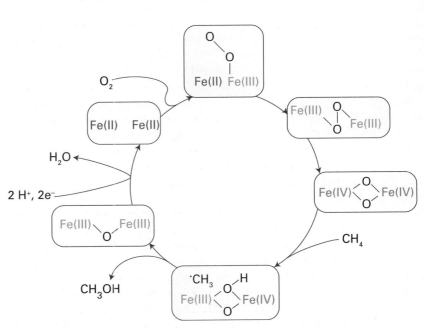

Fig. 26.36 A possible catalytic cycle for methane monooxygenase.

and which can be formed by reacting the Fe(II) or Fe(III) forms with a peroxo acid. The formation of Fe(IV) porphyrins is well established but non-haem Fe(IV) has proved to be extremely elusive. Small models of non-haem Fe(IV) species have been prepared using ligands based on cyclam in acetonitrile solution. Reaction of Fe(II)(TMC)(OTf)$_2$ (where TMC is 1,4,8,11-tetramethyl-1,4,8,11 tetraazacyclotetra-decane) with the oxo-transfer agent iodosylbenzene in acetonitrile at $-40°C$ yields the green cationic complex, symbolically $[(`N_4N')Fe^{IV}=O]^{2+}$ (**37**), which is paramagnetic ($S=1$) and shows a characteristic absorption band in the near-infrared region, at 820 nm. The Fe—O bond length is 165 pm, which is fully consistent with a multiple bond in which the O atom is acting as a π donor.

Tyrosinase and catechol oxidase, two enzymes responsible for producing melanin-type pigments, each contain a strongly coupled binuclear Cu centre that coordinates O$_2$ in a similar manner to haemocyanin. However, unlike in haemocyanin, the ligand$\sigma^*$to-Cu charge transfer is enhanced sufficiently to activate the coordinated O$_2$ for electrophilic attack at a phenolic ring of the substrate. The structure of the active site of catechol oxidase complexed with the inhibitor phenolthiourea (**38**) shows how the phenol ring of the substrate can be oriented in close proximity to a bridging O$_2$. Copper enzymes are also responsible for the production of important neurotransmitters and hormones, such as dopamine and noradrenaline. These enzymes contain two Cu atoms that are well separated in space and uncoupled magnetically.

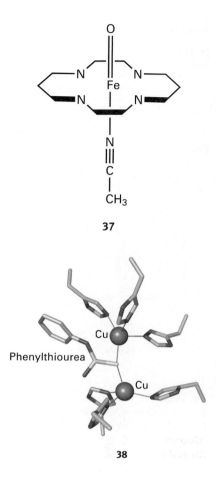

37

Phenylthiourea

38

(c) Photosynthetic O$_2$ production

Key points: Biological solar energy capture by photoactive centres results in the generation of species with sufficiently negative reduction potentials to reduce CO$_2$ to produce organic molecules; in higher plants and cyanobacteria, the electrons are derived from water, which is converted to O$_2$ by a complex catalytic centre containing four Mn atoms and one Ca atom.

Photosynthesis is the production of organic molecules using solar energy. It is conveniently divided into the **light reactions** (the processes by which electromagnetic energy is trapped) and the **dark reactions** (in which the energy acquired in the light reactions is used to convert CO$_2$ and H$_2$O into carbohydrates). We have already mentioned the most important of the dark reactions, the incorporation of CO$_2$ into organic molecules, which is catalysed by rubisco. In this section, we describe some of the roles that metals play in the light reactions.

In photosynthesis by green plants, which occurs in special organelles known as chloroplasts, photons from the Sun excite pigments present in giant membrane-bound proteins known as **photosystems**. The most important pigment, chlorophyll, is a Mg complex that is very similar to a porphyrin (**8**). Most chlorophyll is located in giant proteins known as **light-harvesting antennae**, the name perfectly describing their function, which is to collect photons and funnel their energy to enzymes that convert it into electrochemical energy. This energy conversion uses further chlorophyll complexes that become powerful reducing agents when excited by light. The principle of this energy capture system is shown in Fig. 26.37. The excited state P* reduces a nearby second pigment molecule; the electron travels down a sequence of acceptors, including FeS clusters, and (through the agency of ferredoxin and other redox enzymes) is eventually used to reduce CO$_2$ to carbohydrate. The P$^+$ cation produced from P* is rapidly restored to P by an electron supplied by an electron donor.

Plant chloroplasts have two photosystems, I and II, operating in series, that allow low-energy light (approximately 680–700 nm, <1 eV) to span the large potential range (>1 V) within which water is stable. The arrangement of proteins is depicted in Fig. 26.38. Some of the energy of the photosynthetic electron transfer chain is used to generate a trans-membrane proton gradient which in turn drives the synthesis of ATP, as in mitochondria. Photosystem I lies at the low-potential end, its electron donor is the blue Cu protein plastocyanin that has been reduced using the electrons generated by photosystem II; in turn, the electron donor to photosystem II is H$_2$O. Thus green plants dispose of the oxidizing power by converting H$_2$O into O$_2$. This four-electron reaction is remarkable

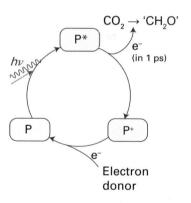

Fig. 26.37 The principle of photosynthetic generation of reducing power for 'fixation' of carbon dioxide.

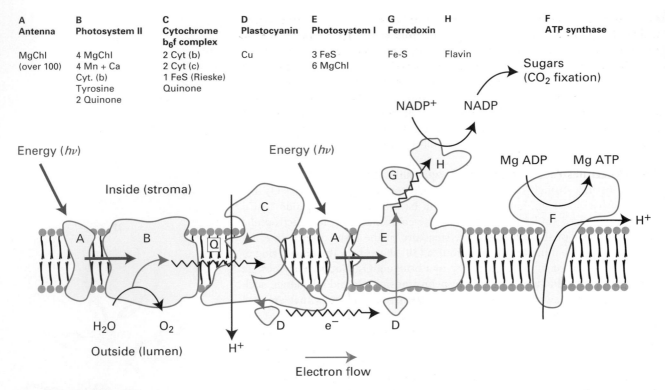

A	B	C	D	E	G	H	F
Antenna	Photosystem II	Cytochrome b_6f complex	Plastocyanin	Photosystem I	Ferredoxin		ATP synthase
MgChl (over 100)	4 MgChl 4 Mn + Ca Cyt. (b) Tyrosine 2 Quinone	2 Cyt (b) 2 Cyt (c) 1 FeS (Rieske) Quinone	Cu	3 FeS 6 MgChl	Fe-S	Flavin	

Fig. 26.38 The arrangement of proteins in the photosynthetic electron-transport chain. **A.** Antenna ('light harvesting') complex. **B.** Photosystem II. **C.** The 'cytochrome b_6f complex' (this is similar to **complex III** in the mitochondrial ET chain). **D.** Plastocyanin (soluble). **E.** Photosystem I. **F.** ATPase. **G.** Ferredoxin (Fe–S). **H.** Ferredoxin-NADP$^+$ reductase (flavin). Blue arrows show transfer of energy. Note how the overall transfer of electrons is from Mn (high potential) to FeS (low potential): this apparently 'uphill' flow reflects the crucial input of energy at each photosystem.

39

because no intermediates are released. The catalyst also has a special significance because its action, commencing over 2 Ga ago, has provided essentially all the O_2 we have in the atmosphere.

The special O_2 evolution catalyst is a metal oxide cluster containing four Mn atoms and one Ca atom that is located in subunit D1 of photosystem II. This subunit has long attracted interest because the cell replaces it at frequent intervals as it quickly becomes worn out by oxidative damage. The 'oxygen evolving centre' exploits the oxidizing ability of Mn(IV) and Mn(III), coupled with that of a nearby tyrosine residue, to oxidize H_2O to O_2. Note that Mn ligands are hard O-atom donors and Mn(III) (d^4) and Mn(IV) (d^3) are hard metal ions. Only recently have X-ray diffraction data been obtained that indicate the positions of the metal atoms in the cluster (**39**).

Based on the available structural evidence, different models have been proposed for the mechanism of O_2 evolution. The principles we have covered in earlier sections of this book allow us to make some proposals. First, as the Mn sites are progressively oxidized, coordinated H_2O molecules become increasingly polarized and lose protons, progressing from H_2O through OH^- to O_2^{2-}. Second, we note that $Mn(IV)=O$ and $Mn(V)=O$ are equivalent to $Mn(II)-[O]$ and $Mn(III)-[O]$, respectively, and if two $[O]$ atoms are close together, an O_2 molecule may form. The presence of the Ca^{2+} is essential, and the only metal ion that can be substituted is Sr^{2+}. A possible role for the Ca is that it provides a site that will remain permanently in the +2 state and provide a fast binding site, whereas were this to be a fifth Mn atom, the latter would certainly become oxidized and this advantage would be lost.

26.11 The reactions of cobalt-containing enzymes

Key points: Coenzyme B_{12} contains a Co—C bond; enzymes containing coenzyme B_{12} catalyse radical-based rearrangements; other cobalamin enzymes catalyse methyl transfer and dehalogenation.

Cobalamins are cobalt macrocycle complexes that are cofactors in many enzymes that catalyse radical-based rearrangements (for example, isomerizations), methyl transfer, and some dehalogenation reactions. The macrocycle is a corrin ring (**9**), which is similar to porphyrin (**8**), except there is less conjugation and it has a smaller ring (15-membered instead of 16-membered). The fifth ligand to the Co is a nitrogen base, usually dimethylbenzimidazole, which is covalently linked to the corrin ring through a nucleotide, but in some enzymes it is a histidine, thus providing a covalent link to the protein. The five-coordinate entity is known as cobalamin. The more elaborate structure known as **coenzyme B_{12}** (**40**) is an important enzyme cofactor for radical rearrangements: the sixth ligand, R, is 5′-deoxyadenosine, which is bonded to the Co atom through the —CH_2— group, making coenzyme B_{12} a rare example of a naturally occurring organometallic compound. The sixth ligand is exchangeable and the complex is ingested in the form of species such as aquacobalamin, hydroxocobalamin, or cyanocobalamin, known

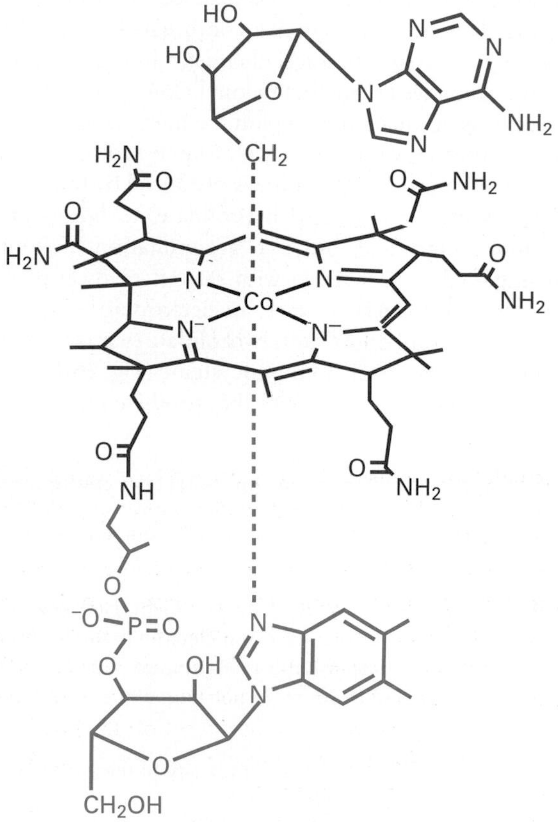

40 Coenzyme B_{12}

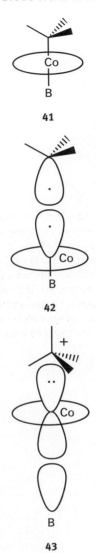

41

42

43

generally as vitamin B_{12}. Cobalamin is essential for higher organisms (the human requirement is only a few milligrams per day) but it is synthesized only by microorganisms.

Cobalamin occurs with the Co atom in three oxidation states, Co(III), Co(II), and Co(I), all of which are low spin. The electronic structure of Co is crucial to its biological activity. As expected, the Co(III) form (d^6) is six-coordinate (**41**). The Co(II) form (**42**) is five-coordinate and has its unpaired electron in the d_{z^2} orbital. The Co(I) form (**43**) is four-coordinate square planar, due to dissociation of the benzimidazole.

Radical-based rearrangements include isomerizations (mutases) and dehydration or deamination (lyases). The generic reaction is

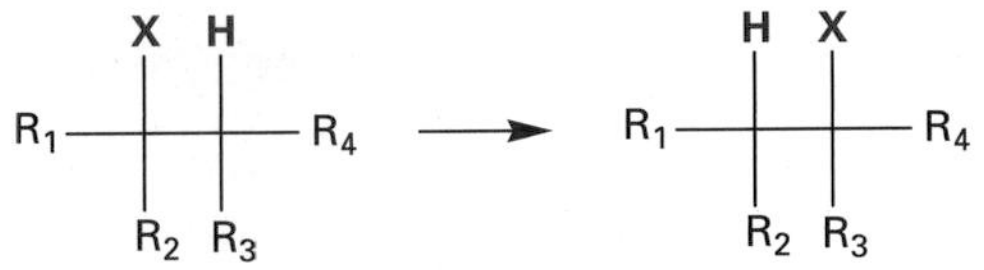

Dehydration and deamination occur after two —OH or —OH and —NH_2 become placed on the same carbon atom and hence are triggered by isomerization:

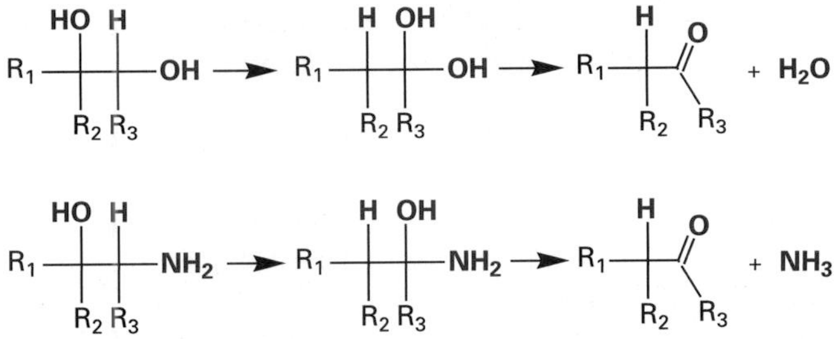

Radical-based rearrangements occur by a mechanism involving initiation of radical formation that begins with enzyme-induced weakening of the Co—C (adenosine) bond. In the free state, the Co—C bond dissociation energy is about 130 kJ mol^{-1}, but when bound in the enzyme the bond is substantially weakened, resulting in homolytic cleavage of the Co—CH_2R bond. This step results in five-coordinate low-spin Co(II) and a ·CH_2R radical, which gives rise to controlled radical chemistry in the enzyme active site pocket (Fig. 26.39). Important examples are methylmalonyl CoA mutase and diol dehydratases.

Methyl transfer reactions of cobalamins exploit the high nucleophility of square-planar Co(I). A particularly important example is methionine synthase, which is responsible for the biosynthesis of methionine. Methionine is produced by transferring a CH_3 group, derived from the methyl carrier N-methyl hydrofolate, to homocysteine. Not only is methionine an essential amino acid, but also accumulation of homocysteine (which occurs if activity is impaired) is associated with serious medical problems. The mechanism involves a cycle in which Co(I) abstracts an electrophilic —CH_3 group (effectively 'CH_3^+') from a quaternary N atom on N^5-tetrahydrofolate to produce methyl cobalamide, Co—CH_3, which then transfers —CH_3 to homocysteine (Fig. 26.40). Another important example occurs in methanogens, the microbes that produce methane.

Example 26.8 Identifying the significance of the d-electron configuration of cobalamin

Why is a Co-based macrocyclic complex (rather than an Fe complex like haem) well suited for radical-based rearrangements?

Answer Radical-based rearrangements depend on homolytic cleavage of the Co—C bond, which generates the adenosine radical and leaves an electron in the d_{z^2} orbital of the Co. This is the stable configuration for a low-spin Co(II) (d^7) complex, but for Fe, this configuration would require the oxidation state Fe(I), which is not normally encountered in coordination complexes.

Self-test 26.8 Provide an explanation for why the toxicity of mercury is greatly increased by the action of enzymes containing cobalamin.

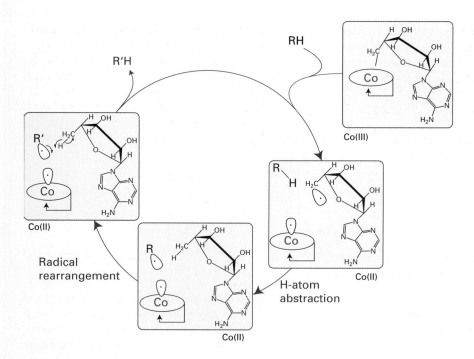

Fig. 26.39 The principle of radical-based rearrangements by coenzyme B_{12}. Homolytic cleavage of the Co—C bond results in low-spin Co(II) ($d_{z^2}^1$) and a carbon radical that abstracts an H atom from the substrate RH. The substrate radical is retained in the active site and undergoes rearrangement before the hydrogen atom transfers back.

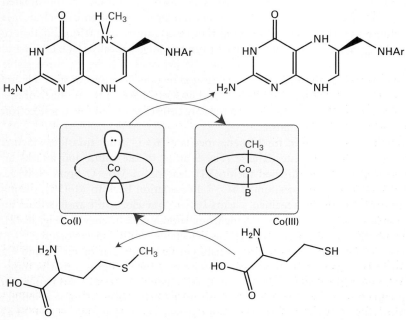

Fig. 26.40 The mechanism of methionine synthase. Co(I) is a strong nucleophile and attacks the electrophilic quaternary-CH_3 group on the methyl carrier tetrahydrofolate. The resulting Co(III) methyl complex transfers CH_3^+ to homocysteine.

26.12 Molybdenum and tungsten enzymes

Key points: Mo is used to catalyse O atom transfer in which the O atom is provided by a water molecule; related chemistry, but in more reducing environments, is displayed by W.

Molybdenum and tungsten are the only heavier elements known so far to have specific functions in biology. Molybdenum is widespread across all life forms, and this section deals with its presence in enzymes other than nitrogenase (Section 26.13). By contrast, tungsten has so far only been found in prokaryotes. Molybdenum enzymes catalyse the oxidation and reduction of small molecules, particularly inorganic species. Reactions include oxidation of sulfite, arsenite, xanthine, aldehydes, and carbon monoxide, and reduction of nitrate and dimethyl sulfoxide (DMSO). Deficiencies of these enzymes (usually genetic) are quite common but often serious, such as the inability of an individual to produce sulfite oxidase (sulfite ions are very toxic).

Fig. 26.41 Oxidation of sulfite to sulfate by sulfite oxidase, illustrating the direct O-atom transfer mechanism for Mo enzymes.

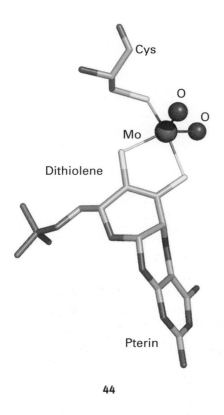

44

Both Mo and W are found in combination with an unusual class of cofactors (**11**), at which the metal is coordinated by a dithiolene group. In higher organisms, Mo is coordinated by one pterin dithiolene along with other ligands that often include cysteine. This is illustrated by the active site of sulfite oxidase, where we see also how the pterin ligand is twisted (**44**). In prokaryotes, Mo enzymes have two pterin cofactors coordinated to the metal and although the role of such an elaborate ligand is not entirely clear, it is potentially redox active and may mediate long-range electron transfer. Coordination of the Mo is usually completed by ligands derived from water, specifically H_2O, OH^-, and O^{2-}. Molybdenum is suited for its role because it provides a series of three stable oxidation states, Mo(IV), Mo(V), and Mo(VI), related by one-electron transfers that are coupled to proton transfer. Typically, Mo(IV) and Mo(VI) differ in the number of oxo groups they contain, and Mo enzymes are commonly considered to couple one-electron transfer reactions with O-atom transfer.

A mechanism often considered for Mo enzymes is direct O atom transfer, which is illustrated in Fig. 26.41 for sulfite oxidase. The S atom of the sulfite ion attacks an electron-deficient O atom coordinated to Mo(VI), leading to Mo—O bond cleavage, formation of Mo(IV), and dissociation of SO_4^{2-}. Reoxidation back to Mo(VI), during which a transferable O atom is regained, occurs by two one-electron transfers from an Fe-porphyrin that is located on a mobile 'cytochrome' domain of the enzyme (Fig. 26.42). The intermediate state containing Mo(V) (d^1) is detectable by EPR spectroscopy.

This kind of oxygenation reaction can be distinguished from that of the Fe and Cu enzymes described previously, because with Mo enzymes the oxo group that is transferred is not derived from molecular O_2 but from water. The Mo(VI)=O unit can transfer an O atom, either directly (inner sphere) or indirectly to reducing (oxophilic) substrates, such as SO_3^{2-} or AsO_3^{2-}, but cannot oxygenate C—H bonds. Figure 26.43 shows the reaction enthalpies for O-atom transfer: we see that the highly oxidizing Fe species formed by reaction with O_2 are able to oxygenate all substrates, whereas Mo(VI) oxo species are limited to more reducing substrates and Mo(IV) is able to extract an O atom from nitrate.

As expected from its position below Mo in Group 6, the lower oxidation states of W are less stable than those of Mo, so W(IV) species are usually potent reducing agents. This is illustrated by the W-containing formate dehydrogenases present in certain primitive organisms, which catalyse the reduction of CO_2 to formate, the first stage in non-photosynthetic carbon assimilation. This reaction does not involve O-atom insertion but rather the formation of a C—H bond. One mechanism that has been proposed is

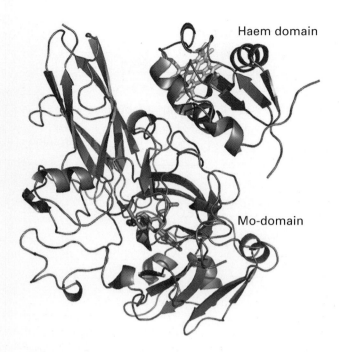

Haem domain

Mo-domain

Fig. 26.42 Structure of sulfite oxidase showing the Mo and haem domains. The region of polypeptide linking the two domains is highly mobile and is not resolved by crystallography.

Biological cycles

Nature is extraordinarily economical and maximizes its use of elements that have been taken up from the non-biological, geological world, often with great difficulty. We have already seen how iron is assimilated by using special ligands, and how this hard-earned resource is stored in ferritin. Thus useful species are recycled rather than returned to the environment. An important example is nitrogen—which is so hard to assimilate from its unreactive gaseous source, N_2, and the elusive gas H_2, rapid cycling of which is achieved by microbes in processes analogous to electrolytic and fuel cells.

26.13 The nitrogen cycle

Key points: The nitrogen cycle involves enzymes containing Fe, Cu, and Mo, often in cofactors having very unusual structures; nitrogenase contains three different kinds of FeS cluster, one of which also contains Mo and a small interstitial atom.

The global biological nitrogen cycle involves organisms of all types and a diverse variety of metalloenzymes (Fig. 26.44). The cycle can be divided into uptake of usable nitrogen (assimilation) from nitrate or N_2 and denitrification (dissimilation). Ammonia is a crucial compound for the biosynthesis of amino acids and NO_3^- is used as an oxidant. Many of the compounds are toxic or environmentally challenging, whereas others, such as NO, are produced in small amounts to serve as hormones. The cycle involves many different organisms and a variety of metal-containing enzymes.

The so-called 'nitrogen-fixing' bacteria found in soil and root nodules of certain plants contain an enzyme called nitrogenase that catalyses the reduction of N_2 to ammonia in a reaction that is coupled to the hydrolysis of 16 molecules of ATP and the production of H_2:

$$N_2 + 8\,H^+ + 8\,e^- + 16\,ATP \rightarrow 2\,NH_3 + H_2 + 16\,ADP + 16\,PO_3(OH)^{2-}$$

'Fixed' nitrogen is essential for the synthesis of amino acids and nucleic acids, so it is central to agricultural production. Industrial production of ammonia by the Haber process involves reaction of N_2 and H_2 at high pressures and high temperatures; by contrast, nitrogenase produces NH_3 under normal conditions, and it is small wonder that it has attracted so much attention. Indeed, the mechanism of activation of the N_2 molecule by nitrogenase has inspired coordination chemists for several decades. The process is very costly in terms of energy for a reaction that is not very unfavourable

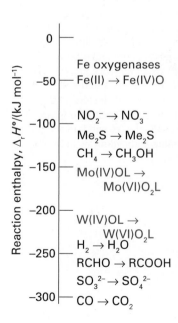

Reaction enthalpy, $\Delta_r H^\ominus/(kJ\,mol^{-1})$

0

−50 — Fe oxygenases
      Fe(II) → Fe(IV)O

−100 — $NO_2^- \rightarrow NO_3^-$
       $Me_2S \rightarrow Me_2S$
       $CH_4 \rightarrow CH_3OH$

−150 — Mo(IV)OL →
             Mo(VI)O$_2$L

−200 — W(IV)OL →
             W(VI)O$_2$L

−250 — $H_2 \rightarrow H_2O$
       RCHO → RCOOH
       $SO_3^{2-} \rightarrow SO_4^{2-}$

−300 — CO → CO$_2$

Fig. 26.43 Scale showing relative enthalpies for oxygen-atom transfer. Fe(IV) oxy species are powerful O-atom donors, whereas Mo(IV) and W(IV) are good O-atom acceptors.

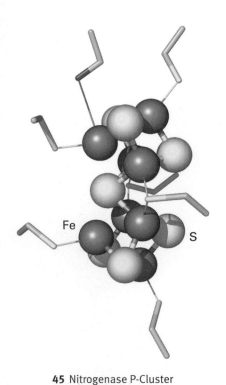

45 Nitrogenase P-Cluster
[8Fe–7S]

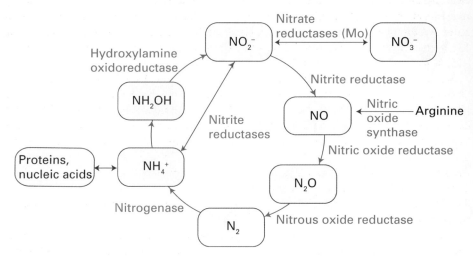

Fig. 26.44 The biological nitrogen cycle.

Fig. 26.45 The structure of nitrogenase showing the Fe protein and the MoFe protein complexed with each other. Positions of the metal centres (black) are indicated. The MoFe protein is an $\alpha_2\beta_2$ of different subunits (red and blue) and it contains two 'P-clusters' and two MoFe cofactors. The Fe protein (green) has a [4Fe–4S] cluster and it is also the site of binding and hydrolysis of Mg-ATP.

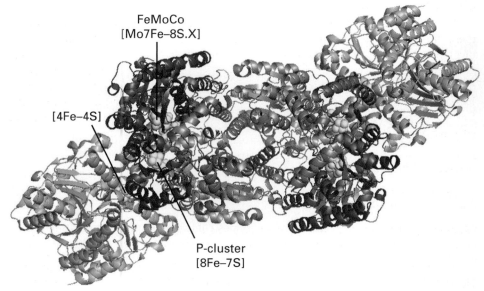

thermodynamically. However, as we saw in Section 14.3, N_2 is an unreactive molecule and energy is required to overcome the high activation barrier for its reduction.

Nitrogenase is a complex enzyme that consists of two types of protein: the larger of the two is called the 'MoFe-protein' and the smaller is the 'Fe-protein' (Fig. 26.45). The Fe-protein contains a single [4Fe–4S] cluster that is coordinated by two cysteine residues from two subunits. The role of the Fe-protein is to transfer electrons to the MoFe-protein in a reaction that is far from understood. Each electron transfer is accompanied by the hydrolysis of two ATP molecule, which are bound to the Fe-protein.

The MoFe protein is an $\alpha_2\beta_2$ tetramer, each $\alpha\beta$ pair of which contains two types of supercluster. The [8Fe–7S] cluster (**45**) is known as the 'P-cluster' and is thought to be an electron transfer centre, whereas the other cluster (**46**), formulated as [Mo7Fe–8S,X] and known as 'FeMoco' (FeMo cofactor), is thought to be the site at which N_2 is reduced to NH_3. The Mo is coordinated also by an imidazole-N from histidine and two O atoms from an exogenous molecule homocitrate. The mechanism of N_2 reduction is still unresolved, the two main questions being whether N_2 is bound and reduced at the Mo atom or at some other site. Here is it significant that nitrogenase is known in which the Mo atom is replaced by vanadium. The cage-like cluster provided by the six central

Fe atoms also coordinates a small central atom ('X') that is either C, N or O. It was nearly ten years after the determination of the structure of the MoFeS cage that this small central atom was discovered, after improvements in resolution and detailed consideration of the X-ray interference characteristics. The six Fe atoms would otherwise be three-coordinate with a flattened trigonal-pyramidal geometry.

Nitrate reductase is another example of a Mo enzyme involved in the transfer of an O atom, in this case catalysing a reduction reaction (the reduction potential for the NO_3^-/NO_2^- couple is $+0.4\,V$ at $pH = 7$; thus NO_3^- is quite strongly oxidizing). The other enzymes in the nitrogen cycle contain either haem or Cu as their active sites. There are two distinct classes of nitrite reductase. One is a multi-haem enzyme that can reduce nitrite all the way to NH_3. The other class contains Cu and carries out one-electron transfer, producing NO: it is a trimer of identical subunits, each of which contains one 'blue' Cu (mediating long-range electron transfer to the electron donor, usually a small 'blue' Cu protein) and a Cu centre with more conventional tetragonal geometry that is thought to be the site of nitrite binding.

The nitrogen cycle is notable for using some of the most unusual redox centres yet encountered as well as some of the strangest reactions. Another unusual cofactor is a [4Cu–S] cluster, named Cu_Z, that is found in N_2O reductase and has the structure shown in (**47**). It is puzzling how this centre is able to bind and activate N_2O, which is a poor ligand. Long-range electron transfer in N_2O reductase is carried out by a Cu_A centre, the same as found for cytochrome c oxidase.

Of particular importance for humans are two enzymes that manipulate NO. One of these, NO synthase, is a haem enzyme that is responsible for producing NO upon receipt of a signal. Its activity is controlled by calmodulin (Section 26.4). The other is an unusual NO-binding haem protein that is found in some bloodsucking parasites, notably 'kissing bugs' (predacious bugs of the family *Reduviidae*). This protein, called nitrophorin, binds NO tightly until it is injected into a victim, where a change in pH causes its release. The free NO causes dilation of the surrounding blood vessels, rendering the victim a more effective blood donor.

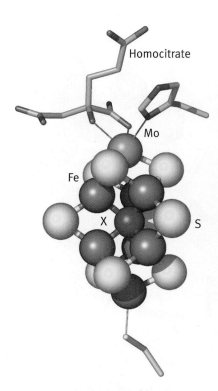

46 Nitrogenase FeMoCo [Mo7Fe–8S, X]

Example 26.9 Identifying intermediates formed during the reduction of N$_2$ to NH$_3$

Suggest likely intermediates formed during the six-electron reduction of an N_2 molecule to NH_3.

Answer By referring to Section 14.6 we would propose diazene (N_2H_2) and hydrazine (N_2H_4) along with their deprotonated conjugate bases.

Self-test 26.9 The MoFe cofactor can be extracted from nitrogenase by using dimethylformamide (DMF), although it is catalytically inactive in this state. Suggest experiments that could establish if the structure of the species in DMF solution is the same as present in the enzyme.

26.14 The hydrogen cycle

Key point: The active sites of hydrogenases contain Fe or Ni, along with CO and CN ligands.

It has been estimated that 99 per cent of all organisms utilize H_2. Even if these species are almost entirely microbes, the fact remains that almost all bacteria and archaea possess extremely active metalloenzymes, known as hydrogenases, that catalyse the interconversion of H_2 and H^+ (as water). The elusive molecule H_2 is produced by some organisms (it is a waste product) and used by others as a fuel, helping to explain why so little H_2 is in fact detected in the atmosphere. Human breath contains measurable amounts of H_2 due to the action of bacteria in the gut.

There are three classes of hydrogenase, based on the structure of the active site. All contain Fe and some also contain Ni. The two best characterized types are known as [FeFe]-hydrogenases and [NiFe]-hydrogenases, and structures of the active sites of two representative enzymes are shown as (**48**) and (**49**). The active sites contain at least one

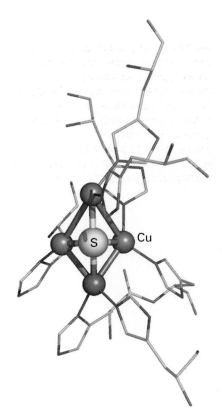

47 Cu_Z cluster [4Cu–S]

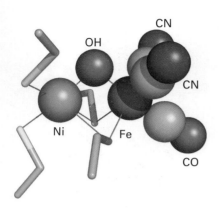

48 [Ni Fe]-hydrogenase

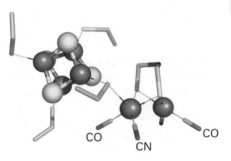

49 [FeFe]-hydrogenase

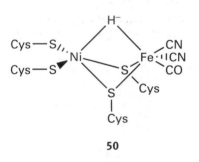

50

CO ligand, and provide further examples of biological organometallic active sites (in addition to coenzyme B_{12}). Further ligation is provided by CN^-, cysteine (and sometimes selenocysteine) and the [FeFe]-enzymes contain an unusual bridging bidentate ligand that is thought to be dithiomethylamine. These fragile active sites are buried deeply within the enzyme, thus necessitating special pores and pathways to convey H_2 and H^+, and a relay of FeS clusters for long-range electron transfer (as shown in Fig. 26.24).

The [FeFe]-enzymes tend to operate in the direction of H_2 production; they are usually found in strictly anaerobic organisms and are very sensitive to O_2. The mechanism of catalysis is uncertain, but it appears that an Fe(I)Fe(I) intermediate is involved. Knowledge is more advanced for the [NiFe]-enzymes for which a catalytically active form assigned as Ni(III)—H^- (a hydrido complex) has been identified (**50**). Single-crystal EPR studies show that the unpaired electron is coupled strongly to an H-atom nucleus lying along an axis that points towards the Fe (that is, the H is in a bridging position).

Sensors

Key points: Cells are able to sense the levels of small molecules such as O_2 and NO, and metal ions such as Fe, and adjust their metabolism accordingly; Cu and Zn are sensed by proteins with binding sites specially tailored to meet the specific coordination preferences of each metal atom.

A number of metalloproteins are used to detect and quantify the presence of small molecules, particularly O_2, NO, and CO. These proteins therefore act as sensors, alerting an organism to an excess or deficit of particular species, and triggering some kind of remedial action. Special proteins are also used to sense the levels of metals such as Cu and Zn that are otherwise always strongly complexed in a cell.

26.15 Iron proteins as sensors

The coordination of an FeS cluster to cysteine ties together different parts of a protein and thus controls its tertiary structure. The sensitivity of the cluster to oxygen, electrochemical potential, or Fe/S concentrations makes it able to be an important sensory device. In the presence of O_2 or other potent oxidizing agents, [4Fe–4S] clusters have a tendency (controlled by the protein) to degrade, producing [3Fe–4S] and [2Fe–2S]

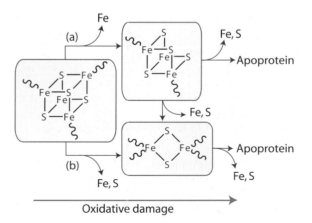

Fig. 26.46 The degradation of Fe–S clusters forms the basis for a sensory system The [4Fe–4S] cluster cannot support a state in which all Fe are Fe(III); thus severe oxidizing conditions, including exposure to O_2, causes their breakdown to [3Fe–4S] or [2Fe–2S] and eventually complete destruction. Degradation to [3Fe–4S] (a) requires only removal of an Fe subsite whereas degradation to [2Fe–2S] (b) may require rearrangement of the ligands (cysteine) and produce a significant protein conformational change. These processes link cluster status to availability of Fe as well as O_2 and other oxidants and provide the basis for sensors and feedback control.

species. The cluster may be removed completely under some conditions (Fig. 26.46). The principle behind an organism's exploitation of FeS clusters as sensors is that the presence or absence of a particular cluster (the structure of which is very sensitive to Fe or oxygen) alters the conformation of the protein and determines its ability to bind to nucleic acids.

In higher organisms, the protein responsible for regulating Fe uptake (transferrin) and storage (ferritin) is an FeS protein known as the **iron regulatory protein** (IRP), which is closely related to aconitase (Section 26.9) but is found in the cytoplasm rather than in mitochondria. It acts by binding to specific regions of messenger RNA (mRNA) that carry the genetic command (transcribed from DNA) to synthesize transferrin receptor or ferritin. The specific interaction regions on the RNA are known as **iron-responsive elements** (IREs). The principle is outlined in Fig. 26.47. When Fe levels are high, a [4Fe–4S] cluster is present, and the protein does not bind to the IRE that controls translation of ferritin. In this case binding would be a 'stop' command, and the cell will respond by synthesizing ferritin. Simultaneously, binding of the [4Fe–4S]-loaded protein to the transferring receptor IRE destabilizes the RNA, so transferrin receptor is not made. When Fe levels are high, the opposite actions occur: a [4Fe–4S] cluster is formed, ferritin synthesis is activated, and transferrin receptor synthesis is switched off (repressed).

The common gut bacterium *E. coli* derives energy either by aerobic respiration (using a terminal oxidase related to cytochrome *c* oxidase, Section 26.10) or by anaerobic respiration with an oxidant such as fumaric acid, or nitrate, using the Mo enzyme nitrate reductase (Section 26.12). The problem the organism faces is how to sense whether O_2 is present at a sufficiently low level to warrant inactivating those genes functioning in aerobic respiration and activate instead the genes producing enzymes necessary for the less efficient anaerobic respiration. This detection is achieved by an FeS protein called **fumarate nitrate regulator** (FNR). The principle is outlined in Fig. 26.48. In the absence of O_2, FNR is a dimeric protein with one [4Fe–4S] cluster per subunit. In this form it binds to specific regions of DNA repressing transcription of the aerobic enzymes and activating transcription of enzymes such as nitrate reductase. When O_2 is present, the [4Fe–4S] cluster is degraded to a [2Fe–2S] cluster and the dimer breaks up so that it cannot bind to DNA. The genes encoding aerobic respiratory enzymes are thus able to be transcribed, whereas those for anaerobic respiration are repressed.

In higher animals, the system that regulates the ability of cells to cope with O_2 shortage involves an Fe oxygenase. In Section 26.9 we mentioned prolyl hydroxylases, which catalyse the hydroxylation of specific proline residues in proteins, thus altering their properties. In higher animals, one such target protein is a transcription factor called **hypoxia inducible factor** (HIF), which mediates the expression of genes responsible for adapting cells to low-O_2 conditions (hypoxia). We should bear in mind here that the internal environment of cells and cell compartments is usually quite reducing, equivalent to an electrode potential below -0.2 V, and even though we regard O_2 as essential for higher organisms, its actual levels may be fairly low. When O_2 levels are above a safe threshold, prolyl hydroxylases catalyse oxygenation of two conserved proline residues of HIF, causing the transcription factor to be recognized by a protein that induces its degradation by proteases. Hence genes such as those ultimately responsible for producing more red blood cells (which will help an individual to cope better when O_2 supply is a problem) are not activated. The principle is outlined in Fig. 26.49.

An underlying principle of haem sensors is that the small molecules being sensed are π acceptors that can bind strongly to the Fe and displace an indigenous ligand. This binding results in a change in conformation that alters catalytic activity or the ability of the protein to bind to DNA. Of the indigenous ligands, two classic examples are NO and CO; although we have long thought of these molecules as toxic, they are becoming well established as hormones. Indeed, there is evidence that the sensing of trace levels of CO is important for controlling circadian rhythms in mammals.

The enzyme guanyl cyclase senses NO, a molecule that is now well established as a hormone that delivers messages between cells. Guanyl cyclase catalyses the conversion of guanidine monophosphate (GMP) to cyclic GMP (cGMP), which is important for activating many cellular processes. The catalytic activity of guanyl cyclase increases

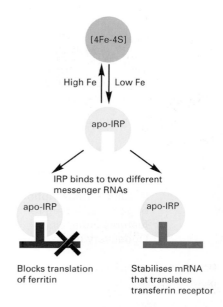

Fig. 26.47 Interactions of iron-regulatory protein with iron-responsive elements on the RNAs responsible for synthesising ferritin or transferrin receptor depend on whether an Fe–S cluster is present and form the basis for regulation of cellular Fe levels.

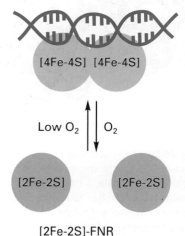

Fig. 26.48 The principle of operation of the fumarate-nitrate regulatory system that controls aerobic vs anaerobic respiration in bacteria.

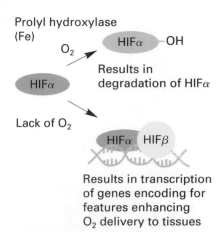

Fig. 26.49 The principle of O₂ sensing by prolyl oxygenases.

greatly (by a factor of 200) when NO binds to the haem, but (as is obviously important) by a factor of only 4 when CO is bound.

An excellent, atomically defined example of CO sensing is provided by a haem-containing transcription factor known as CooA. This protein is found in some photosynthetic bacteria that are able to grow on CO as their sole energy source under anaerobic conditions. Whether growth on CO takes place depends on the ambient CO level, as an organism will not waste its resources synthesizing the necessary enzymes when the essential substrate is not present. CooA is a dimer, each subunit of which contains a single *b*-type cytochrome (the sensor) and a 'helix–turn–helix' protein fold that binds to DNA (Fig. 26.50). In the absence of CO, each Fe(II) is six-coordinate and both its axial ligands are amino acids of the protein, a histidine imidazole and, unusually, the main-chain —NH₂ group of a proline that is also the N-terminal residue of the other subunit. In this form, CooA cannot bind to the specific DNA sequence to transcribe the genes for synthesizing the CO-oxidizing enzymes necessary for existing on CO. When CO is present it binds to the Fe, displacing the distal proline residue and causing CooA to adopt a conformation that will bind to the DNA. The likelihood of NO binding in place of CO to cause a false transcriptional response is prevented because NO not only displaces proline but also results in dissociation of the proximal histidine; the NO complex is thus not recognized.

26.16 Proteins that sense Cu and Zn levels

The levels of Cu in cells are so strictly controlled that almost no uncomplexed Cu is present. An imbalance in Cu levels is associated with serious health problems such as Menkes (Cu deficiency) and Wilson's disease (Cu accumulation). Most of what we know about how Cu levels are sensed and converted to cell signals stems from studies on the *E. coli* system, which involves a transcription factor called CueR (Fig. 26.51). This protein binds Cu(I) with high selectivity, although it also binds Ag(I) and Au(I). Metal coordination causes a conformational change that enables CueR to bind to DNA at a receptor site that controls transcription of an enzyme known as CopA, which is an ATP-driven Cu pump. CopA is located in the cytoplasmic membrane and exports Cu into the periplasm. In CueR, the Cu(I) is coordinated by two cysteine-S atoms arranged in a linear coordination geometry. Titrations using CN^- as a buffer show that Cu^+ is bound with a dissociation constant of approximately 10^{-21}! As may be understood by reference to Section 8.1, this ligand environment leads to remarkably selective binding for d^{10} ions, and measurements with Ag and Au show that these ions are taken up with similar affinities.

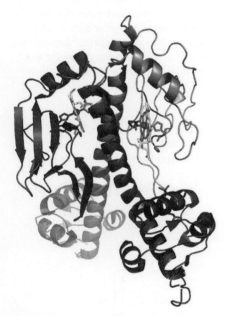

Fig. 26.50 The structure of CooA, a bacterial CO sensor and transcription factor. The molecule, a dimer of two identical subunits, has two haem binding domains and two 'helix–turn–helix' domains that recognize a section of DNA. The protein ligands to the Fe atoms are a histidine from one subunit and an N-terminal proline from the other. The binding of CO and displacement of proline disrupts the assembly and allows CooA to bind to DNA.

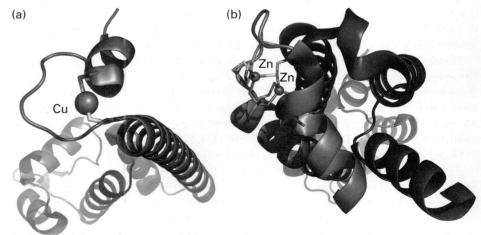

Fig. 26.51 Comparison of (a) the Cu and (b) the Zn binding sites in the respective transcription factors CueR and ZntR. Note how Cu(I) is recognized by a linear binding site (to two cysteines) whereas Zn is recognized by an arrangement of Cys and His ligands that binds two Zn(II) atoms together with a bridging phosphate group.

Most of our current insight about Zn sensing, as for Cu, is provided by studies of bacterial systems. The major difference with respect to Cu is that although Zn is also coordinated (mainly) by cysteine thiolates, the geometry is tetrahedral rather than linear. *E. coli* contains a Zn^{2+}-sensing transcription factor known as ZntR that is closely related to CueR. The factor ZntR contains two Zn binding domains each of which coordinates a pair of Zn using cysteine and histidine ligands. The surrounding protein fold is shown in Fig. 26.51, for comparison with CueR. The extent to which these dynamic Zn binding sites can be identified with zinc fingers remains unclear.

Example 26.10 Identifying links between redox chemistry and metal ion sensing

Suggest a way in which the binding of Cu or Zn to their respective sensor proteins might be linked to the level of cellular O_2.

Answer Pairs of cysteine thiolate ligands can undergo oxidation by O_2 or other oxidants, resulting in the formation of a disulfide bond (cystine). This reaction prevents the cysteine-S atoms from acting as ligands.

Self-test 26.10 Why might Cu sensors be 'designed' to bind Cu(I) rather than Cu(II)?

Biomineralization

Key points: Calcium compounds are used in exoskeletons, bones, teeth, and other devices; some organisms use crystals of magnetite, Fe_3O_4, as a compass; plants produce silica-based protective devices.

Biominerals can be either infinite covalent networks or ionic. The former include the silicates, which occur extensively in the plant world. Leaves, even whole plants, are often covered with silica hairs or spines that offer protection against predatory herbivores. Ionic biominerals are mainly based on calcium salts, and exploit the high lattice energy and low solubilities of these compounds. Calcium carbonate (calcite or aragonite, Section 11.7) is the material present in sea shells and eggshells. These minerals persist long after the organism has died, indeed chalk is a biogenic mineral, a result of the process of **calciferation** by prehistoric organisms. Calcium phosphate (hydroxyapatite, Section 11.7) is the mineral component of bones and teeth, which are particularly good examples of how organisms fabricate 'living' composite materials. Indeed, the different properties of bone found among species (such as stiffness) are produced by varying the amount of organic component, mostly a fibrous protein called collagen, with which hydroxyapatite is associated. High hydroxyapatite/collagen ratios are found for large marine animals, whereas low ratios are found for animals requiring agility and elasticity.

Biomineralization is crystallization under biological control. There are some striking examples that have no counterparts in the laboratory. Perhaps the most familiar example is the exoskeleton of the sea urchin, which comprises large sponge-like plates containing continuous macropores 15 μm in diameter: each of these plates is a single crystal of Mg-rich calcite. Large single crystals support other organisms, such as diatoms and radiolarians, with their skeletons of silica cages (Fig. 26.52).

Biominerals produce some intriguing gadgetry. Figure 26.53 shows crystals of calcite that are part of the gravity sensor device in the inner ear. These crystals are located on a membrane above sensory cells and any change in acceleration that causes them to move results in an electrical signal to the brain. The crystals are uniform in size and spindle-shaped so that they move evenly without becoming hooked together. The same property is seen for crystals of magnetite (Fe_3O_4) that are found in a variety of magnetotactic bacteria (Fig. 26.54). Across different species, there are considerable variations in size and shape, but within any one species, the crystals, formed in magnetosome vesicles (MV) are uniform. Magnetotactic bacteria live in fluid sediment suspensions in marine and freshwater environments, and it is thought that their microcompasses allow them to

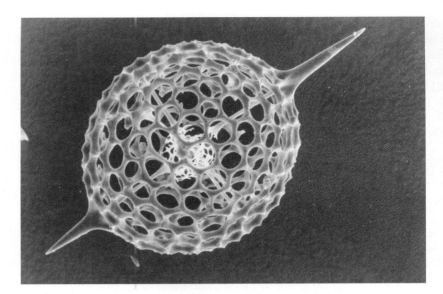

Fig. 26.52 The porous silica structure of a radiolarian microskeleton showing the large radial spines. (Photograph supplied by Professor S. Mann, University of Bristol.)

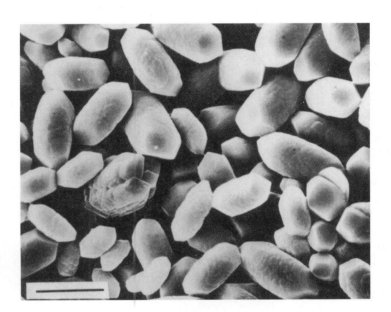

Fig. 26.53 Our gravity sensor. Crystals of biologically formed calcite that are found in the inner ear. (Photograph supplied by Professor S. Mann, University of Bristol.)

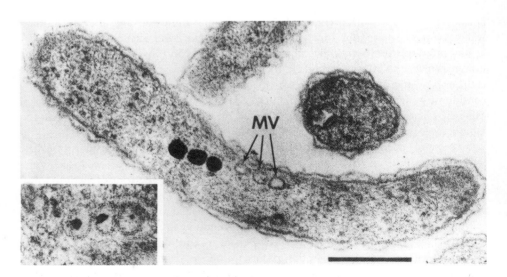

Fig. 26.54 Magnetite crystals in magnetotactic bacteria. These are tiny compasses that guide these organisms to move vertically in river bed sludge. MV indicates empty vesicles (Photograph supplied by Professor S. Mann, University of Bristol.)

swim always in a downwards direction to maintain their chemical environment during turbulent conditions.

There is great interest in how biomaterials are formed, not least because of the inspiration they provide for nanotechnology (Chapter 24). The formation of biominerals involves the following hierarchy of control mechanisms:

1 Chemical control (solubility, supersaturation, nucleation).

2 Spatial control (confinement of crystal growth by boundaries such as cells, subcompartments, and even proteins in the case of ferritin, Section 26.16).

3 Structural control (nucleation is favoured on a specific crystal face).

4 Morphological control (growth of the crystal is limited by boundaries imposed by organic material that grows with time).

5 Constructional control (interweaving inorganic and organic materials to form a higher-order structure, such as bone).

Bone is continually being dissolved and reformed; indeed it functions not only as a structural support but also as the central Ca store. Thus, during pregnancy, bones tend to be raided for their Ca in a process called **demineralization**, which occurs in special cells called **osteoclasts**. Depleted or damaged bones are restored by mineralization, which occurs in cells called **osteoblasts**. These processes involve phosphatases (Section 26.9).

The chemistry of elements in medicine

Serendipity has played an important role in drug discovery, with many effective treatments arising from chance discoveries. There appears to be a special role for compounds containing metals that are not otherwise present in biological systems, for example Pt, Au, Ru, and Bi. A major challenge in pharmacology is to determine the mechanism of action at the molecular level, and we should recognize that the drug that is administered is unlikely to be same molecule that reacts at the target site. This is particularly true for metal complexes, which are usually more susceptible to hydrolysis than organic molecules. In general, the mechanism of action is proposed by extrapolation of *in vitro* studies. Orally administered drugs are highly desirable because they avoid the trauma and potential hazards of injection; however, they may not pass through the gut wall or survive hydrolytic enzyme action. Inorganic compounds are also used in the diagnosis of disease or damage, a particularly interesting example being the use of radioactive technetium.

51 Desferral

26.17 Chelation therapy

Key points: The treatment of Fe overload involves sequestration of Fe by ligands based upon or inspired by siderophores.

Iron overload is the name given to several serious conditions that affect a large proportion of the world's population. Here we recall that, despite its great importance, Fe is potentially a highly toxic element, particularly in its ability to produce harmful radicals by reaction with O_2, and its levels are normally strictly controlled by regulatory systems. In many groups of people, a genetic disorder results in breakdown of this regulation. One kind of Fe overload is caused by an inability for the body to produce sufficient porphyrin. Other problems are caused by faults in the regulation of Fe levels by ferritin or transferrin production. These disorders are treated by **chelation therapy**, the administration of a ligand to sequester Fe and allow it to be excreted. Desferrioxamine ('Desferral', **51**) is a ligand that is similar to the siderophores described in Section 26.6. It is a very successful agent for Fe overload, apart from the trauma of its introduction into the body, which involves being plumbed into an intravenous supply.

A special case of chelation therapy is the treatment of individuals who have been contaminated with plutonium following exposure to nuclear weapons. In its common oxidation states, Pu(IV) and Pu(III) have similar charge densities to Fe(III) and Fe(II). Siderophore-like chelating ligands have been developed, such as 3,4,3-LIMACC (**52**), which contains four catechol groups.

52

26.18 Cancer treatment

Key points: The complex *cis*-[PtCl$_2$(NH$_3$)$_2$] results in the inhibition of DNA replication and prevention of cell division; other drugs cause DNA to be degraded by oxygenation.

'Cancer' is a term that covers a large number of different types of the disease, all characterized by the uncontrolled replication of transformed cells that overwhelm the normal operation of the body. The principle of treatment is to apply drugs that selectively destroy these malignant cells.

The remarkable action of the complex *cis*-[PtCl$_2$(NH$_3$)$_2$] (**53**, known as cisplatin) was discovered in 1964 while examining the effect of an electric field on the growth of bacteria. The behaviour of a colony of bacteria suspended in solution between two Pt electrodes was observed, and it was noted that the cells continued to grow in size, forming long filaments, but stopped replicating. The effect was traced to a complex that was formed electrochemically by dissolution of Pt into the electrolyte, which contained NH$_4$Cl. Since then, cisplatin has been a successful drug for the treatment of many forms of cancer, particularly testicular cancer for which the success rate approaches 100 per cent. The other geometric isomer, *trans*-[PtCl$_2$(NH$_3$)$_2$], is inactive.

The ultimate molecular basis of the chemotherapeutic action of cisplatin and related drugs is thought to be the formation of a stable complex between Pt(II) and DNA. Cisplatin is administered into the bloodstream of the patient, where, because the plasma contains high concentrations of Cl$^-$, it tends to remain as the neutral dichloro species. The electrical neutrality of the dichloro complex facilitates its passage through the cell and nuclear membranes. Once it is subjected to the lower Cl$^-$ concentrations inside the cell (Table 26.1), the chloride ligands are lost from the complex, and the resulting cationic (1+, 2+) species are electrostatically attracted to DNA and inner-sphere complexation occurs, in which the —Pt(NH$_3$)$_2$ fragment becomes coordinated to N atoms of the nuclear bases. Some classic studies have shown that the preferred target is a pair of N atoms on consecutive guanines in the same strand. Complexes of the —Pt(NH$_3$)$_2$ fragment with oligonucleotides have been studied by X-ray crystallography and ^{195}Pt NMR and the structure of one of these is shown in Fig. 26.55. Complexation with Pt causes the helix to bend and partially unwind. It is thought that this renders the DNA incapable of replication or repair, and recognizable by proteins known as high mobility group (HMG) proteins that bind to bent DNA; the cell may thus be targeted for death (a process known as apoptosis).

Despite its efficacy, cisplatin has highly undesirable side-effects, in particular it causes serious damage to the kidneys before it is eventually excreted. Great efforts have been made to find Pt complexes that are effective with fewer side-effects and one example in

Cl
|
H$_3$N—Pt—Cl
|
NH$_3$

53 Cisplatin

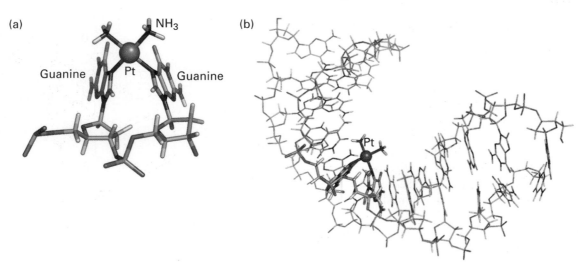

(a)   (b)

Fig. 26.55 Structure of an adduct formed between —Pt(NH$_3$)$_2$ and two adjacent guanine bases on an oligonucleotide. (a) The square-planar ligand arrangement around the Pt atom. (b) Coordination of Pt causes bending of the DNA helix.

clinical use is carboplatin (**54**). Effective drugs may also include trinuclear Pt(II) (**55**) as well as Pt(IV) complexes such as (**56**), which can be administered orally.

Other metal complexes are being discovered that offer improved efficacy over Pt drugs. These include Ru(III) complexes such as *fac*-[RuCl$_3$(NH$_3$)$_3$], which are believed to function by providing a source of Ru(III) that is carried to cancer cells by transferrin, and the complex (**57**), which may be activated by reduction to Ru(II) *in vivo*. Even common Ti compounds such as TiCp$_2$Cl$_2$ have undergone clinical trials.

The challenge with cancer chemotherapy is to identify complexes that will select for malignant cells and leave healthy cells unattacked. Some Ru complexes undergo selective interaction with DNA, and are activated upon irradiation, becoming potent oxidizing agents capable of carrying out cleavage of phosphodiester linkages. This method of treating cancers is known as **phototherapy**. Bleomycin (**58**) is representative of a class of drug that appears to function by binding to DNA and generating, upon reaction with O$_2$, an Fe(IV) (ferryl) species that oxygenates particular sites and leads to degradation.

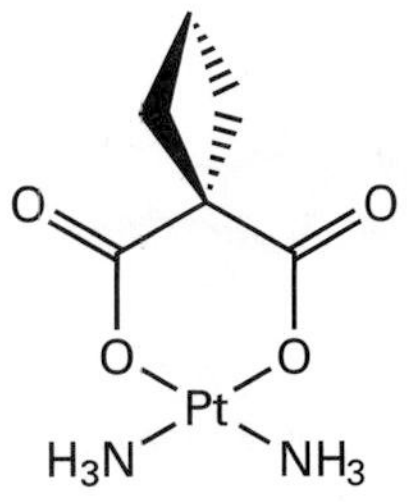

26.19 Anti-arthritis drugs

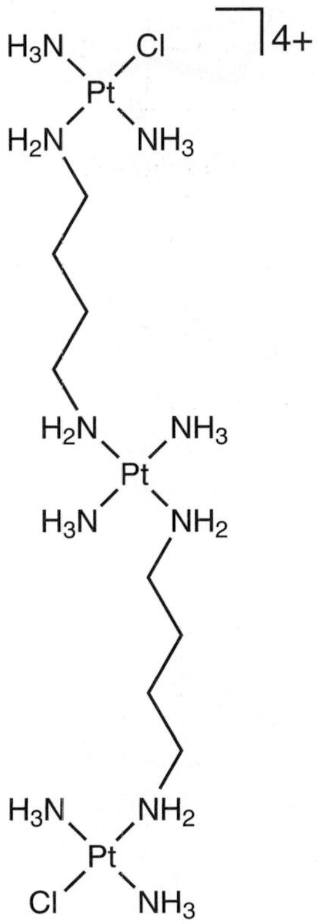

Key point: Complexes of Au are effective against rheumatoid arthritis.

Gold drugs are used in the treatment of rheumatoid arthritis, an inflammatory disease that affects the tissue around joints. The inflammation arises by the action of hydrolytic enzymes in cell compartments known as lysosomes that are associated with the Golgi apparatus (see Fig. 26.1). Although the mechanism of action is not established, it is known that Au accumulates in the lysosomes, and it is therefore possible that Au inhibits these hydrolytic enzymes. Another hypothesis is that Au(I) compounds deactivate singlet O$_2$, a harmful species that can be formed by oxidation of superoxide. The mechanism for this reaction might involve promotion of intersystem crossing by the high spin–orbit coupling constant of the heavy element.

Commonly administered drugs include sodium aurothiomalate (myochrisin, **59**), sodium aurothioglucose (solganol, **60**; the linkage between units is uncertain), and others, all of which feature Au(I) with linear coordination. Many are water-soluble polymers, but they cannot be administered orally due to acid hydrolysis in the stomach. By contrast, the compound known as Auranofin (**61**) can be given orally. Au(I) is chemically soft and it is likely that it targets sulfur groups such as cysteine side chains in proteins. It is much more likely to survive in biological environments than Au(III), which is highly oxidizing. As expected, Au compounds lead to side-effects, which include skin allergies as well as kidney and gastrointestinal problems.

26.20 Imaging agents

Key points: Particular organs and tissues are targeted according to the ligands that are present.

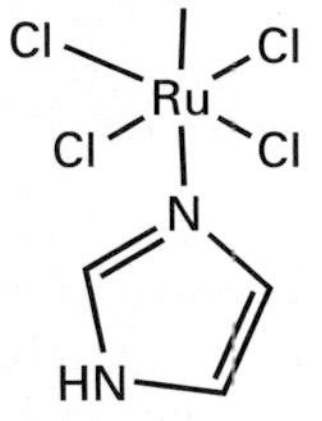

54 Carboplatin

55

56

57

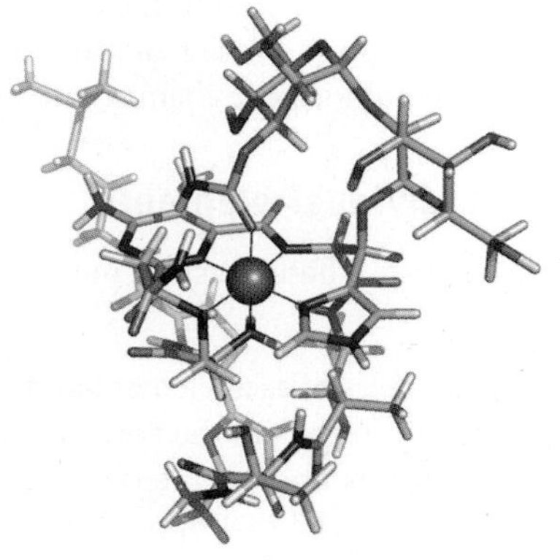

58 Bleomycin

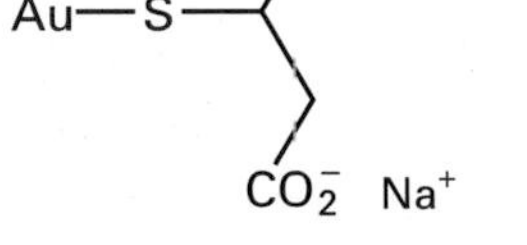

59 Myochrisin

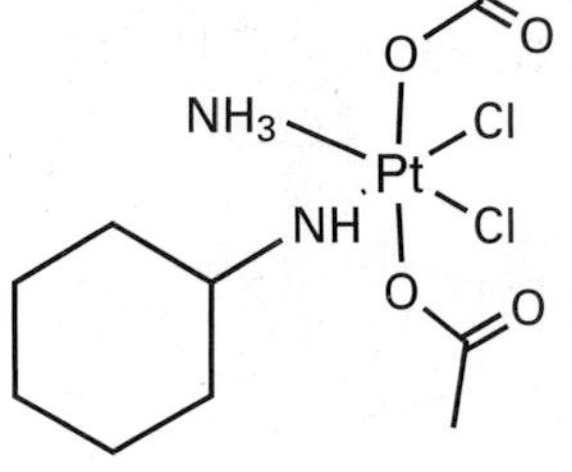

60 Solganol

61 Auranofin

62 Dotarem

63 Cardiolyte

64 Tc-MAG-3

Complexes of gadolinium(III) (f^7) are used in magnetic resonance imaging (MRI), which has become an important technique in medical diagnosis. Through their effect on the relaxation time of ^{1}H NMR resonances, Gd(III) complexes are able to enhance the contrasts between different tissues and highlight details such as the abnormalities of the blood–brain barrier. A number of Gd(III) complexes are approved for clinical use, each exhibiting different degrees of rejection or retention by certain tissues, as well as stability, rates of water exchange, and magnitude of relaxation parameters. All are based on chelating ligands, particularly those having multiple carboxylate groups. One example is the complex (**62**) formed with the macrocyclic aminocarboxylate ligand DOTA, which is known as Dotarem.

Technetium is an artificial element that is produced by a nuclear reaction, but it has found an important use as an imaging agent. The active radionuclide is ^{99m}Tc (m for metastable), which decays by γ-emission and has a half-life of 6 h. Production of ^{99m}Tc involves bombarding ^{98}Mo with neutrons and separating it as it is formed from the unstable product ^{99}Mo:

$$^{98}\text{Mo} \xrightarrow{\text{neutron capture}} {}^{99}\text{Mo} \xrightarrow{\beta\text{-decay, 90 h}} {}^{99m}\text{Tc} \xrightarrow{\gamma\text{-emission, 6 h}}$$

$$^{99}\text{Tc} \xrightarrow{\beta\text{-decay, 200 ka}} {}^{99}\text{Ru}$$

High-energy γ-rays are less harmful to tissue than α- or β-particles. The chemistry of technetium resembles manganese except that higher oxidation states are much less oxidizing.

A variety of substitution-inert Tc complexes can be made that, when injected into the patient, target particular tissues and report on their status. Complexes have been developed that target specific organs such as the heart (revealing tissue damage due to a heart attack), kidney (imaging renal function), or bone (revealing abnormalities and fracture lines). A good basis for organ targeting appears to be the charge on the complex: cationic complexes target the heart, neutral complexes target the brain, and anionic complexes target the bone and kidney. Of the different imaging agents, $[\text{Tc(CNR)}_6]^+$ (**63**) is the best established: known as Cardiolyte, it widely used as a heart imaging agent. The compound of Tc(V) with mercaptoacetyltriglycine (**64**), known as Tc-MAG-3, is used to image kidneys because of its rapid excretion. For imaging bone, complexes of Tc(VII) with hard diphosphonate ligands (**65**) are effective. Brain imaging is carried out with compounds such as Ceretec (**66**).

To produce Tc tracers, radioactive $^{98}\text{MoO}_4^{2-}$ is passed onto an anion exchange column, where it binds tightly until it decays to the pertechnate ion TcO_4^- and the lower charge causes it to be eluted. The eluate is treated with a reducing agent, usually Sn(II), and the ligands required to convert it into the desired imaging agent. The resulting compound is then administered to the patient.

Perspectives

In this final section, we stand back from the material in the chapter and review it from a variety of different perspectives, from the point of view of individual elements and from the point of view of the contribution of bioinorganic chemistry to urgent social problems.

26.21 The contributions of individual elements

Key point: Elements are selected for their inherent useful properties and their availability.

In this section we summarize the major roles of each element and correlate what we have discussed with emphasis on the element rather than the type of reaction that is involved.

Na, K and Li The ions of these elements are characterized by weak binding to hard ligands and their specificity is based on size and hydrophobicity that arises from a lower charge density. Compared to Na$^+$, K$^+$ is more likely to be found coordinated within a protein and is more easily dehydrated. Both Na$^+$ and K$^+$ are important agents in controlling cell structure through osmotic pressure, but whereas Na^+ is ejected from cells, K^+ is accumulated, contributing to a sizeable potential difference across the cell membrane. This differential is maintained by specific enzymes known as ion pumps, in particular the Na^+,K^+-ATPase, also known as the Na-pump. The electrical energy is released by specific gated ion channels, of which the K^+ channel (Section 26.3) is the best studied.

An important related issue is the widespread use of simple Li compounds (particularly Li_2CO_3) as psychotherapeutic agents in the treatment of mental disorders, notably bipolar disorder (manic depression). One possibility is that hydrated Li^+ binds tightly in the Na^+ or K^+ channels in place of the dehydrated ions that are normally transported selectively in the filter region of these proteins. As Li^+ ions are highly labile, we can be confident that the aqua ion itself, or a complex with an abundant ligand, is the active species. As a simple aqua ion, Li^+ is strongly solvated, in fact the solvated radius is higher than that of Na^+.

Mg Magnesium ions are the dominant $2+$ ion in cytoplasm and the only ones to occur above millimolar levels in the free, uncomplexed state. The energy currency for enzyme catalysis, ATP, is always present as its Mg^{2+} complex. Magnesium has a special role in the light-harvesting molecule chlorophyll, because it is a small $2+$ cation that is able to adopt octahedral geometry and can stabilize a structure without promoting energy loss by fluorescence. Mg^{2+} is a weak acid catalyst and is the active metal ion in rubisco, a highly abundant enzyme that is responsible for removing from the atmosphere some 10^{11} t of CO_2 per year. Rubisco is activated by weak binding of Mg^{2+} to two carboxylates and a special carbamate ligand, leaving three exchangeable water molecules. It is spectroscopically challenging.

Ca Calcium ions are important only in eukaryotes. The bulk of biological Ca is used for structural support and devices such as teeth. The choice of Ca for this function is due to the insolubility of Ca salts with carbonate and phosphate. However, a tiny amount of Ca is used as the basis of a sophisticated intracellular signalling system. The principle of this process is that Ca is suited for rapid coordination to hard acid ligands, especially carboxylates from protein side chains, and has no preference for any particular coordination geometry.

Mn Manganese has several oxidation states, most of which are very oxidizing. It is well suited as a redox catalyst for reactions involving positive reduction potentials. One reaction in particular, in which H_2O is used as the electron donor in photosynthesis, is responsible for producing most of the O_2 in the Earth's atmosphere. This involves a special Mn_4Ca cluster. Mn(II) is also used as a weak acid–base catalyst in some enzymes. Spectroscopic detectability varies depending on oxidation state. EPR has been useful for Mn(II) and for particular states of the Mn cluster that is the O_2-evolution catalyst.

Fe Versatile Fe is probably essential to all organisms and was certainly a very early element in biology. Three oxidation states are important, Fe(II), Fe(III), and Fe(IV). Active sites based on Fe catalyse a great variety of redox reactions ranging from electron transfer to oxygenation, as well as acid–base reactions that include reversible O_2 binding, dehydration/hydration and ester hydrolysis. Iron-containing active sites contain ligands ranging from soft donors such as sulfide (as in FeS clusters) to hard donors such as carboxylate. The porphyrin macrocycle is particularly important as a ligand. Fe(II) in various coordination environments is used to bind O_2, either reversibly or as a prerequisite for activation. Fe(III) is a good Lewis acid, whereas the Fe(IV)$=$O (ferryl) group may be considered as biology's way of managing a reactive O atom for insertion into C—H bonds. Cells contain very little uncomplexed Fe(II) and extremely low levels of Fe(III). These ions are toxic, particularly in terms of their reaction with peroxides, which generates the hydroxyl radical. Uptake into organisms poses problems because Fe(III) salts are insoluble at neutral pH. Iron uptake, delivery, and storage are controlled by sophisticated transport systems, including a special storage protein known as ferritin. Spectroscopic detectability is high. Fe porphyrins (cytochromes) show intense UV–visible absorption bands and most active sites with unpaired electrons give rise to characteristic EPR spectra.

Co Cobalt and nickel are among the most ancient biocatalysts. Cobalt is processed only by microorganisms and higher organisms must ingest it as 'vitamin B_{12}'. All Co is complexed by a special macrocycle called corrin. Complexes in which the fifth ligand is a

65

66 Ceretec

benzimidazole that is covalently linked to the corrin ring are known as **cobalamins**. Cobalamins are cofactors in enzymes that catalyse radical-based rearrangements and alkyl transfer reactions. In the special cofactor known as coenzyme B_{12}, the sixth ligand to Co(III) is a carbanion donor atom from deoxyadenosine. Radical-based rearrangements involve the ability of coenzyme B_{12} to undergo facile homolytic cleavage of the Co—C bond, producing stable low-spin Co(II) and a carbon radical that can abstract a hydrogen atom from substrates. Alkyl transfer reactions exploit the high nucelophilicity of Co(I). Spectroscopic detectability is quite good. Cobalamin-containing enzymes show strong UV–visible absorption bands. EPR spectra are observed for Co(II).

Ni Nickel is important in bacterial enzymes, notably hydrogenases, where it also uses the 3+ and 1+ oxidation states, which are rare in conventional chemistry. A particularly remarkable enzyme, coenzyme A synthase, uses Ni to produce CO and then react it with CH_3— (provided by a cobalamin enzyme) to produce a C—C bond in the form of an acetyl ester. Nickel is also found in plants as the active site of urease. Urease was the first enzyme to be crystallized (in 1926), yet it was not until 1976 that it was discovered to contain Ni. EPR spectra are observed for Ni(III) and Ni(I).

Cu Unlike Fe, copper probably became important only after O_2 had become established in the Earth's atmosphere and it became available as soluble Cu(II) salts rather than insoluble sulfides (Cu_2S). The main role of Cu is in electron-transfer reactions at the higher end of the potential scale and catalysis of redox reactions involving O_2. It is also used for reversible O_2 binding. Both Cu(II) and Cu(I) are strongly bound to biological ligands, particularly soft bases. Free Cu ions are highly toxic, and almost absent from cells. Spectroscopic detectability is good for Cu(II) (EPR, and strong visible absorption bands for 'blue' Cu centres) but poor for Cu(I).

Zn Zinc is an excellent Lewis acid, forming stable complexes with ligands such as N and S donors and catalysing reactions such as ester and peptide hydrolysis. The biological importance of Zn stems largely from its lack of redox chemistry, although its common adoption of di-, tri-, and tetrathiolate ligation provides a link to the redox chemistry of cysteine/cystine interconversions. Zinc is used as a structure former in enzymes and proteins that bind to DNA. A major problem has been the lack of good spectroscopic methods for studying this d^{10} ion. In some cases, Zn enzymes have been studied by EPR, after substituting the Zn by Co(II).

Mo and W Molybdenum is an abundant element that is probably used by all organisms as a redox catalyst for the transfer of O atoms derived from H_2O. In these oxo-transfer enzymes the Mo is always part of a larger pterin-containing cofactor in which it is coordinated by a special dithiolene ligand. Interconversion between Mo(IV) and Mo(VI) usually results in a change in the number of terminal oxo ligands, and recovery of the starting material occurs by single-electron transfer reactions with Mo(V) as an intermediate. Aside from oxo-transfer and related reactions, Mo has another intriguing role, that of nitrogen fixation, in which it is part of a special FeS cluster. Use of W is confined to prokaryotes, where it is also used as a redox catalyst, but in reactions where a stronger reducing agent is required. Conventional spectroscopic studies have largely been restricted to Mo(V).

Si Silicon is often neglected among biological elements, yet its turnover in some organisms is comparable to that of carbon. Silica is an important material for the fabrication of the exoskeleton and of prickly defensive armour in plants.

Pt, Au, Bi, and Ru These elements have no known deliberate biological functions and are foreign agents to biological systems, acting under normal conditions as poisons. However, used in controlled procedures, and 'dressed up' by complexation to target a particular site, they are potent drugs, active against a range of diseases and disorders.

26.22 Future directions

Key points: Biological metals and metalloproteins have important futures in medicine, energy production, green synthesis, and nanotechnology.

The pioneering studies of the structures and mechanism of ion channels mentioned in Section 26.3 are providing important new leads in neurophysiology, including the rational design of drugs that can block or modify their action in some way. New functions for Ca are continually emerging, and one intriguing aspect is its role in determining the left–right asymmetry of higher organisms, a prime example being the specific placements of heart and liver in the body cavity. The so-called **Notch signalling pathway** in embryonic cells depends on transient extracellular bursts of Ca^{2+} that are dependent in some way on the activity of an H^+/K^+-ATPase. There is also a growing awareness of the role of Zn and Zn transport proteins in control of cellular activity, and also of neural transmission. Indeed, the term **metalloneurochemistry** has been coined to describe the study of metal ion function in the brain and nervous system at the molecular level. An important challenge is to map out the distribution and flow of Zn in tissue such as brain, and advances are being made in the design of fluorescent ligands that will bind Zn selectively at cellular levels and report on its transport across different zones, for example the synaptic junctions. Metal ions are involved in protein folding, and Cu may have a role in controlling the behaviour of the prions implicated in certain diseases.

In many regions of the world, rice is the staple food but this commodity is low in Fe. Thus transgenic techniques are being used to improve Fe content. The object is to produce better plant siderophores and improve Fe storage (by enhanced expression of the ferritin gene).

In the not so distant future, when fossil fuels have been depleted, it is certain that H_2 will become an extremely important energy carrier, used either directly or indirectly (its conversion into fuels such as alcohol) to power cars and planes. One of the scientific challenges is how to obtain efficient electrolytic production of H_2 from water, given that electricity will be widely available from a variety of sources. This requires catalysts, which are currently based on Pt and other precious metals. These metals are also required for fuel cells. However, Biology has already shown us that rapid hydrogen cycling is possible using just the common metals Fe and Ni. A related challenge is the synthesis of efficient electrocatalysts that can convert water to O_2 without requiring a large overpotential, not because there is a need for O_2 itself, but because it is an essential by-product of electrolytic H_2 production. Once again, we can turn to the biosphere for inspiration, because by elucidating the mechanism of the Mn catalyst, we might synthesize new catalysts that are both cheap and durable.

We have seen the exquisite structures of materials that are produced by organisms. This understanding is now leading to new directions in nanotechnology (Chapter 24). For example, sponge-like single crystals of calcite, having intricate morphological features, have been produced on polymer membranes formed by templating the sketal plates of the sea urchin. Another recent development is the production of Pd nanoclusters by hydrogen oxidizing bacteria, the action of hydrogenases making available controlled electron flow to effect the electroplating of Pd onto microscopic sites.

FURTHER READING

J.J.R. Frausto da Silva and R.J.P. Williams, *The biological chemistry of the elements*. Oxford University Press (2001). This textbook provides critical and informative opinions on the roles of different elements in biology.

K.N. Ferreira, T.M. Iverson, K. Maghlaoui, J. Barber, and S. Iwata, Architecture of the photosynthetic oxygen-evolving center. *Science*, 2004, **303**, 1831. This article describes the complex structure of the catalyst that places O_2 in our atmosphere.

H.B. Gray, B.G. Malmström, and R.J.P. Williams, Copper coordination in blue proteins. *J. Biol. Inorg. Chem.*, 2000, **5**, 551. A definitive article on the nature and properties of blue copper centres.

P.J. Kiley and H. Beinert, The role of Fe-S proteins in sensing and regulation in bacteria. *Curr. Opin. Chem. Biol.*, 2003, **6**, 182. A broad account of how FeS proteins are used to sense Fe and O_2 and link their activities to cell regulation.

S. Mann, *Biomineralisation: principles and concepts in bioinorganic materials chemistry*. Oxford University Press (2001). A clearly written guide to 'solid-state' biological chemistry.

Z. Guo and P.J. Sadler, Metals in medicine. *Angew. Chem., Int. Ed. Engl.*, 1999, **38**, 1512. A survey of the major uses of inorganic compounds in medicine.

J. Reedijk and E. Bouwman (ed.), *Bioinorganic catalysis*. Marcel Dekker, New York and Basel (1999).

M.D. Archer and J. Barber (ed.), *Molecular to global photosynthesis*. Imperial College Press (2004).

EXERCISES

26.1 Calcium-binding proteins can be studied using lanthanoid ions (Ln^{3+}). Compare and contrast the coordination preferences of the two types of metal ion and suggest techniques in which lanthanoid ions would be useful.

26.2 In zinc enzymes, 'spectroscopically silent' Zn(II) can often be replaced by Co(II) with high retention of activity. Explain how this substitution can be exploited to obtain structural and mechanistic information.

26.3 Compare and contrast the acid–base catalytic activities of Zn(II), Fe(III), and Mg(II).

26.4 The structure of the P-cluster in nitrogenase differs significantly between oxidized and reduced states. Comment on this observation in the light of proposals that it functions in long-range electron transfer.

26.5 Microorganisms can synthesize the acetyl group (CH₃CO—) by direct combination of methyl groups with CO. Make some predictions about the metals that are involved.

PROBLEMS

26.1 Apart from direct O-atom transfer (Fig. 26.41), another mechanism proposed for Mo enzymes is indirect O-atom transfer, also known as coupled electron-proton transfer. In this mechanism, as shown in Fig. 26.56 for sulfite oxidase, the O atom that is transferred originates instead from an uncoordinated water molecule. Propose a way to distinguish between direct and indirect O-atom transfer mechanisms.

26.2 A sample of ferredoxin from chloroplasts gives the Mössbauer spectra measured at 77 K shown in Fig. 26.57. Interpret the data.

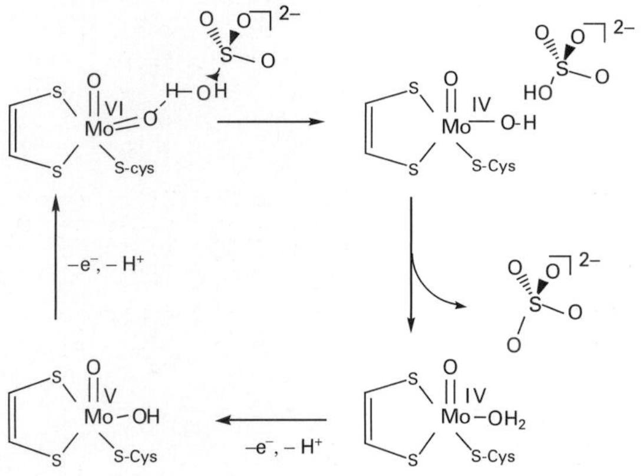

26.3 Comment on the statement that the structure of nitrogenase has not enlightened us as to its mechanism of action and discuss how this view might be valid for enzymes for which a structure is known.

26.4 'In Mo enzymes, the bond between a terminal oxo anion and Mo(VI) is usually written as a double bond, whereas it is more correctly assigned as a triple bond.' Discuss this statement. Suggest how a terminal oxo ligand influences the reactivity of other coordination sites on the Mo atom and explain how a terminal sulfido ligand (as occurs in xanthine oxidase) would alter the properties of the active site.

Fig. 26.56

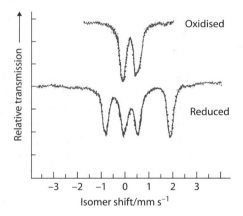

Fig. 26.57

Resource section 1:
Selected ionic radii (in picometres, pm)

Ionic radii are given for the most common oxidation states and coordination geometries. The coordination number is given in parentheses. All *d*-block species are low-spin unless labelled with # in which case values for high-spin are quoted. Most data are taken from R.D. Shannon, *Acta Cryst.*, 1976, **A32**, 751, where values for other coordination geometries can be found. Where Shannon values are not available, Pauling ionic radii are quoted and are indicated by *.

1	2	3	4	5	6	7	8	9	10	11	12	13	14	15	16	17	18
Li^+ 73 (4), 90 (6), 106 (8)	Be^{2+} 41 (4), 59 (6)											B^{3+} 25 (4), 41 (6)	C^{4+} 29 (4), 30 (6)	N^{3-} 132 (4); N^{3+} 30 (6)	O^{2-} 124 (4), 126 (6), 128 (8)	F^- 117 (4), 116 (6)	Ne^+ 112*
Na^+ 113 (4), 116 (6), 132 (8)	Mg^{2+} 71 (4), 86 (6), 103 (8)											Al^{3+} 53 (4), 68 (6)	Si^{4+} 40 (4), 54 (6)	P^{5+} 31 (4), 52 (6); P^{3+} 58 (6)	S^{2-} 170 (6); S^{6+} 26 (4), 43 (6); S^{4+} 51 (6)	Cl^- 167 (6); Cl^{7+} 22 (4), 41 (6)	Ar^+ 154*
K^+ 151 (4), 152 (6), 165 (8)	Ca^{2+} 114 (6), 126 (8)	Sc^{3+} 89 (6), 101 (8)	Ti^{4+} 56 (4), 75 (6), 88 (8); Ti^{3+} 81 (6); Ti^{2+} 100 (6)	V^{5+} 50 (4), 68 (6); V^{4+} 72 (6), 86 (8); V^{3+} 78 (6); V^{2+} 93 (6)	Cr^{6+} 40 (4), 58 (6); Cr^{5+} 63 (6); Cr^{4+} 55 (4), 69 (6); Cr^{3+} 76 (6); Cr^{2+} 87 (6)	Mn^{7+} 39 (4), 60 (6); Mn^{6+} 40 (4); Mn^{5+} 60 (4), 75 (6); Mn^{4+} 53 (4), 67 (6); Mn^{3+} 72 (6); Mn^{2+} 81 (6), 110 (8)	Fe^{6+} 39 (4); Fe^{4+} 73 (6); Fe^{3+} 63 (4)#, 69 (6), 92 (8)#; Fe^{2+} 77 (4)#, 75 (6), 106 (8)#	Co^{4+} 54 (4), 67 (6)#; Co^{3+} 69 (6); Co^{2+} 72 (4)#, 79 (6), 104 (8)	Ni^{4+} 62 (6); Ni^{3+} 70 (6); Ni^{2+} 69 (6), 83 (8)	Cu^{3+} 86 (6); Cu^{2+} 71 (4), 87 (6); Cu^+ 74 (4), 91 (6)	Zn^{2+} 75 (4), 88 (6), 104 (8)	Ga^{3+} 61 (4), 76 (6)	Ge^{4+} 53 (4), 76 (6); Ge^{2+} 87 (6)	As^{5+} 48 (4), 60 (6); As^{3+} 72 (6)	Se^{6+} 42 (4), 56 (6); Se^{4+} 64 (6)	Br^- 182 (6); Br^{7+} 53 (6)	Kr^+ 169*
Rb^+ 166 (6), 175 (8)	Sr^{2+} 132 (6), 140 (8)	Y^{3+} 104 (6), 116 (8)	Zr^{4+} 73 (4), 86 (6), 89 (8)	Nb^{5+} 62 (4), 83 (6), 88 (8); Nb^{4+} 82 (6), 93 (8); Nb^{3+} 86 (6)	Mo^{6+} 55 (4), 83 (6), 88 (8); Mo^{5+} 60 (4), 75 (6); Mo^{4+} 79 (6); Mo^{3+} 83 (6)	Tc^{7+} 51 (4), 73 (6); Tc^{5+} 74 (6); Tc^{4+} 90 (6), 102 (8)	Ru^{8+} 50 (4); Ru^{7+} 74 (6); Ru^{5+} 71 (6); Ru^{4+} 74 (6); Ru^{3+} 82 (6)	Rh^{5+} 69 (6); Rh^{4+} 74 (6); Rh^{3+} 81 (6)	Pd^{4+} 76 (6); Pd^{3+} 90 (6); Pd^{2+} 78 (4), 100 (6); Pd^+ 73 (2)	Ag^{3+} 81 (4), 89 (6); Ag^{2+} 93 (4), 108 (6); Ag^+ 81 (2), 114 (4), 129 (6)	Cd^{2+} 92 (4), 109 (6), 124 (8)	In^{3+} 76 (4), 94 (6), 106 (8)	Sn^{4+} 69 (4), 83 (6), 95 (8)	Sb^{5+} 74 (6); Sb^{3+} 90 (6)	Te^{6+} 57 (4), 60 (6); Te^{4+} 81 (4), 111 (6)	I^- 206 (6); I^{7+} 56 (4), 67 (6)	Xe^+ 190*; Xe^{8+} 54 (4), 62 (6)

1	2	3	4	5	6	7	8	9	10	11	12	13	14	15	16	17	18
Cs^+ 181 (6), 188 (8)	Ba^{2+} 149 (6), 156 (8)	La^{3+} 117 (6), 130 (8)	Hf^{4+} 72 (4), 90 (6), 97 (8)	Ta^{5+} 78 (6), 88 (8)	W^{6+} 56 (4), 74 (6)	Re^{7+} 52 (4), 67 (6)	Os^{8+} 53 (4)	Ir^{5+} 71 (6)	Pt^{5+} 71 (6)	Au^{5+} 71 (6)	Hg^{2+} 110 (4), 116 (6), 128 (8)	Tl^{3+} 89 (4), 103 (6), 112 (8)	Pb^{4+} 79 (4), 92 (6), 108 (8)	Bi^{5+} 90 (6)	Po^{6+} 81 (6)	At^{7+} 76 (6)	
				Ta^{4+} 82 (6)	W^{5+} 76 (6)	Re^{6+} 69 (6)	Os^{7+} 67 (6)	Ir^{4+} 77 (6)	Pt^{4+} 77 (6)	Au^{3+} 82 (4), 99 (6)	Hg^+ 133 (6)	Tl^+ 164 (6), 173 (8)	Pb^{2+} 133 (6), 143 (8)	Bi^{3+} 117 (6), 131 (8)	Po^{4+} 108 (6), 122 (8)		
				Ta^{3+} 86 (6)	W^{4+} 80 (6)	Re^{5+} 72 (6)	Os^{6+} 69 (6)	Ir^{3+} 82 (6)	Pt^{2+} 74 (4), 94 (6)	Au^+ 151 (6)							
						Re^{4+} 77 (6)	Os^{5+} 72 (6)										
							Os^{4+} 77 (6)										
Fr^+ 196 (6)	Ra^{2+} 162 (8)																

Lanthanoids

Ce^{4+} 101 (6), 111 (8)	Pr^{4+} 99 (6), 110 (8)	Nd^{3+} 112 (6), 125 (8)	Pm^{3+} 111 (6), 123 (8)	Sm^{3+} 110 (6), 122 (8)	Eu^{3+} 109 (6), 121 (8)	Gd^{3+} 108 (6), 119 (8)	Tb^{4+} 90 (6), 102 (8)	Dy^{3+} 106 (6), 117 (8)	Ho^{3+} 104 (6), 116 (8)	Er^{3+} 103 (6), 114 (8)	Tm^{3+} 102 (6), 113 (8)	Yb^{3+} 101 (6), 113 (8)	Lu^{3+} 100 (6), 112 (8)
Ce^{3+} 115 (6), 128 (8)	Pr^{3+} 113 (6), 127 (8)	Nd^{2+} 143 (8)		Sm^{2+} 141 (8)	Eu^{2+} 131 (6), 139 (8)		Tb^{3+} 106 (6), 118 (8)	Dy^{2+} 121 (6), 133 (8)			Tm^{2+} 117 (6), 123 (8)	Yb^{2+} 116 (6), 128 (8)	

Actinoids

Th^{4+} 108 (6), 119 (8)	Pa^{5+} 92 (6), 109 (8)	U^{6+} 66 (4), 87 (6), 100 (8)	Np^{7+} 85 (6)	Pu^{6+} 85 (6)	Am^{4+} 99 (6), 109 (8)	Cm^{4+} 99 (6), 109 (8)	Bk^{4+} 97 (6), 107 (8)	Cf^{4+} 96 (6), 106 (8)	Es	Fm	Md	No^{2+} 124 (6)	Lr
	Pa^{4+} 104 (6), 118 (8)	U^{5+} 90 (6)	Np^{6+} 86 (6)	Pu^{5+} 88 (6)	Am^{3+} 112 (6), 123 (8)	Cm^{3+} 111 (6)	Bk^{3+} 110 (6)	Cf^{3+} 109 (6)				
	Pa^{3+} 118 (6)	U^{4+} 103 (6), 114 (8)	Np^{5+} 89 (6)	Pu^{4+} 100 (6), 110 (8)	Am^{2+} 140 (8)							
		U^{3+} 117 (6)	Np^{4+} 101 (4), 112 (6)	Pu^{3+} 114 (6)								
			Np^{3+} 115 (6)									
			Np^{2+} 124 (6)									

Resource section 2:
Electronic properties of the elements

Ground-state electron configurations of atoms are determined experimentally from spectroscopic and magnetic measurements. The results of these determinations are listed below. They can be rationalized in terms of the building-up principle, in which electrons are added to the available orbitals in a specific order in accord with the Pauli exclusion principle. Some variation in order is encountered in the d- and f-block elements to accommodate the effects of electron–electron interaction more faithfully. The closed shell configuration $1s^2$ characteristic of helium is denoted [He] and likewise for the other noble-gas element configurations. The ground-state electron configurations and term symbols listed below have been taken from S. Fraga, J. Karwowski, and K.M.S. Saxena, *Handbook of atomic data*. Elsevier, Amsterdam (1976).

The first three ionization energies of an element E are the energies required for the following processes:

$$I_1 : \quad E(g) \rightarrow E^+(g) + e^-(g)$$
$$I_2 : \quad E^+(g) \rightarrow E^{2+}(g) + e^-(g)$$
$$I_3 : \quad E^{2+}(g) \rightarrow E^{3+}(g) + e^-(g)$$

The electron affinity E_{ea} is the energy *released* when an electron attaches to a gas-phase atom:

$$E_{ea} : \quad E(g) + e^-(g) \rightarrow E^-(g)$$

The values given here are taken from various sources, particularly C.E. Moore, *Atomic energy levels*, NBS Circular 467, Washington (1970) and W.C. Martin, L. Hagan, J. Reader, and J. Sugar, *J. Phys. Chem. Ref. Data*, 1974, **3**, 771. Values for the actinoids are taken from J.J. Katz, G.T. Seaborg, and L.R. Morss (ed.), *The chemistry of the actinide elements*. Chapman & Hall, London (1986). Electron affinities are from H. Hotop and W.C. Lineberger, *J. Phys. Chem. Ref. Data*, 1985, **14**, 731.

For conversions to kilojoules per mole and reciprocal centimetres, see the endpapers.

Atom		Ionization energy/eV			Electron affinity E_{ea}/eV	Atom		Ionization energy/eV			Electron affinity E_{ea}/eV
		I_1	I_2	I_3				I_1	I_2	I_3	
1 H	$1s^1$	13.60			+0.754	19 K	[Ar]$4s^1$	4.340	31.62	45.71	+0.502
2 He	$1s^2$	24.59	54.51		−0.5	20 Ca	[Ar]$4s^2$	6.111	11.87	50.89	+0.02
3 Li	[He]$2s^1$	5.320	75.63	122.4	+0.618	21 Sc	[Ar]$3d^14s^2$	6.54	12.80	24.76	
4 Be	[He]$2s^2$	9.321	18.21	153.85	≤0	22 Ti	[Ar]$3d^24s^2$	6.82	13.58	27.48	
5 B	[He]$2s^22p^1$	8.297	25.15	37.93	+0.277	23 V	[Ar]$3d^34s^2$	6.74	14.65	29.31	
6 C	[He]$2s^22p^2$	11.257	24.38	47.88	+1.263	24 Cr	[Ar]$3d^54s^1$	6.764	16.50	30.96	
7 N	[He]$2s^22p^3$	14.53	29.60	47.44	−0.07	25 Mn	[Ar]$3d^54s^2$	7.435	15.64	33.67	
8 O	[He]$2s^22p^4$	13.62	35.11	54.93	+1.461	26 Fe	[Ar]$3d^64s^2$	7.869	16.18	30.65	
9 F	[He]$2s^22p^5$	17.42	34.97	62.70	+3.399	27 Co	[Ar]$3d^74s^2$	7.876	17.06	33.50	
10 Ne	[He]$2s^22p^6$	21.56	40.96	63.45	−1.2	28 Ni	[Ar]$3d^84s^2$	7.635	18.17	35.16	
11 Na	[Ne]$3s^1$	5.138	47.28	71.63	+0.548	29 Cu	[Ar]$3d^{10}4s^1$	7.725	20.29	36.84	
12 Mg	[Ne]$3s^2$	7.642	15.03	80.14	≤0	30 Zn	[Ar]$3d^{10}4s^2$	9.393	17.96	39.72	
13 Al	[Ne]$3s^23p^1$	5.984	18.83	28.44	+0.441	31 Ga	[Ar]$3d^{10}4s^24p^1$	5.998	20.51	30.71	+0.30
14 Si	[Ne]$3s^23p^2$	8.151	16.34	33.49	+1.385	32 Ge	[Ar]$3d^{10}4s^24p^2$	7.898	15.93	34.22	+1.2
15 P	[Ne]$3s^23p^3$	10.485	19.72	30.18	+0.747	33 As	[Ar]$3d^{10}4s^24p^3$	9.814	18.63	28.34	+0.81
16 S	[Ne]$3s^23p^4$	10.360	23.33	34.83	+2.077	34 Se	[Ar]$3d^{10}4s^24p^4$	9.751	21.18	30.82	+2.021
17 Cl	[Ne]$3s^23p^5$	12.966	23.80	39.65	+3.617	35 Br	[Ar]$3d^{10}4s^24p^5$	11.814	21.80	36.27	+3.365
18 Ar	[Ne]$3s^23p^6$	15.76	27.62	40.71	−1.0	36 Kr	[Ar]$3d^{10}4s^24p^6$	13.998	24.35	36.95	−1.0

Atom		Ionization energy/eV			Electron affinity	Atom		Ionization energy/eV			Electron affinity
		I_1	I_2	I_3	E_{ea}/eV			I_1	I_2	I_3	E_{ea}/eV
37 Rb	$[Kr]5s^1$	4.177	27.28	40.42	+0.486	71 Lu	$[Xe]4f^{14}5d^16s^2$	5.425	13.89	20.96	
38 Sr	$[Kr]5s^2$	5.695	11.03	43.63	+0.05	72 Hf	$[Xe]4f^{14}5d^26s^2$	6.65	14.92	23.32	
39 Y	$[Kr]4d^15s^2$	6.38	12.24	20.52		73 Ta	$[Xe]4f^{14}5d^36s^2$	7.89	15.55	21.76	
40 Zr	$[Kr]4d^15s^2$	6.84	13.13	22.99		74 W	$[Xe]4f^{14}5d^46s^2$	7.89	17.62	23.84	
41 Nb	$[Kr]4d^45s^1$	6.88	14.32	25.04		75 Re	$[Xe]4f^{14}5d^56s^2$	7.88	13.06	26.01	
42 Mo	$[Kr]4d^55s^1$	7.099	16.15	27.16		76 Os	$[Xe]4f^{14}5d^66s^2$	8.71	16.58	24.87	
43 Tc	$[Kr]4d^55s^2$	7.28	15.25	29.54		77 Ir	$[Xe]4f^{14}5d^76s^2$	9.12	17.41	26.95	
44 Ru	$[Kr]4d^75s^1$	7.37	16.76	28.47		78 Pt	$[Xe]4f^{14}5d^96s^1$	9.02	18.56	29.02	
45 Rh	$[Kr]4d^85s^1$	7.46	18.07	31.06		79 Au	$[Xe]4f^{14}5d^{10}6s^1$	9.22	20.52	30.05	
46 Pd	$[Kr]4d^{10}$	8.34	19.43	32.92		80 Hg	$[Xe]4f^{14}5d^{10}6s^2$	10.44	18.76	34.20	
47 Ag	$[Kr]4d^{10}5s^1$	7.576	21.48	34.83		81 Tl	$[Xe]4f^{14}5d^{10}6s^26p^1$	6.107	20.43	29.83	
48 Cd	$[Kr]4d^{10}5s^2$	8.992	16.90	37.47		82 Pb	$[Xe]4f^{14}5d^{10}6s^26p^2$	7.415	15.03	31.94	
49 In	$[Kr]4d^{10}5s^25p^1$	5.786	18.87	28.02	+0.3	83 Bi	$[Xe]4f^{14}5d^{10}6s^26p^3$	7.289	16.69	25.56	
50 Sn	$[Kr]4d^{10}5s^25p^2$	7.344	14.63	30.50	+1.2	84 Po	$[Xe]4f^{14}5d^{10}6s^26p^4$	8.42	18.66	27.98	
51 Sb	$[Kr]4d^{10}5s^25p^3$	8.640	18.59	25.32	+1.07	85 At	$[Xe]4f^{14}5d^{10}6s^26p^5$	9.64	16.58	30.06	
52 Te	$[Kr]4d^{10}5s^25p^4$	9.008	18.60	27.96	+1.971	86 Rn	$[Xe]4f^{14}5d^{10}6s^26p^6$	10.75			
53 I	$[Kr]4d^{10}5s^25p^5$	10.45	19.13	33.16	+3.059	87 Fr	$[Rn]7s^1$	4.15	21.76	32.13	
54 Xe	$[Kr]4d^{10}5s^25p^6$	12.130	21.20	32.10	−0.8	88 Ra	$[Rn]7s^2$	5.278	10.15	34.20	
55 Cs	$[Xe]6s^1$	3.894	25.08	35.24		89 Ac	$[Rn]6d^17s^2$	5.17	11.87	19.69	
56 Ba	$[Xe]6s^2$	5.211	10.00	37.51		90 Th	$[Rn]6d^27s^2$	6.08	11.89	20.50	
57 La	$[Xe]5d^16s^2$	5.577	11.06	19.17		91 Pa	$[Rn]5f^26d^17s^2$	5.89	11.7	18.8	
58 Ce	$[Xe]4f^15d^16s^2$	5.466	10.85	20.20		92 U	$[Rn]5f^36d^17s^2$	6.19	14.9	19.1	
59 Pr	$[Xe]4f^36s^2$	5.421	10.55	21.62		93 Np	$[Rn]5f^46d^17s^2$	6.27	11.7	19.4	
60 Nd	$[Xe]4f^46s^2$	5.489	10.73	20.07		94 Pu	$[Rn]5f^67s^2$	6.06	11.7	21.8	
61 Pm	$[Xe]4f^56s^2$	5.554	10.90	22.28		95 Am	$[Rn]5f^77s^2$	5.99	12.0	22.4	
62 Sm	$[Xe]4f^66s^2$	5.631	11.07	23.42		96 Cm	$[Rn]5f^76d^17s^2$	6.02	12.4	21.2	
63 Eu	$[Xe]4f^76s^2$	5.666	11.24	24.91		97 Bk	$[Rn]5f^97s^2$	6.23	12.3	22.3	
64 Gd	$[Xe]4f^75d^16s^2$	6.140	12.09	20.62		98 Cf	$[Rn]5f^{10}7s^2$	6.30	12.5	23.6	
65 Tb	$[Xe]4f^96s^2$	5.851	11.52	21.91		99 Es	$[Rn]5f^{11}7s^2$	6.42	12.6	24.1	
66 Dy	$[Xe]4f^{10}6s^2$	5.927	11.67	22.80		100 Fm	$[Rn]5f^{12}7s^2$	6.50	12.7	24.4	
67 Ho	$[Xe]4f^{11}6s^2$	6.018	11.80	22.84		101 Md	$[Rn]5f^{13}7s^2$	6.58	12.8	25.4	
68 Er	$[Xe]4f^{12}6s^2$	6.101	11.93	22.74		102 No	$[Rn]5f^{14}7s^2$	6.65	13.0	27.0	
69 Tm	$[Xe]4f^{13}6s^2$	6.184	12.05	23.68		103 Lr	$[Rn]5f^{14}6d^17s^2$	4.6	14.8	23.0	
70 Yb	$[Xe]4f^{14}6s^2$	6.254	12.19	25.03							

Resource section 3:
Standard potentials

The standard potentials quoted here are presented in the form of Latimer diagrams (Section 5.10) and are arranged according to the blocks of the periodic table in the order *s, p, d, f*. Data and species in parentheses are uncertain. Most of the data, together with occasional corrections, come from A.J. Bard, R. Parsons, and J. Jordan (ed.), *Standard potentials in aqueous solution*. Marcel Dekker, New York (1985). Data for the actinoids are from L.R. Morss, *The chemistry of the actinide elements*, Vol. 2 (ed. J.J. Katz, G.T. Seaborg, and L.R. Morss). Chapman & Hall, London (1986). The value for $[Ru(bipy)_3]^{3+/2+}$ is from B. Durham, J.L. Walsh, C.L. Carter, and T.J. Meyer, *Inorg. Chem.*, 1980, **19**, 860. Potentials for carbon species and some *d*-block elements are taken from S.G. Bratsch, *J. Phys. Chem. Ref. Data*, 1989, **18**, 1. For further information on standard potentials of unstable radical species see D.M. Stanbury, *Adv. Inorg. Chem.*, 1989, **33**, 69. Potentials in the literature are occasionally reported relative to the standard calomel electrode (SCE) and may be converted to the H^+/H_2 scale by adding 0.2412 V. For a detailed discussion of other reference electrodes, see D.J.G. Ives and G.J. Janz, *Reference electrodes*. Academic Press, New York (1961).

---

s Block • Group 1

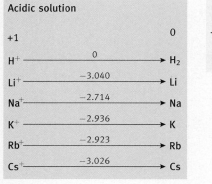

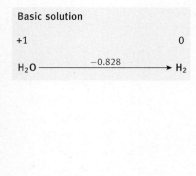

---

s Block • Group 2

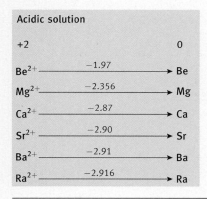

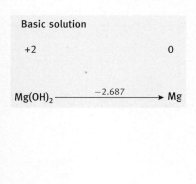

---

p Block • Group 13

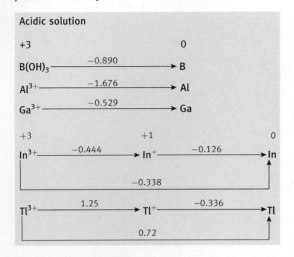

Acidic solution

+3   0

$B(OH)_3 \xrightarrow{-0.890} B$

$Al^{3+} \xrightarrow{-1.676} Al$

$Ga^{3+} \xrightarrow{-0.529} Ga$

+3   +1   0

$In^{3+} \xrightarrow{-0.444} In^+ \xrightarrow{-0.126} In$

$\xrightarrow{-0.338}$

$Tl^{3+} \xrightarrow{1.25} Tl^+ \xrightarrow{-0.336} Tl$

$\xrightarrow{0.72}$

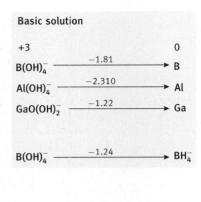

Basic solution

+3   0

$B(OH)_4^- \xrightarrow{-1.81} B$

$Al(OH)_4^- \xrightarrow{-2.310} Al$

$GaO(OH)_2^- \xrightarrow{-1.22} Ga$

$B(OH)_4^- \xrightarrow{-1.24} BH_4^-$

p Block • Group 14

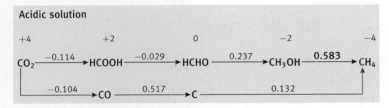

Acidic solution

+4   +2   0   −2   −4

$CO_2 \xrightarrow{-0.114} HCOOH \xrightarrow{-0.029} HCHO \xrightarrow{0.237} CH_3OH \xrightarrow{\mathbf{0.583}} CH_4$

$\xrightarrow{-0.104} CO \xrightarrow{0.517} C \xrightarrow{0.132}$

Basic solution

$CO_3^{2-} \xrightarrow{-0.930} HCO_2^- \xrightarrow{-1.160} HCHO \xrightarrow{-0.591} CH_3OH \xrightarrow{-0.245} CH_4$

$C \xrightarrow{-1.148}$

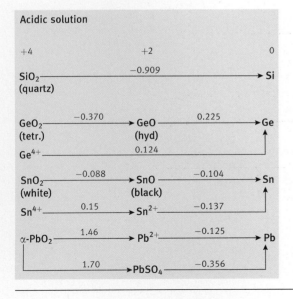

Acidic solution

+4   +2   0

$SiO_2 \xrightarrow{-0.909} Si$
(quartz)

$GeO_2 \xrightarrow{-0.370} GeO \xrightarrow{0.225} Ge$
(tetr.)   (hyd)

$Ge^{4+} \xrightarrow{0.124}$

$SnO_2 \xrightarrow{-0.088} SnO \xrightarrow{-0.104} Sn$
(white)   (black)

$Sn^{4+} \xrightarrow{0.15} Sn^{2+} \xrightarrow{-0.137}$

$\alpha\text{-}PbO_2 \xrightarrow{1.46} Pb^{2+} \xrightarrow{-0.125} Pb$

$\xrightarrow{1.70} PbSO_4 \xrightarrow{-0.356}$

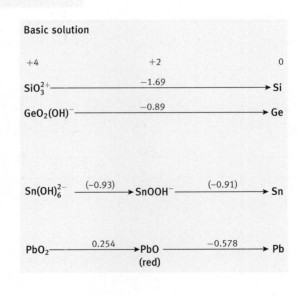

Basic solution

+4   +2   0

$SiO_3^{2+} \xrightarrow{-1.69} Si$

$GeO_2(OH)^- \xrightarrow{-0.89} Ge$

$Sn(OH)_6^{2-} \xrightarrow{(-0.93)} SnOOH^- \xrightarrow{(-0.91)} Sn$

$PbO_2 \xrightarrow{0.254} PbO \xrightarrow{-0.578} Pb$
            (red)

p Block • Group 15

Acidic solution

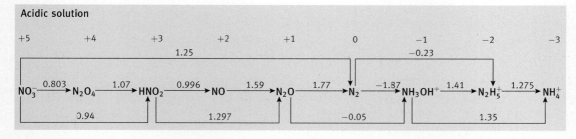

+5   +4   +3   +2   +1   0   −1   −2   −3

NO_3^- —0.803→ N_2O_4 —1.07→ HNO_2 —0.996→ NO —1.59→ N_2O —1.77→ N_2 —−1.87→ NH_3OH^+ —1.41→ $N_2H_5^+$ —1.275→ NH_4^+

1.25   −0.23

0.94   1.297   −0.05   1.35

Basic solution

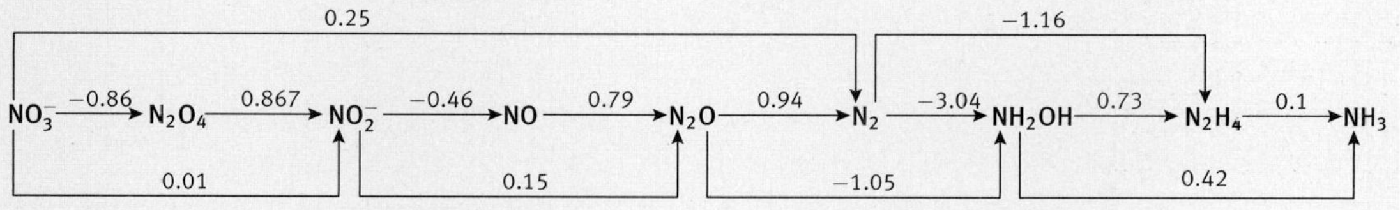

+5   +4   +3   +2   +1   0   −1   −2   −3

NO_3^- —−0.86→ N_2O_4 —0.867→ NO_2^- —−0.46→ NO —0.79→ N_2O —0.94→ N_2 —−3.04→ NH_2OH —0.73→ N_2H_4 —0.1→ NH_3

0.25   −1.16

0.01   0.15   −1.05   0.42

Acidic solution

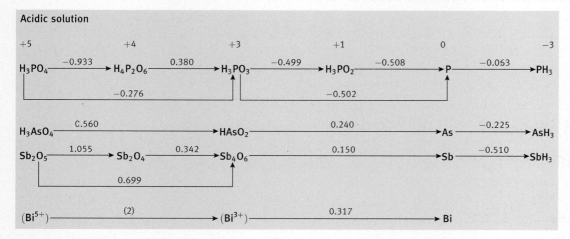

+5   +4   +3   +1   0   −3

H_3PO_4 —−0.933→ $H_4P_2O_6$ —0.380→ H_3PO_3 —−0.499→ H_3PO_2 —−0.508→ P —−0.063→ PH_3

−0.276   −0.502

H_3AsO_4 —0.560→ $HAsO_2$ —0.240→ As —−0.225→ AsH_3

Sb_2O_5 —1.055→ Sb_2O_4 —0.342→ Sb_4O_6 —0.150→ Sb —−0.510→ SbH_3

0.699

(Bi^{5+}) —(2)→ (Bi^{3+}) —0.317→ Bi

Basic solution

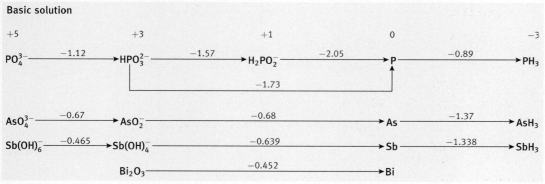

+5   +3   +1   0   −3

PO_4^{3-} —−1.12→ HPO_3^{2-} —−1.57→ $H_2PO_2^-$ —−2.05→ P —−0.89→ PH_3

−1.73

AsO_4^{3-} —−0.67→ AsO_2^- —−0.68→ As —−1.37→ AsH_3

$Sb(OH)_6^-$ —−0.465→ $Sb(OH)_4^-$ —−0.639→ Sb —−1.338→ SbH_3

Bi_2O_3 —−0.452→ Bi

p Block • Group **16**

Acidic solution

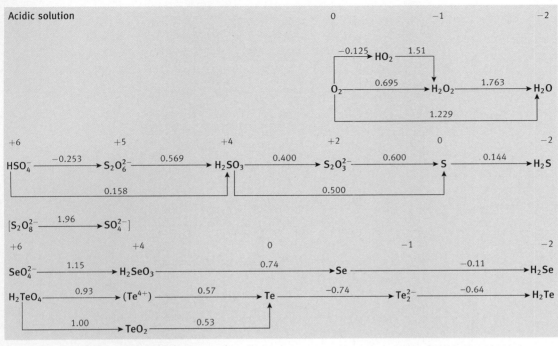

$[S_2O_8^{2-} \xrightarrow{1.96} SO_4^{2-}]$

Basic solution

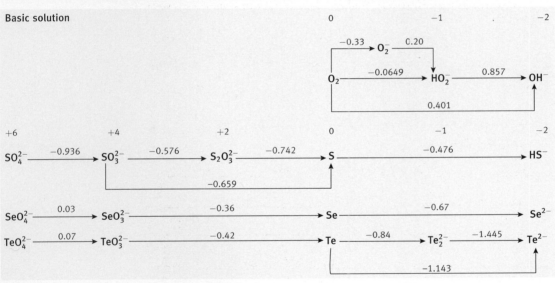

p Block • Group 17

Acidic solution

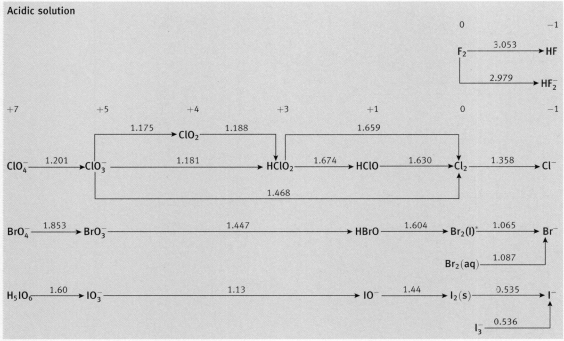

*Bromine is not sufficiently solube in water at room temperature to achieve unit activity. Therefore, the value for a saturated solution in contact with $Br_2(l)$ should be used in all practical calculations.

Basic solution

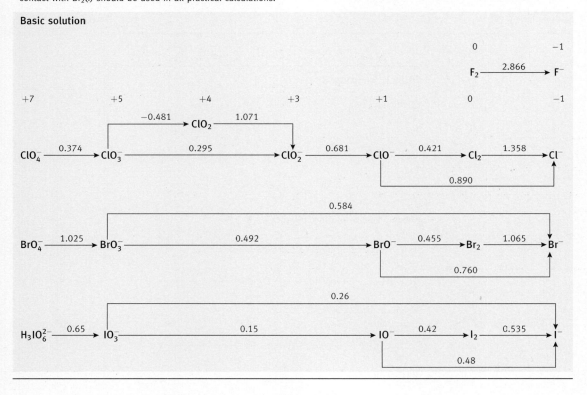

p Block • Group **18**

Acidic solution

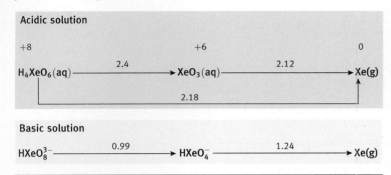

+8   +6   0

H_4XeO_6(aq) ——2.4——→ XeO_3(aq) ——2.12——→ Xe(g)

——————2.18——————

Basic solution

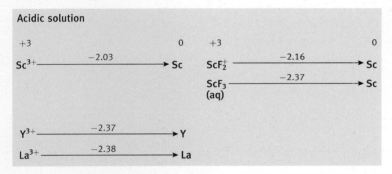

$HXeO_8^{3-}$ ——0.99——→ $HXeO_4^{-}$ ——1.24——→ Xe(g)

d Block • Group **3**

Acidic solution

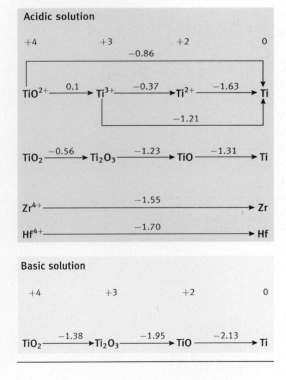

+3   0   +3   0

Sc^{3+} ——−2.03——→ Sc ScF_2^+ ——−2.16——→ Sc

                         ScF_3(aq) ——−2.37——→ Sc

Y^{3+} ——−2.37——→ Y

La^{3+} ——−2.38——→ La

Basic solution

+3   0

$Sc(OH)_3$ ——−2.60——→ Sc

d Block • Group **4**

Acidic solution

+4   +3   +2   0

         ——−0.86——

TiO^{2+} ——0.1——→ Ti^{3+} ——−0.37——→ Ti^{2+} ——−1.63——→ Ti

         ——−1.21——

TiO_2 ——−0.56——→ Ti_2O_3 ——−1.23——→ TiO ——−1.31——→ Ti

Zr^{4+} ——−1.55——→ Zr

Hf^{4+} ——−1.70——→ Hf

Basic solution

+4   +3   +2   0

TiO_2 ——−1.38——→ Ti_2O_3 ——−1.95——→ TiO ——−2.13——→ Ti

d Block • Group **5**

Acidic solution

+5   +4   +3   +2   0

VO_2^+ ——1.000——→ VO^{2+} ——0.337——→ V^{3+} ——−0.255——→ V^{2+} ——−1.13——→ V

         ——0.668——

Weakly acidic solution, pH about 3.0–3.5

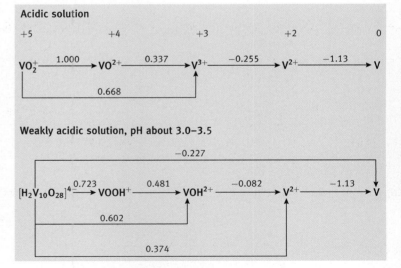

         ——−0.227——

$[H_2V_{10}O_{28}]^{4-}$ ——0.723——→ $VOOH^+$ ——0.481——→ VOH^{2+} ——−0.082——→ V^{2+} ——−1.13——→ V

         ——0.602——

         ——0.374——

Basic solution

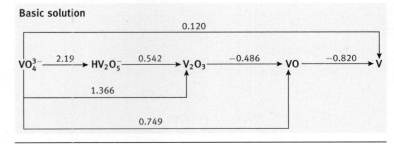

         ——0.120——

VO_4^{3-} ——2.19——→ HV_2O_5 ——0.542——→ V_2O_3 ——−0.486——→ VO ——−0.820——→ V

         ——1.366——

         ——0.749——

d Block • Group 5 (*Continued*)

Acidic solution

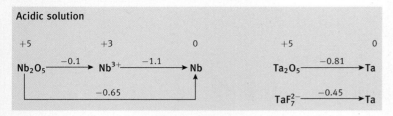

+5   +3   0   +5   0

$Nb_2O_5 \xrightarrow{-0.1} Nb^{3+} \xrightarrow{-1.1} Nb$

$\xrightarrow{-0.65}$

$Ta_2O_5 \xrightarrow{-0.81} Ta$

$TaF_7^{2-} \xrightarrow{-0.45} Ta$

d Block • Group 6

Acidic solution

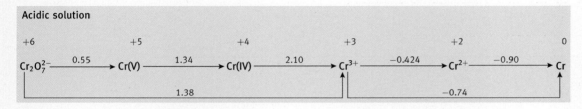

+6   +5   +4   +3   +2   0

$Cr_2O_7^{2-} \xrightarrow{0.55} Cr(V) \xrightarrow{1.34} Cr(IV) \xrightarrow{2.10} Cr^{3+} \xrightarrow{-0.424} Cr^{2+} \xrightarrow{-0.90} Cr$

$\xrightarrow{1.38}$ $\xrightarrow{-0.74}$

Neutral solution

+3   +2

$[Cr(CN)_6]^{3-} \xrightarrow{-1.143} [Cr(CN)_6]^{4-}$

$[Cr(edta)(OH_2)]^- \xrightarrow{-0.99} [Cr(edta)(OH_2)]^{2-}$

Basic solution

+6   +3   0

$CrO_4^{2-} \xrightarrow{-0.11} Cr(OH)_3(s) \xrightarrow{-1.33} Cr$

$\xrightarrow{-0.13} Cr(OH)_4^- \xrightarrow{-1.33}$

Acidic solution

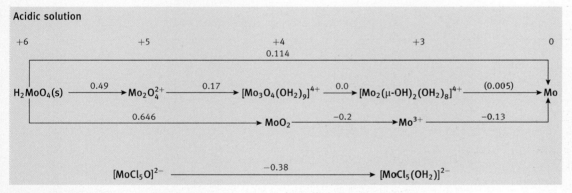

+6   +5   +4   +3   0

                   0.114

$H_2MoO_4(s) \xrightarrow{0.49} Mo_2O_4^{2+} \xrightarrow{0.17} [Mo_3O_4(OH_2)_9]^{4+} \xrightarrow{0.0} [Mo_2(\mu\text{-}OH)_2(OH_2)_8]^{4+} \xrightarrow{(0.005)} Mo$

$\xrightarrow{0.646} MoO_2 \xrightarrow{-0.2} Mo^{3+} \xrightarrow{-0.13}$

$[MoCl_5O]^{2-} \xrightarrow{-0.38} [MoCl_5(OH_2)]^{2-}$

Neutral solution

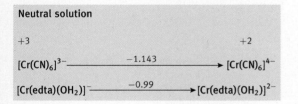

$[Mo(CN)_8]^{3-} \xrightarrow{0.725} [Mo(CN)_8]^{4-}$

Basic solution

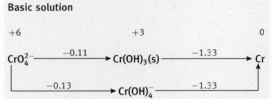

+6   +4   0

$MoO_4^{2-} \xrightarrow{-0.780} MoO_2 \xrightarrow{-0.980} Mo$

$\xrightarrow{-0.913}$

Acidic solution

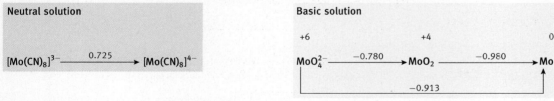

+6   +5   +4   0

$WO_3 \xrightarrow{-0.029} W_2O_5 \xrightarrow{-0.031} WO_2^* \xrightarrow{-0.119} W$

$\xrightarrow{-0.090}$

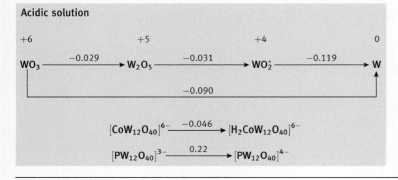

$[CoW_{12}O_{40}]^{6-} \xrightarrow{-0.046} [H_2CoW_{12}O_{40}]^{6-}$

$[PW_{12}O_{40}]^{3-} \xrightarrow{0.22} [PW_{12}O_{40}]^{4-}$

d Block • Group **6** (*Continued*)

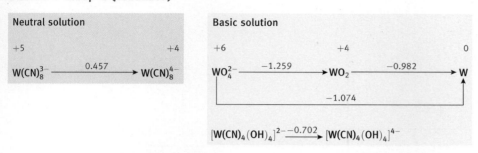

Neutral solution

+5   +4

$W(CN)_8^{3-} \xrightarrow{\;0.457\;} W(CN)_8^{4-}$

Basic solution

+6   +4   0

$WO_4^{2-} \xrightarrow{\;-1.259\;} WO_2 \xrightarrow{\;-0.982\;} W$

-1.074

$[W(CN)_4(OH)_4]^{2-} \xrightarrow{\;-0.702\;} [W(CN)_4(OH)_4]^{4-}$

*Probably $[W_3(\mu_3\text{-O})(\mu\text{-O})_3(OH_2)_9]^{4+}$. See S.P. Gosh and E.S. Gould, *Inorg. Chem.*, 1991, **30**, 3662.

d Block • Group **7**

Acidic solution

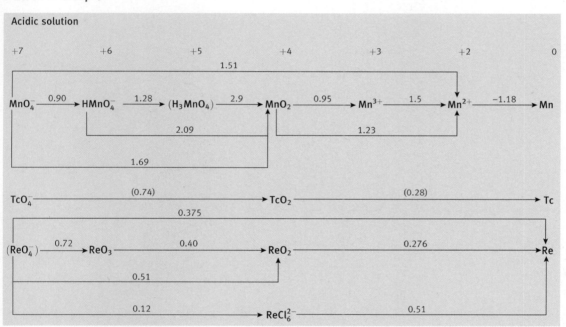

+7   +6   +5   +4   +3   +2   0

1.51

$MnO_4^- \xrightarrow{0.90} HMnO_4^- \xrightarrow{1.28} (H_3MnO_4) \xrightarrow{2.9} MnO_2 \xrightarrow{0.95} Mn^{3+} \xrightarrow{1.5} Mn^{2+} \xrightarrow{-1.18} Mn$

2.09   1.23

1.69

$TcO_4^- \xrightarrow{(0.74)} TcO_2 \xrightarrow{(0.28)} Tc$

0.375

$(ReO_4^-) \xrightarrow{0.72} ReO_3 \xrightarrow{0.40} ReO_2 \xrightarrow{0.276} Re$

0.51

$0.12 \longrightarrow ReCl_6^{2-} \xrightarrow{0.51}$

Basic solution

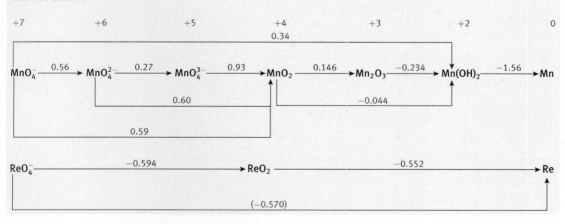

+7   +6   +5   +4   +3   +2   0

0.34

$MnO_4^- \xrightarrow{0.56} MnO_4^{2-} \xrightarrow{0.27} MnO_4^{3-} \xrightarrow{0.93} MnO_2 \xrightarrow{0.146} Mn_2O_3 \xrightarrow{-0.234} Mn(OH)_2 \xrightarrow{-1.56} Mn$

0.60   −0.044

0.59

$ReO_4^- \xrightarrow{-0.594} ReO_2 \xrightarrow{-0.552} Re$

(−0.570)

d Block • Group **8**

Acidic solution

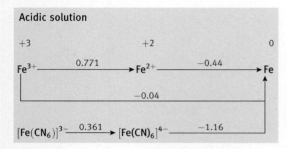

+3   +2   0

$$Fe^{3+} \xrightarrow{0.771} Fe^{2+} \xrightarrow{-0.44} Fe$$

$$\xrightarrow{-0.04}$$

$$[Fe(CN_6)]^{3-} \xrightarrow{0.361} [Fe(CN)_6]^{4-} \xrightarrow{-1.16}$$

Basic solution

+6   +3   +2   0

$$FeO_4^{2-} \xrightarrow{0.81} Fe_2O_3 \xrightarrow{-0.86} Fe(OH)_2 \xrightarrow{-0.89} Fe$$

$$[Fe(ox)_3]^{3-} \xrightarrow{0.005} [Fe(ox)_3]^{4-} \text{ (excess ox}^-\text{)}$$

Acidic solution

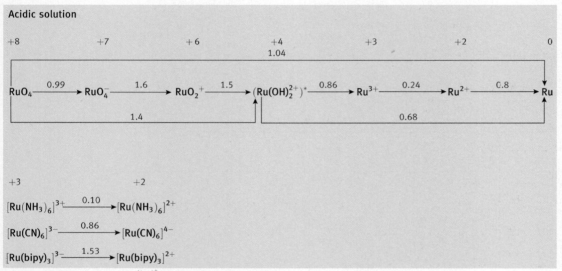

+8   +7   +6   +4   +3   +2   0

1.04

$$RuO_4 \xrightarrow{0.99} RuO_4^- \xrightarrow{1.6} RuO_2^+ \xrightarrow{1.5} (Ru(OH)_2^{2+})^* \xrightarrow{0.86} Ru^{3+} \xrightarrow{0.24} Ru^{2+} \xrightarrow{0.8} Ru$$

1.4   0.68

+3   +2

$$[Ru(NH_3)_6]^{3+} \xrightarrow{0.10} [Ru(NH_3)_6]^{2+}$$

$$[Ru(CN)_6]^{3-} \xrightarrow{0.86} [Ru(CN)_6]^{4-}$$

$$[Ru(bipy)_3]^{3-} \xrightarrow{1.53} [Ru(bipy)_3]^{2+}$$

*Likely to be $H_n[Ru_4O_6(OH_2)_{12}]^{(4+n)+}$. See A. Patel and D.T. Richen, *Inorg. Chem.*, 1991, **30**, 3792.

Acidic solution

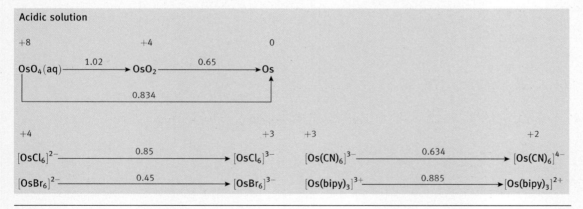

+8   +4   0

$$OsO_4(aq) \xrightarrow{1.02} OsO_2 \xrightarrow{0.65} Os$$

0.834

+4   +3   +3   +2

$$[OsCl_6]^{2-} \xrightarrow{0.85} [OsCl_6]^{3-} \qquad [Os(CN)_6]^{3-} \xrightarrow{0.634} [Os(CN)_6]^{4-}$$

$$[OsBr_6]^{2-} \xrightarrow{0.45} [OsBr_6]^{3-} \qquad [Os(bipy)_3]^{3+} \xrightarrow{0.885} [Os(bipy)_3]^{2+}$$

d Block • Group **9**

Acidic solution

+4   +3   +2   0

$$CoO_2 \xrightarrow{1.4} Co^{3+} \xrightarrow{1.92} Co^{2+} \xrightarrow{-0.282} Co$$

Neutral solution

+3   +2

$$[Co(NH_3)_6]^{3+} \xrightarrow{0.058} [Co(NH_3)_6]^{2+}$$

$$[Co(phen)_3]^{3+} \xrightarrow{0.33} [Co(phen)_3]^{2+}$$

$$[Co(ox)_3]^{3-} \xrightarrow{0.57} [Co(ox)_3]^{4-}$$

Acidic solution

+3   0

$$Rh^{3+} \xrightarrow{0.76} Rh$$

Neutral solution

+3   +2

$$[Rh(CN)_6]^{3-} \xrightarrow{0.9} [Rh(CN)_6]^{4-}$$

Basic solution

+4   +3   +2   0

$$CoO_2 \xrightarrow{0.7} Co(OH)_3 \xrightarrow{0.42} Co(OH)_2 \xrightarrow{-0.733} Co$$

Acidic solution

+4   +3   0

$$IrO_2 \xrightarrow{0.2} (Ir^{3+}) \xrightarrow{1.0} Ir$$

(with 0.8 line from IrO_2 to Ir)

$$[IrCl_6]^{2-} \xrightarrow{0.867} [IrCl_6]^{3-} \xrightarrow{0.86} Ir$$

$$[IrBr_6]^{2-} \xrightarrow{0.805} [IrBr_6]^{3-}$$

$$[IrI_6]^{2-} \xrightarrow{0.49} [IrI_6]^{3-}$$

d Block • Group **10**

Acidic solution

+4   +3   +2   0

$$NiO_2 \xrightarrow{1.5} Ni^{2+} \xrightarrow{-0.257} Ni$$

Basic solution

$$NiO_2^+ \xrightarrow{0.7} NiOOH \xrightarrow{0.52} Ni(OH)_2 \xrightarrow{-0.72} Ni$$

Neutral solution

$$[Ni(NH_3)_6]^{2+} \xrightarrow{-0.49} Ni$$

Acidic solution

+4   +2   0

$$PdO_2 \xrightarrow{1.194} Pd^{2+} \xrightarrow{0.915} Pd$$

$$[PdCl_6]^{2-} \xrightarrow{1.47} [PdCl_4]^{2-} \xrightarrow{0.60} Pd$$

$$[PdBr_4]^{2-} \xrightarrow{0.49} Pd$$

Basic solution

+4   +2   0

$$PdO_2 \xrightarrow{1.47} PdO \xrightarrow{0.897} Pd$$

Acidic solution

$$PtO_2(s) \xrightarrow{1.01} PtO(s) \xrightarrow{0.98} Pt$$

$$[PtCl_6]^{2-} \xrightarrow{0.726} [PtCl_4]^{2-} \xrightarrow{0.758} Pt$$

$$[PtBr_6]^{2-} \xrightarrow{0.631} [PtBr_4]^{2-} \xrightarrow{0.698} Pt$$

$$[PtI_6]^{2-} \xrightarrow{0.329} [PtI_4]^{2-} \xrightarrow{0.40} Pt$$

d Block • Group 11

Acidic solution

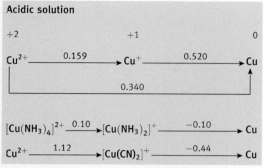

+2   +1   0

Cu^{2+} —0.159→ Cu^+ —0.520→ Cu

—0.340—

$[Cu(NH_3)_4]^{2+}$ —0.10→ $[Cu(NH_3)_2]^+$ —−0.10→ Cu

Cu^{2+} —1.12→ $[Cu(CN)_2]^+$ —−0.44→ Cu

Basic solution

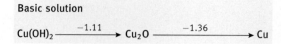

$Cu(OH)_2$ —−1.11→ Cu_2O —−1.36→ Cu

Acidic solution

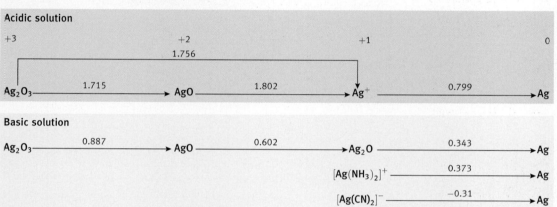

+3   +2   +1   0

1.756

Ag_2O_3 —1.715→ AgO —1.802→ Ag^+ —0.799→ Ag

Basic solution

Ag_2O_3 —0.887→ AgO —0.602→ Ag_2O —0.343→ Ag

$[Ag(NH_3)_2]^+$ —0.373→ Ag

$[Ag(CN)_2]^-$ —−0.31→ Ag

Acidic solution

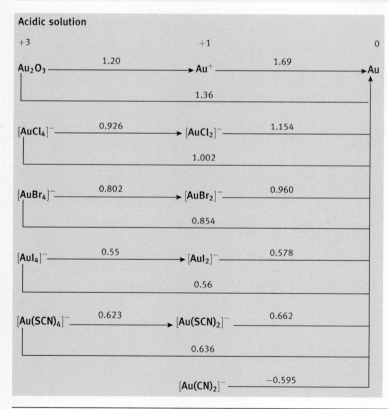

+3   +1   0

Au_2O_3 —1.20→ Au^+ —1.69→ Au

1.36

$[AuCl_4]^-$ —0.926→ $[AuCl_2]^-$ —1.154→

1.002

$[AuBr_4]^-$ —0.802→ $[AuBr_2]^-$ —0.960→

0.854

$[AuI_4]^-$ —0.55→ $[AuI_2]^-$ —0.578→

0.56

$[Au(SCN)_4]^-$ —0.623→ $[Au(SCN)_2]^-$ —0.662→

0.636

$[Au(CN)_2]^-$ —−0.595→

d Block • Group 12

Acidic solution

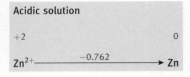

+2		0
Zn^{2+}	$\xrightarrow{-0.762}$	Zn

Basic solution

$[Zn(OH)_4]^{2-} \xrightarrow{-1.199}$ Zn

$Zn(OH)_2 \xrightarrow{-1.246}$ Zn

Acidic solution

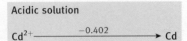

$Cd^{2+} \xrightarrow{-0.402}$ Cd

Basic solution

$Cd(OH)_2(s) \xrightarrow{-0.824}$ Cd

Acidic solution

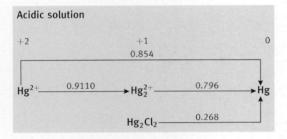

+2		+1		0
		0.854		
Hg^{2+}	$\xrightarrow{0.9110}$	Hg_2^{2+}	$\xrightarrow{0.796}$	Hg
		Hg_2Cl_2	$\xrightarrow{0.268}$	

Basic solution

$HgO \xrightarrow{0.0977}$ Hg

f Block • Lanthanoids

Acidic solution

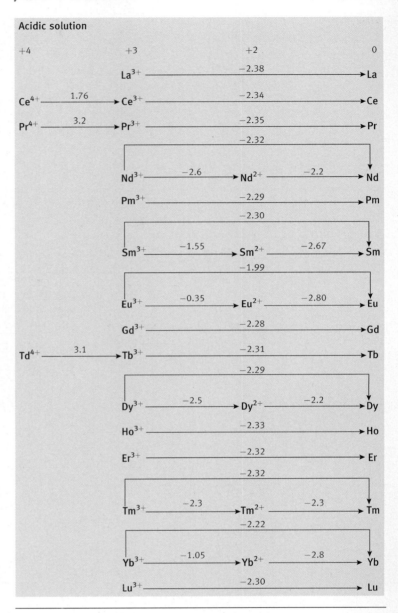

+4		+3		+2		0
		La^{3+}	$\xrightarrow{-2.38}$			La
Ce^{4+}	$\xrightarrow{1.76}$	Ce^{3+}	$\xrightarrow{-2.34}$			Ce
Pr^{4+}	$\xrightarrow{3.2}$	Pr^{3+}	$\xrightarrow{-2.35}$			Pr
				$\xrightarrow{-2.32}$		
		Nd^{3+}	$\xrightarrow{-2.6}$	Nd^{2+}	$\xrightarrow{-2.2}$	Nd
		Pm^{3+}	$\xrightarrow{-2.29}$			Pm
				$\xrightarrow{-2.30}$		
		Sm^{3+}	$\xrightarrow{-1.55}$	Sm^{2+}	$\xrightarrow{-2.67}$	Sm
				$\xrightarrow{-1.99}$		
		Eu^{3+}	$\xrightarrow{-0.35}$	Eu^{2+}	$\xrightarrow{-2.80}$	Eu
		Gd^{3+}	$\xrightarrow{-2.28}$			Gd
Td^{4+}	$\xrightarrow{3.1}$	Tb^{3+}	$\xrightarrow{-2.31}$			Tb
				$\xrightarrow{-2.29}$		
		Dy^{3+}	$\xrightarrow{-2.5}$	Dy^{2+}	$\xrightarrow{-2.2}$	Dy
		Ho^{3+}	$\xrightarrow{-2.33}$			Ho
		Er^{3+}	$\xrightarrow{-2.32}$			Er
				$\xrightarrow{-2.32}$		
		Tm^{3+}	$\xrightarrow{-2.3}$	Tm^{2+}	$\xrightarrow{-2.3}$	Tm
				$\xrightarrow{-2.22}$		
		Yb^{3+}	$\xrightarrow{-1.05}$	Yb^{2+}	$\xrightarrow{-2.8}$	Yb
		Lu^{3+}	$\xrightarrow{-2.30}$			Lu

f Block • Actinoids

Acidic solution

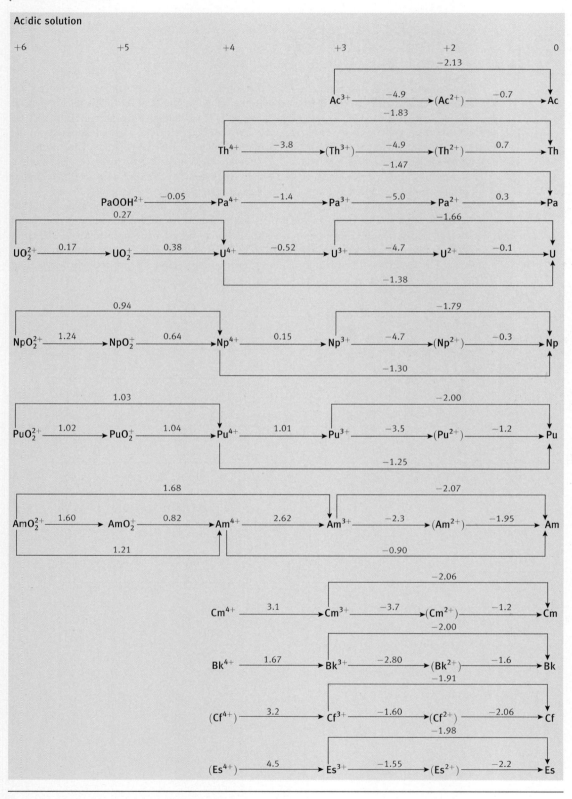

Resource section 4: Character tables

The character tables that follow are for the most common point groups encountered in inorganic chemistry. Each one is labelled with the symbol adopted in the Schoenflies system of nomenclature (such as C_{3v}). Point groups that qualify as crystallographic point groups (because they are also applicable to unit cells) are also labelled with the symbol adopted in the International System (or the Hermann–Mauguin system, such as $2/m$). In the latter system, a number n represents an n-fold axis and a letter m represents a mirror plane. A diagonal line indicates that a mirror plane lies perpendicular to the symmetry axis and a bar over the number indicates that the rotation is combined with an inversion.

The symmetry species of the p and d orbitals are shown on the right of the tables. Thus, in C_{2v}, a p_x orbital (which is proportional to x) has B_1 symmetry. The functions x, y, and z also show the transformation properties of translations and of the electric dipole moment. The set of functions that span a degenerate representation (such as x and y, which jointly span E in C_{3v}) are enclosed in parentheses. The transformation properties of rotation are shown by the letters R on the right of the tables.

The groups C_1, C_s, C_i

C_1 (1)	E	$h=1$
A	1	

$C_s = C_h$ (m)	E	σ_h		$h=2$
A$'$	1	1	x, y, R_z	x^2, y^2, z^2, xy
A$''$	1	-1	z, R_x, R_y	yz, zx

$C_i = S_2$ (1)	E	i		$h=2$
A$_g$	1	1	R_x, R_y, R_z	x^2, y^2, z^2, xy, zx, yz
A$_u$	1	-1	x, y, z	

The groups C_n

C_2 (2)	E	C_2		$h=2$
A	1	1	z, R_z	x^2, y^2, z^2, xy
B	1	-1	x, y, R_x, R_y	yz, zx

C_3 (3)	E	C_3	C_3^2	$\varepsilon = \exp(2\pi i/3)$	$h=3$
A	1	1	1	z, R_z	x^2+y^2, z^2
E	$\left\{\begin{matrix} 1 & \varepsilon & \varepsilon^* \\ 1 & \varepsilon^* & \varepsilon \end{matrix}\right\}$			$(x, y)(R_x, R_y)$	$(x^2-y^2, xy)\,(yz, zx)$

C_4 (4)	E	C_4	C_2	C_4^3		$h=4$
A	1	1	1	1	z, R_z	x^2+y^2, z^2
B	1	-1	1	-1		x^2-y^2, xy
E	$\left\{\begin{matrix} 1 & i & -1 & -i \\ 1 & -i & 1 & i \end{matrix}\right\}$				$(x, y)(R_x, R_y)$	(yz, zx)

The groups C_{nv}

C_{2v} (2mm)	E	C_2	σ_v (xz)	σ_v'(yz)	$h=4$	
A_1	1	1	1	1	z	x^2, y^2, z^2
A_2	1	1	−1	−1	R_z	xy
B_1	1	−1	1	−1	x, R_y	zx
B_2	1	−1	−1	1	y, R_x	yz

C_{3v} (3m)	E	$2C_3$	$3\sigma_v$	$h=6$		
A_1	1	1	1	z		x^2+y^2, z^2
A_2	1	1	−1	R_z		
E	2	−1	0	$(x, y) (R_x, R_y)$		$(x^2-y^2, xy) (zx, yz)$

C_{4v} (4mm)	E	$2C_4$	C_2	$2\sigma_v$	$2\sigma_d$	$h=8$	
A_1	1	1	1	1	1	z	x^2+y^2, z^2
A_2	1	1	1	−1	−1	R_z	
B_1	1	−1	1	1	−1		x^2-y^2
B_2	1	−1	1	−1	1		xy
E	2	0	−2	0	0	$(x, y) (R_x, R_y)$	(zx, yz)

C_{5v}	E	$2C_5$	$2C_5^2$	$5\sigma_v$	$h=10, \alpha=72°$	
A_1	1	1	1	1	z	x^2+y^2, z^2
A_2	1	1	1	−1	R_z	
E_1	2	$2\cos\alpha$	$2\cos 2\alpha$	0	$(x, y)(R_x, R_y)$	(zx, yz)
E_2	2	$2\cos 2\alpha$	$2\cos\alpha$	0		(x^2-y^2, xy)

C_{6v} (6mm)	E	$2C_6$	$2C_3$	C_2	$3\sigma_v$	$3\sigma_d$	$h=12$	
A_1	1	1	1	1	1	1	z	x^2+y^2, z^2
A_2	1	1	1	1	−1	−1	R_z	
B_1	1	−1	1	−1	1	−1		
B_2	1	−1	1	−1	−1	1		
E_1	2	1	−1	−2	0	0	$(x, y) (R_x, R_y)$	(zx, yz)
E_2	2	−1	−1	2	0	0		(x^2-y^2, xy)

$C_{\infty v}$	E	C_2	$2C_\phi$	$\infty\sigma_v$	$h=\infty$	
$A_1 (\Sigma^+)$	1	1	1	1	z	x^2+y^2, z^2
$A_2 (\Sigma^-)$	1	1	1	−1	R_z	
$E_1 (\Pi)$	2	−2	$2\cos\phi$	0	$(x, y) (R_x, R_y)$	(zx, yz)
$E_2 (\Delta)$	2	2	$2\cos 2\phi$	0		(xy, x^2-y^2)
⋮	⋮	⋮	⋮	⋮		

The groups D_n

D_2 (222)	E	$C_2(z)$	$C_2(y)$	$C_2(x)$	$h=4$	
A	1	1	1	1		x^2, y^2, z^2
B_1	1	1	−1	−1	z, R_z	xy
B_2	1	−1	1	−1	y, R_y	zx
B_3	1	−1	−1	1	x, R_x	yz

D_3 (32)	E	$2C_3$	$3C_2$	$h=6$		
A_1	1	1	1			x^2+y^2, z^2
A_2	1	1	−1	z, R_z		
E	2	−1	0	$(x, y) (R_x, R_y)$		$(x^2-y^2, xy) (zx, yz)$

The groups D_{nh}

D_{2h} (mmm)	E	$C_2(z)$	$C_2(y)$	$C_2(x)$	i	$\sigma(xy)$	$\sigma(xz)$	$\sigma(yz)$	$h=8$	
A_g	1	1	1	1	1	1	1	1		x^2, y^2, z^2
B_{1g}	1	1	−1	−1	1	1	−1	−1	R_z	xy
B_{2g}	1	−1	1	−1	1	−1	1	−1	R_y	zx
B_{3g}	1	−1	−1	1	1	−1	−1	1	R_x	yz
A_u	1	1	1	1	−1	−1	−1	−1		
B_{1u}	1	1	−1	−1	−1	−1	1	1	z	
B_{2u}	1	−1	1	−1	−1	1	−1	1	y	
B_{3u}	1	−1	−1	1	−1	1	1	−1	x	

D_{3h} $(6m2)$	E	$2C_3$	$3C_2$	σ_h	$2S_3$	$3\sigma_v$	$h=12$	
A_1'	1	1	1	1	1	1		x^2+y^2, z^2
A_2'	1	1	−1	1	1	−1	R_z	
E'	2	−1	0	2	−1	0	(x, y)	(x^2-y^2, xy)
A_1''	1	1	1	−1	−1	−1		
A_2''	1	1	−1	−1	−1	1	z	
E''	2	−1	0	−2	1	0	(R_x, R_y)	(zx, yz)

D_{4h} $(4/mmm)$	E	$2C_4$	C_2	$2C_2'$	$2C_2''$	i	$2S_4$	σ_h	$2\sigma_v$	$2\sigma_d$	$h=16$	
A_{1g}	1	1	1	1	1	1	1	1	1	1		x^2+y^2, z^2
A_{2g}	1	1	1	−1	−1	1	1	1	−1	−1	R_z	
B_{1g}	1	−1	1	1	−1	1	−1	1	1	−1		x^2-y^2
B_{2g}	1	−1	1	−1	1	1	−1	1	−1	1		xy
E_g	2	0	−2	0	0	2	0	−2	0	0	(R_x, R_y)	(zx, yz)
A_{1u}	1	1	1	1	1	−1	−1	−1	−1	−1		
A_{2u}	1	1	1	−1	−1	−1	−1	−1	1	1	z	
B_{1u}	1	−1	1	1	−1	−1	1	−1	−1	1		
B_{2u}	1	−1	1	−1	1	−1	1	−1	1	−1		
E_u	2	0	−2	0	0	−2	0	2	0	0	(x, y)	

D_{5h}	E	$2C_5$	$2C_5^2$	$5C_2$	σ_h	$2S_5$	$2S_5^2$	$5\sigma_v$	$h=20, \alpha=72°$	
A_1'	1	1	1	1	1	1	1	1		x^2+y^2, z^2
A_2'	1	1	1	−1	1	1	1	−1	R_z	
E_1'	2	$2\cos\alpha$	$2\cos 2\alpha$	0	2	$2\cos\alpha$	$2\cos 2\alpha$	0	(x, y)	
E_2'	2	$2\cos 2\alpha$	$2\cos\alpha$	0	2	$2\cos 2\alpha$	$2\cos\alpha$	0		(x^2-y^2, xy)
A_1''	1	1	1	1	−1	−1	−1	−1		
A_2''	1	1	1	−1	−1	−1	−1	1	z	
E_1''	2	$2\cos\alpha$	$2\cos 2\alpha$	0	−2	$-2\cos\alpha$	$-2\cos 2\alpha$	0	(R_x, R_y)	(zx, yz)
E_2''	2	$2\cos 2\alpha$	$2\cos\alpha$	0	−2	$-2\cos 2\alpha$	$-2\cos\alpha$	0		

The groups $D_{n\text{h}}$ (continued)

$D_{6\text{h}}$ (6/mmm)	E	$2C_6$	$2C_3$	C_2	$3C_2'$	$3C_2''$	i	$2S_3$	$2S_6$	σ_h	$3\sigma_\text{d}$	$3\sigma_\text{v}$	$h = 24$	
A_{1g}	1	1	1	1	1	1	1	1	1	1	1	1		$x^2 + y^2,\ z^2$
A_{2g}	1	1	1	1	−1	−1	1	1	1	1	−1	−1	R_z	
B_{1g}	1	−1	1	−1	1	−1	1	−1	1	−1	1	−1		
B_{2g}	1	−1	1	−1	−1	1	1	−1	1	−1	−1	1		
E_{1g}	2	1	−1	−2	0	0	2	1	−1	−2	0	0	$(R_x,\ R_y)$	$(zx,\ yz)$
E_{2g}	2	−1	−1	2	0	0	2	−1	−1	2	0	0		$(x^2 - y^2,\ xy)$
A_{1u}	1	1	1	1	1	1	−1	−1	−1	−1	−1	−1		
A_{2u}	1	1	1	1	−1	−1	−1	−1	−1	−1	1	1	z	
B_{1u}	1	−1	1	−1	1	−1	−1	1	−1	1	−1	1		
B_{2u}	1	−1	1	−1	−1	1	−1	1	−1	1	1	−1		
E_{1u}	2	1	−1	−2	0	0	−2	−1	1	2	0	0	$(x,\ y)$	
E_{2u}	2	−1	−1	2	0	0	−2	1	1	−2	0	0		

$D_{\infty\text{h}}$	E	$\infty C_2'$	$2C_\phi$	i	$\infty\sigma_\text{v}$	$2S_\phi$	$h = \infty$	
$A_{1g}(\Sigma_g^+)$	1	1	1	1	1	1	$z^2,\ x^2 + y^2$	
$A_{1u}(\Sigma_u^+)$	1	−1	1	−1	1	−1	z	
$A_{2g}(\Sigma_g^-)$	1	−1	1	1	−1	1	R_z	
$A_{2u}(\Sigma_u^-)$	1	1	1	−1	−1	−1		
$E_{1g}(\Pi_g)$	2	0	$2\cos\phi$	2	0	$-2\cos\phi$	$(R_x,\ R_y)$	$(zx,\ yz)$
$E_{1u}(\Pi_u)$	2	0	$2\cos\phi$	−2	0	$2\cos\phi$	$(x,\ y)$	
$E_{2g}(\Delta_g)$	2	0	$2\cos 2\phi$	2	0	$2\cos 2\phi$		$(xy,\ x^2 - y^2)$
$E_{2u}(\Delta_u)$	2	0	$2\cos 2\phi$	−2	0	$-2\cos 2\phi$		
⋮	⋮	⋮	⋮	⋮	⋮	⋮		

The groups $D_{n\text{d}}$

$D_{2\text{d}} = V_\text{d}$ (42m)	E	$2S_4$	C_2	$2C_2'$	$2\sigma_\text{d}$	$h = 8$	
A_1	1	1	1	1	1	$x^2 + y^2,\ z^2$	
A_2	1	1	1	−1	−1	R_z	
B_1	1	−1	1	1	1	$x^2 - y^2$	
B_2	1	−1	1	−1	1	z	xy
E	2	0	−2	0	0	$(x,\ y)$	$(zx,\ yz)$
						$(R_x,\ R_y)$	

$D_{3\text{d}}$ (3m)	E	$2C_3$	$3C_2$	i	$2S_6$	$3\sigma_\text{d}$	$h = 12$	
A_{1g}	1	1	1	1	1	1	$x^2 + y^2,\ z^2$	
A_{2g}	1	1	−1	1	1	−1	R_z	
E_g	2	−1	0	2	−1	0	$(R_x,\ R_y)$	$(x^2 - y^2,\ xy)$
								$(zx,\ yz)$
A_{1u}	1	1	1	−1	−1	−1		
A_{2u}	1	1	−1	−1	−1	1	z	
E_u	2	−1	0	−2	1	0	$(x,\ y)$	

The groups D_{nd} (continued)

D_{4d}	E	$2S_8$	$2C_4$	$2S_8^3$	C_2	$4C_2'$	$4\sigma_d$	$h = 16$	
A_1	1	1	1	1	1	1	1		$x^2 + y^2, z^2$
A_2	1	1	1	1	1	−1	−1	R_z	
B_1	1	−1	1	−1	1	1	−1		
B_2	1	−1	1	−1	1	−1	1	z	
E_1	2	$\sqrt{2}$	0	$-\sqrt{2}$	−2	0	0	(x, y)	
E_2	2	0	−2	0	2	0	0		$(x^2 - y^2, xy)$
E_3	2	$-\sqrt{2}$	0	$\sqrt{2}$	−2	0	0	(R_x, R_y) (zx, yz)	

The cubic groups

T_d $(43m)$	E	$8C_3$	$3C_2$	$6S_4$	$6\sigma_d$	$h = 24$	
A_1	1	1	1	1	1	$x^2 + y^2 + z^2$	
A_2	1	1	1	−1	−1		
E	2	−1	2	0	0	$(2z^2 - x^2 - y^2, x^2 - y^2)$	
T_1	3	0	−1	1	−1	(R_x, R_y, R_z)	
T_2	3	0	−1	−1	1	(x, y, z)	(xy, yz, zx)

O_h $(m3m)$	E	$8C_3$	$6C_2$	$6C_4$	$3C_2(= C_4^2)$	i	$6S_4$	$8S_6$	$3\sigma_h$	$6\sigma_d$	$h = 48$	
A_{1g}	1	1	1	1	1	1	1	1	1	1	$x^2 + y^2 + z^2$	
A_{2g}	1	1	−1	−1	1	1	−1	1	1	−1		
E_g	2	−1	0	0	2	2	0	−1	2	0	$(2z^2 - x^2 - y^2, x^2 - y^2)$	
T_{1g}	3	0	−1	1	−1	3	1	0	−1	−1	(R_x, R_y, R_z)	
T_{2g}	3	0	1	−1	−1	3	−1	0	−1	1		(xy, yz, zx)
A_{1u}	1	1	1	1	1	−1	−1	−1	−1	−1		
A_{2u}	1	1	−1	−1	1	−1	1	−1	−1	1		
E_u	2	−1	0	0	2	−2	0	1	−2	0		
T_{1u}	3	0	−1	1	−1	−3	−1	0	1	1	(x, y, z)	
T_{2u}	3	0	1	−1	−1	−3	1	0	1	−1		

The icosahedral group

I	E	$12C_5$	$12C_5^2$	$20C_3$	$15C_2$	$h = 60$	
A_1	1	1	1	1	1	$x^2 + y^2 + z^2$	
T_1	3	$\frac{1}{2}(1 + \sqrt{5})$	$\frac{1}{2}(1 - \sqrt{5})$	0	−1	(x, y, z)	
						(R_x, R_y, R_z)	
T_2	3	$\frac{1}{2}(1 - \sqrt{5})$	$\frac{1}{2}(1 + \sqrt{5})$	0	−1		
G	4	−1	−1	1	0		
H	5	0	0	−1	1	$(2z^2 - x^2 - y^2, x^2 - y^2, xy, yz, zx)$	

Further information: P.W. Atkins, M.S. Child, and C.S.G. Phillips, *Tables for group theory*. Oxford University Press (1970).

Resource section 5:
Symmetry-adapted orbitals

Table RS5.1 gives the symmetry classes of the s, p, and d orbitals of the central atom of an AB_n molecule of the specified point group. In most cases, the z-axis is the principal axis of the molecule; in C_{2v} the x-axis lies perpendicular to the molecular plane.

The orbital diagrams that follow show the linear combinations of atomic orbitals on the peripheral atoms of AB_n molecules of the specified point groups. Where a view from above is shown, the dot representing the central atom is either in the plane of the paper (for the D groups) or above the plane (for the corresponding C groups). Different phases of the atomic orbitals ($+$ or $-$ amplitudes) are shown by different colours. Where there is a large difference in the magnitudes of the orbital coefficients in a particular combination, the atomic orbitals have been drawn large or small to represent their relative contributions to the linear combination. In the case of degenerate linear combinations (those labelled E or T), any linearly independent combination of the degenerate pair is also of suitable symmetry. In practice, these different linear combinations look like the ones shown here, but their nodes are rotated by an arbitrary axis around the z-axis.

Molecular orbitals are formed by combining an orbital of the central atom (as in Table RS5.1) with a linear combination of the same symmetry.

Table RS5.1 Symmetry species of orbitals on the central atom

	$D_{\infty h}$	C_{2v}	D_{3h}	C_{3v}	D_{4h}	C_{4v}	D_{5h}	C_{5v}	D_{6h}	C_{6v}	T_d	O_h
s	Σ	A_1	A_1'	A_1	A_{1g}	A_1	A_1'	A_1	A_{1g}	A_1	A_1	A_{1g}
p_x	Π	B_1	E'	E	E_u	E	E_1'	E_1	E_{1u}	E_1	T_2	T_{1u}
p_y	Π	B_2	E'	E	E_u	E	E_1'	E_1	E_{1u}	E_1	T_2	T_{1u}
p_z	Σ	A_1	A_2''	A_1	A_{2u}	A_1	A_2''	A_1	A_{2u}	A_1	T_2	T_{1u}
d_{z^2}	Σ	A_1	A_1'	A_1	A_{1g}	A_1	A_1'	A_1	A_{1g}	A_1	E	E_g
$d_{x^2-y^2}$	Δ	A_1	E'	E	B_{1g}	B_1	E_2'	E_2	E_{2g}	E_2	E	E_g
d_{xy}	Δ	A_2	E'	E	B_{2g}	B_2	E_2'	E_2	E_{2g}	E_2	T_2	T_{2g}
d_{yz}	Π	B_2	E''	E	E_g	E	E_1''	E_1	E_{1g}	E_1	T_2	T_{2g}
d_{zx}	Π	B_1	E''	E	E_g	E	E_1''	E_1	E_{1g}	E_1	T_2	T_{2g}

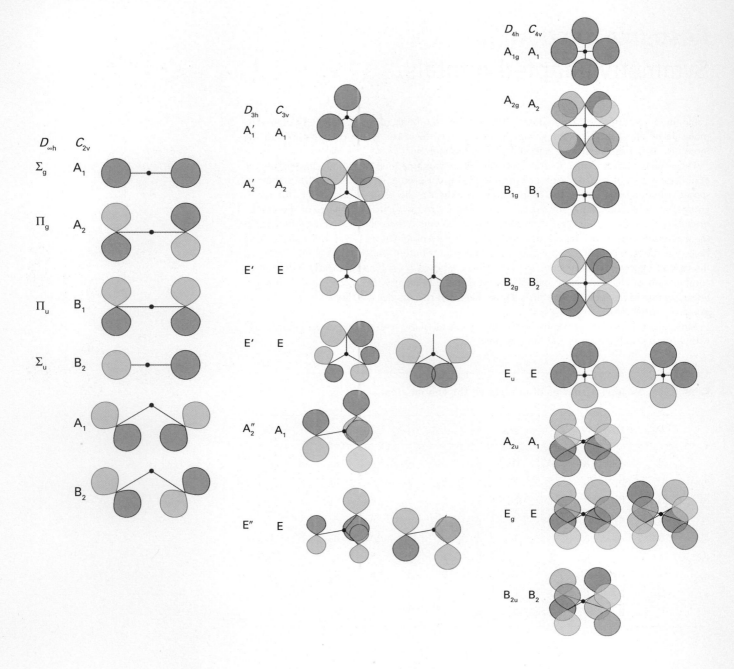

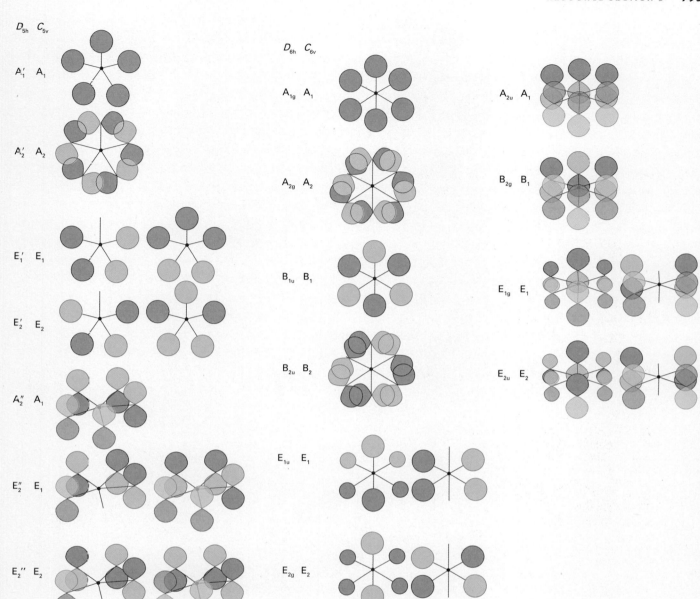

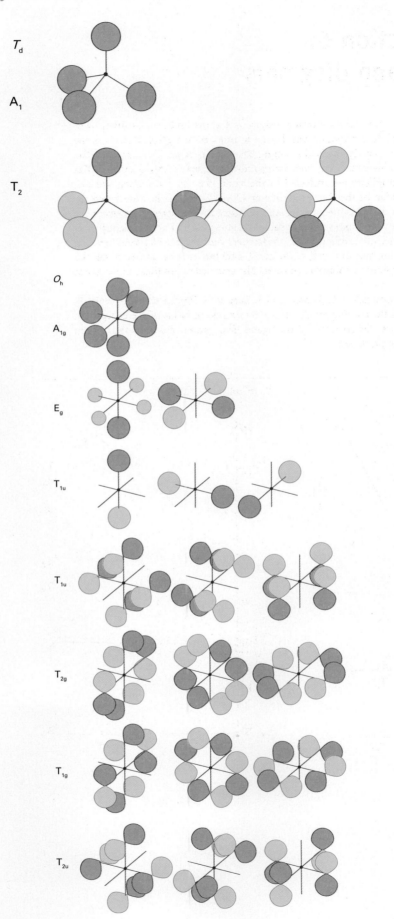

Resource section 6:
Tanabe–Sugano diagrams

This section collects together the Tanabe–Sugano diagrams for octahedral complexes with electron configurations d^2 to d^8. The diagrams, which were introduced in Section 19.4, show the dependence of the term energies on ligand-field strength. The term energies E are expressed as the ratio E/B, where B is a Racah parameter, and the ligand-field splitting Δ_O is expressed as Δ_O/B. Terms of different multiplicity are included in the same diagram by making specific, plausible choices about the value of the Racah parameter C, and these choices are given for each diagram. The term energy is always measured from the lowest energy term, and so there are discontinuities of slope where a low-spin term displaces a high-spin term at sufficiently high ligand-field strengths for d^4 to d^8 configurations. Moreover, the noncrossing rule requires terms of the same symmetry to mix rather than to cross, and this mixing accounts for the curved rather than the straight lines in a number of cases. The term labels are those of the point group O_h.

The diagrams were first introduced by Y. Tanabe and S. Sugano, *J. Phys. Soc. Japan*, 1954, **9**, 753. They may be used to find the parameters Δ_O and B by fitting the ratios of the energies of observed transitions to the lines. Alternatively, if the ligand-field parameters are known, then the ligand-field spectra may be predicted.

1. d^2 with $C = 4.42B$

2. d^3 with $C = 4.5B$

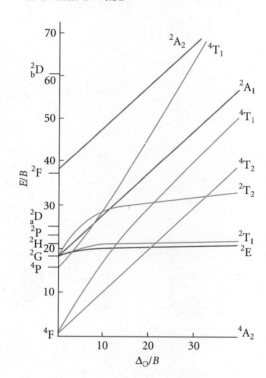

3. d^4 with $C = 4.61B$

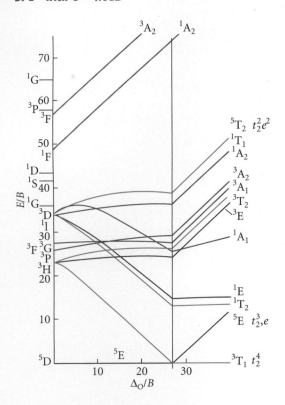

4. d^5 with $C = 4.477B$

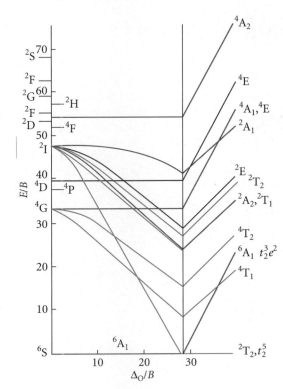

5. d^6 with $C = 4.8B$

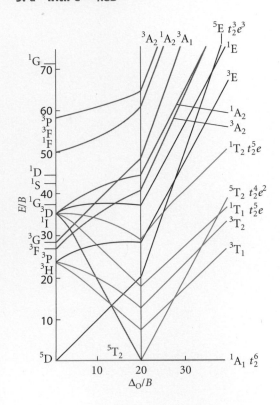

6. d^7 with $C = 4.633B$

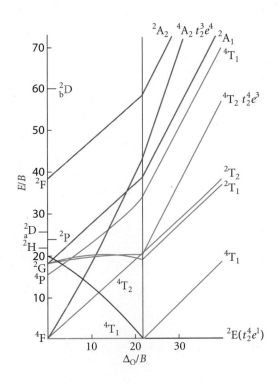

7. d^8 with $C = 4.709B$

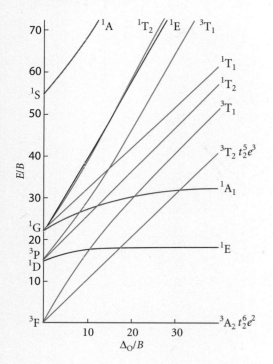

Answers to self-tests and exercises

Chapter 1

Self-tests

1.1 $^{80}_{35}Br + ^{1}_{0}n \rightarrow ^{81}_{35}Br + \gamma$

1.2 $3d$.

1.3 $3p$.

1.4 The added p electron is in a different (p) orbital, so it is less shielded.

1.5 Ni $[Ar]3d^84s^2$, Ni^{2+} $[Ar]3d^8$.

1.6 Period 4, Group 2, s block.

1.7 Going down a group the atomic radius increases and the first ionization energy generally decreases.

1.8 Group 16. The first four electrons are removed with gradually increasing values. Removing the fifth electron requires a large increase in energy, indicating breaking into a complete subshell.

1.9 Adding another electron to C would result in the stable half filled p subshell.

1.10 Cs^+.

Exercises

1.1 (a) $^{14}_{7}N + ^{4}_{2}He \rightarrow ^{17}_{8}O + ^{1}_{1}p + \gamma$

(b) $^{12}_{6}C + ^{1}_{1}p \rightarrow ^{13}_{7}N + \gamma$

(c) $^{14}_{7}N + ^{1}_{0}n \rightarrow ^{12}_{6}C + ^{3}_{1}H$

1.2 $^{22}_{10}Ne + ^{4}_{2}He \rightarrow ^{25}_{12}Mg + ^{1}_{0}n$

1.3 Check inside front cover.

1.4 0.25.

1.5 $-13.2\,V$.

1.6 1524 nm, $1.524 \times 10^4\,cm^{-1}$.

1.7 0 up to $n-\chi$.

1.8 n^2.

1.9

n	l	m_l	Orbital designation	Number of orbitals
2	1	$+1, 0, -1$	$2p$	3
3	2	$+2, +1, \ldots, -2$	$3d$	5
4	0	0	$4s$	1
4	3	$+3, +2, \ldots, -3$	$4f$	7

1.10 See Figs 1.11 through 1.16.

1.11 See Table 1.6 and discussion.

1.12 Table 1.6 shows $Sr > Ba < Ra$. Ra is anomalous because of higher Z_{eff} due to lanthanide contraction.

1.13 Anomalously high value for Cr is associated with the stability of a half filled d shell.

1.14 (a) $[He]2s^22p^2$
(b) $[He]2s^22p^5$
(c) $[Ar]4s^2$
(d) $[Ar]3d^{10}$
(e) $[Xe]4f^{14}5d^{10}6s^23p^3$

(f) $[Xe]4f^{14}5d^{10}6s^2$

1.15 (a) $[Ar]3d^{10}4s^2$
(b) $[Ar]3d^2$
(c) $[Ar]3d^5$
(d) $[Ar]3d^4$
(e) $[Ar]3d^6$
(f) $[Ar]$
(g) $[Ar]3d^{10}4s^1$
(h) $[Xe]4f^7$

1.16 (a) $[Xe]4f^{14}5d^46s^2$
(b) $[Kr]4d^6$
(c) $[Xe]4f^6$
(d) $[Xe]4f^7$
(e) $[Ar]$
(f) $[Kr]4d^2$

1.17 (a) Se
(b) Sr
(c) V
(d) Tc
(e) In
(f) Eu

1.18 (a) I_1 increases across the row except for a dip at S; (b) A_e tends to increase except for Mg (filled subshell), P (half filled subshell), and at Ar (filled shell).

1.19 Radii of Period 4 and 5 d-metals are similar because of lanthanide contraction.

1.20 χ increases steadily in most scales. For χ_M anomalies in I are generally offset by those in A.

1.21 $2s^2$ and $2p^0$.

1.22 With some exceptions, generally associated with half full or full subshells, I and χ increase and r decreases across a period.

Chapter 2

Self-tests

2.1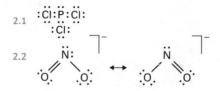

2.2

2.3 $+\frac{1}{2}$, $+5$.

2.4 $24\,kJ\,mol^{-1}$.

2.5 Covalent.

2.6 (a) Angular; (b) square planar.

2.7 Linear.

2.8 S_2^{2-}: $1\sigma_g^2 1\sigma_u^2 2\sigma_g^2 1\pi_u^4 2\pi_g^4$; Cl_2^-: $[S_2^{2-}]2\sigma_u^1$.

2.9 Molecular orbital diagram similar to Fig. 2.24 except Cl $3s$ and $3p$ on left and O $2s$ and $2p$ on right. Same number of electrons as ICl and same electron configuration.

2.10 Bond enthalpy $C\equiv N > C=N > C-N$, bond length $C\equiv N < C=N < C-N$.

2.11 If it contains 4 or fewer electrons.

Exercises

2.1 (a)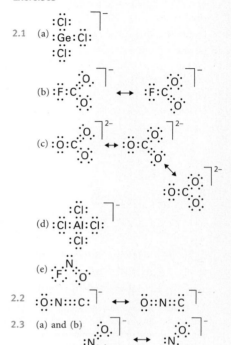

(b)

(c)

(d)

(e)

2.2

2.3 (a) and (b)

(c) N($+3$), O(-2)

(d) (i) formal charge, (ii) oxidation number, (iii) neither.

2.4 (a)

(b)

(c)

(d)

(e)

2.5 (a) Angular; (b) tetrahedral; (c) tetrahedral.

2.6 (a) Trigonal planar; (b) trigonal pyramidal; (c) square pyramidal.

2.7 (a) T-shaped; (b) square planar; (c) linear.

2.8 PCl_4^+ is tetrahedral, PCl_6^- is octahedral.

2.9 (a) 176 pm, (b) 217 pm, (c) 221 pm.

2.10 $2\,(Si-O) = 932\,kJ > Si=O = 640\,kJ$; therefore two Si—O are preferred and SiO_2 should (and does) have four single Si—O bonds.

2.11 Bond enthalpy data (Table 2.5) indicate $2\,N\equiv N \rightarrow N_4$ $\Delta H^{\ominus} \approx +912\,kJ\,mol^{-1}$, $2P_2 \rightarrow P_4$ $\Delta H^{\ominus} \approx -246\,kJ\,mol^{-1}$. Multiple bonds are much stronger for period 2 elements than heavier elements.

2.12 −483 kJ difference is smaller than expected because bond energies are not accurate.

2.13 (a) 0; (b) 205 kJ mol⁻¹.

2.14 Difference in electronegativities are; AB 0.5, AD 2.5, BD 2.0, and AC 1.0. Increasing covalent character AB < AC < BD < AD.

2.15 (a) covalent; (b) ionic; (c) ionic.

2.16 (a) One; (b) one; (c) none; (d) two.

2.17 (a) $1\sigma_g^2 1\sigma_u^2$; (b) $1\sigma_g^2 1\sigma_u^2 1\pi_u^2$; (c) $1\sigma_g^2 1\sigma_u^2 1\pi_u^4 2\sigma_g^1$; (d) $1\sigma_g^2 1\sigma_u^2 2\sigma_g^2 1 \pi_u^4 1\pi_g^3$

2.18 (a) 2; (b) 1; (c) 2.

2.19 (a) +0.5; (b) −0.5; (c) +0.5.

2.20 (a) 4; (b, c)

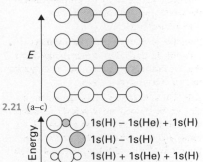

2.21 (a–c)

1s(H) − 1s(He) + 1s(H)

1s(H) − 1s(H)

1s(H) + 1s(He) + 1s(H)

(d) Possibly stable in isolation (only bonding and nonbonding orbitals are filled; not stable in solution because solvents would have higher proton affinity than He.

2.22 1.

2.23 HOMO exclusively F; LUMO mainly S.

2.24 (a) Electron deficient; (b) electron precise.

2.25 (a) NH₃; (b) BH₃; (c) NH₃.

Chapter 3

Self-tests

3.1 See Fig. 3.34.

3.2 Each unit cell contains 1 sphere (equivalent to $8 \times \frac{1}{8}$ spheres on the vertices) in contact along the edges. With a radius r the volume within the sphere is $\frac{4}{3}\pi r^3$ and the volume of the unit cell is $(2r)^3$ so fraction filled is $\frac{4}{3}\pi r^3 / 8r^3 = 0.52$.

3.3 $r_h = ((\frac{3}{2})^{\frac{1}{2}} - 1)r = 0.225\, r$.

3.4 The distorted octahedral site has a minimum vertex to vertex distance of $\sqrt{2}a - 2r$ where a is the bcc unit cell dimension and r is the sphere radius. So the maximum ion radius for this site is $\frac{1}{2}(\sqrt{2}. 4/\sqrt{3}. r - 2r) = 0.63r$.

3.5 Primitive cubic.

3.6 Cs and $(\frac{1}{8} \times 8)$ Cl = 1 Cl.

3.7 Ti CN = 6 (though these are as two slightly different distances it is often described as 4 + 2) and O CN = 3.

3.8 LaInO₃.

3.9 Fluorite.

3.10 6 and 6.

3.11 2421 kJ mol⁻¹.

3.12 621 kJ mol⁻¹.

3.13 Calculations give a large positive heat of formation for CsCl₂ due to the high second ionization energy of Cs. So unlikely to exist.

3.14 2.63 MJ mol⁻¹.

3.15 Order of decomposition temperatures: MgSO₄ < CaSO₄ < SrSO₄ < BaSO₄.

3.16 NaClO₄.

3.17 The $d_{x^2-y^2}$ and d_{z^2} have lobes pointing along the cell edges to the nearest neighbour metals.

3.18 V₂O₅ n-type; CoO p-type (Co(II) easily oxidised to Co(III)).

Exercises

3.1 $a \neq b \neq c$ $\beta = 90°$ $\alpha \neq 90°$ $\gamma \neq 90°$

3.2 Point on the cell corners at (0,0,0), (1,0,0), (0,1,0), (0,0,1), (1,1,0), (1,0,1), (0,1,1), (1,1,1) and in cell faces at $(\frac{1}{2}, \frac{1}{2}, 0)$, $(\frac{1}{2}, 1, \frac{1}{2})$, $(0, \frac{1}{2}, \frac{1}{2})$ $(\frac{1}{2}, \frac{1}{2}, 1)$, $(\frac{1}{2}, 1, \frac{1}{2})$, $(1, \frac{1}{2}, \frac{1}{2})$.

3.3 (c) and (f) are not as they have neighbouring layers of the same position.

3.4 Close-packed W with C in octahedral holes.

3.5 283 pm.

3.6 363 pm.

3.7 CuAu. Primitive 12 carat.

3.8 Zintl phase.

3.9 6:6 and 8:8 (b) CsCl.

3.10 6.

3.11 2 B type and 4 A type—in a distorted octahedral arrangement.

3.12 (a) $\rho = 0.78$ so fluorite, (b) $\rho = 0.94$ so CsCl, (c) $\rho = 0.19$ so ZnS (d) $\rho = 0.46$ so NaCl.

3.13 Lattice enthalpy with di and trivalent ions. Also the bond energy and third electron gain enthalpy for nitrogen with be large.

3.14 4 times the NaCl value so 3144 kJ mol⁻¹.

3.15 (a) 10464 kJ mol⁻¹, (b) 1960 kJ mol⁻¹ (c) 690 kJ mol⁻¹.

3.16 (a) MgSO₄ (b) NaBF₄.

3.17 CsI < RbCl < LiF < CaO < NiO < AlN.

Chapter 4

Self-tests

4.1 (a) Acid/conjugate base HNO₃/NO₃⁻, base/ conjugate acid H₂O/H₃O⁺.
(b) Acid/conjugate base H₂O/OH⁻, base/ conjugate acid HCO₃²⁻/HCO₃⁻.
(c) Acid/conjugate base H₂S/HS⁻, base/ conjugate acid NH₃/NH₄⁺.

4.2 5.55×10^{-6}.

4.3 DMSO and NH₃.

4.4 $[Na(OH_2)_4]^+ < [Mn(OH_2)_6]^{2+} < [Ni(OH_2)_6]^{2+} < [Sc(OH_2)_6]^{3+}$.

4.5 (a) 3; (b) 8; (c) 3.

4.6 Aqueous NH₃ precipitates TiO₂, which dissolves in NaOH.

4.7 Acid/base: (a) FeCl₃/Cl⁻; (b) I₂/I⁻; (c) Mn(CO)₅/SnCl₃⁻.

4.8 (H₃Si)₃N trigonal planar; (H₃C)₃N pyramidal.

4.9 Aluminosilicates: Rb, Cs, Sr; sulfides: Cd, Pd, Pb.

4.10 B coordinated to O; pyramidal C₂OB array.

Exercises

4.1 Check element symbols inside cover. Check acidity and basicity in Fig. 4.4.

4.2 $[Co(NH_3)_5(OH)]^{2+}$; SO₄²⁻; CH₃O⁻; HPO₄²⁻; $[Si(OH)_3O]^-$; S²⁻.

4.3 C₅H₅NH⁺; H₂PO₄⁻; OH⁻; CH₃COOH₂⁺; HCo(CO)₄; HCN.

4.4 5.6×10^{-10}.

4.5 See Table 4.2.

4.6 (a) In water too strong: O²⁻; too weak: ClO₄⁻, NO₃⁻; measurable: CO₃²⁻; (b) too weak: ClO₄⁻; measurable: NO₃⁻; HSO₄⁻.

4.7 CN is electron withdrawing.

4.8 From Pauling's rules, pK_a = 13.

4.9 HClO₄ > HBrO₃ > H₂SO₄ > HNO₂.

4.10 HSiO₄³⁻ < HPO₄²⁻ < HSO₄⁻ < HClO₄.

4.11 (a) $[Fe(OH_2)_6]^{3+}$, higher ξ; (b) $[Al(OH_2)_6]^{3+}$, higher ξ; (c) Si(OH)₄, higher ξ; (d) HClO₄, higher oxidations number or Pauling's rules; (e) HMnO₄, higher oxidation number; (f) H₂SO₄, higher oxidation number or Pauling's rules.

4.12 Cl₂O₇ < SO₃ < CO₂ < B₂O₃ < Al₂O₃ < BaO.

4.13 NH₃ < CH₃GeH < H₄ SiO₄ < HSO₄⁻ < H₃O⁺ < HSO₃F.

4.14 Ag⁺.

4.15 Polyanions: As, Si, Bi, Mo; polyoxocationa: Al, Ti, Cu.

4.16 Reduction of one.

4.17 4PO₄³⁻ + 8H⁺ → P₄O₁₂⁴⁻ + 4H₂O, 2[Fe(OH₂)₆]³⁺ → [(H₂O)₄Fe(μ−OH)₂× Fe(OH₂)₄]⁴⁺ + 2H₃O⁺.

4.18 (a) H₃PO₄ + HPO₄²⁻ → 2H₂PO₄⁻; (b) CO₂ + CaCO₃ + H₂O → Ca²⁺ + 2HCO₃⁻.

4.19 Lewis acids: BF₃, AlCl₃, M(I) or M(II) halides of Ga, In, Td; Cl₂; M(IV) halides of Si, Ge, Sn; PbCl₂; M(V) halides of P, Sb, As; BiCl₃; dioxides of S, Se, Te; Br₂, I₂, IF₇.

4.20 Acids: SO₃; Hg²⁺; SnCl₂; SbF₅; hydrogen bond.

4.21 (a) BBr₃, BCl₃, B(n-Bu)₃; (b) Me₃N, 4-CH₃C₅H₄N.

4.22 K > 1: (b) and (d).

4.23 BH₃ binds to P, BF₃ binds to N.

4.24 Steric repulsions.

4.25 (a) DMSO stronger base to hard and soft acid; (b) SMe₂ stronger base toward soft acids than DMSO.

4.26 SiO₂ + 4HF → 2H₂O + SiF₄.

4.27 Al₂S₃ + 3H₂O → Al₂O₃ + 3H₂S.

4.28 (a) and (b) Hard hydrogen bonding solvent; (c) hard donor e.g. H₂O; (d) softer base than Cl⁻, possibly PR₃.

4.29 I₂⁺ and Se₈⁺ disproportionate on or coordinate to bases; S₄²⁻ and Pb₉⁴⁻ disproportionate in or coordinate to acids.

4.30 Al_2O_3 acid sites coordinate Cl^-.

4.31 HgS soft-soft, ZnO hard-hard.

4.32 (a) $EtOH + HF \rightarrow EtOH_2^+ + F^-$;
(b) $NH_3 + HF \rightarrow NH_4^+ + F^-$;
(c) $PhCOOH + HF \rightarrow PhCOO^- + H_2F^+$.

4.33 Both.

4.34 Hard.

Chapter 5

Self-tests

5.1 $2MnO_4^-(aq) + 5Zn(s) + 16H^+ \rightarrow 5Zn^{2+}(aq) + 2Mn^{2+}(aq) + 8H_2O(l)$.

5.2 No. $E^\ominus$ for the reaction $Cu^{2+}(aq) + 2e^- \rightarrow Cu(s)$, $E^\ominus$ $(Cu^{2+},Cu) = +0.34\,V$ is positive.

5.3 Yes; $E^\ominus$ for reaction with Cl^- is slightly negative ($-0.02\,V$) (but in practice the reaction is very slow).

5.4 $0.85\,V$.

5.5 $1.25\,V$.

5.6 Oxidation by O_2 plus H_2O to H_2SO_4.

5.7 No as $E^\ominus$ is negative.

5.8 $Ni(en)_3^{2+}/Ni$ potential more negative (less favoured).

5.9 $1.43\,V$.

5.10 (a) Pu(IV) disproportionates to Pu(III) and Pu(V) in aqueous solution; (b) Pu(V) doesn't not disproportionate into Pu(VI) and Pu(IV) (though the potentials are so similar that solutions of Pu(IV) and Pu(V) will contain proportions of all Pu oxidation states between Pu(III) and Pu(VI).

5.11 Tl^+ at a minimum (N $E^\ominus = -0.34\,V$), Tl^{3+} the highest point (N $E^\ominus = +2.19\,V$).

5.12 Mn^{2+}.

5.13 Much stronger in acid solution.

5.14 Fe(II) favoured.

5.15 Above approx. $1750°C$.

Exercises

5.1 Thermodynamically suitable in acid solution:
(a) examples include $HClO$, α-PbO_2 etc.;
(b) Fe, Zn, etc.; (c) Al, Fe, Zn etc.; (d) same as (c).

5.2 (a) $4\,Cr^{2+} + O_2 + 4\,H^+ \rightarrow 4\,Cr^{3+} + 2\,H_2O$;
(b) $4\,Fe^{2+} + O_2 + 4\,H^+ \rightarrow 4\,Fe^{3+} + 2\,H_2O$;
(c) No reaction; (d) No reaction;
(e) $2\,Zn + O_2 + 4\,H^+ \rightarrow 2\,Zn^{2+} + 2H_2O$.

5.3 (a) $4Fe^{2+} + O_2 + 4H^+ \rightarrow 4Fe^{3+} + 2H_2O$;
(b) $4Ru^{2+} + O_2 + 4H^+ \rightarrow 4Ru^{3+} + 2H_2O$ and/ or $3Ru^{2+} \rightarrow Ru(s) + 2Ru^{3+}$;
(c) $3HClO_2 + H_2O \rightarrow ClO_3^- + 2HClO + H^+$ (followed by ClO_3^- disproportionation)
$ClO_2 + O_2 \rightarrow ClO_3^-$ (slow due to O_2 over potential).
(d) No reaction

5.4 (a) $E = E^\ominus - (0.059\ V/4)[\log p_{O_2}^{-1} + 4\ pH]$;
(b) $E = E^\ominus - (0.059\ V/6)\ (6\ pH)$.

5.5 For $Cr_2O_4^{2-}$: $\Delta G^\ominus = +31.8\,kJ\,mol^{-1}$,
$K = 2.7 \times 10^{-6}$;
For $[Cu(NH_3)_2]^+$: $\Delta G^\ominus = +9.65\,kJ\,mol^{-1}$,
$K = 2.0 \times 10^{-2}$;

Despite similar $E^\ominus$, $\Delta G^\ominus$ and K differ because of differing n.

5.6 (a) Disproportionation:
$Cl_2 + 2OH^- \rightarrow Cl^- + ClO^- + H_2O$;
(b) very little disproportionation; (c) in acid kinetic; in base thermodynamic.

5.7 (a) $5N_2O + 2OH^- \rightarrow 2NO_3^- + 4N_2 + H_2O$;
(b) $Zn + I_3^- \rightarrow Zn^{2+} + 3I^-$;
(c) $3I_2 + 5ClO_3^- + 3H_2O \rightarrow 6IO_3^- + 5Cl^- + 6H^+$.

5.8 Added acid (lower pH) (a) disfavoured;
(b) favoured; (c) disfavoured; (d) no effect.

5.9 $+1.39\,V$; $2ClO_4^- + 16\ H^+ + 14e^- \rightarrow Cl_2 + 8H_2O$.

5.10 3.7×10^{38}.

5.11 (a) Fe_2O_3; (b) Mn_2O_3; (c) HSO_4^-.

5.12 Approx. $-0.1\,V$.

5.13 Potential less positive.

5.14 $Fe^{2+}/Fe(OH)_2$; $Fe^{3+}/Fe(OH)_3$.

5.15 Above $1400°C$.

Chapter 6

Self-tests

6.1 The ionic radius of Cr(IV), 69 pm, is smaller than that of Ti(IV) 75 pm. The unit cell and d spacings will shrink in line with this. The pattern from CrO_2 will show identical reflections to those from rutile TiO_2 but shifted to higher diffraction angles.

6.2 The hydride resonance couples to three nuclei that are 100% abundant and have $I = \frac{1}{2}$: the two different ^{31}P nuclei and the ^{103}Rh nucleus. Since the three coupling constants are different, the effect is to split the signal into a doublet of doublet of doublets, i.e. 8 lines.

6.3 14% of naturally occurring tungsten is ^{183}W which has $I = \frac{1}{2}$ and therefore splits the signal into 2 lines. This will be superimposed on a non-split signal that arises from the 86% of tungsten that does not have a spin.

6.4 Both chlorine (^{35}Cl, 76% and ^{37}Cl 24%) and bromine (^{79}Br 51% and ^{81}Br 49%) exist as two isotopes. Therefore any compound containing either Cl or Br will have molecular ions 2 u apart. The lightest isotopomer of $ClBr_3$ is $^{35}Cl\ ^{79}Br_3$ at 272 u and the heaviest is $^{37}Cl\ ^{81}Br_3$ at 280 u.

Exercises

6.1 Use powder X-ray diffraction, collect data from the reaction mixture periodically and check for absence of reflections characteristic of reactants.

6.2 A glass has no long range periodicity as required for diffraction.

6.3 $0.5\,\mu m \times 0.5\,\mu m \times 0.5\,\mu m$.

6.4 180 pm. Useful for diffraction as the wavelength is comparable to inter-atomic spacings.

6.5 The bond orders for CN^- and CO are the same but N is lighter than O, hence the reduced mass for CN^- is smaller than for CO. From equation 6.4, CN will have the higher stretching

frequency. The bond order for NO is 2.5 and N is also lighter than C, hence CO has a higher stretching frequency than NO.

6.6 The bond in O_2^+ (bond order 2.5) is stronger than in O_2 (bond order 2). Therefore, O_2^+ has the higher force constant and the stretching frequency is expected to be in the region of $1800\ cm^{-1}$.

6.7 The band at 11 eV is due to the lone pair. This might be expected to be sharp as it is 'non-bonding'. However, the occupancy of the lone-pair orbital is linked strongly to the pyramidal angle. The ionised molecule has greater planarity.

6.8 $N(SiH_3)_3$ is planar and thus N is at the centre of the molecule and does not move in the symmetric stretch. $N(CH_3)_3$ is pyramidal and the N—C symmetric stretch involves displacement of N.

6.9 Even though there are chemically distinct carbonyls in $Co_2(CO)_9$, at room temperature they are exchanging position sufficiently quickly that an average signal is seen.

6.10 XeF_5^- has a pentagonal planar geometry, so all 5 of the F atoms are chemically equivalent and thus the molecule shows a single ^{19}F resonance. Approximately 25% of the Xe is present as ^{129}Xe, which has $I = \frac{1}{2}$, and in this case the ^{19}F resonance is split into a doublet. The final result is a composite: two lines each of 12.5% intensity from the ^{19}F coupled to the ^{129}Xe and one line of 75% intensity for the remainder.

6.11 1.94, 1.74 and 1.56

6.12 NMR, typically 0.5 GHz, responds to slower chemical processes than EPR (typically 10 GHz).

6.13 In aqueous solution at room temperature, molecular tumbling is fast compared to the timescale of the EPR transition (approximately $\mu\beta\beta/h$) and this removes the effect of the g-value anisotropy: we expect a derivative-type spectrum, possibly exhibiting hyperfine structure, that is centred at the average g-value. In frozen solution, g-value anisotropy can be observed because the spectrum records the values of g projected along each of the three axes.

6.14 The total charge on Fe atoms must add up to 18 (3×6). Thus we can have 4Fe(III) and three Fe(II). EPR would detect Fe(III) but not Fe(II). Mössbauer spectroscopy can distinguish between Fe(II) and Fe(III) and between 'complexed' Fe(II) coordinated by $6CN^-$ and 'complexed' Fe(III) coordinated by $6CN^-$. Uncomplexed Fe(II) will have a larger quadrupole splitting than uncomplexed Fe(III), whereas complexed Fe(II) will have a smaller quadrupole splitting than complexed Fe(III).

6.15 From VSEPR theory, SbF_5 is expected to have trigonal bipyramidal geometry, which is lower than cubic and thus subject to an electric field gradient. The absence of quadrupole splitting suggests that the geometry must be closer to cubic; an octahedral geometry can be achieved if each Sb accepts a lone pair from a F-atom on an adjacent molecule.

6.18 The complex undergoes a reversible one-electron reduction with a reduction potential of 0.21 V. Above 0.7 V the complex is oxidised to a species that undergoes a further chemical reaction, and thus is not re-reduced.

6.19 1:3

Chapter 7

Self-tests

7.1 The S_4 axis is best portrayed by separating the components. There are three S_4 axes.

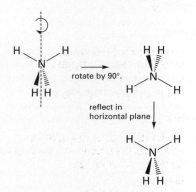

rotate by 90°.

reflect in horizontal plane

7.2 (a) D_{3h} ; (b) T_d.

7.3 No.

7.4 The conformation shown in (14) is chiral. The point group (C_2) does not contain an improper rotation axis. Optically active H_2O_2 might be observable, even transiently, if bound in a chiral host such as the active site of an enzyme (see Chapter 26).

7.5 Triple.

7.6 B_{2g}.

7.7 B_{1g}.

7.8 Yes.

7.9 See the first row in the D_{2h} character table.

7.10 SF_6 has O_h symmetry. Analysis of the stretching vibrations leads to $\Gamma_{str} = A_{1g}$ (Raman, polarised) $+ E_g$ (Raman) $+ T_{1u}$ (IR); SF_5Cl has C_{4v} symmetry. Analysis of the stretching vibrations leads to $\Gamma_{str} = 3A_1$ (IR and Raman, polarised) $+ 2B_1$ (Raman) $+ E$ (IR, Raman). Of the A_1 vibrations, one (S—Cl) will be at much lower frequency than the other two.

Exercises

7.1

(a)

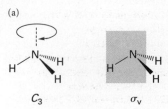

C_3 σ_v

(b)

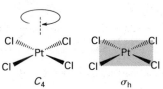

C_4 σ_h

7.2 Centre of inversion: (a) CO_2, (b) C_2H_2; an S_4 axis: (d) SO_4^{2-}

7.3 (a) NH_2Cl E, σ_h, (C_s)
(b) CO_3^{2-} $E, C_3, 3C_2, S_3, 3\sigma_v, \sigma_h$, ($D_{3h}$)
(c) SiF_4 $E, 4C_3, 3C_2, 3S_4, 6\sigma_d$, ($T_d$)
(d) HCN $E, C_\infty, \infty\sigma_v$, ($C_{\infty h}$)
(e) $SiFClBrI$ E, (C_1)
(f) BF_4^- $E, 4C_3, 3C_2, 3S_4, 6\sigma_d$, ($T_d$)

7.4 (a) s-orbital: $E, \infty C_\infty, \infty\sigma_v, \infty S_\infty$
(b) p-orbital: $E, C_\infty, \infty\sigma_v$
(c) d_{xy} orbital: $E, C_2(x), C_2(y), C_2(z), \sigma(xy)$,
       $\sigma(xz), \sigma(yz), I$;
d_{z^2} orbital: $E, C_\infty, \infty C_2, \infty S_\infty, \infty\sigma_v, i, \sigma_h$.

7.5 (a) C_{3v}; (b) 2; (c) $3p_x, 3p_y$.

7.6 (a) D_{3h} (b) 2 (c) $3p_x, 3p_y$.

7.7 SnF_4: T_d; SeF_4: C_{2v}; BrF_4^-: D_{4h}.

7.8 From VSEPR, $AsCl_5$ should be trigonal pyramidal with symmetry D_{3h}.

7.9 (a) In plane: A_1' is symmetric stretch; $2E'$ are modes consisting of (i) mainly asymmetric stretch and (ii) deformation; (b) Perpendicular: A_2'' is the deformation in and out of the plane.

7.10 (a) Γ_{vib} : $A_{1g} + E_g + T_{2g} + 2T_{1u} + T_{2u}$. No modes are both IR and Raman active. (b) Γ_{vib} : A_1' (Raman) $+ 2E'$ (IR and Raman) $+ A_2''$ (IR). The E' modes are active in both IR and Raman.

7.11 From the character table, we see directly that any mode having A_2, B_1 or B_2 symmetry should be inactive in both IR and Raman spectroscopy. Examples of species having C_{6v} symmetry are arene complexes of Group 3 elements, such as $[Tl(benzene)]^+$.

7.12 $A_{1g} + A_{2g} + B_{1g} + B_{2g} + E_g + 2A_{2u} + B_{2u} + 3E_u$.

7.13 (a) B_1; (b) $T_1 + T_2$; (c) $A_{1u} + A_{2u} + B_{1u} + B_{2u}$

7.14 Γ_{vib} : A_1 (Raman) $+ E$ (Raman) $+ 2T_2$ (IR and Raman).

7.15 (a) Planar PF_3 has D_{3h} symmetry. The vibrations are A_1' (Raman, polarised) $+ 2E'$ (IR and Raman) $+ A_2''$ (IR). Pyramidal PF_3 has C_{3v} symmetry. The vibrations are $2A_1$ (IR and Raman, polarised) $+ 2E'$ (IR and Raman). (b) For the planar form of B_2F_4 (D_{2h}), the vibrations are: $3A_g$ (Raman, polarised) $+ 2B_{2g}$ (Raman) $+ B_{3g}$ (Raman) $+ A_u$(inactive) $+ 2B_{1u}$ (IR) $+ B_{2u}$ (IR) $+ 2B_{3u}$ (IR). For the 90°-twisted form of B_2F_4 (D_{2d}) the vibrations are: $3A_1$ (Raman, polarised) $+ B_1$ (Raman) $+ 2B_2$ (IR and Raman) $+ 3E$ (IR and Raman).

Chapter 8

Self-tests

8.1 (a) *trans*, (b) *cis*.

8.2 (a) $[Pt(Cl)_2(OH_2)_2]$, (b) $[Cr(NCS)_4(NH_3)_2]^-$, (c) $[Rh(en)_3]^{3+}$, (d) $[MnBr(CO)_5]$, (e) $[RhCl(PPh_3)_3]$.

8.3 The hydrate isomers $[Cr(NO_2)(H_2O)_5]NO_2 \cdot H_2O$ and $[Cr(H_2O)_6](NO_2)_2$ are possible as are linkage isomers of the NO_2 group. One further isomer is $[Cr(ONO)(H_2O)_5]NO_2 \cdot H_2O$.

8.4

mer   *fac*

8.5 (a) chiral, (b) nonchiral, (c) nonchiral.

Exercises

8.1 (a) Tetrahedral: square planar:
(b) square planar

8.2

a = axial
e = equatorial

Trigonal bipyramid

a = axial
b = basal

Square based pyramid

8.3 (a)

Octahedral   Trigonal prismatic
(b) Trigonal prismatic.

8.4 A monodentate lignd can bond to a metal atom only at a single atom, a bidentate ligand can bond through two atoms, a quadridentate ligand can bond through four atoms.

8.5 Linkage isomerism arises with ambidentate ligands, as in M-NCS and M-SCN.

8.6 Only (b) and (c) can act as chelating ligands.

8.7 (a)

(b)

(c)

(d)

8.8 Ionization isomers.

8.9 No isomers with $[CoCl_2Br_2]^-$ or $[CoCl_2Br(OH_2)]$ but two optical isomers for $[CoClBrI(OH_2)]$:

8.10 One isomer only for $Pt(ox)(NH_3)_2$

Cis and *trans* isomers for $Pd(PEt_3)_2ClBr$

Cis and *trans* isomers for $[IrH(CO)(PR_3)_2]$

Cis and *trans* isomers for $Pd(gly)_2$

8.11 One isomer only for $[Fe(OH_2)_5Cl]^{2+}$

Fac and *mer* isomers for $Ir(PEt_3)_3Cl_3$

Optical isomers for $[Ru(bipy)_3]^{2+}$

Cis, *trans* and optical isomers for $[CoCl_2(en)(NH_3)_2]^+$

Cis and *trans* isomers for $W(CO)_4(py)_2$

8.12 Ignoring optical isomers, nine isomers possible. Including optical isomers, 15 isomers possible.

8.13 (a) Chiral

(b) Nonchiral. Two planes of symmetry: one through the plane of the molecule, and one perpendicular to this plane, bisecting the en and the two chloride ligands.

(c) Nonchiral. Two planes of symmetry: one through the plane defined by the two chloride ligands and the metal atom, and one perpendicular to this plane (through the metal atom and bisecting the the two chloride ligands).

(d) Chiral

(e) Nonchiral. Mirror plane through the metal, bisecting the dien ligand.

(f) Nonchiral. Two mirro planes: one through all 3 coordinating N atoms and the metal atom; the other through all three nitro groups and the metal atom.

8.14 (a) Tetracarbonylnickel(0)

(b) Tetracyanonickelate(II)

(c) Tetrachlorocobaltate(II)

(d) Hexaamminemanganese(II)

8.15 (a) $[CoCl(NH_3)_5]Cl_2$, (b) $[Fe(OH_2)_6](NO_3)_3$, (c) *cis*-$[FeCl_2(en)_2]$, (d) $[Cr(NH_3)_5\mu$-OH-$Cr(NH_3)_5]Cl_5$.

8.16 (a) *cis*-tetraamminedichlorochromium(III), (b) *trans*-diamminetetraisothiocyanato-chromate(III), (c) bis(ethylenediamine)oxalatocobalt(III).

Chapter 9

Self-tests

9.1 (a) $Ca(s) + H_2(g) \rightarrow CaH_2(s)$;
(b) $NH_3(g) + BF_3(g) \rightarrow H_3NBF_3$;
(c) $LiOH(s) + H_2(g) \rightarrow NR$.

9.2 $Et_3SnH + CH_3Br \rightarrow Et_3SnCH_3 + HBr$.

Exercises

9.3 (a) H(+1), S(−2); (b) K(+1), H(−);
(c) Re(+7), H(−); (d) H(+1), S(+6), O(−2);
(e) H(+1), P(+1), O(−2).

9.4 Industry
$CH_4 + H_2O \rightarrow CO + 3H_2$
$CO + H_2O \rightarrow CO_2 + H_2$
$C + H_2O \rightarrow CO_2 + 4H_2$
Laboratory
$2HCl + Zn \rightarrow ZnCl_2 + H_2$
$NaH + CH_3OH \rightarrow NaOCH_3 + H_2$

9.5 (a) Check with Fig. 9.3; (b) Check with Table 9.7; (c) Electron deficient: Group 13, Electron precise: Group 14, Electron rich: Groups 15 through 17.

9.6 Gaseous at STP.

9.7 O—H bond more polar than S—H. Therefore O—H $\oplus$ S is stronger than S—H $\oplus$ O.

9.8 (a) Barium hydride, saline; (b) silane, electron precise, molecular; (c) ammonia, electron rich, molecular; (d) arsane, as ammonia; (d) palladium hydride, metallic; (e) hydrogen iodide, electron rich, molecular.

9.9 (a) $BaH_2 + 2H_2O \rightarrow 2H_2 + Ba(OH)_2$
(b) $HI + NH_3 \rightarrow NH_4I$
(c) $PdH_x \rightarrow PdH_{x-\delta} + \frac{1}{2}\delta H_2$
(d) $NH_3 + BMe_3 \rightarrow H_3NBMe_3$

9.10 Solids: BaH_2, $PdH_{0.9}$; Liquids: none; Gases: SiH_4, NH_3, AsH_3, HI; Conductor: PdH_x.

9.11 H_2Se, bent, C_{2v}; P_2H_4, pyramidal around each P, C_2; H_3O^+, pyramidal, C_{3v}.

9.12 Reaction (b)

9.13 $(CH_3)_3SnH$, weak Sn—H bond.

9.14 (a) $H_2O < H_2S < H_2Se$;
(b) $H_2Se < H_2S < H_2O$

9.15 Direct combination, $Pd + \frac{1}{2}x\, H_2 \rightarrow PdH_x$; protonation of a salt, $NaCl + H_2SO_4 \rightarrow HCl + NaHSO_4$; metathesis, $4BCl_3 + 3LiAlH_4 \rightarrow 2B_2H_6 + LiAlCl_4$.

9.16 $BH_4^- \ll AlH_4^- > GaH_4^-$; AlH_4^- strongest reducing agent;
$GaH_4^- + 4HCl \rightarrow GaCl_4^- + 4H_2$

9.17   Period 2 hydrides volatile except BeH_2; Period 3 CaH_2 and AlH_3 not volatile; Period 2 generally more thermodynamically stable.

9.18   Clathrate hydrates. Krypton contained in hydrogen bonded $(H_2O)_n$ cage.

9.19   See Fig. 9.7.

Chapter 10

Self-tests

10.1   LiF 625 $kJ\,mol^{-1}$, NaF 535 $kJ\,mol^{-1}$; Li^+ smaller than Na^+ giving larger lattice enthalpy.

10.2   (a) Na_2O 2636 $kJ\,mol^{-1}$, Na_2O_2 2309 $kJ\,mol^{-1}$, 327 $kJ\,mol^{-1}$ differenc; (b) Rb_2O 2218 $kJ\,mol^{-1}$, Rb_2O_2 1980 $kJ\,mol^{-1}$, 237 $kJ\,mol^{-1}$ difference.

Exercises

10.1   (a) One valence electron, s^1; (b) all have large positive values of $E^\ominus$; (c) little tendency to act as Lewis acid to accept lone pairs from ligands.

10.2

	$-\Delta_f H^\ominus/$ (kJ mol^{-1})		$-\Delta_f H^\ominus/$ (kJ mol^{-1})
LiF	625	LiCl	470
NaF	535	NaCl	411
KF	564	KCl	466
RbF	548	RbCl	458
CsF	537	CsCl	456

10.3   (a) $C_2H_5Cl + 2Li \rightarrow Li + LiCl$; (b) $(C_2H_5)_2Hg + 2Na \rightarrow 2NaC_2H_5 + Hg$

10.4   (a) Mg^{2+}; (b) K^+.

10.5   A = NaOH, B = Na_2O_2, C = Na_2O, D = $NaNH_2$, E = NaH.

10.6   In LiF and CsI the cation and anion have similar radii. Solubility is lower if large difference between radii of cation and anion, as in CsF and LiI.

10.7   Hydrides decompose to elements. Lattice energy decreases down the group as radius of cation increases. Carbonates decompose to oxides. Difference in lattice energy between carbonate and oxide decreases down the group resulting in increased stability.

10.8   See Fig. 3.22 and Fig. 3.24; 6-coordinate Na^+, 8-coordinate Cs^+; different r^+/r^-.

10.9   Bulky alkyl groups lead to polymerization: methyl produces tetramer or hexamer, ^tBu produces monomer.

10.10   (a) $CH_3Br + Li \rightarrow Li(CH_3) + LiBr$; (b) $MgCl_2 + LiC_2H_5 \rightarrow M(C_2H_5)Br + LiBr$; (c) $C_2H_5Li + C_6H_6 \rightarrow LiC_6H_5 + C_2H_6$.

Chapter 11

Self-tests

11.1   (a) Covalent; (b) ionic.

11.2 Mg^{2+} is smaller than Ba^{2+}. Therefore the lattice enthalpy of MgO_2 is larger than that of BaO_2.

11.3   CaO 3475 $kJ\,mol^{-1}$, CaO_2 3035 $kJ\,mol^{-1}$, trend confirmed.

11.4   pH = 11.3.

11.5 MgF_2: 2991 $kJ\,mol^{-1}$ will offset hydration enthalpy and reduce solubility compared to $MgCl_2$.

Exercises

11.1   Small radius of Be has large polarizing power.

11.2   There is a diagonal relationship between Be and Al arising from their similar atomic radii.

11.3   A = $M(OH)_2$, B = MCO_3, C = MC_2, D = MCl_2.

11.4 $CaCO_3 \cdot MgCO_3(s) \rightarrow 2MgO(s) + 2CaO(s) + 2CO_2(g) \xrightarrow{FeSi} \frac{1}{2}Ca_2SiO_4(s) + Mg(l)$.

11.5 $3BaO(s) + 2Al(s) \rightarrow 3Ba(l) + Al_2O_3(s)$.

11.6 $Mg(OH)_2$ is sparingly soluble and mildly basic, $Ca(OH)_2$ is more soluble an so moderately basic, $Ba(OH)_2$ is soluble and so is strongly basic, also a poison.

11.7   Group 1 hydroxides more soluble so higher OH^- concentrations.

11.8   Salts of divalent ions have low solubility. Hard water due to formation of $CaCO_3$ from $CaHCO_3$ on heating and precipitation of $CaSO_4$.

11.9 $BeCl_2$ covalent; MgI_2 covalent; BaF_2 ionic.

11.10 C_2H_5MgBr will be tetrahedral with two molecules of solvent coordinated to the magnesium. The bulky organo group in 2,4,6-$(CH_3)_3C_6H_2MgBr$ leads to a coordination number of two.

11.11   (a) $MgCl_2 + 2\,LiC_2H_5 \rightarrow 2\,LiCl + Mg(C_2H_5)_2$; (b) $Mg + (C_2H_5)_2Hg \rightarrow Mg(C_2H_5)_2 + Hg$; (c) $Mg + C_2H_5HgCl \rightarrow C_2H_5MgCl + Hg$.

Chapter 12

Self-tests

12.1 $H_3B{:}OEt_2 + CH_2CH_2 \rightarrow H_2BCH_2CH_3 + Et_2O$, $LiBH_4 + NH_4Cl \rightarrow LiCl + H_2 + H_3B{:}NH_3$.

12.2   (a) $BCl_3 + 3EtOH \rightarrow B(Oet)_3 + 3\,HCl$; (b) $BCl_3 + NC_5H_5 \rightarrow Cl_3BNC_5H_5$; (c) $BBr_3 + F_3BNMe_3 \rightarrow Br_3BNMe_3 + BF_3$.

12.3 $3NMeH_2 + 3BCl_3 \rightarrow Me_3N_3B_3Cl_3 + 6\,HCl$; $Me_3N_3B_3Cl_3 + 3MgMeBr \rightarrow Me_3N_3B_3Me_3 + 3MgBrCl$.

12.4   7 pairs.

12.5   14, *arachno*.

12.6 $2[B_{10}H_{13}]^- + Al_2Me_6 \rightarrow 2[B_{10}H_{11}AlMe]^- + 4CH_4$.

12.7 $1,2\text{-}B_{10}C_2H_2 + 2LiBu \rightarrow 1,2\text{-}B_{10}C_2H_{10}Li_2 + 2C_4H_{10}$; $1,2\text{-}B_{10}C_2H_{10}Li_2 + 2SiMe_2Cl_2 \rightarrow 1,2\text{-}B_{10}C_2H_{10}(SiMe_2Cl)_2 + 2LiCl$.

12.8   (a) $(Me)_2SalCl_3 + GaBr_3 \rightarrow Me_2SgaBr_3 + AlCl_3$; (b) $2TlCl_3 + CH_2CO + H_2O \rightarrow TlCl + CO_2 + 4H^+ + 4Cl^-$.

Exercises

12.1 $B_2O_3 + 3Mg \rightarrow 2B + 3MgO$.

12.2   (a) trigonal planar, $F\pi\text{-}B\pi$ bonding removes electron efficiency; (b) dimmer, $2c2e$ bridging bonds; (c) dimer, $3c2e$ bridging bonds.

12.3 $BF_3 < BCl_3 < AlCl_3$; (a) $BF_3N(CH_3)_3 + BCl_3 \rightarrow BCl_3N(CH_3)_3 + BF_3$; (b) $BH_3CO + BBr_3 \rightarrow NR$.

12.4 $Na[TlBr_4]$.

12.5   A = B_2H_6, B = $B(OH)_3$, C = B_2O_3.

12.6   No, $B_2H_6 + 3O_2 \rightarrow B_2O_3 + 3H_2O_2$.

12.7 $BCl_3 + Hg \rightarrow B_2Cl_4 + HgCl_2$ (electric discharge), $B_2Cl_4 + C_2H_4 \rightarrow Cl_2BCH_2CH_2BCl_2$, $3Cl_2BCH_2CH_2BCl_2 + 4AsF_3 \rightarrow 3F_2BCH_2CH_2BF_2 + 4AsCl_3$.

12.8   (a) $2NaBH_4 + 2H_3PO_4 \rightarrow 2NaHPO_4 + B_2H_6$, $B_2H_6 + THF \rightarrow 2H_3BTHF$, $H_3BTHF + 2C_2H_4 \rightarrow 3BEt_3 + THF$; (b) BH_3THF (from above) $+ Et_3N \rightarrow Et_3NBH + THF$.

12.9   Work from Fig. 12.2; two boron atoms are closest to the viewer in this orientation.

12.10 B_6H_{10} *nido* boranes more stable than *arachno* (B_6H_{12}).

12.11   (a) $2B_5H_9(l) + 12O_2(g) \rightarrow 5B_2O_3(s) + 9H_2O(g)$; (b) the solid would deposit inside the engine.

12.12   (a) *nido*; (b) 12 skeletal (framework) electron pairs.

12.13   See synthesis of *closo*-$B_{10}C_2H_{12}$, follow by base cleavage and reaction with $FeCl_2$.

12.14   (a) Hexagonal array of atoms in stacked sheets; stacking B over N in BN, C offset from C in graphite; (b) graphite reacts with Na and Br_2, BN does not; (c) large HOMO-LUMO gap in BN does not provide source and sink for e^-.

12.15   (a) $3PhNH_2 + 3BCl_3 \rightarrow Ph_3N_3B_3Cl_3 + 6HCl$; (b) $Ph_3N_3B_3Cl_3 + 3NaBH_4 \rightarrow Ph_3N_3B_3H_3 + 3NaBH_3Cl$.

12.16   Check against Table12.4 and structure 28.

12.17 $B_2H_6 < B_5H_9 < B_{10}H_{14}$.

Chapter 13

Self-tests

13.1   (a) Electrons are added to the π bond; (b) electrons are withdrawn from the π bond.

13.2 CH_4 !61 $kJ\,mol^{!1}$, $SiH_4 + 39\,kJ\,mol^{!1}$.

13.3 $^{13}CO + 2MnO_2 \rightarrow {}^{13}CO_2 + Mn_2O_3$, $^{12}CO_2 + LiBMe_3H \rightarrow Li[^{13}CHO_2] + BMe_3$. Cyclic $Si_4O_{12}^{n-}$ with one bridging O between Si and two terminal O on each Si.

13.5   Fig. 10.29(a) has 24 vertices; therefore the total number of Si and Al is 24.

Exercises

13.1 All tetrahedral.

13.2 Higher C—F bond enthalpy.

13.3 (a) $[Me_4N]^+$ tetrahedral, $[SiF_5]^!$ trigonal bipyramid: (b) equatorial and axial fluorines.

13.4 A: $SiCl_4$, B: $SiRCl_3$, C: $Rsi(OH)_3$, D: $RsiOsiR$, E: SiR_4, F: SiO_2.

13.5 (a) +4 most stable for lighter elements, +2 most stable for Sn and Pb; (b) (i) $Sn^{2+} + PbO_2 + 4 H^+ \rightarrow Sn^{4+} + Pb^{2+} + H_2O$, (ii) $2Sn^{2+} + O_2 + 4H^+ \rightarrow 2Sn^{4+} + 2H_2O$.

13.6 (i) 1.32 V; (ii) 1.08 V.

13.7 $SiO_2 + 2C \rightarrow Si + 2CO$; $GeO_2 + 2H_2 \rightarrow Ge + 2H_2O$.

13.8 (a) E_g: C > Si > Sn $\approx$ 0; (b) increase.

13.9 Check against Fig. 13.11

13.10 (a) graphite + K(am) $\rightarrow$ KC_8; (b) C heated with Cu $\rightarrow$ CuC_2; (c) $8K(g) + C_{60} \rightarrow K_8C_{60}$.

13.11 $K_2CO_3 + 2HCl \rightarrow 2KCl + H_2O + CO_2$, $Na_2SiO_4 + 2H^+ \rightarrow 2Na^+ + SiO_2.H_2O$ (silica gel).

13.12 Jadeite linear chains of $(SiO_3^{2-})_n$; kaolenite layered see Fig. 10.14.

13.13 (a) 48 bridging O; (b) see Fig. 13.17.

Chapter 14

Self-tests

14.1 According to VSEPR interpretation the 3-coordinate Bi must have a lone pair on Bi.

14.2 The reaction produces very reactive Na metal. This reacts with KNO_3 and SiO_2 to produce unreactive silicates.

14.3 Dimtehylhydrazine ignites spontaneously, it is not a gas and does not need to be liquefied, and it produces less CO_2.

14.4 As electronegativity increases across a period sulphur should be the stronger oxidizing agent.

14.5 The reaction employed to synthezise hydroxylamine probably involves the attack of HSO_3^- (acting as a Lewis base) on NO_2^- (acting as a Lewis acid). Since the formal oxidation state of the N atom changes from +3 to +1 this can also be seen as a redox reaction.

14.6 4 P atoms

14.7 Polydichlorophosphazene can be converted to other polyphosphazenes by nucleophilic substitution.

Exercises

14.1 Check periodic table on inside front cover. Bismuth displays a strong inert-pair effect.

14.2 (a) $2Ca(PO_4)_2 + 6SiO_2 + 10C \rightarrow 6CaSiO_3 + 10CO + P_4$ followed by $P_4 + 5O_2 \rightarrow P_4O_{10}$ then $P_4O_{10} + 6H_2O \rightarrow 4H_3PO_4$; (b) fertilizer grade; (c) high energy cost for preparation of P_4.

14.3 (a) $6Li + N_2 \rightarrow 2Li_3N$, $Li_3N + 3H_2O \rightarrow 3LiOH + NH_3$; (b) $3H_2 + N_2 \rightarrow 2NH_3$; (c) cheaper to produce H_2 than Li.

14.4 NCl_3 highly unstable, no NCl_5; PCl_3 and PCl_5 stable.

14.5 (a) tetrahedral; (b) see-saw; (c) trigonal bipyramidal.

14.6 (a) and (b) see 14.2(a); (c) $2H_3PO_4 + 3CaCl_2 \rightarrow Ca_3(PO_4)_2 + 6HCl$.

14.7 (a) $4NH_3 + 7O_2 \rightarrow 6H_2O + 2NO_2$, $NO_2 + H_2O \rightarrow 2HNO_3 + 2NO_2$; (b) $2NO_2 + 2OH^- \rightarrow NO_2^- + NO_3^- + H_2O$; (c) $NO_3^- + 2HSO_3^- + H_2O \rightarrow NH_3OH^+ + SO_4^{2-}$, $NH_3OH^+ + OH^- \rightarrow NH_2OH + H_2O$; (d) $NaNH_2 + N_2O \rightarrow NaN_3 + NaOH + NH_3$.

14.8 $P_4(s) + 5O_2(g) \rightarrow P_4O_{10}(s)$; P_4 is not the most stable allotrope.

14.9 Check against Fig. 14.6. Bi(V) is readily reduced; Bi(III), P(III), and P(V) oxo species are much less prone to redox.

14.10 Faster: the good electrophile NO^+ is formed and mechanisms proceed via attack of NO^+ on nucleophiles.

14.11 The rate law contains $p(NO)^3$ terms, therefore at low partial pressure NO is not rapidly oxidized.

14.12 (a) $PCl_5 + H_2O \rightarrow POCl_3 + 2HCl$; (b) $PCl_5 + 8H_2O \rightarrow 2H_3PO_4 - 10HCl$; (c) $PCl_5 + AlCl_3 \rightarrow [PCl_4][AlCl_4]$; (d) $3PCl_5 + 3NH_4Cl \rightarrow N_3P_3Cl_6 + 12HCl$.

14.13 $E^{\ominus} = +0.839$ V, H_3PO_4 is a useful oxidizing agent.

14.14 $A = AsCl_3$, $B = AsCl_5$, $C = AsR_3$, $D = AsH_3$.

14.15 $A = NO_2$, $B = HNO_3$, $C = NO$, $D = N_2O_4$, $E = NO$, $F = NH_4^+$.

14.16 cis isomer gives two ^{19}F signals, tran, isomer gives one signal.

14.17 N_2O_4, NO, N_2O, NH_3OH^+, $H_4P_2O_6$, P.

Chapter 15

Self-tests

15.1 Both Br^- and Cl^- are potential catalysts.

15.2 (a) SO_3 see resonance forms for isoelectronic NO_3^-, D_{3h}; (b) SO_3F^-: isoelectronic XO_4^{n-} Table 2.1, C_{3v}.

15.3 All except $Mn(OH)_2$.

15.4 No reaction.

Exercises

15.1 Acidic: CO_2, SO_3, P_2O_5, Neutral: CO, Basic: MgO, K_2O, Amphoteric: Al_2O_3.

15.2 (a) + 1.068 V; (b) Cr^{2+} is not a good catalyst for H_2O_2 disproportionation; (c) + 1.64 V, greater tendency to disproportionate than H_2O_2.

15.3 $O—H \oplus S$ and OH bond most polar.

15.4 (a) en; (b) en.

15.5 Most reducing SO_3^{2-}, SO_4^{2-}, $S_2O_8^{2-}$ most oxidizing.

15.6 (a) $Te(OH)_6$; SO_4^{2-}, (b) The large Te(VII) accommodates more ligands and is less acidic.

15.7 $S_2O_6^{2-}$, $S_2O_3^{2-}$.

15.8 Fe^{3+}, Co^{3+}.

15.9 SF_4^+ trigonal pyramid, BF_4^- tetrahedron.

15.10 (a) $SF_4 + (CH_3)_4NF \rightarrow [(CH_3)_4N]^+[SF_5]^-$; (b) square pyramid; (c) two F environments.

15.11 $A = S_2Cl_2$, $B = S_4N_4$, $C = S_2N_2$, $D = K_2S_2O_3$, $E = S_2O_6^{2-}$, $F = SO_2$.

Chapter 16

Self-tests

16.1 Both SO_2 and Sn^{2+} are thermodynamically suitable, but the oxidation products of the former are easier to separate from the product and S is much cheaper than Sn.

16.2 Pyramidal, C_s.

16.3 I_3^-, IBr_2^-, Br_3^-, central atom hypervalent.

16.4 In acid solution: F_2 and Cl_2. F_2 reaction is fast.

Exercises

16.1 Check with Table 12.1, F_2 light green-yellow, Cl_2 green-yellow, Br_2 brown, I_2 purple in gas or dilute hydrocarbon solution; Noble gases are colourless.

16.2 Br_2 or I_2 by oxidation of halide solution with Cl_2; Cl_2 aqueous electrolysis, F_2 nonaqueous electrolysis. See pp. 407–8.

16.3 Check with Fig. 12.3; $Cl_2 + OH^- \rightarrow HClO + Cl^-$.

16.4

Vacant lobes on each end accept electrons from donor molecules.

16.5 (a) NH_3 is hydrogen bonded; (b) high electronegativity of F decreases basicity of N.

16.6 (a) $NCCN + 2OH^- \rightarrow NCO^- + CN^- + H_2O$; (b) $2SCN^- + MnO_2 + H^+ \rightarrow Mn^{2+} + (SCN)_2 + 2H_2O$; (c) C_{3v}.

16.7 (a) $[(CH_3)_4N][IF_4]$; IF_3 T-shape, Me_4N^+ tetrahedral, IF_4^+ square planar; (c) two, one.

16.8 Trigonal bipyramidal, pyramidal, octahedral.

16.9 $[ClF_4][SbF_6]$

16.10 MCl_4F_2 cis 1 resonance, trans 1 resonance; MCl_3F_3 fac 1 resonance, mer 2 resonances of intensity 1:2.

16.11 (a) Octahedral, pentagonal bipyramid; (b) $IF_7 + SbF_5 \rightarrow [IF_5][SbF_6]$.

16.12 (a) Stronger; (b) no change; (c) weaker.

16.13 (a) No, Sb in maximum oxidation state; (b) explosive CH_3OH is a reducing agent, BrF_5 a strong oxidizing agent; (c) Br_3 can be oxidized, but probably not explosively; (d) explosive, S_2Cl_2 easily oxidized.

16.14 $Br_3^- + I_2 \rightarrow 2IBr + Br^-$.

16.15 Large cations stabilize large unstable anions.

16.16 ClO_2 angular radical; I_2O_5 has oxo bridge between two IO_2.

16.17 $HBrO_4$ is more acidic but less stable than H_5IO_6.

16.18 (a) Reduction potential increases with decreasing pH;

(b) at $E^{\ominus}$ (pH = 7) = 0.78 V vs.
$E^{\ominus}$ (pH = 0) = +1.20 V.

16.19 Reduction of the oxoanion is more favourable because protons are consumed in that half reaction.

16.20 Periodic I—O bonds are more labile.

16.21 (a) ClO_2^-; (b) ClO^-.

16.22 (a) and (d) because they have reducible cations associated with a strongly oxidizing anion, ClO_4^-.

16.23 (a) No; (b) yes; (c) yes; (d) no.

Chapter 17

Self-tests

17.1 $2XeO_4^{2-} \rightarrow XeO_6^{4-} + Xe + O_2$, $2XeO_4^{2-} + 2H_2O \rightarrow 2Xe + 4OH^- + 3O_2$, two independent reactions (disproportionation of xenate and oxidation of water); their ratio will depend on conditions.

Exercises

17.1 He has sufficient speed to escape from the atmosphere.

17.2 (a) He; (b) Xe; (c) Ar.

17.3 (a) $Xe + F_2 \rightarrow XeF_2$ sunlight;
(b) $Xe + 3F_2 \rightarrow XeF_6$ high pressure of F_2, 300°C;
(c) $XeF_6 + 3H_2O \rightarrow XeO_3 + 6HF$.

17.5 (a) XeF_4 (square-planar); (b) XeF_2 linear; (c) XeO_3 pyramidal; (d) IF^+.

17.6 All three share a single electron pair and have an octet around each atom. Formal charges: Cl(0) O(−1); Br(0); Xe(+1), F(0); (b) They are isolobal; (c) All three display some electrophilicity in the order $ClO^- < Br_2 < XeF^+$. Positive formal charge on XeF^+ correlates with its Lewis acidity and electrophilicity.

Chapter 18

Self-tests

18.1 V(V), VO_2^+.

18.2 It has a layered structure with weak van der Waals interaction between layers.

18.3 $Re_3Cl_9(PPh_3)_3$; Re_3 triangle with bridging and terminal Cl and one PPh_3 on each Re.

Exercises

18.1 See periodic table, C: Sc^{3+}, Ti^{4+}, V^{5+}; O: Cr, Mn; N: Fe, Co, Ni, Cu, Zn.

18.2 Metal-metal bonding becomes stronger on descending a group as the d-orbitals become larger. Also, heavier atoms need more energy to vaporize them.

18.3 Descending: higher oxidation states more stable. See Resource section 3.

18.4 (a) $Cr^{2+} + Fe^{3+} \rightarrow Cr^{3+} + Fe^{2+}$ (+2 ox. state more stable to the right of a d-block series);
(b) $2CrO_4^{2-} + 2MoO_2 + 2OH^- \rightarrow Cr_2O_3 +$

$2MoO_4^{2-} + H_2O$ (stability of highest oxidation state increases down a group);
(c) $6MnO_4^- + 10Cr^{3+} + 11H_2O \rightarrow 6Mn^{2+} + 5Cr_2O_7^{2-} + 22H^+$ (maximum ox. state less stable from right to left in $3d$ series).

18.5 (a) Ni^{2+}; (b) for the same oxidation number, metal ions become softer going toward the right of a d series;
(c) $Ni^{2+} + H_2S \rightarrow NiS + 2H^+$.

18.6 (a) See periodic table; (b) TiF_2 through NiF_2; (c) Sc, Y and lanthanides, Zr and Hf, Nb and Ta, Mo and W, Tc and Re; one example is Mo_6Cl_{12}.

18.7 cis-$[Ru^{II}LCl(OH_2)]^+ + OH^- \rightarrow cis$ $[Ru^{III}LCl(OH)]^+ + H_2O + e^-$ (e might be transferred to an electrode which maintains the potential on the system). Higher oxidation state promotes acidity thus $[Fe(OH_2)_6]^{3+}$ is more acidic than $[Fe(OH_2)_6]^{2+}$.

18.8 (a) NR;
(b) $6MoO_4^{2-} + 10H^+ \rightarrow [Mo_6O_{19}]^{2-} + 5H_2O$;
(c) $3ReCl_5 + 2MnO_4^- + 8H_2O \rightarrow 3ReO_4^- + 2MnO_2 + 16H^+$;
(d) $6MoCl_2 + 2HCl + 2H_2O \rightarrow [H_3O]_2[Mo_6Cl_{14}]$;
(e) $2TiO + 6HCl \rightarrow 2Ti^{3+} + H_2 + 2H_2O + 6Cl^-$;
(f) $Hg^{2+} + Cd \rightarrow Hg + Cd^{2+}$.

18.9 $trans$-octahedral (d^2); cis-octahedral (d^0), $trans$-octahedral (d^2), sq. pyramid.

18.10 Ionic or borderline: NiI_2, $NbCl_4$, FeF_2; more covalent PtS; M—M bonded: WCl_2.

18.11 (a) $\sigma^2 \pi^4 \delta^2$, B.O. = 4;
(b) $\sigma^2 \pi^4 \delta^2$, B.O. = 4;
(c) $\sigma^2 \pi^4 \delta^2 \delta^{*2} \pi^{*2} \sigma^{*2}$, B.O. = 0.

18.12 Higher oxidation states become more stable on descending a group.

Chapter 19

Self-tests

19.1 High spin, LFSE = $0.4\Delta_O$ Low spin, LFSE = $2.4\Delta_O$.

19.2 $t_{2g}^3 e_g^2$.

19.3 On the basis of a linear increase in lattice enthalpy from Mn to Zn we would expect FeF_2 to have a lattice enthalpy of 2821 kJ mol^{-1}, CoF_2 2862 kJ mol^{-1}, and NiF_2 2903 kJ mol^{-1}. The observation of larger lattice enthalpies can be accounted for by an LFSEs of $0.4\Delta_O$, $0.8\Delta_O$ and $1.2\Delta_O$ respectively.

19.4 The differences in the 6–8 eV region show that the iron complex has two largely metal based MOs of similar energy (corresponding to the t_{2g} orbitals) whereas the magnesium complex has MOs of widely differing energy: this can be attributed to the lack of d electrons for Mg (II).

19.5 ^1F and ^3F.

19.6 (a) ^3P; (b) ^2D.

19.7 $^3T_{1g}$, $^3T_{2g}$, $^3A_{2g}$ and $^1T_{2g}$, ^1E_g.

19.8 17 500 cm^{-1} and 22 400 cm^{-1}.

19.9 $^4A_{2g}$ to ^2E_g, $^4T_{1g}$, $^4T_{2g}$, ligand.

Exercises

19.1 (a) t_{2g}^6, $2.4\Delta_o$; (b) $t_{2g}^4 e_g^2$, $0.4\Delta_o$; (c) t_{2g}^5, $2\Delta_o$;
(d) t_{2g}^3, $1.2\Delta_o$; (e) t_{2g}^6, $2.4\Delta_o$; (f) $e^3 t_2^3$, $0.6\Delta_T$;
(g) $e^4 t_2^6$, 0.

19.2 No; H$^-$, strong σ donor no π interaction; PPh_3, weak σ donor + π acceptor.

19.3 (a) 0; (b) 4.9; (c) 1.7; (d) 3.9; (e) 0;
(f) 4.9; (g) 0.

19.4 Yellow, pink, and blue, respectively.

19.5 (a) $[Cr(OH_2)_6]^{2+}$; (b) $[Fe(OH_2)_6]^{3+}$;
(c) $[Fe(CN)_6]^{3-}$; (d) $[Ru(CN)_6^{3}]$;
(e) $[CoCl_4]^{2-}$.

19.6 See Fig. 19.6 and associated discussion.

19.7 Perchlorate: square-planar $(d_{xz}, d_{yz})^4 d_{xy}^2 d_{z^2}^2$ diamagnetic; thiocyanate approximately octahedral $t_{2g}^6 e_g^2$ with two unpaired electrons in e_g.

19.8 Distorted octahedral.

19.9 Jahn–Teller distortion in excited state.

19.10 (a) ^6S; (b) ^3F; (c) ^2D; (d) ^3P.

19.11 (a) ^3F; (b) ^5D; (c) ^6S.

19.12 (a) ^2S; (b) ^3P (ground), ^1D, ^1S.

19.13 B$^+$: ^1S; Na: ^2S; Ti^{2+}: ^3F; Ag$^+$: ^1S.

19.14 $B = 861$ cm^{-1}, $C = 3168$ cm^{-1}.

19.15 d^6, $^1A_{1g}$; d^1, $^2T_{2g}$; d^5, $^6A_{1g}$.

19.16 (a) B ≈ 770 cm^{-1}, Δ_O ≈ 8500 cm^{-1};
(b) B ≈ 720 cm^{-1}, Δ_O ≈ 10750 cm^{-1}.

19.17 $^5T_{2g}$.

19.18 Weak: spin forbidden LF; medium: spin-allowed LF transitions; strong: CT transition.

19.19 d^5 $[FeF_3]^{3-}$ has no spin-allowed LF transitions; $[CoF_6]^{3-}$ has $^5E_g \leftarrow ^5T_{2g}$ which is spin-allowed.

19.20 CN$^-$ give a more covalent complex.

19.21 Two spin-allowed LF bands, a spin-forbidden LF band, and a LMCT band respectively.

19.22 High spin d^5 Fe^{3+} has only spin-forbidden LF bands.

19.23 $[Cr(OH_2)_6]^{3+}$: LF bands which are weak; CrO_4^{2-}: allowed LMCT bands.

19.24 a_{1g}; (a) d_{xz}, d_{yz}; (b) d_{z^2}; (c) lowest xy; (xz, yz), z^2, $x^2 - y^2$.

19.25 Mn(VII) is d^0 and has no LF bands.

19.26 The two LMCT bands are to t_{2g} and e_g. The difference is Δ_T i.e. 13 700 cm^{-1}.

19.27 Metal–metal CT.

Chapter 20

Self-tests

20.1 Increased LFSE for the high spin maximally coordinated complex.

20.2 5.1 dm^3 mol^{-1} s^{-1}.

20.3 $[PtCl_4]^{2-}$ reacted with PPh_3 then NH_3 leads to $[PtCl_3PPh_3]^-$ and then $trans$-$[PtCl_2NH_3PPh_3]$; $[PtCl_4]^{2-}$ reacted with

NH$_3$ then PPh$_3$ leads to [PtCl$_3$NH$_3$]$^-$ and then *cis*-[PtCl$_2$NH$_3$PPh$_3$].

20.4 $K_E \approx 1$, $k = 1.2 \times 10^2\,s^{-1}$.

20.5 2 *cis* to 1 *trans*.

20.6 [Ru(NH$_3$)$_5$]$^{3+}$, $k = 2.93 \times 10^3\,dm^3\,mol^{-1}\,s^{-1}$; [Fe(OH$_2$)$_6$]$^{3+}$, $k = 3.12 \times 10^7\,dm^3\,mol^{-1}\,s^{-1}$; [Ru(bipy)$_3$]$^{3+}$, $k = 8.23 \times 10^{15}\,dm^3\,mol^{-1}\,s^{-1}$.

Exercises

20.1 The value of the stepwise formation constants drop from 1 to 4, as expected on statistical ground. However the fifth stepwise formation constant is substantially lower suggesting a change in coordination. In fact what happens is that the square-planar [Cu(NH$_3$)$_4$]$^{2+}$ ion forms, and further coordination does not occur.

20.2 The salient comparison is of log β_2 (7.65) for the ammonia reaction with log K_{f1} (10.72) for the en reacion. The value for the en reaction is substantially higher indicating more favourable complex formation, and can be attributed to the chelate effect. A comparison can also be drawn between the 3rd and 4th stepwise formation constants with ammonia (5.02) and the 2nd one for en (9.31).

20.3 Dissociative mechanism.

20.4 Rate of associative process depends on identity of entering ligand and, therefore, it isn't an inherent property of [M(OH$_2$)$_6$]$^{n+}$.

20.5 *d.*

20.6 Rate = k[Mn(OH$_2$)$_6^{2+}$][X$^-$]/(1 + K_a[X$^-$]); vary X$^-$.

20.7 Stronger covalent bonds to be broken for higher oxidation states and heavier metals.

20.8 The methyl groups provide steric hindrance to nucleophilic attack.

20.9 The rate-controlling step appears to be dissociative.

20.10 Yes, a favourable A intermediate.

20.11 Reacting [PtCl$_4$]$^{2-}$ with NH$_3$ and then NO$_2^-$ would lead to [PtCl$_3$NH$_3$]$^-$ and then *cis*-[PtCl$_2$(NO$_2$)(NH$_3$)]$^-$, whereas reacting [PtCl$_4$]$^{2-}$ with NO$_2^-$ and then NH$_3$ would lead to [PtCl$_3$NO$_2$]$^-$ and then *trans*-[PtCl$_2$(NO$_2$)(NH$_3$)]$^-$.

20.12 (a) Decrease rate; (b) decrease rate; (c) decrease rate; (d) increase rate.

20.13 Conjugate base path with OH$^-$; Brønsted acidity is required for conjugate base path.

20.14 (a) *cis*-[PtCl$_2$(PR$_3$)$_2$]; (b) *trans*-[PtCl$_2$(PR$_3$)$_2$]; (c) *trans*-[PtCl$_2$(NH$_3$)(py)].

20.15 [Ir(NH$_3$)$_6$]$^{3+}$ < [Rh(NH$_3$)$_6$]$^{3+}$ < [Co(NH$_3$)$_6$]$^{3+}$ < [Ni(OH$_2$)$_6$]$^{2+}$ < [Mn(OH$_2$)$_6$]$^{2+}$.

20.16 (a) Decreases; (b) decreases; (c) little effect; (d) decreases.

20.17 Dissociative reaction mechanism.

20.18 Inner sphere reaction would go through N$_3^-$ bridged complex.

20.19 Inner-sphere mechanism.

20.21 [W(CO)$_5$(NEt$_3$)] 0.4; ligand field.

20.20 (a) [Ru(NH$_3$)$_6$]$^{3+}$,

$k = 4.53 \times 10^3\,dm^3\,mol^{-1}\,s^{-1}$; [Co(NH$_3$)$_6$]$^{3+}$, $k = 1.41 \times 10^{-2}\,dm^3\,mol^{-1}\,s^{-1}$; Even though the reduction of the Ru complex is more thermodynamically favoured than the reduction of the Co complex, it is faster.

20.22 250 nm.

20.23 Rate-determining step is SMe$_2$ dissociation.

Chapter 21

Self-tests

21.1 No, it would have 20 electrons.

21.2 Pt(II), 16 electrons.

21.3 Dibromocarbonylmethylbis (triphenylphosphine)iridium(III).

21.4 Fe(CO)$_5$ will have the higher CO stretching frequency and the longer M—C bond.

21.5 (a) The η^6-C$_7$H$_8$ provides 6 electrons, as does the formally neutral Mo atom. Each carbonyl provides another 2 electrons, giving a total of 18. (b) The η^7-C$_7$H$_7^+$, group provides 6 electrons, leaving a formally neutral Mo atom and the 3 carbonyl groups each providing a further 6, giving a total of 18.

21.6 Reductive cleavage of [Mn$_2$(CO)$_{10}$] with Na would give [Mn(CO)$_5$]$^-$; this would react with MeI to give [MnMe(CO)$_5$]. Subsequent reaction with PPh$_3$ would induce a migratory insertion reaction and give the desired product.

21.7 Bridging only.

21.8 A small change only: the electron is lost from the largely non-bonding HOMO.

21.9 CVE = 60; tetrahedral Fe$_4$ with Cp on each vertex, CO on each face.

21.10 P(CH$_3$)$_3$ would be preferred, as it is sterically less demanding.

21.11 Fluorenyl compounds are more reactive than indenyl as they have two aromatic resonance forms when they bond in the η^3 mode.

21.12 The six-coordinate palladium starting material is an 18-electron Pd(IV) species. The four-coordinate product is a 16-electron Pd(II) species. The decrease in both coordination number and oxidation number by 2 identifies the reaction as reductive elimination.

21.13 The ethyl group [Pt(PEt$_3$)$_2$(Et)(Cl)] is prone to β-hydride elimination, whereas the methyl group in [Pt(PEt$_3$)$_2$(Me)(Cl)] is not.

Exercises

21.1 (a) Pentacarbonyliron(0), 18;
(b) decacarbonyldimanganese(0), 18;
(c) hexcarbonylvanadium(0), 17, easily reduced; (d) tetracarbonylferrate(-2), 18;
(e) tris(pentamethylcyclopentadienyl) lanthanum(III), 18;
(f) allyltricarbonylchloroiron(II), 18;
(g) tetracarbonyltriethylphosphineiron(0);
(h) dicarbonylmethyltriphenyl phosphinerhodium(I), 16, sq planar);
(i) chloromethylbis(triphenylphosphine)

palladium(II), 16, sq planar;
(j) cyclopentadienyltetraphenylcyclo-butadinecobalt(I), 18;
(k) dicarbonylcyclopentadienylferrate(0), 18;
(l) benzenecycloheptatrienechromium(0), 18;
(m) trichlorobiscyclopentadineyl-tanatalum(V), 18; (n) cyclopentadineylnitrosylnickel(0), 18.

21.2

M	M
(a)	(b)

21.3 (a) η^2; (b) η^1, η^3, η^5; (c) η^2, η^4, η^6; (d) η^2, η^4; (e) η^2, η^4, η^6, η^8.

21.4 (a) 16e sandwich; (b) 18e; (c) 18e.

21.5 Direct combination, reductive carbonylation. Kinetic.

21.6 Fe + 5CO(high pressure) → Fe(CO)$_5$; diphos + Fe(CO)$_5$ → Fe(CO)$_3$(diphos) + 2CO.

21.7 C_s (3); C_{3v} (2); D_{3h} (1).

21.8 (a) PF$_3$ π-acceptor ligand: increases v(CO) relative to PMe$_3$ (σ donor ligand); (b) Cp* stronger donor, and therefore lower v(CO).

21.9 Ni$_3$ triangle each Ni capped by Cp, CO above and below nickel triangle and collinear; $18\frac{1}{3}$ e per Ni atom, deviations from 18-electron rule are common in Group 10.

21.10 (b) Reacts by an associative process.

21.11 (a) [Fe(CO$_4$)]$^{2-}$ more negative, complex is more basic; (b) [Re(CO)$_5$]$^-$ heavier metal, more basic.

21.12 (a) 3; (b) 2; (c) 12.

21.13 (i) Reduce Mn$_2$(CO)$_{10}$ with Na to give Mn(CO)$_5^-$; react with MeI to give MnMe(CO)$_5$. (ii) Oxidise with Br$_2$. to give MnBr(CO)$_5$; displace the bromide with MeLi to give MnMe(CO)$_5$.

21.14 Mo(CO)$_5$(C(OMe)Ph)

21.15

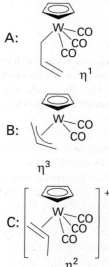

A: η^1

B: η^3

C: η^2

21.16 Compounds formed are TiR$_4$; neither the methyl or trimethylsilyl groups have β

hydrogens, so, unlike the ethyl compound, the low energy β-hydride elimination decomposition reaction is not available to them.

21.17 (a) Reflux $Mo(CO)_6$ with cycloheptatriene to give $[Mo(\eta^6\text{-}C_7H_8)(CO)_3]$; treat with the trityl tetrafluoroborate to abstract a hydride and give $[Mo(\eta^7\text{-}C_7H_7)(CO)_3]BF_4$.
(b) React $[IrCl(CO)(PPh_3)_2]$ with MeCl to give (oxidative addtion) and then expose to CO atmosphere to induce the migration to give $[IrCl_2(COMe)(CO)(PPh_3)_2]$.

21.18 $RuCp_2$ (18e) more stable than $RhCp_2$ (19e)

21.19 (a) $(\eta^5\text{-}C_5H_5)_2Fe + CH_3C(O)Cl \rightarrow$ $(\eta^5\text{-}C_5H_5)(\eta^5\text{-}C_5H_4COCH_3)Fe + HCl$.
(b) $(\eta^5\text{-}C_5H_5)_2Fe + LiBu \rightarrow (\eta^5\text{-}C_5H_5)$ $(\eta^5\text{-}C_5H_4Li)Fe + BuH$, $(\eta^5 - C_5H_5)$ $(\eta^5\text{-}C_5H_4Li)Fe + CO_2 \rightarrow (\eta^5 - C_5H_5)$ $(\eta^5\text{-}C_5H_4CO_2Li)Fe$, $(\eta^5 - C_5H_5)$ $(\eta^5\text{-}C_5H_4CO_2Li)Fe + HCl \rightarrow$ $(\eta^5\text{-}C_5H_5)(\eta^5\text{-}C_5H_4CO_2H)Fe + LiCl$.

21.21 C_5H_6 is a tetrahapto four-electron ligand. Its formation reduces the metal electron count to 18 on Ni; protonation of Fe in $FeCp_2$ leaves the 18e count unchanged.

21.22 (a) Attack of H_2O on the CF_2 carbon followed by elimination of HF; (b) β-H elimination.

21.23 The rate constant for the initial Me migration step.

21.24 (a) Octahedral (86), trigonal prismatic, CVE = 90; (b) no; (c) $[Fe_6C(CO)_{16}]^{2-}$, CVE = 86, octahedral; $[Co_6C(CO)_{16}]^{2-}$, CVE = 92, trigonal prismatic.

21.25 (a) $Si(CH_3)$; (b) I.

21.26 $Co_4(CO)_{12}$ has weaker MM bonds.

Chapter 22

Self-tests

22.1 Use the sterically demanding Cp* group.

22.2 U(IV) is the most stable species, but is readily oxidised by air to give the U(VI) species UO_2^{2+}.

Exercises

22.1 (a) $2Ln(s) + 6H^+ \rightarrow 2Ln^{3+} + 3H_2$. (b) The +3 state is by far the most stable. (c) Cerium (+4) and europium (+2).

22.2 $Ce(IO_3)_4$ and $EuSO_4$ have low solubility compared with the Ln^{3+} salts of these anions.

22.3 First and second IEs would show general increase across the lanthanoids (higher atomic number means higher ionisation energy). With the third, anomalies arise; these are a peak at Eu (gives a f^6 configuration, therefore requires removal of electron from a stable, half-filled f^7) and a trough at Gd (gives the stable f^7 configuration).

22.4 Stable carbonyl compounds need back-bonding from metal orbitals of the appropriate symmetry. With the lanthanoids, the 5d-orbitals are empty, and the 4f-orbitals do not extend beyond the [Xe] core and cannot participate in bonding.

22.5 A simple salt elimination reaction with Li_2COT: $2(Li^+)_2(C_8H_8^{2-}) + UCl_4 \rightarrow$ $U(\eta^8C_8H_8)_2 + 4$ LiCl

22.6 The 4f orbitals in lanthanides do not have significant probability density at the ionic radii. The 5f orbitals are more exposed to ligands.

22.7 Sm^{2+}, f^6, ^7F; Eu^{2+}, f^7, ^8S; Yb^{2+}, f^{14}, ^1S.

22.8 See Box 22.1; (c) ^{90}Sr and (d) ^{144}Ce have high yields and present serious hazards.

Chapter 23

Self-tests

23.1 Vacancies on both anion and cation sites.

23.2 Yes. Mn is easily oxidized from Mn(III) to Mn(IV) and Sr^{2+} and La^{3+} have similar ionic radii.

23.3 Reduction in the thickness of conduction plane impedes the larger cation more.

23.4 Reaction of $6SrO + Mo + 2 MoO_3$ in a sealed tube.

23.5 Cr(III) in octahedral holes provides maximum LFSE.

23.6 1759 pm ($2 \times 569 + 621$) pm.

23.7 $Ca[Al_2Si_2O_8].nH_2O$.

23.9 Make Na_2C_{60} in a similar route to K_3C_{60} and then react with Cs.

Exercises

23.1 NiO (metal vacancies), PbF_2 (Frenkel defects) Fe_2O_3 (oxygen vacancies).

23.2 (a) NiO, (b) Fe_2O_3.

23.3 Forms $(Ni_{1-x}Li_x)O$ which contains Ni(III) an so an electron can hop through the structure from Ni(II) to Ni(III).

23.4 Both. It is a defect when disordered and a new phase when ordered.

23.5 See Fig. 23.17 for the interstitial sites. Triangular bottle necks bounded by three oxide ions—e.g. those at (0,0,0), $(\frac{1}{2}, \frac{1}{2}, 0)$ and $(\frac{1}{2}, 0, \frac{1}{2})$.

23.6 Electron microscopy or X-ray diffraction.

23.7 $SrTiO_3$ [$SrCO_3 + TiO_2$ at 1000°C], Sr_2FeO_3Cl [$SrO, + SrCl_2 + Fe_2O_3$ in correct stoichiometric proportions in a sealed tube], TiN [TiO_2 under NH_3 at 1000 °C].

23.8 $LiNiO_2$, Sr_2WMnO_6 ($=Sr(W_{0.5}Mn_{0.5})O_3$—a perovskite).

23.9 See Fig. 23.22. Yes. Na can be intercalated into ReO_3 and would reside at the cell centre producing a perovskite structure.

23.10 In the inverse spinel structure, Fe[CoFe] with Co^{2+} and Fe^{3+}; three unpaired electrons are expected from the d^7 octahedral Co^{2+}. The structure appears to be close to the inverse spinel formation.

23.11 A strong preference for inverse spinel as found in practice.

23.12 All except $Gd_2Ba_2Ti_2Cu_2O_{11}$ which contains only Cu^{2+}.

23.13 Ferromagnet: domains of parallel spins on atoms/ions, antiferromagnet domains of antiparallel spins on atoms/ions.

23.14 Glass forming BeO, B_2O_3 and to some extent GeO_2.

23.15 e.g. $Zn(II)Cr(III)_2S_4$, Li_2NiF_4.

23.16 Reaction of TaS_2 with BuLi, electrochemical insertion.

23.17 Be, Ga, Zn, and P.

23.18 $(AlP)O_4$, $(BP)O_4$, $(ZnP_2)O_6$.

23.19 The Cu site in Egyptian blue is square planar and thus has a centre of symmetry and d-d transitions are thus symmetry forbidden. In cobalt aluminate blue the site is tetrahedral with no centre of symmetry.

23.20 BN > C(diamond) > AlP > InSb.

23.21 Na_2C_{60} (all tetrahedral holes filled), Na_3C_{60} (all tetrahedral and all octahedral holes filled).

Chapter 24

Self-tests

24.1 Shell formed by controlling the heterogeneous nucleation of shell material on the core particles. Thermodynamic driving force is adjusted to a level that allows for heterogeneous nucleation of the shell material on the core but prevents homogeneous nucleation of the shell material. This heterogeneous nucleation occurs slowly, best results are obtained when growth occurs uniformly on all core particles.

24.2 Radius = 4.21 nm, diameter = 8.42 nm.

24.3 Obtain (100) and (111) perovskite substrates, such $SrTiO_3$. By using an atomic layer epitaxial approach, the (100) superlattices can be grown by alternating the deposition of . . . $[LaO—CrO_2—LaO—FeO_2]$ . . . (100)$SrTiO_3$. The (111) superlattices can be grown by alternating the deposition of . . . $[LaO_3—Cr—LaO_3—Fe]$ . . . on (111)$SrTiO_3$. This leads to distinct type of ordered arrangements of the Cr—O—Fe interactions.

24.4 MCM41 is a better choice as the pore size can range from 2 nm up to 20 nm matching the sizes of QDs. ZSM5 is a poor choice as the pore size is smaller than 0.5 nm.

Exercises

24.1 (a) 3.14×10^2 nm^2 versus 3.14×10^6 nm^2 (a factor of 10^4). (b) Nanoparticle: 10 nm but not 1000 nm particle. (c) Considering a surface plasmon, a nanoparticle should exhibit properties different from a molecule or an extended solid. The particle would have to exhibit a standing wave behaviour of this surface property.

24.2 Exciton Bohr radius, or the average physical separation between an electron-hole pair. If the exciton Bohr radius is on the order of the physical size of the particle, which can occur in ≈ 2 − 10 nm particles, then quantum confinement will occur since the two particles are forced to be closer than they prefer.

24.3 Quantum dots exhibit broadband absorption and single wavelength emission, which can be tailored by engineering the size or surface termination of the quantum dot. Organic fluorophores exhibit narrow-band absorption and single wavelength emission. In general, one light source can be used to excite different quantum dots, but each organic fluorophore requires it own excitation source.

24.4 Energies of the band edges for a QD nanocrystal are more widely separated (i.e., a larger bandgap) than similar bulk semiconductor. Also, the states are localized in quantum dots and are characterized by no linear momentum.

24.5 (a) 'Top-down' approach: 'carve out' nanoscale features from a larger object (e.g. lithography). 'Bottom-up' approach: 'build up' nanoscale features from smaller entities (e.g. thin film deposition of quantum wells or superlattices). (b) Top-down methods allow for precise control over the spatial relationships between nanoscale entities, but they are limited to the design of large nanoscale items. Bottom up methods allow for the precise spatial control over atoms and molecules relative to each other, but the long-range spatial arrangement is often difficult to realize.

24.6 A method to image the microscale (and below) features of a material by scanning a very small probe over the surface and measuring some physical interaction between the tip and material. (b) The local magnetic domains of a materials can be imaged using magnetic force microscopy, where the probe is sensitive to magnetic fields.

24.7 SEM: an electron beam is scanned over a material and an image is typically generated by recording the intensity of secondary or back-scattered electrons. TEM: an electron beam is transmitted through the materials and the image is that is collected is simply the spatial variation in the number of transmitted electrons. Sample preparation for SEM: ensure material is conductive (perhaps requiring a coating), TEM: sample to be made transparent to the electron beam (thin samples).

24.8 (a) Species must first be solvated; stable nuclei of nanometre dimensions must be formed; growth of particles to the final desired size. (b) So that nucleation fixes the total number of particles and growth leads to a controlled size and a narrow size distribution.

24.9 Vapour phase techniques lead to large sizes because it is difficult to control the nucleation/growth regimes and because temperatures are typically higher. (b) Vapour phase synthesis lead to more agglomeration. Because it is difficult to add stabilizers/surfactants to the vapour phase,

agglomeration and excessive growth during nucleation often occur. If temperature is high, agglomerates can sinter into hard-agglomerates.

24.10 (a)

(b) In both cases, one could cause nucleation in one solution and then grow the particle in another solution. (c) In biosensing, the dielectric property of the shell can control the surface plasmon of the core. The shell can be affected by the environment. In drug delivery, the shell could react with a specific location and the core could be used as a treatment (drug).

24.11 Homogeneous nucleation leads to solid formation throughout the vapour phase. Heterogeneous nucleation, solid formation occurs only at specific locations such as a pre-existing solid surface. (b) Heterogeneous nucleation. (c) Homogeneous nucleation.

24.12 Chemical vapour deposition: the atomic species of interest are bound chemically to other species; they are therefore molecular and are stable chemical species. Also, their thermal energies are typically rather low. Physical vapour deposition: the atomic species of interest are typically atomic (or polyatomic), (poly)ionic, radicals, and clusters; they are therefore unstable—the atomic species are most stable. Also, their energies can range from thermal to plasma energies.

24.13 (a) Multiple layers of quantum dots can, if they do not interact, simply increase the intensity of any optical absorption or emission. They can also, if the do interact, be used to form quantum cascade lasers. (b) The limitations come from the requirements on how coherent the interface between the two materials must be and on how easy it is to grow flat layers of one and QDs of the other. In general, it is somewhat strict.

24.14 (a) Epitaxy: a definite crystallographic relationship between each layer between the layers and the substrate. (b) The lines of dots in each layer do line up, implying that there is an epitaxial relationship. (c) Heteroepitaxial: $PrMnO_3$ on $SrTiO_3$, $SrMnO_3$ on $PrMnO_3$. Homoepitaxial: $PrMnO_3$ on $PrMnO_3$.

24.15 (a) Lasers and optical sensors. (b) Because they exhibit properties that are not observed in molecular or traditional solid state materials. (c) Using MBE.

24.16 By developing superlattices of two materials with different elastic constants and that have a large number of interfaces spaced on the nanoscale, much improved mechanical properties (hardness values) can be realized.

24.17 Consider the Ruddlesden-Popper phases (Chapter 23); one could grow phases such as

Sr_2TiO_4 by controlling the growth of each monolayer in a bottom-up approach on an appropriate susbtrate, e.g., ... [SrO—SrO—TiO_2] ... (b) If one were to deposit in an alternating fashion ... [SrO—SrO—TiO_2—SrO—TiO_2—SrO—TiO_2—SrO—TiO_2] ... then one could grow the n = 4 phase which is not known in the bulk (but has been made as a thin film).

24.18 (a) Self assembly offers a route to bridging the bottom-up and top-down approaches to synthesis. (b) Offers a route to assemble nanosized particles/entities into macroscopic structures.

24.19 See section 24.11. Common features: molecular or nanoscale subunits; spontaneous assembly of the subunits; non-covalent interactions between the assembled subunits; and a longer range structures arising from the assembly process.

24.20 Static self-assembly: when a system self-assembles to a stable state, either a local or global equilibrium (e.g. liquid crystal self-assembly). Dynamic self-assembly: when the system is oscillating between states and is dissipating energy (e.g. oscillating chemical reaction).

24.21 Surfactants: molecules that contain both hydrophilic and hydrophobic moieties that influence materials architectures. Our cell membranes are composed of phospholipid bilayers that contain anionic phosphate polar head groups and twin hydrocarbon tails. Like the Au-organothiol films, these self-assembly owing specific regions of hydrophobic and hydrophilic interactions.

24.22 Morphosynthesis: the control of nanoarchitectures in inorganic materials through changes in synthesis parameters (e.g. Ethylenediamine (en) used to control dimensionality and achieve ZnS nanowires).

24.23 PPF/PPF-DA is an injectable, bionanomaterial used for bone tissure engineering. See section 24.16.

24.24 (a) Biomimetics for nanotechnology involves designing nanomaterials that mimmick biological systems. (b) See Box 24.2.

24.25 For biomimetics you need trabecular and cortical bone tissues which have different mechanical properties that combine compressibility and tensile strength. Understanding the chemistry that limits composite interactions is crucial in development of artificial bone. More specifically, hybrid alumoxane nanoparticles dispersed within PPF/PPF-DA have enhanced strength.

24.26 Class I: hybrid materials where no covalent or ionic bonds are present (e.g. block copolymers). Class II:some of the components are linked through strong chemical bonds (e.g. polymer/clay nanocomposites).

24.27 (a) Highly dispersed inorganic nanoparticles lead to increased exposed surface areas that afford control over the interfaces between the inorganic and organic components thus allowing for the tuning of materials properties including stress and strain. (b) The often nonpolar organic polymers (like oil) do not have strong interactions with the polar (like water) or ionic inorganic components leading to problems to inhomogeneity and materials failure.

Chapter 25

Self-tests

25.1 Additional PPh_3 will result in a decrease in reaction rate.

25.2 Single $1465\,cm^{-1}$ band expected.

25.3 No, if it was pure and defect free it would lack acidic sites.

25.4 Bis(cyclopentadienyl) zirconium catalysts have no sterically demanding groups, which could favour one orientation of the incoming propene and therefore give rise to atactic polypropene.

Exercises

25.1 (a) Catalytic; (b) noncatalytic; (c) noncatalytic.

25.2 Sections 25.1 and 25.2

25.3 (a) Homogeneous; (b) heterogeneous; (c) homogeneous.

25.4 Accomplished with existing technology: (d); thermodynamically prohibited: (a) and (b); need research: (c).

25.5 Added PPh_3 suppresses equilibrium concentration of $RhClL_2(sol)$.

25.6 Steric effects reduce equilibrium formation of (D) in Fig. 25.4.

25.7 Formation of (C) or (E) may be rate-limiting with added phosphine. Attack of coordinated OH^- gives one enantiomer, attack of solution OH^- gives the other.

25.8 The difference here is that iodide ions produce MeI and $MeCO_2^-$, and reaction of the ethanoate ion with the acetyl iodide produced leads to the ethanoic anhydride.

25.9 (a) ROMP can result in reduced steric strain, thereby providing a thermodynamic driving force for the reaction. (b) RCM results in the loss of ethene, and by removing this gas, the position of equilibrium can easily be driven over to favour the ring product.

25.10 Coordination of formic acid with loss of L, reductive elimination of H_2, ox. addn. HCO_2H, and CO_2 elim.

25.11 For charge balance an extra proton is required on the surface of the zeolite. (b) Other 3+ ions such as Ga^{3+}, Co^{3+} and Fe^{3+}.

25.12 To achieve a very high surface area.

25.13 Initial chemisorption as $C^*H_2C^*HC(CH_3)_2CH_2CH_3$, exchange of H for D at surface attached carbon atoms (*), dissociation (via reductive elimination).

25.14 CO strongly chemiabsorbs and blocks sites for H_2.

Chapter 26

Self-tests

26.1 Uncomplexed Fe^{2+} is present at low concentrations ($\geq 10^{-7}\,M$). Fe^{2+} is low in the Irving-Williams series. By contrast Fe^{3+} is strongly complexed (relative to Fe^{2+}) by highly specific ligands such as ferritin and by smaller polyanionic ligands, particularly citrate. Free Fe^{2+} is therefore easily oxidised.

26.2 The protein's 3D structure can place any particular atom in a suitable position for axial coordination and prevent water molecules (which would be the natural choice for Mg^{2+}, a hard cation) from occupying that site.

26.3 The need for osmotic balance requires that the saline fluid being administered has the same concentration of solutes as blood plasma. Blood plasma (the external medium with respect to red-blood cells) contains high levels of Na^+ and Cl^-, but low levels of K^+ and other anions: use of NaCl rather than another salt is therefore essential in order to maintain the electrical potential.

26.4 Calmodulin does not bind to the pump unless Ca^{2+} is coordinated. The binding of calmodulin is thus a signal informing the pump that the cytoplasmic Ca^{2+} level has risen above a certain level.

26.5 Without steric protection, a series of reactions leads to inactive Fe(III) oxo products.

26.6 There is greater covalence in blue Cu centres than in simple Cu(II) compounds. The unpaired electron is delocalised onto the cysteine sulphur and 'spends' more time away from the Cu nucleus.

26.7 Cu(III) is d^8. It is likely to be highly oxidising, and probably diamagnetic with square-planar geometry.

26.8 Species such as CH_3Hg^+ and $(CH_3)_2Hg$ are hydrophobic and can penetrate cell membranes.

26.9 Spectroscopic measurements such as EPR which detects unpaired electrons, or EXAFS which reveals distances between Fe or Mo and the sulphur ligands, can be carried out on both enzyme and cofactor dissolved in DMF. See Chapter 6.

26.10 Cu(I) has an almost unique ability to undergo linear coordination by sulphur ligands. The only other metals with this property are Ag(I), Au(I), and Hg(II) but these are not common in biology. Binding as Cu(II) would be less specific because it adopts more common geometries. Also, the cell redox potential is quite low, which favours the presence of Cu(I) rather than Cu(II).

Exercises

26.1 Like Ca^{2+}, Ln^{3+} ions are 'hard' and prefer coordination by anionic oxygen-containing ligands such as carboxylates. Their larger size is compensated for by their higher charge, although they are likely to have higher coordination numbers. Many lanthanoid ions have useful electronic (fluorescence) and magnetic properties (e.g. Gd(III)). 'Ca^{2+}-binding proteins' can thus be studied by fluorescence (which is quenched when Ln^{3+} binds close to certain amino acids, such as tryptophan) and by NMR.

26.2 Co(II) commonly adopts distorted tetrahedral and five-coordinate geometries typical of Zn(II) in enzymes. The d-d transitions that occur for Co(II) in the visible spectral region are quite intense and report on the structure and ligand binding properties of the native Zn sites. Co(II) is also paramagnetic, enabling Zn enzymes to be studied by EPR.

26.3 Acid strengths (coordinated water molecules) lie in the order Fe(III) > Zn(II) > Mg(II). Ligand binding rates are Mg(II) > Zn(II) > Fe(III). Mg is usually six-coordinate so it can accommodate more complex reactions (such as Rubisco).

26.4 Fast and efficient electron transfer requires a low-reorganisation energy. The large difference in structures between oxidised and reduced states is not consistent with this requirement and suggests that the P-clusters perform a function apart from just electron transfer.

26.5 Transfer of CH_3 is expected to involve cobalamin. Transfer and activation of CO is expected to involve an electron-rich metal (Fe(II), Ni(I) Cu(I)).

Index